Construction

Consumer Information

Education

Environment

Finance

(continued on next page)

INDEX OF APPLICATIONS (continued)

(continued on inside back cover)

A Survey of MATHEMATICS with Applications

SEVENTH EDITION

ALLEN R. ANGEL
Monroe Community College

CHRISTINE D. ABBOTT
Monroe Community College

DENNIS C. RUNDE
Manatee Community College

Boston San Francisco New York
London Toronto Sydney Tokyo Singapore Madrid
Mexico City Munich Paris Cape Town Hong Kong Montreal

Publisher: Greg Tobin
Senior Acquisitions Editor: Anne Kelly
Project Editor: Lauren Morse
Editorial Assistant: Cecilia Fleming
Managing Editor: Karen Guardino
Senior Production Supervisor: Peggy McMahon
Senior Manufacturing Buyer: Evelyn Beaton
Marketing Manager: Becky Anderson
Marketing Coordinator: Carolyn Buddeke
Associate Media Producer: Sara Anderson
Software Development: Jan Wann and Marty Wright
Photo Research: Beth Anderson
Rights and Permissions Advisor: Dana Weightman
Senior Designer: Barbara T. Atkinson
Text and Cover Designer: Leslie Haimes
Cover Photo: Cyclists with lavender field in foreground, Tour de France, France © Graham Chadwick / Stone / GettyImages
Composition and Illustration: Nesbitt Graphics, Inc.
Production Services: Nesbitt Graphics, Inc.

For permission to use copyrighted material, grateful acknowledgment is made to the copyright holders in the Credits section beginning on page C-1, which is hereby made part of this copyright page.

Library of Congress Cataloging-in-Publication Data
Angel, Allen R., 1942–
A survey of mathematics with applications.—7th ed. / Allen R. Angel, Christine D. Abbott, Dennis C. Runde.
p. cm.
Includes index.
ISBN 0-321-11250-4
1. Mathematics I. Abbott, Christine D. II. Runde, Dennis C. III. Title.
QA39.3.A54 2004
510—dc22 2003053690

4 5 6 7 8 9 10—QWT—070605

To my wife, Kathy Angel (photo on page 56)
A. R. A.

To my husband, Jason; and children, Matthew and Jake
C. D. A.

To my parents, Bud and Tina Runde
D. C. R.

CONTENTS

MATH: IT'S ALL AROUND US!

We present *A Survey of Mathematics with Applications,* seventh edition, with that vision in mind. Our primary goal in writing this book was to give students a text that they can read, understand, and enjoy while learning how mathematics affects the world around them. Numerous applied examples motivate topics. A variety of interesting applied exercises demonstrate the real-life nature of mathematics and its importance in the students' lives.

The text is intended for students who require a broad-based general overview of mathematics, especially those majoring in the liberal arts, elementary education, the social sciences, business, nursing, and allied health fields. It is particularly suitable for those courses that satisfy the minimum competency requirement in mathematics for graduation or transfer.

EXPANDED EDITION

- The expanded version of *A Survey of Mathematics with Applications,* seventh edition, contains all the material covered in the basic text, with additional chapters on *Graph Theory* and *Voting and Apportionment.* While many wished inclusion of these topics, others did not. Therefore, we have written two versions of the book.

NEW TO THIS EDITION

In this edition we made several important improvements in presentation.

- The interior design has been modified, and many new photographs were added to make the book more inviting and motivational. A new, more open design has been used to make the book easier for the student to read.
- **Recreational Mathematics Exercises** have been added in many sections to help make the book more enjoyable to students. These exercises take a variety of forms, from puzzles to exploring new mathematical ideas and concepts. Often, but not always, the material in these exercises is related to the material covered in the section or chapter.
- **Timely Tips** have been added to provide useful and timely information for students. This information often helps students to understand the concepts, or relates the material to other sections of the book.
- A new feature, **Mathematics Everywhere,** has been added. The information provided in these boxes relates mathematics to students' everyday lives. This material will help students to see a need for and gain an appreciation of mathematics.
- Additional interesting Did You Knows . . . and Profiles in Mathematics have been added throughout the book.
- In various exercise sets, the number and variety of exercises has been increased.

- The exercise sets have been reclassified to include: Concept/Writing Exercises; Practice the Skills Exercises; Problem Solving Exercises; Challenge Problem/Group Activity Exercises; Recreational Mathematics; and Internet/Research Activities.
- Sources have been added; up-to-date tables, graphs, and charts make the material more relevant and encourage students to read graphs and analyze data.
- Approximately 40% of the exercises are new.
- The number of writing exercises was increased. The writing exercises are denoted in magenta in the exercise sets.
- The number of examples was increased throughout the text to promote student understanding.

CONTENT REVISION

In addition, we revised and expanded certain topics to introduce new material and to increase understanding.

CHAPTER 1 "Critical Thinking Skills," was updated with exciting and current examples and exercises.

CHAPTER 2 "Sets," includes more information on countable sets and infinite sets.

CHAPTER 3 "Logic," has more exercises and a greater variety of exercises. Certain material has been rewritten for greater clarity.

CHAPTER 4 "Systems of Numeration," covers the ancient Chinese numeration system in more depth. More material on the hexadecimal system is included.

CHAPTER 5 "Number Theory and the Real Number System," was expanded to include additional coverage of Fibonacci numbers. The most current number theory information (largest prime number, most accurate value of pi, etc.) has been included.

CHAPTER 6 "Algebra, Graphs and Functions," has more and a greater variety of examples and exercises dealing with real-life situations.

CHAPTER 7 "Systems of Linear Equations and Inequalities," now has annotations included, and more detailed explanations of certain topics.

CHAPTER 8 "The Metric System," now includes many examples and interesting photographs of real-life (metric) situations taken from around the world. More information is provided on larger and smaller metric units of measurements.

CHAPTER 9 "Geometry," now includes a new section on Transformational Geometry (transformations, symmetry, and tessellations) and additional information on M. C. Escher and his works.

CHAPTER 10 "Mathematical Systems," has additional exercises and examples. We have tied in this material more with real-life situations.

CHAPTER 11 "Consumer Mathematics," includes current interest rates and updated information on items that may be of interest to students including updated material on sources of credit and mutual funds. There is also additional information on stocks, bonds, and mutual funds.

CHAPTER 12 "Probability," has a greater variety of examples and exercises, more examples and exercises have been added that deal with real-life situations. Certain material has been rewritten for greater clarity.

CHAPTER 13 "Statistics," now includes expanded coverage of measures of location. Also, material related to z-scores has been revised.

CONTINUING FEATURES

Several features appear throughout the book, adding interest and provoking thought.

- **Chapter Openers** Interesting and motivational photo essays introduce each chapter and illustrate the real-world nature of the chapter topics.
- **Problem Solving** Beginning in Chapter 1, students are introduced to problem solving and critical thinking. The theme of problem solving is then continued throughout the text, and special problem-solving exercises are presented in the exercise sets.
- **Critical Thinking Skills** In addition to a focus on problem solving, the book also features sections on *Inductive Reasoning* and the important skills of *Estimation* and *Dimensional Analysis.*
- **Profiles in Mathematics** Brief historical sketches and vignettes present the stories of people who have advanced the discipline of mathematics.
- **Did You Know . . .** These colorful, engaging, and lively boxed features highlight the connection of mathematics to history, to the arts and sciences, to technology, and to a broad variety of disciplines.
- **Group Projects** At the end of each chapter are suggested projects that can be used to have students work together. These projects can also be assigned to individual students if desired.
- **Chapter Summaries, Review Exercises, and Chapter Tests** This end-of-chapter material helps students to review material and to prepare for tests.

INSTRUCTOR'S SUPPLEMENTS

Instructor's Edition ISBN 0-321-20566-9

This special edition of the text includes an answer section consisting of answers to all text exercises.

Instructor's Solutions Manual ISBN 0-321-20594-4

This manual contains solutions to all exercises in the text and answers to Group Projects.

Instructor's Testing Manual ISBN 0-321-20595-2

This manual includes three alternate tests per chapter.

PowerPoint Lecture Presentation

Available through http://www.aw-bc.com/suppscentral or at MyMathLab.

This classroom presentation software covers all-important topics from sections in the text.

TestGen with QuizMaster ISBN 0-321-20593-6

TestGen enables instructors to build, edit, print, and administer tests using a computerized bank of questions developed to cover all the objectives of the text. TestGen is algorithmically based so that multiple, yet equivalent, versions of the same question or test can be generated at the click of a button. Instructors can also modify test bank questions or add new questions by using the built-in question editor, which allows users to create graphs, import graphics, insert math notation, and insert variable numbers or text. Tests can be printed or administered on-line via the Internet or another network. Many questions in TestGen can be expressed in a short-answer or multiple-choice form, giving instructors greater flexibility in their test preparation. TestGen comes packaged with QuizMaster, which allows students to take tests on a local area network. The software is available on a dual-platform Windows/Macintosh CD-ROM.

MathXL® www.mathxl.com

MathXL® is a powerful online homework, tutorial, and assessment system that accompanies your Addison-Wesley textbook in mathematics or statistics. With MathXL, instructors can create, edit, and assign online homework and tests using algorithmically generated exercises correlated at the objective level to your textbook. All student work is tracked in MathXL's online gradebook. Students can take chapter tests in MathXL and receive personalized study plans based on their test results. The study plan diagnoses weaknesses and links students directly to tutorial exercises for the objectives they need to study and retest. Students can also access supplemental animations and video clips directly from selected exercises. MathXL is available to qualified adopters. For more information, visit our website at www.mathxl.com.

MyMathLab® www.mymathlab.com

MyMathLab® is a series of text-specific, easily customizable online courses for Addison-Wesley textbooks in mathematics and statistics. MyMathLab is powered by CourseCompass™—Pearson Education's online teaching and learning environment—and by MathXL®—our online homework, tutorial, and assessment system. MyMathLab gives you the tools you need to deliver all or a portion of your course online, whether your students are in a lab setting or working from home. MyMathLab provides a rich and flexible set of course materials, featuring free-response exercises that are algorithmically generated for unlimited practice and mastery. Students can also use online tools, such as video lectures, animations, and a multimedia textbook, to independently improve their understanding and performance. Instructors can use MyMathLab's homework and test managers to select and assign online exercises correlated directly to the textbook, and they can import TestGen tests into MyMathLab for added flexibility. MyMathLab's online gradebook—designed specifically for mathematics and statistics—automatically tracks students' homework and test results and gives the instructor control over how to calculate final grades.

MyMathLab is available to qualified adopters. For more information, visit our website at www.mymathlab.com.

STUDENT'S SUPPLEMENTS

Student's Solutions Manual ISBN 0-321-20597-9

This for-sale manual contains solutions to all odd-numbered exercises and to all review and chapter test exercises.

Guide to CLAST Mathematical Competency ISBN 0-201-61327-1

This special for-sale study guide for use in Florida offers help in preparing for the College Level Academic Skills Test (CLAST). It includes a review of arithmetic, a CLAST pretest, supplementary exercises, and a CLAST posttest.

Videotape Series ISBN 0-321-20596-0

Videotapes presenting the text's topics are available. A qualified college-level mathematics instructor presents worked-out examples and uses visual aids to reinforce the mathematical concepts.

Digital Video Tutor ISBN 0-321-20598-7

The videotape series for this text is provided on CD-ROM, making it easy and convenient for students to watch video segments from a computer at home or on campus. This complete video set, now affordable and portable for students, is ideal for distance learning or extra instruction.

InterActMath® Tutorial Website

Get practice and tutorial help online! This interactive tutorial website provides algorithmically generated practice exercises that correlate directly to the exercises in the text. A detailed worked-out example and guided solution accompany each practice exercise. The website recognizes student errors and provides feedback.

MathXL will help students succeed in their math course!

MathXL is an online homework, tutorial, and assessment system that uses algorithmically generated exercises correlated to the objectives in your textbook. Students can take chapter tests and receive personalized study plans that diagnose weaknesses and link students to areas they need to study and retest. Students can also work unlimited practice exercises that provide tutorial instruction, and they can access animations and video clips directly from selected exercises. An access code is required. For more information go to www.mathxl.com.

MyMathLab

MyMathLab is a complete online course available with this text and is perfect for a lecture-based, self-paced, or online course. This site offers instructors and students a wide variety of resources from dynamic multimedia—video clips, animations, and more—to course management tools. With *MyMathLab,* instructors can customize their online course and help students increase their comprehension and success!

- The entire textbook is available online and is supplemented by multimedia content, such as videos and animations, which is used to explain concepts. With *MyMathLab,* students can work with tutorial exercises tied directly to those in their textbook.
- *MyMathLab* allows students to do practice work and to complete instructor-assigned tests and homework assignments online. Based on their results, *MyMathLab* automatically builds individual study plans that students can use to improve their skills.

MyMathLab requires a student access code. For more information about *MyMathLab,* go to www.mymathlab.com.

Addison-Wesley Math Tutor Center

The Addison-Wesley Math Tutor Center is staffed by qualified math and statistics instructors who provide students with tutoring on examples and odd-numbered exercises from the textbook. Tutoring is available via toll-free telephone, fax, email, or the Internet and White Board technology allows tutors and students to actually see the problems worked while they "talk" in real time over the Internet during tutoring sessions. An access card is required. For more information, go to www.aw-bc.com/tutorcenter.

ACKNOWLEDGMENTS

We would like to thank our spouses Kathy Angel, Jason Abbott, and Kris Runde for their support and encouragement throughout the project. Our spouses helped us in a great many ways, including proofreading, typing, and offering valuable suggestions. We cannot thank them enough for their wonderful support and understanding while we worked on the book.

We would also like to thank our children: Robert and Steven Angel; Matthew and Jake Abbott; and Alex, Nicholas, and Max Runde. They also gave us support and encouragement. They were very understanding when we could not spend as much time with them as we wished because of book deadlines. Without the support and understanding of our families, this book would not be a reality.

We would like to thank Cathy Ferrer of Valencia Community College and Deana Richmond for accuracy checking of the text and answers. They did a very conscientious job. We would also like to thank Sherry Tornwall of the University of Florida for reading through the book and making many valuable suggestions for improving the content of the book.

There are many people at Addison-Wesley who deserve thanks. We would like to thank all those listed on the Library of Congress categorizing page. In particular, we would like to thank Anne Kelly, Senior Acquisitions Editor; Lauren Morse, Project Editor; Cecilia Fleming, Editorial Assistant; Peggy McMahon, Senior Production Supervisor; Becky Anderson, Marketing Manager; Sara Anderson, Associate Media Producer; Barbara Atkinson, Senior Designer; and Karen Guardino, Managing Editor. We would also like to thank Maria McColligan of Nesbitt Graphics, Inc., for her assistance as Production Manager for this project.

Aimee Calhoun and Dick Stewart of Monroe Community College also deserve our thanks for the excellent work they did on the *Student's Solutions Manual* and the *Instructor's Solutions Manual.*

Finally, we would like to thank the reviewers from all editions of the book and all the students who have offered suggestions for improving the book. A list of reviewers for all editions of this book follows. Thanks to all of you for helping make *A Survey of Mathematics with Applications* the most successful Liberal Arts book in the country.

Allen R. Angel

Christine D. Abbott

Dennis C. Runde

REVIEWERS FOR THIS AND PREVIOUS EDITIONS

Frank Asta, *College of DuPage, IL*

*Robin L. Ayers, *Western Kentucky University*

Hughette Bach, *California State University–Sacramento*

Madeline Bates, *Bronx Community College, NY*

Rebecca Baum, *Lincoln Land Community College, IL*
Vivian Baxter, *Fort Hayes State University, KS*
Una Bray, *Skidmore College, NY*
David H. Buckley, *Polk Community College, FL*
Robert C. Bueker, *Western Kentucky University*
Carl Carlson, *Moorhead State University, MN*
Kent Carlson, *St. Cloud State University, MN*
Donald Catheart, *Salisbury State College, MD*
*Yungchen Cheng, *Southwest Missouri State University*
Joseph Cleary, *Massasoit Community College, MA*
Donald Cohen, *SUNY Ag & Tech College at Cobleskill, NY*
David Dean, *Santa Fe Community College, FL*
Charles Downey, *University of Nebraska*
*Annie Droullard, *Polk Community College, FL*
Ruth Ediden, *Morgan State University, MD*
Lee Erker, *Tri-County Community College, NC*
Karen Estes, *St. Petersburg College, FL*
Teklay Fessahaye, *Santa Fe Community College, FL*
Kurtis Fink, *Northwest Missouri State University*
Raymond Flagg, *McPherson College, KS*
Penelope Fowler, *Tennessee Wesleyan College*
Gilberto Garza, *El Paso Community College, TX*
Judith L. Gersting, *Indiana University–Purdue University at Indianapolis*
Lucille Groenke, *Mesa Community College, AZ*
John Hornsby, *University of New Orleans, LA*
Nancy Johnson, *Broward Community College, FL*
Daniel Kimborowicz, *Massasoit Community College, MA*
Mary Lois King, *Tallahassee Community College, FL*
David Lehmann, *Southwest Missouri State University*
Peter Lindstrom, *North Lake College, TX*
James Magliano, *Union College, NJ*
Yash Manchanda, *East Los Angeles College & Fullerton College, CA*
Don Marsian, *Hillsborough Community College, FL*
Marilyn Mays, *North Lake College, TX*
Robert McGuigan, *Westfield State College, MA*
*Wallace H. Memmer, *Brookdale Community College, NJ*
Maurice Monahan, *South Dakota State University*
Julie Monte, *Daytona Beach Community College, FL*
Karen Mosely, *Alabama Southern Community College*

Edwin Owens, *Pennsylvania College of Technology*
Wing Park, *College of Lake County, IL*
Bettye Parnham, *Daytona Beach Community College, FL*
Joanne Peeples, *El Paso Community College, TX*
Nelson Rich, *Nazareth College, NY*
Kenneth Ross, *University of Oregon*
Ronald Ruemmler, *Middlesex County College, NJ*
Rosa Rusinek, *Queensborough Community College, NY*
Len Ruth, *Sinclair Community College, OH*
John Samoylo, *Delaware County Community College, PA*
Sandra Savage, *Orange Coast College, CA*
Gerald Schultz, *Southern Connecticut State University*
Richard Schwartz, *College of Staten Island, NY*
Kara Shavo, *Mercer County Community College, NJ*
Minnie Shuler, *Chipola Junior College, FL*
*Paula R. Stickles, *University of Southern Indiana*
Kristin Stoley, *Blinn College–Bryan, TX*
Steve Sworder, *Saddleback College, CA*
Shirley Thompson, *Moorhead College, GA*
Alvin D. Tinsley, *Central Missouri State University*
*Sherry Tornwall, *University of Florida*
William Trotter, *University of South Carolina*
Sandra Welch, *Stephen F. Austin State University, TX*
Joyce Wellington, *Southeastern Community College, NC*
Sue Welsch, *Sierra Nevada College*
Robert F. Wheeler, *Northern Illinois University*
Susan Wirth, *Indian River Community College, FL*
James Wooland, *Florida State University*
*Judith B. Wood, *Central Florida Community College*
Jean Woody, *Tulsa Community College, OK*
Michael A. Zwick, *Monroe Community College, NY*
*Denotes reviewers for seventh edition.

TO THE STUDENT

Mathematics is an exciting, living study. It has applications that shape the world around you and influence your everyday life. We hope that as you read through this book you will realize just how important mathematics is and gain an appreciation of both its usefulness and its beauty. We also hope to teach you some practical mathematics that you can use in your everyday life and that will prepare you for further courses in mathematics.

Our primary purpose in writing this text was to provide material that you could read, understand, and enjoy. To this end we have used straightforward language and tried to relate mathematical concepts to everyday experiences. We have also provided many detailed examples for you to follow.

The concepts, definitions, and formulas that deserve special attention have been either boxed or set in boldface type. Within each category the exercises are graded so that the more difficult problems appear at the end. The problems with exercise numbers set in color are writing exercises. At the end of most exercise sets are Challenge Problem/Group Activity exercises that contain challenging or exploratory exercises. At the end of each chapter are Group Projects which reinforce the material learned or provide related material.

Each chapter has a summary, review exercises, and a chapter test. When studying for a test, be sure to read the chapter summary, work the review exercises, and take the chapter test. The answers to the odd-numbered exercises, all review exercises, all chapter test exercises, and selected recreational mathematics exercises appear in the answer section in the back of the text. However, you should use the answers only to check your work.

It is difficult to learn mathematics without becoming involved. To be successful, we suggest you read the text carefully *and work each exercise in each assignment in detail.* Check with your instructor to determine which supplements are available for your use.

We welcome your suggestions and your comments. You may contact us at the following address:

Allen Angel
c/o Marketing
Mathematics & Statistics
Addison-Wesley
75 Arlington St., Ste 300
Boston, MA 02116

or by email at:

math@awl.com
Subject: for Allen Angel

Good luck in your adventure in mathematics!

Allen R. Angel

Christine D. Abbott

Dennis C. Runde

CHAPTER 1

San Francisco, CA

CRITICAL THINKING SKILLS

Life constantly presents new problems. The more sophisticated our society becomes, the more complex the problems. We as individuals are constantly solving problems and do so daily. For example, when we consider ways of reducing our expenses or when planning a trip, we need to make problem-solving decisions. Businesses are constantly trying to solve problems that involve making a profit for the company and keeping customers satisfied.

The goal of this chapter is to help you master the skills of reasoning, estimating, and problem solving. These skills will aid you in solving problems in the remainder of this book as well as problems that you will encounter in everyday life.

Engineers, scientists, and others who worked in the energy industry in California had many problems to solve in the years 2000 and 2001 when the entire state felt the power shortage and experienced many blackouts. Even with Californians cutting back on power, power companies still barely met demands. Under the utilities' rotating system, often during the blackouts, when one group of customers got its electricity back, another group had its power shut off. And when federal energy regulators lifted price caps so that the companies involved could increase their access to electrical supplies, customers' electric bills rose dramatically.

Effective January 2002, California customers were free to choose their energy supplier. Two organizations, the California Power Exchange and the California Independent System Operator, now coordinate the buying, selling, and transmitting of electricity. This system will create many problem-solving opportunities for the employees of these two organizations, as well as for California residents who now can choose electricity suppliers.

1.1 INDUCTIVE REASONING

The goal of this chapter is to help you improve your reasoning and problem-solving skills. This section introduces inductive and deductive reasoning, which are used in problem solving. The next section introduces the concept of estimation. Estimation is a technique that can be used to determine if an answer obtained for a problem or from a calculation is "reasonable." Section 1.3 introduces and applies problem-solving techniques.

Before looking at some examples of inductive reasoning and problem solving, let us first review a few facts about certain numbers. The *natural numbers* or *counting numbers* are the numbers 1, 2, 3, 4, 5, 6, 7, 8, ... The three dots, called an *ellipsis*, mean that 8 is not the last number but that the numbers continue in the same manner. A word that we sometimes use is "divisible." If $a \div b$ has a remainder of zero, then *a is divisible by b*. The counting numbers that are divisible by 2 are 2, 4, 6, 8, ... These are called the *even counting numbers*. The numbers that are not divisible by 2 are 1, 3, 5, 7, 9, ... These are the *odd counting numbers*. When we refer to *odd numbers* or *even numbers*, we mean odd or even counting numbers.

Recognizing patterns is sometimes helpful in solving problems, as Examples 1 and 2 illustrate.

EXAMPLE 1 *The Product of Two Odd Numbers*

If two odd numbers are multiplied together, will the product always be an odd number?

SOLUTION: To answer this question, we will examine the products of several pairs of odd numbers to see if there is a pattern.

$1 \times 3 = 3$	$3 \times 5 = 15$	$5 \times 7 = 35$
$1 \times 5 = 5$	$3 \times 7 = 21$	$5 \times 9 = 45$
$1 \times 7 = 7$	$3 \times 9 = 27$	$5 \times 11 = 55$
$1 \times 9 = 9$	$3 \times 11 = 33$	$5 \times 13 = 65$

All the products are odd numbers. Thus, we might predict from these examples that the product of any two odd numbers is an odd number. ▲

EXAMPLE 2 *The Sum of an Odd Number and an Even Number*

If an odd number and an even number are added, will the sum be an odd or an even number?

SOLUTION: Let's look at a few examples where one number is odd and the other number is even.

$3 + 4 = 7$	$9 + 6 = 15$	$23 + 18 = 41$
$5 + 12 = 17$	$5 + 14 = 19$	$81 + 32 = 113$

All these sums are odd numbers. Therefore, we might predict that the sum of an odd and an even number is an odd number. ▲

In Examples 1 and 2, we cannot conclude that the results are true for all counting numbers. From the patterns developed, however, we can make predictions. This type of reasoning process, arriving at a general conclusion from specific observations or examples, is called *inductive reasoning*, or *induction*.

DID YOU KNOW

The Eyes Tell It All

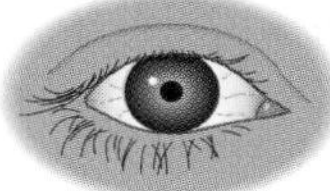

Most of us realize that thus far, no two people have been found to have the same fingerprints or DNA, so fingerprints and DNA have become forms of identification. As technology improves, so do identification techniques. Today, when you purchase a season pass at Disney World, you place your forefinger and middle finger in a device that measures and records your finger size, shape, and density. Then you are given a seasonal pass. Each time you enter Disney World, you put those two fingers in the device, which indicates whether you are allowed or not allowed entry under the seasonal pass that you show.

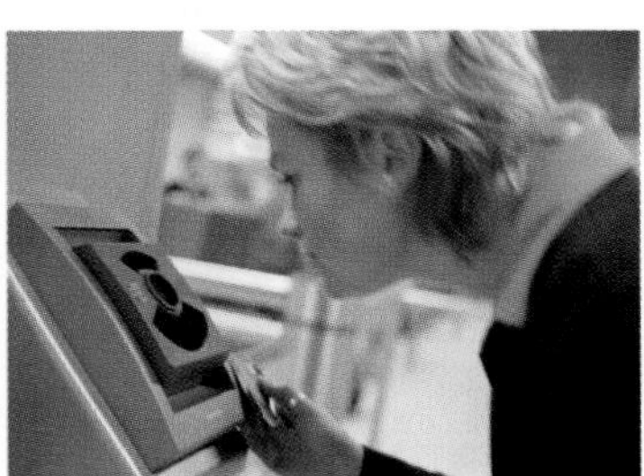

Automated teller machines (ATMs) are now experimenting with determining identity using a person's eye. When you open an account, a photo of your eye is taken and entered into a computer. When you use an ATM, a powerful camera automatically (without you even knowing it) checks the veins in your eye against the computer's files. Shortly, we will no longer need personal identification numbers (PINs) to withdraw money from our accounts. All these techniques are based on inductive reasoning. Based on past experiences, no two people have been found to have the same fingerprints; finger size, shape, and density; DNA; or eyes.

Inductive reasoning is the process of reasoning to a general conclusion through observations of specific cases.

Induction often involves observing a pattern and from that pattern predicting a conclusion. Imagine an endless row of dominoes. You knock down the first, which knocks down the second, which knocks down the third, and so on. Assuming the pattern will continue uninterrupted, you conclude that eventually all the dominoes will fall, even though you may not witness the event.

Inductive reasoning is often used by mathematicians and scientists to predict answers to complicated problems. For this reason, inductive reasoning is part of the *scientific method*. When a scientist or mathematician makes a prediction based on specific observations, it is called a *hypothesis* or *conjecture*. After looking at the products in Example 1, we might conjecture that the product of two odd numbers will be an odd number. After looking at the sums in Example 2, we might conjecture that the sum of an odd number and an even number is an odd number.

Examples 3 and 4 illustrate how we arrive at a conclusion using inductive reasoning.

EXAMPLE 3 *Fingerprints and DNA*

What reasoning process has led to the conclusion that no two people have the same fingerprints or DNA? This conclusion has resulted in fingerprints and DNA being used in courts of law as evidence to convict persons of crimes.

SOLUTION: In millions of tests, no two people have been found to have the same fingerprints or DNA. By induction, then, we believe that fingerprints and DNA provide a unique identification and can therefore be used in a court of law as evidence. Is it possible that, sometime in the future, two people will be found who do have exactly the same fingerprints or DNA? ▲

EXAMPLE 4 *Divisibility by 4*

Consider the conjecture "If the last two digits of a number are divisible by 4, then the number itself is divisible by 4." We will test several numbers to see if the conjecture appears true or false.

SOLUTION: Let's look at some numbers whose last two digits are divisible by 4.

Number	Are the Last Two Digits Divisible by 4?	Is the Number Divisible by 4?
344	yes, $44 \div 4 = 11$	yes, $344 \div 4 = 86$
4312	yes, $12 \div 4 = 3$	yes, $4312 \div 4 = 1078$
10,528	yes, $28 \div 4 = 7$	yes, $10{,}528 \div 4 = 2632$
20,104	yes, $04 \div 4 = 1$	yes, $20{,}104 \div 4 = 5026$

In each case, we find that if the last two digits of a number are divisible by 4, then the number itself is divisible by 4. From these examples, we might be tempted to generalize that the conjecture "If the last two digits of a number are divisible by 4, then the number itself is divisible by 4" is true.* ▲

*This statement is in fact true, as is discussed in Section 5.1.

DID YOU KNOW

An Experiment Revisited

Apollo 15 astronaut David Scott used the moon as his laboratory to show that a heavy object (a hammer) does indeed fall at the same rate as a light object (a feather). Had Galileo dropped a hammer and feather from the Tower of Pisa, the hammer would have fallen more quickly to the ground and he still would have concluded that a heavy object falls faster than a lighter one. If it is not the object's mass that is affecting the outcome, then what is it? The answer is air resistance or friction: Earth has an atmosphere that creates friction on falling objects. The moon does not have an atmosphere; therefore, no friction is created.

EXAMPLE 5 *Pick a Number, Any Number*

Pick any number, multiply the number by 4, add 6 to the product, divide the sum by 2, and subtract 3 from the quotient. Repeat this procedure for several different numbers and then make a conjecture about the relationship between the original number and the final number.

SOLUTION: Let's go through this one together.

Pick a number:	say, 5
Multiply the number by 4:	$4 \times 5 = 20$
Add 6 to the product:	$20 + 6 = 26$
Divide the sum by 2:	$26 \div 2 = 13$
Subtract 3 from the quotient:	$13 - 3 = 10$

Note that we started with the number 5 and finished with the number 10. If you start with the number 2, you will end with the number 4. Starting with 3 would result in a final number of 6, 4 would result in 8, and so on. On the basis of these few examples, we may conjecture that when you follow the given procedure, the number you end with will always be twice the original number. ▲

The result reached by inductive reasoning is often correct for the specific cases studied but not correct for all cases. History has shown that not all conclusions arrived at by inductive reasoning are correct. For example, Aristotle (384–322 B.C.) reasoned inductively that heavy objects fall at a faster rate than light objects. About 2000 years later, Galileo (1564–1642) dropped two pieces of metal—one 10 times heavier than the other—from the Leaning Tower of Pisa in Italy. He found that both hit the ground at exactly the same moment, so they must have traveled at the same rate.

When forming a general conclusion using inductive reasoning, you should test it with several special cases to see whether the conclusion appears correct. If a special case is found that satisfies the conditions of the conjecture but produces a different result, such a case is called a *counterexample*. A counterexample proves that the conjecture is false because only one exception is needed to show that a conjecture is not valid. Galileo's counterexample disproved Aristotle's conjecture. If a counterexample cannot be found, the conjecture is neither proven nor disproven.

A second type of reasoning process is called *deductive reasoning*, or *deduction*. Mathematicians use deductive reasoning to *prove* conjectures true or false.

> **Deductive reasoning** is the process of reasoning to a specific conclusion from a general statement.

EXAMPLE 6 *Pick a Number, n*

Prove, using deductive reasoning, that the procedure in Example 5 will always result in twice the original number selected.

SOLUTION: To use deductive reasoning, we begin with the *general* case rather than specific examples. In Example 5, specific cases were used. Let's select the letter n to represent *any number*.

Pick any number:	n
Multiply the number by 4:	$4n$ ($4n$ means 4 times n)
Add 6 to the product:	$4n + 6$
Divide the sum by 2:	$\dfrac{4n + 6}{2} = \dfrac{\overset{2}{\cancel{4}}n}{\underset{1}{\cancel{2}}} + \dfrac{\overset{3}{\cancel{6}}}{\underset{1}{\cancel{2}}} = 2n + 3$
Subtract 3 from the quotient:	$2n + 3 - 3 = 2n$

Note that, for any number n selected, the result is $2n$, or twice the original number selected. ▲

In Example 5, you may have *conjectured*, using specific examples and inductive reasoning, that the result would be twice the original number selected. In Example 6, we *proved*, using deductive reasoning, that the result will always be twice the original number selected.

SECTION 1.1 EXERCISES

Concept/Writing Exercises

1. **a)** List the natural numbers.
   b) What is another name for the natural numbers?
2. **a)** What does it mean to say, "a is divisible by b," where a and b represent natural numbers?
   b) List three natural numbers that are divisible by 4.
   c) List three natural numbers that are divisible by 9.

In Exercises 3–6, explain your answer in one or two sentences.

3. What is a conjecture?
4. What is inductive reasoning?
5. What is deductive reasoning?
6. What is a counterexample?
7. Which type of reasoning is generally used to arrive at a conjecture?
8. Which type of reasoning is used to prove a conjecture?
9. You have purchased one lottery ticket each week for many months and have not won more than $5.00. You decide, based on your past experience, that you are not going to win the grand prize and so you stop playing the lottery. What type of reasoning did you use? Explain.
10. In the 1950s, doctors noticed that many of their lung cancer patients were also cigarette smokers. Doctors reasoned that cigarette smoking increased a person's chance of getting lung cancer. What type of reasoning did the doctors use? Explain.

Practice the Skills

In Exercises 11–14, use inductive reasoning to predict the next line in the pattern.

11.

```
        1
      1   1
    1   2   1
  1   3   3   1
1   4   6   4   1
```

12.
$10 = 10^1$
$100 = 10^2$
$1000 = 10^3$
$10{,}000 = 10^4$

13.
$1 \times 9 = 9$
$2 \times 9 = 18$
$3 \times 9 = 27$
$4 \times 9 = 36$

14.
$11 \times 11 = 121$
$11 \times 12 = 132$
$11 \times 13 = 143$

In Exercises 15–18, draw the next figure in the pattern (or sequence).

15. . . .

16. . . .

17.  . . .

18. . . .

In Exercises 19–28, use inductive reasoning to predict the next three numbers in the pattern (or sequence).

19. 3, 6, 9, 12, . . .
20. 26, 20, 14, 8, . . .
21. 1, −1, 1, −1, 1, . . .
22. 5, 3, 1, −1, −3, . . .
23. $1, \frac{1}{3}, \frac{1}{9}, \frac{1}{27}, \ldots$
24. 2, −6, 18, −54, . . .
25. 1, 4, 9, 16, 25, . . .
26. 0, 1, 3, 6, 10, 15, . . .
27. 1, 1, 2, 3, 5, 8, 13, 21, . . .
28. $5, -\frac{10}{3}, \frac{20}{9}, -\frac{40}{27}, \ldots$

Problem Solving

29. Find the letter that is the 118th entry in the following sequence. Explain how you determined your answer.

Y, R, R, Y, R, R, Y, R, R, Y, R, R, Y, R, R, . . .

30. **a)** Select a variety of one- and two-digit numbers between 1 and 99 and multiply each by 9. Record your results.
b) Find the sum of the digits in each of your products in part (a). If the sum is not a one-digit number, find the sum of the digits again until you obtain a one-digit number.
c) Make a conjecture about the sum of the digits when a one- or two-digit number is multiplied by 9.

31. *A Square Pattern* The ancient Greeks labeled certain numbers as **square numbers**. The numbers 1, 4, 9, 16, 25, and so on are square numbers.

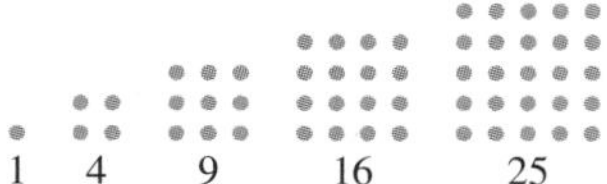

a) Determine the next three square numbers.
b) Describe a procedure to determine the next five square numbers without drawing the figures.
c) Is 72 a square number? Explain how you determined your answer.

32. *A Triangular Pattern* The ancient Greeks labeled certain numbers as **triangular numbers**. The numbers 1, 3, 6, 10, 15, 21, and so on are trianglular numbers.

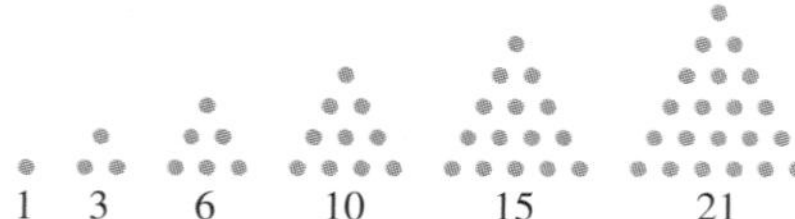

a) Can you determine the next two triangular numbers?
b) Describe a procedure to determine the next five triangular numbers without drawing the figures.
c) Is 72 a triangular number? Explain how you determined your answer.

33. *Quilt Design* The pattern shown above on the right is taken from a quilt design known as a triple Irish chain. Complete the color pattern by indicating the color assigned to each square.

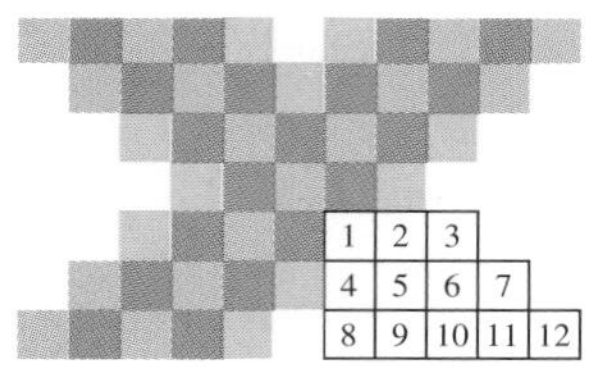

34. *Triangles in a Triangle* Four rows of a triangular figure are shown.

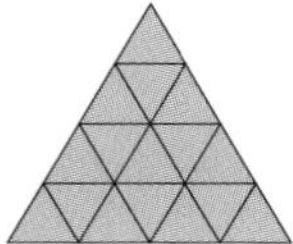

a) If you added six additional rows to the bottom of this triangle, using the same pattern displayed, how many triangles would appear in the 10th row?
b) If the triangles in all 10 rows were added, how many triangles would appear in the entire figure?

35. *Satellite vs. Cable*
a) The graph shows the past and expected future trend in the number of satellite and cable subscribers from 1994 to 2005. If you had to make a prediction for the number of cable subscribers in the year 2010, what would you predict?
b) Using the graph, predict the number of satellite subscribers in 2010.
c) Explain how you are using inductive reasoning in determining your answer.

Satellite TV's Big Gains

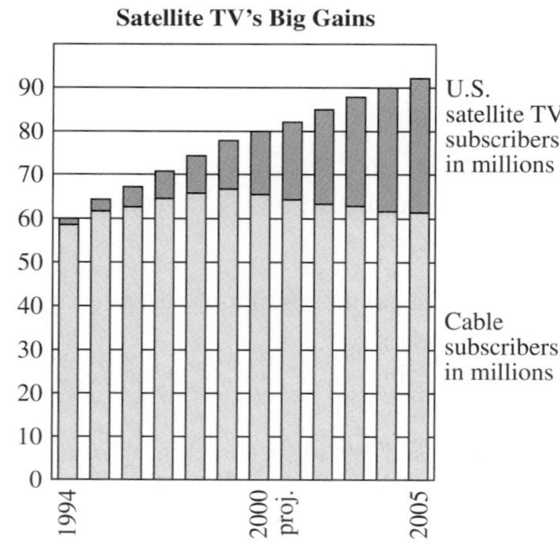

Source: *Fortune*, Feb. 5, 2001

36. *Income and Tuition* The graph on the top of the next page shows the cost of a year's tuition at the average public four-year college, average private four-year college, and the median family income from 1990 to 2000.

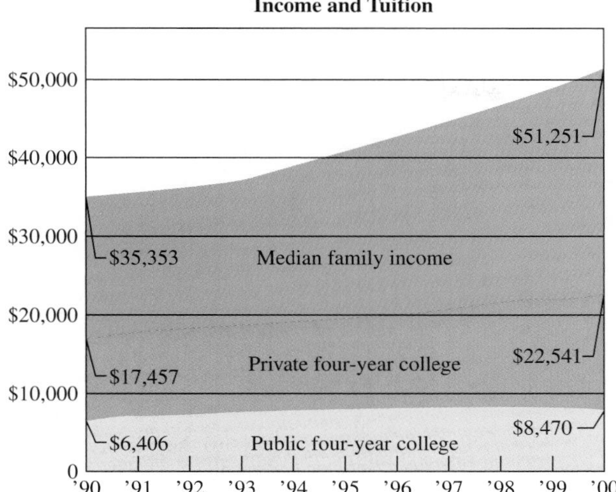

Note: College costs include tuition, room, and board
Source: Census Bureau, College Board
U.S. News and World Report, Oct. 30, 2000

a) Using the graph, make a prediction of the cost of tuition at the average private four-year college in 2005.
b) Using the graph, predict the median family income in 2005.
c) Explain how you are using inductive reasoning in determining your answer.

In Exercises 37 and 38, draw the next diagram in the pattern (or sequence).

37. 38.

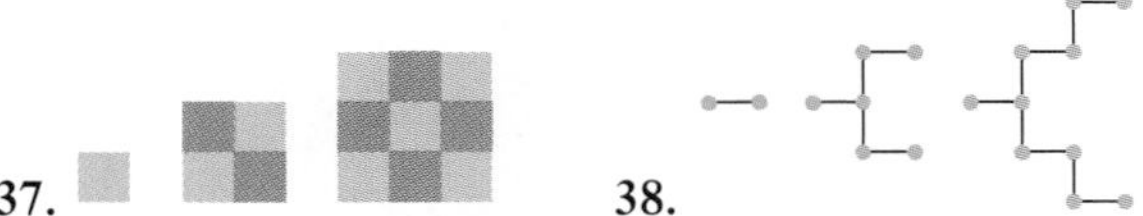

39. Pick any number, multiply the number by 4, add 8 to the product, divide the sum by 4, and subtract 2 from the quotient. See Example 5.
   a) What is the relationship between the number you started with and the final number?
   b) Arbitrarily select some different numbers and repeat the process, recording the original number and the result.
   c) Can you make a conjecture about the relationship between the original number and the final number?
   d) Try to prove, using deductive reasoning, the conjecture you made in part (c). See Example 6.

40. Pick any number and multiply the number by 10. Add 5 to the product. Divide the sum by 5 and subtract 1 from the quotient.
   a) What is the relationship between the number you started with and the final answer?
   b) Arbitrarily select some different numbers and repeat the process, recording the original number and the results.
   c) Can you make a conjecture about the relationship between the original number and the final number?
   d) Try to prove, using deductive reasoning, the conjecture you made in part (c).

41. Pick any number and add 1 to it. Find the sum of the new number and the original number. Add 9 to the sum. Divide the sum by 2 and subtract the original number from the quotient.
   a) What is the final number?
   b) Arbitrarily select some different numbers and repeat the process. Record the results.
   c) Can you make a conjecture about the final number?
   d) Try to prove, using deductive reasoning, the conjecture you made in part (c).

42. Pick any number and add 10 to the number. Divide the sum by 5. Multiply the quotient by 5. Subtract 10 from the product. Then subtract your original number.
   a) What is the result?
   b) Arbitrarily select some different numbers and repeat the process, recording the original number and the result.
   c) Can you make a conjecture regarding the result when this process is followed?
   d) Try to prove, using deductive reasoning, the conjecture you made in part (c).

In Exercises 43–48, find a counterexample to show that each of the statements is incorrect.

43. The product of 2 three-digit numbers is a five-digit number.
44. The sum of 3 two-digit numbers is a three-digit number.
45. When a counting number is added to 3 and the sum is divided by 2, the quotient will be an even number.
46. The product of any two counting numbers is divisible by 2.
47. The difference of any two counting numbers will be a counting number.
48. The sum of any two odd numbers is divisible by 4.

49. *Interior Angles of a Triangle*
   a) Construct a triangle and measure the three interior angles with a protractor. What is the sum of the measures?
   b) Construct three other triangles, measure the angles, and record the sums. Are your answers the same?
   c) Make a conjecture about the sum of the measures of the three interior angles of a triangle.

50. *Interior Angles of a Quadrilateral*
   a) Construct a quadrilateral (a four-sided figure) and measure the four interior angles with a protractor. What is the sum of their angle measures?
   b) Construct three other quadrilaterals, measure the angles, and record the sums. Are your answers the same?
   c) Make a conjecture about the sum of the measures of the four interior angles of a quadrilateral.

Challenge Problems/Group Activities

51. Complete the following square of numbers. Explain how you determined your answer.

1	2	3	4
2	5	10	17
3	10	25	52
4	17	52	?

52. Find the next three numbers in the sequence.

1, 8, 11, 88, 101, 111, 181, 1001, 1111, . . .

Recreational Mathematics

53. Trace out a 13-letter word that has been highlighted in this section by moving along the lines. You will need to use all the letters, and you cannot retrace a line.

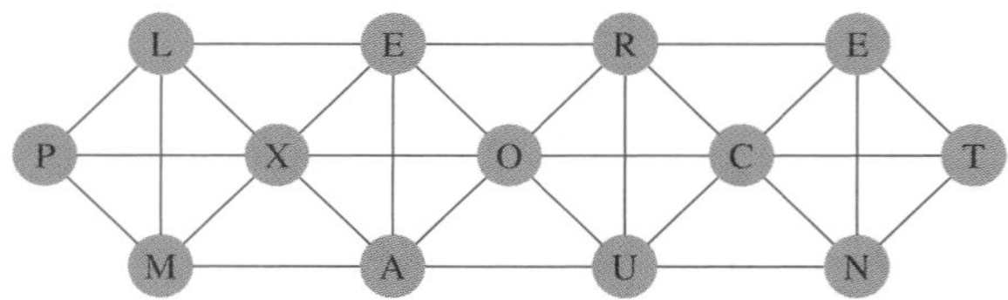

54.

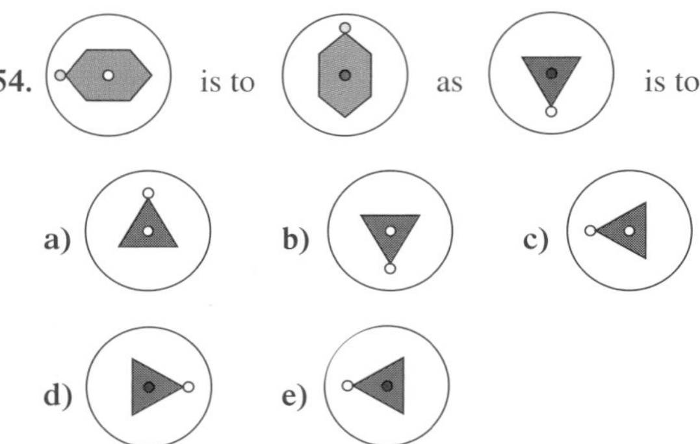

Internet/Research Activities

55. **a)** Using newspapers, the Internet, magazines, and other sources, find examples of conclusions arrived at by inductive reasoning.

b) Explain how inductive reasoning was used in arriving at the conclusion.

56. When a jury decides the guilt or innocence of a defendant, do the jurors collectively use primarily inductive reasoning, deductive reasoning, or an equal amount of each? Write a brief report supporting your answer.

1.2 ESTIMATION

An important step in solving mathematical problems—or, in fact, *any* problem—is to make sure that the answer you've arrived at makes sense. One technique for determining whether an answer is reasonable is to estimate. *Estimation* is the process of arriving at an approximate answer to a question. This section demonstrates several estimation methods.

To estimate, or approximate, an answer, we often round numbers as illustrated in the following examples. The symbol $\approx$ means *is approximately equal to.*

EXAMPLE 1 *Estimating the Cost of Doughnuts*

For a division meeting, Judy Wood decides to treat her colleagues to fresh doughnuts. Estimate her cost if she purchases 38 doughnuts at \$0.49 each.

SOLUTION: We may round the amounts as follows to obtain an estimate.

Number		Number rounded
38	→	40
× $0.49	→	× $0.50
		$20.00

Thus, the 38 doughnuts would cost approximately $20.00, written ≈ $20. ▲

In Example 1, the true cost is $ 0.49 × 38, or $18.62. *Estimates are not meant to give exact values for answers but are a means of determining whether your answer is reasonable.* If you calculated an answer of $18.62 and then did a quick estimate to check it, you would know that the answer is reasonable because it is close to your estimated answer.

EXAMPLE 2 *Two Ways to Estimate*

At a local supermarket, Kaitlyn purchased ice cream for $5.19, lettuce for $1.09, bread for $1.98, a frozen dinner for $4.79, ground beef for $4.26, steaks for $15.37, and a green onion for $0.92. The total bill was $44.08. Use estimation to determine whether this amount is reasonable.

SOLUTION: The most expensive item is $15.37, and the least expensive is $0.92. How should we estimate? We will estimate two different ways. First, we will round the cost of each item to the nearest 10 cents. For the second method, we will round the cost of each item to the nearest dollar. Rounding to the nearest 10 cents is more accurate. To determine whether the total bill is reasonable, however, we may need to round only to the nearest dollar.

	Round to the nearest 10 cents		Round to the nearest dollar	
Ice cream	$5.19 →	$5.20	$5.19 →	$5.00
Lettuce	1.09 →	1.10	1.09 →	1.00
Bread	1.98 →	2.00	1.98 →	2.00
Frozen dinner	4.79 →	4.80	4.79 →	5.00
Ground beef	4.26 →	4.30	4.26 →	4.00
Steaks	15.37 →	15.40	15.37 →	15.00
Onion	0.92 →	0.90	0.92 →	1.00
		$33.70		$33.00

Using either estimate, we find that the bill of $44.08 is quite high. Therefore, Kaitlyn should check the bill carefully before paying it. Adding the prices of all seven items gives the true cost of $33.60. ▲

EXAMPLE 3 *Select the Best Estimate*

The number of bushels of grapes produced at a vineyard are 71,309 Cabernet Sauvignon, 123,879 French Colombard, 106,490 Chenin Blanc, 5960 Charbono, and 12,104 Chardonnay. Select the best estimates of the total number of bushels produced by the vineyard.

a) 500,000 b) 30,000 c) 300,000 d) 5,000,000

SOLUTION: Following are suggested roundings. On the left, the numbers are rounded to thousands. For a less close estimate, round to ten thousands, as illustrated on the right.

Round to the nearest thousand		Round to the nearest ten thousand	
71,309 →	71,000	71,309 →	70,000
123,879 →	124,000	123,879 →	120,000
106,490 →	106,000	106,490 →	110,000
5,960 →	6,000	5,960 →	10,000
12,104 →	12,000	12,104 →	10,000
	319,000		320,000

Either rounding procedure indicates that the best estimate is (c), or 300,000. ▲

EXAMPLE 4 *Using Estimation in Calculations*

The odometer of an automobile reads 48,213.7 miles.

a) If the automobile averaged 22.1 miles per gallon (mpg) for that mileage, estimate the number of gallons of gasoline used.

b) If the cost of the gasoline averaged \$1.59 per gallon, estimate the total cost of the gasoline.

SOLUTION:

a) To estimate the number of gallons, divide the mileage by the number of miles per gallon.

$$\frac{48{,}213.7}{22.1}$$

Round these numbers to obtain an estimate.

$$\frac{50{,}000}{20} = 2500$$

Therefore, the car used approximately 2500 gallons (gal) of gasoline.

b) Rounding the price of the gasoline to \$1.60 per gallon gives the cost of the gasoline as 2500 × \$1.60, or \$4000. ▲

Now let's look at some different types of estimation problems.

EXAMPLE 5 *Estimating Distances on Route 66*

Route 66, also called America's Main Street, or the Mother Road, is probably the most famous road in the United States. Its 75th birthday was officially on November 11, 2001. About 85% of the road is still drivable through 8 states. A map of Route 66 in its heyday, before the opening of the Interstate bypasses, is shown on the next page.

a) Using Route 66, estimate the distance from Springfield, Illinois, to Tulsa, Oklahoma.

b) Using Route 66, estimate the distance from Chicago, Illinois, to Los Angeles, California.

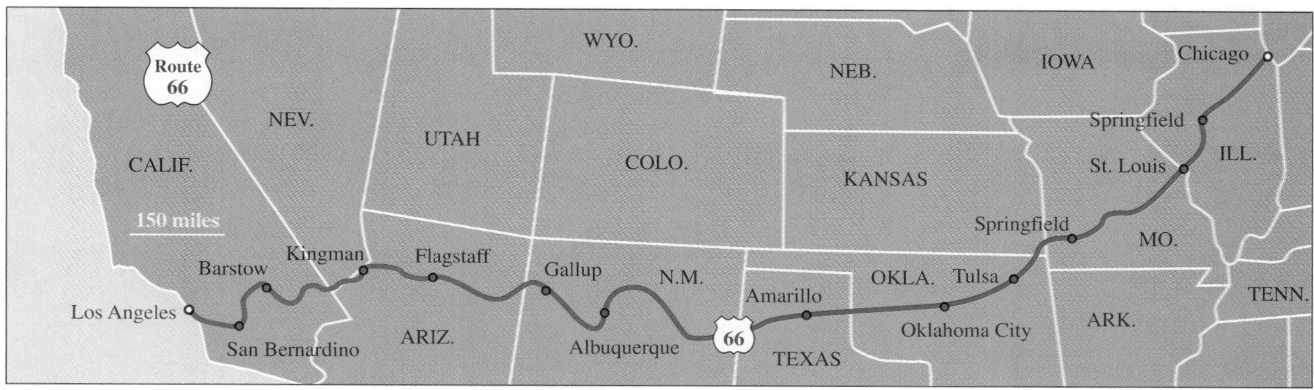

Source: Gannett News Service

SOLUTION:

a) Using a ruler and the scale given on the map, we can determine that approximately $\frac{1}{2}$ inch represents 150 miles (mi). One way to estimate the distance between Springfield, Illinois, and Tulsa, Oklahoma, is to mark off $\frac{1}{2}$-inch intervals along Route 66. If you do so you may obtain about three intervals. Thus, the distance is 3×150 mi, or about 450 mi.

Sometimes on a map like this one, it may be difficult to get an accurate estimate because of the curves on the map. To get a more accurate estimate, you may want to use a piece of string. Place the beginning of the string at Springfield, Illinois, and, using tape or pins, align the string with the road. Indicate on the string where the road ends. Then remove the string and make interval markings on the string (or measure the length of string you have marked off). If, for example, your string from Springfield, Illinois, to Tulsa, Oklahoma, measures $3\frac{1}{8}$ intervals, then the distance is about 3.125×150 mi, or about 469 mi.

b) Using the procedure discussed in part (a), we estimate that the distance from Chicago, Illinois, to Los Angeles, California is about 8 in., or about 16 half-inches. Thus, the distance of Route 66 is about 16×150, or about 2400 mi. ▲

DID YOU KNOW

Estimating Techniques in Medicine

Estimating is one of the diagnostic tools used by the medical profession. Physicians take a small sample of blood, tissue, or body fluids to be representative of the body as a whole. Human blood contains different types of white blood cells that fight infection. When a bacterium or virus gets into the blood, the body responds by producing more of the type of white blood cell whose job it is to destroy that particular invader. Thus, an increased level of white blood cells in a sample of blood not only indicates the presence of an infection but also helps identify its type. A trained medical technician estimates the relative number of each kind of white blood cell found in a count of 100 white blood cells. An increase in any one kind indicates the type of infection present. The accuracy of this diagnostic tool is impressive when you consider that there are normally 5000 to 9000 white blood cells in a dropful of blood (1 cubic millimeter).

Most analysis of blood samples done today is performed using computers.

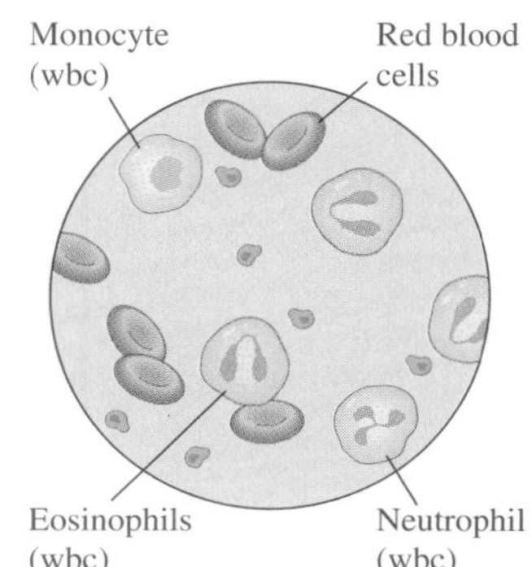

A high count of eosinophils (a particular kind of white blood cell, or wbc) can be an indicator of an allergic reaction.

DID YOU KNOW

Early Estimation

The ancient Egyptians and Greeks used sticks to estimate heights. To measure height, a person held a stick in his or her outstretched arm until it just covered the height of the object. The person then turned the stick through a right angle. Measuring the distance the stick appeared to cover on the ground provided an estimate of the object's height.

EXAMPLE 6 *Estimated Energy Use*

Utility bills sometimes contain graphs illustrating the amount of electricity and gas used. The following graphs show gas and electric use at a specific residence for a period of 13 months, starting in April 2002 and going through April 2003 (the month of the current bill). Also shown, in red, is the bill for the average residential customer for April 2003. Using these graphs, answer the following questions.

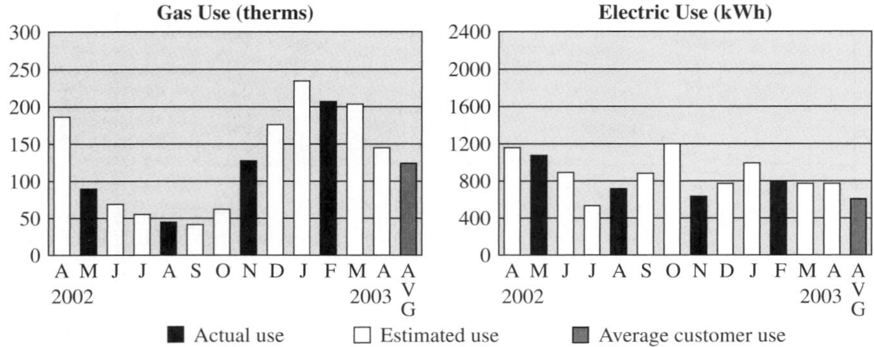

a) How often was an actual gas and electric reading made?
b) Estimate the number of therms of gas used by the average residential customer in April 2003.
c) Estimate the amount of gas used by the resident in April 2003.
d) If the cost of gas is 69.3672 cents per therm, estimate the gas bill in November 2002.
e) In which month was the most electricity used? How many kilowatt hours (kWh) were used in this month?
f) If the cost of electricity is 9.0983 cents per kilowatt hour, estimate the cost of electricity in February 2003.

SOLUTION:

a) Actual readings were made in only four months, May, August, November, and February.
b) Approximately 125 therms were used, as shown by the height of the red bar.
c) Approximately 145 therms were used (slightly less than 150).
d) In November, about 125 therms were used. The rate, 69.3672 cents per therm, is the same as \$0.693672 per therm. To get a rough approximation, round the rate to \$0.70 per therm.

$$0.70 \times 125 = 87.50^*$$

Thus, the cost of gas used was about \$88.

e) The most electricity was used in October. Approximately 1200 kWh were used.
f) In February, about 800 kWh were used. Write 9.0983 cents as \$0.090983. Rounding the rate to \$0.09 per kilowatt hour and multiplying by 800 yields an estimate of \$72.

$$0.09 \times 800 = \$72$$

Thus, the cost of electricity in February 2003 was about \$72. ▲

*The amounts here and in part (f) do not include the basic monthly charge, fuel adjustment, taxes, and other extra charges often included on utility bills.

EXAMPLE 7 *Estimating the Number of Birds in a Photo*

Scientists who are concerned about dwindling animal populations often use aerial photography to make estimates. Estimate the number of birds in the accompanying photograph.

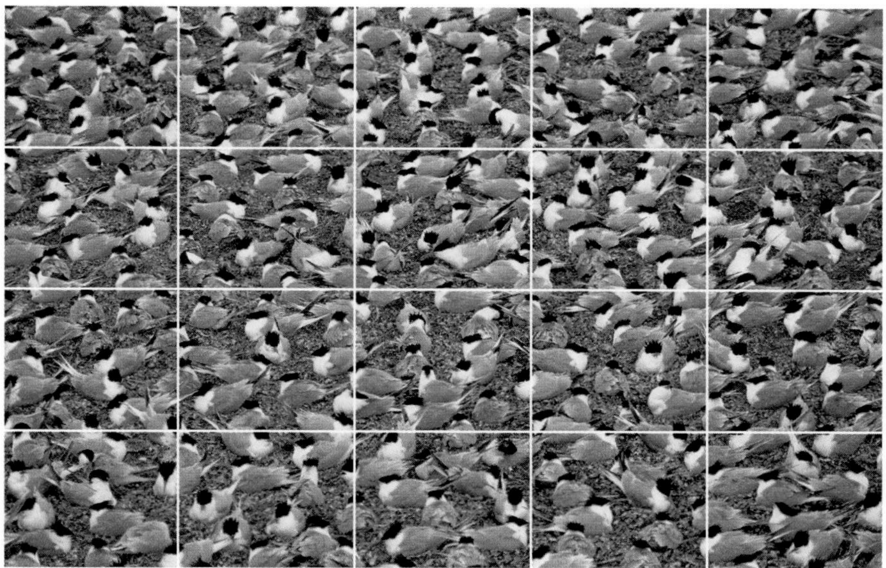

SOLUTION: To estimate the number of birds, we can divide the photograph into rectangles with equal areas, then select one area that appears to be representative of all the areas. Estimate (or count) the number of birds in this single area, and then multiply this number by the number of equal areas.

Let's divide the photo into 20 approximately equal areas. We will select the middle region in the bottom row as the representative region. We enlarge this region and count the birds in it. If half a bird is in the region, we count it (see enlargement). There are 13 birds in this region. Multiplying by 20 gives $13 \times 20 = 260$. Thus, there are about 260 birds in the photo.

▲

In problems similar to that in Example 7, the number of regions or areas into which you choose to divide the total area is arbitrary. Generally, the more regions, the better the approximation, as long as the region selected is representative of the other regions in the map, diagram, or photo.

When you estimate an answer, the amount that your approximation differs from the actual answer will depend on how you round the numbers. Thus, in estimating the product of $196{,}000 \times 0.02520$, using the rounded values $195{,}000 \times 0.025$ would yield an estimate much closer to the true answer than using the rounded values

$200{,}000 \times 0.03$. Without a calculator, however, the product of $195{,}000 \times 0.025$ might be more difficult to find than $200{,}000 \times 0.03$. When estimating, you need to determine the accuracy desired in your estimate and round the numbers accordingly.

SECTION 1.2 EXERCISES

Practice the Skills

In Exercises 1–57, your answers may vary from the answers given in the back of the text, depending on how you round your answers.

In Exercises 1–12, estimate the answer. There is no one correct estimate. Your answer, however, should be something near the answer given.

1. $431 + 327.2 + 73.5 + 20.4 + 315.9$

2. $3.89 + 402.8 + 156.9 + 189 + 0.23 + 416$

3. $297{,}700 \times 4087$

4. 1854×0.0096

5. $\frac{405}{0.049}$

6. $297.521 - 85.964$

7. 0.049×1989

8. 9% of 2164

9. $51{,}608 \times 6981$

10. $\frac{0.0498}{0.00052}$

11. $592 \times 2070 \times 992.62$

12. $296.3 \div 0.0096$

Problem Solving

In Exercises 13–24, estimate the answer.

13. The cost of 52 thirty-seven-cent stamps.

14. The income earned for 32 hours at $7.95 per hour.

15. The total frequent flier mileage awarded for four trips of 1521, 1897, 2324, and 2817 miles.

16. The weight of six buckets of water if each bucket weighs 15.87 pounds (lb).

17. The cost of eight items purchased at a drugstore if the items cost $2.29, $12.16, $4.97, $6.69, $49.76, $0.47, $3.49, and $5.65.

18. The weight of one hamburger in a package of six hamburgers if the weight of the package is 3.12 lb.

19. One fifth of an annual profit of $44,569.

20. The weight of the load of an 18-wheel truck if the weight of the truck when empty is 14,292 pounds and the weight of the truck when loaded is 32,798 lb.

21. The weight of nine identical packages of paper if each package weighs 5.12 lb.

22. An 8% sales tax on a car that cost $14,876 before tax.

23. The average monthly distance traveled if Paul travels 23,663 miles in 1 year.

24. The cost of a pound of grapes if 3.2 lb of grapes cost $10.87.

25. *Time Warner Bill* Dale Gray has both his cable TV and Road Runner internet service through Time Warner Communications. His cable TV costs $29.17 per month and his Road Runner service costs $39.95 per month. Both charges appear on the same monthly bill. Estimate the annual amount Dale pays to Time Warner.

26. *Estimating Weights* In a tug of war, the weight of the members of the two three-person teams is given below. Estimate the difference in the weights of the teams.

Team A	Team B
189	183
172	229
191	167

27. *Estimating the Tip* Ed and Dorothy Ruff go out for dinner and spend $38.60 for their meal. If they want to leave a 15% tip, estimate the amount that they should leave.

28. *Estimating Area* Mrs. Sanchez determines that her lawn contains an average of 3.8 grubs per square foot (ft^2). If her rectangular lawn measures 60 ft by 80.2 ft, estimate the total number of grubs in her lawn.

29. *Currency* Estimate the difference in the value of 100 Mexican pesos and 50 U.S. dollars. Assume that one Mexican peso is about 0.092 U.S. dollar.

30. *The Cost of a Vacation* The Kleins are planning a vacation in the Great Smokey Mountain National Park. Their round-trip airfare from Houston, Texas, to Knoxville, Tennessee, totals $973. Car rental is $41 per day, lodging is a total of $97 per day, and they estimate a total of $90 per day for food, gas, and other miscellaneous items. If they are planning to stay six full days and nights, estimate their total expenses.

For Exercises 31 and 32, refer to the maps below. In the April 23, 2001, issue of U.S. News and World Report, *an article by Marc Silver discusses great vacation drives. The maps that follow were selected from that article.*

31. *A Drive in Mississippi* Using the scale on the map, estimate the distance of the route shown in red starting and ending in Oxford.

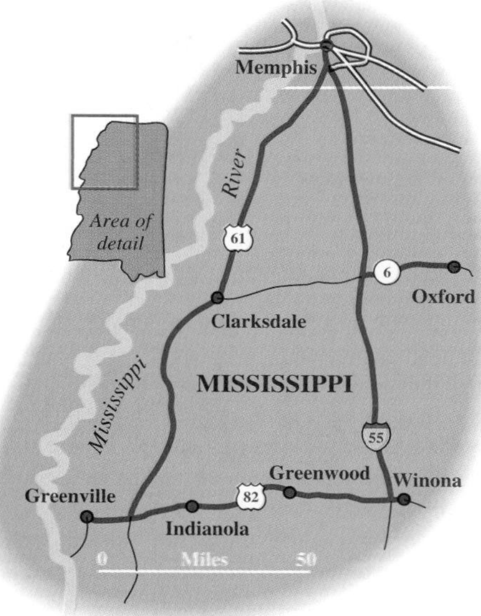

32. *A Drive in Pennsylvania* Using the scale on the map, estimate the distance of the route shown in red by starting and ending in Chadds Ford.

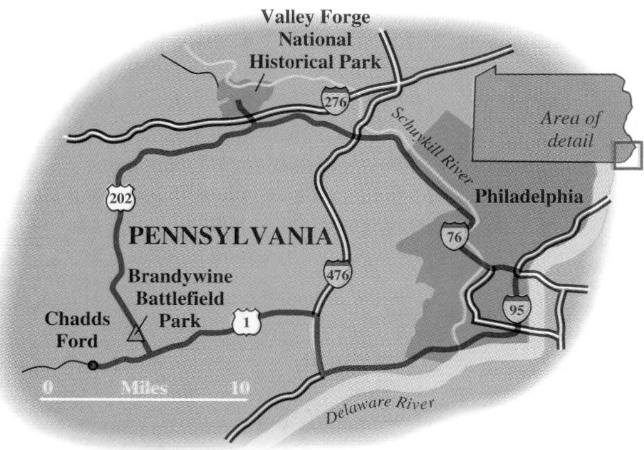

33. *Voting* The circle graphs above and to the right show the methods that Americans used to vote as of November 2000, according to Election Data Services, Inc. There were approximately 105 million Americans from all 3141 counties who voted in the 2000 presidential election.

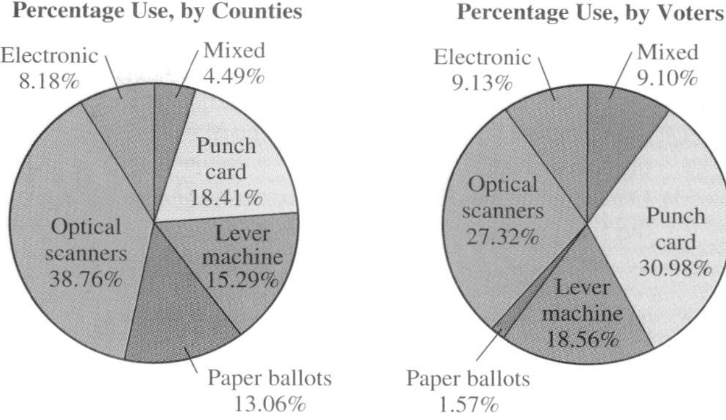

Source: Election Data Services Inc.

a) Using the graphs, estimate the number of people who voted using punch cards.

b) Using the graphs, estimate the number of counties that voted using punch cards.

c) The percent of voters who use punch cards is greater than the percent of counties that use punch cards. Can you offer any suggestions as to how this is possible?

34. *Retirement* The circle graphs show sources of retirement income for retirees with at least $20,000 in annual income for the years 1999 and 2002. Assume that Sheila Abbruzzo had a retirement income of $40,075 from the various sources shown in the figures. Estimate her income from investments/savings in

a) 1999. b) 2002.

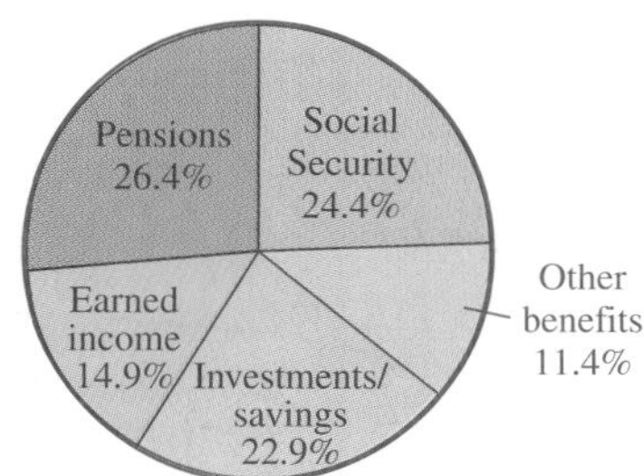

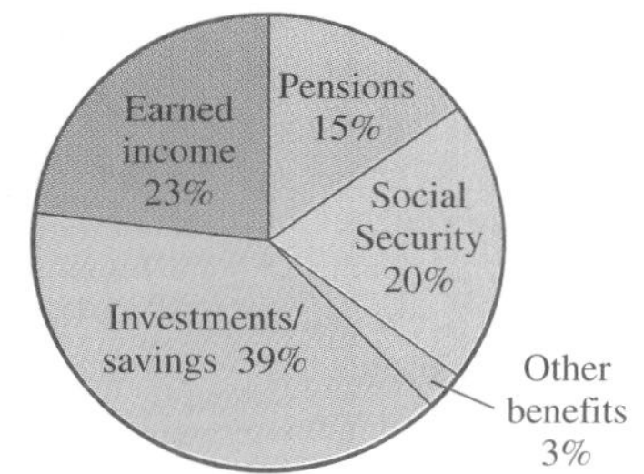

Source: Social Security Administration

35. *An Aging Population* The bar graph shows population figures for 1900 and 2000 and estimated population figures for 2050.
   a) Estimate the number of people 65 and over in 1900.
   b) Estimate the number of people 65 and older in 2050.
   c) Estimate the increase in the number of people 65 and older from 2000 to 2050.
   d) Estimate the total U.S. population in 2000 by adding the five categories.

Population by Age

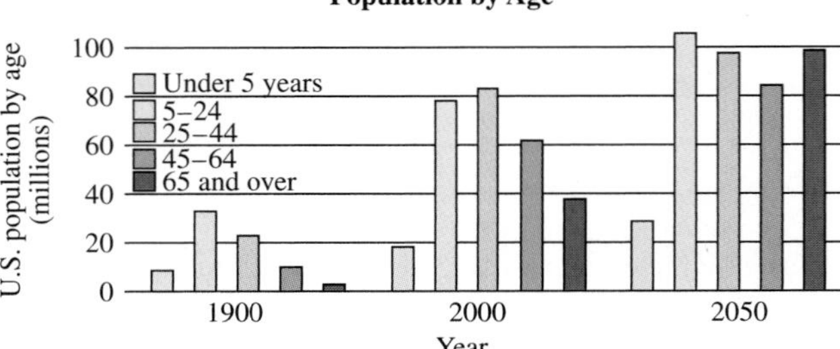

Source: U.S. Census Bureau

36. *Gaining Weight* As the graph shows, as a society we tend to get heavier as we get older. Also, with age, the amount of muscle tends to drop, and fat accounts for a greater percentage of weight.
   a) Estimate the average percent of body fat for a male, age 18 to 25.
   b) Estimate the average percent of body fat for a female, age 56+.
   c) Greg, an average 40-year-old, weighs 179 lb. Estimate the number of pounds of body fat he has.

Getting Older Usually Means Getting Fatter

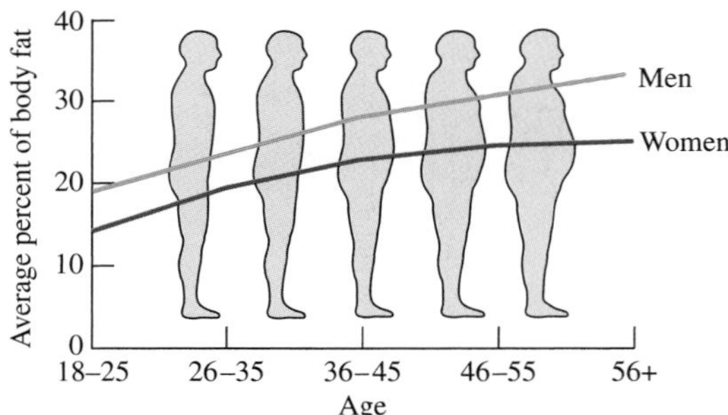

Source: Mayo Clinic Newsletter

37. *Land Owned by the Federal Government* The federal government owns a great deal of land in the United States. The following graph shows the percentage of land owned by the federal government in the 12 states in which it owns the greatest percentage of a state's land.

The Geography of Ownership

Percentage of Land Owned by the Federal Government

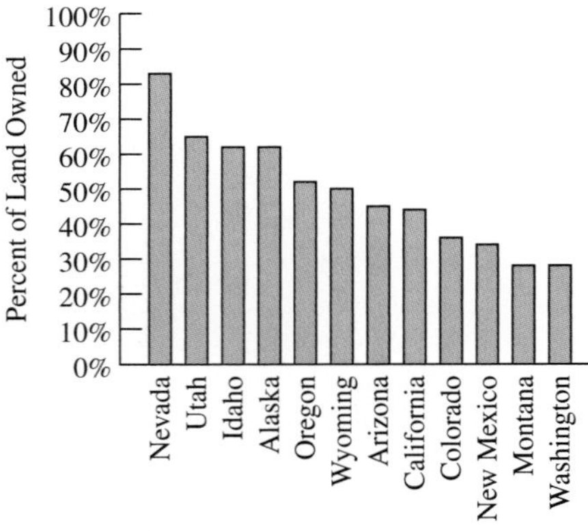

Source: *Time Magazine*, July 16, 2001

   a) Estimate the percent of land in Nevada owned by the federal government.
   b) Estimate the difference in the percent of land owned by the federal government in Utah and in Arizona.
   c) Nevada has a total area of 110,567 square miles. Estimate the number of acres owned by the federal government in Nevada.
   d) By just looking at the graph, is it possible to determine whether the federal government owns more land in Nevada or Washington? Explain.

38. *Calories and Exercise* The chart shows the calories burned per hour for an average person who weighs 150 lb.
   a) Estimate the number of calories Phyllis Nye, who weighs 150 lb, burns in a week if she stair-climbs for 2 hours each week and jogs at 5 miles per hour 4 hours each week.
   b) Estimate the difference in the calories Phyllis will burn each week if she runs for 4 hours at 8 miles per hour rather than does casual bike riding for 4 hours.
   c) Assume Phyllis jogs at 5 miles per hour for 3 hours and bicycles at 13 miles per hour for 3 hours each week. Estimate the number of calories she will burn in a year from these exercises.

Activity	Calories* per hour
Running, 8 mph	920
Bicycling, 13 mph	545
Jogging, 5 mph	545
Air-walking	480
Stair-climbing	410
Weight-lifting	410
Walking, 4 mph	330
Casual bike riding	300

*For a 150 lb person.

In Exercises 39 and 40, estimate the maximum number of smaller figures (at left) that can be placed in the larger figure (at right) without the small figures overlapping.

39.

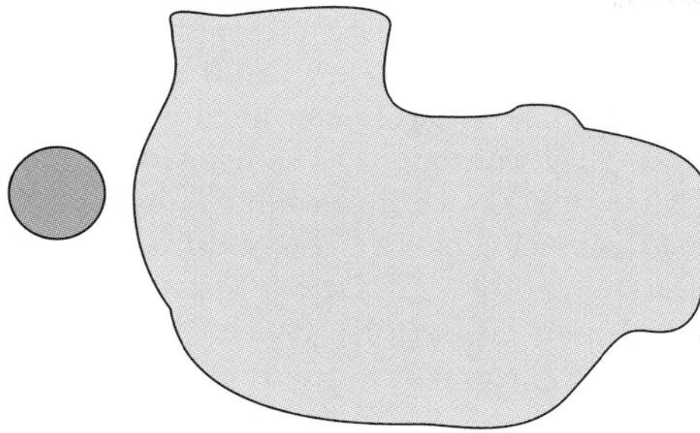

40.

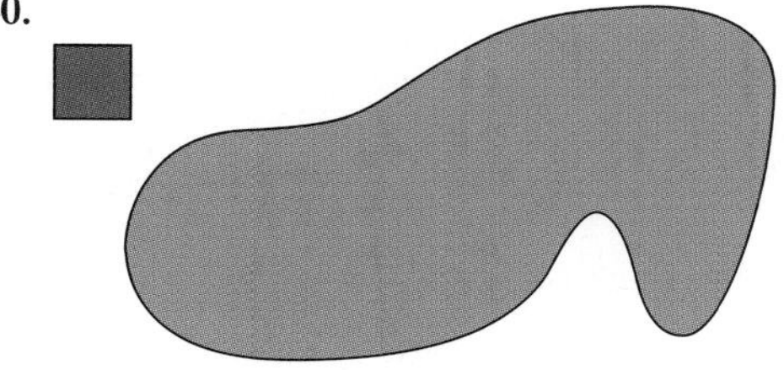

41. Estimate the number of berries shown in the photo.

42. Estimate the number of leaves shown in the photo.

In Exercises 43 and 44, estimate, in degrees, the measure of the angles depicted. For comparison purposes a right angle, ⦜, measures 90°.

43.

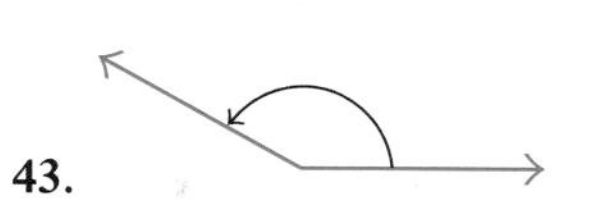

44.

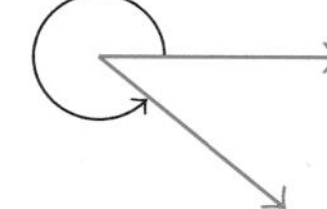

In Exercises 45 and 46, estimate the percent of area that is shaded in the following figures.

45.

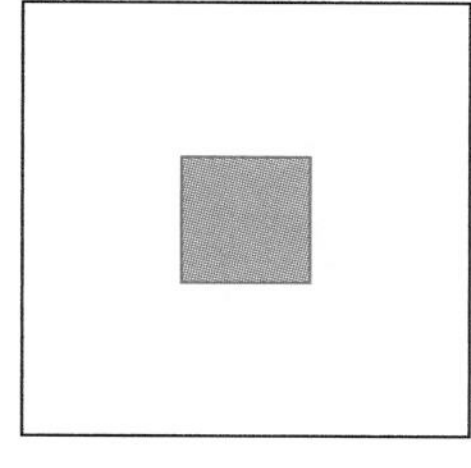

46.

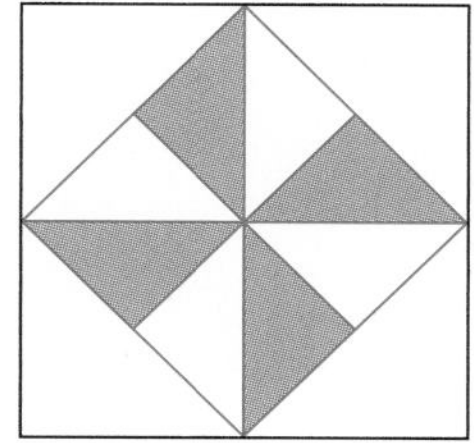

In Exercises 47 and 48, if each square represents one square unit, estimate the area of the shaded figure in square units.

47.

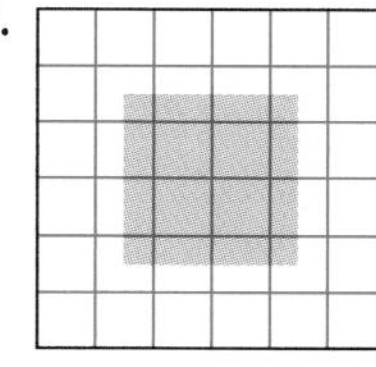

48.

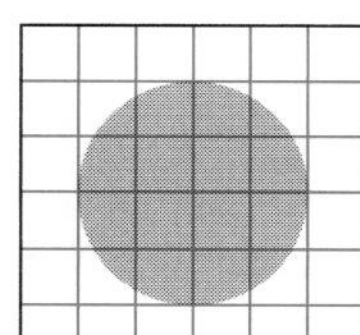

49. *Statue of Liberty* The length of the torch of the Statue of Liberty, from the tip of the flame to the bottom of the torch's baton, is 29 feet. Estimate the height of the Statue of Liberty from the top of the base to the top of the torch (the statue itself).

50. *Estimating Heights* If the height of the middle person in the photo is 62 in. tall, estimate the height of the tree.

51. *Weight* In a bag, place objects that you feel have a total weight of 10 lb. Weigh the bag to determine the accuracy of your estimate.

52. *Distance* Estimate, without a ruler, a distance of 12 in. Measure the distance. How good was your estimate?

53. *Temperature* Fill a glass with water and estimate the water's temperature. Then use a thermometer to measure the temperature and check your estimate.

54. *Phone Call* Estimate the number of times the phone will ring in 1 minute if unanswered. Have a classmate phone you so that you can count the rings and thus test your estimate.

55. *Pennies* Estimate the number of pennies that will fill a 3-ounce (oz) paper cup. Then actually fill a 3 oz paper cup with pennies, counting them to determine the accuracy of your estimate.

56. *Height* Estimate the ratio of your height to your neck size. Then have a friend measure your height and neck size. Determine the stated ratio and check the accuracy of your estimate.

57. *Walking Speed* Estimate how fast you can walk 60 ft. Then mark off a distance of 60 ft and use a watch with a second hand to time yourself walking it. Determine the accuracy of your estimate.

Challenge Problems/Group Activities

58. *Shopping* Make a shopping list of 20 items you use regularly that can be purchased at a supermarket. Beside each item write down what you estimate to be its price. Add these price guesses to estimate the total cost of the 20 items. Next, make a trip to your local supermarket and record the actual price of each item. Add these prices to determine the actual total cost. How close was your estimate? (Don't forget to add tax on the taxable items.)

59. *A Ski Vacation* Two friends, Tiffany Connolly and Ana Pott, are planning a skiing vacation in the Rockies. They plan to purchase round-trip airline tickets from Atlanta, Georgia, to Denver, Colorado. They will fly into Denver on a Friday morning, rent a midsize car, and drive to Aspen that same day. They will stay at the Holiday Inn in Aspen. They will begin skiing at the Buttermilk Ski Area on Saturday, ski up to and including Wednesday, drive back to Denver on Thursday, and fly out of Denver Thursday evening.

a) Estimate the total cost of the vacation for the two friends. Do not forget items such as food, tips, gas, and other incidentals.

b) Using informational sources, including the Internet, determine the airfare cost, hotel cost, cost of ski tickets, cost of a car rental, and so forth. You will need to make an estimate for food and other incidentals.

c) How close was your estimate in part (a) to the amount you found in part (b)? Was your estimate in part (a) lower or higher than the amount obtained in part (b)?

Recreational Mathematics

60. *A Dime* Look at a dime. Around the edge of a dime are many lines. Estimate the number of lines there are around the edge of a dime.

61. *Golf Ball* Look at a golf ball. Estimate the number of dimples (depressed areas) on a golf ball.

62. *A Million Dollars*

a) Estimate the time it would take, in days, to spend \$1 million if you spent \$1 a second until the \$1 million is used up.

b) Calculate the actual time it would take, in days, to spend \$1 million if you spent \$1 a second. How close was your estimate?

63. *The Middle Class* An article in the February 5, 2001, *U.S. News and World Report* indicated that many Americans believe that they are in the middle class (in terms of annual household income; see the chart), although according to U.S. government guidelines they are not. Estimate the annual household income range that you believe is classified as middle class according to the U.S. government.

Middle-Class Perception
By income, households that call themselves middle class

Less than \$15,000	**37%**
\$15,000–\$35,000	**29%**
\$35,000–\$50,000	**42%**
\$50,000–\$75,000	**57%**
\$75,000–\$110,000	**73%**
More than \$110,000	**70%**

Internet/Research Activities

64. *Water Usage*
   a) About how much water does your household use per day? Use the following data to estimate your household's daily water usage.

How much water do you use?

Activity	Typical Use
Running clothes washer	40 gal
Bath	35 gal
5-minute shower	25 gal
Doing dishes in sink, water running	20 gal
Running dishwasher	11 gal
Flushing toilet	4 gal
Brushing teeth, water running	2 gal

Source: U.S. Environmental Protection Agency

   b) Determine from your water department (or company) your household's average daily usage by obtaining the total number of gallons used per year and dividing that amount by 365. How close was your estimate in part (a)?
   c) Current records indicate that the average household uses about 300 gal of water per day (the average daily usage is 110 gal per person). Based on the number of people in your household, do you believe your household uses more or less than the average amount of water? Explain your answer.

65. Develop a monthly budget by estimating your monthly income and your monthly expenditures. Your monthly income should equal your monthly expenditures.

66. Identify three ways that you use estimation in your daily life. Discuss each of them briefly and give examples.

1.3 PROBLEM SOLVING

Solving mathematical puzzles and real-life mathematical problems can be enjoyable. You should work as many exercises in this section as possible. By doing so, you will sample a variety of problem-solving techniques.

You can approach any problem by using a general procedure developed by George Polya. Before learning Polya's problem-solving procedure, let's consider an example.

EXAMPLE 1 *Saving Money When Purchasing Videotapes*

Businesses, to maximize their profits, try to keep their expenses down. We, as individuals, also try to keep our expenses down and we often look for "bargains" or the "best deal."

A video store owned by Roberto Santiago plans to purchase a large number of blank videotapes. One supplier, the Tashanna Miller Company, is selling boxes of 20 tapes for $48 and boxes of 12 tapes for $30. Only complete boxes of tapes are sold.

a) Find the maximum number of tapes that can be purchased for $280 or less. Indicate how many boxes of 20 and how many boxes of 12 will be purchased.
b) If the maximum number of tapes determined in part (a) is purchased in the most economical way, how much will the tapes cost?

SOLUTION:

a) The first thing to do is to read the problem carefully. Read it at least twice and be sure you understand the facts given and what you are being asked to find. Next, make a list of the given facts and determine which are relevant to answering the question asked.

Given information

Store owner: Roberto Santiago

Supplier: Tashanna Miller Company

A box of 20 tapes costs $48.

A box of 12 tapes costs $30.

Only complete boxes of tapes can be purchased.

We need to determine the maximum number of tapes that the video store can purchase for $280 or less. To determine this number, we need to know the number of tapes in each of the boxes and the cost of the boxes. We also need to know that only complete boxes of tapes may be purchased.

Relevant information

A box of 20 tapes costs $48.

A box of 12 tapes costs $30.

Only complete boxes of tapes may be purchased.

The next step is to determine the answer to the question. That is, we need to determine the maximum number of tapes that can be purchased for $280 or less.

We now need a plan for solving the problem. One method is to set up a table or chart to compare costs of different combinations of boxes of tapes. Start by using the maximum number of boxes of 20 tapes. Then reduce the number of boxes of 20 tapes, and add more boxes of 12 tapes. In each case, we need to keep the cost at $280 or less.

Since 1 box of 20 tapes costs $48, we can determine the number of boxes of 20 tapes that can be purchased by dividing 280 by 48. Since the quotient is about 5.83, and since only whole boxes of tapes may be purchased, only 5 boxes of 20 may be purchased. Five boxes would cost $5 \times 48 = 240$. The remaining $40 from the $280 could be used to purchase boxes of 12 tapes. Since each box of 12 tapes costs $30, only one box of 12 tapes could be purchased. Thus, for $280 or less, one option is 5 boxes of 20 tapes and 1 box of 12 tapes. This option is indicated in the first row of the table below. Also given in the table is the cost of this option, which is $270. We complete the other rows of the table in a similar manner.

Boxes of 20 and Boxes of 12 Tapes	Number of Tapes	Cost
5 boxes of 20 and 1 box of 12	$(5 \times 20) + (1 \times 12) = 112$	$270
4 boxes of 20 and 2 boxes of 12	$(4 \times 20) + (2 \times 12) = 104$	$252
3 boxes of 20 and 4 boxes of 12	$(3 \times 20) + (4 \times 12) = 108$	$264
2 boxes of 20 and 6 boxes of 12	$(2 \times 20) + (6 \times 12) = 112$	$276
1 box of 20 and 7 boxes of 12	$(1 \times 20) + (7 \times 12) = 104$	$258
0 boxes of 20 and 9 boxes of 12	$(0 \times 20) + (9 \times 12) = 108$	$270

The question asks us to find the maximum number of tapes that can be purchased for $280 or less. From the second column of the table, we see that the answer is 112 tapes. This result can be done in two different ways: either 5 boxes of 20 and 1 box of 12 tapes, or 2 boxes of 20 and 6 boxes of 12 tapes.

PROFILE IN MATHEMATICS

GEORGE POLYA

George Polya (1887–1985) was educated in Europe and taught at Stanford University. In his book *How to Solve It,* Polya outlines four steps in problem solving. We will use Polya's four steps as guidelines for problem solving.

b) When comparing the two possibilities for purchasing the 112 tapes discussed in part (a), we see that the most economical way to purchase the tapes is to purchase 5 boxes of 20 and 1 box of 12 tapes. The cost is \$270. ▲

Following is a general procedure for problem solving as given by George Polya. Note that Example 1 demonstrates many of these guidelines.

Guidelines for Problem Solving

1. *Understand the problem.*
   - Read the problem *carefully* at least twice. In the first reading, get a general overview of the problem. In the second reading, determine (a) exactly what you are being asked to find and (b) what information the problem provides.
   - Try to make a sketch to illustrate the problem. Label the information given.
   - Make a list of the given facts that are pertinent to the problem.
   - Determine if the information you are given is sufficient to solve the problem.
2. *Devise a plan to solve the problem.*
   - Have you seen the problem or a similar problem before? Are the procedures you used to solve the similar problem applicable to the new problem?
   - Can you express the problem in terms of an algebraic equation? (We explain how to write algebraic equations in Chapter 6.)
   - Look for patterns or relationships in the problem that may help in solving it.
   - Can you express the problem more simply?
   - Can you substitute smaller or simpler numbers to make the problem more understandable?
   - Will listing the information in a table help in solving the problem?
   - Can you make an educated guess at the solution? Sometimes if you know an approximate solution, you can work backward and eventually determine the correct procedure to solve the problem.
3. *Carry out the plan.*
   Use the plan you devised in step 2 to solve the problem.
4. *Check the results.*
   - Ask yourself, "Does the answer make sense?" and "Is the answer reasonable?" If the answer is not reasonable, recheck your method for solving the problem and your calculations.
   - Can you check the solution using the original statement?
   - Is there an alternative method to arrive at the same conclusion?
   - Can the results of this problem be used to solve other problems?

The following examples show how to apply the guidelines for problem solving.

EXAMPLE 2 *Hotel Cost*

At the Courtyard by Marriot Hotel in Irving, Texas (near the Dallas/Fort Worth airport), the room rate is \$139 per day on weekdays and \$55 per day on weekends (Saturday and Sunday). In addition, a 13% sales tax is added to the cost of a room. In-room movies cost \$8.95 plus the 13% sales tax. Robin Ayers stays on the third floor of the Courtyard for four nights (Wednesday, Thursday, Friday, and Saturday) and watches three movies. Determine her hotel bill when she checks out.

DID YOU KNOW

Archimedes Treatise, On Floating Bodies

On October 14, 2000, the Associated Press reported that scientists are using problem-solving techniques and modern technology to unlock ancient secrets. Two teams of scientists, one from Johns Hopkins University and the other from the Rochester Institute of Technology, have deciphered five pages of the only known copy of a 2300-year-old Greek text by the mathematician Archimedes. The scientists hope to complete a translation of this only original copy of the 174-page treatise, *On Floating Bodies*, in the not too distant future. Scholars think the treatise was copied by a scribe in the 10th century from Archimedes' original Greek scrolls, written in the third century B.C. The treatise was erased 200 years later by a monk who reused the parchment for a prayer book, creating a twice-used parchment book.

The scientists are using digital cameras and processing techniques, as well as ultraviolet and infrared filters developed for medicine and space research, to reveal the hidden text. The text and diagrams also contain the roots of modern calculus and gravitational theory.

SOLUTION: We need to find the total cost of the hotel bill. Let's make a list of the information given and mark with an asterisk (*) the information that is pertinent to solving the problem.

*Cost of room per day on Wednesday, Thursday, and Friday = $139 + sales tax
*Cost of room on Saturday = $55 + sales tax
*Days at hotel: one Wednesday, one Thursday, one Friday, one Saturday
*Sales tax is 13%
*Cost per movie = $8.95 + sales tax
*Movies watched: 3
Room on third floor

All the information is needed to solve the problem except for the floor on which Robin stayed.

Let's first determine the cost of the room, before tax, for the four days.

Day	Cost (before tax)
Wednesday	$139
Thursday	$139
Friday	$139
Saturday	$55
	$472

The cost of 3 movies before tax is $3 \times \$8.95 = \$26.85$. Thus, the total amount of the bill before tax is $\$472 + \$26.85 = \$498.85$. Let's determine the sales tax to be added. We could find the sales tax for each individual item and add it to the items separately. However, since the same tax rate applies to each item, it is easier to just determine 13% of the total amount.

$$\begin{aligned}\text{Sales tax} &= 13\% \text{ of } \$498.85\\ &= 0.13(498.85) = \$64.85\end{aligned}$$

The total hotel bill is determined by adding the tax to the pretax amount. Thus, the total hotel bill is $\$498.85 + \$64.85 = \$563.70$.

In Example 2, there are other ways that the total cost could have been determined. We presented the method we believe you would best understand.

EXAMPLE 3 *More Food, Less Water*

We see graphs every day in newpapers and magazines. It is important to you as a consumer to be able to interpret the graphs. The following graph, which shows the top 10 irrigators worldwide and total cropland worldwide, was taken from the February 2001 issue of *Scientific American*. The article mentions that 6000 years ago, farmers in Mesopotamia dug a ditch to divert water from the Euphrates River to form the world's first irrigation-based civilization. Far more people today depend on irrigation than they did in Mesopotamia. Today, about 40% of the world's food grows in irrigated soil. The world's total irrigated cropland is about 670 million acres, and the world's total cropland is about 3722 million acres.

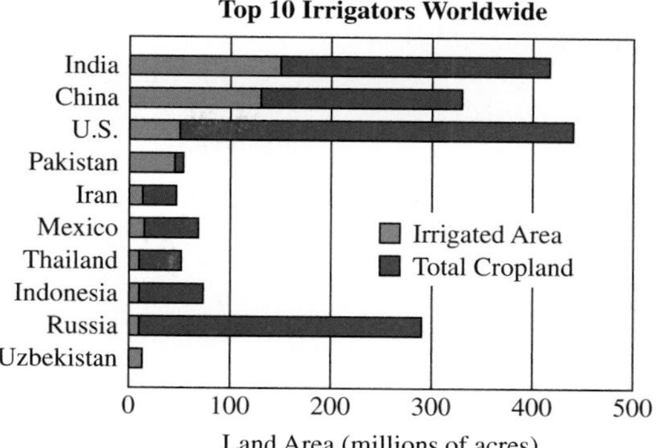

a) Approximately what percent of the world's total irrigated cropland is found in the top four countries indicated on the graph?

b) Determine the total percent of the area of the world's cropland that is irrigated.

c) Estimate the difference in total cropland between the United States and China.

SOLUTION: A great deal of information is provided. The first thing to do is to read the question carefully to make sure you understand what is given and what you are asked to determine. You may need to use different information to answer different parts of the question.

a) To determine the percent we are seeking, we need to divide the sum of the irrigated areas of the top four countries by the world's total irrigated area. From the graph, we determine that the top four countries for irrigated acreage are India, China, United States, and Pakistan. We are given that the world's total irrigated cropland is 670 million acres. Below we estimate the irrigated acreage of the top four countries, in millions of acres.

$$\begin{aligned} \text{India} &\approx 150 \\ \text{China} &\approx 130 \\ \text{United States} &\approx 50 \\ \text{Pakistan} &\approx 45 \\ \hline &\approx 375 \end{aligned}$$

The total acreage of irrigated cropland of these four countries is about 375 million acres.

$$\begin{aligned} \text{Percent of irrigated areas of top four countries} &= \frac{\text{sum of irrigated areas of four countries}}{\text{total world area of irrigated land}} \\ &\approx \frac{375}{670} \approx 0.560 \approx 56\%^{*} \end{aligned}$$

Thus, these four countries account for about 56% of the world's total irrigated cropland.

b) Since not all countries are represented on the graph, we cannot use the graph to obtain our answer. The relevant information given is as follows.

The world's total irrigated cropland is about 670 million acres.

The world's total cropland is about 3722 million acres.

---

*If you have forgotten how to change a decimal number to a percent, see Section 11.1.

DID YOU KNOW

Two Medieval Riddles

A boatman must transport a wolf, a goat, and a bundle of herbs across a river. If the boat is so small he can carry only one of them at a time, how can he make the crossing so that the wolf will not devour the goat and the goat will not eat the herbs?

(Clue: The boatman must make multiple round trips.)

As I was going to St. Ives,
I met a man with seven wives;
Every wife had seven sacks,
Every sack had seven cats,
Every cat had seven kits.
Kits, cats, sacks, and wives,
How many were going to St. Ives?

To find the percent we are seeking, we use the following.

$$\text{Percent of world's cropland that is irrigated} = \frac{\text{total irrigated cropland area}}{\text{total cropland area}}$$

$$\approx \frac{670}{3722} \approx 0.18 \approx 18\%$$

Thus, about 18% of the world's cropland is irrigated.

c) From the graph we estimate the following.

U.S. total cropland is about 440 million acres

China's total cropland is about 330 million acres

$$\text{Difference in cropland between U.S. and China} = \text{U.S. acreage} - \text{China's acreage}$$

$$= 440 - 330 = 110$$

Thus, the difference in total cropland is about 110 million acres. ▲

EXAMPLE 4 *Determining a Tip*

The cost of Sarah Decker's meal before tax is \$23.60.

a) If a $7\frac{1}{2}\%$ sales tax is added to her bill, determine the total cost of the meal including tax.

b) If Sarah wants to leave a 10% tip on the *pretax* cost of the meal, how much should she leave?

c) If she wants to leave a 15% tip on the *pretax* cost of the meal, how much should she leave?

SOLUTION:

a) The sales tax is $7\frac{1}{2}\%$ of \$23.60. To determine the sales tax, first change the $7\frac{1}{2}\%$ to a decimal number. $7\frac{1}{2}\%$ when written as a decimal number is 0.075 (if you have forgotten how to change a percent to a decimal number, read Section 11.1). Next, multiply the decimal number, 0.075, by the amount, \$23.60.

$$\text{Sales tax} = 7\tfrac{1}{2}\% \text{ of } 23.60$$

$$= 0.075(23.60) = 1.77$$

The sales tax is \$1.77. The total bill is the cost of the meal plus the sales tax.

$$\text{Total bill} = \text{cost of meal} + \text{sales tax}$$

$$= 23.60 + 1.77 = 25.37$$

Thus, the bill, including sales tax, is \$25.37.

b) To find 10% of any number, we can multiply the number by 0.10.

$$10\% \text{ of pretax cost} = 0.10(23.60)$$

$$= 2.36$$

Thus, a 10% tip is \$2.36.

A simple way to find 10% of any number is to simply move the decimal point in the number one place to the left. Moving the decimal point in \$23.60 one place to the left gives \$2.36, the same answer we obtained by our calculations.

c) To find 15% of \$23.60, multiply as follows.

$$15\% \text{ of } 23.60 = 0.15(23.60) = 3.54$$

Thus, 15% of \$23.60 is \$3.54. A second method to find a 15% tip is to find 10% of the cost, as in part (b), then add half that amount. Following this procedure we get

$$\$2.36 + \frac{\$2.36}{2} = \$2.36 + \$1.18 = \$3.54$$

In most cases, tips are rounded. If the service is excellent, some people leave a 20% tip. Can you give two methods for determining a 20% tip on \$23.60? Determine the 20% tip now.

EXAMPLE 5 *A Recipe for 6*

The following chart shows the amount of each ingredient recommended to make 2, 4, and 8 servings of Potato Buds. Determine the amount of each ingredient necessary to make 6 servings of Potato Buds by using the following procedures.

a) Multiply the amount for 2 servings by 3.*
b) Add the amounts for 2 servings to the amounts for 4 servings.
c) Find the average of the amounts for 4 servings and for 8 servings.
d) Subtract the amounts for 2 servings from the amounts for 8 servings.
e) Compare the answers for parts (a) through (d). Are they the same? If not, explain why not.
f) Which is the correct procedure for obtaining 6 servings?

Servings	2	4	8
Water	$\frac{2}{3}$ cup	$1\frac{1}{3}$ cups	$2\frac{2}{3}$ cups
Milk	2 tbsp	$\frac{1}{3}$ cup	$\frac{2}{3}$ cup
Butter or margarine	1 tbsp	2 tbsp	4 tbsp
Salt†	$\frac{1}{4}$ tsp	$\frac{1}{2}$ tsp	1 tsp
Potato Buds	$\frac{2}{3}$ cup	$1\frac{1}{3}$ cups	$2\frac{2}{3}$ cups

†Less salt can be used if desired.

SOLUTION:

a) We multiply the amounts for 2 servings by 3.

Water: $3(\frac{2}{3}) = 2$ cups
Milk: $3(2) = 6$ tablespoons (tbsp)
Butter or margarine: $3(1) = 3$ tbsp
Salt: $3(\frac{1}{4}) = \frac{3}{4}$ teaspoon (tsp)
Potato Buds: $3(\frac{2}{3}) = 2$ cups

*Addition, subtraction, multiplication, and division of fractions are discussed in detail in Section 5.3.

MATHEMATICS *Everywhere*

Decisions, Decisions

We deal with problem solving daily. There are aspects of problem solving to almost every decision we make. Can we afford to take that vacation in the mountains? Which vehicle should I purchase? What should I do with my tax return? What shall I major in in college? What college shall I attend? The list of questions we ask ourselves daily goes on and on. To make decisions, we often need to consider and weigh many factors, as well as consider all the possible consequences of our decisions.

Often, without consciously realizing it, different branches of mathematics are involved in our decision-making process. For example, we often use statistical data and consider the probability (or chance) of an event occurring or not occurring when we make decisions. Probability and statistics are two branches of mathematics covered in later chapters of the book.

For many people, problem solving is a recreational activity, as is evident by the great number of crossword puzzles and puzzle books sold daily. This is a very large and expanding market.

Today, as you go about your daily business, keep a record of all the problem-solving decisions you need to make! If you do this conscientiously, you will be amazed at the outcome.

A vacation in the mountains

b) We find the amount of each ingredient by adding the amount for 2 and 4 servings.

Water: $\frac{2}{3}$ cup + $1\frac{1}{3}$ cup = 2 cups
Milk: 2 tbsp + $\frac{1}{3}$ cup

To add these two amounts, we must convert one of them so that they have the same units. By looking in a cookbook or a book of conversion factors, we see that 16 tbsp = 1 cup. The milk in part (a) was given in tablespoons, so we convert $\frac{1}{3}$ cup to tablespoons to compare answers. One third cup equals $\frac{1}{3}(16) = \frac{16}{3}$ or $5\frac{1}{3}$ tbsp. Therefore,

Milk: 2 tbsp + $5\frac{1}{3}$ tbsp = $7\frac{1}{3}$ tbsp

Let's continue with the rest of the ingredients:

Butter: 1 tbsp + 2 tbsp = 3 tbsp
Salt: $\frac{1}{4}$ tsp + $\frac{1}{2}$ tsp = $\frac{3}{4}$ tsp
Potato Buds: $\frac{2}{3}$ cup + $1\frac{1}{3}$ cups = 2 cups

c) We compute the amounts of the ingredients by finding the average of the amounts for 4 and 8 servings. We do so by adding the amounts for each ingredient and dividing the sum by 2.

$$\text{Water: } \frac{1\frac{1}{3}\text{ cups} + 2\frac{2}{3}\text{ cups}}{2} = \frac{4\text{ cups}}{2} = 2\text{ cups}$$

$$\text{Milk: } \frac{\frac{1}{3}\text{ cup} + \frac{2}{3}\text{ cup}}{2} = \frac{1\text{ cup}}{2} = \frac{1}{2}\text{ cup (or 8 tbsp)}$$

$$\text{Butter: } \frac{2\text{ tbsp} + 4\text{ tbsp}}{2} = \frac{6\text{ tbsp}}{2} = 3\text{ tbsp}$$

$$\text{Salt: } \frac{\frac{1}{2}\text{ tsp} + 1\text{ tsp}}{2} = \frac{\frac{3}{2}\text{ tsp}}{2} = \frac{3}{4}\text{ tsp}$$

$$\text{Potato Buds: } \frac{1\frac{1}{3}\text{ cups} + 2\frac{2}{3}\text{ cups}}{2} = \frac{4\text{ cups}}{2} = 2\text{ cups}$$

d) We obtain the amounts of ingredients by subtracting the amounts for 2 servings from the amounts for 8 servings.

Water: $2\frac{2}{3}$ cups − $\frac{2}{3}$ cup = 2 cups

$$\begin{aligned}\text{Milk: } \tfrac{2}{3}\text{ cup} - 2\text{ tbsp} &= \tfrac{2}{3}(16)\text{ tbsp} - 2\text{ tbsp}\\ &= \tfrac{32}{3}\text{ tbsp} - \tfrac{6}{3}\text{ tbsp}\\ &= \tfrac{26}{3}\text{ tbsp, or } 8\tfrac{2}{3}\text{ tbsp}\end{aligned}$$

Butter: 4 tbsp − 1 tbsp = 3 tbsp
Salt: 1 tsp − $\frac{1}{4}$ tsp = $\frac{3}{4}$ tsp
Potato Buds: $2\frac{2}{3}$ cups − $\frac{2}{3}$ cup = 2 cups

e) Comparing the answers in parts (a) through (d), we find that the amounts of all ingredients, except milk, are the same. For milk, we get the following results.

Part (a): Milk = 6 tbsp   Part (c): Milk = 8 tbsp
Part (b): Milk = $7\frac{1}{3}$ tbsp   Part (d): Milk = $8\frac{2}{3}$ tbsp

Why are all these answers different? After rechecking, we find that all our calculations are correct, so we must look deeper. Note that milk is the only ingredient that has different units for 2 servings and 4 servings. Let's check the relationship between 2 tbsp and $\frac{1}{3}$ cup. In going from 2 servings to 4 servings, we would expect that $\frac{1}{3}$ cup should be twice 2 tbsp. We know that 1 cup = 16 tbsp, so

$$\frac{1}{3}\text{ cup} = \frac{1}{3}(16) = \frac{16}{3} = 5\frac{1}{3}\text{ tbsp}$$

Therefore, instead of the 4 tbsp of milk we expected for 4 servings, we get $5\frac{1}{3}$ tbsp. This change causes all our calculations for milk to be different.

f) Which is the correct answer? Because all our calculations for milk are correct, there is no single correct answer. All our answers are correct. Using 8 tbsp instead of $5\frac{1}{3}$ tbsp might make the Potato Buds a little thinner. When we cook, we generally do not add the *exact* amount recommended. We rely on experience to alter the recommended amounts according to individual taste. ▲

Many real-life problems, such as the one in Example 6, can be solved by using proportions. A proportion is a statement of equality between two ratios (or fractions).*

EXAMPLE 6 *Spraying Weed Killer*

The instructions on the Ortho Weed-Be-Gone lawn weed killer indicates that to cover 1000 square feet (ft^2) of lawn, 20 teaspoons (tsp) of the weed killer should be mixed in 5 gallons (gal) of water. Ron Haines wishes to spray his lawn with the weed killer using his pressurized sprayer.

a) How much weed killer should be mixed with 8 gal of water to get the proper strength solution?

b) How much weed killer is needed to cover an area of 2820 ft^2 of lawn?

SOLUTION:

a) Use the information that 20 teaspoons of weed killer is to be mixed with 5 gal of water.

$$\text{Given ratio}\left\{\frac{20\text{ tsp}}{5\text{ gal water}} = \frac{?\text{ tsp}}{8\text{ gal}}\right. \quad \begin{array}{l}\leftarrow \text{Item to be found}\\ \leftarrow \text{Other information given}\end{array}$$

Notice in the proportion that teaspoons and gallons are placed in the same relative positions. Often the unknown quantity is replaced with an x. The proportion may be written as follows and solved using cross multiplication.

$$\frac{20}{5} = \frac{x}{8}$$

$$20(8) = 5x$$

$$160 = 5x$$

$$\frac{160}{5} = \frac{\cancel{5}x}{\cancel{5}} \quad \text{Divide both sides of the equation by 5 to solve for } x.$$

$$32 = x$$

Thus, Ron must mix 32 tsp [or $10\frac{2}{3}$ tablespoons (tbsp) or $\frac{2}{3}$ cup] of the weed killer with 8 gal of water. This answer seems reasonable since we would expect to get an answer greater than 20 tsp.

b) To answer this question, we use the same procedure discussed in part (a). This time we will use the information that 1000 ft^2 requires 20 tsp of weed killer. The areas may be placed either on the top or bottom of the fraction, as long as they are placed in the same relative position.

*Proportions are discussed in greater detail in Section 6.2.

DID YOU KNOW

East Meets West: Magic Squares

A Chinese myth says that in about 2200 B.C., a divine tortoise emerged from the Yellow River. On his back was a special diagram of numbers from which all mathematics was derived. The Chinese called this diagram Lo Shu. The Lo Shu diagram is the first known magic square.

Arab traders brought the Chinese magic square to Europe during the Middle Ages, when the plague was killing millions of people. Magic squares were considered strong talismans against evil, and possession of a magic square was thought to ensure health and wealth.

$$\text{Given ratio} \begin{cases} \dfrac{1000 \text{ sq ft}}{20 \text{ tsp}} = \dfrac{2820 \text{ sq ft}}{? \text{ tsp}} \end{cases}$$

← Other information given
← Item to be found

Now replace the question mark with an x and solve the proportion.

$$\frac{1000}{20} = \frac{2820}{x}$$
$$1000(x) = 20(2820)$$
$$1000x = 56{,}400$$
$$\frac{\cancel{1000}x}{\cancel{1000}} = \frac{56{,}400}{1000}$$

Divide both sides of the equation by 1000 to solve for x.

$$x = 56.4$$

Thus, about 56.4 tsp are needed. This answer is reasonable for we would expect the answer to be more than twice the 20 tsp required for 1000 ft^2. ▲

Most of the problems solved so far have been practical ones. Many people, however, enjoy solving brainteasers. One example of such a puzzle follows.

EXAMPLE 7 *Magic Squares*

A magic square is a square array of numbers such that the numbers in all rows, columns, and diagonals have the same sum. Use the digits 1, 2, 3, 4, 5, 6, 7, 8, and 9 to construct a magic square.

SOLUTION: The first step is to create a figure with nine cells as in Fig. 1.1(a). We must place the nine numbers in the cells so that the same sum is obtained in each row, column, and diagonal. Common sense tells us that 7, 8, and 9 cannot be in the same row, column, or diagonal. We need some small and large numbers in the same row, column, and diagonal. To see a relationship, we list the numbers in order:

1, 2, 3, 4, 5, 6, 7, 8, 9

Note that the middle number is 5 and the smallest and largest numbers are 1 and 9, respectively. The sum of 1, 5, and 9 is 15. If the sum of 2 and 8 is added to 5, the sum is 15. Likewise 3, 5, 7, and 4, 5, 6 have sums of 15. We see that in each group of three numbers the sum is 15 and 5 is a member of the group.

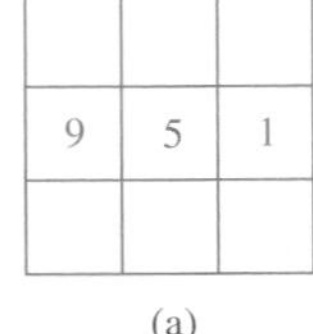

(a)

		8
9	5	1
2		

(b)

4		8
9	5	1
2		6

(c)

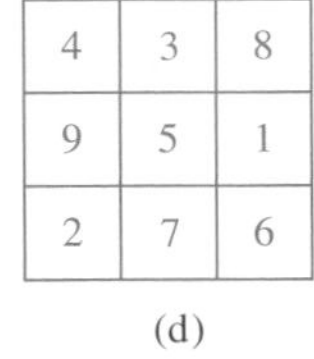

(d)

Figure 1.1

Because 5 is the middle number in the list of numbers, place 5 in the center square. Place 9 and 1 to the left and right of 5 as in Fig. 1.1(a). Now we place the 2 and the 8. The 8 cannot be placed next to 9 because 8 + 9 = 17, which is greater than 15. Place the smaller number 2 next to the larger number 9. We elected to place the 2 in the lower left-hand cell and the 8 in the upper right-hand cell as in Fig. 1.1(b). The sum of 8 and 1 is 9. To arrive at a sum of 15, we place 6 in the lower right-hand cell as in Fig. 1.1(c). The sum of 9 and 2 is 11. To arrive at a sum of 15, we place 4 in the upper left-hand cell as in Fig. 1.1(c). Now the diagonals 2, 5, 8 and 4, 5, 6 have sums of 15. The numbers that remain to be placed in the empty cells are 3 and 7. Using arithmetic, we can see that 3 goes in the top middle cell and 7 goes in the bottom middle cell as in Fig. 1.1(d). A check shows that the sum in all the rows, columns, and diagonals is 15. ▲

The solution to Example 7 is not unique. Other arrangements of the nine numbers in the cells will produce a magic square. Also, other techniques of arriving at a solution for a magic square may be used. In fact, the process described will not work if the number of squares is even, for example, 16 instead of 9. Magic squares are not limited to the operation of addition or to the set of counting numbers.

SECTION 1.3 EXERCISES

Practice the Skills/Problem Solving

1. *Reading a Map* The scale on a map is 1 inch = 50 miles. How long a distance is a route on the map if it measures 3.75 in.?

2. *Blueprints* Chalon Bridges, an architect, is designing a shopping mall. The scale of her plan is 1 in. = 12 ft. If one store in the mall is to have a frontage of 82 ft, how long will the line representing that store's frontage be on the blueprint?

3. *Height of a Tree* At a given time of day, the ratio of the height of an object to the length of its shadow is the same for all objects. If a 3-ft stick in the ground casts a shadow of 1.2 ft, find the length of the shadow of a 48.4 ft tree.

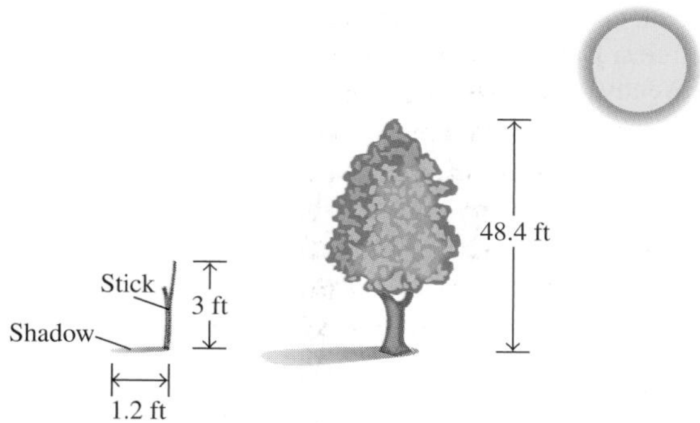

4. *Use of Fertilizer* A bag of fertilizer covers 6000 ft^2. How many bags are needed to cover an area of 26,000 ft^2?

5. *Health Care Cost* According to Hewitt Associates, in 2000 the annual cost to companies that provided health care coverage to their employees averaged $4222. The 2001 average cost per employee for employees' health care was 11.5% greater than the companies' 2000 cost. Determine the annual 2001 cost, per employee, for health care.

6. *New York City Taxi* As of this writing, the fare for a New York City taxi is as follows:
   $2.00 upon entering the cab
   $0.30 for each $\frac{1}{5}$ mile
   $0.30 for each 30 seconds of waiting (or moving less than 8 mph)

   Determine the cost of a 12-mile ride with 2 minutes of standing still in traffic.

7. *Cost of Gas* The bar graph shows the average price for self-serve regular gasoline on August 1, 2001, in the states with the most expensive and least expensive gasoline cost. Assuming that the cost of the gasoline remained constant for the year, how much more would it cost to drive 20,000 miles in Hawaii than it would in South Carolina if your car's mileage was 20 miles per gallon?

A wide variety of pump prices

Costliest States for Gasoline
Self-serve regular, average price per gallon

1. Hawaii	$2.02
2. California	1.73
3. Alaska	1.71
4. Connecticut	1.61
5. New York	1.58

Cheapest States for Gasoline
Self-serve regular, average price per gallon

46. North Carolina	$1.26
47. Missouri	1.25
48. Oklahoma	1.24
49. Georgia	1.22
50. South Carolina	1.22

National average price: $1.40 as of Monday, July 30, 2001
Source: AAA

8. *Hours of Work* An article in the June 11, 2001, issue of *Business Week* indicates that in the United States we tend to put in more hours at work than many other countries, especially European countries, and each year the difference increases. The graph on the top of the next page shows a comparison of annual hours worked by employees in Germany and the United States from 1970 to 2001.

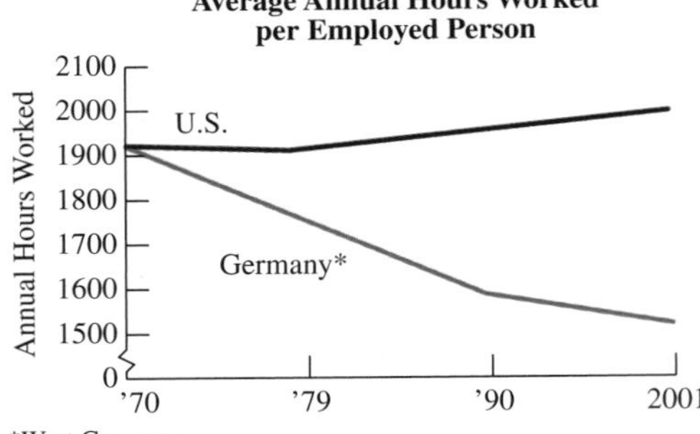

*West Germany
Data: Organization for Economic Cooperation and Development

a) Estimate the increase in annual hours worked by Americans from 1970 to 2001.
b) Estimate the average *weekly* number of hours worked in the United States in 2001.
c) Estimate the average *weekly* number of hours worked in Germany in 2001.

9. *Parking Costs* The Main Street Garage charges $2.50 for the first hour of parking and $1.00 for each additional hour or part thereof. Denise Tomey parks her car in the garage from 9 A.M. to 5 P.M., 5 days a week. How much money does she save by paying a weekly parking rate of $35.00?

10. *Calling Person to Person* From a phone booth, Andrea Sheehan makes a person-to-person call from Houston, Texas, to Seattle, Washington. The call costs $3.75 for the first 3 minutes and $0.50 for each additional minute. How much did Andrea's 21-minute call cost?

11. *Buying a Computer* Emily Falcon wants to purchase a computer that sells for $1250. She can either pay the total amount at the time of purchase, or she can agree to pay the store $120 down and $80 a month for 15 months. How much money can she save by paying the total amount at the time of purchase?

12. *Buying a House* The Suarezes want to purchase a house that costs $120,000. With a $20,000 down payment, their monthly cost would be $699.99 for 30 years. With a $40,000 down payment, their monthly cost would be $559.20 for 30 years. Determine the total amount they would save on the cost of the house over 30 years if they used the $40,000 down payment.

13. *Buying a House* The Browns want to purchase a house that costs $130,000. They plan to take out a $90,000 mortgage on the house and put $40,000 as a down payment. The bank informs them that with a 20-year mortgage their monthly cost would be $752.40, and with a 30-year mortgage their monthly cost would be $660.60. Determine the amount they would save on the cost of the house if they selected the 20-year mortgage rather than the 30-year mortgage.

14. *Getting an 80 Average* On four exams, Wallace Memmer's grades were 77, 93, 90 and 76. What grade must he obtain on his fifth exam to have an 80 average?

15. *Japanese Sizes* The following chart shows men's jacket sizes as would be given in the United States and in Japan.
a) Justin Smith is in Japan and finds a sports jacket that he wishes to buy. He is a size 48 in the United States. Determine the size of the jacket he should try on.
b) Determine a procedure (or a formula) for converting a jacket from a U.S. size to a size in Japan.

	chest								
U.S.		34	36	38	40	42	44	46	48
Japan		86.5	91.5	96.5	101.5	106.5	112	117	?

16. *Playing a Lottery* In one state lottery game, you must select a four-digit number (digits may be repeated). If your number matches exactly the four-digit number selected by the lottery commission, you win.
a) How many different numbers may be chosen?
b) If you purchase one lottery ticket, what is your chance of winning?

17. *Energy Value and Energy Consumption* The table gives the approximate energy values of some foods, in kilojoules (kJ), and the energy requirements of some activities. How soon would you use up the energy from
a) a fried egg by swimming?
b) a hamburger by walking?
c) a piece of strawberry shortcake by cycling?
d) a hamburger and a chocolate milkshake by walking?

Food	Energy Value (kj)	Activity	Energy Consumption (kj/min)
Chocolate milkshake	2200	Walking	25
Fried egg	460	Cycling	35
Hamburger	1550	Swimming	50
Strawberry shortcake	1400	Running	80
Glass of skim milk	350		

18. *Gas Mileage* Wendy Weisner fills her gas tank completely and makes a note that the odometer reads 38,451.4 miles. The next time she stops to put gas in her car, filling the tank takes 12.6 gal, and the odometer reads 38,687.0 miles. Determine the number of miles per gallon that Wendy's car gets on this tank of gas.

19. *Diversity of Doctors* In 2000, the U.S. population was about 273,300,000 and the number of U.S. doctors was about 970,000. Use the graph to determine the number of
a) Hispanics in the United States.
b) Asian doctors in the United States.
c) African-American doctors in the United States.

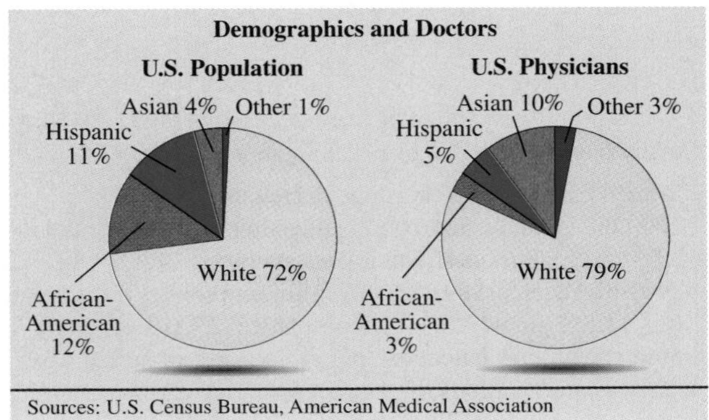

Sources: U.S. Census Bureau, American Medical Association

20. *Saving for a Stereo*
   a) Fernando Diez works 40 hours per week and makes \$8.50 per hour. How much money can he expect to earn in 1 year (52 weeks)?
   b) If he saves all the money that he earns, how long will he have to work to save for a stereo receiver that costs \$1275?

21. *Mail Order Purchase* Mary Liotta purchased 4 tires by mail order. She paid \$52.80 per tire plus \$5.60 per tire for shipping and handling. There is no sales tax on this purchase because they were purchased out of state. She also had to pay \$8.56 per tire for mounting and balancing. At a local tire store, her total for the 4 tires with mounting and balancing would be \$324 plus an 8% sales tax. How much did Mary save by purchasing the tires through the mail?

22. *Sealing a Gym Floor* A gymnasium floor has an area of 2400 square yards. Each gallon of floor sealant covers an area of 350 square feet. How many gallons of sealant are needed to cover the gymnasium floor?

23. *Profit Margins* The following chart shows retail stores' average percent profit margin on certain items.

Product Category	Average Profit Margin (%)
Video equipment	12
Audio components	14
Stereo speakers	20–25
Extended warranties	50–60

Source: *Consumer Reports*

   a) Determine the average profit of a store that has the list price on a camcorder of \$620.
   b) Determine the average profit of a store that has a list price on a pair of speakers for \$1200 (use a 22% profit margin).
   c) If you negotiate with the salesperson and get him or her to sell the speakers for \$1000, find the store's profit (use a 22% profit margin).

24. *Leaking Faucet* A faucet is leaking at a rate of one drop of water per second. If the volume of one drop of water is 0.1 cubic centimeter (0.1 cm^3), find
   a) the volume of water in cubic centimeters lost in a year.
   b) how long it would take, in days, to fill a rectangular basin 30 cm by 20 cm by 20 cm.

25. *Income Taxes* The federal income tax rate schedule for a joint return in 2002 is illustrated in the table. If Steve and Maureen Tomlin paid \$4590 in federal taxes, find the family's adjusted gross income.

Adjusted Gross Income	Taxes
\$0–\$12,000	10% of income
\$12,000–\$46,700	\$1200 + 15% in excess of \$12,000
\$46,700–\$112,850	\$6405 + 27% in excess of \$46,700
\$112,850–\$171,950	\$24,265.50 + 30% in excess of \$112,850
\$171,950–\$307,050	\$41,995.50 + 35% in excess of \$171,950
\$307,050 and above	\$89,280.50 + 38.6% in excess of \$307,050

26. *Wasted Water* A faucet leaks 1 oz of water per minute.
   a) How many gallons of water are wasted in a year? (A gallon contains 128 oz.)
   b) If water costs \$11.20 per 1000 gal, how much additional money is being spent on the water bill?

27. *Water Restrictions* In 2000 and 2001, the entire state of Florida had a water shortage, and restrictions were placed on water usage throughout the state. The number of water restriction ordinance violations for Pinellas County, Florida, from March 2000 to March 2001 is shown in the table below.

Water Restriction Ordinance Violations
March 2000–March 2001

Number of Violations	Number of Customers
7	2
6	0
5	1
4	29
3	201
2	1,408
1	10,352

Source: Pinellas County Utilities

Determine the total number of violations illustrated by the chart.

28. *Tire Pressure* When a car's tire pressure is 30 pounds per square inch (psi), it averages 20.8 mpg of gasoline. If the tire pressure is increased to 35 psi, the car averages 21.6 mpg of gasoline.
   a) If Mr. Levy drives an average of 20,000 mi per year, how many gallons of gasoline will he save in a year by increasing his tire pressure from 30 to 35 psi?
   b) If gasoline costs \$1.60 per gallon, how much will he save in a year?
   c) If we assume that there are about 140 million cars in the United States and that these changes are typical of each car, how many gallons of gasoline would be saved if all drivers increased their cars' tire pressure?

29. *Air Pollution* The graph illustrates the 10 countries that produced the most carbon dioxide, methane, and nitrous oxide in 2000. These gases make up the majority of the greenhouse gases. (The term *per capita* means per person.)

Producers of Major Greenhouse Gases

	TOTAL EMISSIONS (in millions of metric tons)	EMISSIONS PER CAPITA (in metric tons of carbon dioxide equivalent)
UNITED STATES	6503.8	24.3
CHINA	4964.8	4.0
INDIA	2081.7	2.2
RUSSIA	1980.3	13.4
JAPAN	1166.1	9.3
GERMANY	956.0	11.6
BRAZIL	695.7	4.2
SOUTH AFRICA	677.2	15.9
CANADA	634.0	20.9
BRITAIN	618.7	10.5

Source: Environmental Protection Agency, Census Bureau International Data Base

a) By looking at the data provided, is it possible to determine the population of each country? If so, explain how to do so.
b) Using the procedure you gave in part (a), determine the population of the United States.
c) Determine the population of China.

30. *Adjusting for Inflation* Assume that the annual rate of inflation is 6% for each of the next 2 years. What will be the cost of goods 2 years from now, adjusted for inflation, if the goods cost $450.00 today?

31. *Investing* You place $1000 in a mutual fund. The first year the value of the fund increases by 10%. The second year the value of the fund decreases by 10%. Determine the value of the fund at the end of the second year. Is it greater than, less than, or equal to your initial investment?

32. *X-rays* With a certain medical insurance policy, the customer must first pay an annual $100 deductible, then the policy covers 80% of the cost of x-rays. The first insurance claims for a specific year submitted by Yungchen Cheng are for two x-rays. The first x-ray cost $640 and the second x-ray cost $920. How much, in total, will Yungchen need to pay for these x-rays?

33. *A Photo Safari* Kelli Hammer is planning a trip to Africa where she will participate in a photo safari. She is planning on bringing a great deal of film. A photography store is selling 4 packs of film for $17 and 10 packs of the same film for $41.
a) If she wishes to purchase only the 4 packs and 10 packs and wishes to spend a maximum of $200 on film, what is the maximum number of rolls of film she can purchase?
b) What will be the cost?

34. *Buying Film* Erika Gutierrez is planning a vacation to Australia and wishes to bring a large supply of film. At Wal-Mart, 4 packs of 24 exposure film costs $4.08 and 4 packs of the same film with 36 exposures costs $5.76.
a) If she wishes to spend a maximum of $50 on film and get the most exposures, how many 4 packs of 24 exposures and how many 4 packs of 36 exposures should she purchase?
b) How many exposures will she get?
c) What will be the cost? If there is more than one choice in part (a), give the minimum cost.

35. *Making Cream of Wheat* The following amounts of ingredients are recommended to make various servings of Nabisco Instant Cream of Wheat. *Note:* 16 tbsp = 1 cup.

Ingredient	1 Serving	2 Servings	4 Servings
Mix water or milk	1 cup	2 cups	$3\frac{3}{4}$ cups
With salt (optional)	$\frac{1}{8}$ tsp	$\frac{1}{4}$ tsp	$\frac{1}{2}$ tsp
Add Cream of Wheat	3 tbsp	$\frac{1}{2}$ cup	$\frac{3}{4}$ cup

Determine the amount of each ingredient needed to make 3 servings using the following procedures.
a) Multiply the amounts for 1 serving by 3.
b) Find the average of the amounts for 2 and 4 servings.
c) Subtract the amounts for 1 serving from the amounts for 4 servings.
d) Compare the answers obtained in parts (a) through (c) and explain any differences.

36. *Making Rice* Following are the amounts of ingredients recommended to make various servings of Uncle Ben's Original Converted Rice. *Note:* 1 tbsp = 3 tsp.

Ingredient	2 Servings	4 Servings	6 Servings	12 Servings
Rice (cups)	$\frac{1}{2}$	1	$1\frac{1}{2}$	3
Water (cups)	$1\frac{1}{3}$	$2\frac{1}{4}$	$3\frac{1}{3}$	6
Salt (teaspoons)	$\frac{1}{4}$	$\frac{1}{2}$	$\frac{3}{4}$	$1\frac{1}{2}$
Butter or margarine	1 tsp	2 tsp	1 tbsp	2 tbsp

Determine the amount of each ingredient needed to make 8 servings using the following procedures.
a) Multiply the amount for 2 servings by 4.
b) Multiply the amount for 4 servings by 2.
c) Add the amounts for 2 and 6 servings.
d) Subtract the amount for 4 servings from the amount for 12 servings.
e) Compare the answers obtained in parts (a) through (d) and explain any differences.

Solve the following problems.

37. *One Square Foot* How many square inches, 1 in. by 1 in., fit in an area of 1 square foot, 1 ft by 1 ft?

38. *Cubic Inches* How many cubic inches fit in 1 cubic foot?

39. *Rectangle* If the length and width of a rectangle each double, what happens to the area of the rectangle?

40. *Cube* If the length, width, and height of a cube all double, what happens to the volume of the cube?

41. *Positive Numbers* Find two positive numbers that have a one-digit answer when multiplied and a two-digit answer when added.

42. *Buying Candy* How much do 10 pieces of candy cost if 1000 pieces cost $10?

43. *A Balance* On the balance below, where should the one missing block ■ be placed so that the balance would balance on the triangle (the fulcrum). Assume that each block has the same weight.

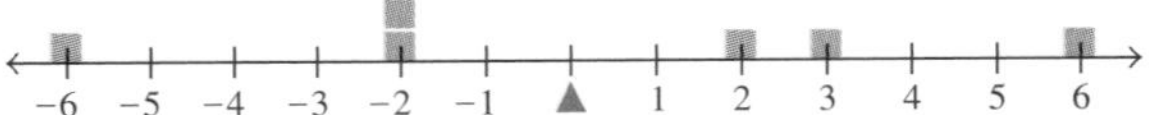

44. *Ties, Ties, Ties* All my ties are red except two. All my ties are blue except two. All my ties are brown except two. How many ties do I have?

45. *Birds and Lizards* A pet store had just received a supply of birds and lizards. Counting heads, the owner got 22. Counting feet, the owner got 68. How many birds and lizards are there?

46. *Palindromes* A *palindrome* is a number (or word) that reads the same forward and backward. The numbers 1991 and 43234 are examples of palindromes. How many palindromes are there between the numbers 2000 and 3000? List them.

47. *Supermarket Display* The figure shows oranges in a supermarket display stacked in a *square pyramid* (the base is a square).

a) How many oranges are in the pyramid shown if the base is 4 oranges by 4 oranges?
b) How many oranges would be in a square pyramid if the base was 7 oranges by 7 oranges?

48. *Balancing a Scale* If you have a balance scale and only the four weights 1 gram (g), 3 g, 9 g, and 27 g, explain how you could show that an object had the following weights.
a) 5 g **b)** 16 g

(*Hint:* Weights must be added to both sides of the balance scale.)

49. *Numbers in Circles* Place the numbers 1 through 6 in the circles below so that the sum along each of the three straight lines is the same. Each number must be used exactly once (*Note:* There is more than one correct answer.)

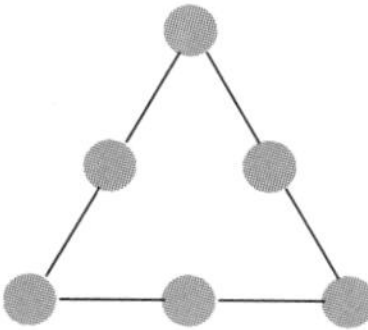

50. *Cuts in Cheese* If you make the three complete cuts in the cheese, as shown, how many pieces of cheese will you have?

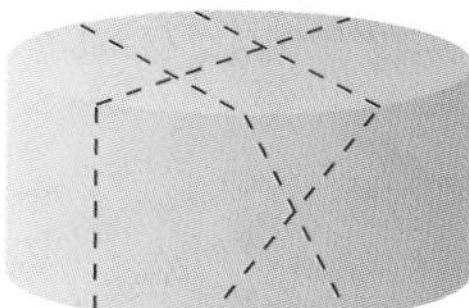

51. *Magic Square* Create a magic square by using the numbers 2, 4, 6, 8, 10, 12, 14, 16, and 18. The sum of the numbers in every column, row, and diagonal must be 30.

52. *Magic Square* Create a magic square by using the numbers 1, 3, 5, 7, 9, 11, 13, 15, and 17. The sum of the numbers in every column, row, and diagonal must be 27.

In Exercises 53–55, use the three magic squares illustrated to obtain the answers.

6	5	10
11	7	3
4	9	8

3	2	7
8	4	0
1	6	5

10	9	14
15	11	7
8	13	12

53. *Magic Square* Examine the 3 by 3 magic squares and find the sum of the four corner entries of each magic square. How can you determine the sum by using a key number in the magic square?

54. *Magic Square* For a 3 by 3 magic square, how can you determine the sum of the numbers in any particular row, column, or diagonal by using a key value in the magic square?

55. *Magic Square* For a 3 by 3 magic square, how can you determine the sum of all the numbers in the square by using a key value in the magic square?

56. *Stack of Cubes* Identical cubes are stacked in the corner of a room, as shown. How many of the cubes are not visible?

57. *Dominos* Consider a domino with six dots, as shown. Two ways of connecting the three dots on the left with the three dots on the right are illustrated. In how many ways can the three dots on the left be connected with the three dots on the right?

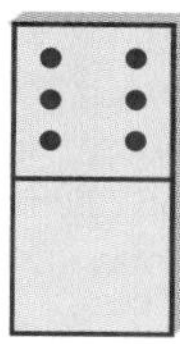 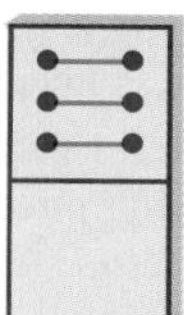 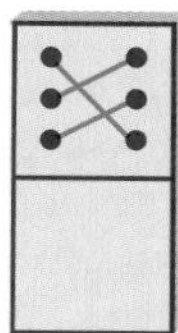

58. *Handshakes All Around* Five salespeople gather for a sales meeting. How many handshakes will each person make if each must shake hands with each of the four others?

59. *Consecutive Digits* Place the digits 1 through 8 in the eight boxes so that each digit is used exactly once and no two consecutive digits touch horizontally, vertically, or diagonally.

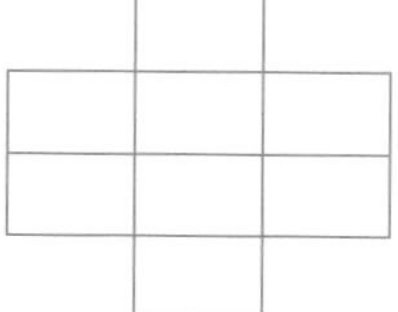

60. *A Digital Clock* Digital clocks display numerals by lighting some or all the seven parts of the pattern shown. If each digit 0 through 9 is displayed once, which of the seven parts is used least often? Which part is used most often?

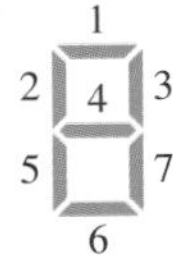

61. *A Grid* Place five 1's, five 2's, five 3's, five 4's, and five 5's in a 5×5 grid so that each digit—that is, 1, 2, 3, 4, 5—appears exactly once in each row and exactly once in each column.

Challenge Problems/Group Activities

62. *Insurance Policies* Ray Kelley owns two cars (a Ford Mustang and a Ford Focus), a house, and a rental apartment. He has auto insurance for both cars, a homeowner's policy, and a policy for the rental property. The costs of the policies are

    Mustang: \$1648 per year
    Focus: \$1530 per year
    Homeowner's: \$640 per year
    Rental property: \$750 per year

    Ray is considering taking out a \$1 million personal umbrella liability policy. The annual cost of the umbrella policy would be \$450. If he has the umbrella policy, he can lower the limits on parts of his auto policies and still have equal or better protection. If Ray purchases the umbrella policy, he can reduce his premium on the Mustang by \$90 per year and his premium on the Focus by 12%. If he purchases the umbrella policy and reduces the amount he pays for auto insurance, what is the net amount he is actually paying for the umbrella policy?

63. *A Sports Puzzle* Peter, Paul, and Mary are three sports professionals. One is a tennis player, one is a golfer, and one is a skier. They live in three adjacent houses on City View Drive. From the following information determine which is the professional skier. (*Hint:* A table may be helpful.)

    Mary does not play tennis.
    Peter skis and plays tennis, but does not golf.
    The golfer and the skier live next to each other.
    Three years ago, Paul broke his leg skiing and has not tried it since.
    Mary lives in the last house.
    The golfer and the tennis player share a common backyard swimming pool.

64. *Counting Triangles* How many triangles are in the figure?

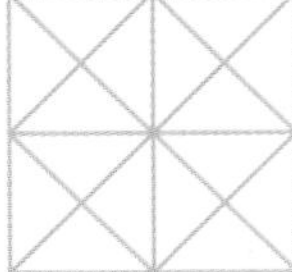

65. *Finding the Area* Rectangle ABCD is made up entirely of squares. The black square has a side of 1 unit. Find the area of ABCD.

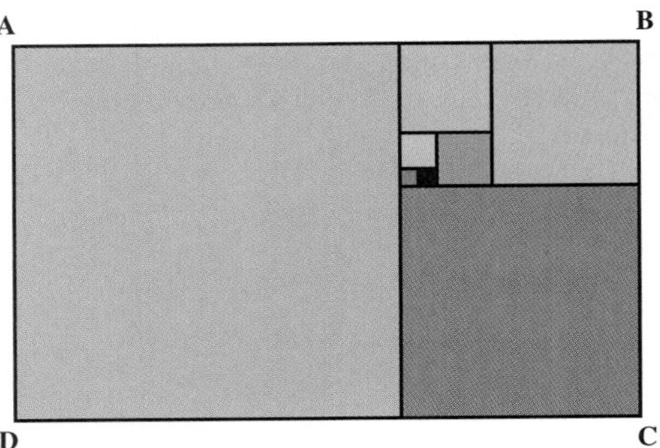

Recreational Math

66. *Ostriches* How many ostriches must replace the question mark to balance the fourth scale? Assume all animals of the same kind have the same weight. That is, all giraffes weigh the same, etc.

Internet/Research Activity

67. *Puzzles* Many fun and interesting puzzle books and magazines are available. Using this chapter and puzzle books as a guide, construct five of your own puzzles and present them to your instructor.

CHAPTER 1 SUMMARY

IMPORTANT FACTS

The **natural numbers** or **counting numbers** are
1, 2, 3, 4, . . .

A **conjecture** is a prediction based on specific observations.

A **counterexample** is a special case that satisfies all the conditions of a conjecture, but proves the conjecture false.

Inductive reasoning is the process of reasoning to a general conclusion through observations of specific cases.

Deductive reasoning is the process of reasoning to a specific conclusion from a general statement.

GUIDELINES FOR PROBLEM SOLVING

1. Understand the problem.
2. Devise a plan to solve the problem.
3. Carry out the plan.
4. Check the results.

CHAPTER 1 REVIEW EXERCISES

1.1*

In Exercises 1–8, use inductive reasoning to predict the next three numbers or figures in the pattern.

1. 3, 8, 13, 18, . . .
2. 1, 4, 9, 16, . . .
3. −3, 6, −12, 24, . . .
4. 5, 7, 10, 14, 19, . . .
5. 30, 29, 27, 24, 20, . . .
6. 6, 3, $\frac{3}{2}$, $\frac{3}{4}$, . . .
7. , , , , . . .
8. , , , , , . . .
9. *Pattern* Examine the following grid for a pattern and then select the answer which completes the pattern. (*Hint:* Think about rotating groups of four squares at a time.)

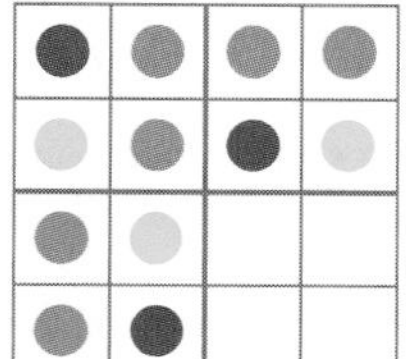

a) 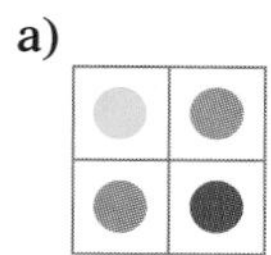

b) 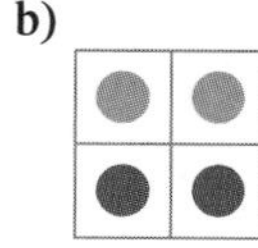

c) 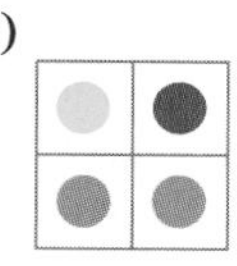

d) 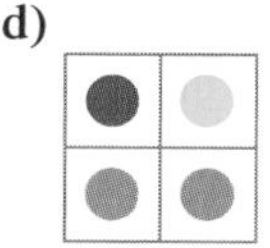

10. Pick any number and multiply the number by 2. Add 10 to the product. Divide the sum by 2. Subtract 5 from the quotient.
    a) What is the relationship between the number you started with and the final number?
    b) Arbitrarily select some different numbers and repeat the process, recording the original number and the results.
    c) Make a conjecture about the original number and the final number.
    d) Prove, using deductive reasoning, the conjecture you made in part (c).
11. Pick any number between 1 and 20. Add 5 to the number. Multiply the sum by 6. Subtract 12 from the product. Divide the difference by 2. Divide the quotient by 3. Subtract the number you started with from the quotient. What is your answer? Try this process with a different number. Make a conjecture as to what your final answer will always be.
12. *Counterexample* Find a counterexample to the statement "The sum of two squares is an even number."

1.2

In Exercises 13–25, estimate the answer. Your answers may vary from those given in the back of the book, depending on how you round to arrive at the answer. However, your answer should be something near the answer given.

13. 210,302 × 1992
14. 346.2 + 96.402 + 1.04 + 897 + 821
15. 21% of 1012
16. *Distance* Estimate the distance from your wrist to your elbow and estimate the length of your foot. Which do you think is greater? With the help of a friend, measure both lengths to determine which is longer.
17. *Cost* Estimate the cost of 82 feet of chain if the cost is 1.09 per foot.

*The number in color indicates the section in which the material is covered.

18. *Sales Tax* Estimate the amount of a 6% sales tax on a coat that costs \$202.

19. *Walking Speed* Estimate your average walking speed in miles per hour if you walked 1.1 mi in 22 min.

20. *Groceries* Estimate the total cost of six grocery items that cost \$2.49, \$0.79, \$1.89, \$0.10, \$2.19, \$6.75.

21. *A Walking Path* The scale of the map is $\frac{1}{4}$ in. = 0.1 mi. Estimate the distance of the walking path indicated in red.

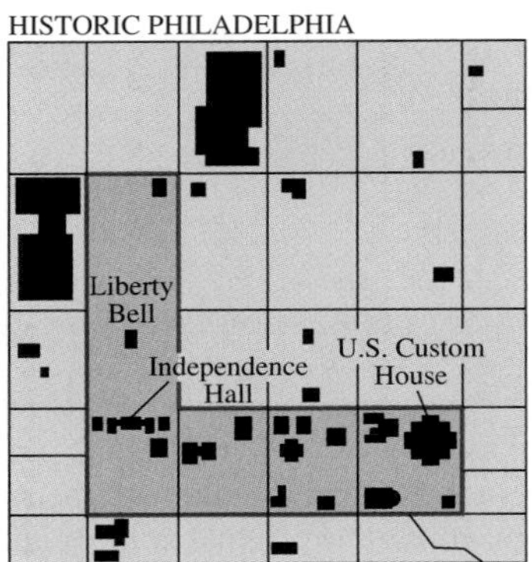

In Exercises 22 and 23, refer to the following graph, which illustrates the percent of people in the different tax brackets for the year 2000.

22. Estimate the percent of tax filers in the 15% tax bracket (the purple area of the circle graph).

23. Estimate the percent of tax filers in the 31% tax bracket (the red area of the circle graph).

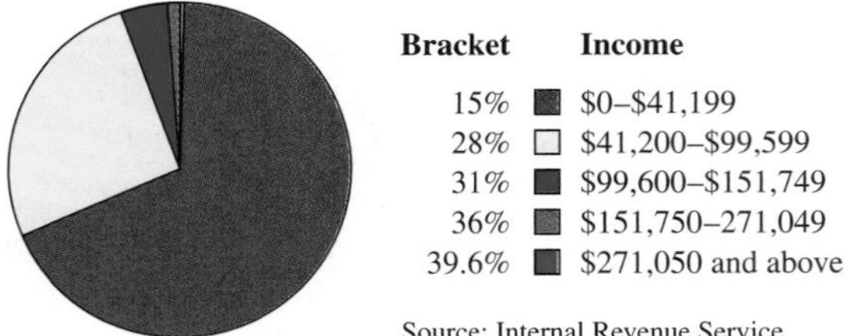

24. *Estimating an Area* If each square represents one square unit, estimate the size of the shaded area.

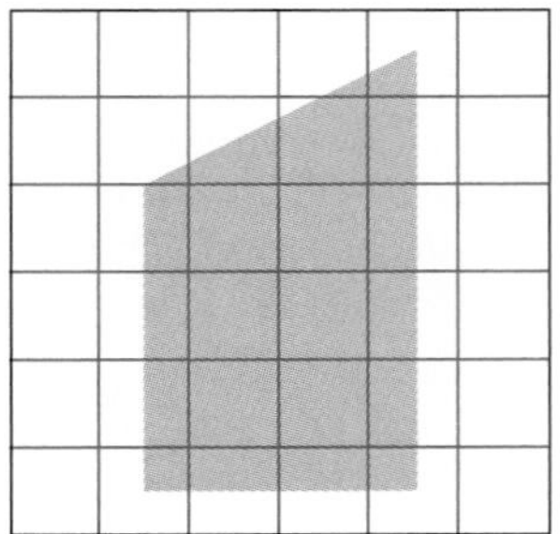

25. *Railroad Car Estimation* The scale of a model railroad is 1 in. = 12.5 ft. Estimate the size of an actual box car if this drawing is the same size as the model box car.

1.3

Solve the following problems.

26. *Change from a Twenty* Jeff Howard parked his car in a lot that charged \$2.00 for the first hour and \$1.50 for each additional hour. He left the car in the lot for 8 hr. How much change did he receive from a \$20 bill?

27. *Buying in Quantity* A six-pack of cola costs \$2.69. A carton of 4 six-packs costs \$9.60. How much will be saved by purchasing the carton rather than 4 individual six-packs?

28. *Jet Ski Rental* The rental cost of a jet ski from Nola Akala's Ski Rental is \$15 per 15 min, and the cost from Jill Berkman's Ski Rental is \$25 per half hour. If you plan to rent the jet ski for 2 hr, which is the better deal, and by how much?

29. *Oscars* In 2001, shortly before the Academy Awards show, many Oscars were lost by the shipping company. Fifty-two of the 55 Oscars were found, before the awards ceremony, in a dumpster by an Illinois man. The man was awarded \$50,000 (and two tickets to the ceremony). The actual cost to have each Oscar produced was \$327. How much more had the man been awarded then the actual cost to produce the 52 Oscars he found (disregarding the cost of the tickets)?

30. *Cab Fare* A taxicab charges \$1.50 for the first 1/5 mi and 30 cents for each additional 1/5 mi. Determine the cost of a 10 mi trip.

31. *Auto Insurance* Most insurance companies reduce premiums by 10% until age 25 for people who successfully pass a driver education course. A particular driver education course costs \$60. Patrick Flanigan, who just turned 18, has auto insurance that costs \$530 per year. By taking the driver education course, how much would he save in auto insurance, including the cost of the course, from the age of 18 until the age of 25?

32. *Pediatric Dosage* If 1.5 milligrams of a medicine is to be given for 10 lb of body weight, how many milligrams should be given to a child who weighs 47 lb?

33. *Qualifying for a Mortgage* Banks will grant an applicant a mortgage if the monthly payments are not greater than 28% of the person's take-home pay. What is the maximum monthly mortgage payment you can make if your gross salary is \$3800 a month and your payroll deductions are 30% of your gross salary?

34. *Flying West* New York City is on eastern standard time, St. Louis is on central standard time (1 hr earlier than eastern standard time), and Las Vegas is on Pacific standard time (3 hr earlier than eastern standard time). A flight leaves New York City at 9 A.M. eastern standard time, stops for 50 min in St. Louis, and arrives in Las Vegas at 1:35 P.M. Pacific time. How long is the plane actually flying?

35. *Crossing Time Zones* The international date line is an imaginary line of longitude (from the North Pole to the South Pole) on Earth's surface between Japan and Hawaii in the Pacific Ocean. Crossing the line east to west adds a day to the present date. Crossing the line west to east subtracts a day. At 3:00 P.M. on July 25 in Hawaii, what is the time and date in Tokyo, Japan, which is four time zones to the west?

36. *Conversions* 1 in. = 2.54 cm.
   a) How many square centimeters are in a square inch?
   b) How many cubic centimeters are in a cubic inch?
   c) How long is a centimeter in terms of inches?

37. *Dot Pattern* If the following pattern is continued, how many dots will be in the hundredth figure?

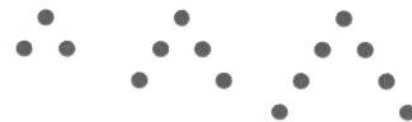

38. *Magic Square* The following magic square uses each number from 6 to 21 exactly once. Complete the magic square by using the unused numbers from 6 through 21 exactly once.

21	7		18
10		15	
14	12	11	17
9	19		

39. *Magic Square* Create a magic square by using the numbers 13, 15, 17, 19, 21, 23, 25, 27, and 29.

40. *Microbes in a Jar* A colony of microbes doubles in number every second. A single microbe is placed in a jar, and in an hour the jar is full. When was the jar half full?

41. *Brothers and Sisters* Jim Carraway has four more brothers than sisters. How many more brothers than sisters does his sister Mary have?

42. *A Missing Dollar* Three friends check into a single room in a motel and pay $10 apiece. The room costs $25 instead of $30, so a clerk is sent to the room to give $5 back. The friends each take back $1, and the clerk is given $2 for his trouble. Now each of the friends paid $9, a total of $27, and the clerk received $2. What happened to the missing dollar?

43. *The Average Weight* Four women in a room have an average weight of 130 lb. A fifth woman who weighs 180 lb enters the room. Find the average weight of all five women.

44. *Change for a Dollar* Could a person have $1.15 worth of change in his pocket and still not be able to give someone change for a dollar bill? If so, what coins might he have?

45. *Volume of a Cube* Here is a flat pattern for a cube to be formed by folding. The sides of each square are 6 cm. Find the volume of the cube.

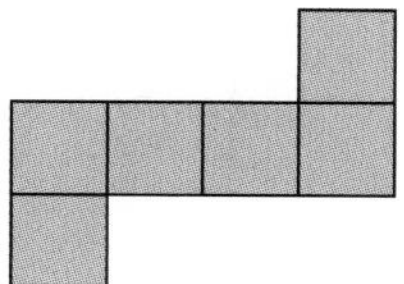

46. *The Heavier Coin* You have 13 coins, which all look alike. Twelve coins weigh exactly the same, but the other one is heavier. You have a pan balance. Tell how to find the heavier coin in just three weighings.

47. *The Sum of Numbers* Find the sum of the first 500 counting numbers. (*Hint:* Group in pairs.)

48. *Balancing a Scale* On a balance scale, three green balls balance six blue balls, two yellow balls balance five blue balls, and six blue balls balance four white balls. How many blue balls are needed to balance four green, two yellow, and two white balls?

49. *Palindromes* How many three-digit numbers greater than 100 are palindromes?

50. *Figures* Describe the fifth figure.

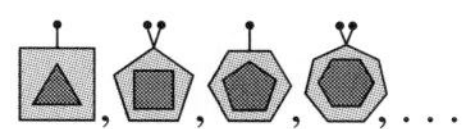

51. *Patterns* How many orange tiles will be required to build the sixth figure in this pattern?

52. *Sum of Numbers* Place the numbers 1 through 12 in the 12 circles so that the sum of the numbers in each of the six rows is 26. Use each number from 1 through 12 exactly once.

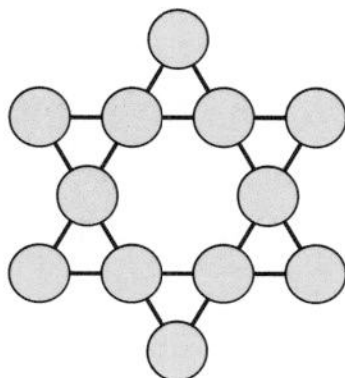

53. *People in a Line* In how many ways can
   a) two people stand in a line?
   b) three people stand in a line?
   c) four people stand in a line?
   d) five people stand in a line?
   e) Using the results from parts (a) through (d), make a conjecture about the number of ways in which n people can stand in a line.

CHAPTER 1 TEST

In Exercises 1 and 2, use inductive reasoning to determine the next three numbers in the pattern.

1. 6, 9, 12, 15, . . .
2. $1, \frac{1}{3}, \frac{1}{9}, \frac{1}{27}, \ldots$
3. Pick any number, multiply the number by 5, and add 10 to the number. Divide the sum by 5. Subtract 1 from the quotient.
   a) What is the relationship between the number you started with and the final answer?
   b) Arbitrarily select some different numbers and repeat the process. Record the original number and the results.
   c) Make a conjecture about the relationship between the original number and the final answer.
   d) Prove, using deductive reasoning, the conjecture made in part (c).

In Exercises 4 and 5, estimate the answers.

4. $0.06 \times 98{,}000$
5. $\dfrac{102{,}000}{0.00302}$
6. *Estimating Area* If each square represents one square unit, estimate the area of the shaded figure.

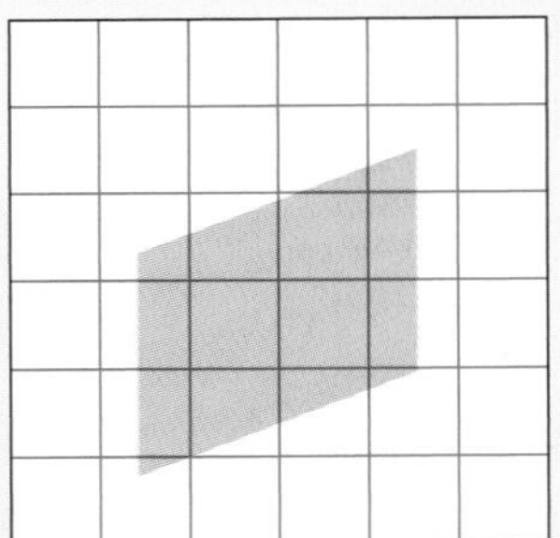

7. *Body Mass Index* During the week of June 12, 2001, the federal government updated its method for determining if a child is overweight. To make this decision, first determine the child's body mass index (BMI). Then compare the BMI with one of the two charts, one for males and one for females, provided by the government. On the right we give the chart for males up to age 20. To determine a child's BMI:
   1) Divide the child's weight (in pounds) by the child's height (in inches).
   2) Divide the results from part 1 by the child's height again.
   3) Multiply the result from part 2 by 703.

   Richard is a 14-year-old male who weighs 130 lb and is 63 in. tall.
   a) Determine his BMI.
   b) Does he appear to be at risk for being overweight, or is he overweight? Explain.

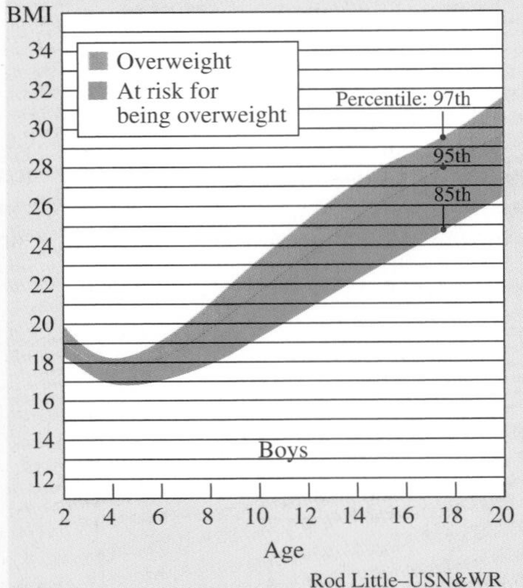

8. *Gas Usage* A gas company charges $9.63 for the basic monthly fee, which includes the first 3 therms of gas used. It charges 72 cents for each additional therm used. If the Smiths' gas bill for December was $122.13, how many therms of gas did they use during that month?
9. *Cans of Soda* At a local store a six-pack of soda costs $2.59 and individual cans cost $0.80. What is the maximum number of cans of soda that can be purchased for $15?
10. *Cutting Wood* How much time does it take Carla Knab, a carpenter, to cut a 10 ft length of wood into four equal pieces, if each cut takes $2\frac{1}{2}$ min?

11. *Determining Size* In this photo of *Sunflowers*, 1889, by Vincent van Gogh, 1 in. equals 15.8 in. on the actual painting. Find the dimensions of the actual painting.

12. *Payment Shortfall* Monica Wilson gets $12.75 per hour with time and a half for any time over 40 hours per week. If she works a 50 hr week and gets paid $652.25, by how much was she underpaid?

13. *Magic Square* Create a magic square by using the numbers 5, 10, 15, 20, 25, 30, 35, 40, and 45. The sum of the numbers in every row, column, and diagonal must be 75.

14. *A Drive to the Beach* Mary Chin drove from her home to the beach that is 30 mi from her house. The first 15 mi she drove at 60 mph, and the next 15 mi she drove at 30 mph. Would the trip take more, less, or the same time if she traveled the entire 30 mi at a steady 45 mph?

15. *Pick Five Numbers* From the six numbers 2, 6, 8, 9, 11, and 13, pick five that, when multiplied, give 11,232.

16. *Jelly Bean Guess* One guess is off by 9, another guess is off by 17, and yet another guess is off by 31. How many jelly beans are in the jar?

17. *Buying Plants* David Mackin wants to purchase nine herb plants. Countryside Nursery has herbs that are on sale at three for $3.99. David has a coupon for 25% off an unlimited number of herb plants at the original price of $1.75 per plant.
    a) Determine the cost of purchasing nine plants at the sale price.
    b) Determine the cost of purchasing nine plants if the coupon is used.
    c) Which is the least expensive way to purchase the nine plants, and by how much?

18. *Setting the Switches* In how many different ways can a panel of four on–off switches be set if no two adjacent switches may be off?

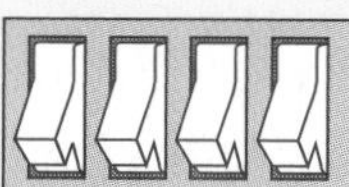

GROUP PROJECTS

Holiday Shopping

1. It is December 1 and John needs to begin his holiday shopping. He intends to purchase gifts for three people: his girlfriend, Melissa; his mother, Ruth; and his father, Don. He doesn't want to spend more than a total of $325, including the 7% sales tax.
    a) If John were to spend the $325 equally among the three people, approximate the amount that would be spent on each person.
    b) If John were to spend the $325 equally among the three people, determine the maximum amount, *before tax*, that he could spend on each person and not exceed the maximum of $325, including tax.
    c) John decides to get a new set of wrenches for his father. He sees the specific set he wants on sale at Sears. He calls four Sears stores to see if they have the set of wrenches in stock. They all reply that the set is out of stock. He decides that calling additional Sears stores is useless, for he believes that

they will also tell him that the set of wrenches is out of stock. What type of reasoning did John use in arriving at his conclusion? Explain.

d) John finds an equivalent set of wrenches at a True Value hardware store. The set he is considering is a combination set that contains both standard U.S. size and metric size wrenches. Its regular price before tax is \$62, but it is selling for 10% off its regular price. He can also purchase the same wrenches by purchasing two separate sets, one for standard U.S. size wrenches and the other for metric sizes. Each of these sets has a regular price, before tax, of \$36, but both are on sale for 20% off their regular prices. Can John purchase the combination set or the two individual sets less expensively?

e) How much will John save, *after tax*, by using the less expensive method?

Going on Vacation

2. Bob and Kristen Williams decide to go on a vacation. They live in San Francisco, California, and plan to drive to New Orleans, Louisiana.

a) Obtain a map that shows routes that they may take from San Francisco to New Orleans. Write directions for them from San Francisco to New Orleans via the shortest distance. Use major highways whenever possible.

b) Use the scale on the map to estimate the one-way distance to New Orleans.

c) If the Williams estimate that they will average 50 mph (including comfort stops), estimate the travel time, in hours, to New Orleans.

d) If the Williams want to travel about 400 miles per day, locate a town in the vicinity of where they will stop each evening.

e) If they begin each segment of the trip each day at 9 A.M., about what time will they look for a hotel each evening?

f) Use the information provided in parts (a) through (e) to estimate the time of day they will arrive in New Orleans.

g) Estimate the mileage of a typical midsized car and the cost per gallon of a gallon of regular unleaded gasoline. Then estimate the cost of gasoline for the Williams' trip.

h) Estimate the cost of a typical breakfast, a typical lunch, and a typical dinner for two adults, and the cost of a typical motel room. Then estimate the total cost, including meals, gas, and lodging, for the Williams' trip from San Francisco to New Orleans (one way).

Problem Solving

3. Four acrobats who bill themselves as the "Tumbling Tumbleweeds" finish up their act with the amazing "Human Pillar," in which the acrobats form a tower, each one standing on the shoulders of the one below. Each acrobat (Ernie, Jed, Tex, and Zeke Tumbleweed) wears a different distinctive item of western garb (chaps, holster, Stetson hat, or leather vest) in the act. Can you identify the members of the "Human Pillar," from top to bottom, by name and apparel?

a) Jed Tumbleweed is not on top, but he is somewhere above the man in the Stetson.

b) Zeke Tumbleweed does not wear the holster.

c) The man in the vest is not on top.

d) The man in the chaps is somewhere above Tex but somewhere below Zeke.

Order	Name	Apparel
________	________	________
________	________	________
________	________	________
________	________	________

Set building is a fundamental learning tool for even the smallest children. As babies, they learn to distinguish "me" from "mom" and "dad." As toddlers, they learn to distinguish and categorize objects as members of a set according to size, color, or shape. The TV show *Sesame Street* teaches children set building in the game "One of these things is not like the other."

SETS

One of the most basic human impulses is to sort and classify things. Consider yourself, for example. How many different sets are you a member of? You might start with some simple categories, such as whether you are male or female, your age group, and the state you live in. Then you might think about your family's ethnic group, socioeconomic group, and nationality. These are but some of the many ways you could describe yourself to other people.

Of what use is this activity of categorization? As you will see in this chapter, putting elements into sets helps you order and arrange your world. It allows you to deal with large quantities of information. Set building is a learning tool that helps answer the question, "What are the characteristics of this group?"

Sets underlie other mathematical topics, such as logic and abstract algebra. In fact, the book *Eléments de Mathématique,* written by a group of French mathematicians under the pseudonym Nicolas Bourbaki, states, "Nowadays it is possible, logically speaking, to derive the whole of known mathematics from a single source, the theory of sets."

2.1 SET CONCEPTS

We encounter sets in many different ways every day of our lives. A *set* is a collection of objects, which are called *elements* or *members* of the set. For example, the United States is a collection or set of 50 states. The 50 individual states are the members or elements of the set that is called the United States.

A set is *well defined* if its contents can be clearly determined. The set of U.S. presidents is a well-defined set because its contents, the presidents, can be named. The set of the three best movies is not a well-defined set because the word *best* is interpreted differently by different people. In this text, we use only well-defined sets.

Three methods are commonly used to indicate a set: (1) description, (2) roster form, and (3) set-builder notation.

The method of indicating a set by *description* is illustrated in Example 1.

PROFILE IN MATHEMATICS

GEORG CANTOR

Georg Cantor (1845–1918), born in St. Petersburg, Russia, is recognized as the founder of set theory. Cantor's creative work in mathematics was nearly lost when his father insisted that he become an engineer rather than a mathematician. His two major books on set theory, *Foundations of General Theory of Aggregates* and *Contributions to the Founding of the Theory of Transfinite Numbers,* were published in 1883 and 1895, respectively.

EXAMPLE 1 *Description of Sets*

Write a description of the set containing the elements Monday, Tuesday, Wednesday, Thursday, Friday, Saturday, Sunday.

SOLUTION: The set is the days of the week. ▲

Listing the elements of a set inside a pair of *braces*, { }, is called *roster form.* The braces are an essential part of the notation because they identify the contents as a set. For example, $\{1, 2, 3\}$ is notation for the set whose elements are 1, 2, and 3, but $(1, 2, 3)$ and $[1, 2, 3]$ are not sets because parentheses and brackets do not indicate a set. For a set written in roster form, commas separate the elements of the set. The order in which the elements are listed is not important.

Sets are generally named with capital letters. For example, the name commonly selected for the set of *natural numbers* or *counting numbers* is N.

Natural Numbers

$N = \{1, 2, 3, 4, 5, \dots\}$

The three dots after the 5, called an *ellipsis,* indicate that the elements in the set continue in the same manner. An ellipsis followed by a last element indicates that the elements continue in the same manner up to and including the last element. This notation is illustrated in Example 2(b).

EXAMPLE 2 *Roster Form of Sets*

Express the following in roster form.

a) Set A is the set of natural numbers less than 6.
b) Set B is the set of natural numbers less than or equal to 50.
c) Set P is the set of planets in Earth's solar system.

SOLUTION:

a) The natural numbers less than 6 are 1, 2, 3, 4, and 5. Thus, set A in roster form is $A = \{1, 2, 3, 4, 5\}$.

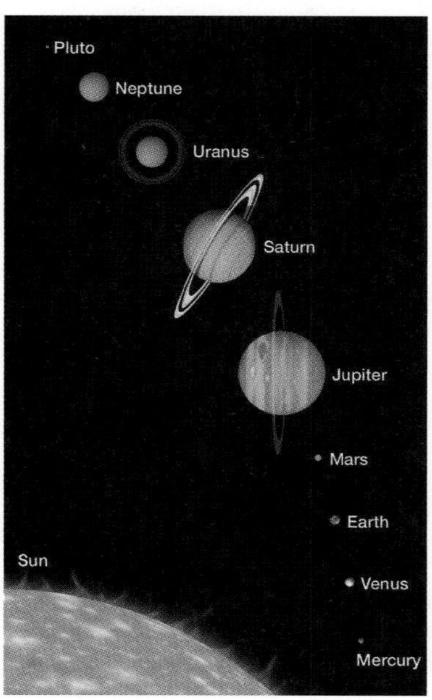

The planets of Earth's solar system.

b) $B = \{1, 2, 3, 4, \dots, 50\}$. The 50 after the ellipsis indicates that the elements continue in the same manner up to and including the number 50.

c) $P =$ {Mercury, Venus, Earth, Mars, Jupiter, Saturn, Uranus, Neptune, Pluto}

EXAMPLE 3 *The Word* Inclusive

Express the following in roster form.

a) The set of natural numbers between 5 and 8.

b) The set of natural numbers between 5 and 8, inclusive.

SOLUTION:

a) $A = \{6, 7\}$

b) $B = \{5, 6, 7, 8\}$. Note that the word *inclusive* indicates that the values of 5 and 8 are included in the set.

The symbol $\in$, read, is an element of, is used to indicate membership in a set. In Example 3, since 6 is an element of set A, we write $6 \in A$. This may also be written $6 \in \{6, 7\}$. We may also write $8 \notin A$, meaning that 8 is not an element of set A.

Set-builder notation (sometimes called *set-generator notation*) may be used to symbolize a set. Set-builder notation is frequently used in algebra. The following example illustrates its form.

D	$=$	$\{$	x	$\vert$	Condition(s) $\}$
↑	↑	↑	↑	↑	↑
Set D	is	the set of	all elements x	such that	the condition(s) x must meet in order to be a member of the set.

Consider $E = \{x \mid x \in N \text{ and } x > 10\}$. The statement is read: "Set E is the set of all the elements x such that x is a natural number and x is greater than 10." The conditions that x must meet to be a member of the set are $x \in N$, which means that x must be a natural number, and $x > 10$, which means that x must be greater than 10. The numbers that meet both conditions are the set of natural numbers greater than 10. Set E in roster form is

$$E = \{11, 12, 13, 14, \dots\}$$

EXAMPLE 4 *Using Set-Builder Notation*

a) Write set $B = \{1, 2, 3, 4, 5\}$ in set-builder notation.

b) Write, in words, how you would read set B in set-builder notation.

SOLUTION:

a) Since set B consists of the natural numbers less than 6, we write

$$B = \{x \mid x \in N \text{ and } x < 6\}$$

Another acceptable answer is $B = \{x \mid x \in N \text{ and } x \leq 5\}$.

b) Set B is the set of all elements x such that x is a natural number and x is less than 6. ▲

EXAMPLE 5 *Roster Form to Set-Builder Notation*

a) Write set $C = \{$North America, South America, Europe, Asia, Australia, Africa, Antarctica$\}$ in set-builder notation.

b) Write in words how you would read set C in set-builder notation.

SOLUTION:

a) $C = \{x \mid x \text{ is a continent}\}$.

b) Set C is the set of all elements x such that x is a continent. ▲

EXAMPLE 6 *Set-Builder Notation to Roster Form*

Write set $A = \{x \mid x \in N \text{ and } 2 \leq x < 8\}$ in roster form.

SOLUTION: $A = \{2, 3, 4, 5, 6, 7\}$ ▲

EXAMPLE 7 *Busiest Ports*

The chart shows the 10 busiest U.S. ports in 2000, ranked by tonnage handled. Also given is a map of Texas and its ports. Let set T be the set of ports in Texas that are among the 10 busiest ports in the United States. Write set T in roster form.

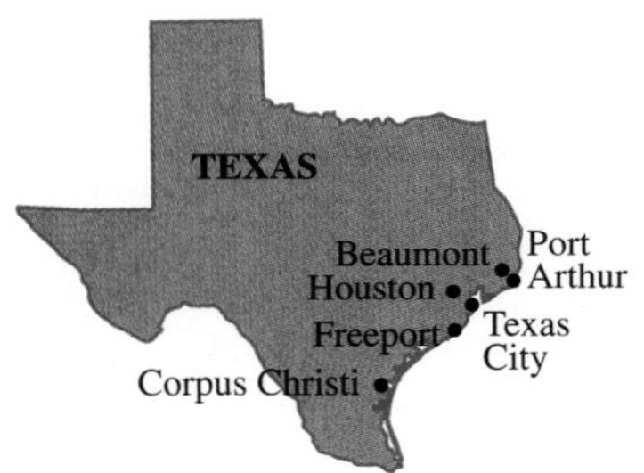

Ten Busiest Ports in the U.S., 2000	Total Tonnage
South Louisiana, LA, Port of	217,756,734
Houston, TX	191,419,265
New York, NY and NJ, Port of	138,669,879
New Orleans, LA	90,768,449
Corpus Christi, TX	83,124,950
Beaumont, TX	82,652,554
Huntington-Tristate, WV	78,867,987
Long Beach, CA	70,149,684
Baton Rouge, LA	65,631,084
Texas City, TX	61,585,891

Source: Corps of Engineers, Department of the U.S. Army, U.S. Department of Defense

SOLUTION: By examining the map and the chart we find that four cites appear on both the map and the chart. They are Beaumont, Corpus Christi, Houston, and Texas City. Thus, set $T = \{$Beaumont, Corpus Christi, Houston, Texas City$\}$. ▲

A set is said to be ***finite*** if it either contains no elements or the number of elements in the set is a natural number. The set $B = \{2, 4, 6, 8, 10\}$ is a finite set because the number of elements in the set is 5, and 5 is a natural number. A set that is not finite is said to be ***infinite***. The set of counting numbers is one example of an infinite set. Infinite sets are discussed in more detail in Section 2.6.

DID YOU KNOW

Creature Cards

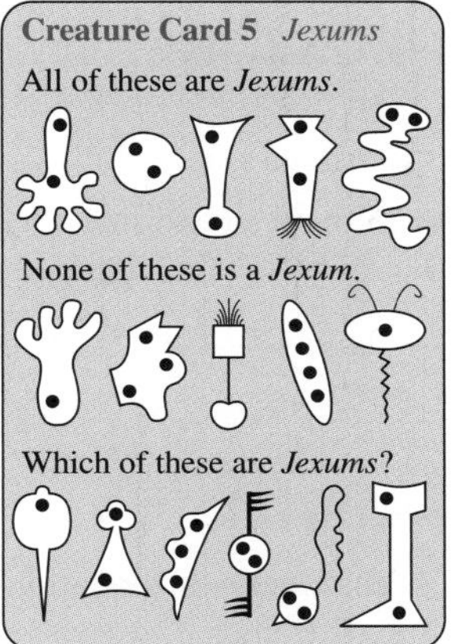

We learn to group objects according to what we see as the relevant distinguishing characteristics. One way used by educators to measure this ability is through visual cues. An example can be seen in this test, called "Creature Cards," offered by the Education Development Center. How would you describe membership in the set of Jexums?

Another important concept is equality of sets.

> Set A is **equal** to set B, symbolized by $A = B$, if and only if set A and set B contain exactly the same elements.

For example, if set $A = \{1, 2, 3\}$ and set $B = \{3, 1, 2\}$, then $A = B$ because they contain exactly the same elements. The order of the elements in the set is not important. If two sets are equal, both must contain the same number of elements. The number of elements in a set is called its *cardinal number.*

> The **cardinal number** of set A, symbolized by $n(A)$, is the number of elements in set A.

Both set $A = \{1, 2, 3\}$ and set $B = \{\text{England, Brazil, Japan}\}$ have a cardinal number of 3; that is, $n(A) = 3$, and $n(B) = 3$. We can say that set A and set B both have a cardinality of 3.

Two sets are said to be *equivalent* if they contain the same number of elements.

> Set A is **equivalent** to set B if and only if $n(A) = n(B)$.

Any sets that are equal must also be equivalent. Not all sets that are equivalent are equal, however. The sets $D = \{\text{a, b, c}\}$ and $E = \{\text{apple, orange, pear}\}$ are equivalent, since both have the same cardinal number, 3. Because the elements differ, however, the sets are not equal.

Two sets that are equivalent or have the same cardinality can be placed in ***one-to-one correspondence***. Set A and set B can be placed in one-to-one correspondence if every element of set A can be matched with exactly one element of set B and every element of set B can be matched with exactly one element of set A. For example, there is a one-to-one correspondence between the student names on a class list and the student identification numbers because we can match each name with a student identification number.

Consider set S, states, and set C, state capitals.

$S = \{$North Carolina, Georgia, South Carolina, Florida$\}$

$C = \{$Columbia, Raleigh, Tallahassee, Atlanta$\}$

Two different one-to-one correspondences for sets S and C follow.

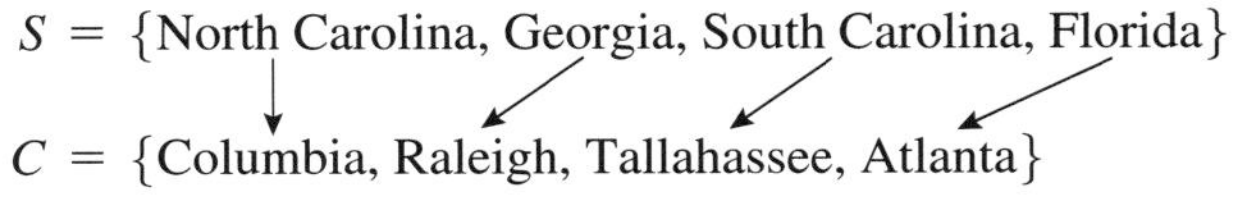

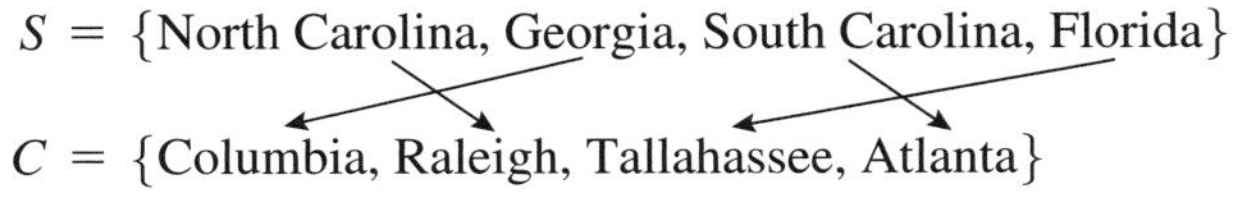

$S = \{$North Carolina, Georgia, South Carolina, Florida$\}$
↓ ↙ ↙ ↙
$C = \{$Columbia, Raleigh, Tallahassee, Atlanta$\}$

$S = \{$North Carolina, Georgia, South Carolina, Florida$\}$
(North Carolina → Raleigh, Georgia → Columbia, South Carolina → Atlanta, Florida → Tallahassee)
$C = \{$Columbia, Raleigh, Tallahassee, Atlanta$\}$

Other one-to-one correspondences between sets S and C are possible. Do you know which capital goes with which state?

Null or Empty Set

Some sets do not contain any elements, such as the set of zebras that are in this room.

The set that contains no elements is called the **empty set** or **null set** and is symbolized by { } or ∅.

Note that {∅} is not the empty set. This set contains the element ∅ and has a cardinality of 1. The set {0} is also not the empty set because it contains the element 0. It also has a cardinality of 1.

EXAMPLE 8 *Natural Number Solutions*

Indicate the set of natural numbers that satisfies the equation $x + 2 = 0$.

SOLUTION: The values that satisfy the equation are those that make the equation a true statement. Only the number -2 satisfies this equation. Because -2 is not a natural number, the solution set of this equation is { } or ∅. ▲

Universal Set

Another important set is a *universal set*.

A **universal set,** symbolized by U, is a set that contains all the elements for any specific discussion.

When a universal set is given, only the elements in the universal set may be considered when working the problem. If, for example, the universal set for a particular problem is defined as $U = \{1, 2, 3, 4, \ldots, 10\}$, then only the natural numbers 1 through 10 may be used in that problem.

SECTION 2.1 EXERCISES

Concept/Writing Exercises

In Exercises 1–12, answer each question with a complete sentence.

1. What is a set?
2. What is an ellipsis, and how is it used?
3. What are the three ways that a set can be written? Give an example of each.
4. What is a finite set?
5. What is an infinite set?
6. What are equal sets?
7. What are equivalent sets?
8. What is the cardinal number of a set?
9. What is the empty set?
10. What are the two ways to indicate the empty set?
11. What does a one-to-one correspondence of two sets mean?
12. What is a universal set?

Practice the Skills

In Exercises 13–18, determine whether each set is well defined.

13. The set of people who own large dogs

14. The set of the best Internet web sites

15. The set of states that have a common border with Colorado
16. The set of the four states in the United States having the largest areas
17. The set of astronauts who walked on the moon
18. The set of the nicest entertainers

In Exercises 19–24, determine whether each set is finite or infinite.

19. $\{1, 3, 5, 7, \dots\}$
20. The set of multiples of 6 between 0 and 90
21. The set of even numbers greater than 19
22. The set of fractions between 1 and 2
23. The set of odd numbers greater than 15
24. The set of cars in the parking lot at the Home Depot store at 770 Jefferson Road, Rochester, New York, on August 1, 2004, at 1:00 P.M.

In Exercises 25–34, express each set in roster form. You may need to use a world almanac or some other reference source.

25. The set of oceans in the world
26. The set of states in the United States whose names begin with the letter I
27. The set of natural numbers between 10 and 178
28. $C = \{x \mid x + 6 = 10\}$
29. $B = \{x \mid x \in N \text{ and } x \text{ is even}\}$
30. The set of states west of the Mississippi River that have a common border with the state of Florida
31. The set of football players over the age of 70 who are still playing in the National Football League
32. The set of states in the United States that have no common border with any other state
33. $E = \{x \mid x \in N \text{ and } 6 \le x < 72\}$
34. The set of professional baseball players in the major leagues who have hit at least 70 home runs in a season prior to 2002

The list above and to the right shows the estimated price of the seven best-selling digital cameras, ranked by market share, in October 2000. In Exercises 35–38, use the list to represent each of the sets in roster form. Let the seven cameras in the list represent the universal set.

35. The set of best-selling digital cameras with an estimated price greater than $500
36. The set of best-selling digital cameras with an estimated price less than $300
37. The set of best-selling digital cameras with an estimated price between $250 and $650
38. The set of best-selling digital cameras with an estimated price between $500 and $800

Camera	Estimated Price
1. Sony Mavica FD-73	$400
2. Olympus D-360L	$290
3. Sony DSC-S50	$550
4. Sony DSC-S70	$750
5. Kodak DC215	$310
6. H-P Photo Smart C315	$300
7. Sony Mavica FD-90	$700

Source: PC Data (Rochester Democrat and Chronicle)

The following graph shows the federal deficit, in billions of dollars, for 2002 and 2003 and the projected federal deficit for the years 2004–2008. In exercises 39–42, use the graph to represent each of the sets in roster form.

Deficit Outlook

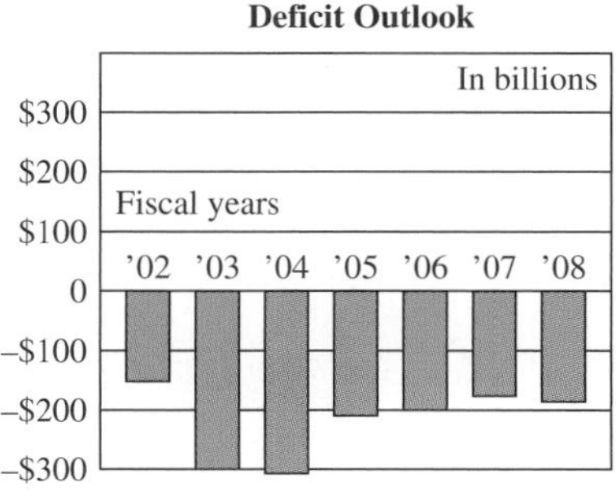

Source: White House Office of Management and Budget

39. The set of years in which the federal deficit or the projected federal deficit is more than $100 billion
40. The set of years in which the federal deficit or the projected federal deficit is between $100 billion and $250 billion
41. The set of years in which the federal deficit or the projected federal deficit is less than $250 billion
42. The set of years in which the federal deficit or the projected federal deficit is more than $250 billion

In Exercises 43–50, express each set in set-builder notation.

43. $B = \{4, 5, 6, 7, 8, 9, 10\}$

44. $A = \{1, 2, 3, 4, 5, 6, 7\}$

45. $C = \{3, 6, 9, 12, \dots\}$

46. $D = \{5, 10, 15, 20, \dots\}$

47. E is the set of odd natural numbers

48. A is the set of national holidays in the United States in September

49. C is the set of months that contain less than 30 days

50. $F = \{15, 16, 17, \dots, 100\}$

In Exercises 51–58, write a description of each set.

51. $A = \{1, 2, 3, 4, 5, 6, 7\}$

52. $D = \{4, 8, 12, 16, 20, \dots\}$

53. $V = \{a, e, i, o, u\}$

54. $S = \{$Bashful, Doc, Dopey, Grumpy, Happy, Sleepy, Sneezy$\}$

55. $C = \{$Casio, Hewlett-Packard, Sharp, Texas Instruments, . . . $\}$

56. $B = \{$Mississippi, Missouri, Yukon, Rio Grande, Arkansas$\}$

57. $B = \{$John Lennon, Ringo Starr, Paul McCartney, George Harrison$\}$

58. $E = \{x \mid x \in N \text{ and } 5 < x \leq 12\}$

The following list shows the top 10 media markets, in order, for advertisements for the 2000 elections of the president and members of Congress, through October 10, 2000. In Exercises 59–62, use the list to represent each of the sets in roster form. Let the 10 markets represent the universal set.

59. $\{x \mid x$ is a city in which the number of advertisements was greater than 18,000$\}$

60. $\{x \mid x$ is a city in which the number of advertisements was less than 10,000$\}$

61. $\{x \mid x$ is a city in which the number of advertisements was between 12,500 and 13,000$\}$

62. $\{x \mid x$ is a city in which the number of advertisements was between 13,000 and 14,000$\}$

Market	Number of Ads
1. St. Louis	18,755
2. Kansas City	14,872
3. Seattle	14,234
4. Detroit	13,490
5. Spokane	13,191
6. Grand Rapids	12,436
7. Flint-Saginaw	11,797
8. Philadelphia	11,006
9. Louisville	10,345
10. Scranton	9016

Source: The Hotline and the Brennan Center for Justice

St. Louis, MO

The following graph shows the advertising revenues for Yahoo, in millions of dollars, for the years 1998–2002. In Exercises 63–66, use the graph to represent each of the sets in roster form.

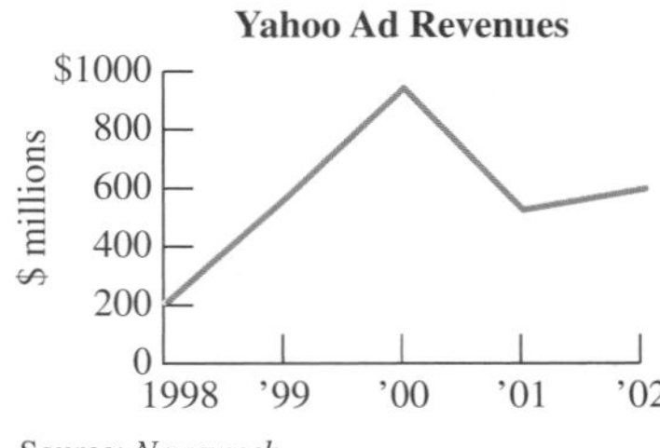

Source: *Newsweek*

63. $\{x \mid x$ is a year in which advertising revenues exceeded \$400 million$\}$

64. $\{x \mid x$ is a year in which advertising revenues were less than \$300 million$\}$

65. $\{x \mid x$ is a year in which advertising revenues exceeded \$500 million but were less than \$800 million$\}$

66. $\{x \mid x$ is a year in which advertising revenues exceeded \$300 million but were less than \$500 million$\}$

In Exercises 67–74, state whether each statement is true or false. If false, give the reason.

67. $\{b\} \in \{a, b, c, d, e, f\}$
68. $b \in \{a, b, c, d, e, f\}$
69. $h \in \{a, b, c, d, e, f\}$
70. Cat in the Hat $\in$ {characters created by Dr. Seuss}
71. $3 \notin \{x \mid x \in N$ and x is odd$\}$
72. Maui $\in$ {capital cities in the United States}
73. *Titanic* $\in$ {top 10 motion pictures with the greatest revenues}
74. $2 \in \{x \mid x$ is an odd natural number$\}$

In Exercises 75–78, for the sets $A = \{2, 4, 6, 8\}$, $B = \{1, 3, 7, 9, 13, 21\}$, $C = \{\ \}$, and $D = \{\#, \&, \%, \square, *\}$, determine

75. $n(A)$.
76. $n(B)$.
77. $n(C)$.
78. $n(D)$.

In Exercises 79–84, determine whether the pairs of sets are equal, equivalent, both, or neither.

79. $A =$ {circle, triangle, square},
 $B =$ {triangle, circle, square}
80. $A = \{7, 9, 10\}$, $B = \{a, b, c\}$
81. $A =$ {grapes, apples, oranges},
 $B =$ {grapes, peaches, apples, oranges}
82. A is the set of collies.
 B is the set of dogs.
83. A is the set of letters in the word *tap.*
 B is the set of letters in the word *ant.*
84. A is the set of states.
 B is the set of state capitals.

Problem Solving

85. Set-builder notation is often more versatile and efficient than listing a set in roster form. This versatility is illustrated with the two sets.

$$A = \{x \mid x \in N \text{ and } x > 2\}$$
$$B = \{x \mid x > 2\}$$

 a) Write a description of set A and set B.
 b) Explain the difference between set A and set B. (*Hint:* Is $4\frac{1}{2} \in A$? Is $4\frac{1}{2} \in B$?)
 c) Write set A in roster form.
 d) Can set B be written in roster form? Explain your answer.

86. Start with sets

$$A = \{x \mid 2 < x \leq 5 \text{ and } x \in N\}$$

and

$$B = \{x \mid 2 < x \leq 5\}$$

 a) Write a description of set A and set B.
 b) Explain the difference between set A and set B.
 c) Write set A in roster form.
 d) Can set B be written in roster form? Explain your answer.

*A cardinal number answers the question "How many?" An **ordinal number** describes the relative position that an element occupies. For example, Molly's desk is the third desk from the aisle.*

In Exercises 87–90, determine whether the number used is a cardinal number or an ordinal number.

87. John Grisham has written 12 books.

John Grisham

88. Study the chart on page 25 in the book.
89. Lincoln was the sixteenth president of the United States.
90. Emily paid \$35 for her new blouse.
91. Describe three sets of which you are a member.
92. Describe three sets that have no members.
93. Write a short paragraph explaining why the universal set and the empty set are necessary in the study of sets.

Challenge Problem/Group Activity

94. a) In a given exercise, a universal set is not specified, but we know that actor Brad Pitt is a member of the universal set. Describe five different possible universal sets of which Brad Pitt is a member.
 b) Write a description of one set that includes all the universal sets in part (a).

Recreational Mathematics

95. *Face to Face* Place the eight squares on the left into the diagram on the right so that two squares with a common border will have the same number on both sides of the border. Do not turn the squares or rearrange the numbers within each square.

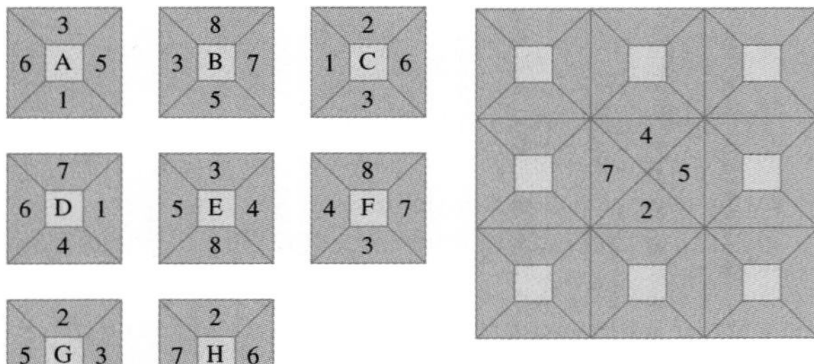

Internet/Research Activity

96. Georg Cantor is recognized as the founder and a leader in the development of set theory. Do research and write a paper on his life and his contributions to set theory and to the field of mathematics. References include history of mathematics books, encyclopedias, and the Internet.

2.2 SUBSETS

In our complex world, we often break larger sets into smaller more manageable sets, called *subsets*. For example, consider the set of people in your class. Suppose we categorize the set of people in your class according to the first letter of their last name (the A's, B's, C's, etc.). When we do this, each of these sets may be considered a subset of the original set. Each of these subsets can be separated further. For example, the set of people whose last name begins with the letter A can be categorized as either male or female or by their age. Each of these collections of people is also a subset. A given set may have many different subsets.

Set A is a **subset** of set B, symbolized by $A \subseteq B$, if and only if all the elements of set A are also elements of set B.

The symbol $A \subseteq B$ indicates that "set A is a subset of set B." The symbol $\not\subseteq$ is used to indicate "is not a subset." Thus, $A \not\subseteq B$ indicates that set A is not a subset of set B. *To show that set A is not a subset of set B, we must find at least one element of set A that is not an element of set B.*

EXAMPLE 1 *A Subset?*

Determine whether set A is a subset of set B.

a) $A = \{$blue jay, robin, cardinal$\}$
   $B = \{$blue jay, robin, chickadee, cardinal$\}$

DID YOU KNOW

Rainbows

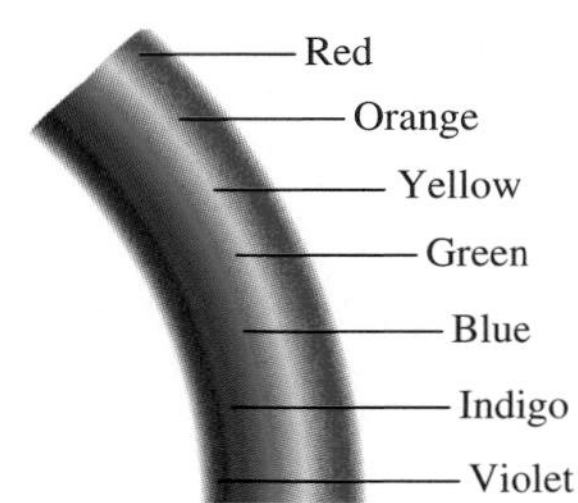

Colors of primary rainbow

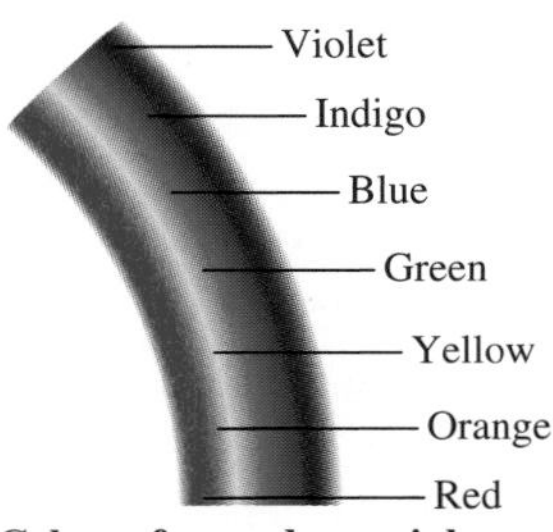

Colors of secondary rainbow

Most rainbows we see are primary rainbows, but there are rare moments when a second, fainter rainbow can be seen behind the first. In this secondary rainbow, the light pattern has been reversed. Both rainbows contain the same set of colors, so each set of colors is a subset of the other.

b) $A = \{2, 3, 4, 5\}$ $B = \{2, 3\}$
c) $A = \{x \mid x \text{ is a yellow fruit}\}$
   $B = \{x \mid x \text{ is a red fruit}\}$
d) $A = \{\text{cassette, compact disc, videotape}\}$
   $B = \{\text{compact disc, videotape, cassette}\}$

SOLUTION:

a) All the elements of set A are contained in set B, so $A \subseteq B$.
b) The elements 4 and 5 are in set A but not in set B, so $A \not\subseteq B$ (A is not a subset of B). In this example, however, all the elements of set B are contained in set A; therefore, $B \subseteq A$.
c) There are fruits, such as bananas, that are in set A that are not in set B, so $A \not\subseteq B$.
d) All the elements of set A are contained in set B, so $A \subseteq B$. Note that set A = set B.

▲

Proper Subsets

Set A is a **proper subset** of set B, symbolized by $A \subset B$, if and only if all the elements of set A are elements of set B and set $A \neq$ set B (that is, set B must contain at least one element not in set A).

Consider the sets $A = \{\text{red, blue, yellow}\}$ and $B = \{\text{red, orange, yellow, green, blue, violet}\}$. Set A is a *subset* of set B, $A \subseteq B$, because every element of set A is also an element of set B. Set A is also a *proper subset* of set B, $A \subset B$, because set A and set B are not equal. Now consider $C = \{\text{car, bus, train}\}$ and $D = \{\text{train, car, bus}\}$. Set C is a subset of set D, $C \subseteq D$, because every element of set C is also an element of set D. Set C, however, is not a proper subset of set D, $C \not\subset D$, because set C and set D are equal sets.

EXAMPLE 2 *A Proper Subset?*

Determine whether set A is a proper subset of set B.

a) $A = \{\text{refrigerator, microwave, dishwasher}\}$
   $B = \{\text{stove, refrigerator, microwave, dishwasher, garbage disposal}\}$
b) $A = \{a, b, c, d\}$ $\quad B = \{a, c, b, d\}$

SOLUTION:

a) All the elements of set A are contained in set B, and sets A and B are not equal; thus, $A \subset B$.
b) Set A = set B, so $A \not\subset B$. (However, $A \subseteq B$.)

▲

Every set is a subset of itself, but no set is a proper subset of itself. For all sets A, $A \subseteq A$, but $A \not\subset A$. For example, if $A = \{1, 2, 3\}$, then $A \subseteq A$ because every element of set A is contained in set A, but $A \not\subset A$ because set A = set A.

Let $A = \{\ \}$ and $B = \{1, 2, 3, 4\}$. Is $A \subseteq B$? To show $A \not\subseteq B$, you must find at least one element of set A that is not an element of set B. As this cannot be done, $A \subseteq B$ must be true. Using the same reasoning, we can show that *the empty set is a subset of every set, including itself.*

EXAMPLE 3 *Element or Subset?*

Determine whether the following are true or false.

a) $3 \in \{3, 4, 5\}$
b) $\{3\} \in \{3, 4, 5\}$
c) $\{3\} \in \{\{3\}, \{4\}, \{5\}\}$
d) $\{3\} \subseteq \{3, 4, 5\}$
e) $3 \subseteq \{3, 4, 5\}$
f) $\{\ \} \subseteq \{3, 4, 5\}$

SOLUTION:

a) $3 \in \{3, 4, 5\}$ is a true statement because 3 is a member of the set $\{3, 4, 5\}$.
b) $\{3\} \in \{3, 4, 5\}$ is a false statement because $\{3\}$ is a set, and the set $\{3\}$ is not an element of the set $\{3, 4, 5\}$.
c) $\{3\} \in \{\{3\}, \{4\}, \{5\}\}$ is a true statement because $\{3\}$ is an element in the set. The elements of the set $\{\{3\}, \{4\}, \{5\}\}$ are themselves sets.
d) $\{3\} \subseteq \{3, 4, 5\}$ is a true statement because every element of the first set is an element of the second set.
e) $3 \subseteq \{3, 4, 5\}$ is a false statement because the 3 is not in braces, so it is not a set and thus cannot be a subset. The 3 is an element of the set as indicated in part (a).
f) $\{\ \} \subseteq \{3, 4, 5\}$ is a true statement because the empty set is a subset of every set. ▲

Number of Subsets

How many distinct subsets can be made from a given set? The empty set has no elements and has exactly one subset, the empty set. A set with one element has two subsets. A set with two elements has four subsets. A set with three elements has eight subsets. This information is illustrated in Table 2.1 on page 53. How many subsets will a set with four elements contain?

By continuing this table with larger and larger sets, we can develop a general formula for finding the number of distinct subsets that can be made from any given set.

The **number of distinct subsets** of a finite set A is 2^n, where n is the number of elements in set A.

TABLE 2.1 Number of Subsets

Set	Subsets	Number of Subsets
{ }	{ }	$1 = 2^0$
$\{a\}$	$\{a\}$	
	{ }	$2 = 2^1$
$\{a, b\}$	$\{a, b\}$	
	$\{a\}, \{b\}$	
	{ }	$4 = 2 \times 2 = 2^2$
$\{a, b, c\}$	$\{a, b, c\}$	
	$\{a, b\}, \{a, c\}, \{b, c\}$	
	$\{a\}, \{b\}, \{c\}$	
	{ }	$8 = 2 \times 2 \times 2 = 2^3$

EXAMPLE 4 *Distinct Subsets*

a) Determine the number of distinct subsets for the set {S, L, E, D}.
b) List all the distinct subsets for the set {S, L, E, D}.
c) How many of the distinct subsets are proper subsets?

SOLUTION:

a) Since the number of elements in the set is 4, the number of distinct subsets is $2^4 = 2 \times 2 \times 2 \times 2 = 16$.

b)

{S, L, E, D}	{S, L, E}	{S, L}	{S}	{ }
	{S, L, D}	{S, E}	{L}	
	{S, E, D}	{S, D}	{E}	
	{L, E, D}	{L, E}	{D}	
		{L, D}		
		{E, D}		

c) There are 15 proper subsets. Every subset except {S, L, E, D} is a proper subset.

EXAMPLE 5 *Variations of Ice Cream*

Shanna Ruben is going to purchase ice cream at Friendly's Restaurant. To her ice cream she can add any of the following toppings: hot fudge, whipped cream, cherries, butterscotch topping, caramel topping, chopped nuts, Reese's Pieces, M & M's, Gummy Bears. How many different variations of the ice cream and toppings can be made?

SOLUTION: Shanna can order the ice cream with no extra toppings, any one topping, any two toppings, any three toppings, and so on, up to all nine toppings. One technique used in problem solving is to consider similar problems that you have solved previously. If you think about this problem, you will realize that this problem is the same as, "How many distinct subsets can be made from a set with nine elements?" The number of different variations of the ice cream is the same as the number of possible subsets of a set that has nine elements. There are 2^9 or 512 possible subsets of a set with nine elements, so there are 512 possible variations of the ice cream and toppings. ▲

DID YOU KNOW

The Ladder of Life

Scientists use sets to classify and categorize knowledge. In biology, the science of classifying all living things is called *taxonomy* and was probably practiced by the earliest cave-dwellers. Over 2000 years ago, Aristotle formalized animal classification with his "ladder of life": higher animals, lower animals, higher plants, lower plants.

Contemporary biologists use a system of classification called the Linnaean system, named after Swedish biologist Carolus Linnaeas (1707–1778). The Linnaean system starts with the smallest unit (member) and assigns it to a specific genus (set) and species (subset).

A zebra, *Equus burchelli*, is a member of the genus *Equus*, as is the horse, *Equus caballus*. Both the zebra and the horse are members of the universal set called the kingdom of animals and the same family, Equidae; they are members of different species (*E. burchelli* and *E. caballus*), however.

Even more general groupings of living things are made according to shared characteristics. The groupings, from most general to most specific are: kingdom, phylum, class, order, family, genus, and species. Each of the groupings is classified into sub groupings. For example, in the sixteenth century, living organisms were classified into two kingdoms, plants and animals. Today, living organisms are classified into six kingdoms called animalia, plantae, archaea, eubacteria, fungi, and protista.

SECTION 2.2 EXERCISES

Concept/Writing Exercises

In Exercises 1–6, answer each question with a complete sentence.

1. What is a subset?
2. What is a proper subset?
3. Explain the difference between a subset and a proper subset.
4. Write the formula for determining the number of distinct subsets for a set with n distinct elements.
5. Write the formula for determining the number of distinct proper subsets for a set with n distinct elements.

6. Can any set be a proper subset of itself? Explain.

Practice the Skills

In Exercises 7–24, answer true or false. If false, give the reason.

7. gold $\subseteq$ {gold, silver, sapphire, emerald}
8. { } $\in$ {knee, ankle, shoulder, hip}
9. { } $\subseteq$ {Tigger, Pooh, Christopher Robin}
10. red $\subset$ {red, green, blue}
11. $5 \notin \{2, 4, 6\}$
12. {Pete, Mike, Amy} $\subseteq$ {Amy, Kaitlyn, Brianna}
13. $\{\ \} = \{\emptyset\}$
14. {engineer} $\subseteq$ {architect, physician, attorney, engineer}
15. $\emptyset = \{\ \}$
16. $0 = \{\ \}$
17. $\{0\} = \emptyset$
18. $\{3, 8, 11\} \subseteq \{3, 8, 11\}$
19. {swimming} $\in$ {sailing, waterskiing, swimming}
20. $\{3, 5, 9\} \not\subseteq \{3, 9, 5\}$
21. $\{\ \} \subseteq \{\ \}$
22. $\{1\} \in \{\{1\}, \{2\}, \{3\}\}$
23. {US Airways, Delta, American} $\subset$ {American, US Airways, Delta}
24. $\{b, a, t\} \subseteq \{t, a, b\}$

In Exercises 25–32, determine whether $A = B$, $A \subseteq B$, $B \subseteq A$, $A \subset B$, $B \subset A$, or none of these. (There may be more than one answer.)

25. A = {Pepsi, Mountain Dew, Coke, Sprite}
    B = {Pepsi, Coke}
26. $A = \{x \mid x \in N \text{ and } x < 6\}$
    $B = \{x \mid x \in N \text{ and } 1 \leq x \leq 5\}$
27. Set A is the set of states east of the Mississippi River. Set B is the set of states that border the Atlantic Ocean.
28. $A = \{1, 3, 5, 7, 9\}$
    $B = \{3, 9, 5, 7, 6\}$
29. $A = \{x \mid x$ is a brand of ice cream$\}$
    B = {Breyers, Ben & Jerry's, Häagen-Dazs}
30. $A = \{x \mid x$ is a sport that uses a ball$\}$
    B = {basketball, soccer, tennis}
31. Set A is the set of natural numbers between 2 and 7. Set B is the set of natural numbers greater than 2 and less than 7.
32. Set A is the set of toys requiring batteries. Set B is the set of toys requiring AA batteries.

In Exercises 33–38, list all the subsets of the sets given.

33. $D = \emptyset$
34. $A = \{\bigcirc\}$
35. B = {pen, pencil}
36. C = {apple, peach, banana}

Problem Solving

37. For set $A = \{a, b, c, d\}$,
    a) list all the subsets of set A.
    b) state which of the subsets in part (a) are not proper subsets of set A.
38. A set contains nine elements.
    a) How many subsets does it have?
    b) How many proper subsets does it have?

In Exercises 39–50, if the statement is true for all sets A and B, write "true." If it is not true for all sets A and B, write "false." Assume that $A \neq \emptyset$, $U \neq \emptyset$, and $A \subset U$.

39. If $A \subseteq B$, then $A \subset B$.
40. If $A \subset B$, then $A \subseteq B$.
41. $A \subseteq A$
42. $A \subset A$
43. $\emptyset \subset A$
44. $\emptyset \subseteq A$
45. $A \subseteq U$
46. $\emptyset \subset \emptyset$
47. $\emptyset \subset U$
48. $U \subseteq \emptyset$
49. $\emptyset \subseteq \emptyset$
50. $U \subset \emptyset$
51. *Building a House* The Jacobsens are planning to build a house in a new development. They can either build the base model offered by the builder or add any of the following options: deck, hot tub, security system, hardwood flooring. How many different variations of the house are possible?
52. *Computer Upgrade* Jason Jackson is considering having his computer upgraded. He can leave the computer as it is, or he can upgrade any of the following set of items: RAM, modem, video card, hard drive, processor, sound card. How many possible options for upgrading does Jason have?

53. *Telephone Features* A customer with Verizon can order telephone service with some, all, or none of the following features: call waiting, call forwarding, caller identification, three-way calling, voice mail, fax line. How many different variations of the set of features are possible?

54. *Hamburger Variations* Customers ordering hamburgers at Vic and Irv's Hamburger stand are always asked, "What do you want on it?" The choices are ketchup, mustard, relish, hot sauce, onions, lettuce, tomato. How many different variations are there for ordering a hamburger?

55. If $E \subseteq F$ and $F \subseteq E$, what other relationship exists between E and F? Explain.

56. How can you determine whether the set of boys is equivalent to the set of girls at a roller-skating rink?

57. For the set $D = \{a, b, c\}$
   a) is a an element of set D? Explain.
   b) is c a subset of set D? Explain.
   c) is $\{a, b\}$ a subset of set D? Explain.

Challenge Problems/Group Activity

58. *Hospital Expansion* A hospital has four members on the board of directors: Arnold, Benitez, Cathy, and Dominique.
   a) When the members vote on whether to add a wing to the hospital, how many different ways can they vote (abstentions are not allowed)? For example, Arnold—yes, Benitez—no, Cathy—no, and Dominique—yes is one of the many possibilities.
   b) Make a listing of all the possible outcomes of the vote. For example, the vote described in part (a) could be represented as (YNNY).
   c) How many of the outcomes given in part (b) would result in a majority supporting the addition of a wing to the hospital? That is, how many of the outcomes have three or more Y's?

Recreational Mathematics

59. How many elements must a set have if the number of proper subsets of the set is $\frac{1}{2}$ of the total number of subsets of the set?

60. If $A \subset B$ and $B \subset C$, must $A \subset C$?

61. If $A \subset B$ and $B \subseteq C$, must $A \subset C$?

62. If $A \subseteq B$ and $B \subseteq C$, must $A \subset C$?

Internet/Research Activity

63. On page 54, we discussed the ladder of life. Do research and indicate all the different classifications in the Linnaean system, from most general to the most specific, in which a koala belongs.

2.3 VENN DIAGRAMS AND SET OPERATIONS

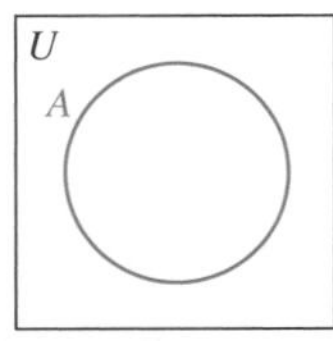

Figure 2.1

A useful technique for picturing set relationships is the Venn diagram, named for the English mathematician John Venn (1834–1923). Venn invented the diagrams and used them to illustrate ideas in his text on symbolic logic, published in 1881.

In a Venn diagram, a rectangle usually represents the universal set, U. The items inside the rectangle may be divided into subsets of the universal set. The subsets are usually represented by circles. In Fig. 2.1, the circle labeled A represents set A, which is a subset of the universal set.

Two sets may be represented in a Venn diagram in any of four different ways (see Fig. 2.2 on page 57). Two sets A and B are *disjoint* when they have no elements in common. Two disjoint sets A and B are illustrated in Fig. 2.2(a). If set A is a proper subset of set B, $A \subset B$, the two sets may be illustrated as in Fig. 2.2(b). If set A contains exactly the same elements as set B, that is, $A = B$, the two sets may be illustrated as in Fig. 2.2(c). Two sets A and B with some elements in common are shown in Fig. 2.2(d), which is regarded as the most general form of a Venn diagram.

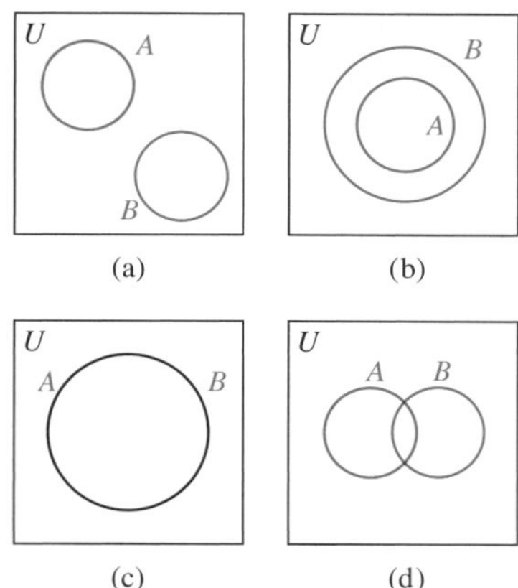

Figure 2.2

If we label the regions of the diagram in Fig. 2.2(d) using I, II, III, and IV, we can illustrate the four possible cases with this one diagram, Fig. 2.3.

CASE 1: DISJOINT SETS When sets A and B are disjoint, they have no elements in common. Therefore, region II of Fig. 2.3 is empty.

CASE 2: SUBSETS When $A \subseteq B$, every element of set A is also an element of set B. Thus, there can be no elements in region I of Fig. 2.3. If $B \subseteq A$, however, then region III of Fig. 2.3 is empty.

CASE 3: EQUAL SETS When set $A =$ set B, all the elements of set A are elements of set B and all the elements of set B are elements of set A. Thus, regions I and III of Fig. 2.3 are empty.

CASE 4: OVERLAPPING SETS When sets A and B have elements in common, those elements are in region II of Fig. 2.3. The elements that belong to set A but not to set B are in region I. The elements that belong to set B but not to set A are in region III.

In each of the four cases, any element not belonging to set A or set B is placed in region IV.

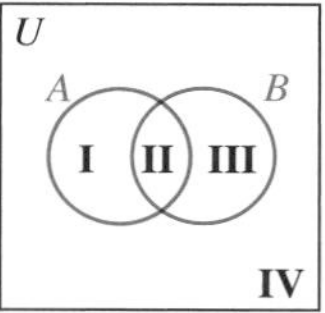

Figure 2.3

Venn diagrams will be helpful in understanding set operations. The basic operations of arithmetic are $+$, $-$, $\times$, and $\div$. When we see these symbols, we know what procedure to follow to determine the answer. Some of the operations in set theory are $'$, $\cup$, and $\cap$. They represent complement, union, and intersection, respectively.

Complement

The **complement** of set A, symbolized by A', is the set of all the elements in the universal set that are not in set A.

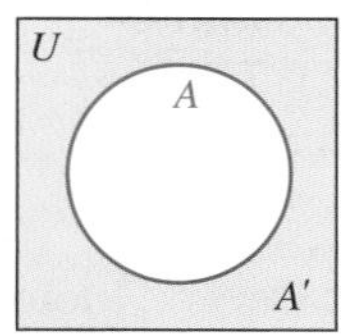

Figure 2.4

In Fig. 2.4, the shaded region outside of set A within the universal set represents the complement of set A, or A'.

EXAMPLE 1 *A Set and Its Complement*

Given

$$U = \{1, 2, 3, 4, 5, 6, 7, 8\} \text{ and } A = \{1, 3, 4\}$$

find A' and illustrate the relationship among sets U, A, and A' in a Venn diagram.

SOLUTION: The elements in U that are not in set A are 2, 5, 6, 7, 8. Thus, $A' = \{2, 5, 6, 7, 8\}$. The Venn diagram is illustrated in Fig. 2.5. ▲

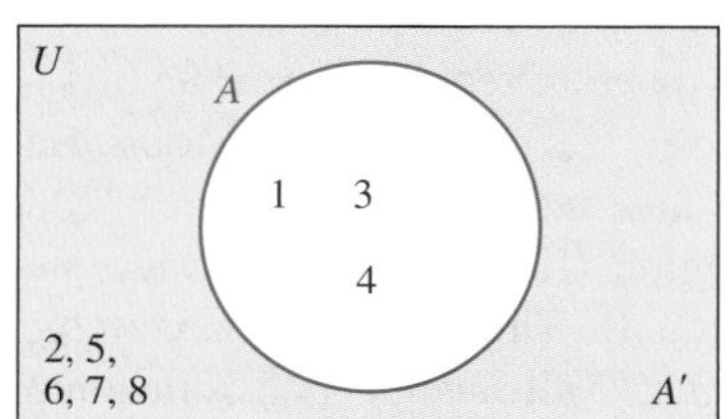

Figure 2.5

Intersection

The word *intersection* brings to mind the area common to two crossing streets. The red car in the figure on the next page is in the intersection of the two streets. The set operation is defined as follows.

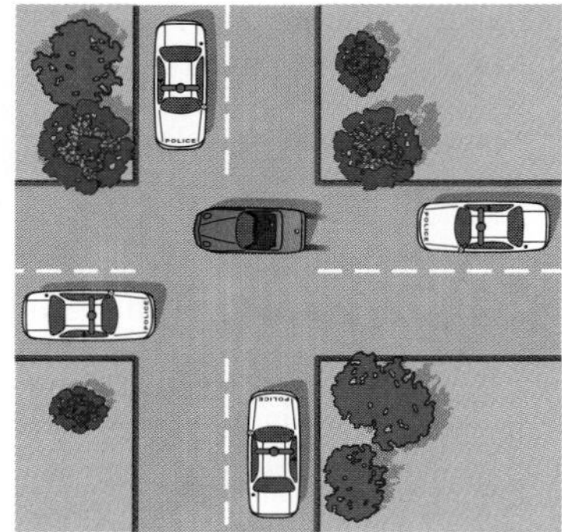

> The **intersection** of sets A and B, symbolized by $A \cap B$, is the set containing all the elements that are common to both set A and set B.

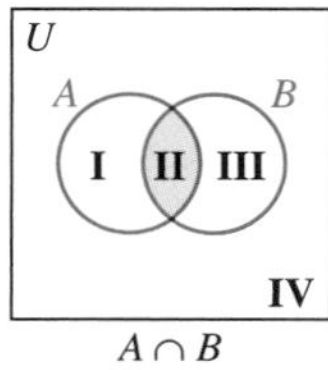

Figure 2.6

The shaded region, region II, in Fig. 2.6 represents the intersection of sets A and B.

EXAMPLE 2 *Sets with Overlapping Regions*

Let the universal set, U, represent the set of all U.S. corporations in 2001. Let set A represent the set of the 10 largest U.S. corporations in 2001, based on revenues. Let set B represent the set of the 10 leading U.S. advertisers in 2001 (see the table). Draw a Venn diagram illustrating the relationship between set A and set B.

Ten Largest U.S. Corporations	Ten Leading U.S. Advertisers
Wal-Mart	General Motors
Exxon Mobil	Procter & Gamble
General Motors	Ford Motor Company
Ford Motor Company	PepsiCo
Enron	Pfizer
General Electric	DaimlerChrysler
Citigroup	AOL Time Warner
ChevronTexaco	Phillip Morris
International Business Machines	Walt Disney
Phillip Morris	Johnson & Johnson

Source: *Fortune*, www.adage.com

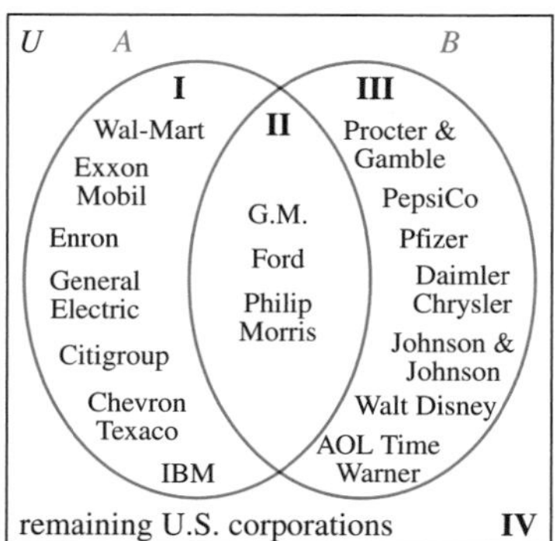

Figure 2.7

SOLUTION: First determine the intersection of sets A and B. General Motors, Ford Motor Company, and Phillip Morris are common to both sets. Therefore,

$$A \cap B = \{\text{General Motors, Ford Motor Company, Phillip Morris}\}$$

Place these elements in region II of Fig. 2.7. Now place in region I the elements in set A that have not been placed in region II. Therefore, Wal-Mart, Exxon Mobil, Enron, General Electric, Citigroup, ChevronTexaco, and International Business Machines (IBM) are placed in region I. Complete region III by determining the elements in set B that have not been placed in region II. Thus, Procter & Gamble, PepsiCo, Pfizer, DaimlerChrysler, AOL Time Warner, Walt Disney, and Johnson & Johnson are placed in region III. Finally, place those elements in U that are not in either set outside both circles. This group includes the remaining U.S. corporations, which are placed in region IV. ▲

EXAMPLE 3 *The Intersection of Sets*

Given

$$U = \{1, 2, 3, 4, 5, 6, 7, 8, 9, 10\}$$
$$A = \{1, 2, 4, 6\}$$
$$B = \{1, 3, 6, 7, 9\}$$
$$C = \{\ \}$$

find

a) $A \cap B$. b) $A \cap C$. c) $A' \cap B$. d) $(A \cap B)'$.

SOLUTION:

a) $A \cap B = \{1, 2, 4, 6\} \cap \{1, 3, 6, 7, 9\} = \{1, 6\}$. The elements common to both set A and set B are 1 and 6.

b) $A \cap C = \{1, 2, 4, 6\} \cap \{\ \} = \{\ \}$. There are no elements common to both set A and set C.

c) $A' = \{3, 5, 7, 8, 9, 10\}$

$A' \cap B = \{3, 5, 7, 8, 9, 10\} \cap \{1, 3, 6, 7, 9\}$

$= \{3, 7, 9\}$

d) To find $(A \cap B)'$, first determine $A \cap B$.

$A \cap B = \{1, 6\}$ from part (a)

$(A \cap B)' = \{1, 6\}' = \{2, 3, 4, 5, 7, 8, 9, 10\}$ ▲

Union

The word *union* means to unite or join together, as in marriage, and that is exactly what is done when we perform the operation of union.

The **union** of set A and set B, symbolized by $A \cup B$, is the set containing all the elements that are members of set A or of set B (or of both sets).

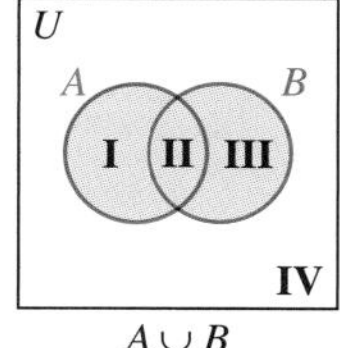

Figure 2.8

The three shaded regions of Fig. 2.8, regions I, II, and III, together represent the union of sets A and B. If an element is common to both sets, it is listed only once in the union of the sets.

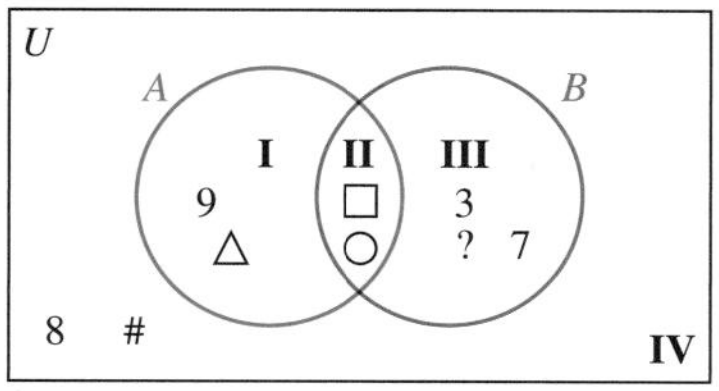

Figure 2.9

EXAMPLE 4 *Determining Sets from a Venn Diagram*

Use the Venn diagram in Fig. 2.9 to determine the following sets.

a) U b) A c) B' d) $A \cap B$

e) $A \cup B$ f) $(A \cup B)'$ g) $n(A \cup B)$

SOLUTION:

a) The universal set consists of all the elements within the rectangle, that is, the elements in regions I, II, III, and IV. Thus, $U = \{9, \triangle, \square, \bigcirc, 3, 7, ?, \#, 8\}$.

b) Set A consists of the elements in regions I and II. Thus, $A = \{9, \triangle, \square, \bigcirc\}$.
c) B' consists of the elements outside set B, or the elements in regions I and IV. Thus, $B' = \{9, \triangle, \#, 8\}$.
d) $A \cap B$ consists of the elements that belong to both set A and set B (region II). Thus, $A \cap B = \{\square, \bigcirc\}$.
e) $A \cup B$ consists of the elements that belong to set A or set B (regions I, II, or III). Thus, $A \cup B = \{9, \triangle, \square, \bigcirc, 3, 7, ?\}$.
f) $(A \cup B)'$ consists of the elements in U that are not in $A \cup B$. Thus, $(A \cup B)' = \{\#, 8\}$.
g) $n(A \cup B)$ represents the *number of elements* in the union of sets A and B. Thus, $n(A \cup B) = 7$, as there are seven elements in the union of sets A and B. ▲

EXAMPLE 5 *The Union of Sets*

Given

$$U = \{1, 2, 3, 4, 5, 6, 7, 8, 9, 10\}$$
$$A = \{1, 2, 4, 6\}$$
$$B = \{1, 3, 6, 7, 9\}$$
$$C = \{\ \}$$

find

a) $A \cup B$. b) $A \cup C$. c) $A' \cup B$. d) $(A \cup B)'$.

SOLUTION:

a) $A \cup B = \{1, 2, 4, 6\} \cup \{1, 3, 6, 7, 9\} = \{1, 2, 3, 4, 6, 7, 9\}$
b) $A \cup C = \{1, 2, 4, 6\} \cup \{\ \} = \{1, 2, 4, 6\}$. Note that $A \cup C = A$.
c) To determine $A' \cup B$, we must determine A'.

$$A' = \{3, 5, 7, 8, 9, 10\}$$
$$A' \cup B = \{3, 5, 7, 8, 9, 10\} \cup \{1, 3, 6, 7, 9\}$$
$$= \{1, 3, 5, 6, 7, 8, 9, 10\}$$

d) Find $(A \cup B)'$ by first determining $A \cup B$, and then find the complement of $A \cup B$.

$$A \cup B = \{1, 2, 3, 4, 6, 7, 9\} \text{ from part (a)}$$
$$(A \cup B)' = \{1, 2, 3, 4, 6, 7, 9\}' = \{5, 8, 10\}$$

▲

EXAMPLE 6 *Union and Intersection*

Given

$$U = \{a, b, c, d, e, f, g\}$$
$$A = \{a, b, e, g\}$$
$$B = \{a, c, d, e\}$$
$$C = \{b, e, f\}$$

find

a) $(A \cup B) \cap (A \cup C)$. b) $(A \cup B) \cap C'$. c) $A' \cap B'$.

SOLUTION:

a) $(A \cup B) \cap (A \cup C) = \{a, b, c, d, e, g\} \cap \{a, b, e, f, g\}$
$= \{a, b, e, g\}$

b) $(A \cup B) \cap C' = \{a, b, c, d, e, g\} \cap \{a, c, d, g\}$
$= \{a, c, d, g\}$

c) $A' \cap B' = \{c, d, f\} \cap \{b, f, g\}$
$= \{f\}$

The Meaning of *and* and *or*

The words *and* and *or* are very important in many areas of mathematics. We use these words in several chapters in this book, including the probability chapter. The word *or* is generally interpreted to mean *union*, whereas *and* is generally interpreted to mean *intersection*. Suppose $A = \{1, 2, 3, 5, 6, 8\}$ and $B = \{1, 3, 4, 7, 9, 10\}$. Then the elements that belong to set A *or* set B are 1, 2, 3, 4, 5, 6, 7, 8, 9, and 10. These are the elements in the union of the sets. The elements that belong to set A *and* set B are 1 and 3. These are the elements in the intersection of the sets.

The Relationship Between $n(A \cup B)$, $n(A)$, $n(B)$, and $n(A \cap B)$

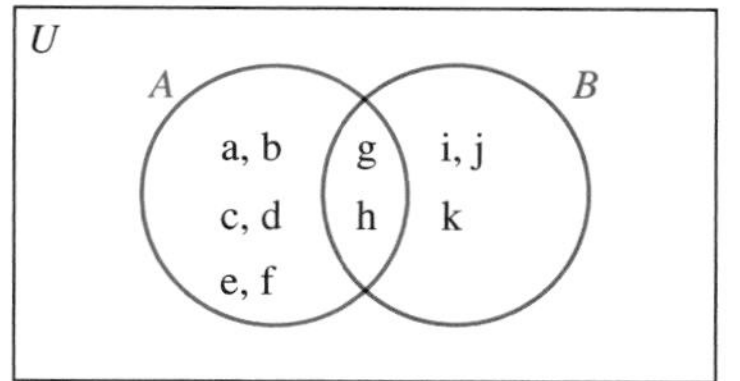

Figure 2.10

Having looked at unions and intersections, we can now determine a relationship between $n(A \cup B)$, $n(A)$, $n(B)$, and $n(A \cap B)$. Suppose set A has eight elements, set B has five elements, and $A \cap B$ has two elements. How many elements are in $A \cup B$? Let's make up some arbitrary sets that meet the criteria specified and draw a Venn diagram. If we let $A = \{a, b, c, d, e, f, g, h\}$, then set B must contain five elements, two of which are also in set A. Let $B = \{g, h, i, j, k\}$. We construct a Venn diagram by filling in the intersection first, as shown in Fig. 2.10. The number of elements in $A \cup B$ is 11. The elements g and h are in both sets, and if we add $n(A) + n(B)$, we are counting these elements twice.

To find the number of elements in the union of sets A and B, we can add the number of elements in sets A and B and then subtract the number of elements common to both sets.

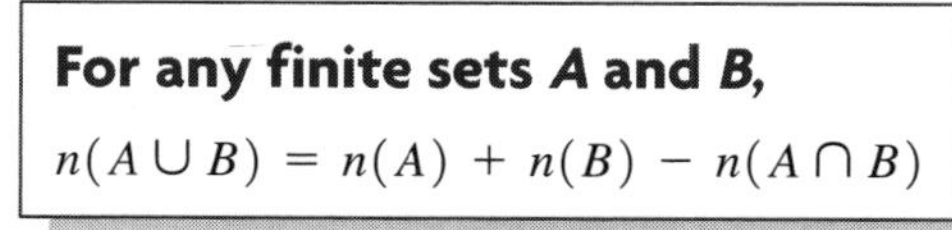

For any finite sets *A* and *B*,

$$n(A \cup B) = n(A) + n(B) - n(A \cap B)$$

EXAMPLE 7 *How Many Visitors Speak Spanish or French?*

The results of a survey of visitors at the Grand Canyon showed that 25 speak Spanish, 14 speak French, and 4 speak both Spanish and French. How many speak Spanish or French?

SOLUTION: If we let set A be the set of visitors who speak Spanish and let set B be the set of visitors who speak French, then we need to determine $n(A \cup B)$. We can use the above formula to find $n(A \cup B)$.

$$n(A \cup B) = n(A) + n(B) - n(A \cap B)$$
$$n(A \cup B) = 25 + 14 - 4$$
$$= 35$$

Thus, 35 of the visitors surveyed speak either Spanish or French. ▲

EXAMPLE 8 *The Number of Elements in Set A or Set B*

Set A contains 6 letters and 5 numbers. Set B contains 4 letters and 9 numbers. Two letters and 1 number are common to both sets A and B. Find the number of elements in set A or set B.

SOLUTION: You are asked to find the number of elements in set A or set B, which is $n(A \cup B)$. Because $n(A \cup B) = n(A) + n(B) - n(A \cap B)$, if you can determine $n(A)$, $n(B)$, and $n(A \cap B)$, you can solve the problem. Set A contains 6 letters and 5 numbers, so $n(A) = 11$. Set B contains 4 letters and 9 numbers, so $n(B) = 13$. Because 2 letters and 1 number are common to both sets, $n(A \cap B) = 3$.

$$n(A \cup B) = n(A) + n(B) - n(A \cap B)$$
$$= 11 + 13 - 3 = 21$$

Thus, the number of elements in set A or set B is 21. ▲

SECTION 2.3 EXERCISES

Concept/Writing Exercises

In Exercises 1–5, use Fig. 2.2 as a guide to draw a Venn diagram that illustrates the situation described.

1. Set A and set B are disjoint sets.
2. $A \subset B$
3. $B \subset A$
4. $A = B$
5. Set A and set B are overlapping sets.
6. If we are given set A, how do we obtain A complement, A'?
7. How do we obtain the union of two sets A and B, $A \cup B$?
8. In a Venn diagram with two overlapping sets, which region(s) represents $A \cup B$?
9. How do we obtain the intersection of two sets A and B, $A \cap B$?
10. In a Venn diagram with two overlapping sets, which region(s) represents $A \cap B$?
11. a) Which set operation is the word *or* generally interpreted to mean?
    b) Which set operation is the word *and* generally interpreted to mean?
12. Give the relationship between $n(A \cup B)$, $n(A)$, $n(B)$, and $n(A \cap B)$.
13. When constructing a Venn diagram with two sets, which region of the diagram do we generally complete first?
14. When constructing a Venn diagram with two sets, which region of the diagram do we generally complete last?

Practice the Skills/Problem Solving

15. *Restaurants* For the sets U, A, and B, construct a Venn diagram and place the elements in the proper regions.

U = {Pizza Hut, Papa John's, McDonald's, Burger King, Wendy's, Roy Rogers, Taco Bell, Subway, Del Taco, Denny's}

A = {Pizza Hut, Papa John's, Wendy's, Roy Rogers, Taco Bell, Denny's}

B = {McDonald's, Burger King, Wendy's, Taco Bell, Subway, Denny's}

16. *Appliances and Electronics* For the sets U, A, and B, construct a Venn diagram and place the elements in the proper regions.

U = {microwave oven, washing machine, dryer, refrigerator, dishwasher, compact disc player, videocassette recorder, computer, camcorder, television}

A = {microwave oven, washing machine, dishwasher, computer, television}

B = {washing machine, dryer, refrigerator, compact disc player, computer, television}

17. *Occupations* The following table shows the fastest-growing occupations based on employment in 2000 and the estimated employment for that profession for 2010. Let the occupations in the table represent the universal set.

Fastest-Growing Occupations, 2000–2010

	Employment (in thousands of jobs)	
Occupation	**2000**	**2010**
Computer software engineers, applications	380	760
Computer support specialists	506	996
Computer software engineers, system software	317	601
Network and computer systems administrators	229	416
Network systems/data communications analysts	119	211
Desktop publishers	38	63
Database administrators	106	176
Personal/home care aides	414	672
Computer systems analysts	431	689
Medical assistants	329	516

Source: U.S. Department of Labor, Bureau of Labor Statistics

Let A = the set of fastest-growing occupations whose 2000 employment was at least 250,000.

Let B = the set of fastest-growing occupations whose estimated employment in 2010 is at least 650,000.

Construct a Venn diagram illustrating the sets.

18. *Basketball Statistics* The table above and to the right shows the number of times certain basketball players were selected as the National Basketball Association's most valuable player and the number of times the player was the league scoring champion. Let these players represent the universal set.

Player	Most Valuable Player	League Scoring Champion
Michael Jordan	4	10
Dominique Wilkins	0	1
Wilt Chamberlain	2	7
Kareem Abdul-Jabbar	4	2
Shaquille O'Neal	1	2
Earvin Johnson	3	0
Kobe Bryant	0	0
Jerry West	0	1
George Gervin	0	4
Alan Iverson	1	1

Source: National Basketball Association

Michael Jordan

Let A = the set of basketball players that were selected most valuable player at least one time.

Let B = the set of basketball players that were league scoring champion at least one time.

Construct a Venn diagram illustrating the sets.

19. Let U represent the set of U.S. colleges and universities. Let A represent the set of U.S. colleges and universities in the state of North Dakota. Describe A'.

20. Let U represent the set of marbles in a box. Let set B represent the set of marbles that contain some blue coloring. Describe B'.

In Exercises 21–26,

U is the set of insurance companies in the U.S.

A is the set of insurance companies that offer life insurance.

B is the set of insurance companies that offer car insurance.

Describe each of the following sets in words.

21. A'

22. B'

23. $A \cup B$

24. $A \cap B$

25. $A \cap B'$

26. $A \cup B'$

In Exercises 27–32,

U is the set of U.S. corporations.

A is the set of U.S. corporations whose headquarters are in the state of New York.

B is the set of U.S. corporations whose chief executive officer is a woman.

C is the set of U.S. corporations that employ at least 100 people.

Describe the following sets.

27. $A \cap B$
28. $A \cup C$
29. $B' \cap C$
30. $A \cap B \cap C$
31. $A \cup B \cup C$
32. $A' \cup C'$

In Exercises 33–40, use the Venn diagram in Fig. 2.11 to list the set of elements in roster form.

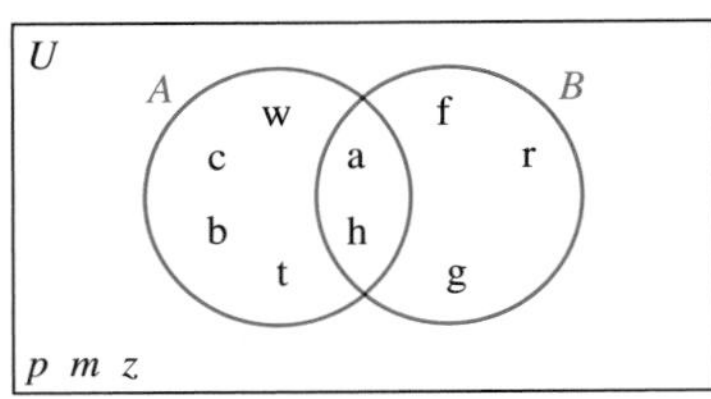

Figure 2.11

33. A
34. B
35. $A \cap B$
36. U
37. $A \cup B$
38. $(A \cup B)'$
39. $A' \cap B'$
40. $(A \cap B)'$

In Exercises 41–48, use the Venn diagram in Fig. 2.12 to list the set of elements in roster form.

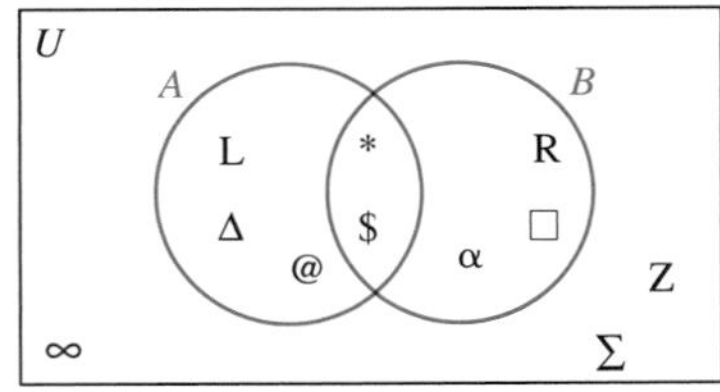

Figure 2.12

41. A
42. B
43. U
44. $A \cup B$
45. $A \cap B$
46. $A \cup B'$
47. $A' \cap B$
48. $(A \cup B)'$

In Exercises 49–58, let

$$\begin{aligned} U &= \{1, 2, 3, 4, 5, 6, 7, 8\} \\ A &= \{1, 2, 4, 5, 8\} \\ B &= \{2, 3, 4, 6\} \end{aligned}$$

Determine the following.

49. $A \cup B$
50. $A \cap B$
51. B'
52. $A \cup B'$
53. $(A \cup B)'$
54. $A' \cap B'$
55. $(A \cup B)' \cap B$
56. $(A \cup B) \cap (A \cup B)'$
57. $(B \cup A)' \cap (B' \cup A')$
58. $A' \cup (A \cap B)$

In Exercises 59–68, let

$$\begin{aligned} U &= \{a, b, c, d, e, f, g, h, i, j, k\} \\ A &= \{a, c, d, f, g, i\} \\ B &= \{b, c, d, f, g\} \\ C &= \{a, b, f, i, j\} \end{aligned}$$

Determine the following.

59. B'
60. $B \cup C$
61. $A \cap C$
62. $A \cup B'$
63. $(A \cap C)'$
64. $(A \cap B) \cup C$
65. $A \cup (C \cap B)'$
66. $A \cup (C' \cup B')$
67. $(A' \cup C) \cup (A \cap B)$
68. $(C \cap B) \cap (A' \cap B)$

Problem Solving

In Exercises 69–82, let

$$\begin{aligned} U &= \{x \mid x \in N \text{ and } x < 10\} \\ A &= \{x \mid x \in N \text{ and } x \text{ is odd and } x < 10\} \\ B &= \{x \mid x \in N \text{ and } x \text{ is even and } x < 10\} \\ C &= \{x \mid x \in N \text{ and } x < 6\} \end{aligned}$$

Determine the following.

69. $A \cap B$
70. $A \cup B$
71. $A' \cup B$
72. $(B \cup C)'$
73. $A \cap C'$
74. $A \cap B'$
75. $(B \cap C)'$
76. $(A \cup C) \cap B$
77. $(C \cap B) \cup A$
78. $(C' \cup A) \cap B$
79. $(A' \cup C) \cap B$
80. $(A \cap B)' \cup C$
81. $(A' \cup B') \cap C$
82. $(A' \cap C) \cup (A \cap B)$
83. When will a set and its complement be disjoint? Explain and give an example.
84. When will $n(A \cap B) = 0$? Explain and give an example.
85. *Visiting California* The results of a survey of visitors in Hollywood, California, showed that 27 visited the Hollywood Bowl, 38 visited Disneyland, and 16 visited both the

Hollywood Bowl and Disneyland. How many people visited either the Hollywood Bowl or Disneyland?

Hollywood Bowl

86. *Chorus and Band* At Henniger High School, 46 students sang in the chorus or played in the stage band, 30 students played in the stage band, and 4 students sang in the chorus and played in the stage band. How many students sang in the chorus?

87. Consider the formula

$$n(A \cup B) = n(A) + n(B) - n(A \cap B)$$

a) Show that this relation holds for $A = \{a, b, c, d\}$ and $B = \{b, d, e, f, g, h\}$.

b) Make up your own sets A and B, each consisting of at least six elements. Using these sets, show that the relation holds.

c) Use a Venn diagram and explain why the relation holds for any two sets A and B.

88. The Venn diagram in Fig. 2.13 shows a technique of labeling the regions to indicate membership of elements in a particular region. Define each of the four regions with a set statement. (*Hint:* $A \cap B'$ defines region I.)

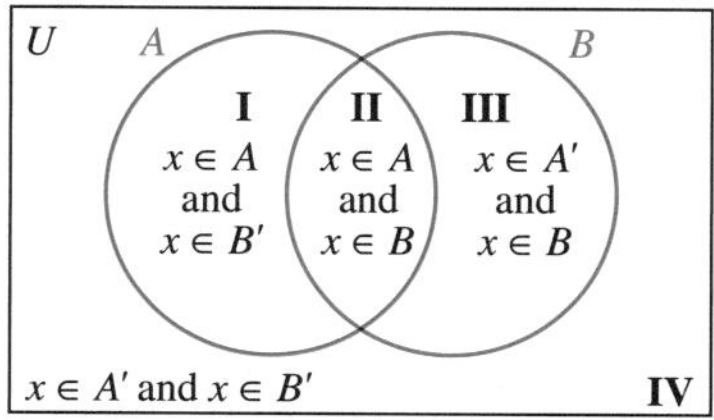

Figure 2.13

In Exercises 89–98, let $U = \{0, 1, 2, 3, 4, 5, \ldots\}$, $A = \{1, 2, 3, 4, \ldots\}$, $B = \{4, 8, 12, 16, \ldots\}$, *and* $C = \{2, 4, 6, 8, \ldots\}$. *Determine the following.*

89. $A \cup B$ **90.** $A \cap B$

91. $B \cap C$ **92.** $B \cup C$

93. $A \cap C$ **94.** $A' \cap C$

95. $B' \cap C$ **96.** $(B \cup C)' \cup C$

97. $(A \cap C) \cap B'$ **98.** $U' \cap (A \cup B)$

Challenge Problems/Group Activities

In Exercises 99–106, determine whether the answer is $\emptyset$, *A, or U. (Assume* $A \neq \emptyset$, $A \neq U$.*)*

99. $A \cup A'$ **100.** $A \cap A'$

101. $A \cup \emptyset$ **102.** $A \cap \emptyset$

103. $A' \cup U$ **104.** $A \cap U$

105. $A \cup U$ **106.** $A \cup U'$

In Exercises 107–112, determine the relationship between set A and set B if

107. $A \cap B = B$. **108.** $A \cup B = B$.

109. $A \cap B = \emptyset$. **110.** $A \cup B = A$.

111. $A \cap B = A$. **112.** $A \cup B = \emptyset$.

Difference of Two Sets Another set operation is the **difference of two sets**. The difference of two sets A and B, symbolized $A - B$, is defined as

$$A - B = \{x \mid x \in A \text{ and } x \notin B\}$$

Thus, $A - B$ is the set of elements that belong to set A but not to set B. For example, if $U = \{1, 2, 3, 4, 5, 6, 7, 8, 9, 10\}$, $A = \{2, 4, 5, 9, 10\}$, and $B = \{1, 3, 4, 5, 6, 7\}$, then $A - B = \{2, 9, 10\}$ and $B - A = \{1, 3, 6, 7\}$.

In Exercises 113–116, let $U = \{a, b, c, d, e, f, g, h, i, j, k\}$, $A = \{b, c, e, f, g, h\}$, *and* $B = \{a, b, c, g, i\}$. *Determine the following.*

113. $A - B$ **114.** $B - A$

115. $A' - B$ **116.** $A - B'$

In Exercises 117–122, let $U = \{1, 2, 3, 4, 5, 6, 7, 8, 9, 10, 11, 12, 13, 14, 15\}$, $A = \{2, 4, 5, 7, 9, 11, 13\}$, *and* $B = \{1, 2, 4, 5, 6, 7, 8, 9, 11\}$. *Determine the following.*

117. $A - B$ **118.** $B - A$

119. $(A - B)'$ **120.** $A - B'$

121. $(B - A)'$ **122.** $A \cap (A - B)$

Recreational Mathematics

123. *What Am I?* The poem below gives clues for a 10-letter word discussed in this section. Use the clues to determine the word.

My first is in card and also cat.
My fourth is in top and also in pat.

My third and my seventh are one and the same—
 you'll find them in math and also in game.
My sixth and my eighth, find one you will find two—
 you will find them in open, you will find them in shoe.
My second is in work and also in told.
My fifth is in label and also in bold.
My ninth is in number and also in change.
My tenth is in table but never in range.
I have more than one meaning, some say I'm "not."
Although you may want to receive me a lot.

124. *Wordgram* Hidden in the box are the following words discussed in this chapter: DIAGRAM, UNION, INTERSECTION, SUBSET. You will find them by going letter to letter either vertically or horizontally. A letter can only be used once when spelling out a word. Find the words and make sure you understand the meaning of each word.

N	O	I	E	L	A
T	C	T	S	E	T
S	E	U	B	P	U
R	D	S	R	I	N
E	I	A	G	O	N
T	N	I	R	A	M

2.4 VENN DIAGRAMS WITH THREE SETS AND VERIFICATION OF EQUALITY OF SETS

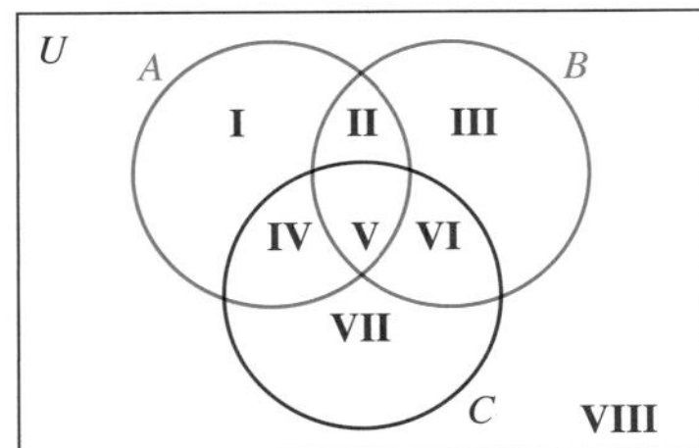

Figure 2.14

Venn diagrams can be used to illustrate three or more sets. For three sets, A, B, and C, the diagram is drawn so the three sets overlap (Fig. 2.14), creating eight regions. The diagrams in Fig. 2.15 emphasize selected regions of three intersecting sets. *When constructing Venn diagrams with three sets, we generally start with region V and work outward*, as explained in the following procedure.

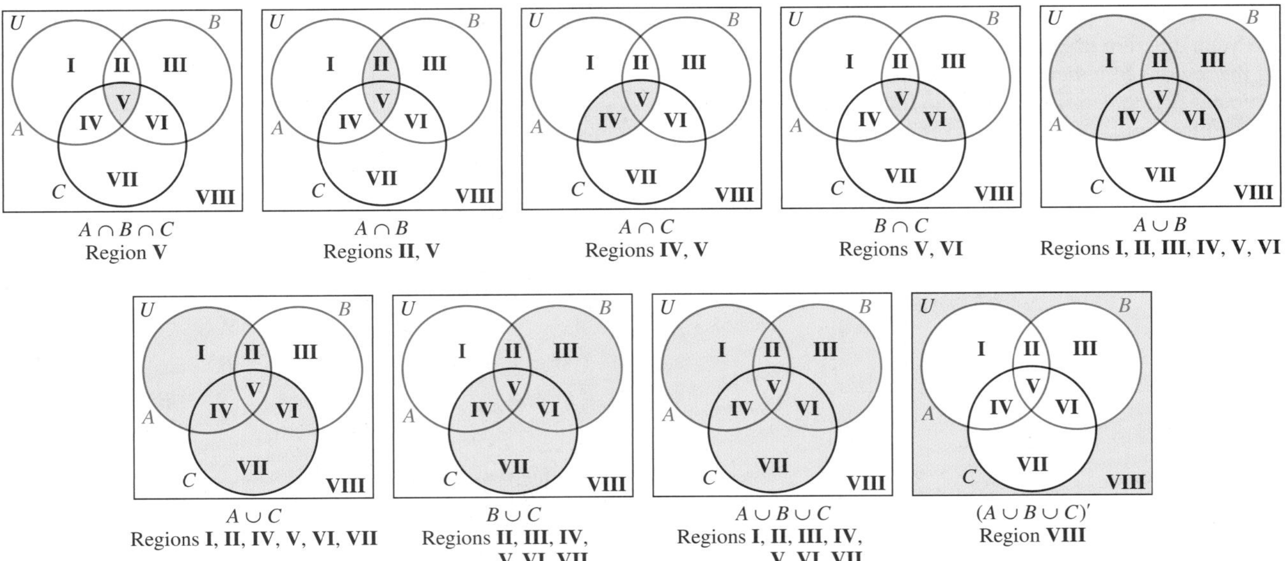

Figure 2.15

General Procedure for Constructing Venn Diagrams with Three Sets, *A*, *B*, and *C*

1. Determine the elements to be placed in region V by finding the elements that are common to all three sets, $A \cap B \cap C$.
2. Determine the elements to be placed in region II. Find the elements in $A \cap B$. The elements in this set belong in regions II and V. Place the elements in the set $A \cap B$ that are not listed in region V in region II. The elements in regions IV and VI are found in a similar manner.
3. Determine the elements to be placed in region I by determining the elements in set A that are not in regions II, IV, and V. The elements in regions III and VII are found in a similar manner.
4. Determine the elements to be placed in region VIII by finding the elements in the universal set that are not in regions I through VII.

Example 1 illustrates the general procedure.

EXAMPLE 1 *Constructing a Venn Diagram for Three Sets*

Construct a Venn diagram illustrating the following sets.

$$U = \{1, 2, 3, 4, 5, 6, 7, 8, 9, 10, 11, 12\}$$
$$A = \{1, 5, 8, 9, 10, 12\}$$
$$B = \{2, 4, 5, 9, 10\}$$
$$C = \{1, 3, 5, 8, 9, 11\}$$

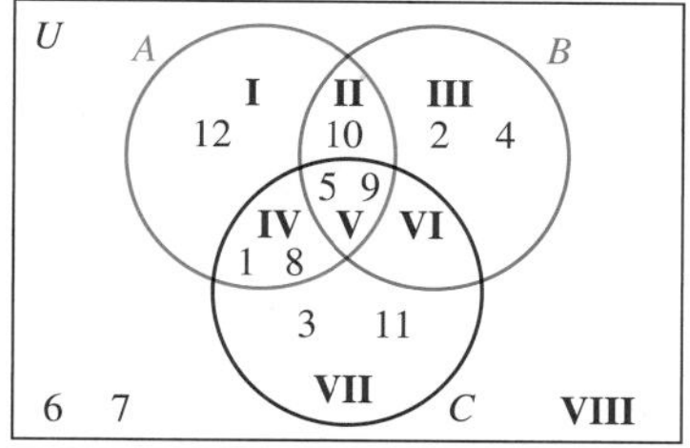

Figure 2.16

SOLUTION: First find the intersection of all three sets. Because the elements 5 and 9 are in all three sets, $A \cap B \cap C = \{5, 9\}$. The elements 5 and 9 are placed in region V in Fig. 2.16. Next complete region II by determining the intersection of sets A and B.

$$A \cap B = \{5, 9, 10\}$$

$A \cap B$ consists of regions II and V. The elements 5 and 9 have already been placed in region V, so 10 must be placed in region II.

Now determine what numbers go in region IV.

$$A \cap C = \{1, 5, 8, 9\}$$

Since 5 and 9 have already been placed in region V, place the 1 and 8 in region IV. Now determine the numbers to go in region VI.

$$B \cap C = \{5, 9\}$$

Since both the 5 and 9 have been placed in region V, there are no numbers to be placed in region VI. Now complete set A. The only element of set A that has not previously been placed in regions II, IV, or V is 12. Therefore, place the element 12 in region I. The element 12 that is placed in region I is only in set A and not in set B or set C. Using set B, complete region III using the same general procedure used to determine the numbers in region I. Using set C, complete region VII by using the same procedure used to complete regions I and III. To determine the elements in region VIII, find the elements in U that have not been placed in regions I–VII. The elements 6 and 7 have not been placed in regions I–VII, so place them in region VIII. ▲

Venn diagrams can be used to illustrate and analyze many everyday problems. One example follows.

Figure 2.17

EXAMPLE 2 ***Blood Types***

Human blood is classified (typed) according to the presence or absence of the specific antigens A, B, and Rh in the red blood cells. Antigens are highly specified proteins and carbohydrates that will trigger the production of antibodies in the blood to fight infection. Blood containing the Rh antigen is labeled positive, +, while blood lacking the Rh antigen is labeled negative, −. Blood lacking both A and B antigens is called type O. Sketch a Venn diagram with three sets A, B, and Rh and place each type of blood listed in the proper region. A person has only one type of blood.

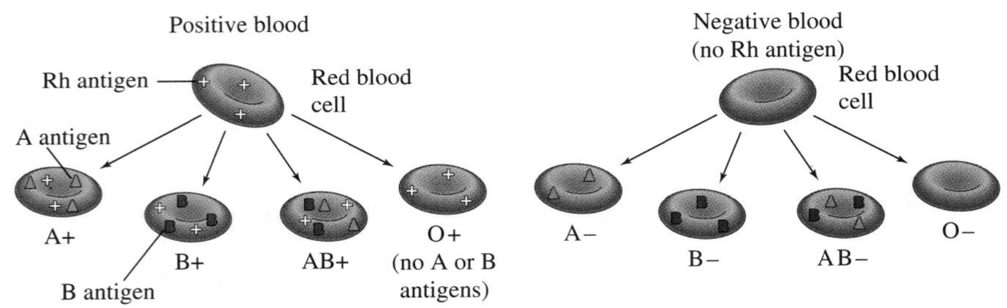

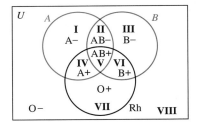

SOLUTION: As illustrated in Chapter 1, the first thing to do is to read the question carefully and make sure you understand what is given and what you are asked to find. There are three antigens A, B, and Rh. Therefore, begin by naming the three circles in a Venn diagram with the three antigens; see Fig. 2.17.

Any blood containing the Rh antigen is positive, and any blood not containing the Rh antigen is negative. Therefore, all blood in the Rh circle is positive, and all blood outside the Rh circle is negative. The intersection of all three sets, region V, is AB+. Region II contains only antigens A and B and is therefore AB−. Region I is A− because it contains only antigen A. Region III is B−, region IV is A+, and region VI is B+. Region VII is O+, containing only the Rh antigen. Region VIII, which lacks all three antigens, is O−. ▲

Verification of Equality of Sets

In this chapter, for clarity we may refer to operations on sets, such as $A \cup B'$ or $A \cap B \cap C$, as *statements involving sets* or simply as *statements.* Now we discuss how to determine if two statements involving sets are equal.

Consider the question: Is $A' \cup B = A' \cap B$ *for all sets A and B*? For the specific sets $U = \{1, 2, 3, 4, 5\}$, $A = \{1, 3\}$, and $B = \{2, 4, 5\}$, is $A' \cup B = A' \cap B$? To answer the question, we do the following.

Find $A' \cup B$	**Find $A' \cap B$**
$A' = \{2, 4, 5\}$	$A' = \{2, 4, 5\}$
$A' \cup B = \{2, 4, 5\}$	$A' \cap B = \{2, 4, 5\}$

For these sets, $A' \cup B = A' \cap B$, because both set statements are equal to $\{2, 4, 5\}$. At this point you may believe that $A' \cup B = A' \cap B$ for all sets A and B.

If we select the sets $U = \{1, 2, 3, 4, 5\}$, $A = \{1, 3, 5\}$, and $B = \{2, 3\}$, we see that $A' \cup B = \{2, 3, 4\}$ and $A' \cap B = \{2\}$. For this case, $A' \cup B \neq A' \cap B$. Thus, we have proved that $A' \cup B \neq A' \cap B$ for all sets A and B by using a *counterexample.* A counterexample, as explained in Chapter 1, is an example that shows a statement is not true.

In Chapter 1, we explained that proofs involve the use of deductive reasoning. Recall that deductive reasoning begins with a general statement and works to a specific conclusion. To verify, or determine whether set statements are equal for any two sets selected, we use deductive reasoning with Venn diagrams. Venn diagrams are used because they can illustrate general cases. To determine if statements that contain sets, such as $(A \cup B)'$ and $A' \cap B'$, are equal for all sets A and B, we use the regions of Venn diagrams. If both statements represent the same regions of the Venn diagram, then the statements are equal for all sets A and B. See Example 3.

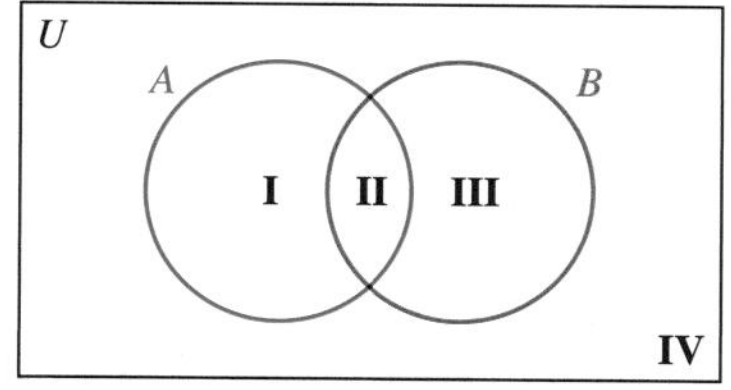

Figure 2.18

EXAMPLE 3 *Equality of Sets*

Determine whether $(A \cap B)' = A' \cup B'$ for all sets A and B.

SOLUTION: Draw a Venn diagram with two sets A and B, as in Fig. 2.18. Label the regions as indicated.

Find $(A \cap B)'$

Set	**Corresponding Regions**
A	I, II
B	II, III
$A \cap B$	II
$(A \cap B)'$	I, III, IV

Find $A' \cup B'$

Set	**Corresponding Regions**
A'	III, IV
B'	I, IV
$A' \cup B'$	I, III, IV

Both statements are represented by the same regions, I, III, and IV, of the Venn diagram. Thus, $(A \cap B)' = A' \cup B'$ for all sets A and B. ▲

In Example 3, when we proved that $(A \cap B)' = A' \cup B'$, we started with two general sets and worked to the specific conclusion that both statements represented

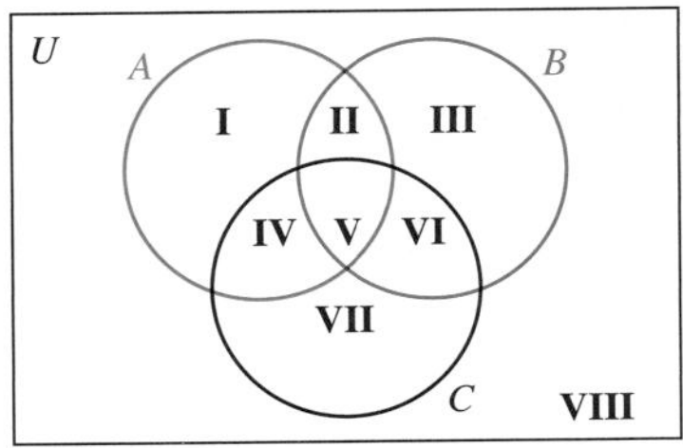

Figure 2.19

the same regions of the Venn diagram. We showed that $(A \cap B)' = A' \cup B'$ *for all sets A and B*. No matter what sets we choose for A and B, this statement will be true. For example, let $U = \{1, 2, 3, 4, 5, 6, 7, 8, 9, 10\}$, $A = \{3, 4, 6, 10\}$, and $B = \{1, 2, 4, 5, 6, 8\}$.

$$
\begin{aligned}
(A \cap B)' &= A' \cup B' \\
\{4, 6\}' &= \{3, 4, 6, 10\}' \cup \{1, 2, 4, 5, 6, 8\}' \\
\{1, 2, 3, 5, 7, 8, 9, 10\} &= \{1, 2, 5, 7, 8, 9\} \cup \{3, 7, 9, 10\} \\
\{1, 2, 3, 5, 7, 8, 9, 10\} &= \{1, 2, 3, 5, 7, 8, 9, 10\}
\end{aligned}
$$

We can also use Venn diagrams to prove statements involving three sets.

MATHEMATICS *Everywhere*

Using Venn Diagrams

A financial planning company uses the following Venn diagram to categorize the financial planning services the company offers. From the diagram, we can see that the company offers financial planning in an "intersection" of the areas investment, retirement, and college planning.

We categorize items on a daily basis, from filing items to planning meals to planning social activities. Children are taught how to categorize items at an early age when they learn how to classify items according to color, shape, and size. Biologists categorize items when they classify organisms according to shared characteristics. A Venn diagram is a very useful tool to help order and arrange items and to picture the relationship between sets.

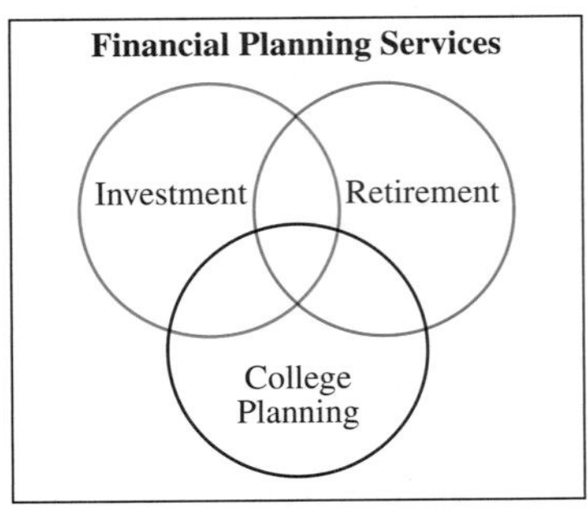

EXAMPLE 4 *Equality of Sets*

Determine whether $A \cap (B \cup C) = (A \cap B) \cup (A \cap C)$ for all sets, A, B, and C.

SOLUTION: Because the statements include three sets, A, B, and C, three circles must be used. The Venn diagram illustrating the eight regions is shown in Fig. 2.19.

First we will find the regions that correspond to $A \cap (B \cup C)$, and then we will find the regions that correspond to $(A \cap B) \cup (A \cap C)$. If both answers are the same, the statements are equal.

Find $A \cap (B \cup C)$

Set	Corresponding Regions
A	I, II, IV, V
$B \cup C$	II, III, IV, V, VI, VII
$A \cap (B \cup C)$	II, IV, V

Find $(A \cap B) \cup (A \cap C)$

Set	Corresponding Regions
$A \cap B$	II, V
$A \cap C$	IV, V
$(A \cap B) \cup (A \cap C)$	II, IV, V

The regions that correspond to $A \cap (B \cup C)$ are II, IV, and V, and the regions that correspond to $(A \cap B) \cup (A \cap C)$ are also II, IV, and V. The results show that both statements are represented by the same regions, namely, II, IV, and V, and therefore $A \cap (B \cup C) = (A \cap B) \cup (A \cap C)$ for all sets A, B, and C. ▲

In Example 4, we proved that $A \cap (B \cup C) = (A \cap B) \cup (A \cap C)$ for all sets A, B, and C. Show that this statement is true for the specific sets $U = \{1, 2, 3, 4, 5, 6, 7, 8, 9, 10\}$, $A = \{1, 2, 3, 7\}$, $B = \{2, 3, 4, 5, 7, 9\}$, and $C = \{1, 4, 7, 8, 10\}$.

De Morgan's Laws

In set theory, logic, and other branches of mathematics, a pair of related theorems known as De Morgan's laws make it possible to transform statements and formulas into alternative and often more convenient forms. In set theory, *De Morgan's laws* are symbolized as follows.

De Morgan's Laws

1. $(A \cup B)' = A' \cap B'$
2. $(A \cap B)' = A' \cup B'$

Law 2 was verified in Example 3. We suggest that you verify law 1 at this time. The laws were expressed verbally by William of Ockham in the fourteenth century. In the nineteenth century, Augustus De Morgan expressed them mathematically. De Morgan's laws will be discussed more thoroughly in Chapter 3, Logic.

SECTION 2.4 EXERCISES

Concept/Writing Exercises

1. How many regions are created when constructing a Venn diagram with three overlapping sets?
2. When constructing a Venn diagram with three overlapping sets, which region do you generally complete first?
3. When constructing a Venn diagram with three overlapping sets, after completing region V, which regions do you generally complete next?
4. A Venn diagram contains three sets, *A*, *B*, and *C*, as in Fig. 2.14. If region V contains 6 elements and there are 10 elements in $A \cap B$, how many elements belong in region II? Explain.
5. A Venn diagram contains three sets, *A*, *B*, and *C*, as in Fig. 2.14. If region V contains 4 elements and there are 12 elements in $B \cap C$, how many elements belong in region VI? Explain.
6. Give De Morgan's laws.
7. a) For $U = \{1, 2, 3, 4, 5\}$, $A = \{1, 4, 5\}$, and $B = \{1, 4, 5\}$, does $A \cup B = A \cap B$?
   b) By observing the answer to part (a), can we conclude that $A \cup B = A \cap B$ for all sets *A* and *B*? Explain.
   c) Using a Venn diagram, determine if $A \cup B = A \cap B$ for all sets *A* and *B*.
8. What type of reasoning do we use when using Venn diagrams to verify or determine whether set statements are equal?

Practice the Skills

9. Construct a Venn diagram illustrating the following sets.

$U = \{a, b, c, d, e, f, g, h, i, j\}$
$A = \{c, d, e, g, h, i\}$
$B = \{a, c, d, g\}$
$C = \{c, f, i, j\}$

10. Construct a Venn diagram illustrating the following sets.

U = {Delaware, Pennsylvania, New Jersey, Georgia, Connecticut, Massachusetts, Maryland, South Carolina, New Hampshire, Virginia, New York, North Carolina, Rhode Island}
A = {New York, New Jersey, Pennsylvania, Massachusetts, New Hampshire}
B = {Delaware, Connecticut, Georgia, Maryland, New York, Rhode Island}
C = {New York, South Carolina, Rhode Island, Massachusetts}

11. Construct a Venn diagram illustrating the following sets.

U = {football, basketball, baseball, gymnastics, lacrosse, soccer, tennis, volleyball, swimming, wrestling, cross-country, track, golf, fencing}
A = {football, basketball, soccer, lacrosse, volleyball}
B = {baseball, lacrosse, tennis, golf, volleyball}
C = {swimming, gymnastics, fencing, basketball, volleyball}

12. Construct a Venn diagram illustrating the following sets.

U = {*The Lion King, Aladdin, Cinderella, Beauty and the Beast, Snow White and the Seven Dwarfs, Toy Story, 101 Dalmatians, The Little Mermaid, Jurassic Park*}
A = {*Aladdin, Toy Story, The Lion King, Snow White and the Seven Dwarfs*}

B = {*Snow White and the Seven Dwarfs, Toy Story, The Lion King, Beauty and the Beast*}

C = {*Snow White and the Seven Dwarfs, Toy Story, Beauty and the Beast, Cinderella, 101 Dalmatians*}

The Lion King

13. Construct a Venn diagram illustrating the following sets.

U = {peach, pear, banana, apple, grape, melon, carrot, corn, orange, spinach}

A = {pear, grape, melon, carrot}

B = {peach, pear, banana, spinach, corn}

C = {pear, banana, apple, grape, melon, spinach}

14. Construct a Venn diagram illustrating the following sets.

U = {Louis Armstrong, Glenn Miller, Stan Kenton, Charlie Parker, Duke Ellington, Benny Goodman, Count Basie, John Coltrane, Dizzy Gillespie, Miles Davis, Thelonius Monk}

A = {Stan Kenton, Count Basie, Dizzy Gillespie, Duke Ellington, Thelonius Monk}

B = {Louis Armstrong, Glenn Miller, Count Basie, Duke Ellington, Miles Davis}

C = {Count Basie, Miles Davis, Stan Kenton, Charlie Parker, Duke Ellington}

15. *Olympic Medals* Consider the chart, which shows countries that won at least 25 medals in the 2000 Summer Olympics. Let the teams shown in the chart represent the universal set.

	Gold	Silver	Bronze	Total
United States	39	25	33	97
Russia	32	28	28	88
China	28	16	15	59
Australia	16	25	17	58
Germany	14	17	26	57
France	13	14	11	38
Italy	13	8	13	34
Cuba	11	11	7	29
Great Britain	11	10	7	28
Korea	8	9	11	28
Romania	11	6	9	26
Netherlands	12	9	4	25

Source: *2001 Time Almanac*

Let A = set of teams that won at least 58 medals.
Let B = set of teams that won at least 20 gold medals.
Let C = set of teams that won at least 10 bronze medals.

Construct a Venn diagram that illustrates the sets A, B, and C.

16. *Popular TV Shows* Let U = {*Friends, CSI, NFL Monday Night Football, Survivor II, E.R., Who Wants to Be a Millionaire—Tues., Who Wants to Be a Millionaire—Wed., Who Wants to Be a Millionaire—Thurs., Who Wants to Be a Millionaire—Sun., Everybody Loves Raymond, 60 Minutes, Law& Order, West Wing*} Set A represents the five most popular prime time shows on television in 2001–2002, set B represents the five most popular prime-time shows on television in 2000–2001, and set C represents the five most popular prime-time shows on television in 1999–2000 (according to Nielsen Media Research). Then

A = {*Friends, CSI, E.R., Everybody Loves Raymond, Law & Order*}

B = {*Survivor II, E.R., Who Wants to Be a Millionaire—Wed., Who Wants to Be a Millionaire—Tues., Friends*}

C = {*Who Wants to Be a Millionaire—Tues., Who Wants to Be a Millionaire—Thurs., Who Wants to Be a Millionaire—Sun., E.R., Friends*}

Construct a Venn diagram illustrating the sets.

Graduate Schools The accompanying chart on page 73 shows the *U.S News and World Report* top 10 rankings of graduate schools in 2002 in the fields of medicine, education, and business. The universal set is the set of all U.S. graduate schools. In Fig. 2.20, the set indicated as Medical represents the 10 highest-rated medical schools, the set indicated as Education represents the 10 highest-rated graduate education schools, and the set listed as Business represents the 10 highest-rated graduate business schools.

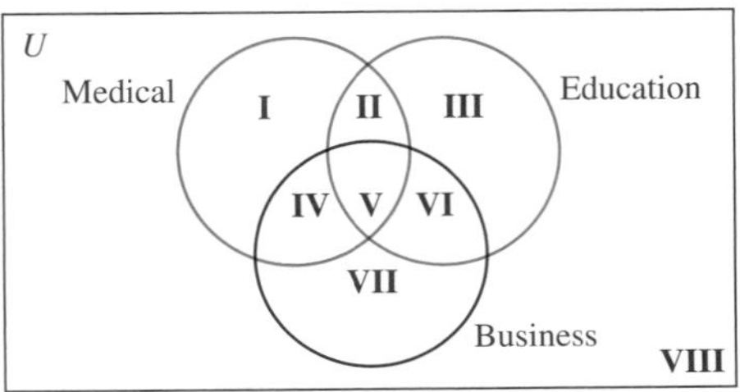

Figure 2.20

Top-Rated Graduate Schools

Medical	Education	Business
1. Harvard	1. Harvard	1. Stanford
2. Johns Hopkins	2. Stanford	2. Harvard
3. Duke	3. Columbia	3. Northwestern
4. University of Pennsylvania	4. UCLA	4. University of Pennsylvania
5. Washington University in St. Louis	5. Vanderbilt	5. M.I.T.
6. Columbia	6. University of California–Berkeley	6. Columbia
7. University of California–San Francisco	7. University of Michigan	7. University of California–Berkeley
8. Yale	8. University of Pennsylvania	8. Duke
9. Stanford	9. University of Wisconsin–Madison	9. University of Chicago
9. University of Michigan (tie)	10. Northwestern	10. University of Michigan

Source: *U.S. News*, www.usnews.com

Indicate in Fig. 2.20 in which region, I–VIII, each of the following belongs.

17. Harvard

18. Yale

19. Boston College

20. University of California–Berkeley

21. Northwestern

22. Duke

Rankings of Metropolitan Areas The table above and to the right, taken from *Places Rated Almanac, Millennium Edition,* shows that almanac's rankings of the overall top 10 metropolitan areas across the United States and Canada. The table also shows the top 10 ranked areas for the categories of transportation and education. The universal set is the set of all metropolitan areas in the United States and Canada. In Fig. 2.21, the set indicated as Overall represents the set of metropolitan areas listed in the table under overall rankings.

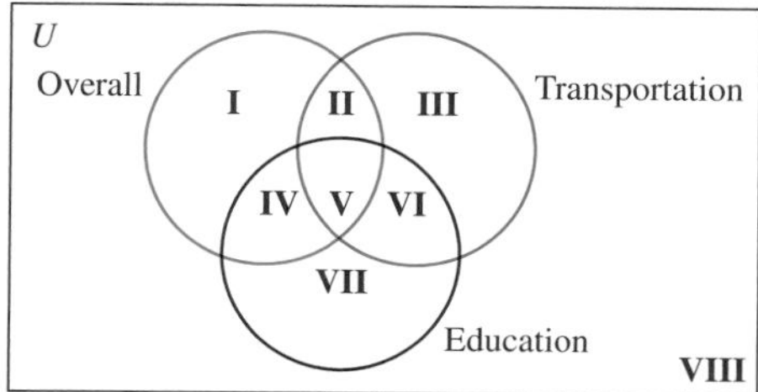

Figure 2.21

The set indicated as Transportation represents the set of metropolitan areas listed in the table under transportation, and the set indicated as Education represents the set of metropolitan areas listed in the table under education.

Rankings of Metropolitan Areas

Overall	Transportation	Education
1. Salt Lake City–Odgen	1. Chicago	1. Raleigh–Durham–Chapel Hill
2. Washington, D.C.	2. Pittsburgh	2. Boston
3. Seattle-Bellevue-Everett	3. New York	3. Albany-Schenectady-Troy
4. Tampa–St. Petersburg–Clearwater	4. Cincinnati	4. St. Louis
5. Denver	5. Detroit	5. Chicago
6. Raleigh–Durham–Chapel Hill	6. Denver	6. Rochester, NY
7. Toronto	7. Atlanta	7. Austin–San Marcos
8. Houston	8. Toronto	8. San Francisco
9. Minneapolis–St. Paul	9. St. Louis	9. Washington, D.C.
10. Phoenix-Mesa	10. Minneapolis–St. Paul	10. Saskatoon

Indicate in Fig. 2.21 in which region, I–VIII, each of the following metropolitan areas belongs.

23. Washington, D.C.

24. Pittsburgh

25. Denver

26. Houston

27. Rochester, NY

28. Chicago

Figures In Exercises 29–40, indicate in Fig. 2.22 the region in which each of the figures would be placed.

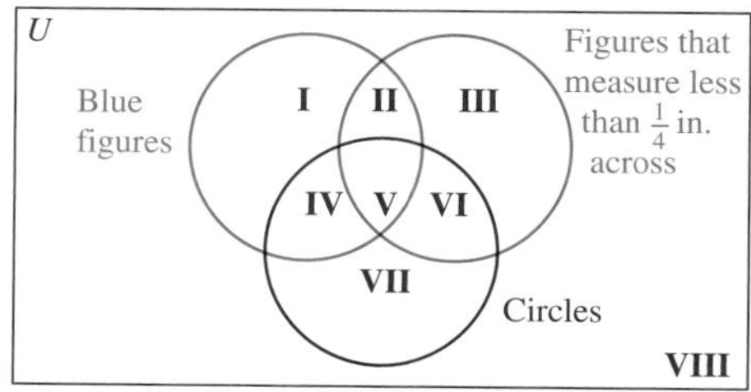

Figure 2.22

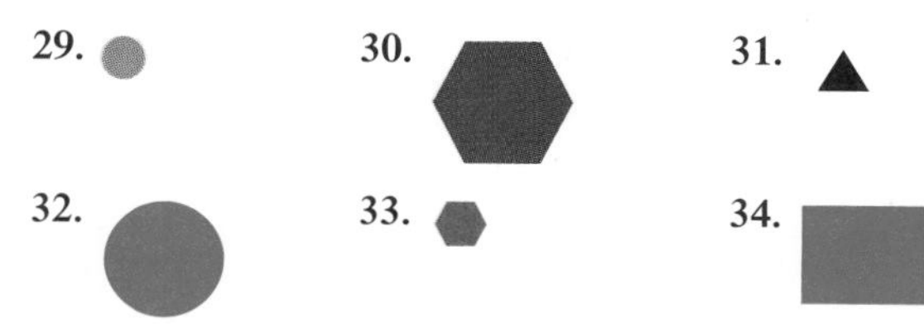

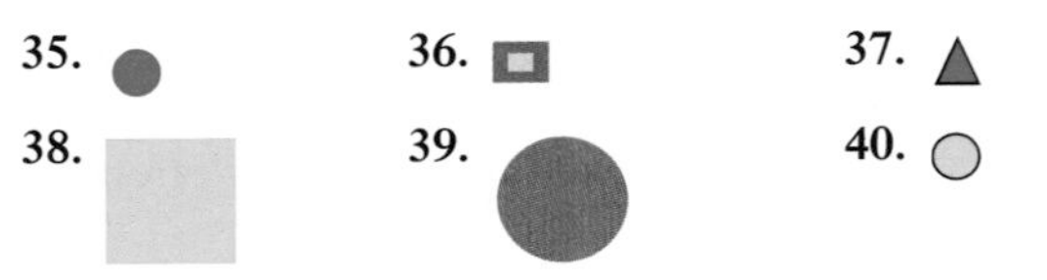

Senate Bills During a session of the U.S. Senate, three bills were voted on. The votes of six senators are shown below the figure. Determine in which region of Fig.2.23 each senator would be placed. The set labeled Bill 1 represents the set of senators who voted yes on Bill 1, and so on.

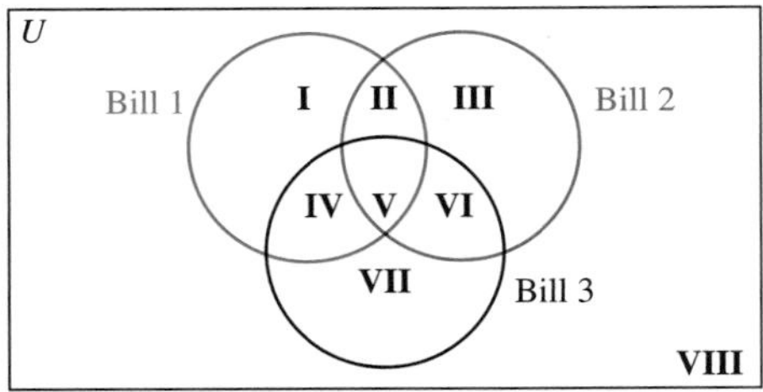

Figure 2.23

Senator	**Bill 1**	**Bill 2**	**Bill 3**
41. Grump	yes	no	no
42. Happi	no	no	yes
43. Turwilliger	no	no	no
44. Dillinger	yes	yes	yes
45. Isaitere	no	yes	yes
46. Smith	no	yes	no

In Exercises 47–60, use the Venn diagram in Fig. 2.24 to list the sets in roster form.

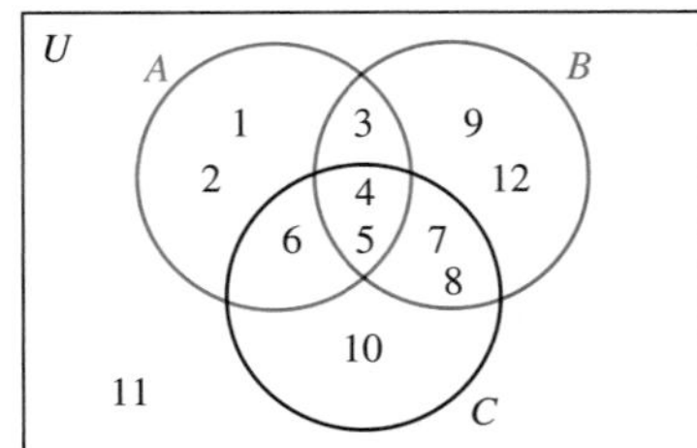

Figure 2.24

47. A **48.** U

49. B **50.** C

51. $A \cap B$ **52.** $A \cap C$

53. $(B \cap C)'$ **54.** $A \cap B \cap C$

55. $A \cup B$ **56.** $B \cup C$

57. $(A \cup C)'$ **58.** $A \cup B \cup C$

59. A' **60.** $(A \cup B \cup C)'$

In Exercises 61–68, use Venn diagrams to determine whether the following statements are equal for all sets A and B.

61. $(A \cup B)'$, $A' \cap B'$

62. $(A \cap B)'$, $A' \cup B$

63. $A' \cup B'$, $A \cap B$

64. $(A \cup B)'$, $(A \cap B)'$

65. $A' \cup B'$, $(A \cup B)'$

66. $A' \cap B'$, $A \cup B'$

67. $(A' \cap B)'$, $A \cup B'$

68. $A' \cap B'$, $(A' \cap B')'$

In Exercises 69–78, use Venn diagrams to determine whether the following statements are equal for all sets A, B, and C.

69. $A \cap (B \cup C)$, $(A \cap B) \cup C$

70. $A \cup (B \cap C)$, $(B \cap C) \cup A$

71. $A \cap (B \cup C)$, $(B \cup C) \cap A$

72. $A \cup (B \cap C)'$, $A' \cap (B \cup C)$

73. $A \cap (B \cup C)$, $(A \cap B) \cup (A \cap C)$

74. $A \cup (B \cap C)$, $(A \cup B) \cap (A \cup C)$

75. $A \cap (B \cup C)'$, $A \cap (B' \cap C')$

76. $(A \cup B) \cap (B \cup C)$, $B \cup (A \cap C)$

77. $(A \cup B)' \cap C$, $(A' \cup C) \cap (B' \cup C)$

78. $(C \cap B)' \cup (A \cap B)'$, $A \cap (B \cap C)$

In Exercises 79–82, use set statements to write a description of the shaded area. Use union, intersection and complement as necessary. More than one answer may be possible.

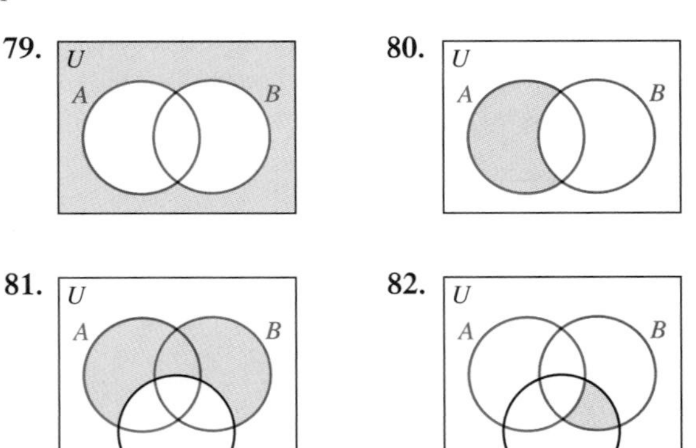

Problem Solving

83. Let

$$U = \{1, 2, 3, 4, 5, 6, 7, 8, 9, 10\}$$
$$A = \{1, 2, 3, 4\}$$
$$B = \{3, 6, 7\}$$
$$C = \{6, 7, 9\}$$

a) Show that $(A \cup B) \cap C = (A \cap C) \cup (B \cap C)$ for these sets.
b) Make up your own sets *A*, *B*, and *C*. Verify that $(A \cup B) \cap C = (A \cap C) \cup (B \cap C)$ for your sets *A*, *B*, and *C*.
c) Use Venn diagrams to verify that $(A \cup B) \cap C = (A \cap C) \cup (B \cap C)$ for all sets *A*, *B*, and *C*.

84. Let

$$U = \{a, b, c, d, e, f, g, h, i\}$$
$$A = \{a, c, d, e, f\}$$
$$B = \{c, d\}$$
$$C = \{a, b, c, d, e\}$$

a) Determine whether $(A \cup C)' \cap B = (A \cap C)' \cap B$ for these sets.
b) Make up your own sets, *A*, *B*, and *C*. Determine whether $(A \cup C)' \cap B = (A \cap C)' \cap B$ for your sets.
c) Determine whether $(A \cup C)' \cap B = (A \cap C)' \cap B$ for all sets *A*, *B*, and *C*.

85. *Blood Types* A hematology text gives the following information on percentages of the different types of blood worldwide.

Type	Positive Blood, %	Negative Blood, %
A	37	6
O	32	6.5
B	11	2
AB	5	0.5

Construct a Venn diagram similar to the one in Example 2 and place the correct percent in each of the eight regions.

86. Define each of the eight regions in Fig. 2.25 using sets *A*, *B*, and *C* and a set operation. (*Hint:* $A \cap B' \cap C'$ defines region I.)

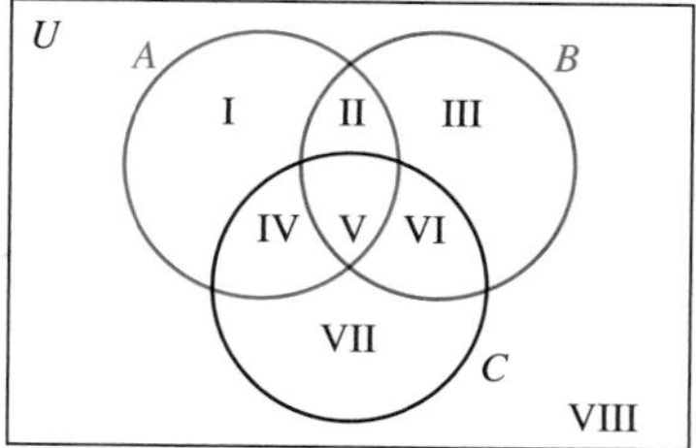

Figure 2.25

87. *Categorizing Contracts* J & C Mechanical Contractors wants to classify its projects. The contractors categorize set *A* as the set of office building construction projects, set *B* as the set of plumbing projects, and set *C* as the set of projects with a budget greater than $300,000.
a) Draw a Venn diagram that can be used to categorize the company projects according to the listed criteria.
b) Determine the region of the diagram that contains office building construction and plumbing projects with a budget greater than $300,000. Describe the region using sets *A*, *B*, and *C* with set operations. Use union, intersection, and complement as necessary.
c) Determine the region of the diagram that contains plumbing projects with a budget greater than $300,000 that are not office building construction projects. Describe the region using sets *A*, *B* and *C* with set operations. Use union, intersection, and complement as necessary.
d) Determine the region of the diagram that contains office building construction and nonplumbing projects whose budget is less than or equal to $300,000. Describe the region using sets *A*, *B*, and *C* with set operations. Use union, intersection, and complement as necessary.

Challenge Problem/Group Activity

88. We were able to determine the number of elements in the union of two sets with the formula

$$n(A \cup B) = n(A) + n(B) - n(A \cap B)$$

Can you determine a formula for finding the number of elements in the union of three sets? In other words, write a formula to determine $n(A \cup B \cup C)$. [*Hint*: The formula will contain each of the following: $n(A)$, $n(B)$, $n(C)$, $n(A \cap B \cap C')$, $n(A \cap B' \cap C)$, $n(A' \cap B \cap C)$, and $2n(A \cap B \cap C)$.]

Recreational Mathematics

89. a) Construct a Venn diagram illustrating four sets, *A*, *B*, *C*, and *D*. (*Hint*: Four circles cannot be used, and you should end up with 16 *distinct* regions.) Have fun!
b) Label each region with a set statement (see Exercise 86). Check all 16 regions to make sure that *each is distinct*.

90. *Triangle Seek* In the word seek on the next page, you are looking for six-letter words that form triangles. You need to trace out a triangle as you move from the first letter to the last letter and back to the first letter. The first letter of the word can be in any position in the triangle. Triangles may overlap other triangles, and the triangles can point up

or down. The word list is below. The word PROPER is shown as an example.

Word List
PROPER
SUBSET
FINITE
NUMBER
ROSTER
PENCIL
REGION

Internet/Research Activity

91. The two Venn diagrams illustrate what happens when colors are added or subtracted. Do research in an art text, an encyclopedia, the Internet, or another source and write a report explaining the creation of the colors in the Venn diagrams, using such terms as union of colors and subtraction (or difference) of colors.

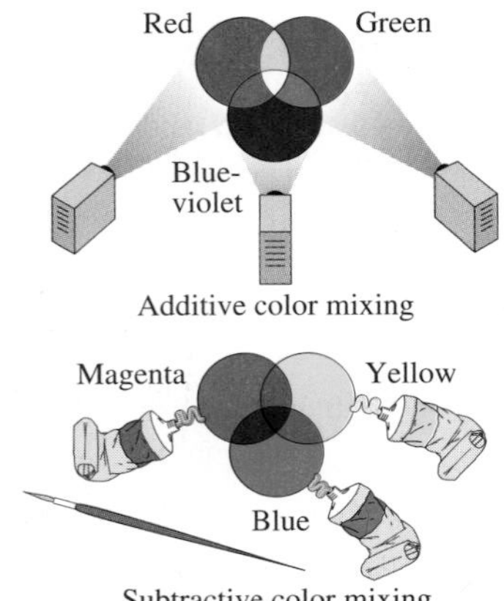

2.5 APPLICATIONS OF SETS

We can solve practical problems involving sets by using the problem-solving process discussed in Chapter 1: Understand the problem, devise a plan, carry out the plan, and then examine and check the results. First determine: What is the problem? or What am I looking for? To devise the plan, list all the facts that are given and how they are related. *Look for key words or phrases* such as "only set *A*," "set *A* and set *B*," "set *A* or set *B*," "set *A* and set *B* and not set *C*." Remember that *and* means intersection, *or* means union, and *not* means complement. The problems we solve in this section contain two or three sets of elements, which can be represented in a Venn diagram. Our plan will generally include drawing a Venn diagram, labeling the diagram, and filling in the regions of the diagram.

Whenever possible, follow the procedure in Section 2.4 for completing the Venn diagram and then answer the questions. Remember, when drawing Venn diagrams, we generally start with the intersection of the sets and work outward.

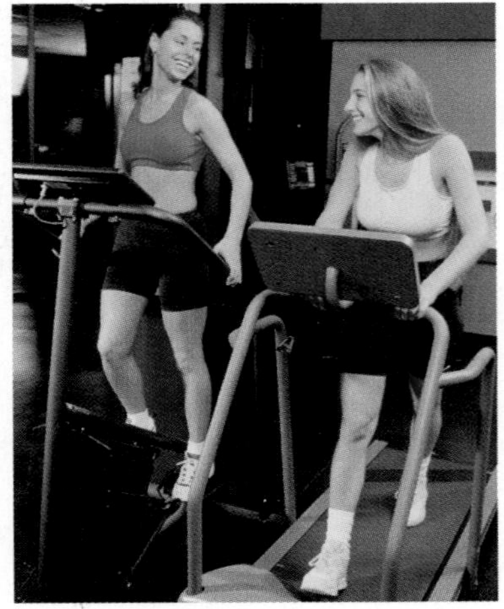

EXAMPLE 1 *Fitness Equipment*

Fitness for Life health club is considering adding additional cardiovascular equipment. It is considering two types of equipment, treadmills (T) and StairMasters (S). The health club surveyed a sample of members and asked which equipment they had used in the previous month. Of 150 members surveyed, it was determined that

102 used the treadmills.
71 used the StairMasters.
40 used both types.

Of those surveyed,

a) how many did not use either the treadmill or the StairMaster?
b) how many used the treadmill but not the StairMaster?
c) how many used the StairMaster but not the treadmill?
d) how many used either the treadmill or StairMaster?

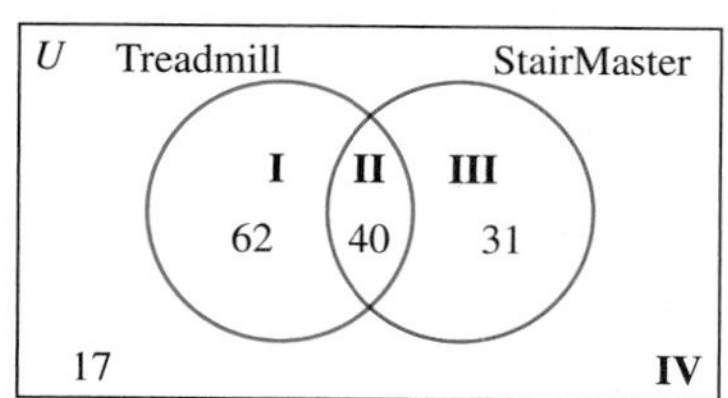

Figure 2.26

SOLUTION: The problem provides the following information.

The number of members surveyed is 150: $n(U)$ is 150.
The number of members surveyed who used the treadmill is 102: $n(T) = 102$.
The number of members surveyed who used the StairMaster is 71: $n(S) = 71$.
The number of members surveyed who used both the treadmill and the StairMaster is 40: $n(T \cap S) = 40$.

We illustrate this information on the Venn diagram shown in Fig. 2.26. We already know that $T \cap S$ corresponds to region II. As $n(T \cap S) = 40$, we write 40 in region II. Set T consists of regions I and II. We know that set T, the members who used the treadmill, contains 102 members. Therefore, region I contains $102 - 40$, or 62 members. We write the number 62 in region I. Set S consists of regions II and III. As $n(S) = 71$, the total in these two regions must be 71. Region II contains 40, leaving $71 - 40$ or 31 for region III. We write 31 in region III.

The total number of members surveyed who used the treadmill or the StairMaster is found by adding the numbers in regions I, II, and III. Therefore, $n(T \cup S) = 62 + 40 + 31 = 133$. The number in region IV is the difference between $n(U)$ and $n(T \cup S)$. There are $150 - 133$, or 17 members in region IV.

a) The members surveyed who did not use either the treadmill or the StairMaster are those members of the universal set who are not contained in set T or set S. The 17 members in region IV did not use the treadmill or StairMaster.
b) The 62 members in region I are those members surveyed who used the treadmill but not the StairMaster.
c) The 31 members in region III are those members surveyed who used the StairMaster but not the treadmill.
d) The members in regions I, II, or III are those members surveyed who used either the treadmill or the StairMaster. Thus, $62 + 40 + 31$ or 133 members surveyed used either the treadmill or the StairMaster. Notice that the 40 members in region II who use both types of equipment are included in those members surveyed who used either the treadmill or the StairMaster. ▲

Similar problems involving three sets can be solved, as illustrated in Example 2.

EXAMPLE 2 *Software Purchases*

CompUSA is considering expanding their computer software department. They are considering additional space for three types of computer software: games, educational software, and utility programs. The following information regarding software purchases was obtained from a survey of 893 customers.

545 purchased games.
497 purchased educational software.
290 purchased utility programs.
297 purchased games and educational software.

196 purchased educational software and utility programs.
205 purchased games and utility programs.
157 purchased all three types of software.

Use a Venn diagram to answer the following questions. How many customers purchased

a) none of these types of software?
b) only games?
c) at least one of these types of software?
d) exactly two of these types of software?

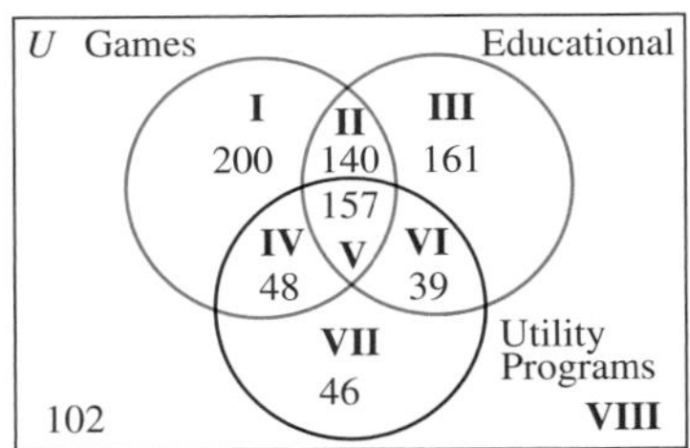

Figure 2.27

SOLUTION: Begin by constructing a Venn diagram with three overlapping circles. One circle represents games, another educational software, and the third utilities. See Fig 2.27. Label the eight regions.

Whenever possible, work from the center of the diagram outwards. First fill in region V. Since 157 customers purchased all three types of software, we place 157 in region V. Next determine the number to be placed in region II. Regions II and V together represent the customers who purchased both games and educational software. Since 297 customers purchased both of these types of software, the sum of the numbers in these regions must be 297. Since 157 have already been placed in region V, $297 - 157 = 140$ must be placed in region II. Now we determine the number to be placed in region IV. Since 205 customers purchased both games and utility programs, the sum of the numbers in regions IV and V must be 205. Since 157 have already been placed in region V, $205 - 157 = 48$ must be placed in region IV. Now determine the number to be placed in region VI. A total of 196 customers purchased educational software and utility programs. The numbers in regions V and VI must total 196. Since 157 have already been placed in region V, the number to be placed in region VI is $196 - 157 = 39$.

Now that we have determined the numbers for regions V, II, IV, and VI, we can determine the numbers to be placed in regions I, III, and VII. We are given that 545 customers purchased games. The sum of the numbers in regions I, II, IV, and V must be 545. To determine the number to be placed in region I, subtract the amounts in regions II, IV, and V from 545. There must be $545 - 140 - 48 - 157 = 200$ in region I. Determine the numbers to be placed in regions III and VII in a similar manner.

$$\text{Region III} = 497 - 140 - 157 - 39 = 161$$
$$\text{Region VII} = 290 - 48 - 157 - 39 = 46$$

Now that we have determined the numbers in regions I through VII, we can determine the number to be placed in region VIII. Adding the numbers in regions I through VII yields a sum of 791. The difference between the total number of customers surveyed, 893, and the sum of the numbers in regions I through VII must be placed in region VIII.

$$\text{Region VIII} = 893 - 791 = 102$$

Now that we have completed the Venn diagram, we can answer the questions.

a) One hundred two customers did not purchase any of these types of software. These customers are indicated in region VIII.

b) Region I represents those customers who purchased only games. Thus, 200 customers purchased only games.

c) The words *at least one* mean "one or more." All those in regions I through VII purchased at least one of the types of software. The sum of the numbers in regions I through VII is 791, so 791 customers purchased at least one of the types of software.

d) The customers in regions II, IV, and VI purchased exactly two of the types of software. Summing the numbers in these regions $140 + 48 + 39$ we find that 227 customers purchased exactly two of these types of software. Notice that we did not include the customers in region V. Those customers purchased all three types of software. ▲

The procedure to work problems like those given in Example 2 is generally the same. Start by completing region V. Next complete regions II, IV, and VI. Then complete regions I, III, and VII. Finally, complete region VIII. When you are constructing Venn diagrams, be sure to check your work carefully.

TIMELY TIP When constructing a Venn diagram, the most common mistake made by students is forgetting to subtract the number in region V from the respective values in determining the numbers to be placed in regions II, IV, and VI.

EXAMPLE 3 *Birds at the Feeders*

In a bird sanctuary, 41 different species of birds are being studied. Three large bird feeders are constructed, each providing a different type of bird feed. One feeder has sunflower seeds. A second feeder has a mixture of seeds, and the third feeder has small pieces of fruit. The following information was obtained.

20 species ate sunflower seeds.
22 species ate the mixture.
11 species ate the fruit.
10 species ate the sunflower seeds and the mixture.
4 species ate the sunflower seeds and the fruit.
3 species ate the mixture and the fruit.
1 species ate all three.

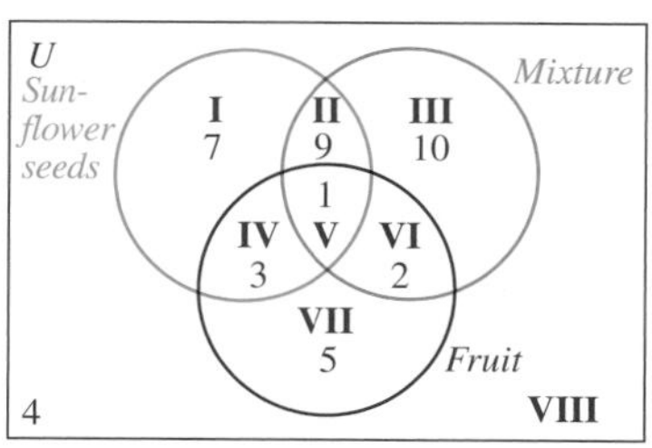

Figure 2.28

Use a Venn diagram to answer the following questions. How many species ate

a) none of the foods?
b) the sunflower seeds, but neither of the other two foods?
c) the mixture *and* the fruit, but not the sunflower seeds?
d) the mixture *or* the fruit, but not the sunflower seeds?
e) exactly one of the foods?

SOLUTION: The Venn diagram is constructed using the procedure we outlined in Example 2. The diagram is illustrated in Fig. 2.28. We suggest you construct the diagram by yourself now and check your diagram with Fig. 2.28.

a) Four species did not eat any of the food (see region VIII).

b) Seven species (see region I) ate the sunflower seeds but neither of the other two foods.

c) Those in region VI ate both the mixture and fruit but not the sunflower seeds. Therefore, two species satisfy the criteria.

d) The word *or* in this type of problem means one or the other or both. All the species in regions II, III, IV, V, VI, and VII ate the mixture or the fruit or both. Those in regions II, IV, and V also ate the sunflower seeds. The species that ate the mixture or fruit, but not the sunflower seeds, are found by adding the numbers in regions III, VI, and VII. There are $10 + 2 + 5 = 17$ species that satisfy the criteria.

e) Those species indicated in regions I, III, and VII ate exactly one of the foods. Therefore, $7 + 10 + 5 = 22$ species ate exactly one of the three types of bird food.

SECTION 2.5 EXERCISES

Practice the Skills/Problem Solving

In Exercises 1–15, draw a Venn diagram to obtain the answers.

1. *Landscape Purchases* Agway Lawn and Garden collected the following information regarding purchases from 130 of its customers.
   74 purchased shrubs.
   70 purchased trees.
   41 purchased both shrubs and trees.

   Of those surveyed,
   a) how many purchased only shrubs?
   b) how many purchased only trees?
   c) how many did not purchase either of these items?

2. *Study Locations* At a local college, a survey was taken to determine where students studied on campus. Of 160 students surveyed, it was determined that
   79 studied in the library.
   65 studied in the student lounge.
   43 studied in both the library and the student lounge.

   Of those interviewed,
   a) how many studied in only the library?
   b) how many studied in only the student lounge?
   c) how many did not study in either location?

3. *Real Estate* The Maiello's are moving to Wilmington, Delaware. Their real estate agent located 83 houses listed for sale, in the Wilmington area, in their price range. Of these houses listed for sale,
   47 had a family room.
   42 had a deck.
   30 had a family room and a deck.

   How many had
   a) a family room but not a deck?
   b) a deck but not a family room?
   c) either a family room or a deck?

4. *Toothpaste Taste Test* A drug company is considering manufacturing a new toothpaste. They are considering two flavors, regular and mint. In a sample of 120 people, it was found that
   74 liked the regular.
   62 liked the mint.
   35 liked both types.

   a) How many liked only the regular?
   b) How many liked only the mint?
   c) How many liked either one or the other or both?

5. *Overnight Delivery Services* In San Diego, California, a sample of 444 businesses was surveyed to determine which overnight mailing services they used. The following information was determined.
   189 used Federal Express.
   205 used United Parcel Service.
   122 used Airborne.
   57 used Federal Express and United Parcel Service.
   34 used Federal Express and Airborne.
   30 used United Parcel Service and Airborne.
   22 used all three.

   How many used
   a) none of these services?
   b) only Airborne?
   c) exactly one of the services?
   d) exactly two of the services?
   e) Federal Express or United Parcel Service?

6. *Professional Teams* Thirty-three U.S. cities with large populations were surveyed to determine whether they had a professional baseball team, a professional football team, or a professional basketball team. The following information was determined.
   16 had baseball.
   17 had football.
   15 had basketball.
   11 had baseball and football.
   7 had baseball and basketball.
   9 had football and basketball.
   5 had all three teams.

   How many had
   a) only a football team?
   b) baseball and football, but not basketball?
   c) baseball or football?
   d) baseball or football, but not basketball?
   e) exactly two teams?

7. *Book Purchases* A survey of 85 customers was taken at Barnes & Noble regarding the types of books purchased. The survey found that
   44 purchased mysteries.
   33 purchased science fiction.
   29 purchased romance novels.
   13 purchased mysteries and science fiction.
   5 purchased science fiction and romance novels.
   11 purchased mysteries and romance novels.
   2 purchased all three types of books.

   How many of the customers surveyed purchased
   a) only mysteries?
   b) mysteries and science fiction, but not romance novels?
   c) mysteries or science fiction?
   d) mysteries or science fiction, but not romance novels?
   e) exactly two types?

8. *Resorts* In a survey of 65 resorts, it was found that
   34 provided refrigerators in the guest rooms.
   30 provided laundry services.
   37 provided child care services.
   15 provided refrigerators in the guest rooms and laundry services.
   17 provided refrigerators in the guest rooms and child care services.
   19 provided laundry services and child care services.
   7 provided all three features.

   How many of the resorts provided
   a) only refrigerators in the guest rooms?
   b) exactly one of the features?
   c) at least one of the features?
   d) exactly two of the features?
   e) none of the features?

9. *Transportation* Seventy businesspeople in Sacramento, California, were asked how they traveled to work during the previous month. The following information was determined.
   28 used public transportation.
   18 rode in a car pool.
   37 drove alone.
   5 used public transportation and rode in a car pool.
   9 used public transportation and drove alone.
   4 rode in a car pool and drove alone.
   3 used all three forms of transportation.

   How many of those surveyed
   a) only used public transportation?
   b) only drove alone?
   c) used public transportation and rode in a car pool, but did not drive alone?
   d) used public transportation or rode in a car pool, but did not drive alone?
   e) used none of these forms of transportation?

10. *Colleges and Universities* In a survey of four-year colleges and universities, it was found that
    356 offered a liberal arts degree.
    293 offered a computer engineering degree.
    285 offered a nursing degree.
    193 offered a liberal arts degree and a computer engineering degree.
    200 offered a liberal arts degree and a nursing degree.
    139 offered a computer engineering degree and a nursing degree.
    68 offered a liberal arts degree, a computer engineering degree, and a nursing degree.
    26 offered none of these degrees.

    a) How many four-year colleges and universities were surveyed?

    Of the four-year colleges and universities surveyed, how many offered
    b) a liberal arts degree and a nursing degree, but not a computer engineering degree?
    c) a computer engineering degree, but neither a liberal arts degree nor a nursing degree?
    d) exactly two of these degrees?
    e) at least one of these degrees?

11. *Fastest-Growing Cities* The following census information was collected regarding the 25 largest U.S. cities based on the 2000 population.

5 cities were in Texas.
13 cities had a population greater than 750,000.
11 cities had a population increase greater than 10% from 1990 to 2000.
3 cities had a population greater than 750,000 and were in Texas.
4 cities had a population increase greater than 10% from 1990 to 2000 and were in Texas.
6 cities had a population greater than 750,000 and a population increase greater than 10% from 1990 to 2000.
3 cities had a population greater than 750,000, a population increase greater than 10% from 1990 to 2000, and were in Texas.

How many of the 25 largest cities
a) with a population greater than 750,000 were not in Texas?
b) were in Texas or had a population greater than 750,000?
c) were in Texas and had a population greater than 750,000, but did not have a population increase greater than 10% from 1990 to 2000?
d) were not in Texas, did not have a population greater than 750,000 and did not have a population increase greater than 10% from 1990 to 2000?

12. *Appetizers Survey* Da Tulio's Restaurant hired Dennis Goldstein to determine what kind of appetizers customers liked. He surveyed 100 people, with the following results: 78 liked shrimp cocktail, 56 liked mozzarella sticks, and 35 liked both shrimp cocktail and mozzarella sticks. Every person interviewed liked one or the other or both kinds of appetizers. Does this result seem correct? Explain your answer.

13. *Discovering an Error* An immigration agent sampled cars going from the United States into Canada. In his report, he indicated that of the 85 cars sampled,
35 cars were driven by women.
53 cars were driven by U.S. citizens.
43 cars had two or more passengers.
27 cars were driven by women who are U.S. citizens.
25 cars were driven by women and had two or more passengers.
20 cars were driven by U.S. citizens and had two or more passengers.
15 cars were driven by women who are U.S. citizens and had two or more passengers.

After his supervisor reads the report, she explains to the agent that he made a mistake. Explain how his supervisor knew that the agent's report contained an error.

Challenge Problems/Group Activities

14. *Parks* A survey of 300 parks showed the following.
15 had only camping.
20 had only hiking trails.
35 had only picnicking.
185 had camping.
140 had camping and hiking trails.
125 had camping and picnicking.
210 had hiking trails.

Find the number of parks that
a) had at least one of these features.
b) had all three features.
c) did not have any of these features.
d) had exactly two of these features.

15. *Surveying Farmers* A survey of 500 farmers in a midwestern state showed the following.
125 grew only wheat.
110 grew only corn.
90 grew only oats.
200 grew wheat.
60 grew wheat and corn.
50 grew wheat and oats.
180 grew corn.

Find the number of farmers who
a) grew at least one of the three.
b) grew all three.
c) did not grow any of the three.
d) grew exactly two of the three.

Recreational Mathematics

16. *Number of Elements* A universal set U consists of 12 elements. If sets A, B, and C are proper subsets of U and $n(U) = 12$, $n(A \cap B) = n(A \cap C) = n(B \cap C) = 6$, $n(A \cap B \cap C) = 4$, and $n(A \cup B \cup C) = 10$ determine
a) $n(A \cup B)$ b) $n(A' \cup C)$ c) $n(A \cap B)'$

2.6 INFINITE SETS

On page 44, we state that a finite set is a set in which the number of elements is zero or the number of elements can be expressed as a natural number. On page 45, we define a one-to-one correspondence. To determine the number of elements in a finite set, we can place the set in a one-to-one correspondence with a subset of the set of counting numbers. For example, the set $A = \{\#, ?, \$\}$ can be placed in one-to-one correspondence with set $B = \{1, 2, 3\}$, a subset of the set of counting numbers.

$$\begin{aligned} A &= \{\#, ?, \$\} \\ & \quad\ \downarrow\ \downarrow\ \downarrow \\ B &= \{1, 2, 3\} \end{aligned}$$

Because the cardinal number of set B is 3, the cardinal number of set A is also 3. Any two sets such as set A and set B that can be placed in a one-to-one correspondence must have the same number of elements (therefore the same cardinality) and must be equivalent sets. Note that $n(A)$ and $n(B)$ both equal 3.

The German mathematician Georg Cantor (1845–1918), known as the father of set theory, thought about sets that were not bounded. He called an unbounded set an *infinite set* and provided the following definition.

> An **infinite set** is a set that can be placed in a one-to-one correspondence with a proper subset of itself.

In Example 1, we use Cantor's definition of an infinite set to show that the set of counting numbers is infinite.

EXAMPLE 1 *The Set of Natural Numbers*

Show that $N = \{1, 2, 3, 4, 5, \ldots, n, \ldots\}$ is an infinite set.

SOLUTION: To show that the set N is infinite, we establish a one-to-one correspondence between the counting numbers and a proper subset of itself. By removing the first element from the set of counting numbers, we get the set $\{2, 3, 4, 5, \ldots\}$, which is a proper subset of the set of counting numbers. Now we establish the one-to-one correspondence.

$$\begin{aligned} \text{Counting numbers} &= \{1, 2, 3, 4, 5, \ldots, \quad n \quad, \ldots\} \\ & \quad\ \downarrow\ \downarrow\ \downarrow\ \downarrow\ \downarrow \qquad\quad \downarrow \\ \text{Proper subset} &= \{2, 3, 4, 5, 6, \ldots, n + 1, \ldots\} \end{aligned}$$

Note that for any number, n, in the set of counting numbers, its corresponding number in the proper subset is one greater, or $n + 1$. We have now shown the desired one-to-one correspondence, and thus the set of counting numbers is infinite. ▲

PROFILE IN MATHEMATICS

LEOPOLD KRONECKER

Mathematician Leopold Kronecker (1823–1891), Cantor's former mentor, ridiculed Cantor's theories and prevented Cantor from gaining a position at the University of Berlin. Although Cantor's work on infinite sets is now considered a masterpiece, it generated heated controversy when originally published. Cantor's claim that the infinite set was unbounded offended the religious views of the time that God had created a complete universe, which could not be wholly comprehended by humans. Eventually Cantor was given the recognition due to him, but by then the criticism had taken its toll on his health. He had several nervous breakdowns and spent his last days in a mental hospital.

GEORG CANTOR

Note in Example 1 that we showed the pairing of the general terms $n \rightarrow (n + 1)$. Showing a one-to-one correspondence of infinite sets requires showing the pairing of the general terms in the two infinite sets.

In the set of counting numbers, n represents the general term. For any other set of numbers, the general term will be different. The general term in any set should be written in terms of n such that when 1 is substituted for n in the general term, we get the first number in the set; when 2 is substituted for n in the general term, we get the second number in the set; when 6 is substituted for n in the general term, we get the sixth number in the set; and so on.

Consider the set $\{4, 9, 14, 19, \ldots\}$. Suppose we want to write the general term for this set (or sequence) or numbers. What would the general term be? The numbers differ by 5, so the general term will be of the form $5n$ plus or minus some number. Substituting 1 for n yields 5(1), or 5. Because the first number in the set is 4, we need to subtract 1 from the 5. Thus, the general term is $5n - 1$. Note that when $n = 1$, the value is $5(1) - 1$ or 4; when $n = 2$, the value is $5(2) - 1$ or 9; when $n = 3$, the value is $5(3) - 1$ or 14; and so on. Therefore, we write the set of numbers with the general term as

$$\{4, 9, 14, 19, \ldots, 5n - 1, \ldots\}$$

Now that you are aware of how to determine the general term of a set of numbers, we can do some more problems involving sets.

EXAMPLE 2 *The Set of Even Numbers*

Show that the set of even counting numbers $\{2, 4, 6, \ldots, 2n, \ldots\}$ is an infinite set.

SOLUTION: First create a proper subset of the set of even counting numbers by removing the first number from the set. Then establish a one-to-one correspondence.

$$\begin{array}{ll} \text{Even counting numbers:} & \{2, 4, 6, 8, \ldots, \quad 2n \quad, \ldots\} \\ & \;\;\downarrow\;\downarrow\;\downarrow\;\downarrow \qquad\quad \downarrow \\ \text{Proper subset:} & \{4, 6, 8, 10, \ldots, 2n + 2, \ldots\} \end{array}$$

A one-to-one correspondence exists between the two sets, so the set of even counting numbers is infinite. ▲

EXAMPLE 3 *The Set of Multiples of Three*

Show that the set $\{3, 6, 9, 12, \ldots, 3n, \ldots\}$ is an infinite set.

SOLUTION:

$$\begin{array}{ll} \text{Given set:} & \{3, 6, 9, 12, 15, \ldots, \quad 3n \quad, \ldots\} \\ & \;\;\downarrow\;\downarrow\;\downarrow\;\downarrow\;\downarrow \qquad\quad \downarrow \\ \text{Proper subset:} & \{6, 9, 12, 15, 18, \ldots, 3n + 3, \ldots\} \end{array}$$

Therefore, the given set is an infinite set. ▲

Countable Sets

In his work with infinite sets, Cantor developed ideas on how to determine the cardinal number of an infinite set. He called the cardinal number of infinite sets "transfinite cardinal numbers" or "transfinite powers." He defined a set as *countable* if it is finite or if it can be placed in a one-to-one correspondence with the set of counting numbers. All infinite sets that can be placed in a one-to-one correspondence with the set of counting numbers have cardinal number, *aleph-null*, symbolized $\aleph_0$ (the first Hebrew letter, aleph, with a zero subscript, read "null").

EXAMPLE 4 *The Cardinal Number of the Set of Even Numbers*

Show that the set of even counting numbers has cardinal number $\aleph_0$.

SOLUTION: In Example 2, we showed that a set of even counting numbers is infinite by setting up a one-to-one correspondence between the set and a proper subset of itself.

Now we will show that it is countable and has cardinality $\aleph_0$ by setting up a one-to-one correspondence between the set of counting numbers and the set of even counting numbers.

$$\begin{array}{lcl} \text{Counting numbers:} & N = & \{1, 2, 3, 4, \ldots, n, \ldots\} \\ & & \;\;\downarrow\;\downarrow\;\downarrow\;\downarrow \qquad\;\; \downarrow \\ \text{Even counting numbers:} & E = & \{2, 4, 6, 8, \ldots, 2n, \ldots\} \end{array}$$

For each number n in the set of counting numbers, its corresponding number is $2n$. Since we found a one-to-one correspondence between the set of counting numbers and the set of even counting numbers, the set of even counting numbers is countable. Thus, the cardinal number of the set of even counting numbers is $\aleph_0$; that is, $n(E) = \aleph_0$. As we mentioned earlier, the set of even counting numbers is an infinite set since it can be placed in a one-to-one correspondence with a proper subset of itself. Therefore, the set of even counting numbers is both infinite and countable. ▲

Any set that can be placed in a one-to-one correspondence with the set of counting numbers has cardinality $\aleph_0$ and is countable.

EXAMPLE 5 *The Cardinal Number of the Set of Odd Numbers*

Show that the set of odd counting numbers has cardinality $\aleph_0$.

SOLUTION: To show that the set of odd counting numbers has cardinality $\aleph_0$, we need to show a one-to-one correspondence between the counting numbers and the odd counting numbers.

$$\begin{array}{lcl} \text{Counting numbers:} & N = & \{1, 2, 3, 4, 5, \ldots, \quad n \quad, \ldots\} \\ & & \;\;\downarrow\;\downarrow\;\downarrow\;\downarrow\;\downarrow \qquad\quad \downarrow \\ \text{Odd counting numbers:} & O = & \{1, 3, 5, 7, 9, \ldots, 2n - 1, \ldots\} \end{array}$$

Since there is a one-to-one correspondence, the odd counting numbers have cardinality $\aleph_0$; that is, $n(O) = \aleph_0$. ▲

We have shown that both the odd and even counting numbers have cardinality $\aleph_0$. Merging the odd counting numbers with the even counting numbers gives the set of counting numbers, and we may reason that

$$\aleph_0 + \aleph_0 = \aleph_0$$

This result may seem strange, but it is true. What could such a statement mean? Well, consider a hotel with infinitely many rooms. If all the rooms are occupied, then the hotel is, of course, full. If more guests appear wanting accommodations, will they be turned away? The answer is *no,* for if the room clerk were to reassign each guest to a new room with a room number twice that of the present room, then all the odd-numbered rooms would become unoccupied and there would be space for more guests!

. . . where there's always room for one more. . .

In Cantor's work, he showed that there are different orders of infinity. Sets that are countable and have cardinal number $\aleph_0$ are the lowest order of infinity. Cantor showed that the set of integers and the set of rational numbers (fractions of the form p/q, where $q \neq 0$) are infinite sets with cardinality $\aleph_0$. He also showed that the set of real numbers (discussed in Chapter 5) could not be placed in a one-to-one correspondence with the set of counting numbers and that they have a higher order of infinity, aleph-one, $\aleph_1$.

SECTION 2.6 EXERCISES

Concept/Writing Exercises

1. What is an infinite set as defined in this section?
2. a) What is a countable set?
   b) How can we determine if a given set has cardinality $\aleph_0$?

Practice the Skills

In Exercises 3–12, show that the set is infinite by placing it in a one-to-one correspondence with a proper subset of itself. Be sure to show the pairing of the general terms in the sets.

3. $\{7, 8, 9, 10, 11, \dots\}$
4. $\{12, 13, 14, 15, 16, \dots\}$
5. $\{3, 5, 7, 9, 11, \dots\}$
6. $\{20, 22, 24, 26, 28, \dots\}$
7. $\{4, 7, 10, 13, 16, \dots\}$
8. $\{4, 8, 12, 16, 20, \dots\}$
9. $\{6, 11, 16, 21, 26, \dots\}$
10. $\{1, \frac{1}{2}, \frac{1}{3}, \frac{1}{4}, \frac{1}{5}, \dots\}$
11. $\{\frac{1}{2}, \frac{1}{4}, \frac{1}{6}, \frac{1}{8}, \frac{1}{10}, \dots\}$
12. $\{\frac{6}{11}, \frac{7}{11}, \frac{8}{11}, \frac{9}{11}, \frac{10}{11}, \dots\}$

In Exercises 13–22, show that the set has cardinal number $\aleph_0$ by establishing a one-to-one correspondence between the set of counting numbers and the given set. Be sure to show the pairing of the general terms in the sets.

13. $\{6, 12, 18, 24, 30, \dots\}$
14. $\{50, 51, 52, 53, 54, \dots\}$
15. $\{4, 6, 8, 10, 12, \dots\}$
16. $\{0, 2, 4, 6, 8, \dots\}$
17. $\{2, 5, 8, 11, 14, \dots\}$
18. $\{4, 9, 14, 19, 24, \dots\}$
19. $\{5, 8, 11, 14, 17, \dots\}$
20. $\{\frac{1}{2}, \frac{1}{4}, \frac{1}{6}, \frac{1}{8}, \dots\}$
21. $\{\frac{1}{3}, \frac{1}{4}, \frac{1}{5}, \frac{1}{6}, \frac{1}{7}, \dots\}$
22. $\{\frac{1}{2}, \frac{2}{3}, \frac{3}{4}, \frac{4}{5}, \frac{5}{6}, \dots\}$

Challenge Problems/Group Activities

In Exercises 23–26, show that the set has cardinality $\aleph_0$ by establishing a one-to-one correspondence between the set of counting numbers and the given set.

23. $\{1, 4, 9, 16, 25, 36, \dots\}$
24. $\{2, 4, 8, 16, 32, \dots\}$
25. $\{3, 9, 27, 81, 243, \dots\}$
26. $\{\frac{1}{3}, \frac{1}{6}, \frac{1}{12}, \frac{1}{24}, \frac{1}{48}, \dots\}$

Recreational Mathematics

In Exercises 27–31, insert the symbol <, >, or = in the shaded area to make a true statement.

27. $\aleph_0$ ▪ $\aleph_0 + \aleph_0$

28. $2\aleph_0$ ▪ $\aleph_0 + \aleph_0$

29. $2\aleph_0$ ▪ $\aleph_0$

30. $\aleph_0 + 5$ ▪ $\aleph_0 - 3$

31. $n(N)$ ▪ $\aleph_0$

32. There are a number of paradoxes (a statement that appears to be true and false at the same time) associated with infinite sets and the concept of infinity. One of these, called *Zeno's Paradox,* is named after the mathematician Zeno, born about 496 BC in Italy. According to Zeno's paradox, suppose Achelles starts out 1 meter behind a tortoise. Also, suppose that Achelles walks 10 times as fast as the tortoise crawls. When Achelles reaches the point where the tortoise started, the tortoise is 1/10 of a meter ahead of Achelles. When Achelles reaches the point where the tortoise was 1/10 of a meter ahead, the tortoise is now 1/100 of a meter ahead. And so on. According to Zeno's Paradox, Achelles gets closer and closer to the tortoise but never catches up to the tortoise.
    a) Do you believe the reasoning process is sound? If not, explain why not.
    b) In actuality, if this were a real situation, would Achelles ever pass the tortoise?

Internet/Research Activities

33. Do research to explain how Cantor proved that the set of rational numbers has cardinal number $\aleph_0$.

34. Do research to explain how it can be shown that the real numbers do not have cardinal number $\aleph_0$.

CHAPTER 2 SUMMARY

IMPORTANT FACTS

Or is generally interpreted to mean *union.*

And is generally interpreted to mean *intersection.*

DE MORGAN'S LAWS

$$(A \cup B)' = A' \cap B'$$
$$(A \cap B)' = A' \cup B'$$

For any sets A and B,

$$n(A \cup B) = n(A) + n(B) - n(A \cap B).$$

Number of distinct subsets of a finite set with n elements is 2^n.

Symbol	Meaning
$\in$	is an element of
$\notin$	is not an element of
$n(A)$	number of elements in set A
$\emptyset$ or $\{\ \}$	the empty set
U	the universal set
$\subseteq$	is a subset of
$\not\subseteq$	is not a subset of
$\subset$	is a proper subset of
$\not\subset$	is not a proper subset of
$'$	complement
$\cup$	union
$\cap$	intersection
$\aleph_0$	aleph-null

CHAPTER 2 REVIEW EXERCISES

2.1, 2.2, 2.3, 2.4, 2.6

In Exercises 1–14, state whether each is true or false. If false, give a reason.

1. The set of counties located in the state of Alabama is a well-defined set.
2. The set of the three best beaches in the United States is a well-defined set.
3. maple $\in$ {oak, elm, maple, sycamore}
4. $\{\ \} \subset \emptyset$
5. $\{3, 6, 9, 12, \ldots\}$ and $\{2, 4, 6, 8, \ldots\}$ are disjoint sets.
6. $\{a, b, c, d, e\}$ is an example of a set in roster form.
7. {computer, calculator, pencil} = {calculator, computer, diskette}
8. {apple, orange, banana, pear} is equivalent to {tomato, corn, spinach, radish}.
9. If $A = \{a, e, i, o, u\}$, then $n(A) = 5$.
10. $A = \{1, 4, 9, 16, \ldots\}$ is a countable set.
11. $A = \{1, 4, 7, 10, \ldots, 31\}$ is a finite set.
12. $\{3, 6, 7\} \subseteq \{7, 6, 3, 5\}$.
13. $\{x \mid x \in N \text{ and } 3 < x \leq 9\}$ is a set in set-builder notation.
14. $\{x \mid x \in N \text{ and } 2 < x \leq 12\} \subseteq \{1, 2, 3, 4, 5, \ldots, 20\}$

In Exercises 15–18, express each set in roster form.

15. Set A is the set of odd natural numbers between 5 and 16.

16. Set B is the set of states that border Nevada.

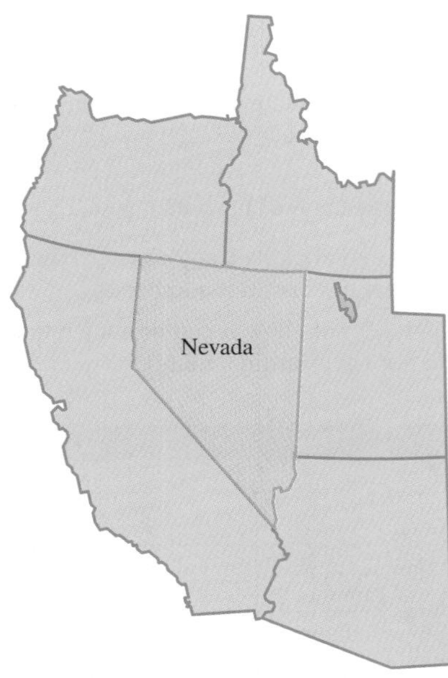

17. $C = \{x \mid x \in N \text{ and } x < 297\}$

18. $D = \{x \mid x \in N \text{ and } 8 < x \leq 96\}$

In Exercises 19–22, express each set in set-builder notation.

19. Set A is the set of natural numbers between 52 and 100.

20. Set B is the set of natural numbers greater than 63.

21. Set C is the set of natural numbers less than 3.

22. Set D is the set of natural numbers between 23 and 41, inclusive.

In Exercises 23–26, express each set with a written description.

23. $A = \{x \mid x \text{ is a letter of the English alphabet from E through M inclusive}\}$

24. $B = \{$penny, nickel, dime, quarter, half-dollar$\}$

25. $C = \{x, y, z\}$

26. $D = \{x \mid 3 \leq x < 9\}$

In Exercises 27–32, let

$$U = \{1, 2, 3, 4, \ldots, 10\}$$
$$A = \{1, 3, 5, 6\}$$
$$B = \{5, 6, 9, 10\}$$
$$C = \{1, 6, 10\}$$

In Exercises 27–32, determine the following.

27. $A \cap B$

28. $A \cup B'$

29. $A' \cap B$

30. $(A \cup B)' \cup C$

31. The number of subsets of set B

32. The number of proper subsets of set A

33. For the following sets, construct a Venn diagram and place the elements in the proper region.

$U = \{$lion, tiger, leopard, cheetah, puma, lynx, panther, jaguar$\}$

$A = \{$tiger, puma, lynx$\}$

$B = \{$lion, tiger, jaguar, panther$\}$

$C = \{$tiger, lynx, cheetah, panther$\}$

In Exercises 34–39, use Fig. 2.29 to determine the sets.

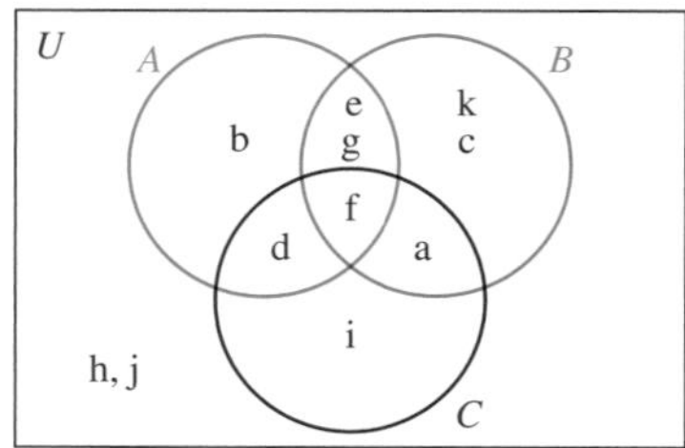

Figure 2.29

34. $A \cup B$

35. $A \cap B'$

36. $A \cup B \cup C$

37. $A \cap B \cap C$

38. $(A \cup B) \cap C$

39. $(A \cap B) \cup C$

Construct a Venn diagram to determine whether the following statements are true for all sets A, B, and C.

40. $(A' \cup B')' = A \cap B$

41. $(A \cup B') \cup (A \cup C') = A \cup (B \cap C)'$

Breakfast Cereals In Exercises 42–47, use the following chart, which shows selected breakfast cereals and some of their nutritional contents for a 1-cup serving. Let the cereals shown represent the universal set.

Cereal	Fat (grams)	Fiber (grams)	Sugar (grams)
Kellogg's Corn Flakes	0	1	2
Kellogg's Product 19	0	1	4
Kellogg's Fruit Loops	1	1	15
Kellogg's All-Bran	2	20	12
Kellogg's Raisin Bran	1.5	8	18
General Mills Wheaties	1	3	4
Kellogg's Special K	0	less than 1	4
General Mills Cinnamon Toast Crunch	3.5	1	10
General Mills Cheerios	2	3	1
General Mills Cookie Crisp	1	0	13

Let A be the set of cereals that contain at least 1 gram of fat.
Let B be the set of cereals that contain at least 3 grams of fiber.
Let C be the set of cereals that contain at least 4 grams of sugar.
Indicate in Fig. 2.30 in which region, I–VIII, each of the following cereals belongs.

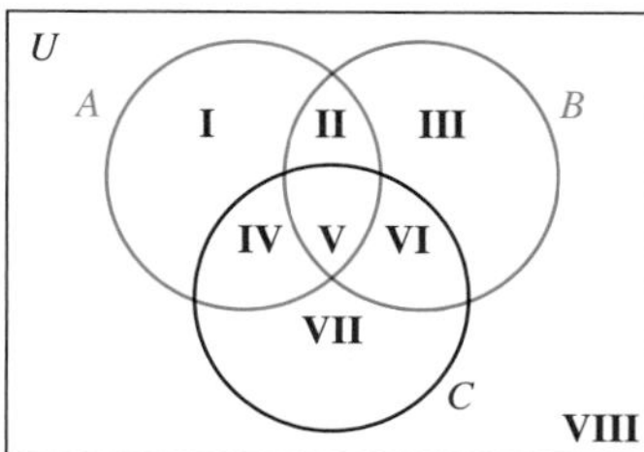

Figure 2.30

42. General Mills Cheerios

43. Kellogg's Raisin Bran

44. Kellogg's Corn Flakes

45. General Mills Cookie Crisp

46. Kellogg's Fruit Loops

47. Kellogg's Special *K*

2.5

48. *Pizza Survey* A pizza chain was willing to pay $1 to each person interviewed about his or her likes and dislikes of types of pizza crust. Of the people interviewed, 200 liked thin crust, 270 liked thick crust, 70 liked both, and 50 did not like pizza at all. What was the total cost of the survey?

49. *Cookie Preferences* The Cookie Shoppe conducted a survey to determine its customers' preferences.
200 people liked chocolate chip cookies.
190 people liked peanut butter cookies.
210 people liked sugar cookies.
100 people liked chocolate chip cookies and peanut butter cookies.
150 people liked peanut butter cookies and sugar cookies.
110 people liked chocolate chip cookies and sugar cookies.
70 people liked all three.
5 people liked none of these cookies.

Draw a Venn diagram, then determine how many people
a) completed the survey.
b) liked only peanut butter cookies.
c) liked peanut butter cookies and chocolate chip cookies, but not sugar cookies.
d) liked peanut butter cookies or sugar cookies, but not chocolate chip cookies.

50. *TV Choices TV Guide* sent a questionnaire to selected subscribers asking which of the following three reality-based shows they watched on a regular basis. The three shows asked about were *Survivor I, Survivor II*, and *Survivor III*. The results of the 510 questionnaires that were returned showed that
175 watched *Survivor I*.
227 watched *Survivor II*.
285 watched *Survivor III*.
100 watched *Survivor I* and *Survivor II*.
87 watched *Survivor II* and *Survivor III*.
96 watched *Survivor I* and *Survivor III*.
59 watched all three shows.

Construct a Venn diagram and determine how many people
a) watched only *Survivor I*.
b) watched exactly one of these shows.
c) watched *Survivor II* and *Survivor III*, but not *Survivor I*.
d) watched *Survivor I* or *Survivor III*, but not *Survivor II*.
e) watched exactly two of these shows.

2.6

In Exercises 51 and 52, show that the sets are infinite by placing each set in a one-to-one correspondence with a proper subset of itself.

51. $\{2, 4, 6, 8, 10, \ldots\}$ **52.** $\{3, 5, 7, 9, 11, \ldots\}$

In Exercises 53 and 54, show that each set has cardinal number $\aleph_0$ by setting up a one-to-one correspondence between the set of counting numbers and the given set.

53. $\{5, 8, 11, 14, 17, \ldots\}$ **54.** $\{4, 9, 14, 19, 24, \ldots\}$

CHAPTER 2 TEST

In Exercises 1–9, state whether each is true or false. If the statement is false, explain why.

1. $[1, y, 8, \$\}$ is equivalent to $\{p, \#, 5, 2\}$.
2. $\{3, 5, 9, h\} = \{9, 5, 3, j\}$
3. $\{\text{star, moon, sun}\} \subset \{\text{star, moon, sun, planet}\}$
4. $\{7\} \subseteq \{x \mid x \in N \text{ and } x < 7\}$
5. $\{\ \} \not\subset \{0\}$
6. $\{p, q, r\}$ has seven subsets.
7. If $A \cap B = \{\ \}$, then A and B are disjoint sets.
8. For any set A, $A \cup A' = \{\ \}$.
9. For any set A, $A \cap U = A$.

In Exercises 10 and 11, use set

$$A = \{x \mid x \in N \text{ and } x < 9\}.$$

10. Write set A in roster form.
11. Write a description of set A.

In Exercises 12–15, use the following information.

$$U = \{3, 5, 7, 9, 11, 13, 15\}$$
$$A = \{3, 5, 7, 9\}$$
$$B = \{7, 9, 11, 13\}$$
$$C = \{3, 11, 15\}$$

Determine the following.

12. $A \cap B$
13. $A \cup C'$
14. $A \cap (B \cap C)'$
15. $n(A \cap B')$
16. Using the sets provided for Exercises 12–15, draw a Venn diagram illustrating the relationship among the sets.
17. Use a Venn diagram to determine whether

$$A \cap (B \cup C') = (A \cap B) \cup (A \cap C')$$

for all sets A, B, and C. Show your work.

18. *Car Accessories* Auto Accessories Unlimited surveyed 155 customers to determine information regarding car accessories their car had. The results of the surveys showed
    76 had keyless entry.
    90 had a sunroof.
    107 had a compact disc player.
    54 had keyless entry and a compact disc player.
    57 had a sunroof and a compact disc player.
    52 had keyless entry and a sunroof.
    35 had all three accessories.

Construct a Venn diagram and then determine how many customers had
a) exactly one of these accessories.
b) none of these accessories.
c) at least two of these accessories.
d) keyless entry and a sunroof, but not a compact disc player.
e) keyless entry or a sunroof, but not a compact disc player.
f) only a compact disc player.

19. Show that the following set is infinite by setting up a one-to-one correspondence between the set and a proper subset of itself.

$$\{7, 8, 9, 10, \dots\}$$

20. Show that the following set has cardinal number $\aleph_0$ by setting up a one-to-one correspondence between the set of counting numbers and the set.

$$\{1, 3, 5, 7, \dots\}$$

GROUP PROJECTS

Selecting a Family Pet

1. The Wilcox family is considering buying a dog. They have established several criteria for the family dog: It must be one of the breeds listed in the table, must not shed, must be less than 16 in. tall, and must be good with children.
   a) Using the information in the table,* construct a Venn diagram in which the universal set is the dogs listed. Indicate the set of dogs to be placed in each region of the Venn diagram.
   b) From the Venn diagram constructed in part (a), determine which dogs will meet the criteria set by the Wilcox family. Explain.

Breed	Sheds	Less than 16 in.	Good with children
Airedale	no	no	no
Basset hound	yes	yes	yes
Beagle	yes	yes	yes
Border terrier	no	yes	yes
Cairn terrier	no	yes	no
Cocker spaniel	yes	yes	yes
Collie	yes	no	yes
Dachshund	yes	yes	no
Poodle, miniature	no	yes	no
Schnauzer, miniature	no	yes	no
Scottish terrier	no	yes	no
Wirehaired fox terrier	no	yes	no

Classification of the Domestic Cat

2. Read the Did You Know feature on page 54. Do research and indicate the name of the following groupings to which the domestic cat belongs.
   a) Kingdom
   b) Phylum
   c) Class
   d) Order
   e) Family
   f) Genus
   g) Species

*The information is a collection of the opinions of an animal psychologist, Dr. Daniel Tortora, and a group of veterinarians.

Who Lives Where

3. On Diplomat Row, an area of Washington, D.C., there are five houses. Each owner is a different nationality, each has a different pet, each has a different favorite food, each has a different favorite drink, and each house is painted a different color.
   The green house is directly to the right of the ivory house.
   The Senegalese has the red house.
   The dog belongs to the Spaniard.
   The Afghanistani drinks tea.
   The person who eats cheese lives next door to the fox.
   The Japanese eats fish.
   Milk is drunk in the middle house.
   Apples are eaten in the house next to the horse.
   Ale is drunk in the green house.
   The Norwegian lives in the first house.
   The peach eater drinks whiskey.
   Apples are eaten in the yellow house.
   The banana eater owns a snail.
   The Norwegian lives next door to the blue house.

   For each house find
   a) the color.
   b) the nationality of the occupant.
   c) the owner's favorite food.
   d) the owner's favorite drink.
   e) the owner's pet.
   f) Finally, the crucial question is: Does the zebra's owner drink vodka or ale?

CHAPTER 3

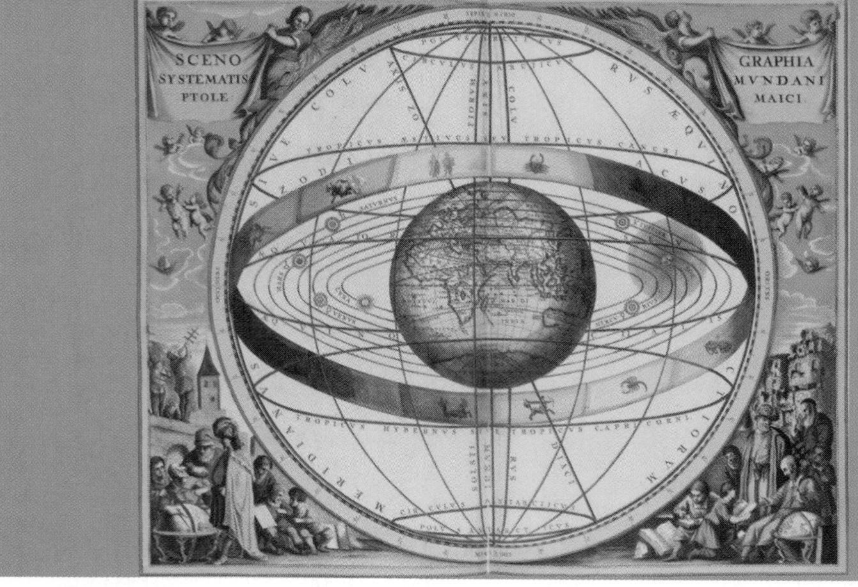

Logical reasoning can tell us whether a conclusion follows from a set of premises, but not whether those premises are true. For example, Greek astronomers, using the assumption that the planets revolved around Earth, correctly predicted the positions of the planets even though their premise was false.

LOGIC

The ancient Greeks were the first people to analyze systematically the way people think and arrive at a conclusion. Aristotle, whose study of logic is presented in a work called *Organon,* is called the father of logic. Since Aristotle's time, the study of logic has been continued by other great mathematicians.

Although most people believe that logic deals with the way people think, it does not. In the study of logic, we use deductive reasoning to analyze complicated situations and come to a reasonable conclusion from a given set of information.

If human thought does not always follow the rules of logic, then why do we study it? Logic enables us to communicate effectively, to make more convincing arguments, and to develop patterns of reasoning for decision making. The study of logic also prepares an individual to better understand other areas of mathematics, computer programming and design, and in general the thought process involved in learning any subject.

3.1 STATEMENTS AND LOGICAL CONNECTIVES

History

The ancient Greeks were the first people to systematically analyze the way humans think and arrive at conclusions. Aristotle (384–322 B.C.) organized the study of logic for the first time in a work called *Organon*. As a result of his work, Aristotle is called the father of logic. The logic from this period, called *Aristotelian logic*, has been taught and studied for more than 2000 years.

Since Aristotle's time, the study of logic has been continued by other great philosophers and mathematicians. Gottfried Wilhelm Leibniz (1646–1716) had a deep conviction that all mathematical and scientific concepts could be derived from logic. As a result, he became the first serious student of *symbolic logic*. One difference between symbolic logic and Aristotelian logic is that in symbolic logic, as its name implies, symbols (usually letters) represent written statements. The forms of the statements in the two types of logic are different. The self-educated English mathematician George Boole (1815–1864) is considered to be the founder of symbolic logic because of his impressive work in this area. Among Boole's publications are *The Mathematical Analysis of Logic* (1847) and *An Investigation of the Law of Thought* (1854). Mathematician Charles Dodgson, better known as Lewis Carroll, incorporated many interesting ideas from logic into his books *Alice's Adventures in Wonderland* and *Through the Looking Glass* and his other children's stories.

Logic has been studied through the ages to exercise the mind's ability to reason. Understanding logic will enable you to think clearly, communicate effectively, make more convincing arguments, and develop patterns of reasoning that will help you in making decisions. It will also help you to detect the fallacies in the reasoning or arguments of others such as advertisers and politicians. Studying logic has other practical applications, such as helping you to understand wills, contracts, and other legal documents.

The study of logic is also good preparation for other areas of mathematics. If you preview Chapter 12, on probability, you will see formulas for the probability of a or b and the probability of a and b, symbolized as $P(A \text{ or } B)$ and $P(A \text{ and } B)$, respectively. Special meanings of common words such as *or* and *and* apply to all areas of mathematics. The meaning of these and other special words is discussed in this chapter.

DID YOU KNOW

Playing on Words

George Boole, Augustus De Morgan, and other mathematicians of the nineteenth century were anxious to make logic an abstract science that would operate like algebra but be applicable to all fields. One of the problems logicians faced was that verbal language could be ambiguous and could easily lead to confusion and contradiction. Comedians Bud Abbott and Lou Costello had fun with the ambiguity of language in their skit about the baseball players: "Who's on first, What's on second, I Don't Know is on third—Yeah, but who's on first?" (in the film *The Naughty Nineties*, 1945).

Logic and the English Language

In reading, writing, and speaking, we use many words such as *and, or,* and *if . . . then . . .* to connect thoughts. In logic we call these words *connectives*. How are these words interpreted in daily communication? A judge announces to a convicted offender, "I hereby sentence you to five months of community service *and* a fine of $100." In this case, we normally interpret the word *and* to indicate that *both* events will take place. That is, the person must do community service and must also pay a fine.

Now suppose a judge states, "I sentence you to six months in prison *or* 10 months of community service." In this case, we interpret the connective *or* as meaning the convicted person must either spend the time in jail or do community service, but not both. The word *or* in this case is the *exclusive or*. When the exclusive or is used, one or the other of the events can take place, but *not both*.

In a restaurant a waiter asks, "May I interest you in a cup of soup or a sandwich?" This question offers three possibilities: You may order soup, you may order a sandwich, or you may order both soup and a sandwich. The *or* in this case is the *inclusive or*. When the inclusive or is used, one or the other, *or both* events can take place. ***In this chapter, when we use the word*** **or** ***in a logic statement, it will mean the*** **inclusive or** ***unless stated otherwise***.

If–then statements are often used to relate two ideas, as in the bank policy statement "If the average daily balance is greater than $500, then there will be no service charge." If–then statements are also used to emphasize a point or add humor, as in the statement "If the Cubs win, then I will be a monkey's uncle."

Now let's look at logic from a mathematical point of view.

Statements and Logical Connectives

The Brooklyn Bridge in New York City

A sentence that can be judged either true or false is called a ***statement***. Labeling a statement true or false is called *assigning a truth value*. Here are some examples of statements.

1. The Brooklyn Bridge goes over San Francisco Bay.
2. Disney World is in Idaho.
3. The Mississippi River is the longest river in the United States.

In each case, we can say that the sentence is either true or false. Statement 1 is false because the Brooklyn Bridge does not go over San Francisco Bay. Statement 2 is false. Disney World is in Florida. By looking at a map or reading an almanac, we can determine that the Mississippi River is the longest river in the United States, and, therefore, statement 3 is true.

The three sentences discussed above are examples of ***simple statements*** because they convey one idea. Sentences combining two or more ideas that can be assigned a truth value are called ***compound statements***. Compound statements are discussed shortly.

Quantifiers

Sometimes it is necessary to change a statement to its opposite meaning. To do so, we use the ***negation*** of a statement. For example, the negation of the statement "Emily is at home" is "Emily is not at home." The negation of a true statement is always a false statement, and the negation of a false statement is always a true statement. We must use special caution when negating statements containing the words ***all, none*** (or ***no***), and ***some***. These words are referred to as ***quantifiers***.

Consider the statement "All lakes contain fresh water." We know this statement is false because the Great Salt Lake in Utah contains salt water. Its negation must therefore be true. We may be tempted to write its negation as "No lake contains fresh water," but this statement is also false because Lake Superior contains fresh water. Therefore, "No lakes contain fresh water" is not the negation of "All lakes contain fresh water." The correct negation of "All lakes contain fresh water" is "Not all lakes contain fresh water" or "At least one lake does not contain fresh water" or "Some lakes do not contain fresh water." These statements all imply that at least one lake does not contain fresh water, which is a true statement.

Now consider the statement "No birds can swim." This statement is false, since at least one bird, the penguin, can swim. Therefore, the negation of this statement must be true. We may be tempted to write the negation as "All birds can swim," but because this statement is also false it cannot be the negation. The correct negation of the statement is "Some birds can swim" or "At least one bird can swim," which are true statements.

Now let's consider statements involving the quantifier *some,* as in "Some students have a driver's license." This is a true statement, meaning that at least one student has a driver's license. The negation of this statement must therefore be false. The negation is "No student has a driver's license," which is a false statement.

Consider the statement "Some students do not ride motorcycles." This statement is true because it means "At least one student does not ride a motorcycle." The negation of this statement must therefore be false. The negation is "All students ride motorcycles," which is a false statement.

The negation of quantified statements is summarized as follows:

Form of statement	Form of negation
All are.	Some are not.
None are.	Some are.
Some are.	None are.
Some are not.	All are.

The following diagram might help you to remember the statements and their negations:

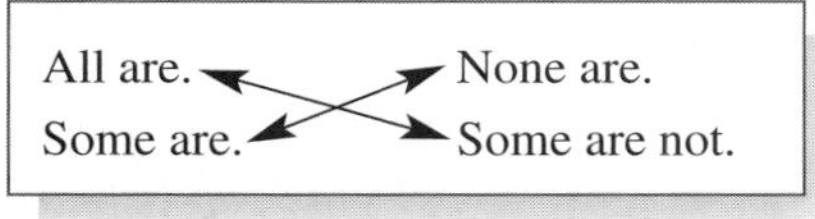

The quantifiers diagonally opposite each other are the negations of each other.

EXAMPLE 1 *Write Negations*

Write the negation of each statement.

a) Some snakes are poisonous.

b) All swimming pools are rectangular.

SOLUTION:

a) Since *some* means "at least one," the statement "Some snakes are poisonous" is the same as "At least one snake is poisonous." Because it is a true statement, its negation must be false. The negation is "No snakes are poisonous," which is a false statement.

b) The statement "All swimming pools are rectangular" is a false statement since some pools are circular, some are oval, and some have other shapes. Its negation must therefore be true. The negation may be written as "Some swimming pools are not rectangular" or "Not all swimming pools are rectangular" or "At least one swimming pool is not rectangular." Each of these statements is true. ▲

Compound Statements

Statements consisting of two or more simple statements are called **compound statements**. The connectives often used to join two simple statements are

and, **or,** **if, . . . then . . . ,** **if and only if**

In addition, we consider a simple statement that has been negated to be a compound statement. The word *not* is generally used to negate a statement.

To reduce the amount of writing in logic, it is common to represent each simple statement with a lowercase letter. For example, suppose we are discussing the simple statement "Leland is a farmer." Instead of writing "Leland is a farmer" over and over again, we can let p represent the statement "Leland is a farmer." Thereafter we can simply refer to the statement with the letter p. It is customary to use the letters p, q, r, and s to represent simple statements, but other letters may be used instead. Let's now look at the connectives used to make compound statements.

Not Statements

The negation is symbolized by $\sim$ and read "not." For example, the negation of the statement "Steve is a college student" is "Steve is not a college student." If p represents the simple statement "Steve is a college student," then $\sim p$ represents the compound statement "Steve is not a college student." For any statement p, $\sim(\sim p) = p$. For example, the negation of the statement "Steve is not a college student" is "Steve is a college student."

Consider the statement "Inga is not at home." This statement contains the word *not,* which indicates that it is a negation. To write this statement symbolically, we let p represent "Inga *is* at home." Then $\sim p$ would be "Inga is not at home." *We will use this convention of letting letters such as p, q, or r represent statements that are not negated. We will represent negated statements with the negation symbol,* $\sim$.

And Statements

The *conjunction* is symbolized by $\wedge$ and read "and." The $\wedge$ looks like an A (for And) with the bar missing. Let p and q represent the simple statements.

p: You will perform 5 months of community service.
q: You will pay a \$100 fine.

Then the following is the conjunction written in symbolic form.

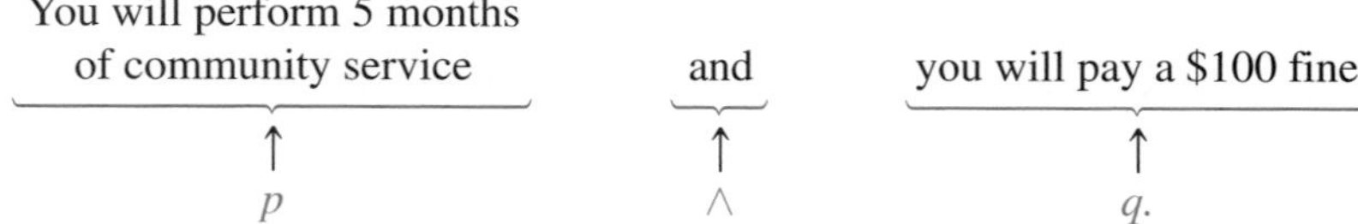

The conjunction is generally expressed as *and*. Other words sometimes used to express a conjunction are *but, however,* and *nevertheless.*

EXAMPLE 2 *Write a Conjunction*

Write the following conjunction in symbolic form. The dish is heavy, but the dish is not hot.

SOLUTION: Let d and h represent the simple statements.

d: The dish is heavy.
h: The dish is hot.

In symbolic form, the compound statement is $d \wedge \sim h$. ▲

In Example 2, the compound statement is "The dish is heavy, but the dish is not hot." This statement could also be repesented as "The dish is heavy, but *it* is not hot." In this problem, it should be clear that the word *it* means *the dish*. Therefore, the statement, "The dish is heavy, but it is not hot" would also be symbolized as $d \wedge \sim h$.

Or Statements

The *disjunction* is symbolized by $\vee$ and read "or." The *or* we use in this book (except where indicated in the exercise sets) is the *inclusive or* described on pages 93 and 94.

EXAMPLE 3 *Write a Disjunction*

Let

p: Maria will go to the circus.
q: Maria will go to the zoo.

Write the following statements in symbolic form.

a) Maria will go to the circus or Maria will go to the zoo.
b) Maria will go to the zoo or Maria will not go to the circus.
c) Maria will not go to the circus or Maria will not go to the zoo.

SOLUTION:

a) $p \vee q$ b) $q \vee \sim p$ c) $\sim p \vee \sim q$ ▲

Because *or* represents the *inclusive or,* the statement "Maria will go to the circus or Maria will go to the zoo" in Example 3(a) may mean that Maria will go to the circus, or that Maria will go to the zoo, or that Maria will go to both the circus *and* the zoo. The statement in Example 3(a) could also be written as "Maria will go to the circus or the zoo."

Geena Davis

When a compound statement contains more that one connective, a comma can be used to indicate which simple statements are to be grouped together. When we write the compound statement symbolically, ***the simple statements on the same side of the comma are to be grouped together within parentheses.***

For example, "Pink is a singer (p) or Geena Davis is an actress (g), and Dallas is in Texas (d)", is written $(p \vee g) \wedge d$. Note that the p and g are both on the same side of the comma in the written statement. They are therefore grouped together within parentheses. The statement "Pink is a singer, or Geena Davis is an actress and Dallas is in Texas" is written $p \vee (g \wedge d)$. In this case, g and d are on the same side of the comma and are therefore grouped together within parentheses.

PROFILE IN MATHEMATICS

GEORGE BOOLE

The self-taught English mathematician George Boole (1815–1864) took the operations of algebra and used them to extend Aristotelian logic. He used symbols such as x and y to represent particular qualities or objects in question. For example, if x represents all butterflies, then $1 - x$ represents everything else except butterflies. If y represents the color yellow, then $(1 - x)(1 - y)$ represents everything except butterflies and things that are yellow, or yellow butterflies. This development added a computational dimension to logic that provided a basis for twentieth-century work in the field of computing.

EXAMPLE 4 *Understand How Commas Are Used to Group Statements*

Let

p: Dinner includes soup.
q: Dinner includes salad.
r: Dinner includes the vegetable of the day.

Write the following statements in symbolic form.

a) Dinner includes soup, and salad or the vegetable of the day.
b) Dinner includes soup and salad, or the vegetable of the day.

SOLUTION:

a) The comma tells us to group the statement "Dinner includes salad" with the statement "Dinner includes the vegetable of the day." Note that both statements are on the same side of the comma. The statement in symbolic form is $p \wedge (q \vee r)$.

In mathematics, we always evaluate the information within the parentheses first. Since the conjunction, $\wedge$, is outside the parentheses and is evaluated *last*, this statement is considered a *conjunction*.

b) The comma tells us to group the statement "Dinner includes soup" with the statement "Dinner includes salad." Note that both statements are on the same side of the comma. The statement in symbolic form is $(p \wedge q) \vee r$. Since the disjunction, $\vee$, is outside the parentheses and is evaluated *last*, this statement is considered a *disjunction*. ▲

The information provided in Example 4 is summarized below.

Statement	Symbolic representation	Type of statement
Dinner includes soup, and salad or the vegetable of the day.	$p \wedge (q \vee r)$	conjunction
Dinner includes soup and salad, or the vegetable of the day.	$(p \wedge q) \vee r$	disjunction

An important point to remember is that a negation symbol has the effect of negating only the statement that directly follows it. To negate a compound statement, we must use parentheses. When a negation symbol is placed in front of a statement in parentheses, it negates the entire statement in parentheses. The negation symbol in this case is read, "It is not true that . . . " or "It is false that . . . "

EXAMPLE 5 *Change Symbolic Statements to Words*

Let

p: Jozsef is making breakfast
q: Arum is setting the table

Write the following symbolic statements in words.

a) $p \wedge \sim q$ b) $\sim p \vee \sim q$ c) $\sim(p \wedge q)$

SOLUTION:

a) Jozsef is making breakfast and Arum is not setting the table.

b) Jozsef is not making breakfast or Arum is not setting the table.

c) It is false that Jozsef is making breakfast and Arum is setting the table. ▲

Recall that the word *but* may also be used in a conjunction. Therefore, Example 5(a) could also be written "Jozsef is making breakfast, *but* Arum is not setting the table."

Part (b) of Example 5 is a disjunction, since it can be written $(\sim p) \vee (\sim q)$. Part (c), which is $\sim(p \wedge q)$, is a negation, since the negation symbol negates the entire statement within parentheses. The similarity of these two statements is discussed in Section 3.4.

Occasionally, we come across a ***neither–nor*** statement, such as "John is neither handsome nor rich." This statement means that John is not handsome *and* John is not rich. If p represents "John is handsome" and q represents "John is rich," this statement is symbolized by $\sim p \wedge \sim q$.

If–Then Statements

The ***conditional*** is symbolized by $\rightarrow$ and is read "if–then." The statement $p \rightarrow q$ is read "If p, then q."* The conditional statement consists of two parts: the part that precedes the arrow is the ***antecedent***, and the part that follows the arrow is the ***consequent***.† In the conditional statement $p \rightarrow q$, the p is the antecedent and the q is the consequent.

In the conditional statement $\sim(p \vee q) \rightarrow (p \wedge q)$, the antecedent is $\sim(p \vee q)$ and the consequent is $(p \wedge q)$. An example of a conditional statement is "If you drink your milk, then you will grow up to be healthy." A conditional symbol may be placed between any two statements even if the statements are not related.

Sometimes the word *then* in a conditional statement is not explicitly stated. For example, the statement "If you pass this course, I will buy you a car" is a conditional statement because it actually means "If you pass this course, then I will buy you a car."

EXAMPLE 6 *Write Conditional Statements*

Let $\quad p$: Jennifer goes to the library.
$\quad\quad\;\; q$: Jennifer will study.

Write the following statements symbolically.

a) If Jennifer goes to the library, then she will study.

b) If Jennifer does not go to the library, then she will not study.

c) It is false that if Jennifer goes to the library then she will study.

SOLUTION:

a) $p \rightarrow q$ b) $\sim p \rightarrow \sim q$ c) $\sim(p \rightarrow q)$ ▲

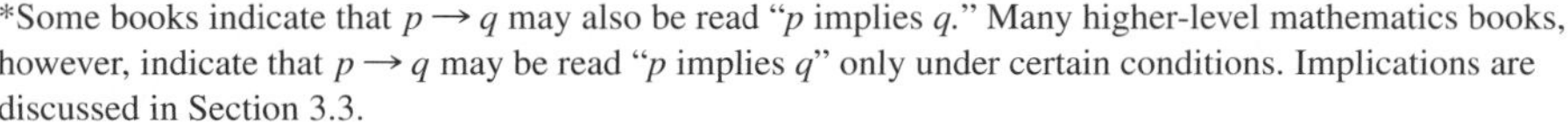

*Some books indicate that $p \rightarrow q$ may also be read "p implies q." Many higher-level mathematics books, however, indicate that $p \rightarrow q$ may be read "p implies q" only under certain conditions. Implications are discussed in Section 3.3.

†Some books refer to the antecedent as the hypothesis or premise and the consequent as the conclusion.

EXAMPLE 7 *Use Commas When Writing a Symbolic Statement in Words*

Let

p: Jorge is enrolled in calculus.
q: Jorge's major is criminal justice.
r: Jorge's major is engineering.

Write the following symbolic statements in words and indicate whether the statement is a negation, conjunction, disjunction, or conditional.

a) $(q \rightarrow \sim p) \vee r$ b) $q \rightarrow (\sim p \vee r)$

SOLUTION: The parentheses indicate where to place the commas in the sentences.

a) "If Jorge's major is criminal justice then Jorge is not enrolled in calculus, or Jorge's major is engineering." This statement is a disjunction because $\vee$ is outside the parentheses.
b) "If Jorge's major is criminal justice, then Jorge is not enrolled in calculus or Jorge's major is engineering." This is a conditional statement because $\rightarrow$ is outside the parentheses. ▲

If and Only if Statements

The *biconditional* is symbolized by $\leftrightarrow$ and is read "if and only if." The phrase *if and only if* is sometimes abbreviated as "iff." The statement $p \leftrightarrow q$ is read "p if and only if q."

EXAMPLE 8 *Write Statements Using the Biconditional*

Let

p: The printer is working.
q: The ink cartridge is correctly inserted.

Write the following symbolic statements in words.

a) $q \leftrightarrow p$ b) $\sim(p \leftrightarrow \sim q)$

SOLUTION:

a) The ink cartridge is correctly inserted if and only if the printer is working.
b) It is false that the printer is working if and only if the ink cartridge is not correctly inserted. ▲

You will learn later that $p \leftrightarrow q$ means the same as $(p \rightarrow q) \wedge (q \rightarrow p)$. Therefore, the statement "I will go to college if and only if I can pay the tuition" has the same logical meaning as "If I go to college then I can pay the tuition, and if I can pay the tuition then I will go to college."

The following is a summary of the connectives discussed in this section.

Formal name	Symbol	Read	Symbolic form
Negation	$\sim$	"Not"	$\sim p$
Conjunction	$\wedge$	"And"	$p \wedge q$
Disjunction	$\vee$	"Or"	$p \vee q$
Conditional	$\rightarrow$	"If-then"	$p \rightarrow q$
Biconditional	$\leftrightarrow$	"If and only if"	$p \leftrightarrow q$

Dominance of Connectives

What is the answer to the problem $2 + 3 \times 4$? Some of you might say 20, but others might say 14. If you evaluate $2 + 3 \times 4$ on a calculator by pressing

$$2 \boxed{+} 3 \boxed{\times} 4$$

some may give you the answer 14, whereas others may give you the answer 20. Which is the correct answer? In mathematics, unless otherwise changed by parentheses or some other grouping symbol, multiplication is *always* performed before addition. Thus,

$$2 + 3 \times 4 = 2 + (3 \times 4) = 14$$

The calculators that gave the incorrect answer of 20 are basic calculators that are not programmed according to the order of operations used in mathematics.

Just as an order of operations exists in the evaluation of arithmetic expressions, a dominance of connectives is used in the evaluation of logic statements. How do we evaluate a symbolic logic statement when no parentheses are used? For example, does $p \vee q \rightarrow r$ mean $(p \vee q) \rightarrow r$, or does it mean $p \vee (q \rightarrow r)$? If we are given a symbolic logic statement for which grouping has not been indicated by parentheses or a written logic statement for which grouping has not been indicated by a comma, then we use the dominance of connectives shown in Table 3.1. Note that *the least dominant connective is the negation and the most dominant is the biconditional.*

TABLE 3.1 Dominance of Connectives

Least dominant	1. Negation, $\sim$	Evaluate first
	2. Conjunction, $\wedge$; disjunction, $\vee$	
	3. Conditional, $\rightarrow$	
Most dominant	4. Biconditional, $\leftrightarrow$	Evaluate last

As indicated in Table 3.1, the conjunction and disjunction have the same level of dominance. Thus, to determine whether the symbolic statement $p \wedge q \vee r$ is a conjunction or a disjunction, we have to use grouping symbols (parentheses). When evaluating a symbolic statement that does not contain parentheses, we *evaluate the least dominant connective first and the most dominant connective last*. For example,

Statement	Most dominant connective used	Statement means	Type of statement
$\sim p \vee q$	$\vee$	$(\sim p) \vee q$	Disjunction
$p \rightarrow q \vee r$	$\rightarrow$	$p \rightarrow (q \vee r)$	Conditional
$p \wedge q \rightarrow r$	$\rightarrow$	$(p \wedge q) \rightarrow r$	Conditional
$p \rightarrow q \leftrightarrow r$	$\leftrightarrow$	$(p \rightarrow q) \leftrightarrow r$	Biconditional
$p \vee r \leftrightarrow r \rightarrow \sim p$	$\leftrightarrow$	$(p \vee r) \leftrightarrow (r \rightarrow \sim p)$	Biconditional
$p \rightarrow r \leftrightarrow s \wedge p$	$\leftrightarrow$	$(p \rightarrow r) \leftrightarrow (s \wedge p)$	Biconditional

EXAMPLE 9 *Use the Dominance of Connectives*

Use the dominance of connectives to add parentheses to each statement. Then indicate whether each statement is a negation, conjunction, disjunction, conditional, or biconditional.

a) $p \rightarrow q \vee r$ b) $\sim p \wedge q \leftrightarrow r \vee p$

SOLUTION:

a) The conditional has greater dominance than the disjunction, so we place parentheses around $q \vee r$, as follows:

$$p \rightarrow (q \vee r)$$

It is a conditional statement because the conditional symbol is outside the parentheses.

b) The biconditional has the greatest dominance, so we place parentheses as follows:

$$(\sim p \wedge q) \leftrightarrow (r \vee p)$$

It is a biconditional statement because the biconditional symbol is outside the parentheses. ▲

EXAMPLE 10 *Identify the Type of Statement*

Use the dominance of connectives and parentheses to write each statement symbolically. Then indicate whether each statement is a negation, conjunction, disjunction, conditional, or biconditional.

a) If you are late in paying your rent or you have damaged the apartment then you may be evicted.

b) You are late in paying your rent, or if you have damaged the apartment then you may be evicted.

SOLUTION:

a) Let

p: You are late in paying your rent.
q: You have damaged the apartment.
r: You may be evicted.

No commas appear in the sentence, so we will evaluate it by using the dominance of connectives. Because the conditional has higher dominance than the disjunction, the conditional statement will be evaluated last. Thus, the statements "You are late in paying your rent" and "You have damaged the apartment" are to be grouped together. The statement written symbolically with parentheses is

$$(p \vee q) \rightarrow r$$

This is a conditional statement.

b) A comma is used in this statement to indicate grouping, just as parentheses do in arithmetic. The placement of the comma indicates that the statements "You have damaged the apartment" and "You may be evicted" are to be grouped together. Therefore, this statement written symbolically is

$$p \vee (q \rightarrow r)$$

This statement is a disjunction. Note that the comma overrides the dominance of connectives and tells us to evaluate the conditional statement before the disjunction. ▲

SECTION 3.1 EXERCISES

Concept/Writing Exercises

1. a) What is a simple statement?
   b) What is a compound statement?
2. List the words identified as quantifiers.
3. Write the general form of the negation for statements of the form
   a) none are.
   b) some are not.
   c) all are.
   d) some are.
4. Represent the statement "The ink is not purple" symbolically. Explain your answer.
5. Draw the symbol used to represent the
   a) conditional.
   b) disjunction.
   c) conjunction.
   d) negation.
   e) biconditional.
6. a) When the *exclusive or* is used as a connective between two events, can both events take place? Explain.
   b) When the *inclusive or* is used as a connective between two events, can both events take place? Explain.
   c) Which *or*, the *inclusive or* or the *exclusive or*, is used in this chapter?
7. Explain how a comma is used to indicate the grouping of simple statements.
8. List the dominance of connectives from the most dominant to the least dominant.

Practice the Skills/Problem Solving

In Exercises 9–22, indicate whether the statement is a simple statement or a compound statement. If it is a compound statement, indicate whether it is a negation, conjunction, disjunction, conditional, or biconditional by using both the word and its appropriate symbol (for example, "a negation," ~).

9. The sun is shining and the air is crisp.
10. The water in the lake is not drinkable.
11. The figure is a quadrilateral if and only if it has four sides.
12. If the electricity goes out then the standard telephone will still work.
13. Joni Burnette is teaching calculus or she is teaching trigonometry.
14. The book was neither a novel nor an autobiography.
15. The hurricane did $400,000 worth of damage to DeSoto County.

16. Inhibor Melendez will be admitted to law school if and only if he earns his bachelor's degree.
17. It is false that Jeffery Hilt is a high school teacher and a grade school teacher.
18. If Cathy Smith walks 4 miles today then she will be sore tomorrow.
19. Mary Jo Woo ran 4 miles today and she lifted weights for 30 minutes.
20. Nancy Wallin went to the game, but she did not eat a hot dog.
21. It is false that if John Wubben fixes your car then you will need to pay him in cash.
22. If Buddy and Evelyn Cordova are residents of Budville, then they must vote for mayor on Tuesday.

In Exercises 23–34, write the negation of the statement.

23. Some picnic tables are portable.
24. No stock mutual funds have guaranteed yields.
25. All chickens fly.
26. All plants contain chlorophyll.
27. Some turtles do not have claws.
28. No teachers made the roster.
29. No bicycles have three wheels.
30. All horses have manes.

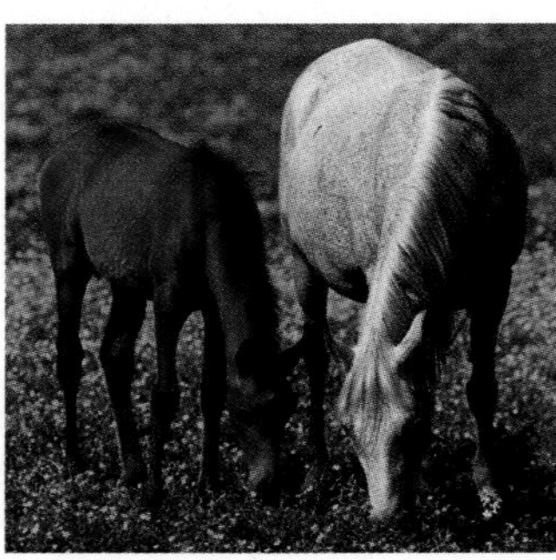

31. Some pine trees do not produce pinecones.
32. No one likes asparagus.
33. Some pedestrians are in the crosswalk.
34. Some dogs with long hair do not get cold.

In Exercises 35–40, write the statement in symbolic form. Let

p: The tent is pitched.
q: The bonfire is burning.

35. The tent is not pitched.
36. The tent is pitched and the bonfire is burning.
37. The bonfire is not burning or the tent is not pitched.
38. The bonfire is not burning if and only if the tent is not pitched.
39. If the tent is not pitched, then the bonfire is not burning.
40. The bonfire is not burning, however the tent is pitched.

In Exercises 41–46, write the statement in symbolic form. Let

p: The charcoal is hot.
q: The chicken is on the grill.

41. If the chicken is not on the grill then the charcoal is not hot.
42. The chicken is not on the grill if and only if the charcoal is not hot.
43. Neither is the charcoal hot nor is the chicken on the grill.
44. The charcoal is not hot, but the chicken is on the grill.
45. It is false that if the chicken is on the grill then the charcoal is not hot.
46. It is false that the charcoal is hot and the chicken is on the grill.

In Exercises 47–56, write the compound statement in words. Let

p: Firemen work hard.
q: Firemen wear red suspenders.

47. $\sim p$
48. $\sim q$
49. $q \vee p$
50. $p \wedge q$
51. $\sim p \leftrightarrow \sim q$
52. $\sim p \rightarrow q$
53. $\sim(q \vee p)$
54. $\sim p \vee \sim q$
55. $\sim p \wedge \sim q$
56. $\sim(p \wedge q)$

In Exercises 57–66, write the statements in symbolic form. Let

p: The temperature is 90°.
q: The air conditioner is working.
r: The apartment is hot.

57. If the temperature is 90° or the air conditioner is not working, then the apartment is hot.

58. The apartment is hot if and only if the temperature is not 90°, or the air conditioner is not working.

59. The temperature is 90° and the air conditioner is working, or the apartment is hot.

60. If the apartment is hot and the air conditioner is working, then the temperature is 90°.

61. If the temperature is 90°, then the air conditioner is working or the apartment is not hot.

62. The temperature is not 90° if and only if the air conditioner is not working, or the apartment is not hot.

63. The apartment is hot if and only if the air conditioner is working, and the temperature is 90°.

64. It is false that if the apartment is hot then the air conditioner is not working.

65. If the air conditioner is working, then the temperature is 90° if and only if the apartment is hot.

66. The apartment is hot or the air conditioner is not working, if and only if the temperature is 90°.

In Exercises 67–76, write each symbolic statement in words. Let

p: The water is 70°.
q: The sun is shining.
r: We go swimming.

67. $(p \vee q) \wedge \sim r$
68. $(p \wedge q) \vee r$
69. $\sim p \wedge (q \vee r)$
70. $(q \rightarrow p) \vee r$
71. $\sim r \rightarrow (q \wedge p)$
72. $(q \wedge r) \rightarrow p$
73. $(q \rightarrow r) \wedge p$
74. $\sim p \rightarrow (q \vee r)$
75. $(q \leftrightarrow p) \wedge r$
76. $q \rightarrow (p \leftrightarrow r)$

Dinner Menu *In Exercises 77–80, use the following information to arrive at your answers. Many restaurant dinner menus include statements such as the following. All dinners are served with a choice of: Soup or Salad, and Potatoes or Pasta, and Carrots or Peas. Which of the following selections are permissible? If a selection is not permissible, explain why. See the discussion of the* exclusive or *on page 93.*

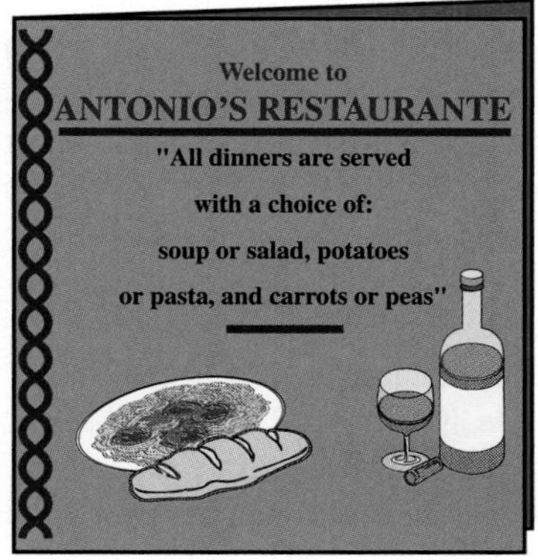

Welcome to
ANTONIO'S RESTAURANTE
"All dinners are served
with a choice of:
soup or salad, potatoes
or pasta, and carrots or peas"

77. Soup, salad, and peas
78. Salad, pasta, and carrots
79. Soup, potatoes, pasta, and peas
80. Soup, pasta, and potatoes

In Exercises 81–94, (a) add parentheses by using the dominance of connectives and (b) indicate whether the statement is a negation, conjunction, disjunction, conditional, or biconditional (see Example 9).

81. $\sim p \rightarrow q$
82. $\sim p \wedge r \leftrightarrow \sim q$
83. $\sim q \wedge \sim r$
84. $\sim p \vee q$
85. $p \vee q \rightarrow r$
86. $q \rightarrow p \wedge \sim r$
87. $r \rightarrow p \vee q$
88. $q \rightarrow p \leftrightarrow p \rightarrow q$
89. $\sim p \leftrightarrow \sim q \rightarrow r$
90. $\sim q \rightarrow r \wedge p$
91. $r \wedge \sim q \rightarrow q \wedge \sim p$
92. $\sim[p \rightarrow q \vee r]$
93. $\sim[p \wedge q \leftrightarrow p \vee r]$
94. $\sim[r \wedge \sim q \rightarrow q \wedge r]$

In Exercises 95–103, (a) select letters to represent the simple statements and write each statement symbolically by using parentheses and (b) indicate whether the statement is a negation, conjunction, disjunction, conditional, or biconditional (see Example 10).

95. Ruth Bignel retired, but she did not start her concrete business.

96. If the water level is up, then we can go canoeing or we can go rafting.

97. It is false that if your speed is below the speed limit then you will not get pulled over.

98. If dinner is ready then we can eat, or we cannot go to the restaurant.

99. If the food has fiber or the food has vitamins, then you will be healthy.

100. If Corliss is teaching then Faye is in the math lab if and only if it is not a weekend.

101. You may take this course if and only if you did not fail the previous course or you passed the placement test.

102. If the car has gas and the battery is charged then the car will start.

103. The classroom is empty if and only if it is the weekend, or it is 7 A.M.

Challenge Problems/Group Activities

104. *An Ancient Question* If Zeus could do anything, could he build a wall that he could not jump over? Explain your answer.

In Exercises 105 and 106, place parentheses in the statement according to the dominance of connectives. Indicate whether the statement is a negation, conjunction, disjunction, conditional, or biconditional.

105. $\sim q \rightarrow r \vee p \leftrightarrow \sim r \wedge q$

106. $\sim[\sim r \rightarrow p \wedge q \leftrightarrow \sim p \vee r]$

107. **a)** We cannot place parentheses in the statement $p \vee q \wedge r$. Explain why.
b) Make up three simple statements and label them p, q, and r. Then write compound statements to represent $(p \vee q) \wedge r$ and $p \vee (q \wedge r)$.
c) Do you think that the statements for $(p \vee q) \wedge r$ and $p \vee (q \wedge r)$ mean the same thing? Explain.

Internet/Research Activities

108. *Legal Documents* Obtain a legal document such as a will or rental agreement and copy one page of the document. Circle every connective used. Then list the number of times each connective appeared. Be sure to include conditional statements from which the word *then* was omitted from the sentence. Give the page and your listing to your instructor.

109. Write a report on the life and accomplishments of George Boole, who was an important contributor to the development of logic. In your report, indicate how his work eventually led to the development of the computer. References include encyclopedias, history of mathematics books, and the Internet.

3.2 TRUTH TABLES FOR NEGATION, CONJUNCTION, AND DISJUNCTION

A *truth table* is a device used to determine when a compound statement is true or false. Five basic truth tables are used in constructing other truth tables. Three are discussed in this section (Tables 3.2, 3.4, and 3.7), and two are discussed in the next section. Section 3.5 uses truth tables in determining whether a logical argument is valid or invalid.

TABLE 3.2 Negation

	p	$\sim p$
Case 1	T	F
Case 2	F	T

TABLE 3.3

	p	q
Case 1	T	T
Case 2	T	F
Case 3	F	T
Case 4	F	F

Negation

The first truth table is for *negation*. If p is a true statement, then the negation of p, "not p," is a false statement. If p is a false statement, then "not p" is a true statement. For example, if the statement "The shirt is blue" is true, then the statement "The shirt is not blue" is false. These relationships are summarized in Table 3.2. For a simple statement, there are exactly two true–false cases, as shown.

If a compound statement consists of two simple statements p and q, there are four possible cases, as illustrated in Table 3.3. Consider the statement "The test is today and the test covers Chapter 5." The simple statement "The test is today" has two possible truth values, true or false. The simple statement "The test covers Chapter 5" also has two truth values, true or false. Thus, for these two simple statements there are four distinct possible true–false arrangements. Whenever we construct a truth table for a compound statement that consists of two simple statements, we begin by listing the four true–false cases shown in Table 3.3.

Conjunction

TABLE 3.4 Conjunction

	p	q	$p \wedge q$
Case 1	T	T	T
Case 2	T	F	F
Case 3	F	T	F
Case 4	F	F	F

To illustrate the conjunction, consider the following situation. You have recently purchased a new house. To decorate it, you ordered a new carpet and new furniture from the same store. You explain to the salesperson that the carpet must be delivered before the furniture. He promises that the carpet will be delivered on Thursday and that the furniture will be delivered on Friday.

To help determine whether the salesperson kept his promise, we assign letters to each simple statement. Let p be "The carpet will be delivered on Thursday" and q be "The furniture will be delivered on Friday." The salesperson's statement written in symbolic form is $p \wedge q$. There are four possible true–false situations to be considered. (Table 3.4).

CASE 1: p is true and q is true. The carpet is delivered on Thursday and the furniture is delivered on Friday. The salesperson has kept his promise and the compound statement is true. Thus, we put a T in the $p \wedge q$ column.

CASE 2: p is true and q is false. The carpet is delivered on Thursday but the furniture is not delivered on Friday. Since the furniture was not delivered as promised, the compound statement is false. Thus, we put an F in the $p \wedge q$ column.

CASE 3: p is false and q is true. The carpet is not delivered on Thursday but the furniture is delivered on Friday. Since the carpet was not delivered on Thursday as promised, the compound statement is false. Thus, we put an F in the $p \wedge q$ column.

CASE 4: p is false and q is false. The carpet is not delivered on Thursday and the furniture is not delivered on Friday. Since the carpet and furniture were not delivered as promised, the compound statement is false. Thus, we put an F in the $p \wedge q$ column.

Examining the four cases, we see that in only one case did the salesperson keep his promise: in case 1. Therefore, case 1 (T, T) is true. In cases 2, 3, and 4, the salesperson did not keep his promise and the compound statement is false. The results are summarized in Table 3.4, the truth table for the conjunction.

The **conjunction** $p \wedge q$ is true only when both p and q are true.

EXAMPLE 1 *Construct a Truth Table*

Construct a truth table for $p \wedge \sim q$.

SOLUTION: Because there are two statements, p and q, construct a truth table with four cases; see Table 3.5(a). Then write the truth values under the p in the compound statement and label this column 1, as in Table 3.5(b). Copy these truth values directly from the p column on the left. Write the corresponding truth values under the q in the compound statement and call this column 2, as in Table 3.5(c). Copy the truth values for column 2 directly from the q column on the left. Now find the truth values of $\sim q$ by negating the truth values in column 2 and call this column 3,

TABLE 3.5

(a)

	p	q	$p \wedge \sim q$
Case 1	T	T	
Case 2	T	F	
Case 3	F	T	
Case 4	F	F	

(b)

p	q	$p \wedge \sim q$
T	T	T
T	F	T
F	T	F
F	F	F
		1

(c)

p	q	p	$\wedge$	$\sim$	q
T	T	T			T
T	F	T			F
F	T	F			T
F	F	F			F
		1			2

(d)

p	q	p	$\wedge$	$\sim$	q
T	T	T		F	T
T	F	T		T	F
F	T	F		F	T
F	F	F		T	F
		1		3	2

(e)

p	q	p	$\wedge$	$\sim$	q
T	T	T	F	F	T
T	F	T	T	T	F
F	T	F	F	F	T
F	F	F	F	T	F
		1	4	3	2

as in Table 3.5(d). Use the conjunction table, Table 3.4, and the entries in columns 1 and 3 to complete column 4, as in Table 3.5(e). The results in column 4 are obtained as follows:

Row 1: T ∧ F is F. Row 2: T ∧ T is T.

Row 3: F ∧ F is F. Row 4: F ∧ T is F.

The answer is always the last column completed. Columns 1, 2, and 3 are only aids in arriving at the answer in column 4. ▲

The statement $p \wedge \sim q$ in Example 1 actually means $p \wedge (\sim q)$. In the future, instead of listing a column for q and a separate column for its negation, we will make one column for $\sim q$, which will have the opposite values of those in the q column on the left. Similarly, when we evaluate $\sim p$, we will use the opposite values of those in the p column on the left. This procedure is illustrated in Example 2.

In Example 1, we spoke about *cases* and also *columns*. Consider Table 3.5(e). This table has four cases indicated by the four different rows of the two left hand (unnumbered) columns. The four *cases* are TT, TF, FT, and FF. In every truth table with two letters, we list the four cases (the first two columns) first. Then we complete the remaining columns in the truth table. In Table 3.5(e), after completing the two left-hand columns, we complete the remaining columns in the order indicated by the numbers below the columns. We will continue to place numbers below the columns to show the order in which the columns are completed.

TIMELY TIP When constructing truth tables it is very important to keep your entries in neat columns and rows. If you are using lined paper, put only one row of the table on each line. If you are not using lined paper, using a straightedge may help you correctly enter the information into the truth table's rows and columns.

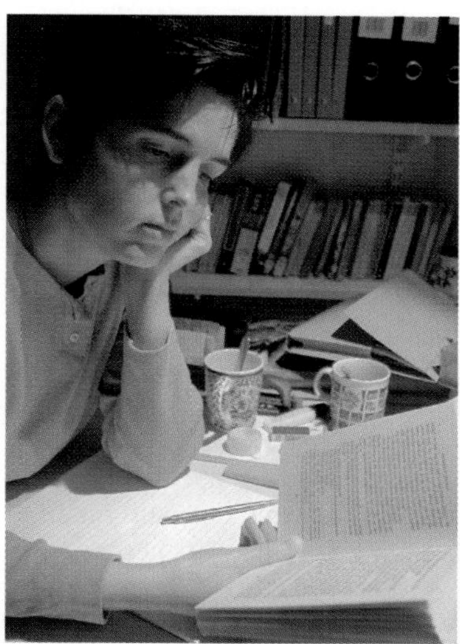

EXAMPLE 2 *Construct and Interpret a Truth Table*

a) Construct a truth table for the following statement:

I have not studied and I am not ready for the test.

b) Under which conditions will the compound statement be true?

c) Suppose "I have studied" is a false statement and "I am ready for the test" is a true statement. Is the compound statement given in part (a) true or false?

SOLUTION:

a) First write the simple statements in symbolic form by using simple nonnegated statements.

Let

p: I have studied.
q: I am ready for the test.

Therefore, the compound statement may be written $\sim p \wedge \sim q$. Now construct a truth table with four cases, as shown in Table 3.6

Fill in the column labeled 1 by negating the truth values under p on the far left. Fill in the column labeled 2 by negating the values under q in the second column from the left. Fill in the column labeled 3 by using the columns labeled 1 and 2 and the definition of conjunction.

In the first row, to determine the entry for column 3, we use false for $\sim p$ and false for $\sim q$. Since false $\wedge$ false is false (see case 4 of Table 3.4), we place an F in column 3, row 1. In the second row, we use false for $\sim p$ and true for $\sim q$. Since false $\wedge$ true is false (see case 3 of Table 3.4), we place an F in column 3, row 2. In the third row, we use true for $\sim p$ and false for $\sim q$. Since true $\wedge$ false is false (see case 2 of Table 3.4), we place an F in column 3, row 3. In the fourth row, we use true for $\sim p$ and true for $\sim q$. Since true $\wedge$ true is true (see case 1 of Table 3.4), we place a T in column 3, row 4.

TABLE 3.6

p	q	~p	∧	~q
T	T	F	F	F
T	F	F	F	T
F	T	T	(F)	F
F	F	T	(T)	T
		1	3	2

b) The compound statement in part (a) will be true only in case 4 (circled in blue) when both simple statements, p and q, are false, that is, when I have not studied and I am not ready for the test.

c) We are told that p, "I have studied," is a false statement and that q, "I am ready for the test," is a true statement. From the truth table (Table 3.6), we can determine that when p is false and q is true, case 3, the compound statement, is false (circled in red).

Disjunction

Consider the job description that contains the following requirements.

> **Civil Technician**
>
> Municipal program for redevelopment seeks on-site technician. **The applicant must have a two-year college degree in civil technology or five years of related experience**. Interested candidates please call 555-1234.

Who qualifies for the job? To help analyze the statement, translate it into symbolic form. Let p be "A requirement for the job is a two-year college degree in civil technology" and q be "A requirement for the job is five years of related experience." The statement in symbolic form is $p \vee q$. For the two simple statements, there are four distinct cases (see Table 3.7).

TABLE 3.7 Disjunction

p	q	$p \vee q$
T	T	T
T	F	T
F	T	T
F	F	F

CASE 1: p is true and q is true. A candidate has a two-year college degree in civil technology and five years of related experience. The candidate has both requirements and qualifies for the job. Consider qualifying for the job as a true statement and not qualifying as a false statement. Since the candidate qualifies for the job, we put a T in the $p \vee q$ column.

CASE 2: p is true and q is false. A candidate has a two-year college degree in civil technology but does not have five years of related experience. The candidate still qualifies for the job with the two-year college degree. Thus, we put a T in the $p \vee q$ column.

CASE 3: p is false and q is true. The candidate does not have a two-year college degree in civil technology but does have five years of related experience. The candidate qualifies for the job with the five years of related experience. Thus, we put a T in the $p \vee q$ column.

CASE 4: p is false and q is false. The candidate does not have a two-year college degree in civil technology and does not have five years of related experience. The candidate does not meet either of the two requirements and therefore does not qualify for the job. Thus, we put an F in the $p \vee q$ column.

In examining the four cases, we see that there is only one case in which the candidate does not qualify for the job: case 4. As this example indicates, an *or* statement will be true in every case, except when both simple statements are false. The results are summarized in Table 3.7, the truth table for the disjunction.

> The **disjunction**, $p \vee q$, is true when either p is true, q is true, or both p and q are true.

The disjunction $p \vee q$ is false only when p and q are both false.

EXAMPLE 3 *Truth Table with a Negation*

Construct a truth table for $\sim(\sim q \wedge p)$.

TABLE 3.8

p	q	$\sim$	$(\sim q$	$\wedge$	$p)$
T	T	T	F	F	T
T	F	F	T	T	T
F	T	T	F	F	F
F	F	T	T	F	F
		4	1	3	2

SOLUTION: First construct the standard truth table listing the four cases. Then work within parentheses. The order to be followed is indicated by the numbers below the columns (see Table 3.8). Under $\sim q$, column 1, write the negation of the q column. Then, in column 2, copy the values from the p column. Next, complete the *and* column, column 3, using columns 1 and 2 and the truth table for the conjunction. The *and* column is true only when both statements are true, as in case 2. Finally, negate the values in the *and* column, column 3, and place these negated values in column 4. By examining the truth table you can see that the compound statement $\sim(\sim q \wedge p)$ is false only in case 2, that is, when p is true and q is false. ▲

A General Procedure for Constructing Truth Tables

1. Study the compound statement and determine whether it is a negation, conjunction, disjunction, conditional, or biconditional statement, as was done in Section 3.1. The answer to the truth table will appear under $\sim$ if the statement is a negation, under $\wedge$ if the statement is a conjunction, under $\vee$ if the statement is a disjunction, under $\rightarrow$ if the statement is a conditional, and under $\leftrightarrow$ if the statement is a biconditional.
2. Complete the columns under the simple statements, p, q, r, and their negations, $\sim p$, $\sim q$, $\sim r$, within parentheses. If there are nested parentheses (one pair of parentheses within another pair), work with the innermost pair first.
3. Complete the column under the connective within the parentheses. You will use the truth values of the connective in determining the final answer in step 5.
4. Complete the column under any remaining statements and their negations.
5. Complete the column under any remaining connectives. Recall that the answer will appear under the column determined in step 1. If the statement is a conjunction, disjunction, conditional, or biconditional, you will obtain the truth values for the connective by using the last column completed on the left side and on the right side of the connective. If the statement is a negation, you will obtain the truth values by negating the truth values of the last column completed within the grouping symbols on the right side of the negation. Be sure to circle or highlight your answer column or number the columns in the order they were completed.

TABLE 3.9

p	q	(~p	∨	q)	∧	~p
T	T	F	T	T	F	F
T	F	F	F	F	F	F
F	T	T	T	T	T	T
F	F	T	T	F	T	T
		1	3	2	5	4

EXAMPLE 4 *Use the General Procedure to Construct a Truth Table*

Construct a truth table for the statement $(\sim p \vee q) \wedge \sim p$.

SOLUTION: We will follow the general procedure outlined in the box. This statement is a conjunction, so the answer will be under the conjunction symbol. Complete columns under $\sim p$ and q within the parentheses and call these columns 1 and 2, respectively (see Table 3.9). Complete the column under the disjunction, $\vee$, using the truth values in columns 1 and 2, and call this column 3. Next complete the column under $\sim p$, and call this column 4. The answer, column 5, is determined from the definition of the conjunction and the truth values in column 3, the last column completed on the left side of the conjunction, and column 4. ▲

TABLE 3.10

	p	q	r
Case 1	T	T	T
Case 2	T	T	F
Case 3	T	F	T
Case 4	T	F	F
Case 5	F	T	T
Case 6	F	T	F
Case 7	F	F	T
Case 8	F	F	F

So far, all the truth tables we have constructed have contained at most two simple statements. Now we will explain how to construct a truth table that consists of three simple statements, such as $(p \wedge q) \wedge r$. When a compound statement consists of three simple statements, there are eight different true–false possibilities, as illustrated in Table 3.10. To begin such a truth table, write four Ts and four Fs in the column under p. Under the second statement, q, pairs of Ts alternate with pairs of Fs. Under the third statement, r, T alternates with F. This technique is not the only way of listing the cases, but it ensures that each case is unique and that no cases are omitted.

EXAMPLE 5 *Construct a Truth Table with Eight Cases*

a) Construct a truth table for the statement "Santana is home and he is not at his desk, or he is sleeping."

b) Suppose that "Santana is home" is a false statement, that "Santana is at his desk" is a true statement, and that "Santana is sleeping" is a true statement. Is the compound statement in part (a) true or false?

SOLUTION:

a) First we will translate the statement into symbolic form.

Let

p: Santana is home.
q: Santana is at his desk.
r: Santana is sleeping.

In symbolic form, the statement is $(p \wedge \sim q) \vee r$.

Since the statement is composed of three simple statements, there are eight cases. Begin by listing the eight cases in the three left-hand columns; see Table 3.11. By examining the statement, you can see that it is a disjunction. Therefore, the answer will be in the $\vee$ column. Fill out the truth table by working in parentheses first. Place values under p, column 1, and $\sim q$, column 2. Then find the conjunctions of columns 1 and 2 to obtain column 3. Place the values of r in column 4. To obtain the answer, column 5, use columns 3 and 4 and your knowledge of the disjunction.

TABLE 3.11

p	q	r	(p	∧	~q)	∨	r
T	T	T	T	F	F	T	T
T	T	F	T	F	F	F	F
T	F	T	T	T	T	T	T
T	F	F	T	T	T	T	F
F	T	T	F	F	F	(T)	T
F	T	F	F	F	F	F	F
F	F	T	F	F	T	T	T
F	F	F	F	F	T	F	F
			1	3	2	5	4

b) We are given the following:

p: Santana is home—false.
q: Santana is at his desk—true.
r: Santana is sleeping—true.

Therefore, we need to find the truth value of the following case: false, true, true. In case 5 of the truth table, p, q, and r are F, T, and T, respectively. Therefore, under these conditions, the original compound statement is true (as circled in the table). ▲

We have learned that a truth table with one simple statement has two cases, a truth table with two simple statements has four cases, and a truth table with three simple statements has eight cases. In general, *the number of distinct cases in a truth table with n distinct simple statements is* 2^n. The compound statement $(p \vee q) \vee (r \wedge \sim s)$ has four simple statements, p, q, r, s. Thus, a truth table for this compound statement would have 2^4, or 16, distinct cases.

When we construct a truth table, we determine the truth values of a compound statement for every possible case. If we want to find the truth value of the compound statement for any specific case when we know the truth values of the simple statements, we do not have to develop the entire table. For example, to determine the truth value for the statement

$$2 + 3 = 5 \quad \text{and} \quad 1 + 1 = 3$$

we let p be $2 + 3 = 5$ and q be $1 + 1 = 3$. Now we can write the compound statement as $p \wedge q$. We know that p is a true statement and q is a false statement. Thus, we can substitute T for p and F for q and evaluate the statement:

$$\begin{array}{c} p \wedge q \\ \text{T} \wedge \text{F} \\ \text{F} \end{array}$$

Therefore, the compound statement $2 + 3 = 5$ and $1 + 1 = 3$ is a false statement.

EXAMPLE 6 *Determine the Truth Value of a Compound Statement*

Determine the truth value for each simple statement. Then, using these truth values, determine the truth value of the compound statement.

a) 15 is less than or equal to 9.

b) George Washington was the first U.S. president or Abraham Lincoln was the second U.S. president, but there has not been a U.S. president born in Antarctica.

SOLUTION:

a) Let

p: 15 is less than 9.
q: 15 is equal to 9.

The statement "15 is less than or equal to 9" means that 15 is less than 9 or 15 is equal to 9. The compound statement can be expressed as $p \vee q$. We know that both p and q are false statements since 15 is greater than 9. Therefore, substitute F for p and F for q and evaluate the statement:

$$\begin{array}{c} p \vee q \\ \text{F} \vee \text{F} \\ \text{F} \end{array}$$

Therefore, the compound statement "15 is less than or equal to 9" is a false statement.

DID YOU KNOW

Applications of Logic

George Boole provided the key that would unlock the door to modern computing. Not until 1938, however, did Claude Shannon, in his master's thesis at MIT, propose uniting the on–off capability of electrical switches with Boole's two-value system of 0's and 1's. The operations AND, OR, and NOT and the rules of logic laid the foundation for computer gates. Such gates determine whether the current will pass. The closed switch (current flow) is represented as 1, and the open switch (no current flow) is represented as 0. For more information, see the group projects on pages 164 and 165.

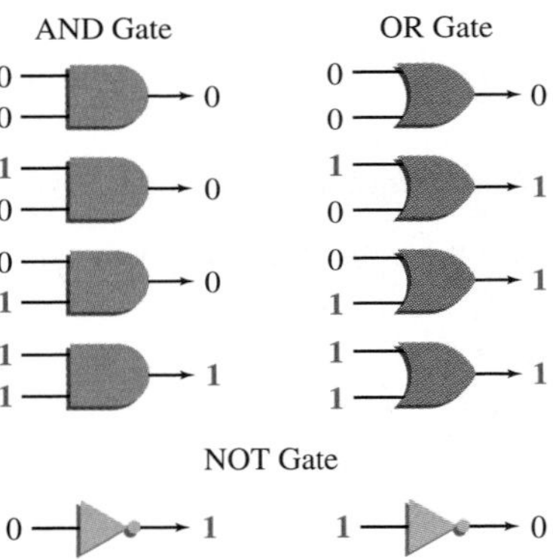

As you can see from the accompanying diagrams, the gates function in essentially the same way as a truth table. The gates shown here are the simplest ones, representing simple statements. There are other gates, such as the NAND and NOR gates, that are combinations of NOT, AND, and OR gates. The microprocessing unit of a computer uses thousands of these switches.

b) Let

p: George Washington was the first U.S. president.

q: Abraham Lincoln was the second U.S. president.

r: There has been a U.S. president who was born in Antarctica.

The compound statement can be written in symbolic form as $(p \vee q) \wedge \sim r$. Recall that *but* is used to express a conjunction. We know that p is a true statement and that q is a false statement. We also know that r is a false statement since all U.S. presidents must be born in the United States. Thus, since r is a false statement, the negation, $\sim r$, is a true statement. So we will substitute T for p, F for q, and T for $\sim r$ and then evaluate the statement:

$$(p \vee q) \wedge \sim r$$
$$(\text{T} \vee \text{F}) \wedge \text{T}$$
$$\text{T} \wedge \text{T}$$
$$\text{T}$$

Therefore, the original compound statement is a true statement. ▲

EXAMPLE 7 *OPEC Oil Production*

The Organization of Petroleum Exporting Countries (OPEC) consists of 11 developing nations whose economies are heavily reliant on oil export revenues. Figure 3.1 shows the percentage of total OPEC oil production produced by each of its member nations in 2002. Use this graph to determine the truth value of the following statement:

Saudi Arabia produces the most oil among OPEC nations and Qatar produces more oil than Venezuela, or Indonesia does not produce the least amount of oil among OPEC nations.

2002 OPEC Oil Production

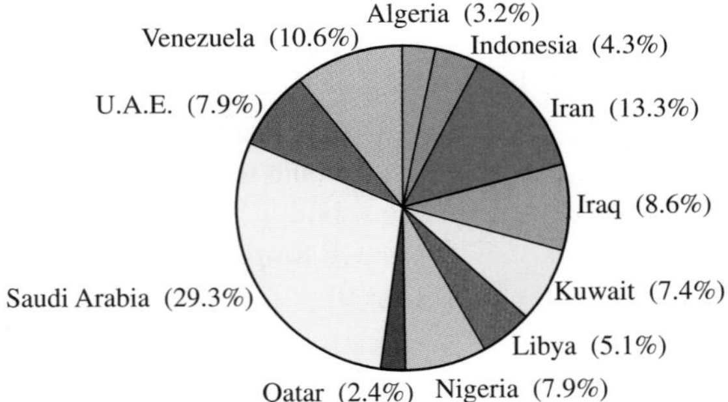

Source: Energy Information Administration

Figure 3.1

SOLUTION:
Let

p: Saudi Arabia produces the most oil among OPEC nations.
q: Qatar produces more oil that Venezuela.
r: Indonesia produces the least amount of oil among OPEC nations.

The given compound statement can be written in symbolic form as $(p \wedge q) \vee \sim r$. From Fig. 3.1, we see that statement p is true: Saudi Arabia does produce the most oil among OPEC nations. We also see that statement q is false: Venezuela actually produces more oil than Qatar. We also see that statement r is false: two OPEC nations produce less oil than Indonesia. Since r is false, its negation, $\sim r$, is true. Therefore, we substitute T for p, F for q, and T for $\sim r$ and get

$$(p \wedge q) \vee \sim r$$
$$(\mathrm{T} \wedge \mathrm{F}) \vee \mathrm{T}$$
$$\mathrm{F} \vee \mathrm{T}$$
$$\mathrm{T}$$

Thus, the original compound statement is true. ▲

SECTION 3.2 EXERCISES

1. a) How many distinct cases must be listed in a truth table that contains two simple statements?
   b) List all the cases.
2. a) How many distinct cases must be listed in a truth table that contains three simple statements?
   b) List all the cases.
3. a) Construct the truth table for the disjunction, $p \vee q$.
   b) Under what circumstances is the *or* table false?
4. a) Construct the truth table for the conjunction, $p \wedge q$.
   b) Under what circumstances is the *and* table true?

In Exercises 5–20, construct a truth table for the statement.

5. $p \vee \sim p$
6. $p \wedge \sim p$
7. $p \wedge \sim q$
8. $q \vee \sim p$
9. $\sim(p \vee \sim q)$
10. $\sim p \vee \sim q$
11. $\sim(p \wedge \sim q)$
12. $\sim(\sim p \wedge \sim q)$
13. $\sim q \vee (p \wedge r)$
14. $(p \vee \sim q) \wedge r$
15. $r \vee (p \wedge \sim q)$
16. $(r \wedge q) \wedge \sim p$
17. $(r \vee \sim p) \wedge \sim q$
18. $\sim p \wedge (q \vee r)$
19. $(\sim q \wedge r) \vee p$
20. $\sim r \vee (\sim p \wedge q)$

In Exercises 21–30, write the statement in symbolic form and construct a truth table.

21. Meetings are dull and teaching is fun.
22. The stadium is enclosed, but it is not air-conditioned.
23. Bob will get a haircut, but he will not shave his beard.
24. It is false that the class must have at least 15 students or the class will be canceled.
25. It is false that Jasper Adams is a tutor and Mark Russo is a secretary.
26. Mike made pizza and Dennis made a chef salad, but Gil burned the lemon squares.
27. The copier is out of toner, or the lens is dirty or the corona wires are broken.
28. I am hungry, and I want to eat a healthy lunch and I want to eat in a hurry.

29. The Congress must act on the bill, and the president must sign the bill or not sign the bill.
30. Gordon Langeneger likes the PowerMac G4 Cube and he likes the iBook, but he does not like the Pentium IV.

In Exercises 31–42, determine the truth value of the statement if

a) p is true, q is false, and r is true.
b) p is false, q is true, and r is true.

31. $\sim p \vee (q \wedge r)$
32. $(\sim p \wedge r) \wedge q$
33. $(\sim q \wedge \sim p) \vee \sim r$
34. $(\sim p \vee \sim q) \vee \sim r$
35. $(p \wedge \sim q) \vee r$
36. $(p \vee \sim q) \wedge \sim(p \wedge \sim r)$
37. $(\sim r \wedge p) \vee q$
38. $\sim q \vee (r \wedge p)$
39. $(\sim q \vee \sim p) \wedge r$
40. $(\sim r \vee \sim p) \vee \sim q$
41. $(\sim p \vee \sim q) \vee (\sim r \vee q)$
42. $(\sim r \wedge \sim q) \wedge (\sim r \vee \sim p)$

In Exercises 43–50, determine the truth value for each simple statement. Then use these truth values to determine the truth value of the compound statement. (You may have to use a reference source such as the Internet or an encyclopedia.)

43. $3 + 5 = 4 + 4$ or $10 - 9 = 9 - 10$
44. $5 < 4$ and $4 < 5$
45. Elvis Presley was a singer or chickens can swim.

Elvis Presley

46. Alaska is the 50th state or Hawaii is a group of islands, and Atlanta is the capital of Alabama.
47. U2 is a rock band and Denzel Washington is an actor, but Jerry Seinfeld is not a comedian.
48. The city of Toronto is in Minnesota or Mexico City is in Texas, and Cairo is in Egypt.
49. Cal Ripken Jr. played football or George Bush was the prime minister of England, and Colin Powell was in the Army.
50. Holstein is a breed of cattle and collie is a breed of dogs, or beagle is not a breed of cats.

Food Consumption In Exercises 51–54, use the chart to determine the truth value of each simple statement. Then determine the truth value of the compound statement.

Annual per capita consumption in pounds:

	1909	2001
Red meat	99	123.5
Poultry	11	66
Fish	11	15
Cheese	4	30
Fats and oils[a]	38	69
Sweeteners[b]	86	154

[a]Added fats and oils
[b]Caloric sweeteners (sugars, honey, corn syrup).
Source: U.S. Department of Agriculture

51. Thirty pounds of cheese were consumed by the average American in 1909, and the average American did not consume 154 pounds of sweeteners in 2001.
52. The per capita consumption of red meat was less for the average American in 2001 than it was in 1909 or the per capita consumption of poultry was greater for the average American in 2001 than it was in 1909.
53. The average American ate approximately the same amount of fish and poultry in 1909, but between 1909 and 2001 the per capita consumption of poultry increased at a rate higher than that of fish.

54. The average American ate approximately nine times as much red meat as fish in 1909, but by 2001 the average American only ate approximately eight times as much red meat as fish.

Sleep Time In Exercises 55–58, use the graph, which shows the number of hours Americans sleep, to determine the truth value of each simple statement. Then determine the truth value of the compound statement.

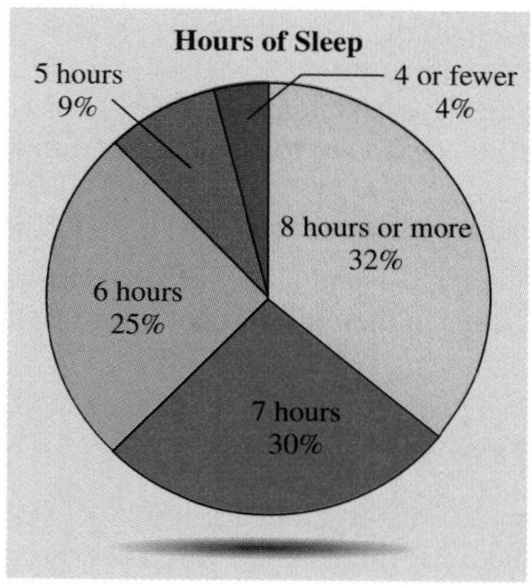

55. It is false that 30% of Americans get 6 hours of sleep each night and 9% get 5 hours of sleep each night.

56. Twenty-five percent of Americans get 6 hours of sleep each night, and 30% get 7 hours of sleep each night or 9% do not get 5 hours of sleep each night.

57. Thirteen percent of Americans get 5 or fewer hours of sleep each night or 32% get 6 or more hours of sleep each night, and 30% get 8 or more hours of sleep each night.

58. Over one-half of all Americans get 7 or fewer hours of sleep each night, and over one-quarter get 6 or fewer hours of sleep each night.

In Exercises 59–62, let

p: Tanisha owns a convertible.
q: Joan owns a Volvo.

Translate each statement into symbols. Then construct a truth table for each and indicate under what conditions the compound statement is true.

59. Tanisha owns a convertible and Joan does not own a Volvo.

60. Tanisha does not own a convertible, but Joan owns a Volvo.

61. Tanisha owns a convertible or Joan does not own a Volvo.

62. Tanisha does not own a convertible or Joan does not own a Volvo.

In Exercises 63–66, let

p: The house is owned by an engineer.
q: The heat is solar generated.
r: The car is run by electric power.

Translate each statement into symbols. Then construct a truth table for each and indicate under what conditions the compound statement is true.

63. The car is run by electric power or the heat is solar generated, but the house is owned by an engineer.

64. The house is owned by an engineer and the heat is solar generated, or the car is run by electric power.

65. The heat is solar generated, or the house is owned by an engineer and the car is not run by electric power.

66. The house is not owned by an engineer, and the car is not run by electric power and the heat is solar generated.

Obtaining a Loan In Exercises 67 and 68, read the requirements and each applicant's qualifications for obtaining a loan.

a) Identify which of the applicants would qualify for the loan.

b) For the applicants who do not qualify for the loan, explain why.

67. To qualify for a loan of $40,000, an applicant must have a gross income of $28,000 if single, $46,000 combined income if married, and assets of at least $6,000.

Mrs. Rusinek, married with three children, makes $42,000 on her job. Mr. Rusinek does not have an income. The Rusineks have assets of $42,000.

Mr. Duncan is not married, works in sales, and earns $31,000. He has assets of $9000.

Mrs. Tuttle and her husband have total assets of $43,000. One earns $35,000, and the other earns $23,500.

68. To qualify for a loan of $45,000, an applicant must have a gross income of $30,000 if single, $50,000 combined income if married, and assets of at least $10,000.

Mr. Argento, married with two children, makes $37,000 on his job. Mrs. Argento earns $15,000 at a part-time job. The Argentos have assets of $25,000.

Ms. McVey, single, has assets of $19,000. She works in a store and earns $25,000.

Mr. Siewert earns $24,000 and Ms. Fox, his wife, earns $28,000. Their assets total $8000.

69. *Airline Special Fares* An airline advertisement states, "To get the special fare you must purchase your tickets between January 1 and February 15 and fly round trip between March 1 and April 1. You must depart on a Monday, Tuesday, or Wednesday, and return on a Tuesday, Wednesday, or Thursday, and stay over at least one Saturday."

a) Determine which of the following individuals will qualify for the special fare.
b) If the person does not qualify for the special fare, explain why.

Wing Park plans to purchase his ticket on January 15, depart on Monday, March 3, and return on Tuesday, March 18.

Gina Vela plans to purchase her ticket on February 1, depart on Wednesday, March 10, and return on Thursday, April 2.

Kara Shavo plans to purchase her ticket on February 14, depart on Tuesday, March 5, and return on Monday, March 18.

Christos Supernaw plans to purchase his ticket on January 4, depart on Monday, March 8, and return on Thursday, March 11.

Alex Chang plans to purchase his ticket on January 1, depart on Monday, March 3, and return on Monday, March 10.

Problem Solving/Group Activities

In Exercises 70 and 71, construct a truth table for the symbolic statement.

70. $\sim[(\sim(p \vee q)) \vee (q \wedge r)]$

71. $[(q \wedge \sim r) \wedge (\sim p \vee \sim q)] \vee (p \vee \sim r)$

72. On page 113, we indicated that a compound statement consisting of n simple statements had 2^n distinct true–false cases.

a) How many distinct true–false cases does a truth table containing simple statements p, q, r, and s have?
b) List all possible true–false cases for a truth table containing the simple statements p, q, r, and s.
c) Use the list in part (b) to construct a truth table for $(q \wedge p) \vee (\sim r \wedge s)$.
d) Construct a truth table for $(\sim r \wedge \sim s) \wedge (\sim p \vee q)$.

73. Must $(p \wedge \sim q) \vee r$ and $(q \wedge \sim r) \vee p$ have the same number of trues in their answer columns? Explain.

Internet/Research Activities

74. Digital computers use gates that work like switches to perform calculations. Information is fed into the gates and information leaves the gates, according to the type of gate. The three basic gates used in computers are the NOT gate, the AND gate, and the OR gate. Do research on the three types of gates.

a) Explain how each gate works.
b) Explain the relationship between each gate and the corresponding logic connectives *not, and,* and *or*.
c) Illustrate how two or more gates can be combined to form a more complex gate.

3.3 TRUTH TABLES FOR THE CONDITIONAL AND BICONDITIONAL

Conditional

In Section 3.1, we mentioned that the statement preceding the conditional symbol is called the *antecedent* and that the statement following the conditional symbol is called the *consequent*. For example, consider $(p \vee q) \rightarrow [\sim(q \wedge r)]$. In this statement, $(p \vee q)$ is the antecedent and $[\sim(q \wedge r)]$ is the consequent.

Now we will look at the truth table for the conditional. Suppose I make the following promise to you: "If you get an A in this class, then I will buy you a car." Consider the statement within the quotation marks. Assume this statement is true except when I have actually broken my promise to you.

Let

p: You get an A.
q: I buy you a car.

Translated into symbolic form, the statement becomes $p \rightarrow q$. Let's examine the four cases shown in Table 3.12.

TABLE 3.12 Conditional

p	q	p→q
T	T	T
T	F	F
F	T	T
F	F	T

CASE 1: (T, T) You get an A, and I buy a car for you. I have met my commitment, and the statement is true.

CASE 2: (T, F) You get an A, and I do not buy a car for you. I have broken my promise, and the statement is false.

What happens if you don't get an A? If you don't get an A, I no longer have a commitment to you, and therefore I cannot break my promise.

CASE 3: (F, T) You do not get an A, and I buy you a car. I have not broken my promise, and therefore the statement is true.

CASE 4: (F, F) You do not get an A, and I don't buy you a car. I have not broken my promise, and therefore the statement is true.

The conditional statement is false when the antecedent is true and the consequent is false. In every other case the conditional statement is true.

The **conditional statement** $p \rightarrow q$ is true in every case except when p is a true statement and q is a false statement.

TABLE 3.13

p	q	~p	→	q
T	T	F	T	T
T	F	F	T	F
F	T	T	T	T
F	F	T	F	F
		1	3	2

EXAMPLE 1 *A Truth Table with a Conditional*

Construct a truth table for the statement $\sim p \rightarrow q$.

SOLUTION: Since this is a conditional statement, the answer will lie under the $\rightarrow$. Fill out the truth table by placing the appropriate values under $\sim p$, column 1, and under q, column 2 (see Table 3.13). Then, using the information given in the truth table for the conditional and the truth values in columns 1 and 2, determine the solution, column 3. In row 1, the antecedent, $\sim p$, is false and the consequent, q, is true. Row 1 is F → T, which according to row 3 of Table 3.12, is T. Likewise, row 2 of Table 3.13 is F → F, which is T. Row 3 is T → T, which is T. Row 4 is T → F, which is F. ▲

TABLE 3.14

p	q	r	p	→	(~q	∧	r)
T	T	T	T	F	F	F	T
T	T	F	T	F	F	F	F
T	F	T	T	T	T	T	T
T	F	F	T	F	T	F	F
F	T	T	F	T	F	F	T
F	T	F	F	T	F	F	F
F	F	T	F	T	T	T	T
F	F	F	F	T	T	F	F
			4	5	1	3	2

EXAMPLE 2 *A Conditional Truth Table with Three Simple Statements*

Construct a truth table for the statement $p \rightarrow (\sim q \wedge r)$.

SOLUTION: Since this is a conditional statement, the answer will lie under the $\rightarrow$. Work within the parentheses first. Place the truth values under $\sim q$, column 1, and r, column 2 (Table 3.14). Then take the conjunction of columns 1 and 2 to obtain column 3. Next, place the truth values under p in column 4. To determine the answer, column 5, use columns 3 and 4 and your knowledge of the conditional statement. Column 4 represents the truth values of the antecedent, and column 3 represents the truth values of the consequent. Remember that the conditional is false only when the antecedent is true and the consequent is false, as in cases (rows) 1, 2, and 4 of column 5. ▲

EXAMPLE 3 *Examining an Advertisement*

An advertisement for Perky Morning coffee makes the following claim: "If you drink Perky Morning coffee, then you will not be sluggish and you will have a great day." Translate the statement into symbolic form and construct a truth table.

SOLUTION: Let

p: You drink Perky Morning coffee.
q: You will be sluggish.
r: You will have a great day.

In symbolic form, the claim is

$$p \rightarrow (\sim q \wedge r)$$

This symbolic statement is identical to the statement in Table 3.14, and the truth tables are the same. Column 3 represents the truth values of $(\sim q \wedge r)$, which corresponds to the statement "You will not be sluggish and you will have a great day." Note that column 3 is true in cases (rows) 3 and 7. In case 3, since p is true, you drank Perky Morning coffee. In case 7, however, since p is false, you did not drink Perky Morning coffee. From this information we can conclude that it is possible for you to not be sluggish and for you to have a great day without drinking Perky Morning coffee. ▲

A truth table alone cannot tell us whether a statement is true or false. It can, however, be used to examine the various possibilities.

Biconditional

The *biconditional statement*, $p \leftrightarrow q$, means that $p \rightarrow q$ and $q \rightarrow p$, or, symbolically, $(p \rightarrow q) \wedge (q \rightarrow p)$. To determine the truth table for $p \leftrightarrow q$, we will construct the truth table for $(p \rightarrow q) \wedge (q \rightarrow p)$ (Table 3.15). Table 3.16 shows the truth values for the biconditional statement.

TABLE 3.15

p	q	(p	→	q)	∧	(q	→	p)
T	T	T	T	T	T	T	T	T
T	F	T	F	F	F	F	T	T
F	T	F	T	T	F	T	F	F
F	F	F	T	F	T	F	T	F
		1	3	2	7	4	6	5

The **biconditional statement,** $p \leftrightarrow q$, is true only when p and q have the same truth value, that is, when both are true or both are false.

TABLE 3.16 Biconditional

p	q	p↔q
T	T	T
T	F	F
F	T	F
F	F	T

EXAMPLE 4 *A Truth Table Using a Biconditional*

Construct a truth table for the statement $p \leftrightarrow (q \rightarrow \sim r)$.

SOLUTION: Since there are three letters, there must be eight cases. The parentheses indicate that the answer must be under the biconditional (Table 3.17). Use columns 3 and 4 to obtain the answer in column 5. When columns 3 and 4 have the same truth values, place a T in column 5. When columns 3 and 4 have different truth values, place an F in column 5.

TABLE 3.17

p	q	r	p	↔	(q	→	~r)
T	T	T	T	F	T	F	F
T	T	F	T	T	T	T	T
T	F	T	T	T	F	T	F
T	F	F	T	T	F	T	T
F	T	T	F	T	T	F	F
F	T	F	F	F	T	T	T
F	F	T	F	F	F	T	F
F	F	F	F	F	F	T	T
			4	5	1	3	2

▲

In the preceding section, we showed that finding the truth value of a compound statement for a specific case does not require constructing an entire truth table. Example 5 illustrates this technique for the conditional and the biconditional.

EXAMPLE 5 *Determine the Truth Value of a Compound Statement*

Determine the truth value of the statement $(q \leftrightarrow r) \rightarrow (\sim p \wedge r)$ when p is true, q is false, and r is true.

SOLUTION: Substitute the truth value for each simple statement:

$$\begin{array}{c} (q \leftrightarrow r) \rightarrow (\sim p \ \wedge \ r) \\ (\mathrm{F} \leftrightarrow \mathrm{T}) \rightarrow (\mathrm{F} \ \wedge \ \mathrm{T}) \\ \mathrm{F} \quad \rightarrow \quad \mathrm{F} \\ \mathrm{T} \end{array}$$

For this specific case, the statement is true. ▲

EXAMPLE 6 *Determine the Truth Value of a Compound Statement*

Determine the truth value for each simple statement. Then use the truth values to determine the truth value of the compound statement.

a) If 15 is an even number, then 29 is an even number.

b) Northwestern University is in Illinois and Marquette University is in Alaska, if and only if Purdue University is in Alabama.

DID YOU KNOW

Satisfiability Problems

Suppose you are hosting a dinner party for seven people: Yasumasa, Marie, Albert, Stephen, Leonhard, Karl, and Emmy. You need to develop a seating plan around your circular dining room table that would satisfy all your guests. Albert and Emmy are great friends and must sit together. Yasumasa and Karl haven't spoken to each other in years and cannot sit by each other. Leonhard must sit by Marie or by Albert, but he cannot sit by Karl. Stephen insists he sit by Albert. Can you come up with a plan that would satisfy all your guests? Now imagine the difficulty of such a problem as the list of guests, and their demands, grows.

Problems such as this are known as *satisfiability* problems. The symbolic logic you are studying in this chapter allows computer scientists to represent these problems with symbols. Once in symbolic form, a problem can be studied with computers using algorithms. An *algorithm* is a general procedure for accomplishing a task. Even with the fastest computers, some satisfiability problems take an enormous amount of time to solve. Other problems may not even have a solution. Computer scientists and mathematicians are continually working to improve the algorithms used to solve satisfiability problems.

One solution to the problem posed above is as follows:

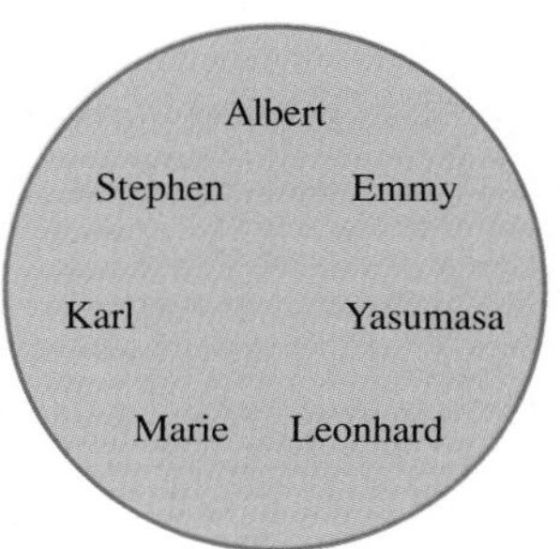

SOLUTION:

a) Let

p: 15 is an even number.
q: 29 is an even number.

Then the statement "If 15 is an even number, then 29 is an even number" can be written $p \rightarrow q$. Since 15 is not an even number, p is a false statement. Also, since 29 is not an even number, q is a false statement. We substitute F for p and F for q and evaluate the statement:

$$\begin{array}{c} p \rightarrow q \\ \text{F} \rightarrow \text{F} \\ \text{T} \end{array}$$

Therefore, "If 15 is an even number, then 29 is an even number" is a true statement.

b) Let

p: Northwestern University is in Illinois.
q: Marquette University is in Alaska.
r: Purdue University is in Alabama.

The original compound statement can be written $(p \wedge q) \leftrightarrow r$. By checking the Internet or other references we can find that Northwestern University is in Illinois, Marquette University is in Wisconsin, and Purdue University is in Indiana. Therefore, p is a true statement, but q and r are false statements. We will substitute T for p, F for q, and F for r and evaluate the compound statement:

$$\begin{array}{c} (p \wedge q) \leftrightarrow r \\ (\text{T} \wedge \text{F}) \leftrightarrow \text{F} \\ \text{F} \quad \leftrightarrow \text{F} \\ \text{T} \end{array}$$

Therefore, the original compound statement is true.

EXAMPLE 7 *Using Real Data in Compound Statements*

The graph in Fig. 3.2 on page 123 represents the U.S. government budget expenditures for fiscal year 2002. Use this graph to determine the truth value of the following compound statements.

a) If social programs account for 17% of the budget then interest on the national debt accounts for 12% of the budget.

b) If physical and community development account for 9% of the budget and social programs account for 37% of the budget, then law enforcement and general government account for 10% of the budget.

SOLUTION:

a) Let

p: Social programs account for 17% of the budget.
q: Interest on the national debt accounts for 12% of the budget.

Then the original compound statement can be written $p \rightarrow q$. We can see from Fig. 3.2 that both p and q are true statements. Substitute T for p and T for q and evaluate the statement:

$$\begin{array}{c} p \rightarrow q \\ \mathrm{T} \rightarrow \mathrm{T} \\ \mathrm{T} \end{array}$$

Therefore, "If social programs account for 17% of the budget then interest on the national debt accounts for 12% of the budget" is a true statement.

2002 United States Federal Government Budget Expenditures

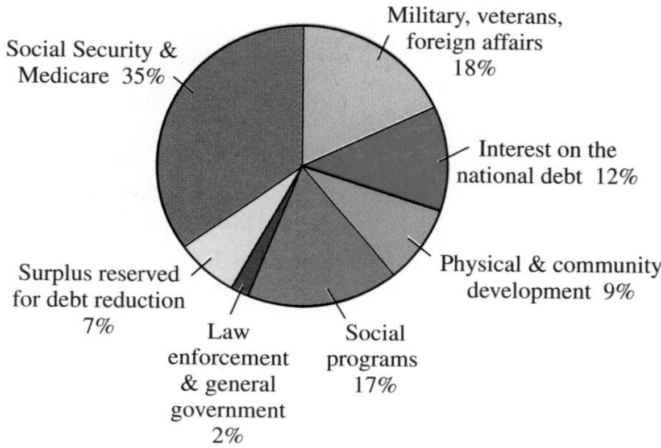

Source: *United States Office of Management and Budget*

Figure 3.2

b) Let

p: Physical and community development account for 9% of the budget.

q: Social programs account for 37% of the budget.

r: Law enforcement and general government account for 10% of the budget.

Then the original compound statement can be written $(p \wedge q) \rightarrow r$. We can see from Fig. 3.2 that p is true, q is false, and r is false. Substitute T for p and F for q and r and evaluate the statement:

$$\begin{array}{c} (p \wedge q) \rightarrow r \\ (\mathrm{T} \wedge \mathrm{F}) \rightarrow \mathrm{F} \\ \mathrm{F} \quad \rightarrow \mathrm{F} \\ \mathrm{T} \end{array}$$

Therefore, the original compound statement, "If physical and community development account for 9% of the budget and social programs account for 37% of the budget, then law enforcement and general government account for 10% of the budget," is true. ▲

Self-Contradictions, Tautologies, and Implications

Two special situations can occur in the truth table of a compound statement: The statement may always be false, or the statement may always be true. We give such statements special names.

> A **self-contradiction** is a compound statement that is always false.

When every truth value in the answer column of the truth table is false, then the statement is a self-contradiction.

TABLE 3.18

p	q	(p ↔ q)	∧	(p	↔	~q)
T	T	T	F	T	F	F
T	F	F	F	T	T	T
F	T	F	F	F	T	F
F	F	T	F	F	F	T
		1	5	2	4	3

EXAMPLE 8 ***All Falses, a Self-Contradiction***

Construct a truth table for the statement $(p \leftrightarrow q) \wedge (p \leftrightarrow \sim q)$.

SOLUTION: See Table 3.18. In this example, the truth values are false in each case of column 5. This statement is an example of a self-contradiction or a *logically false statement.* ▲

> A **tautology** is a compound statement that is always true.

When every truth value in the answer column of the truth table is true, the statement is a tautology.

EXAMPLE 9 ***All Trues, a Tautology***

Construct a truth table for the statement $(p \wedge q) \rightarrow (p \vee r)$.

SOLUTION: The answer is given in column 3 of Table 3.19. The truth values are true in every case. Thus, the statement is an example of a tautology or a *logicallly true statement.*

TABLE 3.19

p	q	r	(p ∧ q)	→	(p ∨ r)
T	T	T	T	T	T
T	T	F	T	T	T
T	F	T	F	T	T
T	F	F	F	T	T
F	T	T	F	T	T
F	T	F	F	T	F
F	F	T	F	T	T
F	F	F	F	T	F
			1	3	2

▲

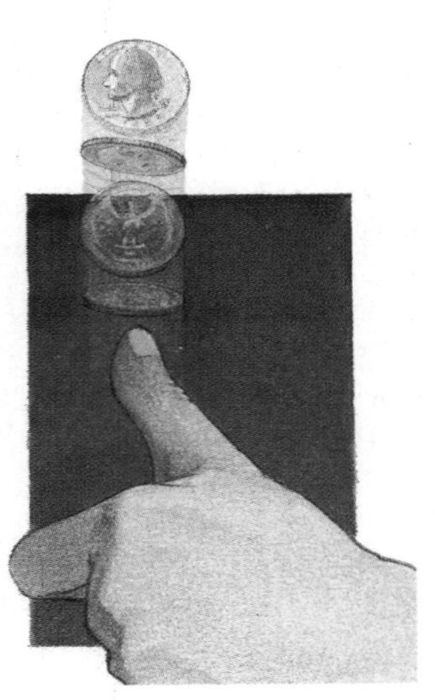

"Heads I win, tails you lose." Do you think that this statement is a tautology, self-contradiction, or neither? See Problem-Solving Exercise 81.

The conditional statement $(p \wedge q) \rightarrow (p \vee r)$ is a tautology. Conditional statements that are tautologies are called *implications*. In Example 9, we can say that $p \wedge q$ implies $p \vee r$.

An **implication** is a conditional statement that is a tautology.

In any implication the antecedent of the conditional statement implies the consequent. In other words, if the antecedent is true, then the consequent must also be true. That is, the consequent will be true whenever the antecedent is true.

TABLE 3.20

p	q	[(p ∧ q)	∧	p]	→	q
T	T	T	T	T	T	T
T	F	F	F	T	T	F
F	T	F	F	F	T	T
F	F	F	F	F	T	F
		1	3	2	5	4

EXAMPLE 10 *An Implication?*

Determine whether the conditional statement $[(p \wedge q) \wedge p] \rightarrow q$ is an implication.

SOLUTION: If the conditional statement is a tautology, the conditional statement is an implication. Because the conditional statement is a tautology (see Table 3.20), the conditional statement is an implication. The antecedent $[(p \wedge q) \wedge p]$ implies the consequent q. Note that the antecedent is true only in case 1 and that the consequent is also true in case 1. ▲

SECTION 3.3 EXERCISES

1. **a)** Construct the truth table for the conditional statement $p \rightarrow q$.
   b) Explain when the conditional statement is true and when it is false.
2. **a)** Construct the truth table for the biconditional statement $p \leftrightarrow q$.
   b) Explain when the biconditional statement is true and when it is false.
3. **a)** Explain the procedure to determine the truth value of a compound statement when specific truth values are provided for the simple statements.
   b) Follow the procedure in part (a) and determine the truth value of the symbolic statement

$$[(p \leftrightarrow q) \vee (\sim r \rightarrow q)] \rightarrow \sim r$$

   when p = true, q = true, and r = false.
4. What is a tautology?
5. What is a self-contradiction?
6. What is an implication?

In Exercises 7–16, construct a truth table for the statement.

7. $\sim q \rightarrow \sim p$
8. $p \rightarrow \sim q$
9. $\sim(q \rightarrow p)$
10. $\sim(p \leftrightarrow q)$
11. $\sim q \leftrightarrow p$
12. $(p \leftrightarrow q) \rightarrow p$
13. $p \leftrightarrow (q \vee p)$
14. $(\sim q \wedge p) \rightarrow \sim q$
15. $q \rightarrow (p \rightarrow \sim q)$
16. $(p \vee q) \leftrightarrow (p \wedge q)$

In Exercises 17–26, construct a truth table for the statement.

17. $r \wedge (\sim q \rightarrow p)$
18. $p \rightarrow (q \vee r)$
19. $(q \leftrightarrow p) \wedge \sim r$
20. $q \leftrightarrow (r \wedge p)$
21. $(q \vee \sim r) \leftrightarrow \sim p$
22. $(p \wedge r) \rightarrow (q \vee r)$
23. $(\sim r \vee \sim q) \rightarrow p$
24. $[r \wedge (q \vee \sim p)] \leftrightarrow \sim p$
25. $(p \rightarrow q) \leftrightarrow (\sim q \rightarrow \sim r)$
26. $(\sim p \leftrightarrow \sim q) \rightarrow (\sim q \leftrightarrow r)$

In Exercises 27–32, write the statement in symbolic form. Then construct a truth table for the symbolic statement.

27. If I drink a glass of water, then I will have a better complexion and I will sleep better.

28. The goalie will make the save if and only if the stopper is in position, or the forward cannot handle the ball.

29. The class has been canceled if and only if the teacher is not here, or we will study together in the library.
30. If the lake rises then we can go canoeing, and if the canoe has a hole in it then we cannot go canoeing.
31. If Mary Andrews does not send me an e-mail then we can call her, or we can write to Mom.
32. It is false that if Eileen Jones went to lunch, then she cannot take a message and we will have to go home.

In Exercises 33–38, determine whether the statement is a tautology, self-contradiction, or neither.

33. $p \rightarrow \sim q$
34. $(p \vee q) \leftrightarrow \sim p$
35. $p \wedge (q \wedge \sim p)$
36. $(p \wedge \sim q) \rightarrow q$
37. $(\sim q \rightarrow p) \vee \sim q$
38. $[(p \rightarrow q) \vee r] \leftrightarrow [(p \wedge q) \rightarrow r]$

In Exercises 39–44, determine whether the statement is an implication.

39. $p \rightarrow (p \wedge q)$
40. $(p \wedge q) \rightarrow (p \vee q)$
41. $(q \wedge p) \rightarrow (p \wedge q)$
42. $(p \vee q) \rightarrow (p \vee \sim r)$
43. $[(p \rightarrow q) \wedge (q \rightarrow p)] \rightarrow (p \leftrightarrow q)$
44. $[(p \vee q) \wedge r] \rightarrow (p \vee q)$

In Exercises 45–56, if p is true, q is false, and r is true, find the truth value of the statement.

45. $p \rightarrow (\sim q \wedge r)$
46. $\sim p \rightarrow (q \vee r)$
47. $(q \wedge \sim p) \leftrightarrow \sim r$
48. $p \leftrightarrow (\sim q \wedge r)$
49. $(\sim p \wedge \sim q) \vee \sim r$
50. $\sim[p \rightarrow (q \wedge r)]$
51. $(p \wedge r) \leftrightarrow (p \vee \sim q)$
52. $(\sim p \vee q) \rightarrow \sim r$
53. $(\sim p \leftrightarrow r) \vee (\sim q \leftrightarrow r)$
54. $(r \rightarrow \sim p) \wedge (q \rightarrow \sim r)$
55. $\sim[(p \vee q) \leftrightarrow (p \rightarrow \sim r)]$
56. $[(\sim r \rightarrow \sim q) \vee (p \wedge \sim r)] \rightarrow q$

In Exercises 57–64, determine the truth value for each simple statement. Then, using the truth values, determine the truth value of the compound statement.

57. If $10 + 5 = 15$, then $56 \div 7 = 8$.
58. If 2 is an even number and 6 is an odd number, then 15 is an odd number.
59. A triangle has four sides or a square has three sides, and a rectangle has four sides.
60. Seattle is in Washington and Portland is in Oregon, or Boise is in California.
61. Dell makes computers, if and only if Gateway makes computers or Canon makes printers.
62. Spike Lee is a movie director, or if Halle Berry is a schoolteacher then George Clooney is a circus clown.

Halle Berry

63. Valentine's Day is in February or President's Day is in March, and Thanksgiving is in November.
64. Honda makes automobiles or Honda makes motorcycles, if and only if Toyota makes cereal.

In Exercises 65–68, use the information provided about the moons for the planets Jupiter and Saturn on page 127 to determine the truth values of the simple statements. Then determine the truth value of the compound statement.

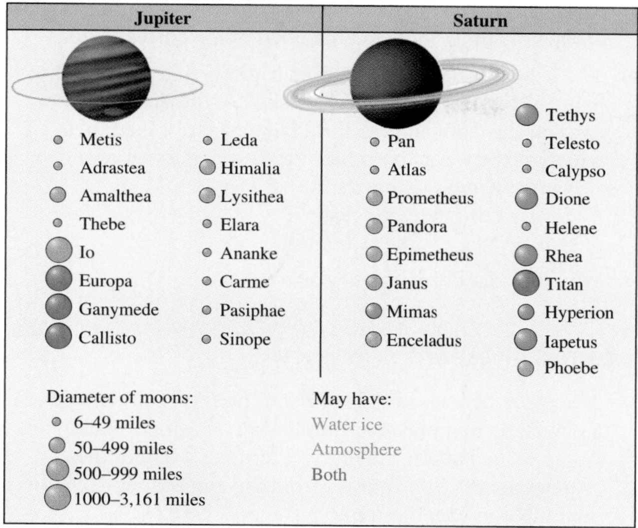

Source: *Time* Magazine

65. *Jupiter's Moons* Io has a diameter of 1000–3161 miles or Thebe may have water, and Io may have atmosphere.
66. *Moons of Saturn* Titan may have water and Titan may have atmosphere, if and only if Janus may have water.
67. *Moon Comparisons* Phoebe has a larger diameter than Rhea if and only if Callisto may have water ice, and Calypso has a diameter of 6–49 miles.
68. *Moon Comparisons* If Jupiter has 16 moons or Saturn does not have 18 moons, then Saturn has 7 moons that may have water ice.

In Exercises 69 and 70, use the graphs to determine the truth values of each simple statement. Then determine the truth value of the compound statement.

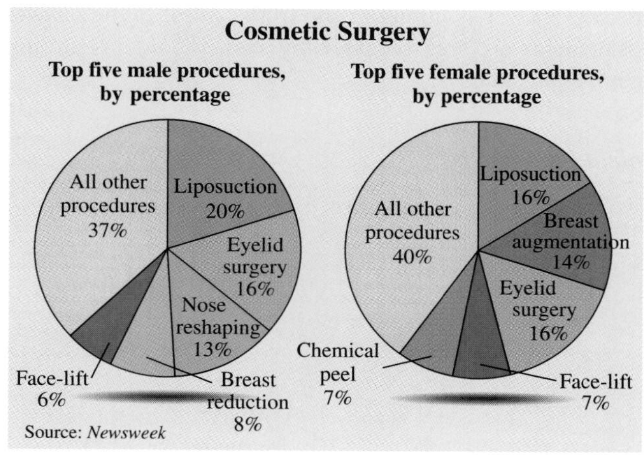

Source: *Newsweek*

69. *Most Common Cosmetic Surgery* The most common cosmetic surgery procedure for females is liposuction or the most common procedure for males is eyelid surgery, and 20% of male cosmetic surgery is for nose reshaping.
70. *Face-lifts and Eyelid Surgeries* 7% of female cosmetic surgeries are for face-lifts and 10% of male cosmetic surgeries are for face-lifts, if and only if males have a higher percent of eyelid surgeries than females.

In Exercises 71–76, suppose both of the following statements are true.

p: Muhundan spoke at the teachers' conference.
q: Muhundan received the outstanding teacher award.

Find the truth values of each compound statement.

71. If Muhundan spoke at the teachers' conference, then Muhundan received the outstanding teacher award.
72. If Muhundan did not speak at the teachers' conference, then Muhundan did not receive the outstanding teacher award.
73. If Muhundan did not speak at the teachers' conference, then Muhundan received the outstanding teacher award.
74. Muhundan did not receive the outstanding teacher award if and only if Muhundan spoke at the teachers' conference.
75. Muhundan received the outstanding teacher award if and only if Muhundan spoke at the teachers' conference.
76. If Muhundan did not receive the outstanding teacher award, then Muhundan did not speak at the teachers' conference.
77. *A New Computer* Your parents make the following statement to your sister, "If you get straight A's this semester, then we will buy you a new computer." At the end of the semester your parents buy your sister a new computer. Can you conclude that your sister got straight A's? Explain.
78. *Job Interview* Consider the statement "If your interview goes well, then you will be offered the job." If you are interviewed and then offered the job, can you conclude that your interview went well? Explain.

Problem Solving/Group Activities

In Exercises 79 and 80, construct truth tables for the symbolic statement. Use the dominance of connectives (see Section 3.1) as needed.

79. $p \vee q \rightarrow \sim r \leftrightarrow p \wedge \sim q$
80. $[(r \rightarrow \sim q) \rightarrow \sim p] \vee (q \leftrightarrow \sim r)$
81. Is the statement "Heads I win, tails you lose" a tautology, a self-contradiction, or neither? Explain your answer.

82. Construct a truth table for
 a) $(p \vee q) \rightarrow (r \wedge s)$.
 b) $(q \rightarrow \sim p) \vee (r \leftrightarrow s)$.

Recreational Mathematics

83. *Cat Puzzle* Solve the following puzzle. The Joneses have four cats. The parents are Tiger and Boots, and the kittens are Sam and Sue. Each cat insists on eating out of its own bowl. To complicate matters, each cat will eat only its own brand of cat food. The colors of the bowls are red, yellow, green, and blue. The different types of cat food are Whiskas, Friskies, Nine Lives, and Meow Mix. Tiger will eat Meow Mix if and only if it is in a yellow bowl. If Boots is to eat her food, then it must be in a yellow bowl. Mrs. Jones knows that the label on the can containing Sam's food is the same color as his bowl. Boots eats Whiskas. Meow Mix and Nine Lives are packaged in a brown paper bag. The color of Sue's bowl is green if and only if she eats Meow Mix. The label on the Friskies can is red. Match each cat with its food and the bowl of the correct color.

84. *The Youngest Triplet* The Barr triplets have an annoying habit: Whenever a question is asked of the three of them, two tell the truth and the third lies. When I asked them which of them was born last, they replied as follows.
 Mary: Katie was born last.
 Katie: I am the youngest.
 Annie: Mary is the youngest.

 Which of the Barr triplets was born last?

Internet/Research Activity

85. Select an advertisement from the Internet, a newspaper, or a magazine that makes or implies a conditional statement. Analyze the advertisement to determine whether the consequent necessarily follows from the antecedent. Explain your answer. (See Example 3.)

3.4 EQUIVALENT STATEMENTS

Equivalent statements are an important concept in the study of logic.

> Two statements are **equivalent,** symbolized $\Leftrightarrow$,* if both statements have exactly the same truth values in the answer columns of the truth tables.

Sometimes the words *logically equivalent* are used in place of the word *equivalent.*

To determine whether two statements are equivalent, construct a truth table for each statement and compare the answer columns of the truth tables. If the answer columns are identical, the statements are equivalent. If the answer columns are not identical, the statements are not equivalent.

EXAMPLE 1 *Equivalent Statements*

Show that the following two statements are equivalent.

$$[p \vee (q \vee r)] \qquad [(p \vee q) \vee r]$$

*The symbol $\equiv$ is also used to indicate equivalent statements.

SOLUTION: Construct a truth table for each statement (see Table 3.21).

TABLE 3.21

p	q	r	[p	∨	(q ∨ r)]	[(p ∨ q)	∨	r]
T	T	T	T	T	T	T	T	T
T	T	F	T	T	T	T	T	F
T	F	T	T	T	T	T	T	T
T	F	F	T	T	F	T	T	F
F	T	T	F	T	T	T	T	T
F	T	F	F	T	T	T	T	F
F	F	T	F	T	T	F	T	T
F	F	F	F	F	F	F	F	F
			1	3	2	1	3	2

Because the truth tables have the same answer (column 3 for both tables), the statements are equivalent. Thus, we can write

$$[p \vee (q \vee r)] \Leftrightarrow [(p \vee q) \vee r].$$

EXAMPLE 2 *Are the Following Equivalent Statements?*

Determine whether the following statements are equivalent.

a) If you work hard and obey all of the rules, then you will succeed in life.

b) If you do not work hard or do not obey all of the rules, then you will not succeed in life.

SOLUTION: First write each statement in symbolic form, then construct a truth table for each statement. If the answer columns of both truth tables are identical, then the statements are equivalent. If the answer columns are not identical, then the statements are not equivalent.

Let

p: You work hard.
q: You obey all of the rules.
r: You will succeed in life.

In symbolic form, the statements are

a) $(p \wedge q) \rightarrow r$. b) $(\sim p \vee \sim q) \rightarrow \sim r$.

The truth tables for these statements are given in Tables 3.22 and 3.23, respectively, on page 130. The answers in the columns labeled 5 are not identical, so the statements are not equivalent.

TABLE 3.22

p	q	r	(p	∧	q)	→	r
T	T	T	T	T	T	T	T
T	T	F	T	T	T	F	F
T	F	T	T	F	F	T	T
T	F	F	T	F	F	T	F
F	T	T	F	F	T	T	T
F	T	F	F	F	T	T	F
F	F	T	F	F	F	T	T
F	F	F	F	F	F	T	F
			1	3	2	5	4

TABLE 3.23

p	q	r	(~p	∨	~q)	→	~r
T	T	T	F	F	F	T	F
T	T	F	F	F	F	T	T
T	F	T	F	T	T	F	F
T	F	F	F	T	T	T	T
F	T	T	T	T	F	F	F
F	T	F	T	T	F	T	T
F	F	T	T	T	T	F	F
F	F	F	T	T	T	T	T
			1	3	2	5	4

EXAMPLE 3 *Which Statements Are Logically Equivalent?*

Determine which statement is logically equivalent to "It is not true that the tire is both out of balance and flat."

a) If the tire is not flat, then the tire is not out of balance.

b) The tire is not out of balance or the tire is not flat.

c) The tire is not flat and the tire is not out of balance.

d) If the tire is not out of balance, then the tire is not flat.

SOLUTION: To determine whether any of the choices are equivalent to the given statement, first write the given statements and the choices in symbolic form. Then construct truth tables and compare the answer columns of the truth tables.

Let

p: The tire is out of balance.

q: The tire is flat.

The given statement may be written "It is not true that the tire is out of balance and the tire is flat." The statement is expressed in symbolic form as $\sim(p \wedge q)$. Using p and q as indicated, choices (a) through (d) may be expressed symbolically as

a) $\sim q \rightarrow \sim p$. b) $\sim p \vee \sim q$. c) $\sim q \wedge \sim p$. d) $\sim p \rightarrow \sim q$.

Now construct a truth table for the given statement (Table 3.24 on page 131) and each possible choice, given in Table 3.25(a) through (d). By examining the truth tables, we see that the given statement, $\sim(p \wedge q)$, is logically equivalent to choice (b), $\sim p \vee \sim q$. Therefore, the correct answer is "The tire is not out of balance or the tire is not flat." This statement is logically equivalent to the statement "It is not true that the tire is both out of balance and flat."

TABLE 3.24

p	q	~	(p	∧	q)
T	T	F	T	T	T
T	F	T	T	F	F
F	T	T	F	F	T
F	F	T	F	F	F
		4	1	3	2

TABLE 3.25

		(a)			(b)			(c)			(d)		
p	q	~q	→	~p	~p	∨	~q	~q	∧	~p	~p	→	~q
T	T	F	T	F	F	F	F	F	F	F	F	T	F
T	F	T	F	F	F	T	T	T	F	F	F	T	T
F	T	F	T	T	T	T	F	F	F	T	T	F	F
F	F	T	T	T	T	T	T	T	T	T	T	T	T

In the preceding section, we showed that $p \leftrightarrow q$ has the same truth table as $(p \rightarrow q) \wedge (q \rightarrow p)$. Therefore, these statements are equivalent, a useful fact for Example 4.

EXAMPLE 4 *Write an Equivalent Biconditional Statement*

Write the following statement as an equivalent biconditional statement: "If the tree produces acorns then the tree is an oak and if the tree is an oak then the tree produces acorns."

SOLUTION: An equivalent statement is "The tree produces acorns if and only if the tree is an oak."

De Morgan's Laws

Example 3 showed that a statement of the form $\sim(p \wedge q)$ is equivalent to a statement of the form $\sim p \vee \sim q$. Thus, we may write $\sim(p \wedge q) \Leftrightarrow \sim p \vee \sim q$. This equivalent statement is one of two special laws called De Morgan's laws. The laws, named after Augustus De Morgan, an English mathematician, were first introduced in Section 2.4, where they applied to sets.

De Morgan's Laws

1. $\sim(p \wedge q) \Leftrightarrow \sim p \vee \sim q$
2. $\sim(p \vee q) \Leftrightarrow \sim p \wedge \sim q$

You can demonstrate that De Morgan's second law is true by constructing and comparing truth tables for $\sim(p \vee q)$ and $\sim p \wedge \sim q$. Do so now.

When using De Morgan's laws, if it becomes necessary to negate an already negated statement, use the fact that $\sim(\sim p)$ is equivalent to p. For example, the negation of the statement "Today is not Monday" is "Today is Monday."

EXAMPLE 5 *Use De Morgan's Laws*

Select the statement that is logically equivalent to "The sun is not shining but it is not raining."

a) It is not true that the sun is shining and it is raining.
b) It is not raining or the sun is not shining.
c) The sun is shining or it is raining.
d) It is not true that the sun is shining or it is raining.

PROFILE IN MATHEMATICS

CHARLES DODGSON

One of the more interesting and well-known students of logic was Charles Dodgson (1832–1898), better known to us as Lewis Carroll, the author of *Alice's Adventures in Wonderland* and *Through the Looking-Glass*. Although the books have a child's point of view, many argue that the audience best equipped to enjoy them is an adult one. Dodgson, a mathematician, logician, and photographer (among other things), uses the naïveté of a 7-year-old girl to show what can happen when the rules of logic are taken to absurd extremes.

"You should say what you mean," the March Hare went on.

"I do," Alice hastily replied; "at least—at least I mean what I say—that's the same thing, you know."

"Not the same thing a bit!" said the Hatter. "You might as well say that 'I see what I eat' is the same thing as 'I eat what I see'!"

SOLUTION: To determine which statement is equivalent, write each statement in symbolic form.

Let

p: The sun is shining.
q: It is raining.

The statement "The sun is not shining but it is not raining" written symbolically is $\sim p \wedge \sim q$. Recall that the word *but* means the same as the word *and*. Now, write parts (a) through (d) symbolically.

a) $\sim(p \wedge q)$ b) $\sim q \vee \sim p$ c) $p \vee q$ d) $\sim(p \vee q)$

De Morgan's law shows that $\sim p \wedge \sim q$ is equivalent to $\sim(p \vee q)$. Therefore, the answer is (d): "It is not true that the sun is shining or it is raining." ▲

EXAMPLE 6 *Using De Morgan's Laws to Write an Equivalent Statement*

Write a statement that is logically equivalent to "It is not true that tomatoes are poisonous or eating peppers cures the common cold."

SOLUTION: Let

p: Tomatoes are poisonous.
q: Eating peppers cures the common cold.

The given statement is of the form $\sim(p \vee q)$. Using the second of De Morgan's laws, we see that an equivalent statement in symbols is $\sim p \wedge \sim q$. Therefore, an equivalent statement in words is "Tomatoes are not poisonous and eating peppers does not cure the common cold." ▲

Consider $\sim(p \wedge q) \Leftrightarrow \sim p \vee \sim q$, one of De Morgan's laws. To go from $\sim(p \wedge q)$ to $\sim p \vee \sim q$, we negate both the p and the q within parentheses; change the conjunction, $\wedge$, to a disjunction, $\vee$; and remove the negation symbol preceding the left parentheses and the parentheses themselves. We can use a similar procedure to obtain equivalent statements. For example,

$$\sim(\sim p \wedge q) \Leftrightarrow p \vee \sim q$$
$$\sim(p \wedge \sim q) \Leftrightarrow \sim p \vee q$$

We can use a similar procedure to obtain equivalent statements when a disjunction is within parentheses. Note that

$$\sim(\sim p \vee q) \Leftrightarrow p \wedge \sim q$$
$$\sim(p \vee \sim q) \Leftrightarrow \sim p \wedge q$$

EXAMPLE 7 *Using De Morgan's Laws to Write an Equivalent Statement*

Use De Morgan's laws to write a statement logically equivalent to "Benjamin Franklin was not a U.S. president, but he signed the Declaration of Independence."

PROFILE IN MATHEMATICS

AUGUSTUS DE MORGAN

Augustus De Morgan (1806–1871), the son of a member of the East India Company, was born in India and educated at Trinity College, Cambridge. One of the great reformers of logic in the nineteenth century, De Morgan made his greatest contribution to the subject by realizing that logic as it had come down from Aristotle was narrow in scope and could be applied to a wider range of arguments. His work laid the foundation for modern, symbolic logic.

SOLUTION: Let

p: Benjamin Franklin was a U.S. president.
q: Benjamin Franklin signed the Declaration of Independence.

The statement written symbolically is $\sim p \wedge q$. Earlier we showed that

$$\sim p \wedge q \Leftrightarrow \sim(p \vee \sim q)$$

Therefore, the statement "It is false that Benjamin Franklin was a U.S. president or Benjamin Franklin did not sign the Declaration of Independence" is logically equivalent to the given statement. ▲

There are strong similarities between the topics of sets and logic. We can see them by examining De Morgan's laws for sets and logic.

De Morgan's laws: set theory	**De Morgan's laws: logic**
$(A \cap B)' = A' \cup B'$	$\sim(p \wedge q) \Leftrightarrow \sim p \vee \sim q$
$(A \cup B)' = A' \cap B'$	$\sim(p \vee q) \Leftrightarrow \sim p \wedge \sim q$

The complement in set theory, $'$, is similar to the negation, $\sim$, in logic. The intersection, $\cap$, is similar to the conjunction, $\wedge$; and the union, $\cup$, is similar to the disjunction, $\vee$. If we were to interchange the set symbols with the logic symbols, De Morgan's laws would remain, but in a different form.

Both $'$ and $\sim$ can be interpreted as *not*.

Both $\cap$ and $\wedge$ can be interpreted as *and*.

Both $\cup$ and $\vee$ can be interpreted as *or*.

For example, the set statement $A' \cup B$ can be written as a statement in logic as $\sim a \vee b$.

Statements containing connectives other than *and* and *or* may have equivalent statements. To illustrate this point, construct truth tables for $p \rightarrow q$ and for $\sim p \vee q$. The truth tables will have the same answer columns and therefore the statements are equivalent. That is,

$$p \rightarrow q \Leftrightarrow \sim p \vee q$$

With these equivalent statements, we can write a conditional statement as a disjunction or a disjunction as a conditional statement. For example, the statement "If the game is polo, then you ride a horse" can be equivalently stated as "The game is not polo or you ride a horse."

To change a conditional statement to a disjunction, negate the antecedent, change the conditional symbol to a disjunction symbol, and keep the consequent the same. To change a disjunction statement to a conditional statement, negate the first statement, change the disjunction symbol to a conditional symbol, and keep the second statement the same.

EXAMPLE 8 ***Rewriting a Disjunction as a Conditional Statement***

Write a conditional statement that is logically equivalent to "The Oregon Ducks will win or the Oregon State Beavers will lose." Assume that the negation of winning is losing.

SOLUTION: Let

p: The Oregon Ducks will win.
q: The Oregon State Beavers will win.

The original statement may be written symbolically as $p \vee \sim q$. To write an equivalent statement, negate the first statement, p, change the disjunction symbol to a conditional symbol, and keep the second statement the same. Symbolically, the equivalent statement is $\sim p \rightarrow \sim q$. The equivalent statement in words is "If the Oregon Ducks lose, then the Oregon State Beavers will lose." ▲

Negation of the Conditional Statement

Now we will discuss how to negate a conditional statement. To negate a statement we use the fact that $p \rightarrow q \Leftrightarrow \sim p \vee q$ and De Morgan's laws. Examples 9 and 10 show the process.

EXAMPLE 9 ***The Negation of a Conditional Statement***

Determine a statement equivalent to $\sim(p \rightarrow q)$.

SOLUTION: Begin with $p \rightarrow q \Leftrightarrow \sim p \vee q$, negate both statements, and use De Morgan's laws.

$$\begin{aligned} p \rightarrow q &\Leftrightarrow \sim p \vee q \\ \sim(p \rightarrow q) &\Leftrightarrow \sim(\sim p \vee q) && \text{Negate both statements} \\ &\Leftrightarrow p \wedge \sim q && \text{De Morgan's laws} \end{aligned}$$

Therefore, $\sim(p \rightarrow q)$ is equivalent to $p \wedge \sim q$. ▲

EXAMPLE 10 ***Write an Equivalent Statement***

Write a statement equivalent to

"It is false that if the dog is snoring then the dog cannot sleep in our bedroom."

SOLUTION: Let

p = the dog is snoring
q = the dog can sleep in our room

Then the given statement can be represented symbolically as $\sim(p \rightarrow \sim q)$. Using the procedure illustrated in Example 9 we can determine that $\sim(p \rightarrow \sim q)$ is equivalent to $p \wedge q$. Verify this yourself now. Therefore, an equivalent statement is "The dog is snoring and the dog can sleep in our bedroom." ▲

DID YOU KNOW

In the Service of Science

Logical reasoning is often used by scientists as they develop theories. Logical reasoning often also plays a part in developing procedures to test theories. Sometimes, their theories are supported by experimental evidence and are accepted. Other times, their theories are contradicted by the evidence and are rejected. For example, Aristotle observed that bugs appeared in spoiling meat and reasoned that life arose spontaneously from nonliving matter. With advances in technology, scientists have had more means at their disposal to test their theories. To refute Aristotle's claim of spontaneous generation of life, Louis Pasteur in 1862 conducted an experiment by isolating some meat broth in a sterile flask to demonstrate that the bugs Aristotle observed grew from microscopic life forms too small to be seen.

Variations of the Conditional Statement

We know that $p \rightarrow q$ is equivalent to $\sim p \vee q$. Are any other statements equivalent to $p \rightarrow q$? Yes, there are many. Now let's look at the variations of the conditional statement to determine whether any are equivalent to the conditional statement. *The variations of the conditional statement are made by switching and/or negating the antecedent and the consequent of a conditional statement.* The variations of the conditional statement are the *converse* of the conditional, the *inverse* of the conditional, and the *contrapositive* of the conditional.

Listed here are the variations of the conditional with their symbolic form and the words we say to read each one.

Variations of the Conditional Statement

Name	Symbolic form	Read
Conditional	$p \rightarrow q$	"If p, then q"
Converse of the conditional	$q \rightarrow p$	"If q, then p"
Inverse of the conditional	$\sim p \rightarrow \sim q$	"If not p, then not q"
Contrapositive of the conditional	$\sim q \rightarrow \sim p$	"If not q, then not p"

To write the converse of the conditional statement, switch the order of the antecedent and the consequent. To write the inverse, negate both the antecedent and the consequent. To write the contrapositive, switch the order of the antecedent and the consequent and then negate both of them.

Are any of the variations of the conditional statement equivalent? To determine the answer, we can construct a truth table for each variation, as shown in Table 3.26. It reveals that the conditional statement is equivalent to the contrapositive statement and that the converse statement is equivalent to the inverse statement.

TABLE 3.26

p	q	Conditional $p \rightarrow q$	Contrapositive $\sim q \rightarrow \sim p$	Converse $q \rightarrow p$	Inverse $\sim p \rightarrow \sim q$
T	T	T	T	T	T
T	F	F	F	T	T
F	T	T	T	F	F
F	F	T	T	T	T

EXAMPLE 11 ***The Converse, Inverse and Contrapositive***

For the conditional statement "If the song contains sitar music, then the song was written by George Harrison," write the

a) converse. b) inverse. c) contrapositive.

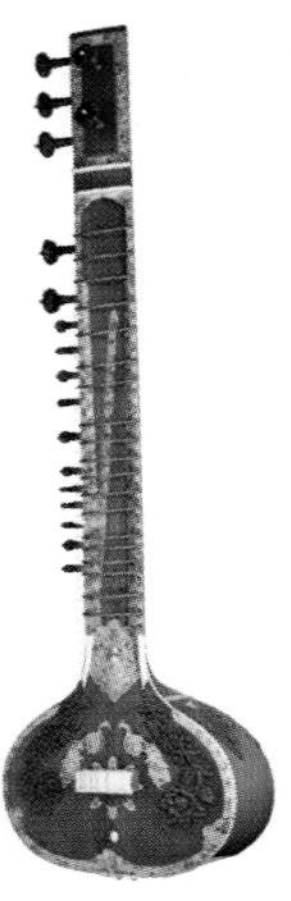

A sitar

SOLUTION:

a) Let

p: The song contains sitar music.
q: The song was written by George Harrison.

The conditional statement is of the form $p \rightarrow q$, so the converse must be of the form $q \rightarrow p$. Therefore, the converse is "If the song was written by George Harrison, then the song contains sitar music."

b) The inverse is of the form $\sim p \rightarrow \sim q$. Therefore, the inverse is "If the song does not contain sitar music, then the song was not written by George Harrison."

c) The contrapositive is of the form $\sim q \rightarrow \sim p$. Therefore, the contrapositive is "If the song was not written by George Harrison, then the song does not contain sitar music." ▲

EXAMPLE 12 *Determine the Truth Values*

Let

p: The number is divisible by 9.
q: The number is divisible by 3.

Write the following statements and determine which are true.

a) The conditional statement, $p \rightarrow q$
b) The converse of $p \rightarrow q$
c) The inverse of $p \rightarrow q$
d) The contrapositive of $p \rightarrow q$

SOLUTION:

a) *Conditional statement:* $(p \rightarrow q)$
If the number is divisible by 9, then the number is divisible by 3. This statement is true. A number divisible by 9 must also be divisible by 3, since 3 is a divisor of 9.

b) *Converse of the conditional:* $(q \rightarrow p)$
If the number is divisible by 3, then the number is divisible by 9. This statement is false. For instance, 6 is divisible by 3, but 6 is not divisible by 9.

c) *Inverse of the conditional:* $(\sim p \rightarrow \sim q)$
If the number is not divisible by 9, then the number is not divisible by 3. This statement is false. For instance, 6 is not divisible by 9, but 6 is divisible by 3.

d) *Contrapositive of the conditional:* $(\sim q \rightarrow \sim p)$
If the number is not divisible by 3, then the number is not divisible by 9. The statement is true, since any number that is divisible by 9 must be divisible by 3. ▲

EXAMPLE 13 *Use the Contrapositive*

Use the contrapositive to write a statement logically equivalent to "If the boat is 24 ft long, then it will not fit into the boathouse."

SOLUTION: Let

p: The boat is 24 ft long.
q: The boat will fit into the boathouse.

The given statement written symbolically is

$$p \rightarrow \sim q$$

The contrapositive of the statement is

$$q \rightarrow \sim p$$

Therefore, an equivalent statement is "If the boat will fit into the boathouse, then the boat is not 24 ft long." ▲

The contrapositive of the conditional is very important in mathematics. Consider the statement "If a^2 is not a whole number, then a is not a whole number." Is this statement true? You may find this question difficult to answer. Writing the statement's contrapositive may enable you to answer the question. The contrapositive is "If a is a whole number, then a^2 is a whole number." Since the contrapositive is a true statement, the original statement must also be true.

EXAMPLE 14 *Which Are Equivalent?*

Determine which, if any, of the following statements are equivalent. You may use De Morgan's laws, the fact that $p \rightarrow q \Leftrightarrow \sim p \vee q$, information from the variations of the conditional, or truth tables.

a) If you leave by 9 A.M., then you will get to your destination on time.
b) You do not leave by 9 A.M. or you will get to your destination on time.
c) It is false that you get to your destination on time or you did not leave by 9 A.M.
d) If you do not get to your destination on time, then you did not leave by 9 A.M.

SOLUTION: Let

p: You leave by 9 A.M.
q: You will get to your destination on time.

In symbolic form, the four statements are

a) $p \rightarrow q$. b) $\sim p \vee q$. c) $\sim(q \vee \sim p)$. d) $\sim q \rightarrow \sim p$.

Which of these statements are equivalent? Earlier in this section, you learned that $p \rightarrow q$ is equivalent to $\sim p \vee q$. Therefore, statements (a) and (b) are equivalent. Statement (d) is the contrapositive of statement (a). Therefore, statement (d) is also equivalent to statement (a) and statement (b). All these statements have the same truth table (Table 3.27 on page 138).

Now let's look at statement (c). If we use De Morgan's laws on statement (c), we get

$$\sim(q \vee \sim p) \Leftrightarrow \sim q \wedge p$$

TABLE 3.27

p	q	(a) $p \to q$	(b) $\sim p \vee q$	(d) $\sim q \to \sim p$
T	T	T	T	T
T	F	F	F	F
F	T	T	T	T
F	F	T	T	T

If $\sim q \wedge p$ was one of the other statements, then $\sim(q \vee \sim p)$ would be equivalent to that statement. Because $\sim q \wedge p$ does not match any of the other choices, it does not necessarily mean that $\sim(q \vee \sim p)$ is not equivalent to the other statements. To determine whether $\sim(q \vee \sim p)$ is equivalent to the other statements, we will construct its truth table (Table 3.28) and compare the answer column with the answer columns in Table 3.27.

TABLE 3.28 (c)

p	q	~	(q	∨	~p)
T	T	F	T	T	F
T	F	T	F	F	F
F	T	F	T	T	T
F	F	F	F	T	T
		4	1	3	2

None of the three answer columns of the truth table in Table 3.27 are the same as the answer column of the truth table in Table 3.28. Therefore $\sim(q \vee \sim p)$ is not equivalent to any of the other statements. Therefore, statements (a), (b), and (d) are equivalent to each other.

DID YOU KNOW

Fuzzy Logic

Modern computers, like truth tables, work with only two values, 1 or 0 (equivalent to true or false in truth tables). This constraint prevents a computer from being able to reason as the human brain can and prevents a computer from being able to evaluate items involving vagueness or value judgments that so often occur in real-world situations. For example, a binary computer will have difficulty evaluating the subjective statement "the air is warm."

Fuzzy logic uses the concept: Everything is a matter of degree. Fuzzy logic manipulates vague concepts such as *bright* and *fast* by assigning values between 0 and 1 to each item. For example, suppose *bright* is assigned a value of 0.80; then *not bright* is assigned a value of $1 - 0.8 = 0.20$. As the value assigned to *bright* changes, so does the value assigned to *not bright*. Not p is always $1 - p$, where $0 < p < 1$. Fuzzy logic is used to operate cameras, air conditioners, subways, and many other devices where the change in one condition changes another condition. For example, when it is bright outside, the camera's lens aperture opens less, and when it is overcast, the camera's lens aperture opens more. How many other devices can you name that may use fuzzy logic? See Problem-Solving Exercises 87 and 88.

SECTION 3.4 EXERCISES

Concept/Writing Exercises

1. What are equivalent statements?
2. Explain how you can determine whether two statements are equivalent.
3. Suppose two statements are connected with the biconditional and the truth table is constructed. If the answer column of the truth table has all trues, what must be true about these two statements? Explain.
4. Write De Morgan's laws for logic.
5. For a statement of the form $p \rightarrow q$, symbolically indicate the form of the
   a) converse.
   b) inverse.
   c) contrapositive.
6. Which of the following are equivalent statements?
   a) The converse
   b) The contrapositive
   c) The inverse
   d) The conditional
7. Write a disjunctive statement that is logically equivalent to $p \rightarrow q$.
8. Write a conjunction involving two conditional statements that is logically equivalent to $p \leftrightarrow q$.

Practice the Skills

In Exercises 9–18, use De Morgan's laws to determine whether the two statements are equivalent.

9. $\sim p \vee \sim q, \sim(p \wedge q)$
10. $\sim(p \vee q), \sim p \wedge \sim q$
11. $\sim(p \wedge q), \sim p \wedge \sim q$
12. $\sim(p \wedge q), \sim p \wedge q$
13. $\sim(p \vee \sim q), \sim p \wedge q$
14. $\sim(p \wedge q), \sim(q \vee \sim p)$
15. $(\sim p \vee \sim q) \rightarrow r, \sim(p \wedge q) \rightarrow r$
16. $q \rightarrow \sim(p \wedge \sim r), q \rightarrow \sim p \vee r$
17. $\sim(p \rightarrow \sim q), p \wedge q$
18. $\sim(\sim p \rightarrow q), \sim p \wedge \sim q$

In Exercises 19–30, use a truth table to determine whether the two statements are equivalent.

19. $p \rightarrow q, \sim p \vee q$
20. $\sim p \rightarrow q, p \wedge q$
21. $(p \wedge q) \wedge r, p \wedge (q \wedge r)$
22. $p \rightarrow q, \sim q \rightarrow \sim p$
23. $(p \vee q) \vee r, p \vee (q \vee r)$
24. $p \vee (q \wedge r), \sim p \rightarrow (q \wedge r)$
25. $p \wedge (q \vee r), (p \wedge q) \vee r$
26. $\sim(q \rightarrow p) \vee r, (p \vee q) \wedge \sim r$
27. $(p \rightarrow q) \wedge (q \rightarrow r), (p \rightarrow q) \rightarrow r$
28. $\sim q \rightarrow (p \wedge r), \sim(p \vee r) \rightarrow q$
29. $(p \rightarrow q) \wedge (q \rightarrow p), (p \leftrightarrow q)$
30. $[\sim(p \rightarrow q)] \wedge [\sim(q \rightarrow p)], \sim(p \leftrightarrow q)$

Problem Solving

In Exercises 31–38, use De Morgan's laws to write an equivalent statement for the sentence.

31. It is false that the Mississippi River runs through Ohio or the Ohio River runs through Mississippi.
32. It is false that the printer is out of toner and the fax machine is out of paper.
33. The snowmobile was neither an Arctic Cat nor was it a Ski-Do.

34. The pot roast is hot, but it is not well done.
35. The hotel does not have a weight room or the conference center does not have an auditorium.
36. Robert Farinelli is an authorized WedgCor dealer or he is not going to work for Prism Construction Company.
37. If we go to Cozumel, then we will go snorkeling or we will not go to Senior Frogs.

38. If Phil Murphy buys us dinner, then we will not go to the top of the CN Tower but we will be able to walk to the Red Bistro Restaurant.

In Exercises 39–44, use the fact that $p \rightarrow q$ is equivalent to $\sim p \vee q$ to write an equivalent form of the given statement.

39. If you drink a glass of orange juice, then you will get a full day supply of folic acid.
40. Nick-at-Nite is showing *Family Ties* or they are showing *The Facts of Life*.
41. Bob the Tomato visited the nursing home or he did not visit the Cub Scout meeting.
42. If John Peden will buy a Harley-Davidson, then he will not buy a Honda.

43. It is false that if the plumbers meet in Kansas City then the Rainmakers will provide the entertainment.
44. Mary Beth Headlee organized the conference or John Waters does not work at Sinclair Community College.

In Exercises 45–48, use the fact that $(p \rightarrow q) \wedge (q \rightarrow p)$ is equivalent to $p \leftrightarrow q$ to write the statement in an equivalent form.

45. If it is cloudy then the front is coming through, and if the front is coming through then it is cloudy.
46. If Model Road is closed then we use Kirkwood Road, and if we use Kirkwood Road then Model Road is closed.
47. The chemistry teacher teaches mathematics if and only if there is a shortage of mathematics teachers.
48. John Deere will hire new workers if and only if the city of Dubuque will pay to retrain the workers.

In Exercises 49–56, write the converse, inverse, and contrapositive of the statement. (For Exercises 55 and 56, use De Morgan's laws.)

49. If the book is interesting, then I will finish the book in 1 week.
50. If the dryer is making a loud noise, then you need to replace the blower fan.
51. If you finish your homework, then you can watch television.
52. If Bob Dylan records a new CD, then he will go on tour.
53. If that annoying paper clip shows up on my computer screen, then I will scream.
54. If the remote control is not within my reach, then I will watch the same channel all night.
55. If the sun is shining, then we will go down to the marina and we will take out the sailboat.

56. If the apple pie is baked, then we will eat a piece of pie and we will save some pie for later.

In Exercises 57–64, write the contrapositive of the statement. Use the contrapositive to determine whether the conditional statement is true or false.

57. If a natural number is not divisible by 5, then the natural number is not divisible by 10.
58. If the opposite sides of the quadrilateral are not parallel, then the quadrilateral is not a parallelogram.
59. If a natural number is divisible by 3, then the natural number is divisible by 6.
60. If $1/n$ is not a natural number, then n is not a natural number.
61. If two lines do not intersect in at least one point, then the two lines are parallel.
62. If $\frac{m \cdot a}{m \cdot b} \neq \frac{a}{b}$, then m is not a counting number.
63. If the sum of the interior angles of a polygon do not measure 360°, then the polygon is not a quadrilateral.
64. If a and b are not both even counting numbers, then the product of a and b is not an even counting number.

In Exercises 65–80, determine which, if any, of the three statements are equivalent (see Example 14).

65. a) Maria has not retired or Maria is still working.
b) If Maria is still working, then Maria has not retired.
c) If Maria has retired, then Maria is not still working.

66. a) If today is Monday, then tomorrow is not Wednesday.
b) It is false that today is Monday and tomorrow is not Wednesday.
c) Today is not Monday or tomorrow is Wednesday.

67. a) The car is not reliable and the car is noisy.
b) If the car is not reliable, then the car is not noisy.
c) It is false that the car is reliable or the car is not noisy.

68. a) The house is not made of wood or the shed is not made of wood.
b) If the house is made of wood, then the shed is not made of wood.
c) It is false that the shed is made of wood and the house is not made of wood.

69. a) Today is not Sunday or the library is open.
b) If today is Sunday, then the library is not open.
c) If the library is open, then today is not Sunday.

70. a) If you are fishing at 1 P.M., then you are driving a car at 1 P.M.
b) You are not fishing at 1 P.M. or you are driving a car at 1 P.M.
c) It is false that you are fishing at 1 P.M. and you are not driving a car at 1 P.M.

71. a) The grass grows and the trees are blooming.
b) If the trees are blooming, then the grass does not grow.
c) The trees are not blooming or the grass does not grow.

72. a) Johnny Patrick is chosen as department chair if and only if he is the only candidate.
b) If Johnny Patrick is chosen as department chair then he is the only candidate, and if Johnny Patrick is the only candidate then he is chosen as department chair.
c) Johnny Patrick is not chosen as department chair and he is not the only candidate.

73. a) It is false that if you do not drink milk then your cholesterol count will be lower.
b) Your cholesterol count will be lower if and only if you drink milk.
c) It is false that if you drink milk then your cholesterol count will not be lower.

74. a) Bruce Springsteen will not go on tour if and only if Clarence Clemmons does not play the saxophone in his band.
b) It is false that Bruce Springsteen will go on tour if and only if Clarence Clemmons does not play the saxophone in his band.
c) If Bruce Springsteen goes on tour, then Clarence Clemmons plays saxophone in his band.

Clarence Clemmons (left) and Bruce Springsteen

75. a) If the pay is good and today is Monday, then I will take the job.
b) If I do not take the job, then it is false that the pay is good or today is Monday.
c) The pay is good and today is Monday, or I will take the job.

76. a) If you are 18 years old and a citizen of the United States, then you can vote in the presidential election.
b) You can vote in the presidential election, if and only if you are a citizen of the United States and you are 18 years old.
c) You cannot vote in the presidential election, or you are 18 years old and you are not a citizen of the United States.

77. a) The package was sent by Federal Express, or the package was not sent by United Parcel Service but the package arrived on time.
b) The package arrived on time, if and only if it was sent by Federal Express or it was not sent by United Parcel Service.
c) If the package was not sent by Federal Express, then the package was not sent by United Parcel Service but the package arrived on time.

78. a) If we put the dog outside or we feed the dog, then the dog will not bark.
b) If the dog barks, then we did not put the dog outside and we did not feed the dog.
c) If the dog barks, then it is false that we put the dog outside or we feed the dog.

79. a) The car needs oil, and the car needs gas or the car is new.
b) The car needs oil, and it is false that the car does not need gas and the car is not new.
c) If the car needs oil, then the car needs gas or the car is not new.

80. a) The mortgage rate went down, if and only if Tim purchased the house and the down payment was 10%.
b) The down payment was 10%, and if Tim purchased the house then the mortgage rate went down.
c) If Tim purchased the house, then the mortgage rate went down and the down payment was not 10%.

81. If p and q represent two simple statements, and if $p \rightarrow q$ is a false statement, what must be the truth value of the converse, $q \rightarrow p$? Explain.

82. If p and q represent two simple statements, and if $p \rightarrow q$ is a false statement, what must be the truth value of the inverse, $\sim p \rightarrow \sim q$? Explain.

83. If p and q represent two simple statements, and if $p \rightarrow q$ is a false statement, what must be the truth value of the contrapositive, $\sim q \rightarrow \sim p$? Explain.

84. If p and q represent two simple statements, and if $p \rightarrow q$ is a true statement, what must be the truth value of the contrapositive, $\sim q \rightarrow \sim p$? Explain.

Challenge Problems/Group Activities

85. We learned that $p \rightarrow q \Leftrightarrow \sim p \vee q$. Determine a conjunctive statement that is equivalent to $p \rightarrow q$. (*Hint:* There are many answers.)

86. Determine whether $\sim[\sim(p \vee \sim q)] \Leftrightarrow p \vee \sim q$. Explain the method(s) you used to determine your answer.

87. In an appliance or device that uses fuzzy logic, a change in one condition causes a change in a second condition. For example, in a camera, if the brightness increases, the lens aperture automatically decreases to get the proper exposure on the film. Name at least 10 appliances or devices that make use of fuzzy logic and explain how fuzzy logic is used in each appliance or device. See the Did You Know on page 138.

88. In symbolic logic, a statement is either true or false (consider true to have a value of 1 and false a value of 0). In fuzzy logic, nothing is true or false, but everything is a matter of degree. For example, consider the statement "The sun is shining." In fuzzy logic, this statement may have a value between 0 and 1 and may be constantly changing. For example, if the sun is partially blocked by clouds, the value of this statement may be 0.25. In fuzzy logic, the values of connective statements are found as follows for statements p and q.

Not p has a truth value of $1 - p$.
$p \wedge q$ has a truth value equal to the lesser of p and q.
$p \vee q$ has a truth value equal to the greater of p and q.
$p \rightarrow q$ has a truth value equal to the lesser of 1 and $1 - p + q$.
$p \leftrightarrow q$ has a truth value equal to $1 - |p - q|$, that is, 1 minus the absolute value* of p minus q.

Suppose the statement "p: The sun is shining" has a truth value of 0.25 and the statement "q: Mary is getting a tan" has a truth value of 0.20. Find the truth value of

a) $\sim p$. b) $\sim q$.
c) $p \wedge q$. d) $p \vee q$.
e) $p \rightarrow q$. f) $p \leftrightarrow q$.

Recreational Mathematics

89. Unscramble the following letters to form the names of five important terms in the study of logic.
a) ACINNLIDOOT
b) DATOONCLIBINI
c) RIENSEV
d) ROCESVEN
e) ARTSOCINVOTEPI

Internet/Research Activities

90. Do research and write a report on fuzzy logic.

91. Read one of Lewis Carroll's books and write a report on how he used logic in the book. Give at least five specific examples.

92. Do research and write a report on the life and achievements of Augustus De Morgan. Indicate in your report his contributions to sets and logic.

3.5 SYMBOLIC ARGUMENTS

In the preceding sections of this chapter, we used symbolic logic to determine the truth value of a compound statement. We now extend those basic ideas to determine whether symbolic arguments are valid or invalid.

*Absolute values are discussed in Section 13.8.

DID YOU KNOW

The Dog Did Nothing

In the case of the disappearance of the racehorse Silver Blaze, Sherlock Holmes demonstrated that sometimes the absence of a clue is itself a clue. The local police inspector asked him, "Is there any point to which you would wish to draw my attention?" Holmes replied, "To the curious incident of the dog in the nighttime." The inspector, confused, asked: "The dog did nothing in the nighttime." "That was the curious incident," remarked Sherlock Holmes. From the lack of the dog's bark, Holmes concluded that the horse had been "stolen" by a stablehand. How did Holmes reach his conclusion?

Consider the statements:

If Jason is a singer, then he is well known.
Jason is a singer.

If you accept these two statements as true, then a conclusion that necessarily follows is that

Jason is well known.

These three statements in the following form constitute a symbolic argument.

Premise 1:	If Jason is a singer, then he is well known.
Premise 2:	Jason is a singer.
Conclusion:	Therefore, Jason is well known.

A *symbolic argument* consists of a set of *premises* and a *conclusion*. It is called a symbolic argument because we generally write it in symbolic form to determine its validity.

> An **argument is valid** when its conclusion necessarily follows from a given set of premises.
>
> An **argument is invalid** or a **fallacy** when the conclusion does not necessarily follow from the given set of premises.

An argument that is not valid is invalid. The argument just presented is an example of a valid argument, as the conclusion necessarily follows from the premises. Now we will discuss a procedure to determine whether an argument is valid or invalid. We begin by writing the argument in symbolic form. To write the argument in symbolic form, we let p and q be

p: Jason is a singer.
q: Jason is well known.

Symbolically, the argument is written

Premise 1:	$p \to q$
Premise 2:	p
Conclusion:	$\therefore q$ (The three-dot triangle is read "therefore.")

Write the argument in the following form.

If	[*premise 1*	**and**	*premise 2*]	**then**	*conclusion*
	$[(p \to q)$	$\wedge$	$p\ ]$	$\to$	q

Then construct a truth table for the statement $[(p \to q) \wedge p] \to q$ (Table 3.29 on page 144). *If the truth table answer column is true in every case, then the statement is a tautology, and the argument is valid. If the truth table is not a tautology, then the argument is invalid.* Since the statement is a tautology (see column 5), the conclusion necessarily follows from the premises and the argument is valid.

TABLE 3.29

p	q	[(p → q)	∧	p]	→	q
T	T	T	T	T	T	T
T	F	F	F	T	T	F
F	T	T	F	F	T	T
F	F	T	F	F	T	F
		1	3	2	5	4

Once we have demonstrated that an argument in a particular form is valid, all arguments with exactly the same form will also be valid. In fact, many of these forms have been assigned names. The argument form just discussed,

$$\begin{array}{l} p \rightarrow q \\ \underline{p \quad\quad} \\ \therefore q \end{array}$$

is called the *law of detachment*, or *modus ponens*.

EXAMPLE 1 *Determining Validity without a Truth Table*

Determine whether the following argument is valid or invalid.

If the water is warm, then the moon is made of cheese.
The water is warm.
∴ The moon is made of cheese.

SOLUTION: Translate the argument into symbolic form.

Let

w: The water is warm.
m: The moon is made of cheese.

In symbolic form the argument is

$$\begin{array}{l} w \rightarrow m \\ \underline{w \quad\quad} \\ \therefore m \end{array}$$

This argument is also the law of detachment, and therefore it is a valid argument. ▲

Note that the argument in Example 1 is valid even though the conclusion, "The moon is made of cheese," is a false statement. It is also possible to have an invalid argument in which the conclusion is a true statement. When an argument is valid, the conclusion necessarily follows from the premises. It is not necessary for the premises or the conclusion to be true statements in an argument.

Procedure to Determine Whether an Argument Is Valid

1. Write the argument in symbolic form.
2. Compare the form of the argument with forms that are known to be valid or invalid. If there are no known forms to compare it with, or you do not remember the forms, go to step 3.
3. If the argument contains two premises, write a conditional statement of the form

   $$[(\text{premise } 1) \wedge (\text{premise } 2)] \rightarrow \text{conclusion}$$

4. Construct a truth table for the statement in step 3.
5. If the answer column of the truth table has all trues, the statement is a tautology, and the argument is valid. If the answer column does not have all trues, the argument is invalid.

Examples 1 through 4 contain two premises. When an argument contains more that two premises, step 3 of the procedure will change slightly, as will be explained shortly.

EXAMPLE 2 *Determining Validity with a Truth Table*

Determine whether the following argument is valid or invalid.

If you score 90% on the final exam, then you will get an A in the course.
You will not get an A in the course.
∴ You do not score 90% on the final exam.

SOLUTION: We first write the argument in symbolic form.

Let

p: You score 90% on the final exam.
q: You will get an A in the course.

In symbolic form, the argument is

$$\begin{array}{l} p \rightarrow q \\ \underline{\sim q} \\ \therefore \sim p \end{array}$$

As we have not tested an argument in this form, we will construct a truth table to determine whether the argument is valid or invalid. We write the argument in the form $[(p \rightarrow q) \wedge \sim q] \rightarrow \sim p$, and construct a truth table (Table 3.30 on page 146). Since the answer, column 5, has all T's, the argument is valid.

DID YOU KNOW

Guilty as Charged?

In U.S. law, a jury must reach unanimous agreement about a verdict. Generally, less than 5% of the 120,000 jury trials conducted each year in the United States result in a "hung jury" in which a jury cannot reach a unanimous agreement. Juries use logical reasoning to determine if there are contradictions in the evidence or testimony when deciding their verdict. Most nations of the world rely solely on a judge's opinion to decide guilt or innocence.

TABLE 3.30

p	q	[(p → q)	∧	~q]	→	~p
T	T	T	F	F	T	F
T	F	F	F	T	T	F
F	T	T	F	F	T	T
F	F	T	T	T	T	T
		1	3	2	5	4

The argument form in Example 2 is an example of the *law of contraposition*, or *modus tollens*.

EXAMPLE 3 *Another Symbolic Argument*

Determine whether the following argument is valid or invalid.

The grass is green or the grass is full of weeds.
The grass is not green.
∴ The grass is full of weeds.

SOLUTION: Let

p: The grass is green.
q: The grass is full of weeds.

In symbolic form, the argument is

$$\begin{array}{l} p \vee q \\ \underline{\sim p} \\ \therefore\ q \end{array}$$

As this form is not one of those we are familiar with, we will construct a truth table. We write the argument in the form $[(p \vee q) \wedge \sim p] \rightarrow q$. Next we construct a truth table, as shown in Table 3.31. The answer to the truth table, column 5, is true in *every case*. Therefore, the statement is a tautology, and the argument is valid.

TABLE 3.31

p	q	[(p ∨ q)	∧	~p]	→	q
T	T	T	F	F	T	T
T	F	T	F	F	T	F
F	T	T	T	T	T	T
F	F	F	F	T	T	F
		1	3	2	5	4

The argument form in Example 3 is an example of *disjunctive syllogism*. Other standard forms of arguments are given in the following chart.

Standard Forms of Arguments

Valid Arguments	*Law of Detachment*	*Law of Contraposition*	*Law of Syllogism*	*Disjunctive Syllogism*
	$p \rightarrow q$	$p \rightarrow q$	$p \rightarrow q$	$p \vee q$
	$\underline{p}$	$\underline{\sim q}$	$\underline{q \rightarrow r}$	$\underline{\sim p}$
	$\therefore q$	$\therefore \sim p$	$\therefore p \rightarrow r$	$\therefore q$
Invalid Arguments	*Fallacy of the Converse*	*Fallacy of the Inverse*		
	$p \rightarrow q$	$p \rightarrow q$		
	$\underline{q}$	$\underline{\sim p}$		
	$\therefore p$	$\therefore \sim q$		

As we saw in Example 1, it is not always necessary to construct a truth table to determine whether or not an argument is valid. The next two examples will show how we can identify an argument as one of the standard arguments given in the chart above.

EXAMPLE 4 *Identifying the Law of Syllogism in an Argument*

Determine whether the following argument is valid or invalid.

> If my laptop battery is dead, then I use my home computer.
> If I use my home computer, then my kids will play outside.
> ∴ If my laptop battery is dead, then my kids will play outside.

SOLUTION: Let

p: My laptop battery is dead.
q: I use my home computer.
r: My kids will play outside.

In symbolic form, the argument is

$$\begin{array}{l} p \rightarrow q \\ \underline{q \rightarrow r} \\ \therefore p \rightarrow r \end{array}$$

The argument is in the form of the law of syllogism. Therefore, the argument is valid, and there is no need to construct a truth table. ▲

MATHEMATICS *Everywhere*

Freedom of Speech or Misleading Advertising?

The first amendment to the United States Constitution guarantees that Americans have the freedom of speech. There are, however, limits to what we may say or write. For example, companies are limited in what they can say or write to advertise their products. The Federal Trade Commission (FTC) states that "ads must be truthful and not misleading; that advertisers must have evidence to back up their assertions; and that ads cannot be unfair." Although most advertisements are truthful and fair, some enter into a "gray area" of truthfulness. Some of these breeches of fairness may be found in logical fallacies either made directly or implicitly by the context of the ad's wording or artwork. One notable FTC action that began in June 2001 was Operation Cure All. This action targeted companies marketing a variety of devices, herbal products, and other dietary supplements to treat or cure cancer, HIV/AIDS, arthritis, hepatitis, Alzheimer's disease, diabetes, and many other diseases. Among the many products for which unfounded claims were being made were a DHEA hormonal supplement, St. John's wort, various multiherbal supplements, colloidal silver, and a variety of electrical therapy devices.

EXAMPLE 5 *Identifying Common Fallacies in Arguments*

Determine whether the following arguments are valid or invalid.

a)

If it is snowing, then we put salt on the driveway.
We put salt on the driveway.
∴ It is snowing.

b)

If it is snowing, then we put salt on the driveway.
It is not snowing.
∴ We do not put salt on the driveway.

SOLUTION:

a) Let

p: It is snowing.
q: We put salt on the driveway.

In symbolic form, the argument is

$$\begin{array}{l} p \rightarrow q \\ \underline{q \quad\quad} \\ \therefore\ p \end{array}$$

This argument is in the form of the fallacy of the converse. Therefore, the argument is a fallacy, or invalid.

b) Using the same symbols defined in the solution to part (a), in symbolic form, the argument is

$$\begin{array}{l} p \rightarrow q \\ \underline{\sim p \quad} \\ \therefore\ \sim q \end{array}$$

This argument is in the form of the fallacy of the inverse. Therefore, the argument is a fallacy, or invalid. ▲

TIMELY TIP If you are not sure whether an argument with two premises is one of the standard forms or if you do not remember the standard forms, you can determine whether a given argument is valid or invalid by using a truth table. To do so, follow the boxed procedure on page 145.

In Example 5b) if you did not recognize that this argument was of the same form as the Fallacy of the Inverse you could construct the truth table for the conditional statement

$$[(p \rightarrow q) \wedge \sim p] \rightarrow \sim q$$

The true-false values under the conditional column, →, would be T, T, F, T. Since the statement is not a tautology, the argument is invalid.

Now we consider an argument that has more than two premises. When an argument contains more than two premises, the statement we test, using a truth table, is formed by taking the conjunction of all the premises as the antecedent and the conclusion as the consequent. For example, if an argument is of the form

$$\begin{array}{l} p_1 \\ p_2 \\ \underline{p_3} \\ \therefore c \end{array}$$

We evaluate the truth table for $[p_1 \wedge p_2 \wedge p_3] \rightarrow c$. When we evaluate $[p_1 \wedge p_2 \wedge p_3]$, it makes no difference whether we evaluate $[(p_1 \wedge p_2) \wedge p_3]$, or $[p_1 \wedge (p_2 \wedge p_3)]$ because both give the same answer. In Example 6, we evaluate $[p_1 \wedge p_2 \wedge p_3]$ from left to right, that is, $[(p_1 \wedge p_2) \wedge p_3]$.

EXAMPLE 6 *An Argument with Three Premises*

Use a truth table to determine whether the following argument is valid or invalid.

If Donna has a pet, then Donna owns a snail.
Donna owns a snail or Donna drives a truck.
Donna drives a truck or Donna has a pet.
∴ Donna has a pet.

SOLUTION: This argument contains three simple statements.

Let

p: Donna has a pet.
q: Donna owns a snail.
r: Donna drives a truck.

In symbolic form, the argument is

$$\begin{array}{l} p \rightarrow q \\ q \vee r \\ \underline{r \vee p} \\ \therefore p \end{array}$$

Write the argument in the form

$$[(p \rightarrow q) \wedge (q \vee r) \wedge (r \vee p)] \rightarrow p.$$

Now construct the truth table (Table 3.32 on page 150). The answer, column 7, is not true in every case. Thus, the argument is a fallacy, or invalid.

TABLE 3.32

p	q	r	[(p → q)	∧	(q ∨ r)	∧	(r ∨ p)]	→	p
T	T	T	T	T	T	T	T	T	T
T	T	F	T	T	T	T	T	T	T
T	F	T	F	F	T	F	T	T	T
T	F	F	F	F	F	F	T	T	T
F	T	T	T	T	T	T	T	F	F
F	T	F	T	T	T	F	F	T	F
F	F	T	T	T	T	T	T	F	F
F	F	F	T	F	F	F	F	T	F
			1	3	2	5	4	7	6

Let's now investigate how we can arrive at a valid conclusion from a given set of premises.

EXAMPLE 7 *Determine a Logical Conclusion*

Determine a logical conclusion that follows from the given statements. "If you own a house, then you will pay property tax. You own a house. Therefore, . . . "

SOLUTION: If you recognize a specific form of an argument, you can use your knowledge of that form to draw a logical conclusion.

Let

p: You own a house.
q: You will pay property tax.

The argument is of the following form.

$$\begin{array}{l} p \rightarrow q \\ \underline{p\quad\quad} \\ \therefore\ ? \end{array}$$

If the question mark is replaced with a q, this argument is of the form of the law of detachment. A logical conclusion is "Therefore, you will pay property tax."

SECTION 3.5 EXERCISES

Concept/Writing Exercises

1. What does it mean when an argument is valid?
2. What does it mean when an argument is a fallacy?
3. Is it possible for an argument to be valid if its conclusion is false? Explain your answer.
4. Is it possible for an argument to be invalid if the premises are all true? Explain your answer.
5. Is it possible for an argument to be valid if the premises are all false? Explain your answer.
6. Explain how to determine whether an argument with premises p_1 and p_2 and conclusion c is a valid or invalid argument.

In Exercises 7–10, (a) indicate the form of the valid argument and (b) write an original argument in words for each form.

7. Disjunctive syllogism
8. Law of contraposition
9. Law of syllogism
10. Law of detachment

In Exercises 11 and 12, (a) indicate the form of the fallacy, and (b) write an original argument in words for each form.

11. Fallacy of the inverse
12. Fallacy of the converse

Practice the Skills

In Exercises 13–32, determine whether the argument is valid or invalid. You may compare the argument to a standard form or use a truth table.

13. $$\begin{array}{l} p \to q \\ \underline{p \qquad} \\ \therefore q \end{array}$$

14. $$\begin{array}{l} p \to q \\ \underline{\sim p \qquad} \\ \therefore q \end{array}$$

15. $$\begin{array}{l} p \wedge \sim q \\ \underline{q \qquad} \\ \therefore \sim p \end{array}$$

16. $$\begin{array}{l} \sim p \vee q \\ \underline{q \qquad} \\ \therefore p \end{array}$$

17. $$\begin{array}{l} \sim p \\ \underline{p \vee q} \\ \therefore \sim q \end{array}$$

18. $$\begin{array}{l} p \to q \\ \underline{\sim q \quad} \\ \therefore \sim p \end{array}$$

19. $$\begin{array}{l} q \to p \\ \underline{\sim q \quad} \\ \therefore \sim p \end{array}$$

20. $$\begin{array}{l} p \vee q \\ \underline{\sim q \quad} \\ \therefore p \end{array}$$

21. $$\begin{array}{l} \sim p \to q \\ \underline{\sim q \qquad} \\ \therefore \sim p \end{array}$$

22. $$\begin{array}{l} q \wedge \sim p \\ \underline{\sim p \qquad} \\ \therefore q \end{array}$$

23. $$\begin{array}{l} p \to q \\ \underline{q \to r \qquad} \\ \therefore p \to r \end{array}$$

24. $$\begin{array}{l} q \wedge p \\ \underline{q \qquad} \\ \therefore \sim p \end{array}$$

25. $$\begin{array}{l} p \leftrightarrow q \\ \underline{q \wedge r \qquad} \\ \therefore p \vee r \end{array}$$

26. $$\begin{array}{l} p \leftrightarrow q \\ \underline{q \to r \qquad\qquad} \\ \therefore \sim r \to \sim p \end{array}$$

27. $$\begin{array}{l} r \leftrightarrow p \\ \underline{\sim p \wedge q \quad} \\ \therefore p \wedge r \end{array}$$

28. $$\begin{array}{l} p \vee q \\ \underline{r \wedge p} \\ \therefore q \end{array}$$

29. $$\begin{array}{l} p \to q \\ q \vee r \\ \underline{r \vee p \quad} \\ \therefore p \end{array}$$

30. $$\begin{array}{l} p \to q \\ q \to r \\ \underline{r \to p \quad} \\ \therefore q \to p \end{array}$$

31. $$\begin{array}{l} p \to q \\ r \to \sim p \\ \underline{p \vee r \qquad} \\ \therefore q \vee \sim p \end{array}$$

32. $$\begin{array}{l} p \leftrightarrow q \\ p \vee r \\ \underline{q \to r \quad} \\ \therefore q \vee r \end{array}$$

Problem Solving

In Exercises 33–50, (a) translate the argument into symbolic form and (b) determine if the argument is valid or invalid. You may compare the argument to a standard form or use a truth table.

33. If Will Smith wins an Academy Award, then he will retire from acting.
Will Smith did not win an Academy Award.
∴ Will Smith will not retire from acting.

Will Smith (see Exercise 33)

34. If the president of the art club resigned, then the vice president becomes president.
The vice president becomes president of the art club.
∴ The president of the art club resigned.

35. If the baby is a boy, then we will name him Alexander Martin.
The baby is a boy.
∴ We will name him Alexander Martin.

36. If I can get my child to preschool by 8:45 A.M., then I can take the 9:00 A.M. class.
If I can take the 9:00 A.M. class, then I can be done by 2:00 P.M.
∴ If I can get my child to preschool by 8:45 A.M., then I can be done by 2:00 P.M.

37. If monkeys can fly, then scarecrows can dance.
Scarecrows cannot dance.
∴ Monkeys cannot fly.

38. Rob Calcatera will go on sabbatical or Frank Cheek will teach logic.
Frank Cheek will not teach logic.
∴ Rob Calcatera will go on sabbatical.

39. If the orange was left on the tree for 1 year, then the orange is ripe.
The orange is ripe.
∴ The orange was left on the tree for 1 year.

40. If you pass general chemistry then you can take organic chemistry.
You pass general chemistry.
∴ You can take organic chemistry.

41. The X-games will be held in San Diego or they will be held in Corpus Christi.
The X-games will not be held in San Diego.
∴ The X-games will be held in Corpus Christi.

42. If Nicholas Thompson teaches this course, then I will get a passing grade.
I did not get a passing grade.
∴ Nicholas Thompson did not teach the course.

43. If it is cold, then graduation will be held indoors.
If graduation is held indoors, then the fireworks will be postponed.
∴ If it is cold, then the fireworks will be postponed.

44. If Miles Davis played with Louis Armstrong, then Charlie Parker played with Dizzy Gillespie.
Miles Davis did not play with Louis Armstrong.
∴ Charlie Parker did not play with Dizzy Gillespie.

45. If the canteen is full, then we can go for a walk.
We can go for a walk and we will not get thirsty.
∴ If we go for a walk, then the canteen is not full.

46. Bryce Canyon National Park is in Utah or Bryce Canyon National Park is in Arizona.
If Bryce Canyon National Park is in Arizona, then it is not in Utah.
∴ Bryce Canyon National Park is not in Arizona.

47. It is snowing and I am going skiing.
If I am going skiing, then I will wear a coat.
∴ If it is snowing, then I will wear a coat.

48. The garden has vegetables or the garden has flowers.
If the garden does not have flowers, then the garden has vegetables.
∴ The garden has flowers or the garden has vegetables.

49. If the house has electric heat, then the Flynns will buy the house.
If the price is not less than $100,000, then the Flynns will not buy the house.
∴ If the house has electric heat, then the price is less than $100,000.

50. If there is an atmosphere, then there is gravity.
If an object has weight, then there is gravity.
∴ If there is an atmosphere, then an object has weight.

In Exercises 51–60, translate the argument into symbolic form. Then determine whether the argument is valid or invalid.

51. If the prescription was called in to Walgreen's, then you can pick it up by 4:00 P.M. You cannot pick it up by 4:00 P.M. Therefore, the prescription was not called in to Walgreen's.

52. The printer has a clogged nozzle or the printer does not have toner. The printer has toner. Therefore, the printer has a clogged nozzle.

53. The television is on or the plug is not plugged in. The plug is plugged in. Therefore, the television is on.

54. If the cat is in the room, then the mice are hiding. The mice are not hiding. Therefore, the cat is not in the room.

55. The test was easy and I received a good grade. The test was not easy or I did not receive a good grade. Therefore, the test was not easy.

56. If Bonnie passes the bar exam, then she will practice law. Bonnie will not practice law. Therefore, Bonnie did not pass the bar exam.

57. The baby is crying but the baby is not hungry. If the baby is hungry then the baby is crying. Therefore, the baby is hungry.

58. If the car is new, then the car has air conditioning. The car is not new and the car has air conditioning. Therefore, the car is not new.

59. If the football team wins the game, then Dave played quarterback. If Dave played quarterback, then the team is not in second place. Therefore, if the football team wins the game, then the team is in second place.

60. The engineering courses are difficult and the chemistry labs are long. If the chemistry labs are long, then the art tests are easy. Therefore, the engineering courses are difficult and the art tests are not easy.

In Exercises 61–67, using the standard forms of arguments and other information you have learned, supply what you believe is a logical conclusion to the argument. Verify that the argument is valid for the conclusion you supplied.

61. If you eat an entire bag of M & M's, your face will break out.
You eat an entire bag of M & M's.
Therefore, . . .

62. If the temperature hits 100°, then we will go swimming.
We did not go swimming.
Therefore, . . .

63. A tick is an insect or a tick is an arachnid.
A tick is not an insect.
Therefore, . . .

64. If Margaret Chang arranges the conference, then many people will attend the conference.
If many people attend the conference, then our picture will be in the paper.
Therefore, . . .

65. If you close the deal, then you will get a commission.
You did not get a commission.
Therefore, . . .

66. If you do not read a lot, then you will not gain knowledge.
You do not read a lot.
Therefore, . . .

67. If you do not pay off your credit card bill, then you will have to pay interest.
If you have to pay interest, then the bank makes money.
Therefore, . . .

Challenge Problems/Group Activities

68. Determine whether the argument is valid or invalid.

If Lynn wins the contest or strikes oil, then she will be rich.
If Lynn is rich, then she will stop working.

∴ If Lynn does not stop working, she did not win the contest.

69. Is it possible for an argument to be invalid if the conjunction of the premises is false in every case of the truth table? Explain your answer.

Recreational Mathematics

70. René Descartes was a seventeenth-century French mathematician and philosopher. One of his most memorable statements is, "I think, therefore, I am." This statement is the basis for the following joke.

Descartes walks into an inn. The innkeeper asks Descartes if he would like something to drink. Descartes replies, "I think not," and promptly vanishes into thin air!

This joke can be summarized in the following argument: If I think, then I am. I think not. Therefore, I am not.

a) Represent this argument symbolically.
b) Is this a valid argument?
c) Explain your answer using either a standard form of argument or using a truth table.

Internet/Research Activities

71. Show how logic is used in advertising. Discuss several advertisements and show how logic is used to persuade the reader.

72. Find examples of valid (or invalid) arguments in printed matter such as newspaper or magazine articles. Explain why the arguments are valid (or invalid).

3.6 EULER DIAGRAMS AND SYLLOGISTIC ARGUMENTS

In the preceding section, we showed how to determine the validity of *symbolic arguments* using truth tables and comparing the arguments to standard forms. This section presents another form of argument called a *syllogistic argument*, better known by the shorter name *syllogism*. The validity of a syllogistic argument is determined by using Euler (pronounced "oiler") diagrams, as is explained shortly.

Syllogistic logic, a deductive process of arriving at a conclusion, was developed by Aristotle in about 350 B.C. Aristotle considered the relationships among the four types of statements that follow.

All ______ are ______.
No ______ are ______.
Some ______ are ______.
Some ______ are not ______.

PROFILE IN MATHEMATICS

LEONHARD EULER

Swiss mathematician Leonhard Euler (1707–1783) produced such a large volume of work that all of it has not yet been compiled. It has been estimated that a complete edition of his work would make up over 60 volumes. Although his greatest contribution was made in the field of modern analysis, his work also explored physics, astronomy, trigonometry, and calculus. Euler also had a great influence on the notation and symbolism of mathematics, and it was through his work that the symbols e and π came into common use. His facility for mental calculation and his uncommon memory allowed him to continue his prolific career even after he lost his vision in both eyes.

Examples of these statements are: *All doctors are tall. No doctors are tall. Some doctors are tall. Some doctors are not tall.* Since Aristotle's time, other types of statements have been added to the study of syllogistic logic, two of which are

_____ is a _____.
_____ is not a _____.

Examples of these statements are: *Maria is a doctor. Maria is not a doctor.*

The difference between a symbolic argument and a syllogistic argument can be seen in the following chart. Symbolic arguments use the connectives *and, or, not, if–then,* and *if and only if.* Syllogistic arguments use the quantifiers *all, some,* and *none,* which were discussed in Section 3.1.

Symbolic Arguments Versus Syllogistic Arguments

	Words or phrases used	*Method of determining validity*
Symbolic argument	and, or, not, if-then, if and only if	Truth tables or by comparison with standard forms of arguments
Syllogistic argument	all are, some are, none are, some are not	Euler diagrams.

As with symbolic logic, the premises and the conclusion together form an argument. An example of a syllogistic argument is

All German shepherds are dogs.
All dogs bark.
$\therefore$ All German shepherds bark.

This is an example of a valid argument. Recall from the previous section that an argument is *valid* when its conclusion necessarily follows from a given set of premises. Recall that an argument in which the conclusion does not necessarily follow from the given premises is said to be an *invalid argument* or a *fallacy.*

Before we give another example of a syllogism, let's review the Venn diagrams discussed in Section 2.3 in relationship with Aristotle's four statements.

All *A*s are *B*s

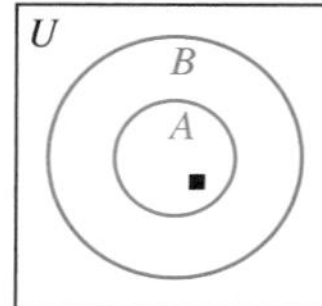

If an element is in set *A*, then it is in set *B*.

No *A*s are *B*s

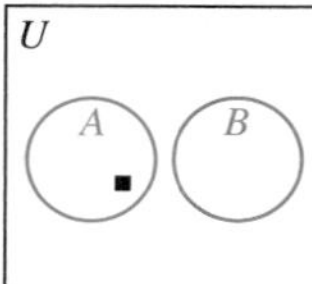

If an element is in set *A*, then it is not in set *B*.

Some *A*s are *B*s

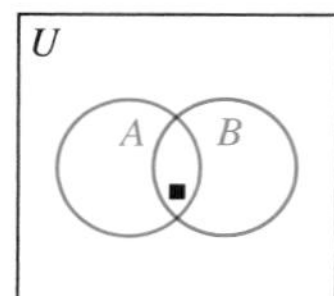

There is at least one element that is in both set *A* and set *B*.

Some *A*s are not *B*s

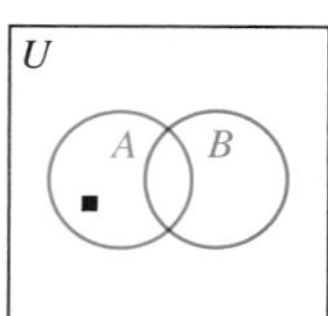

There is at least one element that is in set *A* that is not in set *B*.

One method used to determine whether an argument is valid or is a fallacy is by means of an *Euler diagram*, named after Leonhard Euler (1707–1783), who used circles to represent sets in syllogistic arguments. The technique of using Euler diagrams is illustrated in Example 1.

EXAMPLE 1 *Using a Euler Diagram*

Determine whether the following syllogism is valid or invalid.

All keys are made of brass.
All things made of brass are valuable.
∴ All keys are valuable.

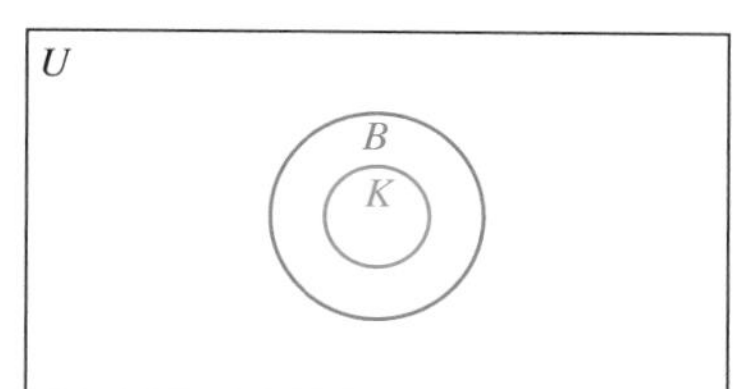

Figure 3.3

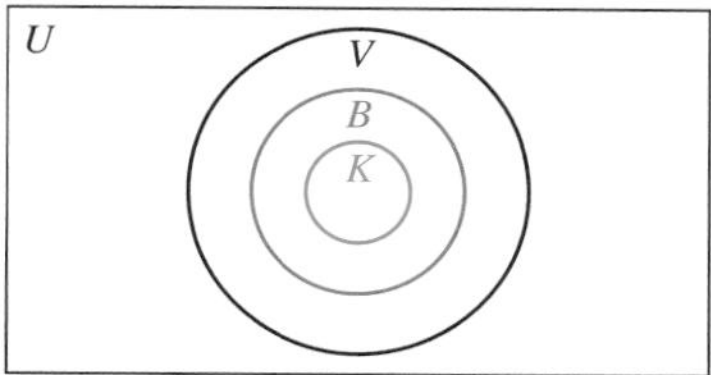

Figure 3.4

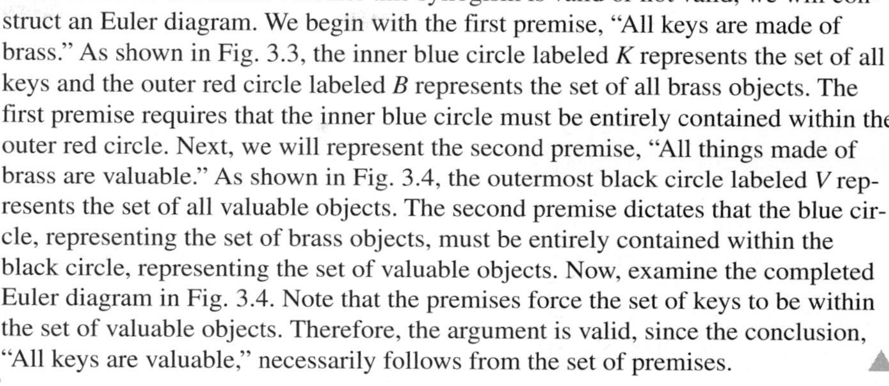

SOLUTION: To determine whether this syllogism is valid or not valid, we will construct an Euler diagram. We begin with the first premise, "All keys are made of brass." As shown in Fig. 3.3, the inner blue circle labeled K represents the set of all keys and the outer red circle labeled B represents the set of all brass objects. The first premise requires that the inner blue circle must be entirely contained within the outer red circle. Next, we will represent the second premise, "All things made of brass are valuable." As shown in Fig. 3.4, the outermost black circle labeled V represents the set of all valuable objects. The second premise dictates that the blue circle, representing the set of brass objects, must be entirely contained within the black circle, representing the set of valuable objects. Now, examine the completed Euler diagram in Fig. 3.4. Note that the premises force the set of keys to be within the set of valuable objects. Therefore, the argument is valid, since the conclusion, "All keys are valuable," necessarily follows from the set of premises. ▲

The syllogism in Example 1 is valid even though the conclusion, "All keys are valuable," is not a true statement. Similarly, a syllogism can be invalid, or a fallacy, even if the conclusion is a true statement.

When we determine the validity of an argument, we are determining whether the conclusion necessarily follows from the premises. When we say that an argument is valid, we are saying that if all the premises are true statements, then the conclusion must also be a true statement.

The form of the argument determines its validity, not the particular statements. For example, consider the syllogism

All Earth people have two heads.
All people with two heads can fly.
∴ All Earth people can fly.

The form of this argument is the same as that of the previous valid argument in Example 1. Therefore, this argument is also valid.

EXAMPLE 2 *Is the Syllogism Valid?*

Determine whether the following syllogism is valid or is invalid.

All pilots have good vision.
Kaitlyn is a pilot.
∴ Kaitlyn has good vision.

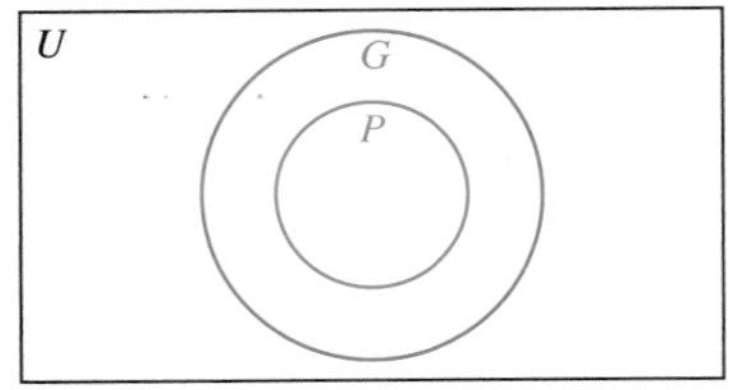

Figure 3.5

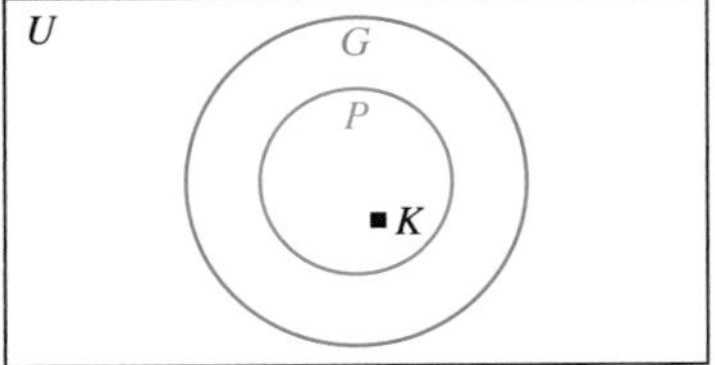

Figure 3.6

SOLUTION: The statement "All pilots have good vision" is illustrated in Fig. 3.5. The second premise, "Kaitlyn is a pilot," tells us that Kaitlyn must be placed in the inner circle (see Fig. 3.6). The Euler diagram illustrates that we must accept the conclusion "Kaitlyn has good vision" as true (when we accept the premises as true). Therefore, the argument is valid. ▲

In both Example 1 and Example 2, we had no choice as to where the second premise was to be placed in the Euler diagram. In Example 1, the set of brass objects had to be placed inside the set of valuable objects. In Example 2, Kaitlyn had to be placed inside the set of people with good vision. Often when determining the truth value of a syllogism, a premise can be placed in more than one area in the diagram. *We always try to draw the Euler diagram so that the conclusion does not necessarily follow from the premises. If that can be done, then the conclusion does not necessarily follow from the premises and the argument is invalid.* If we cannot show that the argument is invalid, only then do we accept the argument as valid. We illustrate this process in Example 3.

DID YOU KNOW

Boolean Algebra

Logic, or Boolean algebra, affects each of us daily, Nowadays, many aspects of our lives are computer related. There are computer chips in our cars, watches, radios, cameras, supermarket checkouts, and most appliances, to name just a few uses. The outcomes of these computer chips are based on the *not, and*, and *or* truth tables we have discussed. Sometimes abstract mathematics is not applied until many years after its development, which is exactly what happened with the abstract Boolean algebra concepts considered in the seventeenth and eighteenth centuries.

EXAMPLE 3 *Ballerinas and Athletes*

Determine whether the following syllogism is valid or is invalid.

All ballerinas are athletic.
Keyshawn is athletic.
∴ Keyshawn is a ballerina.

SOLUTION: The statement "All ballerinas are athletic" is illustrated in Fig. 3.7(a). The next premise "Keyshawn is athletic" tells us that Keyshawn must be placed in the set of athletic people. Two diagrams in which both premises are satisfied are shown in Fig. 3.7(b) and (c). By examining Fig. 3.7(b), however, we see that Keyshawn is not a ballerina. Therefore, the conclusion "Keyshawn is a ballerina" does not necessarily follow from the set of premises. Thus, the argument is invalid, or a fallacy.

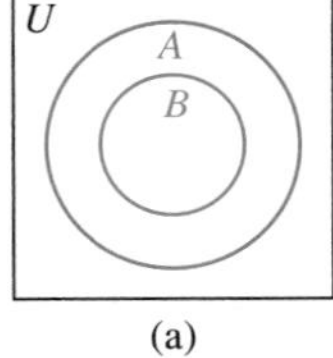

(a)

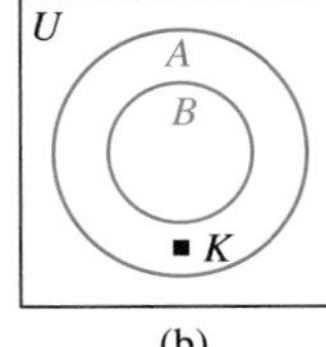

(b)

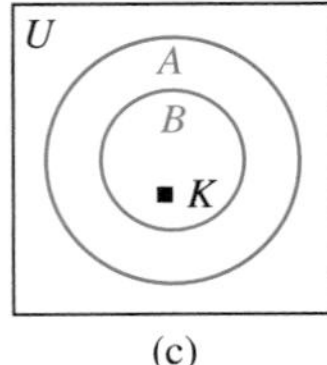

(c)

Figure 3.7 ▲

EXAMPLE 4 *Parrots and Chickens*

Determine whether the following syllogism is valid or invalid.

No parrots eat chicken.
Fletch does not eat chicken.
∴ Fletch is a parrot.

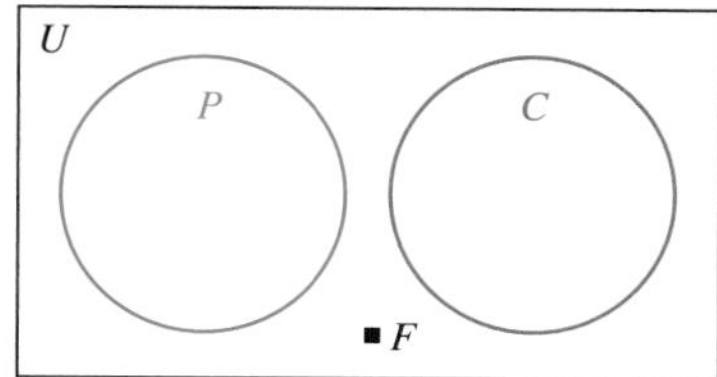

Figure 3.8

SOLUTION: The diagram in Fig. 3.8 satisfies the two given premises and also shows that Fletch is not a parrot. Therefore, the argument is invalid, or is a fallacy. ▲

Note that in Example 4 if we placed Fletch in circle P, the argument would appear to be valid. Remember that *whenever testing the validity of an argument, always try to show that the argument is invalid.* If there is any way of showing that the conclusion does not necessarily follow from the premises, then the argument is invalid.

EXAMPLE 5 *A Syllogism Involving the Word Some*

Determine whether the following syllogism is valid or invalid.

All As are Bs.
Some Bs are Cs.
$\therefore$ Some As are Cs.

SOLUTION: The statement "All As are Bs" is illustrated in Fig. 3.9. The statement "Some Bs are Cs" means that there is at least one B that is a C. We can illustrate this set of premises in four ways, as illustrated in Fig. 3.10.

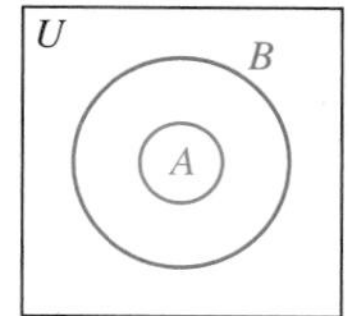

Figure 3.9

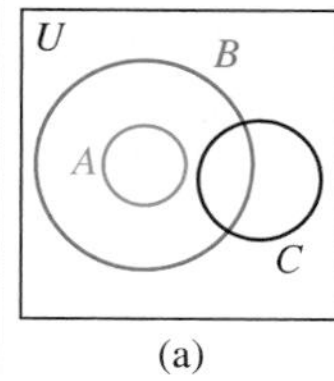

(a)

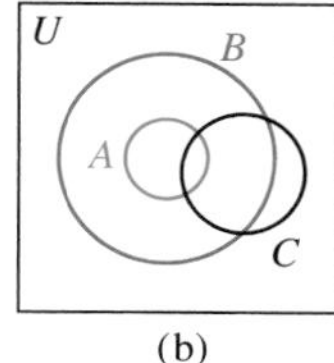

(b)

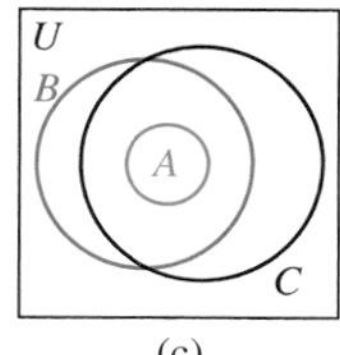

(c)

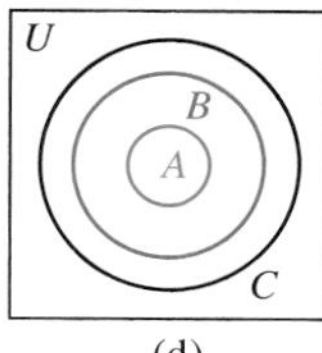

(d)

Figure 3.10

In all four illustrations, we see that (1) all As are Bs and (2) some Bs are Cs. The conclusion is "Some As are Cs." Since at least one of the illustrations, Fig. 3.10(a), shows that the conclusion does not necessarily follow from the given premises, the argument is invalid. ▲

EXAMPLE 6 *Cowboys and Debutantes*

Determine whether the following syllogism is valid or invalid.

No cowboys drink lemonade.
All debutantes drink lemonade.
$\therefore$ No cowboys are debutantes.

SOLUTION: The first premise tells us that cowboys and the people who drink lemonade are disjoint sets—that is, sets that do not intersect—as shown in Fig. 3.11. The second premise tells us that the set of debutantes is a subset of the people who drink lemonade. Therefore, the circle representing the set of debutantes must go within the circle representing the set of people who drink lemonade.

The set of debutantes and the set of cowboys cannot be made to intersect without violating a premise. Thus, no cowboys can be debutantes, and the syllogism is valid. Note that we did not say that this conclusion is true, only that the argument is valid. ▲

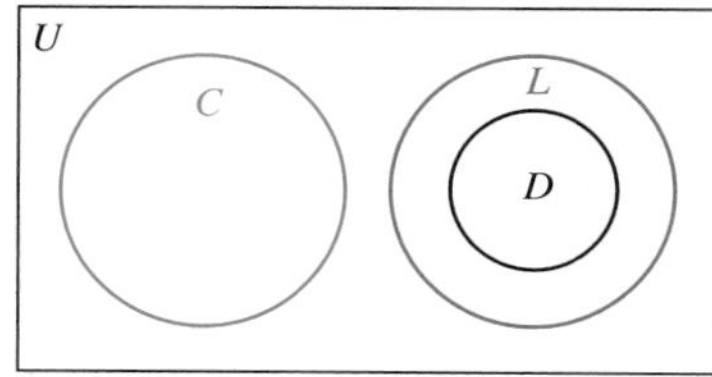

Figure 3.11

SECTION 3.6 EXERCISES

Concept/Writing Exercises

1. If an Euler diagram can be drawn in which the conclusion does not necessarily follow from the premises, what can be said about the syllogistic argument under consideration?
2. If an Euler diagram can only be drawn in a way in which the conclusion necessarily follows from the premises, what can be said about the syllogistic argument under consideration?
3. What does it mean when we determine that an argument is valid?
4. Explain the differences between a symbolic argument and a syllogistic argument.
5. Can an argument be valid if the conclusion is a false statement? Explain your answer.
6. Can an argument be invalid if the conclusion is a true statement? Explain.

Practice the Skills/Problem Solving

In Exercises 7–30, use an Euler diagram to determine whether the syllogism is valid or invalid.

7. All cows give milk.
Mags is a cow.
∴ Mags gives milk.

8. All cordless telephones have antennas.
All things with antennas are insects.
∴ All cordless telephones are insects.

9. No apples are citrus fruits.
All Granny Smiths are apples.
∴ No Granny Smiths are citrus fruits.

10. All dolphins are mammals.
All mammals are vertebrates.
∴ All dolphins are vertebrates.

11. All theme parks have walkways.
Metropolitan Community College has walkways.
∴ Metropolitan Community College is a theme park.

12. All golfers have rain gear.
John Pearse has rain gear.
∴ John Pearse is a golfer.

13. No horses buck.
Palominos are horses.
∴ Palominos do not buck.

14. No jockeys weigh more than 200 pounds.
Deb Otto is not a jockey.
∴ Deb Otto weighs more than 200 pounds.

15. Some mushrooms are poisonous.
A morel is a mushroom.
∴ A morel is poisonous.

16. Some policemen are polite.
Jarod Harshbarger is a policeman.
∴ Jarod Harshbarger is not polite.

17. Some farmers are politicians.
Some politicians are senators.
∴ Some farmers are senators.

18. Some professional golfers give golf lessons.
All people who belong to the PGA are professional golfers.
∴ All people who belong to the PGA give golf lessons.

Tiger Woods

19. No tennis players are wrestlers.
Allison is not a wrestler.
$\therefore$ Allison is a tennis player.

20. Some soaps float.
All things that float are lighter than water.
$\therefore$ Some soaps are lighter than water.

21. Some people love mathematics.
All people who love mathematics love physics.
$\therefore$ Some people love physics.

22. Some desks are made of wood.
All paper is made of wood.
$\therefore$ Some desks are made of paper.

23. No *x*s are *y*s.
No *y*s are *z*s.
$\therefore$ No *x*s are *z*s.

24. All pilots can fly.
All astronauts can fly.
$\therefore$ Some pilots are astronauts.

25. Some dogs wear glasses.
Fido wears glasses.
$\therefore$ Fido is a dog.

26. All rainy days are cloudy.
Today it is cloudy.
$\therefore$ Today is a rainy day.

27. All sweet things taste good.
All things that taste good are fattening.
All things that are fattening put on pounds.
$\therefore$ All sweet things put on pounds.

28. All books have red covers.
All books that have red covers contain 200 pages.
Some books that contain 200 pages are novels.
$\therefore$ All books that contain 200 pages are novels.

29. All country singers play the guitar.
All country singers play the drums.
Some people who play the guitar are rock singers.
$\therefore$ Some country singers are rock singers.

Bonnie Raitt

30. Some hot dogs are made of turkey.
All things made of turkey are edible.
Some things that are made of beef are edible.
$\therefore$ Some hot dogs are made of beef.

Challenge Problem/Group Activity

31. Statements in logic can be translated into set statements: for example, $p \wedge q$ is similar to $P \cap Q$; $p \vee q$ is similar to $P \cup Q$; and $p \rightarrow q$ is equivalent to $\sim p \vee q$, which is similar to $P' \cup Q$. Euler diagrams can also be used to show that arguments similar to those discussed in Section 3.5 are valid or invalid. Use Euler diagrams to show that the symbolic argument is invalid.

$$\begin{array}{l} p \rightarrow q \\ \underline{p \vee q} \\ \therefore \sim p \end{array}$$

Internet/Research Activity

32. Leonhard Euler is considered one of the greatest mathematicians of all time. Do research and write a report on Euler's life. Include information on his contributions to sets and to logic. Also indicate other areas of mathematics in which he made important contributions. References include encyclopedias, history of mathematics books, and the Internet.

CHAPTER 3 SUMMARY

IMPORTANT FACTS

Quantifiers

Form of Statement	Form of Negation
All are.	Some are not.
None are.	Some are.
Some are.	None are.
Some are not.	All are.

Summary of connectives

Formal Name	Symbol	Read	Symbolic Form
Negation	$\sim$	not	$\sim p$
Conjunction	$\wedge$	and	$p \wedge q$
Disjunction	$\vee$	or	$p \vee q$
Conditional	$\rightarrow$	if–then	$p \rightarrow q$
Biconditional	$\leftrightarrow$	if and only if	$p \leftrightarrow q$

Basic truth tables

Negation

p	~p
T	F
F	T

p	q	Conjunction $p \wedge q$	Disjunction $p \vee q$	Conditional $p \rightarrow q$	Biconditional $p \leftrightarrow q$
T	T	T	T	T	T
T	F	F	T	F	F
F	T	F	T	T	F
F	F	F	F	T	T

De Morgan's laws

$$\sim(p \wedge q) \Leftrightarrow \sim p \vee \sim q$$
$$\sim(p \vee q) \Leftrightarrow \sim p \wedge \sim q$$

Other equivalent forms

$$p \rightarrow q \Leftrightarrow \sim p \vee q$$
$$\sim(p \rightarrow q) \Leftrightarrow p \wedge \sim q$$
$$p \leftrightarrow q \Leftrightarrow [(p \rightarrow q) \wedge (q \rightarrow p)]$$

Variations of the conditional statement

Name	Symbolic Form	Read
Conditional	$p \rightarrow q$	If p, then q.
Converse of the conditional	$q \rightarrow p$	If q, then p.
Inverse of the conditional	$\sim p \rightarrow \sim q$	If not p, then not q.
Contrapositive of the conditional	$\sim q \rightarrow \sim p$	If not q, then not p.

Standard forms of arguments

Valid arguments

Law of Detachment	Law of Contra-position	Law of Syllogism	Disjunctive Syllogism
$p \rightarrow q$	$p \rightarrow q$	$p \rightarrow q$	$p \vee q$
p	$\sim q$	$q \rightarrow r$	$\sim p$
$\therefore q$	$\therefore \sim p$	$\therefore p \rightarrow r$	$\therefore q$

Invalid arguments

Fallacy of the Converse	Fallacy of the Inverse
$p \rightarrow q$	$p \rightarrow q$
q	$\sim p$
$\therefore p$	$\therefore \sim q$

Symbolic argument vs. syllogistic argument

	Words or Phrases Used	Method of Determining Validity
Symbolic argument	and, or, not, if–then, if and only if	Truth tables or by comparison with standard forms of arguments
Syllogistic argument	all are, some are, none are, some are not	Euler diagrams

CHAPTER 3 REVIEW EXERCISES

3.1

In Exercises 1–6, write the negation of the statement.

1. Some rock bands play ballads.
2. Some bananas are not ripe.
3. No chickens have lips.
4. All panthers are endangered.

5. All pens use ink.
6. No rabbits wear glasses.

In Exercises 7–12, write each compound statement in words.

p: The coffee is Maxwell House.
q: The coffee is hot.
r: The coffee is strong.

7. $p \vee q$
8. $\sim q \wedge r$
9. $q \rightarrow (r \wedge \sim p)$
10. $p \leftrightarrow \sim r$
11. $\sim p \leftrightarrow (r \wedge \sim q)$
12. $(p \vee \sim q) \wedge \sim r$

3.2

In Exercises 13–18, use the statements for p, q, and r as in Exercises 7–12 to write the statement in symbolic form.

13. The coffee is strong and the coffee is hot.
14. If the coffee is Maxwell House, then it is strong.
15. If the coffee is strong then the coffee is hot, or the coffee is not Maxwell House.
16. The coffee is hot if and only if the coffee is Maxwell House, and the coffee is not strong.
17. The coffee is strong and the coffee is hot, or the coffee is not Maxwell House.
18. It is false that the coffee is strong and the coffee is hot.

In Exercises 19–24, construct a truth table for the statement.

19. $(p \vee q) \wedge \sim p$
20. $q \leftrightarrow (p \vee \sim q)$
21. $(p \vee q) \leftrightarrow (p \vee r)$
22. $p \wedge (\sim q \vee r)$
23. $p \rightarrow (q \wedge \sim r)$
24. $(p \wedge q) \rightarrow \sim r$

3.2, 3.3

In Exercises 25–28, determine the truth value of the statement.

25. If 7 is an odd number, then 11 is an even number.
26. The St. Louis Arch is in St. Louis or Abraham Lincoln is buried in Grant's Tomb.

27. If Oregon borders the Pacific Ocean or California borders the Atlantic Ocean, then Minnesota is south of Texas.
28. $15 - 7 = 22$ or $4 + 9 = 13$, and $9 - 8 = 1$.

In Exercises 29 and 30, the circle graph shows the sources and percentages of Oregon's electricity sources in 2002. Use the graph to determine the truth value of each simple statement. Then determine the truth value of the compound statement.

Oregon's Electricity Sources

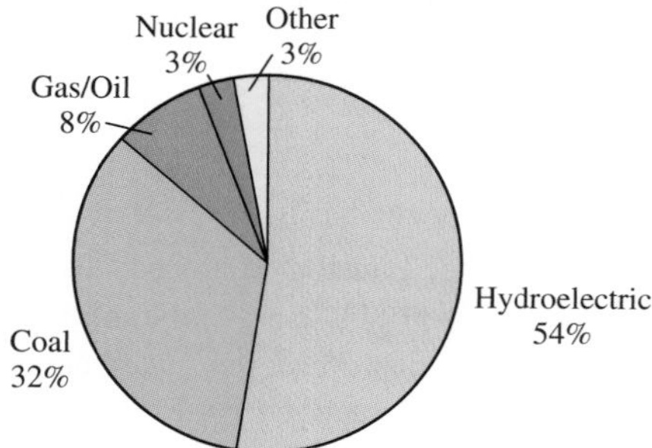

Source: Renewable Northwest Project

29. Thirty-two percent of Oregon's electricity sources are coal if and only if 54% of Oregon's electricity sources are hydroelectric sources, or 38% of Oregon's electricity sources are nuclear.
30. If 3% of Oregon's electricity sources are gas/oil, then 45% of Oregon's electricity sources are coal and 3% of Oregon's electricity comes from other sources.

3.3

In Exercises 31–34, determine the truth value of the statement when p is T, q is F, and r is F.

31. $(p \rightarrow \sim r) \vee (p \wedge q)$
32. $(p \vee q) \leftrightarrow (\sim r \wedge p)$
33. $\sim r \leftrightarrow [(p \vee q) \leftrightarrow \sim p]$
34. $\sim[(q \wedge r) \rightarrow (\sim p \vee r)]$

3.4

In Exercises 35–38, determine whether the pairs of statements are equivalent. You may use De Morgan's laws, the fact that $(p \rightarrow q) \Leftrightarrow (\sim p \vee q)$, truth tables, or equivalent forms of the conditional statement.

35. $\sim p \vee \sim q$ $\quad \sim p \leftrightarrow q$
36. $\sim p \rightarrow \sim q$ $\quad p \vee \sim q$
37. $\sim p \vee (q \wedge r)$ $\quad (\sim p \vee q) \wedge (\sim p \vee r)$
38. $(\sim q \rightarrow p) \wedge p$ $\quad \sim(\sim p \leftrightarrow q) \vee p$

In Exercises 39–43, use De Morgan's laws or the fact that $(p \rightarrow q) \Leftrightarrow (\sim p \vee q)$ to write an equivalent statement for the given statement.

39. Johnny Cash is in the Rock and Roll Hall of Fame and India Arie recorded *Acoustic Soul*.

India Arie

40. Her foot fell asleep or she has injured her ankle.
41. It is not true that Altec Lansing only produces speakers or Harman Kardon only produces stereo receivers.
42. Travis Tritt did not win an Academy Award and Randy Jackson does not do commercials for Milk Bone Dog Biscuits.
43. If the temperature is not above 32°, then we will go ice fishing at O'Leary's Lake.

In Exercises 44–48, write the (a) converse, (b) inverse, and (c) contrapositive for the given statement.

44. If you hear a beautiful songbird today, then you enjoy life.
45. If you followed the correct pattern, then the quilt has a uniform design.
46. If Maureen Gerald is not in attendance, then she is helping at the school.
47. If the desk is made by Winner's Only and the desk is in the Rose catalog, then we will not buy a desk at Miller's Furniture.
48. If you get straight A's on your report card, then I will let you attend the prom.

In Exercises 49–52, determine which, if any, of the three statements are equivalent.

49. a) If the temperature is over 80°, then the air conditioner will come on.
    b) The temperature is not over 80° or the air conditioner will come on.
    c) It is false that the temperature is over 80° and the air conditioner will not come on.
50. a) The screwdriver is on the workbench if and only if the screwdriver is not on the counter.
    b) If the screwdriver is not on the counter, then the screwdriver is not on the workbench.
    c) It is false that the screwdriver is on the counter and the screwdriver is not on the workbench.
51. a) If $2 + 3 = 6$, then $3 + 1 = 5$.
    b) $2 + 3 = 6$ if and only if $3 + 1 \neq 5$.
    c) If $3 + 1 \neq 5$, then $2 + 3 \neq 6$.
52. a) If the sale is on Tuesday and I have money, then I will go to the sale.
    b) If I go to the sale, then the sale is on Tuesday and I have money.
    c) I go to the sale, or the sale is on Tuesday and I have money.

3.5, 3.6

In Exercises 53–58, determine whether the argument is valid or invalid.

53. $p \rightarrow q$
    $\underline{\sim p}$
    $\therefore q$

54. $p \wedge q$
    $\underline{q \rightarrow r}$
    $\therefore p \rightarrow r$

55. Nicole is in the hot tub or she is in the shower.
    Nicole is in the hot tub.
    ∴ Nicole is not in the shower.
56. If the car has a sound system, then Rick will buy the car. If the price is not less than \$18,000, then Rick will not buy the car. Therefore, if the car has a sound system, then the price is less than \$18,000.
57. All plumbers wear overalls.
    Some electricians wear overalls.
    ∴ Some electricians are plumbers.
58. Some submarines are yellow.
    All dandelions are yellow.
    ∴ Some dandelions are submarines.

CHAPTER 3 TEST

In Exercises 1–3, write the statement in symbolic form.

p: Ann is the secretary.

q: Dick is the vice president.

r: Elaine is the president.

1. Ann is the secretary but Elaine is the president, or Dick is not the vice president.
2. If Elaine is the president then Dick is the vice president, or Ann is not the secretary.
3. It is false that Elaine is the president if and only if Dick is not the vice president.

In Exercises 4 and 5, use p, q, and r as above to write each symbolic statement in words.

4. $p \leftrightarrow (q \wedge r)$
5. $\sim(p \rightarrow \sim r)$

In Exercises 6 and 7, construct a truth table for the given statement.

6. $[\sim(p \rightarrow r)] \wedge q$
7. $(q \leftrightarrow \sim r) \vee p$

In Exercises 8 and 9, find the truth value of the statement.

8. $2 + 6 = 8$ or $7 - 12 = 5$.
9. Scissors can cut paper or a dime has the same value as two nickels, if and only if Louisville is a city in Kentucky.

In Exercises 10 and 11, given that p is true, q is false, and r is true, determine the truth value of the statement.

10. $(r \vee q) \leftrightarrow (p \wedge \sim q)$
11. $[\sim(r \rightarrow \sim p)] \wedge (q \rightarrow p)$
12. Determine whether the pair of statements are equivalent.

$$\sim p \vee q, \qquad \sim(p \wedge \sim q)$$

In Exercises 13 and 14, determine which, if any, of the three statements are equivalent.

13. a) If the bird is red, then it is a cardinal.
    b) The bird is not red or it is a cardinal.
    c) If the bird is not red, then it is not a cardinal.
14. a) It is not true that the test is today or the concert is tonight.
    b) The test is not today and the concert is not tonight.
    c) If the test is not today, then the concert is not tonight.
15. *Translate the following argument into symbolic form. Determine whether the argument is valid or invalid by comparing the argument to a recognized form or by using a truth table.*

    If the soccer team wins the game, then Sue played fullback. If Sue played fullback, then the team is in second place. Therefore, if the soccer team wins the game, then the team is in second place.
16. *Use an Euler diagram to determine whether the syllogism is valid or is a fallacy.*

    All cars have engines.
    Some things with engines use gasoline.
    ∴ Some cars use gasoline.

In Exercises 17 and 18, write the negation of the statement.

17. All leopards are spotted.

18. Some jacks-in-the-box are electronic.
19. Write the converse, inverse, and contrapositive of the conditional statement, "If the garbage truck comes, then today is Saturday."
20. Is it possible for an argument to be valid when the conclusion is a false statement? Explain your answer.

GROUP PROJECTS

Switching Circuits

1. An application of logic is *switching circuits*. There are two basic types of electric circuits: *series circuits* and *parallel circuits*. In a series circuit, the current can flow in only one path; see Fig. 3.12. In a parallel circuit the current can flow in more than one path; see Fig. 3.13.

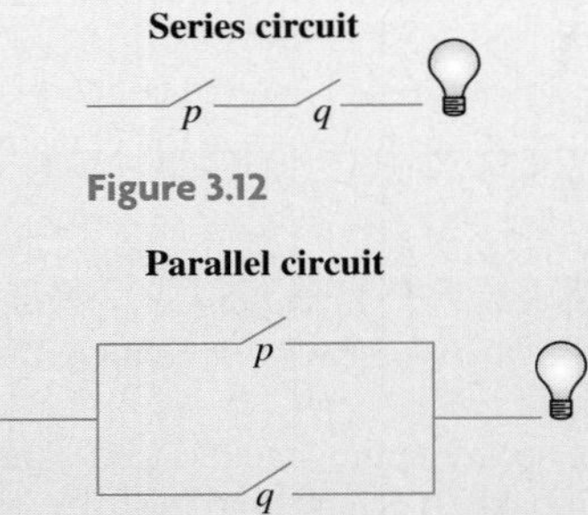

Figure 3.12

Figure 3.13

In Figs. 3.12 and 3.13, the p and q represent switches that may be opened or closed. In the series circuit in Fig. 3.12, if both switches are closed, the current will reach the bulb and the bulb will light. In the parallel circuit in Fig. 3.13, if either switch p or switch q is closed, or if both switches are closed, the current will reach the bulb and the bulb will light.

a) How many different open/closed arrangements of the two switches in Fig. 3.12 are possible? List all the possibilities.

b) Series circuits are represented using conjunctions. The circuit in Fig. 3.12 may be represented as $p \wedge q$. Construct a four-row truth table to represent the series circuit. Construct the table with columns for p, q, and $p \wedge q$. The statement $p \wedge q$ represents the outcome of the circuit (either the bulb lighting or the bulb not lighting). Represent a closed switch with the number 1, an open switch with the number 0, the bulb lighting with the number 1, and the bulb not lighting with the number 0. For example, if both switches are closed, the bulb will light, and so we write the first row of the truth table as

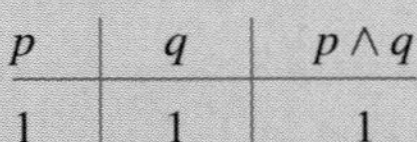

p	q	$p \wedge q$
1	1	1

c) How is the truth table determined in part (b) similar to the truth table for $p \wedge q$ discussed in earlier sections of this chapter?

d) Parallel circuits are represented using disjunctions. The circuit in Fig. 3.13 may be represented as $p \vee q$. Construct a truth table to represent the parallel circuit. Construct the table with columns for p, q, and $p \vee q$. The statement $p \vee q$ represents the outcome of the circuit (either the bulb lighting or the bulb not lighting). Use 1's and 0's as indicated in part (b).

e) How is the truth table determined in part (d) similar to the truth table for $p \vee q$ discussed in earlier sections of this chapter?

f) Represent the following circuit as a symbolic logic statement using parentheses. Explain how you determined your answer.

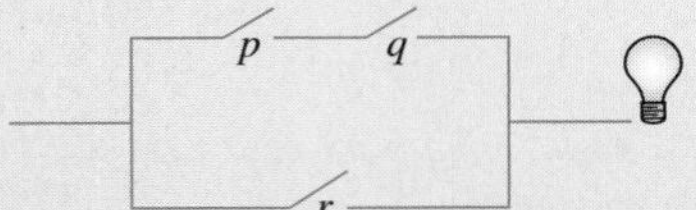

g) Draw a circuit to represent the logic statement $p \wedge (q \vee r)$. Explain how you determined your answer.

Computer Gates

2. Gates in computers work on the same principles as switching circuits. The three basic types of gate are the NOT gate, the AND gate, and the OR gate. Each is illustrated along with a table that indicates current flow entering and exiting the gate. If current flows into a NOT gate, then no current exits, and vice versa. Current exits an AND gate only when both inputs have a current flow. Current exits an OR gate if current flows through either, or both, inputs. In the table, a 1 represents a current flow and a 0 indicates no current flow. For example, in the AND gate, if there is a current flow in input A (I_a has a value of 1) and no current flow in input B (I_b has a value of 0), there is no current flow in the output (O has a value of 0); see row 2 of the AND Gate table.

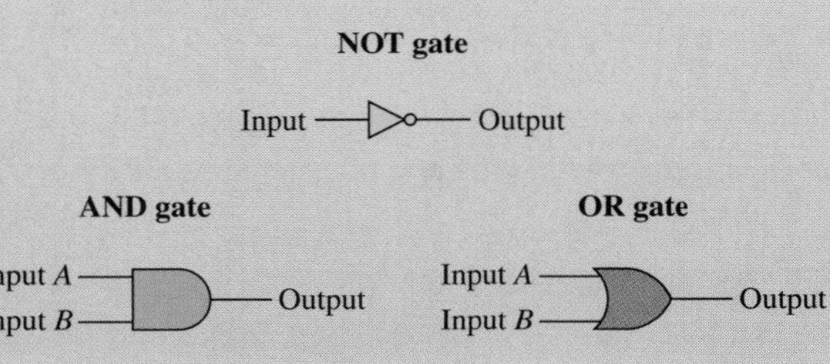

NOT gate

I	O
1	0
0	1

AND gate

I_a	I_b	O
1	1	1
1	0	0
0	1	0
0	0	0

OR gate

I_a	I_b	O
1	1	1
1	0	1
0	1	1
0	0	0

a) If 1 is considered true and 0 is considered false, explain how these tables are similar to the *not, and,* and *or* truth tables.

For the inputs indicated in the following figures determine whether the output is 1 or 0.

b) 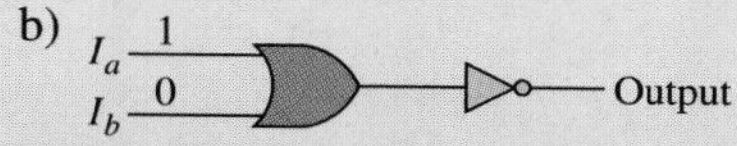

c) 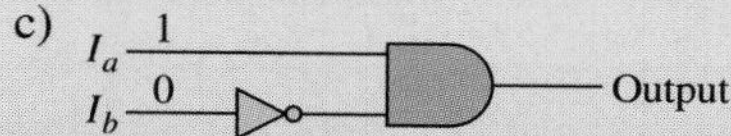

d) 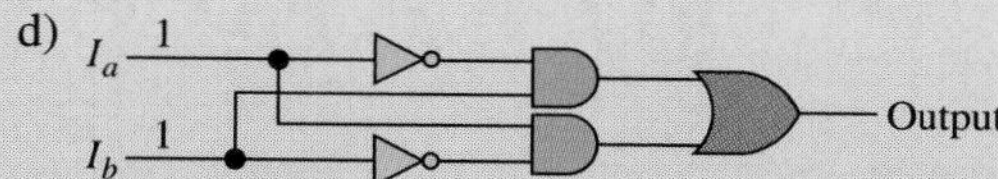

e) What values for I_a and I_b will give an output of 1 in the figure in part (d)? Explain how you determined your answer.

f) Construct a truth table using 1's and 0's for the following gate. Your truth table should have columns I_a, I_b, and O and should indicate the four possible cases for the inputs and each corresponding output.

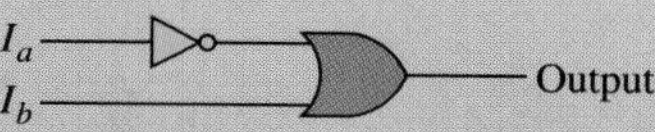

Logic Game

3. a) Shown is a photograph of a logic game at the Ontario Science Centre. There are 12 balls on top of the game board, numbered from left to right, with ball 1 on the extreme left and ball 12 on the extreme right. On the platform in front of the players are 12 buttons, one corresponding to each of the balls. When 6 buttons are pushed, the 6 respective balls are released. When 1 or 2 balls reach an *and* gate or an *or* gate, a single ball may or may not pass through the gate. The object of the game is to select a proper combination of 6 buttons that will allow 1 ball to reach the bottom. Using your knowledge of *and* and *or*, select a combination of 6 buttons that will result in a win. (There is more than one answer.) Explain how you determined your answer.

b) Construct a game similar to this one where 15 balls are at the top and 8 must be selected to allow 1 ball to reach the bottom.

c) Indicate all solutions to the game you constructed in part (c).

One of the earliest reasons that human beings needed numbers was for reckoning time, marking off days in the lunar month, so the seasonal changes that dictated human activity could be anticipated. The Mayans, the Egyptians, and the ancient Britons constructed monumental stone observatories that enabled them to mark the passage of the seasons, especially the summer solstice, using the alignment of the sun as a guide. The photo is Stonehenge in England.

SYSTEMS OF NUMERATION

The number system we use—called the Hindu–Arabic system—seems to be a permanent, unchanging means of communicating quantities. However, just as languages evolve over time, so do numerical symbols that represent numbers.

Mathematics began with the practical problem of counting and record keeping. People had to count their herds, the passage of days, and objects of barter. They used physical objects—stones, shells, fingers—to represent the objects counted.

As primitive cultures grew from villages to cities, the complexity of human activities increased. Now people needed better ways of recording and communicating. It was a revolutionary step when people started using physical objects to represent not only specific objects like sheep and grain, but also the concept of pure quantity.

Through the course of human history, the evolution of numeration systems has expanded our knowledge and abilities for record keeping, communication, and computation. As a society's numeration system changes, so do the capabilities of that society. Without an understanding of the binary number system, the computer as we know it today could not exist. Without the computer, our lifestyle would not be as it is today.

4.1 ADDITIVE, MULTIPLICATIVE, AND CIPHERED SYSTEMS OF NUMERATION

Just as the first attempts to write were made long after the development of speech, the first representation of numbers by symbols came long after people had learned to count. A tally system using physical objects, such as scratch marks in the soil or on a stone, notches on a stick, pebbles, or knots on a vine, was probably the earliest method of recording numbers.

In primitive societies, such a tally system adequately served the limited need for recording livestock, agriculture, or whatever was counted. As civilization developed, however, more efficient and accurate methods of calculating and keeping records were needed. Because tally systems are impractical and inefficient, societies developed symbols to replace them. For example, the Egyptians used the symbol ∩ and the Babylonians used the symbol ⟨ to represent the number we symbolize by 10.

A *number* is a quantity, and it answers the question "How many?" A *numeral* is a symbol such as ∩, ⟨, or 10 used to represent the number. We think a number but write a numeral. The distinction between number and numeral will be made here only if it is helpful to the discussion.

In language, relatively few letters of the alphabet are used to construct a large number of words. Similarly, in arithmetic, a small variety of numerals can be used to represent all numbers. In general, when writing a number, we use as few numerals as possible. One of the greatest accomplishments of humankind has been the development of systems of numeration, whereby all numbers are "created" from a few symbols. Without such systems, mathematics would not have developed to its present level.

> A **system of numeration** consists of a set of numerals and a scheme or rule for combining the numerals to represent numbers.

DID YOU KNOW

An Important Discovery

Archaeologist Denise Schmandt-Besserat made a breakthrough discovery about early systems of numeration. She realized that the little clay geometric objects that had been found in many archaeological sites had actually been used by people to account for their goods. Later in history, these tokens were impressed on a clay tablet to represent quantities, the beginning of writing.

Four types of numeration systems used by different cultures are the topic of this chapter. They are additive (or repetitive), multiplicative, ciphered, and place-value systems. You do not need to memorize all the symbols, but you should understand the principles behind each system. By the end of this chapter, we hope that you better understand the system we use, the *Hindu–Arabic system*, and its relationship to other types of systems.

Additive Systems

An additive system is one in which the number represented by a particular set of numerals is simply the sum of the values of the numerals. The additive system of numeration is one of the oldest and most primitive types of numeration systems. One of the first additive systems, the Egyptian hieroglyphic system, dates back to about 3000 B.C. The Egyptians used symbols for the powers of 10: 10^0 or 1, 10^1 or 10, 10^2 or $10 \cdot 10$, 10^3 or $10 \cdot 10 \cdot 10$, and so on. Table 4.1 on page 168 lists the Egyptian hieroglyphic numerals with the equivalent Hindu–Arabic numerals.

To write the number 600 in Egyptian hieroglyphics, we write the numeral for 100 six times: 999999.

DID YOU KNOW

The Rhind Papyrus

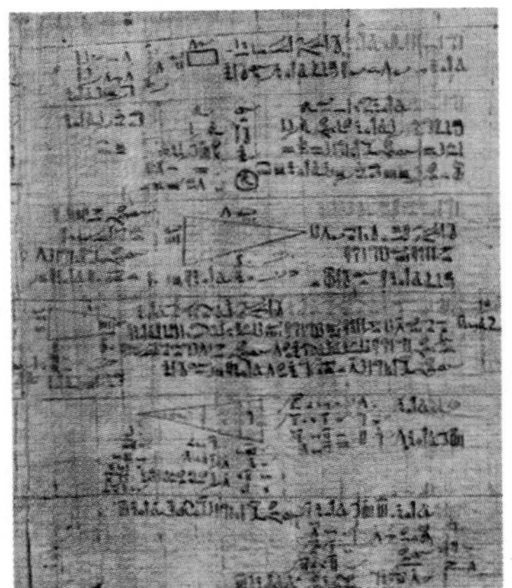

Much of our knowledge of Egyptian mathematics and numeration comes from a roll of papyrus measuring 18 ft by 13 in. It was discovered in a shop in Luxor, Egypt, in 1858 by Henry Rhind, a Scottish lawyer turned archaeologist. Unlike the straightforward accounting of property and events common to Egyptian tombs, the Rhind Papyrus has inscribed on it 85 mathematical problems and solutions involving addition, subtraction, multiplication, division, and geometry.

The key to translation of the papyrus was the Rosetta Stone, which had been discovered some 60 years earlier by one of Napoleon's officers. Both of these treasures now reside at the British Museum in London.

TABLE 4.1 Egyptian Hieroglyphics

Hindu–Arabic Numerals	Egyptian Numerals	Description
1	𓏺	Staff (vertical stroke)
10	𓎆	Heel bone (arch)
100	𓍢	Scroll (coiled rope)
1,000	𓆼	Lotus flower
10,000	𓂭	Pointing finger
100,000	𓆐	Tadpole (or whale)
1,000,000	𓁨	Astonished person

EXAMPLE 1 *From Egyptian to Hindu–Arabic Numerals*

Write the following numeral as a Hindu–Arabic numeral.

𓂭𓂭𓍢𓍢𓍢𓍢𓏺𓏺

SOLUTION:

$$10{,}000 + 10{,}000 + 100 + 100 + 100 + 100 + 1 + 1 = 20{,}402$$ ▲

EXAMPLE 2 *From Hindu–Arabic to Egyptian Numerals*

Write 43,628 as an Egyptian numeral.

SOLUTION:

$$43{,}628 = 40{,}000 + 3000 + 600 + 20 + 8$$

𓂭𓂭𓂭𓂭𓆼𓆼𓆼𓍢𓍢𓍢𓍢𓍢𓍢𓎆𓎆𓏺𓏺𓏺𓏺𓏺𓏺𓏺𓏺 ▲

In this system, the order of the symbols is not important. For example, 𓆐𓏺𓏺𓍢𓍢𓎆 and 𓏺𓏺𓍢𓍢𓆐𓎆 both represent 100,212.

Users of additive systems easily accomplished addition and subtraction by combining or removing symbols. Multiplication and division were more difficult; they were performed by a process called *duplation and mediation* (see Section 4.5). The Egyptians had no symbol for zero, but they did have an understanding of fractions. The symbol 𓂋 was used to take the reciprocal of a number; thus, 𓂋𓏺𓏺𓏺 meant $\frac{1}{3}$ and 𓂋𓎆𓏺 was $\frac{1}{11}$. Writing large numbers in the Egyptian Hieroglyphics system takes longer than in other systems because so many symbols have to be listed. For example, 45 symbols are needed to represent the number 99,999.

The Roman numeration system, a second example of an additive system, was developed later than the Egyptian system. Roman numerals (Table 4.2) were used in most European countries until the eighteenth century. They are still commonly seen on buildings, on clocks, and in books. Roman numerals are selected letters of the Roman alphabet.

TABLE 4.2 Roman Numerals

Roman numerals	I	V	X	L	C	D	M
Hindu–Arabic numerals	1	5	10	50	100	500	1000

DID YOU KNOW

Roman Numerals

Roman numerals remained popular on large clock faces long after their disappearance from daily transactions because they are easier to read from a distance than Hindu–Arabic numerals. The clock shown in this photo is in Vancouver, Canada.

The Roman system has two advantages over the Egyptian system. The first is that it uses the subtraction principle as well as the addition principle. Starting from the left, we add each numeral unless its value is smaller than the value of the numeral to its right. In that case, we subtract its value from the value of the numeral to its right. Only the numbers 1, 10, 100, 1000, . . . can be subtracted, and they can only be subtracted from the next two higher numbers. For example, C (100) can be subtracted only from D (500) or M (1000). The symbol DC represents 500 + 100, or 600, and CD represents 500 − 100, or 400. Similarly, MC represents 1000 + 100, or 1100, and CM represents 1000 − 100, or 900.

EXAMPLE 3 *A Roman Numeral*

Write MMCCCLXII as a Hindu–Arabic numeral.

SOLUTION: Since each numeral is larger than the one on its right, no subtraction is necessary.

$$\begin{aligned}\text{MMCCCLXII} &= 1000 + 1000 + 100 + 100 + 100 + 50 + 10 + 1 + 1\\ &= 2362\end{aligned}$$

▲

EXAMPLE 4 *A Roman Numeral Involving a Subtraction*

Write DCXLVI as a Hindu–Arabic numeral.

SOLUTION: Checking from left to right, we see that X (10) has a smaller value than L (50). Therefore, XL represents 50 − 10, or 40.

$$\text{DCXLVI} = 500 + 100 + (50 - 10) + 5 + 1 = 646$$

▲

EXAMPLE 5 *Writing a Roman Numeral*

Write 289 as a Roman numeral.

SOLUTION:

$$289 = 200 + 80 + 9 = 100 + 100 + 50 + 10 + 10 + 10 + 9$$

(Nine is treated as 10 − 1.)

$$289 = \text{CCLXXXIX}$$

▲

In the Roman numeration system, a symbol does not have to be repeated more than three consecutive times. For example, the number 646 would be written DCXLVI instead of DCXXXXVI.

The second advantage of the Roman numeration system over the Egyptian numeration system is that it makes use of the multiplication principle for numbers over 1000. A bar above a symbol or group of symbols indicates that the symbol or symbols are to be multiplied by 1000. Thus, $\overline{\text{V}} = 5 \times 1000 = 5000$, $\overline{\text{X}} = 10 \times 1000 = 10{,}000$, and $\overline{\text{CD}} = 400 \times 1000 = 400{,}000$. Other examples are, $\overline{\text{VI}} = 6 \times 1000 = 6000$, $\overline{\text{XIX}} = 19 \times 1000 = 19{,}000$, and $\overline{\text{XCIV}} = 94 \times 1000 = 94{,}000$. This greatly reduces the number of symbols needed to write large numbers. Still, it requires 19 symbols, including the bar, to write the number 33,888 in Roman numerals. Write the number 33,888 in Roman numerals now.

Multiplicative Systems

Multiplicative numeration systems are more similar to our Hindu–Arabic system than are additive systems. The number 642 in a multiplicative system might be written (6) (100) (4) (10) (2) or

$$\begin{array}{c} 6 \\ 100 \\ 4 \\ 10 \\ 2 \end{array}$$

Note that no addition signs are needed to represent the number. From this illustration, try to formulate a rule explaining how multiplicative systems work.

The principle example of a multiplicative system is the traditional Chinese system. The numerals used in this system are given in Table 4.3.

TABLE 4.3 Traditional Chinese Numerals

Traditional Chinese numerals	零	一	二	三	四	五	六	七	八	九	十	百	千
Hindu–Arabic numerals	0	1	2	3	4	5	6	7	8	9	10	100	1000

Chinese numerals are always written vertically. The number on top will be a number from 1 to 9 inclusive. This number is to be multiplied by the power of 10 below it. The number 20 is written

二
十 } $2 \times 10 = 20$

The number 400 is written

四
百 } $4 \times 100 = 400$

EXAMPLE 6 ***A Traditional Chinese Numeral***

Write 538 as a Chinese numeral.

SOLUTION:

538 =	500 =	5	五
		100	百
	30 =	3	三
		10	十
	8 =	8	八

▲

Note that in Example 6 the units digit, the 8, is not multiplied by a power of the base.

When writing Chinese numerals, there are some special cases that need to be considered. When writing a number between 11 and 19 it is not necessary to include the 1 before the 10. Thus, the number 18 would be written 十/八 rather than 一/十/八. Another special case involves the use of zero.

When more than one consecutive zero occurs (except at the end of a number) you need to write a zero, but only once for two or more consecutive zeros. Zeros are not included at the end of numbers. The top two illustrations that follow show how zeros are used within a number and the bottom two show that zeros are not used at the end of a number.

406 =
四 百 } $4 \times 100 = 400$
零 } $0 \times 10 = 0$
六 } $6 = 6$

4006 =
四 千 } $4 \times 1000 = 4000$
零 } 0×100, 0×10 $= 0$
六 } $6 = 6$

460 =
四 百 } $4 \times 100 = 400$
六 十 } $6 \times 10 = 60$

4600 =
四 千 } $4 \times 1000 = 4000$
六 百 } $6 \times 100 = 600$

EXAMPLE 7 *Traditional Chinese Numerals*

Write the following as traditional Chinese numerals.

a) 7080 **b)** 7008

SOLUTION: In part (a) there is one zero between the 7 and the 8. In part (b) there are two zeros between the 7 and the 8. As just mentioned, the symbol for zero is used only once in each of those numbers.

a) 7080 =
七 千 } 7×1000
零 } 0×100
八 十 } 8×10

b) 7008 =
七 千 } 7×1000
零 } 0×100, 0×10
八 } 8

▲

TIMELY TIP Notice the difference between our Hindu–Arabic numeration system, which is a positional numeration system, and the Chinese system, which is a multiplicative numeration system. Consider the number 5678. Below we show how that number would be written in Chinese numerals if the Chinese system was a positional value system similar to ours.

Multiplicative		Positional Value	
五	5	五	5
千	1000	六	6
六	6	七	7
百	100	八	8
七	7		
十	10		
八	8		

Note that the multiples of base 10 are removed when writing the number as a positional value number. We will discuss positional value systems in more detail shortly.

> **DID YOU KNOW**
>
> *Hindu-Arabic Numerals*
>
> 
>
>
> The Hindu–Arabic system of numeration is used throughout the world. In many countries, numbers are displayed in Hindu–Arabic numerals as well as the numerals used in the country itself.

Ciphered Systems

A ciphered numeration system is one in which there are numerals for numbers up to and including the base and for multiples of the base. The numbers represented by a particular set of numerals is the sum of the values of the numerals.

Ciphered numeration systems require the memorization of many different symbols but have the advantage that numbers can be written in a compact form. The ciphered numeration system that we discuss is the Ionic Greek system (see Table 4.4). The Ionic Greek system was developed in about 3000 B.C., and it used letters of their alphabet for numerals. Other ciphered systems include the Hebrew, Coptic, Hindu, Brahmin, Syrian, Egyptian Hieratic, and early Arabic systems.

TABLE 4.4 Ionic Greek Numerals

1	α	alpha	60	ξ	xi
2	β	beta	70	o	omicron
3	γ	gamma	80	π	pi
4	δ	delta	90	Q	koph*
5	ϵ	epsilon	100	ρ	rho
6	ϛ	vau*	200	σ	sigma
7	ζ	zeta	300	τ	tau
8	η	eta	400	υ	upsilon
9	θ	theta	500	ϕ	phi
10	ι	iota	600	χ	chi
20	κ	kappa	700	ψ	psi
30	λ	lambda	800	ω	omega
40	μ	mu	900	ͳ	sampi*
50	ν	nu			

*Taken from the Phoenician alphabet.

Since the Greek alphabet contains 24 letters but 27 symbols were needed, the Greeks borrowed the symbols ϛ, Q, and ͳ from the Phoenician alphabet.

The number 24 = 20 + 4. When 24 is written as a Greek numeral, the plus sign is omitted:

$$24 = \kappa\delta$$

The number 996 written as a Greek numeral is ͳQϛ.

When a prime (′) is placed above a number, it multiplies that number by 1000. For example,

$$\beta' = 2 \times 1000 = 2000$$
$$\sigma' = 200 \times 1000 = 200{,}000$$

EXAMPLE 8 ***The Ionic Greek System: A Ciphered System***

Write $\phi\,\nu\,\gamma$ as a Hindu–Arabic numeral.

SOLUTION: $\phi = 500$, $\nu = 50$, and $\gamma = 3$. Adding these numbers gives 553. 

EXAMPLE 9 ***Writing an Ionic Greek Numeral***

Write 9432 as an Ionic Greek numeral.

SOLUTION:

$$\begin{aligned} 9432 &= 9000 + 400 + 30 + 2 \\ &= (9 \times 1000) + 400 + 30 + 2 \\ &= \theta' \qquad\qquad\quad \upsilon \quad\ \ \lambda \quad \beta \\ &= \theta'\upsilon\,\lambda\,\beta \end{aligned}$$

SECTION 4.1 EXERCISES

Concept/Writing Exercises

1. What is the difference between a number and a numeral?
2. List four numerals given in this section that may be used to represent the number ten.
3. What is a system of numeration?
4. List four numerals given in this section that may be used to represent the number one hundred.
5. What is the name of the system of numeration that we presently use?
6. Explain how numbers are represented in an additive numeration system.
7. Explain how numbers are represented in a multiplicative numeration system.
8. Explain how numbers are represented in a ciphered numeration system.

Practice the Skills

In Exercises 9–14, write the numeral as a Hindu–Arabic numeral.

9. 𓍢𓎆𓎆𓎆𓎆𓏤𓏤
10. 𓍢𓍢𓎆𓎆𓏤𓏤
11. 𓆼𓆼𓍢𓍢𓍢𓍢𓎆𓎆𓏤𓏤𓏤
12. 𓂭𓂭𓂭𓂭𓆼𓍢𓍢𓎆
13. 𓆐𓆐𓆐𓂭𓂭𓂭𓆼𓆼𓆼𓆼𓍢𓍢𓎆𓏤𓏤𓏤𓏤
14. 𓁨𓁨𓁨𓆐𓆐𓆐𓍢𓍢𓍢𓍢𓎆𓎆𓎆𓏤

In Exercises 15–20, write the numeral as an Egyptian numeral.

15. 634
16. 752
17. 2045
18. 1812
19. 173,845
20. 3,235,614

In Exercises 21–32, write the numeral as a Hindu–Arabic numeral.

21. XIX
22. XVI
23. DXLVII
24. DLXXV
25. MCDXCII
26. MCMXVIII
27. MMCMXLVI
28. MDCCXLVI
29. $\overline{\text{X}}$MMDCLXVI
30. $\overline{\text{L}}$MCMXLIV
31. $\overline{\text{IX}}$CDLXIV
32. $\overline{\text{V}}$MCCCXXXIII

In Exercises 33–44, write the numeral as a Roman numeral.

33. 59
34. 94
35. 134
36. 269
37. 2005
38. 4285
39. 4793
40. 6274
41. 9999
42. 14,315
43. 20,644
44. 99,999

In Exercises 45–52, write the numeral as a Hindu–Arabic numeral.

45. 七十四
46. 六十二
47. 四千零八十一
48. 三千零二十九

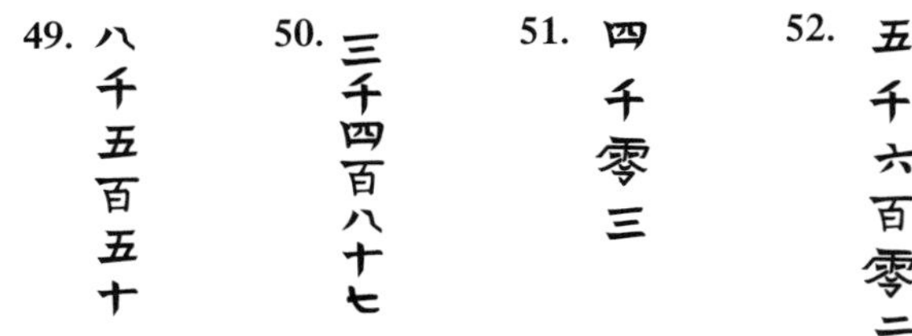

In Exercises 53–60, write the numeral as a traditional Chinese numeral.

53. 53
54. 178
55. 378
56. 2001
57. 4260
58. 6905
59. 7056
60. 3009

In Exercises 61–66, write the numeral as a Hindu–Arabic numeral.

61. τ μ α
62. ψ λ ϛ
63. κ′β′φε
64. ρ′ ν′ω ι γ
65. θ′χ ζ
66. δ′ ϡ Q θ

In Exercises 67–72, write the numeral as an Ionic Greek numeral.

67. 59
68. 178
69. 726
70. 2001
71. 82,704
72. 690,540

In Exercises 73–75, compare the advantages and disadvantages of a ciphered system of numeration with those of the named system.

73. An additive system
74. A multiplicative system
75. The Hindu–Arabic system

In Exercises 76–79, write the numeral as numerals in the indicated systems of numeration.

76. 𓆼𓎆𓎆𓏺 in Hindu–Arabic, Roman, traditional Chinese, and Greek
77. MCMXXXVI in Hindu–Arabic, Egyptian, Greek, and traditional Chinese
78. 五百二十七 in Hindu–Arabic, Egyptian, Roman, and Greek
79. υκβ in Hindu–Arabic, Egyptian, Roman, and traditional Chinese

Challenge Problems/Group Activities

80. Write the Roman numeral for 999,999.
81. Write the Ionic Greek numeral for 999,999.
82. Make up your own additive system of numeration and indicate the symbols and rules used to represent numbers. Using your system of numeration, write
   a) your age.
   b) the year you were born.
   c) the current year.

Recreational Mathematics

83. Without using any type of writing instrument, what can you do to make the following incorrect statement a correct statement?

   XI + I = X

84. Words and numbers that read the same both backward and forward are called *palindromes*. Some examples are the words CIVIC and RACECAR, and the numbers 121 and 32523. Using Roman numerals, list the last year that was a palindrome.
85. Which year in the past 2000 years required the most Roman numerals to write? Write out the year in Roman numerals.

Internet/Research Activity

86. In this section we discussed Egyptian hieroglyphics, Ionic Greek numerals, and other numeration systems. Select either Egypt or Greece.
   a) Give the current numerals used in that country.
   b) Explain how their current system of numeration works.

   If more than one numeration system is used in the country you selected, discuss the system most commonly used.

4.2 PLACE-VALUE OR POSITIONAL-VALUE NUMERATION SYSTEMS

Eighteenth-century mathematician Pierre Simon, Marquis de Laplace, speaking of the positional principle, said: "The idea is so simple that this very simplicity is the reason for our not being sufficiently aware of how much attention it deserves."

Today the most common type of numeration system is the place-value system. The Hindu–Arabic numeration system, used in the United States and many other countries, is an example of a place-value system. In a ***place-value system***, which is also called a ***positional value system***, the value of the symbol depends on its position in the representation of the number. For example, the 2 in 20 represents 2 tens, and the 2 in 200 represents 2 hundreds. A true positional-value system requires a ***base*** and a set of symbols, including a symbol for zero and one for each counting number less than the base. Although any number can be written in any base, the most common positional system is the base 10 system which is called the ***decimal number system***.

The Hindus in India are credited with the invention of zero and the other symbols used in our system. The Arabs, who traded regularly with the Hindus, also adopted the system, thus the name Hindu–Arabic. Not until the middle of the fifteenth century, however, did the Hindu–Arabic numerals take the form we know today.

The Hindu–Arabic numerals and the positional system of numeration revolutionized mathematics by making addition, subtraction, multiplication, and division much easier to learn and very practical to use. Merchants and traders no longer had to depend on the counting board or abacus. The first group of mathematicians, who computed with the Hindu–Arabic system rather than with pebbles or beads on a wire, were known as the "algorists."

DID YOU KNOW

Babylonian Numerals

The form Babylonian numerals took is directly related to their writing materials. Babylonians used a reed (later a stylus) to make their marks in wet clay. The end could be used to make a thin wedge, 𒁹, which represents a unit, or a wider wedge, 𒌋, which represents 10 units. The clay dried quickly, so the writings tended to be short but extremely durable.

In the Hindu–Arabic system, the symbols 0, 1, 2, 3, 4, 5, 6, 7, 8, and 9 are called ***digits***. The base 10 system was developed from counting on fingers, and the word *digit* comes from the Latin word for fingers.

The positional values in the Hindu–Arabic system are

$$\ldots, (10)^5, (10)^4, (10)^3, (10)^2, 10, 1$$

To evaluate a number in the Hindu–Arabic system, we multiply the first digit on the right by 1. We multiply the second digit from the right by the base, 10. We multiply the third digit from the right by the base squared, 10^2 or 100. We multiply the fourth digit from the right by the base cubed, 10^3 or 1000, and so on. In general, we multiply the digit n places from the right by 10^{n-1}. Therefore, we multiply the digit eight places from the right by 10^7. Using the place-value rule, we can write a number in ***expanded form***. The number 1234 written in expanded form is

$$1234 = (1 \times 10^3) + (2 \times 10^2) + (3 \times 10) + (4 \times 1)$$

or

$$(1 \times 1000) + (2 \times 100) + (3 \times 10) + 4$$

The oldest known numeration system that resembled a place-value system was developed by the Babylonians in about 2500 B.C. Their system resembled a place-value system with a base of 60, a sexagesimal system. It was not a true place-value system because it lacked a symbol for zero. The lack of a symbol for zero led to a great deal of ambiguity and confusion. Table 4.5 gives the Babylonian numerals.

TABLE 4.5 Babylonian Numerals

Babylonian Numerals	𒁹	𒌋
Hindu–Arabic numerals	1	10

The positional values in the Babylonian system are

$$\ldots, (60)^3, (60)^2, 60, 1$$

DID YOU KNOW

Counting Boards

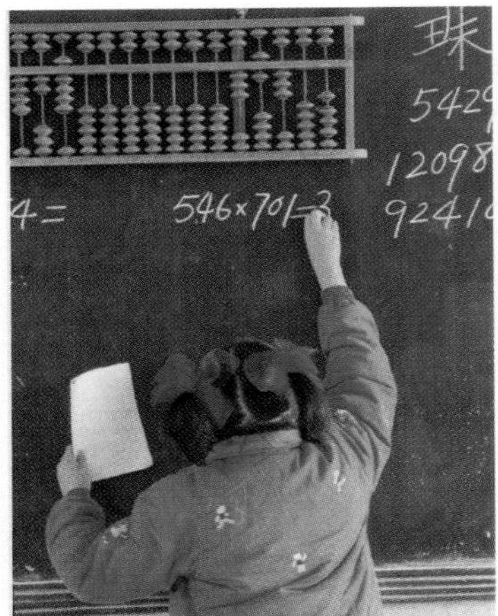

One of the earliest counting devices, used in most ancient civilizations, was the counting board. On such a board, each column represents a positional value. The number of times a value occurs is represented by markers (beads, stones, sticks) in the column. An empty column signifies "no value." The widespread use of counting boards meant that Europeans were already long accustomed to working with positional values when they were introduced to Hindu–Arabic numerals in the fifteenth century. Some people in China, Japan, Russia, Eastern Europe, and the United States still commonly use a type of counting board known as the abacus to perform routine computations.

In a Babylonian numeral, a gap is left between the characters to distinguish between the various place values. From right to left, the sum of the first group of numerals is multiplied by 1. The sum of the second group is multiplied by 60. The sum of the third group is multiplied by $(60)^2$, and so on.

EXAMPLE 1 ***The Babylonian System: A Positional Value System***

Write 𒌋𒌋 𒌋𒌋𒁹𒁹𒁹𒁹 as a Hindu–Arabic numeral.

SOLUTION:

𒌋𒌋	𒌋𒌋𒁹𒁹𒁹𒁹
60's	units
10 + 10	10 + 10 + 1 + 1 + 1 + 1
60's	units

$$(20 \times 60) + (24 \times 1)$$
$$1200 + 24 = 1224$$

▲

The Babylonians used the symbol 𒇲 to indicate subtraction. The numeral 𒌋𒇲𒁹𒁹 represents 10 − 2, or 8. The numeral 𒌋𒌋𒌋𒌋𒇲𒁹𒁹𒁹 represents 40 − 3, or 37 in base 10 or decimal notation.

EXAMPLE 2 ***From Babylonian to Hindu–Arabic Numerals***

Write 𒁹𒁹 𒌋𒁹 𒌋𒌋𒇲𒁹𒁹 as a Hindu–Arabic numeral.

SOLUTION: The place value of these three groups of numerals from left to right is

$$(60)^2, \quad 60, \quad 1$$

or

$$3600, \quad 60, \quad 1$$

The numeral in the group on the right has a value of 20 − 2, or 18. The numeral in the center group has a value of 10 + 1, or 11. The numeral on the left represents 1 + 1, or 2. Multiplying each group by its positional value gives

$$\begin{aligned} & (2 \times 60^2) + (11 \times 60) + (18 \times 1) \\ &= (2 \times 3600) + (11 \times 60) + (18 \times 1) \\ &= 7200 + 660 + 18 \\ &= 7878 \end{aligned}$$

▲

To explain the procedure used to convert from a Hindu–Arabic numeral to a Babylonian numeral, we will consider a length of time. How can we change 9820 seconds into hours, minutes, and seconds? Since there are 3600 seconds in an hour

DID YOU KNOW

On the High Seas

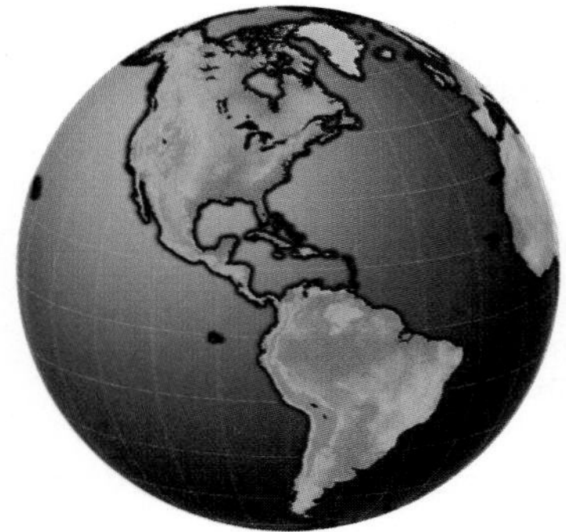

Vestiges of the Babylonian sexagesimal system are still with us today, especially in navigation. Navigators use degrees, minutes, and seconds of longitude and latitude and global positioning systems (GPS) to chart their course. If you look at a globe or world map, you will see that Earth is divided into 360° of latitude and 360° of longitude. Each degree can be divided into 60 minutes, and each minute can be divided into 60 seconds.

Early explorers had to have an easy means of computing angles as they guided their ships by the stars. Base 60 easily divides into halves, thirds, fourths, fifths, sixths, tenths, twelfths, fifteenths, twentieths, and thirtieths, making such computations easier. Hence, the use of a base 60 system became popular.

(60 seconds to a minute and 60 minutes to an hour), we can find the number of hours in 9820 seconds by dividing 9820 by 60^2, or 3600.

```
         2   ← Hours
3600)9820
     7200
     2620    ← Remaining seconds
```

Now we can determine the number of minutes by dividing the remaining seconds by 60, the number of seconds in a minute.

```
     43   ← Minutes
60)2620
   2400
    220
    180
     40   ← Remaining seconds
```

Since the remaining number of seconds, 40, is less than the number of seconds in a minute, our task is complete.

$$9820 \text{ sec} = 2 \text{ hr, } 43 \text{ min, and } 40 \text{ sec}$$

The same procedure is used to convert a decimal (base 10) number to a Babylonian number or any number in a different base.

EXAMPLE 3 *From Hindu–Arabic to Babylonian Numerals*

Write 2519 as a Babylonian numeral.

SOLUTION: The Babylonian numeration system has positional values of

$$\ldots, 60^3, 60^2, 60, 1$$

which can be expressed as

$$\ldots, 216000, 3600, 60, 1$$

The largest positional value less than or equal to 2519 is 60. To determine how many groups of 60 are in 2519, divide 2519 by 60.

```
     41   ← Groups of 60
60)2519
   240
    119
     60
     59   ← Units remaining
```

Thus, $2519 \div 60 = 41$ with remainder 59. There are 41 groups of 60 and 59 units remaining. Because the remainder, 59, is less than the base, 60, no further division is necessary. The remainder represents the number of units when the number is

DID YOU KNOW

Sacred Mayan Glyphs

3

4

In addition to their base 20 numerals the Mayans had a holy numeration system used by priests to create and maintain calendars. They used a special set of hieroglyphs that consisted of pictograms of Mayan gods. For example, the number 3 was represented by the god of wind and rain, the number 4 by the god of sun.

written in expanded form. Therefore, $2519 = (41 \times 60) + (59 \times 1)$. When written as a Babylonian numeral, 2519 is

<<<<▼ <<<<<▼▼ ▲

EXAMPLE 4 *Using Division to Determine a Babylonian Numeral*

Write 6270 as a Babylonian numeral.

SOLUTION: Divide 6270 by the largest positional value less than or equal to 6270. That value is 3600.

$$6270 \div 3600 = 1 \text{ with remainder } 2670$$

There is one group of 3600 in 6270. Next divide the remainder 2670 by 60 to determine the number of groups of 60 in 2670.

$$2670 \div 60 = 44 \text{ with remainder } 30$$

There are 44 groups of 60 and 30 units remaining.

$$6270 = (1 \times 60^2) + (44 \times 60) + (30 \times 1)$$

Thus, 6270 written as a Babylonian numeral is

▼ <<<<▼▼▼▼ <<< ▲

Another place-value system is the Mayan numeration system. The Mayans, who lived on the Yucatan Peninsula in present day Mexico, developed a sophisticated numeration system based on their religious and agricultural calendar. The numbers in this system are written vertically rather than horizontally, with the units position on the bottom. In the Mayan system, the number in the bottom row is to be multiplied by 1. The number in the second row from the bottom is to be multiplied by 20. The number in the third row is to be multiplied by 18×20, or 360. You probably expected the number in the third row to be multiplied by 20^2 rather than 18×20. It is believed that the Mayans used 18×20 so that their numeration system would conform to their calendar of 360 days. The positional values above 18×20 are 18×20^2, 18×20^3, and so on.

Positional values in the Mayan system

... $18 \times (20)^3$,	$18 \times (20)^2$,	18×20,	20,	1
or... 144,000,	7200,	360,	20,	1

The digits 0, 1, 2, 3, ..., 19 of the Mayan systems are formed by a simple grouping of dots and lines, as shown in Table 4.6.

TABLE 4.6 Mayan Numerals

0	1	2	3	4	5	6	7	8	9
⬭	•	••	•••	••••	—	• —	•• —	••• —	•••• —
10	**11**	**12**	**13**	**14**	**15**	**16**	**17**	**18**	**19**
— —	• — —	•• — —	••• — —	•••• — —	— — —	• — — —	•• — — —	••• — — —	•••• — — —

DID YOU KNOW

Numerals of the World

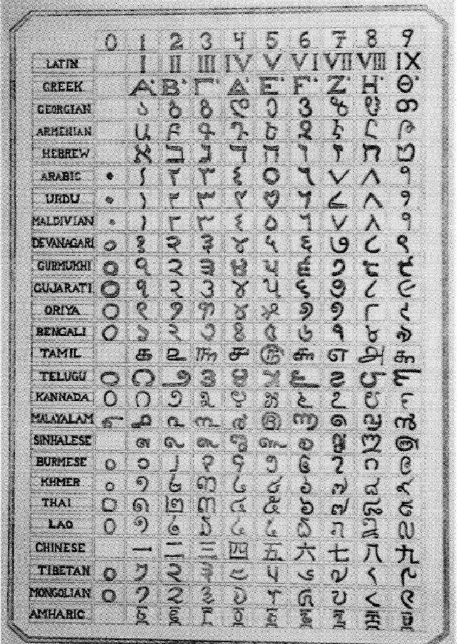

Although most countries presently use a place value (or positional value) numeration system with base 10, the numerals used for the digits differ by country. The photo of the artwork entitled *Numbers* by Jan Fleck shows numerals currently used in many countries of the world. For example, in Burmese, the numeral ၃ has a value of 3.

EXAMPLE 5 ***The Mayan System: A Positional Value System***

Write (•••• over —; •• ; ••• over two bars) as a Hindu–Arabic numeral.

SOLUTION: In the Mayan numeration system, the first three positional values are

$$18 \times 20$$
$$20$$
$$1$$

•••• over —	=	$9 \times (18 \times 20)$	=	3240
••	=	2×20	=	40
••• over two bars	=	13×1	=	13
				3293

▲

EXAMPLE 6 ***From Mayan to Hindu–Arabic Numerals***

Write (••• over —; • over two bars; ••••) as a Hindu–Arabic numeral.

SOLUTION:

••• over —	=	$8 \times (18 \times 20)$	=	2880
• over two bars	=	11×20	=	220
••••	=	4×1	=	4
				3104

▲

EXAMPLE 7 ***From Hindu–Arabic to Mayan Numerals***

Write 4025 as a Mayan numeral.

SOLUTION: To convert from a Hindu–Arabic to a Mayan numeral, we use a procedure similar to the one used to convert to a Babylonian numeral. The Mayan positional values are . . . , 7200, 360, 20, 1. The greatest positional value less than or equal to 4025 is 360. Divide 4025 by 360.

$$4025 \div 360 = 11 \text{ with remainder } 65$$

There are 11 groups of 360 in 4025. Next, divide the remainder, 65, by 20.

$$65 \div 20 = 3 \text{ with remainder } 5$$

There are 3 groups of 20 with five units remaining.

$$4025 = (11 \times 360) + (3 \times 20) + (5 \times 1)$$

4025 written as a Mayan numeral is

11×360	=	• over two bars
3×20		•••
5×1		—

▲

> **TIMELY TIP** Notice that changing a number from the Babylonian or Mayan numeration System *to the Hindu–Arabic* (or decimal or base 10) system involves *multiplication*. Changing a number *from the Hindu–Arabic system* to the Babylonian or Mayan numeration system involves *division*.

SECTION 4.2 EXERCISES

Concept/Writing Exercises

1. What is the most common type of numeration system used in the world today?
2. What is another name for a place-value system?
3. Consider the numbers 40 and 400 in the Hindu–Arabic numeration system. What does the 4 represent in each number?
4. What is the most common base used for a positional value system? Explain why you believe the base you named is the most common base.
5. In a true positional-value system, what symbols are required?
6. a) What is the base in the Hindu–Arabic numeration system?
   b) What are the digits in the Hindu–Arabic numeration system?
7. Explain how to write a number in expanded form in a positional-value numeration system.
8. Why was the Babylonian system not a true place-value system?
9. a) The Babylonian system did not have a symbol for zero. Why did this lead to some confusion?
   b) Write the numbers 133 and 7980 as Babylonian numerals.
10. Consider the Babylonian number represented by <▼. Give two numbers in Hindu–Arabic numerals this number may represent. Explain your answer.
11. List the first five positional values, starting with the units position, for the Mayan numeration system.
12. Describe two ways that the Mayan place-value system differs from the Hindu–Arabic place-value system.

Practice the Skills

In Exercises 13–24, write the Hindu–Arabic numeral in expanded form.

13. 63	**14.** 75	**15.** 359
16. 562	**17.** 897	**18.** 3769
19. 4387	**20.** 23,468	**21.** 16,402
22. 125,678	**23.** 346,861	**24.** 3,765,934

In Exercises 25–30, write the Babylonian numeral as a Hindu–Arabic numeral.

25. <<<<▼▼ **26.** <<<▼▼▼▼▼

27. <▼▼▼ ▼▼▼▼ **28.** <▼ <<▼▼▼▼

29. ▼ <<▼ <▼▼▼ **30.** < <<▼▼▼▼ ▼▼

In Exercises 31–36, write the numeral as a Babylonian numeral.

31. 88	**32.** 97	**33.** 295
34. 512	**35.** 3685	**36.** 3030

In Exercises 37–42, write the Mayan numeral as a Hindu–Arabic numeral.

37.
••••
••
═

38.
═

39.
••
═
⬭
•

40.
••
—
••••
••
—

41.
•
═
••
⬭

42.
••
≡

In Exercises 43–48, write the numeral as a Mayan numeral.

43. 17	**44.** 257	**45.** 297
46. 406	**47.** 2163	**48.** 1978

49. *Comparisons of Systems* Compare the advantages and disadvantages of a place-value system with those of
    a) additive numeration systems.
    b) multiplicative numeration systems.
    c) ciphered numeration systems.
50. *Your Own System* Create your own place-value system. Write 2005 in your system.

In Exercises 51 and 52, write the numeral in the indicated systems of numeration.

51. ◀◀◀ ▼▼▼ in Hindu–Arabic and Mayan

52. ―― in Hindu–Arabic and Babylonian
 ••
 ――
 ••••

In Exercises 53 and 54, suppose a place-value numeration system has base ▢, with digits represented by the symbols △, ϕ, □, and ⊠. Write each expression in expanded form.

53. △□ϕ

54. ⊠△ϕ□

Challenge Problems/Group Activities

55. **a)** Is there a largest number in the Babylonian numeration system? Explain.
 b) Write the Babylonian numeral for 999,999.

56. **a)** Is there a largest number in the Mayan numeration system? Explain.
 b) Write the Mayan numeral for 999,999.

In Exercises 57–60, first convert each numeral to a Hindu–Arabic numeral and then perform the indicated operation. Finally, convert the answer back to a numeral in the original numeration system.

57. ▼▼ ◀◀ ▼▼▼ + ◀◀▼▼▼

58. ▼▼▼ ◀◀◀▼▼▼ − ◀◀◀▼▼

59. (•• / • / ══) + (• / •• / •••)

60. (•• / • / ══) − (• / •• / •••)

Recreational Mathematics

61. Hidden in the box are the names of the four different types of systems of numeration we discussed in this chapter: ADDITIVE, MULTIPLICATIVE, CIPHERED, and PLACE-VALUE (without the hyphen). You find them by going box by box, but the boxes you move between must touch vertically, horizontally, or diagonally. You can use the same box more than once when spelling out a word. Find the names of the numeration systems. Make sure you understand how each system of numeration works.

M	L	P	R	E
A	I	A	D	H
V	C	T	D	P
E	A	L	I	C
A	M	U	E	V

Internet/Research Activities

62. Investigate and write a report on the development of the Hindu–Arabic system of numeration. Start with the earliest records of this system in India.

63. The Arabic numeration system currently in use is a base 10 positional-value system, which uses different symbols than the Hindu–Arabic numeration system. Write the symbols used in the Arabic system of numeration and their equivalent symbols in the Hindu–Arabic numeration system. Write 54, 607, and 2000 in Arabic numerals.

4.3 OTHER BASES

The positional values in the Hindu–Arabic numeration system are

$$\ldots, (10)^4, (10)^3, (10)^2, 10, 1$$

The positional values in the Babylonian numeration system are

$$\ldots, (60)^4, (60)^3, (60)^2, 60, 1$$

The numbers 10 and 60 are called the *bases* of the Hindu–Arabic and Babylonian systems, respectively.

DID YOU KNOW

Beyond Their Fingers

The Kewa people of Papua, New Guinea, have gone well beyond counting on their fingers: They use the entire upper body. Going from the little finger of one hand, down the elbow, to shoulder to head to shoulder, down the other elbow, to the little finger of the opposite hand provides them with a count of 68.

Any counting number greater than 1 may be used as a base for a positional-value numeration system. If a positional-value system has a base b, then its positional values will be

$$\ldots, b^4, b^3, b^2, b, 1$$

The positional values in a base 8 system are

$$\ldots, 8^4, 8^3, 8^2, 8, 1$$

and the positional values in a base 2 system are

$$\ldots, 2^4, 2^3, 2^2, 2, 1$$

As we indicated earlier, the Mayan numeration system is based on the number 20. It is not, however, a true base 20 positional-value system. Why not?

The reason for the almost universal acceptance of base 10 numeration systems is that most human beings have 10 fingers. Even so, there are still some positional-value numeration systems that use bases other than 10. Some societies are still using a base 2 numeration system. They include some groups of people in Australia, New Guinea, Africa, and South America. Bases 3 and 4 are also used in some areas of South America. Base 5 systems were used by some primitive tribes in Bolivia, but the tribes are now extinct. The pure base 6 system occurs only sparsely in Northwest Africa. Base 6 also occurs in other systems in combination with base 12, the *duodecimal system*.

We continue to see remains of other base systems in many countries. For example, there are 12 inches in a foot, 12 months in a year. Base 12 is also evident in the dozen, the 24-hour day, and the gross (12×12). English uses the word *score* to mean 20, as in "Four score and seven years ago." Remains of base 60 are found in measurements of time (60 seconds to a minute, 60 minutes to an hour) and angles (60 seconds to one minute, 60 minutes to one degree).

The base 2, or ***binary system***, has become very important because it is the internal language of the computer. For example, when a grocery store's cash register computer records the price of your groceries by using a scanning device, the bar codes it scans on the packages are in binary form. Computers use a two-digit "alphabet" that consists of the numerals 0 and 1. Every character on a standard keyboard can be represented by a combination of those two numerals. A single numeral such as 0 or 1 is called a ***bit***. Other bases that computers make use of are base 8 and base 16. A group of eight bits is called a ***byte***. In the American Standard Code for Information Interchange (ASCII) code, used in most computers, the byte 01000001 represents the character A, 01100001 represents the character a, 00110000 represents the character 0, and 00110001 represents the character 1.

A place-value system with base b must have b distinct symbols, one for zero and one for each number less than the base. A base 6 system must have symbols for the numbers 0, 1, 2, 3, 4, and 5. All numbers in base 6 are constructed from these 6 symbols. A base 8 system must have symbols for 0, 1, 2, 3, 4, 5, 6, and 7. All numbers in base 8 are constructed from these 8 symbols, and so on.

A number in a base other than base 10 will be indicated by a subscript to the right of the number. Thus, 123_5 represents a number in base 5. The number 123_6 represents a number in base 6. The value of 123_5 is not the same as the value of 123_{10}, and the value of 123_6 is not the same as the value 123_{10}. A base 10 number may be written without a subscript. For example 123 means 123_{10} and 456 means 456_{10}. For clarity in certain problems, we will use the subscript 10 to indicate a number in base 10.

Remember the symbols that represent the base itself, in any base b, are 10_b. For example, in base 5, the symbols 10_5 represent the number 5. Note that $10_5 = 1 \times 5 + 0 \times 1 = 5 + 0 = 5_{10}$, or the number 5 in base 10. The symbols 10_5 mean one group of 5 and no units. In base 6, the symbols 10_6 represent the number 6. The symbols 10_6 represent one group of 6 and no units, and so on.

To change a number in a base other than 10 to a base 10 number, we follow the same procedure we used in Section 4.2 to change the Babylonian and Mayan numbers to base 10 numbers. Multiply each digit in the number by its respective positional value. Then find the sum of the products.

EXAMPLE 1 *Converting from Base 6 to Base 10*

Convert 453_6 to base 10.

SOLUTION: In base 6, the positional values are $\ldots, 6^3, 6^2, 6, 1$. In expanded form,

$$\begin{aligned} 453_6 &= (4 \times 6^2) + (5 \times 6) + (3 \times 1) \\ &= (4 \times 36) + (5 \times 6) + (3 \times 1) \\ &= 144 + 30 + 3 \\ &= 177 \end{aligned}$$

In Example 1, the units digit in 453_6 is 3. Notice that 3_6 has the same value as 3_{10} since both are equal to 3 units. That is, $3_6 = 3_{10}$. If n is a digit less than the base b, and the base b is less than or equal to 10, then $n_b = n_{10}$.

EXAMPLE 2 *Converting from Base 8 to Base 10*

Convert 3615_8 to base 10.

SOLUTION:

$$\begin{aligned} 3615_8 &= (3 \times 8^3) + (6 \times 8^2) + (1 \times 8) + (5 \times 1) \\ &= (3 \times 512) + (6 \times 64) + (1 \times 8) + (5 \times 1) \\ &= 1536 + 384 + 8 + 5 \\ &= 1933 \end{aligned}$$

A base 12 system must have 12 distinct symbols. In this text, we use the symbols 0, 1, 2, 3, 4, 5, 6, 7, 8, 9, T, and E, where T represents ten and E represents eleven. Why will the numerals 10_{12} and 11_{12} have different meanings than 10 and 11? The number 10_{12} represents 1 group of twelve plus 0 units, or twelve. The number 11_{12} represents 1 group of twelve plus 1 unit, or 13.

EXAMPLE 3 *Converting from Base 12 to Base 10*

Convert $12T6_{12}$ to base 10.

SOLUTION:

$$\begin{aligned} 12T6_{12} &= (1 \times 12^3) + (2 \times 12^2) + (T \times 12) + (6 \times 1) \\ &= (1 \times 1728) + (2 \times 144) + (10 \times 12) + (6 \times 1) \\ &= 1728 + 288 + 120 + 6 \\ &= 2142 \end{aligned}$$

EXAMPLE 4 *Converting from Base 2*

Convert 101101_2 to base 10.

SOLUTION:

$$
\begin{aligned}
101101_2 &= (1 \times 2^5) + (0 \times 2^4) + (1 \times 2^3) + (1 \times 2^2) + (0 \times 2) + (1 \times 1) \\
&= 32 + 0 + 8 + 4 + 0 + 1 \\
&= 45
\end{aligned}
$$

▲

To change a number from a base 10 system to a different base, we will use a procedure similar to the one we used to convert base 10 numbers to Babylonian and Mayan numbers, as was explained in Section 4.2. Divide the base 10 number by the highest power of the new base that is less than or equal to the given number. Record this quotient. Then divide the remainder by the next smaller power of the new base and record this quotient. Repeat this procedure until the remainder is a number less than the new base. The answer is the set of quotients listed from left to right, with the remainder on the far right. This procedure is illustrated in Examples 5 through 7.

EXAMPLE 5 *Convert to Base 8*

Convert 486 to base 8.

SOLUTION: We are converting a number in base 10 to a number in base 8. The positional values in the base 8 system are . . . , 8^3, 8^2, 8, 1, or . . . , 512, 64, 8, 1. The highest power of 8 that is less than or equal to 486 is 8^2, or 64. Divide 486 by 64.

First digit in answer
↓
$$486 \div 64 = 7 \text{ with remainder } 38$$

Therefore, there are 7 groups of 8^2 in 486. Next divide the remainder, 38, by 8.

Second digit in answer
↓
$$38 \div 8 = 4 \text{ with remainder } 6$$
↑
Third digit in answer

There are 4 groups of 8 in 38 and 6 units remaining. Since the remainder, 6, is less than the base, 8, no further division is required.

$$
\begin{aligned}
&= (7 \times 64) + (4 \times 8) + (6 \times 1) \\
&= (7 \times 8^2) + (4 \times 8) + (6 \times 1) \\
&= 746_8
\end{aligned}
$$

Notice that we placed the subscript 8 to the right of 746 to show that it is a base 8 number. ▲

EXAMPLE 6 *Convert to Base 3*

Convert 273 to base 3.

SOLUTION: The place values in the base 3 system are . . . , $3^6, 3^5, 3^4, 3^3, 3^2, 3, 1$, or . . . , 729, 243, 81, 27, 9, 3, 1. The highest power of the base that is less than or equal to 273 is 3^5, or 243. Successive divisions by the powers of the base give the following result.

$$\begin{aligned} 273 \div 243 &= 1 \text{ with remainder } 30 \\ 30 \div 81 &= 0 \text{ with remainder } 30 \\ 30 \div 27 &= 1 \text{ with remainder } 3 \\ 3 \div 9 &= 0 \text{ with remainder } 3 \\ 3 \div 3 &= 1 \text{ with remainder } 0 \end{aligned}$$

The remainder, 0, is less than the base, 3, so no further division is necessary. To obtain the answer, list the quotients from top to bottom followed by the remainder in the last division.

The number 273 can be represented as one group of 243, no groups of 81, one group of 27, no groups of 9, one group of 3, and no units.

$$\begin{aligned} 273 &= (1 \times 243) + (0 \times 81) + (1 \times 27) + (0 \times 9) + (1 \times 3) + (0 \times 1) \\ &= (1 \times 3^5) + (0 \times 3^4) + (1 \times 3^3) + (0 \times 3^2) + (1 \times 3) + (0 \times 1) \\ &= 101010_3 \end{aligned}$$

EXAMPLE 7 *Convert to Base 12*

Convert 558 to base 12.

SOLUTION: The place values in base 12 are . . . , $12^3, 12^2, 12, 1$, or . . . , 1728, 144, 12, 1. The highest power of the base that is less than or equal to 558 is 12^2, or 144.

$$\begin{aligned} 558 \div 144 &= 3 \text{ with remainder } 126 \\ 126 \div 12 &= \text{T with remainder } 6 \end{aligned}$$

(Remember that T is used to represent 10 in base 12.)

$$558 = (3 \times 12^2) + (\text{T} \times 12) + (6 \times 1) = 3\text{T}6_{12}$$

TIMELY TIP It is important to remember the following items presented in this section.

- If a number is shown without a base, we assume the number is a base 10 number, and
- When converting a base 10 number to a different base, your answer should never contain a digit greater than or equal to that different base.

You should also remember that changing a number given in a base other than 10 to a number in base 10 involves multiplication. Changing a base 10 number to a number in a different base involves division.

MATHEMATICS *Everywhere*

We use our place-value system daily without thinking of its complexity. The place-value system has come a long way from the first and oldest type of numeration system, the additive numeration system. The changes in numeration systems evolved slowly. Most countries now use a place-value system. For example, the numeration system used in China today is different from the traditional system discussed in this chapter. The present-day system in China is a positional-value system rather than a multiplicative system, and in some areas of China, 0 is used as the numeral for zero. Often when you travel to ethnic areas of cities or to foreign countries, numbers are indicated using both that country's numerals and Hindu–Arabic numerals, as shown in the photo.

It is likely that when new numeration systems are presented they appear abstract to the public. Abstract mathematics often becomes the basis for important discoveries and inventions in the future. That is the case with the binary numeration system. When the binary numeration system was first introduced, nobody could have dreamed that it would form the basis for our computers. Today, computers are found everywhere: in our cars, cameras, watches, calculators, sewing machines; at checkout counters; and in hundreds of other places. When a number or letter is entered into a computer it is converted internally into a binary number. The computations within a computer are made using binary arithmetic, and the binary answer is converted back to a decimal number for us to read. Computers also use octal (base 8) and hexadecimal (base 16) numeration systems.

Computers make use of three numeration systems: the *binary* (base 2), *octal* (base 8), and *hexadecimal* (base 16) numeration systems. Computers and calculators use the binary system to perform their internal computations. The binary number system contains only two digits, 0 and 1. All numbers we enter into a computer are converted internally into a series of 0's and 1's. When a computer performs a calculation it treats 0 as an "off" switch and 1 as an "on" switch. Using these electronic switches, the computer performs calculations using binary numbers, and then the internal result is converted back to a decimal number for us to view.

The octal system is used by computer programmers who work with internal computer codes. In a computer, the central processing unit (CPU) often uses the hexadecimal system to convey information to the printer and other output devices.

We have already given examples of converting numbers given in the binary system and the octal system to numbers in the decimal system, and vice versa. Now let's work an example using the hexadecimal system. Since a hexadecimal system contains 16 symbols, it is treated similarly to a base 12 system in that we need to use additional symbols for numerals. See Examples 3 and 7.

EXAMPLE 8 *Convert to and from Base 16*

In this example, let the numerals 0, 1, 2, 3, 4, 5, 6, 7, 8, 9, A, B, C, D, E, and F represent the numerals in a base 16 system where A through F represent ten through fifteen, respectively.

a) Convert $7DE_{16}$ to base 10. **b)** Convert 6713 to base 16.

SOLUTION:

a) In a base 16 system the positional values are $\ldots, 16^3, 16^2, 16, 1$ or $\ldots$ 4096, 256, 16, 1. Since D has a value of 13 and E has a value of 14, we perform the following calculation.

$$\begin{aligned}7DE_{16} &= (7 \times 16^2) + (D \times 16) + (E \times 1)\\ &= (7 \times 256) + (13 \times 16) + (14 \times 1)\\ &= 1792 + 208 + 14\\ &= 2014\end{aligned}$$

b) The highest power of base 16 less than or equal to 6713 is 16^3, or 4096. If we obtain a quotient greater than nine but less than sixteen, we will use the corresponding letter A through F.

$$\begin{aligned}6713 \div 4096 &= 1 \text{ with remainder } 2617\\ 2617 \div 256 &= A \text{ with remainder } 57\\ 57 \div 16 &= 3 \text{ with remainder } 9\end{aligned}$$

Note that A has a value of ten

Thus, $6713 = 1A39_{16}$.

SECTION 4.3 EXERCISES

Concept/Writing Exercises

1. In your own words, explain how to change a number in a base other than base 10 to base 10.
2. In your own words, explain how to change a number in base 10 to a base other than base 10.

Practice the Skills

In Exercises 3–20, convert the numeral to a numeral in base 10.

3. 5_6 **4.** 60_7 **5.** 42_5
6. 101_2 **7.** 1011_2 **8.** 1101_2
9. 84_{12} **10.** 21021_3 **11.** 565_8
12. 654_7 **13.** 20432_5 **14.** 101111_2
15. 4003_6 **16.** $123E_{12}$ **17.** 123_8
18. 2043_8 **19.** 14705_8 **20.** 67342_9

In Exercises 21–36, convert the base 10 numeral to a numeral in the base indicated.

21. 8 to base 2 **22.** 16 to base 2
23. 23 to base 2 **24.** 243 to base 6
25. 635 to base 6 **26.** 908 to base 4
27. 2061 to base 12 **28.** 200 to base 4
29. 529 to base 8 **30.** 81 to base 3
31. 2867 to base 12 **32.** 4312 to base 6
33. 1011 to base 2 **34.** 1589 to base 7
35. 2307 to base 8 **36.** 13,469 to base 8

In Exercises 37–40, assume that a base 16 positional-value system uses the numerals 0, 1, 2, 3, 4, 5, 6, 7, 8, 9, A, B, C, D, E, *and* F, *where* A *through* F *represent ten through fifteen, respectively. Convert the numeral to a numeral in base 10. See Example 8.*

37. 735_{16} **38.** 581_{16}
39. $6D3B7_{16}$ **40.** $24FEA_{16}$

In Exercises 41–44, convert the numeral to a numeral in base 16. See Example 8.

41. 573 **42.** 349
43. 5478 **44.** 34,721

In Exercises 45–50, convert 2005 to a numeral in the base indicated.

45. 2 **46.** 3 **47.** 5
48. 7 **49.** 12 **50.** 16

In Exercises 51–56, if any numerals are written incorrectly, explain why.

51. 5013_5 **52.** 1203_3
53. 674_8 **54.** 1206_{12}
55. 4086_7 **56.** 3004_5

Problem Solving

In Exercises 57–60, assume the numerals given are in a base 5 numeration system. The numerals in this system and their equivalent Hindu–Arabic numerals are

= 0 = 1 = 2 = 3 = 4

Write the Hindu–Arabic numerals equivalent to each of the following.

57. $_5$ **58.** $_5$ **59.** $_5$ **60.** $_5$

In Exercises 61–64, write the Hindu–Arabic numerals in the numeration system discussed in Exercises 57–60.

61. 19 **62.** 23 **63.** 74 **64.** 85

In Exercises 65–68, suppose colors as indicated below represent numerals in a base 4 numeration system.

= 0 = 1 = 2 = 3

Write the Hindu–Arabic numerals equivalent to each of the following.

65. $_4$ **66.** $_4$ **67.** $_4$ **68.** $_4$

In Exercises 69–72, write the Hindu–Arabic numerals in the base 4 numeration system discussed in Exercises 65–68. You will need to use the colors indicated above to write the answer.

69. 10 **70.** 15 **71.** 60 **72.** 56

73. *Another Conversion Method* There is an alternative method for changing a number in base 10 to a different base. This method will be used to convert 328 to base 5.

Dividing 328 by 5 gives a quotient of 65 and a remainder of 3. Write the quotient below the dividend and the remainder on the right, as shown.

$$\begin{array}{r|l l} 5 & 328 & \text{remainder} \\ \hline & 65 & \quad 3 \end{array}$$

Continue this process of division by 5.

$$\begin{array}{r|r l} 5 & 328 & \text{remainder} \\ 5 & 65 & 3 \\ 5 & 13 & 0 \\ 5 & 2 & 3 \\ & 0 & 2 \end{array}$$

Answer (read upward)

(In the last division, since the dividend, 2, is smaller than the divisor, 5, the quotient is 0 and the remainder is 2.)

Note that the division continues until the quotient is zero. The answer is read from the bottom number to the top number in the remainder column. Thus, $328 = 2303_5$.

a) Explain why this procedure results in the proper answer.

b) Convert 683 to base 5 by this method.

c) Convert 763 to base 8 by this method.

Challenge Problems/Group Activities

74. a) Use the numerals 0, 1, and 2 to write the first 20 numbers in the base 3 numeration system.

b) What is the next number after 222_3?

75. a) *Your Own Numeration System* Make up your own base 20 positional-value numeration system. Indicate the 20 numerals you will use to represent the 20 numbers less than the base.

b) Write the numbers 523 and 5293 in your base 20 numeration system.

76. *Computer Code* The ASCII code used by most computers uses the last seven positions of an eight-bit byte to represent all the characters on a standard keyboard. How many different orderings of 0's and 1's (or how many different characters) can be made by using the last seven positions of an eight-bit byte?

Recreational Mathematics

77. Find b if $111_b = 43$.

78. Find d if $ddd_5 = 124$.

79. Suppose a base 4 place-value system has its digits represented by colors as follows:

● = 0 ● = 1 ● = 2 ● = 3

a) Determine the value of ●●●●●$_4$ in base 10.

b) Write 177 in the base 4 system using only the four colors given in the exercise.

Internet/Research Activities

80. Write a report on how digital computers use the binary number system.

81. We mention at the beginning of this section that some societies still use a base 2 and base 3 numeration system. These societies are in Australia, New Guinea, Africa, and South America. Write a report on these societies, covering the symbols they use and how they combine these symbols to represent numbers in their numeration system.

4.4 COMPUTATION IN OTHER BASES

Addition

When computers perform calculations, they do so in base 2, the binary system. In this section, we explain how to perform calculations in base 2 and other bases.

In a base 2 system, the only digits are 0 and 1, and the place values are

$$\ldots,\ 2^4,\ 2^3,\ 2^2,\ 2,\ 1$$

or

$$\ldots,\ 16,\ 8,\ 4,\ 2,\ 1$$

TABLE 4.7 Base 2 Addition Table

+	0	1
0	0	1
1	1	10

Suppose we want to add $1_2 + 1_2$. The subscript 2 indicates that we are adding in base 2. Remember the answer to $1_2 + 1_2$ must be written using only the digits 0 and 1. The sum of $1_2 + 1_2$ is 10_2, which represents 1 group of two and 0 units in base 2. Recall that 10_2 means $1(2) + 0(1)$.

If we wanted to find the sum of $10_2 + 1_2$, we would add the digits in the right-hand, or units, column. Since $0_2 + 1_2 = 1_2$, the sum of $10_2 + 1_2 = 11_2$.

We are going to work additional examples and exercises in base 2, so rather than performing individual calculations in every problem, we can construct and use an addition table, Table 4.7, for base 2 (just as we used an addition table in base 10 when we first learned to add in base 10).

EXAMPLE 1 *Adding in Base 2*

Add 1101_2
$\underline{\quad 111_2}$

SOLUTION: Begin by adding the numbers in the right-hand, or units, column. From previous discussion, and as can be seen in Table 4.7, $1_2 + 1_2 = 10_2$. Place the 0 under the units column and carry the 1 to the 2's column, the second column from the right.

Place value of columns

$$\begin{array}{cccc} 2^3 & 2^2 & 2 & 1 \\ \downarrow & \downarrow & \downarrow & \downarrow \\ 1 & 1 & {}^{1}0 & 1 \\ & 1 & 1 & 1 \\ \hline & & & 0_2 \end{array}$$

Now add the three digits in the 2's column, $1_2 + 0_2 + 1_2$. Treat this as $(1_2 + 0_2) + 1_2$. Therefore, add $1_2 + 0_2$ to get 1_2, then add $1_2 + 1_2$ to get 10_2. Place the 0 under the 2's column and carry the 1 to the 2^2 column (the third column from the right).

$$\begin{array}{cccc} 1 & {}^{1}1 & {}^{1}0 & 1 \\ & 1 & 1 & 1 \\ \hline & & 0 & 0_2 \end{array}$$

Now add the three 1's in the 2^2 column to get $(1_2 + 1_2) + 1_2 = 10_2 + 1_2 = 11_2$. Place the 1 under the 2^2 column and carry the 1 to the 2^3 column (the fourth column from the right).

$$\begin{array}{cccc} {}^{1}1 & {}^{1}1 & {}^{1}0 & 1 \\ & 1 & 1 & 1 \\ \hline & 1 & 0 & 0_2 \end{array}$$

Now add the two 1's in the 2^3 column, $1_2 + 1_2 = 10_2$. Place the 10 as follows.

$$\begin{array}{ccccc} & {}^{1}1 & {}^{1}1 & {}^{1}0 & 1 \\ & & 1 & 1 & 1 \\ \hline 1 & 0 & 1 & 0 & 0_2 \end{array}$$

Therefore, the sum is 10100_2. ▲

TABLE 4.8 Base 5 Addition Table

+	0	1	2	3	4
0	0	1	2	3	4
1	1	2	3	4	10
2	2	3	4	10	11
3	3	4	10	11	12
4	4	10	11	(12)	13

Let's now look at addition in a base 5 system. In base 5, the only digits are 0, 1, 2, 3, and 4, and the positional values are

$$\ldots,\ 5^4,\ 5^3,\ 5^2,\ 5,\ 1$$
or
$$\ldots,\ 625,\ 125,\ 25,\ 5,\ 1$$

What is the sum of $4_5 + 3_5$? We can consider this to mean $(1 + 1 + 1 + 1) + (1 + 1 + 1)$. We can regroup the seven 1's into one group of five and two units as $(1 + 1 + 1 + 1 + 1) + (1 + 1)$. Thus, the sum of $4_5 + 3_5 = 12_5$ (circled in Table 4.8). Recall that 12_5 means $1(5) + 2(1)$. We can use this same procedure in obtaining the remaining values in the base 5 addition table.

EXAMPLE 2 *Use the Base 5 Addition Table*

Add 42_5
$\underline{33_5}$

SOLUTION: First determine that $2_5 + 3_5$ is 10_5 from Table 4.8. Record the 0 and carry the 1 to the 5's column.

$$\begin{array}{r} {}^{1}4\ 2_5 \\ \underline{3\ 3_5} \\ 0_5 \end{array}$$

Add the numbers in the second column, $(1_5 + 4_5) + 3_5 = 10_5 + 3_5 = 13_5$. Record the 13.

$$\begin{array}{r} {}^{1}4\ 2_5 \\ \underline{3\ 3_5} \\ 1\ 3\ 0_5 \end{array}$$

The sum is 130_5. ▲

EXAMPLE 3 *Add in Base 5*

Add 1234_5
$\underline{2042_5}$

SOLUTION:

$$\begin{array}{r} 1\ {}^{1}2\ {}^{1}3\ 4_5 \\ \underline{2\ \ 0\ \ 4\ 2_5} \\ 3\ \ 3\ \ 3\ 1_5 \end{array}$$

▲

You can develop an addition table for any base and use it to add in that base. As you get more comfortable with addition in other bases, however, you may prefer to add numbers in other bases by using mental arithmetic. To do so, convert the sum of the numbers being added from the given base to base 10 and then convert the base 10 number back into the given base. You must clearly understand how to convert from base 10 to the given base, as discussed in Section 4.3. For example, to add $7_9 + 8_9$, add 7 + 8 in base 10 to get 15_{10} and then mentally convert 15_{10} to 16_9 using the procedure given earlier. Remember, 16_9 when converted to base 10 becomes $1(9) + 6(1)$, or 15. Addition using this procedure is illustrated in Examples 4 and 5.

DID YOU KNOW

Speaking to Machines

GRADE A LOWFAT MILK (1% MILKFAT)
INGREDIENTS: LOWFAT MILK, VITAMIN A PALMITATE, D3
Dist. by MARKET BASKET Tewksbury, MA
PLT. NO. STAMPED ON CONTAINER
0 49705 68773 4
1/2 GALLON (1.89L)

Almost all packaged goods we buy today are marked with a universal product code (UPC), a black-and-white bar code. An optical scanner "reads" the pattern of black and white, thick and thin, and converts it to a binary code that is sent to the scanner's computer, which then calls up the appropriate price and records the sale for inventory purposes.

For the past 600 years, we have used the Hindu–Arabic system of numeration without change. Our base 10 numeration system seems so obvious to us, perhaps because of our 10 fingers and 10 toes, but it would be rash to think that numbers in other bases are not useful. In fact, one of the most significant numeration systems is the binary system, or base 2. This system, with its elemental simplicity, is what is used by computers to process information and "talk" to one another. When a computer receives a command or data, every character in the command or data must first be converted into a binary number for the computer to understand and use it. Because of the ever-expanding number of computers in use, the users of the binary number system may soon outnumber the users of base 10.

Pit Land

On digital video discs (DVDs) and compact discs (CDs), video and sound are digitally encoded on the underside of the disc in a binary system of pits and "lands" (nonpits). To play the disc, a laser beam tracks along the spiral and is reflected when it hits a land (signal sent = 1), but it is not reflected by the pits (no signal = 0). The binary sequence is then converted into video images and music.

EXAMPLE 4 *Adding in Base 10; Converting to Base 3*

Add

$$\begin{array}{r} 1022_3 \\ \underline{2121_3} \end{array}$$

SOLUTION: To solve this problem, make the necessary conversions by using mental arithmetic. $2 + 1 = 3_{10} = 10_3$. Record the 0 and carry the 1.

$$\begin{array}{rrrr} 1 & 0 & {}^{1}2 & 2_3 \\ 2 & 1 & 2 & 1_3 \\ \hline & & & 0_3 \end{array}$$

$1 + 2 + 2 = 5_{10} = 12_3$. Record the 2 and carry the 1.

$$\begin{array}{rrrr} 1 & {}^{1}0 & {}^{1}2 & 2_3 \\ 2 & 1 & 2 & 1_3 \\ \hline & & 2 & 0_3 \end{array}$$

$1 + 0 + 1 = 2_{10} = 2_3$. Record the 2.

$$\begin{array}{rrrr} 1 & {}^{1}0 & {}^{1}2 & 2_3 \\ 2 & 1 & 2 & 1_3 \\ \hline & 2 & 2 & 0_3 \end{array}$$

$1 + 2 = 3_{10} = 10_3$. Record the 10.

$$\begin{array}{r} 1022_3 \\ \underline{2121_3} \\ 10220_3 \end{array}$$

The sum is 10220_3. ▲

EXAMPLE 5 ***Adding in Base 10; Converting to Base 5***

Add
$$\begin{array}{r} 332_5 \\ 344_5 \\ 443_5 \\ \underline{314_5} \end{array}$$

SOLUTION: Adding the digits in the right-hand column gives $2 + 4 + 3 + 4 = 13_{10} = 23_5$. Record the 3 and carry the 2. Adding the 2 with the digits in the next column yields $2 + 3 + 4 + 4 + 1 = 14_{10} = 24_5$. Record the 4 and carry the 2. Adding the 2 with the digits in the left-hand column gives $2 + 3 + 3 + 4 + 3 = 15_{10} = 30_5$. Record both digits. The sum of these four numbers is 3043_5.

$$\begin{array}{rrrr} & {}^{2}3 & {}^{2}3 & 2_5 \\ & 3 & 4 & 4_5 \\ & 4 & 4 & 3_5 \\ & \underline{3} & \underline{1} & \underline{4_5} \\ 3 & 0 & 4 & 3_5 \end{array}$$

▲

Subtraction

Subtraction can also be performed in other bases. Always remember that when you "borrow," you borrow the amount of the base given in the subtraction problem. For example, if subtracting in base 5, when you borrow, you borrow 5. If subtracting in base 12, when you borrow, you borrow 12.

EXAMPLE 6 ***Subtracting in Base 5***

Subtract
$$\begin{array}{r} 3032_5 \\ \underline{-1004_5} \end{array}$$

SOLUTION: We will perform the subtraction in base 10 and convert the results to base 5. Since 4 is greater than 2, we must borrow one group of 5 from the preceding column. This action gives a sum of $5 + 2$, or 7 in base 10. Now we subtract 4 from 7; the difference is 3. We complete the problem in the usual manner. The 3 in the second column becomes a 2, $2 - 0 = 2$, $0 - 0 = 0$, and $3 - 1 = 2$.

$$\begin{array}{r} 3032_5 \\ \underline{-1004_5} \\ 2023_5 \end{array}$$

▲

EXAMPLE 7 *Subtracting in Base 12*

Subtract $\begin{array}{r} 468_{12} \\ -295_{12} \\ \hline \end{array}$

SOLUTION: $8 - 5 = 3$. Next we must subtract 9 from 6. Since 9 is greater than 6, borrowing is necessary. We must borrow one group of 12 from the preceding column. We then have a sum of $12 + 6 = 18$ in base 10. Now we subtract 9 from 18, and the difference is 9. The 4 in the left column becomes 3, and $3 - 2 = 1$.

$$\begin{array}{r} 468_{12} \\ -295_{12} \\ \hline 193_{12} \end{array}$$

Multiplication

Multiplication can also be performed in other bases. Doing so is helped by forming a multiplication table for the base desired. Suppose we want to determine the product of $4_5 \times 3_5$. In base 10, 4×3 means there are four groups of three units. Similarly, in a base 5 system, $4_5 \times 3_5$ means there are four groups of three units, or

$$(1 + 1 + 1) + (1 + 1 + 1) + (1 + 1 + 1) + (1 + 1 + 1)$$

Regrouping the 12 units above into groups of five gives

$$(1 + 1 + 1 + 1 + 1) + (1 + 1 + 1 + 1 + 1) + (1 + 1)$$

or two groups of five, and two units. Thus, $4_5 \times 3_5 = 22_5$.

We can construct other values in the base 5 multiplication table in the same way. You may, however, find it easier to multiply the values in the base 10 system and then change the product to base 5 by using the procedure discussed in Section 4.3. Multiplying 4×3 in base 10 gives 12, and converting 12 from base 10 to base 5 gives 22_5.

The product of $4_5 \times 3_5$ is circled in Table 4.9, the base 5 multiplication table. The other values in the table may be found by either method discussed.

TABLE 4.9 Base 5 Multiplication Table

×	**0**	**1**	**2**	**3**	**4**
0	0	0	0	0	0
1	0	1	2	3	4
2	0	2	4	11	13
3	0	3	11	14	22
4	0	4	13	(22)	31

EXAMPLE 8 *Using the Base 5 Multiplication Table*

Multiply $\begin{array}{r} 13_5 \\ \times\ 3_5 \\ \hline \end{array}$

SOLUTION: Multiply as you would in base 10, but use the base 5 multiplication table to find the products. When the product consists of two digits, record the right digit and carry the left digit. Multiplying gives $3_5 \times 3_5 = 14_5$. Record the 4 and carry the 1.

$$\begin{array}{r} {}^{1}13_5 \\ \times\ \ 3_5 \\ \hline 4 \end{array}$$

$(3_5 \times 1_5) + 1_5 = 4_5$. Record the 4.

$$\begin{array}{r} {}^{1}13_5 \\ \times\ \ 3_5 \\ \hline 44_5 \end{array}$$

The product is 44_5.

Constructing a multiplication table is often tedious, especially when the base is large. To multiply in a given base without the use of a table, multiply in base 10 and convert the products to the appropriate base number before recording them. This procedure is illustrated in Example 9.

EXAMPLE 9 *Multiplying in Base 7*

Multiply $\begin{array}{r} 43_7 \\ \times\ 25_7 \\ \hline \end{array}$

SOLUTION: $5 \times 3 = 15_{10} = 2(7) + 1(1) = 21_7$. Record the 1 and carry the 2.

$$\begin{array}{r} {}^{2}43_7 \\ \times\ 25_7 \\ \hline 1 \end{array}$$

$(5 \times 4) + 2 = 20 + 2 = 22_{10} = 3(7) + 1(1) = 31_7$. Record the 31.

$$\begin{array}{r} {}^{2}43_7 \\ \times\ 25_7 \\ \hline 311 \end{array}$$

$2 \times 3 = 6_{10} = 6_7$. Record the 6.

$$\begin{array}{r} {}^{2}43_7 \\ \times\ 25_7 \\ \hline 311 \\ 6 \end{array}$$

$2 \times 4 = 8_{10} = 1(7) + 1(1) = 11_7$. Record the 11. Now add in base 7 to determine the answer. Remember, in base 7, there are no digits greater than 6.

$$\begin{array}{r} {}^{2}43_7 \\ \times\ 25_7 \\ \hline 311 \\ 116 \\ \hline 1501_7 \end{array}$$

DID YOU KNOW

Ever-Changing Numbers

12th century	1	2	3	4	5	6	7	8	9	0
13th century	1	2	3	4	5	6	7	8	9	0
About 1300	1	2	3	4	5	6	7	8	9	0
About 1429	1	2	3	4	5	6	7	8	9	0
15th century	1	2	3	4	5	6	7	8	9	0
20th century	1	2	3	4	5	6	7	8	9	0
Computer*	1	2	3	4	5	6	7	8	9	0

* Times New Roman font

During the Middle Ages, Western Europeans were reluctant to give up Roman numerals in favor of Hindu–Arabic numerals. The rapid expansion of trade and commerce during the fifteenth century, however, caused the need for quicker systems of calculation. The invention of movable type in 1450 also ensured a certain consistency in the way numerals were depicted, yet we still find ways to alter them.

Division

Division is performed in much the same manner as long division in base 10. A detailed example of a division in base 5 is illustrated in Example 10. The same procedure is used for division in any other base.

EXAMPLE 10 *Dividing in Base 5*

Divide $2_5\overline{)143_5}$.

SOLUTION: Using the multiplication table for base 5, Table 4.9 on page 193, we list the multiples of the divisor, 2.

$$\begin{aligned} 2_5 \times 1_5 &= 2_5 \\ 2_5 \times 2_5 &= 4_5 \\ 2_5 \times 3_5 &= 11_5 \\ 2_5 \times 4_5 &= 13_5 \end{aligned}$$

Since $2_5 \times 4_5 = 13_5$, which is the largest product less than 14_5, 2_5 divides into 14_5 four times.

$$\begin{array}{r} 4 \\ 2_5\overline{)143_5} \\ \underline{13} \\ 1 \end{array}$$

Subtract 13_5 from 14_5. The difference is 1_5. Record the 1. Now bring down the 3 as when dividing in base 10.

$$\begin{array}{r} 4 \\ 2_5\overline{)143_5} \\ \underline{13} \\ 13 \end{array}$$

We see that $2_5 \times 4_5 = 13_5$. Use this information to complete the problem.

$$\begin{array}{r} 44_5 \\ 2_5\overline{)143_5} \\ \underline{13} \\ 13 \\ \underline{13} \\ 0 \end{array}$$

Therefore, $143_5 \div 2_5 = 44_5$ with remainder 0_5.

A division problem can be checked by multiplication. If the division was performed correctly, (quotient × divisor) + remainder = dividend. We can check Example 10 as follows.

$$(44_5 \times 2_5) + 0_5 = 143_5$$

$$\begin{array}{r} 44_5 \\ \times\ \ 2_5 \\ \hline 143_5 \end{array} \quad \text{Check}$$

EXAMPLE 11 *Dividing in Base 6*

Divide $4_6\overline{)2430_6}$.

SOLUTION: The multiples of 4 in base 6 are

$$4_6 \times 1_6 = 4_6$$
$$4_6 \times 2_6 = 12_6$$
$$4_6 \times 3_6 = 20_6$$
$$4_6 \times 4_6 = 24_6$$
$$4_6 \times 5_6 = 32_6$$

$$\begin{array}{r} 404_6 \\ 4_6\overline{)2430_6} \\ \underline{24} \\ 03 \\ \underline{00} \\ 30 \\ \underline{24} \\ 2 \end{array}$$

Thus, the quotient is 404_6, with remainder 2_6.

Be careful when subtracting! When subtracting 4 from 0, you will need to borrow. Remember that you borow 10_6, which is the same as 6 in base 10.

Check: Does $(404_6 \times 4_6) + 2_6 = 2430_6$?

$$\begin{array}{r} 404_6 \\ \times\ \ \ 4_6 \\ \hline 2424_6 \end{array} + 2_6 = 2430_6 \quad \text{True}$$

SECTION 4.4 EXERCISES

Concept/Writing Exercises

1. a) What are the first five positional values, from right to left, in base b?
   b) What are the first five positional values, from right to left, in base 6?
2. In the addition

$$\begin{array}{r} 367_8 \\ +24_8 \\ \hline \end{array}$$

what are the positional values of the first column on the right, the second column from the right, and the third column from the right? Explain how you determined your answer.

3. Suppose you add two base 5 numbers and you obtain an answer of 463_5. Can your answer be correct? Explain.
4. Suppose you add two base 3 numbers and you obtain an answer of 20323_3. Can your answer be correct? Explain.
5. In your own words, explain how to add two numbers in a given base. In your explanation, answer the question, "What happens when the sum of the numbers in a column is greater than the base?"
6. In your own words, explain how to subtract two numbers in a given base. Include in your explanation what you do when, in one column, you must subtract a larger number from a smaller number.

Practice the Skills

In Exercises 7–18, add in the indicated base.

7. $\begin{array}{r} 43_5 \\ 41_5 \\ \hline \end{array}$
8. $\begin{array}{r} 33_8 \\ 65_8 \\ \hline \end{array}$
9. $\begin{array}{r} 2303_4 \\ 232_4 \\ \hline \end{array}$
10. $\begin{array}{r} 101_2 \\ 11_2 \\ \hline \end{array}$
11. $\begin{array}{r} 799_{12} \\ 218_{12} \\ \hline \end{array}$
12. $\begin{array}{r} 222_3 \\ 22_3 \\ \hline \end{array}$
13. $\begin{array}{r} 1112_3 \\ 1011_3 \\ \hline \end{array}$
14. $\begin{array}{r} 470_{12} \\ 347_{12} \\ \hline \end{array}$
15. $\begin{array}{r} 14631_7 \\ 6040_7 \\ \hline \end{array}$
16. $\begin{array}{r} 1341_8 \\ 341_8 \\ \hline \end{array}$
17. $\begin{array}{r} 1110_2 \\ 110_2 \\ \hline \end{array}$
18.* $\begin{array}{r} 43A_{16} \\ 496_{16} \\ \hline \end{array}$

In Exercises 19–30, subtract in the indicated base.

19. $\begin{array}{r} 322_4 \\ -103_4 \\ \hline \end{array}$
20. $\begin{array}{r} 526_7 \\ -145_7 \\ \hline \end{array}$
21. $\begin{array}{r} 2342_5 \\ -1442_5 \\ \hline \end{array}$
22. $\begin{array}{r} 1011_2 \\ -\ 101_2 \\ \hline \end{array}$
23. $\begin{array}{r} 782_{12} \\ -13T_{12} \\ \hline \end{array}$
24. $\begin{array}{r} 1221_3 \\ -\ 202_3 \\ \hline \end{array}$
25. $\begin{array}{r} 1001_2 \\ -\ 110_2 \\ \hline \end{array}$
26. $\begin{array}{r} 2T34_{12} \\ -\ 345_{12} \\ \hline \end{array}$
27. $\begin{array}{r} 4223_7 \\ -\ 304_7 \\ \hline \end{array}$
28. $\begin{array}{r} 4232_5 \\ -2341_5 \\ \hline \end{array}$
29. $\begin{array}{r} 2100_3 \\ -1012_3 \\ \hline \end{array}$
30.* $\begin{array}{r} 4E7_{16} \\ -189_{16} \\ \hline \end{array}$

In Exercises 31–42, multiply in the indicated base.

31. $\begin{array}{r} 33_5 \\ \times\ 2_5 \\ \hline \end{array}$
32. $\begin{array}{r} 323_6 \\ \times\ 4_6 \\ \hline \end{array}$
33. $\begin{array}{r} 342_7 \\ \times\ 5_7 \\ \hline \end{array}$
34. $\begin{array}{r} 101_2 \\ \times\ 11_2 \\ \hline \end{array}$
35. $\begin{array}{r} 512_6 \\ \times\ 23_6 \\ \hline \end{array}$
36. $\begin{array}{r} 124_{12} \\ \times\ 6_{12} \\ \hline \end{array}$
37. $\begin{array}{r} 436_9 \\ \times\ 25_9 \\ \hline \end{array}$
38. $\begin{array}{r} 6T3_{12} \\ \times\ 24_{12} \\ \hline \end{array}$
39. $\begin{array}{r} 111_2 \\ \times\ 101_2 \\ \hline \end{array}$
40. $\begin{array}{r} 584_9 \\ \times\ 24_9 \\ \hline \end{array}$
41. $\begin{array}{r} 316_7 \\ \times\ 16_7 \\ \hline \end{array}$
42. $\begin{array}{r} 8T_{12} \\ \times\ 2T_{12} \\ \hline \end{array}$

In Exercises 43–54, divide in the indicated base.

43. $1_2\overline{)110_2}$
44. $4_6\overline{)231_6}$
45. $3_5\overline{)143_5}$
46. $7_8\overline{)335_8}$
47. $2_4\overline{)312_4}$
48. $6_{12}\overline{)431_{12}}$
49. $2_4\overline{)213_4}$
50. $5_6\overline{)214_6}$
51. $3_5\overline{)224_5}$
52. $4_6\overline{)210_6}$
53. $6_7\overline{)404_7}$
54. $3_7\overline{)2101_7}$

Problem Solving

In Exercises 55–58, the numerals in a base 5 numeration system are as illustrated with their equivalent Hindu–Arabic numerals.

= 0 = 1 = 2 = 3 = 4

Add the following base 5 numbers.

55.

56.

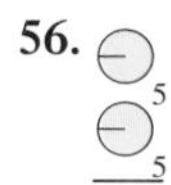

57.

58.

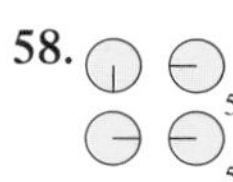

*For Exercises 18 and 30, see Exercises 37–40 in Section 4.3.

In Exercises 59–66, assume the numerals given are in a base 4 numeration system. In this system, suppose colors are used as numerals, as indicated below.

● = 0, ● = 1, ● = 2, ● = 3

Add the following base 4 numbers. Your answers will contain a variety of the colors indicated.

59. $●_4$ + $●_4$

60. 

61. 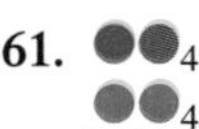

62. 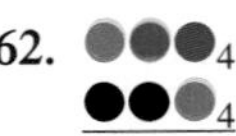

Subtract the following in base 4. Your answer will contain a variety of the colors indicated.

63. $●●_4 - ●●_4$ **64.** $●●_4 - ●●_4$ **65.** $●●●_4 - ●●●_4$ **66.** $●●●_4 - ●●●_4$

For Exercises 67 and 68, study the pattern in the boxes. The number in the bottom row of each box represents the value of each dot in the box directly above it. For example, the following box represents $(3 \times 7^2) + (2 \times 7) + (4 \times 1)$, *or the number* 324_7. *This number in base 10 is 165.*

• • •	• •	• • • •
7^2	7	1

67. Determine the base 5 number represented by the dots in the top row of the boxes. Then convert the base 5 number to a number in base 10.

• •	• • •		• •
5^3	5^2	5	1

68. Fill in the correct amount of dots in the columns above the base values if the number represented by the dots is to equal 327 in base 10.

9^2	9	1

Challenge Problems/Group Activities

Divide in the indicated base.

69. $14_5\overline{)242_5}$ **70.** $20_4\overline{)223_4}$

71. Consider the multiplication

$$\begin{array}{r} 462_8 \\ \times\ 35_8 \\ \hline \end{array}$$

a) Multiply the numbers in base 8.
b) Convert 462_8 and 35_8 to base 10.
c) Multiply the base 10 numbers determined in part (b).
d) Convert the answer obtained in base 8 in part (a) to base 10.
e) Are the answers obtained in parts (c) and (d) the same? Why or why not?

Recreational Mathematics

72. Determine b, by trial and error, if $1304_b = 204$.

73. In a base 4 system, each of the four numerals is represented by one of the following colors:

Determine the value of each color if the following addition is true in base 4.

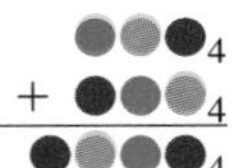

Internet/Research Activities

74. Investigate and write a report on the use of the duodecimal (base 12) system as a system of numeration. You might contact the Dozenal Society (formerly the Duodecimal Society), Nassau Community College, Garden City, NY 11530 or use their website www.polar.sunynassau.edu/~dozenal/ for information.

75. One method used by computers to perform subtraction is the "end around carry method." Do research and write a report explaining, with specific examples, how a computer performs subtraction by using the end around carry method.

4.5 EARLY COMPUTATIONAL METHODS

Our present procedures for multiplying and dividing numbers are the most recent to be developed. Early civilizations used various methods for multiplying and dividing. Multiplication was performed by *duplation and mediation*, by the *galley method*, and by *Napier rods*. Following is an explanation of each method.

Duplation and Mediation

EXAMPLE 1 ***A Pairing Technique for Multiplying***

Multiply 17×30 using duplation and mediation.

SOLUTION: Write 17 and 30 with a dash between to separate them. Divide the number on the left, 17, in half, drop the remainder, and place the quotient, 8, under the 17. Double the number on the right, 30, obtaining 60, and place it under the 30. You will then have the following paired lines.

17—30
8—60

Continue this process, taking one-half the number in the left-hand column, disregarding the remainder, and doubling the number in the right-hand column, as shown below. When a 1 appears in the left-hand column, stop.

17—30
8—60
4—120
2—240
1—480

Cross out all the even numbers in the left-hand column and the corresponding numbers in the right-hand column.

17—30
~~8—60~~
~~4—120~~
~~2—240~~
1—480

Now add the remaining numbers in the right-hand column, obtaining $30 + 480 = 510$, which is the product you want. If you check, you will find that $17 \times 30 = 510$. ▲

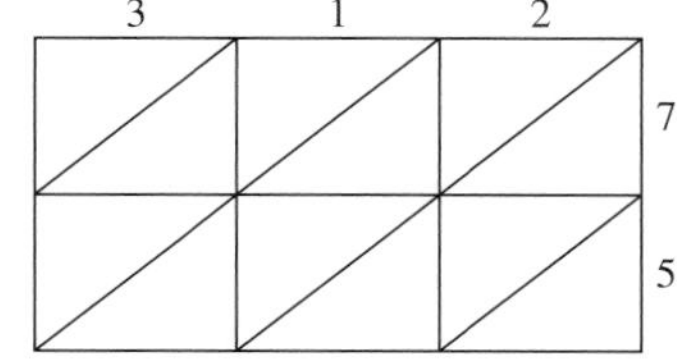

Figure 4.1

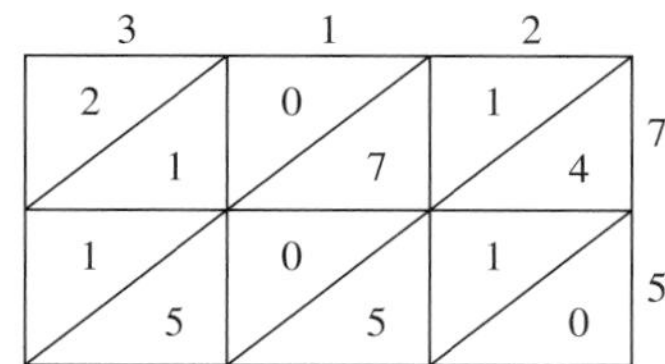

Figure 4.2

The Galley Method

The galley method (sometimes referred to as the Gelosia method) was developed after duplation and mediation. To multiply 312×75 using the galley method, you construct a rectangle consisting of three columns (one for each digit of 312) and two rows (one for each digit of 75).

Place the digits 3, 1, 2 above the boxes and the digits 7, 5 on the right of the boxes, as shown in Fig. 4.1. Then place a diagonal in each box.

Complete each box by multiplying the number on top of the box by the number to the right of the box (Fig. 4.2). Place the units digit of the product below the diagonal and the tens digit of the product above the diagonal.

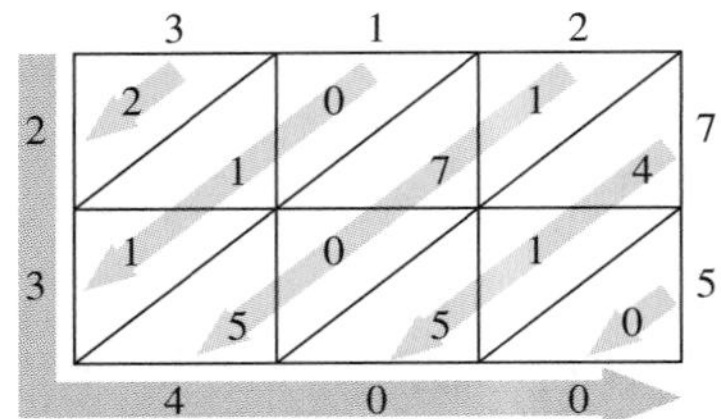

Figure 4.3

Add the numbers along the diagonals, as shown with the blue shaded arrows in Fig. 4.3, starting with the bottom right diagonal. If the sum in a diagonal is 10 or greater, record the units digit below the rectangle and carry the tens digit to the next diagonal to the left.

For example, when adding 4, 1, and 5 (along the second blue diagonal from the right), the sum is 10. Record the 0 below the rectangle and carry the 1 to the next blue diagonal. The sum of $1 + 1 + 7 + 0 + 5$ is 14. Record the 4 and carry the 1. The sum of the numbers in the next blue diagonal is $1 + 0 + 1 + 1$ or 3.

The answer is read down the left-hand column and along the bottom, as shown by the purple arrow in Fig. 4.3. The answer is 23,400.

PROFILE IN MATHEMATICS

JOHN NAPIER

During the seventeenth century, the growth of scientific fields such as astronomy required the ability to perform often unwieldy calculations. The English mathematician John Napier (1550–1617) made great contributions toward solving the problem of computing these numbers. His inventions include simple calculating machines and a device for performing multiplication and division known as Napier rods. Napier also developed the theory of logarithms.

Napier Rods

The third method used to multiply numbers was developed from the galley method by John Napier in the seventeenth century. His method of multiplication, known as Napier rods, proved to be one of the forerunners of the modern-day computer. Napier developed a system of separate rods numbered from 0 through 9 and an additional strip for an index, numbered vertically 1 through 9 (Fig. 4.4). Each rod is divided into 10 blocks. Each block below the first block contains a multiple of the number in the first block, with a diagonal separating the digits. The units digits are placed to the right of the diagonals and the tens digits to the left. Example 2 explains how Napier rods are used to multiply numbers.

INDEX	0	1	2	3	4	5	6	7	8	9
1	0/0	0/1	0/2	0/3	0/4	0/5	0/6	0/7	0/8	0/9
2	0/0	0/2	0/4	0/6	0/8	1/0	1/2	1/4	1/6	1/8
3	0/0	0/3	0/6	0/9	1/2	1/5	1/8	2/1	2/4	2/7
4	0/0	0/4	0/8	1/2	1/6	2/0	2/4	2/8	3/2	3/6
5	0/0	0/5	1/0	1/5	2/0	2/5	3/0	3/5	4/0	4/5
6	0/0	0/6	1/2	1/8	2/4	3/0	3/6	4/2	4/8	5/4
7	0/0	0/7	1/4	2/1	2/8	3/5	4/2	4/9	5/6	6/3
8	0/0	0/8	1/6	2/4	3/2	4/0	4/8	5/6	6/4	7/2
9	0/0	0/9	1/8	2/7	3/6	4/5	5/4	6/3	7/2	8/1

Figure 4.4

EXAMPLE 2 *Using Napier Rods*

Multiply 8×365, using Napier rods.

SOLUTION: To multiply 8×365, line up the rods 3, 6, and 5 to the right of the index, as shown in Fig. 4.5 on page 201. Below the 3, 6, and 5 place the blocks that

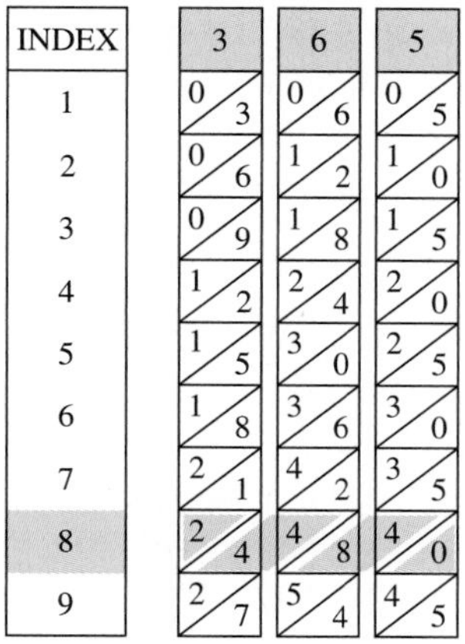

INDEX	3	6	5
1	0/3	0/6	0/5
2	0/6	1/2	1/0
3	0/9	1/8	1/5
4	1/2	2/4	2/0
5	1/5	3/0	2/5
6	1/8	3/6	3/0
7	2/1	4/2	3/5
8	2/4	4/8	4/0
9	2/7	5/4	4/5

Figure 4.5

contain the products of 8 × 3, 8 × 6, and 8 × 5, respectively. To obtain the answer, add along the diagonals as in the galley method.

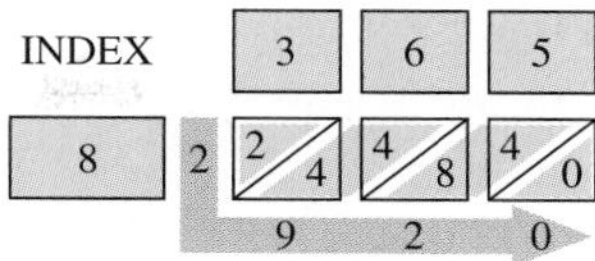

Thus, 8 × 365 = 2920. ▲

Example 3 illustrates the procedure to follow to multiply numbers containing more than one digit, using Napier rods.

EXAMPLE 3 *Using Napier Rods to Multiply Two- and Three-Digit Numbers*

Multiply 48 × 365, using Napier rods.

SOLUTION: 48 × 365 = (40 + 8) × 365

Write (40 + 8) × 365 = (40 × 365) + (8 × 365). To find 40 × 365, determine 4 × 365 and multiply the product by 10. To evaluate 4 × 365, set up Napier rods for 3, 6, and 5 with index 4, and then evaluate along the diagonals, as indicated.

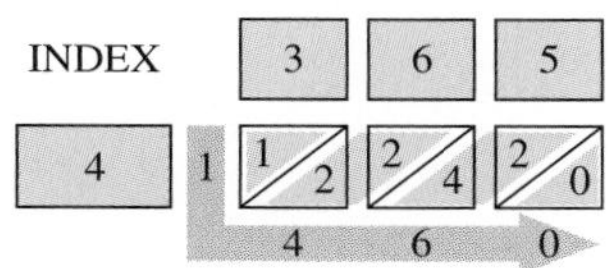

Therefore, 4 × 365 = 1460. Then 40 × 365 = 1460 × 10 = 14,600.

$$
\begin{aligned}
48 \times 365 &= (40 \times 365) + (8 \times 365) \\
&= 14{,}600 + 2920 \\
&= 17{,}520
\end{aligned}
$$

8 × 365 = 2920 from Example 2

▲

SECTION 4.5 EXERCISES

Concept/Writing Exercises

1. What are the three early computational methods discussed in this section?
2. a) Explain in your own words how multiplication by duplation and mediation is performed.
   b) Using the procedure given in part (a), multiply 267 × 193.
3. a) Explain in your own words how multiplication by the galley method is performed.
   b) Using the procedure given in part (a), multiply 362 × 29.
4. a) Explain in your own words how multiplication using Napier rods is performed.
   b) Using the procedure given in part (a), multiply 25 × 6.

Practice the Skills

In Exercises 5–12, multiply using duplation and mediation.

5. 23 × 31
6. 35 × 23
7. 9 × 162
8. 175 × 86
9. 35 × 236
10. 96 × 53
11. 93 × 93
12. 49 × 124

In Exercises 13–20, multiply using the galley method.

13. 6 × 375
14. 8 × 365
15. 4 × 583
16. 7 × 125
17. 75 × 12
18. 47 × 259
19. 314 × 652
20. 634 × 832

In Exercises 21–28, multiply using Napier rods.

21. 8 × 63
22. 7 × 63
23. 7 × 58
24. 7 × 125
25. 5 × 125
26. 75 × 125
27. 9 × 6742
28. 7 × 3456

Problem Solving

In Exercises 29 and 30, we show multiplications using the galley method. (a) Determine the numbers being multiplied. Explain how you determined your answer. (b) Find the product.

29.
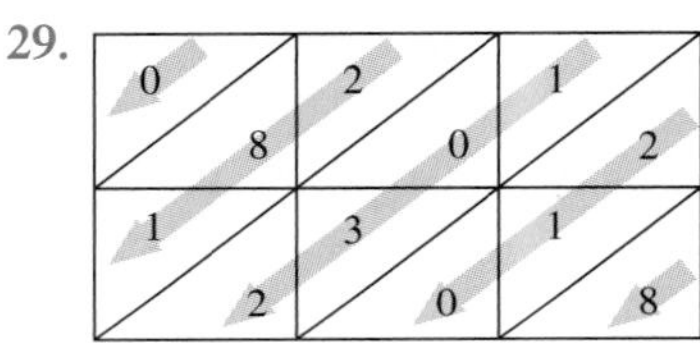

30.
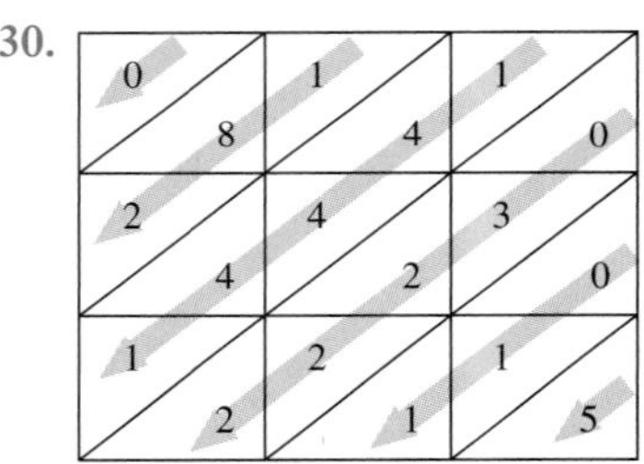

In Exercises 31 and 32, we solve a multiplication problem using Napier rods. (a) Determine the numbers being multiplied. Each empty box contains a single digit. Explain how you determined your answer. (b) Find the product.

31.

32.
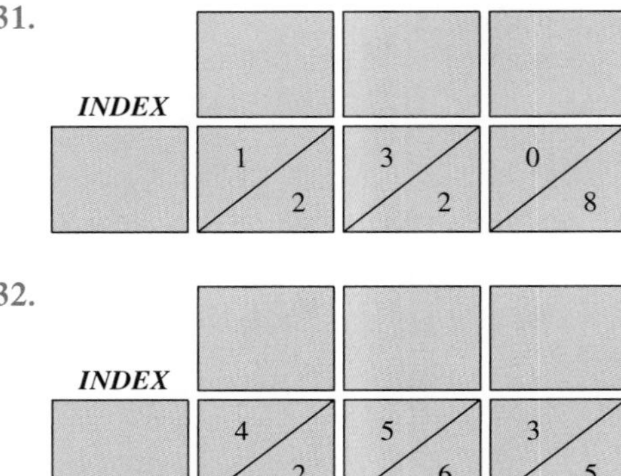

Challenge Problems/Group Activities

In Exercises 33 and 34, use the method of duplation and mediation to perform the multiplication. Write the answer in the numeration system in which the exercise is given.

33. (∩|||) • (∩∩||)
34. (XXVI) · (LXVII)

35. Develop a set of Napier rods that can be used to multiply numbers in base 5. Illustrate how your rods can be used to multiply $3_5 \times 21_5$.

In Exercises 36 and 37, (a) use the galley method to perform the multiplication. (Hint: Be sure not to list any number greater than or equal to the base within the box). Write the answer in the base in which the exercise is given.
(b) Multiply the numbers as explained in Section 4.4. If you do not obtain the results obtained in part (a), explain why.

36. $21_3 \times 21_3$
37. $24_5 \times 234_5$

Recreational Mathematics

38. Obtain a clean U.S. $1 bill. On the back side of the bill is a circle containing a pyramid. On the base of the pyramid is a Roman numeral. a) Determine the value of that Roman numeral in our Hindu–Arabic system.

While you have that bill out, let's consider something else. For many people the number 13 is considered an unlucky number. In fact, many hotels do not have a thirteenth floor because many guests refuse to stay on the thirteenth floor of a hotel. Yet, if you look at the back of a $1 bill, you will find

13 steps on the pyramid.
13 letters in the Latin words *Annuit Coeptis.*
13 letters in "E Pluribus Unum".
13 stars above the eagle.
13 plum feathers on each span of the eagle's wing.
13 bars on the eagle's shield.
13 leaves on the olive branch.
13 fruits.
13 arrows.

In addition, the U.S. flag has 13 stripes, there were 13 signers of the Declaration of Independence, there were 13 original colonies, and don't forget the important Thirteenth Amendment that abolished slavery. b) So why, in your opinion, do we as a society fear the number 13? By the way, the official name for the fear of the number 13 is *triskaidekaphobia*.

Internet/Research Activities

39. In addition to Napier rods, John Napier is credited with making other important contributions to mathematics. Write a report on John Napier and his contributions to mathematics.

40. Write a paper explaining why the duplation and mediation method works.

CHAPTER 4 SUMMARY

IMPORTANT FACTS

Types of Numeration Systems

Additive (Egyptian hieroglyphics, Roman)
Multiplicative (traditional Chinese)
Ciphered (Ionic Greek)
Place-value (Babylonian, Mayan, Hindu–Arabic)

Early Computational Methods

Duplation and mediation
The galley method
Napier rods

CHAPTER 4 REVIEW EXERCISES

4.1, 4.2

In Exercises 1–6, assume an additive numeration system in which $a = 1$, $b = 10$, $c = 100$, and $d = 1000$. Find the value of the numeral.

1. *dddcaaa*
2. *ccbda*
3. *bcccad*
4. *cbdadaaa*
5. *ddcccbaaaa*
6. *ccbaddac*

In Exercises 7–12, assume the same additive numeration system as in Exercises 1–6. Write the numeral in terms of a, b, c, and d.

7. 56
8. 125
9. 293
10. 2005
11. 6851
12. 2314

In Exercises 13–18, assume a multiplicative numeration system in which $a = 1$, $b = 2$, $c = 3$, $d = 4$, $e = 5$, $f = 6$, $g = 7$, $h = 8$, $i = 9$, $x = 10$, $y = 100$, and $z = 1000$. Find the value of the numeral.

13. *dxc*
14. *bxg*
15. *gydxi*
16. *dzfxh*
17. *ezfydxh*
18. *fziye*

In Exercises 19–24, assume the same multiplicative numeration system as in Exercises 13–18. Write the Hindu–Arabic numeral in that system.

19. 82
20. 295
21. 862
22. 3094
23. 6004
24. 2001

In Exercises 25–36, use the following ciphered numeration system.

Decimal	1	2	3	4	5	6	7	8	9
Units	a	b	c	d	e	f	g	h	i
Tens	j	k	l	m	n	o	p	q	r
Hundreds	s	t	u	v	w	x	y	z	A
Thousands	B	C	D	E	F	G	H	I	J
Ten thousands	K	L	M	N	O	P	Q	R	S

Convert the numeral to a Hindu–Arabic numeral.

25. me
26. uh
27. woh
28. NGzqc
29. PEvqa
30. Pwki

Write the numeral in the ciphered numeration system.

31. **85**
32. **372**
33. **493**
34. **1997**
35. **53,467**
36. **75,496**

In Exercises 37–42, convert 1462 to a numeral in the indicated numeration system.

37. **Egyptian**
38. **Roman**
39. **Chinese**
40. **Ionic Greek**
41. **Babylonian**
42. **Mayan**

In Exercises 43–48, convert the numeral to a Hindu–Arabic numeral.

43. 𓆐𓆐𓂭𓂭𓆼𓆼𓎆𓎆𓎆𓏺𓏺𓏺𓏺𓏺

44. 八千二百五十四

45. $\chi\pi\epsilon$

46. MCMXCI

47. <<Y <<YYYY

48. ••
•••
═

4.3

In Exercises 49–54, convert the numeral to a Hindu–Arabic numeral.

49. 47_8
50. 101_2
51. 130_4
52. 3425_7
53. TOE_{12}
54. 20220_3

In Exercises 55–60, convert 463 to a numeral in the base indicated.

55. base 4
56. base 3
57. base 2
58. base 5
59. base 12
60. base 8

4.4

In Exercises 61–66, add in the base indicated.

61. $52_7 + 55_7$
62. $10110_2 + 11001_2$
63. $TE_{12} + 87_{12}$
64. $234_7 + 456_7$
65. $3024_5 + 4023_5$
66. $3407_8 + 7014_8$

In Exercises 67–72, subtract in the base indicated.

67. $4032_7 - 321_7$
68. $1001_2 - 101_2$
69. $3TT_{12} - E7_{12}$
70. $4321_5 - 442_5$
71. $1713_8 - 1243_8$
72. $2021_3 - 212_3$

In Exercises 73–78, multiply in the base indicated.

73. $32_6 \times 4_6$
74. $34_5 \times 21_5$
75. $126_{12} \times 47_{12}$
76. $221_3 \times 22_3$
77. $1011_2 \times 101_2$
78. $476_8 \times 23_8$

In Exercises 79–84, divide in the base indicated.

79. $1_2\overline{)1011_2}$
80. $2_4\overline{)320_4}$
81. $3_5\overline{)130_5}$
82. $4_6\overline{)3020_6}$
83. $3_6\overline{)2034_6}$
84. $6_8\overline{)5072_8}$

4.5

85. Multiply 142 × 24, using the duplation and mediation method.
86. Multiply 142 × 24, using the galley method.
87. Multiply 142 × 24, using Napier rods.

CHAPTER 4 TEST

1. Explain the difference between a numeral and a number.

In Exercises 2–7, convert the numeral to a Hindu–Arabic numeral.

2. MMMDCXLVI

3. <<Y <YYYYY

4. 八
千
零
九
十

5. ••
••
═
••••

6. 𓆼𓍢𓏲𓏲𓎆𓎆𓎆𓎆𓏺𓏺

7. $\theta'\pi$ Ϙ θ

In Exercises 8–12, convert the number written in base 10 to a numeral in the numeration system indicated.

8. 463 to Egyptian
9. 2476 to Ionic Greek
10. 1434 to Mayan
11. 1596 to Babylonian
12. 2378 to Roman

In Exercises 13–16, describe briefly each of the systems of numeration. Explain how each type of numeration system is used to represent numbers.

13. Additive system
14. Multiplicative system
15. Ciphered system
16. Place-value system

In Exercises 17–20, convert the numeral to a numeral in base 10.

17. 56_7

18. 403_5

19. 101101_2

20. 368_9

In Exercises 21–24, convert the base 10 numeral to a numeral in the base indicated.

21. 36 to base 2

22. 93 to base 5

23. 2356 to base 12

24. 2938 to base 7

In Exercises 25–28, perform the indicated operations.

25. $\begin{array}{r} 133_5 \\ +\ 434_5 \\ \hline \end{array}$

26. $\begin{array}{r} 324_6 \\ -\ 142_6 \\ \hline \end{array}$

27. $\begin{array}{r} 45_6 \\ \times\ 23_6 \\ \hline \end{array}$

28. $3_5\overline{)1210_5}$

29. Multiply 35 × 28, using duplation and mediation.

30. Multiply 43 × 196, using the galley method.

GROUP PROJECTS

U.S. Postal Service Bar Codes

Wherever we look nowadays, we see bar codes. We find them on items we buy at grocery stores and department stores and on many pieces of mail we receive. There are various types of bar codes, but each can be considered a type of numeration system. Although bar codes may vary in design, most are made up of a series of long and short bars. (New bar codes now being developed use a variety of shapes.) In this group project, we explain how postal codes are used.

The U.S. Postal Service introduced a bar coding system for zip codes in 1976. The system became known as Postnet (*post*al *n*umeric *e*ncoding *t*echniqe), and it has been refined over the years. Our basic zip code consists of five digits. The post office would like us to use the basic zip code followed by a hyphen and four additional digits. The post office refers to this nine-digit zip code as "zip + 4."

The Postnet bar code uses a series of long and short bars. A bar code may contain either 52 or 62 bars. The code designates the location to which the letter is being sent. The following bar code, with 52 bars, is for an address in Pittsburgh, Pennsylvania.

15250-7406 (Pittsburgh, PA)

In bar codes, each short bar represents 0 and each long bar represents 1. Each code starts and ends with a long bar that is *not* used in determining the zip + 4. If the code contains 52 bars, the code represents the zip + 4 and an extra digit referred to as a check digit. If the code contains 62 bars, it contains the zip + 4, the last two digits of the address number, and a check digit. If the code contains 52 bars, the sum of the zip + 4 and the check digit must equal a number that is divisible by 10. If the code contains 62 bars, the sum of the zip + 4, the last two digits of the address number, and the check digit must equal a number that is divisible by 10. The check digit is added to make each sum divisible by 10.

In a postal bar code, each of the digits 0 through 9 is represented by a series of five digits containing zeros and ones:

11000 (0)	00011 (1)	00101 (2)	00110 (3)	01001 (4)
01010 (5)	01100 (6)	10001 (7)	10010 (8)	10100 (9)

Consider the postal code from Pittsburgh given earlier. If you disregard the bar on the left, the next five bars are ıııll. Since each small bar represents a 0 and each large bar represents a 1, these five bars can be represented as 00011. From the chart, we see that this represents the number 1. The first five bars (after the bar on the far left has been excluded) tell the region of the country in which the address is located on the map shown on the next page.

Notice that Pennsylvania is located, along with New York, in the region marked 1 on the map.

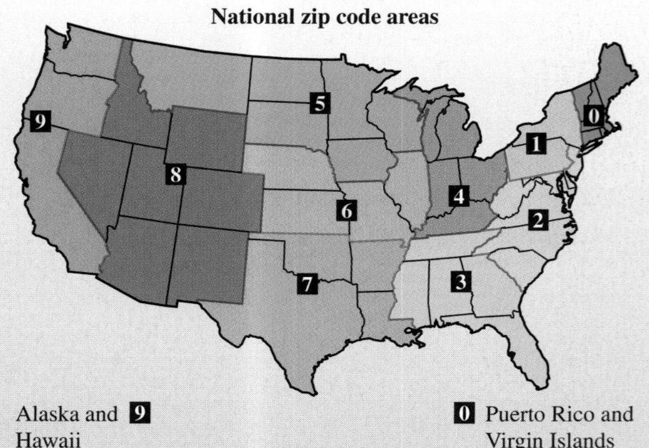

The second five lines in the bar code, ı|ı|ı, represent 01010 and have a value of 5. The other digits in the zip + 4 are determined in a similar manner. This code has 52 bars. The 45 bars, after the first bar, give the zip + 4. If you add the digits in the zip + 4, you get 1 + 5 + 2 + 5 + 0 + 7 + 4 + 0 + 6 = 30. Since 30 is divisible by 10, the five bars to the left of the bar on the very right should be 0. Note that ||ııı is represented as 11000 and has a value of 0. If, for example, the sum of the nine digits in the zip + 4 were 36, then the last five digits would need to represent the number 4, to make the sum of the digits divisible by 10. The five bars to the left of the last bar on the right are always used as a check.

Now let's work some problems.

a) For the Postnet code

|||ıııı||ııı|ıı||ııı|||ııı|ı|ıı|ıı|ıı||ıı|ı|ııııı|||

determine the zip + 4 and the check digit. Then check by adding the zip + 4 and the check digit. Is the sum divisible by 10?

b) For each of the following Postnet codes, determine the zip + 4, the last two numbers of the address number (if applicable), and the check digit.

i) |ı|ı|ıııı||ı|ı|ı|ı|ıııı||ıı|ıı||ıı|ııı||ı|ııı|ı|ı|ı|

ii) |ııı||ı|ıı|ı|ı|ııı|ı||ııı||ıı|ı|ı|ıııı|ı|ı|ıı||ııı|ı|ı|ıı||ıı|

c) Construct the Postnet code of long and short bars for each of the numbers. The numbers represent the zip + 4 and the last two digits of the address number. Do not forget the check digit.

i) 32226-8600-34 **ii)** 20794-1063-50

d) Construct the 52-bar Postnet code for your college's zip + 4. Don't forget to include the check digit.

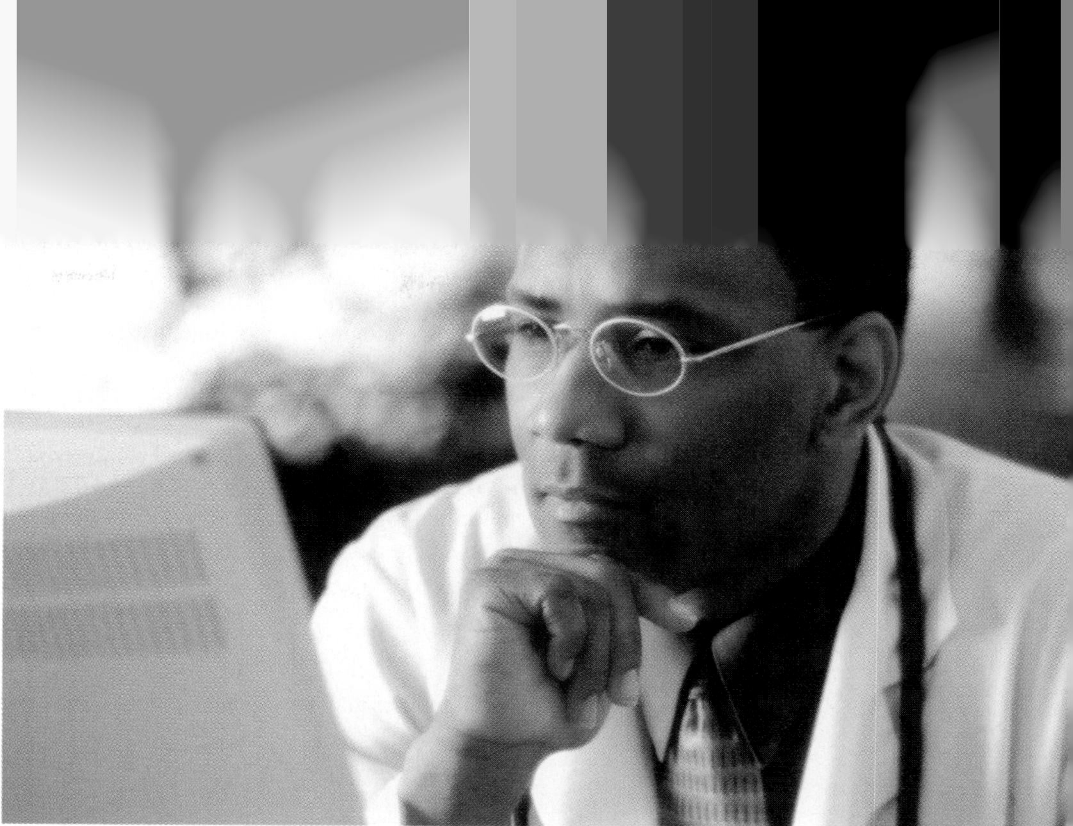

Number Theory plays an important role in computer science, medical research, and satellite technology.

NUMBER THEORY AND THE REAL NUMBER SYSTEM

It is impossible to live in our modern world without encountering numbers on a regular basis. In addition to playing a role in our everyday lives, numbers are used to describe the natural world, to communicate vast quantities of information, and to model problems facing scientists and researchers. *Number theory,* the study of numbers and their properties, makes all these roles possible. Mathematicians and computer scientists use number theory extensively to improve the speed of computers in our homes, businesses, and schools. Research scientists use number theory along with another branch of mathematics known as *knot theory* to conduct DNA research, research new drugs, and study how infectious diseases spread. Engineers use number theory in satellite technology which is used in virtually every modern means of communication.

5.1 NUMBER THEORY

This chapter introduces *number theory*, the study of numbers and their properties. The numbers we use to count are called the *counting numbers* or *natural numbers*. Since we begin counting with the number 1, the set of natural numbers begins with 1. The set of natural numbers is frequently denoted by N:

$$N = \{1, 2, 3, 4, 5, \dots\}$$

Any natural number can be expressed as a product of two or more natural numbers. For example, $8 = 2 \times 4$, $16 = 4 \times 4$, and $19 = 1 \times 19$. The natural numbers that are multiplied together are called factors of the product. For example,

$$\underset{\uparrow}{2} \times \underset{\uparrow}{4} = 8$$

Factors

A natural number may have many factors. For example, what pairs of numbers have a product of 18?

$$1 \cdot 18 = 18$$
$$2 \cdot 9 = 18$$
$$3 \cdot 6 = 18$$

The numbers 1, 2, 3, 6, 9, and 18 are all factors of 18. Each of these numbers divides 18 without a remainder.

If a and b are natural numbers, we say that a is a *divisor* of b or *a divides* b, symbolized $a \mid b$, if the quotient of b divided by a has a remainder of 0. If a divides b, then b is *divisible* by a. For example, 4 divides 12, symbolized $4 \mid 12$, since the quotient of 12 divided by 4 has a remainder of 0. Note that 12 is divisible by 4. The notation $7 \nmid 12$ means that 7 does not divide 12. Note that every factor of a natural number is also a divisor of the natural number. *Caution:* Do not confuse the symbols $a \mid b$ and a / b; $a \mid b$ means "a divides b" and a / b means "a divided by b" $(a \div b)$. The symbols a / b and $a \div b$ indicate that the operation of division is to be performed, and b may or may not be a divisor of a.

Prime and Composite Numbers

Every natural number greater than 1 can be classified as either a prime number or a composite number.

> A **prime number** is a natural number greater than 1 that has exactly two factors (or divisors), itself and 1.

The number 5 is a prime number because it is divisible only by the factors 1 and 5. The first eight prime numbers are 2, 3, 5, 7, 11, 13, 17, and 19. The number 2 is the only even prime number. All other even numbers have at least three divisors: 1, 2, and the number itself.

A **composite number** is a natural number that is divisible by a number other than itself and 1.

Any natural number greater than 1 that is not prime is composite. The first eight composite numbers are 4, 6, 8, 9, 10, 12, 14, and 15.

The number 1 is neither prime nor composite; it is called a *unit*. The number 38 has at least three divisors, 1, 2, and 38, and hence is a composite number. In contrast, the number 23 is a prime number since its only divisors are 1 and 23.

More than 2000 years ago, the ancient Greeks developed a technique for determining which numbers are prime numbers and which are not. This technique is named the *sieve of Eratosthenes*, for the Greek mathematician Eratosthenes of Cyrene who first used it.

~~1~~	(2)	(3)	~~4~~	(5)	~~6~~	(7)	~~8~~	~~9~~	~~10~~
(11)	~~12~~	(13)	~~14~~	~~15~~	~~16~~	(17)	~~18~~	(19)	~~20~~
~~21~~	~~22~~	(23)	~~24~~	~~25~~	~~26~~	~~27~~	~~28~~	(29)	~~30~~
(31)	~~32~~	~~33~~	~~34~~	~~35~~	~~36~~	(37)	~~38~~	~~39~~	~~40~~
(41)	~~42~~	(43)	~~44~~	~~45~~	~~46~~	(47)	~~48~~	~~49~~	~~50~~

Figure 5.1

To find the prime numbers less than or equal to any natural number, say, 50, using this method, list the first 50 counting numbers (Fig. 5.1). Cross out 1 since it is not a prime number. Circle 2, the first prime number. Then cross out all the multiples of 2: 4, 6, 8, . . . , 50. Circle the next prime number, 3. Cross out all multiples of 3 that are not already crossed out. Continue this process until you reach the prime number p, such that $p \cdot p$, or p^2, is greater than the last number listed, in this case 50. Therefore, we next circle 5 and cross out its multiples. Then circle 7 and cross out its multiples. The next prime number is 11, and $11 \cdot 11$, or 121, is greater than 50, so you are done. At this point, circle all the remaining numbers to obtain the prime numbers less than or equal to 50. The prime numbers less than or equal to 50 are 2, 3, 5, 7, 11, 13, 17, 19, 23, 29, 31, 37, 41, 43, and 47.

Now we turn our attention to composite numbers and their factors. The rules of divisibility given in the chart on page 210 are helpful in finding divisors (or factors) of composite numbers.

The test for divisibility by 6 is a particular case of the general statement that the product of two prime divisors of a number is a divisor of the number. Thus, for example, if both 3 and 7 divide a number, then 21 will also divide the number.

Note that the chart does not list rules of divisibility for the number 7. There is a rule for 7, but it is difficult to remember. The easiest way to check divisibility by 7 is just to perform the division.

PROFILE IN MATHEMATICS

ERATOSTHENES OF CYRENE

Eratosthenes of Cyrene (275–195 B.C.) was born in northern Africa near the present-day city of Shahhat, Libya. Eratosthenes is best known for being the first to estimate accurately the diameter of Earth. He is also credited for developing a method of finding prime numbers known as the **sieve of Eratosthenes**. Although he is most known for his work in mathematics, Eratosthenes also was influential in the fields of history, geography, and astronomy. In addition, Eratosthenes served for many years as the director of the famous library in Alexandria, Egypt. Although Eratosthenes was a highly regarded scholar throughout the ancient world, only fragments of his writing remain today. Eratosthenes was near 80 years old when, after going blind, he died from voluntary starvation.

DID YOU KNOW

No More Odd Days!

The date November 19, 1999, was a very special day. It was a rare "odd day." The numerical format is 11-19-1999 (or some may write it as 19-11-1999), which contains only odd digits. The next odd day will be January 1, 3111, which would be written 1-1-3111, and is over a thousand years away—a date we certainly will never see.

Days such as 8-27-2002 have both odd and even digits. Thus, it is neither an odd day nor an even day. The first of many even days in the year 2000 was 2-2-2000, the first one since 8-28-888.

So now you have a reason to celebrate, since you have seen your last odd day on Earth!

Rules of Divisibility

Divisible by	Test	Example
2	The number is even.	924 is divisible by 2 since 924 is even.
3	The sum of the digits of the number is divisible by 3.	924 is divisible by 3 since the sum of the digits, $9 + 2 + 4 = 15$, and 15 is divisible by 3.
4	The number formed by the last two digits of the number is divisible by 4.	924 is divisible by 4 since the number formed by the last two digits, 24, is divisible by 4.
5	The number ends in 0 or 5.	265 is divisible by 5 since the number ends in 5.
6	The number is divisible by both 2 and 3.	924 is divisible by 6 since it is divisible by both 2 and 3.
8	The number formed by the last three digits of the number is divisible by 8.	5824 is divisible by 8 since the number formed by the last three digits, 824, is divisible by 8.
9	The sum of the digits of the number is divisible by 9.	837 is divisible by 9 since the sum of the digits, 18, is divisible by 9.
10	The number ends in 0.	290 is divisible by 10 since the number ends in 0.

EXAMPLE 1 *Using the Divisibility Rules*

Determine whether 145,860 is divisible by

a) 2 b) 3 c) 4 d) 5 e) 6 f) 8 g) 9 h) 10

SOLUTION:

a) Since 145,860 is even, it is divisible by 2.

b) The sum of the digits of 145,860 is $1 + 4 + 5 + 8 + 6 + 0 = 24$. Since 24 is divisible by 3, the number 145,860 is divisible by 3.

c) The number formed by the last two digits is 60. Since 60 is divisible by 4, the number 145,860 is divisible by 4.

d) Since 145,860 has 0 as the last digit, 145,860 is divisible by 5.

e) Since 145,860 is divisible by both 2 and 3, 145,860 is divisible by 6.

f) The number formed by the last three digits is 860. Since 860 is not divisible by 8, the number 145,860 is not divisible by 8.

g) The sum of the digits of 145,860 is 24. Since 24 is not divisible by 9, the number 145,860 is not divisible by 9.

h) Since 145,860 has 0 as the last digit, 145,860 is divisible by 10. ▲

Every composite number can be expressed as a product of prime numbers. The process of breaking a given number down into a product of prime numbers is called *prime factorization*. The prime factorization of 18 is $3 \times 3 \times 2$. No other natural number listed as a product of primes will have the same prime factorization as 18. The *fundamental theorem of arithmetic* states this concept formally. (A *theorem* is a statement or proposition that can be proven true.)

The Fundamental Theorem of Arithmetic
Every composite number can be expressed as a *unique* product of prime numbers.

In writing the prime factorization of a number, the order of the factors is immaterial. For example, we may write the prime factors of 18 as $3 \times 3 \times 2$ or $2 \times 3 \times 3$ or $3 \times 2 \times 3$.

A number of techniques can be used to find the prime factorization of a number. Two methods are illustrated.

Method 1: Branching

To find the prime factorization of a number, select any two numbers whose product is the number to be factored. If the factors are not prime numbers, then continue factoring each composite number until all numbers are prime.

EXAMPLE 2 *Prime Factorization by Branching*

Write 2100 as a product of primes.

SOLUTION: Select any two numbers whose product is 2100. Among the many choices, two possibilities are $21 \cdot 100$ and $30 \cdot 70$. Let us consider $21 \cdot 100$. Since neither 21 nor 100 are prime numbers, find any two numbers whose product is 21 and any two numbers whose product is 100. Continue branching as shown in Fig. 5.2 until the numbers in the last row are all prime numbers. To determine the answer, write the product of all the prime factors. The branching diagram is sometimes called a *factor tree.*

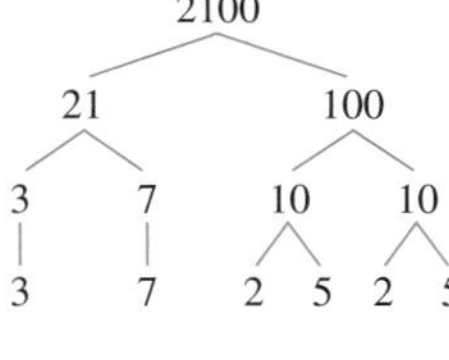

Figure 5.2

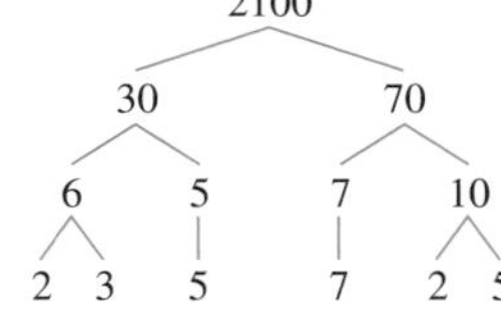

Figure 5.3

We see that the numbers in the last row of factors in Fig. 5.2 are all prime numbers. Thus, the prime factorization of 2100 is $3 \cdot 7 \cdot 2 \cdot 5 \cdot 2 \cdot 5 = 2 \cdot 2 \cdot 3 \cdot 5 \cdot 5 \cdot 7 = 2^2 \cdot 3 \cdot 5^2 \cdot 7$. Note from Fig. 5.3 that had we chosen 30 and 70 as the first pair of factors, the prime factorization would still be $2^2 \cdot 3 \cdot 5^2 \cdot 7$. ▲

Method 2: Division

To obtain the prime factorization of a number by this method, divide the given number by the smallest prime number by which it is divisible. Place the quotient under the given number. Then divide the quotient by the smallest prime number by which it is divisible and again record the quotient. Repeat this process until the quotient is a prime number. The prime factorization is the product of all the prime divisors and the prime (or last) quotient. This procedure is illustrated in Example 3.

DID YOU KNOW

The Long-Lost Factoring Machine

In 1989, Jeffrey Shallit of the University of Waterloo came across an article in a 1920 French journal regarding a machine that was built in 1914 for factoring numbers. Eugene Oliver Carissan, an amateur mathematician who had invented the machine, wrote the article. After reading the article, Shallit wondered whatever became of the factoring machine.

After considerable searching, Shallit found the machine in a drawer of an astronomical observatory in Floirac, France. The machine was in good condition, and it still worked. By rotating the machine by a hand crank at two revolutions per minute, an operator could process 35 to 40 numbers per second. Carissan needed just 10 minutes to prove that 708,158,977 is a prime number, an amazing feat in precomputer times. The machine is now housed at the Conservatoire National des Arts et Métiers in Paris.

PROFILE IN MATHEMATICS

SRINIVASA RAMANUJAN

One of the most interesting mathematicians of modern times is Srinivasa Ramanujan (1887–1920). Born to an impoverished middle-class family in India, he virtually taught himself higher mathematics. He went to England to study with the number theorist G. H. Hardy. Hardy tells the story of a taxicab ride he took to visit Ramanujan. The cab had the license plate number 1729, and he challenged the young Indian to find anything interesting in that. Without hesitating, Ramanujan pointed out that it was the smallest positive integer that could be represented in two different ways as the sum of two cubes: $1^3 + 12^3$ and $9^3 + 10^3$.

EXAMPLE 3 *Prime Factorization by Division*

Write 2100 as a product of prime numbers.

SOLUTION: Because 2100 is an even number, the smallest prime number that divides it is 2. Divide 2100 by 2. Place the quotient, 1050, below the 2100. Repeat this process of dividing each quotient by the smallest prime number that divides it.

2	2100
2	1050
3	525
5	175
5	35
	7

The final quotient, 7, is a prime number, so we stop. The prime factorization of 2100 is

$$2 \cdot 2 \cdot 3 \cdot 5 \cdot 5 \cdot 7 = 2^2 \cdot 3 \cdot 5^2 \cdot 7.$$

Note that, despite the different methods used in Examples 2 and 3, the answer is the same.

Greatest Common Divisor

The discussion in Section 5.3 of how to reduce fractions makes use of the greatest common divisor (GCD). One technique of finding the GCD is to use prime factorization.

The **greatest common divisor (GCD)** of a set of natural numbers is the largest natural number that divides (without remainder) every number in that set.

What is the GCD of 12 and 18? One way to determine it is to make a list of the divisors (or factors) of 12 and 18:

Divisors of 12	{**1, 2, 3**, 4, **6**, 12}
Divisors of 18	{**1, 2, 3, 6**, 9, 18}

The common divisors are 1, 2, 3, and 6. Therefore, the greatest common divisor is 6.

If the numbers are large, this method of finding the GCD is not practical. The GCD can be found more efficiently by using prime factorization.

To Find the Greatest Common Divisor of Two or More Numbers

1. Determine the prime factorization of each number.
2. Find each prime factor with the smallest exponent that appears in each of the prime factorizations.
3. Determine the product of the factors found in step 2.

Example 4 illustrates this procedure.

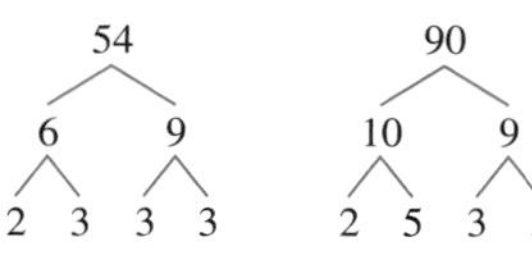

Figure 5.4

EXAMPLE 4 *Using Prime Factorization to Find the GCD*

Find the GCD of 54 and 90.

SOLUTION: The branching method of finding the prime factors of 54 and 90 is illustrated in Fig. 5.4.

a) The prime factorization of 54 is $2 \cdot 3^3$, and the prime factorization of 90 is $2 \cdot 3^2 \cdot 5$.

b) The prime factors with the smallest exponents that appear in each of the factorizations of 54 and 90 are 2 and 3^2.

c) The product of the factors found in step 2 is $2 \cdot 3^2 = 2 \cdot 9 = 18$. The GCD of 54 and 90 is 18. Eighteen is the largest number that divides both 54 and 90. ▲

DID YOU KNOW

Friendly Numbers

The ancient Greeks often thought of numbers as having human qualities. For example, the numbers 220 and 284 were considered "friendly" or "amicable" numbers because each number was the sum of the other number's proper factors. (A proper factor is any factor of a number other than the number itself.) If you sum all the proper factors of 284 $(1 + 2 + 4 + 71 + 142)$, you get the number 220, and if you sum all the proper factors of 220 $(1 + 2 + 4 + 5 + 10 + 11 + 20 + 22 + 44 + 55 + 110)$, you get 284.

EXAMPLE 5 *Finding the GCD*

Find the GCD of 225 and 525.

SOLUTION:

a) The prime factorization of 225 is $3^2 \cdot 5^2$, and the prime factorization of 525 is $3 \cdot 5^2 \cdot 7$ (you should verify these using the branching method or the division method).

b) The prime factors with the smallest exponents that appear in each of the factorizations of 225 and 525 are 3 and 5^2.

c) The product of the factors found in step 2 is $3 \cdot 5^2 = 3 \cdot 25 = 75$. The GCD of 225 and 525 is 75. ▲

Two numbers with a GCD of 1 are said to be *relatively prime*. The numbers 9 and 14 are relatively prime, since the GCD is 1.

Least Common Multiple

To perform addition and subtraction of fractions (Section 5.3), we use the least common multiple (LCM). One technique of finding the LCM is to use prime factorization.

> The **least common multiple (LCM)** of a set of natural numbers is the smallest natural number that is divisible (without remainder) by each element of the set.

What is the least common multiple of 12 and 18? One way to determine the LCM is to make a list of the multiples of each number:

Multiples of 12 {12, 24, **36**, 48, 60, **72**, 84, 96, **108**, 120, 132, **144**, . . .}
Multiples of 18 {18, **36**, 54, **72**, 90, **108**, 126, **144**, 162, . . .}

Some common multiples of 12 and 18 are 36, 72, 108, and 144. The least common multiple, 36, is the smallest number that is divisible by both 12 and 18. Usually, the most efficient method of finding the LCM is to use prime factorization.

DID YOU KNOW

Safer Secrets

Prime numbers have for many years played an essential role in protecting the databases of government and business. From 1977 through 2000, data were *encrypted* to allow access only to legitimate users. The key to entering the database was a large composite number over 300 digits long. This number, n, was the product of two very large prime numbers, $p \cdot q$. The number n was publicly available, so information could be entered into the database. The numbers p and q, however, were given only to those users who had the right to access the database. The system was considered secure because computers of the day could not factor a number as large as n. However, with the rapid advances in computer science and mathematics research, larger and larger numbers had to be found to keep databases secure. By 1997, realizing a better protection system was needed, the National Institute for Standardization and Technology began a global contest to choose a better protection system. In October 2000, the Rijndael Block Cipher, developed by two Belgian cryptographers, was selected as the winner. The new system involves various areas of mathematics, including many that are discussed in this book such as permutations, modular arithmetic, polynomials, matrices, and group theory.

To Find the Least Common Multiple of Two or More Numbers

1. Determine the prime factorization of each number.
2. List each prime factor with the greatest exponent that appears in any of the prime factorizations.
3. Determine the product of the factors found in step 2.

Example 6 illustrates this procedure.

EXAMPLE 6 *Using Prime Factorization to Find the LCM*

Find the LCM of 54 and 90.

SOLUTION:

a) Find the prime factors of each number. In Example 4, we determined that

$$54 = 2 \cdot 3^3 \qquad \text{and} \qquad 90 = 2 \cdot 3^2 \cdot 5$$

b) List each prime factor with the greatest exponent that appears in either of the prime factorizations: 2, 3^3, 5.

c) Determine the product of the factors found in step 2:

$$2 \cdot 3^3 \cdot 5 = 2 \cdot 27 \cdot 5 = 270$$

Thus, 270 is the LCM of 54 and 90. It is the smallest number that is divisible by both 54 and 90. ▲

EXAMPLE 7 *Finding the LCM*

Find the LCM of 225 and 525.

SOLUTION:

a) Find the prime factors of each number. In Example 5, we determined that

$$225 = 3^2 \cdot 5^2 \quad \text{and} \quad 525 = 3 \cdot 5^2 \cdot 7$$

b) List each prime factor with the greatest exponent that appears in either of the prime factorizations: 3^2, 5^2, 7.

c) Determine the product of the factors found in step 2:

$$3^2 \cdot 5^2 \cdot 7 = 9 \cdot 25 \cdot 7 = 1575$$

Thus, 1575 is the least common multiple of 225 and 525. It is the smallest number divisible by both 225 and 525. ▲

MATHEMATICS *Everywhere*

Gimps

George Woltman

The Great Internet Mersenne Prime Search (GIMPS) was started by George Woltman to provide free software and access to a large database to coordinate the efforts of thousands of people interested in seeking new Mersenne prime numbers. Scott Kurowski provided the technology and services needed to make it simple for anyone to join the project via the Internet. GIMPS allows participants to use their computers to search for prime numbers while their computers are idle. A computer communicates, via the specialized software, with the database. Currently, there are over 130,000 worldwide participants in GIMPS. As of July 2003, the largest Mersenne prime was found by the 20-year-old Michael Cameron on November 14, 2001 in conjunction with GIMPS. The number, $2^{13,466,917} - 1$, which is over 4 million digits long, is the 39th Mersenne prime ever found and the fifth discovered by GIMPS participants. Michael's home computer, an AMD T-Bird 880 MHz with 512 Megs of RAM, took 42 days of idle time to conclude that this number was prime.

The Search For Larger Prime Numbers

More than 2000 years ago, the Greek mathematician Euclid proved that there is no largest prime number. Mathematicians, however, continue to strive to find larger and larger prime numbers.

Marin Mersenne (1588–1648), a seventeenth-century monk, found that numbers of the form $2^n - 1$ are often prime numbers when n is a prime number. For example,

$$2^2 - 1 = 4 - 1 = 3 \qquad 2^3 - 1 = 8 - 1 = 7$$
$$2^5 - 1 = 32 - 1 = 31 \qquad 2^7 - 1 = 128 - 1 = 127$$

Numbers of the form $2^n - 1$ that are prime are referred to as *Mersenne primes*. The first 10 Mersenne primes occur when $n = 2, 3, 5, 7, 13, 17, 19, 31, 61, 89$. The first time the expression $2^n - 1$ does not generate a prime number, for prime number n, is when n is 11. The number $2^{11} - 1$ is a composite number (see Exercise 90).

Scientists frequently use Mersenne primes in their search for larger and larger primes. The largest prime number found to date was discovered on November 14, 2001, by 20-year-old Michael Cameron of Owen Sound, Ontario, Canada in conjunction with the Great Internet Mersenne Prime Search (GIMPS). The number is the Mersenne prime $2^{13,466,917} - 1$. This record prime is the 39th known Mersenne prime, and when written out, it is 4,053,946 digits long—over 11 miles long if written using the same size font as this textbook!

More About Prime Numbers

Another mathematician who studied prime numbers was Pierre de Fermat (1601–1665). A lawyer by profession, Fermat became interested in mathematics as a hobby. He became one of the finest mathematicians of the seventeenth century. Fermat conjectured that each number of the form $2^{2^n} + 1$, now referred to as a *Fermat number*, was prime for each natural number n. Recall that a *conjecture* is a supposition that has not been proved nor disproved. In 1732, Leonhard Euler proved that for $n = 5$, $2^{32} + 1$ was a composite number, thus disproving Fermat's conjecture.

Since Euler's time, mathematicians have only been able to evaluate the sixth, seventh, eighth, ninth, tenth, and eleventh Fermat numbers to determine whether they are prime or composite. Each of these numbers has been shown to be composite. The eleventh Fermat number was factored by Richard Brent and François Morain in 1988. The sheer magnitude of the numbers involved makes it difficult to test these numbers, even with supercomputers.

In 1742, Christian Goldbach conjectured in a letter to Euler that every even number greater than or equal to 4 can be represented as the sum of two (not necessarily distinct) prime numbers (for example, $4 = 2 + 2$, $6 = 3 + 3$, $8 = 3 + 5$, $10 = 5 + 5$, $12 = 5 + 7$). This conjecture became known as *Goldbach's conjecture*, and it remains unproven to this day. The *twin prime conjecture* is another famous long-standing conjecture. *Twin primes* are primes of the form p and $p + 2$ (for example, 3 and 5, 5 and 7, 11 and 13). This conjecture states that there are an infinite number of pairs of twin primes. At the time of this writing the largest twin primes are of the form $665,551,035 \cdot 2^{80,025}$ plus or minus 1, which were found by David Underbakke and Phil Carmody on November 28, 2000.

SECTION 5.1 EXERCISES

Concept/Writing Exercises

1. What is number theory?
2. What does "a and b are factors of c" mean?
3. **a)** What does "a divides b" mean?
   b) What does "a is divisible by b" mean?
4. What is a prime number?
5. What is a composite number?
6. What does the fundamental theorem of arithmetic state?
7. **a)** What is the least common multiple of a set of natural numbers?
   b) In your own words, explain how to find the LCM of a set of natural numbers by using prime factorization.
   c) Find the LCM of 16 and 40 by using the procedure given in part (b).
8. **a)** What is the greatest common divisor of a set of natural numbers?
   b) In your own words, explain how to find the GCD of a set of natural numbers by using prime factorization.
   c) Find the GCD of 16 and 40 by using the procedure given in part (b).
9. What are Mersenne primes?
10. What is a conjecture?
11. What is Goldbach's conjecture?
12. What are twin primes?

Practice the Skills

13. Use the sieve of Eratosthenes to find the prime numbers up to 100.

14. Use the sieve of Eratosthenes to find the prime numbers up to 150.

In Exercises 15–26, determine whether the statement is true or false. Modify each false statement to make it a true statement.

15. 9 is a factor of 54.

16. $4|36$.

17. 7 is a multiple of 21.

18. 35 is a divisor of 5.

19. 8 is divisible by 56.

20. 15 is a factor of 45.

21. If a number is not divisible by 5, then it is not divisible by 10.

22. If a number is not divisible by 10, then it is not divisible by 5.

23. If a number is divisible by 3, then every digit of the number is divisible by 3.

24. If every digit of a number is divisible by 3, then the number itself is divisible by 3.

25. If a number is divisible by 2 and 3, then the number is divisible by 6.

26. If a number is divisible by 3 and 4, then the number is divisible by 12.

In Exercises 27–32, determine whether the number is divisible by each of the following numbers: 2, 3, 4, 5, 6, 8, 9, and 10.

27. 10,368 **28.** 19,200
29. 2,763,105 **30.** 3,126,120
31. 1,882,320 **32.** 3,941,221

33. Determine a number that is divisible by 2, 3, 4, 5, and 6.

34. Determine a number that is divisible by 3, 4, 5, 9, and 10.

In Exercises 35–46, find the prime factorization of the number.

35. 45 **36.** 52 **37.** 196
38. 198 **39.** 303 **40.** 400
41. 513 **42.** 663 **43.** 1336
44. 1313 **45.** 2001 **46.** 3190

In Exercises 47–56, find (a) the greatest common divisor (GCD) and (b) the least common multiple (LCM).

47. 6 and 15 **48.** 20 and 36
49. 48 and 54 **50.** 22 and 231
51. 40 and 900 **52.** 120 and 240
53. 96 and 212 **54.** 240 and 285
55. 24, 48, and 128 **56.** 18, 78, and 198

Problem Solving

57. Find the next two sets of twin primes that follow the set 11, 13.

58. The primes 2 and 3 are consecutive natural numbers. Is there another pair of consecutive natural numbers both of which are prime? Explain.

59. For each pair of numbers, determine whether the numbers are relatively prime. Write yes or no as your answer.
   a) 14, 15
   b) 21, 30
   c) 24, 25
   d) 119, 143

60. Find the first three Fermat numbers and determine whether they are prime or composite.

61. Show that Goldbach's conjecture is true for the even numbers 4 through 20.

62. Find the first five Mersenne prime numbers.

63. *Barbie and Ken* Mary Lois King collects Barbie dolls and Ken dolls. She has 350 Barbie dolls and 140 Ken dolls. Mary Lois wishes to display the dolls in groups so that the same number of dolls are in each group and that each doll belongs to one group. If each group is to consist only of Barbie dolls or only of Ken dolls, what is the largest number of dolls Mary Lois can have in each group?

64. *Toy Car Collection* Martha Goshaw collects Matchbox® and HotWheels® toy cars. She has 288 red cars and 192 blue cars. She wants to line up her cars in groups so that each group has the same number of cars and each group contains only red cars or only blue cars. What is the largest number of cars she can have in a group?

65. *Stacking Trading Cards* Desmond Freeman collects trading cards. He has 432 baseball cards and 360 football cards. He wants to make stacks of cards on a table so that each stack contains the same number of cards and each card belongs to one stack. If the baseball and football cards must not be mixed in the stacks, what is the largest number of cards that he can have in a stack?

66. *Tree Rows* Elizabeth Dwyer is the manager at Queen Palm Nursery and is in charge of displaying potted trees in rows. Elizabeth has 150 citrus trees and 180 palm trees. She wants to make rows of trees so that each row has the same number of trees and each tree is in a row. If the citrus trees and the palm trees must not be mixed in the rows, what is the largest number of trees that she can have in a row?

67. *Airport Activity* O'Hare International Airport in Chicago has a flight leaving for New York City every 45 minutes and a flight leaving for Atlanta every 60 minutes. If a flight to New York City and a flight to Atlanta leave at the same time, how many minutes will it be before a flight to New York City and a flight to Atlanta again leave at the same time?

O'Hare International Airport

68. *Car Maintenance* For many sport utility vehicles, it is recommended that the oil be changed every 3500 miles and that the tires be rotated every 6000 miles. If Carmella Gonzalez just had the oil changed and tires rotated on her SUV during the same visit to her mechanic, how many miles will she drive before she has the oil changed and tires rotated again during the same visit?

69. *Work Schedules* Sara Pappas and Harry Kinnan both work the 3:00 P.M. to 11:00 P.M. shift. Sara has every fifth night off and Harry has every sixth night off. If they both have tonight off, how many days will pass before they have the same night off again?

70. *Restaurant Service* Peter Theodus runs a professional accounting service for restaurants. Peter goes to Arturo's Family Restaurant every 15 days, and he goes to Xang's Great Wall Restaurant every 18 days. If on October 1 Peter visits both restaurants, how many days would it be before he visited both restaurants on the same day again?

71. *U.S. Senate Committees* The U.S. Senate consists of 100 members. Senate committees are to be formed so that each of the committees contains the same number of senators and each senator is a member of exactly one committee. The committees are to have more than 2 members but fewer than 50 members. There are various ways that these committees can be formed.
   a) What size committees are possible?
   b) How many committees are there for each size?

72. *Prime Numbers* Consider the first eight prime numbers greater than 3. The numbers are 5, 7, 11, 13, 17, 19, 23, and 29.
   a) Determine which of these prime numbers differs by 1 from a multiple of the number 6.
   b) Use inductive reasoning and the results obtained in part (a) to make a conjecture regarding prime numbers.
   c) Select a few more prime numbers and determine whether your conjecture appears to be correct.

73. State a procedure that defines a divisibility test for 15.

74. State a procedure that defines a divisibility test for 22.

Euclidean Algorithm Another method that can be used to find the greatest common divisor is known as the Euclidean algorithm. We illustrate this procedure by finding the GCD of 60 and 220.

First divide 220 by 60 as shown below. Disregard the quotient 3 and then divide 60 by the remainder 40. Continue this process of dividing the divisors by the remainders until you obtain a remainder of 0. The divisor in the last division, in which the remainder is 0, is the GCD.

$$\begin{array}{r} 3 \\ 60\overline{)220} \\ \underline{180} \\ 40 \end{array} \qquad \begin{array}{r} 1 \\ 40\overline{)60} \\ \underline{40} \\ 20 \end{array} \qquad \begin{array}{r} 2 \\ 20\overline{)40} \\ \underline{40} \\ 0 \end{array}$$

Since 40/20 had a remainder of 0, the GCD is 20.

In Exercises 75–80, use the Euclidean algorithm to find the GCD.

75. 15, 35

76. 16, 28

77. 36, 108

78. 76, 240

79. 150, 180

80. 210, 560

Perfect Numbers A number whose proper factors (factors other than the number itself) add up to the number is called a perfect number. For example, 6 is a perfect number because its proper factors are 1, 2, and 3, and $1 + 2 + 3 = 6$. Determine which, if any, of the following numbers are perfect.

81. 12

82. 28

83. 496

84. 48

Challenge Problems/Group Activities

85. *Number of Factors* The following procedure can be used to determine the *number of factors* (or *divisors*) of a composite number. Write the number in prime factorization form. Examine the exponents on the prime numbers in the prime factorization. Add 1 to each exponent and then find the product of these numbers. This product gives the number of positive divisors of the composite number.

a) Use this procedure to determine the number of divisors of 60.

b) To check your answer, list all the divisors of 60. You should obtain the same number of divisors found in part (a).

86. Recall that if a number is divisible by both 2 and 3, then the number is divisible by 6. If a number is divisible by both 2 and 4, is the number necessarily divisible by 8? Explain your answer.

87. The product of any three consecutive natural numbers is divisible by 6. Explain why.

88. A number in which each digit except 0 appears exactly three times is divisible by 3. For example, 888,444,555 and 714,714,714 are both divisible by 3. Explain why this outcome must be true.

89. Use the fact that if $a|b$ and $a|c$, then $a|(b + c)$ to determine whether 36,018 is divisible by 18. (*Hint:* Write 36,018 as $36{,}000 + 18$.)

90. Show that the $2^n - 1$ is a (Mersenne) prime for $n = 2, 3, 5$, and 7 but composite for $n = 11$.

91. Goldbach also conjectured in his letter to Euler that *every* integer greater than 5 is the sum of three prime numbers. For example, $6 = 2 + 2 + 2$ and $7 = 2 + 2 + 3$. Show that this conjecture is true for integers 8 through 20.

Recreational Mathematics

92. *Country, Animal, Fruit* Select a number from 1 to 10. Multiply your selected number by 9. Add the digits of the product together (if the product has more than one digit). Now subtract 5. Determine which letter of the alphabet corresponds to the number you ended with (for example, 1 = A, 2 = B, 3 = C, and so on). Think of a country whose name begins with that letter. Remember the last letter of the name of that country. Think of an animal whose name begins with that letter. Remember the last letter in the name of that animal. Think of a fruit whose name begins with that letter. a) What country, animal, and fruit did you select? Turn to the answer section to see if your response matches the responses of over 90% of the people who attempt this activity. b) Can you explain why most people select the given answer?

Research Activities

93. Do research and explain what *deficient numbers* and *abundant numbers* are. Give an example of each type of number. References include history of mathematics books, encyclopedias, and the Internet.

94. Conduct an Internet search on the GIMPS project. Write a report describing the history and development of the project. Include a current update of the project's findings.

5.2 THE INTEGERS

DID YOU KNOW

Virgil's Party

The poet Virgil was born in 70 B.C. In 1930, some highly respected scholars decided to celebrate his 2000th birthday. The only problem was that the accounting of time does not include a year 0 between B.C. and A.D. So Virgil would have turned 70 in our year 1 A.D. and, had he lived so long, 2000 in 1931. This fact was pointed out to the celebrants, but only after the party was well under way.

In Section 5.1, we introduced the natural or counting numbers:

$$N = \{1, 2, 3, 4, \dots\}$$

Another important set of numbers, the ***whole numbers***, help to answer the question "How many?"

$$\text{Whole numbers} = \{0, 1, 2, 3, 4, \dots\}$$

Note that the set of whole numbers contains the number 0 but that the set of counting numbers does not. If a farmer were asked how many chickens were in a coop, the answer would be a whole number. If the farmer had no chickens, he or she would answer zero. Although we use the number 0 daily and take it for granted, the number 0 as we know it was not used and accepted until the sixteenth century.

If the temperature is 12°F and drops 20°, the resulting temperature is −8°F. This type of problem shows the need for negative numbers. The set of ***integers*** consists of the negative integers, 0, and the positive integers.

$$\text{Integers} = \{\underbrace{\dots, -4, -3, -2, -1}_{\text{Negative integers}}, 0, \underbrace{1, 2, 3, \dots}_{\text{Positive integers}}\}$$

The term *positive integers* is yet another name for the natural numbers or counting numbers.

An understanding of addition, subtraction, multiplication, and division of the integers is essential in understanding algebra (Chapter 6). To aid in our explanation of addition and subtraction of integers, we introduce the real number line (Fig. 5.5). To construct the real number line, arbitrarily select a point for zero to serve as the starting point. Place the positive integers to the right of 0, equally spaced from one another. Place the negative integers to the left of 0, using the same spacing. The real number line contains the integers and all the other real numbers that are not integers. Some examples of real numbers that are not integers are indicated in Fig. 5.5, namely $-\frac{5}{2}$, $\frac{1}{2}$, $\sqrt{2}$, and π. We discuss real numbers that are not integers in the next two sections.

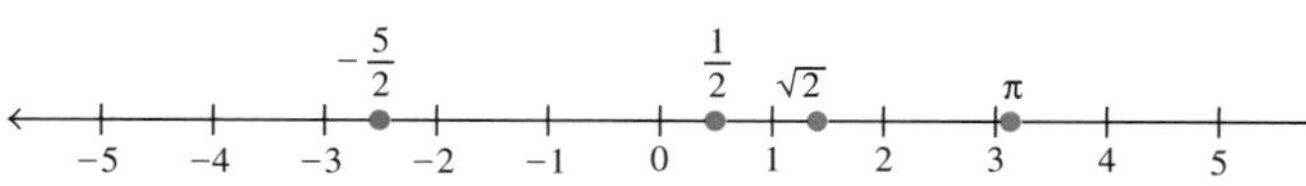

Figure 5.5

The arrows at the ends of the real number line indicate that the line continues indefinitely in both directions. Note that for any natural number, n, on the number line, the *opposite of* that number, $-n$, is also on the number line. This real number line was drawn horizontally, but it could just as well have been drawn vertically. In fact, in the next chapter, we show that the axes of a graph are the union of two number lines, one horizontal and the other vertical.

The number line can be used to determine the greater (or lesser) of two integers. Two *inequality symbols* that we will use in this chapter are $>$ and $<$. The symbol $>$ is read "is greater than," and the symbol $<$ is read "is less than." Expressions that contain an inequality symbol are called *inequalities.* On the number line, the numbers increase from left to right. The number 3 is greater than 2, written $3 > 2$. Observe that 3 is to the right of 2. Similarly, we can see that $0 > -1$ by observing that 0 is to the right of -1 on the number line.

Instead of stating that 3 is greater than 2, we could state that 2 is less than 3, written $2 < 3$. Note that 2 is to the left of 3 on the number line. We can also see that $-1 < 0$ by observing that -1 is to the left of 0. The inequality symbol always points to the smaller of the two numbers when the inequality is true.

EXAMPLE 1 *Writing an Inequality*

Insert either $>$ or $<$ in the shaded area between the paired numbers to make the statement correct.

a) -3 ■ 1 b) -3 ■ -5 c) -6 ■ -4 d) 0 ■ -7

SOLUTION:

a) $-3 < 1$ since -3 is to the left of 1 on the number line.
b) $-3 > -5$ since -3 is to the right of -5 on the number line.
c) $-6 < -4$ since -6 is to the left of -4 on the number line.
d) $0 > -7$ since 0 is to the right of -7 on the number line. ▲

Addition of Integers

Addition of integers can be represented geometrically with a number line. To do so, begin at 0 on the number line. Represent the first addend (the first number to be added) by an arrow starting at 0. Draw the arrow to the right if the addend is positive. If the addend is negative, draw the arrow to the left. From the tip of the first arrow, draw a second arrow to represent the second addend. Draw the second arrow to the right or left, as just explained. The sum of the two integers is found at the tip of the second arrow.

EXAMPLE 2 *Adding Integers*

Evaluate the following using the number line.

a) $3 + (-5)$ b) $-1 + (-4)$ c) $-6 + 4$ d) $3 + (-3)$

SOLUTION:

a)

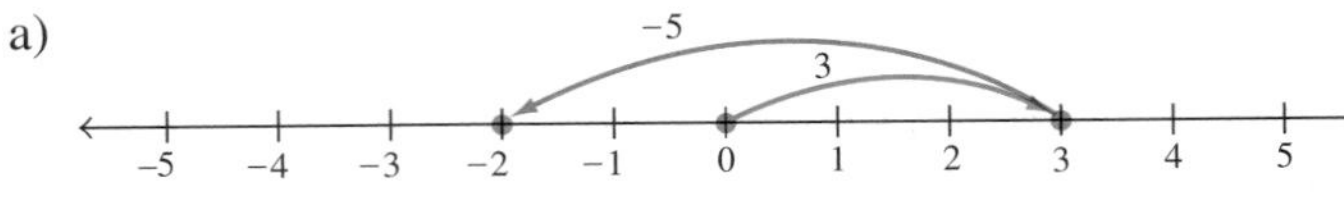

Thus, $3 + (-5) = -2$

b)

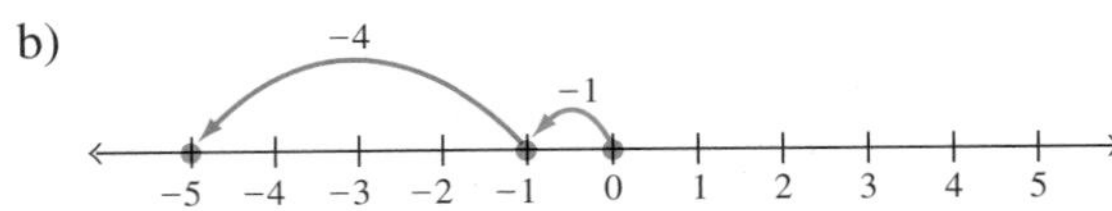

Thus, $-1 + (-4) = -5$

c)

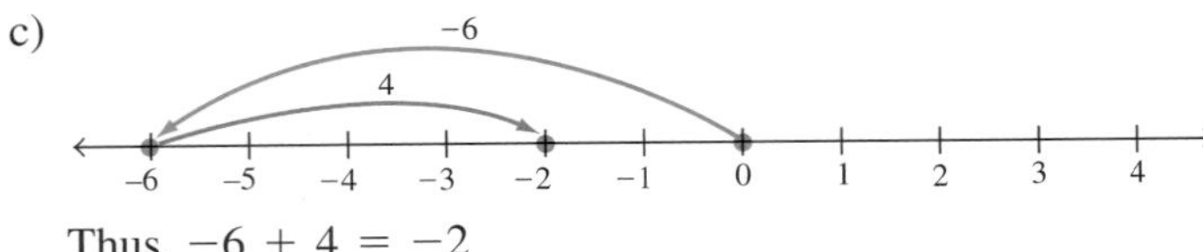

Thus, $-6 + 4 = -2$

d)

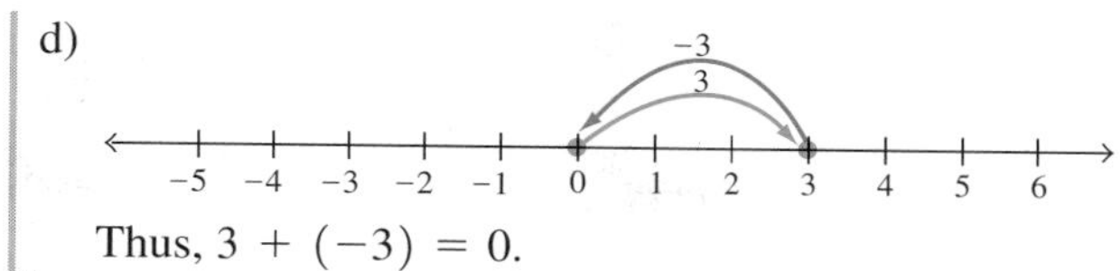

Thus, $3 + (-3) = 0$.

In Example 2(d), the number -3 is said to be the additive inverse of 3 and 3 is the additive inverse of -3, because their sum is 0. In general, the *additive inverse* of the number n is $-n$, since $n + (-n) = 0$. Inverses are discussed more formally in Chapter 10.

Subtraction of Integers

Any subtraction problem can be rewritten as an addition problem. To do so, we use the following definition of subtraction.

Subtraction

$$a - b = a + (-b)$$

The rule for subtraction indicates that to subtract b from a, *add* the additive inverse of b to a. For example,

$$3 - 5 = 3 + (-5)$$

↑ Subtraction ↑ Addition ↑ Additive inverse of 5

Now we can determine the value of $3 + (-5)$.

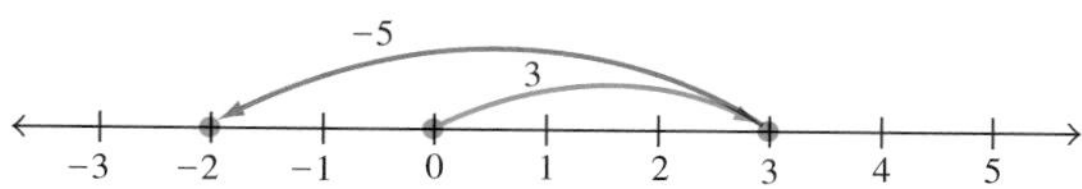

Thus, $3 - 5 = 3 + (-5) = -2$.

EXAMPLE 3 *Subtracting Integers*

Evaluate $-4 - (-3)$ using the number line.

SOLUTION: We are subtracting -3 from -4. The additive inverse of -3 is 3; therefore, we add 3 to -4. We now add $-4 + 3$ on the number line to obtain the answer -1.

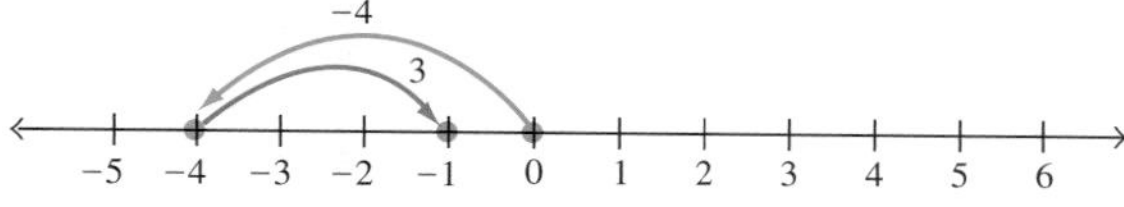

Thus, $-4 - (-3) = -4 + 3 = -1$.

DID YOU KNOW

Sparsely Populated

More people have made it into outer space than have stood on the top of Nepal's 29,035-foot-high Mount Everest. Fewer people still have visited the Pacific Ocean's 36,198-foot-deep Mariana Trench. In 1960, Jacques Piccard and Don Walsh were the first to make the journey in a vessel designed to withstand the immense pressure: 7 tons per square inch. The diving time was $8\frac{1}{2}$ hours, of which only 20 minutes were spent on the ocean floor.

In Example 3, we found that $-4 - (-3) = -4 + 3$. In general, $a - (-b) = a + b$. As you get more proficient in working with integers, you should be able to answer questions involving them without drawing a number line.

EXAMPLE 4 ***Subtracting: Adding the Inverse***

Evaluate

a) $-5 - 2$ b) $-5 - (-2)$ c) $5 - (-2)$ d) $5 - 2$

SOLUTION:

a) $-5 - 2 = -5 + (-2) = -7$ b) $-5 - (-2) = -5 + 2 = -3$

c) $5 - (-2) = 5 + 2 = 7$ d) $5 - 2 = 5 + (-2) = 3$

EXAMPLE 5 ***Elevation Difference***

The highest point on Earth is Mount Everest, in the Himalayas, at a height of 29,035 ft above sea level. The lowest point on Earth is the Mariana Trench, in the Pacific Ocean, at a depth of 36,198 ft below sea level ($-36{,}198$ ft). Find the vertical height difference between Mount Everest and the Mariana Trench.

SOLUTION: We obtain the vertical difference by subtracting the lower elevation from the higher elevation.

$$29{,}035 - (-36{,}198) = 29{,}035 + 36{,}198 = 65{,}233$$

The vertical difference is 65,233 ft.

Multiplication of Integers

The multiplication property of zero is important in our discussion of multiplication of integers. It indicates that the product of 0 and any number is 0.

Multiplication Property of Zero

$$a \cdot 0 = 0 \cdot a = 0$$

We will develop the rules for multiplication of integers using number patterns. The four possible cases are

1. positive integer × positive integer,
2. positive integer × negative integer,
3. negative integer × positive integer, and
4. negative integer × negative integer.

CASE 1: ***POSITIVE INTEGER × POSITIVE INTEGER*** The product of two positive integers can be defined as repeated addition of a positive integer. Thus, $3 \cdot 2$ means $2 + 2 + 2$. This sum will always be positive. Thus, a positive integer times a positive integer is a positive integer.

CASE 2: ***POSITIVE INTEGER × NEGATIVE INTEGER*** Consider the following patterns:

$$3(3) = 9$$
$$3(2) = 6$$
$$3(1) = 3$$

Note that each time the second factor is reduced by 1, the product is reduced by 3. Continuing the process gives

$$3(0) = 0$$

What comes next?

$$3(-1) = -3$$
$$3(-2) = -6$$

The pattern indicates that a positive integer times a negative integer is a negative integer.

We can confirm this result by using the number line. The expression $3(-2)$ means $(-2) + (-2) + (-2)$. Adding $(-2) + (-2) + (-2)$ on the number line, we obtain a sum of -6.

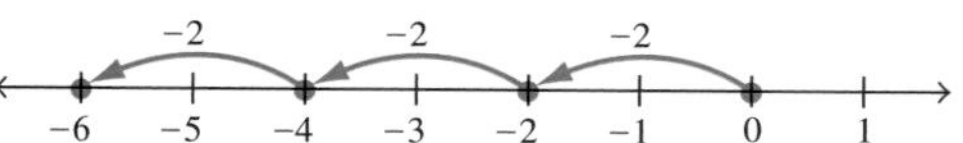

CASE 3: ***NEGATIVE INTEGER × POSITIVE INTEGER*** A procedure similar to that used in case 2 will indicate that a negative integer times a positive integer is a negative integer.

CASE 4: ***NEGATIVE INTEGER × NEGATIVE INTEGER*** We have illustrated that a positive integer times a negative integer is a negative integer. We make use of this fact in the following pattern:

$$4(-4) = -16$$
$$3(-4) = -12$$
$$2(-4) = -8$$
$$1(-4) = -4$$

In this pattern, each time the first term is decreased by 1, the product is increased by 4. Continuing this process gives

$$0(-4) = 0$$
$$(-1)(-4) = 4$$
$$(-2)(-4) = 8$$

This pattern illustrates that a negative integer times a negative integer is a positive integer.

The examples were restricted to integers. The rules for multiplication, however, can be used for any numbers. We summarize them as follows.

Rules for Multiplication

1. The product of two numbers with *like signs* (positive $\times$ positive or negative $\times$ negative) is a *positive number.*
2. The product of two numbers with *unlike signs* (positive $\times$ negative or negative $\times$ positive) is a *negative number.*

EXAMPLE 6 ***Multiplying Integers***

Evaluate

a) $5 \cdot 9$ b) $5 \cdot (-9)$ c) $(-5) \cdot 9$ d) $(-5)(-9)$

SOLUTION:

a) $5 \cdot 9 = 45$ b) $5 \cdot (-9) = -45$ c) $(-5) \cdot 9 = -45$

d) $(-5)(-9) = 45$ ▲

Division of Integers

You may already realize that a relationship exists between multiplication and division.

$$6 \div 2 = 3 \quad \text{means that} \quad 3 \cdot 2 = 6$$

$$\frac{20}{10} = 2 \quad \text{means that} \quad 2 \cdot 10 = 20$$

These examples demonstrate that division is the reverse process of multiplication.

Division

For any a, b, and c where $b \neq 0$, $\frac{a}{b} = c$ means that $c \cdot b = a$.

We discuss the four possible cases for division, which are similar to those for multiplication.

CASE 1: ***POSITIVE INTEGER ÷ POSITIVE INTEGER*** A positive integer divided by a positive integer is positive.

$$\frac{6}{2} = 3 \quad \text{since} \quad 3(2) = 6$$

CASE 2: ***POSITIVE INTEGER ÷ NEGATIVE INTEGER*** A positive integer divided by a negative integer is negative.

$$\frac{6}{-2} = -3 \quad \text{since} \quad (-3)(-2) = 6$$

CASE 3: ***NEGATIVE INTEGER ÷ POSITIVE INTEGER*** A negative integer divided by a positive integer is negative.

$$\frac{-6}{2} = -3 \quad \text{since} \quad (-3)(2) = -6$$

CASE 4: ***NEGATIVE INTEGER ÷ NEGATIVE INTEGER*** A negative integer divided by a negative integer is positive.

$$\frac{-6}{-2} = 3 \quad \text{since} \quad 3(-2) = -6$$

The examples were restricted to integers. The rules for division, however, can be used for any numbers. You should realize that division of integers does not always result in an integer. The rules for division are summarized as follows.

Rules for Division

1. The quotient of two numbers with *like signs* (positive ÷ positive or negative ÷ negative) is a *positive number.*
2. The quotient of two numbers with *unlike signs* (positive ÷ negative or negative ÷ positive) is a *negative number.*

EXAMPLE 7 *Dividing Integers*

Evaluate

a) $\frac{56}{8}$ b) $\frac{-56}{8}$ c) $\frac{56}{-8}$ d) $\frac{-56}{-8}$

SOLUTION:

a) $\frac{56}{8} = 7$ b) $\frac{-56}{8} = -7$ c) $\frac{56}{-8} = -7$ d) $\frac{-56}{-8} = 7$ ▲

In the definition of division, we stated that the denominator could not be 0. Why not? Suppose we are trying to find the quotient $\frac{5}{0}$. Let's say that this quotient is equal to some number x. Then we would have $\frac{5}{0} = x$. If true, this would mean that $5 = x \cdot 0$. The right side of the equation is $x \cdot 0$, which is equal to 0 for any real value of x. This leads us to conclude that $5 = 0$, which is false. Thus, there is no number that can replace x that makes the equation $\frac{5}{0} = x$ true. Therefore, in mathematics, division by 0 is not allowed and we say that a quotient of any number divided by zero is *undefined.*

SECTION 5.2 EXERCISES

Concept/Writing Exercises

1. Explain how to add numbers using a number line.
2. What is the additive inverse of a number n?
3. Explain how to rewrite a subtraction problem as an addition problem.
4. Explain the rule for multiplication of real numbers.
5. Explain the rule for division of real numbers.

6. Explain why the quotient of a number divided by 0 is undefined.

Practice the Skills

In Exercises 7–16, evaluate the expression.

7. $-6 + 9$
8. $4 + (-5)$
9. $(-7) + 9$
10. $(-3) + (-3)$
11. $[6 + (-11)] + 0$
12. $(2 + 5) + (-4)$
13. $[(-3) + (-4)] + 9$
14. $[8 + (-3)] + (-2)$
15. $[(-23) + (-9)] + 11$
16. $[5 + (-13)] + 18$

In Exercises 17–26, evaluate the expression.

17. $3 - 6$
18. $-3 - 7$
19. $-4 - 6$
20. $7 - (-1)$
21. $-5 - (-3)$
22. $-4 - 4$
23. $14 - 20$
24. $8 - (-3)$
25. $[5 + (-3)] - 4$
26. $6 - (8 + 6)$

In Exercises 27–36, evaluate the expression.

27. $-4 \cdot 5$
28. $4(-3)$
29. $(-12)(-12)$
30. $5(-5)$
31. $[(-8)(-2)] \cdot 6$
32. $4(-5)(-6)$
33. $(5 \cdot 6)(-2)$
34. $(-9)(-1)(-2)$
35. $[(-3)(-6)] \cdot [(-5)(8)]$
36. $[(-8 \cdot 4) \cdot 5](-2)$

In Exercises 37–46, evaluate the expression.

37. $-26 \div (-13)$
38. $-56 \div 8$
39. $23 \div (-23)$
40. $-64 \div 16$
41. $56/-8$
42. $-75/15$
43. $-210/14$
44. $186/-6$
45. $144 \div (-3)$
46. $(-900) \div (-4)$

In Exercises 47–56, determine whether the statement is true or false. Modify each false statement to make it a true statement.

47. Every whole number is an integer.

48. Every integer is a whole number.

49. The difference of any two negative integers is a negative integer.

50. The sum of any two negative integers is a negative integer.

51. The product of any two positive integers is a positive integer.

52. The difference of a positive integer and a negative integer is always a negative integer.

53. The quotient of a negative integer and a positive integer is always a negative number.

54. The quotient of any two negative integers is a negative number.

55. The sum of a positive integer and a negative integer is always a positive integer.

56. The product of a positive integer and a negative integer is always a positive integer.

In Exercises 57–66, evaluate the expression.

57. $(5 + 7) \div 2$
58. $(-4) \div [14 \div (-7)]$
59. $[6(-2)] - 5$
60. $[(-5)(-6)] - 3$
61. $(4 - 8)(3)$
62. $[18 \div (-2)](-3)$
63. $[2 + (-17)] \div 3$
64. $(5 - 9) \div (-4)$
65. $[(-22)(-3)] \div (2 - 13)$
66. $[15(-4)] \div (-6)$

In Exercises 67–70, write the numbers in increasing order from left to right.

67. 0, −5, −10, 10, 5, −15
68. −20, 30, −40, 10, 0, −10
69. −5, −2, −3, −1, −4, −6
70. 106, 33, −47, −108, 72, −76

Problem Solving

71. *Extreme Temperatures* The hottest temperature ever recorded in the United States was 134°F, which occurred at Greenland Ranch, California, in Death Valley on July 10, 1913. The coldest temperature ever recorded in the United States was −79.8°F, which occurred at Prospect Creek Camp, Alaska, in the Endicott Mountains on January 23, 1971. Determine the difference between these two temperatures.

72. *NASDAQ Average* On August 28, 2002, the NASDAQ composite average opened at 1347 points. During that day it lost 33 points. On August 29 it gained 22 points, and on August 30 it lost 21 points. What was the closing NASDAQ composite average on August 30, 2002?

73. *Pit Score* While playing the game of Pit, John Pearse began with a score of zero points. He then gained 100 points, lost 40 points, gained 90 points, lost 20 points, and gained 80 points on his next five rounds. What is John's score after five rounds?

74. *Elevation Difference* Mount Whitney, in the Sierra Nevada mountains of California, is the highest point in the contiguous United States. It is 14,495 ft above sea level. Death Valley, in California and Nevada, is the lowest point in the United States, 282 ft below sea level. Find the vertical height difference between Mount Whitney and Death Valley.

75. *Vertical Distance Traveled* A helicopter drops a package from a height of 842 ft above sea level. The package lands in the ocean and settles at a point 927 ft below sea level. What was the vertical distance the package traveled?

76. *Football Yardage* In the first four plays of the game, the Texans gained 8 yd, lost 5 yd, gained 3 yd, and gained 4 yd. What is the total number of yards gained in the first four plays? Did the Texans make a first down? (Ten yards are needed for a first down.)

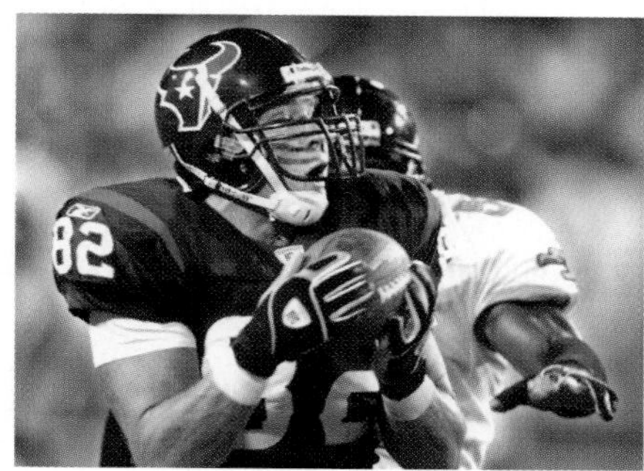

77. *Time Zone Calculations* Part of a World Standard Time Zones chart used by airlines and the United States Navy is shown. The scale along the bottom is just like a number line with the integers $-12, -11, \ldots, 11, 12$ on it.

a) Find the difference in time between Amsterdam (zone $+1$) and Los Angeles (zone -8).

b) Find the difference in time between Boston (zone -5) and Puerto Vallarta (zone -7).

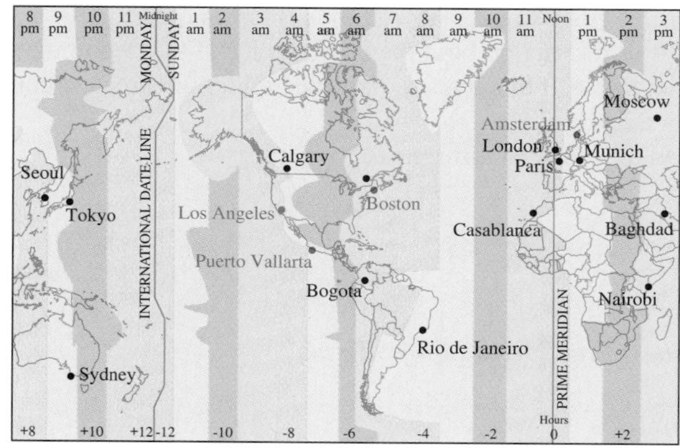

78. Explain why $\dfrac{a}{b} = \dfrac{-a}{-b}$.

Challenge Problems/Group Activities

79. Find the quotient:

$$\frac{-1 + 2 - 3 + 4 - 5 + \cdots - 99 + 100}{1 - 2 + 3 - 4 + 5 - \cdots + 99 - 100}$$

80. *Pentagonal Numbers* Triangular numbers and square numbers were introduced in the Section 1.1 Exercises. There are also **pentagonal numbers**, which were also studied by the Greeks. Four pentagonal numbers are 1, 5, 12, and 22.

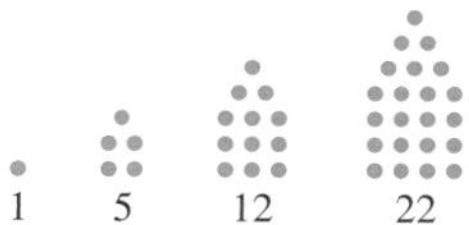

a) Determine the next three pentagonal numbers.

b) Describe a procedure to determine the next five pentagonal numbers without drawing the figures.

c) Is 72 a pentagonal number? Explain how you determined your answer.

81. Place the appropriate plus or minus signs between each digit so that the total will equal 1.

$$0 \quad 1 \quad 2 \quad 3 \quad 4 \quad 5 \quad 6 \quad 7 \quad 8 \quad 9 = 1$$

Recreational Mathematics

82. *Four 4's* The game of *Four 4's* is a challenging way to learn about some of the operations on integers. In this game you must use exactly four 4's along with one or more of the operations of addition, subtraction, multiplication, and division* to write and evaluate expressions. The object of the game is to write expressions that when evaluated will give each whole number from 0 through 9. You may use as many grouping symbols as you wish, but you must use exactly four 4's. For example, one way to obtain 1 is as follows: $(4 + 4) \div (4 + 4) = 8 \div 8 = 1$. One way to obtain 2 is as follows:
$(4 \div 4) + (4 \div 4) = 1 + 1 = 2$.

a) Use the rules as defined above to obtain each whole number 0 through 9.

b) Use the rules as defined above to obtain the following whole numbers: 12, 15, 16, 17, 20.

c) We will now change our rules to allow the number 44 to count as two of the four 4's. Use the number 44 and two other fours to obtain the whole number 10.

Research Activity

83. Do research and write a report on the history of the number 0 in the Hindu–Arabic numeration system.

*This game will be expanded in future exercise sets to include other operations such as exponents and square roots.

5.3 THE RATIONAL NUMBERS

"When you can measure what you are talking about and express it in numbers, you know something about it."
Lord Kelvin

We introduced the number line in Section 5.1 and discussed the integers in Section 5.2. The numbers that fall between the integers on the number line are either rational or irrational numbers. In this section, we discuss the rational numbers, and in Section 5.4, we discuss the irrational numbers.

Any number that can be expressed as a quotient of two integers (denominator not 0) is a rational number.

> The set of **rational numbers**, denoted by Q, is the set of all numbers of the form p/q, where p and q are integers and $q \neq 0$.

The following numbers are examples of rational numbers:

$$\frac{1}{3}, \quad \frac{3}{4}, \quad -\frac{7}{8}, \quad 1\frac{2}{3}, \quad 2, \quad 0, \quad \frac{15}{7}$$

The integers 2 and 0 are rational numbers because each can be expressed as the quotient of two integers: $2 = \frac{2}{1}$ and $0 = \frac{0}{1}$. In fact, every integer n is a rational number, since it can be written in the form of $\frac{n}{1}$.

Numbers such as $\frac{1}{3}$ and $-\frac{7}{8}$ are also called *fractions*. The number above the fraction line is called the *numerator*, and the number below the fraction line is called the *denominator*.

DID YOU KNOW

Triangular Numbers

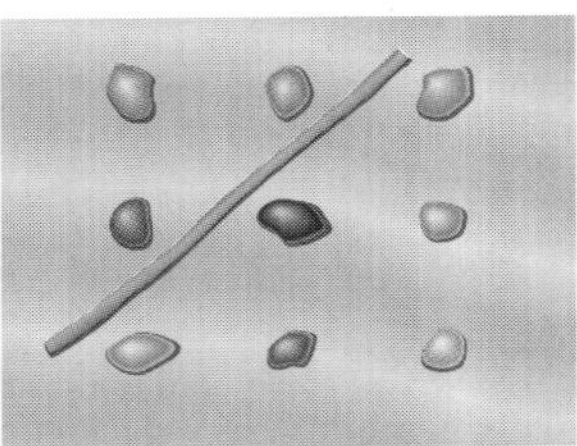

One very simple reason that the ancient Greek mathematicians thought of mathematics in terms of whole numbers and their ratios (the rational numbers) was that they were still working with numbers that were represented by objects, such as squares and triangles, not number symbols. Still, that did not prevent them from drawing some conclusions about number theory. Consider two consecutive triangular numbers, for instance 3 and 6. You can see from the diagram that the sum of the two triangular numbers is equal to the square number 9, the square of a side of the larger triangle.

Reducing Fractions

Sometimes the numerator and denominator in a fraction have a common divisor (or common factor). For example, both the numerator and denominator of the fraction $\frac{6}{10}$ have the common divisor 2. When a numerator and denominator have a common divisor, we can *reduce the fraction to its lowest terms.*

A fraction is said to be in its lowest terms (or reduced) when the numerator and denominator are relatively prime (that is, have no common divisors other than 1). To reduce a fraction to its lowest terms, divide both the numerator and the denominator by the greatest common divisor. Recall that a procedure for finding the greatest common divisor was discussed in Section 5.1.

The fraction $\frac{6}{10}$ is reduced to its lowest terms as follows.

$$\frac{6}{10} = \frac{6 \div 2}{10 \div 2} = \frac{3}{5}$$

EXAMPLE 1 *Reducing a Fraction to Lowest Terms*

Reduce $\frac{54}{90}$ to its lowest terms.

SOLUTION: On page 213 in Example 4 of Section 5.1, we determined that the GCD of 54 and 90 is 18. Divide the numerator and the denominator by GCD, 18.

$$\frac{54}{90} = \frac{54 \div 18}{90 \div 18} = \frac{3}{5}$$

Since there are no common divisors of 3 and 5 other than 1, this fraction is in its lowest terms. ▲

Mixed Numbers and Improper Fractions

Consider the number $2\frac{3}{4}$. It is an example of a *mixed number*. It is called a mixed number because it consists of an integer, 2, and a fraction, $\frac{3}{4}$. The mixed number $2\frac{3}{4}$ means $2 + \frac{3}{4}$. The mixed number $-4\frac{1}{4}$ means $-(4 + \frac{1}{4})$. Rational numbers greater than 1 or less than -1 that are not integers may be represented as mixed numbers, or as *improper fractions*. An improper fraction is a fraction whose numerator is greater than its denominator. An example of an improper fraction is $\frac{8}{5}$. Figure 5.6 shows both mixed numbers and improper fractions indicated on a number line. In this section, we show how to convert mixed numbers to improper fractions and vice versa.

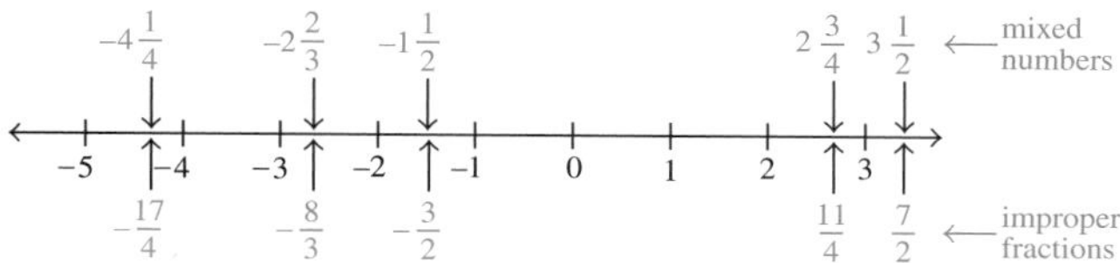

Figure 5.6

We begin by limiting our discussion to positive mixed numbers and positive improper fractions.

Converting a Positive Mixed Number to an Improper Fraction

1. Multiply the denominator of the fraction in the mixed number by the integer preceding it.
2. Add the product obtained in step 1 to the numerator of the fraction in the mixed number. This sum is the numerator of the improper fraction we are seeking. The denominator of the improper fraction we are seeking is the same as the denominator of the fraction in the mixed number.

EXAMPLE 2 *From Mixed Number to Improper Fraction*

Convert the following mixed numbers to improper fractions.

a) $1\frac{7}{8}$ b) $3\frac{5}{6}$

SOLUTION:

a) $1\frac{7}{8} = \frac{8 \cdot 1 + 7}{8} = \frac{8 + 7}{8} = \frac{15}{8}$

b) $3\frac{5}{6} = \frac{6 \cdot 3 + 5}{6} = \frac{18 + 5}{6} = \frac{23}{6}$

Notice that both $\frac{15}{8}$ and $\frac{23}{6}$ have numerators larger than their denominators and that both are improper fractions. ▲

Converting a Positive Improper Fraction to a Mixed Number

1. Divide the numerator by the denominator. Identify the quotient and the remainder.
2. The quotient obtained in step 1 is the integer part of the mixed number. The remainder is the numerator of the fraction in the mixed number. The denominator in the fraction of the mixed number will be the same as the denominator in the original fraction.

EXAMPLE 3 *From Improper Fraction to Mixed Number*

Convert the following improper fractions to mixed numbers.

a) $\frac{8}{5}$ b) $\frac{225}{8}$

SOLUTION:

a) Divide the numerator, 8, by the denominator, 5.

$$\begin{array}{rl} & 1 \quad \leftarrow \text{Quotient} \\ \text{Divisor} \rightarrow & 5\overline{)8} \quad \leftarrow \text{Dividend} \\ & \underline{5} \\ & 3 \quad \leftarrow \text{Remainder} \end{array}$$

Therefore,

$$\frac{8}{5} = \underset{\text{Quotient}}{1}\frac{3}{5} \quad \begin{array}{l}\leftarrow \text{Remainder} \\ \leftarrow \text{Divisor}\end{array}$$

The mixed number is $1\frac{3}{5}$.

b) Divide the numerator, 225, by the denominator, 8.

$$\begin{array}{rl} & 28 \quad \leftarrow \text{Quotient} \\ \text{Divisor} \rightarrow & 8\overline{)225} \quad \leftarrow \text{Dividend} \\ & \underline{16} \\ & 65 \\ & \underline{64} \\ & 1 \quad \leftarrow \text{Remainder} \end{array}$$

Therefore,

$$\frac{225}{8} = \underset{\text{Quotient}}{28}\frac{1}{8} \quad \begin{array}{l}\leftarrow \text{Remainder} \\ \leftarrow \text{Divisor}\end{array}$$

The mixed number is $28\frac{1}{8}$.

Up to this point, we have only worked with positive mixed numbers and positive improper fractions. When converting a negative mixed number to an improper fraction, or a negative improper fraction to a mixed number, it is best to ignore the negative sign temporarily. Perform the calculation as described earlier and then reattach the negative sign.

EXAMPLE 4 ***Negative Mixed Numbers and Improper Fractions***

a) Convert $-1\frac{7}{8}$ to an improper fraction.
b) Convert $-\frac{8}{5}$ to a mixed number.

SOLUTION:

a) First, ignore the negative sign and examine $1\frac{7}{8}$. We learned in Example 2(a) that $1\frac{7}{8} = \frac{15}{8}$. Now to convert $-1\frac{7}{8}$ to an improper fraction, we reattach the negative sign. Thus, $-1\frac{7}{8} = -\frac{15}{8}$.
b) We learned in Example 3(a) that $\frac{8}{5} = 1\frac{3}{5}$. Therefore, $-\frac{8}{5} = -1\frac{3}{5}$. ▲

Terminating or Repeating Decimal Numbers

Note the following important property of the rational numbers.

Every *rational number* when expressed as a decimal number will be either a terminating or a repeating decimal number.

Examples of terminating decimal numbers are 0.5, 0.75, and 4.65. Examples of repeating decimal numbers are 0.333..., 0.2323..., and 8.13456456.... One way to indicate that a number or group of numbers repeat is to place a bar above the number or group of numbers that repeat. Thus, 0.333... may be written $0.\overline{3}$, 0.2323... may be written $0.\overline{23}$, and 8.13456456... may be written $8.13\overline{456}$.

EXAMPLE 5 ***Terminating Decimal Numbers***

Show that the following rational numbers are terminating decimal numbers.

a) $\frac{2}{5}$ b) $-\frac{7}{8}$ c) $\frac{17}{16}$

SOLUTION: To express the rational number in decimal form, divide the numerator by the denominator. If you use a calculator, or use long division, you will obtain the following results.

a) $\frac{2}{5} = 0.4$ b) $-\frac{7}{8} = -0.875$ c) $\frac{17}{16} = 1.0625$ ▲

EXAMPLE 6 ***Repeating Decimal Numbers***

Show that the following rational numbers are repeating decimal numbers.

a) $\frac{2}{3}$ b) $\frac{14}{99}$ c) $1\frac{4}{33}$

SOLUTION: If you use a calculator, or use long division, you will see that each fraction results in a repeating decimal number.

a) $2 \div 3 = 0.6666\ldots$ or $0.\overline{6}$

b) $14 \div 99 = 0.141414\ldots$ or $0.\overline{14}$

c) $1\frac{4}{33} = \frac{37}{33} = 1.121212\ldots$ or $1.\overline{12}$ ▲

Note that in each part of Example 6, the quotient has no final digit and continues indefinitely. Each number is a repeating decimal number.

When a fraction is converted to a decimal number, the maximum number of digits that can repeat is $n - 1$, where n is the denominator of the fraction. For example, when $\frac{2}{7}$ is converted to a decimal number, the maximum number of digits that can repeat is $7 - 1$, or 6.

Converting Decimal Numbers to Fractions

We can convert a terminating or repeating decimal number into a quotient of integers. The explanation of the procedure will refer to the positional values to the right of the decimal point, as illustrated here:

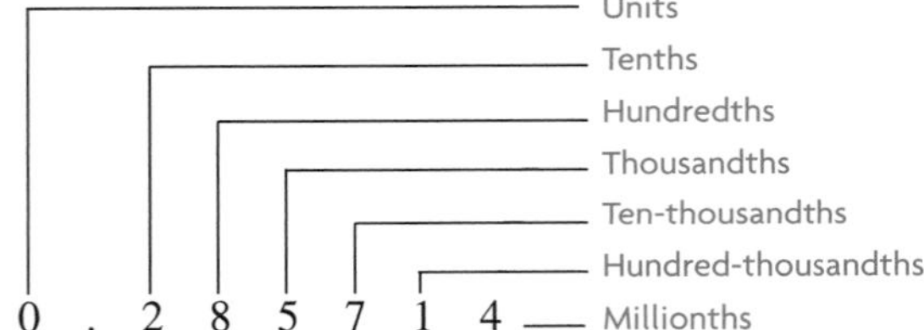

Example 7 demonstrates how to convert from a decimal number to a fraction.

EXAMPLE 7 *Converting a Decimal Number into a Fraction*

Convert the following terminating decimal numbers to a quotient of integers.

a) 0.4 b) 0.62 c) 0.062 d) 1.37

SOLUTION: When converting a terminating decimal number to a quotient of integers, we observe the last digit to the right of the decimal point. The position of this digit will indicate the denominator of the quotient of integers.

a) $0.4 = \frac{4}{10}$ because the 4 is in the tenths position.

b) $0.62 = \frac{62}{100}$ because the last digit on the right, 2, is in the hundredths position.

c) $0.062 = \frac{62}{1000}$ because the last digit on the right, 2, is in the thousandths position.

d) $1.37 = \frac{137}{100}$ because the last digit on the right, 7, is in the ten-thousandths position. Notice the numerator is the decimal number with the decimal point removed. When the decimal number is larger than 1, the numerator of the resulting fraction will be greater than the denominator or an improper fraction. ▲

Converting a repeating decimal number to a quotient of integers is more difficult. To do so, we must "create" another repeating decimal number with the same repeating digits so that when one repeating decimal number is subtracted from the other repeating decimal number, the difference will be a whole number. To create a number with the same repeating digits, multiply the original repeating decimal number by 10 if one digit repeats, by 100 if two digits repeat, by 1000 if three digits repeat, and so on. Examples 8 through 10 demonstrate this procedure.

DID YOU KNOW

Where Do These Fractions Come From?

Until recently, new investors in U.S. stock markets were often puzzled by the practice of quoting share prices in fractions of a dollar instead of in dollars and cents. For example, a stock quote of $20\frac{1}{8}$ meant \$20.125.

In the late 1700s, most stock traders were merchants who were involved in foreign trade and currency exchange. The Spanish dollar coin, or piece of eight, was a widely held and stable currency used for trading both stocks and goods. With a hammer and chisel, these dollars could be divided into halves, quarters, and eighths. Eighths were also known as bits. It is from this practice that the synonym of "two bits" for 25 cents comes.

Up until recently, trading in eighths remained part of the securities industry culture. By 2002, all the stock exchanges in the United States had converted from fractional to decimal representations for share prices.

EXAMPLE 8 *Converting a Repeating Decimal Number into a Fraction*

Convert $0.\overline{3}$ to a quotient of integers.

SOLUTION: $0.\overline{3} = 0.3\overline{3} = 0.33\overline{3}$, and so on.

Let the original repeating decimal number be n; thus, $n = 0.\overline{3}$. Because one digit repeats, we multiply both sides of the equation by 10, which gives $10n = 3.\overline{3}$. Then we subtract.

$$\begin{aligned} 10n &= 3.\overline{3} \\ -\quad n &= 0.\overline{3} \\ \hline 9n &= 3.0 \end{aligned}$$

Note that $10n - n = 9n$ and $3.\overline{3} - 0.\overline{3} = 3.0$.

Next, we solve for n by dividing both sides of the equation by 9.

$$\frac{9n}{9} = \frac{3.0}{9}$$

$$n = \frac{3}{9} = \frac{1}{3}$$

Therefore, $0.\overline{3} = \frac{1}{3}$. Evaluate $1 \div 3$ on a calculator now and see what value you get. ▲

EXAMPLE 9 *Converting a Repeating Decimal Number into a Fraction*

Convert $0.\overline{35}$ to a quotient of integers.

SOLUTION: Let $n = 0.\overline{35}$. Since two digits repeat, multiply both sides of the equation by 100. Thus, $100n = 35.\overline{35}$. Now we subtract n from $100n$.

$$\begin{aligned} 100n &= 35.\overline{35} \\ -\quad n &= 0.\overline{35} \\ \hline 99n &= 35 \end{aligned}$$

Finally, we divide both sides of the equation by 99.

$$\frac{99n}{99} = \frac{35}{99}$$

$$n = \frac{35}{99}$$

Therefore, $0.\overline{35} = \frac{35}{99}$. Evaluate $35 \div 99$ on a calculator now and see what value you get. ▲

EXAMPLE 10 *Converting a Repeating Decimal Number into a Fraction*

Convert $12.14\overline{2}$ to a quotient of integers.

SOLUTION: This problem is different from the two preceding examples in that the repeating digit, 2, is not directly to the right of the decimal point. When this situation arises, move the decimal point to the right until the repeating terms are directly to its right. For each place the decimal point is moved, the number is multiplied by 10. In this example, the decimal point must be moved two places to the right. Thus, the number must be multiplied by 100.

$$\begin{aligned} n &= 12.14\overline{2} \\ 100n &= 100 \times 12.14\overline{2} = 1214.\overline{2} \end{aligned}$$

Now proceed as in the previous two examples. Since one digit repeats, multiply both sides by 10.

$$\begin{aligned} 100n &= 1214.\overline{2} \\ 10 \times 100n &= 10 \times 1214.\overline{2} \\ 1000n &= 12142.\overline{2} \end{aligned}$$

Now subtract $100n$ from $1000n$ so that the repeating part will drop out.

$$\begin{aligned} 1000n &= 12142.\overline{2} \\ -\ 100n &= \ \ 1214.\overline{2} \\ \hline 900n &= 10928 \\ n &= \frac{10{,}928}{900} = \frac{2732}{225} \end{aligned}$$

Therefore, $12.14\overline{2} = \frac{2732}{225}$. Evaluate $2732 \div 225$ on a calculator now and see what value you get. ▲

Multiplication and Division of Fractions

The product of two fractions is found by multiplying the numerators together and multiplying the denominators together.

Multiplication of Fractions

$$\frac{a}{b} \cdot \frac{c}{d} = \frac{a \cdot c}{b \cdot d} = \frac{ac}{bd}, \quad b \neq 0, \quad d \neq 0$$

EXAMPLE 11 *Multiplying Fractions*

Evaluate the following.

a) $\frac{3}{5} \cdot \frac{7}{8}$ b) $\left(\frac{-2}{3}\right)\left(\frac{-4}{9}\right)$ c) $\left(1\frac{7}{8}\right)\left(2\frac{1}{4}\right)$

SOLUTION:

a) $\frac{3}{5} \cdot \frac{7}{8} = \frac{3 \cdot 7}{5 \cdot 8} = \frac{21}{40}$

b) $\left(\frac{-2}{3}\right)\left(\frac{-4}{9}\right) = \frac{(-2)(-4)}{(3)(9)} = \frac{8}{27}$

c) $\left(1\frac{7}{8}\right)\left(2\frac{1}{4}\right) = \frac{15}{8} \cdot \frac{9}{4} = \frac{135}{32} = 4\frac{7}{32}$ ▲

The *reciprocal* of any number is 1 divided by that number. The product of a number and its reciprocal must equal 1. Examples of some numbers and their reciprocals follow.

Number		Reciprocal		Product
3	$\cdot$	$\frac{1}{3}$	$=$	1
$\frac{3}{5}$	$\cdot$	$\frac{5}{3}$	$=$	1
-6	$\cdot$	$-\frac{1}{6}$	$=$	1

To find the quotient of two fractions, multiply the first fraction by the reciprocal of the second fraction.

Division of Fractions

$$\frac{a}{b} \div \frac{c}{d} = \frac{a}{b} \cdot \frac{d}{c} = \frac{ad}{bc}, \quad b \neq 0, \quad d \neq 0, \quad c \neq 0$$

EXAMPLE 12 ***Dividing Fractions***

Evaluate the following.

a) $\frac{2}{3} \div \frac{5}{7}$ b) $\frac{-3}{5} \div \frac{5}{7}$

SOLUTION:

a) $\frac{2}{3} \div \frac{5}{7} = \frac{2}{3} \cdot \frac{7}{5} = \frac{2 \cdot 7}{3 \cdot 5} = \frac{14}{15}$

b) $\frac{-3}{5} \div \frac{5}{7} = \frac{-3}{5} \cdot \frac{7}{5} = \frac{-3 \cdot 7}{5 \cdot 5} = \frac{-21}{25}$ or $-\frac{21}{25}$

Addition and Subtraction of Fractions

Before we can add or subtract fractions, the fractions must have a common denominator. A common denominator is another name for a common multiple of the denominators. The *lowest common denominator (LCD)* is the least common multiple of the denominators.

To add or subtract two fractions with a common denominator, we add or subtract their numerators and retain the common denominator.

Addition and Subtraction of Fractions

$$\frac{a}{c} + \frac{b}{c} = \frac{a+b}{c}, \quad c \neq 0; \qquad \frac{a}{c} - \frac{b}{c} = \frac{a-b}{c}, \quad c \neq 0$$

DID YOU KNOW

Mathematical Music

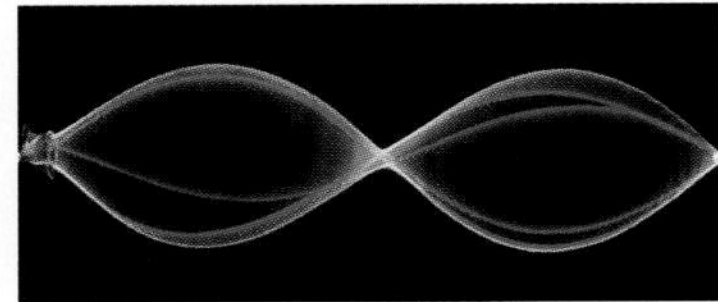

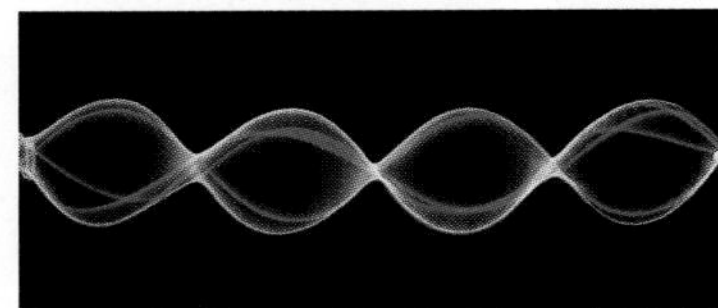

When the string of a musical instrument is plucked or bowed, it moves in a wave-like pattern like the strings shown here. The vibration this creates in the surrounding air is what your eardrums detect as sound.

The ancient Greeks believed that in nature, all harmony and everything of beauty could be explained with rational numbers. This belief was reinforced by the discovery that the sound of plucked strings could be quite pleasing if the strings plucked were in the ratio of 1 to 2 (an octave), 2 to 3 (a fifth), 3 to 4 (a fourth), and so on. Thus, the secret of harmony lies in the rational numbers such as $\frac{1}{2}$, $\frac{2}{3}$, and $\frac{3}{4}$.

The theory of vibrating strings has applications today that go well beyond music. How materials vibrate, and hence the stress they can absorb, is a vital matter in the construction of rockets, buildings, and bridges.

EXAMPLE 13 *Adding and Subtracting Fractions with a Common Denominator*

Evaluate the following.

a) $\frac{3}{8} + \frac{2}{8}$

b) $\frac{15}{16} - \frac{7}{16}$

SOLUTION:

a) $\frac{3}{8} + \frac{2}{8} = \frac{3 + 2}{8} = \frac{5}{8}$

b) $\frac{15}{16} - \frac{7}{16} = \frac{15 - 7}{16} = \frac{8}{16} = \frac{1}{2}$ ▲

Note that in Example 13, the denominators of the fractions being added or subtracted were the same; that is, they have a common denominator. *When adding or subtracting two fractions with unlike denominators, first rewrite each fraction with a common denominator. Then add or subtract the fractions.*

Writing fractions with a common denominator is accomplished with the *fundamental law of rational numbers.*

Fundamental Law of Rational Numbers
If a, b, and c are integers, with $b \neq 0$ and $c \neq 0$, then

$$\frac{a}{b} = \frac{a}{b} \cdot \frac{c}{c} = \frac{a \cdot c}{b \cdot c}$$

The terms $\frac{a}{b}$ and $\frac{a \cdot c}{b \cdot c}$ are called *equivalent fractions*. For example, since $\frac{5}{12} = \frac{5 \cdot 5}{12 \cdot 5} = \frac{25}{60}$, the fractions $\frac{5}{12}$ and $\frac{25}{60}$ are equivalent fractions. We will see the importance of equivalent fractions in the next two examples.

EXAMPLE 14 *Subtracting Fractions with Unlike Denominators*

Evaluate $\frac{5}{12} - \frac{3}{10}$.

SOLUTION: Using prime factorization (Section 5.1), we find that the LCM of 12 and 10 is 60. We will therefore express each fraction as an equivalent fraction with a denominator of 60. Sixty divided by 12 is 5. Therefore, the denominator, 12, must be multiplied by 5 to get 60. If the denominator is multiplied by 5, the numerator must also be multiplied by 5 so that the value of the fraction remains unchanged. Multiplying both numerator and denominator by 5 is the same as multiplying by 1.

We follow the same procedure for the other fraction, $\frac{3}{10}$. Sixty divided by 10 is 6. Therefore, we multiply both the denominator, 10, and the numerator, 3, by 6 to obtain an equivalent fraction with a denominator of 60.

$$\begin{aligned}\frac{5}{12} - \frac{3}{10} &= \left(\frac{5}{12}\cdot\frac{5}{5}\right) - \left(\frac{3}{10}\cdot\frac{6}{6}\right)\\ &= \frac{25}{60} - \frac{18}{60}\\ &= \frac{7}{60}\end{aligned}$$

▲

EXAMPLE 15 *Adding Fractions with Unlike Denominators*

Evaluate $\dfrac{1}{54} + \dfrac{1}{90}$.

SOLUTION: On page 214, in Example 6 of Section 5.1, we determined that the LCM of 54 and 90 is 270. Rewrite each fraction as an equivalent fraction using the LCM as the common denominator.

$$\begin{aligned}\frac{1}{54} + \frac{1}{90} &= \left(\frac{1}{54}\cdot\frac{5}{5}\right) + \left(\frac{1}{90}\cdot\frac{3}{3}\right)\\ &= \frac{5}{270} + \frac{3}{270}\\ &= \frac{8}{270}\end{aligned}$$

Now we reduce $\frac{8}{270}$ by dividing both 8 and 270 by 2, their greatest common factor.

$$\frac{8}{270} = \frac{8 \div 2}{270 \div 2} = \frac{4}{135}$$

▲

EXAMPLE 16 *Rice Preparation*

Following are the instructions given on a box of Minute Rice. Determine the amount of (a) rice and water, (b) salt, and (c) butter or margarine needed to make 3 servings of rice.

Directions

1. Bring water, salt, and butter (or margarine) to a boil.
2. Stir in rice. Cover; remove from heat. Let stand 5 minutes. Fluff with fork.

To Make	Rice & Water (Equal Measures)	Salt	Butter or Margarine (If Desired)
2 servings	$\frac{2}{3}$ cup	$\frac{1}{4}$ tsp	1 tsp
4 servings	$1\frac{1}{3}$ cups	$\frac{1}{2}$ tsp	2 tsp

SOLUTION: Since 3 is halfway between 2 and 4, we can find the amount of each ingredient by finding the average of the amount for 2 and 4 servings. To do so, we add the amounts for 2 servings and 4 servings and divide the sum by 2.

a) Rice and water: $\dfrac{\frac{2}{3} + 1\frac{1}{3}}{2} = \dfrac{\frac{2}{3} + \frac{4}{3}}{2} = \dfrac{\frac{6}{3}}{2} = \dfrac{2}{2} = 1$ cup

b) Salt: $\dfrac{\frac{1}{4} + \frac{1}{2}}{2} = \dfrac{\frac{1}{4} + \frac{2}{4}}{2} = \dfrac{\frac{3}{4}}{2} = \dfrac{3}{4} \cdot \dfrac{1}{2} = \dfrac{3}{8}$ tsp

c) Butter or margarine: $\dfrac{1 + 2}{2} = \dfrac{3}{2}$, or $1\frac{1}{2}$ tsp ▲

The solution to Example 16 can be found in other ways. Suggest two other procedures for solving the same problem.

SECTION 5.3 EXERCISES

Concept/Writing Exercises

1. Describe the set of rational numbers.
2. a) Explain how to write a terminating decimal number as a fraction.
   b) Write 0.397 as a fraction.
3. a) Explain how to reduce a fraction to lowest terms.
   b) Reduce $\frac{15}{27}$ to lowest terms by using the procedure in part (a).
4. Explain how to convert an improper fraction into a mixed number.
5. Explain how to convert a mixed number into an improper fraction.
6. a) Explain how to multiply two fractions.
   b) Multiply $\frac{15}{16} \cdot \frac{24}{25}$ by using the procedure in part (a).
7. a) Explain how to determine the reciprocal of a number.
   b) Using the procedure in part (a), determine the reciprocal of -2.
8. a) Explain how to divide two fractions.
   b) Divide $\frac{4}{15} \div \frac{16}{55}$ by using the procedure in part (a).
9. a) Explain how to add or subtract two fractions having a common denominator.
   b) Add $\frac{11}{36} + \frac{13}{36}$ by using the procedure in part (a).
   c) Subtract $\frac{37}{48} - \frac{13}{48}$ using the procedure in part (a).
10. a) Explain how to add or subtract two fractions having unlike denominators.
    b) Using the procedure in part (a), add $\frac{5}{12} + \frac{4}{9}$.
    c) Subtract $\frac{5}{6} - \frac{2}{15}$ using the procedure in part (a).
11. In your own words, state the fundamental law of rational numbers.
12. Are $\frac{4}{7}$ and $\frac{20}{35}$ equivalent fractions? Explain your answer.

Practice the Skills

In Exercises 13–22, reduce each fraction to lowest terms.

13. $\dfrac{4}{6}$ 14. $\dfrac{21}{35}$ 15. $\dfrac{26}{91}$

16. $\dfrac{36}{56}$ 17. $\dfrac{525}{800}$ 18. $\dfrac{13}{221}$

19. $\dfrac{112}{176}$ 20. $\dfrac{120}{135}$ 21. $\dfrac{45}{495}$

22. $\dfrac{124}{148}$

In Exercises 23–28, convert each mixed number into an improper fraction.

23. $3\dfrac{4}{7}$ 24. $4\dfrac{5}{6}$ 25. $-1\dfrac{15}{16}$

26. $-7\dfrac{1}{5}$ 27. $-4\dfrac{15}{16}$ 28. $11\dfrac{9}{16}$

In Exercises 29–32, write the number of inches indicated by the arrows as an improper fraction.

29. 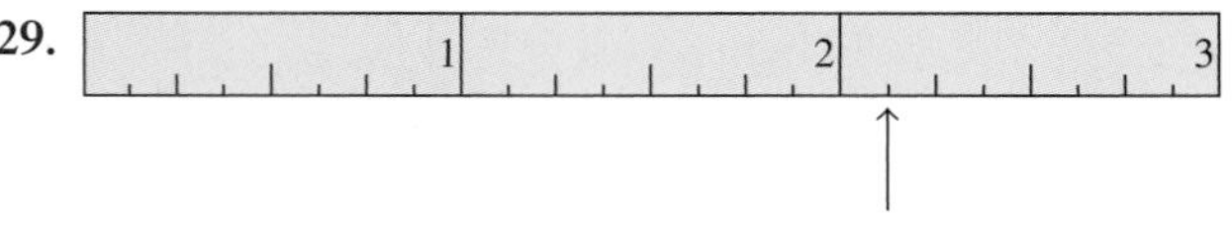

30. 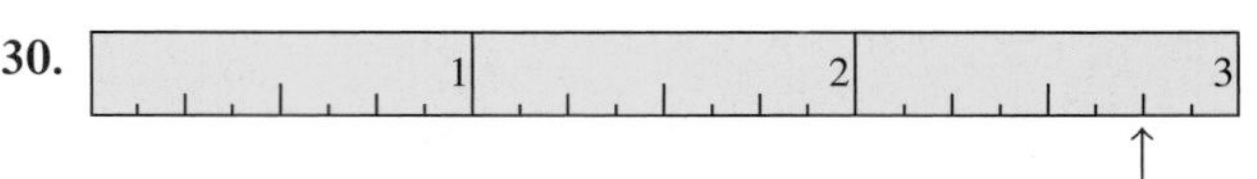

31.

32.

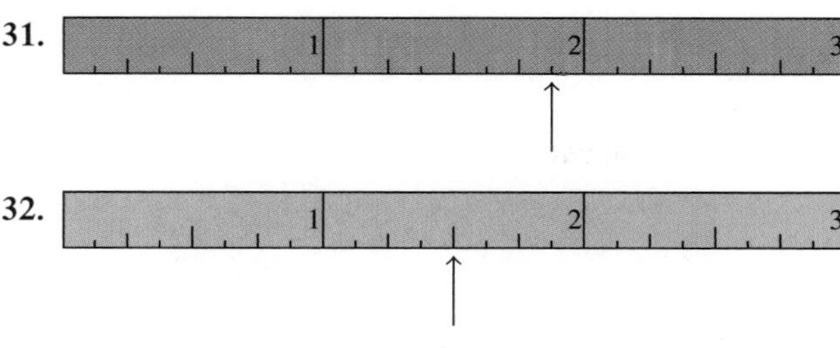

In Exercises 33–38, convert each improper fraction into a mixed number.

33. $\frac{11}{8}$ **34.** $\frac{23}{4}$ **35.** $-\frac{73}{6}$

36. $-\frac{457}{11}$ **37.** $-\frac{878}{15}$ **38.** $\frac{1028}{21}$

In Exercises 39–48, express each rational number as terminating or repeating decimal number.

39. $\frac{3}{5}$ **40.** $\frac{15}{16}$ **41.** $\frac{2}{9}$

42. $\frac{5}{6}$ **43.** $\frac{3}{8}$ **44.** $\frac{23}{7}$

45. $\frac{13}{3}$ **46.** $\frac{115}{15}$ **47.** $\frac{85}{15}$

48. $\frac{1002}{11}$

In Exercises 49–58, express each terminating decimal number as a quotient of two integers.

49. 0.25 **50.** 0.29 **51.** 0.045

52. 0.0125 **53.** 0.2 **54.** 0.251

55. 0.452 **56.** 0.2345 **57.** 0.0001

58. 0.2535

In Exercises 59–68, express each repeating decimal number as a quotient of two integers.

59. $0.\overline{6}$ **60.** $0.\overline{5}$ **61.** $1.\overline{9}$

62. $0.\overline{51}$ **63.** $1.\overline{36}$ **64.** $0.\overline{135}$

65. $1.0\overline{2}$ **66.** $2.4\overline{9}$ **67.** $3.4\overline{78}$

68. $5.2\overline{39}$

In Exercises 69–78, perform the indicated operation and reduce your answer to lowest terms.

69. $\frac{4}{11}\cdot\frac{3}{8}$ **70.** $\frac{3}{5}\div\frac{6}{7}$

71. $\left(\frac{-3}{8}\right)\left(\frac{-16}{15}\right)$ **72.** $\left(-\frac{3}{5}\right)\div\frac{10}{21}$

73. $\frac{7}{8}\div\frac{8}{7}$ **74.** $\frac{3}{7}\div\frac{3}{7}$

75. $\left(\frac{3}{5}\cdot\frac{4}{7}\right)\div\frac{1}{3}$ **76.** $\left(\frac{4}{7}\div\frac{4}{5}\right)\cdot\frac{1}{7}$

77. $\left[\left(\frac{-3}{4}\right)\left(\frac{-2}{7}\right)\right]\div\frac{3}{5}$ **78.** $\left(\frac{3}{8}\cdot\frac{5}{9}\right)\cdot\left(\frac{4}{7}\div\frac{5}{8}\right)$

In Exercises 79–88, perform the indicated operation and reduce your answer to lowest terms.

79. $\frac{2}{3}+\frac{1}{5}$ **80.** $\frac{5}{6}-\frac{1}{8}$

81. $\frac{5}{13}+\frac{11}{26}$ **82.** $\frac{5}{12}+\frac{7}{36}$

83. $\frac{5}{9}-\frac{7}{54}$ **84.** $\frac{13}{30}-\frac{17}{120}$

85. $\frac{1}{12}+\frac{1}{48}+\frac{1}{72}$ **86.** $\frac{3}{5}+\frac{7}{15}+\frac{9}{75}$

87. $\frac{1}{30}-\frac{3}{40}-\frac{7}{50}$ **88.** $\frac{4}{25}-\frac{9}{100}-\frac{7}{40}$

Problem Solving

Alternative methods for adding and subtracting two fractions are shown. These methods may not result in a solution in its lowest terms.

$$\frac{a}{b}+\frac{c}{d}=\frac{ad+bc}{bd} \quad \text{and} \quad \frac{a}{b}-\frac{c}{d}=\frac{ad-bc}{bd}$$

In Exercises 89–94, use one of the two formulas to evaluate the expression.

89. $\frac{2}{5}+\frac{7}{8}$ **90.** $\frac{3}{4}+\frac{2}{9}$

91. $\frac{5}{6}-\frac{7}{8}$ **92.** $\frac{7}{3}-\frac{5}{12}$

93. $\frac{3}{8}+\frac{5}{12}$ **94.** $\left(\frac{2}{3}+\frac{1}{4}\right)-\frac{3}{5}$

In Exercises 95–100, evaluate each expression.

95. $\left(\frac{2}{3}\cdot\frac{9}{10}\right)+\frac{2}{5}$ **96.** $\left(\frac{7}{6}\div\frac{4}{3}\right)-\frac{11}{12}$

97. $\left(\frac{1}{2}+\frac{3}{10}\right)\div\left(\frac{1}{5}+2\right)$

98. $\left(\frac{1}{9}\cdot\frac{3}{5}\right)+\left(\frac{2}{3}\cdot\frac{1}{5}\right)$

99. $\left(3 - \frac{4}{9}\right) \div \left(4 + \frac{2}{3}\right)$

100. $\left(\frac{2}{5} \div \frac{4}{9}\right)\left(\frac{3}{5} \cdot 6\right)$

In Exercises 101–114, write an expression that will solve the problem and then evaluate the expression.

101. *Thistles* Diane Helbing has four different varieties of thistles invading her pasture. She estimates that of these thistles, $\frac{1}{2}$ are Canada thistles, $\frac{1}{4}$ are bull thistles, $\frac{1}{6}$ are plumeless thistles, and the rest are musk thistles. What fraction of the thistles are musk thistles?

Bull Thistle Flower

102. *Height Increase* When David Conway finished his freshman year of high school, his height was $69\frac{7}{8}$ inches. When he returned to school after the summer, David's height was $71\frac{5}{8}$ inches. How much did David's height increase over the summer?

103. *Stairway Height* A stairway consists of 14 stairs, each $8\frac{5}{8}$ inches high. What is the vertical height of the stairway?

104. *Math Team* Julie Cholet is hosting the math team after school and wants to share a $67\frac{5}{8}$ oz bottle of soda. How much soda should she pour into each of six glasses so that each glass contains the same amount of soda?

105. *Alphabet Soup* Margaret Cannata's recipe for alphabet soup calls for (among other items) $\frac{1}{4}$ cup snipped parsley, $\frac{1}{8}$ teaspoon pepper, and $\frac{1}{2}$ cup sliced carrots. Margaret is expecting company and needs to multiply the amounts of the ingredients by $1\frac{1}{2}$ times. Determine the amount of (a) snipped parsley, (b) pepper, and (c) sliced carrots she needs for the soup.

106. *Sprinkler System* To repair his sprinkler system, Tony Gambino needs a total of $20\frac{5}{16}$ inches of PVC pipe. He has on hand pieces that measure $2\frac{1}{4}$ inches, $3\frac{7}{8}$ inches, and $4\frac{1}{4}$ inches in length. If he can combine these pieces and use them in the repair, how long of a piece of PVC pipe will Tony need to purchase to repair his sprinkler system?

107. *Crop Storage* Todd Schroeder has a silo on his farm in which he can store silage made from his various crops. He currently has a silo that is $\frac{1}{4}$ full of corn silage, $\frac{2}{5}$ full of hay silage, and $\frac{1}{3}$ full of oats silage. What fraction of Todd's silo is currently in use?

108. *Department Budget* Jaime Bailey is chair of the humanities department at Santa Fe Community College. Jaime has a budget in which $\frac{1}{2}$ of the money is for photocopying, $\frac{2}{5}$ of the money is for computer-related expenses, and the rest of the money is for student tutors in the foreign languages lab. What fraction of Jaime's budget is for student tutors?

109. *Proofreading a Textbook* To help proofread her new textbook, Chris Mishke assigns three students to proofread $\frac{1}{4}$, $\frac{1}{5}$, and $\frac{1}{2}$ of the book, respectively. She decides to proofread the rest of the book herself. If the book has 540 pages, how many pages must Chris proofread herself?

110. *Art Supplies* Denise Viale teaches kindergarten and is buying supplies for her class to make papier-mâché piggy banks. Each piggy bank to be made requires $1\frac{1}{4}$ cups of flour. If Denise has 15 students who are going to make piggy banks, how much flour does Denise need to purchase?

111. *Height of a Computer Stand* The instructions for assembling a computer stand include a diagram illustrating its dimensions. Find the total height of the stand.

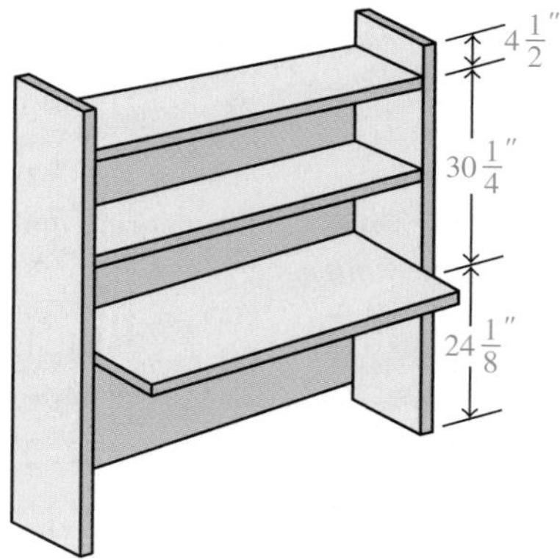

112. *Cutting Lumber* A piece of wood measures $15\frac{3}{8}$ in.

a) How far from one end should you cut the wood if you want to cut the length in half?

b) What is the length of each piece after the cut? You must allow $\frac{1}{8}$ in. for the saw cut.

113. *Width of a Picture* The width of a picture is $24\frac{7}{8}$ in., as shown in the diagram. Find x, the distance from the edge of the frame to the center.

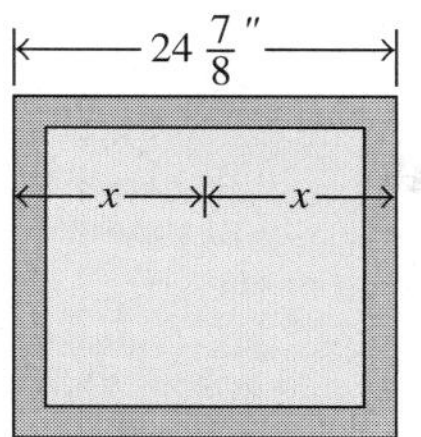

114. *Floor Molding* Rafela Weiss wants to place $\frac{1}{2}$ in. molding along the floor around the perimeter of her room (excluding door openings). She finds that she needs lengths of $26\frac{1}{2}$ in., $105\frac{1}{4}$ in., $53\frac{1}{4}$ in., and $106\frac{5}{16}$ in. How much molding will she need?

Challenge Problems/Group Activities

115. *Cutting Lumber* If a piece of wood $8\frac{3}{4}$ ft long is to be cut into four equal pieces, find the length of each piece. (Allow $\frac{1}{8}$ in. for each saw cut.)

116. *Increasing a Book Size* The dimensions of the cover of a book have been increased from $8\frac{1}{2}$ in. by $9\frac{1}{4}$ in. to $8\frac{1}{2}$ in. by $10\frac{1}{4}$ in. By how many square inches has the surface area increased? Use area = length × width.

117. *Dimensions of a Room* A rectangular room measures 8 ft 3 in. by 10 ft 8 in. by 9 ft 2 in. high.
- **a)** Determine the perimeter of the room in feet.
- **b)** Calculate the area of the floor of the room in square feet.
- **c)** Calculate the volume of the room in cubic feet.

118. *Hanging a Picture* The back of a framed picture that is to be hung is shown. A nail is to be hammered into the wall, and the picture will be hung by the wire on the nail.

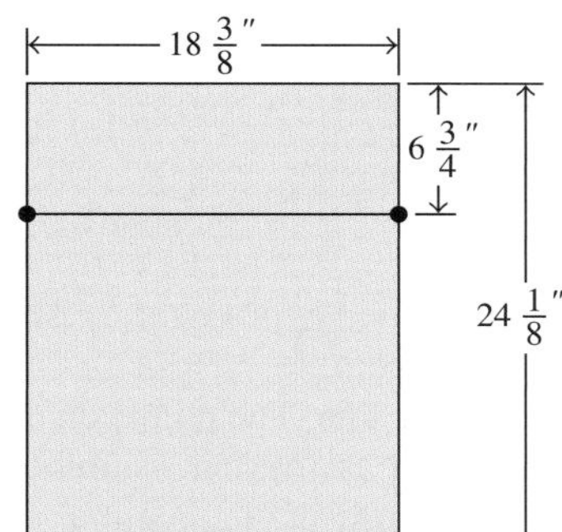

- **a)** If the center of the wire is to rest on the nail and a side of the picture is to be 20 in. from the window, how far from the window should the nail be placed?
- **b)** If the top of the frame is to be $26\frac{1}{4}$ in. from the ceiling, how far from the ceiling should the nail be placed? (Assume the wire will not stretch.)
- **c)** Repeat part (b) if the wire will stretch $\frac{1}{4}$ in. when the picture is hung.

Dense Set of Numbers A set of numbers is said to be a dense set if between any two distinct members of the set there exists a third distinct member of the set. The set of integers is not dense, since between any two consecutive integers, there is not another integer. For example, between 1 and 2 there are no other integers. The set of rational numbers is dense because between any two distinct rational numbers there exists a third distinct rational number. For example, we can find a rational number between 0.243 and 0.244. The number 0.243 can be written as 0.2430, and 0.244 can be written as 0.2440. There are many numbers between these two. Some of them are 0.2431, 0.2435, and 0.243912. In Exercises 119–126, find a rational number between the two numbers in each pair.

119. 0.10 and 0.11

120. 5.03 and 5.003

121. −2.176 and −2.175

122. 1.3457 and 1.34571

123. 3.12345 and 3.123451

124. 0.4105 and 0.4106

125. 4.872 and 4.873

126. −3.7896 and −3.7895

Halfway Between Two Numbers To find a rational number halfway between any two rational numbers given in fraction form, add the two numbers together and divide their sum by 2. In Exercises 127–134, find a rational number halfway between the two fractions in each pair.

127. $\frac{1}{3}$ and $\frac{2}{3}$

128. $\frac{2}{7}$ and $\frac{3}{7}$

129. $\frac{1}{100}$ and $\frac{1}{10}$

130. $\frac{7}{13}$ and $\frac{8}{13}$

131. $\frac{1}{4}$ and $\frac{1}{5}$

132. $\frac{1}{3}$ and $\frac{2}{3}$

133. $\frac{1}{10}$ and $\frac{1}{100}$

134. $\frac{1}{2}$ and $\frac{2}{3}$

135. *Cooking Oatmeal* Following are the instructions given on a box of oatmeal. Determine the amount of water (or milk) and oats needed to make $1\frac{1}{2}$ servings by:
- **a)** Adding the amount of each ingredient needed for 1 serving to the amount needed for 2 servings and dividing by 2.
- **b)** Adding the amount of each ingredient needed for 1 serving to half the amount needed for 1 serving.

Directions

1. Boil water or milk and salt (if desired).
2. Stir in oats.
3. Stirring occasionally, cook over medium heat for 5 minutes.

Servings	1	2
Water (or milk)	1 cup	$1\frac{3}{4}$ cup
Oats	$\frac{1}{2}$ cup	1 cup
Salt (optional)	dash	$\frac{1}{8}$ tsp

136. Consider the rational number $0.\overline{9}$.
   a) Use the method from Example 8 on page 233 to convert $0.\overline{9}$ to a quotient of integers.
   b) Find a number halfway between $0.\overline{9}$ and 1 by adding the two numbers and dividing by 2.
   c) Find $\frac{1}{3} + \frac{2}{3}$. Express $\frac{1}{3}$ and $\frac{2}{3}$ as repeating decimals. Now find the same sum using the repeating decimal representation of $\frac{1}{3}$ and $\frac{2}{3}$.
   d) What conclusion can you draw from parts (a), (b), and (c)?

Recreational Mathematics

137. *Paper Folding* Fold a sheet of paper in half. Now unfold the paper. You will see that this one crease divided the paper into two equal regions. Each of these regions represents $\frac{1}{2}$ of the area of the entire sheet of paper. Next, fold a sheet of paper in half and then fold it in half again. Now unfold this piece of paper. You will see that these two creases divided the paper into four equal regions. Each of these regions can be considered $\frac{1}{4}$ of the area of the sheet of paper. Continue this process and answer the following questions.
   a) If you fold the paper in half three times, each region will be what fraction of the area of the sheet of paper?
   b) If you fold the paper in half four times, each region will be what fraction of the area of the sheet of paper?
   c) How many creases will you need to form regions that are $\frac{1}{32}$ of the area of the sheet of paper?
   d) How many creases will you need to form regions that are $\frac{1}{64}$ of the area of the sheet of paper?

Internet/Research Activity

138. The ancient Greeks are often considered the first true mathematicians. Write a report summarizing the ancient Greeks' contributions to rational numbers. Include in your report what they learned and believed about the rational numbers. References include encyclopedias, history of mathematics books, and Internet websites.

5.4 THE IRRATIONAL NUMBERS AND THE REAL NUMBER SYSTEM

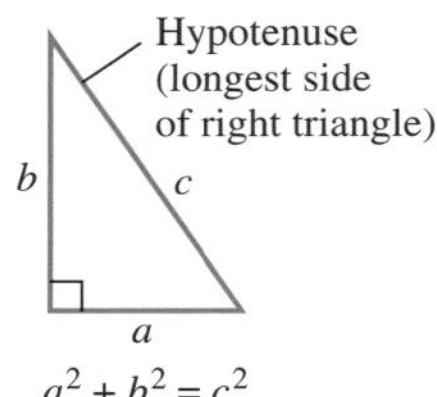

Figure 5.7

Pythagoras (ca. 585–500 B.C.), a Greek mathematician, is credited with providing a written proof that in any *right triangle* (a triangle with a 90° angle; see Fig. 5.7), the square of the length of one side (a^2) added to the square of the length of the other side (b^2) equals the square of the length of the hypotenuse (c^2). The formula $a^2 + b^2 = c^2$ is now known as the *Pythagorean theorem*.* Pythagoras found that the solution of the formula, where $a = 1$ and $b = 1$, is not a rational number.

$$\begin{aligned} a^2 + b^2 &= c^2 \\ 1^2 + 1^2 &= c^2 \\ 1 + 1 &= c^2 \\ 2 &= c^2 \end{aligned}$$

There is no rational number that when squared will equal 2. This prompted a need for a new set of numbers, the irrational numbers.

In Section 5.2, we introduced the real number line. The points on the real number line that are not rational numbers are referred to as irrational numbers. Recall that every rational number is either a terminating or a repeating decimal number. Therefore, irrational numbers, when represented as decimal numbers, will be nonterminating, nonrepeating decimal numbers.

An **irrational number** is a real number whose decimal representation is a nonterminating, nonrepeating decimal number.

*The Pythagorean theorem is discussed in more detail in Section 9.3.

PROFILE IN MATHEMATICS

PYTHAGORAS OF SAMOS

Pythagoras of Samos founded a philosophical and religious school in southern Italy in the sixth century B.C. The scholars at the school, known as Pythagoreans, produced important works of mathematics, astronomy, and theory of music. Although the Pythagoreans are credited with proving the Pythagorean theorem, it was known to the ancient Babylonians 1000 years earlier. The Pythagoreans were a secret society that formed a model for many secret societies in existence today. One practice was that students were to spend their first three years of study in silence, while their master, Pythagoras, spoke to them from behind a curtain. Among other philosophical beliefs was "that at its deepest level, reality is mathematical in nature."

A nonrepeating decimal number such as 5.12639537 . . . can be used to indicate an irrational number. Notice that no number or set of numbers repeat on a continuous basis, and the three dots at the end of the number indicate that the number continues indefinitely. Nonrepeating number patterns can be used to indicate irrational numbers. For example, 6.1011011101111 . . . and 0.525225222 . . . are both irrational numbers.

The expression $\sqrt{2}$ is read "the square root of 2" or "radical 2." The symbol $\sqrt{\ }$ is called the *radical sign*, and the number or expression inside the radical sign is called the *radicand*. In $\sqrt{2}$, 2 is the radicand.

The square roots of some numbers are rational, whereas the square roots of other numbers are irrational. The *principal* (or *positive*) *square root* of a number n, written $\sqrt{n}$, is the positive number that when multiplied by itself, gives n. Whenever we mention the term "square root" in this text, we mean the principal square root. For example,

$$\sqrt{9} = 3 \quad \text{since} \quad 3 \cdot 3 = 9$$
$$\sqrt{36} = 6 \quad \text{since} \quad 6 \cdot 6 = 36$$

Both $\sqrt{9}$ and $\sqrt{36}$ are examples of numbers that are rational numbers because their square roots, 3 and 6 respectively, are terminating decimal numbers.

Returning to the problem faced by Pythagoras: If $c^2 = 2$, then c has a value of $\sqrt{2}$, but what is $\sqrt{2}$ equal to? The $\sqrt{2}$ is an irrational number, and it cannot be expressed as a terminating or repeating decimal number. It can only be approximated by a decimal number: $\sqrt{2}$ is approximately 1.4142135 (to seven decimal places). Later in this section, we will discuss using a calculator to approximate irrational numbers.

Other irrational numbers include $\sqrt{3}$, $\sqrt{5}$, and $\sqrt{37}$. Another important irrational number used to represent the ratio of a circle's circumference to its diameter is pi, symbolized π. Pi is approximately 3.1415926.

We have discussed procedures for performing the arithmetic operations of addition, subtraction, multiplication, and division with rational numbers. We can perform the same operations with the irrational numbers. Before we can proceed, however, we must understand the numbers called perfect squares. Any number that is the square of a natural number is said to be a *perfect square*.

Natural numbers	1,	2,	3,	4,	5,	6, . . .
Squares of the natural numbers	1^2,	2^2,	3^2,	4^2,	5^2,	6^2, . . .
or perfect squares	1,	4,	9,	16,	25,	36, . . .

The numbers 1, 4, 9, 16, 25, and 36 are some of the perfect square numbers. Can you determine the next two perfect square numbers? How many perfect square numbers are there? The square root of a perfect square number will be a natural number. For example, $\sqrt{1} = 1$, $\sqrt{4} = 2$, $\sqrt{9} = 3$, $\sqrt{16} = 4$, $\sqrt{25} = 5$, and so on.

The number that multiplies a radical is called the radical's *coefficient*. For example, in $3\sqrt{5}$, the 3 is the coefficient of the radical.

Some irrational numbers can be simplified by determining whether there are any perfect square factors in the radicand. If there are, the following rule can be used to simplify the radical.

Product Rule for Radicals

$$\sqrt{a \cdot b} = \sqrt{a} \cdot \sqrt{b}, \qquad a \geq 0, \qquad b \geq 0$$

DID YOU KNOW

A Piece of the Pi

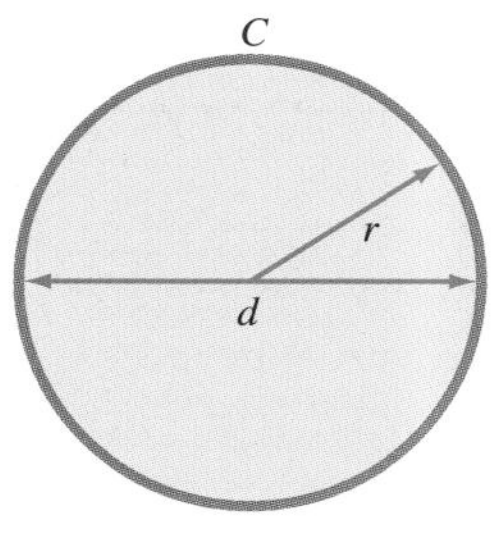

$C = \pi d$
or $\pi = \frac{C}{d}$

The American philosopher and psychologist William James wrote in 1909: "The thousandth decimal of π sleeps there though no one may ever try to compute it." It sleeps no more. We now know π, the ratio of a circle's circumference to its diameter, to over 1.2 trillion digits.

In December 2002, Yasumasa Kanada and others at the University of Tokyo announced that they had calculated π to 1,241,100,000,000 decimal places beating their previous record set in 1999. Their computation of π consumed more than 600 hours of time on a Hitachi SR8000 supercomputer.

This record, like the record for the largest prime number, will most likely be broken in the near future (it might already be broken as you read this). Mathematicians and computer scientists continue to improve both their computers and their methods used to find numbers like the most accurate approximation for π or the largest prime number.

To simplify a radical, write the radical as a product of two radicals. One of the radicals should contain the greatest perfect square that is a factor of the radicand in the original expression. Then simplify the radical containing the perfect square factor.

For example,

$$\sqrt{18} = \sqrt{9 \cdot 2} = \sqrt{9} \cdot \sqrt{2} = 3 \cdot \sqrt{2} = 3\sqrt{2}$$

and

$$\sqrt{75} = \sqrt{25 \cdot 3} = \sqrt{25} \cdot \sqrt{3} = 5 \cdot \sqrt{3} = 5\sqrt{3}$$

EXAMPLE 1 *Simplifying Radicals*

Simplify

a) $\sqrt{28}$ b) $\sqrt{48}$

SOLUTION:

a) Since 4 is a perfect square factor of 28, we write

$$\sqrt{28} = \sqrt{4 \cdot 7} = \sqrt{4} \cdot \sqrt{7} = 2 \cdot \sqrt{7} = 2\sqrt{7}$$

Since 7 has no perfect square factors, $\sqrt{7}$ cannot be simplified.

b) Since 16 is a perfect square factor of 48, we write

$$\sqrt{48} = \sqrt{16 \cdot 3} = \sqrt{16} \cdot \sqrt{3} = 4 \cdot \sqrt{3} = 4\sqrt{3}$$

In Example 1(b), you can obtain the correct answer if you start out factoring differently:

$$\sqrt{48} = \sqrt{4 \cdot 12} = \sqrt{4} \cdot \sqrt{12} = 2\sqrt{12}$$

Note that 12 has 4 as a perfect square factor.

$$2\sqrt{12} = 2\sqrt{4 \cdot 3} = 2 \cdot \sqrt{4} \cdot \sqrt{3} = 2 \cdot 2 \cdot \sqrt{3} = 4\sqrt{3}$$

The second method will eventually give the same answer, but it requires more work. It is best to try to factor out the *largest* perfect square factor from the radicand.

Addition and Subtraction of Irrational Numbers

To add or subtract two or more square roots with the same radicand, add or subtract their coefficients. The answer is the sum or difference of the coefficients multiplied by the common radical.

EXAMPLE 2 *Adding and Subtracting Radicals with the Same Radicand*

Simplify

a) $3\sqrt{2} + 6\sqrt{2}$ b) $4\sqrt{7} + \sqrt{7} - 8\sqrt{7}$

SOLUTION:

a) $3\sqrt{2} + 6\sqrt{2} = (3 + 6)\sqrt{2} = 9\sqrt{2}$

b) $4\sqrt{7} + \sqrt{7} - 8\sqrt{7} = (4 + 1 - 8)\sqrt{7} = -3\sqrt{7}$

Note that $\sqrt{7} = 1\sqrt{7}$.

> **DID YOU KNOW**
>
> *Is the Computer Always Right?*
>
> We often assume so, on the basis of the history and performance of the computer language and programs we use. Obviously, we cannot check calculations that are being made in billionths of a second. We could run a computation twice or on two different systems to be sure we get the same result to a specified number of decimal places. That's what is done with the big supercomputers. Their computation of the value of π often serves as a test of the program's ability to compute accurately.

EXAMPLE 3 *Adding and Subtracting Radicals with Different Radicands*

Simplify $5\sqrt{3} - \sqrt{12}$.

SOLUTION: These radicals cannot be added in their present form because they contain different radicands. When this occurs, determine whether one or more of the radicals can be simplified so that they have the same radicand.

$$\begin{aligned} 5\sqrt{3} - \sqrt{12} &= 5\sqrt{3} - \sqrt{4 \cdot 3} \\ &= 5\sqrt{3} - \sqrt{4} \cdot \sqrt{3} \\ &= 5\sqrt{3} - 2\sqrt{3} \\ &= (5 - 2)\sqrt{3} = 3\sqrt{3} \end{aligned}$$

▲

Multiplication of Irrational Numbers

When multiplying irrational numbers, we again make use of the product rule for radicals. After the radicands are multiplied, simplify the remaining radical when possible.

EXAMPLE 4 *Multiplying Radicals*

Simplify

a) $\sqrt{3} \cdot \sqrt{27}$ b) $\sqrt{3} \cdot \sqrt{7}$ c) $\sqrt{6} \cdot \sqrt{10}$

SOLUTION:

a) $\sqrt{3} \cdot \sqrt{27} = \sqrt{3 \cdot 27} = \sqrt{81} = 9$

b) $\sqrt{3} \cdot \sqrt{7} = \sqrt{3 \cdot 7} = \sqrt{21}$

c) $\sqrt{6} \cdot \sqrt{10} = \sqrt{6 \cdot 10} = \sqrt{60} = \sqrt{4 \cdot 15} = \sqrt{4} \cdot \sqrt{15} = 2\sqrt{15}$ ▲

Division of Irrational Numbers

To divide irrational numbers, use the following rule. After performing the division, simplify when possible.

Quotient Rule for Radicals

$$\frac{\sqrt{a}}{\sqrt{b}} = \sqrt{\frac{a}{b}}, \qquad a \geq 0, \qquad b > 0$$

EXAMPLE 5 *Dividing Radicals*

Divide

a) $\dfrac{\sqrt{8}}{\sqrt{2}}$ b) $\dfrac{\sqrt{96}}{\sqrt{2}}$

SOLUTION:

a) $\dfrac{\sqrt{8}}{\sqrt{2}} = \sqrt{\dfrac{8}{2}} = \sqrt{4} = 2$

b) $\dfrac{\sqrt{96}}{\sqrt{2}} = \sqrt{\dfrac{96}{2}} = \sqrt{48} = \sqrt{16 \cdot 3} = \sqrt{16} \cdot \sqrt{3} = 4\sqrt{3}$ ▲

Rationalizing the Denominator

A denominator is *rationalized* when it contains no radical expressions. To rationalize a denominator that contains only a square root, multiply both the numerator and denominator of the fraction by a number that will result in the radicand in the denominator becoming a perfect square. (This action is the equivalent of multiplying the fraction by 1 because the value of the fraction does not change.) Then simplify the fractions when possible.

EXAMPLE 6 ***Rationalizing the Denominator***

Rationalize the denominator of

a) $\frac{5}{\sqrt{2}}$ b) $\frac{5}{\sqrt{12}}$ c) $\frac{\sqrt{5}}{\sqrt{10}}$

SOLUTION:

a) Multiply the numerator and denominator by a number that will make the radicand a perfect square.

$$\frac{5}{\sqrt{2}} = \frac{5}{\sqrt{2}} \cdot \frac{\sqrt{2}}{\sqrt{2}} = \frac{5\sqrt{2}}{\sqrt{4}} = \frac{5\sqrt{2}}{2}$$

Note that the 2's in the answer cannot be divided out because one 2 is a radicand and the other is not.

b) $\frac{5}{\sqrt{12}} = \frac{5}{\sqrt{12}} \cdot \frac{\sqrt{3}}{\sqrt{3}} = \frac{5\sqrt{3}}{\sqrt{36}} = \frac{5\sqrt{3}}{6}$

You could have obtained the same answer to this problem by multiplying both the numerator and denominator by $\sqrt{12}$ and then simplifying. Try to do so now.

c) Write $\frac{\sqrt{5}}{\sqrt{10}}$ as $\sqrt{\frac{5}{10}}$ and reduce the fraction to obtain $\sqrt{\frac{1}{2}}$. By the quotient rule for radicals, $\sqrt{\frac{1}{2}} = \frac{\sqrt{1}}{\sqrt{2}}$ or $\frac{1}{\sqrt{2}}$. Now rationalize $\frac{1}{\sqrt{2}}$.

$$\frac{1}{\sqrt{2}} = \frac{1}{\sqrt{2}} \cdot \frac{\sqrt{2}}{\sqrt{2}} = \frac{\sqrt{2}}{2}$$

Approximating Square Roots on a Scientific Calculator

Consider the irrational number the square root of two. We use the symbol $\sqrt{2}$ to represent the *exact value* of this number. Although exact values are important, approximations are also important, especially when working with application problems. We can use a scientific calculator to obtain approximations for square roots. Scientific calculators generally have one of the following square root keys:*

[$\sqrt{\ }$] or [$\sqrt{x}$]

*If your calculator has the $\sqrt{\ }$ symbol printed *above* the key instead of on the face of the key, you can access the square root function by first pressing the "2nd" or the "inverse" key.

For simplicity, we will refer to the square root key with the [$\sqrt{\ }$] symbol.

To approximate $\sqrt{2}$, perform the following keystrokes:

2 [$\sqrt{\ }$]

or, depending on your model of calculator, you may have to do the following:

[$\sqrt{\ }$] [2] [ENTER]

The display on your calculator may read 1.414213562. Your calculator may display more or fewer digits. It is important to realize that 1.414213562 is a rational number *approximation* for the irrational number $\sqrt{2}$. The symbol $\approx$ means *is approximately equal to,* and we write

$$\sqrt{2} \approx 1.414213562$$

Exact value (irrational number) ↗ ↖ Approximation (rational number)

EXAMPLE 7 *Approximating Square Roots*

Use a scientific calculator to approximate the following square roots. Round your answers to two decimal places.

a) $\sqrt{5}$ b) $\sqrt{17}$ c) $\sqrt{91}$ d) $\sqrt{237}$

SOLUTION:

a) $\sqrt{5} \approx 2.24$ b) $\sqrt{17} \approx 4.12$ c) $\sqrt{91} \approx 9.54$ d) $\sqrt{237} \approx 15.39$

SECTION 5.4 EXERCISES

Concept/Writing Exercises

1. Explain the difference between a rational number and an irrational number.
2. What is the principal square root of a number?
3. What is a perfect square?
4. a) State the product rule for radicals.
   b) State the quotient rule for radicals.
5. a) Explain how to add or subtract square roots that have the same radicand.
   b) Using the procedure in part (a), add $3\sqrt{6} + 5\sqrt{6} - 9\sqrt{6}$.
6. What does it mean to rationalize the denominator?
7. a) Explain how to rationalize a denominator that contains a square root.
   b) Using the procedure in part (a), rationalize $\frac{7}{\sqrt{3}}$.
8. a) Explain how to approximate square roots on your calculator.
   b) Using the procedure in part (a), approximate $\sqrt{7}$. Round your answer to the nearest hundredth.

Practice the Skills

In Exercises 9–18, determine whether the number is rational or irrational.

9. $\sqrt{36}$
10. $\sqrt{18}$
11. $\frac{2}{3}$
12. 0.212112111...
13. 3.575775777...
14. π
15. $\frac{22}{7}$
16. 3.14159
17. 3.14159...
18. $\frac{\sqrt{5}}{\sqrt{5}}$

In Exercises 19–28, evaluate the expression.

19. $\sqrt{64}$
20. $\sqrt{144}$
21. $\sqrt{100}$
22. $-\sqrt{144}$
23. $-\sqrt{169}$
24. $\sqrt{25}$
25. $-\sqrt{225}$
26. $-\sqrt{36}$
27. $-\sqrt{100}$
28. $\sqrt{256}$

In Exercises 29–38, classify the number as a member of one or more of the following sets: the rational numbers, the integers, the natural numbers, the irrational numbers.

29. 1 **30.** -4

31. $\sqrt{49}$ **32.** $\frac{4}{5}$

33. 0.040040004 **34.** 2.718

35. $-\frac{7}{8}$ **36.** 0.123123123

37. $0.\overline{123}$ **38.** 0.123112311123...

In Exercises 39–48, simplify the radical.

39. $\sqrt{18}$ **40.** $\sqrt{20}$ **41.** $\sqrt{48}$

42. $\sqrt{60}$ **43.** $\sqrt{63}$ **44.** $\sqrt{75}$

45. $\sqrt{80}$ **46.** $\sqrt{90}$ **47.** $\sqrt{162}$

48. $\sqrt{300}$

In Exercises 49–58, perform the indicated operation.

49. $2\sqrt{6} + 5\sqrt{6}$ **50.** $3\sqrt{17} + \sqrt{17}$

51. $5\sqrt{12} - \sqrt{75}$ **52.** $2\sqrt{5} + 3\sqrt{20}$

53. $4\sqrt{12} - 7\sqrt{27}$ **54.** $2\sqrt{7} + 5\sqrt{28}$

55. $5\sqrt{3} + 7\sqrt{12} - 3\sqrt{75}$

56. $13\sqrt{2} + 2\sqrt{18} - 5\sqrt{32}$

57. $\sqrt{8} - 3\sqrt{50} + 9\sqrt{32}$

58. $\sqrt{63} + 13\sqrt{98} - 5\sqrt{112}$

In Exercises 59–68, perform the indicated operation. Simplify the answer when possible.

59. $\sqrt{2}\sqrt{8}$ **60.** $\sqrt{5}\sqrt{15}$ **61.** $\sqrt{6}\sqrt{10}$

62. $\sqrt{3}\sqrt{6}$ **63.** $\sqrt{10}\sqrt{20}$ **64.** $\sqrt{11}\sqrt{33}$

65. $\frac{\sqrt{8}}{\sqrt{4}}$ **66.** $\frac{\sqrt{125}}{\sqrt{5}}$ **67.** $\frac{\sqrt{72}}{\sqrt{8}}$

68. $\frac{\sqrt{136}}{\sqrt{8}}$

In Exercises 69–78, rationalize the denominator.

69. $\frac{1}{\sqrt{2}}$ **70.** $\frac{3}{\sqrt{3}}$ **71.** $\frac{\sqrt{3}}{\sqrt{7}}$

72. $\frac{\sqrt{3}}{\sqrt{10}}$ **73.** $\frac{\sqrt{20}}{\sqrt{3}}$ **74.** $\frac{\sqrt{50}}{\sqrt{14}}$

75. $\frac{\sqrt{9}}{\sqrt{2}}$ **76.** $\frac{\sqrt{15}}{\sqrt{3}}$ **77.** $\frac{\sqrt{10}}{\sqrt{6}}$

78. $\frac{8}{\sqrt{8}}$

Problem Solving

Approximating Radicals *The following diagram shows a 16 in. ruler marked using $\frac{1}{2}$ inches.*

1 2 3 4 5 6 7 8 9 10 11 12 13 14 15

In Exercises 79–84, without using a calculator, indicate between which two adjacent markers each of the following irrational numbers will fall. Explain how you obtained your answer. Support your answer by obtaining an approximation with a calculator.

79. $\sqrt{7}$ in. 80. $\sqrt{37}$ in.

81. $\sqrt{107}$ in. 82. $\sqrt{135}$ in.

83. $\sqrt{170}$ in. 84. $\sqrt{200}$ in.

In Exercises 85–90, determine whether the statement is true or false. Rewrite each false statement to make it a true statement. A false statement can be modified in more than one way to be made a true statement.

85. $\sqrt{p}$ is a rational number for any prime number p.

86. $\sqrt{c}$ is a rational number for any composite number c.

87. The sum of any two rational numbers is always a rational number.

88. The product of any two rational numbers is always a rational number.

89. The product of an irrational and a rational number is always an irrational number.

90. The product of any two irrational numbers is always an irrational number.

In Exercises 91–94, give an example to show that the stated case can occur.

91. The sum of two irrational numbers may be a rational number.

92. The sum of two irrational numbers may be an irrational number.

93. The product of two irrational numbers may be an irrational number.

94. The product of two irrational numbers may be a rational number.

95. Without doing any calculations, determine whether $\sqrt{5} = 2.236$. Explain your answer.

96. Without doing any calculations, determine whether $\sqrt{14} = 3.742$. Explain your answer.

97. The number π is an irrational number. Often the values 3.14 or $\frac{22}{7}$ are used for π. Does π equal either 3.14 or $\frac{22}{7}$? Explain your answer.

98. Give an example to show that $\sqrt{a+b} \neq \sqrt{a} + \sqrt{b}$.

99. Give an example to show that $\sqrt{a \cdot b} = \sqrt{a} \cdot \sqrt{b}$.

100. *A Swinging Pendulum* The time T required for a pendulum to swing back and forth may be found by the formula

$$T = 2\pi\sqrt{\frac{l}{g}}$$

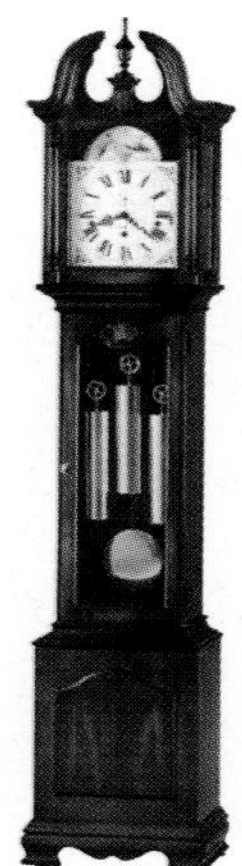

where l is the length of the pendulum and g is the acceleration of gravity. Find the time in seconds if $l = 35$ cm and $g = 980$ cm/sec^2. Round answer to the nearest tenth of a second.

101. *Estimating Speed of a Vehicle* The speed a vehicle was traveling, s, in miles per hour, when the brakes were first applied, can be estimated using the formula $s = \sqrt{\frac{d}{0.04}}$ where d is the length of the vehicle's skid marks, in feet.

a) Determine the speed of a car that made skid marks 4 ft long.
b) Determine the speed of a car that made skid marks 16 ft long.
c) Determine the speed of a car that made skid marks 64 ft long.
d) Determine the speed of a car that made skid marks 256 ft long.

102. *Dropping an Object* The formula $t = \frac{\sqrt{d}}{4}$ can be used to estimate the time, t, in seconds it takes for an object dropped to travel d feet.

a) Determine the time it takes for an object to drop 100 ft.
b) Determine the time it takes for an object to drop 400 ft.
c) Determine the time it takes for an object to drop 900 ft.
d) Determine the time it takes for an object to drop 1600 ft.

Challenge Problems/Group Activities

103. **a)** If a radical expression is evaluated on a calculator, explain how you can determine whether the expression is a rational or irrational number.
b) Is $\sqrt{0.04}$ rational or irrational? Explain.
c) Is $\sqrt{0.7}$ rational or irrational? Explain.

104. One way to find a rational number between two distinct rational numbers is to add the two distinct rational numbers and divide by 2. Do you think that this method will work for finding an irrational number between two distinct irrational numbers? Explain.

Recreational Mathematics

105. *More Four 4's* In Exercise 82 on page 227, we introduced some of the basic rules of the game Four 4's. We now expand our operations to include square roots. For example, one way to obtain the whole number 8 is $\sqrt{4} + \sqrt{4} + \sqrt{4} + \sqrt{4} = 2 + 2 + 2 + 2 = 8$. Using the rules given on page 227, and using at least one square root of 4, $\sqrt{4}$, play Four 4's to obtain the following whole numbers:
a) 11
b) 13
c) 14
d) 18

Internet/Research Activities

In Exercises 106 and 107, references include history of mathematics books, encyclopedias, and Internet web sites.

106. Write a report on the history of the development of the irrational numbers.

107. Write a report on the history of pi. In your report, indicate when the symbol π was first used and list the first 10 digits of π.

5.5 REAL NUMBERS AND THEIR PROPERTIES

Now that we have discussed both the rational and irrational numbers, we can discuss the real numbers and the properties of the real number system. The union of the rational numbers and the irrational numbers is the *set of real numbers*, symbolized by $\mathbb{R}$.

Figure 5.8 illustrates the relationship among various sets of numbers. It shows that the natural numbers are a subset of the whole numbers, the integers, the rational numbers, and the real numbers. For example, since the number 3 is a natural or counting number, it is also a whole number, an integer, a rational number, and a real number. Since the rational number $\frac{1}{4}$ is outside the set of integers, it is not an integer, a whole number, or a natural number. The number $\frac{1}{4}$ is a real number, however, as is the irrational number $\sqrt{2}$. Note that the real numbers are the union of the rational numbers and the irrational numbers.

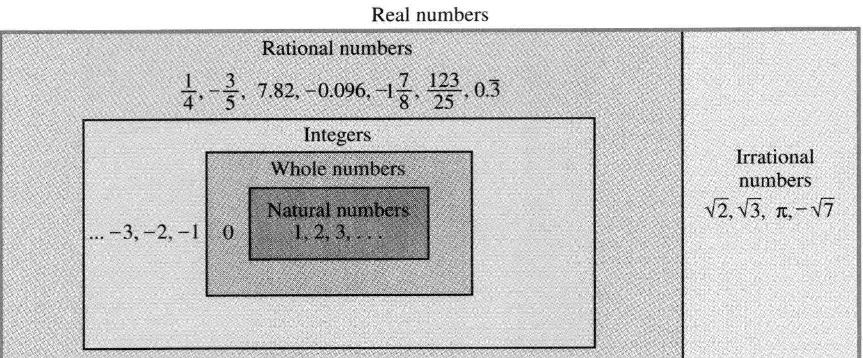

Figure 5.8

The relationship between the various sets of numbers in the real number system can also be illustrated with a tree diagram, as in Fig. 5.9.

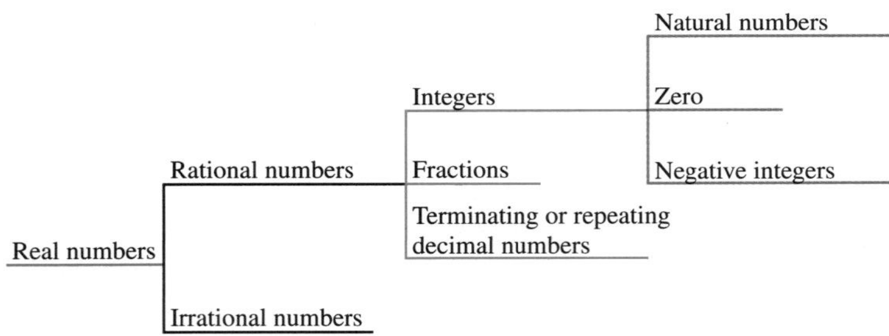

Figure 5.9

Figure 5.9 shows that, for example, the natural numbers are a subset of the integers, the rational numbers, and the real numbers. We can also see, for example, the natural numbers, zero, and the negative integers together form the integers.

Properties of the Real Number System

We are now prepared to consider the properties of the real number system. The first property that we will discuss is *closure.*

> If an operation is performed on any two elements of a set and the result is an element of the set, we say that the set is **closed** under that given operation.

DID YOU KNOW

An Important Number

Some years ago, a group of French mathematicians who worked under the collective pseudonym of "Monsieur Nicholas Bourbaki" embarked on the development of an encyclopedic description of all mathematics. They devoted 200 pages simply to introduce the innocent-looking concept, the number 1.

Is the sum of any two natural numbers a natural number? The answer is yes. Thus, we say that the natural numbers are closed under the operation of addition.

Are the natural numbers closed under the operation of subtraction? If we subtract one natural number from another natural number, must the difference always be a natural number? The answer is no. For example, $3 - 5 = -2$, which is not a natural number. Therefore, the natural numbers are not closed under the operation of subtraction.

EXAMPLE 1 *Closure of Sets*

Determine whether the integers are closed under the operations of (a) multiplication and (b) division.

SOLUTION:

a) If we multiply any two integers, will the product always be an integer? The answer is yes. Thus, the integers are closed under the operation of multiplication.

b) If we divide any two integers, will the quotient always be an integer? The answer is no. For example, $6 \div 5 = \frac{6}{5}$, which is not an integer. Therefore, the integers are not closed under the operation of division. ▲

Next we will discuss three important properties: the commutative property, the associative property, and the distributive property. A knowledge of these properties is essential for the understanding of algebra. We begin with the commutative property.

Commutative Property

ADDITION	MULTIPLICATION
$a + b = b + a$	$a \cdot b = b \cdot a$

for any real numbers a and b.

The commutative property states that the *order* in which two numbers are added or multiplied is not important. For example, $4 + 5 = 5 + 4 = 9$ and $3 \cdot 6 = 6 \cdot 3 = 18$. Note that the commutative property does not hold for the operations of subtraction or division. For example,

$$4 - 7 \neq 7 - 4 \quad \text{and} \quad 9 \div 3 \neq 3 \div 9$$

Now we introduce the associative property.

Associative Property

ADDITION	MULTIPLICATION
$(a + b) + c = a + (b + c)$	$(a \cdot b) \cdot c = a \cdot (b \cdot c)$

for any real numbers a, b, and c.

The associative property states that when adding or multiplying three real numbers, we may place parentheses around any two adjacent numbers. For example,

$$\begin{aligned} (3 + 4) + 5 &= 3 + (4 + 5) \\ 7 + 5 &= 3 + 9 \\ 12 &= 12 \end{aligned} \qquad \begin{aligned} (3 \cdot 4) \cdot 5 &= 3 \cdot (4 \cdot 5) \\ 12 \cdot 5 &= 3 \cdot 20 \\ 60 &= 60 \end{aligned}$$

The associative property does not hold for the operations of subtraction and division. For example,

$$(10 - 6) - 2 \neq 10 - (6 - 2) \quad \text{and} \quad (27 \div 9) \div 3 \neq 27 \div (9 \div 3)$$

Note the difference between the commutative property and the associative property. The commutative property involves a change in *order,* whereas the associative property involves a change in *grouping* (or the *association* of numbers that are grouped together).

Another property of the real numbers is the distributive property of multiplication over addition.

Distributive Property of Multiplication over Addition

$$a \cdot (b + c) = a \cdot b + a \cdot c$$

for any real numbers a, b, and c.

For example, if $a = 3$, $b = 4$, and $c = 5$, then

$$\begin{aligned} 3 \cdot (4 + 5) &= (3 \cdot 4) + (3 \cdot 5) \\ 3 \cdot 9 &= 12 + 15 \\ 27 &= 27 \end{aligned}$$

This result indicates that, when using the distributive property, you may either add first and then multiply or multiply first and then add. Note that the distributive property involves two operations, addition and multiplication. Although positive integers were used in the example, any real numbers could have been used.

We frequently use the commutative, associative, and distributive properties without realizing that we are doing so. To add $13 + 4 + 6$, we may add the $4 + 6$ first to get 10. To this sum we then add 13 to get 23. Here we have done the equivalent of placing parentheses around the $4 + 6$. We can do so because of the associative property of addition.

To multiply 102×11 in our heads, we might multiply $100 \times 11 = 1100$ and $2 \times 11 = 22$ and add these two products to get 1122. We are permitted to do so because of the distributive property.

$$\begin{aligned} 102 \times 11 = (100 + 2) \times 11 &= (100 \times 11) + (2 \times 11) \\ &= 1100 + 22 = 1122 \end{aligned}$$

EXAMPLE 2 *Identifying Properties of Real Numbers*

Name the property illustrated.

a) $2 + 5 = 5 + 2$

b) $(x + 3) + 5 = x + (3 + 5)$

c) $4 \cdot (3 \cdot y) = (4 \cdot 3) \cdot y$

d) $9(w + 3) = 9 \cdot w + 9 \cdot 3$
e) $5 + (z + 3) = 5 + (3 + z)$
f) $(2p) \cdot 5 = 5 \cdot (2p)$

SOLUTION:

a) Commutative property of addition
b) Associative property of addition
c) Associative property of multiplication
d) Distributive property of multiplication over addition
e) The only change between the left and right sides of the equal sign is the order of the z and 3 within the parentheses. The order is changed from $z + 3$ to $3 + z$ using the commutative property of addition.
f) The order of 5 and $(2p)$ is changed by using the commutative property of multiplication. ▲

EXAMPLE 3 *Simplifying by Using the Distributive Property*

Use the distributive property to simplify

a) $2(3 + \sqrt{5})$ b) $\sqrt{2}(7 + \sqrt{3})$

SOLUTION:

a) $2(3 + \sqrt{5}) = (2 \cdot 3) + (2 \cdot \sqrt{5})$
$= 6 + 2\sqrt{5}$

b) $\sqrt{2}(7 + \sqrt{3}) = (\sqrt{2} \cdot 7) + (\sqrt{2} \cdot \sqrt{3})$
$= 7\sqrt{2} + \sqrt{6}$

Note that $\sqrt{2} \cdot 7$ is written $7\sqrt{2}$. ▲

EXAMPLE 4 *Distributive Property*

Use the distributive property to multiply $5(q + 7)$. Then simplify the result.

SOLUTION:

$$5(q + 7) = 5 \cdot q + 5 \cdot 7$$
$$= 5q + 35$$ ▲

We summarize the properties mentioned in this section as follows, where a, b, and c are any real numbers.

Commutative property of addition	$a + b = b + a$
Commutative property of multiplication	$a \cdot b = b \cdot a$
Associative property of addition	$(a + b) + c = a + (b + c)$
Associative property of multiplication	$(a \cdot b) \cdot c = a \cdot (b \cdot c)$
Distributive property of multiplication over addition	$a \cdot (b + c) = a \cdot b + a \cdot c$

SECTION 5.5 EXERCISES

Concept/Writing Exercises

1. What are the real numbers?
2. What symbol is used to represent the set of real numbers?
3. What does it mean if a set is closed under a given operation?
4. Give the commutative property of multiplication, explain what it means, and give an example illustrating it.
5. Give the commutative property of addition, explain what it means, and give an example illustrating it.
6. Give the associative property of addition, explain what it means, and give an example illustrating it.
7. Give the associative property of multiplication, explain what it means, and give an example illustrating it.
8. Give the distributive property of multiplication over addition, explain what it means, and give an example illustrating it.

Practice the Skills

In Exercises 9–12, determine whether the natural numbers are closed under the given operation.

9. Addition
10. Subtraction
11. Division
12. Multiplication

In Exercises 13–16, determine whether the integers are closed under the given operation.

13. Subtraction
14. Addition
15. Division
16. Multiplication

In Exercises 17–20, determine whether the rational numbers are closed under the given operation.

17. Addition
18. Subtraction
19. Multiplication
20. Division

In Exercises 21–24, determine whether the irrational numbers are closed under the given operation.

21. Addition
22. Subtraction
23. Multiplication
24. Division

In Exercises 25–28, determine whether the real numbers are closed under the given operation.

25. Addition
26. Subtraction
27. Division
28. Multiplication

29. Does $x + (3 + 4) = (3 + 4) + x$ illustrate the commutative property or the associative property? Explain your answer.
30. Does $4 + (5 + 6) = 4 + (6 + 5)$ illustrate the commutative property or the associative property? Explain your answer.
31. Give an example to show that the commutative property of multiplication may be true for the negative integers.
32. Give an example to show that the commutative property of addition may be true for the negative integers.
33. Does the commutative property hold for the rational numbers under the operation of division? Give an example to support your answer.
34. Does the commutative property hold for the integers under the operation of subtraction? Give an example to support your answer.
35. Give an example to show that the associative property of multiplication may be true for the negative integers.
36. Give an example to show that the associative property of addition may be true for the negative integers.
37. Does the associative property hold for the integers under the operation of division? Give an example to support your answer.
38. Does the associative property hold for the integers under the operation of subtraction? Give an example to support your answer.
39. Does the associative property hold for the real numbers under the operation of division? Give an example to support your answer.
40. Does $a + (b \cdot c) = (a + b) \cdot (a + c)$? Give an example to support your answer.

In Exercises 41–56, state the name of the property illustrated.

41. $24 + 7 = 7 + 24$
42. $5(x + 3) = 5 \cdot x + 5 \cdot 3$
43. $(7 \cdot 4) \cdot 5 = 7 \cdot (4 \cdot 5)$
44. $v + w = w + v$
45. $(24 + 7) + 3 = 24 + (7 + 3)$
46. $4 \cdot (11 \cdot x) = (4 \cdot 11) \cdot x$
47. $\sqrt{3} \cdot 7 = 7 \cdot \sqrt{3}$
48. $\frac{3}{8} + \left(\frac{1}{8} + \frac{3}{2}\right) = \left(\frac{3}{8} + \frac{1}{8}\right) + \frac{3}{2}$
49. $8 \cdot (7 + \sqrt{2}) = 8 \cdot 7 + 8 \cdot \sqrt{2}$
50. $\sqrt{5} \cdot \frac{2}{3} = \frac{2}{3} \cdot \sqrt{5}$
51. $(1 + 10) + 100 = (10 + 1) + 100$

52. $(r + s) + t = t + (r + s)$
53. $(r + s) \cdot t = (r \cdot t) + (s \cdot t)$
54. $g \cdot (h + i) = (h + i) \cdot g$
55. $(p + q) + (r + s) = (r + s) + (p + q)$
56. $(a \cdot b) + (c \cdot d) = (b \cdot a) + (c \cdot d)$

In Exercises 57–68, use the distributive property to multiply. Then, if possible, simplify the resulting expression.

57. $2(c + 7)$
58. $-3(d - 1)$
59. $\frac{2}{3}(x - 6)$
60. $-\frac{5}{8}(k + 8)$
61. $6\left(\frac{x}{2} + \frac{2}{3}\right)$
62. $24\left(\frac{x}{3} - \frac{1}{8}\right)$
63. $32\left(\frac{1}{16}x - \frac{1}{32}\right)$
64. $15\left(\frac{2}{3}x - \frac{4}{5}\right)$
65. $3(5 - \sqrt{5})$
66. $-7(2 + \sqrt{11})$
67. $\sqrt{2}(\sqrt{2} + \sqrt{3})$
68. $\sqrt{3}(\sqrt{15} + \sqrt{21})$

In Exercises 69–74, name the property used to go from step to step. You only need to name the properties indicated by an a), b), c), or d). For example, in Exercise 69, you need to supply an answer for part a), and an answer to go from the second step of part a) to part b).

69. a) $7(x + 2) + 3 = (7 \cdot x + 7 \cdot 2) + 3$
$= (7x + 14) + 3$
b) $= 7x + (14 + 3)$
$= 7x + 17$

70. a) $3(n + 5) + 6 = (3 \cdot n + 3 \cdot 5) + 6$
$= (3n + 15) + 6$
b) $= 3n + (15 + 6)$
$= 3n + 21$

71. a) $7(k + 1) + 2k = (7 \cdot k + 7 \cdot 1) + 2k$
$= (7k + 7) + 2k$
b) $= 7k + (7 + 2k)$
c) $= 7k + (2k + 7)$
d) $= (7k + 2k) + 7$
$= 9k + 7$

72. a) $11(h + 6) + 5h = (11 \cdot h + 11 \cdot 6) + 5h$
$= (11h + 66) + 5h$
b) $= 11h + (66 + 5h)$
c) $= 11h + (5h + 66)$
d) $= (11h + 5h) + 66$
$= 16h + 66$

73. a) $9 + 2(t + 3) + 7t = 9 + (2 \cdot t + 2 \cdot 3) + 7t$
$= 9 + (2t + 6) + 7t$
b) $= 9 + (6 + 2t) + 7t$
c) $= (9 + 6) + 2t + 7t$
$= 15 + 9t$
d) $= 9t + 15$

74. a) $7 + 5(s + 4) + 3s = 7 + (5 \cdot s + 5 \cdot 4) + 3s$
$= 7 + (5s + 20) + 3s$
b) $= 7 + (20 + 5s) + 3s$
c) $= (7 + 20) + 5s + 3s$
$= 27 + 8s$
d) $= 8s + 27$

In Exercises 75–80, determine whether the activity can be used to illustrate the commutative property. For the property to hold, the end result must be identical, regardless of the order in which the actions are performed.

75. Putting on your seat belt and locking your car door

76. Putting on your left shoe and putting on your right shoe
77. Washing clothes and drying clothes
78. Turning on a computer and typing a term paper on the computer
79. Filling your car with gasoline and washing the windshield
80. Turning on a lamp and reading a book

In Exercises 81–88, determine whether the activity can be used to illustrate the associative property. For the property to hold, doing the first two actions followed by the third would produce the same end result as doing the second and third actions followed by the first.

81. Washing the exterior of your car, vacuuming out the interior, and checking the oil
82. Reading a novel, writing a book report on the novel, and making a presentation to your class about your book report
83. Sending a holiday card to your grandmother, sending one to your parents, and sending one to your teacher
84. Mowing the lawn, trimming the bushes, and removing dead limbs from trees

85. Brushing your teeth, washing your face, and combing your hair

86. Cracking an egg, pouring out the egg, and cooking the egg

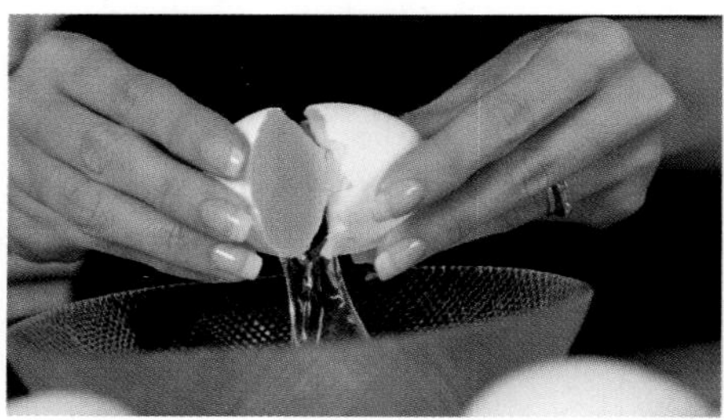

87. Taking a bath, brushing your teeth, and taking your vitamins

88. While making meatloaf, mixing in the milk, mixing in the spices, and mixing in the bread crumbs

Challenge Problems/Group Activities

89. Describe two other activities that can be used to illustrate the commutative property (see Exercises 75–80).

90. Describe three other activities that can be used to illustrate the associative property (see Exercises 81–88).

91. Does $0 \div a = a \div 0$ (assume $a \neq 0$)? Explain.

Recreational Mathematics

92. **a)** Consider the three words *man eating tiger.* Does (*man eating*) *tiger* mean the same as *man* (*eating tiger*)?
    b) Does (*horse riding*) *monkey* mean the same as *horse* (*riding monkey*)?
    c) Can you find three other nonassociative word triples?

Internet/Research Activity

93. A set of numbers that was not discussed in this chapter is the set of *complex numbers.* Write a report on complex numbers. Include their relationship to the real numbers.

5.6 RULES OF EXPONENTS AND SCIENTIFIC NOTATION

An understanding of exponents is important in solving problems in algebra. In the expression 5^2, the 2 is referred to as the *exponent* and the 5 is referred to as the *base*. We read 5^2 as 5 to the second power, or 5 squared, which means

$$5^2 = \underbrace{5 \cdot 5}_{\text{2 factors of 5}}$$

The number 5 to the third power, or 5 cubed, written 5^3, means

$$5^3 = \underbrace{5 \cdot 5 \cdot 5}_{\text{3 factors of 5}}$$

In general, the number b to the nth power, written b^n, means

$$b^n = \underbrace{b \cdot b \cdot b \cdot \cdots \cdot b}_{n \text{ factors of } b}$$

EXAMPLE 1 ***Evaluating the Power of a Number***

Evaluate the following.

a) 4^2 b) $(-5)^2$ c) 5^3 d) 1^{1000} e) 7^1

SOLUTION:

a) $4^2 = 4 \cdot 4 = 16$

b) $(-5)^2 = (-5)(-5) = 25$

c) $5^3 = 5 \cdot 5 \cdot 5 = 125$

d) $1^{1000} = 1$. (The number 1 times itself any number of times equals 1.)

e) $7^1 = 7$. (Any number with an exponent of 1 equals the number itself.) ▲

EXAMPLE 2 *The Importance of Parentheses*

Evaluate the following.

a) $(-2)^4$ b) -2^4 c) $(-2)^5$ d) -2^5

SOLUTION:

a) $(-2)^4 = (-2)(-2)(-2)(-2) = 4(-2)(-2) = -8(-2) = 16$

b) -2^4 means take the opposite of 2^4 or $-1 \cdot 2^4$.
$-1 \cdot 2^4 = -1 \cdot 2 \cdot 2 \cdot 2 \cdot 2 = -1 \cdot 16 = -16$

c) $(-2)^5 = (-2)(-2)(-2)(-2)(-2) = 4(-2)(-2)(-2) = -8(-2)(-2)$
$= 16(-2) = -32$

d) $-2^5 = -1 \cdot 2^5 = -1 \cdot 32 = -32$ ▲

From Example 2, we can see that $(-x)^n \neq -x^n$, where n is an even natural number.

DID YOU KNOW

A Very Large Number

170,141,183,460,469,231,731,687, 303,715,884,105,727 is a very large number. How would you read this number? Take a breath.

170 undecillion, 141 decillion, 183 nonillion, 460 octillion, 469 septillion, 231 sextillion, 731 quintillion, 687 quadrillion, 303 trillion, 715 billion, 884 million, 105 thousand, 727.

Often numbers this large can be represented with an approximation involving scientific notation. This number, however, is a prime number, and its *exact* representation is very important. Mathematicians frequently represent such large numbers by using exponents. Here, another exact, and more efficient, representation of this number is $2^{127} - 1$.

Rules of Exponents

Now that we know how to evaluate powers of numbers we can discuss the rules of exponents. Consider

$$2^2 \cdot 2^3 = \underbrace{2 \cdot 2}_{\text{2 factors}} \cdot \underbrace{2 \cdot 2 \cdot 2}_{\text{3 factors}} = 2^5$$

This example illustrates the product rule for exponents.

Product Rule for Exponents

$$a^m \cdot a^n = a^{m+n}$$

Therefore, by using the product rule, $2^2 \cdot 2^3 = 2^{2+3} = 2^5$.

EXAMPLE 3 *Using the Product Rule for Exponents*

Use the product rule to simplify.

a) $3^4 \cdot 3^5$ b) $7^2 \cdot 7^6$

SOLUTION:

a) $3^4 \cdot 3^5 = 3^{4+5} = 3^9$ b) $7^2 \cdot 7^6 = 7^{2+6} = 7^8$ ▲

Consider

$$\frac{2^5}{2^2} = \frac{2\cdot 2\cdot 2\cdot \cancel{2}\cdot \cancel{2}}{\cancel{2}\cdot \cancel{2}} = 2\cdot 2\cdot 2 = 2^3$$

This example illustrates the quotient rule for exponents.

Quotient Rule for Exponents

$$\frac{a^m}{a^n} = a^{m-n}, \qquad a \neq 0$$

Therefore, $\frac{2^5}{2^2} = 2^{5-2} = 2^3$.

EXAMPLE 4 ***Using the Quotient Rule for Exponents***

Use the quotient rule to simplify.

a) $\frac{5^8}{5^5}$ b) $\frac{8^{12}}{8^5}$

SOLUTION:

a) $\frac{5^8}{5^5} = 5^{8-5} = 5^3$ b) $\frac{8^{12}}{8^5} = 8^{12-5} = 8^7$ ▲

Consider $2^3 \div 2^3$. The quotient rule gives

$$\frac{2^3}{2^3} = 2^{3-3} = 2^0$$

But $\frac{2^3}{2^3} = \frac{8}{8} = 1$. Therefore, 2^0 must equal 1. This example illustrates the zero exponent rule.

Zero Exponent Rule

$$a^0 = 1, \qquad a \neq 0$$

Note that 0^0 is not defined by the zero exponent rule.

EXAMPLE 5 ***The Zero Power***

Use the zero exponent rule to simplify.

a) 2^0 b) $(-2)^0$ c) -2^0 d) $(5x)^0$ e) $5x^0$

SOLUTION:

a) $2^0 = 1$ b) $(-2)^0 = 1$ c) $-2^0 = -1\cdot 2^0 = -1\cdot 1 = -1$

d) $(5x)^0 = 1$ e) $5x^0 = 5\cdot x^0 = 5\cdot 1 = 5$ ▲

Consider $2^3 \div 2^5$. The quotient rule yields

$$\frac{2^3}{2^5} = 2^{3-5} = 2^{-2}$$

But $\frac{2^3}{2^5} = \frac{\cancel{2}\cdot\cancel{2}\cdot\cancel{2}}{\cancel{2}\cdot\cancel{2}\cdot\cancel{2}\cdot 2\cdot 2} = \frac{1}{2^2}$. Since $\frac{2^3}{2^5}$ equals both 2^{-2} and $\frac{1}{2^2}$, then 2^{-2} must equal $\frac{1}{2^2}$. This example illustrates the negative exponent rule.

Negative Exponent Rule

$$a^{-m} = \frac{1}{a^m}, \qquad a \neq 0$$

EXAMPLE 6 *Using the Negative Exponent Rule*

Use the negative exponent rule to simplify.

a) 5^{-2} b) 8^{-1}

SOLUTION:

a) $5^{-2} = \frac{1}{5^2} = \frac{1}{25}$ b) $8^{-1} = \frac{1}{8^1} = \frac{1}{8}$ ▲

Consider $(2^3)^2$.

$$(2^3)^2 = (2^3)(2^3) = 2^{3+3} = 2^6$$

This example illustrates the power rule for exponents.

Power Rule for Exponents

$$(a^m)^n = a^{m\cdot n}$$

Thus, $(2^3)^2 = 2^{3\cdot 2} = 2^6$.

EXAMPLE 7 *Evaluating a Power Raised to Another Power*

Use the power rule to simplify.

a) $(5^4)^3$ b) $(7^2)^5$

SOLUTION:

a) $(5^4)^3 = 5^{4\cdot 3} = 5^{12}$ b) $(7^2)^5 = 7^{2\cdot 5} = 7^{10}$ ▲

DID YOU KNOW

Large and Small Numbers

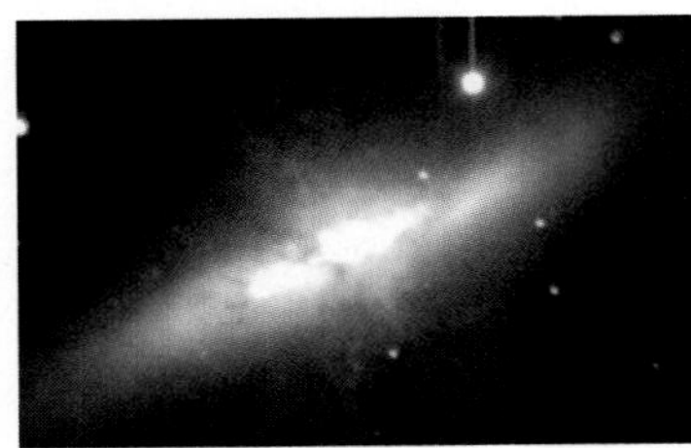

Diameter of a galaxy may be 1×10^5 light-years

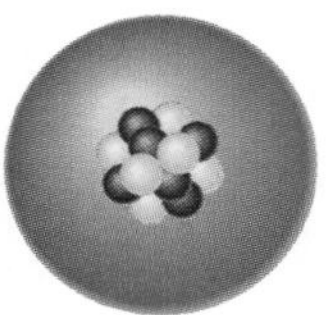

Diameter of an atom may be 1×10^{-10} meter

Our everyday activities don't require us to deal with quantities much above those in the thousands: \$6.95 for lunch, 100 meters to a lap, a \$15,000 car loan, and so on. Yet as modern technology has developed, so has our ability to study all aspects of the universe we live in, from the very large to the very small. Modern technology can be used with the rules of exponents and scientific notation to study everything from the diameter of a galaxy to the diameter of an atom.

Summary of the Rules of Exponents

$a^m \cdot a^n = a^{m+n}$	Product rule for exponents
$\dfrac{a^m}{a^n} = a^{m-n}, \quad a \neq 0$	Quotient rule for exponents
$a^0 = 1, \quad a \neq 0$	Zero exponent rule
$a^{-m} = \dfrac{1}{a^m}, \quad a \neq 0$	Negative exponent rule
$(a^m)^n = a^{m \cdot n}$	Power rule for exponents

Scientific Notation

Often scientific problems deal with very large and very small numbers. For example, the distance from Earth to the sun is about 93,000,000 miles. The wavelength of a yellow color of light is about 0.0000006 meter. Because working with many zeros is difficult, scientists developed a notation that expresses such numbers with exponents. For example, consider the distance from Earth to the sun, 93,000,000 miles.

$$\begin{aligned} 93{,}000{,}000 &= 9.3 \times 10{,}000{,}000 \\ &= 9.3 \times 10^7 \end{aligned}$$

The wavelength of a yellow color of light is about 0.0000006 meter.

$$\begin{aligned} 0.0000006 &= 6.0 \times 0.0000001 \\ &= 6.0 \times 10^{-7} \end{aligned}$$

The numbers 9.3×10^7 and 6.0×10^{-7} are written in a form called *scientific notation*. Each number written in scientific notation is written as a number greater than or equal to 1 and less than 10 multiplied by some power of 10.

Some examples of numbers in scientific notation are

$$3.7 \times 10^3, \qquad 2.05 \times 10^{-3}, \qquad 5.6 \times 10^8, \qquad \text{and} \qquad 1.00 \times 10^{-5}$$

The following is a procedure for writing a number in scientific notation.

To Write a Number in Scientific Notation

1. Move the decimal point in the original number to the right or left until you obtain a number greater than or equal to 1 and less than 10.
2. Count the number of places you have moved the decimal point to obtain the number in step 1. If the decimal point was moved to the left, the count is to be considered positive. If the decimal point was moved to the right, the count is to be considered negative.
3. Multiply the number obtained in step 1 by 10 raised to the count found in step 2. (Note that the count determined in step 2 is the exponent on the base 10.)

DID YOU KNOW

Catalan's Conjecture

Eugène Catalan Preda Mihailescu

Consider all the squares of positive integers:

$1^2 = 1, \quad 2^2 = 4, \quad 3^2 = 9,$
$4^2 = 16,$ and so on

Now consider all the cubes of positive integers:

$1^3 = 1, \quad 2^3 = 8, \quad 3^3 = 27,$
$4^3 = 64,$ and so on

Next, consider all the fourth powers of positive integers:

$1^4 = 1, \quad 2^4 = 16, \quad 3^4 = 81,$
$4^4 = 256,$ and so on

And so on. If we were to put *all* powers of the positive integers into one set, the set would begin as follows: $\{1, 4, 8, 9, 16, 25, 27, 32, 36, 49, 64, \ldots\}$. In 1844, Belgian mathematician Eugène Catalan proposed that of all the numbers in this infinite set, only 8 and 9 are consecutive integers. Although this conjecture was easily stated, the formal proof of it eluded mathematicians for over 150 years. In April 2002, Preda Mihailescu announced that he had proved the Catalan conjecture.

EXAMPLE 8 *Converting from Decimal Notation to Scientific Notation*

Write each number in scientific notation.

a) In 2002, the population of the United States was about 288,000,000.
b) In 2002, the population of China was about 1,283,000,000.
c) In 2002, the population of the world was about 6,251,000,000.
d) The diameter of a hydrogen atom nucleus is about 0.0000000000011 millimeter.
e) The wavelength of an x-ray is about 0.000000000492 meter.

SOLUTION

a) $288{,}000{,}000 = 2.88 \times 10^8$
b) $1{,}283{,}000{,}000 = 1.283 \times 10^9$
c) $6{,}251{,}000{,}000 = 6.251 \times 10^9$
d) $0.0000000000011 = 1.1 \times 10^{-12}$
e) $0.000000000492 = 4.92 \times 10^{-10}$ ▲

To convert from a number given in scientific notation to decimal notation we reverse the procedure.

To Change a Number in Scientific Notation to Decimal Notation

1. Observe the exponent on the 10.
2. a) If the exponent is positive, move the decimal point in the number to the right the same number of places as the exponent. Adding zeros to the number might be necessary.
   b) If the exponent is negative, move the decimal point in the number to the left the same number of places as the exponent. Adding zeros might be necessary.

EXAMPLE 9 *Converting from Scientific Notation to Decimal Notation*

Write each number in decimal notation.

a) The average distance from Earth to the sun is about 9.3×10^7 miles.
b) The half-life of uranium 235 is about 4.5×10^9 years.
c) The average grain size in siltstone is 1.35×10^{-3} inch.
d) A *millimicron* is a unit of measure used for very small distances. One millimicron is about 3.94×10^{-8} inch.

SOLUTION:

a) $9.3 \times 10^7 = 93{,}000{,}000$
b) $4.5 \times 10^9 = 4{,}500{,}000{,}000$
c) $1.35 \times 10^{-3} = 0.00135$
d) $3.94 \times 10^{-8} = 0.0000000394$ ▲

In scientific journals and books, we occasionally see numbers like 10^{15} and 10^{-6}. We interpret these numbers as 1×10^{15} and 1×10^{-6}, respectively, when converting the numbers to decimal form.

EXAMPLE 10 ***Multiplying Numbers in Scientific Notation***

Multiply $(2.1 \times 10^5)(9 \times 10^{-3})$. Write the answer in scientific notation and in decimal notation.

SOLUTION:

$$\begin{aligned}(2.1 \times 10^5)(9 \times 10^{-3}) &= (2.1 \times 9)(10^5 \times 10^{-3}) \\ &= 18.9 \times 10^2 \\ &= 1.89 \times 10^3 \quad \text{Scientific notation} \\ &= 1.890 \quad \text{Decimal notation}\end{aligned}$$

▲

EXAMPLE 11 ***Dividing Numbers Using Scientific Notation***

Divide $\frac{0.000000000048}{24{,}000{,}000{,}000}$. Write the answer in scientific notation.

SOLUTION: First write each number in scientific notation.

$$\begin{aligned}\frac{0.000000000048}{24{,}000{,}000{,}000} = \frac{4.8 \times 10^{-11}}{2.4 \times 10^{10}} &= \left(\frac{4.8}{2.4}\right)\left(\frac{10^{-11}}{10^{10}}\right) \\ &= 2.0 \times 10^{-11-10} \\ &= 2.0 \times 10^{-21}\end{aligned}$$

▲

Scientific Notation on the Scientific Calculator

One of the advantages of using scientific notation when working with very large and very small numbers is the ease with which you can perform operations. Performing these operations is even easier with the use of a scientific calculator. Most scientific calculators have a scientific notation key labeled "Exp," "EXP," or "EE." We will refer to the scientific notation key as [EXP]. The following keystrokes can be used to enter the number 4.3×10^6

Keystroke(s)	**Calculator display**
4.3	4.3
[EXP]	4.3^{00}
6	4.3^{06}

Your calculator may have some slight variations to the display shown here. The display 4.3^{06} means 4.3×10^6. We now will use our calculators to perform some computations using scientific notation.

EXAMPLE 12 ***Use Scientific Notation on a Calculator to Find a Product***

Multiply $(4.3 \times 10^6)(2 \times 10^{-4})$ using a scientific calculator. Write the answer in decimal notation.

DID YOU KNOW

What's the Difference Between Debt and Deficit?

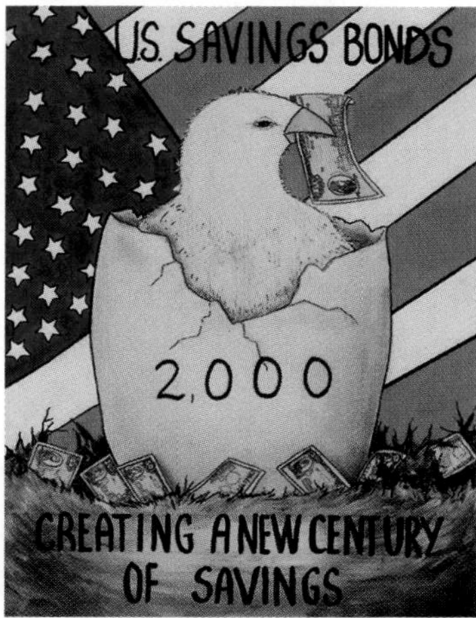

We often hear economists and politicians talk about such things as revenue, expenditures, deficit, surplus, and national debt. Revenue is the money the government collects annually, mostly through taxes. Expenditures are the money the government spends annually. If revenue exceeds expenditures, a surplus occurs; if expenditures exceed revenue, a deficit occurs. The national debt is the total of all the budget deficits and (the few) surpluses encountered by the federal government for over 200 years. How big is the national debt? Search the Internet for the latest figures. As of July 30, 2003, the national debt was \$6,737,879,490,363.90. This amount is owed to investors worldwide—perhaps including you—who own U.S. government bonds.

SOLUTION: Our sequence of keystrokes is as follows:

Keystroke(s)	Display	
4.3	4.3	
EXP	4.3^{00}	
6	4.3^{06}	
×	4300000	4.3×10^6 is now entered in the calculator. Most calculators will convert to decimal notation (if the display can show the number)
2	2.	
EXP	$2.^{00}$	
4	$2.^{04}$	Enter the positive form of the exponent
+/−	$2.^{-04}$	Make the exponent negative*
=	860.	Press = to obtain the answer of 860†

▲

EXAMPLE 13 *Use Scientific Notation on a Calculator to Find a Quotient*

Divide $\dfrac{0.000000000048}{24{,}000{,}000{,}000}$ using a scientific calculator. Write the answer in scientific notation.

SOLUTION: We first rewrite the numerator and denominator using scientific notation. See Example 11.

$$\frac{0.000000000048}{24{,}000{,}000{,}000} = \frac{4.8 \times 10^{-11}}{2.4 \times 10^{10}}$$

Next we use a scientific calculator to perform the computation. The keystrokes are as follows

4.8 [EXP] 11 [+/−] [÷] 2.4 [EXP] 10 [=]

The display on the calculator is $2.^{-21}$, which means 2.0×10^{-21}. ▲

EXAMPLE 14 *U.S. Debt per Person*

On July 30, 2003, the U.S. Department of the Treasury estimated the U.S. federal debt to be about \$6.74 trillion. On this same day, the U.S. Bureau of the Census estimated the U.S. population to be about 292 million people. Determine the average debt, per person, by dividing the U.S. federal debt by the U.S. population.

---

*Some calculators will require you to enter the negative sign before entering the exponent.

†Some calculators will display the answer in scientific notation. In this case, the display will show $8.6^{\ 02}$, which means 8.6×10^2 and equals 860.

SOLUTION: First, we will write the numbers involved using decimal notation and then convert them to scientific notation.

$$6.74 \text{ trillion} = 6{,}740{,}000{,}000{,}000 = 6.74 \times 10^{12}$$
$$292 \text{ million} = 292{,}000{,}000 = 2.92 \times 10^{8}$$

Now we will divide 6.74×10^{12} by 2.92×10^{8} using a scientific calculator. The keystrokes are

6.74 [EXP] 12 [÷] 2.92 [EXP] 8 [=]

The display shows 23,082.1918. This number indicates that on July 30, 2003, the U.S. government owed about $23,082.19 per man, woman, and child living in the United States.

SECTION 5.6 EXERCISES

Concept/Writing Exercises

1. In the expression 2^3, what is the name given to the 2, and what is the name given to the 3?
2. Explain the meaning of b^n.
3. a) Explain the product rule for exponents.
   b) Use the product rule to simplify $2^3 \cdot 2^4$.
4. a) Explain the quotient rule for exponents.
   b) Use the quotient rule to simplify $\frac{5^6}{5^4}$.
5. a) Explain the zero exponent rule.
   b) Use the zero exponent rule to simplify 7^0.
6. a) Explain the negative exponent rule.
   b) Use the negative exponent rule to simplify 2^{-3}.
7. a) Explain the power rule for exponents.
   b) Use the power rule to simplify $(3^2)^4$.
8. Explain how you can simplify the expression 1^{500}.
9. Explain how you can simplify the following expressions and then simplify the expression.
   a) -1^{500}
   b) $(-1)^{500}$
   c) -1^{501}
   d) $(-1)^{501}$
10. a) In your own words, explain how to change a number in decimal notation to scientific notation.
    b) Using the procedure in part (a), change 0.000426 to scientific notation.
11. a) In your own words, explain how to change a number in scientific notation to decimal notation.
    b) Using the procedure in part (a), change 5.76×10^{-4} to decimal notation.
12. A number is given in scientific notation. What does it indicate about the number when the exponent on the 10 is (a) positive, (b) zero, and (c) negative?

Practice the Skills

In Exercises 13–44, evaluate the expression.

13. 5^2
14. 3^4
15. $(-2)^4$
16. -2^4
17. -3^2
18. $(-3)^2$
19. $\left(\frac{2}{3}\right)^2$
20. $\left(-\frac{7}{8}\right)^2$
21. $(-5)^2$
22. -5^2
23. $2^3 \cdot 3^2$
24. $\frac{15^2}{3^2}$
25. $\frac{5^7}{5^5}$
26. $3^3 \cdot 3^4$
27. $\frac{7}{7^3}$
28. $3^4 \cdot 7^0$
29. $(-13)^0$
30. $(-3)^4$
31. 3^4
32. -3^4
33. 3^{-2}
34. 3^{-3}
35. $(2^3)^4$
36. $(1^{12})^{13}$
37. $\frac{11^{25}}{11^{23}}$
38. $5^2 \cdot 5$
39. $(-4)^2$
40. 4^{-2}
41. -4^2
42. $(4^3)^2$
43. $(2^2)^{-3}$
44. $3^{-3} \cdot 3$

In Exercises 45–60, express the number in scientific notation.

45. 231,000
46. 297,000,000
47. 15
48. 0.000034
49. 0.56
50. 0.00467

51. 19,000 **52.** 1,260,000,000 **53.** 0.000186
54. 0.0003 **55.** 0.00000423 **56.** 54,000
57. 711 **58.** 0.02 **59.** 0.153
60. 416,000

In Exercises 61–76, express the number in decimal notation.

61. 2.3×10^3 **62.** 4.78×10^5 **63.** 3.901×10^{-3}
64. 1.764×10^7 **65.** 8.62×10^{-5} **66.** 2.19×10^{-4}
67. 3.12×10^{-1} **68.** 4.6×10^1 **69.** 9×10^6
70. 7.3×10^4 **71.** 2.31×10^2 **72.** 1.04×10^{-2}
73. 3.5×10^4 **74.** 2.17×10^{-6} **75.** 1×10^4
76. 1×10^{-3}

In Exercises 77–86, (a) perform the indicated operation without the use of a calculator and express each answer in decimal notation. (b) Confirm your answer from part (a) by using a scientific calculator to perform the operations. If the calculator displays the answer in scientific notation, convert the answer to decimal notation.

77. $(2 \times 10^3)(4 \times 10^2)$ **78.** $(4.1 \times 10^{-3})(2 \times 10^3)$
79. $(5.1 \times 10^1)(3 \times 10^{-4})$
80. $(1.6 \times 10^{-2})(4 \times 10^{-3})$
81. $\dfrac{6.4 \times 10^5}{2 \times 10^3}$ **82.** $\dfrac{8 \times 10^{-3}}{2 \times 10^1}$
83. $\dfrac{8.4 \times 10^{-6}}{4 \times 10^{-3}}$ **84.** $\dfrac{25 \times 10^3}{5 \times 10^{-2}}$
85. $\dfrac{4 \times 10^5}{2 \times 10^4}$ **86.** $\dfrac{16 \times 10^3}{8 \times 10^{-3}}$

In Exercises 87–96, (a) perform the indicated operation without the use of a calculator and express each answer in scientific notation. (b) Confirm your answer from part (a) by using a scientific calculator to perform the operations. If the calculator displays the answer in decimal notation, convert the answer to scientific notation.

87. (300,000)(2,000,000) **88.** (0.000041)(3000)
89. (0.003)(0.00015) **90.** (230,000)(3000)
91. $\dfrac{1{,}400{,}000}{700}$ **92.** $\dfrac{20{,}000}{0.0005}$
93. $\dfrac{0.00004}{200}$ **94.** $\dfrac{0.0012}{0.000006}$
95. $\dfrac{150{,}000}{0.0005}$ **96.** $\dfrac{24{,}000}{8{,}000{,}000}$

Problem Solving

In Exercises 97–100, list the numbers from smallest to largest.

97. 5.8×10^5; 3.2×10^{-1}; 4.6; 8.3×10^{-4}
98. 8.5×10^{-5}; 8.2×10^3; 1.3×10^{-1}; 6.2×10^4
99. 40,000; 4.1×10^3; 0.00079; 8.3×10^{-5}
100. 267,000,000; 3.14×10^7; 1,962,000; 4.79×10^6

In Exercises 101–107, express your answer (a) using decimal notation and (b) using scientific notation. You may use a scientific calculator to perform the necessary operations.

101. *Gross Domestic Product* The gross domestic product (GDP) of a country is the total national output of goods and services produced within that country. In 2001, the GDP of the United States was about $10.1432 trillion and the U.S. population was about 285 million people. Determine the U.S. GDP per person by dividing the GDP by the population.

102. *Japan's GDP* In 2001, the GDP (see Exercise 101) of Japan was about $4.1468 trillion and the population of Japan was about 127 million people. Determine Japan's GDP per person by dividing the GDP by the population.

103. *Computer Speed* On April 20, 2002, the NEC Earth Simulator computer—located in Yokohama, Japan—broke the world's record for being the fastest supercomputer. This computer is capable of performing 36.6 trillion calculations per second. At this rate, how long would it take to perform a task requiring 7.69×10^{33} calculations?

104. *World Population* According to the U.S. Bureau of the Census, the population of the world in 2002 was approximately 6.251×10^9 and the population of China was about 1.283×10^9. How many people lived outside China?

105. *Traveling to Jupiter* The distance from Earth to the planet Jupiter is approximately 4.5×10^8 mi. If a spacecraft traveled at a speed of 25,000 mph, how many hours would the spacecraft need to travel from Earth to Jupiter? Use distance = rate × time.

106. *Traveling to the Moon* The distance from Earth to the moon is approximately 239,000 mi. If a spacecraft travels at a speed of 20,000 mph, how many hours would the spacecraft need to travel from Earth to the moon? Use distance = rate × time.

107. *Bucket Full of Molecules* A drop of water contains about 40 billion molecules. If a bucket has half a million drops of water in it, how many molecules of water are in the bucket?

108. *Blood Cells in a Cubic Millimeter* If a cubic millimeter of blood contains 5,800,000 red blood cells, how many red blood cells are contained in 50 cubic millimeters of blood?

109. *Radioactive Isotopes* The half-life of a radioactive isotope is the time required for half the quantity of the isotope to decompose. The half-life of uranium 238 is 4.5×10^9 years, and the half-life of uranium 234 is 2.5×10^5 years. How many times greater is the half-life of uranium 238 than uranium 234?

110. *1950 Niagara Treaty* The 1950 Niagara Treaty between the United States and Canada requires that during the tourist season a minimum of 100,000 cubic feet of water per second (ft^3/sec) flow over Niagara Falls (another 130,000–160,000 ft^3/sec are diverted for power generation). Find the minimum amount of water that will flow over the falls in a 24-hour period during the tourist season.

111. *U.S. Debt per Person: 1993 versus 2003* In Example 14, the U.S. government debt was discussed. It was found that the debt in 2003 was \$23,082.19 per person. In 1993, the U.S. government debt was about \$4.41 trillion, and the population of the United States was about 258 million people.

a) Determine the amount of debt per person in the United States in 1993.
b) How much more per person did the U.S. government owe in 2003 than in 1993?

112. *Disposable Diaper Quantity* Laid end to end, the 18 billion disposable diapers thrown away in the United States each year would reach the moon and back seven times.

a) Write 18 billion in scientific notation.
b) If the distance from Earth to the moon is 2.38×10^5 miles, what is the length of all these diapers placed end to end? Write the answer in decimal notation.

113. *Mutual Fund Manager* Lauri Mackey is the fund manager for the Mackey Mutual Fund. This mutual fund has total assets of \$1.2 billion. Lauri wants to maintain the investments in this fund according to the following pie chart.

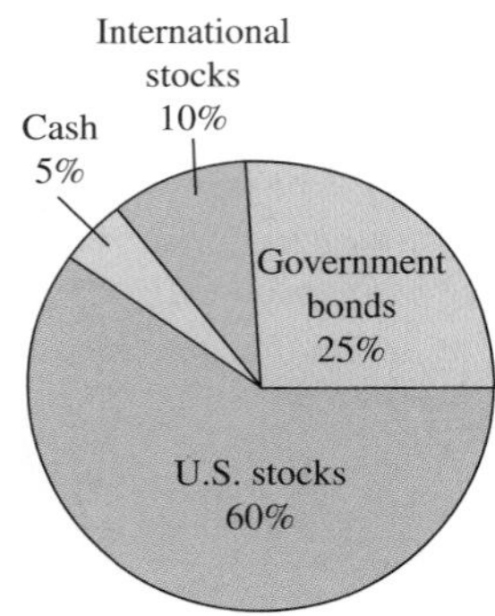

Source: Ibbotson Associates

a) How much of the \$1.2 billion should be invested in U.S. stocks?
b) How much should be invested in government bonds?
c) How much should be invested in international stocks?
d) How much should remain in cash?

114. *Another Mutual Fund Manager* Susan Dratch is the fund manager for the Dratch Mutual Fund. This mutual fund has total assets of \$3.4 billion. Susan wants to maintain the investments in this fund according to the following pie chart.

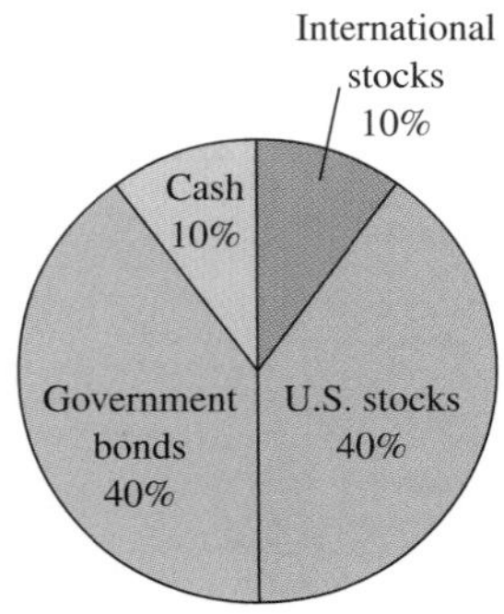

Source: Ibbotson Associates

a) How much of the \$3.4 billion should be invested in U.S. stocks?

b) How much should be invested in government bonds?
c) How much should be invested in international stocks?
d) How much should remain in cash?

115. *Metric System Comparison* In the metric system, 1 meter $= 10^3$ millimeters. How many times greater is a meter than a millimeter? Explain how you determined your answer.

116. In the metric system, 1 gram $= 10^3$ millimeters and 1 gram $= 10^{-3}$ kilogram. What is the relationship between milligrams and kilograms? Explain how you determined your answer.

117. *Earth to Sun Comparison* The mass of the sun is approximately 2×10^{30} kilograms, and the mass of Earth is approximately 6×10^{24} kilograms. How many times greater is the mass of the sun than the mass of Earth? Write your answer in decimal notation.

118. *The Day of Six Billion* The United Nations declared October 12, 1999, the *Day of Six Billion*. On this day, Earth's population was estimated to reach 6 billion. Currently, Earth's population is doubling about every 35 years.
a) Using this figure, estimate the world's population in the year 2034.
b) Assuming 365 days in a year, estimate the average number of additional people added to Earth's population each day between 1999 and 2034.

119. *Computer Calculation Speed* The IBM Blue Pacific computer is capable of operating at a peak speed of about 3.9 trillion (3,900,000,000,000) calculations per second. At this rate, how long would it take to perform a task requiring 897 quadrillion (897,000,000,000,000,000) calculations?

Challenge Problems/Group Activities

120. *Comparing a Million to a Billion* Many people have no idea of the difference in size between a million (1,000,000), a billion (1,000,000,000), and a trillion (1,000,000,000,000).
a) Write a million, a billion, and a trillion in scientific notation.
b) Determine how long it would take to spend a million dollars if you spent \$1000 a day.
c) Repeat part (b) for a billion dollars.
d) Repeat part (b) for a trillion dollars.
e) How many times greater is a billion dollars than a million dollars?

121. *Speed of Light*
a) Light travels at a speed of 1.86×10^5 mi/sec. A *light-year* is the distance that light travels in 1 year. Determine the number of miles in a light year.
b) Earth is approximately 93,000,000 mi from the sun. How long does it take light from the sun to reach Earth?

122. *Bacteria in a Culture* The exponential function $E(t) = 2^{10} \cdot 2^t$ approximates the number of bacteria in a certain culture after t hours.
a) The initial number of bacteria is determined when $t = 0$. What is the initial number of bacteria?
b) How many bacteria are there after $\frac{1}{2}$ hour?

Internet/Research Activities

123. John Allen Paulos of Temple University has written many entertaining books about mathematics for nonmathematicians. Included among these are *Mathematics and Humor* (1980); *I Think, Therefore I Laugh* (1985); *Innumeracy—Mathematical Illiteracy and Its Consequences* (1989); *Beyond Numeracy—Ruminations of a Numbers Man* (1991); *A Mathematician Reads the Newspaper* (1995); *Once Upon a Number* (1998); *A Mathematician Plays the Stock Market* (2003); and *Music and Gender* (2003). Read one of Paulos's books and write a 500-word report on it.

John Allen Paulos

124. Obtain data from the U.S. Department of the Treasury and from the U.S. Bureau of the Census Internet web sites to calculate the current U.S. government debt per person. Write a report in which you compare your figure with those obtained in Exercise 111 and Example 14. Include in your report definitions of the following terms: revenues, expenditures, deficit, and surplus.

125. Find an article in a newspaper or magazine that contains scientific notation. Write a paragraph explaining how scientific notation was used. Attach a copy of the article to your report.

5.7 ARITHMETIC AND GEOMETRIC SEQUENCES

Now that you can recognize the various sets of real numbers and know how to add, subtract, multiply, and divide real numbers, we can discuss sequences. A *sequence* is a list of numbers that are related to each other by a rule. The numbers that form the sequence are called its *terms*. If your salary increases or decreases by a fixed amount over a period of time, the listing of the amounts, over time, would form an arithmetic sequence. When interest in a savings account is compounded at regular intervals, the listing of the amounts in the account over time will be a geometric sequence.

Arithmetic Sequences

A sequence in which each term after the first term differs from the preceding term by a constant amount is called an *arithmetic sequence*. The amount by which each pair of successive terms differs is called the *common difference*, d. The common difference can be found by subtracting any term from the term that directly follows it.

Examples of arithmetic sequences	Common differences
$1, 5, 9, 13, 17, \ldots$	$d = 5 - 1 = 4$
$-7, -5, -3, -1, 1, \ldots$	$d = -5 - (-7) = -5 + 7 = 2$
$\frac{5}{2}, \frac{3}{2}, \frac{1}{2}, -\frac{1}{2}, \ldots$	$d = \frac{3}{2} - \frac{5}{2} = -\frac{2}{2} = -1$

EXAMPLE 1 *The First Five Terms of an Arithmetic Sequence*

Write the first five terms of the arithmetic sequence with first term 5 and a common difference of 4.

SOLUTION: The first term is 5. The second term is $5 + 4$ or 9. The third term is $9 + 4$ or 13. The fourth term is $13 + 4$ or 17. The fifth term is $17 + 4$ or 21. Thus, the first five terms of the sequence are 5, 9, 13, 17, 21. ▲

EXAMPLE 2 *An Arithmetic Sequence with a Negative Difference*

Write the first five terms of the arithmetic sequence with first term 10 and common difference of -3.

SOLUTION: The sequence is

$$10, 7, 4, 1, -2$$ ▲

When discussing a sequence, we often represent the first term as a_1 (read "a sub 1"), the second term as a_2, the fifteenth term as a_{15}, and so on. We use the notation a_n to represent the general or nth term of a sequence. Thus a sequence may be symbolized as

$$a_1, a_2, a_3, a_4, \ldots, a_n, \ldots$$

For example, in the sequence 2, 5, 8, 11, 14, ..., we have

$$a_1 = 2, a_2 = 5, a_3 = 8, a_4 = 11, a_5 = 14, \ldots.$$

PROFILE IN MATHEMATICS

CARL FRIEDRICH GAUSS

Carl Friedrich Gauss (1777–1855), often called the "Prince of Mathematicians," made significant contributions to the fields of algebra, geometry, and number theory. Gauss was only 22 years old when he proved the fundamental theorem of algebra for his doctoral dissertation.

When Gauss was only 10, his mathematics teacher gave him the problem of finding the sum of the first 100 natural numbers, thinking that this would keep him busy for a while. Gauss recognized a pattern in the sequence of numbers when he considered the sum of the following numbers.

$$\begin{array}{l} 1 + \ \ 2 + \ \ 3 + \cdots + \ 99 + 100 \\ \underline{100 + \ 99 + \ 98 + \cdots + \ \ 2 + \ \ \ 1} \\ 101 + 101 + 101 + \cdots + 101 + 101 \end{array}$$

He had the required answer in no time at all. When he added, he had one hundred 101's. Therefore, the sum is $\frac{1}{2}(100)(101) = 5050$.

When we know the first term of an arithmetic sequence and the common difference, we can use the following formula to find the value of any specific term.

General or nth Term of an Arithmetic Sequence

$$a_n = a_1 + (n - 1)d$$

EXAMPLE 3 *Finding the Seventh Term of an Arithmetic Sequence*

Find the seventh term of the arithmetic sequence whose first term is 3 and whose common difference is −6.

SOLUTION: To find the seventh term, or a_7, replace n in the formula with 7, a_1 with 3, and d with −6.

$$\begin{aligned} a_n &= a_1 + (n - 1)d \\ a_7 &= 3 + (7 - 1)(-6) \\ &= 3 + (6)(-6) \\ &= 3 - 36 \\ &= -33 \end{aligned}$$

The seventh term is −33. As a check, we have listed the first seven terms of the sequence: 3, −3, −9, −15, −21, −27, −33. ▲

EXAMPLE 4 *Finding the nth Term of an Arithmetic Sequence*

Write an expression for the general or nth term, a_n, for the sequence 1, 6, 11, 16, ...

SOLUTION: In this sequence, the first term a_1, is 1, and the common difference, d, is 5. We substitute these values into $a_n = a_1 + (n - 1)d$ to obtain an expression for the nth term, a_n.

$$\begin{aligned} a_n &= a_1 + (n - 1)d \\ &= 1 + (n - 1)5 \\ &= 1 + 5n - 5 \\ &= 5n - 4 \end{aligned}$$

Note that when $n = 1$, the first term is $5(1) - 4 = 1$. When $n = 2$, the second term is $5(2) - 4 = 6$, and so on. ▲

We can use the following formula to find the sum of the first n terms in an arithmetic sequence.

Sum of the First n Terms in an Arithmetic Sequence

$$s_n = \frac{n(a_1 + a_n)}{2}$$

In this formula, s_n represents the sum of the first n terms, a_1 is the first term, a_n is the nth term, and n is the number of terms in the sequence from a_1 to a_n.

EXAMPLE 5 ***Finding the Sum of a Sequence***

Find the sum of the first 25 natural numbers.

SOLUTION: The sequence we are discussing is

$$1, 2, 3, 4, 5, \ldots, 25$$

In this sequence, $a_1 = 1$, $a_{25} = 25$, and $n = 25$. Thus, the sum of the first 25 terms is

$$s_n = \frac{n(a_1 + a_n)}{2}$$
$$s_{25} = \frac{25(1 + 25)}{2}$$
$$= \frac{25(26)}{2} = 325$$

Thus, the sum of the terms $1 + 2 + 3 + 4 + \cdots + 25$ is 325. ▲

DID YOU KNOW

A Heavenly Sequence

The sequence 0, 3, 6, 12, 24, 48, 96, 192, . . . , which resembles a geometric sequence, is known collectively as the **Titius–Bode law**. The sequence, discovered in 1766 by the two German astronomers, was of great importance to astronomy in the eighteenth and nineteenth centuries. When 4 was added to each term and the result was divided by 10, the sequence closely corresponded with the observed mean distance from the sun to the known principal planets of the solar system (in astronomical units, or au): Mercury (0.4), Venus (0.7), Earth (1.0), Mars (1.6), missing planet (2.8), Jupiter (5.2), Saturn (10.0), missing planet (19.6). The exciting discovery of Uranus in 1781 with a mean distance of 19.2 astronomical units was in near agreement with the Titius–Bode law and stimulated the search for an undiscovered planet at a predicted 2.8 astronomical units. This effort led to the discovery of Ceres and other members of the asteroid belt. Although the Titius–Bode law broke down after discoveries of Neptune (30.1 au) and Pluto (39.5 au), many scientists still believe that other applications of the Titius–Bode law will emerge in the future.

Geometric Sequences

The next type of sequence we will discuss is the geometric sequence. A *geometric sequence* is one in which the ratio of any term to the term that directly precedes it is a constant. This constant is called the *common ratio*. The common ratio, r, can be found by taking any term except the first and dividing that term by the preceding term.

Examples of geometric sequences	Common ratios
$2, 4, 8, 16, 32, \ldots$	$r = 4 \div 2 = 2$
$-3, 6, -12, 24, -48, \ldots$	$r = 6 \div (-3) = -2$
$\frac{2}{3}, \frac{2}{9}, \frac{2}{27}, \frac{2}{81}, \ldots$	$r = \frac{2}{9} \div \frac{2}{3} = \left(\frac{2}{9}\right)\left(\frac{3}{2}\right) = \frac{1}{3}$

To construct a geometric sequence when the first term, a_1, and common ratio are known, multiply the first term by the common ratio to get the second term. Then multiply the second term by the common ratio to get the third term, and so on.

EXAMPLE 6 ***The First Five Terms of a Geometric Sequence***

Write the first five terms of the geometric sequence whose first term, a_1, is 3 and whose common ratio, r, is 4.

SOLUTION: The first term is 3. The second term, found by multiplying the first term by 4, is $3 \cdot 4$ or 12. The third term is $12 \cdot 4$ or 48. The fourth term is $48 \cdot 4$ or 192. The fifth term is $192 \cdot 4$ or 768. Thus, the first five terms of the sequence are 3, 12, 48, 192, 768. ▲

When we know the first term of a geometric sequence and the common ratio, we can use the following formula to find the value of the general or nth term, a_n.

General or nth Term of a Geometric Sequence

$$a_n = a_1 r^{n-1}$$

EXAMPLE 7 *Finding the Seventh Term of a Geometric Sequence*

Find the seventh term of the geometric sequence whose first term is -3 and whose common ratio is -2.

SOLUTION: In this sequence, $a_1 = -3$, $r = -2$, and $n = 7$. Substituting the values, we obtain

$$\begin{aligned} a_n &= a_1 r^{n-1} \\ a_7 &= -3(-2)^{7-1} \\ &= -3(-2)^6 \\ &= -3(64) \\ &= -192 \end{aligned}$$

As a check, we have listed the first seven terms of the sequence: $-3, 6, -12, 24, -48, 96, -192$.

EXAMPLE 8 *Finding the nth Term of a Geometric Sequence*

Write an expression for the general or nth term, a_n, of the sequence 2, 6, 18, 54, ...

SOLUTION: In this sequence, $a_1 = 2$ and $r = 3$. We substitute these values into $a_n = a_1 r^{n-1}$ to obtain an expression for the nth term, a_n.

$$\begin{aligned} a_n &= a_1 r^{n-1} \\ &= 2(3)^{n-1} \end{aligned}$$

Note than when $n = 1$, $a_1 = 2(3)^0 = 2(1) = 2$. When $n = 2$, $a_2 = 2(3)^1 = 6$, and so on.

We can use the following formula to find the sum of the first n terms of a geometric sequence.

Sum of the First n Terms of a Geometric Sequence

$$s_n = \frac{a_1(1 - r^n)}{1 - r}, \qquad r \neq 1$$

EXAMPLE 9 *Adding the First n Terms of a Geometric Sequence*

Find the sum of the first five terms in the geometric sequence whose first term is 4 and whose common ratio is 2.

SOLUTION: In this sequence, $a_1 = 4$, $r = 2$, and $n = 5$. Substituting these values into the formula, we get

$$s_n = \frac{a_1(1 - r^n)}{1 - r}$$

$$s_5 = \frac{4[1 - (2)^5]}{1 - 2}$$

$$= \frac{4(1 - 32)}{-1}$$

$$= \frac{4(-31)}{-1} = \frac{-124}{-1} = 124$$

The sum of the first five terms of the sequence is 124. The first five terms of the sequence are 4, 8, 16, 32, 64. If you add these five numbers, you will obtain the sum 124.

EXAMPLE 10 *Pounds and Pounds of Silver*

As a reward for saving his kingdom from a band of thieves, a king offered a knight one of two options. The knight's first option was to be paid 100,000 pounds of silver all at once. The second option was to be paid over the course of a month. On the first day, he would receive one pound of silver. On the second day, he would receive two pounds of silver. On the third day, he would receive four pounds of silver, and so on, each day receiving double the amount given on the previous day. Assuming the month is 30 days, which option would pay the knight more silver?

SOLUTION: The first option pays the knight 100,000 pounds of silver. The second option pays according to the geometric sequence 1, 2, 4, 8, 16, ... In this sequence, $a_1 = 1$, $r = 2$, and $n = 30$. The sum of this sequence can be found by substituting these values into the formula to obtain

$$s_n = \frac{a_1(1 - r^n)}{1 - r}$$

$$s_{30} = \frac{1(1 - 2^{30})}{1 - 2}$$

$$= \frac{1 - 1{,}073{,}741{,}824}{-1}$$

$$= \frac{-1{,}073{,}741{,}823}{-1}$$

$$= 1{,}073{,}741{,}823$$

Thus, the knight would get paid 1,073,741,823 pounds of silver with the second option. The second option pays 1,073,641,823 more pounds of silver than the first option.

SECTION 5.7 EXERCISES

Concept/Writing Exercises

1. State the definition of *sequence* and give an example.
2. What are the numbers that make up a sequence called?
3. a) State the definition of *arithmetic sequence* and give an example.
   b) State the definition of *geometric sequence* and give an example.
4. a) In the arithmetic sequence 2, 5, 8, 11, 14, . . . , state the common difference, d.
   b) In the geometric sequence 3, 6, 12, 24, 48, . . . , state the common ratio, r.
5. For an arithmetic sequence, state the meaning of each of the following symbols.
   a) a_n b) a_1 c) d d) s_n
6. For a geometric sequence, state the meaning of each of the following symbols.
   a) a_n b) a_1 c) r d) s_n

Practice the Skills

In Exercises 7–14, write the first five terms of the arithmetic sequence with the first term, a_1, and common difference, d.

7. $a_1 = 3, d = 2$
8. $a_1 = 1, d = 3$
9. $a_1 = -5, d = 3$
10. $a_1 = -11, d = 5$
11. $a_1 = 5, d = -2$
12. $a_1 = -3, d = -4$
13. $a_1 = \frac{1}{2}, d = \frac{1}{2}$
14. $a_1 = \frac{5}{2}, d = -\frac{3}{2}$

In Exercises 15–22, find the indicated term for the arithmetic sequence with the first term, a_1, and common difference, d.

15. Find a_6 when $a_1 = 2, d = 3$.
16. Find a_9 when $a_1 = 3$ and $d = -2$.
17. Find a_{10} when $a_1 = -5, d = 2$.
18. Find a_{12} when $a_1 = 7, d = -3$.
19. Find a_{20} when $a_1 = \frac{4}{5}, d = -1$.
20. Find a_{15} when $a_1 = -\frac{1}{2}, d = -2$.
21. Find a_{11} when $a_1 = 4, d = \frac{1}{2}$.
22. Find a_{15} when $a_1 = \frac{4}{3}, d = \frac{1}{3}$.

In Exercises 23–30, write an expression for the general or nth term, a_n, for the arithmetic sequence.

23. 1, 2, 3, 4, . . .
24. 1, 3, 5, 7, . . .
25. 2, 4, 6, 8, . . .
26. 3, 1, −1, −3, . . .
27. $-\frac{5}{3}, -\frac{4}{3}, -1, -\frac{2}{3}, \ldots$
28. −15, −10, −5, 0, . . .
29. $-3, -\frac{3}{2}, 0, \frac{3}{2}, \ldots$
30. −5, −2, 1, 4, . . .

In Exercises 31–38, find the sum of the terms of the arithmetic sequence. The number of terms, n, is given.

31. $1, 2, 3, 4, \ldots, 50; n = 50$
32. $2, 4, 6, 8, \ldots, 100; n = 50$
33. $1, 3, 5, 7, \ldots, 99; n = 50$
34. $-4, -7, -10, -13, \ldots, -28; n = 9$
35. $11, 6, 1, -4, \ldots, -24; n = 8$
36. $-9, -\frac{17}{2}, -8, -\frac{15}{2}, \ldots, -\frac{1}{2}; n = 18$
37. $\frac{1}{2}, \frac{5}{2}, \frac{9}{2}, \frac{13}{2}, \ldots, \frac{29}{2}; n = 8$
38. $\frac{3}{5}, \frac{4}{5}, 1, \frac{6}{5}, \ldots, 4; n = 18$

In Exercises 39–46, write the first five terms of the geometric sequence with the first term, a_1, and common ratio, r.

39. $a_1 = 3, r = 2$
40. $a_1 = 6, r = 3$
41. $a_1 = 2, r = -2$
42. $a_1 = 8, r = \frac{1}{2}$
43. $a_1 = -3, r = -1$
44. $a_1 = -6, r = -2$
45. $a_1 = -16, r = -\frac{1}{2}$
46. $a_1 = 5, r = \frac{3}{5}$

In Exercises 47–54, find the indicated term for the geometric sequence with the first term, a_1, and common ratio, r.

47. Find a_6 when $a_1 = 3, r = 4$.
48. Find a_5 when $a_1 = 2, r = 2$.
49. Find a_3 when $a_1 = 3, r = \frac{1}{2}$.
50. Find a_7 when $a_1 = -3, r = -3$.
51. Find a_5 when $a_1 = \frac{1}{2}, r = 2$.
52. Find a_{25} when $a_1 = 1, r = 2$.
53. Find a_{10} when $a_1 = -2, r = 3$.
54. Find a_{18} when $a_1 = -5, r = -2$.

In Exercises 55–62, write an expression for the general or nth term, a_n, for the geometric sequence.

55. 1, 2, 4, 8, . . .
56. 3, 6, 12, 24, . . .
57. 3, −3, 3, −3, . . .
58. −16, −8, −4, −2, . . .
59. $\frac{1}{4}, \frac{1}{2}, 1, 2, \ldots$
60. −3, 6, −12, 24, . . .
61. $9, 3, 1, \frac{1}{3}, \frac{1}{9}, \ldots$
62. $-4, -\frac{8}{3}, -\frac{16}{9}, -\frac{32}{27}, \ldots$

In Exercises 63–70, find the sum of the first n terms of the geometric sequence for the values of a_1 and r.

63. $n = 4, a_1 = 3, r = 2$
64. $n = 5, a_1 = 2, r = 3$
65. $n = 7, a_1 = 5, r = 4$

66. $n = 9, a_1 = -3, r = 5$

67. $n = 11, a_1 = -7, r = 3$

68. $n = 15, a_1 = -1, r = 2$

69. $n = 15, a_1 = -1, r = -2$

70. $n = 10, a_1 = 512, r = \frac{1}{2}$

Problem Solving

71. Find the sum of the first 100 natural numbers.

72. Find the sum of the first 100 even natural numbers.

73. Find the sum of the first 100 odd natural numbers.

74. Find the sum of the first 50 multiples of 3.

75. *Annual Pay Raises* Rita Fernandez is given a starting salary of $20,200 and promised a $1200 raise per year after each of the next eight years.
   a) Determine her salary during her eighth year of work.
   b) Determine the total salary she received over the 8 years.

76. *Pendulum Movement* Each swing of a pendulum (from far left to far right) is 3 in. shorter than the preceding swing. The first swing is 8 ft.
   a) Find the length of the twelfth swing.
   b) Determine the total distance traveled by the pendulum during the first 12 swings.

77. *A Bouncing Ball* Each time a ball bounces, the height attained by the ball is 6 in. less than the previous height attained. If on the first bounce the ball reaches a height of 6 ft, find the height attained on the eleventh bounce.

78. *Clock Strikes* A clock strikes once at 1 o'clock, twice at 2 o'clock, and so on. How many times does it strike over a 12 hr period?

79. *Squirrels and Pinecones* A tree squirrel cut down 1 pinecone on the first day of October, 2 pinecones on the second day, 3 pinecones on the third day, and so on. How many pinecones did this squirrel cut down during the month of October, which contains 31 days?

80. *Enrollment Increase* The enrollment at Loras College in 2001 was 8000 students. If the enrollment increases by 8% per year, determine the enrollment 10 years later.

81. *Decomposing Substance* A certain substance decomposes and loses 20% of its weight each hour. If there are originally 200 g of the substance, how much remains after 6 hr?

82. *Samurai Sword Construction* While making a traditional Japanese samurai sword, the master sword maker prepares the blade by heating a bar of iron until it is white hot. He then folds it over and pounds it smooth. Therefore, after each folding, the number of layers of steel is doubled. Assuming the sword maker starts with a bar of one layer and folds it 15 times, how many layers of steel will the finished sword contain?

83. *Salary Increase* If your salary were to increase at a rate of 6% per year, find your salary during your 15th year if your original salary is $20,000.

84. *A Bouncing Ball* When dropped, a ball rebounds to four-fifths of its original height. How high will the ball rebound after the fourth bounce if it is dropped from a height of 30 ft?

85. *Value of a Stock* Ten years ago, Nancy Hart purchased $2,000 worth of shares in RCF, Inc. Since then, the price of the stock has roughly tripled every two years. Approximately how much are Nancy's shares worth today?

86. *A Baseball Game* During a baseball game, the visiting team scored 1 run in the first inning, 2 runs in the second inning, 3 runs in the third inning, 4 runs in the fourth inning, and so on. The home team scored 1 run in the first inning, 2 runs in the second inning, 4 runs in the third inning, 8 runs in the fourth inning, and so on. What is the score of the game after eight innings?

Inning	1	2	3	4	5	6	7	8	9
Visitors	1	2	3	4					
Home	1	2	4	8					

Challenge Problems/Group Activities

87. A geometric sequence has $a_1 = 82$ and $r = \frac{1}{2}$; find s_6.

88. *Sums of Interior Angles* The sums of the interior angles of a triangle, a quadrilateral, a pentagon, and a sextagon are 180°, 360°, 540°, and 720°, respectively. Use this pattern to

find a formula for the general term, a_n, where a_n represents the sum of the interior angles of an n-sided quadrilateral.

89. *Divisibility by 6* Determine how many numbers between 7 and 1610 are divisible by 6.

90. Find r and a_1 for the geometric sequence with $a_2 = 24$ and $a_5 = 648$.

91. *Total Distance Traveled by a Bouncing Ball* A ball is dropped from a height of 30 ft. On each bounce it attains a height four-fifths of its original height (or of the previous bounce). Find the total vertical distance traveled by the ball after it has completed its fifth bounce (therefore has hit the ground six times).

Recreational Mathematics

92. *A Wagering Strategy* The following is a strategy used by some people involved in games of chance. A player begins by betting a standard bet, say \$1. If the player wins, the player again bets \$1 in the next round. If the player loses, the player bets \$2 in the next round. Next, if the player wins, the player again bets \$1; if the player loses, the player now bets \$4 in the next round. The process continues as long as the player keeps playing, betting \$1 after a win or doubling the previous bet after a loss.

a) Assume a player is using a \$1 standard bet and loses five times in a row. How much money should the player bet in the sixth round? How much money has the player lost at the end of the fifth round?

b) Assume a player is using a \$10 standard bet and loses five times in a row. How much money should the player bet in the sixth round? How much money has the player lost at the end of the fifth round?

c) Assume a player is using a \$1 standard bet and loses 10 times in a row. How much money should the player bet in the 11th round? How much money has the player lost at the end of the 10th round?

d) Assume a player is using a \$10 standard bet and loses 10 times in a row. How much money should the player bet in the 11th round? How much money has the player lost at the end of the 10th round?

e) Why is this a dangerous strategy?

Internet/Research Activity

93. A topic generally associated with sequences is *series.*

a) Research *series* and explain what a series is and how it differs from a sequence. Also write a formal definition of series. Give examples of different kinds of series.

b) Write the arithmetic series associated with the arithmetic sequence 1, 4, 7, 10, 13, ...

c) Write the geometric series associated with the geometric sequence 3, 6, 12, 24, 48, ...

d) What is an infinite geometric series?

e) Find the sum of the terms of the infinite geometric series $1 + \frac{1}{2} + \frac{1}{4} + \frac{1}{8} + \frac{1}{16} + \cdots$.

5.8 FIBONACCI SEQUENCE

Our discussion of sequences would not be complete without mentioning a sequence known as the ***Fibonacci sequence***. The sequence is named after Leonardo of Pisa, also known as Fibonacci. He was one of the most distinguished mathematicians of the Middle Ages. This sequence is first mentioned in his book *Liber Abacci* (Book of the Abacus), which contained many interesting problems, such as: "A certain man put a pair of rabbits in a place surrounded on all sides by a wall. How many pairs of rabbits can be produced from that pair in a year if it is assumed that every month each pair begets a new pair which from the second month becomes productive?"

The solution to this problem (Fig. 5.10 on page 276) led to the development of the sequence that bears its author's name: the Fibonacci sequence. The sequence is shown in Table 5.1 on page 276. The numbers in the columns titled *Pairs of Adults* form the Fibonacci sequence.

PROFILE IN MATHEMATICS

FIBONACCI

Leonardo of Pisa (1170–1250) is considered one of the most distinguished mathematicians of the Middle Ages. He was born in Italy and was sent by his father to study mathematics with an Arab master. When he began writing, he referred to himself as Fibonacci, or "son of Bonacci," the name by which he is known today. In addition to the famous sequence bearing his name, Fibonacci is also credited with introducing the Hindu–Arabic number system into Europe. His 1202 book, *Liber Abacci* (Book of the Abacus), explained the use of this number system and emphasized the importance of the number zero.

TABLE 5.1

Month	Pairs of Adults	Pairs of Babies	Total Pairs
1	1	0	1
2	1	1	2
3	2	1	3
4	3	2	5
5	5	3	8
6	8	5	13
7	13	8	21
8	21	13	34
9	34	21	55
10	55	34	89
11	89	55	144
12	144	89	233

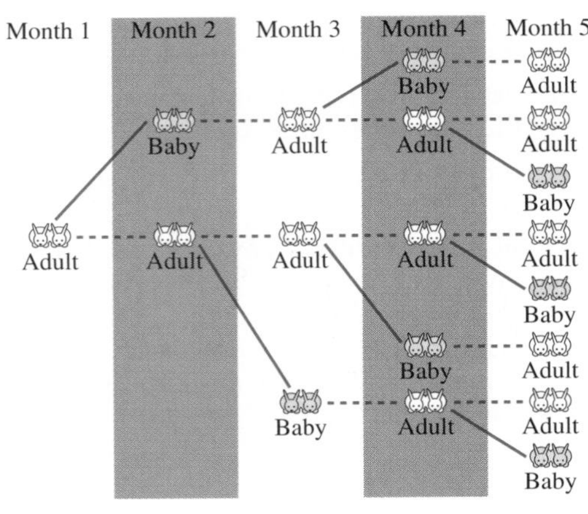

Figure 5.10

Fibonacci Sequence

1, 1, 2, 3, 5, 8, 13, 21, . . .

In the Fibonacci sequence, the first and second terms are 1. The sum of these two terms is the third term. The sum of the second and third terms is the fourth term, and so on.

In the middle of the nineteenth century, mathematicians made a serious study of this sequence and found strong similarities between it and many natural phenomena. Fibonacci numbers appear in the seed arrangement of many species of plants and in the petal counts of various flowers. For example, when the flowering head of the sunflower matures to seed, the seeds' spiral arrangement becomes clearly visible. A typical count of these spirals may give 89 steeply curving to the right, 55 curving more shallowly to the left, and 34 again shallowly to the right. The largest known specimen to be examined had spiral counts of 144 right, 89 left, and 55 right. These numbers, like the other three mentioned, are consecutive terms of the Fibonacci sequence.

The head of a sunflower

On the heads of many flowers, petals surrounding the central disk generally yield a Fibonacci number. For example, some daisies contain 21 petals, and others contain 34, 55, or 89 petals. (People who use a daisy to play the "love me, love me not" game will likely pluck 21, 34, 55, or 89 petals before arriving at an answer.)

Fibonacci numbers are also observed in the structure of pinecones and pineapples. The tablike or scalelike structures called bracts that make up the main body of the pinecone form a set of spirals that start from the cone's attachment to the branch. Two sets of oppositely directed spirals can be observed, one steep and the other more gradual. A count on the steep spiral will reveal a Fibonacci number, and a count on the gradual one will be the adjacent smaller Fibonacci number, or if not, the next smaller

TABLE 5.2

Numbers	Ratio
1, 1	$\frac{1}{1} = 1$
1, 2	$\frac{2}{1} = 2$
2, 3	$\frac{3}{2} = 1.5$
3, 5	$\frac{5}{3} = 1.666\ldots$
5, 8	$\frac{8}{5} = 1.6$
8, 13	$\frac{13}{8} = 1.625$
13, 21	$\frac{21}{13} \approx 1.615$
21, 34	$\frac{34}{21} \approx 1.619$
34, 55	$\frac{55}{34} \approx 1.618$
55, 89	$\frac{89}{55} \approx 1.618$

Fibonacci number. One investigation of 4290 pinecones from 10 species of pine trees found in California revealed that only 74 cones, or 1.7%, deviated from this Fibonacci pattern.

Like pinecone bracts, pineapple scales are patterned into spirals, and because they are roughly hexagonal in shape, three distinct sets of spirals can be counted.

Fibonacci Numbers and Divine Proportions

In 1753, while studying the Fibonacci sequence, Robert Simson, a mathematician at the University of Glasgow, noticed that when he took the ratio of any term to the term that immediately preceded it, the value he obtained remained in the vicinity of one specific number. To illustrate this, we indicate in Table 5.2 the ratio of various pairs of sequential Fibonacci numbers.

The ratio of the 50th term to the 49th term is 1.6180. Simson proved that the ratio of the $(n + 1)$ term to the nth term as n gets larger and larger is the irrational number $(\sqrt{5} + 1)/2$, which begins 1.61803. . . . This number was already well known to mathematicians at that time as the *golden number*.

Many years earlier, the Bavarian astronomer and mathematician Johannes Kepler wrote that for him the golden number symbolized the Creator's intention "to create like from like." The golden number $(\sqrt{5} + 1)/2$ is frequently referred to as "phi," symbolized by the Greek letter Φ.

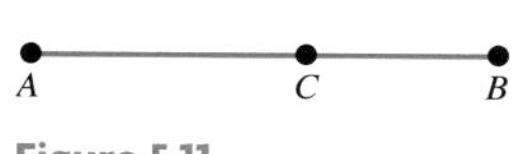

Figure 5.11

The Great Pyramid of Gizeh

The ancient Greeks, in about the sixth century B.C., sought unifying principles of beauty and perfection, which they believed could be described by using mathematics. In their study of beauty, the Greeks used the term *golden ratio.* To understand the golden ratio, let's consider the line segment *AB* in Fig. 5.11. When this line segment is divided at a point *C*, such that the ratio of the whole, *AB*, to the larger part, *AC*, is equal to the ratio of the larger part, *AC*, to the smaller part, *CB*, then each ratio *AB/AC* and *AC/CB* is referred to as a *golden ratio*. The proportion they form, $AB/AC = AC/CB$, is called the *golden proportion*. Furthermore, each ratio in the proportion will have a value equal to the golden number, $(\sqrt{5} + 1)/2$.

$$\frac{AB}{AC} = \frac{AC}{CB} = \frac{\sqrt{5} + 1}{2} \approx 1.618$$

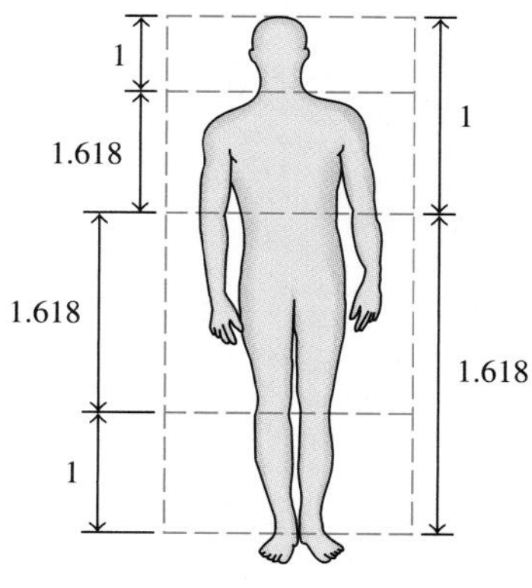

Figure 5.12

The Great Pyramid of Gizeh in Egypt, built about 2600 B.C., is the earliest known example of use of the golden ratio in architecture. The ratio of any of its sides of the square base (775.75 ft) to its altitude (481.4 ft) is about 1.611. Other evidence of the use of the golden ratio appears in other Egyptian buildings and tombs.

In medieval times, people referred to the golden proportion as the *divine proportion*, reflecting their belief in its relationship to the will of God.

The twentieth-century architect Le Corbusier developed a scale of proportions for the human body that he called the Modulor (Fig. 5.12). Note that the navel separates the entire body into golden proportions, as does the neck and knee.

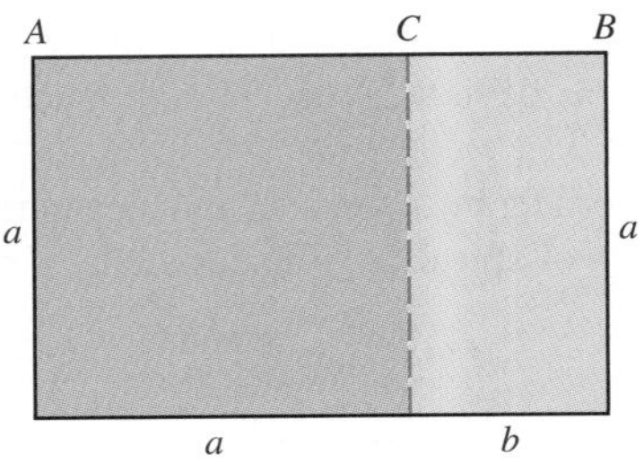

Figure 5.13

From the golden proportion, the *golden rectangle* can be formed, as shown in Fig. 5.13.

$$\frac{\text{Length}}{\text{Width}} = \frac{a+b}{a} = \frac{a}{b} = \frac{\sqrt{5}+1}{2}$$

Note that when a square is cut off one end of a golden rectangle, as in Fig. 5.13, the remaining rectangle has the same properties as the original golden rectangle (creating "like from like" as Johannes Kepler had written) and is therefore itself a golden rectangle. Interestingly, the curve derived from a succession of diminishing golden rectangles, as shown in Fig. 5.14, is the same as the spiral curve of the chambered nautilus. The same curve appears on the horns of rams and some other animals. It is the same curve that is observed in the plant structures mentioned earlier—sunflowers, other flower heads, pinecones, and pineapples. You will recall that Fibonacci numbers were observed in each of these plant structures. The curve shown in Fig. 5.14 closely approximates what mathematicians call a *logarithmic spiral.*

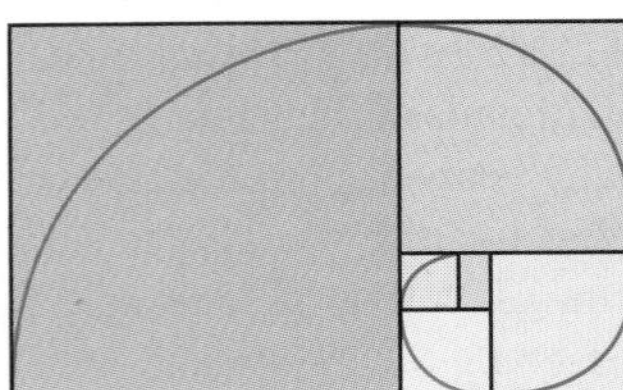

Figure 5.14

The Parthenon

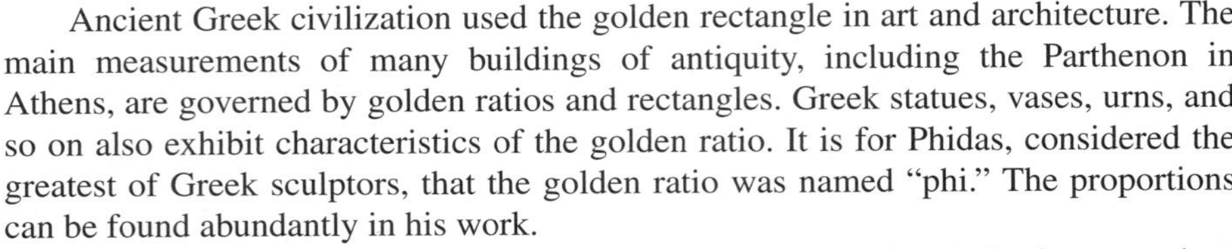

Ancient Greek civilization used the golden rectangle in art and architecture. The main measurements of many buildings of antiquity, including the Parthenon in Athens, are governed by golden ratios and rectangles. Greek statues, vases, urns, and so on also exhibit characteristics of the golden ratio. It is for Phidas, considered the greatest of Greek sculptors, that the golden ratio was named "phi." The proportions can be found abundantly in his work.

The proportions of the golden rectangle can be found in the work of many artists, from the old masters to the moderns. For example, the golden rectangle can be seen in the painting *Invitation to the Sideshow (La Parade de Cirque),* 1887, by George Seurat, a French neoimpressionist artist.

Fibonacci's Garden by Caryl Bryer Fallert (see page 279)

Invitation to the Sideshow (La Parade de Cirque), 1887, by George Seurat

DID YOU KNOW

Fibonacci and the Male Bee's Ancestors

The most frequent example given to introduce the Fibonacci sequence involves rabbits producing offspring, two at a time. Although this makes for a nice introduction to Fibonacci sequences, it is not at all realistic. A much better example comes from the breeding practices of bees. Female or worker bees are produced when the queen bee mates with a male bee. Male bees are produced from the queen's unfertilized eggs. In essence, then, female bees have two parents, whereas male bees only have one parent. The family tree of a male bee would look like this:

F M F M F F M F
M F F M F
F M F
M F
F
(M)

From this tree, we can see that the **1** male bee (circled) has **1** parent, **2** grandparents, **3** great-grandparents, **5** great-great-grandparents, **8** great-great-great-grandparents, and so on. We see the Fibonacci sequence as we move back through the male bees' generations.

In addition to using the golden rectangle in art, several artists have used Fibonacci numbers in art. One contemporary example is the 1995 work by Caryl Bryer Fallert called *Fibonacci's Garden* (see page 278). This artwork is a quilt constructed from two separate fabrics that are put together in a pattern based on the Fibonacci sequence.

Fibonacci numbers are also found in another form of art, namely music. Perhaps the most obvious link between Fibonacci numbers and music can be found on the piano keyboard. An octave (Fig. 5.15) on a keyboard has 13 keys: 8 white keys and 5 black keys (the 5 black keys are in one group of 2 and one group of 3).

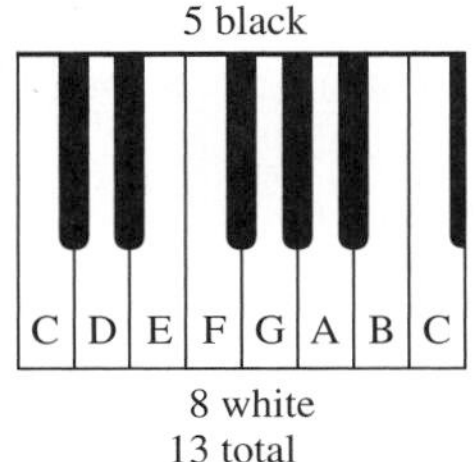

Figure 5.15

In Western music, the most complete scale, the chromatic scale, consists of 13 notes (from C to the next higher C). Its predecessor, the diatonic scale, contains 8 notes (an octave). The diatonic scale was preceded by a 5-note pentatonic scale (*penta* is Greek for "five"). Each number is a Fibonacci number.

The visual arts deal with what is pleasing to the eye, whereas musical composition deals with what is pleasing to the ear. Whereas art achieves some of its goals by using division of planes and area, music achieves some of its goals by a similar division of time, using notes of various duration and spacing. The musical intervals considered by many to be the most pleasing to the ear are the major sixth and minor sixth. A major sixth, for example, consists of the note C, vibrating at about 264* vibrations per second, and note A, vibrating at about 440 vibrations per second. The ratio of 440 to 264 reduces to 5 to 3, or $\frac{5}{3}$, a ratio of two consecutive Fibonacci numbers. An example of a minor sixth is E (about 330 vibrations per second) and C (about 528 vibrations per second). The ratio 528 to 330 reduces to 8 to 5, or $\frac{8}{5}$, the next ratio of two consecutive Fibonacci numbers. The vibrations of any sixth interval reduce to a similar ratio.

Patterns that can be expressed mathematically in terms of Fibonacci relationships have been found in Gregorian chants and works of many composers, including Bach, Beethoven, and Bartók. A number of twentieth-century musical works, including Ernst Krenek's *Fibonacci Mobile*, have been deliberately structured by using Fibonacci proportions.

A number of studies have tried to explain why the Fibonacci sequence and related items are linked to so many real-life situations. It appears that the Fibonacci numbers are a part of natural harmony that is pleasing to both the eye and the ear. In the nineteenth century, German physicist and psychologist Gustav Fechner tried to determine which dimensions were most pleasing to the eye. Fechner, along with psychologist Wilhelm Wundt, found that most people do unconsciously favor golden dimensions when purchasing greeting cards, mirrors, and other rectangular objects. This discovery has been widely used by commercial manufacturers in their packaging and labeling designs, by retailers in their store displays, and in other areas of business and advertising.

*Frequencies of notes vary in different parts of the world and change over time.

SECTION 5.8 EXERCISES

Concept/Writing Exercises

1. Explain how to construct the Fibonacci sequence.
2. a) Write out the first ten terms of the Fibonacci sequence.
   b) Divide the ninth term by the eighth term, rounding to the nearest thousandth.
   c) Divide the tenth term by the ninth term, rounding to the nearest thousandth.
   d) Try a few more divisions and then make a conjecture about the result.
3. a) What is the value of the golden number?
   b) What is the golden ratio?
   c) What is the golden proportion?
   d) What is the golden rectangle?
4. In your own words, explain the relationship between the golden number, golden ratio, golden proportion, and golden rectangle.
5. Describe three examples of where the golden ratio can be found
   a) in nature.
   b) in manufactured items.
6. Describe three examples of where Fibonacci numbers can be found
   a) in nature.
   b) in manufactured items.

Practice the Skills/Problem Solving

7. a) To what decimal value is $(\sqrt{5} + 1)/2$ aproximately equal?
   b) To what decimal value is $(\sqrt{5} - 1)/2$ approximately equal?
   c) By how much do the results in parts (a) and (b) differ?
8. The eleventh Fibonacci number is 89. Examine the first six digits in the decimal expression of its reciprocal, $\frac{1}{89}$. What do you find?
9. Find the ratio of the second to the first term of the Fibonacci sequence. Then find the ratio of the third to the second term of the sequence and determine whether this ratio was an increase or decrease from the first ratio. Continue this process for 10 ratios and then make a conjecture regarding the increasing or decreasing values in consecutive ratios.
10. A musical composition is described as follows. Explain why this piece is based on the golden ratio.

Entire Composition

34 measures	55 measures	21 measures	34 measures
Theme	Fast, Loud	Slow	Repeat of theme

11. The greatest common factor of any two consecutive Fibonacci numbers is 1. Show this is true for the first 15 Fibonacci numbers.
12. The sum of any 10 consecutive Fibonacci numbers is always divisible by 11. Select any 10 consecutive Fibonacci numbers and show that for your selection this is true.
13. Twice any Fibonacci number minus the next Fibonacci number equals the second number preceding the original number. Select a number in the Fibonacci sequence and show that this pattern holds for the number selected.
14. For any four consecutive Fibonacci numbers, the difference of the squares of the middle two numbers equals the product of the smallest and largest numbers. Select four consecutive Fibonacci numbers and show that this pattern holds for the numbers you selected.
15. Determine the ratio of the length to width of various photographs and compare these ratios to Φ.
16. Determine the ratio of the length to the width of a 6 inch by 4 inch standard index card, and compare the ratio to Φ.
17. Determine the ratio of the length to width of several picture frames and compare these ratios to Φ.
18. Determine the ratio of the length to the width of your television screen and compare this ratio to Φ.
19. Determine the ratio of the length to width of a desktop in your classroom and compare this ratio to Φ.
20. Determine the ratio of the length to the width of this textbook and compare this ratio to Φ.
21. Determine the ratio of the length to the width of a computer screen and compare this ratio to Φ.
22. Find three physical objects whose dimensions are very close to a golden rectangle.
    a) List the articles and record the dimensions.
    b) Compute the ratios of their lengths to their widths.
    c) Find the difference between the golden ratio and the ratio you obtain in part (b)—to the nearest tenth—for each object.

In Exercises 23–30, determine whether the sequence is a Fibonacci-type sequence (each term is the sum of the two preceding terms). If it is, determine the next two terms of the sequence.

23. 1, 3, 4, 7, 11, 18, . . .
24. 1, 1, 2, 2, 3, 3, . . .
25. 1, 4, 9, 16, 25, 36, . . .
26. $-1, 1, 0, 1, 1, 2, \ldots$
27. 5, 10, 15, 25, 40, 65, . . .
28. $\frac{1}{4}, \frac{1}{4}, \frac{1}{2}, \frac{3}{4}, 1\frac{1}{4}, 2, \ldots$
29. $-5, 3, -2, 1, -1, 0, \ldots$
30. $-4, 5, 1, 6, 7, 13, \ldots$
31. a) Select any two nonzero digits and add them to obtain a third digit. Continue adding the two previous terms to get a Fibonacci-type sequence.

b) Form ratios of successive terms to show how they will eventually approach the golden number.

32. Repeat Exercise 31 for two different nonzero numbers.

33. **a)** Select any three consecutive terms of a Fibonacci sequence. Subtract the product of the terms on each side of the middle term from the square of the middle term. What is the difference?

b) Repeat part (a) with three different consecutive terms of the sequence.

c) Make a conjecture about what will happen when you repeat this process for any three consecutive terms of a Fibonacci sequence.

34. *Pascal's Triangle* One of the most famous number patterns involves *Pascal's triangle*. The Fibonacci sequence can be found by using Pascal's triangle. Can you explain how that can be done? A hint is shown.

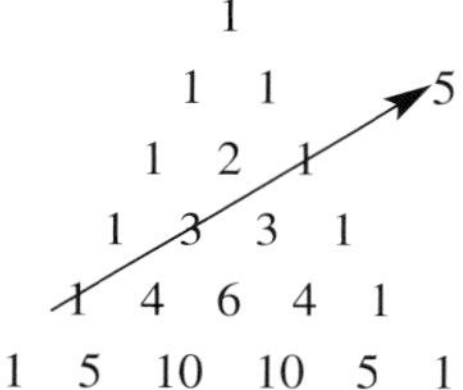

35. *Lucas Sequence* **a)** A sequence related to the Fibonacci sequence is the *Lucas sequence*. The Lucas sequence is formed in a manner similar to the Fibonacci sequence. The first two numbers of the Lucas sequence are 1 and 3. Write the first eight terms of the Lucas sequence.

b) Complete the next two lines of the following chart.

$$1 + 2 = 3$$
$$1 + 3 = 4$$
$$2 + 5 = 7$$
$$3 + 8 = 11$$
$$5 + 13 = 18$$

c) What do you observe about the first column in the chart in part (b)?

Challenge Problems/Group Activities

36. *Fibonacci-Type Sequence* The following sequence represents a Fibonacci-type sequence (each term is the sum of the two preceding terms). Here x represents any natural number from 1 to 10:

$$-10, x, -10 + x, -10 + 2x, -20 + 3x, -30 + 5x, \ldots$$

For example, if $x = 2$, the first 10 terms of the sequence would be $-10, 2, -8, -6, -14, -20, -34, -54, -88, -142$.

Write out the first 10 terms of this Fibonacci-type sequence for x equal to

a) 4. **b)** 5. **c)** 6. **d)** 7. **e)** 8.

f) For values of x of 4, 5, and 6, you should have found that each term after the seventh term in the sequence was a negative number. For values of x of 7 and 8, you should have found that each term after the seventh term in the sequence was a positive number. Do you believe that for any value of x greater than or equal to 7, each term after the seventh term of the sequence will always be a positive number? Explain.

37. The divine proportion is $(a + b)/a = a/b$ (see Fig. 5.13), which can be written $1 + (b/a) = a/b$. Now let $x = a/b$, which gives $1 + (1/x) = x$. Multiply both sides of this equation by x to get a quadratic equation and then use the quadratic formula (Section 6.8) to show that one answer is $x = (1 + \sqrt{5})/2$ (the golden ratio).

38. Draw a line of length 5 in. Determine and mark the point on the line that will create the golden ratio. Explain how you determined your answer.

39. *Pythagorean Triples* A Pythagorean triple is a set of three whole numbers, $\{a, b, c\}$, such that $a^2 + b^2 = c^2$. For example, since $6^2 + 8^2 = (10)^2$, $\{6, 8, 10\}$ is a Pythagorean triple. The following steps show how to find Pythagorean triples using any four consecutive Fibonacci numbers. Here we will demonstrate the process with the Fibonacci numbers 3, 5, 8, and 13.

1. Determine the product of 2 and the two inner Fibonacci numbers. We have $2(5)(8) = 80$, which is the first number in the Pythagorean triple. So $a = 80$.
2. Determine the product of the two outer numbers. We have $3(13) = 39$, which is the second number in the Pythagorean triple. So $b = 39$.
3. Determine the sum of the squares of the inner two numbers. We have $5^2 + 8^2 = 25 + 64 = 89$, which is the third number in the Pythagorean triple. So $c = 89$.

This process has produced the Pythagorean triple, $\{80, 39, 89\}$. To verify,

$$(80)^2 + (39)^2 = (89)^2$$
$$6400 + 1521 = 7921$$
$$7921 = 7921$$

Use this process to produce four other Pythagorean triples.

40. *Reflections* When two panes of glass are placed face to face, four interior reflective surfaces exist labeled 1, 2, 3, and 4. If light is not reflected, it has just one path through the glass (see the figure below). If it has one reflection, it can be reflected in two ways. If it has two reflections, it can be reflected in three ways. Use this information to answer parts (a) through (c).

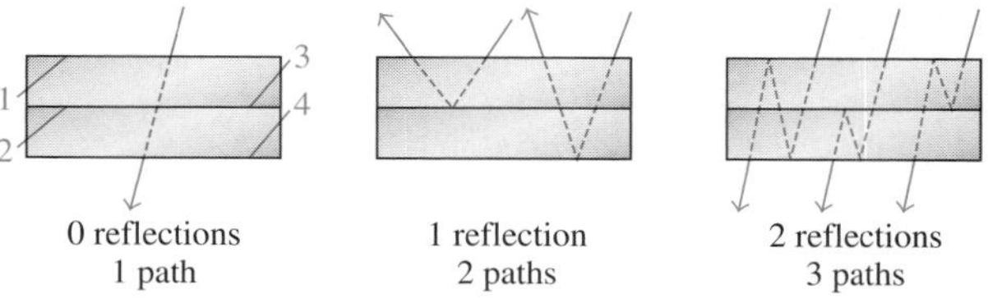

a) If a ray is reflected three times, there are five paths it can follow. Show the paths.

b) If a ray is reflected four times, there are eight paths it can follow. Show the paths.

c) How many paths can a ray follow if it is reflected five times? Explain how you determined your answer.

Internet/Research Activities

41. The digits 1 through 9 have evolved considerably since they appeared in Fibonacci's book *Liber Abacci.* Write a report tracing the history of the evolution of the digits 1 through 9 since Fibonacci's time.

42. Write a report on the history and mathematical contributions of Fibonacci.

43. Write a report indicating where the golden ratio and golden rectangle have been used in art and architecture. You may wish to include information on art and architecture related to the golden ratio and Fibonacci sequences.

CHAPTER 5 SUMMARY

IMPORTANT FACTS

Fundamental theorem of arithmetic

Every composite number can be expressed as a unique product of prime numbers.

Sets of numbers

Natural or counting numbers: $\{1, 2, 3, 4, \ldots\}$

Whole numbers: $\{0, 1, 2, 3, 4, \ldots\}$

Integers: $\{\ldots, -3, -2, -1, 0, 1, 2, 3, \ldots\}$

Rational numbers: Numbers of the form p/q, where p and q are integers, $q \neq 0$. Every rational number when expressed as a decimal number will be either a terminating or repeating decimal number.

Irrational number: A real number whose representation is a nonterminating, nonrepeating decimal number (not a rational number).

Definition of subtraction

$$a - b = a + (-b)$$

Fundamental law of rational numbers

$$\frac{a}{b} = \frac{a}{b} \cdot \frac{c}{c} = \frac{ac}{bc}, \qquad b \neq 0, \qquad c \neq 0$$

Rules of radicals

Product rule for radicals:

$$\sqrt{a \cdot b} = \sqrt{a} \cdot \sqrt{b}, \qquad a \geq 0, \qquad b \geq 0$$

Quotient rule for radicals:

$$\frac{\sqrt{a}}{\sqrt{b}} = \sqrt{\frac{a}{b}}, \qquad a \geq 0, \qquad b > 0$$

Properties of real numbers

Commutative property of addition: $a + b = b + a$

Commutative property of multiplication: $a \cdot b = b \cdot a$

Associative property of addition:

$$(a + b) + c = a + (b + c)$$

Associative property of multiplication:

$$(a \cdot b) \cdot c = a \cdot (b \cdot c)$$

Distributive property: $a \cdot (b + c) = ab + ac$

Rules of exponents

Product rule for exponents: $a^m \cdot a^n = a^{m+n}$

Quotient rule for exponents: $\frac{a^m}{a^n} = a^{m-n}, \qquad a \neq 0$

Zero exponent rule: $a^0 = 1, \qquad a \neq 0$

Negative exponent rule: $a^{-m} = \frac{1}{a^m}, \qquad a \neq 0$

Power rule: $(a^m)^n = a^{m \cdot n}$

Arithmetic sequence

$$a_n = a_1 + (n - 1)d$$

$$s_n = \frac{n(a_1 + a_n)}{2}$$

Geometric sequence

$$a_n = a_1 r^{n-1}$$

$$s_n = \frac{a_1(1 - r^n)}{1 - r}, \qquad r \neq 1$$

Fibonacci sequence

$$1, 1, 2, 3, 5, 8, 13, 21, \ldots$$

Golden number

$$\frac{\sqrt{5} + 1}{2} \approx 1.618$$

Golden proportion

$$\frac{a + b}{a} = \frac{a}{b}$$

CHAPTER 5 REVIEW EXERCISES

5.1

In Exercises 1 and 2, determine whether the number is divisible by each of the following numbers: 2, 3, 4, 5, 6, 8, 9, and 10.

1. 894,348 **2.** 400,644

In Exercises 3–7, find the prime factorization of the number.

3. 252 **4.** 385 **5.** 840
6. 882 **7.** 1452

In Exercises 8–13, find the GCD and LCM of the numbers.

8. 15, 60 **9.** 63, 108 **10.** 45, 250
11. 840, 320 **12.** 60, 40, 96 **13.** 36, 108, 144

14. *Train Stops* From 1912 to 1971, the Milwaukee Road Railroad Company had a train stop every 15 days in Dubuque, Iowa. During this same period, the same train also stopped in Des Moines, Iowa, every 9 days. If on April 18, 1964, the train made a stop in Dubuque and a stop in Des Moines, how many days was it until the train again stopped in both cities on the same day?

5.2

In Exercises 15–22, use a number line to evaluate the expression.

15. $-2 + 5$ **16.** $4 + (-7)$
17. $4 - 8$ **18.** $-2 + (-4)$
19. $-5 - 4$ **20.** $-3 - (-6)$
21. $(-3 + 7) - 4$ **22.** $-1 + (9 - 4)$

In Exercises 23–30, evaluate the expression.

23. $(-3)(-11)$ **24.** $-4(9)$
25. $(14)(-4)$ **26.** $\frac{-35}{-7}$
27. $\frac{12}{-6}$ **28.** $[8 \div (-4)](-3)$
29. $[(-4)(-3)] \div 2$ **30.** $[(-30) \div (10)] \div (-1)$

5.3

In Exercises 31–39, express the fraction as a terminating or repeating decimal.

31. $\frac{3}{10}$ **32.** $\frac{3}{5}$ **33.** $\frac{15}{40}$
34. $\frac{13}{4}$ **35.** $\frac{3}{7}$ **36.** $\frac{7}{12}$
37. $\frac{3}{8}$ **38.** $\frac{7}{8}$ **39.** $\frac{5}{7}$

In Exercises 40–46, express the decimal number as a quotient of two integers.

40. 0.225 **41.** 4.5 **42.** $0.\overline{6}$ **43.** $2.\overline{37}$
44. 0.083 **45.** 0.0042 **46.** $2.3\overline{4}$

In Exercises 47–50, express each mixed number as an improper fraction.

47. $2\frac{5}{7}$ **48.** $4\frac{1}{6}$ **49.** $-3\frac{1}{4}$ **50.** $-35\frac{3}{8}$

In Exercises 51–54, express each improper fraction as a mixed number.

51. $\frac{11}{5}$ **52.** $\frac{27}{15}$ **53.** $-\frac{12}{7}$ **54.** $-\frac{136}{5}$

In Exercises 55–63, perform the indicated operation and reduce your answer to lowest terms.

55. $\frac{1}{2} + \frac{4}{5}$ **56.** $\frac{7}{8} - \frac{3}{4}$
57. $\frac{1}{6} + \frac{5}{4}$ **58.** $\frac{4}{5} \cdot \frac{15}{16}$
59. $\frac{5}{9} \div \frac{6}{7}$ **60.** $\left(\frac{4}{5} + \frac{5}{7}\right) \div \frac{4}{5}$
61. $\left(\frac{2}{3} \cdot \frac{1}{7}\right) \div \frac{4}{7}$ **62.** $\left(\frac{1}{5} + \frac{2}{3}\right)\left(\frac{3}{8}\right)$
63. $\left(\frac{1}{5} \cdot \frac{2}{3}\right) + \left(\frac{1}{5} \div \frac{1}{2}\right)$

64. *Cajun Turkey* A recipe for Roasted Cajun Turkey calls for $\frac{1}{8}$ teaspoon of cayenne pepper per pound of turkey. If

Jennifer Thornton is preparing a turkey that weighs $17\frac{3}{4}$ pounds, how much cayenne pepper does she need?

5.4

In Exercises 65–80, simplify the expression. Rationalize the denominator when necessary.

65. $\sqrt{50}$
66. $\sqrt{200}$
67. $\sqrt{5} + 7\sqrt{5}$
68. $\sqrt{3} - 4\sqrt{3}$
69. $\sqrt{8} + 6\sqrt{2}$
70. $\sqrt{3} - 7\sqrt{27}$
71. $\sqrt{75} + \sqrt{27}$
72. $\sqrt{3} \cdot \sqrt{6}$
73. $\sqrt{8} \cdot \sqrt{6}$
74. $\frac{\sqrt{18}}{\sqrt{2}}$
75. $\frac{\sqrt{56}}{\sqrt{2}}$
76. $\frac{4}{\sqrt{3}}$
77. $\frac{\sqrt{3}}{\sqrt{5}}$
78. $3(2 + \sqrt{7})$
79. $\sqrt{3}(4 + \sqrt{6})$
80. $\sqrt{3}(\sqrt{6} + \sqrt{15})$

5.5

In Exercises 81–90, state the name of the property illustrated.

81. $x + 2 = 2 + x$
82. $5 \cdot m = m \cdot 5$
83. $(1 + 2) + 3 = 1 + (2 + 3)$
84. $4(y + 3) = 4 \cdot y + 4 \cdot 3$
85. $(1 + 2) + 3 = 3 + (1 + 2)$
86. $(3 + 5) + (4 + 3) = (4 + 3) + (3 + 5)$
87. $(3 \cdot a) \cdot b = 3 \cdot (a \cdot b)$
88. $a \cdot (2 + 3) = (2 + 3) \cdot a$
89. $2(x + 3) = (2 \cdot x) + (2 \cdot 3)$
90. $x \cdot 2 + 6 = 2 \cdot x + 6$

In Exercises 91–96, determine whether the set of numbers is closed under the given operation.

91. Natural numbers, addition
92. Whole numbers, subtraction
93. Integers, division
94. Real numbers, subtraction
95. Irrational numbers, multiplication
96. Rational numbers, division

5.6

In Exercises 97–104, evaluate each expression.

97. 3^2
98. 3^{-2}
99. $\frac{9^5}{9^3}$
100. $5^2 \cdot 5$
101. 7^0
102. 4^{-3}
103. $(2^3)^2$
104. $(3^2)^2$

In Exercises 105–108, write each number in scientific notation.

105. 230,000,000
106. 0.0000158
107. 0.00275
108. 4,950,000

In Exercises 109–112, express each number in decimal notation.

109. 4.3×10^7
110. 1.39×10^{-4}
111. 1.75×10^{-4}
112. 1×10^5

In Exercises 113–116, (a) perform the indicated operation and write your answer in scientific notation. (b) Confirm the result found in part (a) by performing the calculation on a scientific calculator.

113. $(7 \times 10^3)(2 \times 10^{-5})$
114. $(4 \times 10^2)(2.5 \times 10^2)$
115. $\frac{8.4 \times 10^3}{4 \times 10^2}$
116. $\frac{1.5 \times 10^{-3}}{5 \times 10^{-4}}$

In Exercises 117–121, (a) perform the indicated calculation by first converting each number to scientific notation. Write your answer in decimal notation. (b) Confirm the result found in part (a) by performing the calculation on a scientific calculator.

117. (4,000,000)(2,000)
118. (35,000)(0.00002)
119. $\frac{9{,}600{,}000}{3000}$
120. $\frac{0.000002}{0.0000004}$
121. *Space Distances* The distance from Earth to the sun is about 1.49×10^{11} meters. The distance from Earth to the moon is about 3.84×10^8 meters. The distance from Earth to the sun is how many times larger than the distance from Earth to the moon? Use a scientific calculator and round your answer to the nearest whole number.

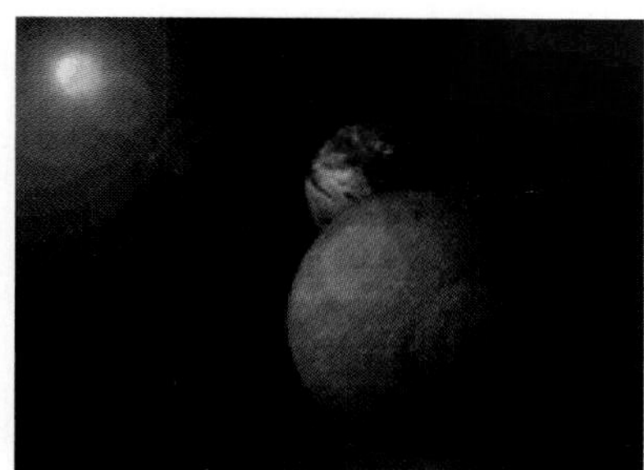

See Exercise 121

122. *Outstanding Debt* As a result of a recent water and sewer system improvement, the city of Galena, Illinois, has an outstanding debt of \$20,000,000. If the population of Galena is 3600 people, how much would each person have to contribute to pay off the outstanding debt?

5.7

In Exercises 123–128, determine whether the sequence is arithmetic or geometric. Then determine the next two terms of the sequence.

123. $2, 5, 8, 11, \ldots$
124. $\frac{1}{2}, 1, 2, 4, \ldots$
125. $-3, -6, -9, -12, \ldots$
126. $\frac{1}{2}, \frac{1}{4}, \frac{1}{8}, \frac{1}{16}, \ldots$
127. $1, 4, 7, 10, 13, \ldots$
128. $2, -2, 2, -2, 2, \ldots$

In Exercises 129–134, find the indicated term of the sequence with the given first term, a_1, and common difference, d, or common ratio, r.

129. Find a_4 when $a_1 = 3, d = 4$.
130. Find a_8 when $a_1 = -6, d = -4$.
131. Find a_{10} when $a_1 = -20, d = 5$.
132. Find a_5 when $a_1 = 3, r = 2$.
133. Find a_5 when $a_1 = 4, r = \frac{1}{2}$.
134. Find a_4 when $a_1 = -6, r = 2$.

In Exercises 135–138, find the sum of the arithmetic sequence. The number of terms, n, is given.

135. $2, 5, 8, 11, \ldots, 89; n = 30$
136. $-4, -3\frac{3}{4}, -3\frac{1}{2}, -3\frac{1}{4}, \ldots, -2\frac{1}{4}; n = 8$
137. $100, 94, 88, 82, \ldots, 58; n = 8$
138. $0.5, 0.75, 1.00, 1.25, \ldots, 5.25; n = 20$

In Exercises 139–142, find the sum of the first n terms of the geometric sequence for the values of a_1 and r.

139. $n = 4, a_1 = 5, r = 3$
140. $n = 4, a_1 = 2, r = 3$
141. $n = 5, a_1 = 3, r = -2$
142. $n = 6, a_1 = 1, r = -2$

In Exercises 143–148, first determine whether the sequence is arithmetic or geometric; then write an expression for the general or nth term, a_n.

143. $7, 4, 1, -2, \ldots$
144. $3, 6, 9, 12, \ldots$
145. $4, \frac{5}{2}, 1, -\frac{1}{2}, \ldots$
146. $3, 6, 12, 24, \ldots$
147. $2, -2, 2, -2, \ldots$
148. $5, \frac{5}{3}, \frac{5}{9}, \frac{5}{27}, \ldots$

5.8

In Exercises 149–152, determine whether the sequence is a Fibonacci-type sequence. If so, determine the next two terms.

149. $0, 1, 1, 2, 3, 5, 8, \ldots$
150. $-3, 4, 1, 5, 6, 11, \ldots$
151. $1, 4, 3, -1, -4, -5, \ldots$
152. $-10, 10, 0, 10, 20, \ldots$

CHAPTER 5 TEST

1. Which of the numbers 2, 3, 4, 5, 6, 8, 9, and 10 divide 38,610?
2. Find the prime factorization of 840.
3. Evaluate $[(-6) + (-9)] + 8$.
4. Evaluate $-7 - 13$.
5. Evaluate $[(-70)(-5)] \div (8 - 10)$.
6. Convert $4\frac{5}{8}$ to an improper fraction.
7. Convert $\frac{176}{9}$ to a mixed number.
8. Write $\frac{5}{8}$ as a terminating or repeating decimal.
9. Express 6.45 as a quotient of two integers.
10. Evalute $(\frac{5}{16} \div 3) + (\frac{4}{5} \cdot \frac{1}{2})$.
11. Perform the operation and reduce the answer to lowest terms: $\frac{11}{12} - \frac{3}{8}$.
12. Simplify $\sqrt{75} + \sqrt{48}$.
13. Rationalize $\dfrac{\sqrt{2}}{\sqrt{7}}$.
14. Determine whether the integers are closed under the operation of multiplication. Explain your answer.

Name the property illustrated.

15. $(4 + y) + 5 = 4 + (y + 5)$

16. $3(x + y) = 3x + 3y$

Evaluate.

17. $\frac{4^5}{4^2}$ **18.** $4^3 \cdot 4^2$ **19.** 3^{-4}

20. Perform the operation by first converting the numerator and denominator to scientific notation. Write the answer in scientific notation.

$$\frac{7{,}200{,}000}{0.000009}$$

21. Write an expression for the general or nth term, a_n, of the sequence $-2, -6, -10, -14, \ldots$.

22. Find the sum of the terms of the arithmetic sequence. The number of terms, n, is given.

$$-2, -5, -8, -11, \ldots, -32;\ n = 11$$

23. Find a_5 when $a_1 = 3$ and $r = 3$.

24. Find the sum of the first five terms of the sequence when $a_1 = 3$ and $r = 4$.

25. Write an expression for the general or nth term, a_n, of the sequence $3, 6, 12, 24, \ldots$.

26. Write the first 10 terms of the Fibonacci sequence.

GROUP PROJECTS

1. *Making Rice* The amount of ingredients needed to make 3 and 5 servings of rice are:

To Make	Rice and Water	Salt	Butter
3 servings	1 cup	$\frac{3}{8}$ tsp	$1\frac{1}{2}$ tsp
5 servings	$1\frac{2}{3}$ cup	$\frac{5}{8}$ tsp	$2\frac{1}{2}$ tsp

Find the amount of each ingredient needed to make (a) 2 servings, (b) 1 serving, and (c) 29 servings. Explain how you determined your answers.

2. *Finding Areas*

a) Determine the area of the trapezoid shown by finding the area of the three parts indicated and finding the sum of the three areas. The necessary geometric formulas are given in Chapter 9.

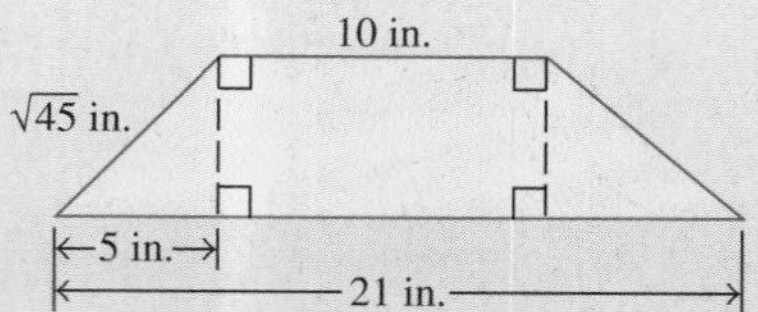

b) Determine the area of the trapezoid by using the formula for the area of a trapezoid given in Chapter 9.

c) Compare your answers from parts (a) and (b). Are they the same? If not, explain why they are different.

3. *Medical Insurance* On a medical insurance policy (such as Blue Cross/Blue Shield), the policyholder may need to make copayments for prescription drugs, office visits, and procedures until the total of all copayments reaches a specified amount. Suppose on the Gattelaro's medical policy that the copayment for prescription drugs is 50% of the cost; the copayment for office visits is \$10; and the copayment for all medical tests, x-rays, and other procedures is 20% of the cost. After the family's copayment totals \$500 in a calendar year, all medical and prescription bills are paid in full by the insurance company. The Gattelaros had the following medical expenses from January 1 through April 30.

Date	Reason	Cost before Copayment
January 10	Office visit	\$40
	Prescription	\$44
February 27	Office visit	\$40
	Medical tests	\$188
April 19	Office visit	\$40
	X-rays	\$348
	Prescription	\$76

a) How much had the Gattelaros paid in copayments from January 1 through April 30?

b) How much had the medical insurance company paid?

c) What is the remaining copayment that must be paid by the Gattelaros before the \$500 copayment limit is reached?

4. *A Branching Plant* A plant grows for two months and then adds a new branch. Each new branch grows for two months and then adds another branch. After the second month, each branch adds a new branch every month. Assume the growth begins in January.

a) How many branches will there be in February?

b) How many branches will there be in May?

c) How many branches will there be after 12 months?

d) How is this problem similar to the problem involving rabbits that appeared in Fibonacci's book *Liber Abacci* (see pages 275 and 276)?

When planning a trip, knowledge of a coordinate system is helpful.

ALGEBRA, GRAPHS, AND FUNCTIONS

Algebra, and, in particular, word problems: The very mention of them is enough to frighten many people, and yet algebra is one of the most practical tools for solving everyday problems. You probably use algebra in your daily life without realizing it.

For example, you use a coordinate system when you consult your car map to find directions to a new destination. You solve simple equations when you change a recipe to increase or decrease the number of servings. To evaluate how much interest you will earn on a savings account or to figure out how long it will take you to travel a given distance, you use common formulas that are algebraic equations.

The symbolic language of algebra makes it an excellent tool for solving problems. Symbolism has three advantages. First, it allows us to write lengthy expressions in compact form. Second, symbolic language is clear—each symbol has a precise meaning. Finally, symbolism allows us to consider a large or infinite number of separate cases with a common property.

The English philosopher Alfred North Whitehead explained the power of algebra when he stated, "By relieving the brain of all unnecessary work, a good notation sets the mind free to concentrate on more advanced problems and in effect increases the mental power of the race."

6.1 ORDER OF OPERATIONS

Algebra is a generalized form of arithmetic. The word *algebra* is derived from the Arabic word *al-jabr* (meaning "reunion of broken parts"), which was the title of a book written by the mathematician Muhammed ibn-Musa al Khwarizmi in about A.D. 825.

Why study algebra? You can solve many problems in everyday life by using arithmetic or by trial and error, but with a knowledge of algebra you can find the solutions with less effort. You can solve other problems, like some we will present in this chapter, only by using algebra.

Algebra uses letters of the alphabet called *variables* to represent numbers. Often the letters x and y are used to represent variables. However, any letter may be used as a variable. A symbol that represents a specific quantity is called a *constant*.

Multiplication of numbers and variables may be represented in several different ways in algebra. Since the "times" sign might be confused with the variable x, a dot between two numbers or variables indicates multiplication. Thus, $3 \cdot 4$ means 3 times 4, and $x \cdot y$ means x times y. Placing two letters or a number and a letter next to one another, with or without parentheses, also indicates multiplication. Thus, $3x$ means 3 times x, xy means x times y, and $(x)(y)$ means x times y.

An *algebraic expression* (or simply an *expression*) is a collection of variables, numbers, parentheses, and operation symbols. Some examples of algebraic expressions are

$$x, \qquad x + 2, \qquad 3(2x + 3), \qquad \frac{3x + 1}{2x - 3}, \qquad \text{and} \qquad x^2 + 7x + 3$$

Two algebraic expressions joined by an equal sign form an *equation*. Some examples of equations are

$$x + 2 = 4, \qquad 3x + 4 = 1, \qquad \text{and} \qquad x + 3 = 2x$$

The *solution to an equation* is the number or numbers that replace the variable to make the equation a true statement. For example, the solution to the equation $x + 3 = 4$ is $x = 1$. When we find the solution to an equation, we *solve the equation*.

We can determine if any number is a solution to an equation by *checking the solution*. To check the solution, we substitute the number for the variable in the equation. If the resulting statement is a true statement, that number is a solution to the equation. If the resulting statement is a false statement, the number is not a solution to the equation. To check the number $x = 1$ in the equation $x + 3 = 4$, we do the following.

$$x + 3 = 4$$
$$1 + 3 = 4 \qquad \text{Substitute 1 for } x$$
$$4 = 4 \qquad \text{True}$$

The same number is obtained on both sides of the equal sign, so the solution is correct. For the equation $x + 3 = 4$, the only solution is $x = 1$. Any other value of x would result in the check being a false statement.

DID YOU KNOW

Broken Bones

Imagine yourself walking down a street in Spain during the Middle Ages. You see a sign over a door: "*Algebrista y Sangrador.*" Inside, you would find a person more ready to give you a haircut than help you with your algebra. The sign translates into "Bonesetter and Bloodletter," relatively simple medical treatments administered in barbershops of the day.

The root word *al-jabr*, which the Muslims (Moors) brought to Spain along with some concepts of algebra, suggests the restoring of broken parts. The parts might be bones, or they might be mathematical expressions that are broken into separate parts and the parts moved from one side of an equation to the other and reunited in such a way as to make a solution more obvious.

To *evaluate an expression* means to find the value of the expression for a given value of the variable. To evaluate expressions and solve equations, you must have an understanding of exponents. Exponents (Section 5.6) are used to abbreviate repeated multiplication. For example, the expressions 5^2 means $5 \cdot 5$. The 2 in the expression 5^2 is the *exponent*, and the 5 is the *base*. We read 5^2 as "5 to the second power" or "5 squared," and 5^2 means $5 \cdot 5$ or 25.

In general, the number b to the nth power, written b^n, means

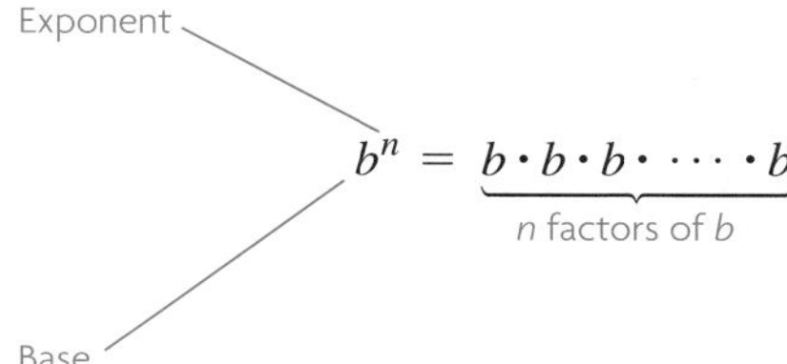

An exponent refers only to its base. In the expression -5^2, the base is 5. In the expression $(-5)^2$, the base is -5.

$$-5^2 = -(5)^2 = -1(5)^2 = -1(5)(5) = -25$$
$$(-5)^2 = (-5)(-5) = 25$$

Note that $-5^2 \neq (-5)^2$ since $-25 \neq 25$.

Order of Operations

To evaluate an expression or to check the solution to an equation, we need to know the *order of operations* to follow. For example, suppose we want to evaluate the expression $2 + 3x$ when $x = 4$. Substituting 4 for x, we obtain $2 + 3 \cdot 4$. What is the value of $2 + 3 \cdot 4$? Does it equal 20, or does it equal 14? Some standard rules, called the order of operations, have been developed to ensure that there is only one correct answer. In mathematics, unless parentheses indicate otherwise, always perform multiplication before addition. Thus, the correct answer is 14.

$$2 + 3 \cdot 4 = 2 + (3 \cdot 4) = 2 + 12 = 14$$

The order of operations for evaluating an expression is as follows.

Order of Operations

1. First, perform all operations within parentheses or other grouping symbols (according to the following order).
2. Next, perform all exponential operations (that is, raising to powers or finding roots).
3. Next, perform all multiplications and divisions from left to right.
4. Finally, perform all additions and subtractions from left to right.

DID YOU KNOW

An Important Breakthrough

This painting by Charles Demuth, called *I Saw the Figure 5 in Gold,* depicts the abstract nature of numbers. Mathematician and philosopher Bertrand Russell observed in 1919 that it must have required "many ages to discover that a brace of pheasants and a couple of days were both instances of the number 2." The discovery that numbers could be used not only to count objects such as the number of birds but also to represent abstract quantities represented a breakthrough in the development of algebra.

TIMELY TIP Some students use the phrase, "Please Excuse My Dear Aunt Sally," or the word "PEMDAS" (Parentheses, Exponents, Multiplication, Division, Addition, Subtraction) to remind them of the order of operations. Remember: Multiplication and division are of the same order, and addition and subtraction are of the same order.

EXAMPLE 1 *Evaluating an Expression*

Evaluate the expression $-x^2 + 3x + 20$ for $x = 4$.

SOLUTION: Substitute 4 for each x and use the order of operations to evaluate the expression.

$$\begin{aligned} &-x^2 + 3x + 20 \\ &= -(4)^2 + 3(4) + 20 \\ &= -16 + 12 + 20 \\ &= -4 + 20 \\ &= 16 \end{aligned}$$

EXAMPLE 2 *Finding the Height*

A ball is thrown upward off a bridge 40 feet above ground. Its height, in feet, above ground, t seconds after it is thrown, can be determined by the expression $-16t^2 + 30t + 40$. Find the height of the ball, above ground, 2 seconds after it is thrown.

SOLUTION: Substitute 2 for each t.

$$\begin{aligned} &-16t^2 + 30t + 40 \\ &= -16(2)^2 + 30(2) + 40 \\ &= -16(4) + 30(2) + 40 \\ &= -64 + 60 + 40 \\ &= -4 + 40 \\ &= 36 \end{aligned}$$

The ball is 36 feet above ground 2 seconds after it is thrown.

EXAMPLE 3 *Substituting for Two Variables*

Evaluate $-3x^2 + 2xy - 2y^2$ when $x = 2$ and $y = 3$.

SOLUTION: Substitute 2 for each x and 3 for each y; then evaluate using the order of operations.

$$\begin{aligned} &-3x^2 + 2xy - 2y^2 \\ &= -3(2)^2 + 2(2)(3) - 2(3)^2 \\ &= -3(4) + 2(2)(3) - 2(9) \\ &= -12 + 12 - 18 \\ &= 0 - 18 \\ &= -18 \end{aligned}$$

EXAMPLE 4 *Is 3 a Solution?*

Determine whether 3 is a solution to the equation $2x^2 + 4x - 9 = 21$.

SOLUTION: To determine whether 3 is a solution to the equation, substitute 3 for each x in the equation. Then evaluate the left-hand side of the equation using the order of operations. If this leads to a 21 on the left-hand side of the equal sign, then both sides of the equation have the same value, and 3 is a solution. In checking the solution, we use $\stackrel{?}{=}$ which means we are not sure if the statement is true.

$$\begin{aligned} 2x^2 + 4x - 9 &\stackrel{?}{=} 21 \\ 2(3)^2 + 4(3) - 9 &\stackrel{?}{=} 21 \\ 2(9) + 12 - 9 &\stackrel{?}{=} 21 \\ 18 + 12 - 9 &\stackrel{?}{=} 21 \\ 30 - 9 &\stackrel{?}{=} 21 \\ 21 &= 21 \quad \text{True} \end{aligned}$$

Because 3 makes the equation a true statement, 3 is a solution to the equation.

SECTION 6.1 EXERCISES

Concept/Writing Exercises

1. What is a *variable*?
2. What is a *constant*?
3. What does it mean when we state that *a number is a solution to an equation*?
4. What is an algebraic expression? Illustrate an algebraic expression with an example.
5. **a)** For the term 4^5, identify the base and the exponent. **b)** In your own words, explain how to evaluate 4^5.
6. In your own words, explain the order of operations.
7. Evaluate $8 + 16 \div 4$ using the order of operations.
8. Evaluate $9 + 6 \cdot 3$ using the order of operations.

Practice the Skills

In Exercises 9–28, evaluate the expression for the given value(s) of the variable(s).

9. $x^2, \quad x = 7$

10. $x^2, \quad x = -8$

11. $-x^2, \quad x = -3$

12. $-x^2, \quad x = -5$

13. $-2x^3, \quad x = -7$

14. $-x^3, \quad x = -4$

15. $x - 7, \quad x = 4$

16. $8x - 3, \quad x = \frac{5}{2}$

17. $-7x + 4, \quad x = -2$

18. $x^2 - 3x + 8, \quad x = 5$

19. $-x^2 + 5x - 13, \quad x = -2$

20. $5x^2 + 7x - 11, \quad x = -1$

21. $\frac{1}{2}x^2 - 5x + 2, \quad x = \frac{2}{3}$

22. $\frac{2}{3}x^2 + x - 1, \quad x = \frac{1}{2}$

23. $8x^3 - 4x^2 + 7, \quad x = \frac{1}{2}$

24. $-x^2 + 4xy, \quad x = 2, \quad y = 3$

25. $2x^2 + xy + 3y^2, \quad x = -2, \quad y = 1$

26. $3x^2 + \frac{2}{5}xy - \frac{1}{5}y^2, \quad x = 2, \quad y = 5$

27. $4x^2 - 12xy + 9y^2, \quad x = 3, \quad y = 2$

28. $(x + 3y)^2, \quad x = 4, \quad y = -3$

In Exercises 29–38, determine whether the value(s) is (are) a solution to the equation.

29. $7x + 3 = 23, \quad x = 3$

30. $5x - 7 = -27, \quad x = -4$

31. $x - 3y = 0, \quad x = 6, \quad y = 3$

32. $4x + 2y = -2, \quad x = -2, \quad y = 3$

33. $x^2 + 3x - 4 = 5, \quad x = 2$

34. $2x^2 - x - 5 = 0, \quad x = 3$

35. $2x^2 + x = 28, \quad x = -4$

36. $y = x^2 + 3x - 5, \quad x = 1, \quad y = -1$

37. $y = -x^2 + 3x - 1, \quad x = 3, \quad y = -1$

38. $y = x^3 - 3x^2 + 1, \quad x = 2, \quad y = -3$

Problem Solving

39. *Sales Tax* If the sales tax on an item is 7%, the sales tax, in dollars, on an item costing d dollars can be found by using the expression $0.07d$. Determine the sales tax on a telescope costing \$175.

40. *Radius of a Circle* A pebble is dropped into a calm pond causing ripples in the form of concentric circles (one circle inside another circle). The radius of the outer circle, in feet, can be determined by the expression $0.5t$, where t is the time in seconds after the pebble strikes the water. Find the radius of the outer circle 3 sec after the pebble strikes the water.

41. *Cost of a Tour* The cost, in dollars, for Crescent City Tours to provide a tour for x people can be determined by the expression $220 + 2.75x$. Determine the cost for Crescent City Tours to provide a tour for 75 people.

42. *Orange Orchard* The number of baskets of oranges that are produced by x trees in a small orchard can be approximated by the expression $25x - 0.2x^2$ (assuming x is no more than 100). Find the number of baskets of oranges produced by 60 trees.

43. *8 Trillion Calculations* If a computer can do a calculation in 0.000002 sec, the time required to do n calculations can be determined by the expression $0.000002n$. Determine the number of seconds needed for the computer to do 8 trillion (8,000,000,000,000) calculations.

44. *Drying Time* The time, in minutes, needed for clothes hanging on a line outdoors to dry, at a specific temperature and wind speed, depends on the humidity, h. The time can be approximated by the expression $2h^2 + 80h + 40$, where h is the percent humidity expressed as a decimal number. Find the length of time required for clothing to dry if there is 60% humidity.

45. *Grass Growth* The rate of growth of grass in inches per week depends on a number of factors, including rainfall and temperature. For a certain area, this can be approximated by the expression $0.2R^2 + 0.003RT + 0.0001T^2$, where R is the weekly rainfall, in inches, and T is the average weekly temperature, in degrees Fahrenheit. Find the amount of growth of grass for a week in which the rainfall is 2 in. and the average temperature is 70°F.

Challenge Problems/Group Activities

46. Explain why $(-1)^n = 1$ for any even number n.

47. Does $(x + y)^2 = x^2 + y^2$? Complete the table and state your conclusion.

x	y	$(x + y)^2$	$x^2 + y^2$
2	3		
−2	−3		
−2	3		
2	−3		

48. Suppose n represents any natural number. Explain why 1^n equals 1?

Internet/Research Activity

49. When were exponents first used? Write a paper explaining how exponents were first used and when mathematicians began writing them in the present form.

6.2 LINEAR EQUATIONS IN ONE VARIABLE

In Section 6.1, we stated that two algebraic expressions joined by an equal sign form an equation. The solution to some equations, such as $x + 3 = 4$, can be found easily by trial and error. However, solving more complex equations, such as $2x - 3 = 4(x + 3)$, requires understanding the meaning of like terms and learning four basic properties.

The parts that are added or subtracted in an algebraic expression are called ***terms***. The expression $4x - 3y - 5$ contains three terms, namely $4x$, $-3y$, and -5. The + and − signs that break the expression into terms are a part of the terms. When listing the terms of an expression, however, it is not necessary to include the + sign at the beginning of the term.

The numerical part of a term is called its ***numerical coefficient*** or, simply, its ***coefficient***. In the term $4x$, the 4 is the numerical coefficient. In the term $-4y$, the -4 is the numerical coefficient.

Like terms are terms that have the same variables with the same exponents on the variables. ***Unlike terms*** have different variables or different exponents on the variables.

Like Terms	Unlike Terms
$2x$, $7x$ (same variable, x)	$2x$, 9 (only first term has a variable)
$-8y$, $3y$ (same variable, y)	$5x$, $6y$ (different variables)
-4, 10 (both constants)	x, 8 (only first term has a variable)
$-5x^2$, $6x^2$ (same variable with same exponent)	$2x^3$, $3x^2$ (different exponents)

To ***simplify an expression*** means to combine like terms by using the commutative, associative, and distributive properties discussed in Chapter 5. For convenience, we list these properties below.

Properties of the Real Numbers

$a(b + c) = ab + ac$	Distributive property
$a + b = b + a$	Commutative property of addition
$ab = ba$	Commutative property of multiplication
$(a + b) + c = a + (b + c)$	Associative property of addition
$(ab)c = a(bc)$	Associative property of multiplication

EXAMPLE 1 *Combining Like Terms*

Combine like terms in each expression.

a) $7x + 3x$ b) $6y - 2y$

c) $x + 12 - 3x + 7$ d) $-2x + 4 - 6y - 11 - 5y + 3x$

SOLUTION:

a) We use the distributive property (in reverse) to combine like terms.

$$\begin{aligned} 7x + 3x &= (7 + 3)x && \text{Distributive property} \\ &= 10x \end{aligned}$$

b) $6y - 2y = (6 - 2)y = 4y$

c) $x + 12 - 3x + 7 = x - 3x + 12 + 7$ Rearrange terms, place like terms together.

$= -2x + 19$ Combine like terms.

d) $-2x + 4 - 6y - 11 - 5y + 3x$

$= -2x + 3x - 6y - 5y + 4 - 11$ Rearrange terms, place like terms together.

$= x - 11y - 7$ Combine like terms. ▲

DID YOU KNOW

In the Beginning

Greek philosopher Diophantus of Alexandria (A.D. 250), who invented notations for powers of a number and for multiplication and division of simple quantities, is thought to have made the first attempts at algebra. But not until the sixteenth century did French mathematician François Viète (1540–1603) use symbols to represent numbers, the foundation of symbolic algebra. However, the work of René Descartes (1596–1660) is considered to be the starting point of modern-day algebra. In 1707, Sir Isaac Newton (1643–1727) gave symbolic mathematics the name *universal arithmetic*.

We are able to rearrange the terms of an expression, as was done in Example 1(c) and (d) by the commutative and associative properties that were discussed in Section 5.5.

The order of the terms in an expression is not crucial. However, when listing the terms of an expression we generally list the terms in alphabetical order with the constant, the term without a variable, last.

Solving Equations

Recall that to solve an equation means to find the value or values for the variable that make(s) the equation true. In this section, we discuss solving *linear (or first degree) equations*. A linear equation in one variable is one in which the exponent on the variable is 1. Examples of linear equations are $5x - 1 = 3$ and $2x + 4 = 6x - 5$.

Equivalent equations are equations that have the same solution. The equations $2x - 5 = 1$, $2x = 6$, and $x = 3$ are all equivalent equations since they all have the same solution, 3. When we solve an equation, we write the given equation as a series of simpler equivalent equations until we obtain an equation of the form $x = c$, where c is some real number.

To solve any equation, we have to *isolate the variable*. That means getting the variable by itself on one side of the equal sign. The four properties of equality that we are about to discuss are used to isolate the variable. The first is the addition property.

Addition Property of Equality
If $a = b$, then $a + c = b + c$ for all real numbers a, b, and c.

The addition property of equality indicates that the same number can be added to both sides of an equation without changing the solution.

EXAMPLE 2 *Using the Addition Property of Equality*

Find the solution to the equation $x - 7 = 10$.

SOLUTION: To isolate the variable, add 7 to both sides of the equation.

$$x - 7 = 10$$
$$x - 7 + 7 = 10 + 7$$
$$x + 0 = 17$$
$$x = 17$$

Check:

$$x - 7 \stackrel{?}{=} 10$$
$$17 - 7 \stackrel{?}{=} 10 \quad \text{Substitute 17 for } x.$$
$$10 = 10 \quad \text{True}$$

In Example 2, we showed the step $x + 0 = 17$. Generally this step is done mentally, and the step is not listed.

Subtraction Property of Equality
If $a = b$, then $a - c = b - c$ for all real numbers a, b, and c.

The subtraction property of equality indicates that the same number can be subtracted from both sides of an equation without changing the solution.

EXAMPLE 3 ***Using the Subtraction Property of Equality***

Find the solution to the equation $x + 7 = 15$.

SOLUTION: To isolate the variable, subtract 7 from both sides of the equation.

$$\begin{aligned} x + 7 &= 15 \\ x + 7 - 7 &= 15 - 7 \\ x &= 8 \end{aligned}$$

Note that we did not subtract 15 from both sides of the equation, since this would not result in getting x on one side of the equal sign by itself.

Multiplication Property of Equality
If $a = b$, then $a \cdot c = b \cdot c$ for all real numbers a, b, and c, where $c \neq 0$.

The multiplication property of equality indicates that both sides of the equation can be multiplied by the same nonzero number without changing the solution.

EXAMPLE 4 ***Using the Multiplication Property of Equality***

Find the solution to $\frac{x}{6} = 3$.

SOLUTION: To solve this equation, multiply both sides of the equation by 6.

$$\begin{aligned} \frac{x}{6} &= 3 \\ 6\left(\frac{x}{6}\right) &= 6(3) \\ \frac{\overset{1}{\cancel{6}}x}{\underset{1}{\cancel{6}}} &= 18 \\ 1x &= 18 \\ x &= 18 \end{aligned}$$

In Example 4, we showed the steps $\frac{6x}{6} = 18$ and $1x = 18$. Generally, we will not illustrate these steps.

Division Property of Equality

If $a = b$, then $\frac{a}{c} = \frac{b}{c}$ for all real numbers a, b, and c, $c \neq 0$.

The division property of equality indicates that both sides of an equation can be divided by the same nonzero number without changing the solution. Note that the divisor, c, cannot be 0 because division by 0 is not permitted.

EXAMPLE 5 *Using the Division Property of Equality*

Find the solution to the equation $3x = 15$.

SOLUTION: To solve this equation, divide both sides of the equation by 3.

$$3x = 15$$
$$\frac{3x}{3} = \frac{15}{3}$$
$$x = 5$$

An *algorithm* is a general procedure for accomplishing a task. The following general procedure is an algorithm for solving linear (or first-degree) equations. Sometimes the solution to an equation may be found more easily by using a variation of this general procedure. Remember that the primary objective in solving any equation is to isolate the variable.

A General Procedure for Solving Linear Equations

1. If the equation contains fractions, multiply both sides of the equation by the lowest common denominator (or least common multiple). This step will eliminate all fractions from the equation.
2. Use the distributive property to remove parentheses when necessary.
3. Combine like terms on the same side of the equal sign when possible.
4. Use the addition or subtraction property to collect all terms with a variable on one side of the equal sign and all constants on the other side of the equal sign. It may be necessary to use the addition or subtraction property more than once. This process will eventually result in an equation of the form $ax = b$, where a and b are real numbers.
5. Solve for the variable using the division or multiplication property. This will result in an answer in the form $x = c$, where c is a real number.

EXAMPLE 6 *Using the General Procedure*

Solve the equation $2x - 9 = 19$ and check your solution.

SOLUTION: Our goal is to isolate the variable; therefore, we start by getting the term $2x$ by itself on one side of the equation.

$$2x - 9 = 19$$
$$2x - 9 + 9 = 19 + 9$$ Add 9 to both sides of the equation (addition property) (step 4).
$$2x = 28$$
$$\frac{2x}{2} = \frac{28}{2}$$ Divide both sides of the equation by 2 (division property) (step 5).
$$x = 14$$

A check will show that 14 is the solution to $2x - 9 = 19$.

DID YOU KNOW

A New Concept

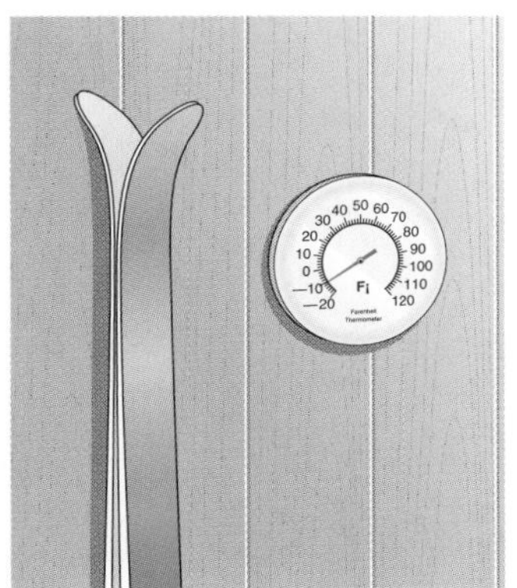

Well into the sixteenth century, mathematicians found it difficult to accept the idea that the solution to a problem (such as Example 7) could be a negative number because negative numbers could not be accepted as physically real. In the early days of algebra, someone working a problem did not isolate a variable by subtracting like terms. Instead, a problem would be put into a form that allowed only positive coefficients and answers. Albert Girard (1595–1637) contributed to the evolution of a correct understanding of negative quantities.

EXAMPLE 7 *Solving a Linear Equation*

Solve the equation $4 = 5 + 2(t + 1)$ for t.

SOLUTION: Our goal is to isolate the variable t. To do so, follow the general procedure for solving equations.

$$4 = 5 + 2(t + 1)$$

$$4 = 5 + 2t + 2 \qquad \text{Distributive property (step 2)}$$

$$4 = 2t + 7 \qquad \text{Combine like terms (step 3)}$$

$$4 - 7 = 2t + 7 - 7 \qquad \text{Subtraction property (step 4)}$$

$$-3 = 2t$$

$$-\frac{3}{2} = \frac{2t}{2} \qquad \text{Division property (step 5)}$$

$$-\frac{3}{2} = t$$

EXAMPLE 8 *Solving an Equation Containing Fractions*

Solve the equation $\frac{2x}{3} + \frac{1}{3} = \frac{3}{4}$.

SOLUTION: When an equation contains fractions, we generally begin by multiplying each term of the equation by the lowest common denominator, LCD (see Chapter 5). In this example, the LCD is 12, since 12 is the smallest number that is divisible by both 3 and 4.

$$12\left(\frac{2x}{3} + \frac{1}{3}\right) = 12\left(\frac{3}{4}\right) \qquad \text{Multiply both sides of the equation by the LCD (step 1).}$$

$$12\left(\frac{2x}{3}\right) + 12\left(\frac{1}{3}\right) = 12\left(\frac{3}{4}\right) \qquad \text{Distributive property (step 2)}$$

$$\overset{4}{\cancel{12}}\left(\frac{2x}{\underset{1}{\cancel{3}}}\right) + \overset{4}{\cancel{12}}\left(\frac{1}{\underset{1}{\cancel{3}}}\right) = \overset{3}{\cancel{12}}\left(\frac{3}{\underset{1}{\cancel{4}}}\right) \qquad \text{Divide out common factors.}$$

$$8x + 4 = 9$$

$$8x + 4 - 4 = 9 - 4 \qquad \text{Subtraction property (step 4)}$$

$$8x = 5$$

$$\frac{8x}{8} = \frac{5}{8} \qquad \text{Division property (step 5)}$$

$$x = \frac{5}{8}$$

A check will show that $\frac{5}{8}$ is the solution to the equation. You could have worked the problem without first multiplying both sides of the equation by the LCD. Try it!

EXAMPLE 9 *Variables on Both Sides of the Equation*

Solve the equation $6x + 8 = 10x + 12$.

SOLUTION: Note that the equation has an x on both sides of the equal sign. In equations of this type, you might wonder what to do first. It really does not matter as long as you do not forget the goal of isolating the variable x. Let's collect the terms containing a variable on the left-hand side of the equation.

$$6x + 8 = 10x + 12$$
$$6x + 8 - 8 = 10x + 12 - 8 \quad \text{Subtraction property (step 4)}$$
$$6x = 10x + 4$$
$$6x - 10x = 10x - 10x + 4 \quad \text{Subtraction property (step 4)}$$
$$-4x = 4$$
$$\frac{-4x}{-4} = \frac{4}{-4} \quad \text{Division property (step 5)}$$
$$x = -1$$

▲

In the solution to Example 9, the terms containing the variable were collected on the left-hand side of the equal sign. Now work Example 9, collecting the terms with the variable on the right-hand side of the equal sign. If you do so correctly, you will get the same result.

EXAMPLE 10 *Solving an Equation Containing Decimals*

Solve the equation $4x - 0.48 = 0.8x + 4$ and check your solution.

SOLUTION: This problem may be solved with the decimals, or you may multiply each term by 100 and eliminate the decimals. We will solve the problem with the decimals.

$$4x - 0.48 = 0.8x + 4$$
$$4x - 0.48 + 0.48 = 0.8x + 4 + 0.48 \quad \text{Addition property}$$
$$4x = 0.8x + 4.48$$
$$4x - 0.8x = 0.8x - 0.8x + 4.48 \quad \text{Subtraction property}$$
$$3.2x = 4.48$$
$$\frac{3.2x}{3.2} = \frac{4.48}{3.2} \quad \text{Division property}$$
$$x = 1.4$$

Check:

$$4x - 0.48 = 0.8x + 4$$
$$4(1.4) - 0.48 = 0.8(1.4) + 4 \quad \text{Substitute 1.4 for each } x \text{ in the equation.}$$
$$5.6 - 0.48 = 1.12 + 4$$
$$5.12 = 5.12 \quad \text{True}$$

▲

In Chapter 5, we explained that $a - b$ can be expressed as $a + (-b)$. We use this principle in Example 11.

EXAMPLE 11 *Using the Definition of Subtraction*

Solve $10 = -5 + 3(p - 4)$ for p.

SOLUTION: Our goal is to isolate the variable p. To do so, follow the general procedure for solving equations.

$$
\begin{aligned}
10 &= -5 + 3(p - 4) \\
10 &= -5 + 3[p + (-4)] && \text{Definition of subtraction} \\
10 &= -5 + 3(p) + 3(-4) && \text{Distributive property} \\
10 &= -5 + 3p - 12 \\
10 &= 3p - 17 && \text{Combine like terms.} \\
10 + 17 &= 3p - 17 + 17 && \text{Addition property} \\
27 &= 3p && \text{Combine like terms.} \\
\frac{27}{3} &= \frac{3p}{3} && \text{Division property} \\
9 &= p
\end{aligned}
$$

TIMELY TIP Remember that the goal in solving an equation is to get the variable alone on one side of the equal sign.

So far, every equation has had exactly one solution. Some equations, however, have no solution and others have more than one solution. Example 12 illustrates an equation that has no solution, and Example 13 illustrates an equation that has an infinite number of solutions.

EXAMPLE 12 *An Equation with No Solution*

Solve $3(x - 4) + x + 2 = 6x - 2(x + 3)$.

SOLUTION:

$$
\begin{aligned}
3(x - 4) + x + 2 &= 6x - 2(x + 3) \\
3x - 12 + x + 2 &= 6x - 2x - 6 && \text{Distributive property} \\
4x - 10 &= 4x - 6 && \text{Combine like terms.} \\
4x - 4x - 10 &= 4x - 4x - 6 && \text{Subtraction property} \\
-10 &= -6 && \text{False}
\end{aligned}
$$

During the process of solving an equation, if you obtain a false statement like $-10 = -6$, or $-4 = 0$, the equation has ***no solution***. An equation that has no solution is called an ***inconsistent equation***. The equation $3(x - 4) + x + 2 = 6x - 2(x + 3)$ is inconsistent and thus has no solution.

EXAMPLE 13 *An Equation with Infinitely Many Solutions*

Solve $3(x + 2) - 5(x - 3) = -2x + 21$.

SOLUTION:

$$
\begin{aligned}
3(x + 2) - 5(x - 3) &= -2x + 21 \\
3x + 6 - 5x + 15 &= -2x + 21 && \text{Distributive property} \\
-2x + 21 &= -2x + 21 && \text{Combine like terms.}
\end{aligned}
$$

Note that at this point both sides of the equation are the same. Every real number will satisfy this equation. This equation has an infinite number of solutions. An equation of this type is called an *identity*. When solving an equation, if you notice that the same expression appears on both sides of the equal sign, the equation is an identity. The solution to any linear equation that is an identity is ***all real numbers***. If you continue to solve an equation that is an identity, you will end up with $0 = 0$, as follows.

$$
\begin{aligned}
-2x + 21 &= -2x + 21 \\
-2x + 2x + 21 &= -2x + 2x + 21 && \text{Addition property} \\
21 &= 21 && \text{Combine like terms.} \\
21 - 21 &= 21 - 21 && \text{Subtraction property} \\
0 &= 0 && \text{True for any value of } x
\end{aligned}
$$

Proportions

A *ratio* is a quotient of two quantities. An example is the ratio of 2 to 5, which can be written 2 : 5 or $\frac{2}{5}$ or 2/5.

A **proportion** is a statement of equality between two ratios.

An example of a proportion is $\frac{a}{b} = \frac{c}{d}$. Consider the proportion

$$\frac{x + 2}{5} = \frac{x + 5}{8}$$

We can solve this proportion by first multiplying both sides of the equation by the least common denominator, 40.

$$
\begin{aligned}
\frac{x + 2}{5} &= \frac{x + 5}{8} \\
\overset{8}{\cancel{40}}\left(\frac{x + 2}{\cancel{5}}\right) &= \overset{5}{\cancel{40}}\left(\frac{x + 5}{\cancel{8}}\right) && \text{Multiplication property} \\
8(x + 2) &= 5(x + 5) \\
8x + 16 &= 5x + 25 \\
3x + 16 &= 25 \\
3x &= 9 \\
x &= 3
\end{aligned}
$$

A check will show that 3 is the solution.

Proportions can often be solved more easily by using cross multiplication.

Cross Multiplication

If $\frac{a}{b} = \frac{c}{d}$, then $ad = bc$, $\quad b \neq 0, d \neq 0$.

Let's use cross multiplication to solve the proportion $\frac{x+2}{5} = \frac{x+5}{8}$.

$$\begin{aligned} \frac{x+2}{5} &= \frac{x+5}{8} \\ 8(x+2) &= 5(x+5) \quad \text{Cross multiplication} \\ 8x + 16 &= 5x + 25 \\ 3x + 16 &= 25 \\ 3x &= 9 \\ x &= 3 \end{aligned}$$

Many practical application problems can be solved using proportions.

To Solve Application Problems Using Proportions

1. Represent the unknown quantity by a variable.
2. Set up the proportion by listing the given ratio on the left-hand side of the equal sign and the unknown and other given quantity on the right-hand side of the equal sign. When setting up the right-hand side of the proportion, the same respective quantities should occupy the same respective positions on the left and right. For example, an acceptable proportion might be

$$\frac{\text{miles}}{\text{hour}} = \frac{\text{miles}}{\text{hour}}$$

3. Once the proportion is properly written, drop the units and use cross multiplication to solve the equation.
4. Answer the question or questions asked.

EXAMPLE 14 *Water Usage*

The cost for water in Orange County is \$1.42 per 750 gallons (gal) of water used. What is the water bill if 30,000 gallons are used?

SOLUTION: This problem may be solved by setting up a proportion. One proportion that can be used is

$$\frac{\text{cost of 750 gal}}{\text{750 gal}} = \frac{\text{cost of 30,000 gal}}{\text{30,000 gal}}$$

We want to find the cost for 30,000 gallons of water, so we will call this quantity x. The proportion then becomes

$$\text{Given ratio} \left\{ \frac{1.42}{750} \right. = \frac{x}{30{,}000}$$

Now we solve for x.

$$\begin{aligned} (1.42)(30{,}000) &= 750x \\ 42{,}600 &= 750x \\ \frac{42{,}600}{750} &= \frac{750x}{750} \\ \$56.80 &= x \end{aligned}$$

The cost of 30,000 gallons of water is \$56.80.

EXAMPLE 15 *Determining the Amount of Insulin*

Insulin comes in 10 cubic centimeter (cc) vials labeled in the number of units of insulin per cubic centimeter of fluid. A vial of insulin marked U40 has 40 units of insulin per cubic centimeter of fluid. If a patient needs 30 units of insulin, how much fluid should be drawn into the syringe from the U40 vial?

SOLUTION: The unknown quantity, x, is the number of cubic centimeters of fluid to be drawn into the syringe. Following is one proportion that can be used to find that quantity.

Given ratio
$$\frac{40 \text{ units}}{1 \text{ cc}} = \frac{30 \text{ units}}{x \text{ cc}}$$
$$40x = 30(1)$$
$$40x = 30$$
$$x = \frac{30}{40} = 0.75$$

The nurse or doctor putting the insulin in the syringe should draw 0.75 cc of the fluid.

SECTION 6.2 EXERCISES

Concept/Writing Exercises

1. Define and give an example of a *term.*
2. Define and give an example of *like terms.*
3. Define and give an example of a *numerical coefficient.*
4. Define and give an example of a *linear equation.*
5. Explain how to simplify an expression. Give an example.
6. State the addition property of equality. Give an example.
7. State the subtraction property of equality. Give an example.
8. State the multiplication property of equality. Give an example.
9. State the division property of equality. Give an example.
10. Define *algorithm.*
11. Define and give an example of a *ratio.*
12. Define and give an example of a *proportion.*
13. Are $3x$ and $\frac{1}{2}x$ like terms? Explain.
14. Are $4x$ and $4y$ like terms? Explain.

Practice the Skills

In Exercises 15–38, combine like terms.

15. $2x + 9x$
16. $-4x - 7x$
17. $5x - 3x + 12$
18. $-6x + 3x + 21$
19. $7x + 3y - 4x + 8y$
20. $x - 4x + 3$
21. $-3x + 2 - 5x$
22. $-3x + 4x - 2 + 5$
23. $2 - 3x - 2x + 1$
24. $-0.2x + 1.7x - 4$
25. $6.2x - 8.3 + 7.1x$
26. $\frac{2}{3}x + \frac{1}{6}x - 5$
27. $\frac{1}{5}x - \frac{1}{3}x - 4$
28. $7t + 5s + 9 - 3t - 2s - 12$
29. $5x - 4y - 3y + 8x + 3$
30. $3(p + 2) - 4(p + 3)$
31. $2(s + 3) + 6(s - 4) + 1$
32. $6(r - 3) - 2(r + 5) + 10$
33. $0.3(x + 2) + 1.2(x - 4)$
34. $\frac{1}{5}(x + 2) - \frac{1}{10}x$
35. $\frac{2}{3}x + \frac{3}{7} - \frac{1}{4}x$
36. $n - \frac{3}{4} + \frac{5}{9}n - \frac{1}{6}$
37. $0.5(2.6x - 4) + 2.3(1.4x - 5)$
38. $\frac{2}{3}(3x + 9) - \frac{1}{4}(2x + 5)$

In Exercises 39–64, solve the equation.

39. $y + 8 = 13$
40. $2y - 7 = 17$
41. $9 = 12 - 3x$
42. $14 = 3x + 5$
43. $\frac{3}{x} = \frac{7}{8}$
44. $\frac{x - 1}{5} = \frac{x + 5}{15}$
45. $\frac{1}{2}x + \frac{1}{3} = \frac{2}{3}$
46. $\frac{1}{2}y + \frac{1}{3} = \frac{1}{4}$
47. $0.7x - 0.3 = 1.8$
48. $5x + 0.050 = -0.732$

49. $6t - 8 = 4t - 2$

50. $\frac{x}{4} + 2x = \frac{1}{3}$

51. $\frac{x-3}{2} = \frac{x+4}{3}$

52. $\frac{x-5}{4} = \frac{x-9}{3}$

53. $6t - 7 = 8t + 9$

54. $12x - 1.2 = 3x + 1.5$

55. $2(x + 3) - 4 = 2(x - 4)$

56. $3(x + 2) + 2(x - 1) = 5x - 7$

57. $4(x - 4) + 12 = 4(x - 1)$

58. $\frac{y}{3} + 4 = \frac{2y}{5} - 6$

59. $\frac{1}{4}(x + 4) = \frac{2}{5}(x + 2)$

60. $\frac{2}{3}(x + 5) = \frac{1}{4}(x + 2)$

61. $3x + 2 - 6x = -x - 15 + 8 - 5x$

62. $6x + 8 - 22x = 28 + 14x - 10 + 12x$

63. $2(t - 3) + 2 = 2(2t - 6)$

64. $5.7x - 3.1(x + 5) = 7.3$

Problem Solving

In Exercises 65 and 66, use the DeKalb County water rate of \$2.05 per 1000 gallons of water used.

65. *Water Bill* What is the water bill if a resident uses 35,300 gal?

66. *Limiting the Cost* How many gallons of water can the customer use if the water bill is not to exceed \$40.68?

67. *Dial Bodywash* A bottle of Dial Bodywash contains 354 milliliters (ml) of soap. If Tony Vaszquez uses 6 ml for each shower, how many times can he shower using one bottle of Dial Bodywash?

68. *Fajitas* A recipe for six servings of beef fajitas requires 16 oz of beef sirloin.
   a) If the recipe were to be made for nine servings, how many ounces of beef sirloin would be needed?
   b) How many servings of beef fajitas can be made with 32 oz of beef sirloin?

69. *Watching Television* Nielson Media Research determines the number of people who watch a television show. One rating point means that about 1,022,000 households watched the show. The top-rated television show for the week of September 23, 2002, was *Friends*, with a rating of 20.3. About how many households watched *Friends* that week?

70. *Grass Seed Coverage* A 20 lb bag of grass seed will cover an area of 10,000 ft^2.
   a) How many pounds are needed to cover an area of 140,000 ft^2?
   b) How many bags of grass seed must be purchased to cover an area of 140,000 ft^2?

71. *Speed Limit* When Jacob Abbott crossed over from Niagara Falls, New York, to Niagara Falls, Canada, he saw a sign that said 50 miles per hour (mph) is equal to 80 kilometers per hour (kph).

   a) How many kilometers per hour are equal to 1 mph?
   b) On a stretch of the Queen Elizabeth Way, the speed limit is 90 kph. What is the speed limit in miles per hour?

72. *The Proper Dosage* A doctor asks a nurse to give a patient 250 milligrams (mg) of the drug Simethicone. The drug is available only in a solution whose concentration is 40 mg Simethicone per 0.6 millimeter (mm) of solution. How many millimeters of solution should the nurse give the patient?

Amount of Insulin In Exercises 73 and 74, how much insulin (in cc) would be given for the following doses? (Refer to Example 15 on page 302.)

73. 12 units of insulin from a vial marked U40

74. 35 units of insulin from a vial marked U40

75. a) In your own words, summarize the procedure to use to solve an equation.
   b) Solve the equation $2(x + 3) = 4x + 3 - 5x$ with the procedure you outlined in part (a).

76. a) What is an identity?
   b) When solving an equation, how will you know if the equation is an identity?

77. a) What is an inconsistent equation?
   b) When solving an equation, how will you know if the equation is inconsistent?

Challenge Problems/Group Activities

78. *Depth of a Submarine* The pressure, P, in pounds per square inch (psi), exerted on an object x ft below the sea is given by the formula $P = 14.70 + 0.43x$. The 14.70 represents the weight in pounds of the column of air (from sea level to the top of the atmosphere) standing over a 1 in. by 1 in. square of seawater. The $0.43x$ represents the weight in pounds of a column of water 1 in. by 1 in. by x ft (see Fig. 6.1).

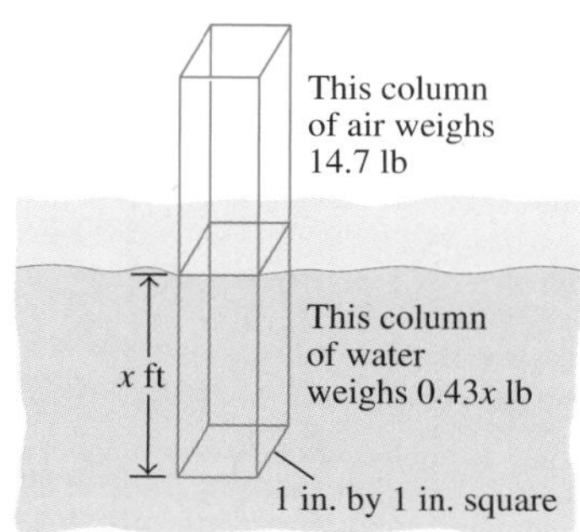

Figure 6.1

a) A submarine is built to withstand a pressure of 148 psi. How deep can that submarine go?

b) If the pressure gauge in the submarine registers a pressure of 128.65 psi, how deep is the submarine?

79. **a)** *Gender Ratios* If the ratio of males to females in a class is 2 : 3, what is the ratio of males to all the students in the class? Explain your answer.

b) If the ratio of males to females in a class is $m : n$, what is the ratio of males to all the students in the class?

Internet/Research Activities

80. Ratio and proportion are used in many different ways in everyday life. Submit two articles from newspapers, magazines, or the Internet in which ratios and/or proportions are used. Write a brief summary of each article explaining how ratio and/or proportion were used.

81. Write a report explaining how the ancient Egyptians used equations. Include in your discussion the forms of the equations used.

6.3 FORMULAS

A *formula* is an equation that typically has a real-life application. To *evaluate a formula*, substitute the given values for their respective variables and then evaluate using the order of operations given in Section 6.1. Many of the formulas given in this section are discussed in greater detail in other parts of the book.

EXAMPLE 1 *Simple Interest*

The simple interest formula*, interest = principal × rate × time, or $i = prt$, is used to find the interest you must pay on a simple interest loan when you borrow principal, p, at simple interest rate, r, in decimal form, for time, t. Chris Campbell borrows $3000 at a simple interest rate of 9% for 3 years.

a) How much will Chris Campbell pay in interest at the end of 3 years?

b) What is the total amount he will repay the bank at the end of 3 years?

SOLUTION:

a) Substitute the values of p, r, and t into the formula; then evaluate.

$$\begin{aligned} i &= prt \\ &= 3000(0.09)(3) \\ &= 810 \end{aligned}$$

Thus, Chris must pay $810 interest.

b) The total he must pay at the end of 3 years is the principal, $3000, plus the $810 interest, for a total of $3810. ▲

EXAMPLE 2 *Volume of a Cereal Box*

The formula for the volume of a box* is volume = length × width × height or $V = lwh$. Use the formula $V = lwh$ to find the width of a Sweet Treats cereal box if $l = 7.5$ in., $h = 10.5$ in., and $V = 196.875$ in.^{3}.

*The simple interest formula is discussed in Section 10.2. The volume formula is discussed in Section 9.4.

SOLUTION: We substitute the appropriate values into the volume formula and solve for the desired quantity, w.

$$\begin{aligned} V &= lwh \\ 196.875 &= (7.5)w(10.5) \\ 196.875 &= 78.75w \\ \frac{196.875}{78.75} &= w \\ 2.5 &= w \end{aligned}$$

Therefore, the width of a Sweet Treats cereal box is 2.5 in. ▲

In Example 1, we used the formula $i = prt$. In Example 2, we used the formula $V = lwh$. In these examples, we used a mathematical equation to represent real phenomena. When we represent real phenomena, such as finding simple interest, mathematically we say we have created a *mathematical model* or simply a *model* to represent the situation. A model may be a single formula, or equation, or a system of many equations. By using models we gain insight into real-life situations, such as how much interest you will accumulate in your savings account. We will use mathematical models throughout this chapter and elsewhere in the text. In some exercises in this and the next chapter, when you are asked to determine an equation to represent a real-life situation, we will sometimes write the word *model* in the instructions.

Many formulas contain Greek letters, such as μ (mu), σ (sigma), Σ (capital sigma), δ (delta), ϵ (epsilon), π (pi), θ (theta), and λ (lambda). Example 3 makes use of Greek letters.

EXAMPLE 3 *A Statistics Formula*

A formula used in the study of statistics to find the standard score (or z-score) is

$$z = \frac{\bar{x} - \mu}{\frac{\sigma}{\sqrt{n}}}$$

Find the value of z when $\bar{x}$ (read "x bar") $= 120$, $\mu = 100$, $\sigma = 16$, and $n = 4$.

SOLUTION:

$$z = \frac{\bar{x} - \mu}{\frac{\sigma}{\sqrt{n}}} = \frac{120 - 100}{\frac{16}{\sqrt{4}}} = \frac{20}{\frac{16}{2}} = \frac{20}{8} = 2.5$$ ▲

Some formulas contain *subscripts*. Subscripts are numbers (or letters) placed below and to the right of variables. They are used to help clarify a formula. For example, if two different amounts are used in a problem, they may be symbolized as A and A_0, or A_1 and A_2. Subscripts are read using the word *sub*; for example, A_0 is read "A sub zero" and A_1 is read "A sub one."

Exponential Equations

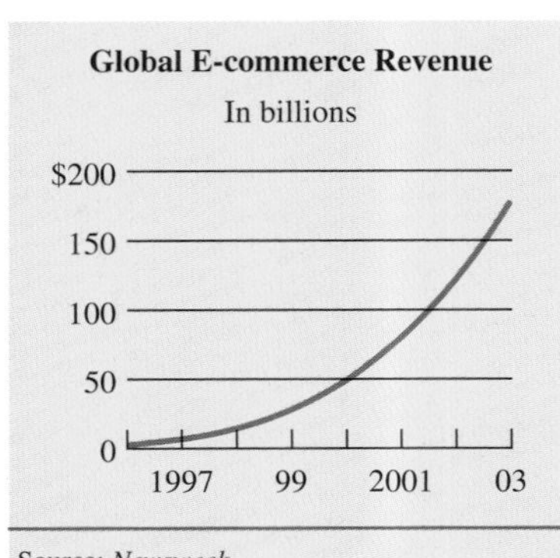

Figure 6.2

Many real-life problems, including population growth, growth of bacteria, and decay of radioactive substances, increase or decrease at a very rapid rate. For example, in Fig. 6.2, which shows global electronic business revenue, in billions of dollars, from 1996 through 2003, the graph is increasing rapidly. This is an example where the graph is increasing *exponentially*. The equation of a graph that increases or decreases exponentially is called an *exponential equation* (or *exponential formula*). An exponential equation is of the form $y = a^x, a > 0, a \neq 1$. We often use exponential equations to model real-life problems. In Section 6.10, we will discuss exponential equations (and exponential formulas) in more detail.

In an exponential formula, letters other than x and y may be used to represent the variables. The following equations are examples of exponential formulas: $y = 2^x$, $A = (\frac{1}{2})^x$, and $P = 2.3^t$. Note in the exponential formula that the variable is the exponent of some positive constant that is not equal to 1. In many real-life applications, the variable t will be used to represent time. Problems involving exponential formulas can be evaluated much more easily if you use a calculator containing a $\boxed{y^x}$, $\boxed{x^y}$, or $\boxed{\wedge}$ key.

The following formula, referred to as the *exponential growth* or *decay formula*, is used to solve many real-life problems.

$$P = P_0 a^{kt}, \qquad a > 0, \quad a \neq 1$$

In the formula, P_0 represents the original amount present, P represents the amount present after t years, and a and k are constants.

When $k > 0$, P increases as t increases and we have exponential growth. When $k < 0$, P decreases as t increases and we have exponential decay.

EXAMPLE 4 *Using an Exponential Decay Formula*

Carbon dating is used by scientists to find the age of fossils, bones, and other items. The formula used in carbon dating is

$$P = P_0 2^{-t/5600}$$

where P_0 represents the original amount of carbon 14 (C_{14}) present and P represents the amount of C_{14} present after t years. If 10 mg of C_{14} is present in an animal bone recently excavated, how many milligrams will be present in 3000 years?

SOLUTION: Substituting the values in the formula gives

$$P = P_0 2^{-t/5600}$$
$$P = 10(2)^{-3000/5600}$$
$$P \approx 10(2)^{-0.54} \qquad \text{(Recall that } \approx \text{ means "is approximately equal to")}$$
$$P \approx 10(0.69)$$
$$P \approx 6.9 \text{ mg}$$

Thus, in 3000 years, approximately 6.9 mg of the original 10 mg of C_{14} will remain. ▲

In Example 4, we used a calculator to evaluate $(2)^{-3000/5600}$. The steps used to find this quantity on a calculator with a [y^x] key are

2 [y^x] [(] 3000 [+/−] [÷] 5600 [)] [=] .6898170602

After the [=] key is pressed, the calculator displays the answer 0.689817. To evaluate $10(2)^{-3000/5600}$ on a scientific calculator, we can press the following keys.

10 [×] 2 [y^x] [(] 3000 [+/−] [÷] 5600 [)] [=] 6.898170602

Notice that the answer obtained using the calculator steps shown above is a little more accurate than the answer we gave when we rounded the values before the final answer in Example 4.

When the a in the formula $P = P_0a^{kt}$ is replaced with the very special letter e, we get the ***natural exponential formula***

$$P = P_0e^{kt}$$

The letter e represents an irrational number whose value is approximately 2.7183. The number e plays an important role in mathematics and is used in finding the solution to many application problems.

To evaluate $e^{(0.04)5}$ on a calculator, as will be needed in Example 5, press*

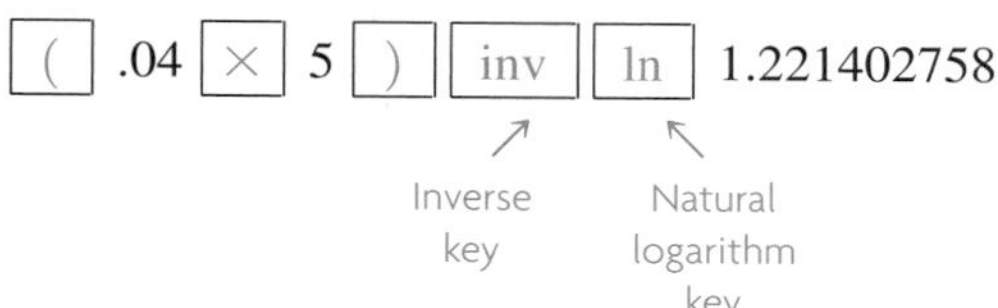

After the [ln] key is pressed, the calculator displays the answer 1.221402758.

To evaluate $10{,}000e^{(0.04)5}$ on a calculator, press

10000 [×] [(] .04 [×] 5 [)] [inv] [ln] [=] 12214.02758

In this calculation, after the [=] key is pressed, the calculator displays the answer 12214.02758.

EXAMPLE 5 *Using an Exponential Growth Formula*

Banks often credit compound interest continuously. When that is done, the principal amount in the account, P, at any time t can be calculated by the natural exponential formula $P = P_0e^{kt}$, where P_0 is the initial principal invested, k is the interest rate in decimal form, and t is the time.

*Keys to press may vary on some calculators.

Suppose \$10,000 is invested in a savings account at a 4% interest rate compounded continuously. What will be the balance (or principal) in the account in 5 years?

SOLUTION:

$$\begin{aligned} P &= P_0e^{kt} \\ &= 10{,}000e^{(0.04)5} \\ &= 10{,}000e^{(0.20)} \\ &\approx 10{,}000(1.221402758) \\ &\approx 12{,}214 \end{aligned}$$

Thus, after 5 years, the account's value will have grown from \$10,000 to about \$12,214, an increase of about \$2214. ▲

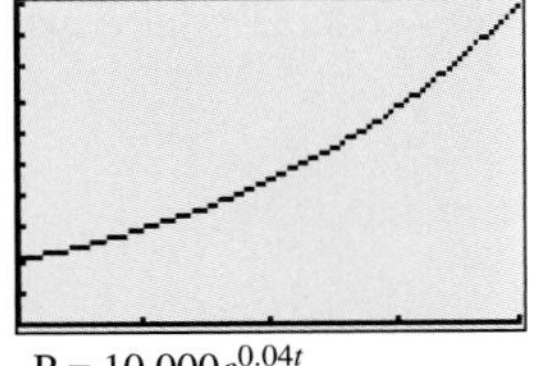

$P = 10{,}000e^{0.04t}$

Figure 6.3

Graphing calculators are a tool that can be used to graph equations. Figure 6.3 shows the graph of $P = 10{,}000e^{0.04t}$ as it appears on the screen (or window) of a Texas Instrument TI-83 Plus graphing calculator. To obtain this screen, the domain (or the x-values) and range (or the y-values) of the window need to be set to selected values. We will speak a little more about graphing calculators shortly. In Section 6.10, we will explain how to graph exponential equations by plotting points.

EXAMPLE 6 *Population of Nevada*

The population of Nevada, which was the fastest growing state in every decade of the twentieth century except for the 1950s, is continuing to grow exponentially at the rate of about 5.10% per year. In 2000, the population of Nevada was 1,998,257. Nevada's expected population, t years after 2000 is given by the formula $P = 1{,}998{,}257e^{0.0510t}$. Find the expected population of Nevada in the year 2010.

SOLUTION: Since 2010 is 10 years after 2000, $t = 10$ years.

$$\begin{aligned} P &= 1{,}998{,}257e^{0.0510t} \\ &= 1{,}998{,}257e^{0.0510(10)} && \text{Substitute 10 for } t. \\ &= 1{,}998{,}257e^{0.510} \\ &\approx 1{,}998{,}257(1.665291195) \\ &\approx 3{,}327{,}679.787 \end{aligned}$$

Thus, in the year 2010, the population of Nevada is expected to be about 3,327,697 people. ▲

TIMELY TIP When doing calculations on the calculator, do not round any value before obtaining the final answer. By not rounding, you will obtain a more accurate answer. For example, if we work Example 6 on a calculator, and rounded $e^{0.510}$ to 1.67, we would determine that the population of Nevada in 2010 is expected to be about 3,337,089 which is a less accurate answer.

DID YOU KNOW

$L = mc^2$
$E = mc^2$

Many of you recognize the formula $E = mc^2$ used by Albert Einstein in 1912 to describe his groundbreaking theory of relativity. In the formula, E is energy, m is mass, and c is the speed of light. In his theory, Einstein hypothesized that time was not absolute and that mass and energy were related.

Einstein's 72-page handwritten manuscript, which went on display in 1999 at Jerusalem's Israel Museum, shows that Einstein toyed with using an L rather than an E to represent energy. He wrote $L = mc^2$, and then scratched out the L and replaced it with an E to produce the equation $E = mc^2$. Yisrail Galili at Jerusalem's Hebrew University said the L might have been a lambda, a Greek letter sometimes used in functions related to energy. He said it might also be related to Joseph Lagrange, an eighteenth-century Italian mathematician.

Solving for a Variable in a Formula or Equation

Often in mathematics and science courses, you are given a formula or an equation expressed in terms of one variable and asked to express it in terms of a different variable. For example, you may be given the formula $P = i^2r$ and asked to solve the formula for r. To do so, treat each of the variables, except the one you are solving for, as if it were a constant. Then solve for the variable desired, using the properties previously discussed. Examples 7 through 9 show how to do this task.

When graphing equations in Section 6.7, you will sometimes have to solve the equation for the variable y as is done in Example 7.

EXAMPLE 7 *Solving an Equation Containing More Than One Variable*

Solve the equation $2x + 5y - 10 = 0$ for y.

SOLUTION: We need to isolate the term containing the variable y. Begin by moving the constant, -10, and the term $2x$ to the right-hand side of the equation.

$$2x + 5y - 10 = 0$$

$$2x + 5y - 10 + 10 = 0 + 10 \quad \text{Addition property}$$

$$2x + 5y = 10$$

$$-2x + 2x + 5y = -2x + 10 \quad \text{Subtraction property}$$

$$5y = -2x + 10$$

$$\frac{5y}{5} = \frac{-2x + 10}{5} \quad \text{Division property}$$

$$y = \frac{-2x + 10}{5}$$

$$y = -\frac{2x}{5} + \frac{10}{5}$$

$$y = -\frac{2}{5}x + 2$$

▲

Note that once you have found $y = \frac{-2x + 10}{5}$, you have solved the equation for y. The solution can also be expressed in the form $y = -\frac{2}{5}x + 2$. This form of the equation is convenient for graphing equations, as will be explained in Section 6.7. Example 7 can also be solved by moving the y term to the right-hand side of the equal sign. Do so now and note that you obtain the same answer.

EXAMPLE 8 *Solving for a Variable in a Formula*

An important formula used in statistics is

$$z = \frac{x - \mu}{\sigma}$$

Solve this formula for x.

SOLUTION: To isolate the term x, use the general procedure for solving linear equations given in Section 6.2. Treat each letter, except x, as if it were a constant.

$$z = \frac{x - \mu}{\sigma}$$

$$z \cdot \sigma = \frac{x - \mu}{\cancel{\sigma}} \cdot \cancel{\sigma} \qquad \text{Multiplication property}$$

$$z\sigma = x - \mu$$

$$z\sigma + \mu = x - \mu + \mu \qquad \text{Add } \mu \text{ to both sides of the equation.}$$

$$z\sigma + \mu = x$$

$$\text{or} \qquad x = z\sigma + \mu$$

▲

EXAMPLE 9 *The Tax Free Yield Formula*

A formula that may be important to you now or sometime in the future is the tax-free yield formula, $T_f = T_a(1 - F)$. This formula can be used to convert a taxable yield, T_a, into its equivalent tax-free yield, T_f, where F is the federal income tax bracket of the individual. A taxable yield is an interest rate for which income tax is paid on the interest made. A tax-free yield is an interest rate for which income tax does not have to be paid on the interest made.

PROFILE IN MATHEMATICS

SOPHIE GERMAIN

Because she was a woman, Sophie Germain (1776–1831) was denied admission to the École Polytechnic, the French academy of mathematics and science. Not to be stopped, she obtained lecture notes from courses in which she had an interest, including one taught by Joseph-Louis Lagrange. Under the pen name M. LeBlanc, she submitted a paper on analysis to Lagrange, who was so impressed with the report that he wanted to meet the author and personally congratulate "him." When he found out that the author was a woman, he became a great help and encouragement to her. Lagrange introduced Germain to many of the French scientists of the time.

In 1801, Germain wrote the great German mathematician Carl Friedrich Gauss to discuss Fermat's equation, $x^n + y^n = z^n$. He commended her for showing "the noblest courage, quite extraordinary talents and a superior genius." Germain's interests included work in number theory and mathematical physics. She would have received an honorary doctorate from the University of Göttingen, based on Gauss's recommendation, but died before the honorary doctorate could be awarded.

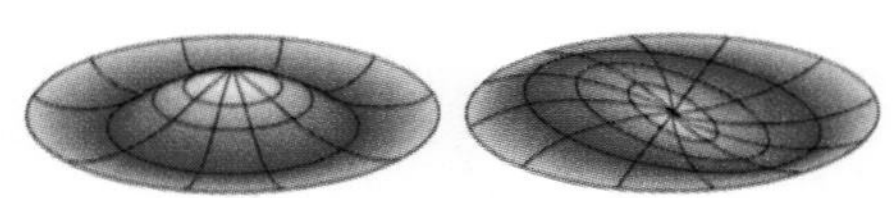

Germain was the first person to devise a formula describing elastic motion. The study of the equations for the elasticity of different materials aided the development of acoustical diaphragms in loudspeakers and telephones.

a) For someone in a 25% tax bracket, find the equivalent tax-free yield of a 4% taxable investment.

b) Solve this formula for T_a. That is, write a formula for taxable yield in terms of tax-free yield.

SOLUTION:

a) $T_f = T_a(1 - F)$

$= 0.04(1 - 0.25) = 0.04(0.75) = 0.03, \quad \text{or} \quad 3\%$

Thus, a taxable investment of 4% is equivalent to a tax-free investment of 3% for a person in a 25% income tax bracket.

b) $T_f = T_a(1 - F)$

$$\frac{T_f}{1 - F} = \frac{T_a\cancel{(1 - F)}}{\cancel{1 - F}}$$ Divide both sides of the equation by $1 - F$.

$$\frac{T_f}{1 - F} = T_a, \quad \text{or} \quad T_a = \frac{T_f}{1 - F}$$

SECTION 6.3 EXERCISES

Concept/Writing Exercises

1. What is a formula?
2. Explain how to evaluate a formula.
3. What are subscripts?
4. What is the simple interest formula?
5. What is an exponential equation?
6. a) In an exponential equation of the form $y = a^x$, what are the restrictions on a?
   b) In an exponential equation of the form $y = P_0a^{kt}$, what does P_0 represent?

Practice the Skills

In Exercises 7–38, use the formula to find the value of the indicated variable for the values given. Use a calculator when one is needed. When necessary, round answers to the nearest hundredth.

7. $P = 4s$; find P when $s = 5$ (geometry).
8. $P = a + b + c$; find P when $a = 25$, $b = 53$ and $c = 32$ (geometry).
9. $P = 2l + 2w$; find P when $l = 12$ and $w = 16$ (geometry).
10. $F = ma$; find m when $F = 40$ and $a = 5$ (physics).
11. $E = mc^2$; find m when $E = 400$ and $c = 4$ (physics).
12. $p = i^2r$; find r when $p = 62{,}500$ and $i = 5$ (electronics).
13. $A = \pi(R^2 - r^2)$; find A when $R = 6$, $\pi = 3.14$, and $r = 4$ (geometry).
14. $B = \dfrac{703w}{h^2}$; find B when $w = 130$ and $h = 67$ (for finding body mass index).
15. $z = \dfrac{x - \mu}{\sigma}$; find μ when $z = 2.5$, $x = 42.1$, and $\sigma = 2$ (statistics).
16. $S = B + \frac{1}{2}Ps$; find P when $s = 10$, $S = 300$, and $B = 100$ (geometry).
17. $T = \dfrac{PV}{k}$; find P when $T = 80$, $V = 20$, and $k = 0.5$ (physics).
18. $m = \dfrac{a + b + c}{3}$; find a when $m = 70$, $b = 60$, and $c = 90$ (statistics).
19. $A = P(1 + rt)$; find P when $A = 3600$, $r = 0.04$, and $t = 5$ (economics).
20. $m = \dfrac{a + b}{2}$; find a when $m = 70$ and $b = 77$ (statistics).
21. $v = \frac{1}{2}at^2$; find a when $v = 576$ and $t = 12$ (physics).
22. $F = \frac{9}{5}C + 32$; find F when $C = 7$ (temperature conversion).
23. $C = \frac{5}{9}(F - 32)$; find C when $F = 77$ (temperature conversion).

24. $K = \frac{F - 32}{1.8} + 273.1$; find K when F = 100 (chemistry).

25. $m = \frac{y_2 - y_1}{x_2 - x_1}$; find m when $y_2 = 8$, $y_1 = -4$, $x_2 = -3$, and $x_1 = -5$ (mathematics).

26. $z = \frac{\bar{x} - \mu}{\frac{\sigma}{\sqrt{n}}}$; find z when $\bar{x} = 66$, $\mu = 60$, $\sigma = 15$, and $n = 25$ (statistics).

27. $S = R - rR$; find R when $S = 186$ and $r = 0.07$ (for determining sale price when an item is discounted).

28. $S = C + rC$; find C when $S = 115$ and $r = 0.15$ (for determining selling price when an item is marked up).

29. $E = a_1p_1 + a_2p_2 + a_3p_3$; find E when $a_1 = 5$, $p_1 = 0.2$, $a_2 = 7$, $p_2 = 0.6$, $a_3 = 10$, and $p_3 = 0.2$ (probability).

30. $x = \frac{-b + \sqrt{b^2 - 4ac}}{2a}$; find x when $a = 2$, $b = -5$, and $c = -12$ (mathematics).

31. $s = -16t^2 + v_0t + s_0$; find s when $t = 4$, $v_0 = 30$, and $s_0 = 150$ (physics).

32. $R = O + (V - D)r$; find O when $R = 670$, $V = 100$, $D = 10$, and $r = 4$ (economics).

33. $P = \frac{f}{1 + i}$; find f when $i = 0.08$ and $P = 3000$ (investment banking).

34. $c = \sqrt{a^2 + b^2}$; find c when $a = 5$ and $b = 12$ (geometry).

35. $F = \frac{Gm_1m_2}{r^2}$; find G when $F = 625$, $m_1 = 100$, $m_2 = 200$, and $r = 4$ (physics).

36. $P = \frac{nRT}{V}$; find V if $P = 12$, $n = 10$, $R = 60$, and $T = 8$ (chemistry).

37. $S_n = \frac{a_1(1 - r^n)}{1 - r}$; find S_n when $a_1 = 8$, $r = \frac{1}{2}$, and $n = 3$ (mathematics).

38. $A = P\left(1 + \frac{r}{n}\right)^{nt}$; find A when $P = 100$, $r = 6\%$, $n = 1$, and $t = 3$ (banking).

In Exercises 39–48, solve the equation for y.

39. $10x - 4y = 13$
40. $8x - 6y = 21$
41. $4x + 7y = 14$
42. $-2x + 4y = 9$
43. $2x - 3y + 6 = 0$
44. $3x + 4y = 0$
45. $-2x + 3y + z = 15$
46. $5x + 3y - 2z = 22$
47. $9x + 4z = 7 + 8y$
48. $2x - 3y + 5z = 0$

In Exercises 49–68, solve for the variable indicated.

49. $E = IR$ for R
50. $p = irt$ for t
51. $p = a + b + c$ for a
52. $p = a + b + s_1 + s_2$ for s_1
53. $V = \frac{1}{3}Bh$ for B
54. $V = \pi r^2h$ for h
55. $C = 2\pi r$ for r
56. $r = \frac{2gm}{c^2}$ for m
57. $y = mx + b$ for b
58. $y = mx + b$ for m
59. $P = 2l + 2w$ for w
60. $A = \frac{d_1d_2}{2}$ for d_2
61. $A = \frac{a + b + c}{3}$ for c
62. $A = \frac{1}{2}bh$ for b
63. $P = \frac{KT}{V}$ for T
64. $\frac{P_1V_1}{T_1} = \frac{P_2V_2}{T_2}$ for V_2
65. $F = \frac{9}{5}C + 32$ for C
66. $C = \frac{5}{9}(F - 32)$ for F
67. $S = \pi r^2 + \pi rs$ for s
68. $A = \frac{1}{2}h(b_1 + b_2)$ for b_2

Problem Solving

69. *Refund Check* Joel and Patti Karpel received a \$600 income tax refund check from the federal government and decided to deposit the check in a money market account that paid 2% simple interest per year. Determine
  a) how much interest was added to their account at the end of 1 year.
  b) the balance in their account at the end of 1 year.

70. *Interest on a Loan* Jeff Hubbard borrowed \$800 from his brother for 2 years. At the end of 2 years, he repaid the \$800 plus \$128 in interest. What simple interest rate did he pay?

71. *Volume in a Soup Can* Determine the volume of a cylindrical soup can if its diameter is 2.5 in. and its height is 3.75 in. (The formula for the volume of a cylinder is $V = \pi r^2h$. Use your pi key, $\boxed{\pi}$, on your calculator, or 3.14 for π if your calculator does not have a $\boxed{\pi}$ key.) Round your answer to the nearest tenth.

72. *Body Mass Index* A person's body mass index (BMI) is found by the formula $B = \frac{703w}{h^2}$, where w is the person's

weight, in pounds, and h is the person's height, in inches. Lance Bass is 6 ft tall and weighs 200 lb.

a) Determine his BMI.

b) If Lance would like to have a BMI of 26, how much weight would he need to gain or lose?

73. *Bacteria* The number of a certain type of bacteria, y, present in a culture is determined by the equation $y = 2000(3)^x$, where x is the number of days the culture has been growing. Find the number of bacteria present after 5 days.

74. *Adjusting for Inflation* If P is the price of an item today, the price of the same item n years from today, P_n, is $P_n = P(1 + r)^n$, where r is the constant rate of inflation. Determine the price of a movie ticket 10 years from today if the price today is \$8.00 and the annual rate of inflation is constant at 3%.

75. *Value of New York City* Assume the value of the island of Manhattan has grown at an exponential rate of 8% per year since 1626 when Peter Minuit of the Dutch West India Company purchased the island for \$24. The value of the island, V, at any time, t, in years after 1626, can be found by the formula $V = 24e^{0.08t}$. What is the value of the island in 2003, 377 years after Minuit purchased it?

76. *Radioactive Decay* Strontium 90 is a radioactive isotope that decays exponentially at a rate of 2.8% per year. The amount of strontium 90, S, remaining after t years can be found by the formula $S = S_0e^{-0.028t}$, where S_0 is the original amount present. If there are originally 1000 g of strontium 90, find the amount of strontium 90 remaining after 30 years.

Challenge Problem/Group Activity

77. Determine the volume of the block shown in Fig. 6.4, excluding the hole.

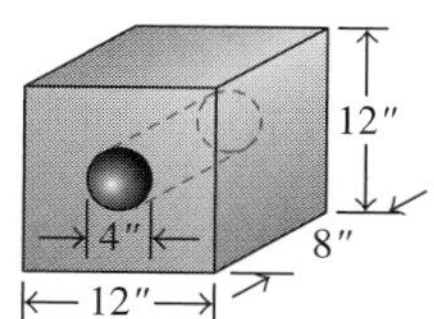

Figure 6.4

Recreational Mathematics

78. *Triangle Seek* In this word seek, you are looking for six-letter words that form triangles. The first letter of the words can be in any position in the triangle; the remaining letters of the words are in order moving clockwise or counterclockwise. Triangles may overlap other triangles and the triangles can point up or down. For example, the word LINEAR is indicated below. The word list is below on the right.

S F A R R L I D G R T P
C C J O A N Y E B O L F
R I E D N E N D O M S O
T E E E W T O M E Y I H
F S E R G H T L Y Z M T
A P J N R W C I N C U D
M D P P R O L R N A T A
G R O I V E U U F R O U

Word List
LINEAR
SYMBOL
DEGREE
FACTOR
GROWTH
NEWTON

6.4 APPLICATIONS OF LINEAR EQUATIONS IN ONE VARIABLE

One of the main reasons for studying algebra is that it can be used to solve everyday problems. In this section, we will do two things: (1) show how to translate a written problem into a mathematical equation and (2) show how linear equations can be used in solving everyday problems. We begin by illustrating how English phrases can be written as mathematical expressions. When writing a mathematical expression, we may use any letter to represent the variable. In the following illustrations, we use the letter x.

Phrase	Mathematical expression
Six more than a number	$x + 6$
A number increased by 3	$x + 3$
Four less than a number	$x - 4$
A number decreased by 9	$x - 9$
Twice a number	$2x$
Four times a number	$4x$
3 decreased by a number	$3 - x$
The difference between a number and 5	$x - 5$

DID YOU KNOW

Survival of the Fittest

Symbols come and symbols go; the ones that find the greatest acceptance are the ones that survive. The Egyptians used pictorial symbols: a pair of legs walking forward for addition or backward for subtraction. Robert Recorde (1510–1558) used two parallel lines to represent "equals" because "no 2 thynges can be moore equalle." Some symbols evolved from abbreviations, such as the "+" sign, which comes from the Latin *et* meaning "and." The evolution of others is less clear. The invention of the printing press in the fifteenth century led to a greater standardization of symbols already in use.

Sometimes the phrase that must be converted to a mathematical expression involves more than one operation.

Phrase	Mathematical expression
Four less than 3 times a number	$3x - 4$
Ten more than twice a number	$2x + 10$
The sum of 5 times a number and 3	$5x + 3$
Eight times a number decreased by 7	$8x - 7$

The word *is* often represents the equal sign.

Phrase	Mathematical equation
Six more than a number is 10.	$x + 6 = 10$
Five less than a number is 20.	$x - 5 = 20$
Twice a number decreased by 6 is 12.	$2x - 6 = 12$
A number decreased by 13 is 6 times the number.	$x - 13 = 6x$

The following is a general procedure for solving word problems.

To Solve a Word Problem

1. Read the problem carefully at least twice to be sure that you understand it.
2. If possible, draw a sketch to help visualize the problem.
3. Determine which quantity you are being asked to find. Choose a letter to represent this unknown quantity. Write down exactly what this letter represents.
4. Write the word problem as an equation.
5. Solve the equation for the unknown quantity.
6. Answer the question or questions asked.
7. Check the solution.

This general procedure for solving word problems is illustrated in Examples 1 through 4. In these examples, the equations we obtain are mathematical models of the given situations.

EXAMPLE 1 *How Much Can You Purchase?*

Roberto Raynor spent $339.95 on textbooks at the bookstore. In addition to textbooks, he wanted to purchase as many notebooks as possible, but he only has a total of $350 to spend. If a notebook, including tax, costs $0.99, how many notebooks can he purchase?

SOLUTION: In this problem, the unknown quantity is the number of notebooks Roberto can purchase. Let's select n to represent the number of notebooks he can purchase. Then we construct an equation using the given information that will allow us to solve for n.

DID YOU KNOW

Aha!

Aha:

or

Heap:

The Egyptians as far back as 1650 B.C. had a knowledge of linear equations. They used the words *aha* or *heap* in place of the variable. Problems involving linear equations can be found in the Rhind Papyrus (See Chapter 4, p. 168).

Let

$$n = \text{number of notebooks Roberto can purchase}$$

Then

$$\$0.99n = \text{cost for } n \text{ notebooks at } \$0.99 \text{ per notebook}$$

$$\begin{array}{ccccc} \text{Cost of textbooks} & + & \text{cost of notebooks} & = & \text{total amount to spend} \\ \$339.95 & + & \$0.99n & = & \$350 \end{array}$$

Now solve the equation.

$$\begin{aligned} 339.95 + 0.99n &= 350 \\ 339.95 - 339.95 + 0.99n &= 350 - 339.95 \\ 0.99n &= 10.05 \\ \frac{0.99n}{0.99} &= \frac{10.05}{0.99} \\ n &\approx 10.15 \end{aligned}$$

Therefore, Roberto can purchase 10 notebooks. When we solve the equation, we obtain $n = 10.15$ (to the nearest hundredth). Since he cannot purchase a part of a notebook, only 10 notebooks can be purchased.

Check: The check is made with the information given in the original problem.

$$\begin{aligned} \text{Total amount to spend} &= \text{cost of textbooks} + \text{cost of notebooks} \\ &= 339.95 + 0.99(10) \\ &= 339.95 + 9.90 \\ &= 349.85 \end{aligned}$$

This result would leave 15 cents change from the \$350 he has to spend, which is not enough to purchase another notebook. Therefore, this answer checks. ▲

EXAMPLE 2

Forty hours of overtime must be split among three workers. One worker will be assigned twice the number of hours as each of the other two. How many hours of overtime will be assigned to each worker?

SOLUTION: Two workers receive the same amount of overtime, and the third worker receives twice that amount.

Let

$$\begin{aligned} x &= \text{number of hours of overtime for the first worker} \\ x &= \text{number of hours of overtime for the second worker} \\ 2x &= \text{number of hours of overtime for the third worker} \end{aligned}$$

Then

$$\begin{aligned} x + x + 2x &= \text{total amount of overtime} \\ x + x + 2x &= 40 \\ 4x &= 40 \\ x &= 10 \end{aligned}$$

Thus, two workers are assigned 10 hours of overtime, and the third worker is assigned 2(10), or 20, hours of overtime. A check in the original problem will verify that this answer is correct.

EXAMPLE 3 *Dimensions of a Fenced in Area*

Robert Koch wants to fence in a rectangular region in his backyard for his poodle. He only has 56 ft of fencing to use for the perimeter of the region. What should the dimensions of the region be if he wants the length to be 4 ft greater than the width?

SOLUTION: The formula for finding the perimeter of a rectangle is $P = 2l + 2w$, where P is the perimeter, l is the length, and w is the width. A diagram, such as the one shown below, is often helpful in solving problems of this type.

Let w equal the width of the region. The length is 4 ft more than the width, so $l = w + 4$. The total distance around the region P, is 56 ft.

Substitute the known quantities in the formula.

$$\begin{aligned} P &= 2w + 2l \\ 56 &= 2w + 2(w + 4) \\ 56 &= 2w + 2w + 8 \\ 56 &= 4w + 8 \\ 48 &= 4w \\ 12 &= w \end{aligned}$$

The width of the region is 12 ft, and the length of the region is $12 + 4$ or 16 ft.

In shopping and other daily activities, we are occasionally asked to solve problems using percents. The word *percent* means "per hundred." Thus, for example, 7% means 7 per hundred, or $\frac{7}{100}$. When $\frac{7}{100}$ is converted to a decimal number, we obtain 0.07. Thus, $7\% = 0.07$.

Let's look at one example involving percent. (See Section 11.1 for a more detailed discussion of percent.)

EXAMPLE 4 *Art Show*

Peggy McMahon is planning to sell her original paintings at an art show. Determine the cost of a painting before tax if the total cost of a painting, including an 8% sales tax, is to be \$145.80.

SOLUTION: We are asked to find the cost of a painting before sales tax.

Let

$$x = \text{cost of a painting before sales tax.}$$

Then

$$0.08x = 8\% \text{ of the cost of the painting (the sales tax)}$$

$$\begin{aligned}
\text{Cost of a painting before tax} + \text{tax on a painting} &= 145.80 \\
x + 0.08x &= 145.80 \\
1.08x &= 145.80 \\
\frac{1.08x}{1.08} &= \frac{145.80}{1.08} \\
x &= \frac{145.80}{1.08} \\
x &= 135
\end{aligned}$$

Thus, the cost of a painting before tax is \$135. ▲

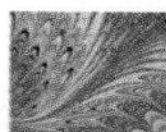

SECTION 6.4 EXERCISES

Concept/Writing Exercises

1. What is the difference between a mathematical expression and an equation?
2. Give an example of a mathematical expression and an example of a mathematical equation.

Practice the Skills

In Exercises 3–14, write the phrase as a mathematical expression.

3. 4 increased by 3 times x
4. 6 times x decreased by 2
5. 5 more than 6 times r
6. 10 times s decreased by 13
7. 15 decreased by twice r
8. 6 more than x
9. 2 times m increased by 9
10. 8 increased by 5 times x
11. 18 decreased by s, divided by 4
12. The sum of 8 and t, divided by 2
13. 6 less than the product of 5 times y, increased by 3
14. The quotient of 8 and y, decreased by 3 times x

In Exercises 15–26, write an equation and solve.

15. A number decreased by 6 is 5.
16. The sum of a number and 7 is 15.
17. The difference between a number and 4 is 20.
18. A number multiplied by 7 is 42.
19. Twelve increased by 5 times a number is 47.
20. Four times a number decreased by 10 is 42.
21. Sixteen more than 8 times a number is 88.
22. Six more than five times a number is 7 times the number decreased by 18.
23. A number increased by 11 is 1 more than 3 times the number.
24. A number divided by 3 is 4 less than the number.

25. A number increased by 10 is 2 times the sum of the number and 3.

26. The product of 2 and a number decreased by 3 is 4 more than the number.

Problem Solving

In Exercises 27–46, set up an equation that can be used to solve the problem. Solve the equation and find the desired value(s).

27. MODELING - *Ticket Sales* New Hyde Park High School sold 600 tickets to the play *The Wiz*. The number of tickets sold to students was three times the number of tickets sold to nonstudents. How many tickets were sold to students and to nonstudents?

28. MODELING - *New Clothing* Miguel Garcia purchases two new pairs of pants at The Gap for \$60. If one pair was \$10 more than the other, how much was the more expensive pair?

29. MODELING - *Income Tax* From 1999 to 2000, there was an 11.6% increase in the number of taxpayers filing their taxes electronically. If 34.20 million taxpayers filed their taxes electronically in 2000, how many million taxpayers filed their taxes electronically in 1999?

30. MODELING - *Sales Commission* Vinny Raineri receives a weekly salary of \$400 at Abbott's Appliances. He also receives a 6% commission on the total dollar amount of all sales he makes. What must his total sales be in a week if he is to make a total of \$790?

31. MODELING - *Pet Supplies* PetSmart has a sale offering 10% off of all pet supplies. If Amanda Miller spent \$15.72 on pet supplies before tax, what was the price of the pet supplies he purchased before the discount?

32. MODELING - *Copying* Ronnie McNeil pays 8¢ to make a copy of a page at a copy shop. She is considering purchasing a photocopy machine that is on sale for \$250, including tax. How many copies would Ronnie have to make in the copy shop for her cost to equal the purchase price of the photocopy machine she is considering buying?

33. MODELING - *Number of CD's* Samantha Silverstone and Josie Appleton receive 12 free compact discs by joining a compact disc club. How many CD's will each receive if Josie is to have three times as many as Samantha?

34. MODELING - *Scholarship Donation* Each year, Andrea Choi donates a total of \$1000 for scholarships at Mercer County Community College. This year she wants the amount she donates for scholarships for liberal arts to be three times the amount she donates for scholarships for business. Determine the amount she will donate for each type of scholarship.

35. MODELING - *Homeowners Association* Cross Creek Townhouses Homeowners Association needs to charge each homeowner a supplemental assessment to help pay for some unexpected repairs to the townhouses. The association has \$2000 in its reserve fund that it will use to help pay for the repairs. How much must the association charge each of the 50 homeowners if the total cost for the repairs is \$13,350?

36. MODELING - *Dimensions of a Deck* Jim Yuhas is building a rectangular deck and wants the length to be 3 ft greater than the width. What will be the dimensions of the deck if the perimeter is to be 54 ft?

37. MODELING - *Floor Area* The total floor space in three barns is 45,000 ft^2. The two smaller ones have the same area, and the largest one is three times the area of the smaller ones.

a) Determine the floor space for each barn.

b) Can merchandise that takes up 8500 ft^2 of floor space fit into either of the smaller barns?

38. MODELING - *Average Salary* According to the Bureau of Economic Analysis, the average per capita income by state in 2000 was highest in Connecticut and lowest in Mississippi. The average per capita income in Connecticut was \$1346 less than two times the per capita income in Mississippi. If the sum of the average per capita income in Connecticut and Mississippi is \$61,663, what is the average per capita income in each state?

39. MODELING - *Vacation Days* According to the World Tourism Organization, the average number of vacation days for employees in Italy is 3 more than 3 times the average number of vacation days for employees in the United States. The sum of the average number of vacation days in Italy and in the United States is 55. Determine the average number of vacation days in each country.

40. MODELING - *Car Purchase* The Gilberts purchased a car. If the total cost, including a 5% sales tax, was \$14,512, find the cost of the car before tax.

41. MODELING - *Enclosing Two Pens* Chuck Salvador has 140 ft of fencing in which he wants to build two connecting, adjacent square pens (see the figure). What will be the dimensions if the length of the entire enclosed region is to be twice the width?

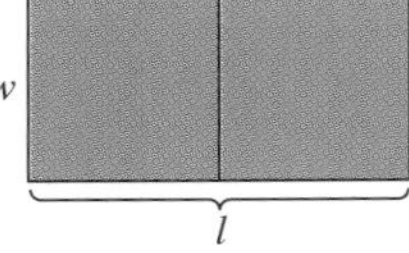

42. **MODELING** - *Dimensions of a Bookcase* A bookcase with three shelves is to be built by a woodworking student. If the height of the bookcase is to be 2 ft longer than the length of a shelf and the total amount of wood to be used is 32 ft, find the dimensions of the bookcase.

43. **MODELING** - *Laundry Cost* The cost of doing the family laundry for a month at a local laundromat is \$70. A new washer and dryer cost a total of \$760. How many months would it take for the cost of doing the laundry at the laundromat to equal the cost of a new washer and dryer?

44. **MODELING** - *Health Club Cost* A health club is offering two new membership plans. Plan A costs \$56 per month for unlimited use. Plan B costs \$20 per month plus \$3 for every visit. How many visits to the health club must Doug Jones make per month for Plan A to result in the same cost as Plan B?

45. **MODELING** - *Airfare* Rachel James has been told that with her half-off airfare coupon, her airfare from New York to San Diego will be \$257.00. The \$257.00 includes a 7% tax *on the regular fare.* On the way to the airport, Rachel realizes that she has lost her coupon. What will her regular fare be before tax?

Hotel del Coronado, San Diego, CA

46. **MODELING** - *Truck Rentals* The cost of renting a small truck at the U-Haul rental agency is \$35 per day plus 20¢ a mile. The cost of renting the same truck at the Ryder rental agency is \$25 per day plus 32¢ a mile. How far would you have to drive in one day for the cost of renting from U-Haul to equal the cost of renting from Ryder?

Challenge Problems/Group Activities

47. *Income Tax* Some states allow a husband and wife to file individual tax returns (on a single form) even though they have filed a joint federal tax return. It is usually to the taxpayers' advantage to do so when both husband and wife work. The smallest amount of tax owed (or the largest refund) will occur when the husband's and wife's taxable incomes are the same.

Mr. McAdams's 2003 taxable income was \$24,200, and Mrs. McAdams's taxable income for that year was \$26,400. The McAdams's total tax deduction for the year was \$3640. This deduction can be divided between Mr. and Mrs. McAdams any way they wish. How should the \$3640 be divided between them to result in each individual's having the same taxable income and therefore the greatest tax refund?

48. Write each equation as a sentence. There are many correct answers.

a) $x + 3 = 13$ **b)** $3x + 5 = 8$
c) $3x - 8 = 7$

49. Show that the sum of any three consecutive integers is 3 less than 3 times the largest.

50. *Auto Insurance* A driver education course at the East Lake School of Driving costs \$45 but saves those under 25 years of age 10% of their annual insurance premiums until they are 25. Dan has just turned 18, and his insurance costs \$600.00 per year.

a) When will the amount saved from insurance equal the price of the course?

b) Including the cost of the course, when Dan turns 25, how much will he have saved?

Recreational Mathematics

51. The relationship between Fahrenheit temperature (F) and Celsius temperature (C) is shown by the formula $F = \frac{9}{5}C + 32$. At what temperature will a Fahrenheit thermometer read the same as a Celsius thermometer?

6.5 VARIATION

In Sections 6.3 and 6.4, we presented many applications of algebra. In this section, we introduce variation, which is an important tool in solving applied problems.

Direct Variation

Many scientific formulas are expressed as variations. A *variation* is an equation that relates one variable to one or more other variables through the operations of multiplication or division (or both operations). Essentially there are four types of variation problems: direct, inverse, joint, and combined variation.

In *direct variation*, the values of the two related variables increase together or decrease together; that is, as one increases so does the other, and as one decreases so does the other.

Consider a car traveling at 40 miles an hour. The car travels 40 miles in 1 hour, 80 miles in 2 hours, and 120 miles in 3 hours. Note that, as the time increases, the distance traveled increases, and, as the time decreases, the distance traveled decreases.

The formula used to calculate distance traveled is

$$\text{Distance} = \text{rate} \cdot \text{time}$$

Since the rate is a constant 40 miles per hour, the formula can be written

$$d = 40t$$

We say that distance *varies directly* as time or that distance is *directly proportional* to time.

The preceding equation is an example of direct variation.

Direct Variation

If a variable y varies directly with a variable x, then

$$y = kx$$

where k is the **constant of proportionality** (or the variation constant).

Examples 1 through 4 illustrate direct variation.

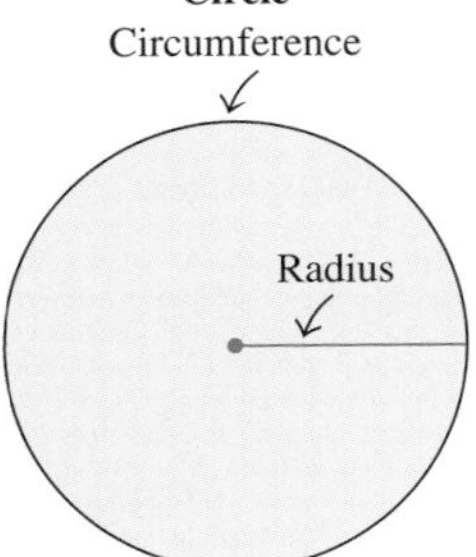

Figure 6.5

EXAMPLE 1 *Direct Variation in Geometry*

The circumference of a circle, C, is directly proportional to (or varies directly as) its radius, r; see Fig. 6.5. Write the equation for the circumference of a circle if the constant of proportionality, k, is 2π.

SOLUTION:

$C = kr$ — C varies directly as r.

$C = 2\pi r$ — Constant of proportionality is 2π.

EXAMPLE 2 *Direct Variation in Medicine*

The recommended dosage, d, of the antibiotic drug vancomycin is directly proportional to a person's weight, w.

a) Write this variation as an equation.

b) Find the recommended dosage, in milligrams, for Doug Kulzer, who weighs 192 lb. Assume the constant of proportionality for the dosage is 18.

SOLUTION:

a) $d = kw$

b) $d = 18(192) = 3456$

The recommended dosage for Doug Kulzer is 3456 mg. ▲

In certain variation problems, the constant of proportionality, k, may not be known. In such cases, we can often find it by substituting the given values in the variation formula and solving for k.

EXAMPLE 3 *Finding the Constant of Proportionality*

Suppose w varies directly as the square of y. If w is 60 when y is 20, find the constant of proportionality.

SOLUTION: Since w varies directly as the *square of* y, we begin with the formula $w = ky^2$. Since the constant of proportionality is not given, we must find k using the given information. Substitute 60 for w and 20 for y.

$$\begin{aligned} w &= ky^2 \\ 60 &= k(20)^2 \\ 60 &= 400k \\ \frac{60}{400} &= \frac{400k}{400} \\ 0.15 &= k \end{aligned}$$

Thus, the constant of proportionality is 0.15. ▲

EXAMPLE 4 *Using the Constant of Proportionality*

The length that a spring will stretch, S, varies directly with the force (or weight), F, attached to the spring. If a spring stretches 4.2 in. when a 60-lb weight is attached, how far will it stretch when a 30-lb weight is attached?

SOLUTION: We begin with the formula $S = kF$. Since the constant of proportionality is not given, we must find k using the given information.

$$\begin{aligned} S &= kF \\ 4.2 &= k(60) \\ \frac{4.2}{60} &= k \\ 0.07 &= k \end{aligned}$$

We now use $k = 0.07$ to find S when $F = 30$.

$$S = kF$$
$$S = 0.07F$$
$$S = 0.07(30)$$
$$S = 2.1 \text{ in.}$$

Thus, a spring will stretch 2.1 in. when a force of 30 lb is attached. ▲

Inverse Variation

A second type of variation is *inverse variation*. When two quantities vary inversely, as one quantity increases, the other quantity decreases, and vice versa.

To explain inverse variation, we use the formula, distance = rate · time. If we solve for time, we get time = distance/rate. Assume the distance is fixed at 100 miles; then

$$\text{Time} = \frac{100}{\text{rate}}$$

At 100 miles per hour it would take 1 hour to cover this distance. At 50 miles an hour, it would take 2 hours. At 25 miles an hour, it would take 4 hours. Note that as the rate (or speed) decreases, the time increases and vice versa.

The preceding equation can be written

$$t = \frac{100}{r}$$

This equation is an example of an inverse variation. The time and rate are inversely proportional. The constant of proportionality in this case is 100.

Inverse Variation

If a variable y varies inversely with a variable x, then

$$y = \frac{k}{x}$$

where k is the constant of proportionality.

Two quantities *vary inversely*, or are *inversely proportional*, when as one quantity increases the other quantity decreases and vice versa. Examples 5 and 6 illustrate inverse variation.

EXAMPLE 5 *Inverse Variation in Astronomy*

The velocity, v, of a meteorite approaching Earth varies inversely as the square root of its distance from the center of Earth. Assuming the velocity is 2 miles per second at a distance of 6400 miles from the center of Earth, find the equation that expresses the relationship between the velocity of a meteorite and its distance from the center of Earth.

SOLUTION: Since the velocity of the meteorite varies inversely as the *square root* of its distance from the center of Earth, the general form of the equation is

$$v = \frac{k}{\sqrt{d}}$$

To find k, we substitute the given values for v and d.

$$\begin{aligned} 2 &= \frac{k}{\sqrt{6400}} \\ 2 &= \frac{k}{80} \\ (2)(80) &= k \\ 160 &= k \end{aligned}$$

Thus, the formula is $v = \dfrac{160}{\sqrt{d}}$. ▲

EXAMPLE 6 ***Using the Constant of Proportionality***

Suppose y varies inversely as x. If $y = 8$ when $x = 15$, find y when $x = 18$.

SOLUTION: First write the inverse variation, then solve for k.

$$\begin{aligned} y &= \frac{k}{x} \\ 8 &= \frac{k}{15} \\ 120 &= k \end{aligned}$$

Now substitute 120 for k in $y = \frac{k}{x}$ and find y when $x = 18$.

$$y = \frac{120}{x} = \frac{120}{18} = 6.7 \quad \text{(to the nearest tenth)}$$ ▲

Joint Variation

One quantity may vary directly as a product of two or more other quantities. This type of variation is called *joint variation*.

Joint Variation
The general form of a joint variation, where y varies directly as x and z, is

$$y = kxz$$

where k is the constant of proportionality.

EXAMPLE 7 *Joint Variation in Geometry*

The area, A, of a triangle varies jointly as its base, b, and height, h. If the area of a triangle is 48 in.^{2} when its base is 12 in. and its height is 8 in., find the area of a triangle whose base is 15 in. and whose height is 20 in.

SOLUTION: First write the joint variation, then substitute the known values and solve for k.

$$\begin{aligned} A &= kbh \\ 48 &= k(12)(8) \\ 48 &= k(96) \\ \frac{48}{96} &= k \\ \tfrac{1}{2} &= k \end{aligned}$$

Now solve for the area of the given triangle.

$$\begin{aligned} A &= kbh \\ &= \tfrac{1}{2}(15)(20) \\ &= 150 \text{ in.}^2 \end{aligned}$$

▲

Summary of Variations

Direct	Inverse	Joint
$y = kx$	$y = \dfrac{k}{x}$	$y = kxz$

Combined Variation

Often in real-life situations, one variable varies as a combination of variables. The following examples illustrate the use of *combined variations*.

EXAMPLE 8 *Combined Variation in Engineering*

The load, L, that a horizontal beam can safely support varies jointly as the width, w, and the square of the depth, d, and inversely as the length, l. Express L in terms of w, d, l, and the constant of proportionality, k.

SOLUTION:

$$L = \frac{kwd^2}{l}$$

▲

EXAMPLE 9 *Pretzel Price, Combined Variation*

The owners of the Colonel Mustard Pretzel Shop find that their weekly sales of pretzels, S, vary directly with their advertising budget, A, and inversely with their pretzel price, P. When their advertising budget is \$600 and the price is \$1.20, they sell 6500 pretzels.

a) Write an equation of variation expressing S in terms of A and P. Include the value of the proportionality constant.

b) Find the expected sales if the advertising budget is \$900 and the pretzel price is \$1.50.

SOLUTION:

a) Since S varies directly as A and inversely as P, we begin with the equation

$$S = \frac{kA}{P}$$

We now find k using the known values.

$$6500 = \frac{k(600)}{1.20}$$

$$6500 = 500k$$

$$13 = k$$

Therefore, the equation for the sales of pretzels is $S = \frac{13A}{P}$.

b) $S = \frac{13A}{P}$

$= \frac{13(900)}{1.50} = 7800$

They can expect to sell 7800 pretzels.

EXAMPLE 10 *Combined Variation*

A varies jointly as B and C and inversely as the square of D. If $A = 1$ when $B = 9$, $C = 4$, and $D = 6$, find A when $B = 8$, $C = 12$, and $D = 5$.

SOLUTION: We begin with the equation

$$A = \frac{kBC}{D^2}$$

We must first find the constant of proportionality, k, by substituting the known values for A, B, C, and D and solving for k.

$$1 = \frac{k(9)(4)}{6^2}$$

$$1 = \frac{36k}{36}$$

$$1 = k$$

Thus, the constant of proportionality equals 1. Now we find A for the corresponding values of B, C, and D.

$$A = \frac{kBC}{D^2}$$

$$A = \frac{(1)(8)(12)}{5^2} = \frac{96}{25} = 3.84$$

SECTION 6.5 EXERCISES

Concept/Writing Exercises

In Exercises 1–4, use complete sentences to answer the question.

1. Describe inverse variation.
2. Describe direct variation.
3. Describe joint variation.
4. Describe combined variation.

In Exercises 5–20, use your intuition to determine whether the variation between the indicated quantities is direct or inverse.

5. The distance between two cities on a map and the actual distance between the two cities
6. The time required to fill a pool with a hose and the volume of water coming from the hose

Allen R. Angel, author

7. The time required to boil water on a burner and the temperature of the burner
8. The time a person spends walking on a treadmill and the number of calories the person burns
9. The interest earned on an investment and the interest rate
10. The volume of a balloon and its radius
11. A person's speed and the time needed for the person to complete the race
12. The time required to cool a room and the temperature of the room
13. The number of painters hired to paint a house and the time required to paint the house
14. The number of calories eaten and the amount of exercise required to burn off those calories
15. The time required to defrost frozen hamburger in a room and the temperature of the room
16. On Earth, the weight and mass of an object
17. The number of people in the cashier line at the bookstore and the time required to stand in line
18. The number of books that can be placed upright on a shelf 3 ft long and the width of the books
19. The displacement, in liters, and the horsepower of an engine
20. The speed of a rider lawn mower and the time it takes to cut a lawn

In Exercises 21 and 22, use Exercises 5–20 as a guide.

21. Name two items that have not been mentioned in this section that have a direct variation.
22. Name two items that have not been mentioned in this section that have an inverse variation.

Practice the Skills

In Exercises 23–40, (a) write the variation and (b) find the quantity indicated.

23. y varies directly as x. Find y when $x = 5$ and $k = 3$.
24. x varies inversely as y. Find x when $y = 12$ and $k = 15$.

25. m varies inversely as the square of n. Find m when $n = 8$ and $k = 16$.

26. r varies directly as the square of s. Find r when $s = 2$ and $k = 13$.

27. R varies inversely as W. Find R when $W = 160$ and $k = 8$.

28. D varies directly as J and inversely as C. Find D when $J = 10$, $C = 25$, and $k = 5$.

29. F varies jointly as D and E. Find F when $D = 3$, $E = 10$, and $k = 7$.

30. A varies jointly as R_1 and R_2 and inversely as the square of L. Find A when $R_1 = 120$, $R_2 = 8$, $L = 5$, and $k = \frac{3}{2}$.

31. t varies directly as the square of d and inversely as f. If $t = 192$ when $d = 8$ and $f = 4$, find t when $d = 10$ and $f = 6$.

32. y varies directly as the square root of t and inversely as s. If $y = 12$ when $t = 36$ and $s = 2$, find y when $t = 81$ and $s = 4$.

33. Z varies jointly as W and Y. If $Z = 12$ when $W = 9$ and $Y = 4$, find Z when $W = 50$ and $Y = 6$.

34. y varies directly as the square of R. If $y = 4$ when $R = 4$, find y when $R = 8$.

35. H varies directly as L. If $H = 15$ when $L = 50$, find H when $L = 10$.

36. C varies inversely as J. If $C = 7$ when $J = 0.7$, find C when $J = 12$.

37. A varies directly as the square of B. If $A = 245$ when $B = 7$, find A when $B = 12$.

38. F varies jointly as M_1 and M_2 and inversely as the square of d. If $F = 20$ when $M_1 = 5$, $M_2 = 10$, and $d = 0.2$, find F when $M_1 = 10$, $M_2 = 20$, and $d = 0.4$.

39. F varies jointly as q_1 and q_2 and inversely as the square of d. If $F = 8$ when $q_1 = 2$, $q_2 = 8$, and $d = 4$, find F when $q_1 = 28$, $q_2 = 12$, and $d = 2$.

40. S varies jointly as I and the square of T. If $S = 8$ when $I = 20$ and $T = 4$, find S when $I = 2$ and $T = 2$.

Problem Solving

In Exercises 41–49, (a) write the variation and (b) find the quantity indicated.

41. *Resistance* The resistance, R, of a wire varies directly as its length, L. If the resistance of a 30 ft length of wire is 0.24 ohm, find the resistance of a 40 ft length of wire.

42. *Finding Interest* The amount of interest earned on an investment, I, varies directly as the interest rate, r. If the interest earned is \$40 when the interest rate is 4%, find the amount of interest earned when the interest rate is 6%.

43. *Speaker Loudness* The loudness of a stereo speaker, l, measured in decibels (dB), is inversely proportional to the square of the distance, d, of the listener from the speaker. If the loudness is 20 dB when the listener is 6 ft from the speaker, find the loudness when the listener is 3 ft from the speaker.

44. *Building a Deck* The time, t, it takes to build a deck for a specific house is inversely proportional to the number, n, of workers building the deck. If it takes two workers 16 hours to build the deck, how many hours will it take for four workers to build the deck?

45. *Video Rentals* The weekly videotape rentals, R, at Busterblock Video vary directly with their advertising budget, A, and inversely with the daily rental price, P. When the video store's advertising budget is \$600 and the rental price is \$3 per day, it rents 4800 tapes per week. How many tapes would it rent per week if the store increased their advertising budget to \$700 and raised its rental price to \$3.50?

46. *Area and Projection* The area, a, of a projected picture on a movie screen varies directly as the square of the distance, d, from the projector to the screen. If a projector at a distance of 25 feet projects a picture with an area of 100 square feet, what is the area of the projected picture when the projector is at a distance of 40 feet?

47. *Strength of a Beam* The strength, s, of a rectangular beam varies jointly as its width, w, and the square of its depth, d. If the strength of a beam 2 inches wide and 10 inches deep is 2250 pounds per square inch, find the strength of a beam 4 inches wide and 12 inches deep.

48. *Electric Resistance* The electrical resistance of a wire, R, varies directly as its length, L, and inversely as its cross-sectional area, A. If the resistance of a wire is 0.2 ohm when the length is 200 ft and its cross-sectional area is 0.05 in.^{2}, find the resistance of a wire whose length is 5000 ft with a cross-sectional area of 0.01 in^2.

49. *Phone Calls* The number of phone calls between two cities during a given time period, N, varies directly as the populations p_1 and p_2 of the two cities and inversely to the distance, d, between them. If 100,000 calls are made between two cities 300 mi apart and the populations of the cities are 60,000 and 200,000, how many calls are made between two cities with populations of 125,000 and 175,000 that are 450 mi apart?

50. **a)** If y varies directly as x and the constant of proportionality is 2, does x vary directly or inversely as y? Explain.
    b) Give the new constant of proportionality for x as a variation of y.

51. **a)** If y varies inversely as x and the constant of proportionality is 0.3, does x vary directly or inversely as y? Explain.
    b) Give the new constant of proportionality for x as a variation of y.

Challenge Problems/Group Activities

52. *Photography* An article in the magazine *Outdoor and Travel Photography* states, "If a surface is illuminated by a point-source of light, the intensity of illumination produced is inversely proportional to the square of the distance separating them. In practical terms, this means that foreground objects will be grossly overexposed if your background subject is properly exposed with a flash. Thus direct flash will not offer pleasing results if there are any intervening objects between the foreground and the subject."

If the subject you are photographing is 4 ft from the flash and the illumination on this subject is $\frac{1}{16}$ of the light of the flash, what is the intensity of illumination on an intervening object that is 3 ft from the flash?

53. *Water Cost* In a specific region of the country, the amount of a customer's water bill, W, is directly proportional to the average daily temperature for the month, T, the lawn area, A, and the square root of F, where F is the family size, and inversely proportional to the number of inches of rain, R.

In one month, the average daily temperature is 78°F and the number of inches of rain is 5.6. If the average family of four who has a thousand square feet of lawn pays \$72.00 for water for that month, estimate the water bill in the same month for the average family of six who has 1500 ft^2 of lawn.

6.6 LINEAR INEQUALITIES

The first four sections of this chapter have dealt with equations. However, we often encounter statements of inequality. The symbols of inequality are as follows.

Symbols of Inequality

$a < b$ means that a is less than b.
$a \leq b$ means that a is less than or equal to b.
$a > b$ means that a is greater than b.
$a \geq b$ means that a is greater than or equal to b.

An *inequality* consists of two (or more) expressions joined by an inequality sign.

Examples of inequalities

$$3 < 5, \qquad x < 2, \qquad 3x - 2 \geq 5$$

A statement of inequality can be used to indicate a set of real numbers. For example, $x < 2$ represents the set of all real numbers less than 2. Listing all these numbers is impossible, but some are $-2, -1.234, -1, -\frac{1}{2}, 0, \frac{97}{163}, 5, 9$.

A method of picturing all real numbers less than 2 is to graph the solution on the number line. The number line was discussed in Chapter 5.

To indicate the solution set of $x < 2$ on the number line, we draw an open circle at 2 and a line to the left of 2 with an arrow at its end. This technique indicates that all points to the left of 2 are part of the solution set. The open circle indicates that the solution set does not include the number 2.

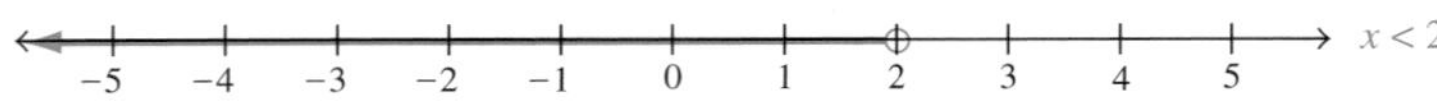

To indicate the solution set of $x \leq 2$ on the number line, we draw a closed (or darkened) circle at 2 and a line to the left of 2 with an arrow at its end. The closed circle indicates that the 2 is part of the solution.

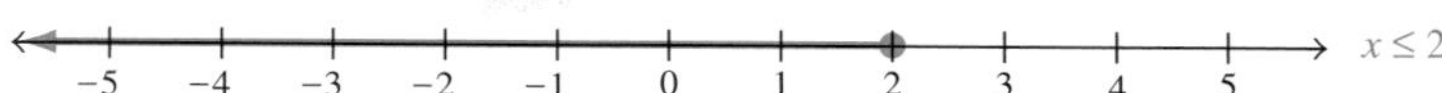

EXAMPLE 1 *Graphing a Less Than or Equal to Inequality*

Graph the solution set of $x \leq -2$, where x is a real number, on the number line.

SOLUTION: The numbers less than or equal to -2 are all the points on the number line to the left of -2 and -2 itself. The closed circle at -2 shows that -2 is included in the solution set.

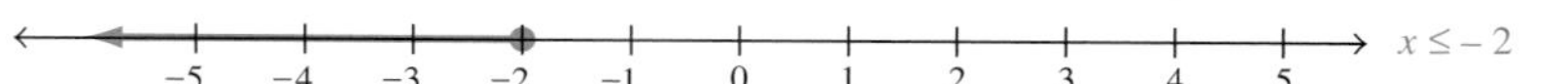

▲

The inequality statements $x < 2$ and $2 > x$ have the same meaning. Note that the inequality symbol points to the x in both cases. Thus, one inequality may be written in place of the other. Likewise, $x > 2$ and $2 < x$ have the same meaning. Note that the inequality symbol points to the 2 in both cases. We make use of this fact in Example 2.

EXAMPLE 2 *Graphing a Less Than Inequality*

Graph the solution set of $3 < x$, where x is a real number, on the number line.

SOLUTION: We can restate $3 < x$ as $x > 3$. Both statements have identical solutions. Any number that is greater than 3 satisfies the inequality $x > 3$. The graph includes all the points to the right of 3 on the number line. To indicate that 3 is not part of the solution set, we place an open circle at 3.

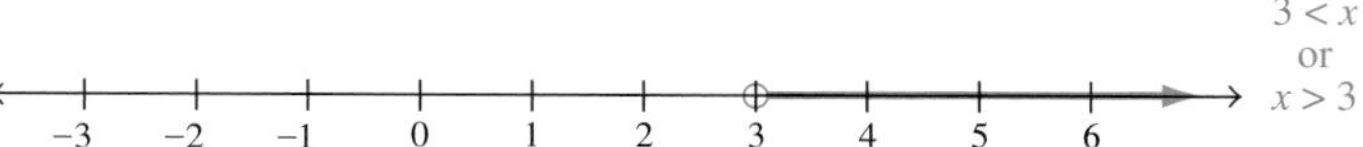

▲

We can find the solution to an inequality by adding, subtracting, multiplying, or dividing both sides of the inequality by the same number or expression. We use the procedure discussed in Section 6.2 to isolate the variable, with one important exception: *When both sides of an inequality are multiplied or divided by a negative number, the direction of the inequality symbol is reversed.*

EXAMPLE 3 *Multiplying by a Negative Number*

Solve the inequality $-x > 3$ and graph the solution set on the number line.

SOLUTION: To solve this inequality, we must eliminate the negative sign in front of the x. To do so, we multiply both sides of the inequality by -1 and change the direction of the inequality symbol.

$$-x > 3$$

$$-1(-x) < -1(3)$$ Multiply both sides of the inequality by -1 and change the direction of the inequality symbol.

$$x < -3$$

DID YOU KNOW

What a Bore

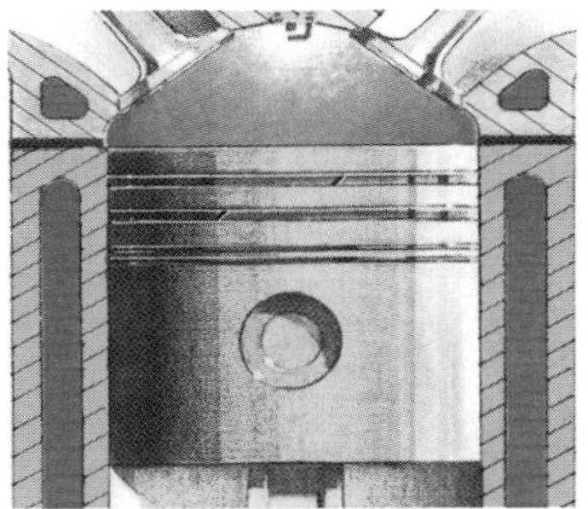

In the production of machine parts, engineers must allow a certain tolerance in the way parts fit together. For example, the boring machines that grind cylindrical openings in an automobile's engine block must create a cylinder that allows the piston to move freely up and down, but still fit tightly enough to ensure that compression and combustion are complete. The allowable tolerance between parts can be expressed as an inequality. For example, the diameter of a cylinder may need to be no less than 3.383 in. and no greater than 3.387 in. We can represent the allowable tolerance as $3.383 \le t \le 3.387$.

The solution set is graphed on the number line as follows.

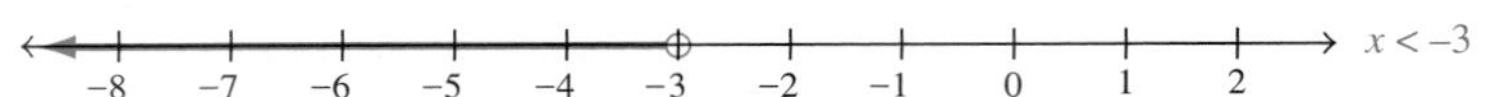

EXAMPLE 4 ***Dividing by a Negative Number***

Solve the inequality $-4x < 16$ and graph the solution set on the number line.

SOLUTION: Solving the inequality requires making the coefficient of the x term 1. To do so, divide both sides of the inequality by -4 and change the direction of the inequality symbol.

$$-4x < 16$$

$$\frac{-4x}{-4} > \frac{16}{-4} \quad \text{Divide both sides of the inequality by } -4 \text{ and change the direction of the inequality symbol.}$$

$$x > -4$$

The solution set is graphed on the number line as follows.

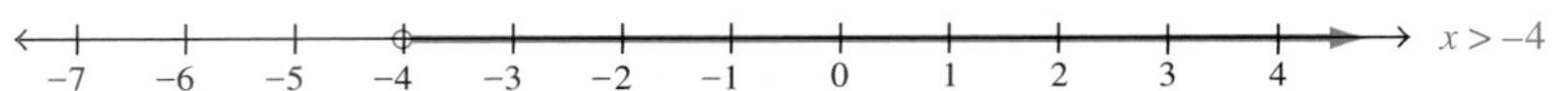

EXAMPLE 5 ***Solving an Inequality***

Solve the inequality $3x - 5 > 13$ and graph the solution set on the number line.

SOLUTION: To find the solution set, isolate x on one side of the inequality symbol.

$$3x - 5 > 13$$

$$3x - 5 + 5 > 13 + 5 \quad \text{Add 5 to both sides of the inequality.}$$

$$3x > 18$$

$$\frac{3x}{3} > \frac{18}{3} \quad \text{Divide both sides of the inequality by 3.}$$

$$x > 6$$

Thus, the solution set to $3x - 5 > 13$ is all real numbers greater than 6.

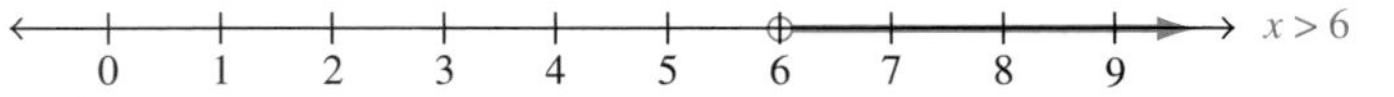

Note that in Example 5, the *direction of the inequality symbol* did not change when both sides of the inequality were divided by the positive number 3.

EXAMPLE 6 ***A Solution of Only Integers***

Solve the inequality $x + 4 < 7$, where x is an integer, and graph the solution set on the number line.

SOLUTION:

$$x + 4 < 7$$

$$x + 4 - 4 < 7 - 4$$

$$x < 3$$

Since x is an integer and is less than 3, the solution set is the set of integers less than 3, or $\{\ldots -3, -2, -1, 0, 1, 2\}$. To graph the solution set, we make solid dots at the corresponding points on the number line. The three smaller dots to the left of -3 indicate that all the integers to the left of -3 are included.

An inequality of the form $a < x < b$ is called a *compound inequality*. Consider the compound inequality $-3 < x \leq 2$, which means that $-3 < x$ *and* $x \leq 2$.

EXAMPLE 7 *A Compound Inequality*

Graph the solution set of the inequality $-3 < x \leq 2$

a) where x is an integer.

b) where x is a real number.

SOLUTION:

a) The solution set is all the integers between -3 and 2, including the 2 but not including the -3, or $\{-2, -1, 0, 1, 2\}$.

b) The solution set consists of all the real numbers between -3 and 2, including the 2 but not including the -3.

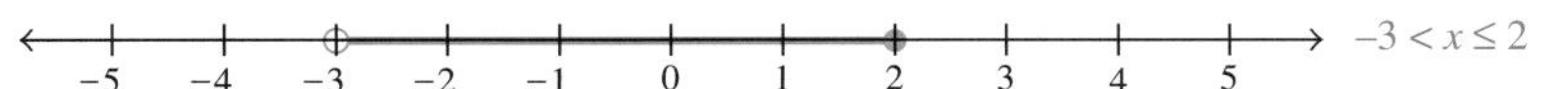

EXAMPLE 8 *Solving a Compound Inequality*

Solve the compound inequality for x and graph the solution set.

$$-4 < \frac{x+3}{2} \leq 5$$

SOLUTION: To solve a compound inequality, we must isolate the x as the middle term. To do so, we use the same principles used to solve inequalities.

$$-4 < \frac{x+3}{2} \leq 5$$

$$2(-4) < 2\left(\frac{x+3}{2}\right) \leq 2(5) \quad \text{Multiply each part of the inequality by 2.}$$

$$-8 < x + 3 \leq 10$$

$$-8 - 3 < x + 3 - 3 \leq 10 - 3 \quad \text{Subtract 3 from each part of the inequality.}$$

$$-11 < x \leq 7$$

The solution set is graphed on the number line as follows.

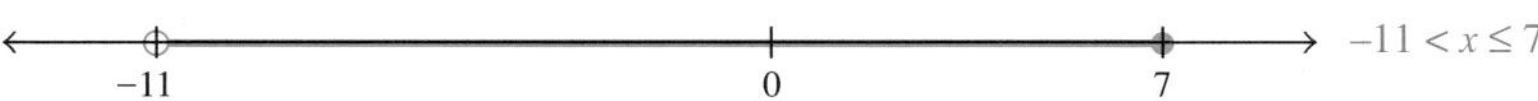

"In mathematics the art of posing problems is easier than that of solving them."

Georg Cantor

EXAMPLE 9 *Average Grade*

A student must have an average (the mean) on five tests that is greater than or equal to 80% but less than 90% to receive a final grade of B. Devon's grades on the first four tests were 98%, 76%, 86%, and 92%. What range of grades on the fifth test would give him a B in the course?

SOLUTION: The unknown quantity is the range of grades on the fifth test. First construct an inequality that can be used to find the range of grades on the fifth exam. The average (mean) is found by adding the grades and dividing the sum by the number of exams.

Let x = the fifth grade. Then

$$\text{Average} = \frac{98 + 76 + 86 + 92 + x}{5}$$

For Devon to obtain a B, his average must be greater than or equal to 80 but less than 90.

$$80 \leq \frac{98 + 76 + 86 + 92 + x}{5} < 90$$

$$80 \leq \frac{352 + x}{5} < 90$$

$$5(80) \leq \cancel{5}\left(\frac{352 + x}{\cancel{5}}\right) < 5(90) \qquad \text{Multiply the three terms of the inequality by 5.}$$

$$400 \leq 352 + x < 450$$

$$400 - 352 \leq 352 - 352 + x < 450 - 352 \qquad \text{Subtract 352 from all three terms.}$$

$$48 \leq x < 98$$

Thus, a grade of 48% up to but not including a grade of 98% on the fifth test will result in a grade of B. ▲

TIMELY TIP Remember to change the direction of the inequality symbol when multiplying or dividing both sides of an inequality by a negative number.

SECTION 6.6 EXERCISES

Concept/Writing Exercises

1. Give the four inequality symbols we use in this section and indicate how each is read.
2. **a)** What is an inequality?
   b) Give an example of three inequalities.
3. When solving an inequality, under what conditions do you need to change the direction of the inequality symbol?
4. Does $x < 2$ have the same meaning as $2 > x$? Explain.
5. Does $x > -3$ have the same meaning as $-3 < x$? Explain.
6. When graphing the solution set to an inequality on the number line, when should you use an open circle and when should you use a closed circle?

Practice the Skills

In Exercises 7–24, graph the solution set of the inequality, where x is a real number, on the number line.

7. $x > 6$
8. $x \leq 9$
9. $x + 4 \geq 7$
10. $3x > 9$
11. $-3x \leq 18$
12. $-4x < 12$
13. $\frac{x}{6} < -2$
14. $\frac{x}{2} > 4$
15. $\frac{-x}{3} \geq 3$
16. $\frac{x}{2} \geq -4$
17. $2x + 6 \geq 14$
18. $3x + 12 < 5x + 14$
19. $4(x - 1) < 6$
20. $-5(x + 1) + 2x > -3x + 6$
21. $3(x + 4) - 2 < 3x + 10$
22. $-2 \leq x \leq 1$
23. $3 < x - 7 \leq 6$
24. $\frac{1}{2} < \frac{x + 4}{2} \leq 4$

In Exercises 25–44, graph the solution set of the inequality, where x is an integer, on the number line.

25. $x \geq 2$
26. $-3 < x$
27. $-3x \leq 27$
28. $3x \geq 27$
29. $x - 2 < 4$
30. $-5x \leq 15$
31. $\frac{x}{3} \leq -2$
32. $\frac{x}{4} \geq -3$
33. $-\frac{x}{6} \geq 3$
34. $\frac{2x}{3} \leq 4$
35. $-11 < -5x + 4$
36. $2x + 5 < -3 + 6x$
37. $3(x + 4) \geq 4x + 13$
38. $-2(x - 1) < 3(x - 4) + 5$
39. $5(x + 4) - 6 \leq 2x + 8$
40. $-3 \leq x < 5$
41. $1 > -x > -5$
42. $-2 < 2x + 3 < 6$
43. $0.2 \leq \frac{x - 4}{10} \leq 0.4$
44. $-\frac{1}{3} < \frac{x - 2}{12} \leq \frac{1}{4}$

Problem Solving

45. *Health Care Costs* The following chart shows the national average for employers for annual health care costs per employee for the years 1997 through 2000 and projected for 2001.

Annual Health Care Costs per Employee, National Average

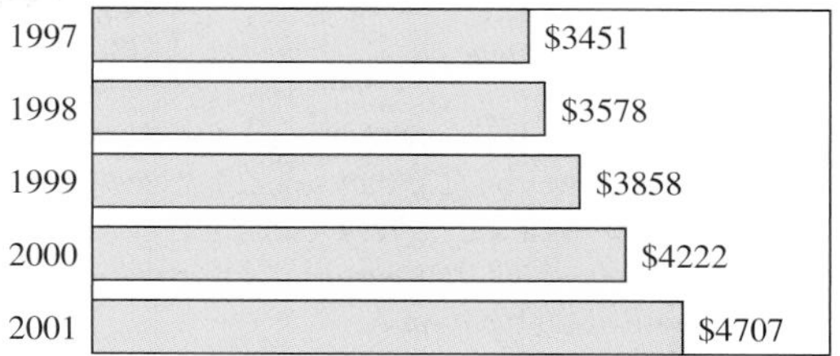

Source: Hewitt Associates

a) In which years was the national average for annual health care costs per employee $> \$4000$?
b) In which years was the national average for annual health care costs per employee $< \$3858$?
c) In which years was the national average for annual health care costs per employee $\leq \$4222$?
d) In which years was the national average for annual health care costs per employee $\geq \$3578$?

46. *U.S. Population* The following bar graph shows the U.S. foreign-born population, in millions, for selected years.

U.S. Foreign-born Population in Millions

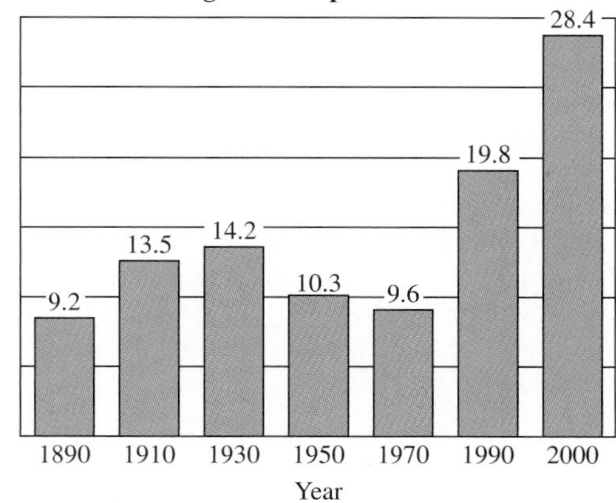

Source: U.S. Census Bureau

a) In which of the years listed on the bar graph was the U.S. foreign-born population >14.2 million?
b) In which of the years listed on the bar graph was the U.S. foreign-born population ≤ 13.5 million?
c) In which of the years listed on the bar graph was the U.S. foreign-born population ≥ 28.4 million?
d) In which of the years listed on the bar graph was the U.S. foreign-born population >19.8 million?

47. *Video Rental* Movie Mania offers two rental plans. One has an annual fee and the other has no annual fee. The annual membership fee and the daily charge per video for each plan are shown in the table on page 334. Determine

the maximum number of videos that can be rented for the no fee plan to cost less than the annual fee plan.

Rental Plan	Yearly Fee	Daily Charge per Video
Annual fee	\$30	\$1.49
No fee	None	\$2.99

48. *Salary Plans* Bobby Exler recently accepted a sales position in Portland, Oregon. He can select between the two salary plans shown in the table. Determine the dollar amount of weekly sales that would result in Bobby earning more by Plan B than by Plan A.

Salary Plan	Weekly Salary	Commission on Sales
Plan A	\$500	6%
Plan B	\$400	8%

49. *Van Rental* The Berrys need to rent a van for their family vacation. They can rent a van from Jason's Auto Rentals for \$200 per week with no charge for mileage or from Fred's Fine Autos for \$110 per week plus \$0.25 per mile. Determine the distances the Berrys can drive in the van if the cost of renting from Fred's is to be less than the cost of renting from Jason's.

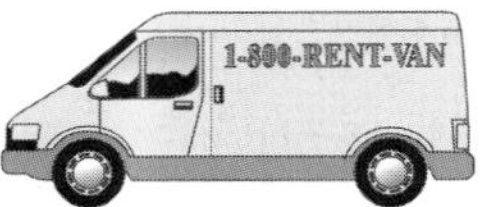

50. *Moving Boxes* A janitor must move a large shipment of books from the first floor to the fifth floor. Each box of books weighs 60 lb, and the janitor weighs 180 lb. The sign on the elevator reads, "Maximum weight 1200 lb."

a) Write a statement of inequality to determine the maximum number of boxes of books the janitor can place on the elevator at one time. (The janitor must ride in the elevator with the books.)

b) Determine the maximum number of boxes that can be moved in one trip.

51. *Price of a Meal* After Mrs. Franklin is seated in a restaurant, she realizes that she has only \$19.00. If she must pay 7% tax and wants to leave a 15% tip on the price of the meal before tax, what is the price range of meals that she can order?

52. *Making a Profit* For a business to realize a profit, its revenue, R, must be greater than its costs, C; that is, a profit will only result if $R > C$ (the company breaks even when $R = C$). A book publishing company has a weekly cost equation of $C = 2x + 2000$ and a weekly revenue equation of $R = 12x$, where x is the number of books produced and sold in a week. How many books must be sold weekly for the company to make a profit?

53. *Finding Velocity* The velocity, v, in feet per second, t sec after a tennis ball is projected directly upward is given by the formula $v = 84 - 32t$. How many seconds after being projected upward will the velocity be between 36 ft/sec and 68 ft/sec?

54. *Speed Limit* The minimum speed for vehicles on a highway is 40 mph, and the maximum speed is 55 mph. If Philip Rowe has been driving nonstop along the highway for 4 hr, what range in miles could he have legally traveled?

55. *A Grade of B* In Example 9 on page 332, what range of grades on the fifth test would result in Devon receiving a grade of B if his grades on the first four tests were 78%, 64%, 88%, and 76%?

56. *Tent Rental* The Cuhayoga Community College Planning Committee wants to rent tents for the spring job fair. Rent-a-Tent charges \$325 for setup and delivery of its tents. This fee is charged regardless of the number of tents delivered and set up. In addition, Rent-a-Tent charges \$125 for each tent rented. If the minimum amount the planning committee wishes to spend is \$950 and the maximum amount they wish to spend is \$1200, determine the minimum and maximum number of tents they can rent.

Challenge Problems/Group Activities

57. *Painting a House* J. B. Davis is painting the exterior of his house. The instructions on the paint can indicate that 1 gal covers from 250 to 400 ft^2. The total surface of the house to be painted is 2750 ft^2. Determine the number of gallons of paint he could use and express the answer as an inequality.

58. *Final Exam* Teresa's five test grades for the semester are 86%, 74%, 68%, 96%, and 72%. Her final exam counts one-third of her final grade. What range of grades on her final exam would result in Teresa receiving a final grade of B in the course? (See Example 9.)

59. A student multiplied both sides of the inequality $-\frac{1}{3}x \le 4$ by -3 and forgot to reverse the direction of the inequality symbol. What is the relation between the student's incorrect solution set and the correct solution set? Is any number in both the correct solution set and the student's incorrect solution set? If so, what is it?

Internet/Research Activity

60. Find a newspaper or a magazine article that contains the mathematical concept of inequality.

a) From the information in the article write a statement of inequality.

b) Summarize the article and explain how you arrived at the inequality statement in part (a).

6.7 GRAPHING LINEAR EQUATIONS

MATHEMATICS Everywhere

Hollywood's Four Quadrants

Mike Myers, the voice of *Shrek*

Hollywood studio executives use four "quadrants" to divide up the movie-going audience. Men 25 years and older, men younger than 25, women 25 years and older, and women younger than 25 are the age groups represented by Hollywood's four quadrants. If a studio produces a movie that appeals to all four quadrants, it is sure to have a hit movie. If the movie appeals to none of the four quadrants, the movie is sure to fail. A challenge for studio executives is to determine the core quadrant and then try to make sure no other movie geared toward the same quadrant debuts at the same time. One of the biggest hit movies of the summer of 2001 was *Shrek*, a movie that appealed to all four quadrants. On the other hand, the target or core audience for the movie *A.I.* Artificial Intelligence, also released in the summer of 2000, was unclear. The logo and previews for *A.I.*, featured the child star, whereas everything else suggested a movie more appropriate for adults. As a result, the marketing campaign confused both audiences, and *A.I.* was much more a critical success than a box office success.

In Section 6.2, we solved equations with a single variable. Real-world problems, however, often involve two or more unknowns. For example, the profit, p, of a company may depend on the amount of sales, s; or the cost, c, of mailing a package may depend on the weight, w, of the package. Thus, it is helpful to be able to work with equations with two variables (for example, $x + 2y = 6$). Doing so requires understanding the *Cartesian* (or *rectangular*) *coordinate system*, named after the French mathematician René Descartes (1596–1650).

The rectangular coordinate system consists of two perpendicular number lines (Fig. 6.6). The horizontal line is the *x-axis*, and the vertical line is the *y-axis*. The point of intersection of the x-axis and y-axis is called the *origin*. The numbers on the axes to the right and above the origin are positive. The numbers on the axes on the left and below the origin are negative. The axes divide the plane into four parts: the first, second, third, and fourth *quadrants*.

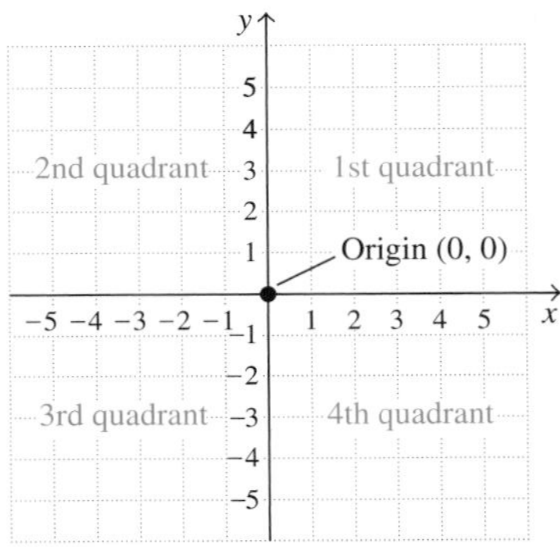

Figure 6.6

We indicate the location of a point in the rectangular coordinate system by means of an *ordered pair* of the form (x, y). The x-coordinate is always placed first and the y-coordinate is always placed second in the ordered pair. Consider the point illustrated in Fig. 6.7. Since the x-coordinate of the point is 5 and the y-coordinate is 3, the ordered pair that represents this point is (5, 3).

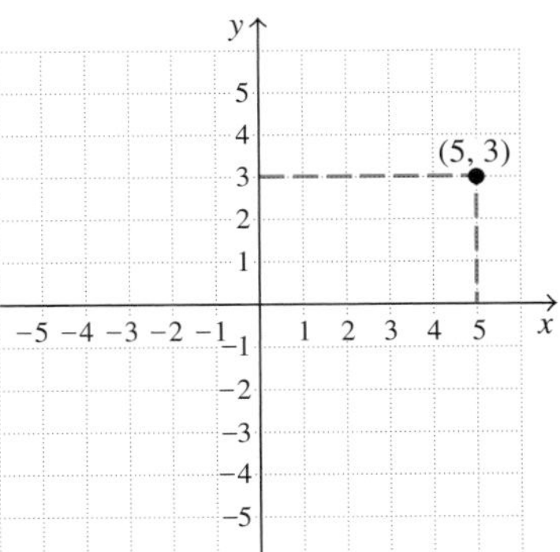

Figure 6.7

The origin is represented by the ordered pair (0, 0). Every point on the plane can be represented by one and only one ordered pair (x, y), and every ordered pair (x, y) represents one and only one point on the plane.

PROFILE IN MATHEMATICS

RENÉ DESCARTES

The Mathematician and the Fly

According to legend, the French mathematician and philosopher René Descartes (1596–1650) did some of his best thinking in bed. He was a sickly child, and so the Jesuits who undertook his education allowed him to stay in bed each morning as long as he liked. This practice he carried into adulthood, seldom getting up before noon. One morning as he watched a fly crawl about the ceiling, near the corner of his room, he was struck with the idea that the fly's position could best be described by the connecting distances from it to the two adjacent walls. These became the coordinates of his rectangular coordinate system and were appropriately named after him (Cartesian coordinates) and not the fly.

EXAMPLE 1 *Plotting Points*

Plot the points $A(-2, 4)$, $B(3, -4)$, $C(6, 0)$, $D(4, 1)$, and $E(0, 3)$.

SOLUTION: Point A has an x-coordinate of -2 and a y-coordinate of 4. Project a vertical line up from -2 on the x-axis and a horizontal line to the left from 4 on the y-axis. The two lines intersect at the point denoted A (Fig. 6.8). The other points are plotted in a similar manner.

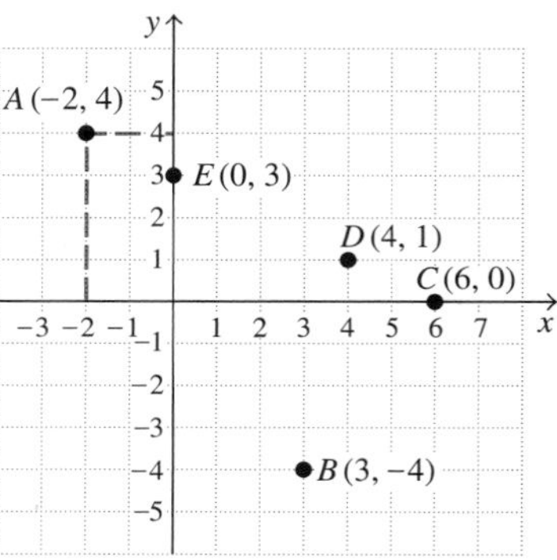

Figure 6.8

EXAMPLE 2 *A Parallelogram*

The points, A, B, and C are three vertices of a parallelogram with two sides parallel to the x-axis. Plot the three points below and determine the coordinates of the fourth vertex, D.

$$A(1, 2) \qquad B(2, 4) \qquad C(7, 4)$$

SOLUTION: A parallelogram is a figure that has opposite sides that are of equal length and are parallel. (Parallel lines are two lines in the same plane that do not intersect.) The horizontal distance between points B and C is 5 units (see Fig. 6.9). Therefore, the horizontal distance between points A and D must also be 5 units. This problem has two possible solutions, as illustrated in Fig. 6.9. In each figure, we have indicated the given points in red.

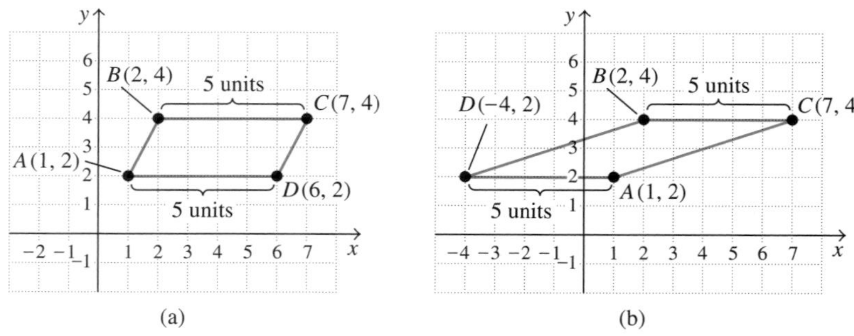

Figure 6.9

The solutions are the points (6, 2) and $(-4, 2)$.

Graphing Linear Equations by Plotting Points

Consider the following equation in two variables: $y = x + 1$. Every ordered pair that makes the equation a true statement is a solution to, or satisfies, the equation. We can mentally find some ordered pairs that satisfy the equation $y = x + 1$ by picking some values of x and solving the equation for y. For example, suppose we let $x = 1$; then $y = 1 + 1 = 2$. The ordered pair (1, 2) is a solution to the equation $y = x + 1$. We can make a chart of other ordered pairs that are solutions to the equation.

x	y	**Ordered Pair**
1	2	(1, 2)
2	3	(2, 3)
3	4	(3, 4)
4.5	5.5	(4.5, 5.5)
−3	−2	(−3, −2)

How many other ordered pairs satisfy the equation? Infinitely many ordered pairs satisfy the equation. Since we cannot list all the solutions, we show them by means of a graph. A *graph* is an illustration of all the points whose coordinates satisfy an equation.

The points (1, 2), (2, 3), (3, 4), (4.5, 5.5), and (−3, −2) are plotted in Fig. 6.10. With a straightedge we can draw one line that contains all these points. This line, when extended indefinitely in either direction, passes through all the points in the plane that satisfy the equation $y = x + 1$. The arrows on the ends of the line indicate that the line extends indefinitely.

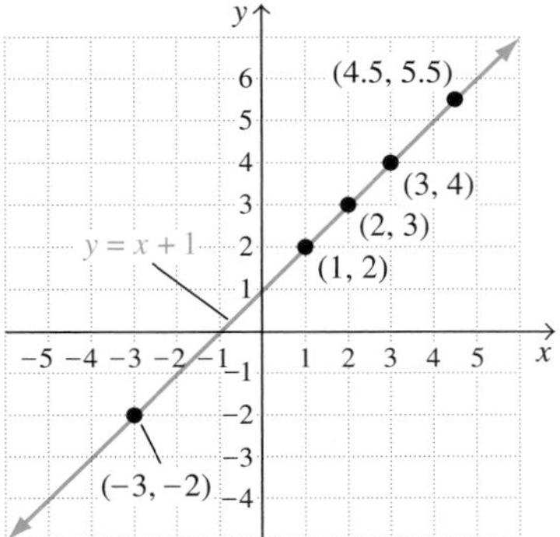

Figure 6.10

All equations of the form $ax + by = c, a \neq 0, b \neq 0$, will be straight lines when graphed. Thus, such equations are called *linear equations in two variables*. The exponents on the variables x and y must be 1 for the equation to be linear. Since only two points are needed to draw a line, only two points are needed to graph a linear equation. It is always a good idea to plot a third point as a checkpoint. If no error has been made, all three points will be in a line, or *collinear*. One method that can be used to obtain points is to solve the equation for y, substitute values for x, and find the corresponding values of y.

DID YOU KNOW

Graphing Calculators

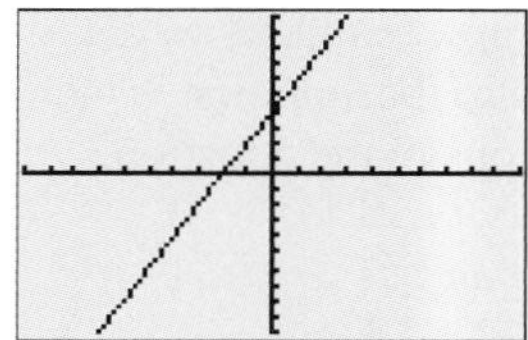

Figure 6.12

On page 308, we illustrated the graph of an exponential equation we obtained from the screen (or window) of a TI-83 Plus graphing calculator. Figure 6.12 shows the graph of the equation $y = 2x + 4$ as illustrated on the screen of that calculator.

Graphing calculators are being used more and more in many mathematics courses. Other makers of graphing calculators include Casio, Sharp, and Hewlett-Packard. Graphing calculators can do a great many more things than a scientific calculator can. If you plan on taking additional mathematics courses, you may find that a graphing calculator is required for those courses.

EXAMPLE 3 *Graphing an Equation by Plotting Points*

Graph $y = 2x + 4$.

SOLUTION: Since the equation is already solved for y, select values for x and find the corresponding values for y. The table indicates values arbitrarily selected for x and the corresponding values for y. The ordered pairs are (0, 4), (1, 6), and $(-2, 0)$. The graph is shown in Fig. 6.11.

x	y
0	4
1	6
−2	0

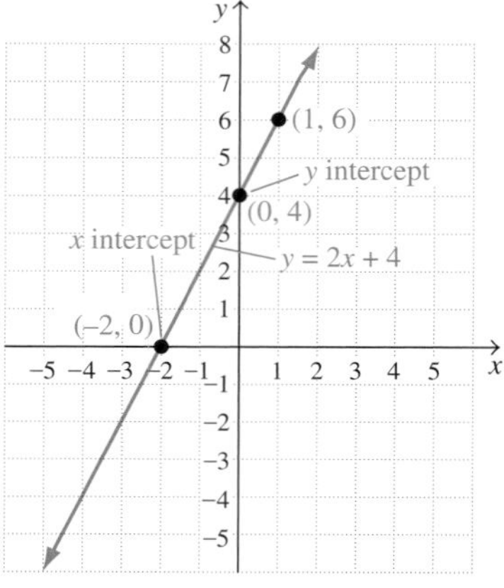

Figure 6.11

To Graph Equations by Plotting Points

1. Solve the equation for y.
2. Select at least three values for x and find their corresponding values of y.
3. Plot the points.
4. The points should be in a straight line. Draw a line through the set of points and place arrow tips at both ends of the line.

In step 4 of the procedure, if the points are not in a straight line, recheck your calculations and find your error.

Graphing by Using Intercepts

Example 3 contained two special points on the graph, $(-2, 0)$ and (0, 4). At these points, the line crosses the x-axis and the y-axis, respectively. The ordered pairs $(-2, 0)$ and (0, 4) represent the *x-intercept* and the *y-intercept*, respectively. Another method that can be used to graph linear equations is to find the x- and y-intercepts of the graph.

Finding the x- and y-Intercepts

To find the x-intercept, set $y = 0$ and solve the equation for x.
To find the y-intercept, set $x = 0$ and solve the equation for y.

DID YOU KNOW

Use of Grids

Grids have long been used in mapping. In archaeological digs, a rectangular coordinate system may be used to chart the location of each find.

An equation may be graphed by finding the x- and y-intercepts, plotting the intercepts, and drawing a straight line through the intercepts. When graphing by this method, you should always plot a checkpoint before drawing your graph. To obtain a checkpoint, select a nonzero value for x and find the corresponding value of y. The checkpoint should be collinear with the x- and y-intercepts.

EXAMPLE 4 *Graphing Using Intercepts*

Graph $2x - 4y = 8$ by using the x- and y-intercepts.

SOLUTION: To find the x-intercept, set $y = 0$ and solve for x.

$$\begin{aligned} 2x - 4y &= 8 \\ 2x - 4(0) &= 8 \\ 2x &= 8 \\ x &= 4 \end{aligned}$$

The x-intercept is $(4, 0)$. To find the y-intercept, set $x = 0$ and solve for y.

$$\begin{aligned} 2x - 4y &= 8 \\ 2(0) - 4y &= 8 \\ -4y &= 8 \\ y &= -2 \end{aligned}$$

The y-intercept is $(0, -2)$. As a checkpoint, try $x = 2$ and find the corresponding value for y.

$$\begin{aligned} 2x - 4y &= 8 \\ 2(2) - 4y &= 8 \\ 4 - 4y &= 8 \\ -4y &= 4 \\ y &= -1 \end{aligned}$$

The checkpoint is the ordered pair $(2, -1)$.

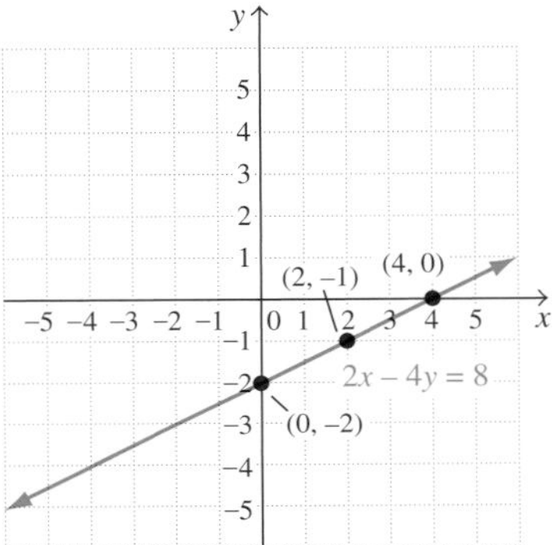

Figure 6.13

Since all three points in Fig. 6.13 are collinear, draw a line through the three points to obtain the graph.

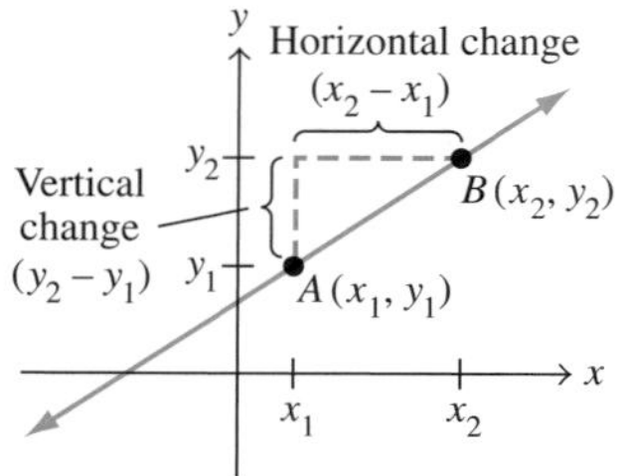

Figure 6.14

Slope

Another useful concept when you are working with straight lines is slope, which is a measure of the "steepness" of a line. The *slope of a line* is a ratio of the vertical change to the horizontal change for any two points on the line. Consider Fig. 6.14. Point A has coordinates (x_1, y_1), and point B has coordinates (x_2, y_2). The vertical change between points A and B is $y_2 - y_1$, and the horizontal change between points A and B is $x_2 - x_1$. Thus, the slope, which is often symbolized with the letter m, can be found as follows.

$$\textbf{Slope} = \frac{\text{vertical change}}{\text{horizontal change}}$$

$$m = \frac{y_2 - y_1}{x_2 - x_1}$$

The Greek capital letter delta, Δ, is often used to represent the words "the change in." Therefore, slope may be defined as

$$m = \frac{\Delta y}{\Delta x}$$

A line may have a positive slope, a negative slope, zero slope, or the slope may be undefined, as indicated in Fig. 6.15. A line with a positive slope rises from left to right, as shown in Fig. 6.15(a). A line with a negative slope falls from left to right, as shown in Fig. 6.15(b). A horizontal line, which neither rises nor falls, has a slope of zero, as shown in Fig. 6.15(c). Since a vertical line does not have any horizontal change (the x value remains constant) and since we cannot divide by 0, the slope of a vertical line is undefined, as shown in Fig. 6.15(d).

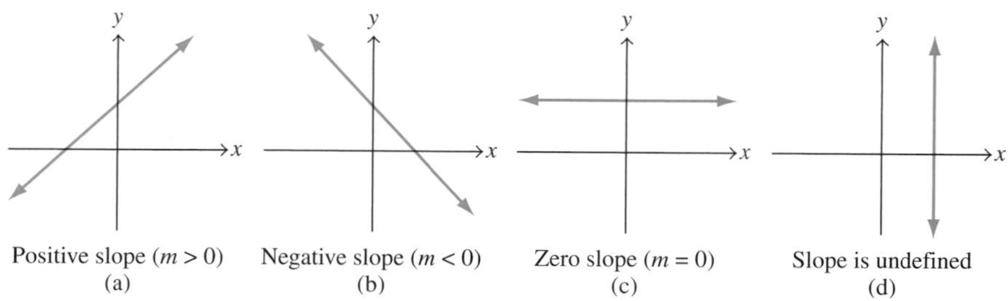

Figure 6.15

EXAMPLE 5 *Finding the Slope of a Line*

Determine the slope of the line that passes through the points $(-1, -3)$ and $(1, 5)$.

SOLUTION: Let's begin by drawing a sketch, illustrating the points and the line. See Fig. 6.16(a) on page 341.

We will let (x_1, y_1) be $(-1, -3)$ and (x_2, y_2) be $(1, 5)$. Then

$$\text{Slope} = \frac{y_2 - y_1}{x_2 - x_1} = \frac{5 - (-3)}{1 - (-1)} = \frac{5 + 3}{1 + 1} = \frac{8}{2} = \frac{4}{1} = 4$$

DID YOU KNOW

Up Up and Away

Simon Ammann

Although we may not think much about it, the slope of a line is something we are altogether familiar with. You confront it every time you run up the stairs, late for class, moving 8 inches horizontally for every 6 inches up. The 2002 Olympic gold medalist Simon Ammann of Switzerland is familiar with the concept of slope. He speeds down a steep 120 meters of a ski ramp at speeds of over 60 mph before he takes flight.

The slope of 4 means that there is a vertical change of 4 units for each horizontal change of 1 unit; see Fig. 6.16(b). The slope is positive, and the line rises from left to right. Note that we would have obtained the same results if we let (x_1, y_1) be $(1, 5)$ and (x_2, y_2) be $(-1, -3)$. Try this now and see.

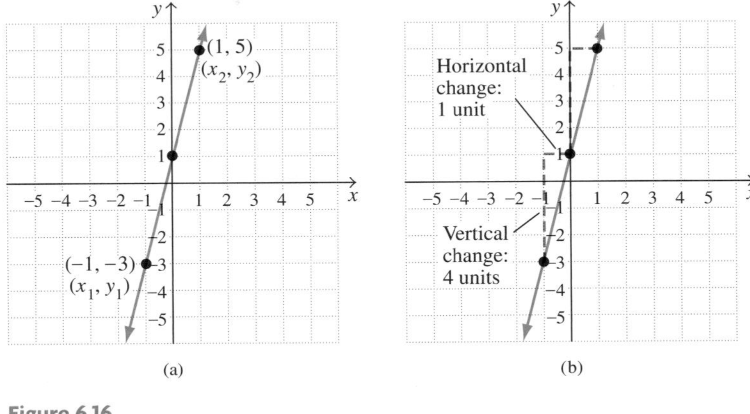

Figure 6.16

Graphing Equations by Using the Slope and *y*-Intercept

A linear equation given in the form $y = mx + b$ is said to be in slope–intercept form.

Slope–Intercept Form of the Equation of a Line

$$y = mx + b$$

where m is the slope of the line and $(0, b)$ is the y-intercept of the line.

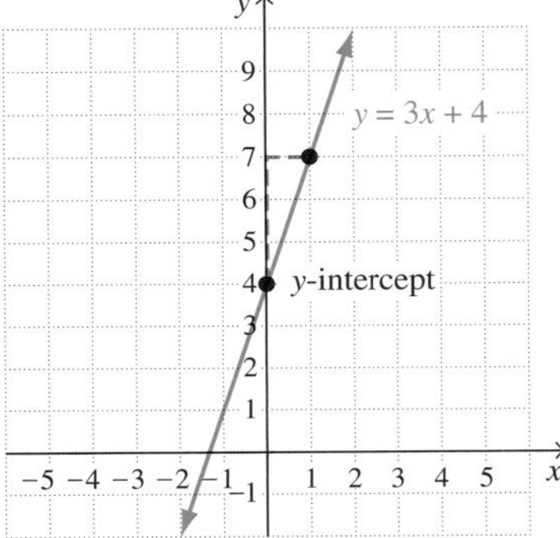

Figure 6.17

In the equation $y = mx + b$, b represents the value of y where the graph of the equation $y = mx + b$ crosses the y-axis.

Consider the graph of the equation $y = 3x + 4$, which appears in Fig. 6.17. By examining the graph, we can see that the y-intercept is $(0, 4)$. We can also see that the graph has a positive slope, since it rises from left to right. Since the vertical change is 3 units for every 1 unit of horizontal change, the slope must be $\frac{3}{1}$ or 3.

We could graph this equation by marking the y-intercept at $(0, 4)$ and then moving *up* 3 units and to the *right* 1 unit to get another point. If the slope were -3, which means $\frac{-3}{1}$, we could start at the y-intercept and move *down* 3 units and to the *right* 1 unit. Thus, if we know the slope and y-intercept of a line, we can graph the line.

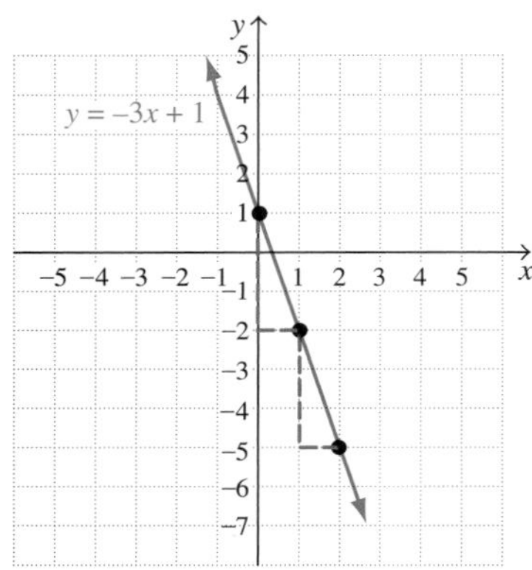

Figure 6.18

To Graph Equations by Using the Slope and y-Intercept

1. Solve the equation for y to place the equation in slope–intercept form.
2. Determine the slope and y-intercept from the equation.
3. Plot the y-intercept.
4. Obtain a second point using the slope.
5. Draw a straight line through the points.

EXAMPLE 6 *Graphing an Equation Using the Slope and y-Intercept*

Graph $y = -3x + 1$ using the slope and y-intercept.

SOLUTION: The slope is -3 or $\frac{-3}{1}$ and the y-intercept is (0, 1). Plot (0, 1) on the y-axis. Then plot the next point by moving *down* 3 units and to the *right* 1 unit (see Fig. 6.18). A third point has been plotted in the same way. The graph of $y = -3x + 1$ is the line drawn through these three points.

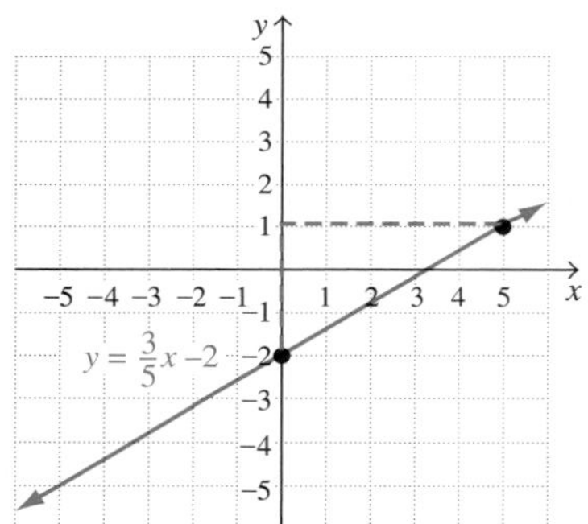

Figure 6.19

EXAMPLE 7 *Write an Equation in Slope–Intercept Form*

a) Write $3x - 5y = 10$ in slope–intercept form.
b) Graph the equation.

SOLUTION:

a) To write $3x - 5y = 10$ in slope–intercept form, we solve the given equation for y.

$$3x - 5y = 10$$
$$3x - 3x - 5y = -3x + 10$$
$$-5y = -3x + 10$$
$$\frac{-5y}{-5} = \frac{-3x + 10}{-5}$$
$$y = \frac{-3x}{-5} + \frac{10}{-5} \quad \text{or} \quad y = \frac{3}{5}x - 2$$

Thus, in slope–intercept form, the equation is $y = \frac{3}{5}x - 2$.

b) The y-intercept is $(0, -2)$ and the slope is $\frac{3}{5}$. Plot a point at $(0, -2)$ on the y-axis, then move *up* 3 units and to the *right* 5 units to obtain the second point (see Fig. 6.19). Draw a line through the two points.

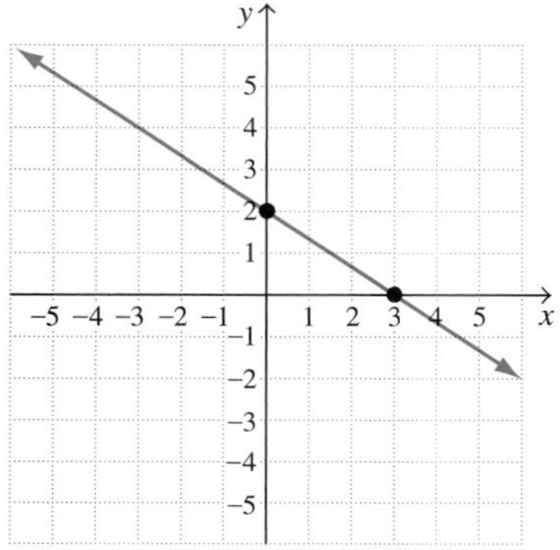
Figure 6.20

EXAMPLE 8 *Determine the Equation of a Line from Its Graph*

Determine the equation of the line in Fig. 6.20.

SOLUTION: If we determine the slope and the y-intercept of the line, then we can write the equation using slope–intercept form, $y = mx + b$. We see from the graph that the y-intercept is (0, 2); thus, $b = 2$. The slope of the line is negative because the graph falls from left to right. The change in y is 2 units for every 3-unit change in x. Thus, m, the slope of the line, is $-\frac{2}{3}$.

$$y = mx + b$$

$$y = -\frac{2}{3}x + 2$$

The equation of the line is $y = -\frac{2}{3}x + 2$.

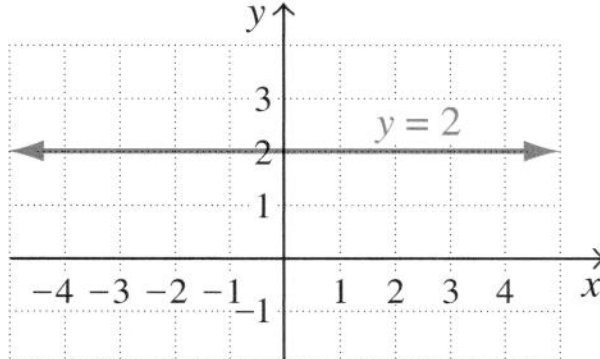

Figure 6.21

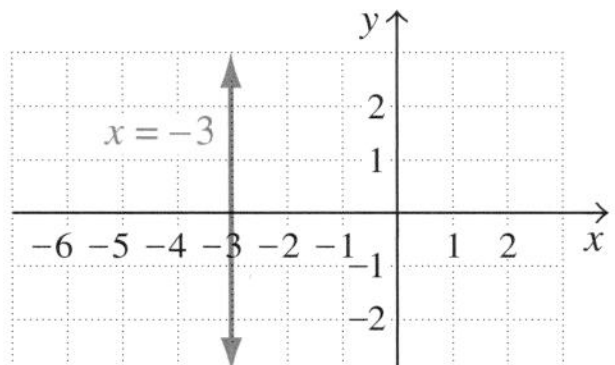

Figure 6.22

EXAMPLE 9 *Horizontal and Vertical Lines*

In the Cartesian coordinate system, graph (a) $y = 2$ and (b) $x = -3$.

SOLUTION:

a) For any value of x, the value of y is 2. Therefore, the graph will be a horizontal line through $y = 2$ (Fig. 6.21).

b) For any value of y, the value of x is -3. Therefore, the graph will be a vertical line through $x = -3$ (Fig. 6.22).

Note that the graph of $y = 2$ has a slope of 0. The slope of the graph of $x = -3$ is undefined.

In graphing the equations in this section, we labeled the horizontal axis the x-axis and the vertical axis the y-axis. For each equation, we can determine values for y by substituting values for x. Since the value of y depends on the value of x, we refer to y as the *dependent variable* and x as the *independent variable*. We label the *vertical axis* with the *dependent variable* and the *horizontal axis* with the *independent variable*. For the equation $C = 3n + 5$, the C is the dependent variable and n is the independent variable. Thus, to graph this equation, we label the vertical axis C and the horizontal axis n.

In many graphs, the values to be plotted on one axis are much greater than the values to be plotted on the other axis. When that occurs, we can use different scales on the horizontal and the vertical axes, as illustrated in Examples 10 and 11. The next two examples illustrate applications of graphing.

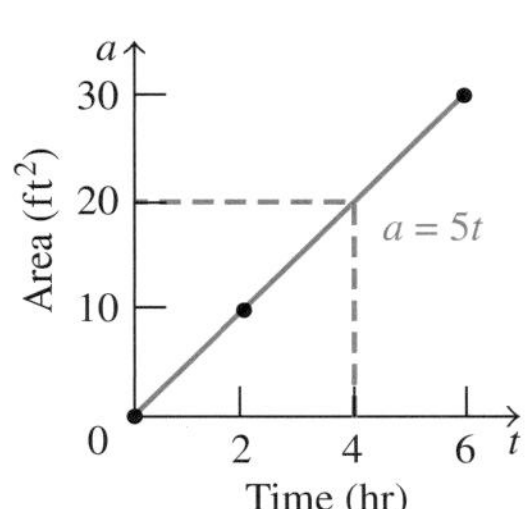

Figure 6.23

EXAMPLE 10 *Using a Graph to Determine Area*

The Professional Patio Company installs brick patios. The area of brick, a, in square feet, the company can install in t hours can be approximated by the formula $a = 5t$.

a) Graph $a = 5t$, for $t \leq 6$.

b) Use the graph to estimate the area of brick the company can install in 4 hours.

SOLUTION:

a) Since $a = 5t$ is a linear equation, its graph will be a straight line. Select three values for t, find the corresponding values for a, and then draw the graph (Fig. 6.23).

$$a = 5t$$

Let $t = 0$, $\quad a = 5(0) = 0$

Let $t = 2$, $\quad a = 5(2) = 10$

Let $t = 6$, $\quad a = 5(6) = 30$

t	a
0	0
2	10
6	30

b) By drawing a vertical line from $t = 4$ on the time axis up to the graph and then drawing a horizontal line across to the area axis, we can determine that the area installed in 4 hours is 20 ft^2.

EXAMPLE 11 *Using a Graph to Determine Profits*

Jonathan Cwirko owns a small business that manufactures compact discs. He believes that the profit (or loss) from the compact discs produced can be estimated by the formula $P = 3.5S - 200{,}000$, where S is the number of compact discs sold.

a) Graph $P = 3.5S - 200{,}000$, for $S \leq 500{,}000$ compact discs.

b) From the graph, estimate the number of compact discs that must be sold for the company to break even.

c) If the profit from selling compact discs is \$1 million, estimate the number of compact discs sold.

SOLUTION:

a) Select values for S and find the corresponding values of P.

S	P
0	−200,000
100,000	150,000
500,000	1,550,000

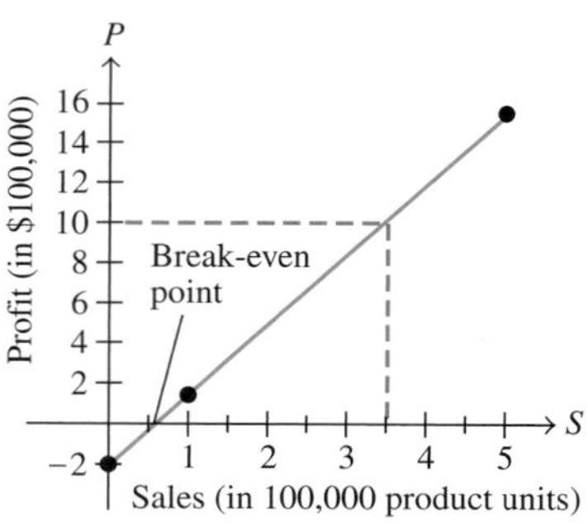

Figure 6.24

b) On the graph (Fig. 6.24), note that the break-even point is about 0.6, or 60,000 compact discs.

c) We can obtain the answer by drawing a horizontal line from 10 on the profit axis. Since the horizontal line cuts the graph at about 3.4 on the S axis, approximately 340,000 compact discs were sold.

SECTION 6.7 EXERCISES

Concept/Writing Exercises

1. What is a graph?
2. Explain how to find the x-intercept of a linear equation.
3. Explain how to find the y-intercept of a linear equation.
4. What is the slope of a line?
5. **a)** Explain in your own words how to find the slope of a line between two points.
   b) Based on your explanation in part (a), find the slope of the line through the points (6, 2) and (−3, 5).
6. Describe the three methods used to graph a linear equation in this section.

7. a) In which quadrant is the point (1, 4) located?
b) In which quadrant is the point (−2, 5) located?

8. What is the minimum number of points needed to graph a linear equation?

Practice the Skills

In Exercises 9–16, plot all the points on the same axes.

9. (−3, 2) **10.** (2, 3) **11.** (−5, −1)
12. (4, 0) **13.** (0, 2) **14.** (0, 0)
15. (0, −5) **16.** $(3\frac{1}{2}, 4\frac{1}{2})$

In Exercises 17–24, plot all the points on the same axes.

17. $(5, 1)$ **18.** $(0, -3)$ **19.** $(-6, -1)$
20. $(1, 0)$ **21.** $(-3, 0)$ **22.** $(-3, 1)$
23. $(4, -1)$ **24.** $(4.5, 3.5)$

In Exercises 25–34 (indicated on Fig. 6.25), write the coordinates of the corresponding point.

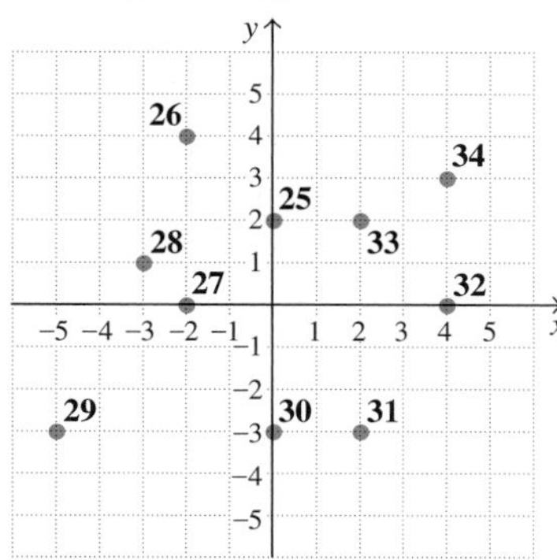

Figure 6.25

In Exercises 35–42, determine which ordered pairs satisfy the given equation.

35. $3x + y = 7$ $(1, 3)$ $(1, 4)$ $(-1, 10)$
36. $4x - y = 4$ $(0, -4)$ $(1, 0)$ $(2, -3)$
37. $2x - 3y = 10$ $(5, 0)$ $(0, 3)$ $(0, -\frac{10}{3})$
38. $3y = 4x + 2$ $(2, 1)$ $(1, 2)$ $(0, \frac{2}{3})$
39. $7y = 3x - 5$ $(1, -1)$, $(-3, -2)$, $(2, 5)$
40. $\frac{x}{2} + 3y = 4$ $(0, \frac{4}{3})$, $(8, 0)$, $(10, -2)$
41. $\frac{x}{2} + \frac{3y}{4} = 2$ $(0, \frac{8}{3})$, $(1, \frac{11}{4})$, $(4, 0)$
42. $2x - 5y = -7$ $(2, 1)$, $(-1, 1)$, $(4, 3)$

In Exercises 43–46, graph the equation and state the slope of the line if the slope exists (see Example 9).

43. $x = 4$ **44.** $x = -2$ **45.** $y = 3$ **46.** $y = -5$

In Exercises 47–56, graph the equation by plotting points, as in Example 3.

47. $y = x + 3$ **48.** $y = x - 2$
49. $y = 2x - 1$ **50.** $y = -x + 4$
51. $y + 3x = 6$ **52.** $y - 4x = 8$
53. $y = \frac{1}{2}x + 4$ **54.** $3y = 2x - 3$
55. $2y = -x + 6$ **56.** $y = -\frac{3}{4}x$

In Exercises 57–66, graph the equation, using the x- and y-intercepts, as in Example 4.

57. $x + y = 1$ **58.** $-x + y = 7$
59. $3x + y = 6$ **60.** $4x - 2y = 12$
61. $2x = -4y - 8$ **62.** $y = 4x + 4$
63. $y = 3x + 5$ **64.** $3x + 6y = 9$
65. $3x - y = 5$ **66.** $4y = 2x + 12$

In Exercises 67–76, find the slope of the line through the given points. If the slope is undefined, so state.

67. $(3, 7)$ and $(10, 21)$ **68.** $(4, 1)$ and $(1, 4)$
69. $(2, 6)$ and $(-5, -9)$ **70.** $(-5, 6)$ and $(7, -9)$
71. $(5, 2)$ and $(-3, 2)$ **72.** $(-3, -5)$ and $(-1, -2)$
73. $(8, -3)$ and $(8, 3)$ **74.** $(2, 6)$ and $(2, -3)$
75. $(-2, 3)$ and $(1, -1)$ **76.** $(-7, -5)$ and $(5, -6)$

In Exercises 77–86, graph the equation using the slope and y-intercept, as in Examples 6 and 7.

77. $y = x + 3$ **78.** $y = 3x + 2$
79. $y = -x - 4$ **80.** $y = -2x + 1$
81. $y = -\frac{1}{3}x + 2$ **82.** $y = -x - 2$
83. $7y = 4x - 7$ **84.** $3x + 2y = 6$
85. $3x - 2y + 6 = 0$ **86.** $3x + 4y - 8 = 0$

In Exercises 87–90, determine the equation of the graph.

87.

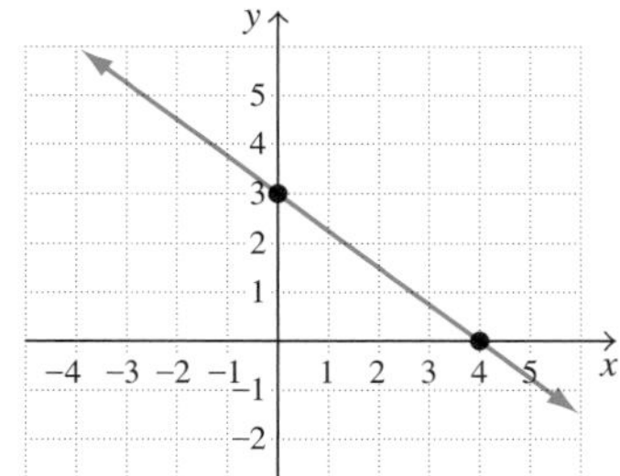

88.

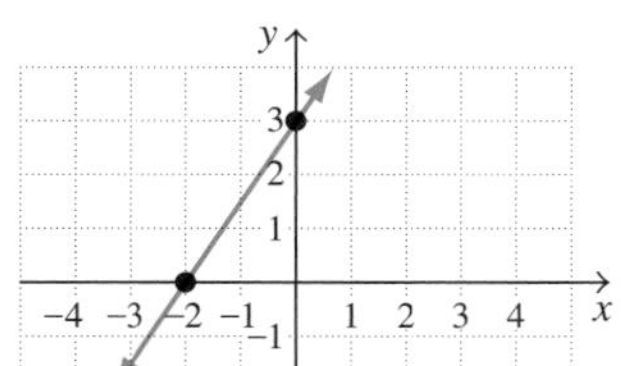

89.

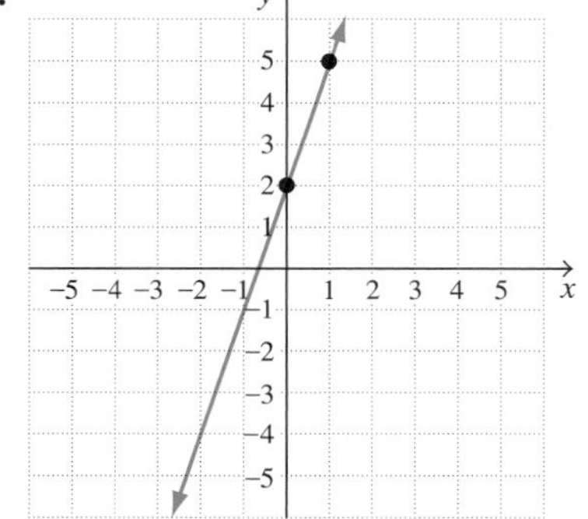

90.

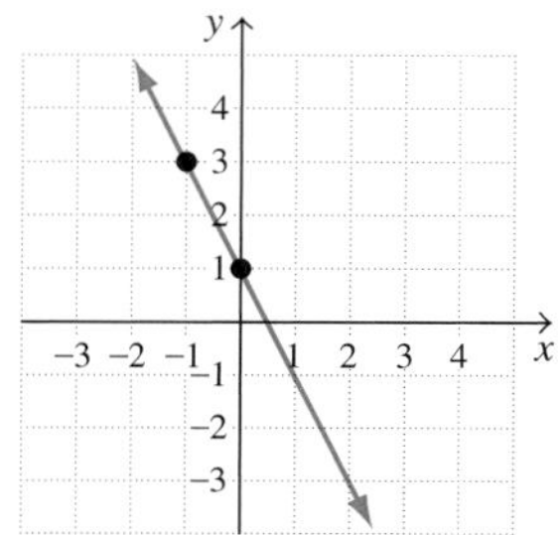

Problem Solving

In Exercises 91 and 92, points A, B, and C are three vertices of a rectangle (the points where two sides meet). Plot the three points. (a) Find the coordinates of the fourth point, D, to complete the rectangle. (b) Find the area of the rectangle; use $A = lw$.

91. $A(-1, 4)$, $B(4, 4)$, $C(4, 2)$

92. $A(-4, 2)$, $B(7, 2)$, $C(7, 8)$

In Exercises 93 and 94, points A, B, and C are three vertices of a parallelogram with sides parallel to the x-axis. Plot the three points. Find the coordinates of the fourth point, D, to complete the parallelogram. Note: There are two possible answers for point D.

93. $A(3, 2)$, $B(5, 5)$, $C(9, 5)$

94. $A(-2, 2)$, $B(3, 2)$, $C(6, -1)$

In Exercises 95–98, for what value of b will the line joining the points P and Q be parallel to the indicated axis?

95. $P(-1, 3)$, $Q(4, b)$; x-axis

96. $P(5, 6)$, $Q(b, -2)$; y-axis

97. $P(3b - 1, 5)$, $Q(8, 4)$; y-axis

98. $P(-6, 2b + 3)$, $Q(7, -1)$; x-axis

99. *Selling Chocolates* Ryan Stewart sells chocolate on the Internet. His monthly profit, p, in dollars, can be estimated by $p = 15n - 300$, where n is the number of dozens of chocolates he sells in a month.

a) Graph $p = 15n - 300$, for $n \leq 60$.

b) From the graph, estimate his profit if he sells 40 dozen chocolates in a month.

c) How many dozens of chocolates must he sell in a month to break even?

100. *Hanging Wallpaper* Tanisha Vizquez owns a wallpaper hanging business. Her charge, C, for hanging wallpaper is \$40 plus \$0.30 per square foot of wallpaper she hangs, or $C = 40 + 0.30s$, where s is the number of square feet of wallpaper she hangs.

a) Graph $C = 40 + 0.30s$, for $s \leq 500$.

b) From the graph, estimate her charge if she hangs 300 square feet of wallpaper.

c) If her charge is \$70, use the equation for C to determine how many square feet of wallpaper she hung.

101. *Photo Processing* The charge, C, for processing a roll of 35-millimeter (mm) film onto a picture compact disc at Costco's 1 Hour Photo is \$8.95 plus \$0.33 per picture, or $C = 8.95 + 0.33n$, where n is the number of pictures printed.

a) Draw a graph of the cost of processing film for up to and including 36 pictures.

b) From the graph, estimate the cost of processing a roll of 35 mm film containing 20 pictures.

c) If the total cost of processing a roll of 35 mm film is \$20.83, estimate the number of pictures.

102. *Earning Simple Interest* When \$1000 is invested in a savings account paying simple interest for a year, the interest, i in dollars, earned can be found by the formula $i = 1000r$, where r is the rate in decimal form.

a) Graph $i = 1000r$, for r up to and including a rate of 15%.

b) If the rate is 4%, what is the simple interest?

c) If the rate is 6%, what is the simple interest?

In Exercises 103 and 104, a set of points is plotted. Also shown is a straight line through the set of points that is called the line of best fit *(or a* regression line, *as will be discussed in Chapter 13, Statistics.)*

103. *Determining a Test Grade* The graph shows the hours studied and the test grades on a biology test for six students. (The two points indicated on the line do not represent any of the six students.) The line of best fit, the red line on the graph, can be used to approximate the test grade the average student receives for the number of hours he or she studies.

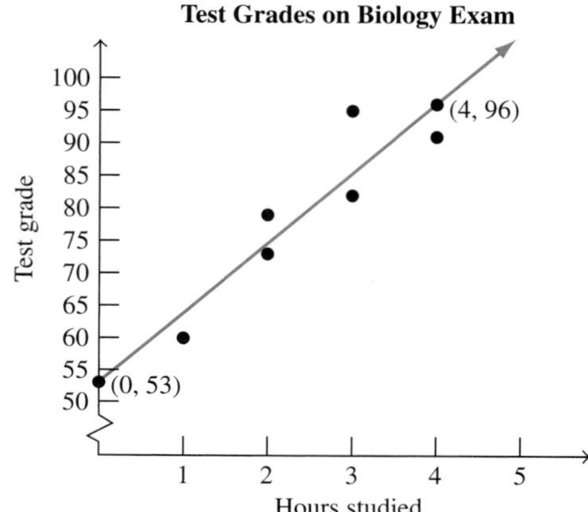

a) Determine the slope of the line of best fit using the two points indicated.

b) Using the slope determined in part (a) and the y-intercept, (0, 53), determine the equation of the line of best fit.

c) Using the equation you determined in part (b), determine the approximate test grade for a student who studied for 3 hours.

d) Using the equation you determined in part (b), determine the amount of time a student would need to study to receive a grade of 80 on the biology test.

104. *Determining the Number of Defects* The graph shows the daily number of workers absent from the assembly line at J. B. Davis Corporation and the number of defects coming off the assembly line for 8 days. (The two points indicated on the line do not represent any of the 8 days.) The line of best fit, the blue line on the graph, can be used to approximate the number of defects coming off the assembly line per day for a given number of workers absent.

a) Determine the slope of the line of best fit using the two points indicated.

b) Using the slope determined in part (a) and the y-intercept, (0, 9), determine the equation of the line of best fit.

c) Using the equation you determined in part (b), determine the approximate number of defects for a day if 3 workers are absent.

d) Using the equation you determined in part (b), approximate the number of workers absent for a day if there are 17 defects that day.

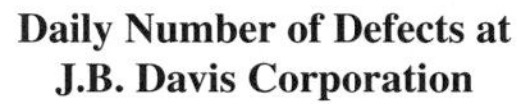

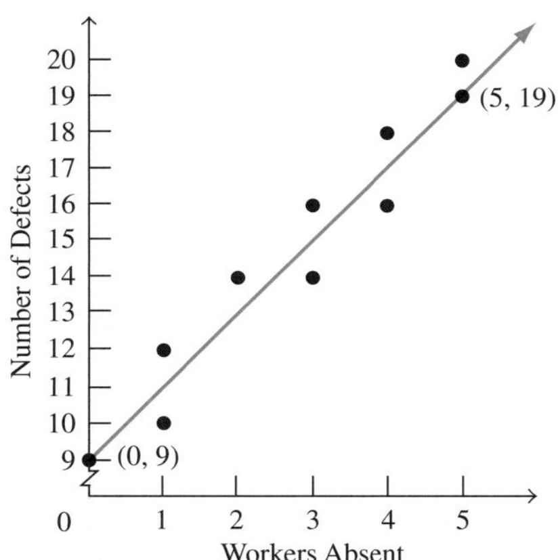

105. *Married Householders with Children* The blue graph shows the percent of married householders with children. The red dashed straight line can be used to approximate the percent of married householders with children. If we let 0 represent 1970, 10 represent 1980, and so on, then 2000 would be represented by 30. Using the ordered pairs (0, 40) and (30, 24),

a) determine the slope of the dashed line.

b) determine the equation of the dashed line using (0, 40) as the y-intercept.

c) Using the equation you determined in part (b), determine the percent of married householders with children in 1985, which would be represented by 15.

d) Using the equation you determined in part (b), determine the year in which the percent of married householders with children was 30.

Married House-holders with Kids

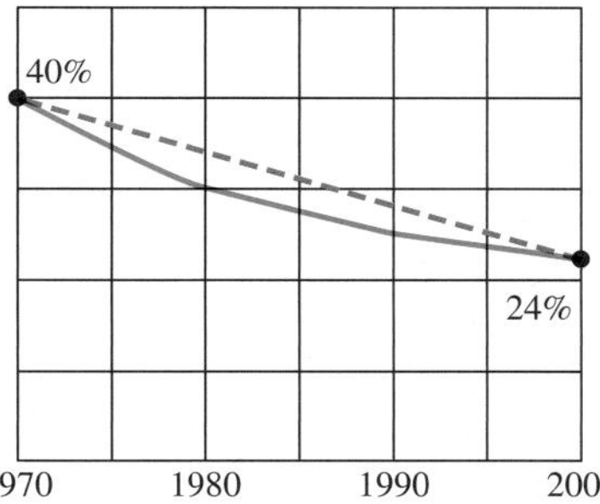

Source: *Newsweek*

106. *Book Sales* The green graph shows book publishers' net sales, in millions of dollars, for trade, mass market, professional, educational, and university press publishers. The red dashed straight line can be used to approximate the book publishers' net dollar sales. If we let 0 represent 1994, 1 represent 1995, 2 represent 1996, and so on, then 2003 would be represented by 9. Using the ordered pairs (0, 17,000) and (9, 25,000),

a) determine the slope of the dashed line.

b) determine the equation of the dashed line using (0, 17,000) as the y-intercept of the graph.

c) Using the equation you determined in part (b), determine the net dollar sales in 1998, which would be represented with year 4.

d) Using the equation you determined in part (b), determine the year that net sales were \$20,000 (in millions).

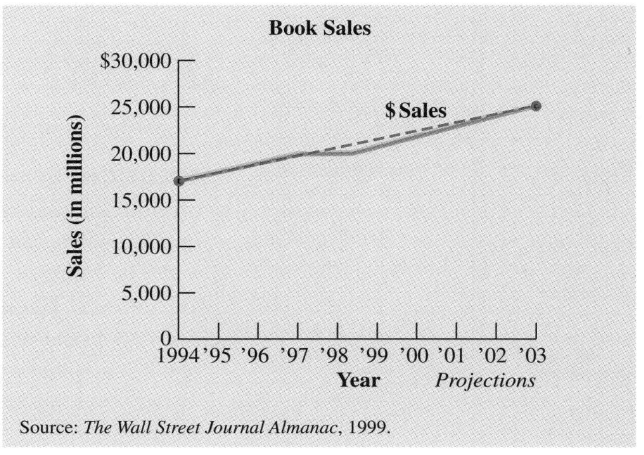

Source: *The Wall Street Journal Almanac*, 1999.

Challenge Problems/Group Activities

107. a) Two lines are parallel when they do not intersect no matter how far they are extended. Explain how you can determine, without graphing the equations, whether two equations will be parallel lines when graphed.
    b) Determine whether the graphs of the equations $2x - 3y = 6$ and $4x = 6y + 6$ are parallel lines.

108. In which quadrants will the set of points that satisfy the equation $x + y = 1$ lie? Explain.

Internet/Research Activity

109. René Descartes is known for his contributions to algebra. Write a paper on his life and his contributions to algebra.

6.8 LINEAR INEQUALITIES IN TWO VARIABLES

In Section 6.6, we introduced linear inequalities in one variable. Now we will introduce linear inequalities in two variables. Some examples of linear inequalities in two variables are $2x + 3y \leq 7$, $x + 7y \geq 5$, and $x - 3y < 6$.

The solution set of a linear inequality in one variable may be indicated on a number line. The solution set of a linear inequality in two variables is indicated on a coordinate plane.

An inequality that is strictly less than ($<$) or greater than ($>$) will have as its solution set a *half-plane*. A half-plane is the set of all the points on one side of a line. An inequality that is less than or equal to ($\leq$) or greater than or equal to ($\geq$) will have as its solution set the set of points that consists of a half-plane and a line. To indicate that the line is part of the solution set, we draw a solid line. To indicate that the line is not part of the solution set, we draw a dashed line.

To Graph Inequalities in Two Variables

1. Mentally substitute the equal sign for the inequality sign and plot points as if you were graphing the equation.
2. If the inequality is $<$ or $>$, draw a dashed line through the points. If the inequality is $\leq$ or $\geq$, draw a solid line through the points.
3. Select a test point not on the line and substitute the x- and y-coordinates into the inequality. If the substitution results in a true statement, shade in the area on the same side of the line as the test point. If the test point results in a false statement, shade in the area on the opposite side of the line as the test point.

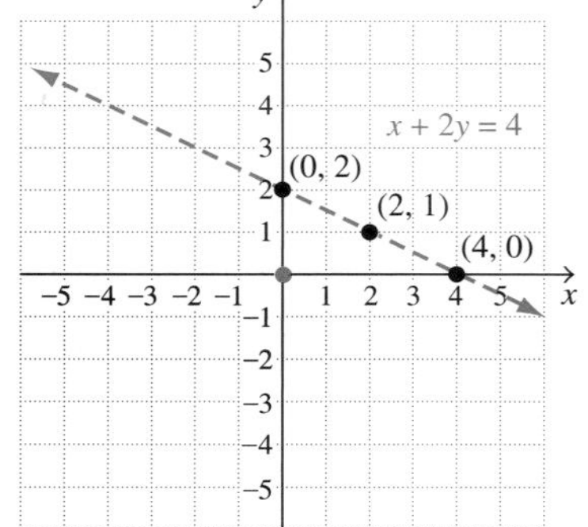

Figure 6.26

EXAMPLE 1 *Graphing an Inequality*

Draw the graph of $x + 2y < 4$.

SOLUTION: To obtain the solution set, start by graphing $x + 2y = 4$. Since the original inequality is strictly "less than," draw a dashed line (Fig 6.26). The dashed line indicates that the points on the line are not part of the solution set.

The line $x + 2y = 4$ divides the plane into three parts, the line itself and two *half-planes*. The line is the boundary between the two half-planes. The points in one half-plane will satisfy the inequality $x + 2y < 4$. The points in the other half-plane will satisfy the inequality $x + 2y > 4$.

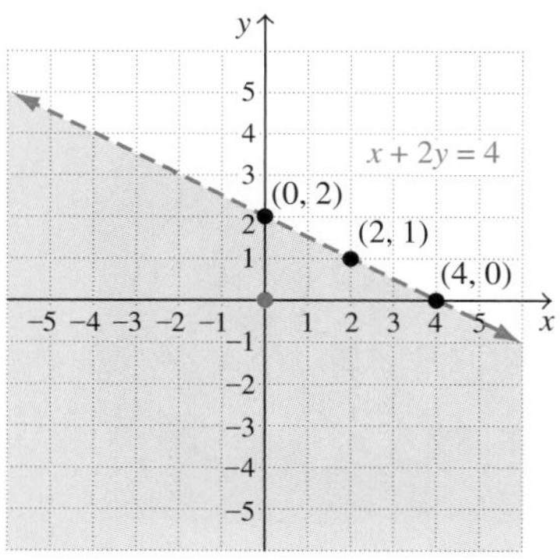

Figure 6.27

To determine the solution set to the inequality $x + 2y < 4$, pick any point on the plane that is not on the line. The simplest point to work with is the origin, (0, 0). Substitute $x = 0$ and $y = 0$ into $x + 2y < 4$.

$$x + 2y < 4$$
$$\text{Is } 0 + 2(0) < 4?$$
$$0 + 0 < 4$$
$$0 < 4 \quad \text{True}$$

Since 0 is less than 4, the point (0, 0) is part of the solution set. All the points on the same side of the graph of $x + 2y = 4$ as the point (0, 0) are members of the solution set. We indicate this by shading the half-plane that contains (0, 0). The graph is shown in Fig. 6.27.

EXAMPLE 2 *Graphing an Inequality*

Draw the graph of $4x - 2y \geq 12$.

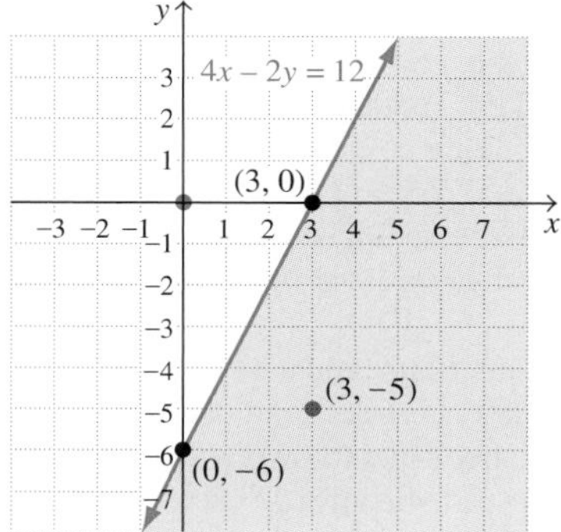

Figure 6.28

SOLUTION: First draw the graph of the equation $4x - 2y = 12$. Use a solid line because the points on the boundary line are included in the solution set. Now pick a point that is not on the line. Take (0, 0) as the test point.

$$4x - 2y \geq 12$$
$$\text{Is } 4(0) - 2(0) \geq 12?$$
$$0 \geq 12 \quad \text{False}$$

Since 0 is not greater than or equal to 12 ($0 \ngeq 12$), the solution set is the line and the half-plane that does not contain the point (0, 0). The graph is shown in Fig. 6.28.

If you had arbitrarily selected the test point $(3, -5)$ from the other half-plane, you would have found that the inequality would be true: $4(3) - 2(-5) \geq 12$, or $22 \geq 12$. Thus, the point $(3, -5)$ would be in the half-plane containing the solution set.

EXAMPLE 3 *Graphing an Inequality*

Draw the graph of $y < x$.

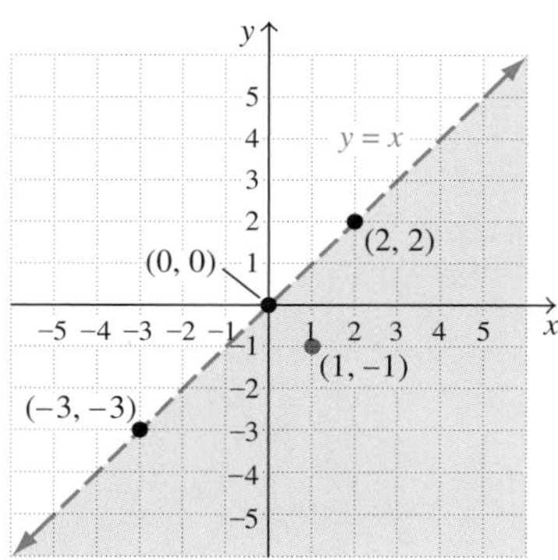

Figure 6.29

SOLUTION: The inequality is strictly "less than," so the boundary line is not part of the solution set. In graphing the equation $y = x$, draw a dashed line (Fig. 6.29). Since (0, 0) is *on* the line, it cannot serve as a test point. Let's pick the point $(1, -1)$.

$$y < x$$
$$-1 < 1 \quad \text{True}$$

Since $-1 < 1$ is true, the solution set is the half-plane containing the point $(1, -1)$.

SECTION 6.8 EXERCISES

Concept/Writing Exercises

1. Outline the procedure used to graph inequalities in two variables.
2. Explain why we use a solid line when graphing an inequality containing $\le$ or $\ge$ and we use a dashed line when graphing an inequality containing $<$ or $>$.

Practice the Skills

In Exercises 3–24, draw the graph of the inequality.

3. $x \le 1$
4. $y \ge -2$
5. $y > x + 3$
6. $y < x - 5$
7. $y \ge 2x - 6$
8. $y < -2x + 2$
9. $3x - 4y > 12$
10. $x + 2y > 4$
11. $3x - 4y \le 9$
12. $4y - 3x \ge 9$
13. $3x + 2y < 6$
14. $-x + 2y < 2$
15. $x + y > 0$
16. $x + 2y \le 0$
17. $5x - 2y \le 10$
18. $y \ge -2x + 1$
19. $3x + 2y > 12$
20. $y \le 3x - 4$
21. $\frac{2}{5}x - \frac{1}{2}y \le 1$
22. $0.1x + 0.3y \le 0.4$
23. $0.2x + 0.5y \le 0.3$
24. $\frac{1}{3}x + \frac{3}{4}y \ge 1$

Problem Solving

25. *Gas Grills* A manufacturer of gas grills must produce and ship x gas grills to one outlet and y gas grills to a second outlet. The maximum number of gas grills the manufacturer can produce and ship is 300.
   a) Write an inequality in two variables that represents this problem.
   b) Graph the inequality.
26. *Flower Garden* Jim Lawler has 40 ft of landscape edging to place around a new rectangular flower garden.

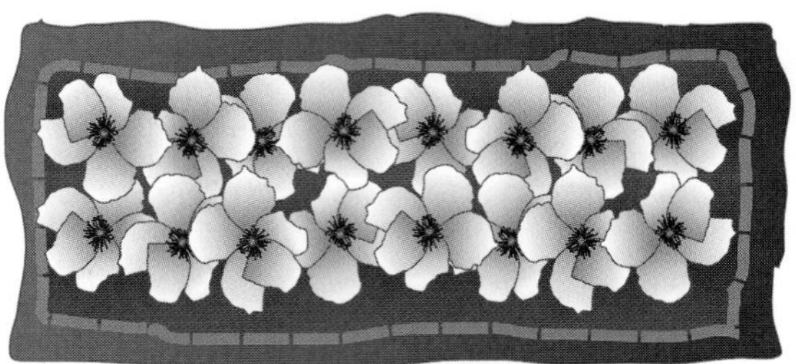

   a) Write an inequality illustrating all possible dimensions of the rectangular garden. $P = 2l + 2w$ is the formula for the perimeter of a rectangle.
   b) Graph the inequality.

Challenge Problems

27. *Building a House* Yolanda Vega has \$150,000 to spend on purchasing land and building a new house in the country. She wants at least 1 acre of land but less than 10 acres. If land costs \$1500 per acre and building costs are \$75 per square foot, the inequality $1500x + 75y \le 150{,}000$, where $1 \le x < 10$, describes the restriction on her purchase.
   a) What quantities do x and y represent in the inequality?
   b) Graph the inequality.
   c) If Yolanda decides that her house must be at least 1950 ft^2 in size, how many acres of land can she buy?
   d) If Yolanda decides that she wants to own at least 5 acres of land, what size house can she afford?
28. *Men's Shirts* The Tommy Hilfiger Company must ship x men's shirts to one outlet and y men's shirts to a second outlet. The maximum number of shirts the manufacturer can produce and ship is 250. We can represent this situation with the inequality $x + y \le 250$.
   a) Can x or y be negative? Explain.
   b) Graph the inequality.
   c) Write one or two paragraphs interpreting the information that the graph provides.
29. Which of the following inequalities have the same graph? Explain how you determined your answer.
   a) $3x - y < 6$ b) $-3x + y > -6$
   c) $3x - 2y < 12$ d) $y > 3x - 6$

6.9 SOLVING QUADRATIC EQUATIONS BY USING FACTORING AND BY USING THE QUADRATIC FORMULA

We begin this section by discussing multiplication of binomials and factoring trinomials. After we discuss factoring trinomials, we will explain how to solve quadratic equations using factoring.

DID YOU KNOW

Algorithm

Recall that the series of steps taken to solve a certain type of equation is called an *algorithm*. Because mathematical procedures to solve equations can be generalized, it is possible to program computers to solve problems. For example, the equation $4x^2 + 7x + 3 = 0$ is an equation of the form $ax^2 + bx + c = 0$. If you can find a procedure to solve an equation of the general form, it can be used to solve all equations of that form. One of the first computer languages, FORTRAN (for FORmula TRANslation), was developed specifically to handle scientific and mathematical applications.

We will now look at the *FOIL method* of multiplying two binomials. A *binomial* is an expression that contains two terms, in which each exponent that appears on a variable is a whole number.

Examples of binomials

$x + 3$	$x - 5$
$3x + 5$	$4x - 2$

To multiply two binomials, we can use the FOIL method. The name of the method, FOIL, is an acronym to help its users remember it as a method that obtains the products of the First, Outer, Inner, and Last terms of the binomials.

$$(a + b)(c + d) = \underbrace{a \cdot c}_{\text{First}} + \underbrace{a \cdot d}_{\text{Outer}} + \underbrace{b \cdot c}_{\text{Inner}} + \underbrace{b \cdot d}_{\text{Last}}$$

After multiplying the first, outer, inner, and last terms, combine all like terms.

EXAMPLE 1 *Multiplying Binomials*

Multiply $(x + 3)(x + 5)$.

SOLUTION: The FOIL method of multiplication yields

$$\begin{aligned}(x + 3)(x + 5) &= \overset{F}{x \cdot x} + \overset{O}{x \cdot 5} + \overset{I}{3 \cdot x} + \overset{L}{3 \cdot 5}\\ &= x^2 + 5x + 3x + 15\\ &= x^2 + 8x + 15\end{aligned}$$

Note that the like terms $5x$ and $3x$ were combined to get $8x$.

EXAMPLE 2 *Multiplying Binomials*

Multiply $(2x - 1)(x + 4)$.

SOLUTION:

$$\begin{aligned}(2x - 1)(x + 4) &= \overset{F}{2x \cdot x} + \overset{O}{2x \cdot 4} + \overset{I}{(-1) \cdot x} + \overset{L}{(-1) \cdot 4}\\ &= 2x^2 + 8x - x - 4\\ &= 2x^2 + 7x - 4\end{aligned}$$

Factoring Trinomials of the Form $x^2 + bx + c$

The expression $x^2 + 8x + 15$ is an example of a trinomial. A *trinomial* is an expression containing three terms in which each exponent that appears on a variable is a whole number.

Example 1 showed that

$$(x + 3)(x + 5) = x^2 + 8x + 15$$

Since the product of $x + 3$ and $x + 5$ is $x^2 + 8x + 15$, we say that $x + 3$ and $x + 5$ are *factors* of $x^2 + 8x + 15$. To *factor* an expression means to write the expression as a product of its factors. For example, to factor $x^2 + 8x + 15$ we write

$$x^2 + 8x + 15 = (x + 3)(x + 5)$$

Let's look at the factors more closely.

$$3 + 5 = 8$$
$$3 \cdot 5 = 15$$
$$x^2 + 8x + 15 = (x + 3)(x + 5)$$

Note that the sum of the two numbers in the factors is $3 + 5$ or 8. The 8 is the coefficient of the x-term. Also note that the product of the numbers in the two factors is $3 \cdot 5$, or 15. The 15 is the constant in the trinomial. In general, when factoring an expression of the form $x^2 + bx + c$, we need to find two numbers whose product is c and whose sum is b. When we determine the two numbers, the factors will be of the form

$$(x + \square)\ (x + \square)$$

One number / Other number

EXAMPLE 3 *Factoring a Trinomial*

Factor $x^2 + 5x + 6$.

SOLUTION: We need to find two numbers whose product is 6 and whose sum is 5. Since the product is $+6$, the two numbers must both be positive or both be negative. Because the coefficient of the x-term is positive, only the positive factors of 6 need to be considered. Can you explain why? We begin by listing the positive numbers whose product is 6.

Factors of 6	Sum of Factors
1(6)	$1 + 6 = 7$
2(3)	$2 + 3 = 5$

Since $2 \cdot 3 = 6$ and $2 + 3 = 5$, 2 and 3 are the numbers we are seeking. Thus, we write

$$x^2 + 5x + 6 = (x + 2)(x + 3)$$

Note that $(x + 3)(x + 2)$ is also an acceptable answer.

To Factor Trinomial Expressions of the Form $x^2 + bx + c$

1. Find two numbers whose product is c and whose sum is b.
2. Write factors in the form

$$(x + \square)\,(x + \square)$$

↑ One number from step 1 ↑ Other number from step 1

3. Check your answer by multiplying the factors using the FOIL method.

If, for example, the numbers found in step 1 of the above procedure were 6 and -4, the factors would be written $(x + 6)(x - 4)$.

EXAMPLE 4 *Factoring a Trinomial*

Factor $x^2 - 6x - 16$.

SOLUTION: We must find two numbers whose product is -16 and whose sum is -6. Begin by listing the factors of -16.

Factors of −16	Sum of Factors
$-16(1)$	$-16 + 1 = -15$
$-8(2)$	$-8 + 2 = -6$
$-4(4)$	$-4 + 4 = 0$
$-2(8)$	$-2 + 8 = 6$
$-1(16)$	$-1 + 16 = 15$

The table lists all the factors of -16. The only factors listed whose product is -16 and whose sum is -6 are -8 and 2. We listed all factors in this example so that you could see, for example, that $-8(2)$ is a different set of factors than $-2(8)$. Once you find the factors you are looking for, there is no need to go any further. The trinomial can be written in factored form as

$$x^2 - 6x - 16 = (x - 8)(x + 2)$$

Factoring Trinomials of the Form $ax^2 + bx + c, a \neq 1$

Now we discuss how to factor an expression of the form $ax^2 + bx + c$, where a, the coefficient of the squared term, is not equal to 1.

Consider the multiplication problem $(2x + 1)(x + 3)$.

$$\begin{aligned}(2x + 1)(x + 3) &= 2x \cdot x + 2x \cdot 3 + 1 \cdot x + 1 \cdot 3 \\ &= 2x^2 + 6x + x + 3 \\ &= 2x^2 + 7x + 3\end{aligned}$$

Since $(2x + 1)(x + 3) = 2x^2 + 7x + 3$, the factors of $2x^2 + 7x + 3$ are $2x + 1$ and $x + 3$.

Let's study the coefficients more closely.

$$\begin{array}{ccc} \text{F} & \text{O + I} & \text{L} \\ \downarrow & \downarrow & \downarrow \\ 2x^2 & +\ 7x & +\ 3 \end{array} \qquad (2x + 1)(1x + 3)$$

F L I O

$$\mathbf{F} = 2 \cdot 1 = 2 \qquad \mathbf{O + I} = (2 \cdot 3) + (1 \cdot 1) = 7 \qquad \mathbf{L} = 1 \cdot 3 = 3$$

Note that the product of the coefficient of the first terms in the multiplication of the binomials equals 2, the coefficient of the squared term. The sum of the products of the coefficients of the outer and inner terms equals 7, the coefficient of the x-term. The product of the last terms equals 3, the constant.

A procedure to factor expressions of the form $ax^2 + bx + c$, $a \neq 1$, follows.

To Factor Trinomial Expressions of the Form $ax^2 + bx + c, a \neq 1$

1. Write all pairs of factors of the coefficient of the squared term, a.
2. Write all pairs of factors of the constant, c.
3. Try various combinations of these factors until the sum of the products of the outer and inner terms is bx.
4. Check your answer by multiplying the factors using the FOIL method.

EXAMPLE 5 *Factoring a Trinomial, $a \neq 1$*

Factor $3x^2 + 17x + 10$.

SOLUTION: The only positive factors of 3 are 3 and 1. Therefore, we write

$$3x^2 + 17x + 10 = (3x \qquad)(x \qquad)$$

The number 10 has both positive and negative factors. However, since both the constant, 10, and the sum of the products of the outer and inner terms, 17, are positive, the two factors must be positive. Why? The positive factors of 10 are 1(10) and 2(5). The following is a list of the possible factors.

Possible Factors	Sum of Products of Outer and Inner Terms
$(3x + 1)(x + 10)$	$31x$
$(3x + 10)(x + 1)$	$13x$
$(3x + 2)(x + 5)$	$17x$ ← Correct middle term
$(3x + 5)(x + 2)$	$11x$

Thus, $3x^2 + 17x + 10 = (3x + 2)(x + 5)$.

Note that factoring problems of this type may be checked by using the FOIL method of multiplication. We will check the results to Example 5:

$$\begin{aligned}(3x + 2)(x + 5) &= 3x \cdot x + 3x \cdot 5 + 2 \cdot x + 2 \cdot 5\\ &= 3x^2 + 15x + 2x + 10\\ &= 3x^2 + 17x + 10\end{aligned}$$

Since we obtained the expression we started with, our factoring is correct.

EXAMPLE 6 *Factoring a Trinomial, $a \neq 1$*

Factor $6x^2 - 11x - 10$.

SOLUTION: The factors of 6 will be either $6 \cdot 1$ or $2 \cdot 3$. Therefore, the factors may be of the form $(6x \quad)(x \quad)$ or $(2x \quad)(3x \quad)$. When there is more than one set of factors for the first term, we generally try the medium-sized factors first. If that does not work, we try the other factors. Thus, we write

$$6x^2 - 11x - 10 = (2x \quad)(3x \quad)$$

The factors of -10 are $(-1)(10)$, $(1)(-10)$, $(-2)(5)$, and $(2)(-5)$. There will be eight different pairs of possible factors of the trinomial $6x^2 - 11x - 10$. Can you list them?

The correct factoring is $6x^2 - 11x - 10 = (2x - 5)(3x + 2)$. ▲

Note that in Example 6 we first tried factors of the form $(2x \quad)(3x \quad)$. If we had not found the correct factors using them, we would have tried $(6x \quad)(x \quad)$.

Solving Quadratic Equations by Factoring

In Section 6.2, we solved linear, or first-degree, equations. In those equations, the exponent on all variables was 1. Now we deal with the *quadratic equation*. The standard form of a quadratic equation in one variable is shown in the box.

Standard Form of a Quadratic Equation

$$ax^2 + bx + c = 0, \qquad a \neq 0$$

Note that in the standard form of a quadratic equation, the greatest exponent on x is 2 and the right side of the equation is equal to zero. *To solve a quadratic equation* means to find the value or values that make the equation true. In this section, we will solve quadratic equations by factoring and by the quadratic formula.

To solve a quadratic equation by factoring, set one side of the equation equal to 0 and then use the *zero-factor* property.

Zero-Factor Property

If $a \cdot b = 0$, then $a = 0$ or $b = 0$.

The zero-factor property indicates that, if the product of two factors is 0, then one (or both) of the factors must have a value of 0.

EXAMPLE 7 *Using the Zero-Factor Property*

Solve the equation $(x + 3)(x - 6) = 0$.

SOLUTION: When we use the zero-factor property, either $(x + 3)$ or $(x - 6)$ must equal 0 for the product to equal 0. Thus, we set each individual factor equal to 0 and solve each resulting equation for x.

$$(x + 3)(x - 6) = 0$$

$$x + 3 = 0 \quad \text{or} \quad x - 6 = 0$$

$$x = -3 \qquad\qquad x = 6$$

Thus, the solutions are -3 and 6.

Check: $x = -3$ $\qquad$ $x = 6$

$$\begin{aligned}(x + 3)(x - 6) &= 0\\ (-3 + 3)(-3 - 6) &= 0\\ 0(-9) &= 0\\ 0 &= 0 \quad \text{True}\end{aligned} \qquad \begin{aligned}(x + 3)(x - 6) &= 0\\ (6 + 3)(6 - 6) &= 0\\ 9(0) &= 0\\ 0 &= 0 \quad \text{True}\end{aligned}$$

To Solve a Quadratic Equation by Factoring

1. Use the addition or subtraction property to make one side of the equation equal to 0.
2. Factor the side of the equation not equal to 0.
3. Use the zero-factor property to solve the equation.

Examples 8 and 9 illustrate this procedure.

EXAMPLE 8 *Solving a Quadratic Equation by Factoring*

Solve the equation $x^2 - 8x = -15$.

SOLUTION: First add 15 to both sides of the equation to make the right side of the equation equal to 0.

$$\begin{aligned}x^2 - 8x &= -15\\ x^2 - 8x + 15 &= -15 + 15\\ x^2 - 8x + 15 &= 0\end{aligned}$$

Factor the left side of the equation. The object is to find two numbers whose product is 15 and whose sum is -8. Since the product of the numbers is positive and the sum of the numbers is negative, the two numbers must both be negative. The numbers are -3 and -5. Note that $(-3)(-5) = 15$ and $-3 + (-5) = -8$.

$$\begin{aligned}x^2 - 8x + 15 &= 0\\ (x - 3)(x - 5) &= 0\end{aligned}$$

Now use the zero-factor property to find the solution.

$$x - 3 = 0 \quad \text{or} \quad x - 5 = 0$$
$$x = 3 \qquad\qquad x = 5$$

The solutions are 3 and 5. ▲

EXAMPLE 9 ***Solve a Quadratic Equation by Factoring***

Solve the equation $3x^2 - 13x + 4 = 0$.

SOLUTION: $3x^2 - 13x + 4$ factors into $(3x - 1)(x - 4)$. Thus, we write

$$3x^2 - 13x + 4 = 0$$
$$(3x - 1)(x - 4) = 0$$
$$3x - 1 = 0 \quad \text{or} \quad x - 4 = 0$$
$$3x = 1 \qquad\qquad x = 4$$
$$x = \frac{1}{3}$$

The solutions are $\frac{1}{3}$ and 4. ▲

TIMELY TIP As stated on page 355, every factoring problem can be checked by multiplying the factors. If you have factored correctly, the product of the factors will be identical to the original expression that was factored. If we wished to check the factoring of Example 9, we would multiply $(3x - 1)(x - 4)$. Since the product of the factors is $3x^2 - 13x + 4$, the expression we started with, our factoring is correct.

Solving Quadratic Equations by Using the Quadratic Formula

Not all quadratic equations can be solved by factoring. When a quadratic equation cannot be easily solved by factoring, we can solve the equation with the *quadratic formula*. The quadratic formula can be used to solve any quadratic equation.

Quadratic Formula

For a quadratic equation in standard form, $ax^2 + bx + c = 0$, $a \neq 0$, the quadratic formula is

$$x = \frac{-b \pm \sqrt{b^2 - 4ac}}{2a}$$

In the quadratic formula, the plus or minus symbol, $\pm$, is used. If, for example, $x = 2 \pm 3$, then $x = 2 + 3 = 5$ or $x = 2 - 3 = -1$.

DID YOU KNOW

The Mathematics of Motion

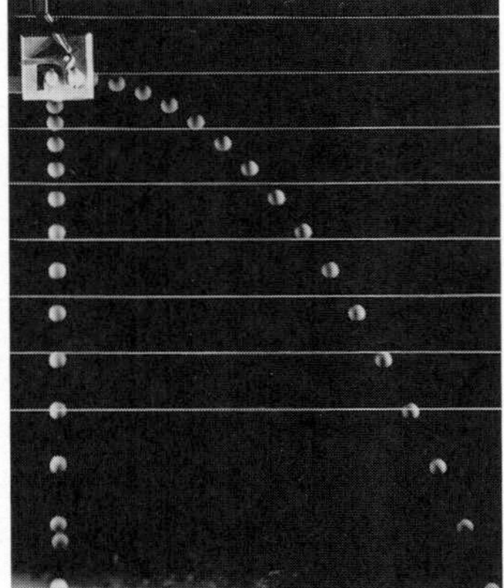

The free fall of an object is something that has interested scientists and mathematicians for centuries. It is described by a quadratic equation. Shown here is a time-lapse photo that shows the free fall of a ball in equal-time intervals. What you see can be described verbally this way: The rate of change in velocity in each interval is the same; therefore, velocity is continuously increasing and acceleration is constant.

It is possible for a quadratic equation to have no real solution. In solving an equation, if the radicand (the expression inside the square root) is a negative number, then the quadratic equation has *no real solution*.

To use the quadratic formula, first write the quadratic equation in standard form. Then determine the values for a (the coefficient of the squared term), b (the coefficient of the x-term), and c (the constant). Finally, substitute the values of a, b, and c into the quadratic formula and evaluate the expression.

EXAMPLE 10 *Solve a Quadratic Equation Using the Quadratic Formula*

Solve the equation $x^2 + 2x - 15 = 0$ using the quadratic formula.

SOLUTION: In this equation, $a = 1$, $b = 2$, and $c = -15$.

$$x = \frac{-b \pm \sqrt{b^2 - 4ac}}{2a} = \frac{-2 \pm \sqrt{2^2 - 4(1)(-15)}}{2(1)}$$

$$= \frac{-2 \pm \sqrt{4 + 60}}{2}$$

$$= \frac{-2 \pm \sqrt{64}}{2}$$

$$= \frac{-2 \pm 8}{2}$$

$$\frac{-2 + 8}{2} = \frac{6}{2} = 3 \quad \text{or} \quad \frac{-2 - 8}{2} = \frac{-10}{2} = -5$$

The solutions are 3 and -5. ▲

Note that Example 10 can also be solved by factoring. We suggest that you do so now.

EXAMPLE 11 *Irrational Solutions to a Quadratic Equation*

Solve $4x^2 - 8x = -1$ using the quadratic formula.

SOLUTION: Begin by writing the equation in standard form by adding 1 to both sides of the equation, which gives the following.

$$4x^2 - 8x + 1 = 0$$

$$a = 4, \qquad b = -8, \qquad c = 1$$

$$x = \frac{-b \pm \sqrt{b^2 - 4ac}}{2a} = \frac{-(-8) \pm \sqrt{(-8)^2 - 4(4)(1)}}{2(4)}$$

$$= \frac{8 \pm \sqrt{64 - 16}}{8}$$

$$= \frac{8 \pm \sqrt{48}}{8}$$

Since $\sqrt{48} = \sqrt{16}\sqrt{3} = 4\sqrt{3}$ (see Section 5.4), we write

$$\frac{8 \pm \sqrt{48}}{8} = \frac{8 \pm 4\sqrt{3}}{8} = \frac{\overset{1}{\cancel{4}}(2 \pm \sqrt{3})}{\underset{2}{\cancel{8}}} = \frac{2 \pm \sqrt{3}}{2}$$

The solutions are $\frac{2 + \sqrt{3}}{2}$ and $\frac{2 - \sqrt{3}}{2}$. ▲

Note that the solutions to Example 11 are irrational numbers.

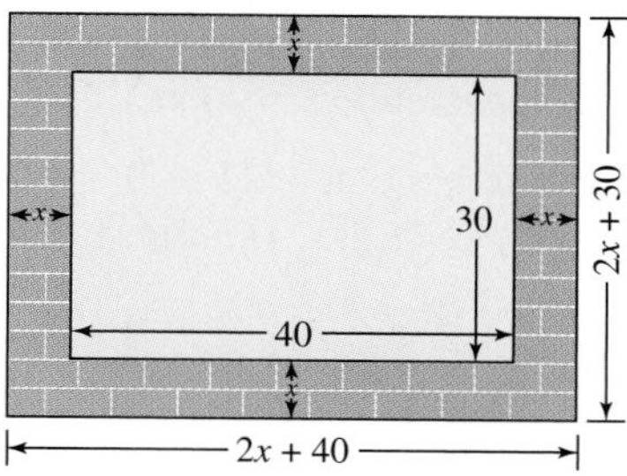

Figure 6.30

EXAMPLE 12 *Brick Border*

Diane Cecero and her husband recently installed an inground rectangular swimming pool measuring 40 ft by 30 ft. They want to add a brick border of uniform width around all sides of the pool. How wide can they make the brick border if they purchased enough brick to cover 296 ft^2?

SOLUTION: Let's make a diagram of the pool and the brick border (Fig. 6.30) Let x = the uniform width of the brick border. Then the total length of the larger rectangular area, the pool plus the border, is $2x + 40$. The total width of the larger rectangular area is $2x + 30$.

The area of the brick border can be found by subtracting the area of the pool from the area of the pool plus the brick border.

$$\begin{aligned} \text{Area of pool} &= l \cdot w = (40)(30) = 1200 \text{ ft}^2 \\ \text{Area of pool plus brick border} &= l \cdot w = (2x + 40)(2x + 30) \\ &= 4x^2 + 140x + 1200 \\ \text{Area of the brick border} &= \text{area of pool plus brick border} - \text{area of pool} \\ &= (4x^2 + 140x + 1200) - 1200 \\ &= 4x^2 + 140x \end{aligned}$$

The total brick border must be 296 ft^2. Therefore,

$$296 = 4x^2 + 140x$$

or

$$\begin{aligned} 4x^2 + 140x - 296 &= 0 \\ 4(x^2 + 35x - 74) &= 0 && \text{Factor out 4 from each term.} \\ \frac{4}{4}(x^2 + 35x - 74) &= \frac{0}{4} && \text{Divide both sides of the equation by 4.} \\ x^2 + 35x - 74 &= 0 \\ (x + 37)(x - 2) &= 0 && \text{Factor trinomial.} \end{aligned}$$

$$x + 37 = 0 \quad \text{or} \quad x - 2 = 0$$
$$x = -37 \qquad\qquad x = 2$$

Since lengths are positive, the only possible answer is $x = 2$. Thus, they can make a brick border 2 ft wide all around the pool. ▲

SECTION 6.9 EXERCISES

Concept/Writing Exercises

1. What is a *binomial*? Give three examples of binomials.
2. What is a *trinomial*? Give three examples of trinomials.
3. In your own words, explain the FOIL method used to multiply two binomials.
4. In your own words, state the zero-factor property.
5. Give the standard form of a quadratic equation.
6. Have you memorized the quadratic formula? If not, you need to do so. Without looking at the book, write the quadratic formula.

Practice the Skills

In Exercises 7–22, factor the trinomial. If the trinomial cannot be factored, so state.

7. $x^2 + 9x + 18$
8. $x^2 + 5x + 4$
9. $x^2 - x - 6$
10. $x^2 + x - 6$
11. $x^2 + 2x - 24$
12. $x^2 - 6x + 8$
13. $x^2 - 2x - 3$
14. $x^2 - 5x - 6$
15. $x^2 - 10x + 21$
16. $x^2 - 81$
17. $x^2 - 25$
18. $x^2 - x - 20$
19. $x^2 + 3x - 28$
20. $x^2 + 4x - 32$
21. $x^2 + 2x - 63$
22. $x^2 - 2x - 48$

In Exercises 23–34, factor the trinomial. If the trinomial cannot be factored, so state.

23. $2x^2 - x - 10$
24. $3x^2 - 2x - 5$
25. $4x^2 + 13x + 3$
26. $2x^2 - 11x - 21$
27. $5x^2 + 12x + 4$
28. $2x^2 - 9x + 10$
29. $4x^2 + 11x + 6$
30. $4x^2 + 20x + 21$
31. $4x^2 - 11x + 6$
32. $6x^2 - 11x + 4$
33. $3x^2 - 14x - 24$
34. $6x^2 + 5x + 1$

In Exercises 35–38, solve each equation, using the zero-factor property.

35. $(x - 1)(x + 2) = 0$
36. $(2x + 5)(x - 1) = 0$
37. $(3x + 4)(2x - 1) = 0$
38. $(x - 6)(5x - 4) = 0$

In Exercises 39–58, solve each equation by factoring.

39. $x^2 + 10x + 21 = 0$
40. $x^2 + 4x - 5 = 0$
41. $x^2 - 4x + 3 = 0$
42. $x^2 - 5x - 24 = 0$
43. $x^2 - 15 = 2x$
44. $x^2 - 7x = -6$
45. $x^2 = 4x - 3$
46. $x^2 - 13x + 40 = 0$
47. $x^2 - 81 = 0$
48. $x^2 - 64 = 0$
49. $x^2 + 5x - 36 = 0$
50. $x^2 + 12x + 20 = 0$
51. $3x^2 + 10x = 8$
52. $3x^2 - 5x = 2$
53. $5x^2 + 11x = -2$
54. $2x^2 = -5x + 3$
55. $3x^2 - 4x = -1$
56. $5x^2 + 16x + 12 = 0$
57. $4x^2 - 9x + 2 = 0$
58. $6x^2 + x - 2 = 0$

In Exercises 59–78, solve the equation, using the quadratic formula. If the equation has no real solution, so state.

59. $x^2 + 2x - 15 = 0$
60. $x^2 + 12x + 27 = 0$
61. $x^2 - 3x - 18 = 0$
62. $x^2 - 6x - 16 = 0$
63. $x^2 - 8x = 9$
64. $x^2 = -8x + 15$
65. $x^2 - 2x + 3 = 0$
66. $2x^2 - x - 3 = 0$
67. $x^2 - 4x + 2 = 0$
68. $2x^2 - 5x - 2 = 0$
69. $3x^2 - 8x + 1 = 0$
70. $2x^2 + 4x + 1 = 0$
71. $4x^2 - x - 1 = 0$
72. $4x^2 - 5x - 3 = 0$
73. $2x^2 + 7x + 5 = 0$
74. $3x^2 = 9x - 5$
75. $3x^2 - 10x + 7 = 0$
76. $4x^2 + 7x - 1 = 0$
77. $4x^2 - 11x + 13 = 0$
78. $5x^2 + 9x - 2 = 0$

Challenge Problems/Group Activities

79. *Flower Garden* Karen and Kurt Ohliger's backyard has a width of 20 meters and a length of 30 meters. Karen and Kurt want to put a flower garden in the middle of the backyard leaving a strip of grass of uniform width around all sides of the flower garden. If they want to have 336 square meters of grass, what will be the width and length of the garden?
80. *Air Conditioning* The yearly profit p of Arnold's Air Conditioning is given by $p = x^2 + 15x - 100$, where x is the number of air conditioners produced and sold. How many air conditioners must be produced and sold to have a yearly profit of \$45,000?

81. **a)** Explain why solving $(x - 4)(x - 7) = 6$ by setting each factor equal to 6 is not correct.
b) Determine the correct solution to $(x - 4)(x - 7) = 6$.

82. The radicand in the quadratic formula, $b^2 - 4ac$, is called the **discriminant.** How many real number solutions will the quadratic equation have if the discriminant is (a) greater than 0, (b) equal to 0, or (c) less than zero? Explain your answer.

83. Write an equation that has solutions −1 and 3.

Recreational Mathematics

84. Hidden in the grid above and to the right are the following words discussed in this chapter: ALGEBRA, FORMULA, SOLUTION, VARIATION, BINOMIAL. You will find these words by going letter to letter. As you move from letter to letter, you may move vertically or horizontally. A letter can be used only once when spelling out each particular word. Find the words listed above in the grid. One example, ALGEBRA, is shown.

R	I	A	A	S	N	E	Y
A	V	T	R	R	I	U	P
U	I	I	O	N	R	A	B
R	R	L	U	A	L	G	D
U	S	O	T	N	A	E	C
A	L	U	I	O	R	B	U
X	S	M	S	E	I	I	N
F	O	R	M	M	O	N	S
C	G	E	M	I	A	L	P

Internet/Research Activity

85. Italian mathematician Girolamo Cardano (1501–1576) is recognized for his skill in solving equations. Write a paper about his life and his contributions to mathematics, in particular his contribution to solving equations.

86. Chinese mathematician Foo Ling Awong, who lived during the Pong dynasty, developed a technique, other than trial and error, to factor trinomials of the form $ax^2 + bx + c$, $a \neq 1$. Write a paper about his life and his contributions to mathematics, in particular his technique for factoring trinomials in the form $ax^2 + bx + c$, $a \neq 1$. (References include history of mathematics books, encyclopedias, and the Internet.)

6.10 FUNCTIONS AND THEIR GRAPHS

The concepts of relations and functions are extremely important in mathematics. A *relation* is any set of ordered pairs. Therefore, every graph will be a relation. A function is a special type of relation. Suppose you are purchasing oranges at a supermarket where each orange costs \$0.20. Then one orange would cost \$0.20, two oranges $2 \times \$0.20 = \$0.40$, three oranges \$0.60, and so on. We can indicate this relation in a table of values.

Number of Oranges	Cost
0	0.00
1	0.20
2	0.40
3	0.60
⋮	⋮
10	2.00
⋮	⋮

In general, the cost for purchasing n oranges will be 20 cents times the number of oranges, or $0.20n$. We can represent the cost, c, of n items by the equation $c = 0.20n$.

Since the value of c depends on the value of n, we refer to c as the *dependent variable* and n as the *independent variable. Note for each value of the independent variable, n, there is one and only one value of the dependent variable, c.* Such an equation is called a *function*. In the equation $c = 0.20n$, the value of c depends on the value of n, so we say that "c is a function of n."

A **function** is a special type of relation where each value of the independent variable corresponds to a unique value of the dependent variable.

The set of values that can be used for the independent variable is called the *domain* of the function, and the resulting set of values obtained for the dependent variable is called the *range*. The domain and range for the function $c = 0.20n$ are illustrated in Fig. 6.31.

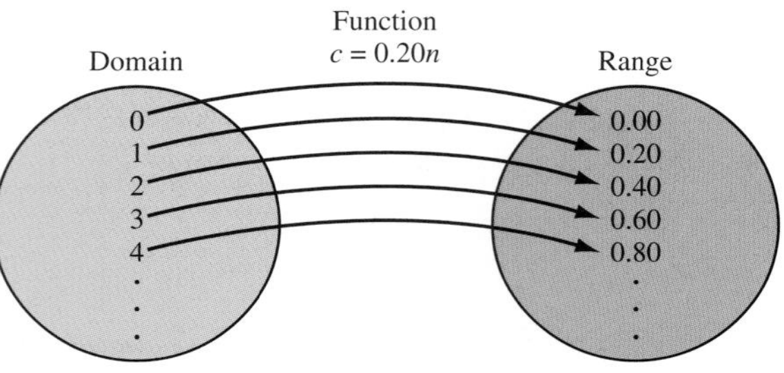

Figure 6.31

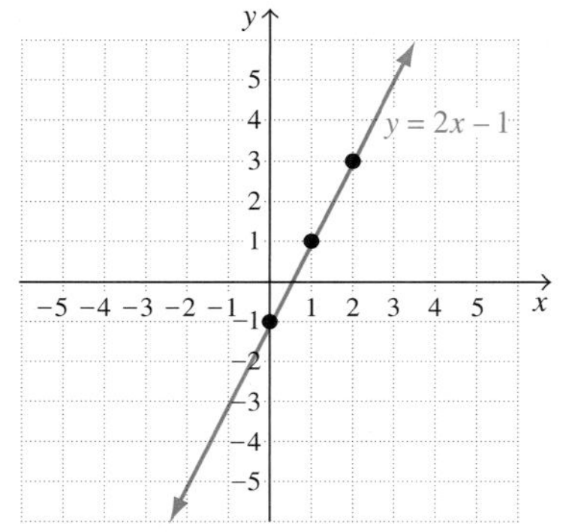

Figure 6.32

When we graphed equations of the form $ax + by = c$ in Section 6.7, we found that they were straight lines. For example, the graph of $y = 2x - 1$ is illustrated in Fig. 6.32.

Is the equation $y = 2x - 1$ a function? To answer this question, we must ask, "Does each value of x correspond to a unique value of y?" The answer is yes; therefore, this equation is a function.

For the equation $y = 2x - 1$, we say that "y is a function of x" and write $y = f(x)$. The notation $f(x)$ is read "f of x." When we are given an equation that is a function, we may replace the y in the equation with $f(x)$, since $f(x)$ represents y. Thus, $y = 2x - 1$ may be written $f(x) = 2x - 1$.

To evaluate a function for a specific value of x, replace each x in the function with the given value, then evaluate. For example, to evaluate $f(x) = 2x - 1$ when $x = 8$, we do the following.

$$f(x) = 2x - 1$$
$$f(8) = 2(8) - 1 = 16 - 1 = 15$$

Thus, $f(8) = 15$. Since $f(x) = y$, when $x = 8$, $y = 15$. What is the domain and range of $f(x) = 2x - 1$? Because x can be any real number, the domain is the set of real numbers, symbolized $\mathbb{R}$. The range is also $\mathbb{R}$.

We can determine whether a graph represents a function by using the *vertical line test*: If a vertical line can be drawn so that it intersects the graph at more than one point, then each x does not have a unique y and the graph does not represent a function. If a vertical line cannot be made to intersect the graph in at least two different places, then the graph represents a function.

EXAMPLE 1 ***Using the Vertical Line Test***

Use the vertical line test to determine which of the graphs in Figure 6.33 represent functions.

a) 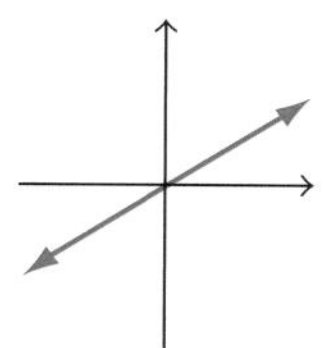 b) 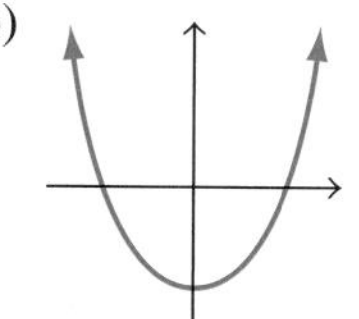 c) 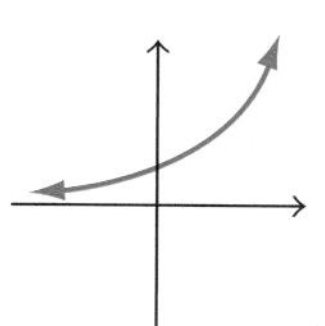 d) 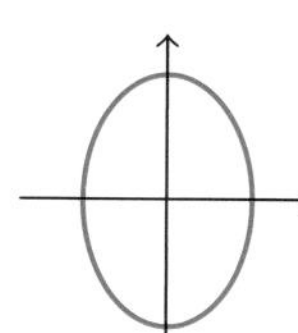

Figure 6.33

SOLUTION: (a), (b), and (c) represent functions, but (d) does not.

a) 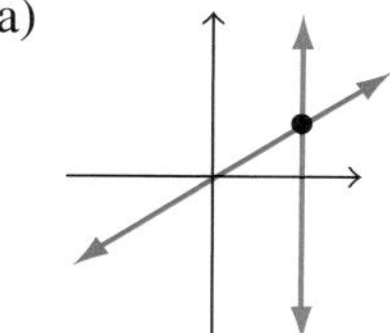 b) 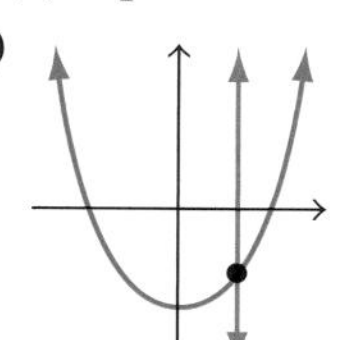 c) d) 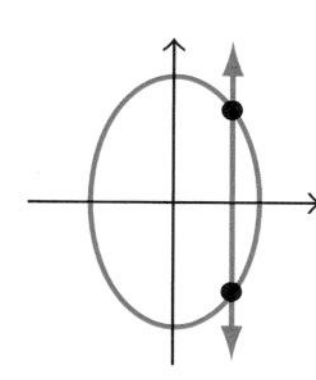

There are many real-life applications of functions. In fact, all the applications illustrated in Sections 6.2 through 6.4 are functions.

In this section, we will discuss three types of functions: linear functions, quadratic functions, and exponential functions.

Linear Functions

In Section 6.7, we graphed linear equations. The graph of any linear equation of the form $y = ax + b$ will pass the vertical line test, and so equations of the form $y = ax + b$ are *linear functions*. If we wished, we could write the linear function as $f(x) = ax + b$ since $f(x)$ means the same as y.

EXAMPLE 2 ***Cost as a Linear Function***

Adam Finiteri's weekly cost of operating a taxi, c, is given by the function $c(m) = 52 + 0.18m$, where m is the number of miles driven per week. What is his weekly cost if he drives 200 miles in a week?

SOLUTION: Substitute 200 for m in the function.

$$c(m) = 52 + 0.18m$$
$$c(200) = 52 + 0.18(200)$$
$$c(200) = 52 + 36 = 88$$

Thus, if Adam drives 200 miles in a week, his weekly cost will be \$88.

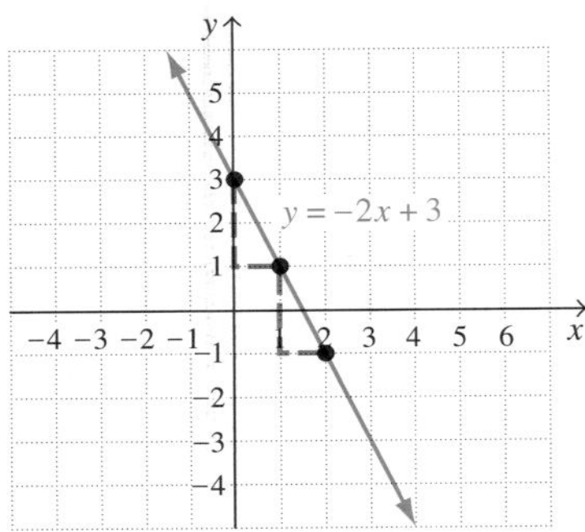

Figure 6.34

Graphs of Linear Functions

The graphs of linear functions are straight lines that will pass the vertical line test. In Section 6.7, we discussed how to graph linear equations. Linear functions can be graphed by plotting points, by using intercepts, or by using the slope and y-intercept.

EXAMPLE 3 *Graphing a Linear Function*

Graph $f(x) = -2x + 3$ by using the slope and y-intercept.

SOLUTION: Since $f(x)$ means the same as y, we can rewrite this function as $y = -2x + 3$. From Section 6.7, we know that the slope is -2 and the y-intercept is (0, 3). Plot (0, 3) on the y-axis. Then plot the next point by moving *down* 2 units and to the *right* 1 unit (see Fig. 6.34). A third point has been plotted in the same way. The graph of $f(x) = -2x + 3$ is the line drawn through these three points. ▲

Quadratic Functions

The standard form of a quadratic equation is $y = ax^2 + bx + c, a \neq 0$. We will learn shortly that graphs of equations of this form always pass the vertical line test and are functions. Therefore, equations of the form $y = ax^2 + bx + c, a \neq 0$, may be referred to as *quadratic functions*. We may express quadratic functions using function notation as $f(x) = ax^2 + bx + c$. Two examples of quadratic functions are $y = 2x^2 + 5x - 7$ and $y = -\frac{1}{2}x^2 + 4$.

EXAMPLE 4 *Landing on the Moon*

On July 20, 1969, Neil Armstrong became the first person to walk on the moon. The velocity, v, of his spacecraft, the *Eagle,* in meters per second, was a function of time before touchdown, t, given by

$$v = f(t) = 3.2t + 0.45$$

The height of the spacecraft, h, above the moon's surface, in meters, was also a function of time before touchdown, given by

$$h = g(t) = 1.6t^2 + 0.45t$$

What was the velocity of the spacecraft and its distance from the surface of the moon

a) at 3 seconds before touchdown? b) at touchdown (0 seconds)?

SOLUTION:

a)
$$\begin{aligned} v = f(t) &= 3.2t + 0.45, \\ f(3) &= 3.2(3) + 0.45 \\ &= 9.6 + 0.45 \\ &= 10.05 \end{aligned}$$

$$\begin{aligned} h = g(t) &= 1.6t^2 + 0.45t \\ g(3) &= 1.6(3)^2 + 0.45(3) \\ &= 1.6(9) + 1.35 \\ &= 14.4 + 1.35 \\ &= 15.75 \end{aligned}$$

The velocity 3 seconds before touchdown was 10.05 meters per second and the height 3 seconds before touchdown was 15.75 meters.

DID YOU KNOW

Sea of Tranquillity

Apollo II touched down at Mare Tranqillitatis, the Sea of Tranquillity. The rock samples taken there placed the age of the rocks at 3.5 billion years old, as old as the oldest known Earth rocks.

$$\begin{aligned} \text{b) } v = f(t) &= 3.2t + 0.45, \\ f(0) &= 3.2(0) + 0.45 \\ &= 0 + 0.45 \\ &= 0.45 \end{aligned} \qquad \begin{aligned} h = g(t) &= 1.6t^2 + 0.45t \\ g(0) &= 1.6(0)^2 + 0.45(0) \\ &= 0 + 0 \\ &= 0 \end{aligned}$$

The touchdown velocity was 0.45 meter per second. At touchdown, the *Eagle* is on the moon, and therefore the distance from the surface of the moon is 0 meter. ▲

DID YOU KNOW

Computing Reality with Mathematical Models

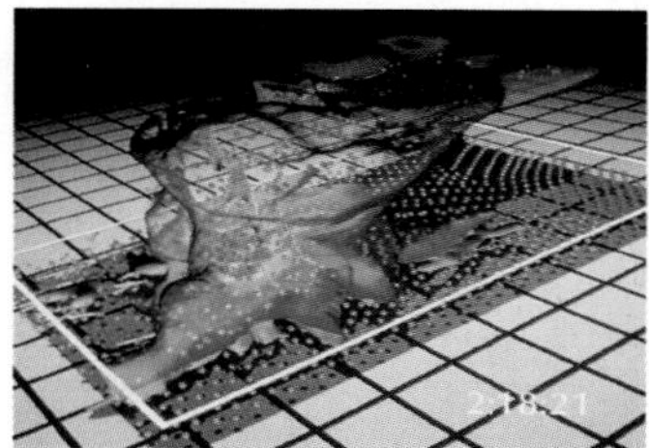

It is a relatively easy matter for scientists and mathematicians to describe and predict a simple motion like that of a falling object. When the phenomenon is complicated, such as making an accurate prediction of the weather, the mathematics becomes much more difficult. The National Weather Service has devised an algorithm that takes the temperature, pressure, moisture content, and wind velocity of more than 250,000 points in Earth's atmosphere and applies a set of equations that, they believe, will reasonably predict what will happen at each point over time. Researchers at the National Center for Supercomputing Applications are working on a computer model that simulates thunderstorms. Researchers want to know why some thunderstorms can turn severe, even deadly.

Graphs of Quadratic Functions

The graph of every quadratic function is a parabola. Two parabolas are illustrated in Fig. 6.35. Note that both graphs represent functions since they pass the vertical line test. A parabola opens upward when the coefficient of the squared term, a, is greater than 0, as shown in Figure 6.35(a). A parabola opens downward when the coefficient of the squared term, a, is less than 0, as shown in Fig. 6.35(b).

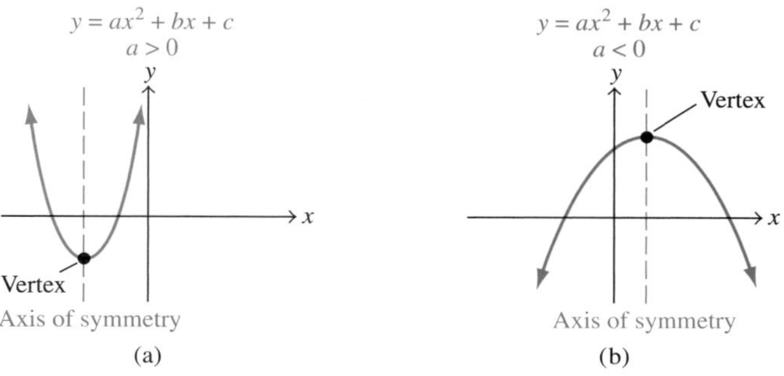

Figure 6.35

The *vertex* of a parabola is the lowest point on a parabola that opens upward and the highest point on a parabola that opens downward. Every parabola is *symmetric* with respect to a vertical line through its vertex. This line is called the *axis of symmetry* of the parabola. The x-coordinate of the vertex and the equation of the axis of symmetry can be found by using the following equation.

Axis of Symmetry of a Parabola

$$x = \frac{-b}{2a}$$

Once the x-coordinate of the vertex has been determined, the y-coordinate can be found by substituting the value found for the x-coordinate into the quadratic equation and evaluating the equation. This procedure is illustrated in Example 5.

EXAMPLE 5 *Describing the Graph of a Quadratic Equation*

Consider the equation $y = -2x^2 + 8x + 1$.

a) Determine whether the graph will be a parabola that opens upward or downward.

DID YOU KNOW

Gravity and the Parabola

Any golf player knows that a golf ball will arch in the same path going down as it did going up. Early gunners knew it too. To hit a distant target, the cannon barrel was pointed skyward, not directly at the target. A cannonball fired at a 45° angle will travel the greatest horizontal distance. What the golfer and gunner alike were allowing for is the effect of gravity on a projectile. Projectile motion follows a parabolic path. Galileo was neither a gunner nor a golfer, but he gave us the formula that effectively describes that motion and the distance traveled by an object if it is projected at a specific angle with a specific initial velocity.

b) Determine the equation of the axis of symmetry of the parabola.

c) Determine the vertex of the parabola.

SOLUTION:

a) Since $a = -2$, which is less than 0, the parabola opens downward.

b) To find the axis of symmetry, we use the equation $x = -\frac{b}{2a}$. In the equation $y = -2x^2 + 8x + 1$, $a = -2$, $b = 8$, and $c = 1$, so

$$x = \frac{-b}{2a} = \frac{-(8)}{2(-2)} = \frac{-8}{-4} = 2$$

The equation of the axis of symmetry is $x = 2$.

c) The x-coordinate of the vertex is 2, from part (b). To find the y-coordinate, we substitute 2 for x in the equation and then evaluate.

$$\begin{aligned} y &= -2x^2 + 8x + 1 \\ &= -2(2)^2 + 8(2) + 1 \\ &= -2(4) + 16 + 1 \\ &= -8 + 16 + 1 \\ &= 9 \end{aligned}$$

Therefore, the vertex of the parabola is located at the point (2, 9) on the graph. ▲

General Procedure to Sketch the Graph of a Quadratic Equation

1. Determine whether the parabola opens upward or downward.
2. Determine the equation of the axis of symmetry.
3. Determine the vertex of the parabola.
4. Determine the y-intercept by substituting $x = 0$ into the equation.
5. Determine the x-intercepts (if they exist) by substituting $y = 0$ into the equation and solving for x.
6. Draw the graph, making use of the information gained in steps 1 through 5. Remember the parabola will be symmetric with respect to the axis of symmetry.

In step 5, to determine the x-intercepts, you may use either factoring or the quadratic formula.

EXAMPLE 6 *Graphing a Quadratic Equation*

Sketch the graph of the equation $y = x^2 - 6x + 8$.

SOLUTION: We follow the steps outlined in the general procedure.

1. Since $a = 1$, which is greater than 0, the parabola opens upward.
2. Axis of symmetry: $x = \frac{-b}{2a} = \frac{-(-6)}{2(1)} = \frac{6}{2} = 3$

   Thus, the axis of symmetry is $x = 3$.
3. y-coordinate of vertex: $y = x^2 - 6x + 8$

   $y = (3)^2 - 6(3) + 8 = 9 - 18 + 8 = -1$

   Thus, the vertex is at $(3, -1)$.

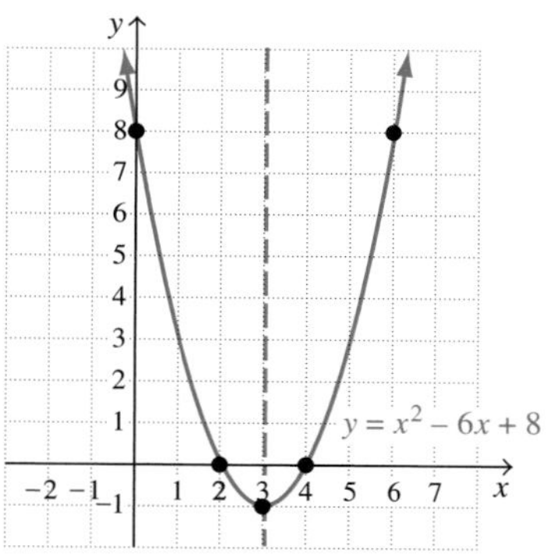

Figure 6.36

4. y-intercept: $y = x^2 - 6x + 8$
   $y = 0^2 - 6(0) + 8 = 8$
   Thus, the y-intercept is at (0, 8).
5. x-intercepts: $0 = x^2 - 6x + 8$, or $x^2 - 6x + 8 = 0$
   We can solve this equation by factoring.

$$x^2 - 6x + 8 = 0$$
$$(x - 4)(x - 2) = 0$$
$$x - 4 = 0 \quad \text{or} \quad x - 2 = 0$$
$$x = 4 \qquad\qquad x = 2$$

   Thus, the x-intercepts are (4, 0) and (2, 0).
6. Plot the vertex (3, −1), the y-intercept (0, 8), and the x-intercepts (4, 0) and (2, 0). Then sketch the graph (Fig. 6.36).

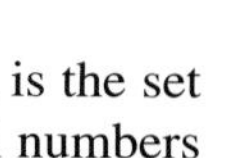

Note that the domain of the graph in Example 6, the possible x-values, is the set of all real numbers, $\mathbb{R}$. The range, the possible y-values, is the set of all real numbers greater than or equal to −1. When graphing parabolas, if you feel that you need additional points to graph the equation, you can always substitute values for x and find the corresponding values of y and plot those points. For example, if you substituted 1 for x, the corresponding value of y is 3. Thus, you could plot the point (1, 3).

EXAMPLE 7 *Domain and Range of a Quadratic Function*

a) Sketch the graph of the function $f(x) = -2x^2 + 3x + 4$.
b) Determine the domain and range of the function.

SOLUTION:

a) Since $f(x)$ means y, we can replace $f(x)$ with y to obtain $y = -2x^2 + 3x + 4$. Now graph $y = -2x^2 + 3x + 4$ using the steps outlined in the general procedure.

1. Since $a = -2$, which is less than 0, the parabola opens downward.
2. Axis of symmetry: $x = \dfrac{-b}{2a} = \dfrac{-(3)}{2(-2)} = \dfrac{-3}{-4} = \dfrac{3}{4}$

   Thus, the axis of symmetry is $x = \dfrac{3}{4}$.

3. y-coordinate of vertex:

$$\begin{aligned} y &= -2x^2 + 3x + 4 \\ &= -2\left(\frac{3}{4}\right)^2 + 3\left(\frac{3}{4}\right) + 4 \\ &= -2\left(\frac{9}{16}\right) + \frac{9}{4} + 4 \\ &= -\frac{9}{8} + \frac{9}{4} + 4 \\ &= -\frac{9}{8} + \frac{18}{8} + \frac{32}{8} = \frac{41}{8} \quad \text{or} \quad 5\frac{1}{8} \end{aligned}$$

   Thus, the vertex is at $\left(\frac{3}{4}, 5\frac{1}{8}\right)$.

4. y-intercept: $y = -2x^2 + 3x + 4$

$$= -2(0)^2 + 3(0) + 4 = 4$$

Thus, the y-intercept is (0, 4).

5. x-intercepts: $y = -2x^2 + 3x + 4$

$$0 = -2x^2 + 3x + 4 \qquad \text{or} \qquad -2x^2 + 3x + 4 = 0$$

This equation cannot be factored, so we will use the quadratic formula to solve it.

$$a = -2, \qquad b = 3, \qquad c = 4$$

$$x = \frac{-b \pm \sqrt{b^2 - 4ac}}{2a}$$

$$= \frac{-3 \pm \sqrt{3^2 - 4(-2)(4)}}{2(-2)}$$

$$= \frac{-3 \pm \sqrt{9 + 32}}{-4}$$

$$= \frac{-3 \pm \sqrt{41}}{-4}$$

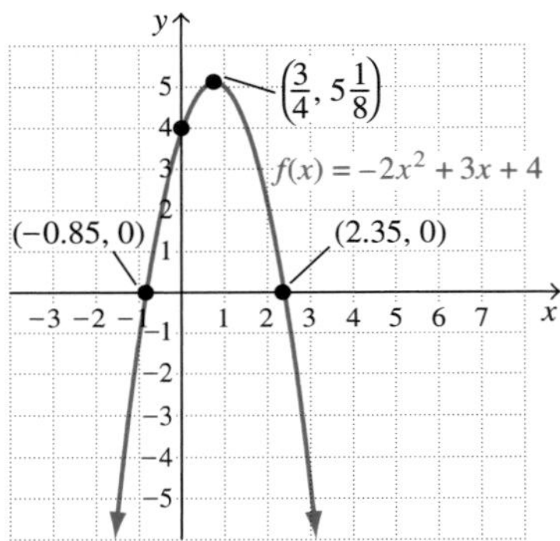

Figure 6.37

Since $\sqrt{41} \approx 6.4$,

$$x \approx \frac{-3 + 6.4}{-4} \approx \frac{3.4}{-4} \approx -0.85 \qquad \text{or} \qquad x \approx \frac{-3 - 6.4}{-4} \approx \frac{-9.4}{-4} \approx 2.35$$

Thus, the x-intercepts are $(-0.85, 0)$ and $(2.35, 0)$.

6. Plot the vertex $(\frac{3}{4}, 5\frac{1}{8})$, the y-intercept (0, 4), and the x-intercepts $(-0.85, 0)$ and (2.35, 0). Then sketch the graph (Fig. 6.37).

b) The domain, the values that can be used for x, is the set of all real numbers, $\mathbb{R}$. The range, the values of y, is $y \leq 5\frac{1}{8}$. ▲

When we use the quadratic formula to find the x-intercepts of a graph, if the radicand, $b^2 - 4ac$, is a negative number, the graph has no x-intercepts. The graph will lie totally above or below the x-axis.

Exponential Functions

In Section 6.3, we discussed exponential equations. Recall that exponential equations are of the form $y = a^x$, $a > 0$, $a \neq 1$. The graph of every exponential equation will pass the vertical line test, and so every exponential equation is also an exponential function. Exponential functions may be written as $f(x) = a^x$, $a > 0$, $a \neq 1$.

In Section 6.3, we also introduced the natural exponential formula $P = P_0e^{kt}$. We can write this formula in function notation as $P(t) = P_0e^{kt}$. This expression is referred to as the ***natural exponential function***. In Example 8, we use the natural exponential function.

EXAMPLE 8 *Evaluating an Exponential Decay Function*

The power supply of a satellite is a radioisotope. The power output, p, in watts remaining in the power supply is a function of the time the satellite is in space. If there are originally 100 grams of the radioisotope, the power remaining after t days is $p(t) = 100e^{-0.001t}$. What will be the remaining power after 1 year (or 365 days) in space?

SOLUTION: Substitute 365 days for t in the function, and then evaluate using a calculator as described in Section 6.3.

$$\begin{aligned} p(t) &= 100e^{-0.001t} \\ p(365) &= 100e^{-0.001(365)} \\ &= 100e^{-0.365} \\ &\approx 100(0.6941966509) \\ &\approx 69.4 \text{ watts} \end{aligned}$$

Thus, after 365 days, the power remaining will be about 69.4 watts. ▲

EXAMPLE 9 *Evaluating an Exponential Decay Function*

Carbon 14 is used by scientists to find the age of fossils and other artifacts. If an object originally had 25 grams of carbon 14, the amount present after t years is $f(t) = 25e^{-0.00012010t}$. How much carbon 14 will be found after 350 years?

SOLUTION: Substitute 350 for t in the function, and then evaluate using a calculator as described in Section 6.3.

$$\begin{aligned} f(t) &= 25e^{-0.00012010t} \\ f(350) &= 25e^{-0.00012010(350)} \\ &= 25e^{-0.042035} \\ &\approx 25(0.9588362207) \\ &\approx 23.97090552 \\ &\approx 24 \text{ grams} \end{aligned}$$

Thus, after 350 years, about 24 grams of carbon 14 will remain. ▲

Graphs of Exponential Functions

What does the graph of an ***exponential function*** of the form $y = a^x$, $a > 0$, $a \neq 1$, look like? Examples 10 and 11 illustrate graphs of exponential functions.

EXAMPLE 10 *Graphing an Exponential Function, $a > 1$*

a) Graph $y = 2^x$.

b) Determine the domain and range of the function.

DID YOU KNOW

Population Growth

Population growth during certain time periods can be described by an exponential function. Whether it is a population of bacteria, fish, flowers, or people, the same general trend emerges: a period of rapid (exponential) growth, which is then followed by a leveling-off period.

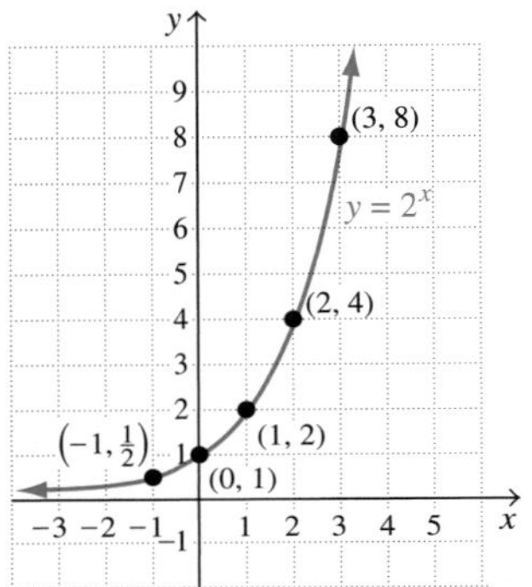

Figure 6.38

SOLUTION:

a) Substitute values for x and find the corresponding values of y. The graph is shown in Fig. 6.38.

$$y = 2^x$$

$$x = -3, \quad y = 2^{-3} = \frac{1}{2^3} = \frac{1}{8}$$

$$x = -2, \quad y = 2^{-2} = \frac{1}{2^2} = \frac{1}{4}$$

$$x = -1, \quad y = 2^{-1} = \frac{1}{2^1} = \frac{1}{2}$$

$$x = 0, \quad y = 2^0 = 1$$

$$x = 1, \quad y = 2^1 = 2$$

$$x = 2, \quad y = 2^2 = 4$$

$$x = 3, \quad y = 2^3 = 8$$

x	y
-3	$\frac{1}{8}$
-2	$\frac{1}{4}$
-1	$\frac{1}{2}$
0	1
1	2
2	4
3	8

b) The domain is all real numbers, $\mathbb{R}$. The range is $y > 0$. Note that y can never have a value of 0.

All exponential functions of the form $y = a^x$, $a > 1$, will have the general shape of the graph illustrated in Fig. 6.38. Since $f(x)$ is the same as y, the graphs of functions of the form $f(x) = a^x$, $a > 1$, will also have the general shape of the graph illustrated in Fig. 6.38. Can you now predict the shape of the graph of $y = e^x$? Remember: e has a value of about 2.7183.

EXAMPLE 11 *Graphing an Exponential Function,* $0 < a < 1$

a) Graph $y = (\frac{1}{2})^x$.

b) Determine the domain and range of the function.

SOLUTION:

a) We begin by substituting values for x and calculating values for y. We then plot the ordered pairs and use these points to sketch the graph. To evaluate a fraction with a negative exponent, we use the fact that

$$\left(\frac{a}{b}\right)^{-x} = \left(\frac{b}{a}\right)^{x}$$

For example,

$$\left(\frac{1}{2}\right)^{-3} = \left(\frac{2}{1}\right)^{3} = 8$$

DID YOU KNOW

The Electronic Superhighway

Electronic mail, or e-mail, is fast becoming one of the most popular and easiest methods of communicating. The number of e-mail messages is increasing exponentially. Although 45% of the U.S. population now uses e-mail, according to Jupiter Research, 93% of people online frequently use e-mail. Researchers expect this percentage to increase to 98% by 2007. Nearly one-quarter of e-mail users maintain three or more personal e-mail accounts. Thirty-five percent of e-mail messages received is considered "spam mail" or unsolicited e-mail, whereas e-mail from friends and family makes up 34% of messages received. E-mail is also becoming a valuable marketing tool. According to the Kelsey Group, annual spending by U.S. small business on localized e-mail marketing will exceed $2.2 billion by 2005.

Then

$$y = \left(\frac{1}{2}\right)^x$$

$$x = -3, \quad y = \left(\frac{1}{2}\right)^{-3} = 2^3 = 8$$

$$x = -2, \quad y = \left(\frac{1}{2}\right)^{-2} = 2^2 = 4$$

$$x = -1, \quad y = \left(\frac{1}{2}\right)^{-1} = 2^1 = 2$$

$$x = 0, \quad y = \left(\frac{1}{2}\right)^{0} = 1$$

$$x = 1, \quad y = \left(\frac{1}{2}\right)^{1} = \frac{1}{2}$$

$$x = 2, \quad y = \left(\frac{1}{2}\right)^{2} = \frac{1}{4}$$

$$x = 3, \quad y = \left(\frac{1}{2}\right)^{3} = \frac{1}{8}$$

x	y
-3	8
-2	4
-1	2
0	1
1	$\frac{1}{2}$
2	$\frac{1}{4}$
3	$\frac{1}{8}$

The graph is illustrated in Fig. 6.39.

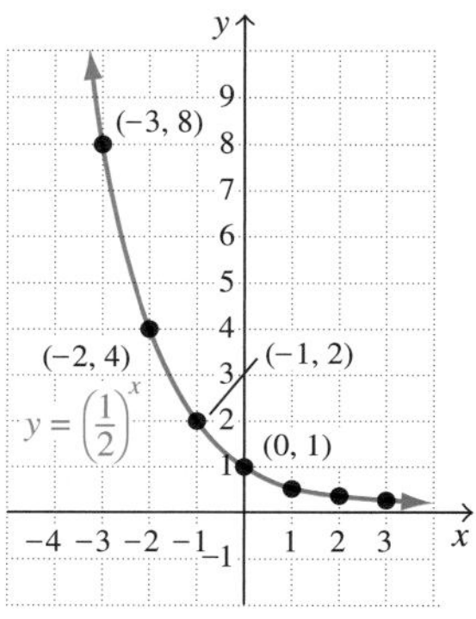

Figure 6.39

b) The domain is the set of all real numbers, $\mathbb{R}$. The range is $y > 0$. ▲

All exponential functions of the form $y = a^x$ or $f(x) = a^x, 0 < a < 1$, will have the general shape of the graph illustrated in Fig. 6.39.

EXAMPLE 12 *Is the Growth Exponential?*

Sales of Botox, a drug injected under the skin to smooth out wrinkles, have increased tremendously since the late 1990s. The graph on page 372 shows the sales of Botox from 1999 through 2001 and projected though 2005.

a) Does the graph approximate the graph of an exponential function?
b) Estimate the sales of Botox in 2004.

SOLUTION:

a) Yes, the graph has the approximate shape of an exponential function. A function that increases rapidly with this general shape, is, or approximates, an exponential function.

b) From the graph, we see that in 2004, sales of Botox were about $800 million.

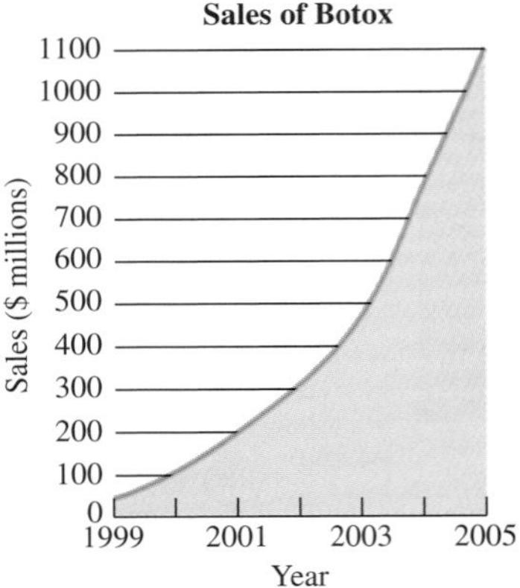

Source: American Society of Plastic Surgeons, Allergan

SECTION 6.10 EXERCISES

Concept/Writing Exercises

1. What is a function?
2. What is a relation?
3. What is the domain of a function?
4. What is the range of a function?
5. Explain how and why the vertical line test can be used to determine whether a graph is a function.
6. Give three examples of one quantity being a function of another quantity.

Practice the Skills

In Exercises 7–24, determine whether the graph represents a function. If it does represent a function, give its domain and range.

7.

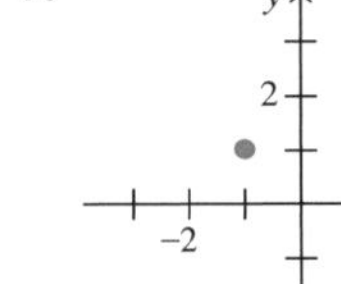

8.

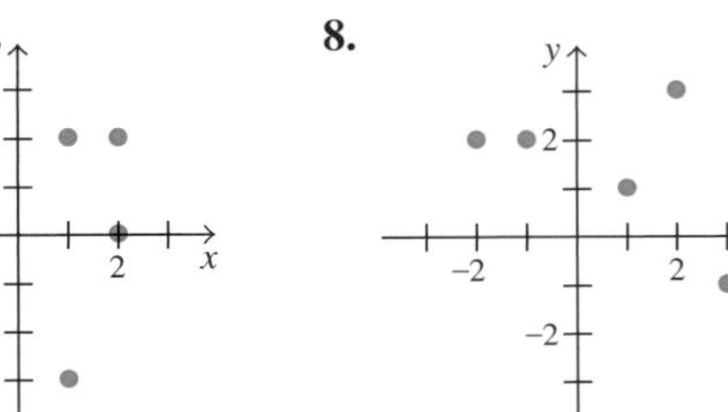

9.

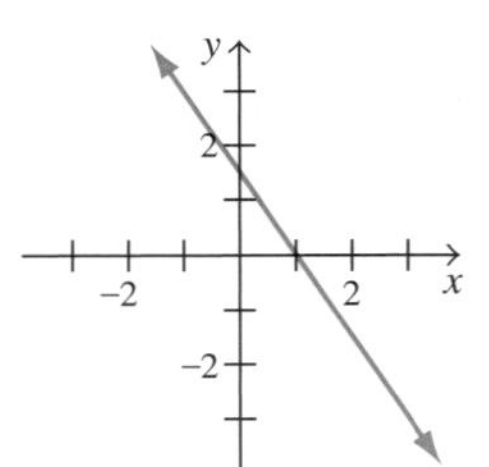

10.

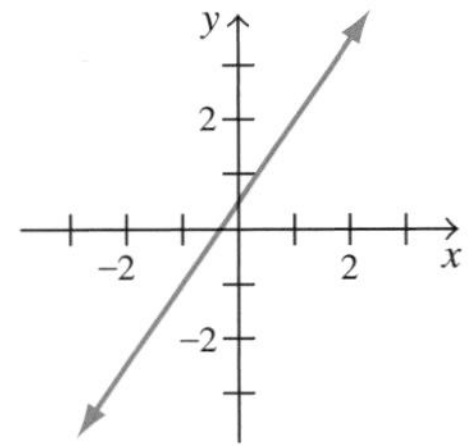

11.

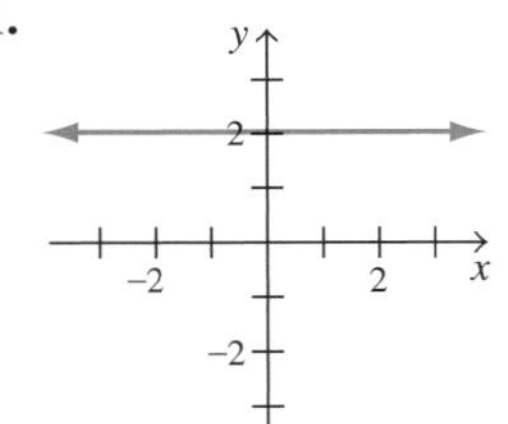

12.

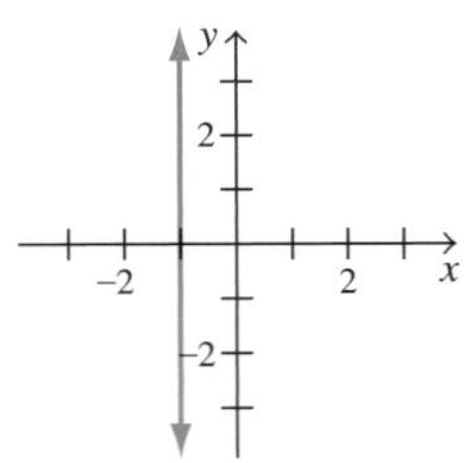

13.

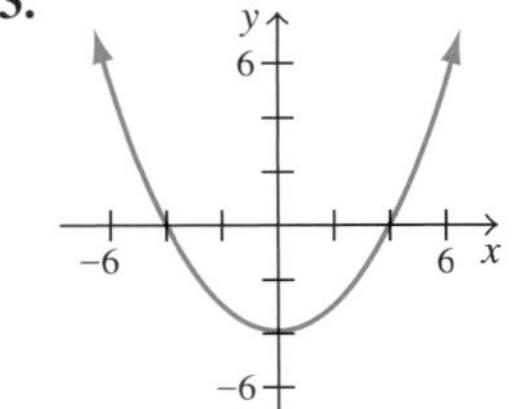

14.

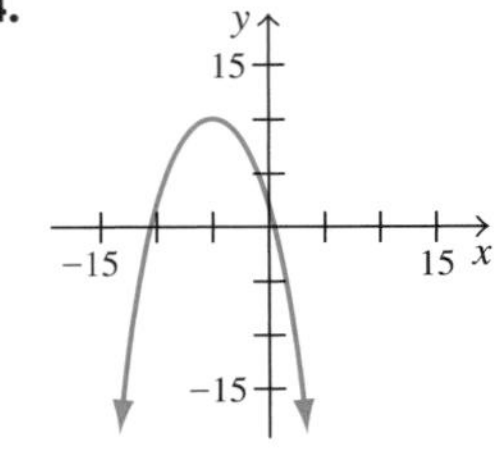

15.
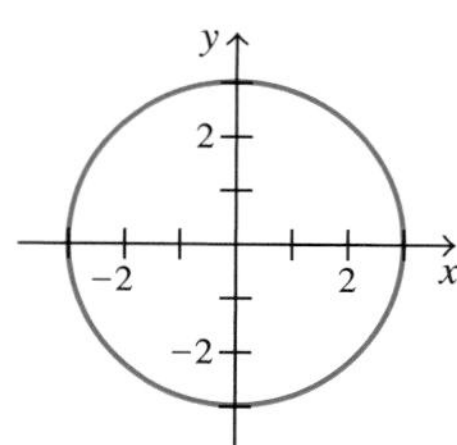

16.
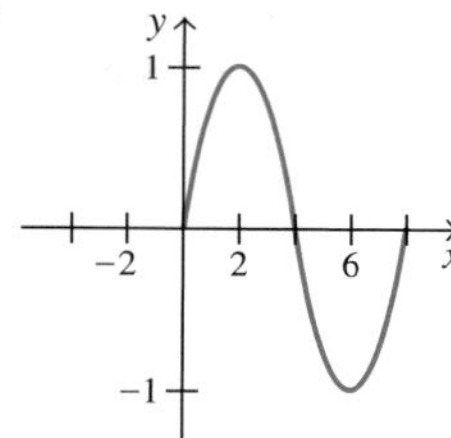

17.
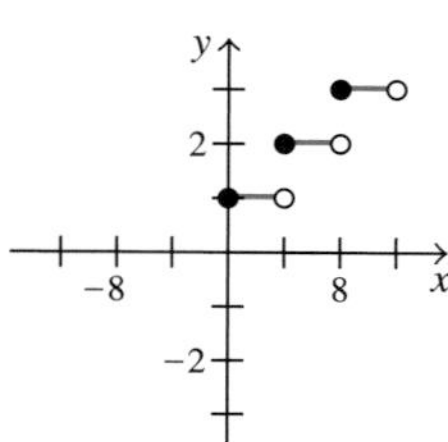

18.
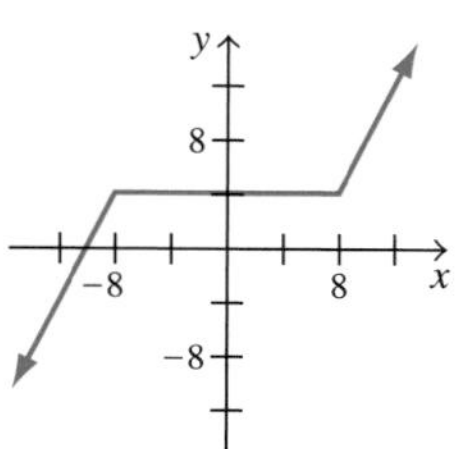

19.
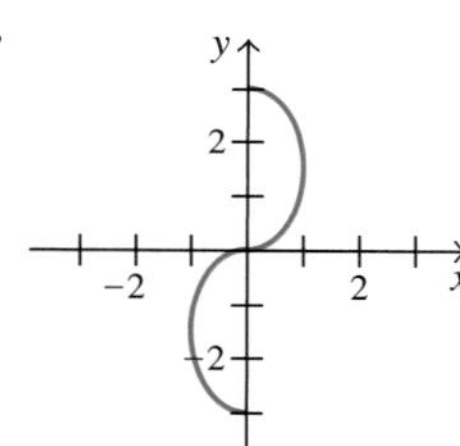

20.
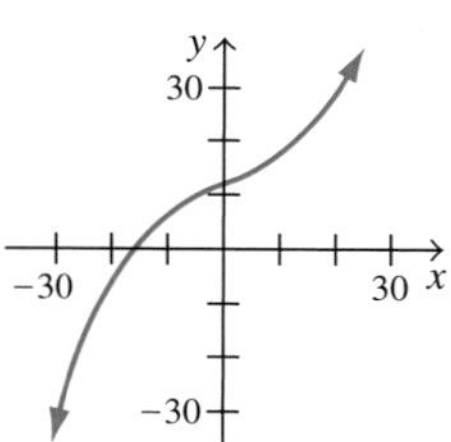

21.
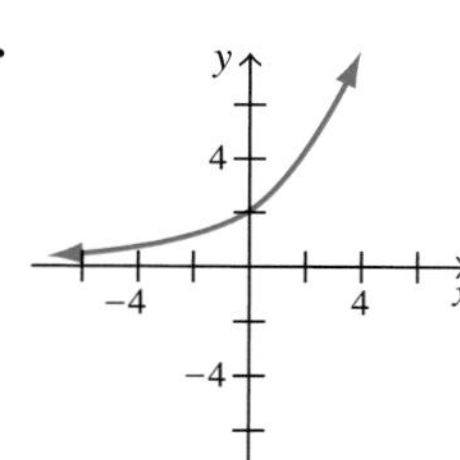

22.
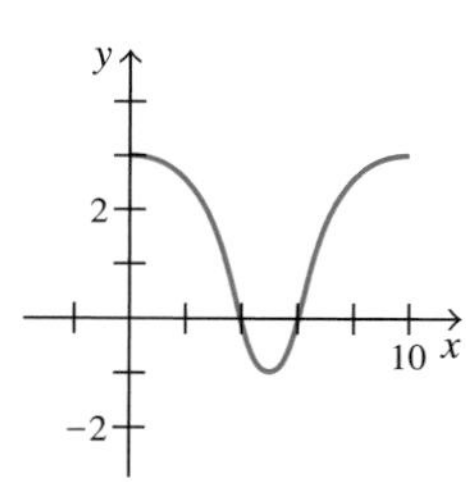

23.
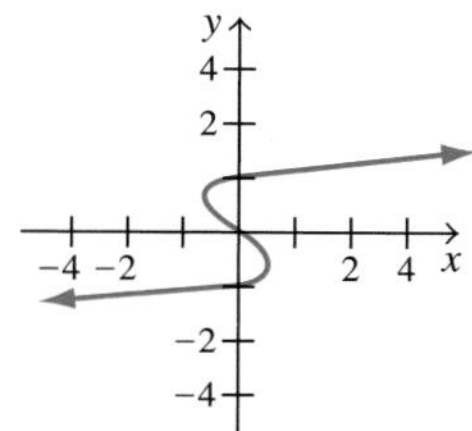

24.
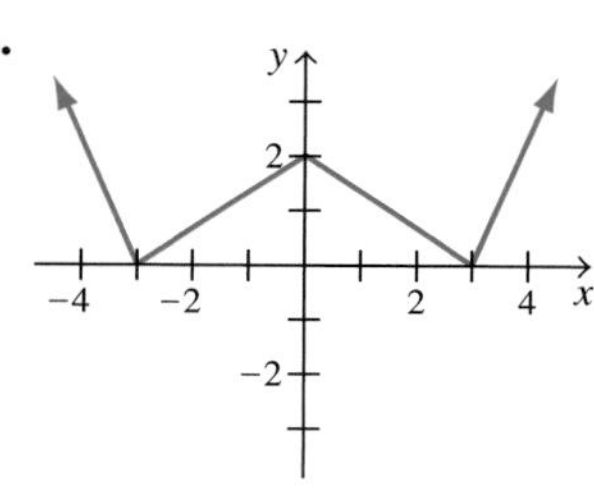

In Exercises 25–30, determine whether the set of ordered pairs is a function.

25. $\{(2, 9), (3, 6), (7, 11), (9, 15)\}$
26. $\{(1, -3), (2, -5), (3, -7), (4, -9)\}$
27. $\{(4, 4), (5, 1), (4, 0)\}$
28. $\{(3, 1), (3, 3), (3, 5)\}$
29. $\{(7, 1), (6, 1), (5, 1)\}$
30. $\{(1, 7), (1, 6), (1, 5)\}$

In Exercises 31–44, evaluate the function for the given value of x.

31. $f(x) = x + 3, \quad x = 2$
32. $f(x) = 2x + 5, \quad x = 4$
33. $f(x) = -2x - 7, \quad x = -4$
34. $f(x) = -5x + 3, \quad x = -1$
35. $f(x) = 10x - 6, \quad x = 0$
36. $f(x) = 7x - 6, \quad x = 4$
37. $f(x) = x^2 - 3x + 1, \quad x = 4$
38. $f(x) = x^2 - 5, \quad x = 7$
39. $f(x) = 2x^2 - 2x - 8, \quad x = -2$
40. $f(x) = -x^2 + 3x + 7, \quad x = 2$
41. $f(x) = -3x^2 + 5x + 4, \quad x = -3$
42. $f(x) = 5x^2 + 2x + 5, \quad x = 4$
43. $f(x) = -5x^2 + 3x - 9, \quad x = -1$
44. $f(x) = -3x^2 - 6x + 10, \quad x = -2$

In Exercises 45–50, graph the function by using the slope and y-intercept.

45. $f(x) = 2x - 1$
46. $f(x) = -x + 3$
47. $f(x) = -4x + 2$
48. $f(x) = 2x + 5$
49. $f(x) = \frac{3}{2}x - 1$
50. $f(x) = -\frac{1}{2}x + 3$

In Exercises 51–66,
a) *determine whether the parabola will open upward or downward.*
b) *find the equation of the axis of symmetry.*
c) *find the vertex.*
d) *find the y-intercept.*
e) *find the x-intercepts if they exist.*
f) *sketch the graph.*
g) *find the domain and range of the function.*

51. $y = x^2 - 16$
52. $y = x^2 - 9$
53. $y = -x^2 + 4$
54. $y = -x^2 + 16$
55. $f(x) = -x^2 - 4$
56. $y = -2x^2 - 8$
57. $y = 2x^2 - 3$
58. $f(x) = -3x^2 - 6$
59. $f(x) = x^2 + 2x + 6$
60. $y = x^2 - 8x + 1$

61. $y = x^2 + 5x + 6$

62. $y = x^2 - 7x - 8$

63. $y = -x^2 + 4x - 6$

64. $y = -x^2 + 8x - 8$

65. $y = -3x^2 + 14x - 8$

66. $y = 2x^2 - x - 6$

In Exercises 67–78, draw the graph of the function and state the domain and range.

67. $y = 3^x$

68. $f(x) = 4^x$

69. $y = (\frac{1}{3})^x$

70. $y = (\frac{1}{4})^x$

71. $f(x) = 2^x + 1$

72. $y = 3^x - 1$

73. $y = 4^x + 1$

74. $y = 2^x - 1$

75. $y = 3^{x-1}$

76. $y = 3^{x+1}$

77. $f(x) = 4^{x+1}$

78. $y = 4^{x-1}$

Problem Solving

79. *Monthly Salary* Chet Rogalski is part owner of a newly opened hardware store. Chet's monthly salary is given by the function $m(s) = 300 + 0.10s$, where s is the store's monthly sales in dollars. If sales for the month of July are \$20,000, determine Chet's monthly salary for July.

80. *Finding Distances* The distance a car travels, $d(t)$, at a constant 60 mph is given by the function $d(t) = 60t$, where t is the time in hours. Find the distance traveled in

a) 3 hours.

b) 7 hours.

81. The following graph indicates the percent of first-time California State University freshmen who entered college with college-level mathematics proficiency for the years 1992 through 2001. The function $f(x) = 0.56x^2 - 5.43x + 59.83$ can be used to estimate the percent of first-time California State University freshmen who entered college with college-level mathematics proficiency, $f(x)$, where x is the number of years since 1992 and $0 \le x \le 9$.

a) Use the function $f(x)$ to estimate the percent of first-time California State University freshmen who entered college with college-level mathematics proficiency in 2000.

b) Use the graph to determine the year in which the percent of first-time California State University freshmen who entered college with college-level mathematics proficiency was a minimum.

c) Determine the x-coordinate of the vertex, then use this value in the function $f(x)$ to estimate the minimum percentage of first-time California State University freshmen who entered college with college-level mathematics proficiency.

Percentage of First-time California State University Freshmen Entering with College-level Mathematics Proficiency

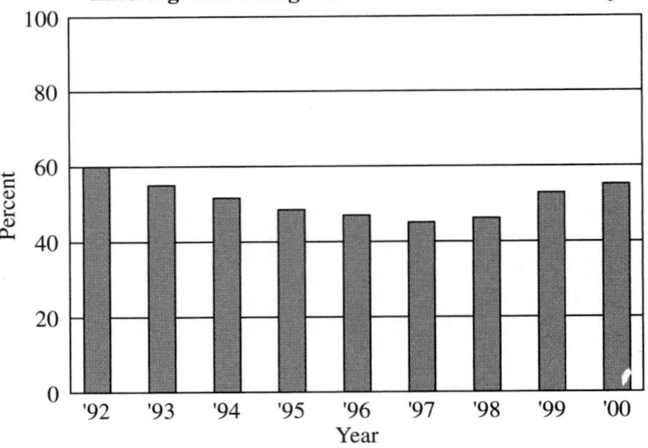

Source: California State University Board of Trustees report

82. *Free Meals* The following graph indicates the number of free lunches, in thousands, served in the Rochester, NY, Summer Meals Program from 1994 through 2000. The function $l(x) = -4.25x^2 + 30.32x + 150.14$ can be used to estimate the number of free lunches served, $l(x)$, where x is the number of years since 1994 and $0 \le x \le 6$.

a) Use the function $l(x)$ to estimate the number of free lunches served in 1999.

b) Use the graph to determine the year in which the number of free lunches served was a maximum.

c) Determine the x-coordinate of the vertex, then use this value in the function $l(x)$ to estimate the maximum number of free lunches served.

Free Lunches

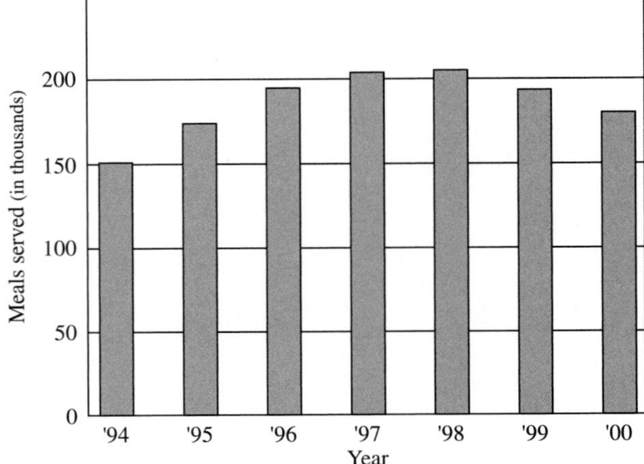

Source: YWCA

83. ***Expected Growth*** The town of Lockport currently has 4000 residents. The expected future population can be approximated by the function $P(x) = 4000(1.3)^{0.1x}$, where x is the number of years in the future. Find the expected population of Lockport in
a) 10 years.
b) 50 years.

84. ***Decay of Plutonium*** Plutonium, a radioactive material used in most nuclear reactors, decays exponentially at a rate of 0.003% per year. The amount of plutonium, P, left after t years can be found by the formula $P = P_0 e^{-0.00003t}$, where P_0 is the original amount of plutonium present. If there are originally 2000 grams of plutonium, find the amount of plutonium left after 50 years.

85. ***Scooter Injuries*** The number of scooter injuries rose rapidly during the summer months in 2000. The graph below shows the number of scooter injuries, by month, in 2000.
a) Does the graph approximate the graph of an exponential function from May through September 2000?
b) Estimate the number of scooter injuries in August 2000.

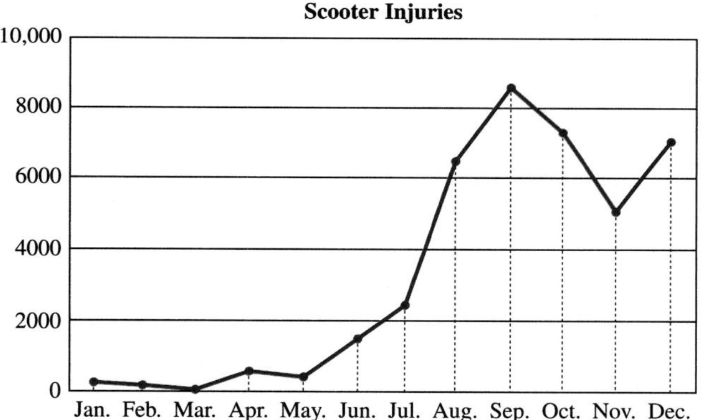

Source: Consumer Product Safety Commission

86. ***Cost of a PC*** The graph shows the U.S. average cost of a personal computer (PC), in thousands of dollars, from 1995 through 2001 projected through 2003.

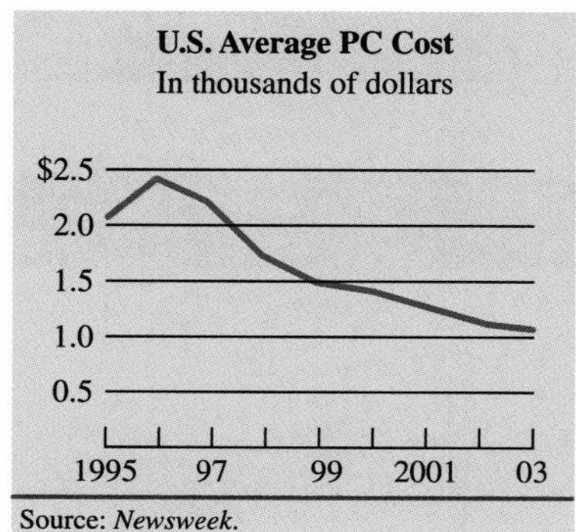

Source: *Newsweek.*

a) From 1995 through 2001 projected through 2003, does the graph approximate the graph of an exponential function? Explain.
b) Estimate the U.S. average cost of a PC in 2002.

87. The spacing of the frets on the neck of a classical guitar is determined from the equation $d = (21.9)(2)^{(20-x)/12}$, where x = the fret number and d = the distance in centimeters of the xth fret from the bridge.

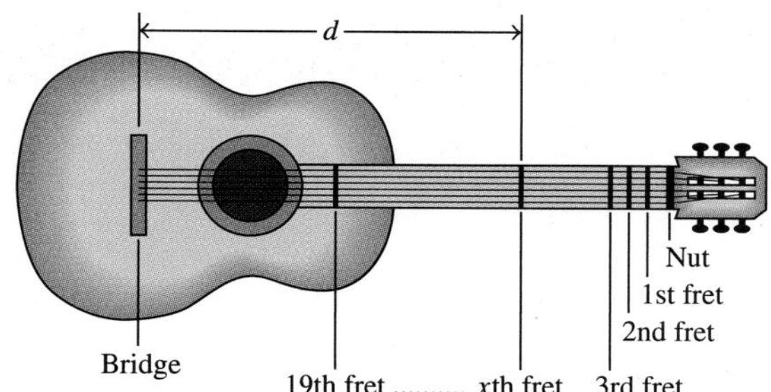

a) Determine how far the 19th fret should be from the bridge (rounded to one decimal place).
b) Determine how far the 4th fret should be from the bridge (rounded to one decimal place).
c) The distance of the nut from the bridge can be found by letting $x = 0$ in the given exponential equation. Find the distance from the nut to the bridge (rounded to one decimal place).

Challenge Problems/Group Activities

88. ***Appreciation of a House*** A house initially cost \$85,000. The value, V, of the house after n years if it appreciates at a constant rate of 4% per year can be determined by the function $V = f(n) = \$85{,}000(1.04)^n$.
a) Determine $f(8)$ and explain its meaning.
b) After how many years is the value of the house greater than \$153,000? (Find by trial and error.)

89. ***Target Heart Rate*** While exercising, a person's recommended target heart rate is a function of age. The recommended number of beats per minute, y, is given by the function $y = f(x) = -0.85x + 187$, where x represents a person's age in years. Determine the number of recommended heart beats per minute for the following ages and explain the results.
a) 20
b) 30
c) 50
d) 60
e) What is the age of a person with a recommended target heart rate of 85?

90. ***Speed of Light*** Light travels at about 186,000 miles per second through space. The distance, d, in miles that light travels in t seconds can be determined by the function $d(t) = 186{,}000t$.

a) Light reaches the moon from Earth in about 1.3 sec. Determine the approximate distance from Earth to the moon.
b) Express the distance in miles, d, traveled by light in t minutes as a function of time, t.
c) Light travels from the sun to Earth in about 8.3 min. Determine the approximate distance from the sun to Earth.

Internet/Research Activity

91. The idea of using variables in algebraic equations was introduced by the French mathematician François Viète (1540–1603). Write a paper about his life and his contributions to mathematics. In particular, discuss his work with algebra equations. (References include history of mathematics books, encyclopedias, and the Internet.)

CHAPTER 6 SUMMARY

IMPORTANT FACTS

Properties used to solve equations

Addition property of equality
If $a = b$, then $a + c = b + c$

Subtraction property of equality
If $a = b$, then $a - c = b - c$

Multiplication property of equality
If $a = b$, then $ac = bc$

Division property of equality
If $a = b$, then $a/c = b/c$, $c \neq 0$

Variation

Direct: $y = kx$

Inverse: $y = \dfrac{k}{x}$

Joint: $y = kxz$

Inequality symbols

$a < b$ means that a is less than b.

$a \leq b$ means that a is less than or equal to b.

$a > b$ means that a is greater then b.

$a \geq b$ means that a is greater than or equal to b.

Intercepts

To find the x-intercept, set $y = 0$ and solve the resulting equation for x.

To find the y-intercept, set $x = 0$ and solve the resulting equation for y.

Slope

Slope (m): $m = \dfrac{y_2 - y_1}{x_2 - x_1}$

Equations and formulas

Linear equation in two variables:
$$ax + by = c, \quad a \neq 0, \quad b \neq 0$$

Quadratic equation in one variable:
$$ax^2 + bx + c = 0, \quad a \neq 0$$

Quadratic equation (or function) in two variables:
$$y = ax^2 + bx + c, \quad a \neq 0$$

Exponential equation (or function):
$$y = a^x, \quad a \neq 1, \quad a > 0$$

Exponential growth or decay formula:
$$P = P_0a^{kt}, \quad a \neq 1, \quad a > 0$$

Quadratic formula:
$$x = \frac{-b \pm \sqrt{b^2 - 4ac}}{2a}$$

Slope–intercept form of a line:
$$y = mx + b$$

Axis of symmetry of a parabola:
$$x = \frac{-b}{2a}$$

Zero-factor property

If $a \cdot b = 0$, then $a = 0$ or $b = 0$.

CHAPTER 6 REVIEW EXERCISES

6.1

In Exercises 1–6, evaluate the expression for the given value(s) of the variable.

1. $x^2 + 12, \quad x = 3$
2. $-x^2 - 9, \quad x = -1$
3. $4x^2 - 2x + 5, \quad x = 2$
4. $-x^2 + 7x - 3, \quad x = \frac{1}{2}$
5. $4x^3 - 7x^2 + 3x + 1, \quad x = -2$
6. $3x^2 - xy + 2y^2, \quad x = 1, \quad y = -2$

6.2

In Exercises 7–9, combine like terms.

7. $3x - 4 + x + 5$
8. $3x + 4(x - 2) + 6x$
9. $4(x - 1) + \frac{1}{3}(9x + 3)$

In Exercises 10–14, solve the equation for the given variable.

10. $4s + 10 = -30$
11. $3t + 8 = 6t - 13$
12. $\frac{x + 5}{6} = \frac{x - 3}{3}$
13. $4(x - 2) = 3 + 5(x + 4)$
14. $\frac{x}{4} + \frac{3}{5} = 7$
15. *Making Oatmeal* A recipe for Hot Oats Cereal calls for 2 cups of water and for $\frac{1}{3}$ cup of dry oats. How many cups of dry oats would be used with 3 cups of water?
16. *Laying Blocks* A mason lays 120 blocks in 1 hr 40 min. How long will it take her to lay 300 blocks?

6.3

In Exercises 17–20, use the formula to find the value of the indicated variable for the values given.

17. $A = bh$

    Find A when $b = 12$ and $h = 4$ (geometry).
18. $V = 2\pi R^2 r^2$

    Find V when $R = 3$, $r = 1\frac{3}{4}$, and $\pi = 3.14$ (geometry).
19. $z = \dfrac{\overline{x} - \mu}{\dfrac{\sigma}{\sqrt{n}}}$

    Find $\overline{x}$ when $z = 2$, $\mu = 100$, $\sigma = 3$, and $n = 16$ (statistics).
20. $K = \frac{1}{2}mv^2$

    Find m when $v = 30$ and $k = 4500$ (physics).

In Exercises 21–24, solve for y.

21. $3x - 9y = 18$
22. $2x + 5y = 12$
23. $2x - 3y + 52 = 30$
24. $-3x - 4y + 5z = 4$

In Exercises 25–28, solve for the variable indicated.

25. $A = lw$, for w
26. $P = 2l + 2w$, for w
27. $L = 2(wh + lh)$, for l
28. $a_n = a_1 + (n - 1)d$, for d

6.4

In Exercises 29–32, write the phrase in mathematical terms.

29. 8 increased by 2 times x
30. 3 times y decreased by 7
31. 10 increased by 3 times r
32. The difference between 8 divided by q and 11

In Exercises 33–36, write an equation that can be used to solve the problems. Solve the equation and find the desired value(s).

33. Four increased by 3 times a number is 22.
34. The product of 3 and a number, increased by 8, is 6 less than the number.
35. Five times the difference of a number and 4 is 45.
36. Fourteen more than 10 times a number is 8 times the sum of the number and 12.

In Exercises 37–40, write the equation and then find the solution.

37. **MODELING** - *Investing* Jim Lawton received an inheritance of \$15,000. If he wants to invest twice as much money in mutual funds as in bonds, how much should he invest in mutual funds?
38. **MODELING** - *Lawn Chairs* Larry's Lawn Chair Company has fixed costs of \$15,000 per month and variable costs of \$9.50 per lawn chair manufactured. The company has \$95,000 available to meet its total monthly expenditures. What is the maximum number of lawn chairs the company can manufacture in a month? (Fixed costs, such as rent and insurance, are those that occur regardless of the level of production. Variable costs, such as those for materials, depend on the level of production.)
39. **MODELING** - *Zoo Animals* According to the 2003 *World Almanac*, the number of species at the San Diego Zoo is 140 more than two times the number of species at the Philadelphia Zoo. The sum of the species at the San Diego Zoo and the Philadelphia Zoo is 1130. Determine the number of species at each zoo.

40. **MODELING** - *Restaurant Profit* John Smith owns two restaurants. His profit for a year at restaurant A is \$12,000 greater than his profit at restaurant B. The total profit from both restaurants is \$68,000. Determine the profit at each restaurant.

6.5

In Exercises 41–44, find the quantity indicated.

41. s is inversely proportional to t. If $s = 10$ when $t = 3$, find s when $t = 5$.

42. J is directly proportional to the square of A. If $J = 32$ when $A = 4$, find J when $A = 7$.

43. W is directly proportional to L and inversely proportional to A. If $W = 80$ when $L = 100$ and $A = 20$, find W when $L = 50$ and $A = 40$.

44. z is jointly proportional to x and y and inversely proportional to the square of r. If $z = 12$ when $x = 20$, $y = 8$, and $r = 8$, find z when $x = 10$, $y = 80$, and $r = 3$.

45. ***Buying Fertilizer***
a) A 30 lb bag of fertilizer will cover an area of 2500 ft^2. How many pounds of fertilizer are needed to cover an area of 12,500 ft^2?
b) How many bags of fertilizer are needed?

46. ***Map Reading*** The scale of a map is 1 in. to 30 mi. What distance on the map represents 120 mi?

47. ***Electric Bill*** An electric company charges \$0.162 per kilowatt-hour (kWh). What is the electric bill if 740 kWh are used in a month?

48. ***A Falling Object*** The distance, d, an object drops in free fall is directly proportional to the square of the time, t. If an object falls 16 ft in 1 sec, how far will an object fall in 5 sec?

6.6

In Exercises 49–52, graph the solution set for the set of real numbers.

49. $5 + 9x \leq 7x - 7$ **50.** $2x + 8 \geq 4x + 10$

51. $3(x + 9) \leq 4x + 11$ **52.** $-3 \leq x + 1 < 7$

In Exercises 53–56, graph the solution set for the set of integers.

53. $2 + 5x > -8$ **54.** $5x + 13 \geq -22$

55. $-1 < x \leq 9$ **56.** $-8 \leq x + 2 \leq 7$

6.7

In Exercises 57–60, graph the ordered pair in the Cartesian coordinate system.

57. $(1, 4)$ **58.** $(-2, 5)$ **59.** $(-4, 3)$ **60.** $(5, -3)$

In Exercises 61 and 62, points A, B, and C are vertices of a rectangle. Plot the points. Find the coordinates of the fourth point, D, to complete the rectangle. Find the area of the rectangle.

61. $A(-3, 3)$, $B(2, 3)$, $C(2, -1)$

62. $A(-3, 1)$, $B(-3, -2)$, $C(4, -2)$

In Exercises 63–66, graph the equation by plotting points.

63. $x - y = 4$ **64.** $2x + 3y = 12$

65. $x = y$ **66.** $x = 3$

In Exercises 67–70, graph the equation, using the x- and y-intercepts.

67. $x - 2y = 6$ **68.** $x + 3y = 6$

69. $4x - 3y = 12$ **70.** $2x + 3y = 9$

In Exercises 71–74, find the slope of the line through the given points.

71. $(1, 3), (6, 5)$ **72.** $(3, -1), (5, -4)$

73. $(-1, -4), (2, 3)$ **74.** $(6, 2), (6, -2)$

In Exercises 75–78, graph the equation by plotting the y-intercept and then plotting a second point by making use of the slope.

75. $y = 2x - 5$ **76.** $2y - 4 = 3x$

77. $2y + x = 8$ **78.** $y = -x - 1$

In Exercises 79 and 80, determine the equation of the graph.

79.

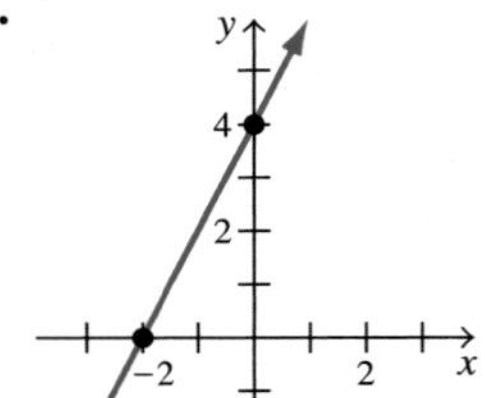

80.

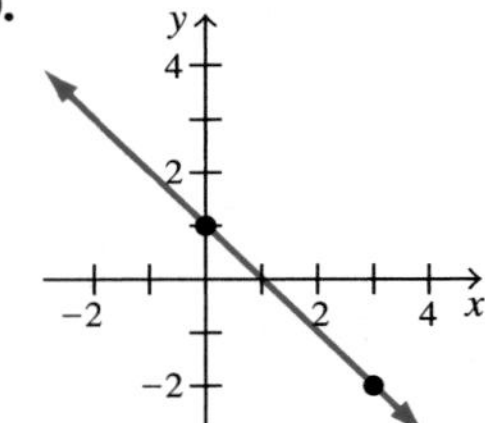

81. ***Disability Income*** The monthly disability income, I, that Nadja Muhidin receives is $I = 460 - 0.5m$, where m is her monthly earnings for her part-time job for the previous month.
a) Draw a graph of disability income versus earnings for earnings up to and including \$920.
b) If Nadja earns \$600 in January, how much disability income will she receive in February?
c) If she received \$380 disability income in November, how much did she earn in October?

82. ***Business Space Rental*** The monthly rental cost, C, in dollars, for space in the Galleria Mall can be approximated by the equation $C = 1.70A + 3000$, where A is the area, in square feet, of space rented.
a) Draw a graph of monthly rental cost versus square feet for up to and including 12,000 ft^2.
b) Determine the monthly rental cost if 2000 ft^2 are rented.
c) If the rental cost is \$10,000 per month, how many square feet are rented?

6.8

In Exercises 83–86, graph the inequality.

83. $4x + 3y \leq 12$

84. $3x + 2y \geq 12$

85. $2x - 3y > 12$

86. $-7x - 2y < 14$

6.9

In Exercises 87–92, factor the trinomial. If the trinomial cannot be factored, so state.

87. $x^2 + 9x + 18$

88. $x^2 + x - 20$

89. $x^2 - 10x + 24$

90. $x^2 - 9x + 20$

91. $6x^2 + 7x - 3$

92. $2x^2 + 13x - 7$

In Exercises 93–96, solve the equation by factoring.

93. $x^2 + 3x + 2 = 0$

94. $x^2 - 5x = -4$

95. $3x^2 - 17x + 10 = 0$

96. $3x^2 = -7x - 2$

In Exercises 97–100, solve the equation, using the quadratic formula. If the equation has no real solution, so state.

97. $x^2 - 4x - 1 = 0$

98. $x^2 - 3x + 2 = 0$

99. $2x^2 - 3x + 4 = 0$

100. $2x^2 - x - 3 = 0$

6.10

In Exercises 101–104, determine whether the graph represents a function. If it does represent a function, give its domain and range.

101.

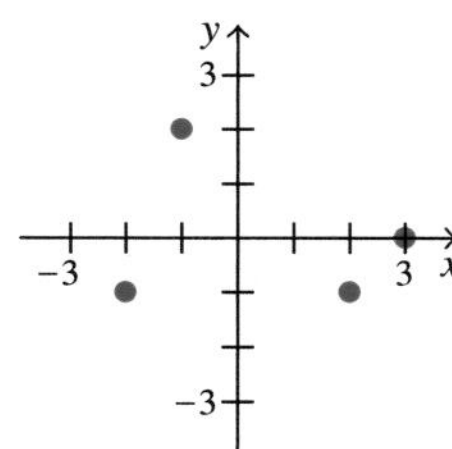

102.

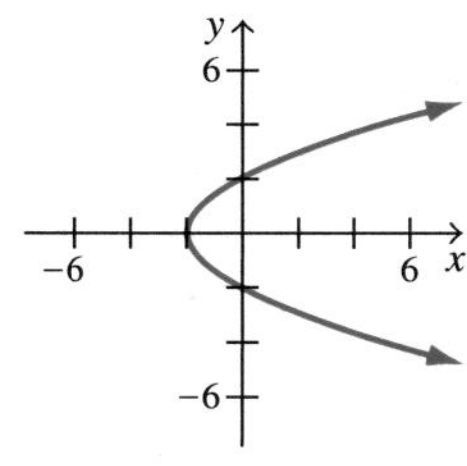

103.

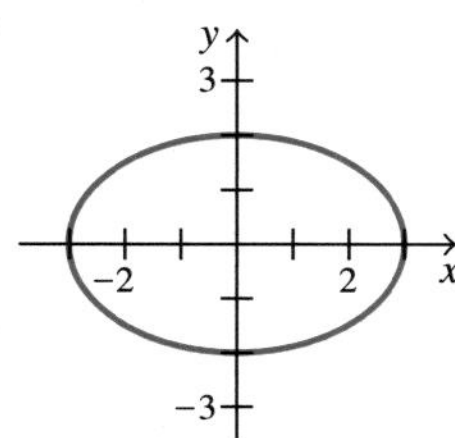

104.

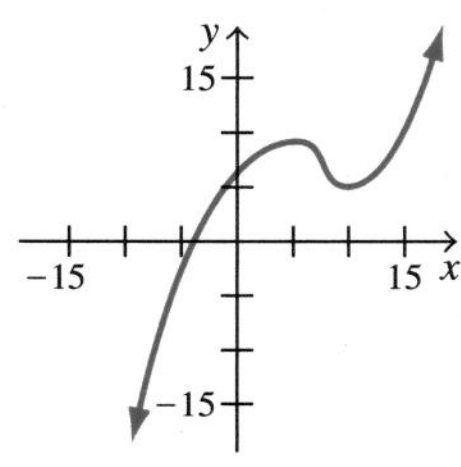

In Exercises 105–108, evaluate f(x) for the given value of x.

105. $f(x) = 5x - 2, \quad x = 4$

106. $f(x) = -2x + 7, \quad x = -3$

107. $f(x) = 2x^2 - 3x + 4, \quad x = 5$

108. $f(x) = -4x^2 + 7x + 9, \quad x = 4$

In Exercises 109 and 110, for each function

a) *determine whether the parabola will open upward or downward.*
b) *find the equation of the axis of symmetry.*
c) *find the vertex.*
d) *find the y-intercept.*
e) *find the x-intercepts if they exist.*
f) *sketch the graph.*
g) *find the domain and range.*

109. $y = -x^2 - 4x + 21$

110. $f(x) = 2x^2 - 8x + 10$

In Exercises 111 and 112, draw the graph of the function and state the domain and range.

111. $y = 2^{2x}$

112. $y = (\frac{1}{2})^x$

6.2, 6.3, 6.10

113. *Gas Mileage* The gas mileage, m, of a specific car can be estimated by the equation (or function)

$$m = 30 - 0.002n^2, \qquad 20 \leq n \leq 80$$

where n is the speed of the car in miles per hour. Estimate the gas mileage when the car travels at 60 mph.

114. *Auto Accidents* The approximate number of accidents in one month, n, involving drivers between 16 and 30 years of age inclusive can be approximated by the equation

$$n = 2a^2 - 80a + 5000, \qquad 16 \leq a \leq 30$$

where a is the age of the driver. Approximate the number of accidents in one month that involved

a) 18-year-olds
b) 25-year-olds.

115. *Filtered Light* The percent of light filtering through Swan Lake, P, can be approximated by the function $P(x) = 100(0.92)^x$, where x is the depth in feet. Find the percent of light filtering through at a depth of 4.5 ft.

CHAPTER 6 TEST

1. Evaluate $3x^2 + 4x - 1$, when $x = -2$.

In Exercises 2 and 3, solve the equation.

2. $3x + 5 = 2(4x - 7)$

3. $-2(x - 3) + 6x = 2x + 3(x - 4)$

In Exercises 4 and 5, write an equation to represent the problem. Then solve the equation.

4. The product of a number and 2, increased by 7 is 25.

5. *Buying a Car* The cost of a car including a 7% sales tax is \$26,750. Determine the cost of the car before tax.

6. Evaluate $L = ah + bh + ch$ when $a = 3, b = 4, c = 5$, and $h = 7$.

7. Solve $3x + 5y = 11$ for y.

8. L varies jointly as M and N and inversely as P. If $L = 12$ when $M = 8$, $N = 3$, and $P = 2$, find L when $M = 10$, $N = 5$, and $P = 15$.

9. For a constant area, the length, l, of a rectangle varies inversely as the width, w. If $l = 15$ ft when $w = 9$ ft, find the length of a rectangle with the same area if the width is 20 ft.

10. Graph the solution set of $-3x + 11 \leq 5x + 35$ on the real number line.

11. Determine the slope of the line through the points $(-3, 5)$ and $(7, 12)$.

In Exercises 12 and 13, graph the equation.

12. $y = 2x - 4$

13. $2x - 3y = 15$

14. Graph the inequality $3y \geq 5x - 12$.

15. Solve the equation $x^2 - 3x = 28$ by factoring.

16. Solve the equation $3x^2 + 2x = 8$ by using the quadratic formula.

17. Determine whether the graph is a function. Explain your answer.

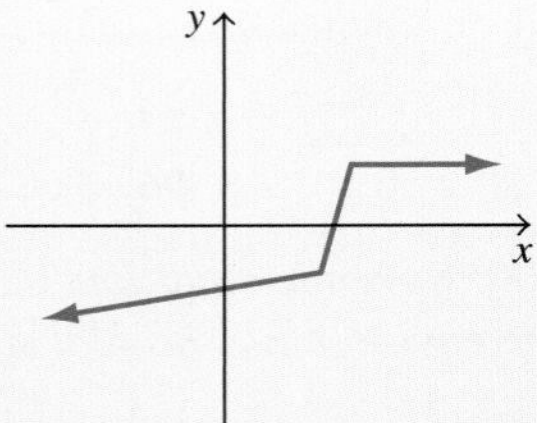

18. Evaluate $f(x) = -4x^2 - 11x + 5$ when $x = -2$.

19. For the equation $y = x^2 - 2x + 4$,

a) determine whether the parabola will open upward or downward.

b) find the equation of the axis of symmetry.

c) find the vertex.

d) find the y-intercept.

e) find the x-intercepts if they exist.

f) sketch the graph.

g) find the domain and range of the function.

GROUP PROJECTS

Archeology: Gaining Information from Bones

1. Archeologists have developed formulas to predict the height and, in some cases, the age at death of the deceased by knowing the lengths of certain bones in the body. The long bones of the body grow at approximately the same rate. Thus, a linear relationship exists between the length of the bones and the person's height. If the length of one of these major bones—the femur (F), the tibia (T), the humerus (H), and the radius (R)— is known, the height, h, of a person can be calculated with one of the following formulas. The relationship between bone length and height is different for males and females.

Male	Female
$h = 2.24F + 69.09$	$h = 2.23F + 61.41$
$h = 2.39T + 81.68$	$h = 2.53T + 72.57$
$h = 2.97H + 73.57$	$h = 3.14H + 64.98$
$h = 3.65R + 80.41$	$h = 3.88R + 73.51$

All measurements are in centimeters.

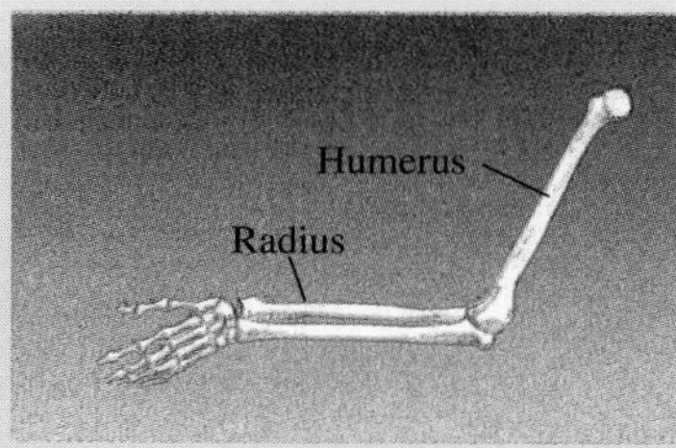

a) Measure your humerus and use the appropriate formula to predict your height in centimeters. How close is this predicted height to your actual height? (The result is an approximation because measuring a bone covered with flesh and muscle is difficult.)

b) Determine and describe where the femur and tibia bones are located.

c) Dr. Juarez, an archeologist, had one female humerus that was 29.42 cm in length. He concluded that the height of the entire skeleton would have been 157.36 cm. Was his conclusion correct?

d) If a 21-year-old woman is 167.64 cm tall, about how long should her tibia be?

e) Sometimes the age of a person may be determined by using the fact that the height of a person, and the length of his or her long bones, decreases at the rate of 0.06 cm per year after the age of 30.

i. At age 30, Jolene is 168 cm tall. Estimate the length of her humerus.

ii. Estimate the length of Jolene's humerus when she is 60 years old.

f) Select six people of the same gender and measure their height and one of the bones for which an equation is given (the same bone on each person). Each measurement should be made to the nearest 0.5 cm. For each person, you will have two measurements, which can be considered an ordered pair (bone length, height). Plot the ordered pairs on a piece of graph paper, with the bone length on the horizontal axis and the height on the vertical axis. Start the scale on both axes at zero. Draw a straight line that you feel is the best approximation, or best fit, through these points. Determine where the line crosses the y-axis and the slope of the line. Your y-intercept and slope should be close to the values in the given equation for that bone. (Reference: M. Trotter and G. C. Gleser, "Estimation of Stature from Long Bones of American Whites and Negroes," *American Journal of Physical Anthropology,* 1952, 10:463–514.)

Graphing Calculator

2. The functions that we graphed in this chapter can be easily graphed with a graphing calculator (or grapher). If you do not have a graphing calculator, borrow one from your instructor or a friend.

a) Explain how you would set the domain and range. The calculator key to set the domain and range may be labeled *range* or *window*. Set the grapher with the following range or window settings:
Xmin = −12, Xmax = 12, Xscl = 1,
Ymin = −13, Ymax = 6, and Yscl = 1.

b) Explain how to enter a function in the graphing calculator. Enter the function $y = 3x^2 - 7x - 8$ in the calculator.

c) Graph the function you entered in part (b).

d) Learn how to use the *trace feature.* Then use it to estimate the x-intercepts. Record the estimated values for the x-intercepts.

e) Learn how to use the *zoom feature* to obtain a better approximation of the x-intercepts. Use the zoom feature twice and record the x-intercepts each time.

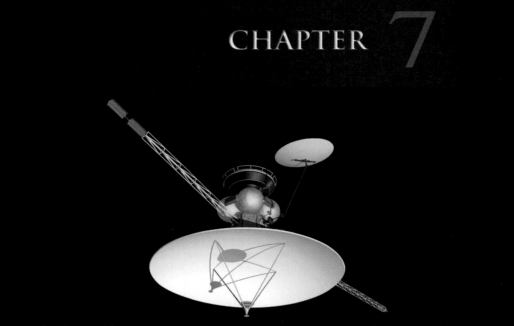

CHAPTER 7

Linear programming provides businesses and governments with a mathematical form of decision making that makes the most efficient use of time and resources. Telecommunications companies use it to route calls through satellites, like this one being tested, so that few of their customers will reach a "no circuits available" message.

SYSTEMS OF LINEAR EQUATIONS AND INEQUALITIES

Imagine that you and a friend have a great idea for a new T-shirt. First you make a few to give away to friends, using your own money. Then other students see the shirt, and soon everyone on campus wants one. Suddenly, you have entered the T-shirt business. To make a profit in your business, you need to keep track of the cost of your materials, the quantity of shirts sold, and the price at which you sell them, a relatively straightforward calculation.

Now suppose you come up with three other designs, and you want to put them on sweatshirts as well as T-shirts, and you want to offer a variety of colors: black, white, blue, and maroon. The equation for finding the profitability of your venture becomes more complicated because there are more variables. To track your profits, you may need to develop and solve systems of equations.

For most business owners, numerous factors must be considered to determine not only whether the business is profitable, but also how much they should charge their customers, which production method is most efficient, what return they can expect by placing advertisements, and so on. Many small-business owners routinely make these calculations based on their own experience, mathematics, and sometimes computer programs. Larger companies often employ inventory analysts, quality control engineers, and efficiency experts to help them, along with computers, keep track of vast quantities of data.

7.1 SYSTEMS OF LINEAR EQUATIONS

In Chapter 6, we discussed linear equations in two variables. In algebra, it is often necessary to find the common solution to two or more such equations. We refer to the equations in this type of problem as a *system of linear equations* or as *simultaneous linear equations*. A *solution to a system of equations* is the ordered pair or pairs that satisfy *all* equations in the system. A system of linear equations may have exactly one solution, no solution, or infinitely many solutions.

The solution to a system of linear equations may be found by a number of different techniques. In this section, we illustrate how a system of linear equations may be solved by graphing. In Section 7.2, we illustrate two algebraic methods, the substitution method and the addition method, for solving a system of linear equations.

EXAMPLE 1 *Is the Ordered Pair a Solution?*

Determine which of the ordered pairs is a solution to the following system of linear equations.

$$\begin{aligned} 3x + y &= -6 \\ 2x - y &= -4 \end{aligned}$$

a) $(1, -9)$ b) $(-2, 0)$ c) $(2, 8)$

SOLUTION: For the ordered pair to be a solution to the system, it must satisfy each equation in the system.

a)

$$\begin{aligned} 3x + y &= -6 \\ 3(1) + (-9) &= -6 \\ -6 &= -6 \quad \text{True} \end{aligned} \qquad \begin{aligned} 2x - y &= -4 \\ 2(1) - (-9) &= -4 \\ 11 &= -4 \quad \text{False} \end{aligned}$$

Since $(1, -9)$ does not satisfy both equations, it is not a solution to the system.

b)

$$\begin{aligned} 3x + y &= -6 \\ 3(-2) + 0 &= -6 \\ -6 &= -6 \quad \text{True} \end{aligned} \qquad \begin{aligned} 2x - y &= -4 \\ 2(-2) - 0 &= -4 \\ -4 &= -4 \quad \text{True} \end{aligned}$$

Since $(-2, 0)$ satisfies both equations, it is a solution to the system.

c)

$$\begin{aligned} 3x + y &= -6 \\ 3(2) + 8 &= -6 \\ 14 &= -6 \quad \text{False} \end{aligned} \qquad \begin{aligned} 2x - y &= -4 \\ 2(2) - 8 &= -4 \\ -4 &= -4 \quad \text{True} \end{aligned}$$

Since $(2, 8)$ does not satisfy both equations, it is not a solution to the system. ▲

To find the solution to a system of linear equations graphically, we graph both of the equations on the same axes. The coordinates of the point or points of intersection of the graphs are the solution or solutions to the system of equations.

$x + y = 4$

x	y
0	4
1	3
4	0

$2x - y = -1$

x	y
0	1
1	3
−2	−3

Procedure for Solving a System of Equations by Graphing

1. Determine three ordered pairs that satisfy each equation.
2. Plot the ordered pairs and sketch the graphs of both equations on the same axes.
3. The coordinates of the point or points of intersection of the graphs are the solution or solutions to the system of equations.

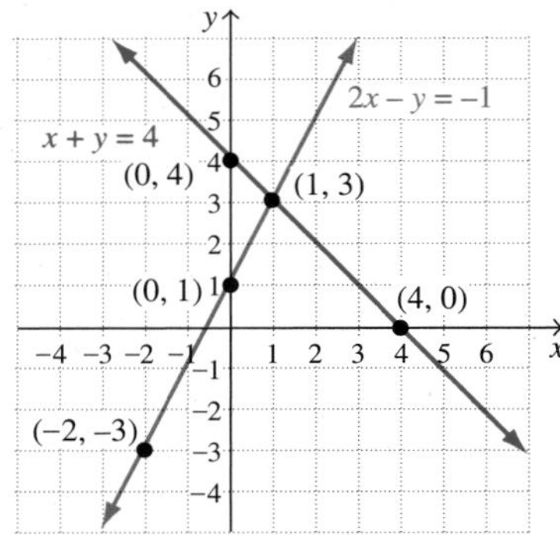

Figure 7.1

When two linear equations are graphed, three situations are possible. The two lines may intersect at one point, as in Example 2; or the two lines may be parallel and not intersect, as in Example 3; or the two equations may represent the same line, as in Example 4.

Since the solution to a system of equations may not be integer values, you may not be able to obtain the exact solution by graphing.

EXAMPLE 2 *A System with One Solution*

Find the solution to the following system of equations graphically.

$$x + y = 4$$
$$2x - y = -1$$

SOLUTION: To find the solution, graph both $x + y = 4$ and $2x - y = -1$ on the same axes (Fig. 7.1). Three points that satisfy each equation are shown in the tables above Fig. 7.1. Figure 7.2 shows the system $x + y = 4$ and $2x - y = -1$ graphed on a Texas Instrument TI-83 Plus graphing calculator.

The graphs intersect at (1, 3), which is the solution. This point is the only point that satisfies *both* equations.

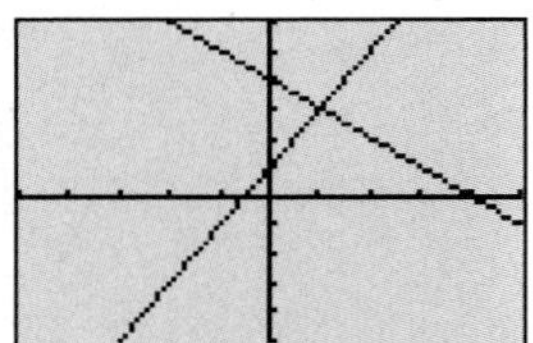

Figure 7.2

Check:

$$x + y = 4 \qquad 2x - y = -1$$
$$1 + 3 = 4 \qquad 2(1) - 3 = -1$$
$$4 = 4 \quad \text{True} \qquad 2 - 3 = -1$$
$$-1 = -1 \quad \text{True}$$

The system of equations in Example 2 is an example of a ***consistent system of equations***. A consistent system of equations is one that has a solution.

EXAMPLE 3 *A System with No Solution*

Find the solution to the following system of equations graphically.

$$2x + y = 3$$
$$2x + y + 5 = 0$$

SOLUTION: Three ordered pairs that satisfy the equation $2x + y = 3$ are (0, 3), $(\frac{3}{2}, 0)$, and $(-1, 5)$. Three ordered pairs that satisfy the equation $2x + y + 5 = 0$ are $(0, -5)$, $(-\frac{5}{2}, 0)$, and $(1, -7)$. The graphs of both equations are given in Fig. 7.3. Since the two lines are parallel, they do not intersect; therefore, the system has *no solution*.

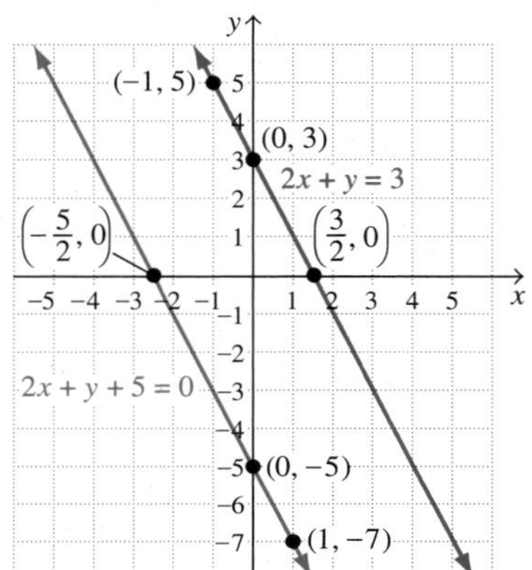

Figure 7.3

The system of equations in Example 3 has no solution. A system of equations that has no solution is called an ***inconsistent system***.

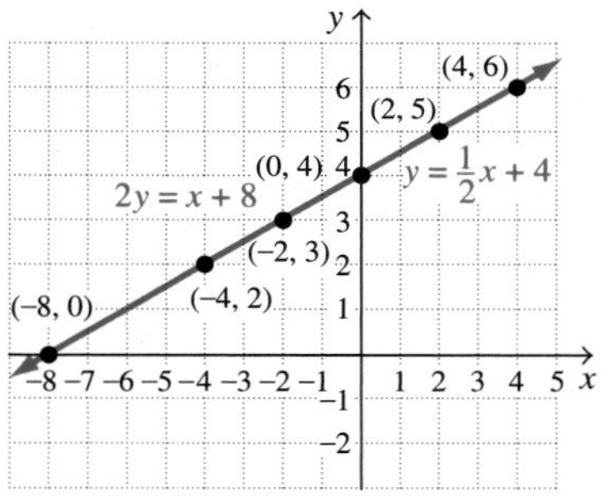

Figure 7.4

EXAMPLE 4 *A System with an Infinite Number of Solutions*

Find the solution to the following system of equations graphically.

$$y = \frac{1}{2}x + 4$$
$$2y = x + 8$$

SOLUTION: Three ordered pairs that satisfy the equation $y = \frac{1}{2}x + 4$ are (0, 4), (2, 5), and $(-2, 3)$. Three ordered pairs that satisfy the equation $2y = x + 8$ are $(-8, 0)$, (4, 6), and $(-4, 2)$. Graph the equations on the same axes (Fig. 7.4). Because all six points are on the same line, the two equations represent the same line. Therefore, every ordered pair that is a solution to one equation is also a solution to the other equation. Every point on the line satisfies both equations; thus, this system has an *infinite number of solutions*. Solving the second equation for y reveals that the equations are equivalent. ▲

When a system of equations has an infinite number of solutions, as in Example 4, it is called a ***dependent system***. Note that a dependent system is also a consistent system, since it has a solution.

Figure 7.5 summarizes the three possibilities for a system of linear equations.

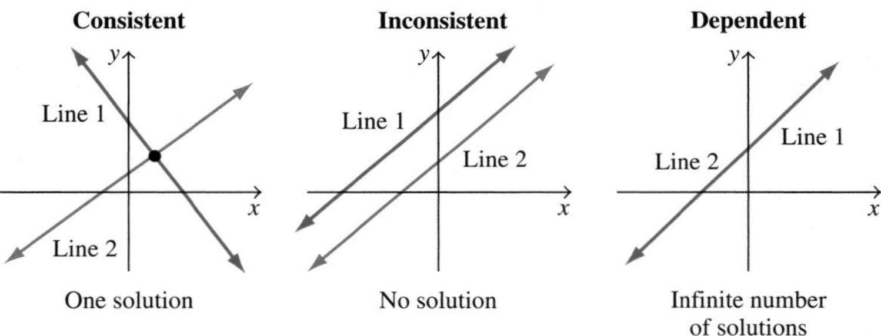

Figure 7.5

In Chapter 6 we introduced *modeling*. Recall that a *mathematical model* is an equation or system of equations that represents a real-life situation. In Examples 5 and 6 we develop equations that model a real-life situation.

EXAMPLE 5 MODELING - *A Landscape Service Application*

Tom's Tree and Landscape Service charges a consultation fee of \$200 plus \$50 per hour for labor for landscaping. Lawn Perfect Landscape Service charges a consultation fee of \$300 plus \$25 per hour for labor for landscaping.

a) Write a system of equations to represent the cost, C, of the two landscaping services, each with h hours of labor.
b) Graph both equations on the same axes and determine the number of hours needed for both services to have the same cost.
c) If the Johnsons need 7 hours of landscaping service done at their home, which service is less expensive?

SOLUTION: Let h = the number of hours of labor. The total cost of each service is the consultation fee plus the cost of the labor.

a) Tom's Tree and Landscape Service: $C = 200 + 50h$

Lawn Perfect Landscape Service: $C = 300 + 25h$

b) We graphed the cost, C, versus the number of hours of labor, h, for 0 to 10 hours (Fig. 7.6). On the graph, the lines intersect at the point (4, 400). Thus, for 4 hours of service, both services would have the same cost, \$400.

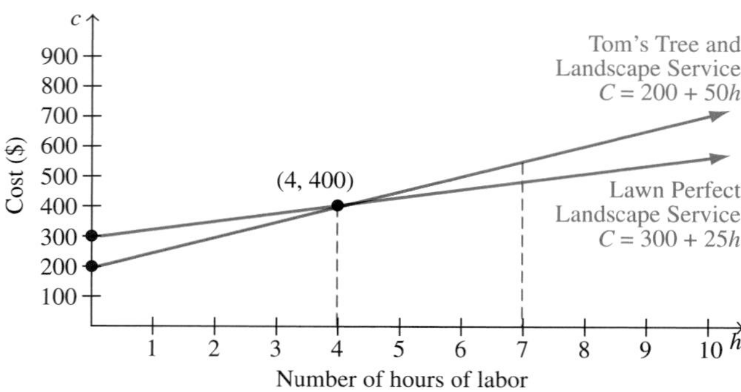

Figure 7.6

c) The graph shows that for more than 4 hours, Lawn Perfect is the least expensive service. Thus, for 7 hours, Lawn Perfect is less expensive than Tom's Tree and Landscape Service.

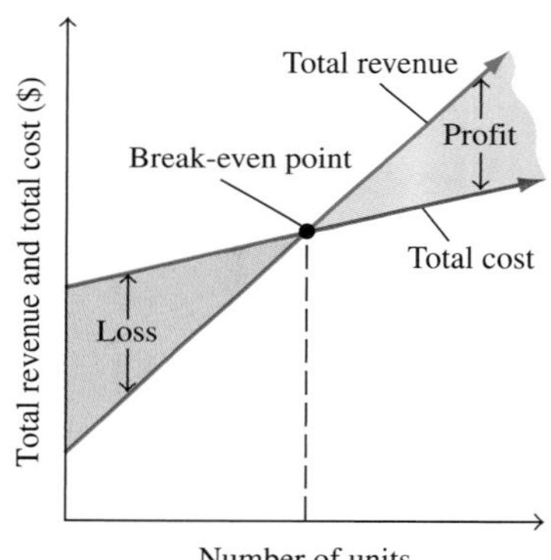

Figure 7.7

Manufacturers use a technique called ***break-even analysis*** to determine how many units of an item must be sold for the business to "break even," that is, for its total revenue to equal its total cost. Suppose we let the horizontal axis represent the number of units manufactured and sold and the vertical axis represent dollars. Then linear equations for cost, C, and revenue, R, can both be sketched on the same axes (Fig. 7.7). Both C and R are expressed in dollars and both are a function of the number of units.

Initially, the cost graph is higher than the revenue graph because of fixed (overhead) costs such as rent and utilities. During low levels of production the manufacturer suffers a loss (the cost graph is greater). During higher levels of production the manufacturer realizes a profit (the profit graph is greater). The point at which the two graphs intersect is called the ***break-even point***. At that number of units sold revenue equals cost, and the manufacturer breaks even.

EXAMPLE 6 MODELING - *Profit and Loss in Business*

At a collectibles show, Richard Lane can sell model cars for \$25. The costs for making the cars are a fixed cost of \$150 and a production cost of \$10 apiece.

a) How many model cars must Richard sell to break even?

b) Determine whether Richard makes a profit if he sells 12 model cars. What is the profit or loss?

c) How many model cars must Richard sell to make a profit of \$450?

SOLUTION:

a) Let x denote the number of model cars made and sold. The revenue is given by the equation

$$R = 25x \quad (\$25 \text{ times the number of units})$$

and the cost is given by the equation

$$C = 150 + 10x \quad (\$150 \text{ plus } \$10 \text{ times the number of units})$$

The break-even point is the point at which the revenue and cost graphs intersect. In Fig. 7.8, the graphs intersect at the point (10, 250), which is the break-even point. Thus, for Richard to break even, he must sell 10 model cars. When 10 model cars are made and sold the cost and revenue are both $250.

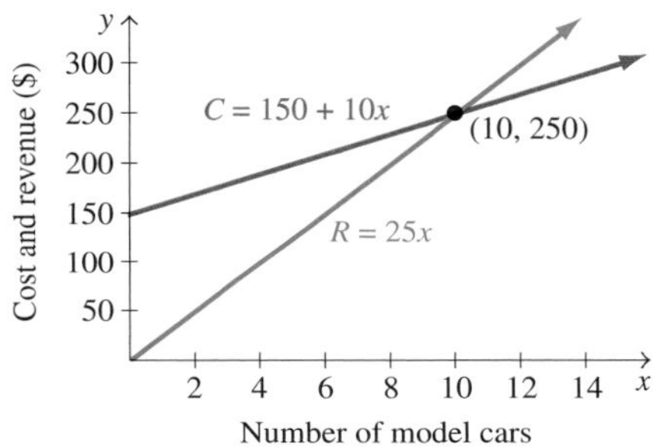

Figure 7.8

b) Examining the graph we can see that if Richard sells 12 model cars he will have a profit, P, which is the revenue minus the cost. The profit formula is

$$\begin{aligned} P &= R - C \\ &= 25x - (150 + 10x) \\ &= 25x - 150 - 10x \\ &= 15x - 150 \end{aligned}$$

Thus, for 12 model cars,

$$\begin{aligned} P &= 15x - 150 \\ &= 15(12) - 150 = 30 \end{aligned}$$

Richard has a profit of $30 if he sells 12 model cars.

c) We can determine the number of model cars that Richard must sell to have a profit of $450 by using the profit formula. Substituting 450 for P we have

$$\begin{aligned} P &= 15x - 150 \\ 450 &= 15x - 150 \\ 600 &= 15x \\ 40 &= x \end{aligned}$$

Thus, Richard must sell 40 model cars to make a profit of $450. ▲

TIMELY TIP Following is a summary of the different types of systems of linear equations.

- A *consistent system of equations* is one that has a solution.
- An *inconsistent system of equations* is one that has no solution.
- A *dependent system of equations* is one that has an infinite number of solutions.

SECTION 7.1 EXERCISES

Concept/Writing Exercises

1. What is a system of linear equations?
2. What is the solution to a system of linear equations?
3. Define a *consistent system of equations.*
4. Define a *dependent system of equations.*
5. Define an *inconsistent system of equations.*
6. Outline the procedure for solving a system of equations by graphing.
7. If a system of linear equations has no solution, what does that mean about the graphs of the equations in the system?
8. If a system of linear equations has one solution, what does that mean about the graphs of the equations in the system?
9. If a system of linear equations has an infinite number of solutions, what does that mean about the graphs of the equations in the system?
10. Can a system of linear equations have exactly two solutions? Explain.

Practice the Skills

In Exercises 11 and 12, determine which ordered pairs are solutions to the given system.

11. $y = 2x - 6$ $(3, 0)$ $(2, -2)$ $(1, 2)$
$y = -x + 3$

12. $x + 2y = 6$ $(-2, 4)$ $(2, 2)$ $(3, -9)$
$x - y = -6$

In Exercises 13–16, solve the system of equations graphically.

13. $x = 1$, $y = 4$

14. $x = -3$, $y = 3$

15. $x = 4$, $y = -3$

16. $x = -5$, $y = -3$

In Exercises 17–32, solve the system of equations graphically. If the system does not have a single ordered pair as a solution, state whether the system is inconsistent or dependent.

17. $x = 2$, $y = -x - 3$

18. $y = 1$, $y = x + 2$

19. $y = 3x - 6$, $y = -x + 6$

20. $x + y = 4$, $-x + y = 2$

21. $x + 2y = 8$, $2x - 3y = 2$

22. $3x - y = 1$, $4x - 3y = 3$

23. $2x + y = 3$, $2y = 6 - 4x$

24. $y = 2x - 4$, $2x + y = 0$

25. $y = x + 3$, $y = -1$

26. $x = 1$, $x + y + 3 = 0$

27. $2x - y = -3$, $2x + y = -9$

28. $3x + 2y = 6$, $6x + 4y = 12$

29. $2x - 3y = 12$, $3y - 2x = 9$

30. $y = \frac{1}{3}x - 4$, $3y - x = 4$

31. $y = -\frac{1}{3}x + 2$, $2x - 2y = 4$

32. $2(x - 1) + 2y = 0$, $3x + 2(y + 2) = 0$

33. **a)** If the two lines in a system of equations have different slopes, how many solutions will the system have? Explain your answer.
b) If the two lines in a system of equations have the same slope but different y-intercepts, how many solutions will the system have? Explain.
c) If the two lines in a system of equations have the same slope and the same y-intercept, how many solutions will the system have? Explain.

34. Indicate whether the graph shown represents a consistent, inconsistent, or dependent system. Explain your answer.

a)

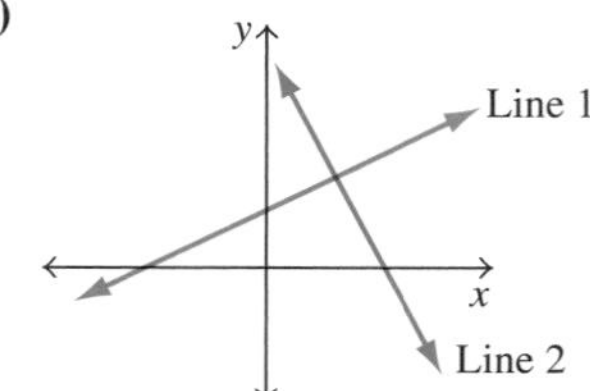

b)

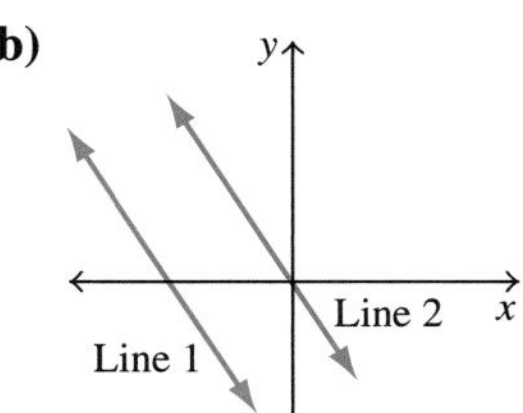

c)

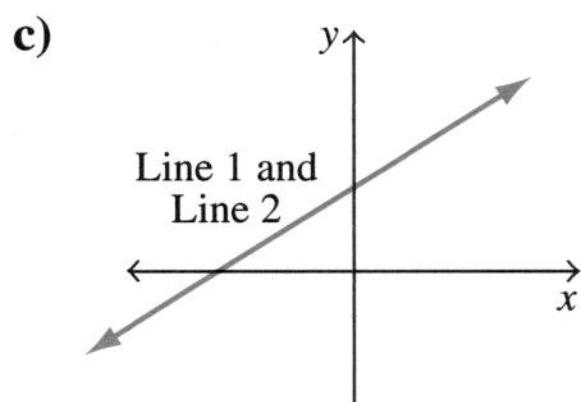

In Exercises 35–46, determine without graphing whether the system of equations has exactly one solution, no solution, or an infinite number of solutions. (Consider your answers to Exercise 33.)

35. $2x - y = 6$
$y = 2x - 6$

36. $3x + 4y = 8$
$8y = -6x + 4$

37. $3x - 4y = 5$
$x - 8y = 10$

38. $x + 3y = 6$
$3x + y = 4$

39. $3x + y = 7$
$y = -3x + 9$

40. $x + 4y = 12$
$x = 4y + 3$

41. $2x - 3y = 6$
$x - \frac{3}{2}y = 3$

42. $x - 2y = 6$
$x + 2y = 4$

43. $3x = 6y + 5$
$y = \frac{1}{2}x - 3$

44. $3y = 6x + 4$
$-2x + y = \frac{4}{3}$

45. $12x - 5y = 4$
$3x + 4y = 6$

46. $4x + 7y = 2$
$4x = 6 + 7y$

Problem Solving

Two lines are perpendicular when they meet at a right angle (a 90° angle). Two lines are perpendicular to each other when their slopes are negative reciprocals of each other. The negative reciprocal of 2 is $-\frac{1}{2}$, the negative reciprocal of $\frac{3}{5}$ is $-1/(\frac{3}{5})$ or $-\frac{5}{3}$, and so on. If a *represents any real number, except 0, its negative reciprocal is $-1/a$. Note that the product of a number and its negative reciprocal is -1. In Exercises 47–50, determine, by finding the slope of each line, whether the lines will be perpendicular to each other when graphed.*

47. $5y - 2x = 15$
$2y - 5x = 2$

48. $4y - x = 6$
$y = x + 8$

49. $2x + y = 3$
$2y - x = 5$

50. $6x + 5y = 3$
$-10x = 2 + 12y$

In Exercises 51–55, part of the question involves determining a system of equations that models the situation.

51. MODELING - *Landscaping Revisited* In Example 5, assume that Tom's Tree and Landscape Service charges \$200 for a consultation fee plus \$60 per hour for labor and that Lawn Perfect Landscape Service charges \$305 for a consultation fee plus \$25 per hour for labor.

a) Write the system of equations to represent the cost of the two landscaping services.
b) Graph both equations for 0 to 10 hours on the same axes.
c) Determine the number of hours of landscaping that must be used for both services to have the same cost.

52. MODELING - *Security Systems* Tamika Dixon plans to install a security system in her house. She is considering two security companies: ABC Security and SafeHomes Security. ABC's system costs \$3380 to install and their monitoring fee is \$18 per month. SafeHomes equivalent system costs only \$2302 to install but their monitoring fee is \$29 per month.
a) Write a system of equations to represent the cost of each system.
b) Graph both equations (for up to and including 180 months) on the same axes.
c) Determine the number of months the service must be used for both companies to have the same cost.
d) If both companies guarantee not to raise monthly fees for 10 years, and if Tamika plans to use the system for 10 years, which system would be less expensive?

53. MODELING - *Selling Backpacks* Benjamin's Backpacks can sell backpacks for \$25 per backpack. The costs for making the backpacks are a fixed cost of \$400 and a production cost of \$15 per backpack (see Example 6 for an example of cost and revenue equations).
a) Write the cost and revenue equations.

b) Graph both equations, for 0 to 50 backpacks, on the same axes.
c) How many backpacks must Benjamin's Backpacks sell to break even?
d) Write the profit formula.
e) Determine whether Benjamin's Backpacks makes a profit or loss if it sells 30 backpacks. What is the profit or loss?
f) How many backpacks must Benjamin's Backpacks sell to realize a profit of $1000?

54. **MODELING** - *Purchasing Stocks* When buying or selling stock for a customer, the Mark Demo Agency charges $40 plus 8 cents per share of stock purchased or sold. Andy Harris and Associates charges $15 plus 18 cents per share of stock purchased or sold.
a) Write a system of equations to represent the cost of purchasing or selling stock with each company.
b) Graph both equations (for up to and including 350 shares of stock) on the same axes.
c) Determine the number of shares of stock that must be purchased or sold for the total cost to be the same.
d) If 300 shares of stock are to be purchased, which firm would be less expensive?

55. **MODELING** - *Manufacturing DVD Players* A manufacturer sells a certain DVD player for $225 per unit. Manufacturing costs consist of a fixed cost of $8400 and a production cost of $155 per unit.
a) Write the cost and revenue equations.
b) Graph both equations (for up to and including 150 units) on the same axes.
c) How many units must the manufacturer sell to break even?
d) Write the profit formula.
e) What is the manufacturer's profit or loss if 100 units are sold?
f) How many units must the manufacturer sell to make a profit of $1260?

56. Explain how you can determine whether a system of two linear equations will be consistent, dependent, or inconsistent without graphing the equations.

Challenge Problems/Group Activities

57. **MODELING** - *Job Offers* Hubert Hotchkiss had two job offers for sales positions. One pays a salary of $300 per week plus a 15% commission on his dollar sales volume. The second position pays a salary of $450 per week with no commission.
a) For each offer, write an equation that expresses the weekly pay.
b) Graph the system of equations and determine the solution.
c) For what dollar sales volume will the two offers result in the same pay?

58. **MODELING**- *Long Distance Calling*
a) In March 2002, an AT&T One Rate Plan charged 7 cents per minute for long-distance calls with a monthly fee of $3.95. The Sprint Nickel Anytime Plan charged 5 cents per minute for long-distance calls with a monthly fee of $8.95. Write an equation to determine the monthly cost for long-distance service with the AT&T One Rate Plan, and write an equation to determine the monthly cost for long-distance service with the Sprint Nickel Anytime Plan.
b) Graph the system of equations and determine the solution.
c) After how many minutes will the cost for the two long-distance service plans be the same?

59. *Points of Intersection* **a)** If two lines have different slopes, what is the maximum possible number of points of intersection?
b) If three lines all have different slopes, what is the maximum possible number of points of intersection?
c) If four lines all have different slopes, what is the maximum possible number of points of intersection?
d) If five lines all have different slopes, what is the maximum possible number of points of intersection?
e) Is there a pattern in the number of points of intersection? If so, explain the pattern. Use the pattern to determine the maximum possible number of points of intersection for six lines.

Recreational Mathematics

60. Connect all the following points using exactly four straight line segments. Do not lift your pencil off the paper.

•   •   •

•   •   •

•   •   •

Internet/Research Activity

61. The Rhind Papyrus indicates that the early Egyptians used linear equations. Do research and write a paper on the symbols used in linear equations and the use of the linear equations by the early Egyptians. (References include history of mathematics books, encyclopedias, and the Internet.)

7.2 SOLVING SYSTEMS OF EQUATIONS BY THE SUBSTITUTION AND ADDITION METHODS

Having solved systems of equations by graphing in Section 7.1, we are now ready to learn two other methods used to solve systems of linear equations: the substitution method and the addition method. We now discuss the substitution method.

Substitution Method

Procedure for Solving a System of Equations Using the Substitution Method

1. Solve one of the equations for one of the variables. If possible, solve for a variable with a numerical coefficient of 1. By doing so, you may avoid working with fractions.
2. Substitute the expression found in step 1 into the other equation. This step yields an equation in terms of a single variable.
3. Solve the equation found in step 2 for the variable.
4. Substitute the value found in step 3 into the equation you rewrote in step 1 and solve for the remaining variable.

Examples 1, 2, and 3 illustrate the *substitution method*. These systems of equations are the same as in Examples 2, 3, and 4 in Section 7.1.

EXAMPLE 1 *A Single Solution, by the Substitution Method*

Solve the following system of equations by substitution.

$$x + y = 4$$
$$2x - y = -1$$

SOLUTION: The numerical coefficients of the x and y terms in the equation $x + y = 4$ are both 1. Thus, we can solve this equation for either x or y. Let's solve for x in the first equation.

STEP 1.

$$x + y = 4$$
$$x + y - y = 4 - y \quad \text{Subtract } y \text{ from both sides of the equation.}$$
$$x = 4 - y$$

STEP 2. Substitute $4 - y$ for x in the other equation.

$$2x - y = -1$$
$$2(4 - y) - y = -1$$

STEP 3. Now solve the equation for y.

$$8 - 2y - y = -1 \quad \text{Distributive property}$$
$$8 - 3y = -1$$
$$8 - 8 - 3y = -1 - 8 \quad \text{Subtract 8 from both sides of the equation.}$$
$$-3y = -9$$
$$\frac{-3y}{-3} = \frac{-9}{-3} \quad \text{Divide both sides of the equation by } -3.$$
$$y = 3$$

STEP 4. Substitute $y = 3$ in the equation solved for x and determine the value of x.

$$x = 4 - y$$
$$x = 4 - 3$$
$$x = 1$$

Thus, the solution is the ordered pair (1, 3). This answer checks with the solution obtained graphically in Section 7.1, Example 2. ▲

TIMELY TIP When solving a system of equations, once you successfully solve for one of the variables, make sure you solve for the other variable. Remember that a solution to a system of equations must contain a numerical value for each variable in the system.

EXAMPLE 2 *No Solution, by the Substitution Method*

Solve the following system of equations by substitution.

$$2x + y = 3$$
$$2x + y + 5 = 0$$

SOLUTION: Solve for y in the first equation.

$$2x + y = 3$$
$$2x - 2x + y = 3 - 2x \quad \text{Subtract } 2x \text{ from both sides of the equation.}$$
$$y = 3 - 2x$$

Now substitute $3 - 2x$ in place of y in the second equation.

$$2x + y + 5 = 0$$
$$2x + (3 - 2x) + 5 = 0$$
$$2x + 3 - 2x + 5 = 0$$
$$8 = 0 \quad \text{False}$$

Since 8 cannot be equal to 0, there is no solution to the system of equations. Thus, the system of equations is inconsistent. This answer checks with the solution obtained graphically in Section 7.1, Example 3. ▲

When solving Example 2, we obtained $8 = 0$ and indicated that the system was inconsistent and that there was no solution. When solving a system of equations, if you obtain a false statement, such as $4 = 0$ or $-2 = 0$, the system is *inconsistent* and has *no solution.*

EXAMPLE 3 *An Infinite Number of Solutions, by the Substitution Method*

Solve the following system of equations by substitution.

$$y = \frac{1}{2}x + 4$$
$$2y = x + 8$$

SOLUTION: The first equation $y = \frac{1}{2}x + 4$ is already solved for y, so we will substitute $\frac{1}{2}x + 4$ for y in the second equation.

$$2y = x + 8$$
$$2\left(\frac{1}{2}x + 4\right) = x + 8$$
$$x + 8 = x + 8 \quad \text{Distributive property}$$
$$x - x + 8 = x - x + 8 \quad \text{Add 8 to both sides of equation}$$
$$8 = 8 \quad \text{True}$$

Since 8 equals 8, the system has an infinite number of solutions. Thus, the system of equations is dependent. This answer checks with the solution obtained in Section 7.1, Example 4.

When solving Example 3, we obtained $8 = 8$ and indicated that the system was dependent and had an infinite number of solutions. When solving a system of equations, if you obtain a true statement, such as $0 = 0$ or $8 = 8$, the system is *dependent* and has an *infinite number of solutions.*

Addition Method

If neither of the equations in a system of linear equations has a variable with a coefficient of 1, it is generally easier to solve the system by using the ***addition*** (or ***elimination***) ***method***.

To solve a system of linear equations by the addition method, it is necessary to obtain two equations whose sum will be a single equation containing only one variable. To achieve this goal, we rewrite the system of equations as two equations where the coefficients of one of the variables are opposites (or additive inverses) of each other. For example, if one equation has a term of $2x$, we might rewrite the other equation so that its x term will be $-2x$. To obtain the desired equations, it might be necessary to multiply one or both equations in the original system by a number. When an equation is to be multiplied by a number, we will place brackets around the equation and place the number that is to multiply the equation before the brackets. For example, $4[2x + 3y = 6]$ means that each term on both sides of the equal sign in the equation $2x + 3y = 6$ is to be multiplied by 4:

$$4[2x + 3y = 6] \quad \text{gives} \quad 8x + 12y = 24.$$

This notation will make our explanations much more efficient and easier for you to follow.

DID YOU KNOW

How to Succeed in Business

Economics, a science dependent on mathematics, dates back to just before the Industrial Revolution of the eighteenth century. Technologies were being invented and applied to the manufacture of cloth, iron, transportation, and agriculture. These new technologies led to the development of mathematically based economic models. French economist Jules Dupuit (1804–1866) suggested a method to calculate the value of railroad bridges; Irish economist Dionysis Larder (1793–1859) showed railroad companies how to structure their rates so as to increase their profits.

Procedure for Solving a System of Equations by the Addition Method

1. If necessary, rewrite the equations so that the variables appear on one side of the equal sign and the constants appear on the other side of the equal sign.
2. If necessary, multiply one or both equations by a constant(s) so that when you add the equations, the result will be an equation containing only one variable.
3. Add the equations to obtain a single equation in one variable.
4. Solve the equation in step 3 for the variable.
5. Substitute the value found in step 4 into either of the original equations and solve for the other variable.

EXAMPLE 4 *Eliminating a Variable by the Addition Method*

Solve the following system of equations by the addition method.

$$\begin{aligned} x + y &= 5 \\ 2x - y &= 7 \end{aligned}$$

SOLUTION: Since the coefficients of the y terms, 1 and -1, are additive inverses, the sum of the y terms will be zero when the equations are added. Thus, the sum of the two equations will contain only one variable, x. Add the two equations to obtain one equation in one variable. Then solve for the remaining variable.

$$\begin{aligned} x + y &= 5 \\ \underline{2x - y} &\underline{= 7} \\ 3x \quad\;\; &= 12 \\ x &= 4 \end{aligned}$$

Now substitute 4 for x in either of the original equations to find the value of y.

$$\begin{aligned} x + y &= 5 \\ 4 + y &= 5 \\ y &= 1 \end{aligned}$$

The solution to the system is (4, 1). ▲

EXAMPLE 5 *Multiplying by -1 in the Addition Method*

Solve the following system of equations by the addition method.

$$\begin{aligned} x + 3y &= 9 \\ x + 2y &= 5 \end{aligned}$$

SOLUTION: We want the sum of the two equations to have only one variable. We can eliminate the variable x by multiplying either equation by -1 and then adding. We will multiply the first equation by -1.

$$\begin{array}{rcr} -1[x + 3y = 9] & \text{gives} & -x - 3y = -9 \\ x + 2y = 5 & & x + 2y = 5 \end{array}$$

We now have a system of equations equivalent to the original system.

Now add the two equations.

$$\begin{aligned} -x - 3y &= -9 \\ x + 2y &= 5 \\ \hline -y &= -4 \\ y &= 4 \end{aligned}$$

Now we solve for x by substituting 4 for y in either of the original equations.

$$\begin{aligned} x + 3y &= 9 \\ x + 3(4) &= 9 \\ x + 12 &= 9 \\ x &= -3 \end{aligned}$$

The solution is $(-3, 4)$. ▲

EXAMPLE 6 ***Multiplying One Equation in the Addition Method***

Solve the following system of equations by the addition method.

$$\begin{aligned} 4x + y &= 6 \\ 3x + 2y &= 7 \end{aligned}$$

SOLUTION: We can multiply the top equation by -2 and then add the two equations to eliminate the variable y.

$$\begin{aligned} -2[4x + y = 6] \quad &\text{gives} \quad -8x - 2y = -12 \\ 3x + 2y = 7 \quad &\qquad\quad\;\; 3x + 2y = 7 \end{aligned}$$

$$\begin{aligned} -8x - 2y &= -12 \\ 3x + 2y &= 7 \\ \hline -5x \qquad &= -5 \\ x &= 1 \end{aligned}$$

Now we find y by substituting 1 for x in either of the original equations.

$$\begin{aligned} 4x + y &= 6 \\ 4(1) + y &= 6 \\ 4 + y &= 6 \\ y &= 2 \end{aligned}$$

The solution is $(1, 2)$. ▲

Note that in Example 6 we could have eliminated the variable x by multiplying the top equation by 3 and the bottom equation by -4, then adding. Try this method now.

EXAMPLE 7 ***Multiplying Both Equations***

Solve the following system of equations by the addition method.

$$3x - 4y = 8$$
$$2x + 3y = 9$$

SOLUTION: In this system, we cannot eliminate a variable by multiplying only one equation by an integer value and then adding. To eliminate a variable, we can multiply each equation by a different number. To eliminate the variable x, we can multiply the top equation by 2 and the bottom by -3 (or the top by -2 and the bottom by 3) and then add the two equations. If we want, we can instead eliminate the variable y by multiplying the top equation by 3 and the bottom by 4 and then adding the two equations. Let's eliminate the variable x.

$$2[3x - 4y = 8] \quad \text{gives} \quad 6x - 8y = 16$$
$$-3[2x + 3y = 9] \quad \text{gives} \quad -6x - 9y = -27$$

$$\begin{array}{r} 6x - 8y = 16 \\ \underline{-6x - 9y = -27} \\ -17y = -11 \\ y = \dfrac{11}{17} \end{array}$$

We could now find x by substituting $\frac{11}{17}$ for y in either of the original equations. Although it can be done, it gets messy. Instead, let's solve for x by eliminating the variable y from the two original equations. To do so, we multiply the first equation by 3 and the second equation by 4.

$$3[3x - 4y = 8] \quad \text{gives} \quad 9x - 12y = 24$$
$$4[2x + 3y = 9] \quad \text{gives} \quad 8x + 12y = 36$$

$$\begin{array}{r} 9x - 12y = 24 \\ \underline{8x + 12y = 36} \\ 17x \qquad = 60 \\ x = \dfrac{60}{17} \end{array}$$

The solution to the system is $(\frac{60}{17}, \frac{11}{17})$. ▲

When solving a system of linear equations by either the substitution or the addition method, if you obtain the equation $0 = 0$ it indicates that the system is *dependent* (both equations represent the same line; see Fig. 7.4 on page 385), and there are an infinite number of solutions. When solving, if you obtain an equation such as $0 = 6$, or any other equation that is false, it means that the system is *inconsistent* (the two equations represent parallel lines; see Fig. 7.5 on page 385), and there is no solution.

EXAMPLE 8 **MODELING** - *When Are Repair Costs the Same?*

Melinda Melendez needs to purchase a new radiator for her car and have it installed by a mechanic. She is considering two garages: Steve's Repair and Greg's Garage.

At Steve's Repair, the parts cost \$200 and the labor cost is \$50 per hour. At Greg's Garage, the parts cost \$375 and the labor cost is \$25 per hour. How many hours would the repair need to take for the total cost at each garage to be the same?

SOLUTION: We are asked to find the number of hours the repair would need to take for each garage to have the same total cost, C. First write a system of equations to represent the total cost for each of the garages. The total cost consists of the cost of the parts and the labor cost. The labor cost depends on the number of hours of labor.

Let x = the number of hours of labor.

$$\text{Total cost} = \text{cost of parts} + \text{labor cost}$$
$$\text{Steve's Repair: } C = 200 + 50x$$
$$\text{Greg's Garage: } C = 375 + 25x$$

We want to determine when the cost will be the same, so we set the two costs equal to each other (substitution method) and solve the resulting equation.

$$200 + 50x = 375 + 25x$$
$$200 - 200 + 50x = 375 - 200 + 25x \quad \text{Subtract 200 from both sides of the equation.}$$
$$50x = 175 + 25x$$
$$50x - 25x = 175 + 25x - 25x \quad \text{Subtract } 25x \text{ from both sides of the equation.}$$
$$25x = 175$$
$$\frac{25x}{25} = \frac{175}{25} \quad \text{Divide both sides of the equation by 25.}$$
$$x = 7$$

Thus, for 7 hours of labor, the cost at both garages would be the same. If we construct a graph (Fig. 7.9) of the two cost equations, the point of intersection is (7, 550). If the repair were to require 7 hours of labor, the total cost at either garage would be \$550.

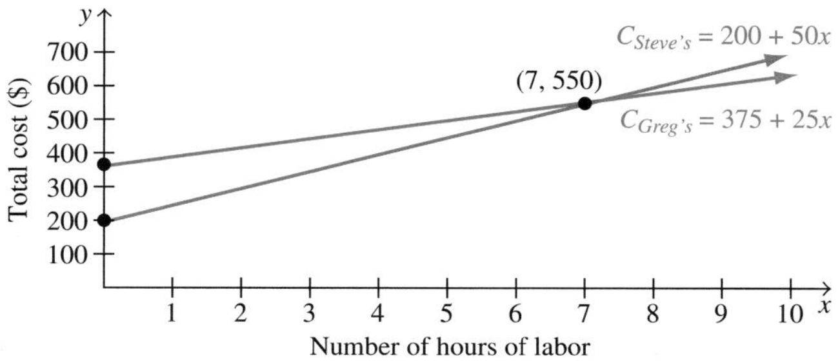

Figure 7.9

EXAMPLE 9 MODELING - *A Mixture Problem*

Karen Guardino, a pharmacist, needs 500 milliliters (mℓ) of a 10% phenobarbital solution. She has only a 5% solution and a 25% solution available. How many milliliters of each solution should she mix to obtain the desired solution?

SOLUTION: First we set up a system of equations. The unknown quantities are the amount of 5% solution and the amount of the 25% solution that must be used. Let

$$x = \text{number of m}\ell\text{ of 5\% solution}$$
$$y = \text{number of m}\ell\text{ of 25\% solution}$$

We know that 500 mℓ of solution are needed. Thus,

$$x + y = 500$$

The total amount of phenobarbital in a solution is determined by multiplying the percent of phenobarbital by the number of milliliters of solution. The second equation comes from the fact that

$$\begin{pmatrix}\text{Total amount of}\\ \text{phenobarbital in}\\ \text{5\% solution}\end{pmatrix} + \begin{pmatrix}\text{total amount of}\\ \text{phenobarbital in}\\ \text{25\% solution}\end{pmatrix} = \begin{pmatrix}\text{total amount of}\\ \text{phenobarbital}\\ \text{in 10\% mixture}\end{pmatrix}$$
$$0.05x \quad + \quad 0.25y \quad = \quad 0.10(500)$$
$$\text{or} \qquad 0.05x + 0.25y = 50$$

The system of equations is

$$x + y = 500$$
$$0.05x + 0.25y = 50$$

Let's solve this system of equations by using the addition method. There are various ways of eliminating one variable. To obtain integer values in the second equation, we can multiply both sides of the equation by 100. That will result in an x-term of $5x$. If we multiply both sides of the first equation by -5, that will result in an x-term of $-5x$. By following this process, we can eliminate the x-terms from the system.

$$-5[x + y = 500] \quad \text{gives} \quad -5x - 5y = -2500$$
$$100[0.05x + 0.25y = 50] \quad \text{gives} \quad 5x + 25y = 5000$$

$$\begin{aligned} -5x - 5y &= -2500 \\ 5x + 25y &= 5000 \\ \hline 20y &= 2500 \\ \frac{20y}{20} &= \frac{2500}{20} \\ y &= 125 \end{aligned}$$

Now we determine x.

$$\begin{aligned} x + y &= 500 \\ x + 125 &= 500 \\ x &= 375 \end{aligned}$$

Therefore, 375 mℓ of a 5% phenobarbital solution must be mixed with 125 mℓ of a 25% phenobarbital solution to obtain 500 mℓ of a 10% phenobarbital solution. ▲

Example 9 can also be solved by using substitution. Try to do so now.

SECTION 7.2 EXERCISES

Concept/Writing Exercises

1. In your own words, explain how to solve a system of linear equations by using the addition method.
2. In your own words, explain how to solve a system of linear equations by using the substitution method.
3. How will you know, when solving a system of linear equations by either the substitution or the addition method, whether the system is dependent?
4. How will you know, when solving a system of linear equations by either the substitution or the addition method, whether the system is inconsistent?
5. When solving the following system of equations by the substitution method, which variable, in which equation, would you choose to solve for in order to make the solution easier? Explain your answer. Do not solve the system.

$$x + 3y = 3$$
$$3x + 4y = -1$$

6. When solving the following system of equations by the addition method, what will your first step be in solving the system? Explain your answer. Do not solve the system.

$$2x + y = 6$$
$$3x + 3y = 9$$

Practice the Skills

In Exercises 7–24, solve the system of equations by the substitution method. If the system does not have a single ordered pair as a solution, state whether the system is inconsistent or dependent.

7. $y = x + 7$, $y = -x + 5$
8. $y = 3x + 7$, $y = -2x - 3$
9. $2x + 4y = 8$, $2x - y = -2$
10. $y + 3x = 7$, $2x + 3y = 14$
11. $y - x = 4$, $x - y = 3$
12. $x + y = 3$, $y + x = 5$
13. $3y + 2x = 4$, $3y = 6 - x$
14. $x = 5y - 12$, $x - y = 0$
15. $y - 2x = 3$, $2y = 4x + 6$
16. $y = 2$, $y + x + 3 = 0$
17. $x = y + 3$, $x = -3$
18. $x + 2y = 6$, $y = 2x + 3$
19. $y + 3x - 4 = 0$, $2x - y = 7$
20. $x + 4y = 7$, $2x + 3y = 5$
21. $x = 2y + 3$, $y = 3x - 1$
22. $x + 4y = 9$, $2x - y - 6 = 0$
23. $y = -2x + 3$, $4x + 2y = 12$
24. $2x + y = 12$, $x = -\frac{1}{2}y + 6$

In Exercises 25–40, solve the system of equations by the addition method. If the system does not have a single ordered pair as a solution, state whether the system is inconsistent or dependent.

25. $3x + y = 10$, $4x - y = 4$
26. $x + 2y = 9$, $x - 2y = -3$
27. $x + y = 10$, $x - 2y = -2$
28. $3x + y = 10$, $-3x + 2y = -16$
29. $2x - y = -4$, $-3x - y = 6$
30. $x + y = 6$, $-2x + y = -3$
31. $4x + 3y = -1$, $2x - y = -13$
32. $2x + y = 6$, $3x + y = 5$
33. $2x + y = 11$, $x + 3y = 18$
34. $5x - 2y = 11$, $-3x + 2y = 1$
35. $3x - 4y = 11$, $3x + 5y = -7$
36. $4x - 2y = 6$, $4y = 8x - 12$
37. $4x + y = 6$, $-8x - 2y = 13$
38. $2x + 3y = 6$, $5x - 4y = -8$
39. $3x - 4y = 10$, $5x + 3y = 7$
40. $6x + 3y = 7$, $5x + 2y = 9$

Problem Solving

In Exercises 41–52, write a system of equations that can be used to solve the problem. Then solve the system and determine the answer.

41. **MODELING** - *Owning a Business* Sosena Milion can join a small business as a full partner and receive a salary of \$12,000 per year plus 15% of the year's profit, or she can join as a sales manager with a salary of \$27,000 per year plus 5% of the year's profit. What must the year's profit be for her total earnings to be the same whether she joins as a full partner or as a sales manager?
42. **MODELING** - *Mortgage Refinancing* In January 2003, mortgage rates were very low so Wayne Morganstein considered refinancing his mortgage. The cost of refinancing his mortgage would include a one-time charge of \$1600. With the reduced mortgage rate, his monthly interest and principal payments would be \$780. At the higher interest

rate he currently has, his interest and principal payments are \$980 per month.

a) Determine how many months it will take until both mortgage plans would have the same total cost.

b) If Wayne plans to remain in his house for exactly 6 years, which mortgage plan would result in a lower total cost?

43. **MODELING** - *Pizza Orders* Pizza Corner sells medium and large specialty pizzas. A medium Meat Lovers pizza costs \$10.95, and a large Meat Lovers pizza costs \$14.95. One Saturday a total of 50 Meat Lovers pizzas were sold, and the receipts from the Meat Lovers pizzas were \$663.50. How many medium and how many large Meat Lovers pizzas were sold?

44. **MODELING** - *Basketball Game* The University of Tennessee women's basketball team made 45 field goals in a recent game; some were 2-pointers and some were 3-pointers. How many 2-point baskets were made and how many 3-point baskets were made if Tennessee scored 101 points?

45. **MODELING** - *Chemical Mixture* Antonio Gonzalez is a chemist and needs 10 liters (ℓ) of a 40% hydrochloric acid solution. He discovers he is out of the 40% hydrochloric acid solution and does not have sufficient time to reorder. He checks his supply shelf and finds he has a large supply of both 25% and 50% hydrochloric acid solutions. He decides to use the 25% and 50% solutions to make 10 ℓ of a 40% solution. How many liters of the 25% solution and of the 50% solution should he mix?

46. **MODELING** - *A Milk Mixture* The Guidas own a dairy. They have milk that is 5% butterfat and skim milk without butterfat. How much of the 5% milk and how much of the skim milk should they mix to make 100 gal of milk that is 3.5% butterfat?

47. **MODELING** - *Choosing a Copy Service* Lori Lanier recently purchased a high-speed copier for her home office and wants to purchase a service contract on the copier. She is considering two sources for the contract. The Economy Sales and Service Company charges \$18 a month plus 2 cents per copy. Office Superstore charges \$24 a month but only 1.5 cents per copy. How many copies would Lori need to make for the monthly costs of both plans to be the same?

48. **MODELING** - *Cellular Phone Plans* Rich Gratien is considering two cellular phone plans. Both plans offer 300 free minutes each month. Cingular Home 300 Plan charges \$30 per month plus 45 cents for each additional minute after 300 minutes. Verizon America's Choice Plan charges \$35 per month plus 20 cents for each additional minute after 300 minutes.

a) In addition to the 300 free minutes, how long would Rich have to talk on the phone, in a month, for the two plans to have the same total cost?

b) If Rich talks for 350 minutes a month, which plan would be less expensive for him?

49. **MODELING** - *Nut and Pretzel Mix* Dave Chwalik wants to purchase 20 pounds of party mix for a total of \$30. To obtain the mixture, he will mix nuts that cost \$3 per pound with pretzels that cost \$1 per pound. How many pounds of each type of mix should he use?

50. **MODELING** - *Laboratory Research* Animals in an experiment are to be kept on a strict diet. Each animal is to receive, among other things, 20 g of protein and 6 g of carbohydrates. The scientist has only two food mixes of the following compositions available.

	Protein (%)	Carbohydrates (%)
Mix A	10	6
Mix *B*	20	2

How many grams of each mix should she use to obtain the right diet for a single animal?

51. **MODELING** - *School Play Tickets* Jefferson High School sold 250 tickets to its annual school play. Student tickets cost \$2 per ticket and nonstudent tickets cost \$5 per ticket. If \$950 in ticket sales is collected, how many tickets of each type were sold?

52. **MODELING** - *Golf Club Membership* Membership in Oakwood Country Club costs \$3000 per year and entitles a member to play a round of golf for greens fee of \$18. At Pinecrest Country Club, membership costs \$2500 per year and the greens fee is \$20.

a) How many rounds must a golfer play in a year for the costs at the two clubs to be the same?

b) If Sally Sestini planned to play 30 rounds of golf in a year, which club would be the least expensive?

53. **MODELING** - *College Applications* As the following graph shows, from 1981 through 2000 the percentage of high school students who applied to exactly one college decreased while the percentage of high school students who applied to exactly five colleges increased. The percentage of high school students who applied to one college (the blue curve) can be approximated by the linear equation $y = -0.58x + 31$, where x is the number of years since 1981. The percentage of high school students who applied to five colleges (the red curve) can be approximated by the linear equation $y = 0.32x + 7$. Assuming the present trend continues, use the substitution method to approximate when the percentage of high school students applying to one college will equal the percentage of high school students applying to five colleges.

Percentage of Students Who Applied to:

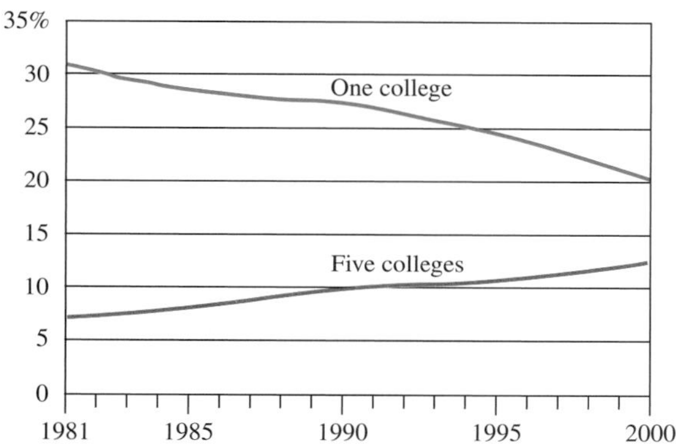

Source: *U.S. News and World Report*

54. **MODELING** - *Drinking Coffee* The following graph shows how coffee consumption compared with consumption of other beverages from 1992 through 2000. The number of gallons of coffee consumed per capita per year (the blue curve) can be approximated by the linear equation $y = -1.13x + 27$, where x is the number of years since 1992. The number of gallons of bottled water consumed per capita per year (the red curve) can be approximated by the linear equation $y = 0.38x + 9$. Assuming the present trend continues, use the substitution method to approximate when the number of gallons per capita of coffee consumed will equal the number of gallons per capita of bottled water consumed.

Beverage Consumption in the U.S.

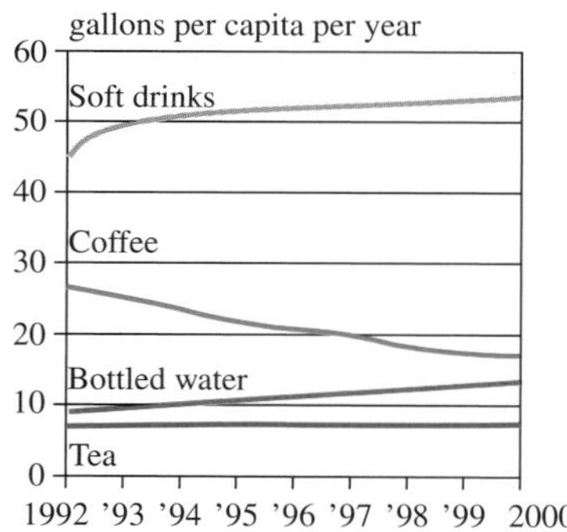

Source: U.S. Department of Agriculture

Challenge Problems/Group Activities

55. Solve the following system of equations for u and v by first substituting x for $\frac{1}{u}$ and y for $\frac{1}{v}$.

$$\frac{1}{u} + \frac{2}{v} = 8$$
$$\frac{3}{u} - \frac{1}{v} = 3$$

56. Develop a system of equations that has (6, 5) as its solution. Explain how you developed your system of equations.

57. The substitution or addition methods can also be used to solve a system of three equations in three variables. Consider the following system.

$$\begin{aligned} x + y + z &= 7 \\ x - y + 2z &= 9 \\ -x + 2y + z &= 4 \end{aligned}$$

The *ordered triple* (x, y, z) is the solution to the system if it satisfies all three equations.

a) Show that the ordered triple (2, 1, 4) is a solution to the system.

b) Use the substitution or addition method to determine the solution to the system. (*Hint:* Eliminate one variable by using two equations. Then eliminate the same variable by using two different equations.)

58. Construct a system of two equations that has no solution. Explain how you know the system has no solution.

59. Construct a system of two equations that has an infinite number of solutions. Explain how you know the system has an infinite number of solutions.

Recreational Mathematics

60. In parts a)–d) make up a system of linear equations whose solution will be the ordered pair given. *Hint:* It may be helpful to visualize possible graphs that have the given solution. There are many possible answers for each part.
a) (0, 0) **b)** (1, 0) **c)** (0, 1) **d)** (1, 1)

7.3 MATRICES

DID YOU KNOW

Matrices Are Everywhere

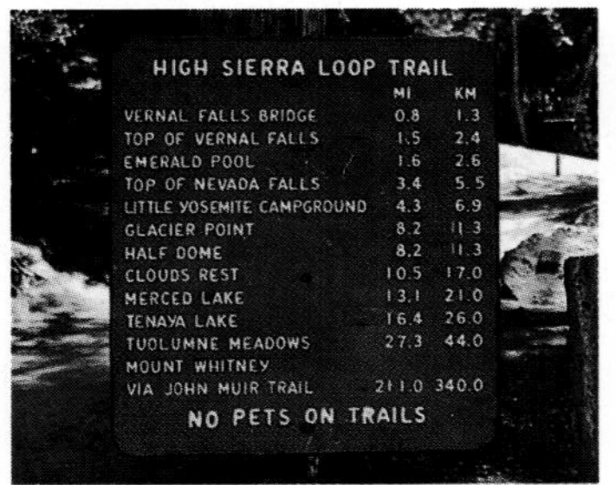

You are already familiar with matrices, although you may not be aware of it. The matrix is a good way to display numerical data, as illustrated on this trail sign in Yosemite National Park.

We have discussed solving systems of equations by graphing, using substitution, and using the addition method. In Section 7.4, we will discuss solving systems of linear equations by using matrices. So that you will become familiar with matrices, in this section, we explain how to add, subtract, and multiply matrices. We also explain how to multiply a matrix by a real number. Matrix techniques are easily adapted to computers.

A ***matrix*** is a rectangular array of elements. An array is a systematic arrangement of numbers or symbols in rows and columns. Matrices (the plural of matrix) may be used to display information and to solve systems of linear equations. The following matrix displays the responses from a survey of 500 students at Morgan State University. The students were asked if they were in favor of an increase in their student fees to pay for building a new meeting room for student clubs.

Rows ↓ / Columns →	Freshmen	Sophomores	Juniors	Seniors
In favor	102	93	22	35
Opposed	82	94	23	49

The numbers in the rows and columns of a matrix are called the ***elements*** of the matrix. The matrix given above contains eight elements. The ***dimensions*** of a matrix may be indicated with the notation $r \times s$, where r is the number of rows and s is the number of columns in the matrix. Because the matrix given above has 2 rows and 4 columns, it is a 2 by 4, written 2×4, matrix. In this text, from this point onward, we use brackets, [ ], to indicate a matrix. Consider the two matrices below. A matrix that contains the same number of rows and columns is called a ***square matrix***. Following is an example of a 2×2 square matrix and a 3×3 square matrix.

$$\begin{bmatrix} 2 & 3 \\ 5 & 2 \end{bmatrix} \qquad \begin{bmatrix} 4 & 6 & -1 \\ 2 & 3 & 0 \\ 5 & 2 & 1 \end{bmatrix}$$

Two matrices are equal if and only if they have the same elements in the same relative positions.

EXAMPLE 1 *Equal Matrices*

Given $A = B$, find x and y.

$$A = \begin{bmatrix} 2 & 5 \\ 1 & 8 \end{bmatrix}, \qquad B = \begin{bmatrix} x & 5 \\ 1 & y \end{bmatrix}$$

SOLUTION: Since the matrices are equal, the corresponding elements must be the same, so $x = 2$ and $y = 8$. ▲

Addition of Matrices

Two matrices can be added only if they have the same dimensions (same number of rows and same number of columns). To obtain the sum of two matrices with the same dimensions, add the corresponding elements of the two matrices.

EXAMPLE 2 *Adding Matrices*

$$A = \begin{bmatrix} 3 & 4 \\ -1 & 7 \end{bmatrix}, \quad B = \begin{bmatrix} 2 & 8 \\ 4 & 0 \end{bmatrix}. \quad \text{Find } A + B.$$

SOLUTION:
$$A + B = \begin{bmatrix} 3 & 4 \\ -1 & 7 \end{bmatrix} + \begin{bmatrix} 2 & 8 \\ 4 & 0 \end{bmatrix}$$
$$= \begin{bmatrix} 3+2 & 4+8 \\ -1+4 & 7+0 \end{bmatrix} = \begin{bmatrix} 5 & 12 \\ 3 & 7 \end{bmatrix}$$

▲

EXAMPLE 3 **MODELING-** *Sales of Bicycles*

Peddler's Bicycle Corporation owns and operates two stores, one in Pennsylvania and one in New Jersey. The number of mountain bicycles, MB, and racing bicycles, RB, sold in each store during January through June and during July through December are indicated in the matrices that follow. We will call the matrices A and B.

	Pennsylvania			*New Jersey*		
	MB	*RB*		*MB*	*RB*	
Jan.–June	515	425	$= A$	520	350	$= B$
July–Dec.	290	250		180	271	

Find the total number of each type of bicycle sold by the corporation during each time period.

SOLUTION: To solve the problem, we add matrices A and B.

	MB	*RB*		*MB*	*RB*
Jan.–June	515 + 520	425 + 350	=	1035	775
July–Dec.	290 + 180	250 + 271		470	521

We can see from the sum matrix that during the period from January through June, a total of 1035 mountain bicycles and 775 racing bicycles were sold. During the period from July through December, a total of 470 mountain bicycles and 521 racing bicycles were sold. ▲

Subtraction of Matrices

Only matrices with the same dimension may be subtracted. To do so, we subtract each entry in one matrix from the corresponding entry in the other matrix.

EXAMPLE 4 *Subtracting Matrices*

Find $A - B$ if

$$A = \begin{bmatrix} 2 & 6 \\ 3 & -1 \end{bmatrix} \quad \text{and} \quad B = \begin{bmatrix} 3 & -4 \\ 7 & -3 \end{bmatrix}$$

SOLUTION:

$$A - B = \begin{bmatrix} 2 & 6 \\ 3 & -1 \end{bmatrix} - \begin{bmatrix} 3 & -4 \\ 7 & -3 \end{bmatrix}$$
$$= \begin{bmatrix} 2-3 & 6-(-4) \\ 3-7 & -1-(-3) \end{bmatrix} = \begin{bmatrix} -1 & 10 \\ -4 & 2 \end{bmatrix}$$

▲

Multiplying a Matrix by a Real Number

A matrix may be multiplied by a real number by multiplying each entry in the matrix by the real number. Sometimes when we multiply a matrix by a real number, we call that real number a *scalar*.

EXAMPLE 5 ***Multiplying a Matrix by a Scalar***

For matrices A and B, find (a) $3A$ and (b) $3A - 2B$.

$$A = \begin{bmatrix} 1 & 4 \\ -3 & 5 \end{bmatrix}, \qquad B = \begin{bmatrix} -1 & 3 \\ 5 & 6 \end{bmatrix}$$

SOLUTION:

a) $3A = 3\begin{bmatrix} 1 & 4 \\ -3 & 5 \end{bmatrix} = \begin{bmatrix} 3(1) & 3(4) \\ 3(-3) & 3(5) \end{bmatrix} = \begin{bmatrix} 3 & 12 \\ -9 & 15 \end{bmatrix}$

b) We found $3A$ in part (a). Now we find $2B$.

$$2B = 2\begin{bmatrix} -1 & 3 \\ 5 & 6 \end{bmatrix} = \begin{bmatrix} 2(-1) & 2(3) \\ 2(5) & 2(6) \end{bmatrix} = \begin{bmatrix} -2 & 6 \\ 10 & 12 \end{bmatrix}$$

$$3A - 2B = \begin{bmatrix} 3 & 12 \\ -9 & 15 \end{bmatrix} - \begin{bmatrix} -2 & 6 \\ 10 & 12 \end{bmatrix}$$

$$= \begin{bmatrix} 3 - (-2) & 12 - 6 \\ -9 - 10 & 15 - 12 \end{bmatrix} = \begin{bmatrix} 5 & 6 \\ -19 & 3 \end{bmatrix}$$

Multiplication of Matrices

Multiplication of matrices is slightly more difficult than addition of matrices. Multiplication of matrices is possible only when the number of *columns* of the first matrix, A, is the same as the number of *rows* of the second matrix, B. We use the notation

$$\underset{3 \times 4}{A}$$

to indicate that matrix A has three rows and four columns. Suppose matrix A is a 3×4 matrix and matrix B is a 4×5 matrix. Then

$$\underset{3 \times 4}{A} \qquad \underset{4 \times 5}{B}$$

Same

Product matrix 3×5

This notation indicates that matrix A has four columns and matrix B has four rows. Therefore, we can multiply these two matrices. The product matrix will have the same number of rows as matrix A and the same number of columns as matrix B. Thus, the dimensions of the product matrix are 3×5.

EXAMPLE 6 ***Can These Matrices Be Multiplied?***

Determine which of the following pairs of matrices can be multiplied.

a) $A = \begin{bmatrix} 3 & 2 \\ 5 & 7 \end{bmatrix}, \qquad B = \begin{bmatrix} 0 & 6 \\ 4 & 1 \end{bmatrix}$

DID YOU KNOW

Spreadsheets

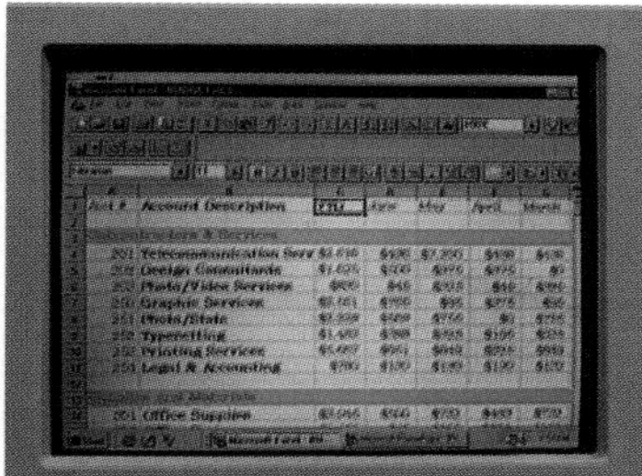

One of the most useful applications of computer software is the spreadsheet. On a computer screen, it looks very much like an accountant's ledger, but there the similarities end. A spreadsheet uses a matrix that assigns an identifying number and letter, in maplike fashion, to each "cell" in the matrix. The data may be dates, numbers, or sums of money. You instruct the computer what operation to perform in each column and row. You can change an entry to see what effect this has on the rest of the spreadsheet. The computer will automatically recompute all affected cells. The computer has the capacity to store information in thousands of cells and can reformat the data in graph form.

b) $A = \begin{bmatrix} 2 & 3 \\ 5 & 6 \end{bmatrix}, \quad B = \begin{bmatrix} 2 & 4 & -1 \\ 6 & 8 & 0 \end{bmatrix}$

c) $A = \begin{bmatrix} 2 & 1 & 4 \\ 3 & 2 & 8 \end{bmatrix}, \quad B = \begin{bmatrix} 2 & 1 & 3 \\ 1 & 0 & -2 \end{bmatrix}$

SOLUTION:

a)

A	B
2×2	2×2

Same

Because matrix A has two columns and matrix B has two rows, the two matrices can be multiplied. The product is a 2×2 matrix.

b)

A	B
2×2	2×3

Same

Because matrix A has two columns and matrix B has two rows, the two matrices can be multiplied. The product is a 2×3 matrix.

c)

A	B
2×3	2×3

Not Same

Because matrix A has three columns and matrix B has two rows, the two matrices cannot be multiplied. ▲

To explain matrix multiplication let's use matrices A and B that follow.

$$A = \begin{bmatrix} 3 & 2 \\ 5 & 7 \end{bmatrix} \quad \text{and} \quad B = \begin{bmatrix} 0 & 6 \\ 4 & 1 \end{bmatrix}$$

Since A contains two rows and B contains two columns, the product matrix will contain two rows and two columns. To multiply two matrices, we use a row–column scheme of multiplying. The numbers in the *first row* of matrix A are multiplied by the numbers in the *first column* of matrix B. These products are then added to determine the entry in the product matrix.

$$A \times B = \begin{bmatrix} 3 & 2 \\ 5 & 7 \end{bmatrix} \begin{bmatrix} 0 & 6 \\ 4 & 1 \end{bmatrix}$$

First row — First column

$$\begin{bmatrix} 3 & 2 \\ 5 & 7 \end{bmatrix} \quad \begin{bmatrix} 0 & 6 \\ 4 & 1 \end{bmatrix}$$

$$(3 \times 0) + (2 \times 4) = 0 + 8 = 8$$

PROFILE IN MATHEMATICS

JAMES SYLVESTER, WILLIAM ROWAN HAMILTON, AND ARTHUR CAYLEY

James Sylvester

William Rowan Hamilton

Arthur Cayley

Three mathematicians played important roles in the development of matrix theory: James Sylvester (1814–1897), William Rowan Hamilton (1805–1865), and Arthur Cayley (1821–1895). Sylvester and Cayley cooperated in developing matrix theory. Sylvester was the first to use the term *matrix*. Hamilton, a noted physicist, astronomer, and mathematician, also used what was essentially the algebra of matrices under the name *linear and vector functions*. The mathematical concept of *vector space* grew out of Hamilton's work on the algebra of vectors.

The 8 is placed in the first-row, first-column position of the product matrix. The other numbers in the product matrix are obtained similarly, as illustrated in the matrix that follows.

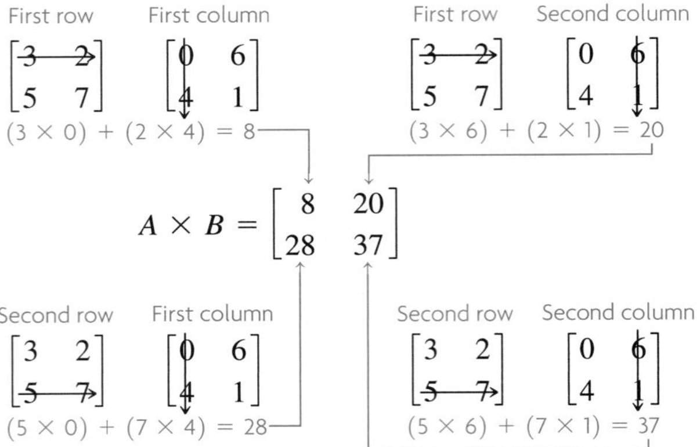

We can shorten the procedure as follows.

$$A \times B = \begin{bmatrix} 3 & 2 \\ 5 & 7 \end{bmatrix}\begin{bmatrix} 0 & 6 \\ 4 & 1 \end{bmatrix}$$

$$= \begin{bmatrix} 3(0) + 2(4) & 3(6) + 2(1) \\ 5(0) + 7(4) & 5(6) + 7(1) \end{bmatrix}$$

$$= \begin{bmatrix} 8 & 20 \\ 28 & 37 \end{bmatrix}$$

In general, if

$$A = \begin{bmatrix} a & b \\ c & d \end{bmatrix} \quad \text{and} \quad B = \begin{bmatrix} e & f \\ g & h \end{bmatrix},$$

then

$$A \times B = \begin{bmatrix} a & b \\ c & d \end{bmatrix}\begin{bmatrix} e & f \\ g & h \end{bmatrix} = \begin{bmatrix} ae + bg & af + bh \\ ce + dg & cf + dh \end{bmatrix}.$$

Let's do one more multiplication of matrices.

EXAMPLE 7 *Multiplying Matrices*

Find $A \times B$, given

$$A = \begin{bmatrix} 2 & 1 \\ 3 & 6 \end{bmatrix} \quad \text{and} \quad B = \begin{bmatrix} 6 & -1 & 3 \\ 2 & 8 & 0 \end{bmatrix}$$

SOLUTION: Matrix A contains two columns, and matrix B contains two rows. Thus, the matrices can be multiplied. Since A contains two rows and B contains three columns, the product matrix will contain two rows and three columns.

$$A \times B = \begin{bmatrix} 2 & 1 \\ 3 & 6 \end{bmatrix} \begin{bmatrix} 6 & -1 & 3 \\ 2 & 8 & 0 \end{bmatrix}$$

$$= \begin{bmatrix} 2(6) + 1(2) & 2(-1) + 1(8) & 2(3) + 1(0) \\ 3(6) + 6(2) & 3(-1) + 6(8) & 3(3) + 6(0) \end{bmatrix}$$

$$= \begin{bmatrix} 14 & 6 & 6 \\ 30 & 45 & 9 \end{bmatrix}$$

It should be noted that multiplication of matrices *is not* commutative; that is, $A \times B \neq B \times A$, except in special instances.

Square matrices have a *multiplicative identity matrix*. The multiplicative identity matrices for a 2×2 and a 3×3 matrix, denoted I, follow. Note that in any multiplicative identity matrix, 1's go diagonally from top left to bottom right and all other elements in the matrix are 0's.

$$I = \begin{bmatrix} 1 & 0 \\ 0 & 1 \end{bmatrix} \qquad I = \begin{bmatrix} 1 & 0 & 0 \\ 0 & 1 & 0 \\ 0 & 0 & 1 \end{bmatrix}$$

For any square matrix, A, $A \times I = I \times A = A$.

EXAMPLE 8 *Using the Identity Matrix in Multiplication*

Use the multiplicative identity matrix for a 2×2 matrix and matrix A to show that $A \times I = A$.

$$A = \begin{bmatrix} 4 & 3 \\ 2 & 1 \end{bmatrix}$$

SOLUTION: The identity matrix is $I = \begin{bmatrix} 1 & 0 \\ 0 & 1 \end{bmatrix}$.

$$A \times I = \begin{bmatrix} 4 & 3 \\ 2 & 1 \end{bmatrix} \begin{bmatrix} 1 & 0 \\ 0 & 1 \end{bmatrix}$$

$$= \begin{bmatrix} 4(1) + 3(0) & 4(0) + 3(1) \\ 2(1) + 1(0) & 2(0) + 1(1) \end{bmatrix}$$

$$= \begin{bmatrix} 4 & 3 \\ 2 & 1 \end{bmatrix} = A$$

Example 9 illustrates an application of multiplication of matrices.

EXAMPLE 9 *A Manufacturing Application*

The Fancy Frock Company manufactures three types of women's outfits: a dress, a two-piece suit (skirt and jacket), and a three-piece suit (skirt, jacket, and a vest). On a particular day, the firm produces 20 dresses, 30 two-piece suits, and 50 three-piece suits. Each dress requires 4 units of material and 1 hour of work to produce, each two-piece suit requires 5 units of material and 2 hours of work to produce, and each three-piece suit requires 6 units of material and 3 hours to produce. Use matrix multiplication to determine the total number of units of material and the total number of hours needed for that day's production.

DID YOU KNOW

The Prisoner's Dilemma

When two parties pursue conflicting interests, the situation can sometimes be described and modeled in a matrix under a branch of mathematics known as game theory. Consider a famous problem called "the prisoner's dilemma." A pair of criminal suspects, A and B, are being held in separate jail cells and cannot communicate with each other. Each one is told that there are four possible outcomes: If both confess, each receives a 3-year sentence. If A confesses and B does not, A receives a 1-year sentence, whereas B receives a 10-year sentence. If B confesses and A does not, B receives a 1-year sentence and A receives a 10-year sentence. Finally, if neither confesses, each will be imprisoned for 2 years. (Try arranging this situation in a matrix.)

If neither prisoner knows whether the other will confess, what should each prisoner do? A study of game theory shows that it is in each prisoner's best interest to confess to the crime.

SOLUTION: Let matrix A represent the number of each type of women's outfits produced.

$$A = \begin{matrix} \text{Dress} & \text{Two piece} & \text{Three piece} \\ [\,20 & 30 & 50\,] \end{matrix}$$

The units of material and time requirements for each type are indicated in matrix B.

$$B = \begin{matrix} \text{Material} & \text{Hours} & \\ \end{matrix}$$
$$B = \begin{bmatrix} 4 & 1 \\ 5 & 2 \\ 6 & 3 \end{bmatrix} \begin{matrix} \text{Dress} \\ \text{Two piece} \\ \text{Three piece} \end{matrix}$$

The product of A and B, or $A \times B$, will give the total number of units of material and the total number of hours of work needed for that day's production.

$$\begin{aligned} A \times B &= [20 \quad 30 \quad 50]\begin{bmatrix} 4 & 1 \\ 5 & 2 \\ 6 & 3 \end{bmatrix} \\ &= [20(4) + 30(5) + 50(6) \qquad 20(1) + 30(2) + 50(3)] \\ &= [530 \quad 230] \end{aligned}$$

Thus, a total of 530 units of material and a total of 230 hours of work are needed that day.

TIMELY TIP Matrices can only be added or subtracted if they have the same dimensions.

Matrices can only be multiplied if the number of *columns* in the first matrix is the same as the number of *rows* in the second matrix.

SECTION 7.3 EXERCISES

Concept/Writing Exercises

1. What is a matrix?
2. Explain how to determine the dimensions of a matrix.
3. What is a square matrix?
4. How many rows does a 4×3 matrix have?
5. How many columns does a 3×2 matrix have?
6. To add or subtract two matrices, what must be true about the dimensions of those matrices?
7. a) In your own words, explain the procedure used to add matrices.
   b) Use the procedure given in part (a) to add

$$\begin{bmatrix} 1 & 4 & -1 \\ 3 & 2 & 5 \end{bmatrix} \quad \text{and} \quad \begin{bmatrix} 3 & 5 & -6 \\ -1 & 2 & 4 \end{bmatrix}$$

8. a) In your own words, explain the procedure used to subtract matrices.
   b) Use the procedure given in part (a) to subtract

$$\begin{bmatrix} 8 & 4 & 2 \\ 0 & -2 & 4 \end{bmatrix} \quad \text{from} \quad \begin{bmatrix} 3 & -5 & 6 \\ -2 & 3 & 4 \end{bmatrix}$$

9. a) To multiply two matrices, what must be true about the dimensions of those matrices?
   b) What will be the dimensions of the product matrix when multiplying a 2×2 matrix with a 2×3 matrix?
10. a) In your own words, explain the procedure used to multiply matrices.
    b) Use the procedure given in part (a) to multiply

$$\begin{bmatrix} 6 & -1 \\ 5 & 0 \end{bmatrix} \quad \text{by} \quad \begin{bmatrix} 2 & -3 \\ 1 & -4 \end{bmatrix}$$

11. a) What is the multiplicative identity matrix for a 2×2 matrix?
    b) What is the multiplicative identity matrix for a 3×3 matrix?

12. A company has three offices: East, West, and Central. Each office has five divisions. The number of employees in each division of the three offices is as follows:
East: 110, 232, 103, 190, 212
West: 107, 250, 135, 203, 189
Central: 115, 218, 122, 192, 210
Express this information in the form of a 3 × 5 matrix.

Practice the Skills

In Exercises 13–16, determine A + B.

13. $A = \begin{bmatrix} 1 & 3 \\ 5 & 7 \end{bmatrix}, \quad B = \begin{bmatrix} -5 & -1 \\ 7 & 2 \end{bmatrix}$

14. $A = \begin{bmatrix} 2 & 3 & -7 \\ 4 & 0 & -1 \end{bmatrix}, \quad B = \begin{bmatrix} -4 & -3 & 8 \\ 6 & 5 & 0 \end{bmatrix}$

15. $A = \begin{bmatrix} 3 & 1 \\ 0 & 4 \\ 6 & 0 \end{bmatrix}, \quad B = \begin{bmatrix} -3 & 3 \\ 4 & 0 \\ -1 & -1 \end{bmatrix}$

16. $A = \begin{bmatrix} 2 & 6 & 3 \\ -1 & -6 & 4 \\ 3 & 0 & 5 \end{bmatrix}, \quad B = \begin{bmatrix} -1 & 3 & 1 \\ 7 & -2 & 1 \\ 2 & 3 & 8 \end{bmatrix}$

In Exercises 17–20, determine A − B.

17. $A = \begin{bmatrix} 4 & -2 \\ -3 & 5 \end{bmatrix}, \quad B = \begin{bmatrix} -2 & 5 \\ 9 & 1 \end{bmatrix}$

18. $A = \begin{bmatrix} 8 & 1 \\ 0 & 2 \\ -3 & -9 \end{bmatrix}, \quad B = \begin{bmatrix} 3 & 3 \\ -4 & 5 \\ -2 & 6 \end{bmatrix}$

19. $A = \begin{bmatrix} -4 & 3 \\ 6 & 2 \\ 1 & -5 \end{bmatrix}, \quad B = \begin{bmatrix} -6 & -8 \\ -10 & -11 \\ 3 & -7 \end{bmatrix}$

20. $A = \begin{bmatrix} 5 & 3 & -1 \\ 7 & 4 & 2 \\ 6 & -1 & -5 \end{bmatrix}, \quad B = \begin{bmatrix} 4 & 3 & 6 \\ -2 & -4 & 9 \\ 0 & -2 & 4 \end{bmatrix}$

In Exercises 21–26,

$$A = \begin{bmatrix} 1 & 2 \\ 0 & 5 \end{bmatrix}, \quad B = \begin{bmatrix} 3 & 2 \\ 5 & 0 \end{bmatrix}, \quad \text{and} \quad C = \begin{bmatrix} -2 & 3 \\ 4 & 0 \end{bmatrix}.$$

Determine the following.

21. $2B$

22. $-3B$

23. $2B + 3C$

24. $2B + 3A$

25. $3B - 2C$

26. $4C - 2A$

In Exercises 27–32, determine A × B.

27. $A = \begin{bmatrix} 2 & 0 \\ 3 & 1 \end{bmatrix}, \quad B = \begin{bmatrix} 2 & 6 \\ 8 & 4 \end{bmatrix}$

28. $A = \begin{bmatrix} 1 & -1 \\ 2 & 6 \end{bmatrix}, \quad B = \begin{bmatrix} 4 & -2 \\ -3 & -2 \end{bmatrix}$

29. $A = \begin{bmatrix} 2 & 3 & -1 \\ 0 & 4 & 6 \end{bmatrix}, \quad B = \begin{bmatrix} 2 \\ 4 \\ 1 \end{bmatrix}$

30. $A = \begin{bmatrix} 1 & 1 \\ 1 & 1 \end{bmatrix}, \quad B = \begin{bmatrix} 1 & -1 \\ -1 & 2 \end{bmatrix}$

31. $A = \begin{bmatrix} 4 & 7 & 6 \\ -2 & 3 & 1 \\ 5 & 1 & 2 \end{bmatrix}, \quad B = \begin{bmatrix} 1 & 0 & 0 \\ 0 & 1 & 0 \\ 0 & 0 & 1 \end{bmatrix}$

32. $A = \begin{bmatrix} -3 & 1 \\ 2 & 7 \end{bmatrix}, \quad B = \begin{bmatrix} 4 & 0 \\ 1 & 6 \end{bmatrix}$

In Exercises 33–38, determine A + B and A × B. If an operation cannot be performed, explain why.

33. $A = \begin{bmatrix} 1 & 3 & -2 \\ 4 & 0 & 3 \end{bmatrix}, \quad B = \begin{bmatrix} 5 & -1 & 3 \\ 2 & -2 & 1 \end{bmatrix}$

34. $A = \begin{bmatrix} 6 & 4 & -1 \\ 2 & 3 & 4 \end{bmatrix}, \quad B = \begin{bmatrix} 1 & 0 \\ 4 & -1 \end{bmatrix}$

35. $A = \begin{bmatrix} 4 & 5 & 3 \\ 6 & 2 & 1 \end{bmatrix}, \quad B = \begin{bmatrix} 3 & 2 \\ 4 & 6 \\ -2 & 0 \end{bmatrix}$

36. $A = \begin{bmatrix} 1 & 2 \\ 3 & 4 \\ 5 & 6 \end{bmatrix}, \quad B = \begin{bmatrix} 1 & 2 \\ 3 & 4 \\ 5 & 6 \end{bmatrix}$

37. $A = \begin{bmatrix} 1 & 2 \\ 3 & 4 \end{bmatrix}, \quad B = \begin{bmatrix} -3 \\ 2 \end{bmatrix}$

38. $A = \begin{bmatrix} 5 & -1 \\ 6 & -2 \end{bmatrix}, \quad B = \begin{bmatrix} 1 & 2 \\ 3 & 4 \end{bmatrix}$

In Exercises 39–41, show the commutative property of addition, A + B = B + A, holds for matrices A and B.

39. $A = \begin{bmatrix} 1 & 2 \\ 2 & -3 \end{bmatrix}, \quad B = \begin{bmatrix} 4 & 5 \\ 6 & 7 \end{bmatrix}$

40. $A = \begin{bmatrix} 9 & 4 \\ 1 & 7 \end{bmatrix}, \quad B = \begin{bmatrix} 0 & 6 \\ -1 & 5 \end{bmatrix}$

41. $A = \begin{bmatrix} 0 & -1 \\ 3 & -4 \end{bmatrix}, \quad B = \begin{bmatrix} 8 & 1 \\ 3 & -4 \end{bmatrix}$

42. Make up two matrices with the same dimensions, A and B, and show that $A + B = B + A$.

In Exercises 43–45, show that the associative property of addition, $(A + B) + C = A + (B + C)$, holds for the matrices given.

43. $A = \begin{bmatrix} 5 & 2 \\ 3 & 6 \end{bmatrix}$, $B = \begin{bmatrix} 3 & 4 \\ -2 & 7 \end{bmatrix}$, $C = \begin{bmatrix} -1 & 4 \\ 5 & 0 \end{bmatrix}$

44. $A = \begin{bmatrix} 4 & 1 \\ 6 & 7 \end{bmatrix}$, $B = \begin{bmatrix} -9 & 1 \\ -7 & 2 \end{bmatrix}$, $C = \begin{bmatrix} -6 & -3 \\ 3 & 6 \end{bmatrix}$

45. $A = \begin{bmatrix} 7 & 4 \\ 9 & -36 \end{bmatrix}$, $B = \begin{bmatrix} 5 & 6 \\ -1 & -4 \end{bmatrix}$, $C = \begin{bmatrix} -7 & -5 \\ -1 & 3 \end{bmatrix}$

46. Make up three matrices with the same dimensions, A, B, and C, and show that $(A + B) + C = A + (B + C)$.

In Exercises 47–51, determine whether the commutative property of multiplication, $A \times B = B \times A$, holds for the matrices given.

47. $A = \begin{bmatrix} 1 & -2 \\ 4 & -3 \end{bmatrix}$, $B = \begin{bmatrix} -1 & -3 \\ 2 & 4 \end{bmatrix}$

48. $A = \begin{bmatrix} 3 & 1 \\ 6 & 6 \end{bmatrix}$, $B = \begin{bmatrix} 1 & 0 \\ 0 & 1 \end{bmatrix}$

49. $A = \begin{bmatrix} 4 & 2 \\ 1 & -3 \end{bmatrix}$, $B = \begin{bmatrix} 2 & 4 \\ -3 & 1 \end{bmatrix}$

50. $A = \begin{bmatrix} -3 & 2 \\ 6 & -5 \end{bmatrix}$, $B = \begin{bmatrix} -\frac{5}{3} & -\frac{2}{3} \\ -2 & -1 \end{bmatrix}$

51. $A = \begin{bmatrix} 3 & 2 & 1 \\ 4 & 2 & 0 \\ 0 & -2 & 5 \end{bmatrix}$, $B = \begin{bmatrix} 1 & 0 & 0 \\ 0 & 1 & 0 \\ 0 & 0 & 1 \end{bmatrix}$

52. Make up two square matrices A and B with the same dimensions, and determine whether $A \times B = B \times A$.

In Exercises 53–57, show that the associative property of multiplication, $(A \times B) \times C = A \times (B \times C)$, holds for the matrices given.

53. $A = \begin{bmatrix} 1 & 3 \\ 4 & 0 \end{bmatrix}$, $B = \begin{bmatrix} 4 & 2 \\ 3 & 1 \end{bmatrix}$, $C = \begin{bmatrix} 2 & 1 \\ 3 & 0 \end{bmatrix}$

54. $A = \begin{bmatrix} -2 & 3 \\ 0 & 4 \end{bmatrix}$, $B = \begin{bmatrix} 4 & 0 \\ 3 & 5 \end{bmatrix}$, $C = \begin{bmatrix} 3 & 4 \\ -2 & 5 \end{bmatrix}$

55. $A = \begin{bmatrix} 4 & 3 \\ -6 & 2 \end{bmatrix}$, $B = \begin{bmatrix} 1 & 2 \\ 0 & 1 \end{bmatrix}$, $C = \begin{bmatrix} 4 & 3 \\ 0 & -2 \end{bmatrix}$

56. $A = \begin{bmatrix} -1 & -2 \\ -3 & -4 \end{bmatrix}$, $B = \begin{bmatrix} 1 & 0 \\ 0 & 1 \end{bmatrix}$, $C = \begin{bmatrix} 0 & 0 \\ 0 & 0 \end{bmatrix}$

57. $A = \begin{bmatrix} 3 & 4 \\ -1 & -2 \end{bmatrix}$, $B = \begin{bmatrix} 0 & 1 \\ 1 & 0 \end{bmatrix}$, $C = \begin{bmatrix} 2 & 0 \\ 3 & 0 \end{bmatrix}$

58. Make up three matrices, A, B, and C, and show that $(A \times B) \times C = A \times (B \times C)$.

Problem Solving

59. **MODELING** - *Cookie Company Costs* The Original Cookie Factory bakes and sells four types of cookies: chocolate chip, sugar, molasses, and peanut butter. Matrix A shows the number of units of various ingredients used in baking a dozen of each type of cookie.

	Sugar	Flour	Milk	Eggs
Chocolate chip	2	2	$\frac{1}{2}$	1
Sugar	3	2	1	2
Molasses	0	1	0	3
Peanut butter	$\frac{1}{2}$	1	0	0

$$A = \begin{bmatrix} 2 & 2 & \frac{1}{2} & 1 \\ 3 & 2 & 1 & 2 \\ 0 & 1 & 0 & 3 \\ \frac{1}{2} & 1 & 0 & 0 \end{bmatrix}$$

The cost, in cents per cup or per egg, for each ingredient when purchased in small quantities and in large quantities is given in matrix B.

	Large quantities	Small quantities
Sugar	10	12
Flour	5	8
Milk	8	8
Eggs	4	6

$$B = \begin{bmatrix} 10 & 12 \\ 5 & 8 \\ 8 & 8 \\ 4 & 6 \end{bmatrix}$$

Use matrix multiplication to find a matrix representing the comparative cost per item for small and large quantities purchased.

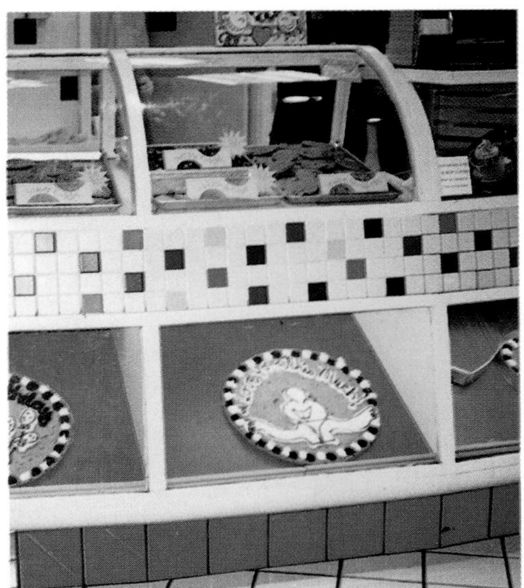

In Exercises 60 and 61, use the information given in Exercise 59. Suppose a typical day's order consists of 40 dozen chocolate chip cookies, 30 dozen sugar cookies, 12 dozen molasses cookies, and 20 dozen peanut butter cookies.

60. a) Express these orders as a 1×4 matrix.

b) Use matrix multiplication to determine the amount of each ingredient needed to fill the day's order.

61. Use matrix multiplication to determine the cost under the two purchase options (small and large quantities) to fill the day's order.

62. **MODELING** - *Food Prices* To raise money for a local charity, the Spanish Club at Montclair High School sold hot dogs, soft drinks, and candy bars for 3 days in the student lounge. The sales for the 3 days are summarized in matrix A.

$$A = \begin{array}{c} \begin{array}{ccc} \text{Hot dogs} & \text{Soft drinks} & \text{Candy bars} \end{array} \\ \begin{bmatrix} 52 & 50 & 75 \\ 48 & 43 & 60 \\ 62 & 57 & 81 \end{bmatrix} \end{array} \begin{array}{l} \\ \text{Day 1} \\ \text{Day 2} \\ \text{Day 3} \end{array}$$

The cost and revenue (in dollars) for hot dogs, soft drinks, and candy are summarized in matrix B.

$$B = \begin{array}{c} \begin{array}{cc} \text{Cost} & \text{Revenue} \end{array} \\ \begin{bmatrix} 0.30 & 0.75 \\ 0.25 & 0.50 \\ 0.15 & 0.45 \end{bmatrix} \end{array} \begin{array}{l} \\ \text{Hot dogs} \\ \text{Soft drinks} \\ \text{Candy bars} \end{array}$$

Multiply the two matrices to form a 3×2 matrix that shows the total cost and revenue for each item.

In Exercises 63 and 64, there are many acceptable answers.

63. a) Construct two matrices A and B whose product is a 3×1 matrix. Explain how you determined your answer.

b) For your matrices, determine $A \times B$.

64. a) Construct two matrices A and B whose product is a 4×1 matrix. Explain how you determined your answer.

b) For your matrices, determine $A \times B$.

Two matrices whose product is the multiplicative identity matrix are said to be **multiplicative inverses**. *That is, if $A \times B = B \times A = I$, where I is the multiplicative identity matrix, then A and B are multiplicative inverses. In Exercises 65 and 66, determine whether A and B are multiplicative inverses.*

65. $A = \begin{bmatrix} 5 & -2 \\ -2 & 1 \end{bmatrix}, \quad B = \begin{bmatrix} 1 & 2 \\ 2 & 5 \end{bmatrix}$

66. $A = \begin{bmatrix} 7 & 3 \\ 2 & 1 \end{bmatrix}, \quad B = \begin{bmatrix} 1 & -3 \\ -2 & 7 \end{bmatrix}$

Challenge Problems/Group Activities

In Exercises 67 and 68, determine whether the statement is true or false. Give an example to support your answer.

67. $A - B = B - A$, where A and B are any matrices.

68. For scalar a and matrices B and C,

$$a(B + C) = aB + aC.$$

69. **MODELING** - *Sofa Manufacturing Costs* The number of hours of labor required to manufacture one sofa of various sizes is summarized in matrix L.

$$L = \begin{array}{c} \overbrace{\begin{array}{ccc} \text{Cutting} & \text{Assembly} & \text{Packing} \end{array}}^{\text{Department}} \\ \begin{bmatrix} 1.4\text{ hr} & 0.7\text{ hr} & 0.3\text{ hr} \\ 1.8\text{ hr} & 1.4\text{ hr} & 0.3\text{ hr} \\ 2.7\text{ hr} & 2.8\text{ hr} & 0.5\text{ hr} \end{bmatrix} \end{array} \left. \begin{array}{l} \text{Small} \\ \text{Medium} \\ \text{Large} \end{array} \right\} \text{Sofa size}$$

The hourly labor rates for cutting, assembly, and packing at the Ames City Plant and at the Bay City Plant are given in matrix C.

$$C = \begin{array}{c} \overbrace{\begin{array}{cc} \text{Ames City} & \text{Bay City} \end{array}}^{\text{Plant}} \\ \begin{bmatrix} \$14 & \$12 \\ \$10 & \$9 \\ \$7 & \$5 \end{bmatrix} \end{array} \left. \begin{array}{l} \text{Cutting} \\ \text{Assembly} \\ \text{Packaging} \end{array} \right\} \text{Department}$$

a) What is the total labor cost for manufacturing a small-sized sofa at the Ames City plant?

b) What is the total cost for manufacturing a large-sized sofa at the Bay City plant?

c) Determine the product $L \times C$ and explain the meaning of the results.

70. Is it possible that two matrices could be added but not multiplied? If so, give an example.

71. Is it possible that two matrices could be multiplied but not added? If so, give an example.

Recreational Mathematics

72. Make up two matrices A and B such that

$$A + B = \begin{bmatrix} 1 & 0 \\ 0 & 1 \end{bmatrix} \text{ and } A \times B = \begin{bmatrix} 0 & 0 \\ 0 & 0 \end{bmatrix}.$$

Internet/Research Activities

73. Find an article that shows information illustrated in matrix form. Write a short paper explaining how to interpret the information provided by the matrix. Include the article with your report.

74. *Messages* The study of encoding and decoding messages is called *cryptography*. Do research on current real-life uses of cryptography and write a paper on how matrix multiplication is used to encode and decode messages. In your paper, include current real-life uses of cryptography.

75. *Graphing Calculator* Some graphing calculators are able to perform matrix operations such as addition, subtraction, and multiplication. Read the instruction manual for your graphing calculator and then use your graphing calculator to evaluate Exercises 13, 15, 17, and 19 and Exercises 33, 35, and 37. Verify that your answers are the same as the answers you obtained without the calculator.

7.4 SOLVING SYSTEMS OF EQUATIONS BY USING MATRICES

In Section 7.3, we introduced matrices. Now we will discuss the procedure to solve a system of linear equations using matrices. We will illustrate how to solve a system of two equations and two unknowns. Systems of equations containing three equations and three unknowns (called third-order systems) and higher-order systems can also be solved by using matrices.

The first step in solving a system of equations using matrices is to represent the system of equations with an *augmented matrix*. An augmented matrix consists of two smaller matrices, one for the coefficients of the variables in the equations and one for the constants in the equations. To determine the augmented matrix, first write each equation in standard form, $ax + by = c$. For the system of equations below, its augmented matrix is shown to its right.

System of equations / **Augmented matrix**

$$\begin{aligned} a_1x + b_1y &= c_1 \\ a_2x + b_2y &= c_2 \end{aligned} \qquad \left[\begin{array}{cc|c} a_1 & b_1 & c_1 \\ a_2 & b_2 & c_2 \end{array}\right]$$

Following is another example.

System of equations / **Augmented matrix**

$$\begin{aligned} x + 2y &= 8 \\ 3x - y &= 7 \end{aligned} \qquad \left[\begin{array}{cc|c} 1 & 2 & 8 \\ 3 & -1 & 7 \end{array}\right]$$

Note that the bar in the augmented matrix separates the numerical coefficients from the constants. The matrix is just a shortened way of writing the system of equations. Thus, we can solve a system of equations by using matrices in a manner very similar to solving a system of equations with the addition method.

To solve a system of equations by using matrices, we use *row transformations* to obtain new matrices that have the same solution as the original system. We will discuss three row transformation procedures.

DID YOU KNOW

Early Matrices

The earliest known use of a matrix to solve linear equations appeared in the ancient Chinese mathematical classic *Jiuzhang Suanshu (Nine Chapters on the Mathematical Art)* in about 200 B.C. The use of matrices to solve problems did not appear in the West until the nineteenth century. Perhaps the fact that the Chinese used a counting board that took the form of a grid made it easier for them to make the leap to the development and use of matrices. The image above shows the same type of counting board as appeared in a book dated 1795.

Procedures for Row Transformations

1. Any two rows of a matrix may be interchanged (which is the same as interchanging any two equations in the system of equations).
2. All the numbers in any row may be multiplied by any nonzero real number (which is the same as multiplying both sides of an equation by any nonzero real number).
3. All the numbers in any row may be multiplied by any nonzero real number, and these products may be added to the corresponding numbers in any other row of numbers.

We use row transformations to obtain an augmented matrix whose numbers to the left of the vertical bar are the same as in the *multiplicative identity matrix*. From this type of augmented matrix, we can determine the solution to the system of equations. For example, if we get

$$\left[\begin{array}{cc|c} 1 & 0 & 3 \\ 0 & 1 & -2 \end{array}\right]$$

it tells us that $1x + 0y = 3$ or $x = 3$, and $0x + 1y = -2$ or $y = -2$. Thus, the solution to the system of equations that yielded this augmented matrix is $(3, -2)$. Now let's work an example.

EXAMPLE 1 *Using Row Transformations*

Solve the following system of equations by using matrices.

$$\begin{aligned} x + 2y &= 5 \\ 3x - y &= 8 \end{aligned}$$

SOLUTION: First we write the augmented matrix.

$$\left[\begin{array}{cc|c} 1 & 2 & 5 \\ 3 & -1 & 8 \end{array}\right]$$

Our goal is to obtain a matrix of the form

$$\left[\begin{array}{cc|c} 1 & 0 & c_1 \\ 0 & 1 & c_2 \end{array}\right]$$

where c_1 and c_2 may represent any real numbers. It is generally easier to work by columns. Therefore, we will try to get the first column of the augmented matrix to be $\begin{smallmatrix} 1 \\ 0 \end{smallmatrix}$ and the second column to be $\begin{smallmatrix} 0 \\ 1 \end{smallmatrix}$. Since the element in the top left position is already a 1, we must work to change the 3 in the first column, second row, into a 0. We use row transformation procedure 3 to change the 3 into a 0. If we multiply the top row of numbers by -3 and add these products to the second row of numbers, the element in the first column, second row will become a 0:

$$\left[\begin{array}{cc|c} 1 & 2 & 5 \\ 3 & -1 & 8 \end{array}\right] \quad \text{Original augmented matrix}$$

The top row of numbers multiplied by -3 gives

$$1(-3), \qquad 2(-3), \qquad \text{and} \qquad 5(-3)$$

Now add these products to their respective numbers in row 2.

$$\left[\begin{array}{cc|c} 1 & 2 & 5 \\ 3 + 1(-3) & -1 + 2(-3) & 8 + 5(-3) \end{array}\right] = \left[\begin{array}{cc|c} 1 & 2 & 5 \\ 0 & -7 & -7 \end{array}\right]$$

MATHEMATICS *Everywhere*

Matrices and Computer Graphics

Woody from *Toy Story*

Did you ever wonder how the characters in computer-generated movies such as *Shrek* and *Toy Story* are created? To create the graphics for computer-generated films, an object is first represented as a collection of geometric figures. Each geometric figure is then represented as points in the Cartesian coordinate plane. The coordinates of the points representing the object are written in columns in a matrix. Matrix operations are used if an object needs to be transformed by scaling, stretching, reflecting, or translating (see transformational geometry in Section 9.5). For example, matrix multiplication is used to translate the character Woody, from the movies *Toy Story* and *Toy Story 2*, from one position to another.

To add a shadow to an object, graphic artists perform several transformations on the matrix representing the object. Each transformation is created using matrix multiplication. The computer graphics used to create special effects in video games and used in areas such as medical imaging, architectural engineering, and weather forecasting are all based on a mathematical object called a matrix.

The next step is to obtain a 1 in the second column, second row. At present, -7 is in this position. To change the -7 to a 1, we use row transformation procedure 2. If we multiply -7 by $-\frac{1}{7}$, the product will be 1. Therefore, we multiply all the numbers in the second row by $-\frac{1}{7}$ to get

$$\left[\begin{array}{cc|c} 1 & 2 & 5 \\ 0(-\frac{1}{7}) & -7(-\frac{1}{7}) & -7(-\frac{1}{7}) \end{array}\right] = \left[\begin{array}{cc|c} 1 & 2 & 5 \\ 0 & 1 & 1 \end{array}\right]$$

The next step is to obtain a 0 in the second column, first row. At present, a 2 is in this position. Multiplying the numbers in the second row by -2 and adding the products to the corresponding numbers in the first row gives a 0 in the desired position.

$$\left[\begin{array}{cc|c} 1 + 0(-2) & 2 + 1(-2) & 5 + 1(-2) \\ 0 & 1 & 1 \end{array}\right] = \left[\begin{array}{cc|c} 1 & 0 & 3 \\ 0 & 1 & 1 \end{array}\right]$$

We now have the desired augmented matrix:

$$\left[\begin{array}{cc|c} 1 & 0 & 3 \\ 0 & 1 & 1 \end{array}\right]$$

With this matrix, we see that $1x + 0y = 3$, or $x = 3$, and $0x + 1y = 1$, or $y = 1$. The solution to the system is (3, 1).

Check:

$$\begin{aligned} x + 2y &= 5 \\ 3 + 2(1) &= 5 \\ 5 &= 5 \quad \text{True} \end{aligned} \qquad \begin{aligned} 3x - y &= 8 \\ 3(3) - 1 &= 8 \\ 8 &= 8 \quad \text{True} \end{aligned}$$

Now we give a general procedure to change an augmented matrix to the desired form.

To Change an Augmented Matrix to the Form $\left[\begin{array}{cc|c} 1 & 0 & c_1 \\ 0 & 1 & c_2 \end{array}\right]$

Use row transformations to:

1. Change the element in the first column, first row, to a 1.
2. Change the element in the first column, second row, to a 0.
3. Change the element in the second column, second row, to a 1.
4. Change the element in the second column, first row, to a 0.

Generally, when changing an element in the augmented matrix to a 1, *we use step* 2 *in the row transformation box on page* 412. *When changing an element to a* 0, *we use step* 3 *in the row transformation box.*

EXAMPLE 2 *Using Matrices to Solve a System of Equations*

Solve the following system of equations using matrices.

$$\begin{aligned} 2x + 4y &= 6 \\ 4x - 2y &= -8 \end{aligned}$$

SOLUTION: First write the augmented matrix.

$$\left[\begin{array}{cc|c} 2 & 4 & 6 \\ 4 & -2 & -8 \end{array}\right]$$

To obtain a 1 in the first column, first row, multiply the numbers in the first row by $\frac{1}{2}$.

$$\left[\begin{array}{cc|c} 1 & 2 & 3 \\ 4 & -2 & -8 \end{array}\right]$$

To obtain a 0 in the first column, second row, multiply the numbers in the first row by -4 and add the products to the corresponding numbers in the second row.

$$\left[\begin{array}{cc|c} 1 & 2 & 3 \\ 4 + 1(-4) & -2 + 2(-4) & -8 + 3(-4) \end{array}\right] = \left[\begin{array}{cc|c} 1 & 2 & 3 \\ 0 & -10 & -20 \end{array}\right]$$

To obtain a 1 in the second column, second row, multiply the numbers in the second row by $-\frac{1}{10}$.

$$\left[\begin{array}{cc|c} 1 & 2 & 3 \\ 0(-\frac{1}{10}) & -10(-\frac{1}{10}) & -20(-\frac{1}{10}) \end{array}\right] = \left[\begin{array}{cc|c} 1 & 2 & 3 \\ 0 & 1 & 2 \end{array}\right]$$

To obtain a 0 in the second column, first row, multiply the numbers in the second row by -2 and add the products to the corresponding numbers in the first row.

$$\left[\begin{array}{cc|c} 1 + (0)(-2) & 2 + (1)(-2) & 3 + (2)(-2) \\ 0 & 1 & 2 \end{array}\right] = \left[\begin{array}{cc|c} 1 & 0 & -1 \\ 0 & 1 & 2 \end{array}\right]$$

The solution to the system of equations is $(-1, 2)$. ▲

Inconsistent and Dependent Systems

Assume that you solve a system of two equations and obtain an augmented matrix in which one row of numbers on the left side of the vertical line are all zeroes but a zero does not appear in the same row on the right side of the vertical line. This situation indicates that the system is inconsistent and has no solution. For example, a system of equations that yields the following augmented matrix is an inconsistent system.

$$\left[\begin{array}{cc|c} 1 & 2 & 5 \\ 0 & 0 & 4 \end{array}\right] \quad \text{Inconsistent system}$$

The second row of the matrix represents the equation

$$0x + 0y = 4 \qquad \text{or} \qquad 0 = 4$$

which is never true.

If you obtain a matrix in which a 0 appears across an entire row, the system of equations is dependent. For example, a system of equations that yields the following matrix is a dependent system.

$$\left[\begin{array}{cc|c} 1 & 5 & -6 \\ 0 & 0 & 0 \end{array}\right] \quad \text{Dependent system}$$

The second row of the matrix represents the equation

$$0x + 0y = 0 \qquad \text{or} \qquad 0 = 0$$

which is always true.

Triangularization Method

Another procedure to solve a system of two equations is to use row transformation procedures to obtain an augmented matrix of the form

$$\left[\begin{array}{cc|c} 1 & a & b \\ 0 & 1 & c \end{array}\right]$$

where a, b, and c represent real numbers. This procedure is called the ***triangularization method***, because the ones and zeroes form a triangle.

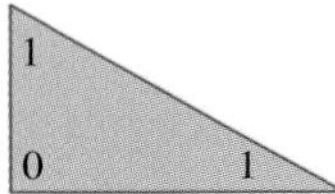

When the matrix is in this form, we can write the following system of equations.

$$\begin{aligned} 1x + ay &= b \\ 0x + 1y &= c \end{aligned} \qquad \text{or} \qquad \begin{aligned} x + ay &= b \\ y &= c \end{aligned}$$

Using substitution, we can solve the system.

For example, in Example 2, in the process of solving the system we obtained the augmented matrix

$$\left[\begin{array}{cc|c} 1 & 2 & 3 \\ 0 & 1 & 2 \end{array}\right]$$

This matrix represents the following system of equations.

$$\begin{aligned} x + 2y &= 3 \\ y &= 2 \end{aligned}$$

To solve for x, we substitute 2 for y in the equation

$$\begin{aligned} x + 2y &= 3 \\ x + 2(2) &= 3 \\ x + 4 &= 3 \\ x &= -1 \end{aligned}$$

Thus, the solution to the system is $(-1, 2)$, as was obtained in Example 2. You may use either method when solving a system of equations with matrices unless your instructor specifies otherwise.

SECTION 7.4 EXERCISES

Concept/Writing Exercises

1. a) What is an augmented matrix?
   b) Determine the augmented matrix for the following system.

$$x + 3y = 7$$
$$2x - y = 4$$

2. In your own words, write the three row transformation procedures.
3. How will you know, when solving a system of equations by using matrices, whether the system is inconsistent?
4. How will you know, when solving a system of equations by using matrices, whether the system is dependent?
5. If you obtained the following augmented matrix when solving a system of equations, what would be your next step in completing the process? Explain your answer.

$$\left[\begin{array}{rr|r} 1 & 3 & 5 \\ 0 & -2 & 1 \end{array}\right]$$

6. If you obtained the following augmented matrix when solving a system of equations, what would be your next step in completing the process? Explain your answer.

$$\left[\begin{array}{rr|r} 1 & -2 & 1 \\ 0 & 1 & 3 \end{array}\right]$$

Practice the Skills

In Exercises 7–20, use matrices to solve the system of equations.

7. $x + 3y = 3$
   $-x + y = -3$
8. $x - y = 5$
   $2x - y = 6$
9. $x - 2y = -1$
   $2x + y = 8$
10. $x + y = -1$
    $2x + 3y = -5$
11. $2x - 5y = -6$
    $-4x + 10y = 12$
12. $x + y = 5$
    $3x - y = 3$
13. $2x - 3y = 10$
    $2x + 2y = 5$
14. $x + 3y = 1$
    $-2x + y = 5$
15. $4x + 2y = -10$
    $-2x + y = -7$
16. $4x + 2y = 6$
    $5x + 4y = 9$
17. $-3x + 6y = 5$
    $2x - 4y = 8$
18. $2x - 5y = 10$
    $3x + y = 15$
19. $2x + y = 11$
    $x + 3y = 18$
20. $4x - 3y = 7$
    $-2x + 5y = 14$

Problem Solving

In Exercises 21–24, use matrices to solve the problem.

21. **MODELING** - *Selling Flags* Michael's Arts and Crafts sells small flags for \$4 and large flags for \$6. Recently, they sold a total of 55 flags in a single day. If flag receipts for the day totaled \$290, how many of each type of flag were sold?

22. **MODELING** - *TV Dimensions* Vita Gunta just purchased a new high-definition television. She noticed that the perimeter of the screen is 124 in. The width of the screen is 8 in. greater than its height. Find the dimensions of the screen.
23. **MODELING** - *On the Job* Peoplepower, Inc., a daily employment agency, charges \$10 per hour for a truck driver and \$8 per hour for a laborer. On a certain job, the laborer worked two more hours than the truck driver, and together they cost \$144. How many hours did each work?
24. **MODELING** - *Sweets* If Meg Cohn buys 2 lb of chocolate-covered cherries and 3 lb of chocolate-covered mints, her total cost is \$23. If she buys 1 lb of chocolate-covered cherries and 2 lb of chocolate-covered mints, her total cost is \$14. Find the cost of 1 lb of chocolate-covered cherries and 1 lb of chocolate-covered mints.

Recreational Mathematics

25. **MODELING** - *Fill in the Missing Information* Pencil World sells two types of mechanical pencils to stationery stores. The nonrefillable pencil sells for \$1.50 each, and the refillable pencil sells for \$2.00 each. Pencil World received an order for 200 pencils and a check for \$337.50 for the pencils. When placing the order, the stationery store clerk failed to specify the number of each type of pencil being ordered. Can Pencil World fill the order with the information given? If so, determine the number of nonrefillable and the number of refillable pencils the clerk ordered.

Internet/Research Activities

26. Do research and write a paper on the development of matrices. In your paper, cover the contributions of James Joseph Sylvester, Arthur Cayley, and William Rowan Hamilton. (References include history of mathematics books, encyclopedias, and the Internet.)

7.5 SYSTEMS OF LINEAR INEQUALITIES

In earlier sections, we showed how to find the solution to a system of linear equations in two variables. Now we are going to explore the techniques of finding the solution set to a system of linear inequalities in two variables.

The solution set of a system of linear inequalities is the set of points that satisfy all inequalities in the system. The solution set of a system of linear inequalities may consist of infinitely many ordered pairs. To determine the solution set to a system of linear inequalities, graph each inequality on the same axes. The ordered pairs common to all the inequalities are the solution set to the system.

Procedure for Solving a System of Linear Inequalities

1. Select one of the inequalities. Replace the inequality symbol with an equal sign and draw the graph of the equation. Draw the graph with a dashed line if the inequality is $<$ or $>$ and with a solid line if the inequality is $\leq$ or $\geq$.
2. Select a test point on one side of the line and determine whether the point is a solution to the inequality. If so, shade the area on the side of the line containing the point. If the point is not a solution, shade the area on the other side of the line.
3. Repeat steps 1 and 2 for the other inequality.
4. The intersection of the two shaded areas and any solid line common to both inequalities form the solution set to the system of inequalities.

EXAMPLE 1 *Solving a System of Inequalities*

Graph the following system of inequalities and indicate the solution set.

$$x + y < 2$$
$$x - y < 4$$

SOLUTION: Graph both inequalities on the same axes. First draw the graph of $x + y < 2$. When drawing the graph, remember to use a dashed line, since the inequality is "less than" (see Fig. 7.10a on page 419). If you have forgotten how to graph inequalities, review Section 6.8.

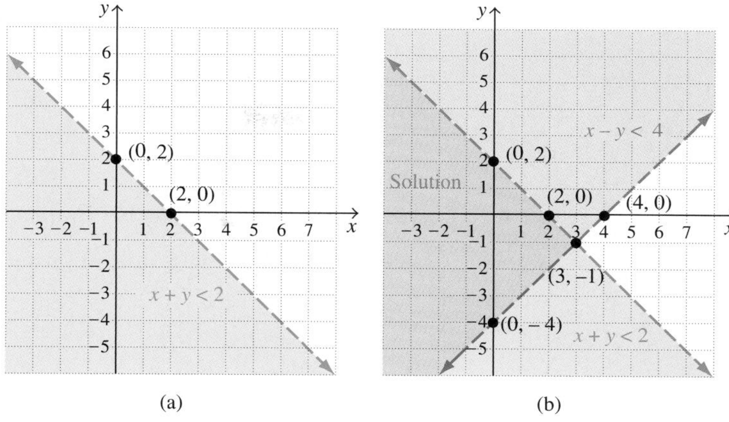

Figure 7.10

Now, on the same axes, shade the half-plane determined by the inequality $x - y < 4$ (see Fig. 7.10b). The solution set consists of all the points common to the two shaded half-planes. These are the points in the region on the graph containing both color shadings. Figure 7.10(b) shows that the two lines intersect at $(3, -1)$. This ordered pair can also be found by any of the algebraic methods discussed in Sections 7.2 and 7.3.

EXAMPLE 2 *Solving a System of Linear Inequalities*

Graph the following system of inequalities and indicate the solution set.

$$4x - 2y \geq 8$$
$$2x + 3y < 6$$

SOLUTION: Graph the inequality $4x - 2y \geq 8$. Remember to use a solid line, because the inequality is "greater than or equal to"; see Fig. 7.11(a).

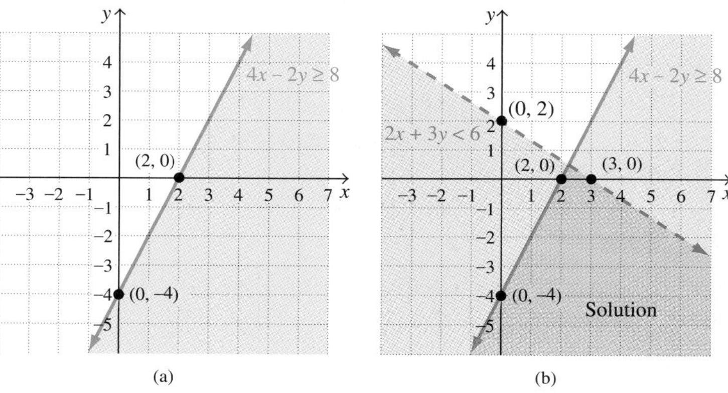

Figure 7.11

On the same set of axes, draw the graph of $2x + 3y < 6$. Use a dashed line, since the inequality is "less than"; see Fig. 7.11(b) on page 419. The solution is the region of the graph that contains both color shadings and the part of the solid line that satisfies the inequality $2x + 3y < 6$. Note that the point of intersection of the two lines is not a part of the solution set.

EXAMPLE 3 *Another System of Inequalities*

Graph the following system of inequalities and indicate the solution set.

$$x \geq -2$$
$$y < 3$$

SOLUTION: Graph the inequality $x \geq -2$; see Fig. 7.12(a). On the same axes, graph the inequality $y < 3$; see Fig. 7.12(b). The solution set is that region of the graph that is shaded in both colors and the part of the solid line that satisfies the inequality $y < 3$. The point of intersection of the two lines, $(-2, 3)$, is not part of the solution because it does not satisfy the inequality $y < 3$.

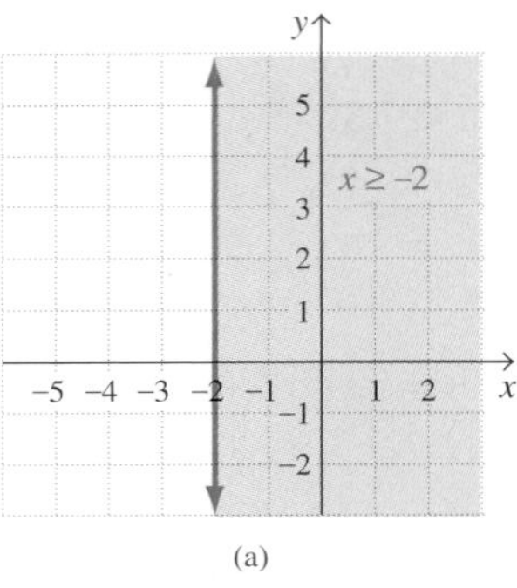

(a)

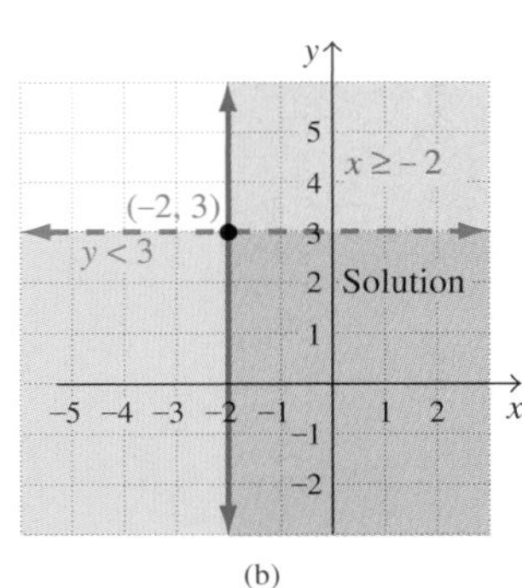

(b)

Figure 7.12

SECTION 7.5 EXERCISES

Concept/Writing Exercises

1. What is the solution set of a system of linear inequalities?
2. In your own words, give the four-step procedure for solving a system of linear inequalities.

Practice the Skills

In Exercises 3–18, graph the system of linear inequalities and indicate the solution set.

3. $y > x + 3$
   $y > 2x$
4. $x + y \geq 1$
   $x - y > 3$
5. $y \leq x - 4$
   $y < -2x + 4$
6. $x + y < 4$
   $3x + 2y \geq 6$
7. $x - y < 4$
   $x + y < 5$
8. $3x - y \leq 6$
   $x - y > 4$
9. $x + 2y \geq 4$
   $3x - y \geq -6$
10. $x - 3y \leq 3$
    $x + 2y \geq 4$
11. $y \leq 3x$
    $x \geq 3y$
12. $y \leq 4$
    $x - y < 1$
13. $x \geq 1$
    $y \leq 1$
14. $x \leq 0$
    $y \leq 0$
15. $4x + 2y > 8$
    $x \geq y - 1$
16. $5y > 3x + 10$
    $3y < -2x - 3$
17. $3x \geq 2y + 10$
    $x \leq y + 8$
18. $3x - 5y < 7$
    $x > 2y + 1$

Challenge Problems/Group Activities

19. **MODELING** - *Camcorder Sales* Best Buy sells two models of a certain brand of camcorder. Based on demand, it is necessary to stock at least twice as many units of Panasonic as Sony. The costs to the store for the two models are \$600 and \$900, respectively. The management wants at least 10 Panasonic camcorders and five Sony camcorders in inventory at all times and does not want more than \$18,000 in camcorder inventory at any one time.
    a) Translate the problem into a system of linear inequalities.
    b) Solve the system graphically. Graph inventory for Panasonic on the horizontal axis and inventory for Sony on the vertical axis.

    c) Select a point in the solution set and determine the inventory cost for the two models that corresponds to that point.
20. Write a system of linear inequalities whose solution is the second quadrant, including the axes.
21. a) Do all systems of linear inequalities have solutions? Explain.
    b) Write a system of inequalities that has no solution.
22. Can a system of linear inequalities have a solution set consisting of a single point? Explain.
23. Can a system of linear inequalities have as its solution set all the points on the coordinate plane? Explain your answer, giving an example to support it.

Recreational Mathematics

24. Write a system of linear inequalities that has the ordered pair (0, 0) as its only solution. There are many possible answers.
25. Write a system of linear inequalities that has the following ordered pairs as some of its solutions. There are many possible answers.

$$\ldots(-3, -3), (-2, -2), (-1, -1), (0, 0), (1, 1), (2, 2), (3, 3), \ldots$$

7.6 LINEAR PROGRAMMING

Government, business, and industry often require decision makers to find cost-effective solutions to a variety of problems. Linear programming often serves as a method of expressing the relationships in many of these problems and uses systems of linear inequalities.

The typical linear programming problem has many variables and is generally so lengthy that it is solved on a computer by a technique called the *simplex method*. The simplex method was developed in the 1940s by George B. Dantzig. Linear programming is used to solve problems in the social sciences, health care, land development, nutrition, military, and many other fields.

We will not discuss the simplex method in this textbook. We will merely give a brief introduction to how linear programming works. You can find a detailed explanation in books on finite mathematics.

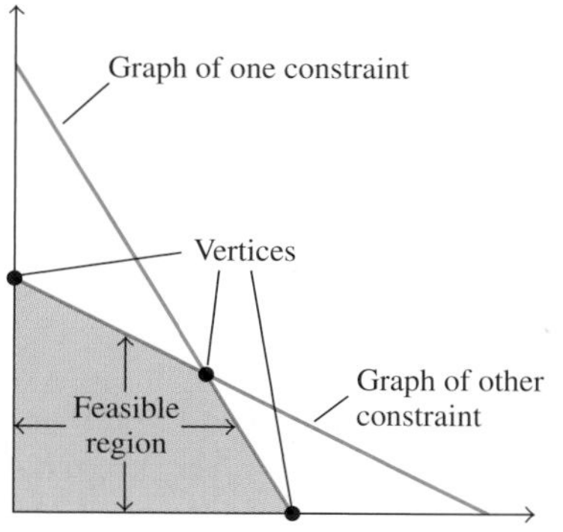

Figure 7.13

In a linear programming problem, there are restrictions called *constraints*. Each constraint is represented as a linear inequality. The list of constraints forms a system of linear inequalities. When the system of inequalities is graphed, we often obtain a region bounded on all sides by line segments (Fig. 7.13). This region is called the *feasible region*. The points where two or more boundaries intersect are called the *vertices* of the feasible region. The points on the boundary of the region and the points inside the feasible region are the solution set for the system of inequalities.

For each linear programming problem, we will obtain a formula of the form $K = Ax + By$, called the *objective function*. The objective function is the formula for the quantity K (or some other variable) that we want to maximize or minimize. The values we substitute for x and y determine the value of K. From the information given in the problem, we determine the real number constants A and B. In a particular

DID YOU KNOW

The Logistics of D-Day

Linear programming was first used to deal with the age-old military problem of logistics: obtaining, maintaining, and transporting military equipment and personnel. George Dantzig developed the simplex method for the Allies of World War II to do just that. Consider the logistics of the Allied invasion of Normandy. Meteorologic experts had settled on three possible dates in June 1944. It had to be a day when low tide and first light would coincide, when the winds should not exceed 8 to 13 mph, and when visibility was not less than 3 miles. A force of 170,000 assault troops was to be assembled and moved to 22 airfields in England where 1200 air transports and 700 gliders would then take them to the coast of France to converge with 5000 ships of the D-day armada. The code name for the invasion was Operation Overlord, but it is known to most as D-day.

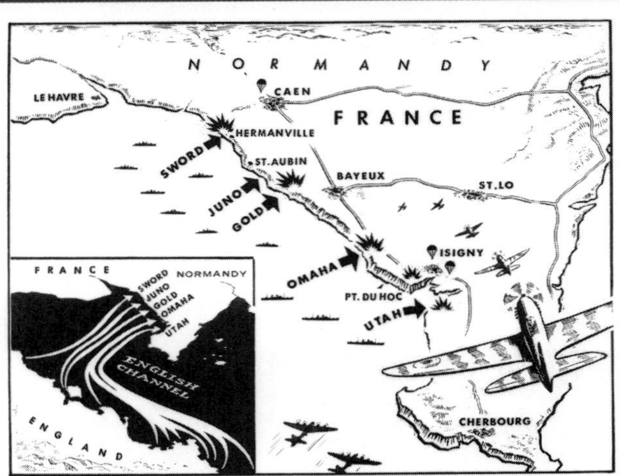

Winston Churchill called the invasion of Normandy "the most difficult and complicated operation that has ever taken place."

linear programming problem, a typical equation that might be used to find the maximum profit, P, is $P = 3x + 7y$. We would find the maximum profit by substituting the ordered pairs (x, y) of the vertices of the feasible region into the formula $P = 3x + 7y$ to see which ordered pair yields the greatest value of P and therefore the maximum profit. The ordered pair that yields the smallest value of P determines the minimum profit.

Linear programming is used to determine which ordered pair will yield the maximum (or minimum) value of the variable that is being maximized (or minimized). The fundamental principle of linear programming provides a rule for finding those maximum and minimum values.

Fundamental Principle of Linear Programming
If the objective function, $K = Ax + By$, is evaluated at each point in a feasible region, the maximum and minimum values of the equation occur at vertices of the region.

Linear programming is a powerful tool for finding the maximum and minimum values of an objective function. Using the fundamental principle of linear programming, we are quickly able to determine the maximum and minimum values of an objective function by using just a few of many points in the feasible region.

Example 1 illustrates how the fundamental principle is used to solve a linear programming problem.

EXAMPLE 1 MODELING - *Using the Fundamental Principle of Linear Programming*

The Ric Shaw Chair company makes two types of rocking chairs, a plain chair and a fancy chair. Each rocking chair must be assembled and then finished. The plain chair takes 4 hours to assemble and 4 hours to finish. The fancy chair takes 8 hours to assemble and 12 hours to finish. The company can provide at most 160 worker-hours of assembling and 180 worker-hours of finishing a day. If the profit on a plain chair is \$25 and the profit on a fancy chair is \$40, how many rocking chairs of each type should the company make per day to maximize profits? What is the maximum profit?

SOLUTION: From the information given, we know the following facts.

	Assembly Time (hr)	**Finishing Time (hr)**	**Profit (\$)**
Plain chair	4	4	25.00
Fancy chair	8	12	40.00

Let

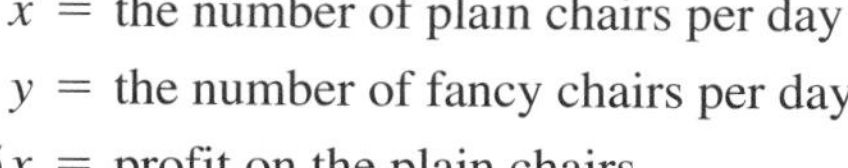

$$
\begin{aligned}
x &= \text{the number of plain chairs per day} \\
y &= \text{the number of fancy chairs per day} \\
25x &= \text{profit on the plain chairs} \\
40y &= \text{profit on the fancy chairs} \\
P &= \text{the total profit}
\end{aligned}
$$

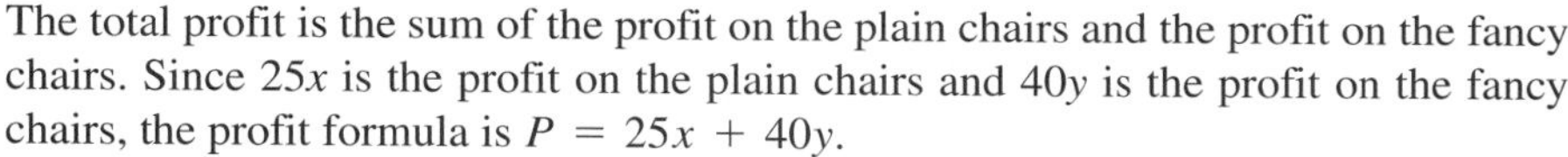

The total profit is the sum of the profit on the plain chairs and the profit on the fancy chairs. Since $25x$ is the profit on the plain chairs and $40y$ is the profit on the fancy chairs, the profit formula is $P = 25x + 40y$.

The maximum profit, P, is dependent on several conditions, called *constraints*. The number of chairs manufactured each day cannot be a negative amount. This condition gives us the constraints $x \geq 0$ and $y \geq 0$. Another constraint is determined by the total number of hours allocated for assembling. Four hours are needed to assemble the plain chair, so the total number of hours per day to assemble x plain chairs is $4x$. Eight hours are required to assemble a fancy chair, so the total number of hours needed to assemble y fancy chairs is $8y$. The maximum number of hours allocated for assembling is 160 per day. Thus, the third constraint is $4x + 8y \leq 160$. The final constraint is determined by the number of hours allotted for finishing. Finishing a plain chair takes 4 hours, or $4x$ hours to finish x plain chairs. Finishing a fancy chair takes 12 hours, or $12y$ hours to finish y fancy chairs. The total number of hours allotted for finishing is 180 per day. Therefore, the fourth constraint is $4x + 12y \leq 180$. Thus, the four constraints are

$$
\begin{aligned}
x &\geq 0 \\
y &\geq 0 \\
4x + 8y &\leq 160 \\
4x + 12y &\leq 180
\end{aligned}
$$

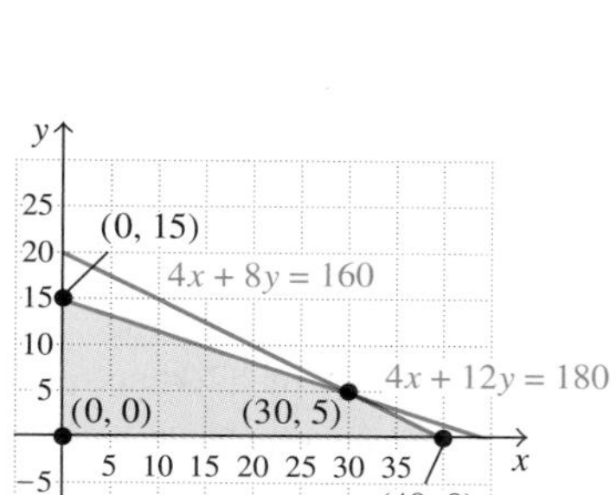

Figure 7.14

The list of constraints is a system of linear inequalities in two variables. The solution to the system of inequalities is the set of ordered pairs that satisfies all the constraints. These points are plotted in Fig. 7.14. Note that the solution to the system consists of the colored region and the solid boundaries. The points (0, 0), (0, 15), (30, 5), and (40, 0) are the points at which the boundaries intersect. These points can also be found by the addition or substitution method described in Section 7.2.

The goal in this example is to maximize the profit. The objective function is given by the profit formula $P = 25x + 40y$. According to the fundamental principle, the maximum profit will be found at one of the vertices of the feasible region.

Calculate P for each one of the vertices.

$$
\begin{aligned}
& P = 25x + 40y \\
\text{At } (0, 0), \quad & P = 25(0) + 40(0) = 0 \\
\text{At } (0, 15), \quad & P = 25(0) + 40(15) = 600 \\
\text{At } (30, 5), \quad & P = 25(30) + 40(5) = 950 \\
\text{At } (40, 0), \quad & P = 25(40) + 40(0) = 1000
\end{aligned}
$$

The maximum profit is at (40, 0), which means that the company should manufacture 40 plain rocking chairs and no fancy rocking chairs. The maximum profit would be \$1000. The minimum profit would be at (0, 0), when no rocking chairs of either style were manufactured.

A variation of the problem in Example 1 could be that the company knows that it cannot sell more than 15 plain rocking chairs per day. With this additional constraint, we now have the following set of constraints.

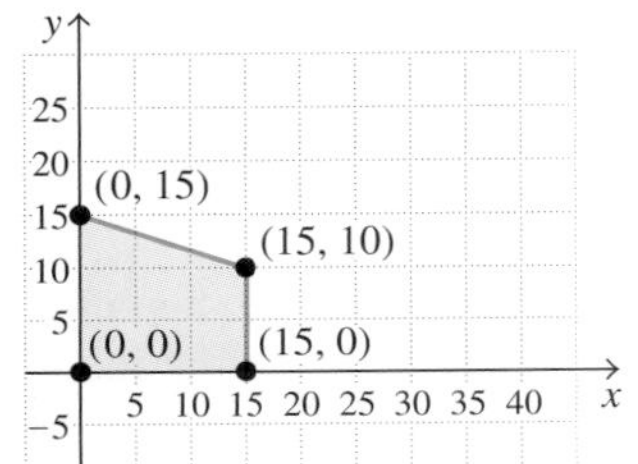

Figure 7.15

$$
\begin{aligned}
x &\ge 0 \\
x &\le 15 \\
y &\ge 0 \\
4x + 8y &\le 160 \\
4x + 12y &\le 180
\end{aligned}
$$

The graph of these constraints is shown in Fig. 7.15.

The vertices of the feasible region are (0, 0), (0, 15), (15, 10), and (15, 0). To determine the maximum profit, we calculate P for each of these vertices:

$$
\begin{aligned}
& P = 25x + 40y \\
\text{At } (0, 0), \quad & P = 25(0) + 40(0) = 0 \\
\text{At } (0, 15), \quad & P = 25(0) + 40(15) = 600 \\
\text{At } (15, 10), \quad & P = 25(15) + 40(10) = 775 \\
\text{At } (15, 0), \quad & P = 25(15) + 40(0) = 375
\end{aligned}
$$

This set of constraints gives the maximum profit of \$775 when the company manufactures 15 plain rocking chairs and 10 fancy rocking chairs.

EXAMPLE 2 **MODELING** - *Washers and Dryers, Maximizing Profit*

The Alexander Appliance Company makes washers and dryers. The company must manufacture at least one washer per day to ship to one of its customers. No more than 6 washers can be manufactured due to production restrictions. The number of dryers cannot exceed 7 per day. Also, the number of washers cannot exceed the number of dryers manufactured per day. If the profit on each washer is \$20 and the profit on each dryer is \$30, how many of each appliance should the company make per day to maximize profits? What is the maximum profit?

SOLUTION: Let

$$\begin{aligned} x &= \text{the number of washers manufactured per day} \\ y &= \text{the number of dryers manufactured per day} \\ 20x &= \text{the profit on washers} \\ 30x &= \text{the profit on dryers} \\ P &= \text{the total profit} \end{aligned}$$

The maximum profit is dependent on several constraints. The number of appliances manufactured each day cannot be a negative amount. This condition gives us the constraints $x \geq 0$ and $y \geq 0$. The company must manufacture at least one washer per day; therefore, $x \geq 1$. No more than 6 washers can be manufactured per day; therefore, $x \leq 6$. No more than 7 dryers can be manufactured per day; therefore, $y \leq 7$. The number of washers cannot exceed the number of dryers manufactured per day; therefore, $x \leq y$. Thus, the six constraints are

$$x \geq 0,\ y \geq 0,\ x \geq 1,\ x \leq 6,\ y \leq 7,\ x \leq y$$

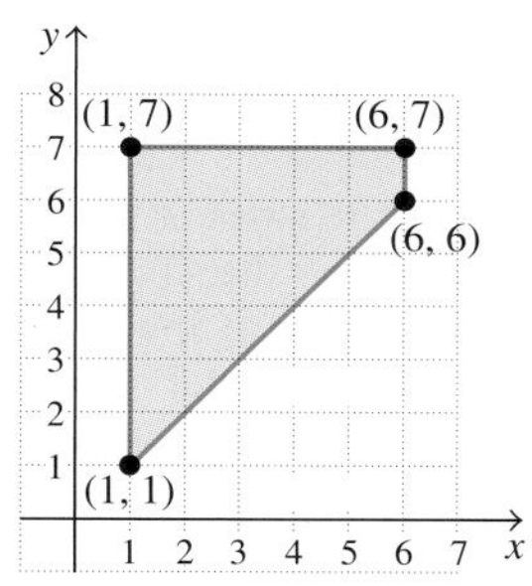

Figure 7.16

Since $20x$ is the profit on x washers and $30y$ is the profit on y dryers, the objective function, the profit formula, is $P = 20x + 30y$. Figure 7.16 shows the feasible region. The feasible region consists of the shaded region and the boundaries. The vertices of the feasible region are the points (1, 1), (1, 7), (6, 7), and (6, 6).

Next we calculate the value of the objective function, P, at each one of the vertices.

$$P = 20x + 30y$$

At (1, 1), $P = 20(1) + 30(1) = 50$

At (1, 7), $P = 20(1) + 30(7) = 230$

At (6, 7), $P = 20(6) + 30(7) = 330$

At (6, 6), $P = 20(6) + 30(6) = 300$

The maximum profit is at (6, 7). This means the company should manufacture 6 washers and 7 dryers to maximize their profit. The maximum profit is $330.

Use the following steps to solve a linear programming problem.

Solving a Linear Programming Problem

1. Determine all necessary constraints.
2. Determine the objective function.
3. Graph the constraints and determine the feasible region.
4. Determine the vertices of the feasible region.
5. Determine the value of the objective function at each vertex.

The solution is determined by the vertex that yields the maximum or minimum value of the objective function.

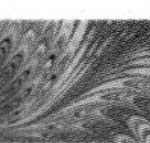

SECTION 7.6 EXERCISES

Concept/Writing Exercises

1. What are constraints in a linear programming problem? How are they represented?
2. In a linear programming problem, how is a feasible region formed?
3. What are the points of intersection of the boundaries of the feasible region called?
4. What is the general form of the objective function?
5. In your own words, state the fundamental principle of linear programming.
6. A profit function is $P = 4x + 6y$ and the vertices of the feasible region are (1, 1), (1, 4), (5, 1), and (7, 1). Determine the maximum profit. Explain how you determined your answer.

Practice the Skills

Exercises 7 and 8 show a feasible region and its vertices. Find the maximum and minimum values of the given objective function.

7. $K = 6x + 4y$

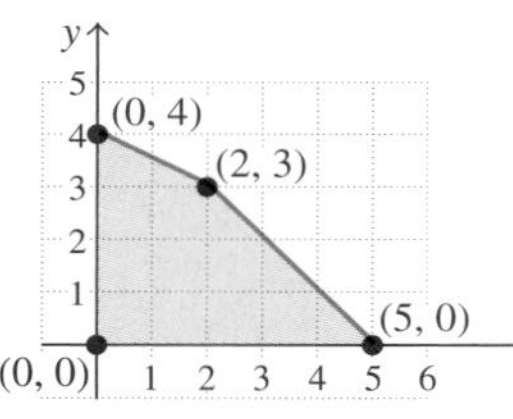

8. $K = 2x + 3y$

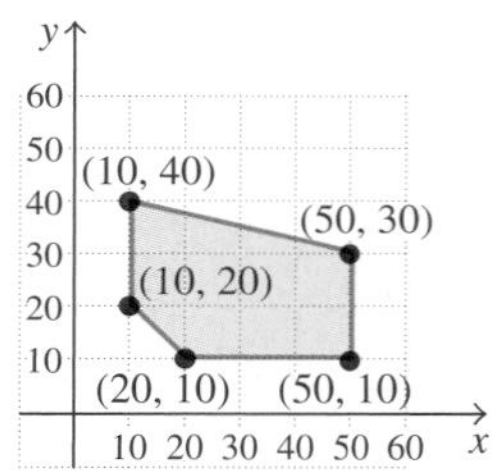

In Exercises 9–14, a set of constraints and a profit formula are given.

a) Draw the graph of the constraints and find the vertices of the feasible region.

b) Use the vertices as obtained in part (a) to determine the maximum and minimum profit.

9. $x + y \le 5$
$2x + y \le 8$
$x \ge 0$
$y \ge 0$
$P = 5x + 4y$

10. $2x + 3y \le 12$
$2x + y \le 8$
$x \ge 0$
$y \ge 0$
$P = 2x + 4y$

11. $x + y \le 4$
$x + 3y \le 6$
$x \ge 0$
$y \ge 0$
$P = 7x + 6y$

12. $x + y \le 50$
$x + 3y \le 90$
$x \ge 0$
$y \ge 0$
$P = 20x + 40y$

13. $4x + 3y \ge 12$
$3x + 4y \le 36$
$x \ge 2$
$y \le 5$
$y \ge 1$
$P = 2.20x + 1.65y$

14. $x + 2y \le 14$
$7x + 4y \ge 28$
$x \ge 2$
$x \le 10$
$y \ge 1$
$P = 15.13x + 9.35y$

Problem Solving

15. **MODELING** - *Stocking Film* A camera store stocks two brands of film, Kodak and Fuji. The manager does not want to keep more than 24 rolls of film on hand. She wants to stock at least twice as many rolls of Kodak film as Fuji film. She also wants to stock at least 4 rolls of Fuji film. Assume that she makes a profit of \$0.35 on Kodak film and a profit of \$0.50 on Fuji film.
    a) List the constraints.
    b) Determine the objective function.
    c) Graph the set of constraints.
    d) Determine the vertices of the feasible region.
    e) How many rolls of each brand of film should she stock to maximize her profit?
    f) Determine the maximum profit.

16. **MODELING** - *On Wheels* The Boards and Blades Company manufactures skateboards and in-line skates. The company can produce a maximum of 20 skateboards and pairs of in-line skates per day. It makes a profit of \$25 on a skateboard and a profit of \$20 on a pair of in-line skates. The company's planners want to make at least 3 skateboards but not more than 6 skateboards per day. To keep customers happy, they must make at least 2 pairs of in-line skates per day.
    a) List the constraints.
    b) Determine the objective function.
    c) Graph the set of constraints.
    d) Determine the vertices of the feasible region.
    e) How many skateboards and pairs of in-line skates should be made to maximize the profit?
    f) Find the maximum profit.

Central Park, New York City

17. **MODELING** - *Paint Production* A paint supplier has two machines that produce both indoor paint and outdoor paint. To meet one of its contractual obligations, the company must produce at least 60 gal of indoor paint and 100 gal of outdoor paint. Machine I makes 3 gal of indoor paint and 10 gal of outdoor paint per hour. Machine II makes 4 gal of indoor paint and 5 gal of outdoor paint per hour. It costs $28 per hour to run machine I and $33 per hour to run machine II.

a) List the constraints.
b) Determine the objective function.
c) Graph the set of constraints.
d) Determine the vertices of the feasible region.
e) How many hours should each machine be operated to fulfill the contract at a minimum cost?
f) Determine the minimum cost.

Challenge Problems/Group Activities

18. **MODELING** - *Hot Dog Profits* To make one package of all-beef hot dogs, a manufacturer uses 1 lb of beef; to make one package of regular hot dogs, the manufacturer uses $\frac{1}{2}$ lb each of beef and pork. The profit on the all-beef hot dogs is 40 cents per pack and the profit on regular hot dogs is 30 cents per pack. If there are 200 lb of beef and 150 lb of pork available, how many packs of all-beef and regular hot dogs should the manufacturer make to maximize the profit? What is the profit?

19. **MODELING** - *Car Seats and Strollers* A company makes car seats and strollers. Each car seat and stroller passes through three processes: assembly, safety testing, and packaging. A car seat requires 1 hr in assembly, 2 hr in safety testing, and 1 hr in packaging. A stroller requires 3 hr in assembly, 1 hr in safety testing, and 1 hr in packaging. Employee work schedules allow for 24 hr per day for assembly, 16 hr per day for safety testing, and 10 hr per day for packaging. The profit for each car seat is $25 and the profit for each stroller is $35. How many units of each type should the company make per day to maximize the profit? What is the maximum profit?

Internet/Research Activity

20. *Operations research* draws on several disciplines, including mathematics, probability theory, statistics, and economics. George Dantzig was one of the key people in developing operations research. Write a paper on Dantzig and his contributions to operations research and linear programming.

CHAPTER 7 SUMMARY

IMPORTANT FACTS

Systems of equations

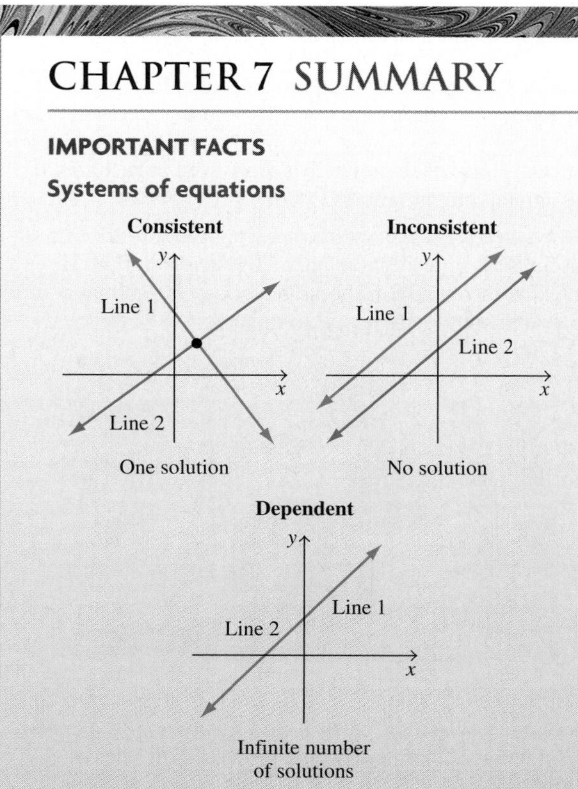

Methods of solving systems of equations

1. Graphing
2. Substitution
3. Addition (or elimination) method

Multiplicative identity matrix

$$\begin{bmatrix} 1 & 0 \\ 0 & 1 \end{bmatrix}$$

Fundamental principle of linear programming

If the objective function $K = Ax + By$ is evaluated at each point in a feasible region, the maximum and minimum values of the equation occur at vertices of the region.

CHAPTER 7 REVIEW EXERCISES

7.1

In Exercises 1–4, solve the system of equations graphically. If the system does not have a single ordered pair as a solution, state whether the system is inconsistent or dependent.

1. $x = 3$
$y = 4$

2. $2x + y = 5$
$-3x + y = 5$

3. $x = 3$
$x + y = 5$

4. $x + 2y = 5$
$2x + 4y = 4$

In Exercises 5–8, determine without graphing whether the system of equations has exactly one solution, no solution, or an infinite number of solutions.

5. $y = \frac{2}{3}x + 5$
$3y - 2x = 15$

6. $2y - 4x = 12$
$2y = 4x + 15$

7. $6y - 2x = 20$
$4y + 2x = 10$

8. $2x - 4y = 8$
$-2x + y = 6$

7.2

In Exercises 9–12, solve the system of equations by the substitution method. If the system does not have a single ordered pair as a solution, state whether the system is inconsistent or dependent.

9. $-x + y = 12$
$x + 2y = -3$

10. $x - 2y = 9$
$y = 2x - 3$

11. $2x - y = 4$
$3x - y = 2$

12. $3x + y = 1$
$3y = -9x - 4$

In Exercises 13–18, solve the system of equations by the addition method. If the system does not have a single ordered pair as a solution, state whether the system is inconsistent or dependent.

13. $x - 2y = 1$
$2x + y = 7$

14. $2x + y = 2$
$-3x - y = 5$

15. $x + y = 2$
$x + 3y = -2$

16. $4x - 8y = 16$
$x - 2y = 4$

17. $3x + 5y = 15$
$2x + 4y = 0$

18. $3x + 4y = 6$
$2x - 3y = 4$

7.3

Given $A = \begin{bmatrix} 1 & -3 \\ 2 & 4 \end{bmatrix}$ *and* $B = \begin{bmatrix} -2 & -5 \\ 6 & 3 \end{bmatrix}$, *determine the following.*

19. $A + B$

20. $A - B$

21. $2A$

22. $2A - 3B$

23. $A \times B$

24. $B \times A$

7.4

In Exercises 25–30, use matrices to solve the system of equations.

25. $x + 2y = 6$
$x + y = 4$

26. $-x + y = 4$
$x + 2y = 2$

27. $2x + y = 3$
$3x - y = 12$

28. $2x + 3y = 2$
$4x - 9y = 4$

29. $x + 3y = 3$
$3x - 2y = 2$

30. $3x - 6y = -9$
$4x + 5y = 14$

7.1–7.4

31. **MODELING** - *Borrowing Money* A company borrows \$600,000 for 1 year to expand its product line. Some of the money was borrowed at an 8% simple interest rate and the rest of the money was borrowed at a 10% simple interest rate. How much money was borrowed at each rate if the annual interest was \$53,000?

32. **MODELING** - *Chemistry* In chemistry class, Mark Damon has an 80% acid solution and a 50% acid solution. How much of each solution should he mix to get 100 liters of a 75% acid solution?

33. **MODELING** - *A Salesman's Earnings* Tom Northrup, an electronics salesman, earns a weekly salary plus a commission on sales. In one week, his salary on sales of \$4000 was \$660. The next week, his salary on sales of

$6000 was $740. Determine his weekly salary and his commission rate.

34. **MODELING** - *Cool Air* Emily Richelieu needs to purchase a new air conditioner for the office. Model 1600A costs $950 to purchase and $32 per month to operate. Model 6070B, a more efficient unit, costs $1275 to purchase and $22 per month to operate.
   a) After how many months will the total cost of both units be equal?
   b) Which model will be the more cost effective if the life of both units is guaranteed for 10 years?

35. **MODELING** - *Minimizing Parking Costs* The cost of parking in All-Day parking lot is $5 for the first hour and $0.50 for each additional hour. Sav-a-Lot parking lot costs $4.25 for the first hour and $0.75 for each additional hour.
   a) In how many hours would the total cost of parking at All-Day and Sav-a-Lot be the same?
   b) If Mark McMahon needed to park his car for 5 hr, which parking lot would be less expensive?

7.5

In Exercises 36–39, graph the system of linear inequalities and indicate the solution set.

36. $y \leq 3x - 1$
    $y > -2x + 1$

37. $2x + y < 8$
    $y \geq 2x - 1$

38. $x + 3y \leq 6$
    $2x - 7y \geq 14$

39. $x - y > 5$
    $6x + 5y \leq 30$

7.6

40. For the following set of constraints and profit formula, graph the constraints and find the vertices. Use the vertices to determine the maximum profit.

$$\begin{aligned} x + y &\leq 10 \\ 2x + 1.8y &\leq 18 \\ x &\geq 0 \\ y &\geq 0 \\ P &= 6x + 3y \end{aligned}$$

CHAPTER 7 TEST

1. From a graph, explain how you would identify a consistent system of equations, an inconsistent system of equations, and a dependent system of equations.

2. Solve the system of equations graphically.

$$\begin{aligned} y &= -2x - 3 \\ -2x + y &= -11 \end{aligned}$$

3. Determine without graphing whether the system of equations has exactly one solution, no solution, or an infinite number of solutions.

$$\begin{aligned} 4x + 5y &= 6 \\ -3x + 5y &= 13 \end{aligned}$$

Solve the system of equations by the method indicated.

4. $x - y = 5$
   $2x + 3y = -5$
   (substitution)

5. $y = 5x + 7$
   $y = 2x + 1$
   (substitution)

6. $x - y = 4$
   $2x + y = 5$
   (addition)

7. $4x + 3y = 5$
   $2x + 4y = 10$
   (addition)

8. $3x + 4y = 6$
   $2x - 3y = 4$
   (addition)

9. $x + 3y = 4$
   $5x + 7y = 4$
   (matrices)

In Exercises 10–12, for $A = \begin{bmatrix} 2 & -5 \\ 1 & 3 \end{bmatrix}$ *and* $B = \begin{bmatrix} -1 & -3 \\ 5 & 2 \end{bmatrix}$, *determine the following.*

10. $A + B$

11. $3A - B$

12. $A \times B$

13. Graph the system of linear inequalities and indicate the solution set.

$$\begin{aligned} y &< -2x + 2 \\ y &> 3x + 2 \end{aligned}$$

Solve Exercises 14 and 15 by using a system of equations.

14. **MODELING** - *A Coffee Blend* Louis DiMento plans to mix coffee that sells for $6 per pound with coffee that sells for $7.50 per pound to get a 30 lb blend that sells for $7 per pound. How many pounds of each type should Louis use?

15. **MODELING** - *Checking Accounts* The charge for maintaining a checking account at Union Bank is \$6 per month plus 10 cents for each check that is written. The charge at Citrus Bank is \$2 per month and 20 cents per check.
    a) How many checks would a customer have to write in a month for the total charges to be the same at both banks?
    b) If Brent Pickett planned to write 14 checks per month, which bank would be the least expensive?

16. The set of constraints and profit formula for a linear programming problem are

$$\begin{aligned} x + 3y &\le 6 \\ 4x + 3y &\le 15 \\ x &\ge 0 \\ y &\ge 0 \\ P &= 5x + 3y \end{aligned}$$

    a) Draw the graph of the constraints and determine the vertices of the feasible region.
    b) Use the vertices to determine the maximum and minimum profit.

GROUP PROJECTS

1. Make up three different systems of equations that have (1, 4) as a solution. Explain how you determined your systems.

Linear Programming

2. **MODELING** - *Profit from Bookcases* The Bookholder Company manufactures two types of bookcases out of oak and walnut. Model 01 requires 5 board feet of oak and 2 board feet of walnut. Model 02 requires 4 board feet of oak and 3 board feet of walnut. A profit of \$75 is made on each Model 01 bookcase and a profit of \$125 is made on each Model 02 bookcase. The company has a supply of 1000 board feet of oak and 600 board feet of walnut. The company has orders for 40 Model 01 bookcases and 50 Model 02 bookcases. These orders indicate the minimum number the company must manufacture of each model.
    a) Write the set of constraints.
    b) Write the objective function.
    c) Graph the set of constraints.
    d) Determine the number of bookcases of each type the company should manufacture in order to maximize profits.
    e) Determine the maximum profit.

Create Your Own Word Problem

3. **a)** Write a word problem that can be solved by using a system of two equations with two unknowns.
    b) For the problem in part (a), write the system of equations and find the answer.
    c) Explain how you developed the problem in part (a).

The 2000–2004 Chevrolet Corvettes have a standard 5.7-liter engine.

THE METRIC SYSTEM

On a trip to Italy or Spain, you might have a difficult time asking for and understanding directions if you do not speak the language of that country. Purchasing clothing or gasoline may also be difficult in those countries because the measurement system is different than ours. Our neighbors in Mexico and Canada, like almost all countries around the world, also use a different measurement system than we use in the United States. Most countries of the world use the *Système international d'unités*, or abbreviated, the *SI system*. In the United States, the SI system is most commonly referred to as the metric system.

The system of measurement most commonly used in the United States is called the U.S. customary system, but metric units are used in many ways in the United States. You can purchase metric tools at your local hardware store. Some clothing is measured in metric units, as are car tires. Drinks are sold in liter bottles, and vitamins and medicines are labeled in milligrams.

As you study this chapter, you will see the many advantages of the metric system. You eventually may even support the movement to change from the U.S. customary system to the metric system.

8.1 BASIC TERMS AND CONVERSIONS WITHIN THE METRIC SYSTEM

Most countries of the world use the *Système international d'unités* or *SI system*. The SI system is generally referred to as the *metric system* in the United States. The metric system was named for the Greek word *metron,* meaning "measure." The standard units in the metric system have gone through many changes since the system was first developed in France during the French Revolution. For example, one unit of measure, the meter, was first defined as one ten-millionth of the distance between the North Pole and the equator. Later, the meter was defined as 1,650,763.73 wavelengths of the orange–red line of krypton 86. Since 1893, the meter has been defined as the distance traveled by light in a vacuum in $\frac{1}{299{,}792{,}458}$ of a second.

Two systems of weights and measures exist side by side in the United States today, the *U.S. customary system* and the metric system. The metric system is used predominantly in the automotive, construction, farm equipment, computer, and bottling industries and in health-related professions. Furthermore, almost every industry that ships internationally uses at least some metric measures.

In this chapter, we will discuss metric measurements of length, area, volume, mass, and temperature. Using the metric system has many advantages. Some of them are summarized here.

1. The metric system is the worldwide accepted standard measurement system. All industrial nations that trade internationally, except the United States, use the metric system as the official system of measurement.
2. There is only one basic unit of measurement for each physical quantity. In the U.S. customary system, many units are often used to represent the same physical quantity. For example, when discussing length, we use inches, feet, yards, miles, and so on. Converting from one of these units to the other is often a tedious task (consider changing 12 miles to inches). In the metric system, we can make many conversions by simply moving the decimal point.
3. The SI system is based on the number 10, and there is less need for fractions, because most quantities can be expressed as decimals.

DID YOU KNOW

Lost in Space

In September 1999, the United States lost the *Mars Climate Orbiter* as it approached Mars. The loss of the $125 million spacecraft was due to scientists confusing U.S. customary units and metric units. Two spacecraft teams, one at NASA's Jet Propulsion Laboratory (JPL) in Pasadena, California, and the other at a Lockheed Martin facility in Colorado, where the spacecraft was built, were unknowingly exchanging some vital information in different measurement units.

The spacecraft team in Colorado used U.S. customary units of *pounds* of force to describe small forces needed to adjust the spacecraft's orbit. The data were shipped via computer, without units, to the JPL, where the navigation team was expecting to receive the information in newtons, a metric measure of force. The mix-up in units led to the JPL scientists giving the spacecraft's computer wrong information, which threw the spacecraft off course. That, in turn, led to the spacecraft entering the Martian atmosphere, where it burned up. NASA has taken steps to prevent this error from ever happening again.

The missing *Mars Climate Orbiter*

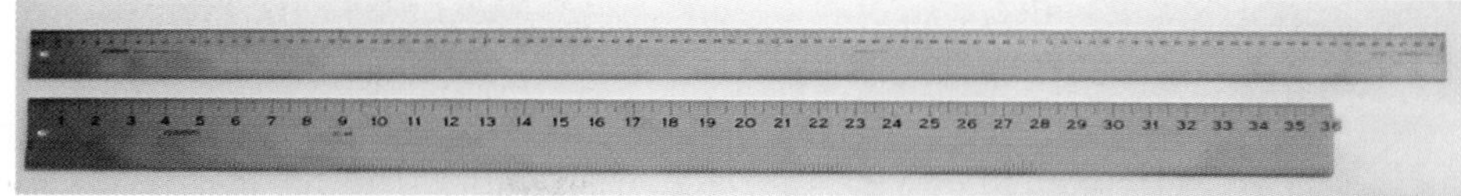

A meter is a little longer than a yard.

Basic Terms

One- and two-liter bottles

Because the official definitions of many metric terms are quite technical, we present them informally.

The *meter* (m) is commonly used to measure *length* in the metric system. One meter is a little more than a yard (see the photo above). A door is about 2 meters high.

The *kilogram* (kg) is commonly used to measure *mass*. (The difference between mass and weight is discussed in Section 8.3.) One kilogram is about 2.2 pounds. A newborn baby may have a mass of about 3 kilograms. The gram (g), a unit of mass derived from the kilogram, is used to measure small amounts. A nickel has a mass of about 5 grams.

The *liter* (ℓ) is commonly used to measure *volume*. One liter is a little more than a quart. The gas tank of a compact car may hold 50 liters of gasoline.

Thus,

$$1 \text{ m} \approx 1 \text{ yd}$$
$$1 \text{ kg} \approx 2.2 \text{ lb}$$
$$1\ \ell \approx 1 \text{ qt}$$

The term *degree Celsius* (°C) is used to measure temperature. The freezing point of water is 0°C, and the boiling point of water is 100°C. The temperature on a warm day may be 30°C.

$$0°\text{C} = 32°\text{F} \quad \text{Water freezes}$$
$$22°\text{C} = 71.6°\text{F} \quad \text{Comfortable room temperature}$$
$$37°\text{C} = 98.6°\text{F} \quad \text{Body temperature}$$
$$100°\text{C} = 212°\text{F} \quad \text{Water boils}$$

Prefixes

The metric system is based on the number 10 and therefore is a decimal system. Prefixes are used to denote a multiple or part of a base unit. Table 8.1 on page 434 summarizes the more commonly used prefixes and their meanings. In the table, where we mention "base units" we mean metric units without prefixes, such as meter, gram, or liter. From Table 8.1 on page 434, we can determine that a *deka*meter represents 10 meters, and a *centi*meter represents $\frac{1}{100}$ of a meter. Also, 1 kiloliter = 1000 liters, 1 kilogram = 1000 grams, and 1 milliliter = $\frac{1}{1000}$ liter.

In the metric system, as used outside the United States, groups of three digits in large numbers are separated by a space, not a comma. For example, the number for thirty thousand is 30 000, and the number for nine million is 9 000 000. Groups of three digits to the right of the decimal point are also separated by spaces. Commas are not used in the SI system because many countries use the comma as we use the decimal point. For example, 16 millionths is written 0,000 016 in many countries of the world. We will use the decimal point in this book and write 0.000 016. In this section, we will separate groups of three digits using spaces as done outside the United States. Note, however, that the space between groups of three digits is usually omitted if there

The computer's hard drive can store 80 gigabytes (80 GB) of information.

TABLE 8.1 Metric Prefixes

Prefix	Symbol	Meaning
kilo	k	$1000 \times$ base unit
hecto	h	$100 \times$ base unit
deka	da	$10 \times$ base unit
—	—	base unit
deci	d	$\frac{1}{10}$ of base unit
centi	c	$\frac{1}{100}$ of base unit
milli	m	$\frac{1}{1000}$ of base unit

are only four digits to the left or right of the decimal point. Thus, we will write three thousand as 3000 and five ten-thousandths as 0.0005.

For scientific work, which involves very large and very small quantities, the following prefixes are also used: *mega* (M) is one million times the base unit, *giga* (G) is one billion times the base unit, *tera* (T) is one trillion times the base unit, *micro* (μ, the Greek letter mu) is one millionth of the base unit, *nano* (n) is one billionth of the base unit, and *pico* (p) is one trillionth of the base unit.

In this book, the abbreviations or symbols for units of measure are not pluralized, but full names are. For example, 5 milliliters is symbolized as 5 mℓ, not 5 mℓs. Some countries that use the metric system do not use an "s" in their abbreviations, whereas others do.

Conversions within the Metric System

We will use Table 8.2 to help demonstrate how to change from one metric unit to another metric unit (meters to kilometers and so on).

The meters in Table 8.2 can be replaced by grams, liters, or any other base unit of the metric system. Regardless of which unit we choose, the procedure is the same. For purposes of explanation, we have used the meter.

TABLE 8.2 Changing Metric Units

Measure of length	kilometer	hectometer	dekameter	meter	decimeter	centimeter	millimeter
Symbol	km	hm	dam	m	dm	cm	mm
Number of meters	1000 m	100 m	10 m	1 m	0.1 m	0.01 m	0.001 m

Table 8.2 shows that 1 hectometer equals 100 meters and 1 millimeter is 0.001 (or $\frac{1}{1000}$) meter. The millimeter is the smallest unit in the table. A centimeter is 10 times as large as a millimeter, a decimeter is 10 times as large as a centimeter, a meter is 10 times as large as a decimeter, and so on. Because each unit is 10 times as large as the unit on its right, converting from one unit to another is simply a matter of multiplying or dividing by powers of 10.

Our neighbors in Canada (and also Mexico) use the metric system. As you will learn shortly, the distance to the Botanical Gardens is about 0.6 mile and the distance to Niagara-on-the-Lake is about 9 miles from the sign.

Changing Units within the Metric System

1. To change from a smaller unit to a larger unit (for example, from meters to kilometers), move the decimal point in the original quantity one place to the left for each larger unit of measurement until you obtain the desired unit of measurement.
2. To change from a larger unit to a smaller unit (for example from kilometers to meters), move the decimal point in the original quantity one place to the right for each smaller unit of measurement until you obtain the desired unit of measurement.

DID YOU KNOW

Additional Metric Prefaces

There are many publications about the metric system available free from the U.S. government. You may contact the National Institute of Standards and Technology (under the U.S. Department of Customs), through their Web site at www.nist.gov/, or you may write to their office (Gaithersburg, MD 20899). Two worthwhile publications are *Metric Style Guide for the News Media* and *A Brief History of Measurement Systems*. The following interesting chart was selected and modified from the latter.

*Some countries use D for deka.

METRIC PREFIXES

Multiples and Submultiples	Prefixes	Symbols
1 000 000 000 000 000 000 000 000 = 10^{24}	yotta	Y
1 000 000 000 000 000 000 000 = 10^{21}	zetta	Z
1 000 000 000 000 000 000 = 10^{18}	exa	E
1 000 000 000 000 000 = 10^{15}	peta	P
1 000 000 000 000 = 10^{12}	tera	T
1 000 000 000 = 10^{9}	giga	G
1 000 000 = 10^{6}	mega	M
1000 = 10^{3}	kilo	k
100 = 10^{2}	hecto	h
10 = 10^{1}	deka	da*
1 = 10^{0}		
0.1 = 10^{-1}	deci	d
0.01 = 10^{-2}	centi	c
0.001 = 10^{-3}	milli	m
0.000 001 = 10^{-6}	micro	µ
0.000 000 001 = 10^{-9}	nano	n
0.000 000 000 001 = 10^{-12}	pico	p
0.000 000 000 000 001 = 10^{-15}	femto	f
0.000 000 000 000 000 001 = 10^{-18}	atto	a
0.000 000 000 000 000 000 001 = 10^{-21}	zepto	z
0.000 000 000 000 000 000 000 001 = 10^{-24}	yocto	y

EXAMPLE 1 *Changing Units*

a) Convert 462.3 m to km.

b) Convert 14 g to cg.

c) Convert 0.76 ℓ to mℓ.

d) Convert 240 daℓ to kℓ.

SOLUTION:

a) Table 8.2 shows that dekameters, hectometers, and kilometers are all larger units of measurements than meters. Kilometers appear three places to the left of meters in the table. Therefore, to change a measure from meters to kilometers, we must move the decimal point in the given number three places to the left, or

$$462.3 \text{ m} = 0.4623 \text{ km}$$

Note that since we are changing from a smaller unit of measurement (meter) to a larger unit of measurement (kilometer), the answer will be a smaller number of units.

b) Grams are a larger unit of measurement than centigrams. To convert grams to centigrams, we move the decimal point two places to the right, or

$$14 \text{ g} = 1400 \text{ cg}$$

Note that since we are changing from a larger unit of measurement (gram) to a smaller unit of measurement (centigram), the answer will be a larger number of units.

c) 0.76 ℓ = 760 mℓ

d) 240 daℓ = 2.40 kℓ ▲

EXAMPLE 2 *Two More Conversions*

a) Convert 305 mm to hectometers.

b) Convert 6.34 dam to decimeters.

SOLUTION:

a) Table 8.2 shows that hectometers are five places to the left of millimeters. Therefore, to make the conversion, we must move the decimal point in the given number five places to the left, or

$$305 \text{ mm} = 0.003\,05 \text{ hm}$$

b) Table 8.2 shows that decimeters are two places to the right of dekameters. Therefore, to make the conversion, we must move the decimal point in the given number two places to the right, or

$$6.34 \text{ dam} = 634 \text{ dm}$$

▲

EXAMPLE 3 *A Metric Road Sign*

The sign in the photo, from Cancun, Mexico, shows that there is a truck crossing 200 meters ahead. Notice that the sign uses "MTS" for meters.

a) Determine the distance in kilometers.

b) Determine the distance in centimeters.

SOLUTION:

a) We must move the decimal point three places to the left to change from meters to kilometers. Therefore,

$$200 \text{ m} = 0.2 \text{ km}$$

b) We must move the decimal point two places to the right to change from meters to centimeters. Therefore,

$$200 \text{ m} = 20\,000 \text{ cm}$$

▲

EXAMPLE 4 *Comparing Lengths*

Arrange in order from the smallest to largest length: 3.4 m, 3421 mm, and 104 cm.

SOLUTION:
To be compared, these lengths should all be in the same units of measure. Let's convert all the measures to millimeters, the smallest units of the lengths being compared.

$$3.4 \text{ m} = 3400 \text{ mm} \qquad 3421 \text{ mm} \qquad 104 \text{ cm} = 1040 \text{ mm}$$

Since the lengths, in millimeters, from smallest to largest are 1040, 3400, 3421, the lengths arranged in order from smallest to largest are 104 cm, 3.4 m, and 3421 mm.

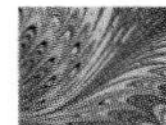

SECTION 8.1 EXERCISES

Concept/Writing Exercises

1. What is the name commonly used for the Système international d'unités in the United States?
2. What is the name of the system of measurement primarily used in the United States today?
3. List three advantages of the metric system.
4. What metric unit is commonly used to measure
   a) length?
   b) mass?
   c) volume?
   d) temperature?
5. **a)** Explain how to convert from one metric unit of length to a different metric unit of length. Then use this procedure in parts (b) and (c).
   b) Convert 714.6 cm to kilometers.
   c) Convert 30.8 hm to decimeters.
6. What is the name of the prefix that is
   a) a million times the basic unit?
   b) one millionth of the base unit?
7. Without referring to any table, name as many of the metric system prefixes as you can and give their meanings. If you don't already know all the prefixes in Table 8.1, memorize them now.
8. **a)** How many times greater is 1 hectometer than 1 centimeter?
   b) Convert 1 hm to centimeters.
   c) Convert 1 cm to hectometers.
9. **a)** How many times greater is 1 dam than 1 dm?
   b) Convert 1 dam to decimeters.
   c) Convert 1 dm to dekameters.
10. **a)** What is the freezing temperature of water in the metric system?
    b) What is the boiling point of water in the metric system?
    c) What is normal human body temperature in the metric system?

Practice the Skills

In Exercises 11–16, fill in the blank.

11. One kilogram is a little more than _____ pounds.
12. One meter is a little longer than a _____.
13. One nickel has a mass of about _____ grams.
14. The temperature on a warm day may be _____ °C.
15. A comfortable room temperature may be _____ °C.
16. A door may be _____ meters high.

In Exercises 17–22, match the prefix with the one letter, a–f, that gives the meaning of the prefix.

17. Kilo	**a)** $\frac{1}{100}$ of base unit
18. Milli	**b)** $\frac{1}{1000}$ of base unit
19. Hecto	**c)** 100 times base unit
20. Deka	**d)** 1000 times base unit
21. Deci	**e)** 10 times base unit
22. Centi	**f)** $\frac{1}{10}$ of base unit

23. Complete the following.
    a) 1 dekaliter = ____ liters
    b) 1 centiliter = ____ liter
    c) 1 milliliter = ____ liter
    d) 1 deciliter = ____ liter
    e) 1 kiloliter = ____ liters
    f) 1 hectoliter = ____ liters
24. Complete the following.
    a) 1 hectogram = ____ grams
    b) 1 milligram = ____ gram
    c) 1 kilogram = ____ grams
    d) 1 centigram = ____ gram
    e) 1 dekagram = ____ grams
    f) 1 decigram = ____ gram

In Exercises 25–30, without referring to any of the tables or your notes, give the symbol and the equivalent in grams for the unit.

25. Milligram **26.** Centigram **27.** Decigram
28. Dekagram **29.** Hectogram **30.** Kilogram

Aerial Tram Load In Exercises 31 and 32, use the photo, which shows the maximum load for an aerial tram in Switzerland. (Notice that in some countries, an "s" is used on the metric abbreviations.)

31. What is the maximum load in grams?

32. What is the maximum load in milligrams?

In Exercises 33–42, fill in the missing values.

33. 2 m = ____ cm
34. 35.7 hg = ____ g
35. 0.095 hℓ = ____ ℓ
36. 7 dam = ____ m
37. 242.6 cm = ____ hm
38. 1.34 mℓ = ____ ℓ
39. 4036 mg = ____ hg
40. 14.27 kℓ = ____ ℓ
41. 1.34 hm = ____ cm
42. 0.000 062 kg = ____ mg

In Exercises 43–50, convert the given unit to the unit indicated.

43. 92.5 kg to grams
44. 7.3 m to millimeters
45. 895 ℓ to milliliters
46. 24 dm to kilometers
47. 240 cg to hectograms
48. 6049 mm to meters
49. 40 302 mℓ to dekaliters
50. 0.034 mℓ to liters

In Exercises 51–56, arrange the quantities in order from smallest to largest.

51. 5.1 dam, 0.47 km, 590 cm
52. 514 hm, 62 km, 680 m
53. 2.2 kg, 2400 g, 24 300 dg
54. 4.3 ℓ, 420 cℓ, 0.045 kℓ
55. 2.6 km, 203 000 mm, 52.6 hm
56. 0.032 kℓ, 460 dℓ, 48 000 cℓ

Problem Solving

57. *Who Ran Faster* Jim ran 100 m, and Bob ran 100 yd in the same length of time. Who ran faster? Explain.

58. *Walking* Would you be walking faster if you walked 1 dam in 10 min or 1 hm in 10 min? Explain.

59. *Water Removal* One pump removes 1 daℓ of water in 1 min, and another pump removes 1 dℓ of water in 1 min. Which pump removes water faster? Explain.

60. *Balance* If 5 kg are placed on one side of a balance and a 15 lb weight is placed on the other side, which way would the balance tip? Explain.

61. *Framing a Masterpiece* The painting by Picasso, including the frame, measures 74 cm by 99 cm.

a) How many centimeters of framing were needed to frame the painting?
b) How many millimeters of framing were needed to frame the painting?

62. *Calcium Tablets* Dr. Driscoll recommends that Sean take two 250 mg chewable calcium tablets each day.
a) How many milligrams of calcium will Sean take in a week?
b) How many grams of calcium will Sean take in a week?

63. *A Home Run* A baseball diamond is a square whose sides are about 27 m in length.
   a) How many meters does a batter run if he hits a home run?
   b) How many kilometers?
   c) How many millimeters?

64. *Gas Consumption* Dale Ewen drove 1200 km and used 187 ℓ of gasoline. What was his average rate of gas use for the trip
   a) in kilometers per liter?
   b) in meters per liter?

65. *Track and Field* The high school has a 400 m oval track. If Patty Burgess runs around the track eight times, how many kilometers has she traveled?

66. *Aerating an Aquarium* The filter pump on an aquarium circulates 360 mℓ of water every minute. If the aquarium holds 30 ℓ of water, how long will it take to circulate all the water?

67. *Liters of Soda* A bottle of soda contains 360 mℓ.
   a) How many milliliters are contained in a six-bottle carton?
   b) How many liters does the amount in part (a) equal?
   c) At $2.45 for the carton of soda, what is its cost per liter?

68. *Fill 'er Up* In Europe, gas may cost the equivalent of about $1.03 (American) per liter. What will be the cost of filling the gas tank of a car that has a capacity of 37.7 ℓ?

69. *Turkey Dinner* After a turkey is cooked it weighs 6.9 kg.
   a) What is its weight in grams?
   b) If Marie Sinclair cuts off one-third of the turkey and places it in the freezer, how many decigrams of turkey has she placed in the freezer?

70. *Road Signs* A road sign indicated that you are 750 km from the city of Kingston, Canada. A while later, while riding on the same road, a different road sign indicated that you are 325 hm from Kingston. How many kilometers have you traveled?

Challenge Exercises/Group Activities

In Exercises 71–74, fill in the blank to make a true statement.

71. 1 gigameter = ____ megameters
72. 1 nanogram = ____ micrograms
73. 1 teraliter = ____ picoliters
74. 1 megagram = ____ nanograms

Calcium The recommended daily amount of calcium for an American adult is 0.8 g. In Exercises 75–78, how much of the food indicated must an adult eat to satisfy the entire daily allowance using only that food?

75. Eggs: 1 egg contains 27 mg calcium.
76. Milk: 1 cup contains 288 mg calcium.
77. Broccoli: 1 cup (cooked) contains 195 mg calcium.
78. Raisin bran: 49 g contains 1.6 mg calcium.

Large and Small Numbers One advantage of the metric system is that by using the proper prefix, you can write large and small numbers without large groups of zeroes. In Exercises 79–84, write an equivalent measurement with an amount greater than one that does not contain any zeroes. For example, you can write 3000 m without zeroes as 3 km and 0.0003 hm as 3 cm.

79. 5000 cm
80. 2000 mm
81. 0.000 06 hg
82. 3000 dm
83. 0.02 kℓ
84. 500 cm

Recreational Mathematics

In Exercises 85–94, unscramble the word to make a metric unit of measurement.

85. magr
86. migradec
87. rteli
88. raktileed
89. terem
90. leritililm
91. reketolim
92. timenceret
93. greseed sulesic
94. togmeharc

Internet/Research Activities

95. Write a report on the development of the metric system in Europe. Indicate which individual people had the most influence in its development.

96. Write a report on why you believe many Americans oppose switching to the metric system. Give your opinion about whether the United States will eventually switch to the metric system and, if so, when it might do so.

8.2 LENGTH, AREA, AND VOLUME

This section and the next section are designed to help you *think metric*, that is, to become acquainted with day-to-day usage of metric units. In this section, we consider length, area, and volume.

Length

The basic unit of length is the meter. In all English-speaking countries except the United States, *meter* is spelled "metre." Until 1960, the meter was officially defined by the length of a platinum bar kept in a vault in France. The modern definition of the meter is based on the speed of light, a constant that has been defined with great precision. Other commonly used units of length are the kilometer, centimeter, and millimeter. The meter, which is a little longer than 1 yard, is used to measure things that we normally measure in yards and feet. A man whose height is about 2 meters is a tall man. A tractor trailer unit (an 18-wheeler) is about 18 meters long.

The kilometer is used to measure what we normally measure in miles. For example, the distance from New York to Seattle is about 5120 kilometers. One kilometer is about 0.6 mile, and 1 mile is about 1.6 kilometers.

Centimeters and millimeters are used to measure what we normally measure in inches. The centimeter is a little less than $\frac{1}{2}$ inch (see Fig. 8.1), and the millimeter is a little less than $\frac{1}{20}$ inch. A millimeter is about the thickness of a dime. A book may measure 20 cm by 25 cm with a thickness of about 3 cm. Millimeters are often used in scientific work and other areas in which small quantities must be measured. The length of a small insect may be measured in millimeters.

DID YOU KNOW

1 Yard = 3 Feet = 36 Inches = . . .

In the U.S. Customary system, 27 different units of length are used. How many of them can you name? Don't forget rod, mil, paris line, toise, cubit, and light-year. The different units can be found in the *CRC Handbook of Chemistry and Physics* or at the website www.hbcpnetbase.com.

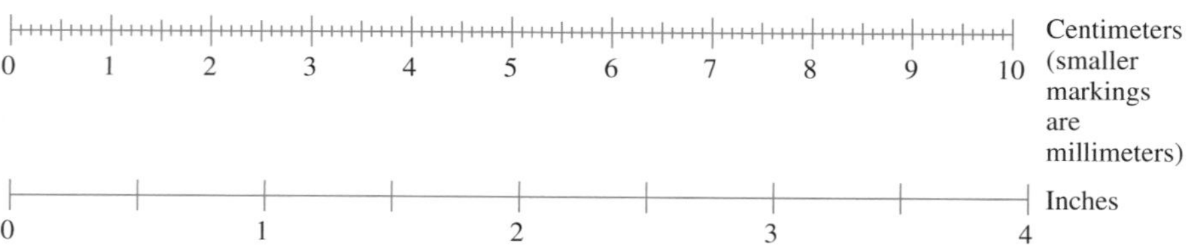

Figure 8.1

EXAMPLE 1 ***Choosing an Appropriate Unit of Length***

Determine which metric unit of length you would use to express the following.

a) The length of a guitar
b) The length of your shoe
c) The height of the Sears Tower in Chicago
d) The length of an ant
e) The diameter of a half-dollar
f) The distance between Lexington, Kentucky, and Houston, Texas.
g) The diameter of a round wastepaper basket.
h) The diameter of a pencil
i) Your waist size
j) Your height

SOLUTION:

a) Meters or centimeters
b) Centimeters
c) Meters
d) Millimeters
e) Centimeters or millimeters
f) Kilometers
g) Centimeters
h) Millimeters
i) Centimeters
j) Meters or centimeters

In some parts of this solution, more than one possible answer is listed. Measurements can often be made by using more than one unit. For example, if someone asks your height, you might answer $5\frac{1}{2}$ feet or 66 inches. Both answers are correct.

Area

The area enclosed in a square with 1-centimeter sides (Fig. 8.2) is $1 \text{ cm} \times 1 \text{ cm} = 1 \text{ cm}^2$. A square whose sides are 2 cm (Fig. 8.3) has an area of $2 \text{ cm} \times 2 \text{ cm} = 2^2 \text{ cm}^2 = 4 \text{ cm}^2$.

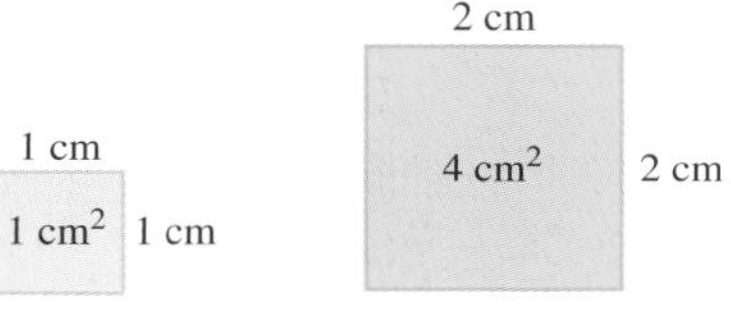

Figure 8.2 **Figure 8.3**

Areas are always expressed in square units, such as square centimeters, square kilometers, or square meters. When finding areas, be careful that all the numbers being multiplied are expressed in the same units.

In the metric system, the square centimeter replaces the square inch. The square meter replaces the square foot and square yard. In the future, you might purchase carpet or other floor covering by the square meter instead of by the square foot.

For measuring large land areas, the metric system uses a square unit 100 meters on each side (a square hectometer). This unit is called a *hectare* (pronounced "hectair" and symbolized ha). A hectare is about 2.5 acres. One square mile of land contains about 260 hectares. Very large units of area are measured in square kilometers. One square kilometer is about $\frac{4}{10}$ square mile.

The Grand Canyon National Park, Arizona, see Example 2a)

EXAMPLE 2 *Choosing an Appropriate Unit of Area*

Determine which metric unit of area you would use to measure the area of the following.

a) The Grand Canyon National Park
b) The top of a kitchen table
c) The floor of the classroom
d) A person's property with an average-sized lot
e) The cover of this book
f) A football field
g) An ice-skating rink
h) A dime
i) A lens in eyeglasses
j) A dollar bill

SOLUTION:

a) Square kilometers or hectares
b) Square meters
c) Square meters
d) Square meters or hectares
e) Square centimeters
f) Hectares or square meters
g) Square meters
h) Square millimeters or square centimeters
i) Square centimeters
j) Square centimeters

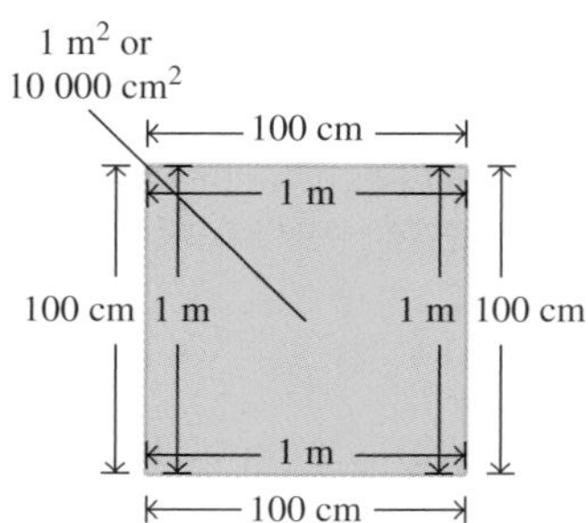

Figure 8.4

EXAMPLE 3 *Same Area, Different Units*

A square meter is how many times as large as a square centimeter?

SOLUTION: Since 1 m equals 100 cm, we can replace 1 m with 100 cm (see Fig. 8.4). The area of 1 m^2 = 1 m × 1 m = 100 cm × 100 cm = 10 000 cm^2. Thus, the area of one square meter is 10 000 times the area of one square centimeter. This technique can be used to convert from any square unit to a different square unit.

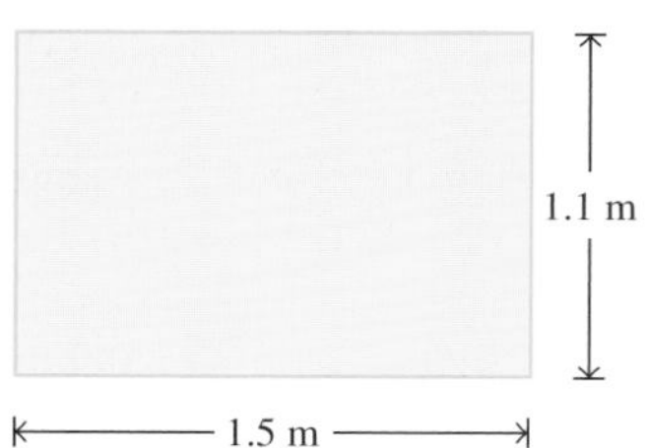

Figure 8.5

EXAMPLE 4 *Table Top*

Find the area of a rectangular table top if its length is 1.5 m and its width is 1.1 m (see Fig. 8.5).

SOLUTION: To find the area, we use the formula

$$\text{Area} = \text{length} \times \text{width}$$

or

$$A = l \times w$$

Substituting values for l and w, we have

$$\begin{aligned} A &= 1.5 \text{ m} \times 1.1 \text{ m} \\ &= 1.65 \text{ m}^2 \end{aligned}$$

Notice that the area is measured in square meters.

Figure 8.6

EXAMPLE 5 *A Quarter*

A quarter has a diameter of about 2.4 cm (Fig. 8.6). Find the surface area of one side of a quarter.

SOLUTION: The formula for the area of a circle is $A = \pi r^2$, where π is *approximately* 3.14. The radius, r, is one-half the diameter. Since the diameter is about 2.4 cm, the radius is about 1.2 cm. Substituting values for π and r, we get the following.

$$\begin{aligned} A &= \pi r^2 \\ &\approx 3.14(1.2 \text{ cm})^2 \\ &\approx 4.52 \text{ cm}^2 \end{aligned}$$

Thus, the area is approximately 4.52 square centimeters. Recall from earlier chapters that the symbol ≈ means "is approximately equal to."

TIMELY TIP Many calculators contain a $\boxed{\pi}$ key. If your calculator contains a $\boxed{\pi}$ key, you should use that key to input the value of π. If you do so, you will get a more accurate answer than if you used 3.14 for pi.

Volume

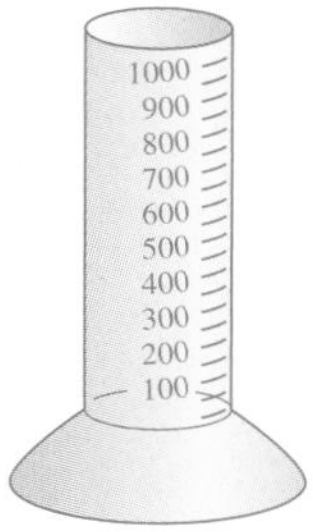

Figure 8.7

When a figure has only two dimensions—length and width—we can find its area. When a figure has three dimensions—length, width, and height—we can find its volume. The volume of an item can be considered the space occupied by the item.

In the metric system, volume may be expressed in terms of liters or cubic meters, depending on what is being measured. In all English-speaking countries except the United States, *liter* is spelled "litre."

The volume of liquids is expressed in liters. A liter is a little larger than a quart. Liters are used in place of pints, quarts, and gallons. A liter can be divided into 1000 equal parts, each of which is called a milliliter. Figure 8.7 illustrates a type of liter container (a 1000 mℓ graduated cylinder) that is often used in chemistry. Milliliters are used to express the volume of very small amounts of liquid. Drug dosages are often expressed in milliliters. An 8 oz cup will hold about 240 mℓ of liquid.

The kiloliter, 1000 liters, is used to represent the volume of large amounts of liquid. Tank trucks carrying gasoline to service stations hold about 10.5 kℓ of gasoline.

Cubic meters are used to express the volume of large amounts of solid material. The volume of a dump truck's load of topsoil is measured in cubic meters. The volume of natural gas used to heat a house may soon be measured in cubic meters instead of cubic feet.

The liquid in a liter container will fit exactly in a cubic decimeter (Fig. 8.8). Note that $1\ \ell = 1000\ \text{m}\ell$ and that $1\ \text{dm}^3 = 1000\ \text{cm}^3$. Because $1\ \ell = 1\ \text{dm}^3$, *1 mℓ must equal 1 cm^3*. Other useful facts are illustrated in Table 8.3. Thus, within the metric system, conversions are much simpler than in the U.S. customary system. For example, how would you change cubic feet of water into gallons of water?

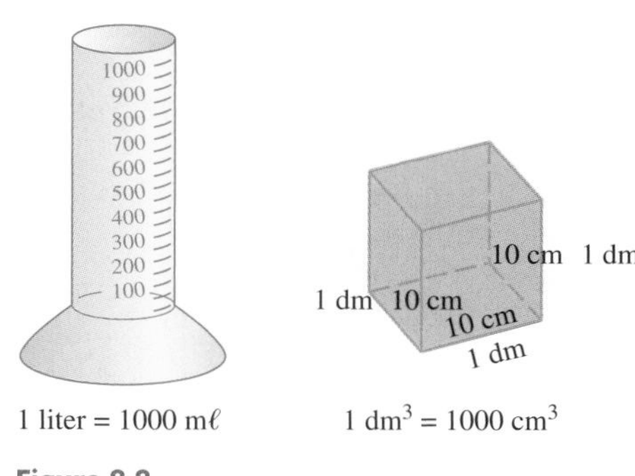

1 liter = 1000 mℓ 1 dm^3 = 1000 cm^3

Figure 8.8

TABLE 8.3

Volume in Cubic Units		Volume in Liters
1 cm^3	=	1 mℓ
1 dm^3	=	1 ℓ
1 m^3	=	1 kℓ

The Alaskan pipeline

EXAMPLE 6 *Choosing an Appropriate Unit of Volume*

Determine which metric unit of volume you would use to measure the volume of the following.

a) The oil that flows through the Alaskan pipeline in a day
b) A carton of milk
c) A truckload of topsoil
d) A drug dosage
e) Sand in a paper cup
f) A dime
g) Water in a drinking glass
h) Water in a water bed
i) Space available in a refrigerator
j) Concrete used to lay the foundation for a basement

SOLUTION:

a) Kiloliters
b) Liters
c) Cubic meters
d) Milliliters
e) Cubic centimeters
f) Cubic milliliters
g) Milliliters
h) Liters or kiloliters
i) Cubic meters
j) Cubic meters

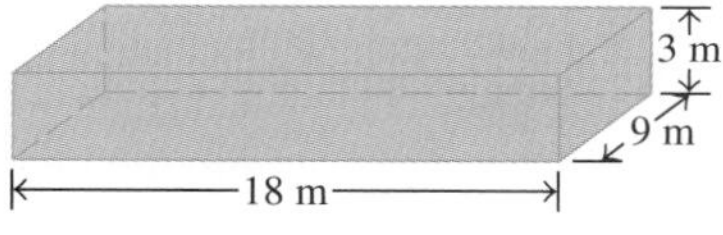

Figure 8.9

EXAMPLE 7 *Swimming Pool Volume*

A swimming pool is 18 m long and 9 m wide, and it has a uniform depth of 3 m (Fig. 8.9). Find (a) the volume of the pool in cubic meters and (b) the volume of water in the pool in kiloliters.

SOLUTION:

a) To find the volume in cubic meters, we use the formula

$$V = l \times w \times h$$

Substituting values for l, w, and h we have

$$\begin{aligned} V &= 18 \text{ m} \times 9 \text{ m} \times 3 \text{ m} \\ &= 486 \text{ m}^3 \end{aligned}$$

b) Since $1 \text{ m}^3 = 1 \text{ k}\ell$, the pool will hold 486 $\text{k}\ell$ of water.

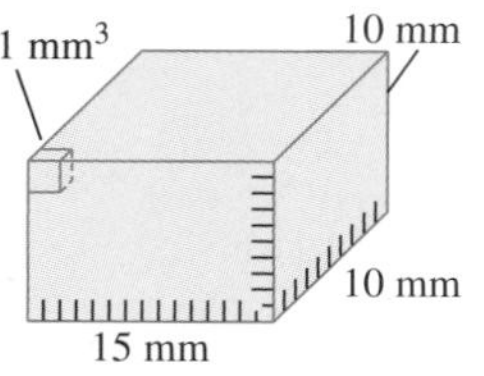

Figure 8.10

EXAMPLE 8 *Choose an Appropriate Unit*

Select the most appropriate answer. The volume of a shoe box is approximately

a) 1500 mm^3. b) 6500 ℓ. c) 6500 cm^3.

SOLUTION: A shoe box is not a liquid, so its volume is not expressed in liters. Thus, (b) is not the answer. The volume of the rectangular solid in Fig. 8.10 is approximately 1500 mm^3, so (a) is not an appropriate answer. A shoe box may measure about 33 cm × 18 cm × 11 cm, or 6534 cm^3. Therefore, 6500 cm^3 or (c) is the most appropriate answer.

When the volume of a liquid is measured, the abbreviation cc is often used instead of cm^3 to represent cubic centimeters. For example, a nurse may give a patient an injection of 3 cc or 3 mℓ of the drug ampicillin.

EXAMPLE 9 *Measuring Medicine*

A nurse must give a patient 3 cc of the drug gentamicin mixed in 100 cc of a normal saline solution.

a) How many milliliters of the drug will the nurse administer?

b) What is the total volume of the drug and saline solution in milliliters?

SOLUTION:

a) Because 1 cc is equal in volume to 1 mℓ, the nurse will administer 3 mℓ of the drug.

b) The total volume is 3 + 100 or 103 cc, which is equal to 103 mℓ. ▲

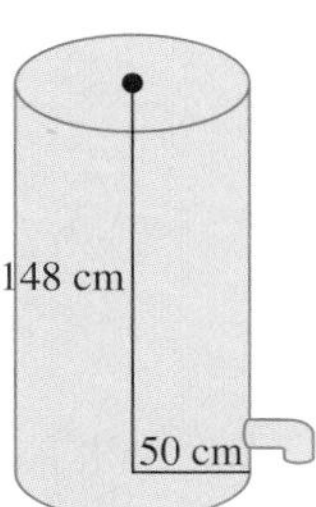

Figure 8.11

EXAMPLE 10 *A Hot-Water Heater*

A hot-water heater, in the shape of a right circular cylinder, has a radius of 50 cm and a height of 148 cm. What is the capacity, in liters, of the hot-water heater?

SOLUTION: The hot-water heater is illustrated in Fig. 8.11. The formula for the volume of a right circular cylinder is $V = \pi r^2 h$, where π is approximately 3.14. If we express all the measurements in meters, the volume will be given in cubic meters. Thus, 50 cm = 0.5 m, and 148 cm = 1.48 m.

$$\begin{aligned} V &= \pi r^2 h \\ &\approx 3.14(0.5)^2(1.48) \\ &\approx 3.14(0.25)(1.48) \approx 1.1618 \text{ m}^3 \end{aligned}$$

We want the volume in liters, so we must change the answer from cubic meters to liters.

$$1 \text{ m}^3 = 1000 \ \ell$$

So,

$$1.1618 \text{ m}^3 = 1.1618 \times 1000 = 1161.8 \ \ell$$

▲

EXAMPLE 11 *Comparing Volume Units*

a) How many times larger is a cubic meter than a cubic centimeter?

b) How many times larger is a cubic dekameter than a cubic meter?

SOLUTION:

a) The procedure used to determine the answer is similar to that used in Example 3 in this section. First we draw a cubic meter, which is a cube 1 m long by 1 m wide by 1 m high. In Fig. 8.12 on page 446, we represent each meter as 100 centimeters. The volume of the cube is its length times its width times its height, or

$$\begin{aligned} V &= l \times w \times h \\ &= 100 \text{ cm} \times 100 \text{ cm} \times 100 \text{ cm} = 1\,000\,000 \text{ cm}^3 \end{aligned}$$

DID YOU KNOW

Metrics and Medicine

Both milliliters and cubic centimeters are commonly used in medicine. In the United States, cubic centimeters are commonly denoted cc rather than the cm^3 used in the metric system. A patient's intake and output of fluids and intravenous injections are commonly measured in cubic centimeters. Drug dosage is measured in milliliters.

The following question is from a nursing exam. Can you determine the correct answer?

In caring for a patient after delivery, you are to give 12 units of Pitocin (in 1000 cc of intravenous fluid). The ampule is labeled 10 units per 0.5 mℓ. How much of the solution would you draw and give?

a) 0.6 cc

b) 1.2 cc

c) 6.0 cc

d) 9.6 cc

Answer: (a)

Since 1 m^3 = 1 000 000 cm^3, a cubic meter is one million times larger than a cubic centimeter.

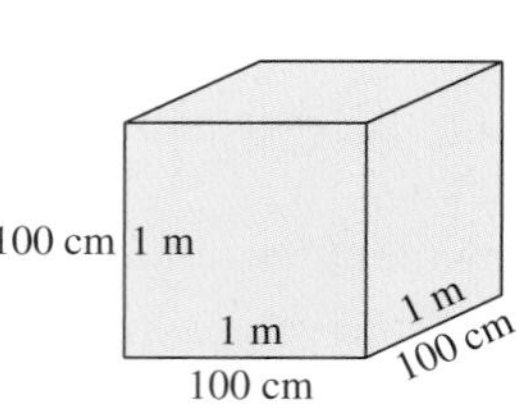

Figure 8.12

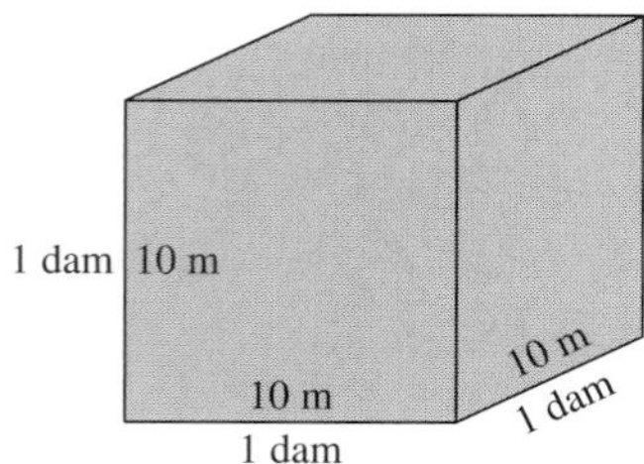

Figure 8.13

b) Work part (b) in a similar manner (Fig. 8.13).

$$V = l \times w \times h$$
$$= 10 \text{ m} \times 10 \text{ m} \times 10 \text{ m} = 1000 \text{ m}^3$$

Since 1 dam^3 = 1000 m^3, a cubic dekameter is one thousand times larger than a cubic meter.

SECTION 8.2 EXERCISES

Concept/Writing Exercises

In Exercises 1–12, an object has been measured and the measurement has been written with the unit indicated. Indicate what was measured: length, area, or volume.

1. m^3	**2.** mm	**3.** ha	**4.** m
5. cc	**6.** ℓ	**7.** cm^3	**8.** kℓ
9. m^2	**10.** dℓ	**11.** cm	**12.** cm^2

13. Estimate your height in (a) centimeters and (b) meters.

14. Estimate, in centimeters, the length of this book.

15. Estimate, in square centimeters, the surface area of this book.

16. Estimate, in meters, the length of the classroom in which your mathematics course is held.

17. Estimate, in centimeters, the length of your arm.

18. Estimate, in square centimeters, the surface area of a dollar bill.

19. One liter of liquid has the equivalent volume of which of the following: a cubic centimeter, a cubic decimeter, or a cubic meter?

20. One cubic meter has the equivalent volume of which of the following liquid measures: a liter, a milliliter, or a kiloliter?

21. One milliliter of liquid has the equivalent volume of which of the following: a cubic centimeter, a cubic decimeter, or a cubic meter?

22. Which metric measurement is used to measure very large areas of land?

23. Is the hectare a measure of length, area, or volume?

24. A hectare has an area of about how many acres: 2.5, 25, or 250?

Practice the Skills

In Exercises 25–36, indicate the metric unit of measurement that you would use to express the following.

25. The length of a calculator

26. The distance between cities

27. The length of a paper clip

28. The width of a Frisbee

29. The length of a newborn infant

30. The diameter of a pencil

31. The diameter of a jump rope

32. The width of an Olympic-size swimming pool

33. The length of a photograph

34. The length of a butterfly

35. The distance to the moon

36. The height of an adult male

In Exercises 37–44, choose the best answer.

37. The distance between home plate and first base is about how long?
a) 27 km **b)** 27 cm **c)** 27 m

38. A U.S. postage stamp is about how wide and how long?
a) 2 cm × 3 cm **b)** 2 mm × 3 mm
c) 2 hm × 3 hm

39. The distance between freeway exits could be how long?
a) 5 mm **b)** 5 m **c)** 5 km

40. A grown woman is about how tall?
a) 160 cm **b)** 160 mm **c)** 160 dm

41. The width of a piece of adhesive tape is about how wide?
a) 2 cm **b)** 2 mm **c)** 2 dm

42. The diameter of a coffee cup is about which of the following?
a) 8 mm **b)** 8 cm **c)** 8 dm

43. The length of the New River Gorge Bridge near Fayetteville, West Virginia is about how long?
a) 1000 dam **b)** 1000 m **c)** 1000 cm

44. The Sears Tower in Chicago is about how tall?
a) 375 cm **b)** 375 km **c)** 375 m

In Exercises 45–50, (a) estimate the item in metric units and (b) measure it with a metric ruler. Record your result.

45. The width of a card from a deck of cards

46. The width of a classroom door

47. The length of a car

48. The diameter of a can of soda

49. The height of a milk carton

50. The thickness of 10 sheets of paper.

In Exercises 51–56, replace the customary measure (shown in parentheses) with the appropriate metric measure.

51. Give him a _____ (inch), and he will take a _____ (mile).

52. There was a crooked man and he walked a crooked _____ (mile).

53. One hundred _____ (yard) dash.

54. I wouldn't touch a skunk with a 10-_____ (foot) pole.

55. I found a _____ (inch) worm.

56. This is a _____ (mile)stone in my life.

In Exercises 57–66, indicate the metric unit of measurement you would use to express the area of the following.

57. A computer monitor screen

58. The city of San Francisco

59. The floor of your classroom

60. The face of a dime

61. A building lot for a house

62. A baseball field

63. A postage stamp

64. A ceiling tile

65. Death Valley National Park

66. A professional basketball court

In Exercises 67–74, choose the best answer.

67. The area of a U.S. flag is about
a) 2.2 cm^2. **b)** 2.2 m^2. **c)** 2.2 km^2.

68. A U.S. postage stamp has an area of about
a) 5 cm^2. **b)** 5 mm^2. **c)** 5 dm^2.

69. The area of a city lot is about
a) 800 m^2. **b)** 800 hm^2. **c)** 800 cm^2.

70. The area of a city lot is about
a) $\frac{1}{8}$ m^2. **b)** $\frac{1}{8}$ ha. **c)** $\frac{1}{8}$ km^2.

71. The area of a ceiling tile is about
a) 360 m^2. **b)** 360 km^2. **c)** 360 cm^2.

72. The area of the face of a dime is about
a) 2.5 cm^2. **b)** 2.5 m^2. **c)** 2.5 mm^2.

73. The area of the screen of a table top TV is about
a) 1200 dm^2. **b)** 1200 mm^2. **c)** 1200 cm^2.

74. The area of the Grand Canyon National Park is about
a) 4900 m^2 **b)** 4900 cm^2 **c)** 4900 km^2

In Exercises 75–80, (a) estimate the area of the item in metric units and (b) measure it in metric units and compute its area.

75. A typical photograph

76. The cover of this book

77. A $5 bill

78. The top of your teacher's desk

79. The bottom of a 12 oz soda can

80. The face of a penny

In Exercises 81–90, determine the metric unit that would best be used to measure the volume of the following.

81. Water flowing over Niagara Falls per minute

Niagara Falls

82. Water in a hot-water heater

83. Liquid in an eye dropper

84. Air in a basketball

85. Oil needed to change the oil in your car

86. A bag of topsoil

87. A truckload of ready-mix concrete

88. Asphalt needed to pave a driveway

89. Soda in a bottle of soda

90. Air in a hot air balloon

In Exercises 91–98, choose the best answer to indicate the volume of the following.

91. A shoe box
a) 7780 mm^3 **b)** 7780 dm^3 **c)** 7780 cm^3

92. A quarter
a) 0.5 cm^3 **b)** 0.5 mm^3 **c)** 0.5 dm^3

93. Water in a 24-ft-diameter above-ground circular swimming pool
a) 55 ℓ **b)** 55 mℓ **c)** 55 kℓ

94. Soda in a can of soda
a) 355 ℓ **b)** 355 mℓ **c)** 355 m^3

95. A can of vegetables
a) 550 cm^3 **b)** 550 mm^3 **c)** 550 dm^3

96. Juice that can be squeezed out of an orange
a) 120 kℓ **b)** 120 mℓ **c)** 120 ℓ

97. Air in a balloon with a diameter of 4 meters
a) 30 m^3 **b)** 30 cm^3 **c)** 30 km^3

98. Air in a basketball
a) 14 000 m^3 **b)** 14 000 cm^3 **c)** 14 000 mm^3

In Exercises 99–102, (a) estimate the volume in metric units and (b) compute the actual volume of the item.

99. Air in a cardboard box that is 61 cm long, 61 cm wide, and 41 cm tall (Use $V = lwh$.)

100. Water in a water bed that is 2 m long, 1.5 m wide, and 25 cm deep

101. Oil in a barrel that has a height of 1 m and a diameter of 0.5 m (Use $V = \pi r^2 h$.)

102. Water in a cylindrical tank that is 40 cm in diameter and 2 m high

Problem Solving

103. *Area* Use a metric ruler to measure the length and width of the sides of the rectangle. Then compute the area of the rectangle. Give your answers in metric units.

104. *Area* Use a metric ruler to find the radius of the circle. Then compute the area of the circle. Give your answers in metric units.

105. *A Mat for a Picture* A framed picture is shown. Find the matted area.

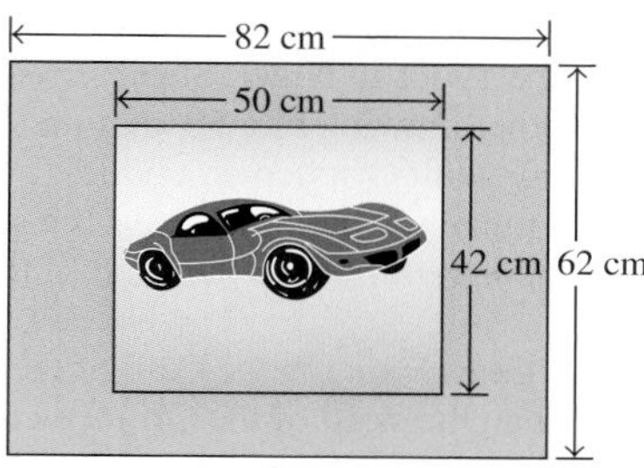

106. *A Walkway* A rectangular building 50 m by 70 m is surrounded by a walk 1.5 m wide.

a) Find the area of the region covered by the building and the walk.

b) Find the area of the walk.

107. *Farmland* Mrs. Manecki has purchased a farm that is in the shape of a rectangle. The dimensions of the piece of land are 1.4 km by 3.75 km.

a) How many square kilometers of land did she purchase?

b) If 1 km^2 equals 100 ha, determine the amount of land she purchased in hectares.

108. *Area of a Garden* Mr. Baumgarten's garden is 22.5 m by 18.3 m.

a) How large is his garden in square meters?

b) If 1 m^2 equals 0.0001 ha, determine the area of his garden in hectares.

109. *Volume of Water* **a)** What is the volume of water in a swimming pool that is 18 m long and 10 m wide and has an average depth of 2.5 m? Give your answer in cubic meters.

b) How many kiloliters of water will the pool hold?

110. *Cost of Paint* The first coat of paint for the outside of a building requires 1 ℓ of paint for each 10 m^2. The second coat requires 1 ℓ for every 15 m^2. If the paint costs $4.75 per liter, what will be the cost of two coats of paint for the four outside walls of a building 20 m long, 12 m wide, and 6 m high?

111. *Fish Tank Volume* A rectangular fish tank is 70 cm long, 40 cm wide, and 20 cm high.

a) How many cubic centimeters of water will the tank hold?

b) How many milliliters of water will the tank hold?

c) How many liters of water will the tank hold?

112. *How Much Soup?* A can of Campbell's Home Cookin' chicken vegetable soup has a diameter of 8.0 cm and a height of 12.5 cm. Determine the volume of soup in the can (assume that the can is filled with soup).

113. How many times larger is a square dekameter than a square meter?

114. How many times larger is a square kilometer than a square dekameter?

115. How many times larger is a cubic meter than a cubic decimeter?

116. How many times larger is a cubic centimeter than a cubic millimeter?

In Exercises 117–124, replace the question mark with the appropriate value.

117. 1 m^2 = ? mm^2

118. 1 hm^2 = ? cm^2

119. 1 km^2 = ? hm^2

120. 1 cm^2 = ? m^2

121. 1 mm^3 = ? cm^3

122. 1 dm^3 = ? mm^3

123. 1 m^3 = ? cm^3

124. 1 hm^3 = ? km^3

In Exercises 125–128, fill in the blank.

125. 435 cm^3 = ____ mℓ

126. 435 cm^3 = ____ ℓ

127. 76 kℓ = ____ m^3

128. 4.2 ℓ = ____ cm^3

Glacier In Exercises 129 and 130, assume that a part of a glacier that contains 60 cubic meters of ice calves (or breaks) off and falls into the ocean.

A Glacier in Alaska

129. When the ice that has fallen into the ocean melts, determine the approximate amount of water, in deciliters, obtained from the ice.

130. When the ice melts, determine the approximate amount of water, in cubic centimeters, obtained from the ice.

Challenge Problems/Group Activities

131. Starting with a straight piece of wood of sufficient size, construct a meter stick. Indicate decimeters, centimeters, and millimeters on the meter stick. Use the centimeter measure in Fig. 8.1 as a guide.

132. Construct a metric tape measure from a piece of tape or rope and then determine your waist measurement.

In Exercises 133 and 134, fill in the blank to make a true statement.

133. 6.7 kℓ = ____ dm^3 **134.** 1.4 ha = ____ cm^2

135. *Conversions* In Example 3, we illustrated how to change an area in a metric unit to an area measured with a different metric unit.

a) Using Example 3 as a guide, change 1 square mile to square inches.

b) Is converting from one unit of area to a different unit of area generally easier in the metric system or the U.S. customary system? Explain.

136. *Conversions* In Example 11, we illustrated how to change a volume in one metric unit to a volume measured with a different metric unit.

a) Using Example 11 as a guide, change 6 yd^3 (a volume 1 yard by 2 yards by 3 yards) into cubic inches.

b) Is converting from one unit of volume to a different unit of volume generally easier in the metric system or the U.S. customary system? Explain.

Recreational Mathematics

137. *Find the Words* In the box below, the following words are spelled out: METER, MILLIMETER, CENTIMETER, LITER, HECTARE, SQUARE METER, MILLILITER. You can find these words by moving from square to square, vertically, horizontally, or diagonally (either up and down or forward or backward). You may leave a square and then return to that square to use that letter again. You can use the same squares to make the different words. There is no space left in the box between the words square meter. How many of the words can you find?

C	E	H	M	I
T	U	Q	S	L
A	R	E	I	L
T	I	M	E	T
N	E	C	E	R

138. *Crocodiles* The following drawing shows a complete 1.5-meter-long fossil skull of an estimated 110-million-year-old crocodile called *Sarcosuchus imperator*, which was found in the 1960s in Niger. Superimposed on the drawing is another drawing of a 50-centimeter-long skull of a modern-day adult Orinoco crocodile.

a) How much longer, in centimeters, is the skull of the *Sarcosuchus imperator* than the skull of the Orinoco crocodile?

b) How many times longer is the skull of the *Sarcosuchus imperator* than the skull of the Orinoco crocodile?

c) Does this photo of the two skulls give a true perspective of the relative sizes of the two skulls? Explain.

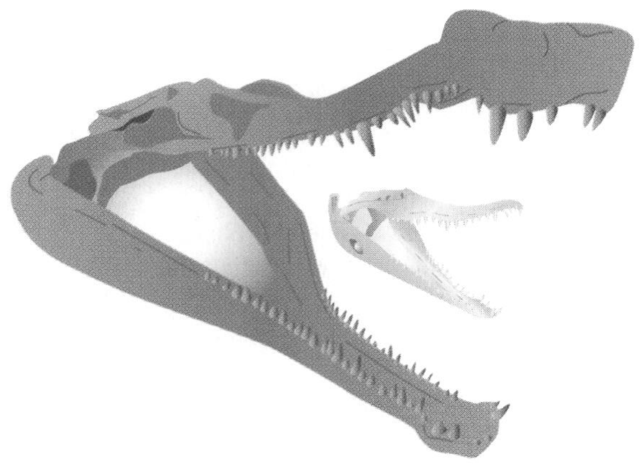

139. *Water Usage* **a)** How much water do we use daily? On the average, people in the United States use more water than people anywhere else in the world. Take a guess at the number of liters of water used per day per person in the United States.

b) Now take a guess at the number of liters used per day per person in the United Kingdom.
Compare your answers to those given in the answer section.

Internet/Research Activities

140. *The Meter* The definition of the meter has changed several times throughout history. Write a one- to two-page report on the history of the meter, from when it was first named to the present.

8.3 MASS AND TEMPERATURE

In this section, we discuss the metric measurements of mass and temperature. As with Section 8.2, the focus of this section is on thinking metric.

Mass

Weight and mass are not the same. *Mass* is a measure of the amount of matter in an object. It is determined by the molecular structure of the object, and it will not change

DID YOU KNOW

The Kilogram

Since 1889, a single platinum–iridium bar has been sealed in an airtight jar in the International Bureau of Weights and Measures in Sèvres, France.

Nicknamed "Le Grand K," this bar constitutes the one and only true kilogram. Of all the standard international units of measure, the kilogram remains the only one whose definition relies on a physical artifact. All the other units have their definitions rooted in constants of nature, such as the speed of light or atomic vibrations.

As part of an international effort, researchers at the U.S. National Institute of Standards and Technology in Washington, D.C., want to redefine the kilogram in a way that will make the standard absolute, unchanging, and accessible to anyone, anywhere.

One problem is that the current standard tends to drift a bit. The kilogram has varied by as much as 0.05 part per million in the last 100 years. The cause of that variance remains unknown.

from place to place. Weight is a measure of the gravitational pull on an object. For example, the gravitational pull of Earth is about six times as great as the gravitational pull of the moon. Thus, a person on the moon weighs about $\frac{1}{6}$ as much as on Earth, even though the person's mass remains the same. In space, where there is no gravity, a person has no weight.

Even on Earth, the gravitational pull varies from point to point. The closer you are to Earth's center, the greater the gravitational pull. Thus, a person weighs very slightly less on a mountain than in a nearby valley. Because the mass of an object does not vary with location, scientists generally use mass rather than weight.

Although weight and mass are not the same, on Earth they are proportional to each other (the greater the weight, the greater the mass). Therefore, for our purposes, we can treat weight and mass as the same.

The *kilogram* is the basic unit of mass in the metric system. It is a little more than 2 lb. The official kilogram is a cylinder of platinum–iridium alloy kept by the International Bureau of Weights and Measures, located in Sèvres, near Paris. (See the Did You Know in the margin.)

Items that we normally measure in pounds are usually measured in kilograms in other parts of the world. For example, an average-sized man has a mass of about 75 kg.

The *gram* (a unit that is 0.001 kg) is relatively small and is used in place of the ounce. A nickel has a mass of about 5 g, a cube of sugar has a mass of about 2 g, and a large paper clip has a mass of about 1 g.

The *milligram* is used extensively in the medical and scientific fields as well as in the pharmaceutical industry. Nearly all bottles of tablets are now labeled in either milligrams or grams.

The *metric tonne* (t) is used to express the mass of heavy items. One metric tonne equals 1000 kg. It is a little larger than our customary ton of 2000 lb. The mass of a large truck may be expressed in metric tonnes.

EXAMPLE 1 *Choosing the Appropriate Unit*

Determine which metric unit you would use to express the mass of the following.

a) An orca (or killer whale)
b) A newborn child
c) A teaspoon of sugar
d) A box of cereal
e) A quarter
f) A fly
g) A frog
h) A refrigerator

SOLUTION:

a) Metric tonnes
b) Kilograms
c) Grams
d) Grams
e) Grams
f) Milligrams
g) Grams
h) Kilograms

Orcas (or killer whales) at Sea World.

One kilogram of water has a volume of exactly 1 liter. In fact, a liter is defined to be the volume of 1 kilogram of water at a specified temperature and pressure. Thus, mass and volume are easily interchangeable in the metric system. Converting from weight to volume is not nearly as convenient in the U.S. customary system. For example, how would you change pounds of water to cubic feet or gallons of water in our customary system?

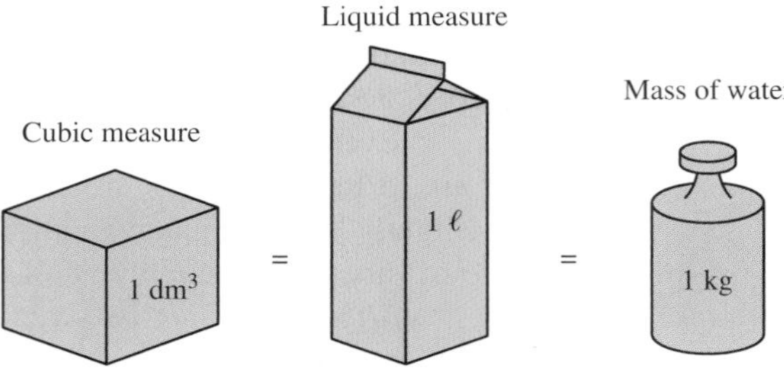

Figure 8.14

Since 1 dm^3 = 1000 cm^3, 1ℓ = 1000 mℓ, and 1kg = 1000g, we have the following relationship.

$$\begin{aligned} 1000 \text{ cm}^3 &= 1000 \text{ m}\ell = 1000 \text{ g} \\ \text{or,} \quad 1 \text{ cm}^3 &= 1 \text{ m}\ell = 1 \text{ g} \end{aligned}$$

Figure 8.14 illustrates the relationship between volume of water in cubic decimeters, in liters, and mass in kilograms. Table 8.4 expands on this relationship between the volume and mass of water.

TABLE 8.4 Volume and Mass of Water

Volume in Cubic Units		Volume in Liters		Mass of Water
1 cm^3	=	1 mℓ	=	1 g
1 dm^3	=	1 ℓ	=	1 kg
1 m^3	=	1 kℓ	=	1 t (1000 kg)

TIMELY TIP In Chapter 9, we provide formulas and discuss procedures for finding the area and volume of many geometric figures. The procedures and formulas for finding area and volume are the same regardless of whether the units are metric units or customary units. When finding areas and volumes, each side of the figure must be given in (or converted to) the same unit.

EXAMPLE 2 *A Fish Tank's Capacity*

A fish tank is 1 m long, 50 cm high, and 250 mm wide (Fig. 8.15).

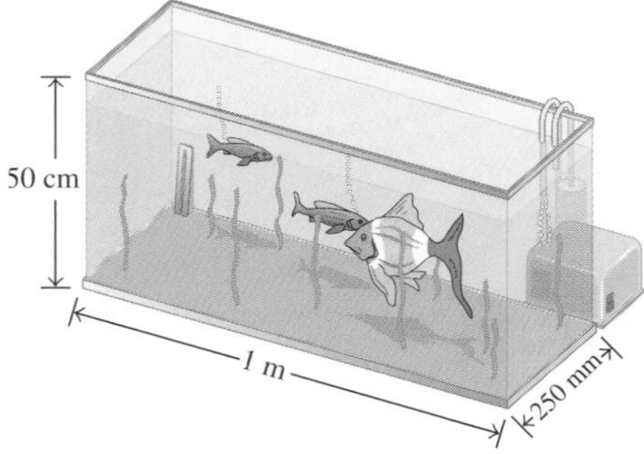

Figure 8.15

DID YOU KNOW

Penny-wise, Kilo-foolish

Steven Thoburn

Steven Thoburn, 36, a British grocer, has been charged in the northern England town of Sunderland with violating British law. His crime: He used a nonmetric scale to sell bananas to undercover officers. This year-old law was instituted to bring Britain in harmony with European Union standards. If convicted, Thoburn could have faced a fine of £7300 (about $11,500) and a six-month jail sentence.

Thoburn pleaded not guilty and did not dispute the facts. His defense argued that the laws that required metrification of scales were invalid. A district judge disagreed and in April 2002 handed Thoburn a six-month conditional discharge (suspended sentence). The trial has already cost Thoburn in excess of £35,000 (about $55,000), but he has indicated that he will appeal the ruling. For further information, see www.metricmartyrs.com.

a) Determine the number of liters of water the tank holds.
b) What is the mass of the water in kilograms?

SOLUTION:

a) We must convert all the measurements to the same units. Let's convert them all to meters: 50 cm is 0.5 m, and 250 mm is 0.25 m.

$$\begin{aligned} V &= l \times w \times h \\ &= 1 \times 0.25 \times 0.5 \\ &= 0.125 \text{ m}^3 \end{aligned}$$

Since 1 m^3 of water = 1 kℓ of water,

$$0.125 \text{ m}^3 = 0.125 \text{ k}\ell\text{, or } 125\ \ell \text{ of water}$$

b) Since 1 ℓ = 1 kg, 125 ℓ = 125 kg of water. ▲

To convince yourself of the advantages of the metric system, do a similar problem involving the customary system of measurement, such as Challenge Problems/Group Activities Exercise 73 at the end of this section.

Temperature

The Celsius scale is used to measure temperatures in the metric system. Figure 8.16 on page 454 shows a thermometer with the Fahrenheit scale on the left and the Celsius scale on the right.

The Celsius scale was named for the Swedish astronomer Anders Celsius (1701–1744), who first devised it in 1742. On the Celsius scale, water freezes at 0°C and boils at 100°C. In the past, the Celsius thermometer was called a "centigrade thermometer." Recall that *centi* means $\frac{1}{100}$, and there are 100 degrees between the freezing point of water and the boiling point of water. Thus, 1°C is $\frac{1}{100}$ of this interval. Table 8.5 gives some common temperatures in both degrees Celsius (°C) and degrees Fahrenheit (°F).

TABLE 8.5

Celsius Temperature		Fahrenheit Temperature
−18°C	A very cold day	0°F
0°C	Freezing point of water	32°F
10°C	A warm winter day	50°F
20°C	A mild spring day	68°F
30°C	A warm summer day	86°F
37°C	Body temperature	98.6°F
100°C	Boiling point of water	212°F
177°C	Oven temperature for baking	350°F

At a temperature of −40° the Celsius and Fahrenheit temperatures are the same. That is, −40°C = −40°F. See Exercise 72.

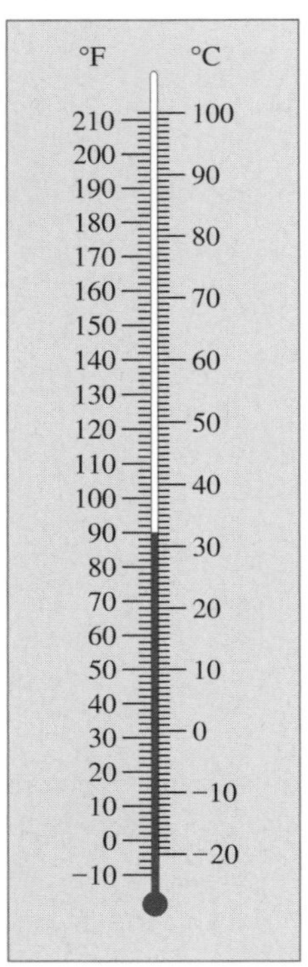

Figure 8.16

EXAMPLE 3 *Think Metric Temperatures*

Choose the best answer. (Refer to the dual-scale thermometer in Fig. 8.16.)

a) Buffalo, New York, on New Year's Day might have a temperature of
   i) −15°C. ii) 15°C. iii) 40°C.

b) Washington, D.C., on July 4 might have a temperature of
   i) 15°C. ii) 30°C. iii) 40°C.

c) The oven temperature for baking a cake might be
   i) 60°C. ii) 100°C. iii) 175°C.

SOLUTION:

a) A temperature of 15°C is possible if it is a very mild winter, but 40°C is much too hot. The best answer for a normal winter is −15°C.

b) The best estimate is 30°C. A temperature of 15°C is too chilly, and 40°C is too hot for July 4.

c) A cake bakes at temperatures well above boiling, so the only reasonable answer is 175°C. ▲

Comparing the temperature in Table 8.5, we see that the Celsius scale has 100° from the boiling point of water to the freezing point of water and the Fahrenheit scale has 180° from the boiling point of water to the freezing point of water. Therefore, one Celsius degree represents a greater change in temperature than one Fahrenheit degree does. In fact, one Celsius degree is the same as $\frac{180}{100}$, or $\frac{9}{5}$ Fahrenheit degrees. When converting from one system to the other system, use the following formulas.

From Celsius to Fahrenheit

$$F = \frac{9}{5}C + 32$$

From Fahrenheit to Celsius

$$C = \frac{5}{9}(F - 32)$$

EXAMPLE 4 *Convert to °C*

A typical setting for home thermostats is 72°F. What is the equivalent temperature on the Celsius thermometer?

SOLUTION: We use the formula $C = \frac{5}{9}(F - 32)$ to convert from °F to °C. Substituting $F = 72$ gives

$$\begin{aligned} C &= \frac{5}{9}(72 - 32) \\ &= \frac{5}{9}(40) \\ &\approx 22.2 \end{aligned}$$

Thus, the equivalent temperature of 72°F is about 22.2°C. ▲

EXAMPLE 5 *Convert to °F*

If the temperature outdoors is 28°C, will you need to wear a sweater if going outdoors?

SOLUTION: We use the formula $F = \frac{9}{5}C + 32$ to convert from °C to °F. Substituting $C = 28$ yields

$$\begin{aligned} F &= \frac{9}{5}(28) + 32 \\ &= 50.4 + 32 \\ &= 82.4 \end{aligned}$$

Since the temperature is about 82.4°F, you will not need to wear a sweater. ▲

DID YOU KNOW

It's a Metric World

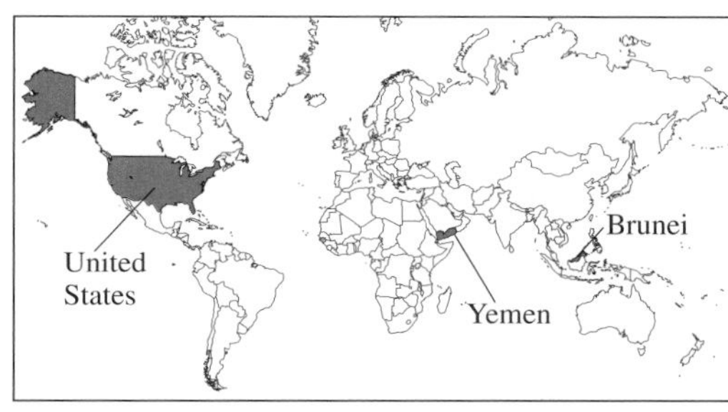

Figure 8.17

The United States is the only westernized country not currently using the metric system as its primary system of measurement. The only countries in the world besides the United States not using or committed to using the metric system are Yemen, Brunei, and a few small islands; see Fig. 8.17.

The European Union (EU) adopted a directive that requires all exporters to EU nations to indicate the dimensions of their products in metric units. Currently, U.S. manufacturers who export goods are doing so. Little by little, the United States is becoming more metric. For example, soft drinks come in liter bottles and prescription drug dosages are given in metric units. Maybe in the not too distant future gasoline will be measured in liters, not gallons, as it is in Canada and Mexico.

SECTION 8.3 EXERCISES

Concept/Writing Exercises

1. What is the basic unit of mass in the metric system?
2. The mass of a nickel is about how many grams?
3. One kilogram is a little more than how many pounds?
4. What unit of mass is used to express the mass of very heavy items?
5. Give an estimate of the temperature, in degrees Celsius, in Florida in August.
6. Give an estimate of the temperature, in degrees Celsius, in North Dakota in February.
7. Give an estimate, in degrees Celsius, of what you would consider an ideal outdoor temperature.
8. **a)** Is a person's mass the same in space as on Earth? Explain.
   b) Is a person's weight the same in space as on Earth? Explain.

Practice the Skills

In Exercises 9–18, indicate the metric unit of measurement that would best express the mass of the following.

9. A quarter
10. A man
11. A pair of eyeglasses
12. A box of cereal
13. A new pencil
14. An SUV
15. A refrigerator
16. A mosquito
17. A roll of paper towels
18. A calculator

In Exercises 19–24, select the best answer.

19. The mass of a 5 lb bag of flour is about how much?
    a) 2.26 g **b)** 2.26 kg **c)** 2.26 dag
20. The mass of a dime is about how much?
    a) 9.1 mg **b)** 9.1 kg **c)** 9.1 g

21. The mass of a child's pail filled with sand is about how much?
a) 1.4 mg **b)** 1.4 kg **c)** 1.4 g

22. The mass of a box of cornflakes is about how much?
a) 0.45 t **b)** 0.45 g **c)** 0.45 kg

23. The mass of a full-grown elephant is about how much?
a) 2800 g **b)** 2800 kg **c)** 2800 dag

24. The mass of a full-size car is about how much?
a) 1 962 000 hg **b)** 380 kg **c)** 1.6 t

In Exercises 25–28, estimate the mass of the item. If a scale with metric measure is available, find the mass.

25. Your body
26. A telephone book
27. A gallon of water
28. A tomato

In Exercises 29–38, choose the best answer. Use Table 8.5 and Fig. 8.16 to help select your answers.

29. Freezing rain is most likely to occur at a temperature of
a) −25°C. **b)** 32°C. **c)** 0°C.

Rochester, New York

30. A cup of coffee might have a temperature of
a) 15°C. **b)** 50°C. **c)** 90°C.

31. The thermostat for an air conditioner was set for 80°F. This setting is closest to
a) 2°C. **b)** 27°C. **c)** 57°C.

32. The temperature of the water in a certain lake is 5°C. You could
a) ice fish.
b) dress warmly and walk along the lake.
c) swim in the lake.

33. What might be the temperature at which a refrigerator is set?
a) 30°C **b)** 5°C **c)** 0°C

34. The weather forecast calls for a high of 32°C. You should plan to wear
a) a down-lined jacket.
b) a sweater.
c) a bathing suit.

35. What might be the temperature of an apple pie baking in the oven?
a) 90°C **b)** 100°C **c)** 177°C

36. The temperature of the water in a car's radiator when the car's engine is operating at its normal temperature might be
a) 70°C. **b)** 300°C. **c)** 110°C.

37. What might be the temperature at which a home freezer is set?
a) −20°C **b)** −7°C **c)** 2°C

38. The temperature of water in a hot tub might be
a) 30°C. **b)** 50°C. **c)** 40°C.

In Exercises 39–52, convert each temperature as indicated. Give your answer to the nearest tenth of a degree.

39. 30°C = ____°F
40. −5°C = ____°F
41. 92°F = ____°C
42. −10°F = ____°C
43. 180°F = ____°C
44. 98°F = ____°C
45. 37°C = ____°F
46. −4°C = ____°F
47. 13°F = ____°C
48. 75°F = ____°C
49. 45°C = ____°F
50. 60°C = ____°F
51. −20°F = ____°C
52. 425°F = ____°C

In Exercises 53–58, use the following graph, which shows the daily low and high temperatures, in degrees Celsius, for the week in a country that uses the metric system. The week illustrated was unseasonably warm. Determine the following temperatures in degrees Fahrenheit.

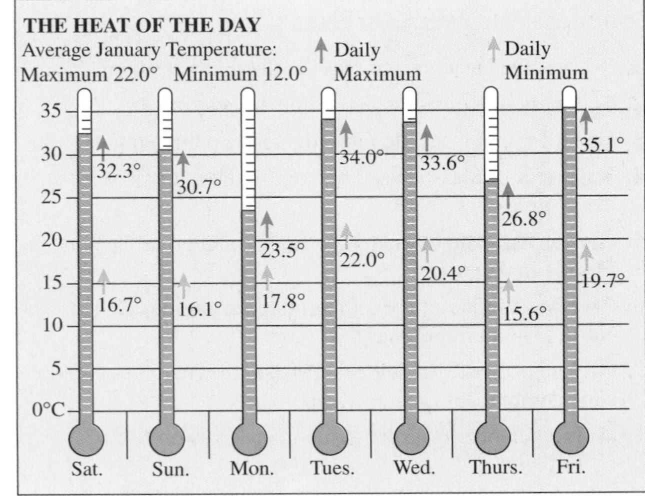

53. The average January maximum temperature

54. The maximum temperature for the week

55. The maximum temperature on Friday

56. The maximum temperature on Saturday

57. The range of temperatures on Monday

58. The range of temperatures on Tuesday

Problem Solving

The photo shows the cost of Crest Grower Crumbles and corn at a farm market in Fiji. Use the information provided in the photo to answer Exercises 59 and 60. In the photo, kilo is an abbreviation for kilogram.

59. *Crest Grower Crumbles* Determine the cost, in dollars, of 6.2 kg of Crest Grower Crumbles. (Grower Crumbles is a poultry feed.)

60. *Corn* Determine the cost, in dollars, of 1.3 kg of corn.

61. *Salt and Soda* A mixture of 45 g of salt and 29 g of baking soda is poured into 370 mℓ of water. What is the total mass of the mixture in grams?

62. *Jet Fuel* A jet can travel about 1 km on 17 kg of fuel. How many metric tonnes of fuel will the jet use flying nonstop between Baltimore and Los Angeles, a distance of about 4320 km?

63. *A Storage Tank* The dimensions of a storage tank are length 16 m, width 12 m, and height 12 m. If the tank is filled with water, determine
   a) the volume of water in the tank in cubic meters.
   b) the number of kiloliters of water the tank will hold.
   c) the mass of the water in metric tonnes.

64. *A Water Heater* A hot-water heater in the shape of a right circular cylinder has a radius of 50 cm and a height of 150 cm. If the tank is filled with water, determine
   a) the volume of water in the tank in cubic meters.
   b) the number of liters of water the tank will hold.
   c) the mass of the water in kilograms.

65. *What's the Problem?* A temperature display at a bank flashes the temperature in degrees Fahrenheit and then flashes the temperature in degrees Celsius. If it flashes 78°F, then 20°C, is there a problem? Explain.

66. *Fever or Chills?* Maria's body temperature is 38.2°C. Should she take an aspirin or put on a sweater? Explain.

In Exercises 67–70, convert as indicated

67. 4.2 kg = ____ t

68. 9.52 t = ____ kg

69. 17.4 t = ____ g

70. 1 460 000 mg = ____ t

Challenge Problems/Group Activities

71. *Gatorade* Gatorade is poured into a plastic bottle that holds 1.2 ℓ of liquid. The bottle is then placed in a freezer. When the bottle is removed from the freezer, the plastic is cut away, leaving just the frozen Gatorade.
   a) What is the approximate mass of the frozen Gatorade in grams?
   b) What is the approximate volume of the frozen Gatorade in cubic centimeters?

72. Show that −40°C = −40°F.

In Example 2, we showed how to find the volume and mass of water in a fish tank. Exercise 73 demonstrates how much more complicated solving a similar problem is in the U.S. customary system.

73. *Fish Tank* A fish tank is 1 yd long by 1.5 ft high by 15 in. wide.
   a) Determine the volume of water in the fish tank in cubic feet.
   b) Determine the weight of the water in pounds. One cubic foot of water weighs about 62.5 lb.
   c) If 1 gal of water weighs about 8.3 lb, how many gallons will the tank hold?

Recreational Mathematics

74. *Balance the Scale* Determine the quantity to replace the question mark to make the scale balance. The weight times the distance on both sides of the fulcum (the triangle) must be the same to make the scale balance.

75. *Interesting Facts* The 2003 Guinness Book of World Records provides some interesting facts.
   a) The lowest temperature ever recorded in the United States was −62.11°C on January 23, 1971, in Prospect Creek, Alaska. What is this temperature in degrees Fahrenheit?

b) International Falls, Minnesota, has the lowest annual mean temperature in the United States (including Alaska). Its mean annual temperature is about 2.5°C. What is this temperature in degrees Fahrenheit?

c) The highest temperature produced in a laboratory was about 918,000,000°F. What is this temperature in degrees Celsius?

Ice Box Day, International Falls, MN

Internet/Research Activity

76. Do industries in your area export goods? If so, are they training employees to use and understand the metric system? Contact local industries that export goods and write a report on your findings.

8.4 DIMENSIONAL ANALYSIS AND CONVERSIONS TO AND FROM THE METRIC SYSTEM

You may sometimes need to change units of measurement in the metric system to equivalent units in the U.S. customary system. To do so, use *dimensional analysis*, which is a procedure used to convert from one unit of measurement to a different unit of measurement. To perform dimensional analysis, you must first understand what is meant by a unit fraction. A *unit fraction* is any fraction in which the numerator and denominator contain different units and the value of the fraction is 1. From Table 8.6, we can obtain many unit fractions involving U.S. customary units.

TABLE 8.6 U.S. Customary Units

1 foot = 12 inches
1 yard = 3 feet
1 mile = 5280 feet
1 pound = 16 ounces
1 ton = 2000 pounds
1 cup (liquid) = 8 fluid ounces
1 pint = 2 cups
1 quart = 2 pints
1 gallon = 4 quarts
1 minute = 60 seconds
1 hour = 60 minutes
1 day = 24 hours
1 year = 365 days

Examples of Unit Fractions

$$\frac{12\text{ in.}}{1\text{ ft}} \quad \frac{1\text{ ft}}{12\text{ in.}} \quad \frac{16\text{ oz}}{1\text{ lb}} \quad \frac{1\text{ lb}}{16\text{ oz}} \quad \frac{60\text{ min}}{1\text{ hr}} \quad \frac{1\text{ hr}}{60\text{ min}}$$

In each of these examples, the numerator equals the denominator, so the value of the fraction is 1.

To convert an expression from one unit of measurement to a different unit, multiply the given expression by the unit fraction (or fractions) that will result in the answer having the units you are seeking. When two fractions are being multiplied and the same unit appears in the numerator of one fraction and the denominator of the other fraction, then that common unit may be divided out. For example, suppose we want to convert 30 inches to feet. We consider the following:

$$30 \text{ in.} = ? \text{ ft}$$

Since inches are given, we will need to eliminate them. Thus, inches will need to appear in the denominator of the unit fraction. We need to convert to feet, so feet will need to appear in the numerator of the unit fraction. If we multiply a quantity in inches by a unit fraction containing feet/inches, the inches will divide out as follows, leaving feet. In the following illustration we have omitted the numbers in the unit fraction so we can concentrate on the units.

$$(\cancel{\text{in.}})\left(\frac{\text{ft}}{\cancel{\text{in.}}}\right) = \text{ft}$$

Thus, to convert 30 inches to feet, we do the following.

$$30 \text{ in.} = (30 \cancel{\text{in.}})\left(\frac{1 \text{ ft}}{12 \cancel{\text{in.}}}\right) = \frac{30}{12}\text{ft} = 2.5 \text{ ft}$$

In Examples 1 through 3, we will give examples that do not involve the metric system. After that, we will use dimensional analysis to make conversions to and from the metric system.

EXAMPLE 1 *Using Dimensional Analysis*

A container contains 26 ounces of salt. Convert 26 ounces to pounds.

SOLUTION: One pound is 16 ounces. Therefore, we write

$$26 \text{ oz} = (26 \cancel{\text{oz}})\left(\frac{1 \text{ lb}}{16 \cancel{\text{oz}}}\right) = \frac{26}{16}\text{lb} = 1.625 \text{ lb}$$

Thus, 26 oz equals 1.625 lb. ▲

EXAMPLE 2 *Euro Dollars*

On January 1, 2002, the euro became the common currency in use in 12 European countries (Austria, Belgium, Finland, France, Germany, Greece, Ireland, Italy, Luxembourg, the Netherlands, Portugal, and Spain). On May 1, 2003, $1 U.S. could be exchanged for about 0.91 euros. What was the amount in U.S. dollars of 1260 euros?

SOLUTION:

$$1260 \text{ euros} = 1260 \cancel{\text{euros}}\left(\frac{\$1.00}{0.91 \cancel{\text{euros}}}\right) = \$\frac{1260}{0.91} \approx \$1384.62$$

Thus, 1260 euros had a value of about $1384.62. ▲

DID YOU KNOW

The Importance of Measurements

John Quincy Adams, 6th president of the United States

We all realize how important the abilities to read and to write are. Yet do we realize how important an understanding of measurements is to our society and our daily lives? As John Quincy Adams wrote in his report to the U.S. Congress in 1821:

"Weights and measures may be ranked among the necessities of life to every individual of human society. They enter into the economical arrangements and daily concerns of every family. They are necessary to every occupation of human industry; to the distribution and security of every species of property; to every transaction of trade and commerce; to the labors of the husbandman; to the ingenuity of the artificer; to the studies of the philosopher; to the researches of the antiquarian; to the navigation of the mariner, and the marches of the soldier; to all the exchanges of peace, and all the operations of war. The knowledge of them, as in established use, is among the first elements of education, and is often learned by those who learn nothing else, not even to read and write. This knowledge is riveted in the memory by the habitual application of it to the employments of men throughout life."

If more than one unit needs to be changed, more than one multiplication may be needed, as illustrated in Example 3.

EXAMPLE 3 ***Using Several Unit Fractions***

Convert 60 miles per hour to feet per second.

SOLUTION: Let's consider the units given and where we want to end up. We are given $\frac{\text{mi}}{\text{hr}}$ and wish to end with $\frac{\text{ft}}{\text{sec}}$. Thus, we need to change miles into feet and hours into seconds. Because two units need to be changed, we will need to multiply the given quantity by two unit fractions, one for each conversion. First we show how to convert the units of measurement from miles per hour to feet per second:

$$\left(\frac{\cancel{\text{mi}}}{\cancel{\text{hr}}}\right)\left(\frac{\text{ft}}{\cancel{\text{mi}}}\right)\left(\frac{\cancel{\text{hr}}}{\text{sec}}\right) \quad \text{gives an answer in} \quad \frac{\text{ft}}{\text{sec}}$$

Now we multiply the given quantity by the appropriate unit fractions to obtain the answer:

$$60\frac{\text{mi}}{\text{hr}} = \left(60\frac{\cancel{\text{mi}}}{\cancel{\text{hr}}}\right)\left(\frac{5280\text{ ft}}{1\ \cancel{\text{mi}}}\right)\left(\frac{1\ \cancel{\text{hr}}}{3600\text{ sec}}\right) = \frac{(60)(5280)}{(1)(3600)}\frac{\text{ft}}{\text{sec}}$$

$$= 88\frac{\text{ft}}{\text{sec}}$$

Note that $\left(60\frac{\text{mi}}{\text{hr}}\right)\left(\frac{1\text{ hr}}{3600\text{ sec}}\right)\left(\frac{5280\text{ ft}}{1\text{ mi}}\right)$ will give the same answer. ▲

Conversions to and from the Metric System

Now we will apply dimensional analysis to the metric system.

Table 8.7 on page 461 is used in making conversions to and from the metric system. The values given in Table 8.7 are often approximations. A more exact table of conversion factors may be found in many science books at your college's library or on the Internet. However, we can use this table to obtain many unit fractions.

Table 8.7 shows that 1 in. = 2.54 cm. From this equality, we can write the two unit fractions

$$\frac{1\text{ in.}}{2.54\text{ cm}} \quad \text{or} \quad \frac{2.54\text{ cm}}{1\text{ in.}}$$

Examples of other unit fractions from Table 8.7 are

$$\frac{1\text{ yd}}{0.9\text{ m}}, \quad \frac{0.9\text{ m}}{1\text{ yd}}, \quad \frac{1\text{ gal}}{3.8\ \ell}, \quad \frac{3.8\ \ell}{1\text{ gal}}, \quad \frac{1\text{ lb}}{0.45\text{ kg}}, \quad \text{and} \quad \frac{0.45\text{ kg}}{1\text{ lb}}$$

To change from a metric unit to a customary unit or vice versa, multiply the given quantity by the unit fraction whose product will result in the units you are seeking.

TABLE 8.7 Conversions Table

Length		
1 inch (in.) =	2.54	centimeters (cm)
1 foot (ft) =	30	centimeters (cm)
1 yard (yd) =	0.9	meter (m)
1 mile (mi) =	1.6	kilometers (km)
Area		
1 square inch ($in.^2$) =	6.5	square centimeters (cm^2)
1 square foot (ft^2) =	0.09	square meter (m^2)
1 square yard (yd^2) =	0.8	square meter (m^2)
1 square mile (mi^2) =	2.6	square kilometers (km^2)
1 acre =	0.4	hectare (ha)
Volume		
1 teaspoon (tsp) =	5	milliliters (mℓ)
1 tablespoon (tbsp) =	15	milliliters (mℓ)
1 fluid ounce (fl oz) =	30	milliliters (mℓ)
1 cup (c) =	0.24	liter (ℓ)
1 pint (pt) =	0.47	liter (ℓ)
1 quart (qt) =	0.95	liter (ℓ)
1 gallon (gal) =	3.8	liters (ℓ)
1 cubic foot (ft^3) =	0.03	cubic meter (m^3)
1 cubic yard (yd^3) =	0.76	cubic meter (m^3)
Weight (Mass)		
1 ounce (oz) =	28	grams (g)
1 pound (lb) =	0.45	kilogram (kg)
1 ton (T) =	0.9	tonne (t)

For example, to convert 5 in. to centimeters, multiply 5 in. by a unit fraction with centimeters in the numerator and inches in the denominator.

$$\begin{aligned} 5 \text{ in.} &= (5 \cancel{\text{in.}})\left(\frac{2.54 \text{ cm}}{1 \cancel{\text{in.}}}\right) \\ &= 5(2.54) \text{ cm} \\ &= 12.7 \text{ cm} \end{aligned}$$

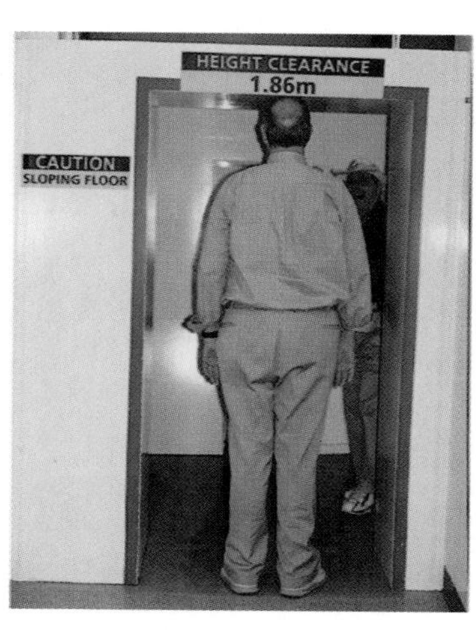

EXAMPLE 4 *Volume and Length Conversions*

a) A recipe for chicken soup requires $4\frac{1}{2}$ cups of water. How many liters does this amount equal?

b) A man measures 1.86 m (see photo). What is his height in feet?

SOLUTION:

a) In Table 8.7, under the heading of volume, we see that 1 cup = 0.24 ℓ. Thus, the unit fractions involving cups and liters are

$$\frac{1 \text{ cup}}{0.24\ \ell} \quad \text{or} \quad \frac{0.24\ \ell}{1 \text{ cup}}$$

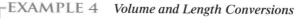

We need to convert from cups to liters. Since $4\frac{1}{2}$ is 4.5 in decimal form, we write

$$4.5 \text{ cups} = 4.5 \cancel{\text{cups}}\left(\frac{0.24\ \ell}{1\ \cancel{\text{cup}}}\right) = (4.5)(0.24)\ \ell = 1.08\ \ell$$

b) In Table 8.7, under the heading of length, we see that there is no conversion given from meters to feet. There are a number of ways this example could be worked. One method is to convert meters to yards and then convert yards to feet. The procedure is shown below.

$$\begin{aligned} 1.86 \text{ m} &= (1.86\ \cancel{\text{m}})\left(\frac{1\ \cancel{\text{yd}}}{0.9\ \cancel{\text{m}}}\right)\left(\frac{3 \text{ ft}}{1\ \cancel{\text{yd}}}\right) \\ &= \frac{(1.86)(3)}{0.9} \text{ ft} \\ &\approx 6.2 \text{ ft} \end{aligned}$$

Thus, the man is about 6.2 ft. ▲

Land for sale in Fiji, measured in hectares.

EXAMPLE 5 *Area Conversion*

The photo shows an area of 31.46 hectares for sale. Find the area in acres.

SOLUTION: From Table 8.7, we determine that 1 acre = 0.4 ha. Thus,

$$31.46 \text{ ha} = (31.46\ \cancel{\text{ha}})\left(\frac{1 \text{ acre}}{0.4\ \cancel{\text{ha}}}\right) = \frac{31.46}{0.4} \text{ acres} = 78.65 \text{ acres}$$ ▲

EXAMPLE 6 *Weight (Mass) Conversion*

The photo shows that tangelos cost \$2.45 per kilogram. Determine the cost per pound for the tangelos.

SOLUTION: First, we will determine the number of pounds that is equivalent to 1 kilogram. From Table 8.7, we obtain the unit fraction

$$\frac{0.45 \text{ kg}}{1 \text{ lb}}$$

We use this unit fraction to determine the tangelos's cost per pound.

$$\begin{aligned} \frac{\$2.45}{1 \text{ kg}} &= \left(\frac{\$2.45}{1\ \cancel{\text{kg}}}\right)\left(\frac{0.45\ \cancel{\text{kg}}}{1 \text{ lb}}\right) \\ &= \$2.45(0.45) \text{ per pound} \\ &\approx \$1.10 \text{ per pound} \end{aligned}$$

Therefore, the tangelos cost about \$1.10 per pound. ▲

EXAMPLE 7 ***Administering a Medicine***

A nurse must administer 4 cc of codeine elixir to a patient.

a) How many milliliters of the drug will be administered?

b) How many ounces is this dosage equivalent to?

SOLUTION:

a) Since 1 cc = 1 mℓ, the nurse will administer 4 mℓ of the drug.

b) Since 1 fl oz = 30 mℓ,

$$4 \text{ m}\ell = (4 \cancel{\text{m}\ell})\left(\frac{1 \text{ fl oz}}{30 \cancel{\text{m}\ell}}\right) = \frac{4}{30}\text{fl oz} \approx 0.13 \text{ fl oz}$$

Suppose we want to convert 150 millimeters to inches. Table 8.7 does not have a conversion factor from millimeters to inches, but it does have one for inches to centimeters. Because 1 inch = 2.54 centimeters and 1 centimeter = 10 millimeters, we can reason that 1 inch = 25.4 millimeters. Therefore, unit fractions we may use are as follows.

$$\frac{1 \text{ in.}}{25.4 \text{ mm}} \quad \text{or} \quad \frac{25.4 \text{ mm}}{1 \text{ in.}}$$

We can solve the problem as follows.

$$150 \text{ mm} = (150 \cancel{\text{mm}})\left(\frac{1 \text{ in.}}{25.4 \cancel{\text{mm}}}\right) = \frac{150}{25.4}\text{in.}$$
$$\approx 5.91 \text{ in.}$$

If we wish, we can use dimensional analysis using two unit fractions to make the conversion. The procedure follows:

$$150 \text{ mm} = (150 \cancel{\text{mm}})\left(\frac{1 \cancel{\text{cm}}}{10 \cancel{\text{mm}}}\right)\left(\frac{1 \text{ in.}}{2.54 \cancel{\text{cm}}}\right) = \frac{150}{(10)(2.54)}\text{ in.}$$
$$\approx 5.91 \text{ in.}$$

EXAMPLE 8 ***Converting a Speed***

The photo shows that a road in Cancun, Mexico, has a speed limit of 50 kilometers per hour (kph). Determine the speed limit in miles per hour.

SOLUTION: In kilometers per hour and miles per hour, the time unit, hour, is the same. Therefore, we just need to convert 50 kilometers to miles. From Table 8.7, we find unit fractions

$$\frac{1 \text{ mi}}{1.6 \text{ km}} \quad \text{or} \quad \frac{1.6 \text{ km}}{1 \text{ mi}}$$

$$50 \text{ km} = (50 \cancel{\text{km}})\left(\frac{1 \text{ mi}}{1.6 \cancel{\text{km}}}\right) = \frac{50}{1.6}\text{mi} = 31.25 \text{ mi}$$

Since 50 km equals 31.25 mi, 50 kph is equivalent to 31.25 mph.

EXAMPLE 9 *Understanding the Label*

The label on a bottle of Vicks Formula 44D Cough Syrup indicates that the active ingredient is dextromethorphan hydrobromide and that 5 mℓ (1 teaspoon) contains 10 mg of this ingredient. If the recommended dosage for adults is 3 teaspoons, determine the following.

a) How many milliliters of cough medicine should be taken?

b) How many milligrams of the active ingredient should be taken?

c) If the bottle contains 8 fluid ounces of medicine, how many milligrams of the active ingredient are in the bottle?

SOLUTION:

a) Since each teaspoon contains 5 mℓ and 3 teaspoons should be taken, 15 mℓ of the cough medicine should be taken.

$$3 \text{ tsp} = (3 \cancel{\text{tsp}})\left(\frac{5 \text{ m}\ell}{1 \cancel{\text{tsp}}}\right) = 15 \text{ m}\ell$$

b) Since each teaspoon contains 10 mg of the active ingredient, 30 mg of the active ingredient should be taken.

$$3 \text{ tsp} = (3 \cancel{\text{tsp}})\left(\frac{10 \text{ mg}}{1 \cancel{\text{tsp}}}\right) = 30 \text{ mg}$$

c) Table 8.7 shows that each fluid ounce contains 30 mℓ. Since each 5 mℓ contains 10 mg of the active ingredient, we can work the problem as follows.

$$8 \text{ fl oz} = (8 \cancel{\text{fl oz}})\left(\frac{30 \cancel{\text{m}\ell}}{1 \cancel{\text{fl oz}}}\right)\left(\frac{10 \text{ mg}}{5 \cancel{\text{m}\ell}}\right) = \frac{8(30)(10)}{5} \text{ mg} = 480 \text{ mg}$$

Therefore, there are 480 mg (or 0.48 g) of the active ingredient in the bottle of cough syrup. ▲

EXAMPLE 10 *Determining Dosage by Weight*

Drug dosage is often administered according to a patient's weight. For example, 30 mg of the drug vancomicin is to be given for each kilogram of a person's weight. If Martha Greene, who weighs 136 lb, is to be given the drug, what dosage should she be given?

SOLUTION: First we need to convert Martha's weight into kilograms. From Table 8.7, we see that 1 lb = 0.45 kg. We obtain our unit fraction from this information. Next, we need to determine the number of milligrams of the drug for Martha's weight in kilograms. To do so, write the given ratio of 30 mg of the drug for each kilogram as $\frac{30 \text{ mg}}{1 \text{ kg}}$. Note that this ratio is not a unit fraction since the numerator and denominator are not equivalent. The answer may be found as follows.

$$136 \text{ lb} = (136 \cancel{\text{lb}})\left(\frac{0.45 \cancel{\text{kg}}}{1 \cancel{\text{lb}}}\right)\left(\frac{30 \text{ mg}}{1 \cancel{\text{kg}}}\right) = (136)(0.45)(30) \text{ mg} = 1836 \text{ mg}$$

Thus, 1836 mg, or 1.836 g, of the drug should be given. ▲

MATHEMATICS *Everywhere*

The Seven Base Units

We all realize how important measurements are to daily life. The Système international d'unités (SI), the modern version of the metric system, provides a logical and interconnected framework for all measurements in science, industry, and commerce. The SI is built upon a foundation of seven base units, as explained below. All other SI units are derived from these units. The base units for time, electric current, amount of substance, and luminous intensity are the same in both the metric system and the U.S. customary system.

Length : Meter The meter is the length of the path traveled by light in a vacuum during a time interval of $\frac{1}{299{,}792{,}458}$ second. Thus, the speed of light in a vacuum is 299 792 458 meters per second.

Time : Second The second is the duration of 9,192,631,770 cycles of the radiation associated with a specific transition of the cesium 133 atom.

Electric Current : Ampere The ampere is the current that, if maintained in each of two infinitely long parallel wires separated by 1 m in free space, would produce a force between the two wires (due to their magnetic fields) of 2×10^{-7} newton for each meter of length. The electrical terms *volt, watt, and ohm* are derived using amperes.

Luminous Intensity : Candela The candela is the luminous intensity, in a given direction, of a source that emits monochromatic radiation of frequency 540×10^{12} hertz (Hz) and that has a radiant intensity in that direction of $\frac{1}{683}$ watt per steradian.

Temperature : Kelvin A kelvin is the fraction $\frac{1}{273.16}$ of the thermodynamic temperature of the triple point of water. The temperature 0 K is commonly referred to as "absolute zero." In the widely used Celsius temperature, 0°C corresponds to 273.15 K. Thus, water freezes at 273.15 K.

Mass : Kilogram The kilogram is a cylinder of platinum–iridium alloy kept by the International Bureau of Weights and Measures in Sèvres, France. A duplicate in the custody of the U.S. National Institute of Standards and Technology serves as the mass standard for the United States. The kilogram is the only base unit still defined by an artifact.

Amount of Substance : Mole The mole is the amount of substance of a system that contains as many elementary entities as there are atoms in 0.012 kilogram of carbon 12. When the mole is used, the elementary entities must be specified and may be atoms, molecules, ions, electrons, other particles, or specific groups of such particles. The SI unit of concentration (of amount of substance) is the *mole per cubic meter* (mol/m^3).

Some of these definitions provided here are quite complex, but they form the basis for all measurements in the metric system. For more complete definitions of unknown terms and for additional information, contact the U.S. Department of Commerce, National Institute of Standards and Technology (www.nist.gov).

Temperature Measurement Systems

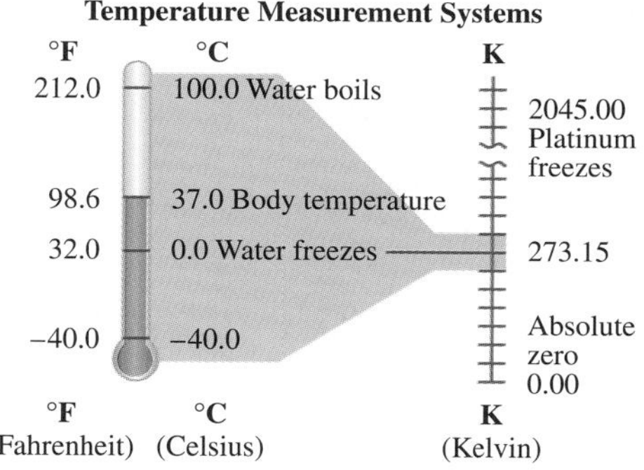

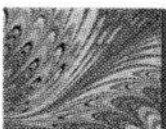

SECTION 8.4 EXERCISES

Concept/Writing Exercises

1. What is dimensional analysis?
2. What is a unit fraction?
3. Give a unit fraction that relates seconds and minutes. Explain how you determined the unit fraction.
4. Give a unit fraction that relates feet and yards. Explain how you determined the unit fraction.
5. When converting from centimeters to feet, which unit fraction would you use? Explain.

$$\frac{1 \text{ ft}}{30 \text{ cm}} \quad \text{or} \quad \frac{30 \text{ cm}}{1 \text{ ft}}$$

6. When converting from kilograms to pounds, which unit fraction would you use? Explain.

$$\frac{1 \text{ lb}}{0.45 \text{ kg}} \quad \text{or} \quad \frac{0.45 \text{ kg}}{1 \text{ lb}}$$

7. When converting from gallons to liters, which unit fraction would you use? Explain.

$$\frac{1 \text{ gal}}{3.8\ \ell} \quad \text{or} \quad \frac{3.8\ \ell}{1 \text{ gal}}$$

8. When converting from square yards to square meters, which unit fraction would you use? Explain.

$$\frac{1 \text{ yd}^2}{0.8 \text{ m}^2} \quad \text{or} \quad \frac{0.8 \text{ m}^2}{1 \text{ yd}^2}$$

Practice the Skills

In Exercises 9–24, convert the quantity to the indicated units.

9. 52 in. to centimeters

10. 9 lb to kilograms

11. 4.2 ft to meters

12. 427 g to ounces

13. 15 yd^2 to square meters

14. 160 kg to pounds

15. 39 mi to kilometers

16. 765 mm to inches

17. 675 ha to acres

18. 192 oz to grams

19. 15.6 ℓ to pints

20. 4 T to tonnes

21. 45.6 mℓ to fluid ounces

22. 1.6 km^2 to square miles

23. 120 lb to kilograms

24. 6.2 acres to hectares

In Exercises 25–32, replace the measurement(s) indicated in blue with an equivalent metric measure(s). For example, a foot could be replaced with 30 cm.

25. More bounce to the *ounce*.

26. An *ounce* of prevention is worth a *pound* of cure.

27. He demanded his *pound* of flesh.

28. *Five foot two* and eyes of blue.

29. Give him an *inch* and he'll take a *mile*.

30. A miss is as good as a *mile*.

31. First down and *10 yards* to go.

32. The longest *yard*.

In Exercises 33–36, use the part of the scorecard, which shows the distance in meters for the first four holes of the Millbrook Resort Golf Course in Queenstown, New Zealand. Determine the distances indicated.

HOLE	BLACK Tees	BLUE Tees	HANDICAP	PAR					WHITE Tees	RED Tees
1	505	505	3	5					466	414
2	185	175	15	3					137	91
3	366	357	11	4					344	287
4	396	376	7	4					376	303

33. Hole 1, black tees, in yards

34. Hole 2, blue tees, in yards

35. Hole 3, white tees, in feet

36. Hole 4, red tees, in feet

Problem Solving

37. *Speed Limit* The speed limit for the sharp curve shown in the photo is 85 kph. Determine the speed in miles per hour.

38. *How Far?* Carol Ann Harle's new car traveled 105 mi on 5 gal of gasoline. How many kilometers can Carol Ann's car travel with the same amount of gasoline?

39. *Buying Carpet* Victoria Montoya is buying outdoor carpet for her lanai, which is 6 yd by 9 yd. The carpeting is sold in square meters. How many square meters of carpeting will she need?

40. *Cincinnati to Columbus* The distance from Cincinnati, Ohio, to Columbus, Ohio, is about 110 mi. What is the distance in kilometers?

41. *Cornflakes* A box of cornflakes purchased in Canada indicates that it contains 400 grams of cornflakes. How many ounces of cornflakes are contained in the box?

42. *The QEW* Part of the Queen Elizabeth Way in Canada has a speed limit of 80 kph. What is the speed in miles per hour?

43. *Milliliters in a Glass* A glass holds 8 fl oz. How many milliliters will it hold?

44. *Swimming Pool* A swimming pool holds 12,500 gal of water. What is this volume in kiloliters?

45. *Building a Basement* A basement is to be 50 ft long, 30 ft wide, and 8 ft high. How much dirt will have to be removed when this basement is built? Answer in cubic meters.

46. *Area of Yosemite National Park* Yosemite National Park has an area of 1189 mi^2. What is its area in square kilometers?

47. *Cost of Rice* If rice costs $1.10 per kilogram, determine the cost of a pound of rice.

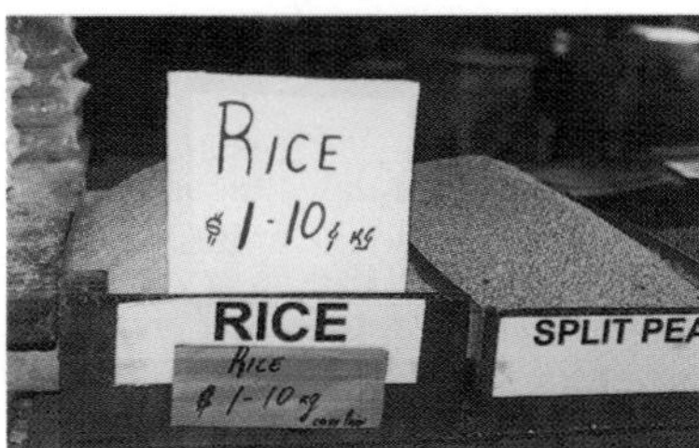

48. *Weight of a Car* A German-made car has a weight of 1.3 t.
 a) How many tons does this weight equal?
 b) How many pounds?

49. *Capacity of a Tank Truck* A tank truck holds 34.5 kℓ of gasoline. How many gallons does it hold?

50. *Cost per Gram* A 0.25 oz bottle of Chanel perfume costs $80. What is the cost per gram?

51. *A Weight in Stones* Some scales in Europe measure a person's weight both in kilograms and in stones. From the photo, we see that a weight of 70 kg is equal to about 11 stones.

 a) Using a unit fraction, determine the weight, in kilograms, of a person who weighs 8 stones.
 b) Determine the person's weight in pounds.

52. *A Precious Stone* One gram is the same as five carats. David Erich's new ring contains a precious stone that is $\frac{1}{8}$ carat. Find the weight of the stone in grams.

53. *Death Valley Elevation* The lowest elevation in the United States is −282 ft at Badwater in Death Valley, California. Determine this elevation in
 a) centimeters.
 b) meters.

54. *Car's Engine* A specific car's engine has a capacity of 5.7 ℓ of oil. How many quarts of oil does the engine have?

55. *Square Meters to Square Feet* One meter is about 3.3 ft. Use this information to determine
 a) the equivalent of one square meter in square feet.
 b) the equivalent of one cubic meter in cubic feet.

56. *Square Feet to Square Centimeters* One foot is about 30 cm. Use this information to determine
 a) the equivalent of one square foot in square centimeters.
 b) the equivalent of one cubic foot in cubic centimeters.

57. *Dosage for a Child* The recommended dosage of the drug codeine for pediatric patients is 1 mg per kilogram of a child's weight. What dosage of codeine should be given to April Adam, who weighs 56 lb?

58. *Dosage for a Man* For each kilogram of a person's weight, 1.5 mg of the antibiotic drug gentamicin is to be administered. If Ron Gigliotti weighs 170 lb, how much of the drug should he receive?

59. *Ampicillin* The recommended dosage of the drug ampicillin for pediatric patients is 200 mg per kilogram of a patient's weight. If Janine Baker weighs 76 lb, how much ampicillin should she receive?

60. *Medicine for a Dog* For each kilogram of weight of a dog, 5 mg of the drug bretylium is to be given. If Blaster, an Irish setter, weighs 82 lb, how much of the drug should be given?

61. *Active Ingredients* The label on the bottle of Triaminic expectorant indicates that each teaspoon (5 mℓ) contains 12.5 mg of the active ingredient phenylpropanolamine hydrochloride.
 a) Determine the amount of the active ingredient in the recommended adult dosage of 2 teaspoons.
 b) Determine the quantity of the active ingredient in a 12 oz bottle.

62. *Stomach Ache Remedy* The label on the bottle of Maximum Strength Pepto-Bismol indicates that each tablespoon contains 236 mg of the active ingredient bismuth subsalicyate.
 a) Determine the amount of the active ingredient in the recommended dosage of 2 tablespoons.
 b) If the bottle contains 8 fl oz, determine the quantity of the active ingredient in the bottle.

63. *Disney Magic* The Disney Magic Cruise Ship is 964 feet long, has a weight of 85,000 tons, and can travel 28 mph.

 a) Determine the length of the ship in meters.
 b) Determine the weight in tonnes.
 c) Determine the speed in kilometers per hour.

64. *Making Cookies* Change all the measurements in the cookie recipe to metric units. Do not forget pan size, temperature, and size of cookies.

Magic Cookie Bar

$\frac{1}{2}$c graham cracker crumbs
12 oz nuts
8 oz chocolate pieces
$1\frac{1}{3}$c flaked coconut
$1\frac{1}{3}$c condensed milk

Coat the bottom of a 9 in. × 13 in. pan with melted margarine. Add rest of ingredients one by one: crumbs, nuts, chocolate, and coconut. Pour condensed milk over all. Bake at 350°F for 25 minutes. Allow to cool 15 minutes before cutting. Makes about two dozen $1\frac{1}{2}$in. by 3 in. bars.

65. *The Space Shuttle* Write each of the metric units, labeled (*a*) through (*n*), in U.S. customary units.

The first human flight, December 17, 1903, was (*a*) *37 m.* Just 66 years later, Neil Armstrong stepped on the moon after journeying (*b*) *370 140 km.* On April 12, 1981, a new era in space flight began when the space shuttle embarked on its maiden voyage.

Here are some characteristics of and facts about the space shuttle. The two solid rocket boosters are jettisoned at (*c*) *44 km.* During reentry, portions of the orbiter's exterior reach temperatures up to (*d*) *1260°C.* The orbiter lands at a speed of (*e*) *335 kph.* It can deliver to orbit up to (*f*) *29 484 kg* of payload in its huge (*g*) *4.5 m × 18 m* cargo bay. Propellants can be supplied to the engines at a rate of about (*h*) *171 396 ℓ*/min of hydrogen and (*i*) *63 588 ℓ*/min of oxygen. The external tank is (*j*) *46.89 m* long and (*k*) *8.4 m* in diameter. When fully loaded, the tank contains (*l*) *632 772 kg* of liquid oxygen and (*m*) *106 142 kg* of cold liquid hydrogen at about (*n*) *−251°C.*

Challenge Problems/Group Activities

66. *Nursing Question* The following question was selected from a nursing exam. Can you answer it?

In caring for a patient after delivery, you are to give 0.2 mg Ergotrate Maleate. The ampule is labeled $\frac{1}{300}$ grain/mℓ. How much would you draw and give? (60 mg = 1 grain)

a) 15 cc **b)** 1.0 cc **c)** 0.5 cc **d)** 0.01 cc

67. *How Much Beef* Paul Gosse is planning a picnic and plans on purchasing 0.18 kg of ground beef for each 100 lb of weight of guests who will be in attendance. If he expects 15 people whose average weight is 130 lb, how many pounds of beef should he purchase?

68. *An Auto Engine* The displacement of automobile engines is measured in liters. A 2003 Ford Explorer has a 4.0 ℓ engine.

a) Determine the displacement of the engine in cubic centimeters.

b) Determine the displacement of the engine in cubic inches.

Recreational Mathematics

In Exercises 69–74, answer the question, What metric unit am I?

69. I am a length greater than a yard, but less than a kitchen tabletop.

70. I am a weight greater than a calculator, but less than a wooden chair.

71. I am an area greater than an acre, but less than a square kilometer.

72. I am a liquid volume greater than a quart, but less than a gallon.

73. I am a weight greater than a ton, but less than a full-grown elephant.

74. I am a length greater than an inch, but less than a yard.

In Exercises 75–84, try to solve the puzzle. What is

75. 2000 pounds of Chinese soup?

76. 1 millionth of a mouthwash?

77. 1000 aches?

78. 448 grahams of cake?

79. 1 million phones?

80. 1 million bicycles?

81. 2000 mockingbirds?

82. 10 cards?

83. 10 rations?

84. 1 millionth of a fish?

CHAPTER 8 SUMMARY

IMPORTANT FACTS

Metric Units

Prefix	Symbol	Meaning
kilo	k	1000 × base unit
hecto	h	100 × base unit
deka	da	10 × base unit
		base unit
deci	d	$\frac{1}{10}$ of base unit
centi	c	$\frac{1}{100}$ of base unit
milli	m	$\frac{1}{1000}$ of base unit

Water

Volume in Cubic Units		Volume in Liters		Mass of Water
1 cm^3	=	1 mℓ	=	1g
1 dm^3	=	1 ℓ	=	1 kg
1 m^3	=	1 kℓ	=	1 t (1000 kg)

Temperature

$$°C = \frac{5}{9}(°F - 32)$$

$$°F = \frac{9}{5}°C + 32$$

CHAPTER 8 REVIEW EXERCISES

8.1

In Exercises 1–6, indicate the meaning of the prefix.

1. Centi
2. Kilo
3. Milli
4. Hecto
5. Deka
6. Deci

In Exercises 7–12, change the given quantity to that indicated.

7. 20 cg to grams
8. 3.2 ℓ to centiliters
9. 0.0004 cm to millimeters
10. 1 000 000 mg to kilograms
11. 4.62 kℓ to liters
12. 192.6 dag to decigrams

In Exercises 13 and 14, arrange the quantities from smallest to largest.

13. 2.67 kℓ, 3000 mℓ, 14 630 cℓ
14. 0.047 km, 4700 m, 47 000 cm

8.2, 8.3

In Exercises 15–24, indicate the metric unit of measurement that would best express the following.

15. The length of a telephone
16. The mass of a cellular telephone
17. The temperature of the sun's surface
18. The diameter of a quarter
19. The area of a room of a house
20. The volume of a glass of milk
21. The length of an ant
22. The mass of a car
23. The distance from Philadelphia, Pennsylvania, to Irvine, Texas
24. The height a dolphin can jump

In Exercises 25 and 26, (a) first estimate the following in metric units and then (b) measure with a metric ruler. Record your results.

25. Your height
26. The length of a new pencil

In Exercises 27–32, select the best answer.

27. The length of the distance between Los Angeles and San Francisco is about
    a) 8000 m. b) 2000 km. c) 650 km.
28. The mass of a full-grown border collie is about
    a) 600 g. b) 20 kg. c) 100 kg.

29. The volume of a gallon of orange juice is about
a) 0.1 kℓ. **b)** 0.5 ℓ. **c)** 4 ℓ.

30. The area of a large vegetable garden in a person's yard may be
a) 200 m^2. **b)** 0.5 ha. **c)** 0.02 km^2.

31. The temperature on a hot summer day in Georgia may be
a) 34°C. **b)** 55°C. **c)** 25°C.

32. The height of a giant sequoia tree is about
a) 300 m. **b)** 3000 cm. **c)** 0.3 m.

33. Convert 2500 kg to tonnes.

34. Convert 6.3 t to grams.

35. If the temperature outside is 18°C, what is the Fahrenheit temperature?

36. If the room temperature is 68°F, what is the Celsius temperature?

37. If your outdoor thermometer shows a temperature of −6°F, what is the Celsius temperature?

38. If Lynn Colgin's body temperature is 39°C, what is her Fahrenheit temperature?

39. Measure, in centimeters, each of the line segments, then compute the area of the figure.

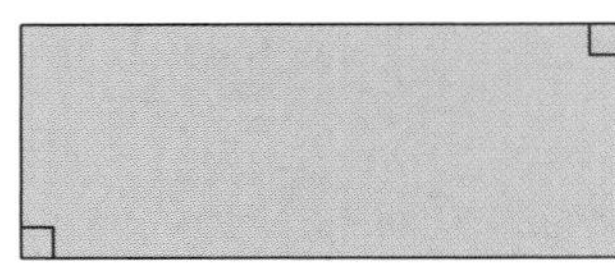

40. Measure, in centimeters, the radius of the circle, then compute the area of the circle.

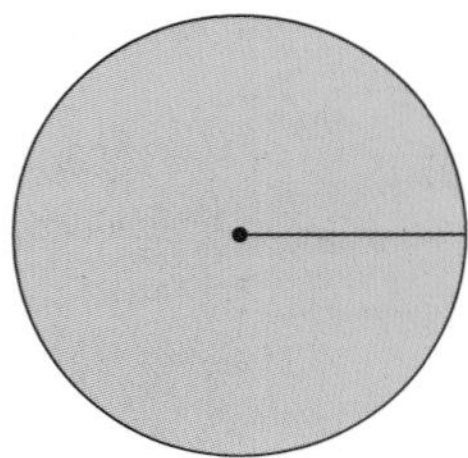

41. a) *A Swimming Pool's Volume* What is the volume of water in a full rectangular swimming pool that is 10 m long and 4 m wide and has an average depth of 2 m? Answer in cubic meters.
b) What is the mass of the water in kilograms?

42. *Area* A rectangular lot measures 22 m by 30 m. Determine
a) the area in square meters.
b) the area in square kilometers.

43. *Volume of a Fish Tank* A small fish tank measures 80 cm long, 40 cm wide, and 30 cm high.
a) What is its volume in cubic centimeters?
b) What is its volume in cubic meters?
c) How many milliliters of water will the tank hold?
d) How many kiloliters of water will the tank hold?

44. A square kilometer is a square with length and width both 1 km. How many times larger is a square kilometer than a square dekameter?

8.4

In Exercises 45–58, change the given quantity to the indicated quantity. When appropriate, round answers to the nearest hundredth.

45. 20 cm = ____ in.
46. 105 kg = ____ lb
47. 83 yd = ____ m
48. 100 m = ____ yd
49. 45 mph = ____ kph
50. 40 ℓ = ____ qt
51. 15 gal = ____ ℓ
52. 40 m^3 = ____ yd^3
53. 83 cm^2 = ____ $in.^2$
54. 4 qt = ____ ℓ
55. 15 yd^3 = ____ m^3
56. 62 mi = ____ km
57. 27 cm = ____ ft
58. $3\frac{1}{4}$ in. = ____ mm

59. *Building a Chimney* Anne Kelly bought 700 bricks to build a chimney. Each brick has a mass of 1.5 kg.
a) What is the total mass of the bricks in kilograms?
b) What is the total weight of the bricks in pounds?

60. *Carpeting a Room* Patricia Burgess is buying new carpet for her family room. The room is 15 ft wide and 24 ft long. The carpeting is sold only in square meters. How many square meters of carpeting will she need? Round your answer to the nearest square meter.

61. *Milk Tank* A cylindrical milk tank can store 50,000 gal of milk.
a) Determine the volume in kiloliters.
b) Estimate the weight of the milk in kilograms. Assume that milk has the same weight as water.

62. *The Speed Limit* The speed limit on a certain road is 35 mph. What is the speed limit in
a) kilometers per hour?
b) meters per hour?

63. *A Water Tank* A rectangular tank used to test leaks in tires is 90 cm by 70 cm by 40 cm deep.
a) Determine the number of liters of water the tank holds.
b) What is the mass of the water in kilograms?

64. *Oranges* If the cost of oranges is $3.50 per kilogram, determine the cost of 1 lb of oranges.

CHAPTER 8 TEST

1. Change 204 cℓ to daℓ.

2. Change 123 km to mm.

3. How many times greater is a kilometer than a dekameter?

4. *Jogging* A high school track is an oval that measures 400 m around. If Dave Camp jogs around the track six times, how many kilometers has he gone?

In Exercises 5–9, choose the best answer.

5. The length of this page is about
a) 10 cm.
b) 25 cm.
c) 60 cm.

6. The surface area of the top of a kitchen table is about
a) 2 m^2.
b) 200 cm^2.
c) 2000 cm^2.

7. The amount of gasoline that an automobile's gas tank can hold is about
a) 200 ℓ.
b) 20 ℓ.
c) 75 ℓ.

8. The mass of a cellular telephone is about
a) 0.1 t.
b) 2 kg.
c) 150 g.

9. The outside temperature on a snowy day is about
a) 18°C.
b) −2°C.
c) −40°C.

10. How many times greater is a square meter than a square centimeter?

11. How many times greater is a cubic meter than a cubic millimeter?

12. Convert 452 in. to centimeters.

13. How far, in yards, is the Marriott from the sign?

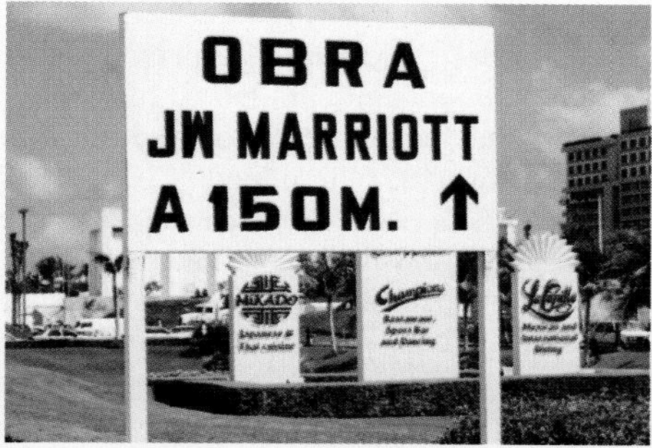

Cancun, Mexico

14. Change −10°F to degrees Celsius.

15. Change 20°C to degrees Fahrenheit.

16. *Giraffe* A giraffe may be 12 ft tall. How many centimeters is this?

17. *At the Aquarium* A fish tank at an aquarium is 20 m long by 20 m wide by 8 m deep.
a) Determine the volume of the tank in cubic meters.
b) Determine the number of liters of water the tank holds.
c) Determine the weight of the water in kilograms.

18. *Cost of Paint* The first coat of paint for the outside walls of a building requires 1 ℓ of paint for each 10 m^2 of wall surface. The second coat requires 1 ℓ for every 15 m^2. If the paint costs $3.50 per liter, what will be the cost of two coats of paint for the four outside walls of a building 20 m long, 15 m wide, and 6 m high?

GROUP PROJECTS

Health and Medicine

Throughout this chapter, we have shown the importance of the metric system in the medical professions. The following two questions involve applications of the metric system to medicine.

1. a) Twenty milligrams of the drug lincomycin is to be given for each kilogram of a person's weight. The drug is to be mixed with 250 cc of a normal saline solution, and the mixture is to be administered intravenously over a 1 hr period. Clyde Dexter, who weighs 196 lb, is to be given the drug. Determine the dosage of the drug he will be given.
   b) At what rate per minute should the 250 cc solution be administered?
2. a) At a pharmacy, a parent asks a pharmacist why her child needs such a small dosage of a certain medicine. The pharmacist explains that a general formula may be used to estimate a child's dosage of certain medicines. The formula is

$$\text{Child's dose} = \frac{\left(\begin{array}{c}\text{child's weight}\\ \text{in kilograms}\end{array}\right)}{67.5 \text{ kg}} \times \text{adult dose}$$

   What is the amount of medicine you would give a 60 lb child if the adult dosage of the medicine is 70 mg?
   b) At what weight, in pounds, would the child receive an adult dose?

Traveling to Other Countries

3. Dale Pollinger is a buyer at General Motors and travels frequently on business to foreign countries. He always plans ahead and does his holiday shopping overseas where he can purchase items not easily found in the United States.
   a) On a trip to Tokyo, he decides to buy a kimono for his sister, Kathy. To determine the length of a kimono, one measures, in centimeters, the distance from the bottom of a person's neck to 5 cm above the floor. If the distance from the bottom of Dale's sister's neck to the floor is 5 ft 2 in., calculate the length of the kimono that Dale should purchase.
   b) If the conversion rate at the time is 1 U.S. dollar = 118.25 yen and the kimono cost 8695.5 yen, determine the cost of the kimono in U.S. dollars.
   c) On a trip to Mexico City, Mexico, Dale finds a small replica of a Mayan castle that he wants to purchase for his wife, Sue. He is going directly from Mexico to Rome, so he wants to mail the castle back to the United States. The mailing rate from Mexico to the United States is 10 pesos per hundred grams. Determine the mailing cost, in U.S. dollars, if the castle weighs 6 lb and the exchange rate is 1 peso = 0.095 U.S. dollar.
   d) This question has three parts. While traveling to New Zealand, Dale finds that unleaded plus gasoline cost $0.929 per liter in New Zealand dollars.

   How much will it cost him, in New Zealand dollars, to fill the 53 ℓ gas tank of his rented car? If the exchange rate is $1 New Zealand = $0.584 U.S., what will it cost in U.S. dollars to fill the tank? What is the cost, in U.S. dollars, of a gallon of gasoline at this gas station?

The geometry of fractals provides mathematicians with a means for describing objects in nature in which a pattern endlessly repeats itself in smaller and smaller versions. This mountain scene is a computer-generated fractal.

GEOMETRY

Many objects in our lives can be described in terms of geometry. The spherical basketball you dribble down a rectangular court, the cylindrical can of soda you drink, and even the rectangular solid shape of this book are all examples of geometric objects that affect our lives. Throughout human history, geometry has played an important role in education, technology, and commerce. This role continues in modern times.

Albert Einstein's use of non-Euclidean geometry in his theory of relativity has enabled mathematicians and scientists to model the universe more accurately. Benoit Mandelbrot's work in fractal geometry has led scientists to discover ways to describe such intricate and detailed objects as weather systems, air passages in our lungs, and earthquake frequency patterns. A recently discovered form of pure carbon naturally forms molecules whose structure involves hexagons and pentagons in a pattern similar to those found on a soccer ball. As the human mind continues to uncover and interact with nature's secrets, geometry undoubtedly will continue to play a vital role.

9.1 POINTS, LINES, PLANES, AND ANGLES

PROFILE IN MATHEMATICS

EUCLID

Euclid (320–275 B.C.) lived in Alexandria, Egypt, and was a teacher and scholar at Alexandria's school called the *Museum*. It was here that Euclid collected and arranged many of the mathematical results known at the time. This collection of works became his 13-volume masterpiece known as *Elements*. Beginning with a list of definitions, postulates, and axioms, Euclid proved one theorem after another, using only previously proven results. This method of proof became a model of mathematical and scientific investigation that survives today. Remarkably, the geometry in *Elements* does not rely on making exact geometric measurements using a ruler or protractor. Rather, the work is developed using only an unmarked straightedge and a drawing compass. Next to the Bible, Euclid's *Elements* may be the most translated, published, and studied of all the books produced in the Western world.

Human beings recognized shapes, sizes, and physical forms long before geometry was developed. Geometry as a science is said to have begun in the Nile Valley of ancient Egypt. The Egyptians used geometry to measure land and to build pyramids and other structures.

The word *geometry* is derived from two Greek words, *ge*, meaning earth, and *metron*, meaning measure. Thus geometry means "earth measure" or "measurement of the earth."

Unlike the Egyptians, the Greeks were interested in more than just the applied aspects of geometry. The Greeks attempted to apply their knowledge of logic to geometry. In about 600 B.C., Thales of Miletus was the first to be credited with using deductive methods to develop geometric concepts. Another outstanding Greek geometer, Pythagoras, continued the systematic development of geometry that Thales had begun.

In about 300 B.C., Euclid collected and summarized much of the Greek mathematics of his time. In a set of 13 books called *Elements*, Euclid laid the foundation for plane geometry, which is also called *Euclidean geometry*.

Euclid is credited with being the first mathematician to use the *axiomatic method* in developing a branch of mathematics. First, Euclid introduced *undefined terms* such as point, line, plane, and angle. He related these to physical space by such statements as "A line is length without breadth" so that we may intuitively understand them. Because such statements play no further role in his system, they constitute primitive or undefined terms.

Second, Euclid introduced certain *definitions*. The definitions are introduced when needed and are often based on the undefined terms. Some terms that Euclid introduced and defined include triangle, right angle, and hypotenuse.

Third, Euclid stated certain primitive propositions called *postulates* (now called *axioms**) about the undefined terms and definitions. The reader is asked to accept these statements as true on the basis of their "obviousness" and their relationship with the physical world. For example, the Greeks accepted all right angles as being equal, which is Euclid's fourth postulate.

Fourth, Euclid proved, using deductive reasoning (see Section 1.1), other propositions called *theorems*. One theorem that Euclid proved is known as the Pythagorean theorem: "The sum of the areas of the squares constructed on the arms of a right triangle is equal to the area of the square constructed on the hypotenuse." He also proved that the sum of the angles of a triangle is 180°.

Using only 10 axioms, Euclid deduced 465 propositions (or theorems) in plane and solid geometry, number theory, and Greek geometric algebra.

Point and Line

Three basic terms in geometry are *point*, *line*, and *plane*. These three terms are not given a formal definition, but we recognize points, lines, and planes when we see them.

*The concept of the axiom has changed significantly since Euclid's time. Now any statement may be designated as an axiom, whether it is self-evident or not. All axioms are *accepted* as true. A set of axioms forms the foundation for a mathematical system.

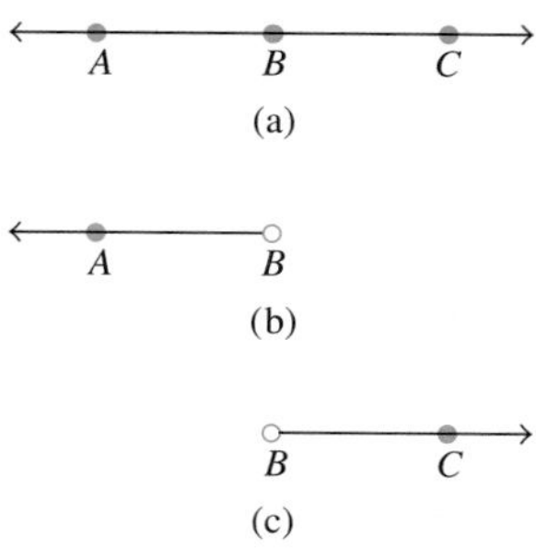

Figure 9.1

Let's consider some properties of a line. Assume that a line means a straight line unless otherwise stated.

1. A line is a set of points. Each point is on the line and the line passes through each point. When we wish to refer to a specific point, we will label it with a single capital letter. For example, in Figure 9.1(a) three points are labled *A*, *B*, and *C*, respectively.
2. Any two distinct points determine a unique line. Figure 9.1(a) illustrates a line. The arrows at both ends of the line indicate that the line continues in each direction. The line in Fig. 9.1(a) may be symbolized with any two points on the line by placing a line with a double-sided arrow above the letters that correspond to the points—for example, $\overleftrightarrow{AB}$, $\overleftrightarrow{BA}$, $\overleftrightarrow{AC}$, $\overleftrightarrow{CA}$, $\overleftrightarrow{BC}$, or $\overleftrightarrow{CB}$.
3. Any point on a line separates the line into three parts: the point itself and two *half lines* (neither of which includes the point). For example, in Fig. 9.1(a) point *B* separates the line into the point *B* and two half lines. Half line *BA*, symbolized $\overset{\circ}{\overrightarrow{BA}}$, is illustrated in Fig. 9.1(b). The open circle above the *B* indicates that point *B* is not included in the half line. Figure 9.1(c) illustrates half line *BC*, symbolized $\overset{\circ}{\overrightarrow{BC}}$.

Look at the half line $\overset{\circ}{\overrightarrow{AB}}$ in Fig. 9.2(b). If the *end point*, *A*, is included with the set of points on the half line, the result is called a *ray*. Ray *AB*, symbolized $\overrightarrow{AB}$, is illustrated in Fig. 9.2(c). Ray *BA*, symbolized $\overrightarrow{BA}$, is illustrated in Fig. 9.2(d).

A *line segment* is that part of a line between two points, including the end points. Line segment *AB*, symbolized $\overline{AB}$, is illustrated in Fig. 9.2(e).

Description	Diagram	Symbol
(a) Line *AB*	A B	$\overleftrightarrow{AB}$
(b) Half line *AB*	A B	$\overset{\circ}{\overrightarrow{AB}}$
(c) Ray *AB*	A B	$\overrightarrow{AB}$
(d) Ray *BA*	A B	$\overrightarrow{BA}$
(e) Line segment *AB*	A B	$\overline{AB}$
(f) Open line segment *AB*	A B	$\overset{\circ\ \ \circ}{\overline{AB}}$
(g) Half open line segments *AB*	A B A B	$\overset{\ \ \circ}{\overline{AB}}$ $\overset{\circ\ \ }{\overline{AB}}$

Figure 9.2

An open line segment is the set of points on a line between two points, excluding the end points. Open line segment *AB*, symbolized $\overset{\circ\ \ \circ}{\overline{AB}}$, is illustrated in Fig. 9.2(f).

Figure 9.2(g) illustrates two half open line segments, symbolized $\overset{\ \ \circ}{\overline{AB}}$ and $\overset{\circ\ \ }{\overline{AB}}$.

DID YOU KNOW

Compass and Straightedge Constructions

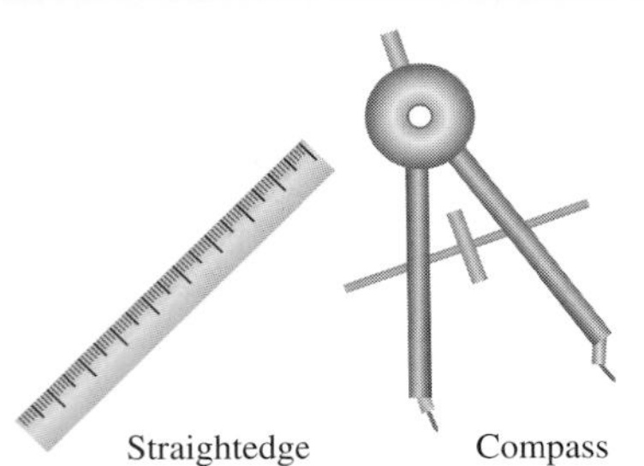

Geometric constructions were central to ancient Greek mathematics. Although these constructions are often referred to as *Euclidean constructions*, they were used centuries before Euclid wrote his classic work, *Elements*. The tools *allowed* in geometric constructions are a pencil, an unmarked straightedge, and a drawing compass. The straightedge is used to draw line segments, and the compass is used to draw circles and arcs. One example of a construction using these tools is shown below. The Internet has many sites devoted to classic geometric constructions.

To construct a triangle with sides of equal length (i.e., an equilateral triangle) do the following:

1. Use the straightedge to draw a line segment of any length and label the end points A and B.
2. Place one end of the compass at point A and the other end on point B and draw an arc as shown.
3. Now turn the compass around and draw another arc as shown. Label the point of intersection of the two arcs C.
4. Draw line segments AC and BC. This completes the construction of equilateral triangle ABC.

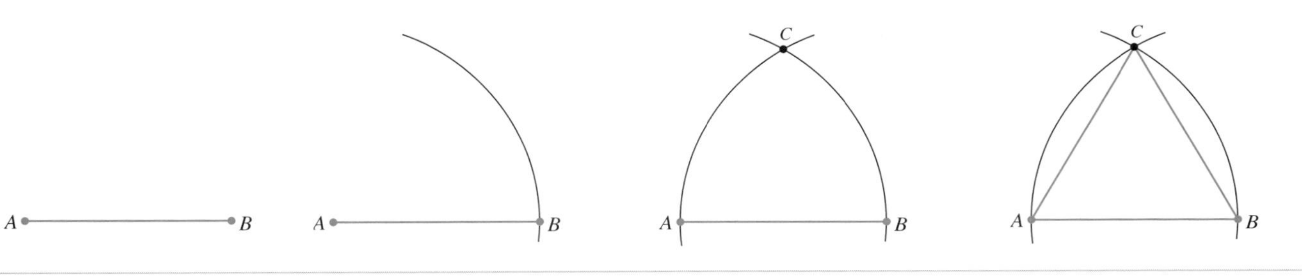

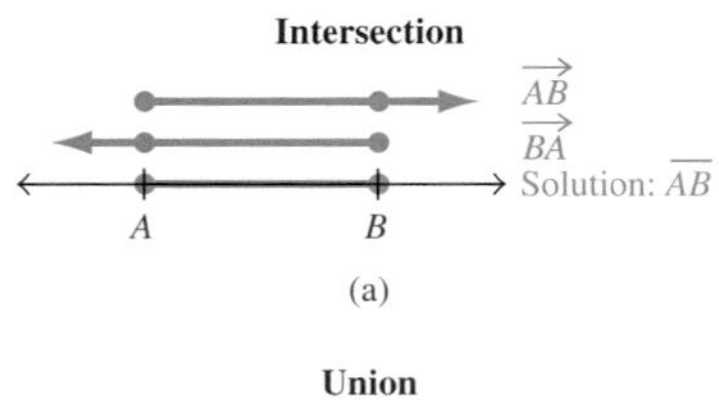

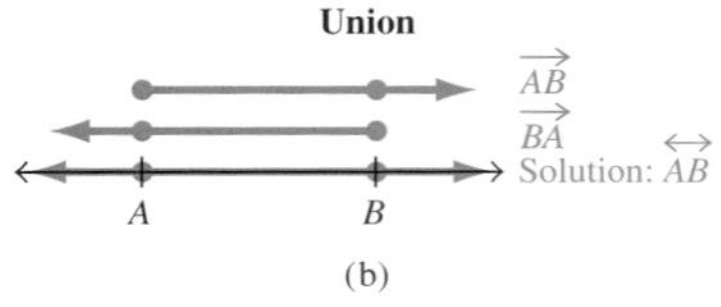

Figure 9.3

In Chapter 2 we discussed intersection of sets. Recall that the intersection (symbolized $\cap$) of two sets is the set of elements (points in this case) common to both sets.

Consider the rays $\overrightarrow{AB}$ and $\overrightarrow{BA}$ in Fig. 9.3(a). The intersection of $\overrightarrow{AB}$ and $\overrightarrow{BA}$ is $\overline{AB}$. Thus, $\overrightarrow{AB} \cap \overrightarrow{BA} = \overline{AB}$.

We also discussed the union of two sets in Chapter 2. The union (symbolized $\cup$) of two sets is the set of elements (points in this case) that belong to either of the sets or both sets. The union of $\overrightarrow{AB}$ and $\overrightarrow{BA}$ is $\overleftrightarrow{AB}$ (Fig. 9.3b). Thus, $\overrightarrow{AB} \cup \overrightarrow{BA} = \overleftrightarrow{AB}$.

EXAMPLE 1 *Unions and Intersections of Parts of a Line*

Using line AD, determine the solution to each part.

a) $\overrightarrow{AB} \cap \overrightarrow{DC}$ b) $\overrightarrow{AB} \cup \overrightarrow{DC}$ c) $\overline{AB} \cap \overline{CD}$ d) $\overline{AD} \cup \overset{\circ}{\overrightarrow{CA}}$

DID YOU KNOW

Let There Be Light

A Kodak Camera advertisement appeared in the first issue of the *Photographic Herald and Amateur Sportsman,* in November 1889. The slogan "You press the button, we do the rest" summed up George Eastman's groundbreaking snapshot camera system.

Sunlight travels in straight lines, or rays. When these rays pass through a small hole of a darkened room, they converge and then spread out to form an inverted image of the sun on the opposite wall. In the sixteenth century, artists used lenses to focus images on paper, which could then be traced to create a reproduction. In the eighteenth century, chemists discovered that certain salts of silver would darken when exposed to light, effectively creating a negative image. For the population at large, it all came together in 1888 when 23-year-old George Eastman, using basic geometric concepts, developed the first modern camera intended for the "casual millions," which sold for $25 apiece.

SOLUTION:

a) $\overrightarrow{AB} \cap \overrightarrow{DC}$

Ray *AB* and ray *DC* are shown below. The intersection of these two rays is that part of line *AD* that is a part of *both* ray *AB and* ray *DC*. The intersection of ray *AB* and ray *DC* is line segment *AD*.

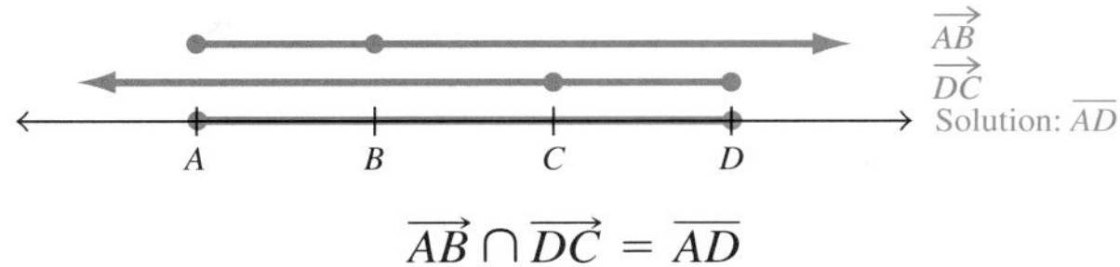

$$\overrightarrow{AB} \cap \overrightarrow{DC} = \overline{AD}$$

b) $\overrightarrow{AB} \cup \overrightarrow{DC}$

Once again ray *AB* and ray *DC* are shown below. The union of these two rays is that part of line *AD* that is part of *either* ray *AB or* ray *DC*. The union of ray *AB* and ray *DC* is the entire line *AD*.

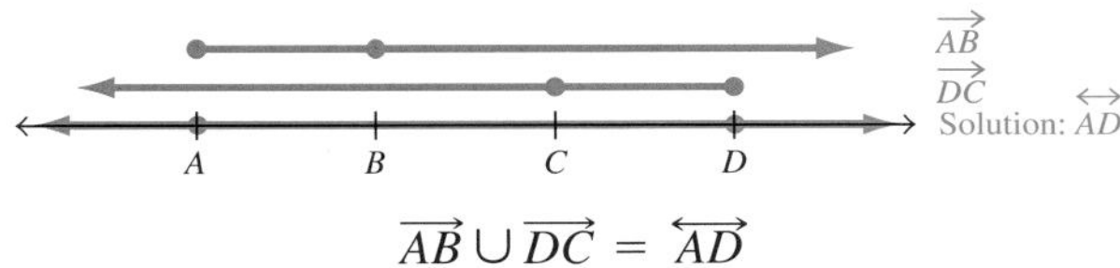

$$\overrightarrow{AB} \cup \overrightarrow{DC} = \overleftrightarrow{AD}$$

c) $\overline{AB} \cap \overrightarrow{CD}$

Line segment *AB* and ray *CD* have no points in common, so their intersection is empty.

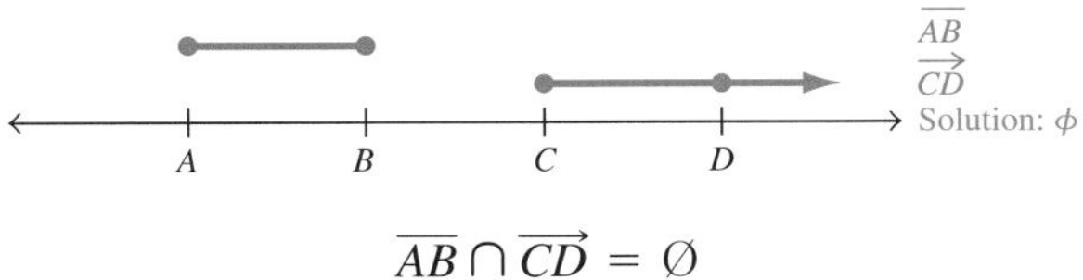

$$\overline{AB} \cap \overrightarrow{CD} = \emptyset$$

d) $\overline{AD} \cup \overset{\circ}{\overrightarrow{CA}}$

The union of line segment *AD* and half line *CA* is ray *DA* (or $\overrightarrow{DB}$ or $\overrightarrow{DC}$).

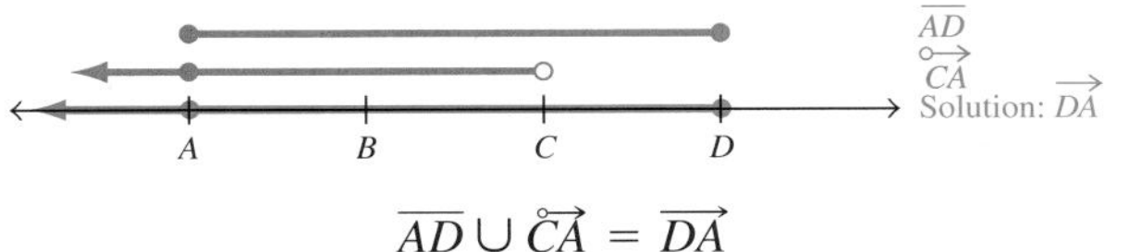

$$\overline{AD} \cup \overset{\circ}{\overrightarrow{CA}} = \overrightarrow{DA}$$

Plane

The term *plane* is one of Euclid's undefined terms. For our purposes, we can think of a plane as a two-dimensional surface that extends infinitely in both directions, like an infinitely large blackboard. Euclidean geometry is called *plane geometry* because it is the study of two-dimensional figures in a plane.

Two lines in the same plane that do not intersect are called *parallel lines*. Figure 9.4(a) on page 478 illustrates two parallel lines in a plane ($\overleftrightarrow{AB}$ is parallel to $\overleftrightarrow{CD}$).

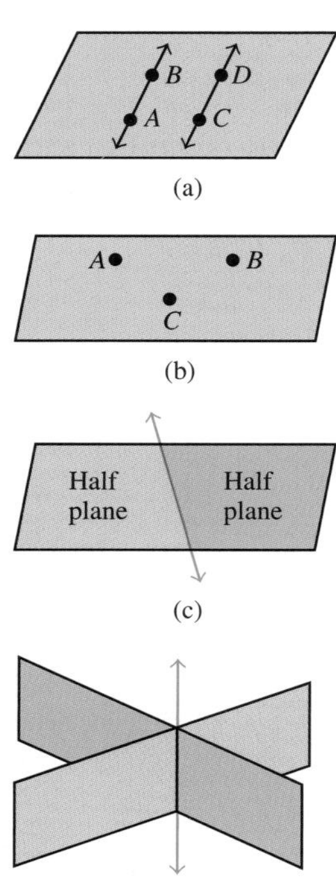

Figure 9.4

Properties of planes include the following:

1. Any three points that are not on the same line (noncollinear points) determine a unique plane (Fig. 9.4b).
2. A line in a plane divides the plane into three parts, the line and two half planes (Fig. 9.4c).
3. Any line and a point not on the line determine a unique plane.
4. The intersection of two planes is a line (Fig. 9.4d).

Two planes that do not intersect are said to be *parallel planes*. For example, in Fig. 9.5 plane *ABE* is parallel to plane *GHF*.

Two lines that do not lie in the same plane and do not intersect are called *skewed lines*. Figure 9.5 illustrates many skewed lines (for example, $\overleftrightarrow{AB}$ and $\overleftrightarrow{CD}$).

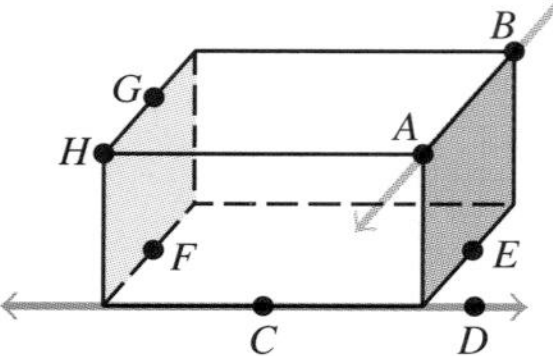

Figure 9.5

Angles

An *angle*, denoted $\measuredangle$, is the union of two rays with a common end point (Fig. 9.6):

$$\overrightarrow{BA} \cup \overrightarrow{BC} = \measuredangle ABC \text{ (or } \measuredangle CBA)$$

An angle can be formed by the rotation of a ray about a point. An angle has an initial side and a terminal side. The initial side indicates the position of the ray prior to rotation; the terminal side indicates the position of the ray after rotation. The point common to both rays is called the *vertex* of the angle. The letter designating the vertex is always the middle one of the three letters designating an angle. The rays that make up the angle are called its *sides*.

There are several ways to name an angle. The angle in Fig. 9.6 may be denoted

$$\measuredangle ABC, \quad \measuredangle CBA, \quad \text{or} \quad \measuredangle B$$

An angle divides a plane into three distinct parts: the angle itself, its interior, and its exterior. In Fig. 9.6 the angle is represented by the blue lines, the interior of the angle is shaded pink, and the exterior is shaded green.

The *measure of an angle*, symbolized *m*, is the amount of rotation from its initial side to its terminal side. In Fig. 9.6, the letter *x* represents the measure of $\measuredangle ABC$; therefore, we may write $m\measuredangle ABC = x$.

Angles can be measured in *degrees*, radians, or gradients. In this text we will discuss only the degree unit of measurement. The symbol for degrees is the same as the symbol for temperature degrees. An angle of 45 degrees is written 45°. A *protractor* is used to measure angles. The angle shown being measured by the protractor in Fig. 9.8 on page 479 is 50°.

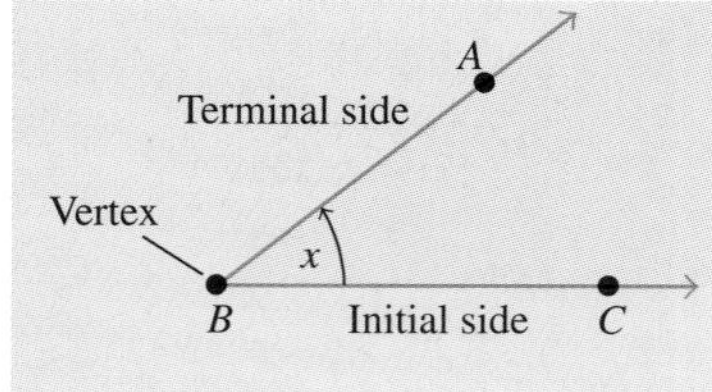

Figure 9.6

EXAMPLE 2 ***Union and Intersection***

Refer to Fig. 9.7. Determine the following.

a) $\overrightarrow{BG} \cup \overrightarrow{BF}$ b) $\measuredangle ABG \cap \measuredangle DBC$ c) $\overleftrightarrow{DE} \cap \measuredangle CBE$ d) $\overrightarrow{BD} \cup \overrightarrow{BC}$

SOLUTION:

a) $\overrightarrow{BG} \cup \overrightarrow{BF} = \overleftrightarrow{GF}$
b) $\measuredangle ABG \cap \measuredangle DBC = \{B\}$
c) $\overleftrightarrow{DE} \cap \measuredangle CBE = \overrightarrow{BE}$
d) $\overrightarrow{BD} \cup \overrightarrow{BC} = \measuredangle DBC \text{ (or } \measuredangle CBD)$

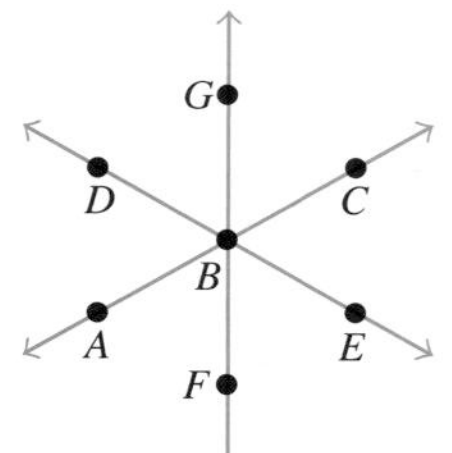

Figure 9.7

DID YOU KNOW

Classical Geometry Problems

Have you ever spent several hours trying to solve a difficult homework problem? Can you imagine studying the same problem for your entire life? Entire generations of mathematicians throughout history spent their entire lives studying three geometry construction problems that originated in ancient Greece. The three problems are (1) *trisecting an angle*, (2) *squaring the circle*, and (3) *doubling the cube*. All three problems involve using only an unmarked straightedge and a compass to construct geometric figures. Trisecting an angle refers to dividing a given angle into three equal angles. Squaring the circle refers to constructing a square that has the exact same area as the area of a given circle. Doubling the cube refers to constructing a cube that has exactly double the volume of a given cube. Although these problems are easily stated, they puzzled mathematicians for thousands of years. The solutions were never found, but in the process of studying these problems, mathematicians were able to expand their knowledge about mathematics. Eventually, mathematicians came to realize that if one is limited to using only an unmarked straightedge and a compass, then these constructions are impossible. Finally, in 1837 Pierre Wantzel proved the impossibility of the first two constructions, and in 1882 Carl Lindemann proved the impossibility of the third construction.

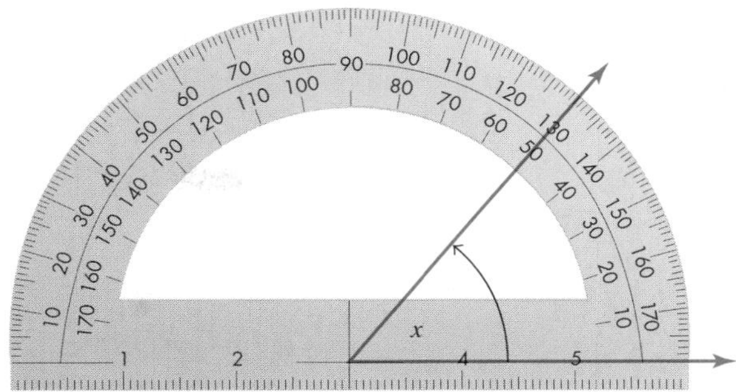

Figure 9.8

Consider a circle whose circumference is divided into 360 equal parts. If we draw a line from each mark on the circumference to the center of the circle, we get 360 wedge-shaped pieces. The measure of an angle formed by the straight sides of each wedge-shaped piece is defined to be 1°.

Angles are classified by their degree measurement, as shown in the following summary. A *right angle* is 90°, an *acute angle* is less than 90°, an *obtuse angle* is greater than 90° but less than 180°, and a *straight angle* is 180°.

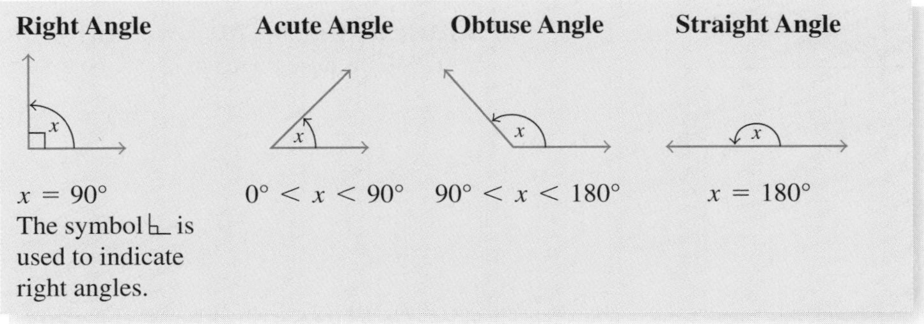

Two angles in the same plane are *adjacent angles* when they have a common vertex and a common side but no common interior points. In Fig. 9.9, $\measuredangle DBC$ and $\measuredangle CBA$ are adjacent angles, but $\measuredangle DBA$ and $\measuredangle CBA$ are not adjacent angles.

Two angles are called *complementary angles* if the sum of their measures is 90°. Two angles are called *supplementary angles* if the sum of their measures is 180°.

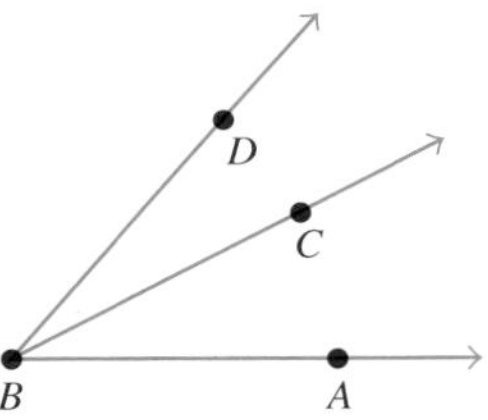

Figure 9.9

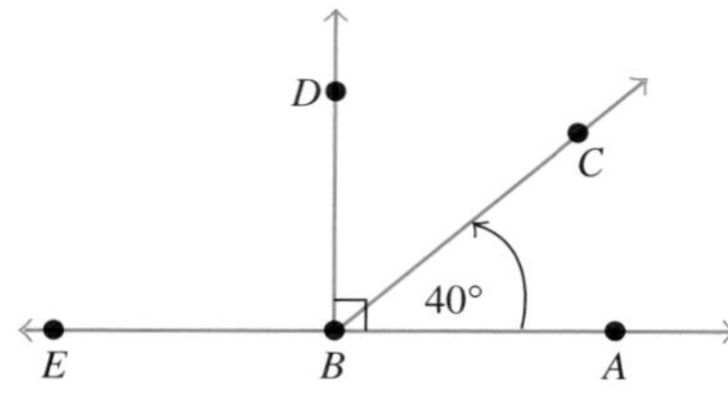

Figure 9.10

EXAMPLE 3 *Determining Complementary and Supplementary Angles*

In Fig. 9.10 we see that $m\measuredangle ABC = 40°$.

a) $\measuredangle ABC$ and $\measuredangle CBD$ are complementary angles. Determine $m\measuredangle CBD$.

b) $\measuredangle ABC$ and $\measuredangle CBE$ are supplementary angles. Determine $m\measuredangle CBE$.

SOLUTION:

a) The sum of two complementary angles must be 90°, so

$$\begin{aligned} m\measuredangle ABC + m\measuredangle CBD &= 90° \\ 40° + m\measuredangle CBD &= 90° \\ m\measuredangle CBD &= 90° - 40° = 50° \end{aligned}$$

Subtract 40° from each side of the equation.

b) The sum of two supplementary angles must be 180°, so

$$\begin{aligned} m\measuredangle ABC + m\measuredangle CBE &= 180° \\ 40° + m\measuredangle CBE &= 180° \\ m\measuredangle CBE &= 180° - 40° = 140° \end{aligned}$$

Subtract 40° from each side of the equation. ▲

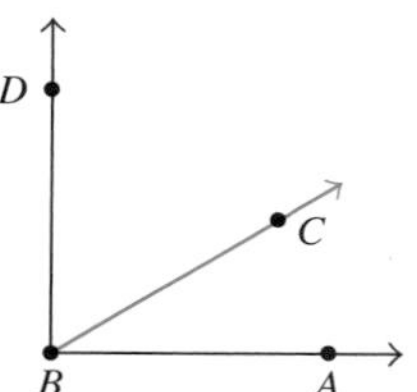

Figure 9.11

EXAMPLE 4 *Determining Complementary Angles*

If $\measuredangle ABC$ and $\measuredangle CBD$ are complementary angles and $m\measuredangle ABC$ is 26° less than $m\measuredangle CBD$, determine the measure of each angle (Fig. 9.11).

SOLUTION: Let $m\measuredangle CBD = x$. Then $m\measuredangle ABC = x - 26$ since it is 26° less than $m\measuredangle CBD$. Because these angles are complementary, we have

$$\begin{aligned} m\measuredangle CBD + m\measuredangle ABC &= 90 \\ x + (x - 26) &= 90 \\ 2x - 26 &= 90 \\ 2x &= 116 \\ x &= 58 \end{aligned}$$

Therefore, $m\measuredangle CBD = 58°$ and $m\measuredangle ABC = 58° - 26°$, or 32°. Note that $58° + 32° = 90°$, which is what we expected. ▲

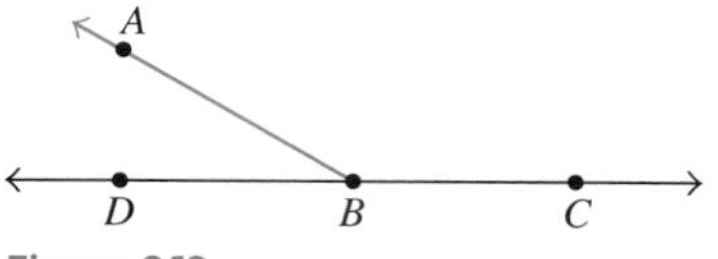

Figure 9.12

EXAMPLE 5 *Determining Supplementary Angles*

If $\measuredangle ABC$ and $\measuredangle ABD$ are supplementary and $m\measuredangle ABC$ is five times larger than $m\measuredangle ABD$, determine $m\measuredangle ABC$ and $m\measuredangle ABD$ (Fig. 9.12).

SOLUTION: Let $m\measuredangle ABD = x$, then $m\measuredangle ABC = 5x$. Since these angles are supplementary, we have

$$\begin{aligned} m\measuredangle ABC + m\measuredangle ABD &= 180° \\ 5x + x &= 180° \\ 6x &= 180° \\ x &= 30° \end{aligned}$$

Thus, $m\measuredangle ABD = 30°$ and $m\measuredangle ABC = 5(30°) = 150°$. Note that $30° + 150° = 180°$, which is what we expected. ▲

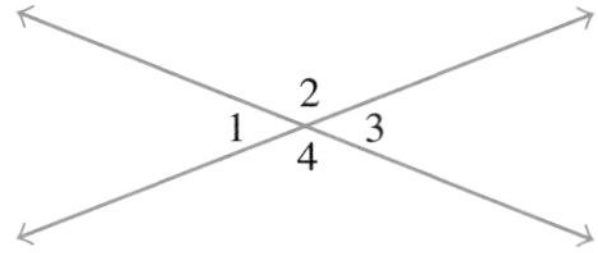

Figure 9.13

When two straight lines intersect, the nonadjacent angles formed are called *vertical angles*. In Fig. 9.13, ∡1 and ∡3 are vertical angles, and ∡2 and ∡4 are vertical angles. We can show that vertical angles have the same measure, that is, they are equal. For example, Fig. 9.13 shows that

$$m\measuredangle 1 + m\measuredangle 2 = 180°. \qquad \text{Why?}$$
$$m\measuredangle 2 + m\measuredangle 3 = 180°. \qquad \text{Why?}$$

Since ∡2 has the same measure in both cases, $m\measuredangle 1$ must equal $m\measuredangle 3$.

Vertical angles have the same measure.

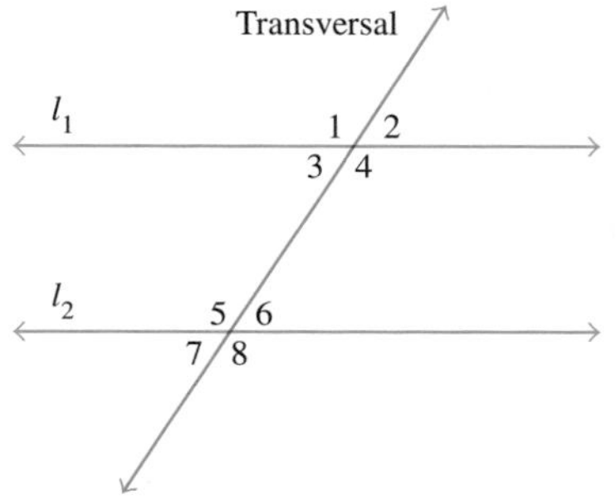

Figure 9.14

A line that intersects two different lines, l_1 and l_2, at two different points is called a *transversal*. Figure 9.14 illustrates that when two parallel lines are cut by a transversal, eight angles are formed. Angles 3, 4, 5, and 6 are called *interior angles*, and angles 1, 2, 7, and 8 are called *exterior angles*. Eight pairs of supplementary angles are formed. Can you list them?

Special names are given to the angles formed by a transversal crossing two parallel lines.

Name	Description	Illustration	Pairs of Angles Meeting Criteria
Alternate interior angles	Interior angles on opposite sides of the transversal	1 2 / 3 4 / 5 6 / 7 8	∡3 and ∡6 ∡4 and ∡5
Alternate exterior angles	Exterior angles on opposite sides of the transversal	1 2 / 3 4 / 5 6 / 7 8	∡1 and ∡8 ∡2 and ∡7
Corresponding angles	One interior and one exterior angle on the same side of the transversal	1 2 / 3 4 / 5 6 / 7 8	∡1 and ∡5 ∡2 and ∡6 ∡3 and ∡7 ∡4 and ∡8

When two parallel lines are cut by a transversal

1. alternate interior angles have the same measure.
2. alternate exterior angles have the same measure.
3. corresponding angles have the same measure.

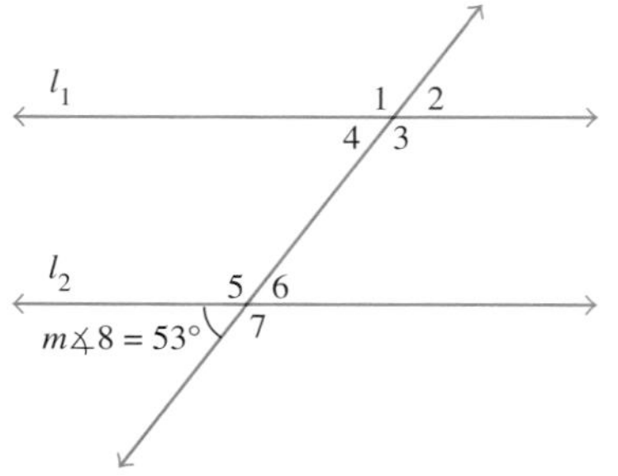

Figure 9.15

EXAMPLE 6 *Finding Angle Measures*

Figure 9.15 shows two parallel lines cut by a transversal. Determine the measure of $\measuredangle 1$ through $\measuredangle 7$.

SOLUTION:

$m\measuredangle 6 = 53°$	$\measuredangle 8$ and $\measuredangle 6$ are vertical angles.
$m\measuredangle 5 = 127°$	$\measuredangle 8$ and $\measuredangle 5$ are supplementary angles.
$m\measuredangle 7 = 127°$	$\measuredangle 5$ and $\measuredangle 7$ are vertical angles.
$m\measuredangle 1 = 127°$	$\measuredangle 1$ and $\measuredangle 7$ are alternate exterior angles.
$m\measuredangle 4 = 53°$	$\measuredangle 4$ and $\measuredangle 6$ are alternate interior angles.
$m\measuredangle 2 = 53°$	$\measuredangle 6$ and $\measuredangle 2$ are corresponding angles.
$m\measuredangle 3 = 127°$	$\measuredangle 3$ and $\measuredangle 1$ are vertical angles.

In Example 6, the angles could have been determined in alternate ways. For Example, we mentioned $m\measuredangle 1 = 127°$ because $\measuredangle 1$ and $\measuredangle 7$ are alternate exterior angles. We could have also stated that $m\measuredangle 1 = 127°$ because $\measuredangle 1$ and $\measuredangle 5$ are corresponding angles.

SECTION 9.1 EXERCISES

Concept/Writing Exercises

1. **a)** What are the four key parts in the axiomatic method used by Euclid?
   b) Discuss each of the four parts.
2. What is the difference between an axiom and a theorem?
3. What are parallel lines?
4. What are skewed lines?
5. What are adjacent angles?
6. What are supplementary angles?
7. What are complementary angles?
8. What is a straight angle?
9. What is an obtuse angle?
10. What is an acute angle?
11. What is a right angle?
12. Draw two intersecting lines. Identify the two pairs of vertical angles.

Practice the Skills

In Exercises 13–20, identify the figure as a line, half line, ray, line segment, open line segment, or half open line segment. Denote it by its appropriate symbol.

13. **14.**

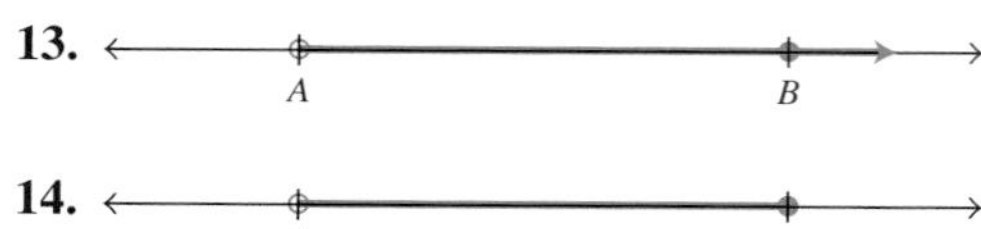

15. **16.** **17.** **18.** **19.** **20.**

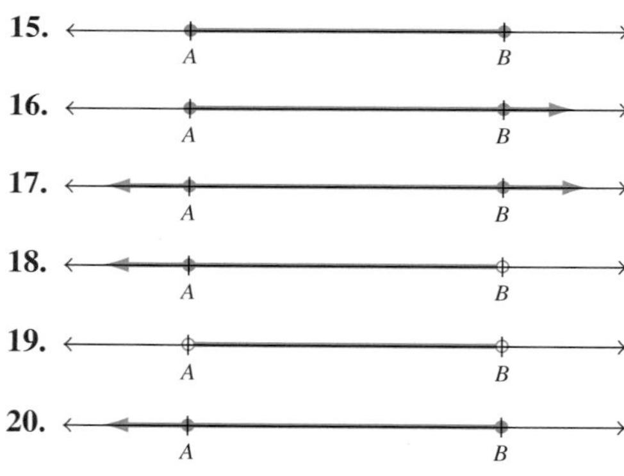

In Exercises 21–32, use the figure to find the following:

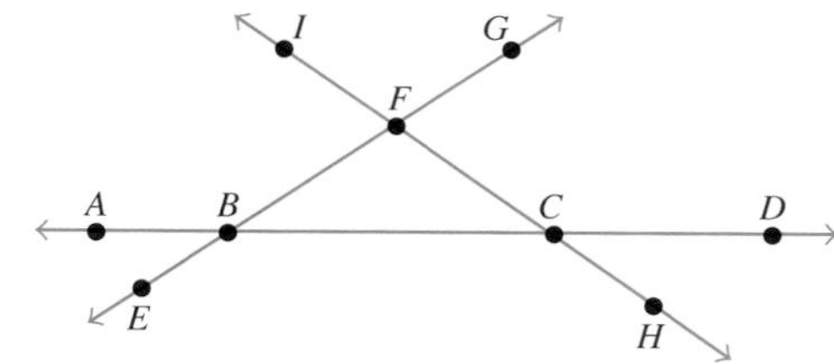

21. $\measuredangle GBD \cap \overrightarrow{AD}$

22. $\overset{\circ\longrightarrow}{FE} \cup \overrightarrow{FG}$

23. $\overset{\longrightarrow\circ}{BC} \cup \overset{\longrightarrow\circ}{CD}$

24. $\overleftrightarrow{AD} \cup \overline{BC}$

25. $\measuredangle ICA \cap \overleftrightarrow{EG}$

26. $\measuredangle HCD \cap \measuredangle ACF$

27. $\overset{\circ\longrightarrow}{AB} \cap \overrightarrow{HC}$

28. $\overset{\circ\longrightarrow}{BD} \cap \overset{\circ\longrightarrow}{CB}$

29. $\overleftrightarrow{AD} \cap \overline{BC}$

30. $\overline{BC} \cup \overline{CF} \cup \overline{FB}$

31. $\overset{\circ\longrightarrow}{BD} \cup \overset{\circ\longrightarrow}{CB}$

32. $\{C\} \cap \overset{\circ\longrightarrow}{CH}$

In Exercises 33–44, use the figure to find each of the following:

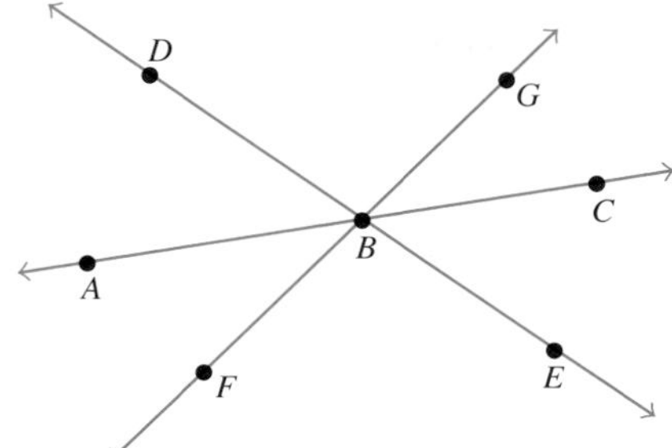

33. $\overset{\circ\!\rightarrow}{BF} \cap \overrightarrow{BE}$ **34.** $\overrightarrow{BD} \cup \overset{\;\;\circ}{\overline{BE}}$

35. $\measuredangle GBC \cap \measuredangle CBE$ **36.** $\overleftrightarrow{DE} \cup \overset{\circ\!\rightarrow}{BE}$

37. $\measuredangle ABE \cup \overset{\circ\!-\!\circ}{AB}$ **38.** $\overset{\circ\!\rightarrow}{BF} \cup \overrightarrow{BE}$

39. $\measuredangle CBE \cap \measuredangle EBC$ **40.** $\{B\} \cap \overrightarrow{BA}$

41. $\overset{\circ\!-\!\circ}{AC} \cap \overline{AC}$ **42.** $\overset{\circ\!-\!\circ}{AC} \cap \overset{\circ\!-\!\circ}{BE}$

43. $\overrightarrow{EB} \cap \overset{\;\;\circ}{\overline{BE}}$ **44.** $\overline{GF} \cap \overline{AB}$

In Exercises 45–52, classify the angle as acute, right, straight, obtuse, or none of these.

45. 

46. 

47. 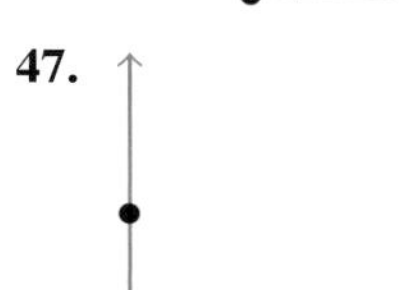

48. 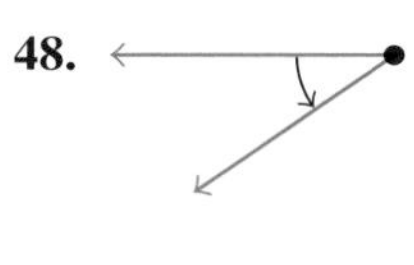

49. 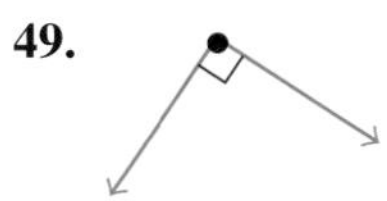

50. 

51. 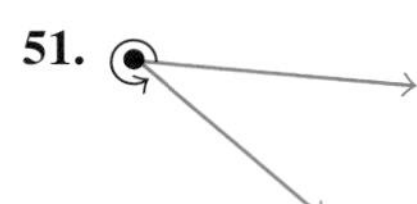

52. 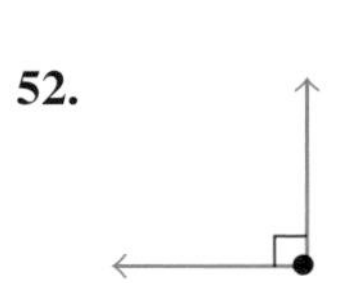

In Exercises 53–58, find the complementary angle of the given angle.

53. 19° **54.** 89° **55.** $32\frac{3}{4}°$

56. $43\frac{1}{3}°$ **57.** 64.7° **58.** 0.01°

In Exercises 59–64, find the supplementary angle of the given angle.

59. 91° **60.** 8° **61.** 20.5°

62. 179.99° **63.** $43\frac{5}{7}°$ **64.** $64\frac{7}{16}°$

In Exercises 65–70, match the names of the angles with the corresponding figure in parts (a)–(f).

65. Corresponding angles **66.** Vertical angles

67. Supplementary angles **68.** Complementary angles

69. Alternate interior angles **70.** Alternate exterior angles

a) 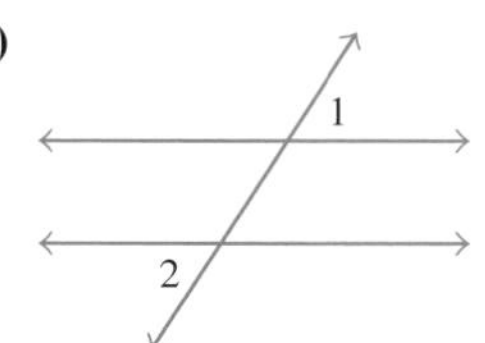

b) 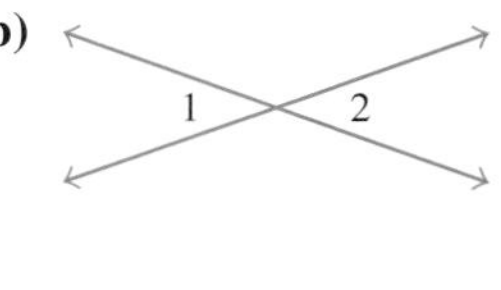

c) 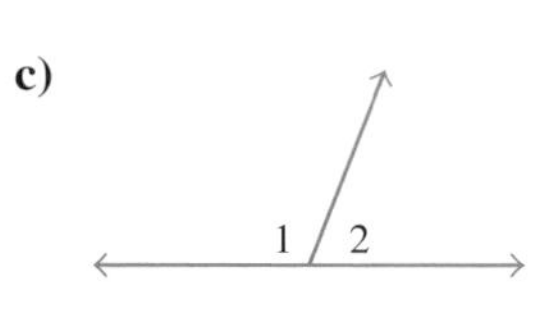

d) 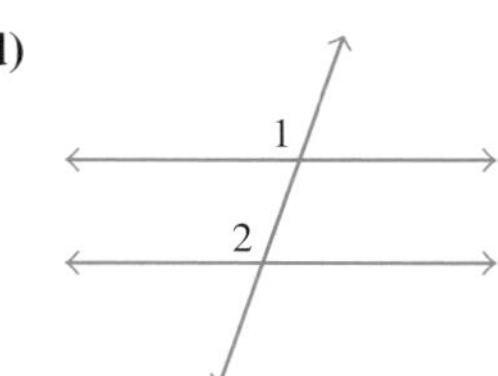

e) 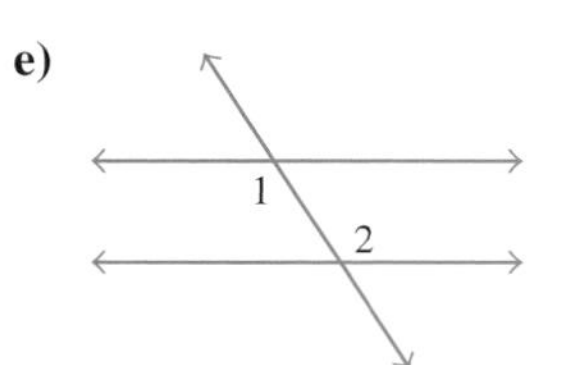

f) 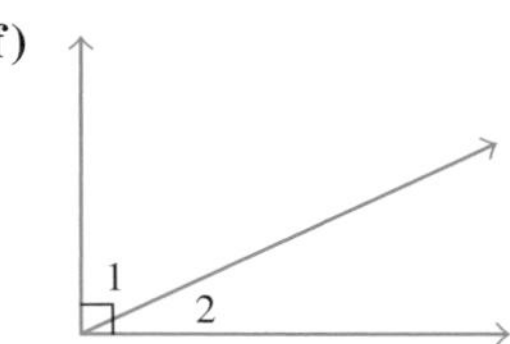

Problem Solving

71. MODELING - *Complementary Angles* If $\measuredangle 1$ and $\measuredangle 2$ are complementary angles and if the measure of $\measuredangle 1$ is four more than the measure of $\measuredangle 2$, determine the measures of $\measuredangle 1$ and $\measuredangle 2$.

72. MODELING - *Complementary Angles* The difference between the measures of two complementary angles is 62°. Determine the measures of the two angles.

73. MODELING - *Supplementary Angles* The difference between the measures of two supplementary angles is 88°. Determine the measures of the two angles.

74. MODELING - *Supplementary Angles* If $\measuredangle 1$ and $\measuredangle 2$ are supplementary angles and if the measure of $\measuredangle 2$ is 17 times the measure of $\measuredangle 1$, determine the measures of the two angles.

In Exercises 75–78, parallel lines are cut by the transversal shown. Determine the measures of $\measuredangle 1$ through $\measuredangle 7$.

75. 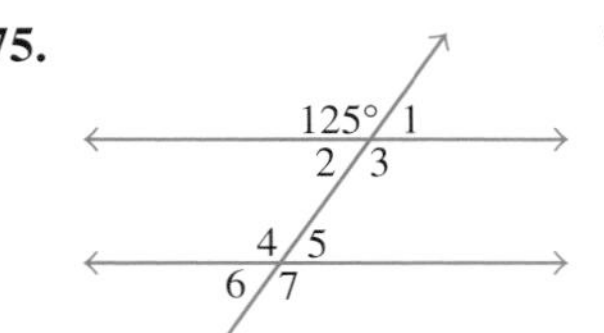

76. 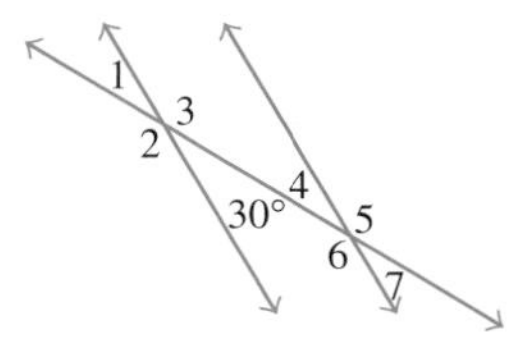

77.

78.

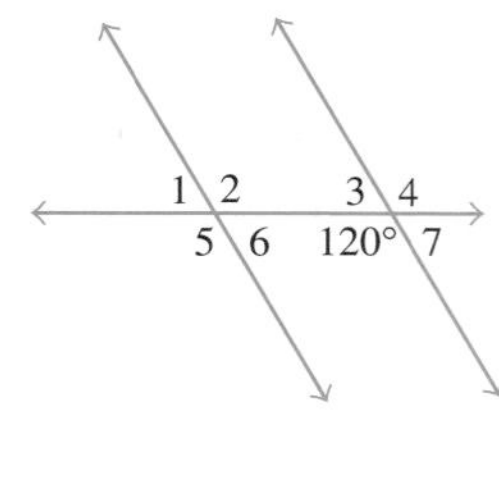

In Exercises 79–82, the angles are complementary angles. Find the measures of ∡1 and ∡2.

79.

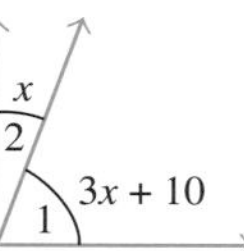

80.

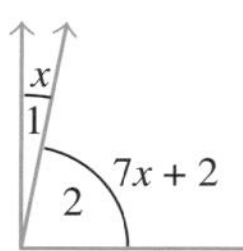

81.

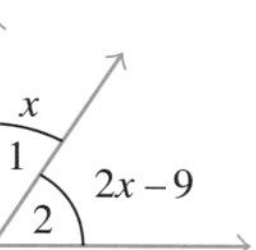

82.

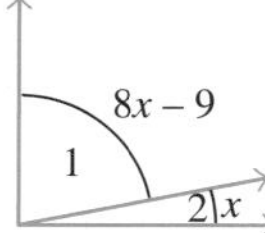

In Exercises 83–86, the angles are supplementary angles. Find the measures of ∡1 and ∡2.

83.

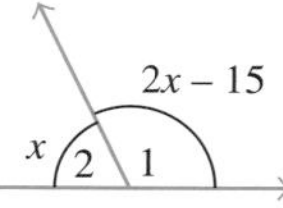

84.

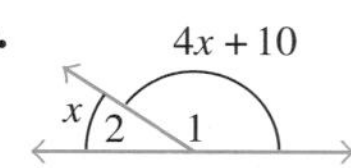

85.

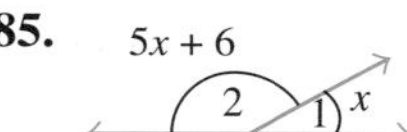

86.

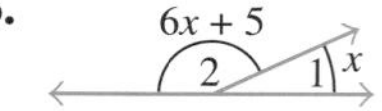

87. a) How many lines can be drawn through a given point?
b) How many planes can be drawn through a given point?

88. What is the intersection of two distinct nonparallel planes?

89. How many planes can be drawn through a given line?

90. **a)** Will three noncollinear points A, B, and C always determine a plane? Explain.
b) Is it possible to determine more than one plane with three noncollinear points? Explain.
c) How many planes can be constructed through three collinear points?

The figure suggests a number of lines and planes. The lines may be described by naming two points, and the planes may be described by naming three points. In Exercises 91–98, use the figure to name the following:

91. Two parallel planes

92. Two parallel lines

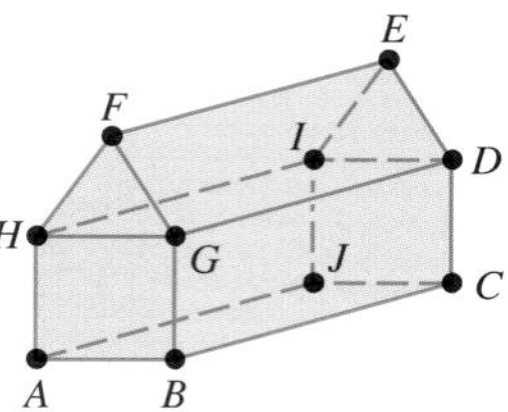

93. Two lines that intersect at right angles

94. Two planes that intersect at right angles

95. Three planes whose intersection is a single point

96. Three planes whose intersection is a line

97. A line and a plane whose intersection is a point

98. A line and a plane whose intersection is a line

In Exercises 99–104, determine whether the statement is always true, sometimes true, or never true. Explain your answer.

99. Two lines that are both parallel to a third line must be parallel to each other.

100. A triangle contains two acute angles.

101. Vertical angles are complementary angles.

102. Alternate exterior angles are supplementary angles.

103. Alternate interior angles are complementary angles.

104. A triangle contains two obtuse angles.

Challenge Problems/Group Activities

105. If lines l and m are parallel lines and if lines l and n are skewed lines, is it true that lines m and n must also be skewed? (*Hint:* Look at Fig. 9.5 on page 478.) Explain your answer and include a sketch to support your answer.

106. Two lines are *perpendicular* if they intersect at right angles. If lines l and m are perpendicular and if lines m and n are perpendicular, is it true that lines l and n must also be perpendicular? Explain your answer and include a sketch to support your answer.

107. Suppose you have three distinct lines, all lying in the same plane. Find all the possible ways in which the three lines can be related. Sketch each case (four cases).

Recreational Mathematics

108. If two straight lines intersect at a point, determine the sum of the measures of the 4 angles formed.

109. ∡ABC and ∡CBD are complementary and m∡CBD is twice the m∡ABC. ∡ABD and ∡DBE are supplementary angles.
a) Draw a sketch illustrating ∡ABC, ∡CBD, and ∡DBE.
b) Determine m∡ABC.
c) Determine m∡CBD.
d) Determine m∡DBE.

Internet/Research Activities

110. Using the Internet and other sources, write a research paper on Euclid's contributions to geometry.

111. Using the Internet and other sources, write a research paper on the three classic geometry problems of Greek antiquity (see Did You Know on page 479).

112. Search the Internet or other sources such as a geometry textbook to study the geometric constructions that use a straightedge and a compass only. Prepare a poster demonstrating five of these basic constructions.

9.2 POLYGONS

A *polygon* is a closed figure in a plane determined by three or more straight line segments. Examples of polygons are given in Fig. 9.16.

The straight line segments that form the polygon are called its *sides* and a point where two sides meet is called a *vertex* (plural *vertices*). The union of the sides of a polygon and its interior is called a *polygonal region*. A *regular polygon* is one whose sides are all the same length and whose interior angles all have the same measure. Figures 9.16(b) and (d) are regular polygons.

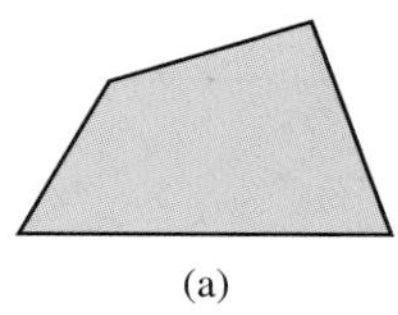
(a)

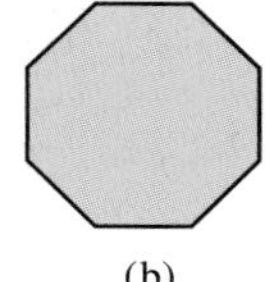
(b)

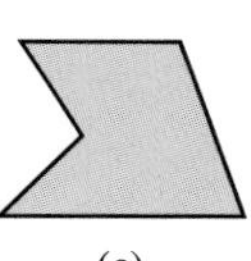
(c)

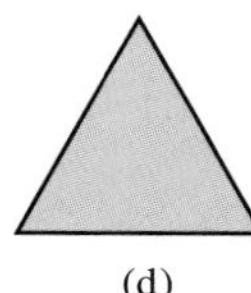
(d)

Figure 9.16

Polygons are named according to their number of sides. The names of some polygons are given in Table 9.1.

TABLE 9.1

Number of Sides	Name	Number of Sides	Name
3	Triangle	8	Octagon
4	Quadrilateral	9	Nonagon
5	Pentagon	10	Decagon
6	Hexagon	12	Dodecagon
7	Heptagon	20	Icosagon

One of the most important polygons is the triangle. The sum of the measures of the interior angles of a triangle is 180°. To illustrate, consider triangle ABC given in Fig. 9.17. The triangle is formed by drawing two transversals through two parallel lines l_1 and l_2 with the two transversals intersecting at a point on l_1.

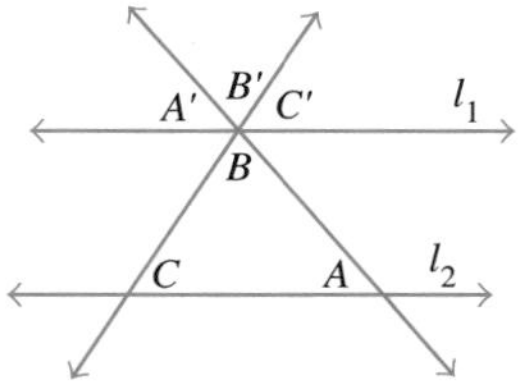

Figure 9.17

MATHEMATICS *Everywhere*

Making Movies Come Alive

Mathematics plays a key role in the animation you see in movies such as those in the *Jurassic Park* series. The images you see on the movie screen are created using software that combines pixels (the smallest piece of a screen image) into geometric shapes including polygons. These shapes are then stored in a computer and manipulated using various mathematical techniques so that the new shapes formed (from the original geometric shapes) approximate curves. Each movie frame has over 2 million pixels and can have over 40 million polygons. With such a huge amount of data, computers are used to carry out the mathematics needed to create animation. One computer animation specialist stated that "it's all controlled by math. . . . All those little X's, Y's, and Z's that you had in school—oh my gosh, suddenly they all apply."

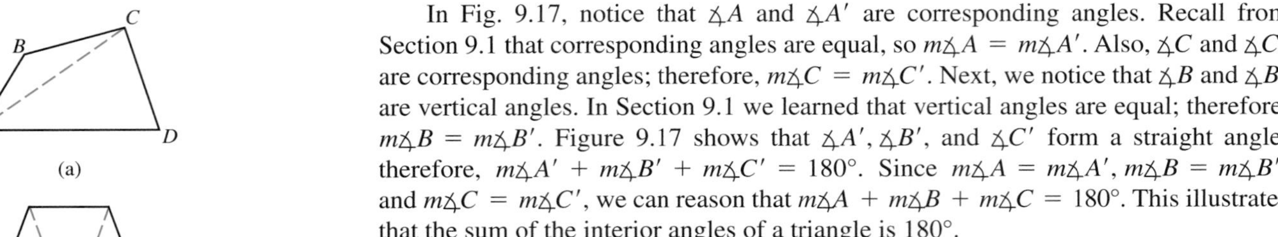

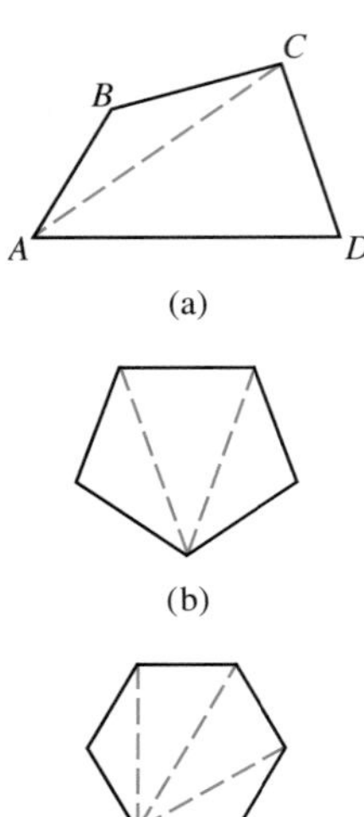

Figure 9.18

In Fig. 9.17, notice that $\measuredangle A$ and $\measuredangle A'$ are corresponding angles. Recall from Section 9.1 that corresponding angles are equal, so $m\measuredangle A = m\measuredangle A'$. Also, $\measuredangle C$ and $\measuredangle C'$ are corresponding angles; therefore, $m\measuredangle C = m\measuredangle C'$. Next, we notice that $\measuredangle B$ and $\measuredangle B'$ are vertical angles. In Section 9.1 we learned that vertical angles are equal; therefore, $m\measuredangle B = m\measuredangle B'$. Figure 9.17 shows that $\measuredangle A'$, $\measuredangle B'$, and $\measuredangle C'$ form a straight angle; therefore, $m\measuredangle A' + m\measuredangle B' + m\measuredangle C' = 180°$. Since $m\measuredangle A = m\measuredangle A'$, $m\measuredangle B = m\measuredangle B'$, and $m\measuredangle C = m\measuredangle C'$, we can reason that $m\measuredangle A + m\measuredangle B + m\measuredangle C = 180°$. This illustrates that the sum of the interior angles of a triangle is 180°.

Consider the quadrilateral $ABCD$ (Fig. 9.18a). Drawing a straight line segment between any two vertices forms two triangles. Since the sum of the measures of the angles of a triangle is 180°, the sum of the measures of the interior angles of a quadrilateral is $2 \cdot 180°$, or 360°.

Now let's examine a pentagon (Fig. 9.18b). We can draw two straight line segments to form three triangles. Thus, the sum of the measures of the interior angles of a five-sided figure is $3 \cdot 180°$, or 540°. Figure 9.18(c) shows that four triangles can be drawn in a six-sided figure. Table 9.2 summarizes this information.

TABLE 9.2

Sides	Triangles	Sum of the Measures of the Interior Angles
3	1	$1(180°) = 180°$
4	2	$2(180°) = 360°$
5	3	$3(180°) = 540°$
6	4	$4(180°) = 720°$

If we continue this procedure, we can see that for an n-sided polygon the sum of the measures of the interior angles is $(n - 2)180°$.

The **sum** of the measures of the interior angles of an n-sided polygon is $(n - 2)180°$.

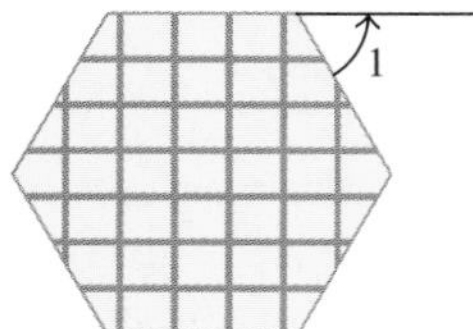

Figure 9.19

EXAMPLE 1 *Angles of a Hexagon*

Crispy Hexagons is a breakfast cereal whose pieces are in the shape of regular hexagons. A regular hexagon is a six-sided figure with all the sides the same length and all interior angles with the same measure. See Fig. 9.19. Determine

a) the measure of an interior angle.

b) the measure of exterior $\measuredangle 1$.

SOLUTION:

a) Using the formula $(n - 2)180°$, we can determine the sum of the measures of the interior angles of a hexagon as follows.

$$\begin{aligned} \text{Sum} &= (6 - 2)180° \\ &= 4(180°) \\ &= 720° \end{aligned}$$

The measure of an interior angle of a regular polygon can be determined by dividing the sum of the interior angles by the number of angles.

The measure of an interior angle of a regular hexagon is determined as follows:

$$\text{Measure} = \frac{720^\circ}{6} = 120^\circ$$

b) Since $\measuredangle 1$ is the supplement of an interior angle,

$$m\measuredangle 1 = 180^\circ - 120^\circ = 60^\circ$$

In order to discuss area in the next section, we must be able to identify various types of triangles and quadrilaterals. The following is a summary of certain types of triangles and their characteristics.

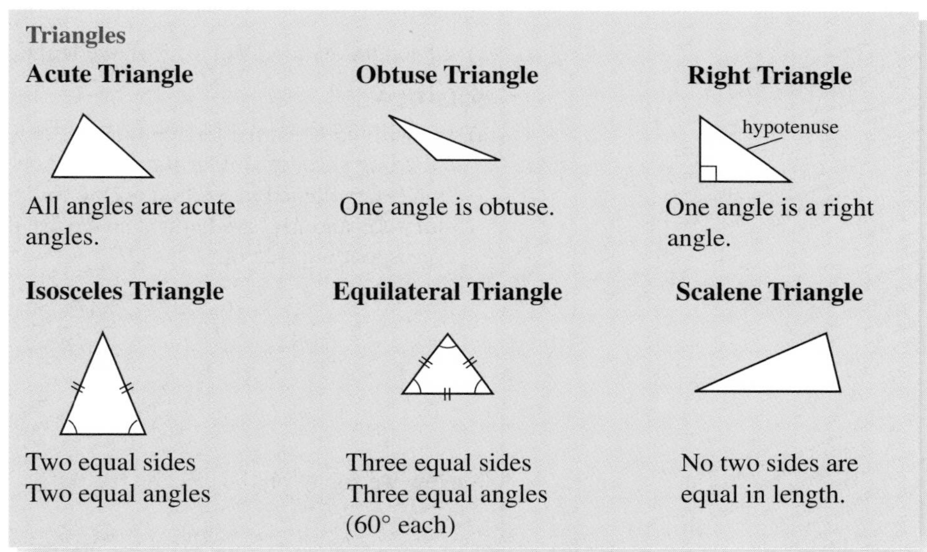

Triangles
Acute Triangle
All angles are acute angles.
Obtuse Triangle
One angle is obtuse.
Right Triangle
hypotenuse
One angle is a right angle.
Isosceles Triangle
Two equal sides
Two equal angles
Equilateral Triangle
Three equal sides
Three equal angles (60° each)
Scalene Triangle
No two sides are equal in length.

Similar Figures

In everyday living we often have to deal with geometric figures that have the "same shape" but are of different sizes. For example, an architect will make a small-scale drawing of a floor plan or a photographer will make an enlargement of a photograph. Figures that have the same shape but may be of different sizes are called *similar figures*. Two similar figures are illustrated in Fig. 9.20.

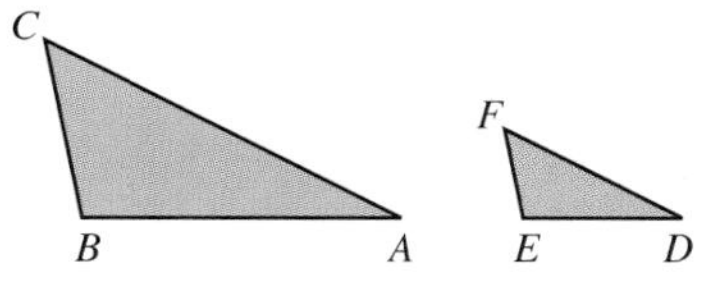

Figure 9.20

Similar figures have *corresponding angles* and *corresponding sides*. In Fig. 9.20 triangle ABC has angles A, B, and C. Their respective corresponding angles in triangle DEF are angles D, E, and F. Sides AB, BC, and AC in triangle ABC have corresponding sides DE, EF, and DF, respectively, in triangle DEF.

Two polygons are **similar** if their corresponding angles have the same measure and their corresponding sides are in proportion.

In Figure 9.20, $\measuredangle A$ and $\measuredangle D$ have the same measure, $\measuredangle B$ and $\measuredangle E$ have the same measure, and $\measuredangle C$ and $\measuredangle F$ have the same measure. Also, the corresponding sides of similar triangles are in proportion:

$$\frac{\overline{AB}}{\overline{DE}} = \frac{\overline{BC}}{\overline{EF}} = \frac{\overline{AC}}{\overline{DF}}.$$

EXAMPLE 2 *Similar Figures*

Consider the similar figures in Fig. 9.21.

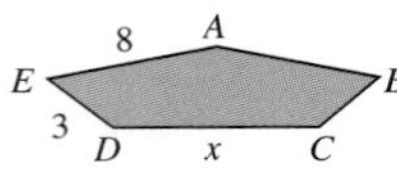

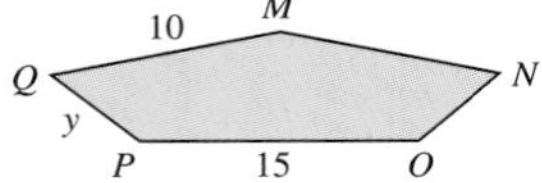

Figure 9.21

Determine

a) the length of side $\overline{CD}$. b) the length of side $\overline{PQ}$.

SOLUTION:

a) We will represent the length of side $\overline{CD}$ with the variable x. Because the corresponding sides of similar figures must be in proportion, we can write a proportion (as explained in Section 6.2) to find the length of side $\overline{CD}$. Corresponding sides $\overline{AE}$ and $\overline{MQ}$ are known, so we use them as one ratio in the proportion. The corresponding side of $\overline{CD}$ is $\overline{OP}$.

$$\frac{\overline{AE}}{\overline{MQ}} = \frac{\overline{CD}}{\overline{OP}}$$

$$\frac{8}{10} = \frac{x}{15}$$

Now we solve for x.

$$8 \cdot 15 = 10 \cdot x$$

$$120 = 10x$$

$$12 = x$$

Thus, the length of side $\overline{CD}$ is 12 units.

b) We will represent the length of side $\overline{PQ}$ with the variable y. We will work part (b) in a similar manner to part (a).

$$\frac{\overline{AE}}{\overline{MQ}} = \frac{\overline{DE}}{\overline{PQ}}$$

$$\frac{8}{10} = \frac{3}{y}$$

$$8 \cdot y = 10 \cdot 3$$

$$8y = 30$$

$$y = \frac{30}{8} = \frac{15}{4} = 3.75$$

Thus, the length of side $\overline{PQ}$ is $\frac{15}{4}$ or 3.75.

EXAMPLE 3 ***Using Similar Triangles to Find the Height of a Tree***

Saraniti Walker plans to remove a tree from her back yard. She needs to know the height of the tree. Saraniti is 5 ft tall and determines that when her shadow is 8 ft long, the shadow of the tree is 50 ft long (see Fig. 9.22). How tall is the tree?

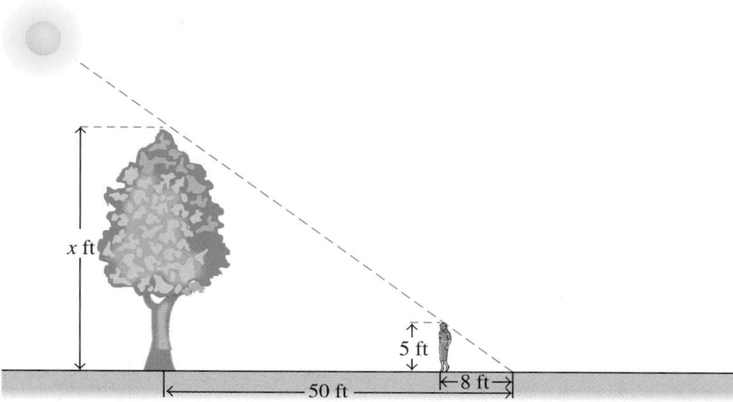

Figure 9.22

SOLUTION: We will let x represent the height of the tree. From Fig. 9.22 we can see that the triangle formed by the sun's rays, Saraniti, and her shadow is similar to the triangle formed by the sun's rays, the tree, and its shadow. To find the height of the tree we will set up and solve the following proportion:

$$\frac{\text{Height of the tree}}{\text{Height of Saraniti}} = \frac{\text{length of tree's shadow}}{\text{length of Saraniti's shadow}}$$

$$\frac{x}{5} = \frac{50}{8}$$

$$8x = 250$$

$$x = 31.25$$

Therefore, the tree is 31.25 ft tall.

Congruent Figures

If the corresponding sides of two similar figures are the same length, the figures are called *congruent figures*. Corresponding angles of congruent figures have the same measure, and the corresponding sides are equal in length. Two congruent figures coincide when placed one upon the other.

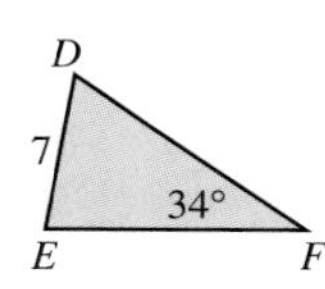

Figure 9.23

EXAMPLE 4 ***Congruent Triangles***

Triangles ABC and DEF in Fig. 9.23 are congruent. Determine

a) the length of side $\overline{DF}$.

b) the length of side $\overline{AB}$.

c) $m\measuredangle FDE$.

d) $m\measuredangle ACB$.

e) $m\measuredangle ABC$.

SOLUTION: Because ΔABC is congruent to ΔDEF, we know that the corresponding sides and angles are equal.

a) $\overline{DF} = \overline{AC} = 12$

b) $\overline{AB} = \overline{DE} = 7$

c) $m\measuredangle FDE = m\measuredangle CAB = 65°$

d) $m\measuredangle ACB = m\measuredangle DFE = 34°$

e) The sum of the angles of a triangle is 180°. Since $m\measuredangle BAC = 65°$ and $m\measuredangle ACB = 34°$, $m\measuredangle ABC = 180° - 65° - 34° = 81°$. ▲

Earlier we learned that *quadrilaterals* are four-sided polygons, the sum of whose interior angles is 360°. Quadrilaterals may be classified according to their characteristics, as illustrated in the summary box below.

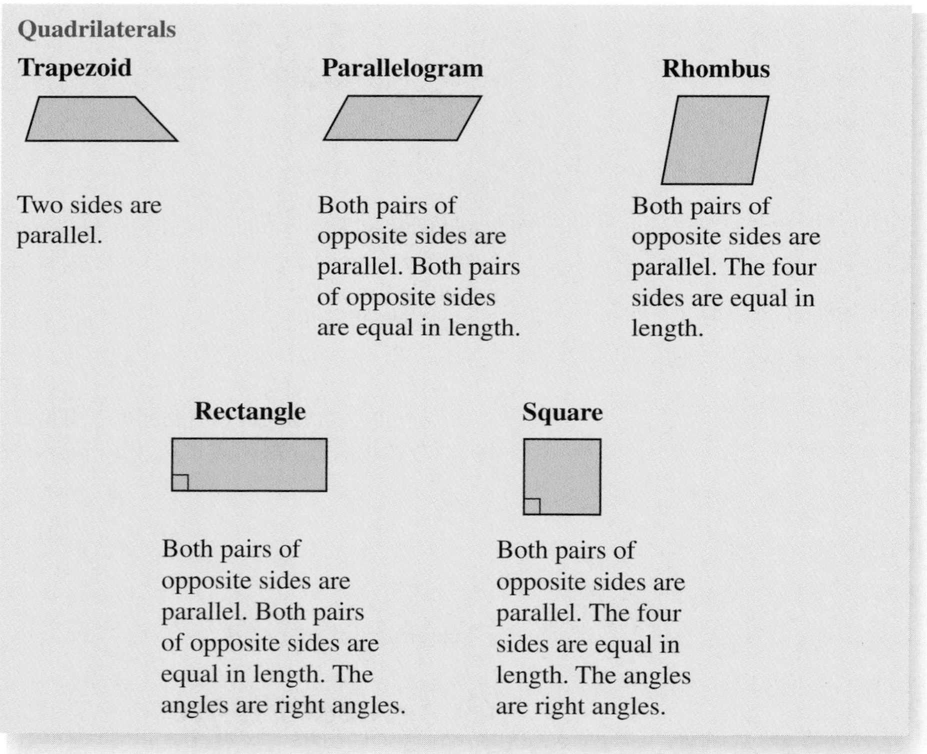

Quadrilaterals
Trapezoid
Two sides are parallel.
Parallelogram
Both pairs of opposite sides are parallel. Both pairs of opposite sides are equal in length.
Rhombus
Both pairs of opposite sides are parallel. The four sides are equal in length.
Rectangle
Both pairs of opposite sides are parallel. Both pairs of opposite sides are equal in length. The angles are right angles.
Square
Both pairs of opposite sides are parallel. The four sides are equal in length. The angles are right angles.

Figure 9.24

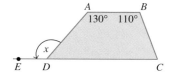

EXAMPLE 5 *Angles of a Trapezoid*

Find the measure of the exterior angle, x, of the trapezoid in Fig. 9.24.

SOLUTION: By the definition of a trapezoid, sides AB and CD must be parallel. Therefore, side AD may be considered a transversal and $\measuredangle BAD$ and $\measuredangle ADE$ are alternate interior angles. Recall from Section 9.1, that alternate interior angles are equal. Thus, $m\measuredangle BAD = m\measuredangle ADE$ and $m\measuredangle x = 130°$. ▲

DID YOU KNOW

Buckyballs

The molecular structure of C_{60} resembles the patterns found on a soccer ball.

Buckminsterfullerenes, also known as fullerenes and affectionately known as buckyballs, are a class of pure carbon molecules. Along with graphite and diamond, buckyballs are the only naturally occurring forms of pure carbon. Named after American architect-engineer F. Buckminster Fuller, who designed hemispherical geodesic domes from hexagonal and pentagonal faces, fullerenes are the most spherical molecules known. Discovered in 1985 by Robert Curl, Harold Kroto, and Richard Smalley at Rice University, buckyballs are only now beginning to see a wide range of applications. Used primarily as microscopic lubricant, buckyballs have potential applications in molecular medical engineering, electrical superconductivity, and computer chip design. The most common form of buckminsterfullerene contains 60 carbon atoms and has the chemical symbol C_{60}. The molecular structure of C_{60} contains 12 pentagons and 20 hexagons arranged in a pattern similar to that found on a soccer ball.

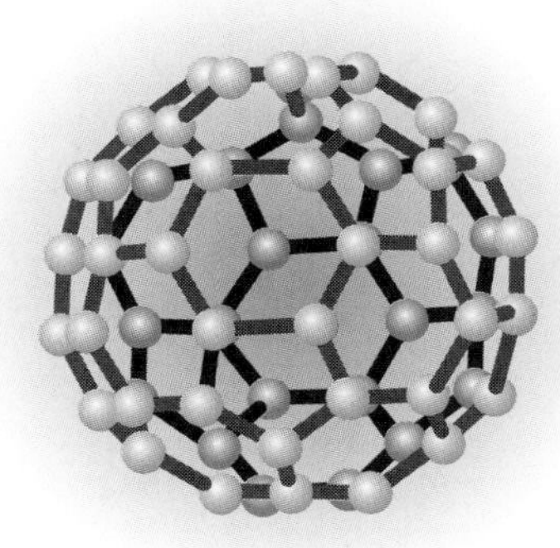

Sketch of a C_{60} molecule (also known as a buckyball)

SECTION 9.2 EXERCISES

Concept/Writing Exercises

1. What is a polygon?
2. What distinguishes regular polygons from other polygons?
3. List six different types of triangles and in your own words describe the characteristics of each.
4. List five different types of quadrilaterals and in your own words describe the characteristics of each.
5. What are congruent figures?
6. What are similar figures?

In Exercises 7–14, (a) name the polygon. If the polygon is a quadrilateral, give its specific name. (b) State whether or not the polygon is a regular polygon.

7. 

8. 

9. 

10. 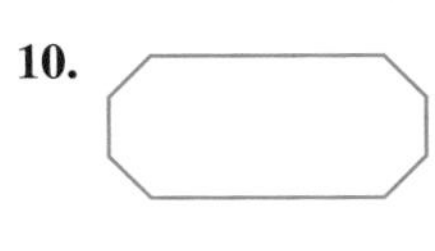

11. 

12. 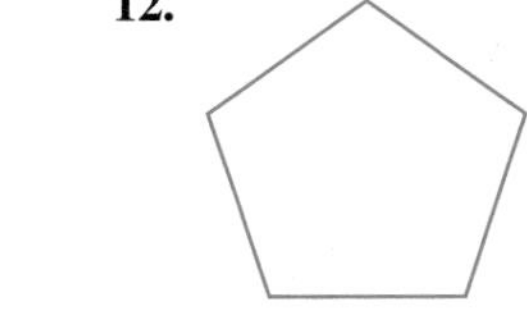

13. 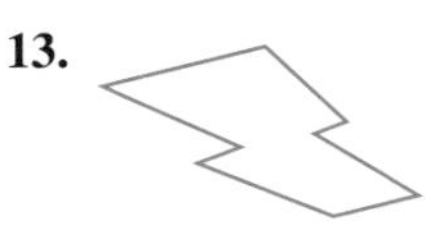

14. 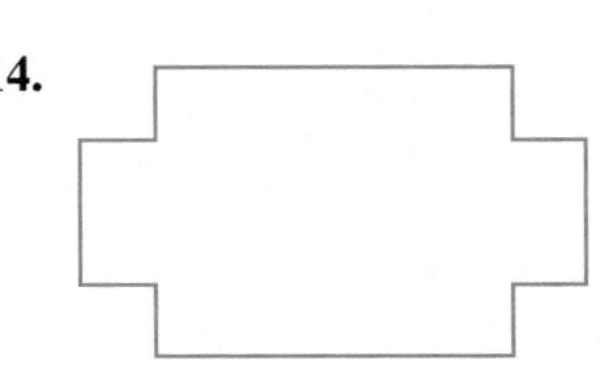

In Exercises 15–22, identify the triangle as (a) scalene, isosceles, or equilateral and as (b) acute, obtuse, or right. The parallel markings (the two small parallel lines) on two or more sides indicate that the marked sides are of equal length.

15. 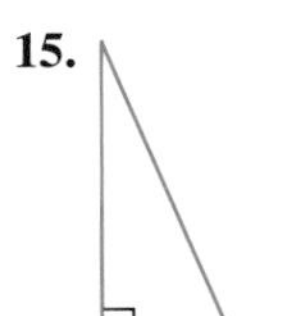

16. 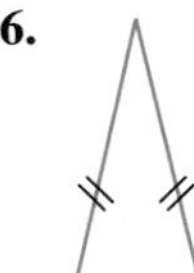

17.

18.

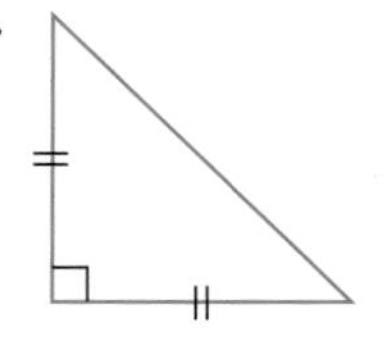

19.

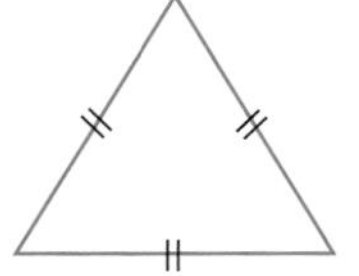

20.

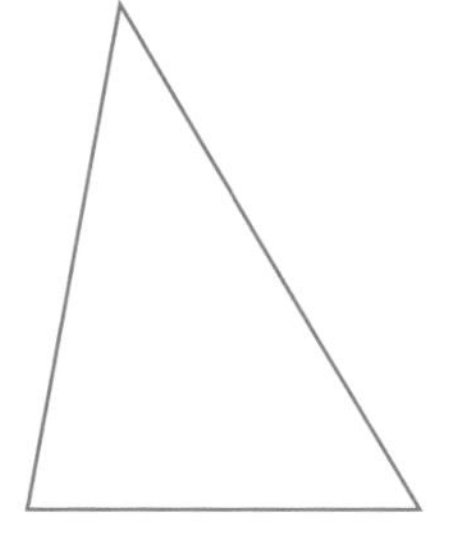

21.

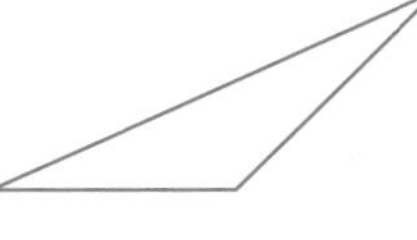

22.

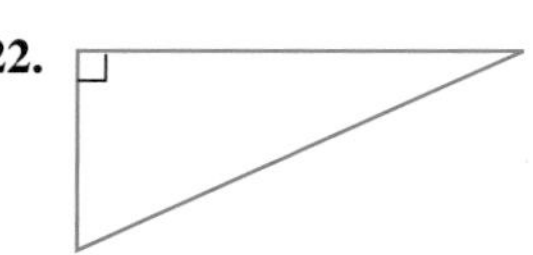

In Exercises 23–28, identify the quadrilateral.

23.

24.

25.

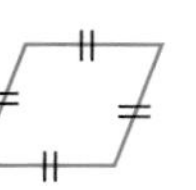

26.

27.

28.

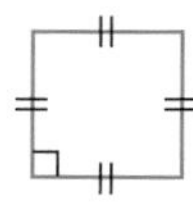

In Exercises 29–32, find the measure of $\measuredangle x$.

29.

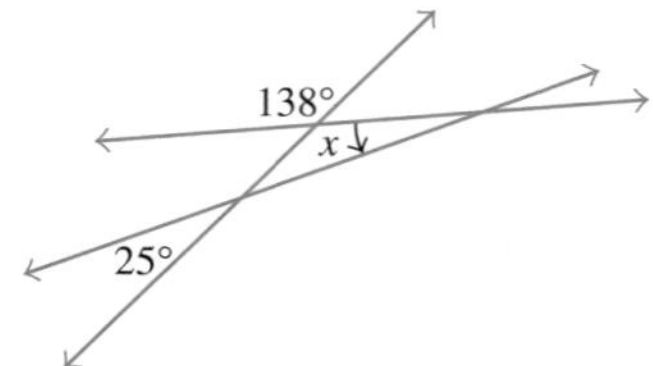

30.

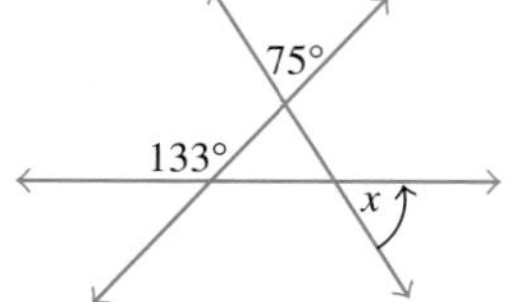

31.

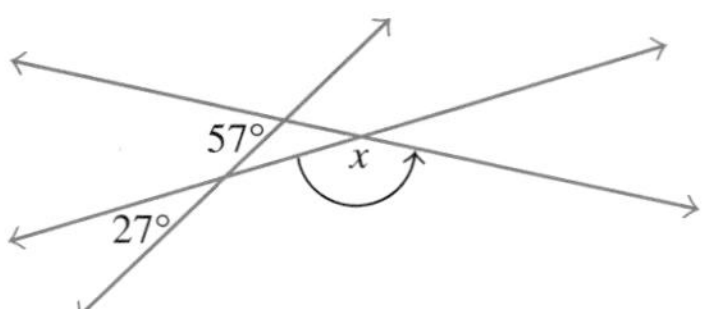

32.

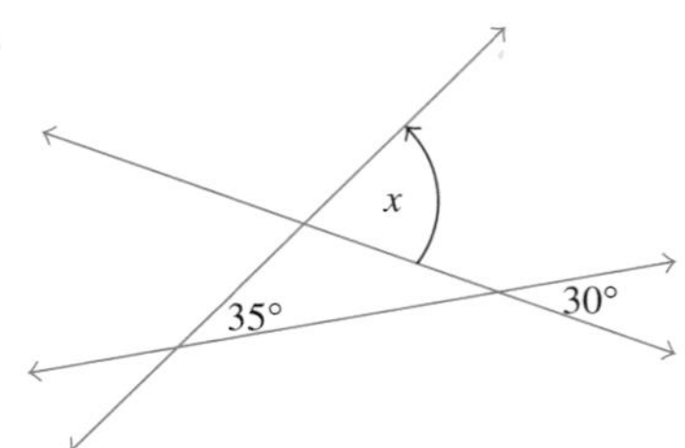

In Exercises 33–34, lines l_1 *and* l_2 *are parallel. Determine the measures of* $\measuredangle 1$ *through* $\measuredangle 12$.

33.

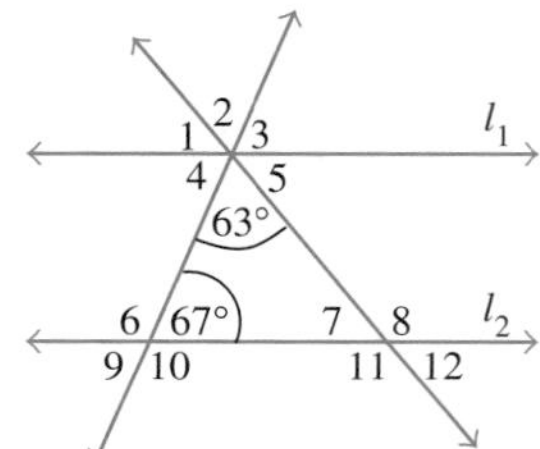

34.

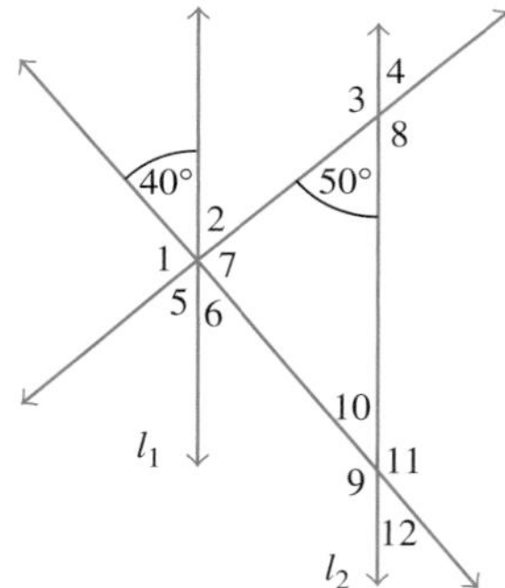

In Exercises 35–40, determine the sum of the measures of the interior angles of the indicated polygon.

35. Pentagon

36. Nonagon

37. Hexagon

38. Decagon

39. Icosagon

40. Dodecagon

In Exercises 41–46, (a) determine the measure of an interior angle of the named regular polygon. (b) If a side of the polygon is extended, determine the supplementary angle of an interior angle. See Example 1.

41. Triangle

42. Quadrilateral

43. Octagon

44. Nonagon

45. Dodecagon

46. Icosagon

In Exercises 47–52, the figures are similar. Find the length of side x and side y.

47.

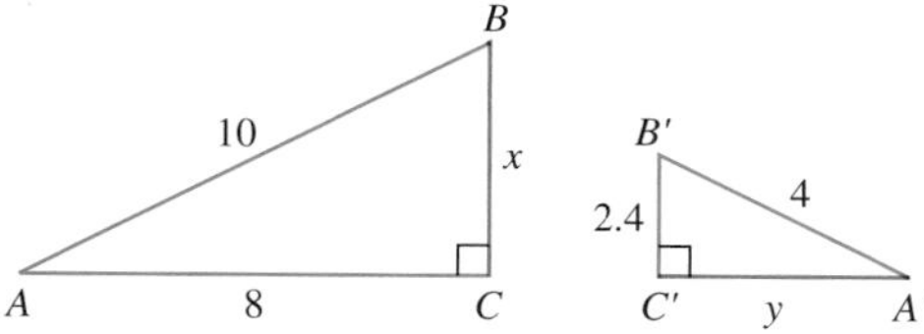

48.

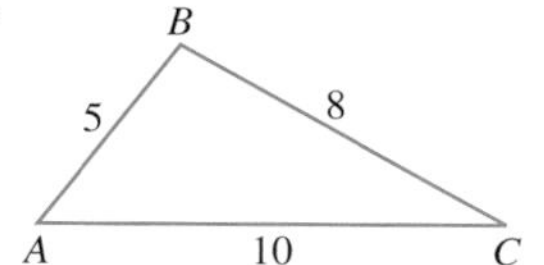

49.

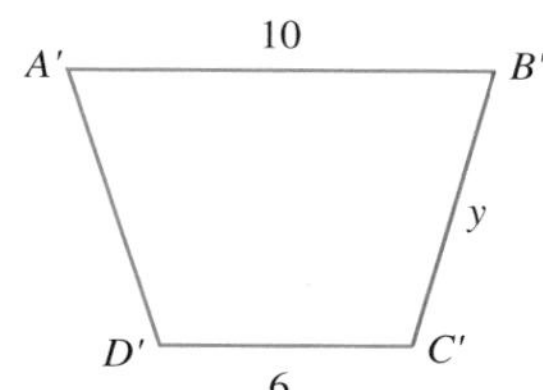

50.

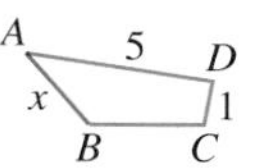

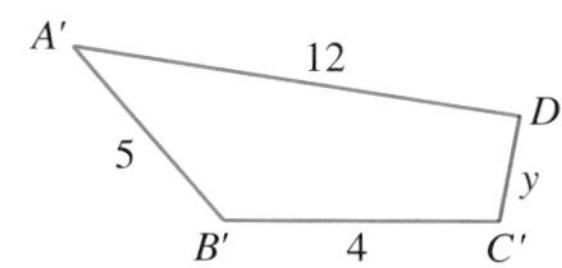

51.

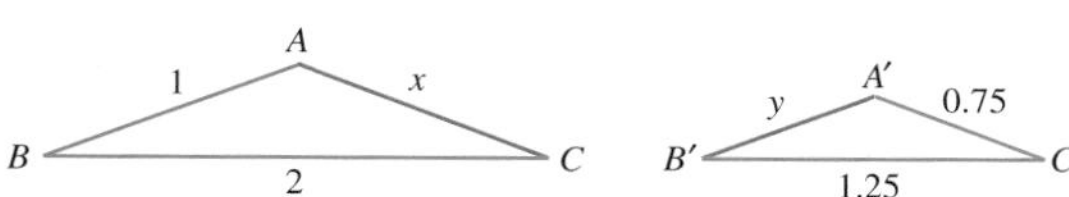

52.

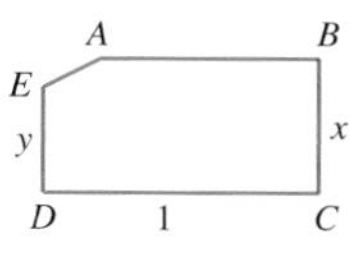

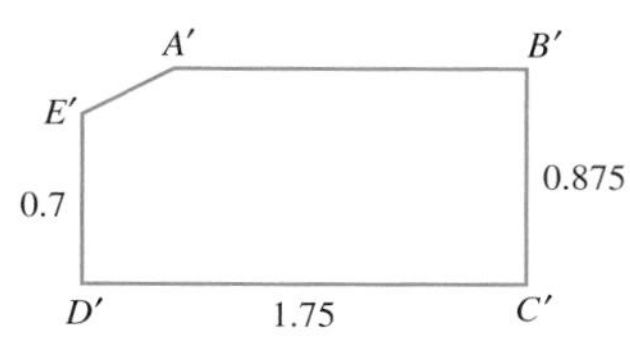

In Exercises 53–56, triangles ABC and DEC are similar figures. Find the length of

53. side $\overline{BC}$.

54. side $\overline{DC}$.

55. side $\overline{AD}$.

56. side $\overline{BE}$.

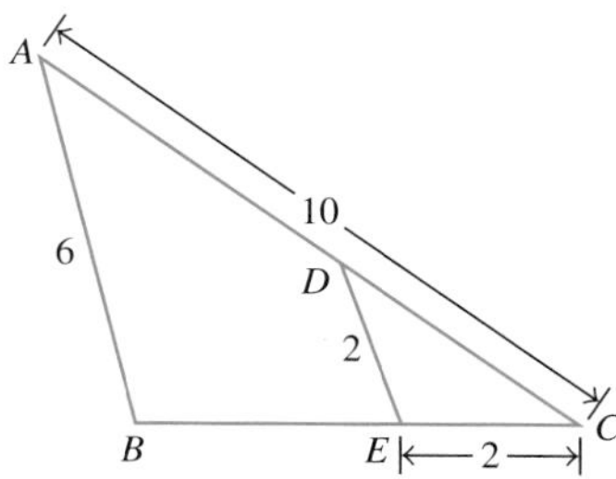

In Exercises 57–62, find the length of the sides and the measures of the angles for the congruent triangles ABC and A′B′C′.

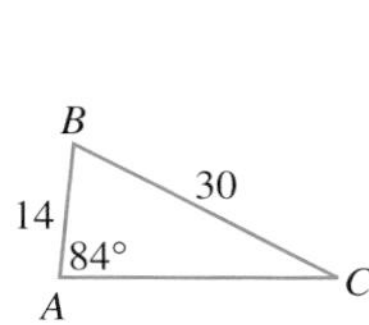

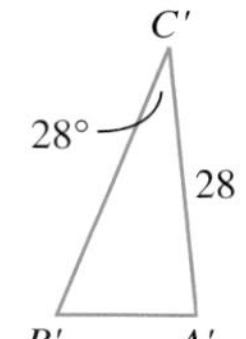

57. The length of side $\overline{A'B'}$

58. The length of side $\overline{B'C'}$

59. The length of side $\overline{AC}$

60. $\measuredangle B'A'C'$

61. $\measuredangle ACB$

62. $\measuredangle ABC$

In Exercises 63–68, find the length of the sides and the measures of the angles for the congruent quadrilaterals ABCD and A′B′C′D′.

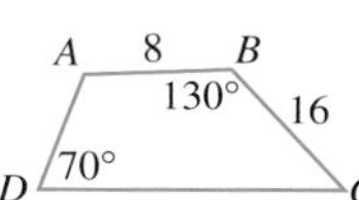

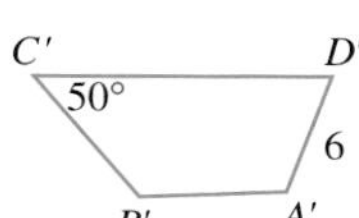

63. The length of side $\overline{A'B'}$

64. The length of side $\overline{AD}$

65. The length of side $\overline{B'C'}$

66. $\measuredangle BCD$

67. $\measuredangle A'D'C'$

68. $\measuredangle DAB$

Problem Solving

In Exercises 69–72, determine the measure of the angle. In the figure, $\measuredangle ABC$ makes an angle of 125° with the floor and l_1 and l_2 are parallel.

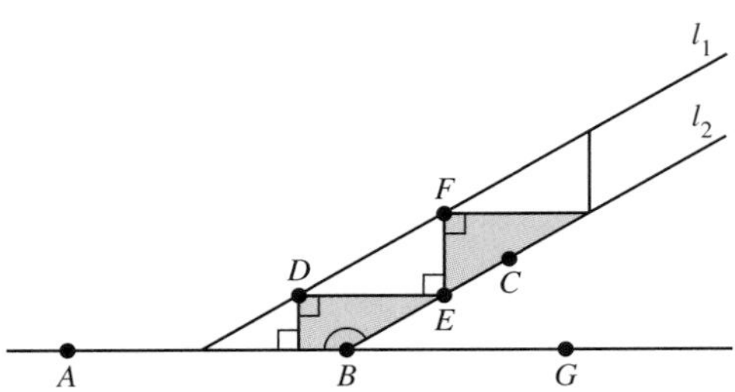

69. $\measuredangle GBC$ **70.** $\measuredangle EDF$

71. $\measuredangle DFE$ **72.** $\measuredangle DEC$

73. *Height of a Silo* Steve Runde is buying a farm and needs to determine the height of a silo on the farm. Steve, who is 6 ft tall, notices that when his shadow is 9 ft long, the shadow of the silo is 105 ft long (see diagram). How tall is the silo? Note that the diagram is not to scale.

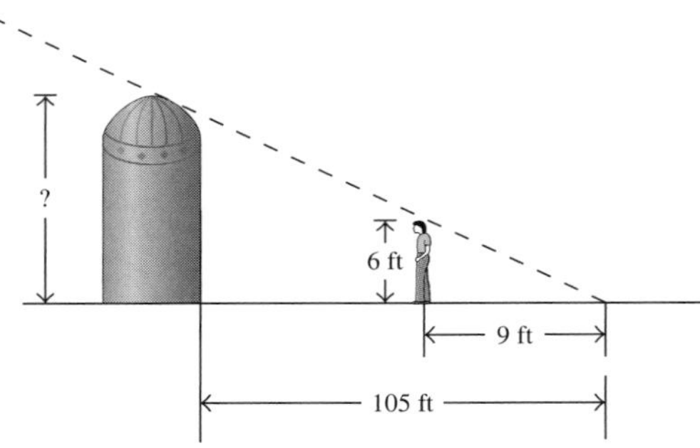

74. *Angles on a Picnic Table* The legs of a picnic table form an isosceles triangle as indicated in the figure. If $\measuredangle ABC = 80°$, determine $m\measuredangle x$ and $m\measuredangle y$ so that the top of the table will be parallel to the ground.

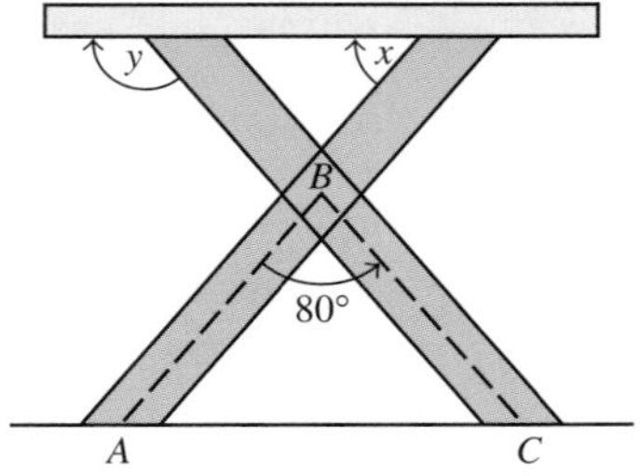

75. *Distances in Texas* A triangle can be formed by drawing line segments on a map of Texas connecting the cities of Dallas, Houston, and San Antonio (see figure). If the actual distance from San Antonio to Houston is approximately 197 miles, use the lengths of the line segments indicated in the figure along with similar triangles to approximate

a) the actual distance from Dallas to Houston.

b) the actual distance from Dallas to San Antonio.

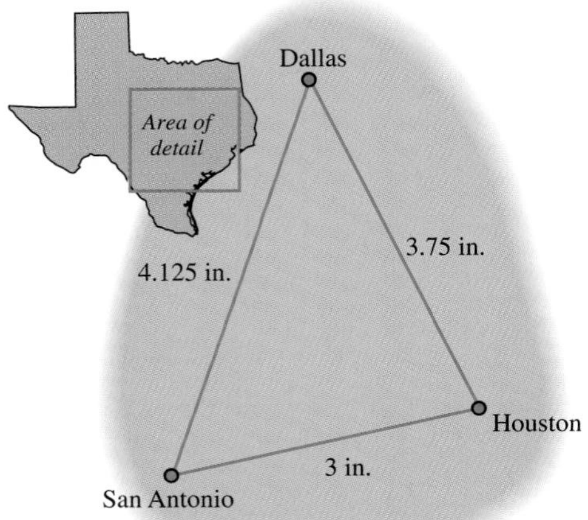

76. *Distances in Minnesota* A triangle can be formed by drawing line segments on a map of Minnesota connecting the cities of Austin, Rochester, and St. Paul (see figure). If the actual distance from Austin to Rochester is approximately 44 miles, use the lengths of the line segments indicated in the figure along with similar triangles to approximate

a) the actual distance from St. Paul to Austin.

b) the actual distance from St. Paul to Rochester.

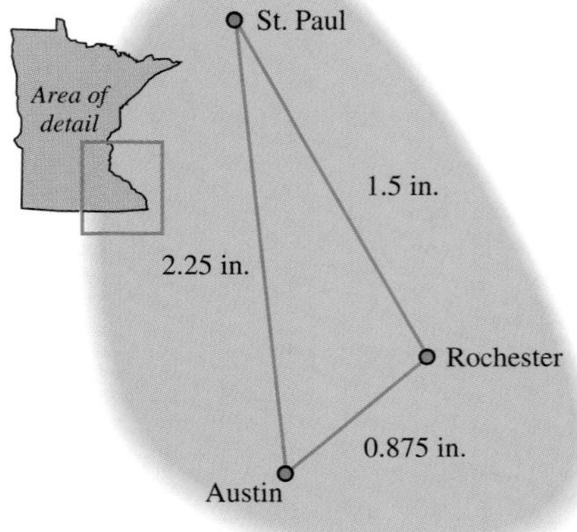

Challenge Problems/Group Activities

Scaling Factor Examine the similar triangles ABC and A′B′C′ in the figure below.

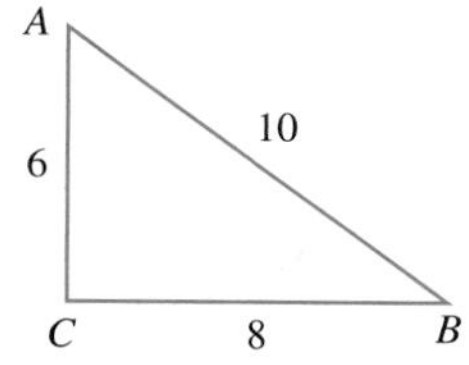

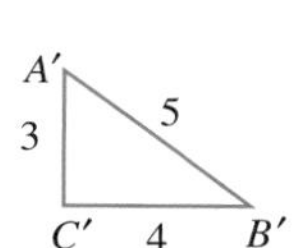

If we calculate the ratios $\frac{\overline{AB}}{\overline{A'B'}}, \frac{\overline{BC}}{\overline{B'C'}}$, *and* $\frac{\overline{CA}}{\overline{C'A'}}$, *we see that each of these ratios is equal to 2. We call this common ratio the* ***scaling factor*** *of* ΔABC *with respect to* $\Delta A'B'C'$. *If we calculate the reciprocal ratios* $\frac{\overline{A'B'}}{\overline{AB}}, \frac{\overline{B'C'}}{\overline{BC}}$, *and* $\frac{\overline{C'A'}}{\overline{CA}}$, *we see that each of these ratios is equal to* $\frac{1}{2}$. *We call this common ratio the scaling factor of* $\Delta A'B'C'$ *with respect to* ΔABC. *Every pair of similar figures has two scaling factors that show the relationship between the corresponding side lengths. Notice that the length of each side of* ΔABC *is two times the length of the corresponding side in* $\Delta A'B'C'$. *We can also state that the length of each side of* $\Delta A'B'C'$ *is one-half the length of the corresponding side of* ΔABC.

77. In the figure, ΔDEF is similar to $\Delta D'E'F'$. The length of the sides of ΔDEF is shown in the figure. If the scaling factor of ΔDEF with respect to $\Delta D'E'F'$ is 3, determine the length of the sides of triangle $\Delta D'E'F'$.

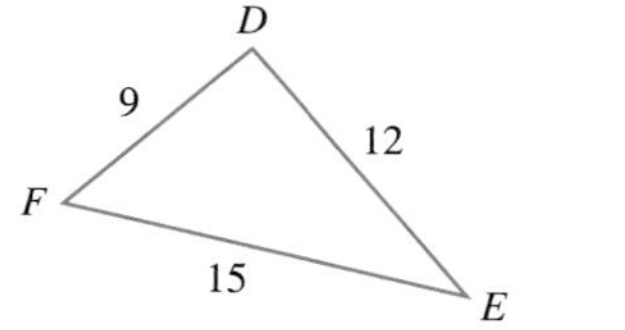

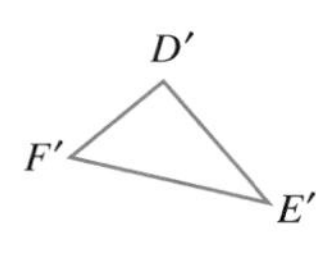

78. In the figure, quadrilateral *EFGH* is similar to quadrilateral $E'F'G'H'$. The length of the sides of quadrilateral *EFGH* is shown in the figure. If the scaling factor of quadrilateral $E'F'G'H'$ with respect to quadrilateral *EFGH* is $\frac{1}{3}$, determine the length of the sides of quadrilateral $E'F'G'H'$.

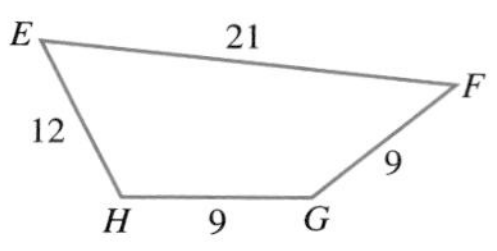

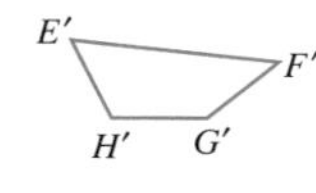

79. *Height of a Wall* You are asked to measure the height of an inside wall of a warehouse. No ladder tall enough to measure the height is available. You borrow a mirror from a salesclerk and place it on the floor. You then move away from the mirror until you can see the reflection of the top of the wall in it, as shown in the figure.
   a) Explain why triangle *HFM* is similar to triangle *TBM*. (*Hint:* In the reflection of light the angle of incidence equals the angle of reflection. Thus, $\measuredangle HMF = \measuredangle TMB$.)
   b) If your eyes are $5\frac{1}{2}$ ft above the floor, you are $2\frac{1}{2}$ ft from the mirror, and the mirror is 20 ft from the wall, how high is the wall?

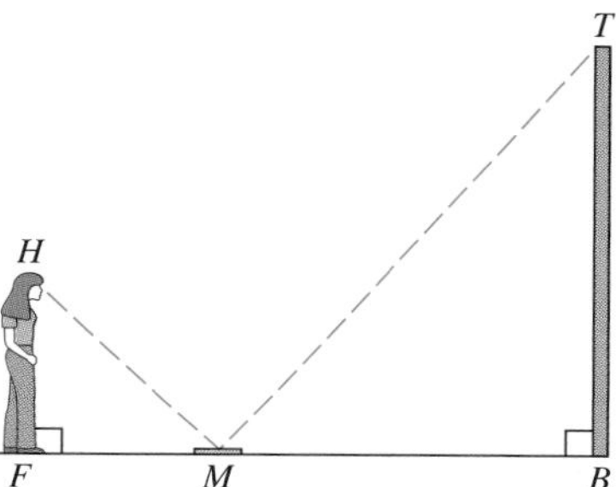

Recreational Mathematics

80. *Distance Across a Lake*
   a) In the figure $m\measuredangle CED = m\measuredangle ABC$. Explain why triangles *ABC* and *DEC* must be similar.
   b) Determine the distance across the lake, $\overline{DE}$.

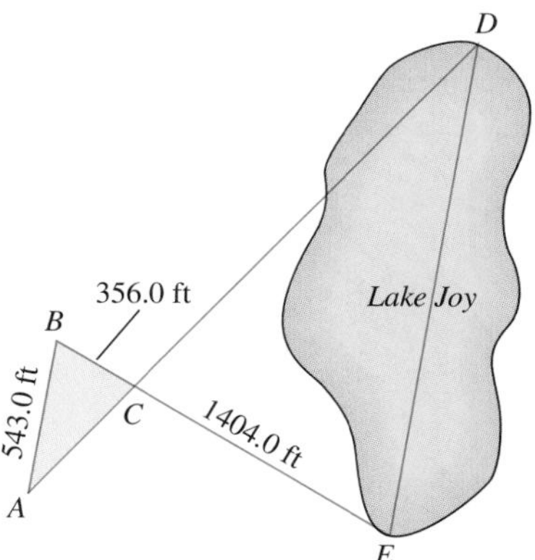

Internet/Research Activities

81. Using the Internet and history of mathematics texts, write a paper on the history and use of the theodolite, a surveying instrument.

82. Using the Internet and other sources, write a paper on the use of geometry in the photographic process. Include discussions on the use of similar figures.

9.3 PERIMETER AND AREA

Perimeter and Area

Figure 9.25

Geometric shapes abound in the natural world and the world made by human beings. For example, a basketball court is a rectangle, a basketball is a sphere, a can of food is a cylinder, and a ream of paper is a rectangular solid.

The *perimeter*, P, of a two-dimensional figure is the sum of the lengths of the sides of the figure. In Figs. 9.25 and 9.26 the sums of the red lines are the perimeters. Perimeters are measured in the same units as the sides. For example, if the sides of a figure are measured in feet, the perimeter will be measured in feet.

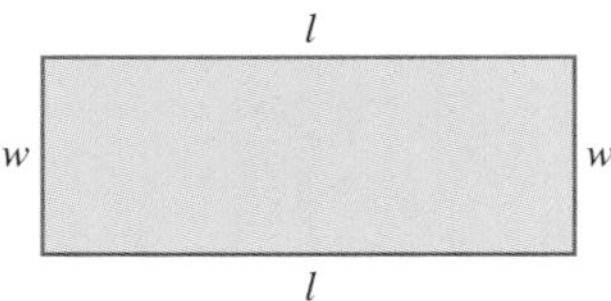

Figure 9.26

The *area*, A, is the region within the boundaries of the figure. The blue color in Figs. 9.25 and 9.26 indicates the areas of the figures. Area is measured in square units. For example, if the sides of a figure are measured in inches, the area of the figure will be measured in square inches (in.^{2}). (See Table 8.7 on page 461 for common units of area in the U.S. customary and metric systems.)

Consider the rectangle in Fig. 9.26. Two sides of the rectangle have length l, and two sides of the rectangle have width w. Thus, if we add the lengths of the four sides to get the perimeter, we find $P = l + w + l + w = 2l + 2w$.

Perimeter of a Rectangle

$$P = 2l + 2w$$

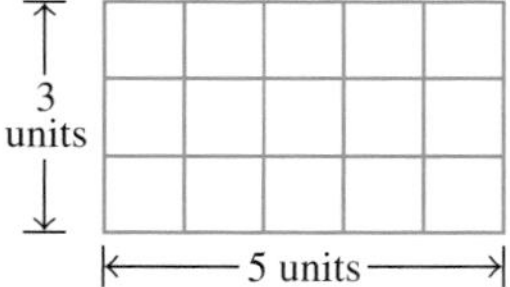

Figure 9.27

Consider a rectangle of length 5 units and width 3 units (Fig. 9.27). Counting the number of 1-unit by 1-unit squares within the figure we obtain the area of the rectangle, 15 square units. The area can also be obtained by multiplying the number of units of length by the number of units of width, or 5 units $\times$ 3 units $=$ 15 square units. We can find the area of a rectangle by the formula area $=$ length $\times$ width.

Area of a Rectangle

$$A = l \times w$$

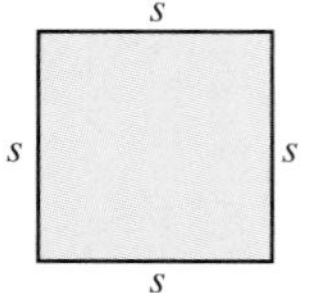

Figure 9.28

Using the formula for the area of a rectangle, we can determine the formulas for the areas of other figures.

A square (Fig. 9.28) is a rectangle that contains four equal sides. Therefore, the length equals the width. If we call both the length and the width of the square s, then

$$A = l \times w, \qquad \text{so} \qquad A = s \times s = s^2$$

Area of a Square

$$A = s^2$$

A parallelogram with height h and base b is shown in Fig. 9.29(a).

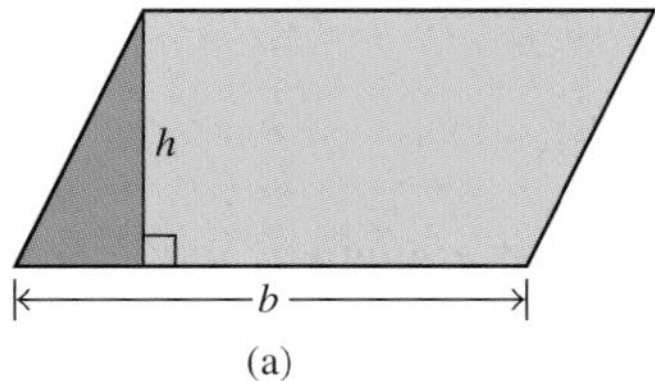

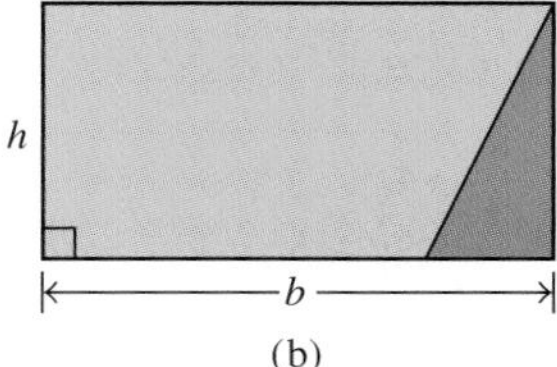

Figure 9.29

If we were to cut off the red portion of the parallelogram on the left, Fig. 9.29(a), and attach it to the right side of the figure, the resulting figure would be a rectangle, Fig. 9.29(b). Since the area of the rectangle is $b \times h$, the area of the parallelogram is also $b \times h$.

Area of a Parallelogram

$$A = b \times h$$

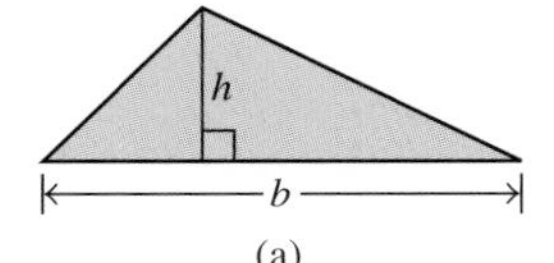

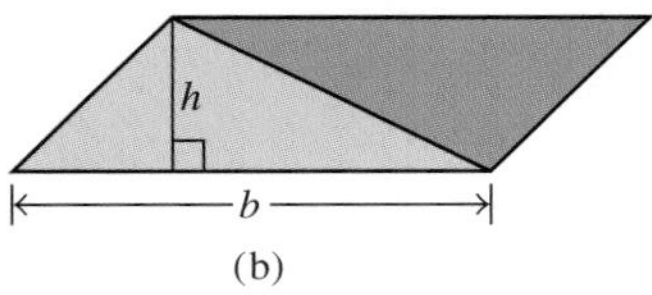

Figure 9.30

Consider the triangle with height, h, and base, b, shown in Fig. 9.30(a). Using this triangle and a second identical triangle, we can construct a parallelogram, Fig. 9.30(b). The area of the parallelogram is bh. The area of the triangle is one-half that of the parallelogram. Therefore, the area of the triangle is $\frac{1}{2}(\text{base})(\text{height})$.

Area of a Triangle

$$A = \frac{1}{2}bh$$

Now consider the trapezoid shown in Fig. 9.31(a). We can partition the trapezoid into two triangles by drawing diagonal $\overline{DB}$, as in Fig. 9.31(b). One triangle has base $\overline{AB}$ (called b_2) with height $\overline{DE}$, and the other triangle has base $\overline{DC}$ (called b_1) with height $\overline{FB}$. Note that the line used to measure the height of the triangle need not be inside the triangle. Because heights $\overline{DE}$ and $\overline{FB}$ are equal, both triangles have the same height, h. The area of triangles DCB and ADB are $\frac{1}{2}b_1h$ and $\frac{1}{2}b_2h$, respectively. The area of the trapezoid is the sum of the areas of the triangles, $\frac{1}{2}b_1h + \frac{1}{2}b_2h$, which can be written $\frac{1}{2}h(b_1 + b_2)$.

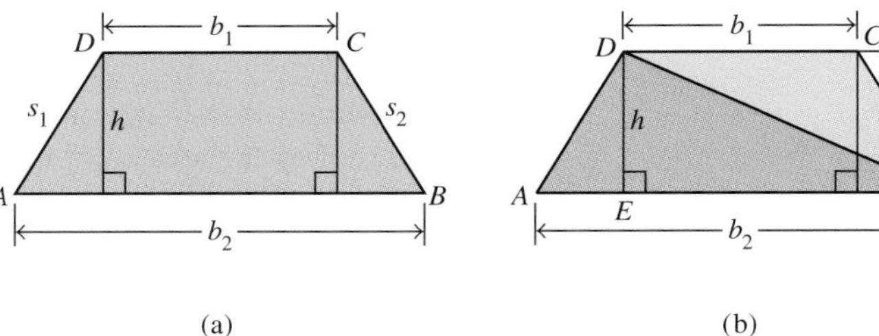

Figure 9.31

> **DID YOU KNOW**
>
> *Area versus Perimeter*
>
> 
>
>
> 
>
> New Mexico
>
> Which geographic area, Italy or New Mexico, do you think has the larger land area? As a matter of fact, the land areas of Italy and New Mexico are very similar: Italy's area is 116,304 mi^2 and New Mexico's area is 121,593 mi^2. Italy's perimeter, 5812 mi, is almost five times greater than New Mexico's 1200 mi, however.

Area of a Trapezoid

$$A = \tfrac{1}{2}h(b_1 + b_2)$$

Following is a summary of the perimeters and areas of selected figures.

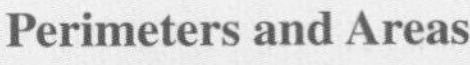

Perimeters and Areas

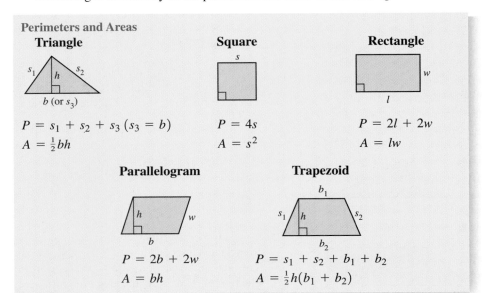

Triangle

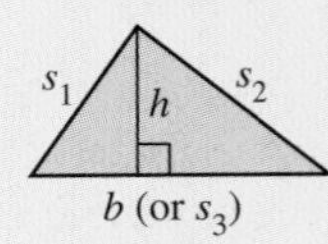

$P = s_1 + s_2 + s_3\ (s_3 = b)$

$A = \tfrac{1}{2}bh$

Square

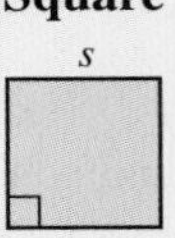

$P = 4s$

$A = s^2$

Rectangle

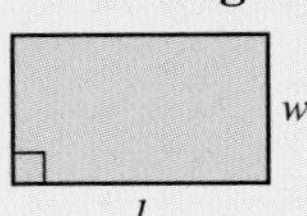

$P = 2l + 2w$

$A = lw$

Parallelogram

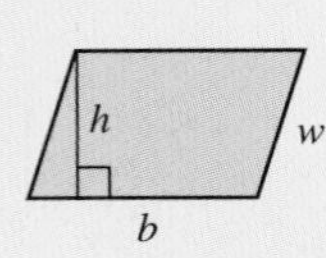

$P = 2b + 2w$

$A = bh$

Trapezoid

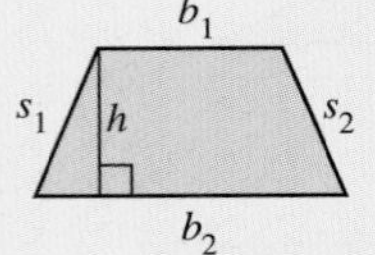

$P = s_1 + s_2 + b_1 + b_2$

$A = \tfrac{1}{2}h(b_1 + b_2)$

EXAMPLE 1 *Resurfacing a Restaurant Roof*

Richard McMenamy recently purchased the Lazy Lobster Restaurant and needs to resurface the roof of the sunporch with aluminum roof coating. One can of Kopper's Aluminum roof coating costs \$10.99 and covers 330 ft^2. If the roof of the sunporch is 40 ft long and 16 ft wide, determine

a) the area of the roof.

b) how many cans of roof coating he needs.

c) the cost of resurfacing the roof of the sunporch.

SOLUTION:

a) The area of the roof is

$$A = l \cdot w = 40 \cdot 16 = 640 \text{ ft}^2$$

The area of the roof is in square feet because both the length and width are measured in feet.

b) To determine the number of cans of roof coating Richard needs, divide the area of the roof by the area covered by one can of roof coating.

$$\frac{\text{Area of roof}}{\text{Area covered by one can}} = \frac{640}{330} = 1\frac{31}{33} \approx 1.94$$

The roof coating must be purchased in whole cans, so Richard needs to buy two cans of roof coating.

c) The cost of two cans of the roof coating is $2 \times \$10.99$, or \$21.98. 

MATHEMATICS *Everywhere*

Pythagorean Theorem

The Pythagorean theorem is one of the most famous theorems of all time. One book, *Pythagorean Propositions*, contains 370 different proofs of the Pythagorean theorem. U.S. President James A. Garfield gave one notable proof. The Pythagorean theorem has found its way into popular culture as well. In the movie *The Wizard of Oz*, the scarecrow incorrectly recites the Pythagorean theorem once the wizard grants him his diploma. In the play *The Pirates of Penzance*, the Major General refers to the Pythagorean theorem when he sings "I'm teeming with a lot o' news, with many cheerful facts about the square of the hypotenuse." Lewis Carroll, author of *Through the Looking Glass* and *Alice's Adventures in Wonderland*, stated that the Pythagorean theorem "is as dazzlingly beautiful now as it was in the day when Pythagoras discovered it." Actually, the Pythagorean theorem was known to the ancient Babylonians in about 1600 B.C., 1000 years before Pythagoras, and it continues to play a huge role in mathematics.

Pythagorean Theorem

We introduced the Pythagorean theorem in Chapter 5. Because this theorem is an important tool for finding the perimeter and area of triangles, we restate it here.

Pythagorean Theorem

The sum of the squares of the lengths of the legs of a right triangle equals the square of the length of the hypotenuse.

$$\text{leg}^2 + \text{leg}^2 = \text{hypotenuse}^2$$

Symbolically, if a and b represent the lengths of the legs and c represents the length of the hypotenuse (the side opposite the right angle), then

$$a^2 + b^2 = c^2$$

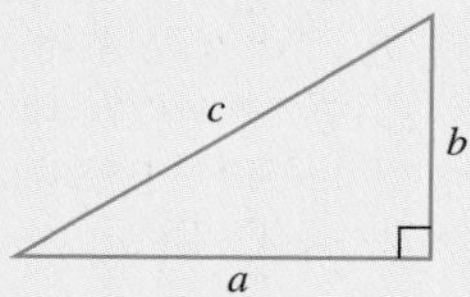

EXAMPLE 2 ***Crossing a Moat***

The moat surrounding a castle is 18 ft wide and the wall by the moat of the castle is 24 ft high (see Figure 9.32). If an invading army wishes to use a ladder to cross the moat and reach the top of the wall, how long must the ladder be?

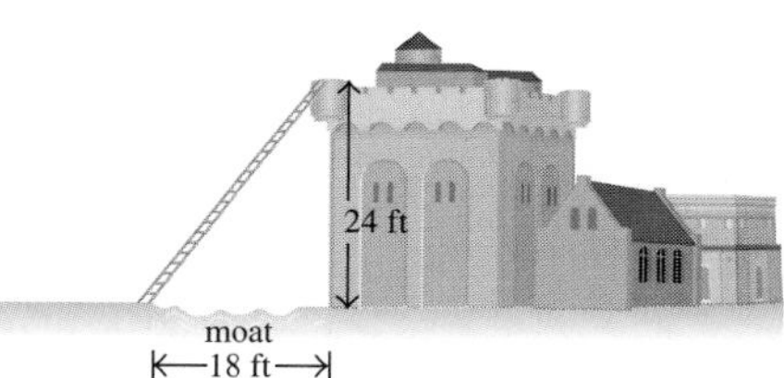

Figure 9.32

SOLUTION: The moat, the castle wall, and the ladder form a right triangle. The moat and the castle wall form the legs of the triangle (sides a and b), and the ladder forms the hypotenuse (side c). By the Pythagorean theorem,

$$\begin{aligned} c^2 &= a^2 + b^2 \\ c^2 &= (18)^2 + (24)^2 \\ c^2 &= 324 + 576 \\ c^2 &= 900 \\ \sqrt{c^2} &= \sqrt{900} \\ c &= 30 \end{aligned}$$

Take the square root of both sides of the equation.

Therefore, the ladder would need to be at least 30 ft long.

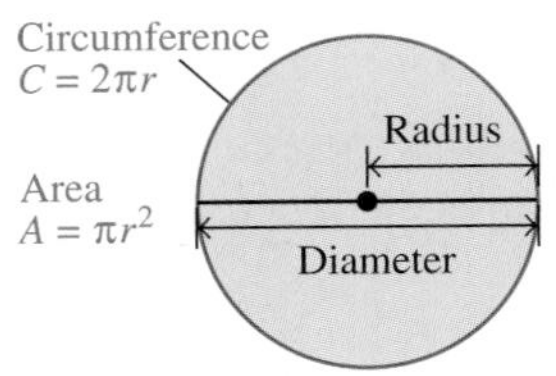

Figure 9.33

Circles

A commonly used plane figure that is not a polygon is a *circle*. A *circle* is a set of points equidistant from a fixed point called the center. A *radius*, r, of a circle is a line segment from the center of the circle to any point on the circle (Fig. 9.33). A *diameter*, d, of a circle is a line segment through the center of a circle with both end points on the circle. Note that the diameter of the circle is twice its radius. The *circumference* is the length of the simple closed curve that forms the circle. The formulas for the area and circumference of a circle are given in Fig. 9.33. The symbol pi, π, was introduced in Chapter 5. Recall that π is approximately 3.14. If your calculator contains a $\boxed{\pi}$ key, you should use that key when working calculations involving pi.

EXAMPLE 3 *Comparing Pizzas*

Victoria Montoya wishes to order a large cheese pizza. She can choose among three pizza parlors in town: Antonio's, Brett's, and Dorsey's. Antonio's large cheese pizza is a round 16-in.-diameter pizza that sells for \$15. Brett's large cheese pizza is a round 14-in.-diameter pizza that sells for \$12. Dorsey's large cheese pizza is a square 12-in. by 12-in. pizza that sells for \$10. All three pizzas have the same thickness. To get the most for her money, from which pizza parlor should Victoria order her pizza?

SOLUTION: To determine the best value, we will calculate the cost per square inch of pizza for each of the three pizzas. To do so we will divide the cost of each pizza by its area. The areas of the two round pizzas can be determined using the formula for the area of a circle, $A = \pi r^2$. Since the radius is half the diameter, we will use $r = 8$ and $r = 7$ for Antonio's and Brett's large pizzas, respectively. The area for the square pizza can be determined using the formula for the area of a square, $A = s^2$. We will use $s = 12$.

$$\text{Area of Antonio's pizza} = \pi r^2 \approx (3.14)(8)^2 \approx 3.14(64) \approx 200.96 \text{ in.}^{2*}$$
$$\text{Area of Brett's pizza} = \pi r^2 \approx (3.14)(7)^2 \approx 3.14(49) \approx 153.86 \text{ in.}^{2*}$$
$$\text{Area of Dorsey's pizza} = s^2 = (12)^2 = 144 \text{ in.}^2$$

Now, to find the cost per square inch of pizza, we will divide the cost of the pizza by the area of the pizza.

$$\text{Cost per square inch of Antonio's pizza} \approx \frac{\$15}{200.96 \text{ in.}^2} \approx \$0.0746$$

Thus, Antonio's pizza costs about \$0.0746, or about 7.5 cents, per square inch.

$$\text{Cost per square inch of Brett's pizza} \approx \frac{\$12}{153.86 \text{ in.}^2} \approx \$0.0780$$

Thus, Brett's pizza costs about \$0.0780, or about 7.8 cents, per square inch.

$$\text{Cost per square inch of Dorsey's pizza} = \frac{\$10}{144 \text{ in.}^2} \approx \$0.0694$$

*If you use the $\boxed{\pi}$ key on your calculator, your answers will be slightly more accurate.

DID YOU KNOW

Fermat's Last Theorem

Pierre de Fermat

In 1637 Pierre de Fermat, an amateur French mathematician, scribbled a note in the margin of the book *Arithmetica* by Diophantus. The note would haunt mathematicians for centuries. Fermat stated that the generalized form of the Pythagorean theorem, $a^n + b^n = c^n$, has no positive integer solutions where $n > 2$. Fermat's note concluded, "I have a truly marvelous demonstration of this proposition, which this margin is too narrow to contain." This conjecture became known as Fermat's last theorem. A formal proof of this conjecture escaped mathematicians until on September 19, 1994, Andrew J. Wiles of Princeton University announced he had found a proof. It took Wiles over 8 years of work—including fixing a flaw in an earlier announced solution—to accomplish the task. Wiles was awarded the Wolfskehl prize at Göttingen University in Germany in acknowledgement of his achievement.

Andrew J. Wiles

Thus, Dorsey's pizza costs about \$0.0694, or about 6.9 cents, per square inch.

Since the cost per square inch of pizza is the lowest for Dorsey's pizza, Victoria would get the most pizza for her money by ordering her pizza from Dorsey's. ▲

EXAMPLE 4 *Applying Lawn Fertilizer*

Steve May plans to fertilize his lawn. The shapes and dimensions of his lot, house, driveway, pool, and rose garden are shown in Fig. 9.34. One bag of fertilizer costs \$29.95 and covers 5000 ft^2. Determine how many bags of fertilizer Steve needs and the total cost of the fertilizer.

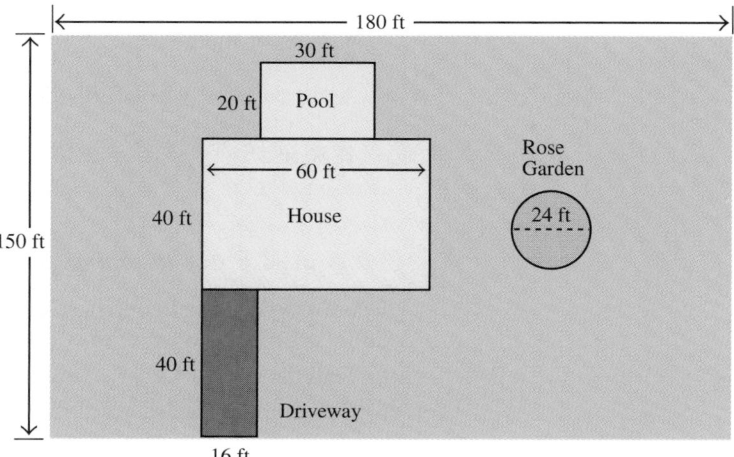

Figure 9.34

DID YOU KNOW

City Planning

The Roman poet Virgil tells the story of Queen Dido, who fled to Africa after her brother murdered her husband. There, she begged for some land from King Iarbus, telling him she only needed as much land as the hide of an ox would enclose. Being very clever, she decided that the greatest area would be enclosed if she tore the hide into thin strips and formed the strips into a circle. On this land she founded the city of Byrsa (the Greek word for "hide"), later known as Carthage in present-day Tunisia.

SOLUTION: The total area of the lot is $150 \cdot 180$, or 27,000 ft^2. To determine the area to be fertilized, subtract the area of house, driveway, pool, and rose garden from the total area.

$$\begin{aligned} \text{Area of house} &= 60 \cdot 40 = 2400 \text{ ft}^2 \\ \text{Area of driveway} &= 40 \cdot 16 = 640 \text{ ft}^2 \\ \text{Area of pool} &= 20 \cdot 30 = 600 \text{ ft}^2 \end{aligned}$$

The diameter of the rose garden is 24 ft, so its radius is 12 ft.

$$\text{Area of rose garden} = \pi r^2 = \pi(12)^2 \approx 3.14(144) \approx 452.16 \text{ ft}^2$$

The total area of the house, driveway, pool, and rose garden is approximately $2400 + 640 + 600 + 452.16$, or 4092.16 ft^2. The area to be fertilized is $27{,}000 - 4092.16$ ft^2, or 22,907.84 ft^2. The number of bags of fertilizer is found by dividing the total area to be fertilized by the number of square feet covered per bag.

The number of bags of fertilizer is $\frac{22{,}907.84}{5000}$, or about 4.58 bags. Therefore, Steve needs five bags. At \$29.95 per bag, the total cost is $5 \times \$29.95$, or \$149.75. ▲

EXAMPLE 5 *Converting between Square Feet and Square Inches*

a) Convert 1 ft^2 to square inches.

b) Convert 86 ft^2 to square inches.

c) Convert 288 in.^{2} to square feet.

d) Convert 1836 in.^{2} to square feet.

SOLUTION:

a) 1 ft = 12 in. Therefore, 1 ft^2 = 12 in. × 12 in. = 144 in.^{2}

b) From part (a) we know that 1 ft^2 = 144 in.^{2} Therefore, 86 ft^2 = 86 × 144 in.^{2} = 12,384 in.^{2}

c) In part (b) we converted from ft^2 to in.^{2} by *multiplying* the number of square feet by 144. Now, to convert from square inches to square feet we will *divide* the number of square inches by 144. Therefore, 288 in.$^2 = \frac{288}{144}$ ft^2 = 2 ft^2.

d) As in part (c), we will divide the number of square inches by 144. Therefore, 1836 in.$^2 = \frac{1836}{144}$ ft^2 = 12.75 ft^2. ▲

EXAMPLE 6 *Installing Ceramic Tile*

Debra Levy wishes to purchase ceramic tile for her family room, which measures 30 ft × 27 ft. The cost of the tile, including installation, is \$21 per square yard.

a) Find the area of Debra's family room in square *yards*.

b) Determine Debra's cost of the ceramic tile for her family room.

SOLUTION:

a) The area of the family room in square feet is $30 \cdot 27 = 810$ ft^2.

Since 1 yd = 3 ft, 1 $\text{yd}^2 = 3 \text{ ft} \times 3 \text{ ft} = 9 \text{ ft}^2$. To find the area of the family room in square yards, divide the area in square feet by 9 ft^2.

$$\text{Area in square yards} = \frac{810}{9} = 90$$

Therefore, the area is 90 yd^2.

b) The cost of 90 yd^2 of ceramic tile, including installation, is $90 \cdot \$21 = \$1890$.

When multiplying units of length, be sure that the units are the same. You can multiply feet by feet to get square feet or yards by yards to get square yards. However, you cannot get a valid answer if you multiply numbers expressed in feet by numbers expressed in yards.

SECTION 9.3 EXERCISES

Concept/Writing Exercises

1. a) Describe in your own words how to determine the *perimeter* of a two-dimensional figure.
   b) Describe in your own words how to determine the *area* of a two-dimensional figure.
   c) Draw a rectangle with a length of 6 units and a width of 2 units. Determine the area and perimeter of this rectangle.
2. What is the relationship between the *radius* and the *diameter* of a circle?
3. a) How do you convert an area from square feet into square inches?
   b) How do you convert an area from square inches into square feet?
4. a) How do you convert an area from square yards into square feet?
   b) How do you convert an area from square feet into square yards?

Practice the Skills

In Exercises 5–8, find the area of the triangle.

5. 7 in. 10 in.

6.

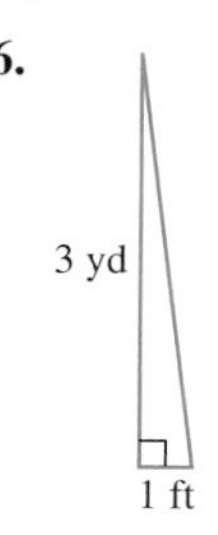

7.

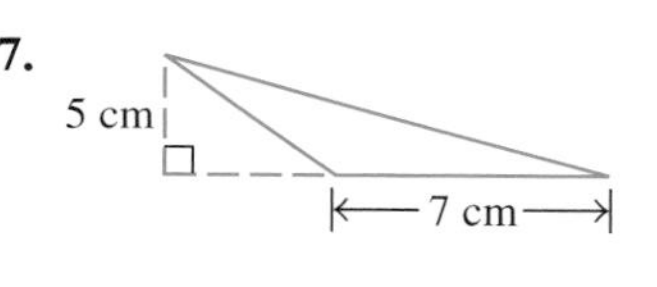

8.

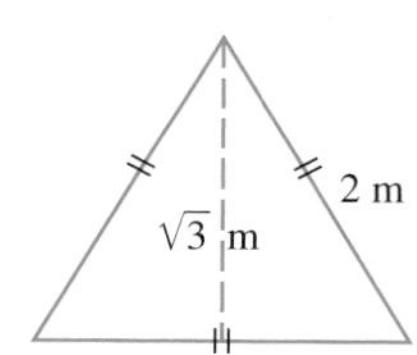

In Exercises 9–14, find the area and perimeter of the quadrilateral.

9.

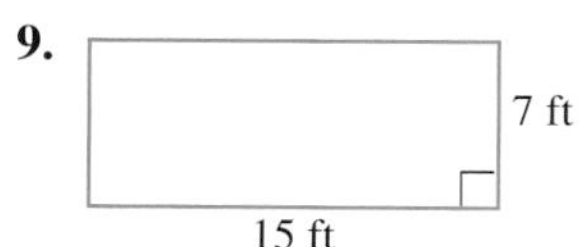

10.

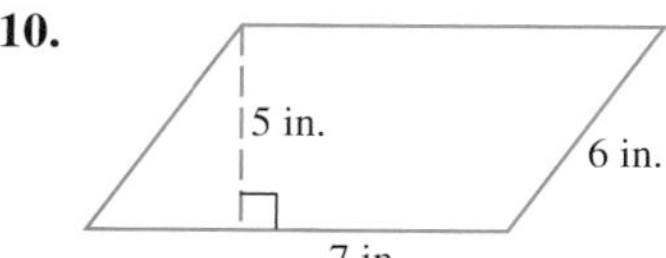

11. 12.

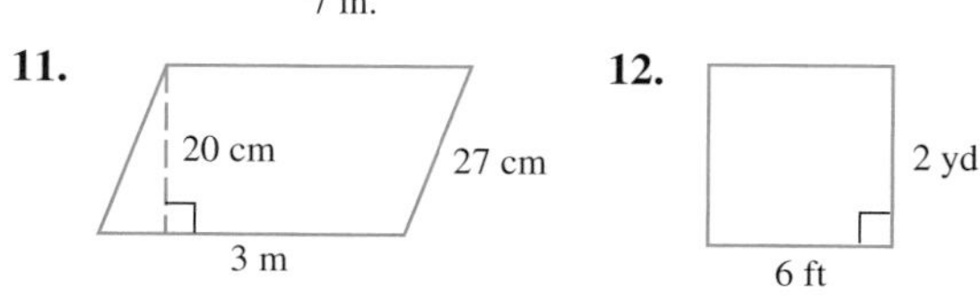

13. 14.

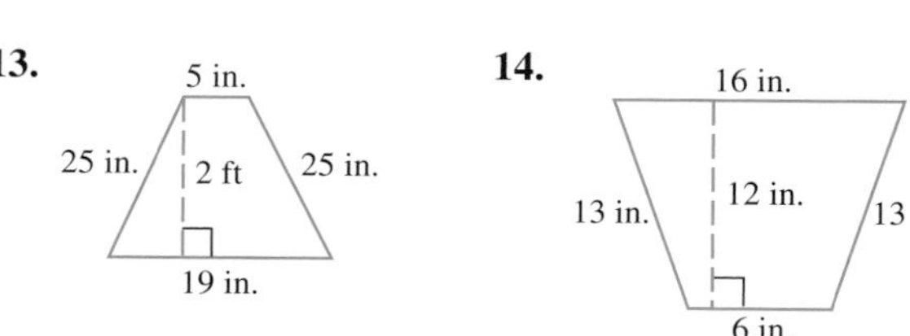

In Exercises 15–18, find the area and circumference of the circle.

15.

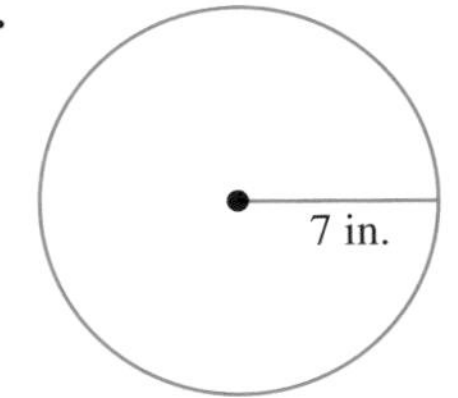

16.

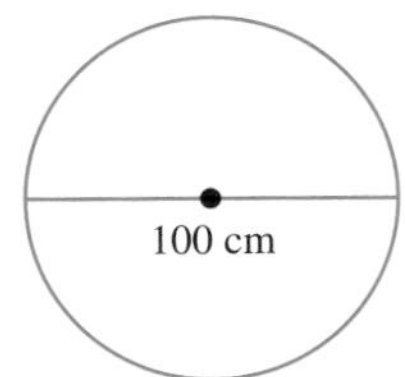

17.

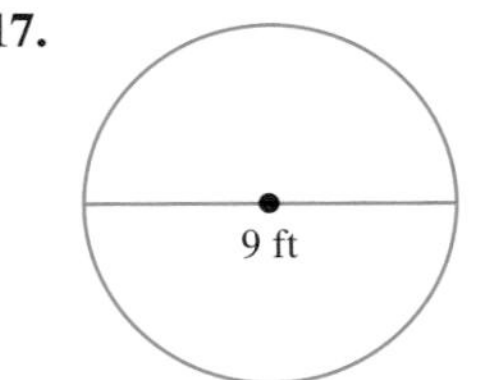

18.

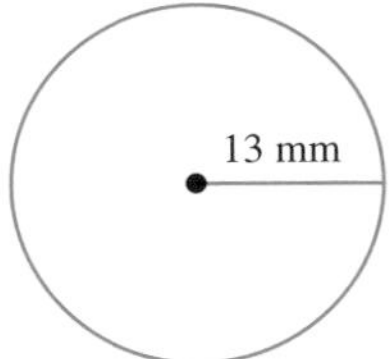

In Exercises 19–22, (a) use the Pythagorean theorem to determine the length of the unknown side of the triangle, (b) determine the perimeter of the triangle, and (c) determine the area of the triangle.

19.

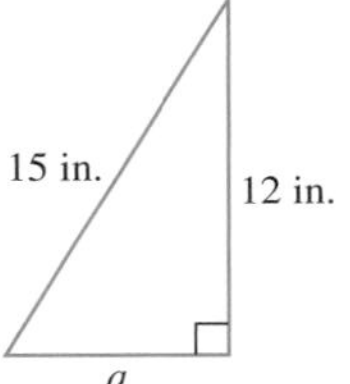

20.

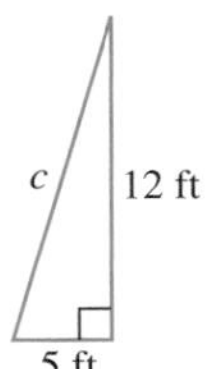

21.

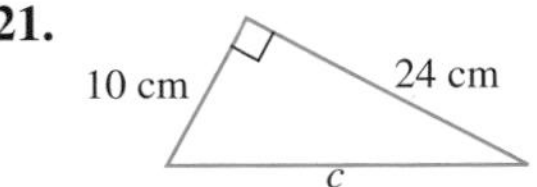

22.

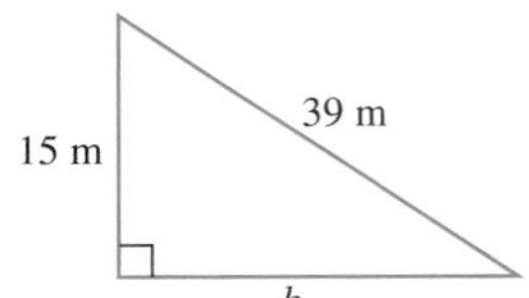

Problem Solving

In Exercises 23–32, find the shaded area. Round your answers to hundredths.

23.

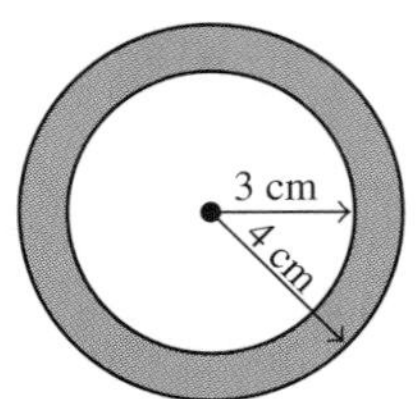

24.

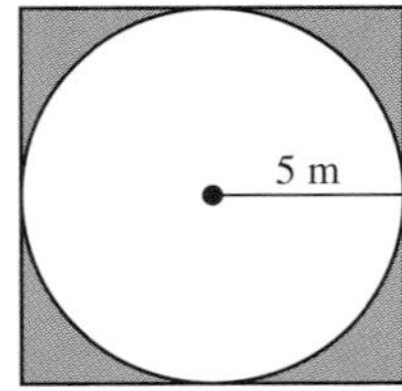

25.

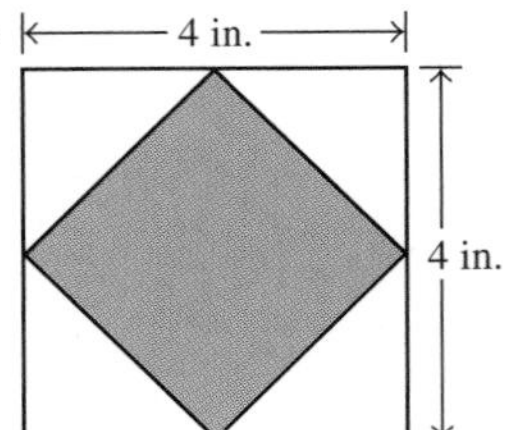

26.

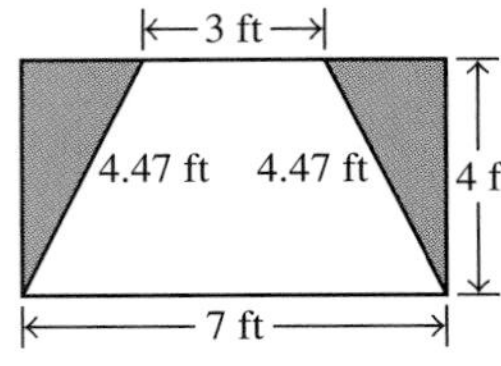

27.

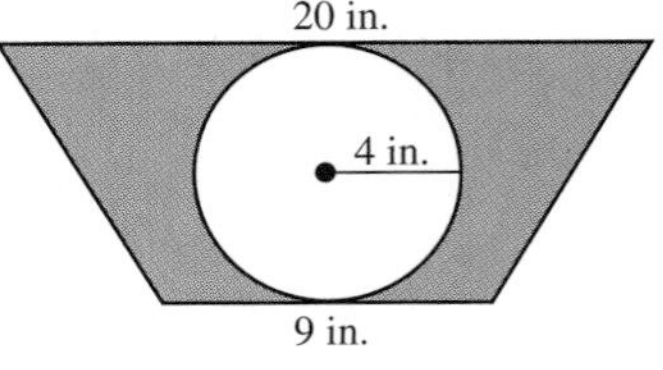

28.

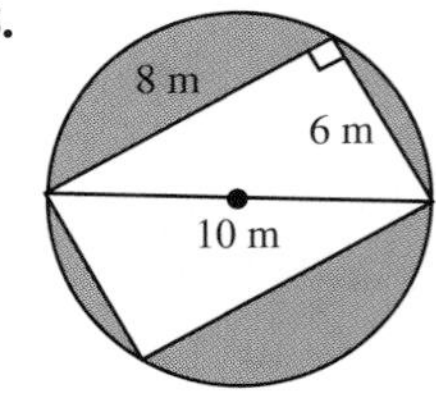

29.

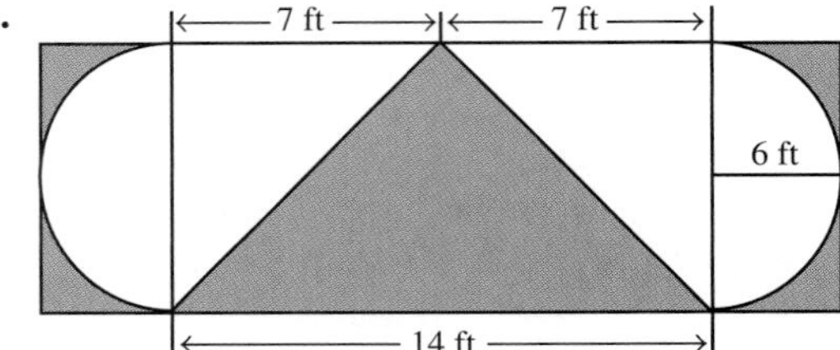

30.

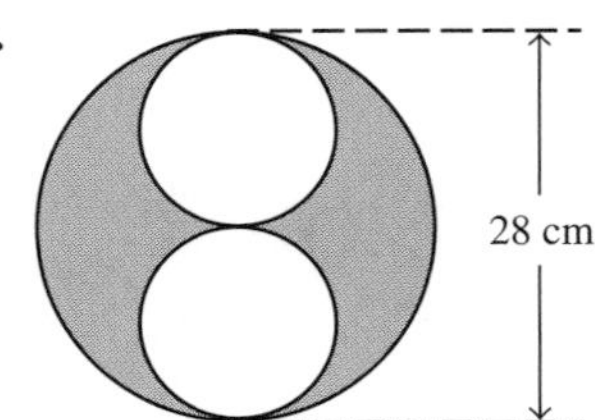

31.

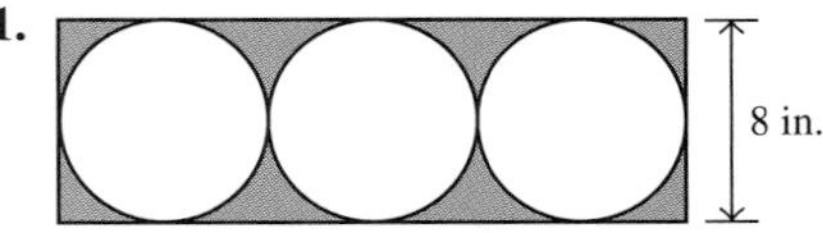

32.

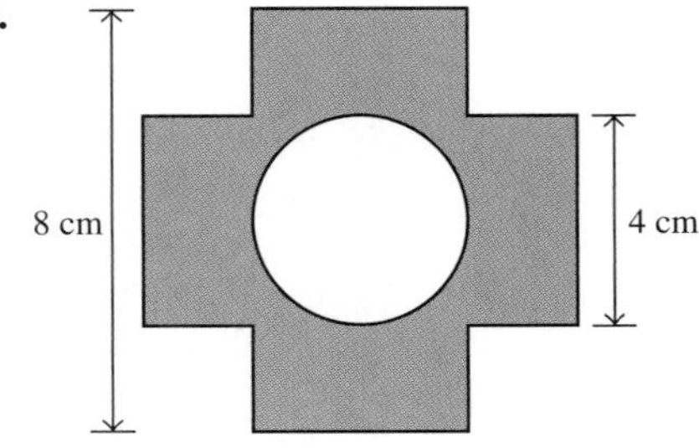

One square yard equals 9 ft^2. Use this information to convert the following.

33. 107 ft^2 to square yards **34.** 15.2 ft^2 to square yards

35. 14.7 yd^2 to square feet **36.** 18.3 yd^2 to square feet

One square meter equals 10,000 cm^2. Use this information to convert the following.

37. 23.4 m^2 to square centimeters **38.** 14.7 m^2 to square centimeters

39. 1075 cm^2 to square meters **40.** 608 cm^2 to square meters

Nancy Wallin has just purchased a new house that is in need of new flooring. In Exercises 41–46, use the measurements given on the floor plans of Nancy's house to obtain the answer.

41. *Cost of Laminate Flooring* The cost of Pergo Select Helsinki Birch laminate flooring is \$5.89 per square foot if Nancy installs the flooring herself or \$8.89 per square foot if she has the flooring installed by the flooring company. Determine the cost for the flooring in the living/dining room if

a) Nancy installs it herself.

b) Nancy has it installed by the flooring company.

42. *Cost of Hardwood Flooring* The cost of Mannington Chestnut hardwood flooring is \$10.86 per square foot if Nancy installs the flooring herself or \$13.86 per square foot if she has the flooring installed by the flooring company. Determine the cost for hardwood flooring in the living/dining room if

a) Nancy installs it herself.

b) Nancy has it installed by the flooring company.

43. *Cost of Linoleum* The cost of Armstrong Solarian Woodcut linoleum is \$5.00 per square foot. This price includes the cost of installation. Determine the cost for Nancy to have this linoleum installed in the kitchen and in both bathrooms.

44. *Cost of Ceramic Tile* The cost of Mohawk Porcelain ceramic tile is \$8.50 per square foot. This price includes the cost of installation. Determine the cost for Nancy to have this ceramic tile installed in the kitchen and in both bathrooms.

45. *Cost of Berber Carpeting* The cost of Bigelow Commodore Berber carpeting is \$6.06 per square foot. This price includes the cost of installation. Determine the cost for Nancy to have this carpeting installed in all three bedrooms.

46. *Cost of Saxony Carpeting* The cost of DuPont Stainmaster Saxony carpeting is \$5.56 per square foot. This price includes the cost of installation. Determine the cost for Nancy to have this carpeting installed in all three bedrooms.

47. *Cost of a Lawn Service* Jim and Wendy Scott's home lot is illustrated here. The Scotts wish to hire a lawn service to cut their lawn. M&M Lawn Service charges \$0.02 per square yard of lawn. How much will it cost the Scotts to have their lawn cut?

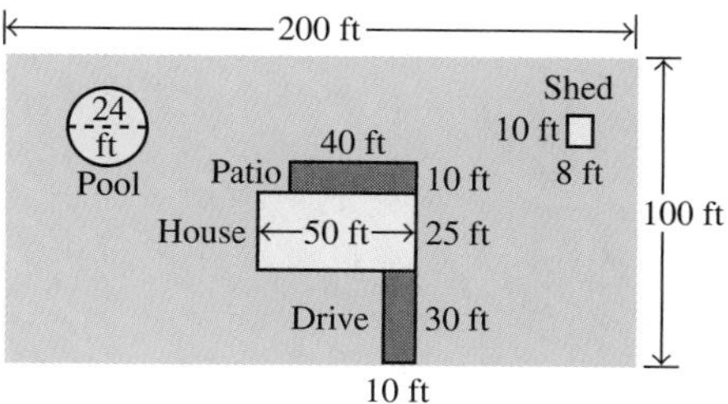

48. *Cost of a Lawn Service* Clarence and Rose Cohen's home lot is illustrated here. Clarence and Rose wish to hire Picture Perfect Lawn Service to cut their lawn. How much will it cost Clarence and Rose to have their lawn cut if Picture Perfect charges \$0.02 per square yard?

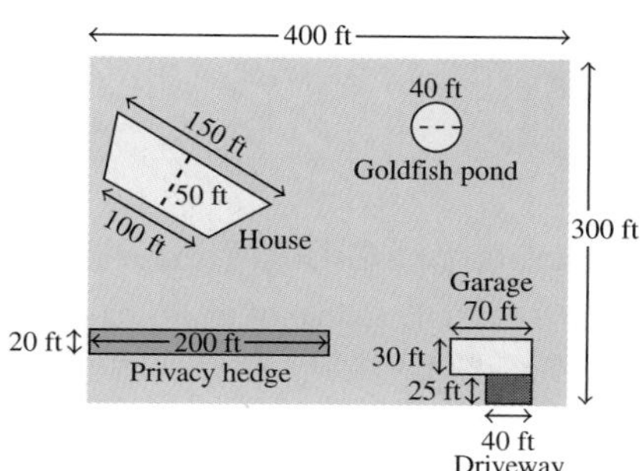

49. *Area of a Garden* Gaetano Cannata's rectangular garden is 11.5 m by 15.4 m.

a) How large is his garden in square meters?

b) If 1 hectare (a measurement of area in the metric system) equals 10,000 m^2, how large is his garden in hectares?

50. *Hamburger Comparison* Which hamburger has the larger surface area: a square hamburger 3 in. on a side from Wendy's or a $3\frac{1}{2}$-in.-diameter round hamburger from Burger King? Explain your answer and give the difference in their surface areas.

51. *Anchoring a Radio Signal Tower* A 100 ft radio signal tower is being constructed. To steady the tower, guy wires are attached to the tower. One end of the highest guy wire is attached to the tower at a point 90 ft above the ground (see figure). The other end is anchored into the ground at a point 52 ft from the base of the tower. How long is this guy wire?

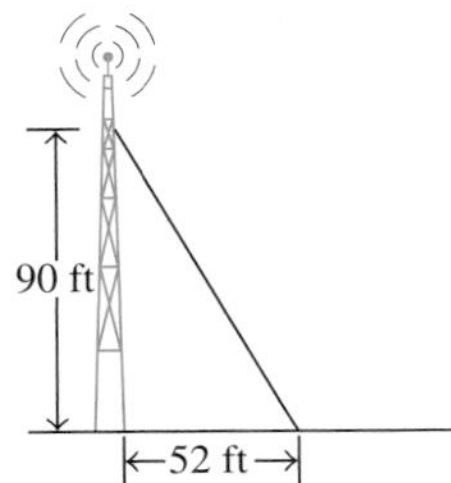

52. *Ladder on a Wall* Lorrie Morgan places a 29 ft ladder against the side of a building with the bottom of the ladder 20 ft away from the building (see figure). How high up on the wall does the ladder reach?

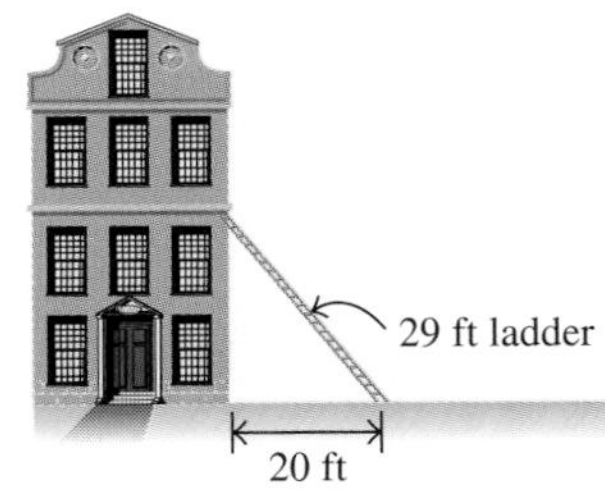

53. *Docking a Boat* Brian Murphy is bringing his boat into a dock that is 9 ft above the water level (see figure). If a 41 ft rope is attached to the dock on one side and to the boat on the other side, determine the horizontal distance from the dock to the boat.

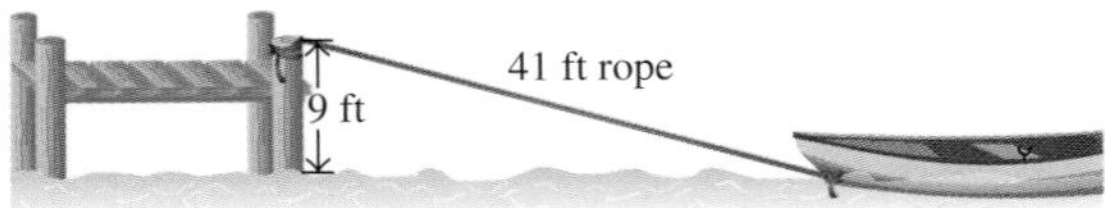

Challenge Problems/Group Activities

54. *Plasma Television* The screen of a plasma television is in the shape of a rectangle with a diagonal of length 43 in. If the height of the screen is 21 in., determine the width of the screen.

55. *Doubling the Sides of a Square* In the figure below, an original square with sides of length s is shown. Also shown is a larger square with sides double in length, or $2s$.

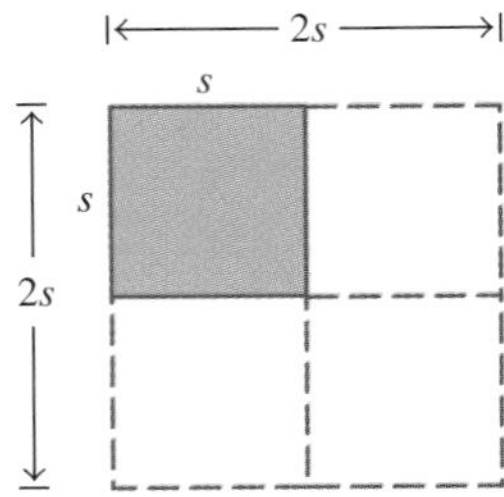

a) Express the area of the original square in terms of s.

b) Express the area of the larger square in terms of s.

c) How many times larger is the area of the square in part (b) than the area of the square in part (a)?

56. *Doubling the Sides of a Parallelogram* In the figure below, an original parallelogram with base b and height h is shown. Also shown is a larger parallelogram with base and height double in length, or $2b$ and $2h$, respectively.

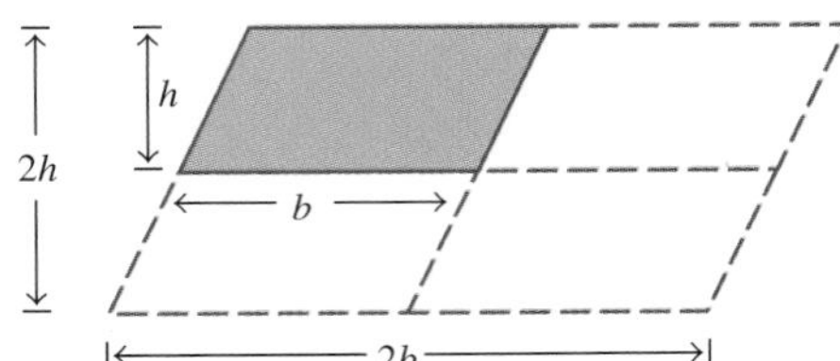

a) Express the area of the original parallelogram in terms of b and h.

b) Express the area of the larger parallelogram in terms of b and h.

c) How many times larger is the area of the parallelogram in part (b) than the area of the parallelogram in part (a)?

57. *Heron's Formula* A second formula for determining the area of a triangle (called Heron's formula) is

$$A = \sqrt{s(s-a)(s-b)(s-c)}$$

where $s = \frac{1}{2}(a + b + c)$ and a, b, and c are the lengths of the sides of the triangle. Use Heron's formula to determine the area of right triangle ABC and check your answer using the formula $A = \frac{1}{2}ab$.

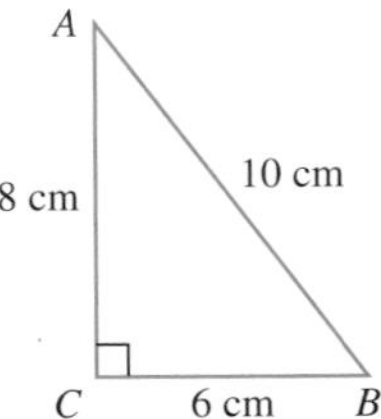

58. *Expansion of $(a + b)^2$* In the figure on page 507, one side of the largest square has length $a + b$. Therefore, the area of the largest square is $(a + b)^2$. Answer the following questions to find a formula for the expansion of $(a + b)^2$.

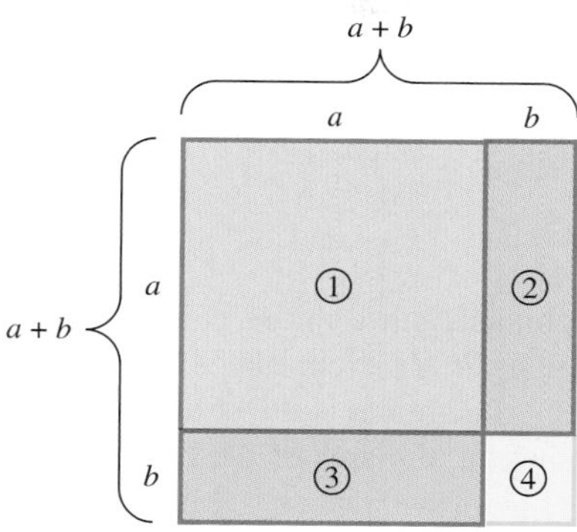

a) What is the area of the square marked ①?
b) What is the area of the rectangle marked ②?
c) What is the area of the rectangle marked ③?
d) What is the area of the square marked ④?
e) Add the four areas found in parts (a) through (d) to write a formula for the expansion of $(a + b)^2$.

Recreational Mathematics

59. *Scarecrow's Error* In the movie *The Wizard of Oz,* once the scarecrow gets his diploma he states the following: "In an isosceles triangle, the sum of the square roots of the two equal sides is equal to the square root of the third side." Discuss why this statement is incorrect.

Internet/Research Activities

For Exercises 60–62, references include the Internet, history of mathematics textbooks, and encyclopedias.

60. Research the proof of the Pythagorean theorem provided by President James Garfield. Write a brief paper and make a poster of this proof and the associated diagrams.
61. The early Babylonians and Egyptians did not know about π and had to devise techniques to approximate the area of a circle. Do research and write a paper on the techniques these societies used to approximate the area of a circle.
62. Write a paper on the contributions of Heron of Alexandria to geometry.

9.4 VOLUME

When discussing a one-dimensional figure, such as a line, we can find its length. When discussing a two-dimensional figure, such as a rectangle, we can find its area. When discussing a three-dimensional figure, such as a sphere, we can find its volume. *Volume* is a measure of the capacity of a figure. The measure of volume may be confusing because we use different units to measure different types of volumes. For example, water and other liquids may be measured in ounces, quarts, or gallons. A volume of topsoil may be measured in cubic yards. In the metric system, liquid may be measured in liters or milliliters, and topsoil may be measured in cubic meters.

Solid geometry is the study of three-dimensional solid figures (also called space figures). Volumes of three-dimensional figures are measured in cubic units such as cubic feet or cubic meters.

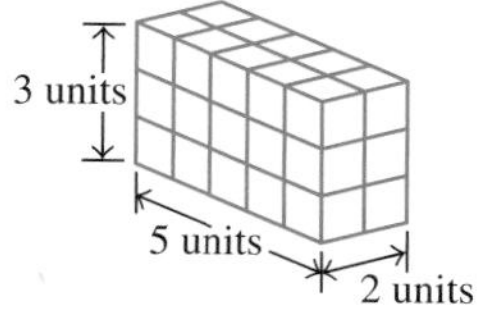

Figure 9.35

We will begin our discussion with the *rectangular solid.* If the length of the solid is 5 units, the width is 2 units, and the height is 3 units, the total number of cubes is 30 (Fig. 9.35). Thus, the volume is 30 cubic units. The volume of a rectangular solid can also be found by multiplying its length times width times height; in this case, 5 units $\times$ 2 units $\times$ 3 units $=$ 30 cubic units. In general, the volume of any rectangular solid (shown in part a in the box on page 508), is $V = l \times w \times h$.

Volume of a Rectangular Solid

$$V = l \times w \times h$$

A *cube* is a rectangular solid with the same length, width, and height (part b in the box below). If we call the side of a cube s and use the formula for a rectangular solid, substituting s for l, w, and h, we obtain $V = s \cdot s \cdot s = s^3$.

Volume of a Cube

$$V = s^3$$

Now consider the right circular cylinder (part c in the box below). The base is a circle with area πr^2. When we add height, h, the figure becomes a cylinder. For the same circular base, the greater the height, the greater is the volume. The volume of the right circular cylinder is found by multiplying the area of the base, πr^2, by the height h. In this book, when we use the term cylinder we mean a right circular cylinder.

Volume of a Cylinder

$$V = \pi r^2 h$$

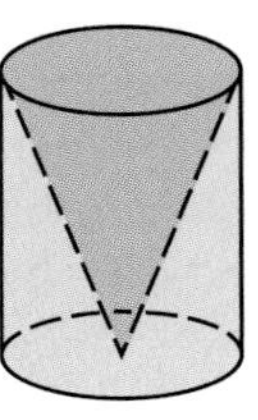

Figure 9.36

A cone is illustrated in part (d) in the box below. Imagine a cone inside a cylinder, sharing the same circle as the base. The volume of the cone is less than the volume of the cylinder that has the same base and the same height (Fig. 9.36). In fact, the volume of the cone is one-third the volume of the cylinder.

Volume of a Cone

$$V = \frac{1}{3}\pi r^2 h$$

The next shape we will discuss in this section is the sphere (part e in the box below). Basketballs, golf balls, and so on have the shape of a sphere. The formula for the volume of a sphere is as follows.

Volume of a Sphere

$$V = \frac{4}{3}\pi r^3$$

The following is a summary of the volumes of selected three-dimensional figures:

Volumes

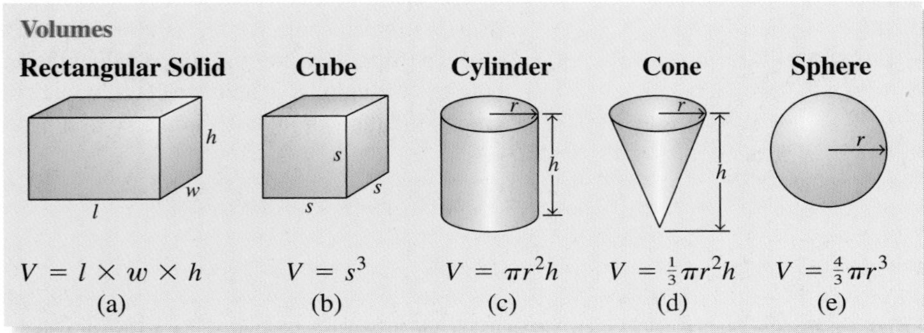

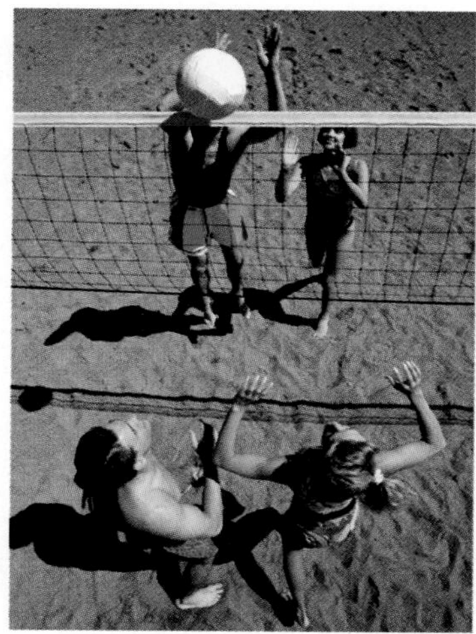

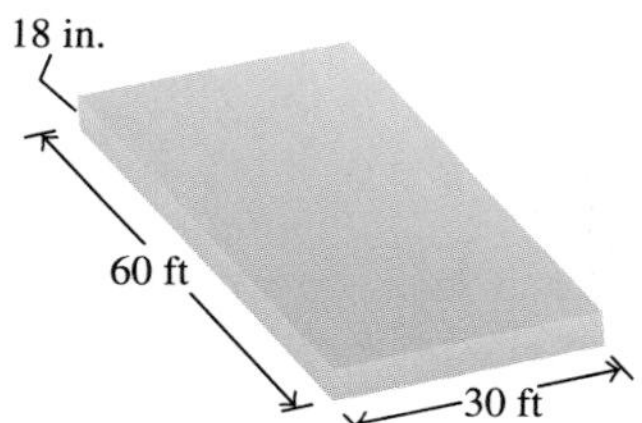

Figure 9.37

EXAMPLE 1 *Replacing a Sand Volleyball Court*

Linda Nelson is the manager at the Colony Apartments and needs to replace the sand in the rectangular sand volleyball court. The court is 30 ft wide by 60 ft long, and the sand has a uniform depth of 18 in., see Fig. 9.37. Volleyball court sand sells for \$15 per cubic yard.

a) How many cubic yards of sand does Linda need?

b) How much will the sand cost?

SOLUTION:

a) Since we are asked to find the volume in cubic yards, we will convert each measurement to yards. There are 3 ft in a yard. Thus, 30 ft equals $\frac{30}{3}$ or 10 yd, and 60 ft equals $\frac{60}{3}$ or 20 yd. There are 36 in. in a yard, so 18 in. equals $\frac{18}{36}$ or $\frac{1}{2}$ yd. The amount of sand needed is determined using the formula for the volume of a rectangular solid, $V = l \cdot w \cdot h$. In this case the height of the rectangular solid can be considered the depth of the sand.

$$V = l \cdot w \cdot h = 10 \cdot 20 \cdot \tfrac{1}{2} = 100 \text{ yd}^3$$

Note that since the measurements for length, width, and height are each in terms of yards, the answer is in terms of cubic yards.

b) One cubic yard of sand costs \$15, so 100 yd^3 will cost 100 × \$15, or \$1500.

EXAMPLE 2 *Silage Storage*

Gordon Langeneger has three silos on his farm. The silos are each in the shape of a right circular cylinder (see Fig. 9.38). One silo has a 12 ft diameter and is 40 ft tall. The second silo has a 14 ft diameter and is 50 ft tall. The third silo has an 18 ft diameter and is 60 ft tall.

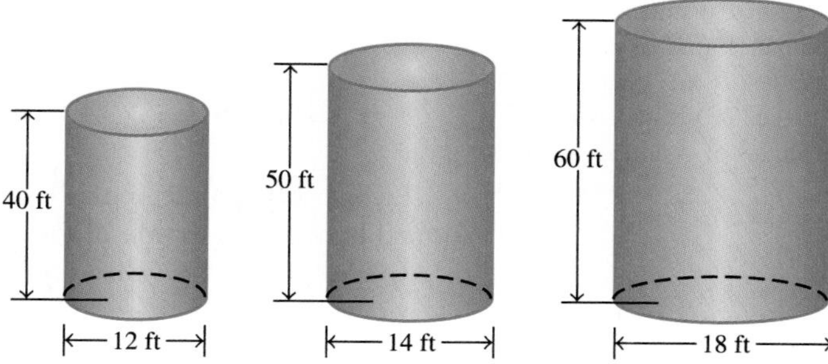

Figure 9.38

a) What is the total capacity of the three silos in cubic feet?

b) If Gordon fills all three of his silos and then feeds his cattle 150 ft^3 of silage per day, in how many days will all three silos be empty?

SOLUTION:

a) The capacity of each silo can be determined using the formula for the volume of a right circular cylinder, $V = \pi r^2 h$. Since the radius is half of the diameter, the

DID YOU KNOW

Fractured Volumes

Picasso, Pablo. *Les Demoiselles d'Avignon*. Paris (June–July 1907). Oil on canvas, 8 ft × 7 ft 8 in. (243.9 cm × 233.7 cm). The Museum of Modern Art, New York. Acquired through the Lillie P. Bliss Bequest. Photograph © 1996 The Museum of Modern Art, New York. © Estate of Pablo Picasso.

The cubist painters of the early twentieth century sought to analyze forms as geometric shapes. They were greatly influenced by Paul Cézanne who, in a famous letter, wrote of treating nature in terms of the cylinder, sphere, and cone. In one of the first paintings of the cubist period, *Les Demoiselles d'Avignon*, Pablo Picasso dismissed the idea of the human figure as a dynamic unity and instead fractured volumes and planes to present them from multiple angles of vision at one time.

radii for the three silos are 6 ft, 7 ft, and 9 ft, respectively. Now let's determine the volumes.

$$\begin{aligned}
\text{Volume of the first silo} &= \pi r^2 h = \pi \cdot 6^2 \cdot 40 \\
&\approx 3.14 \cdot 36 \cdot 40 \approx 4521.6 \text{ ft}^3 \\
\text{Volume of the second silo} &= \pi r^2 h = \pi \cdot 7^2 \cdot 50 \\
&\approx 3.14 \cdot 49 \cdot 50 \approx 7693.0 \text{ ft}^3 \\
\text{Volume of the third silo} &= \pi r^2 h = \pi \cdot 9^2 \cdot 60 \\
&\approx 3.14 \cdot 81 \cdot 60 \approx 15{,}260.4 \text{ ft}^3
\end{aligned}$$

Therefore, the total capacity of all three silos is about $4521.6 + 7693.0 + 15{,}260.4 \approx 27{,}475.0 \text{ ft}^3$.

b) To find how long it takes to empty all three silos, we will divide the total capacity by 150 ft^3, the amount fed to Gordon's cattle every day.

$$\frac{27{,}475}{150} \approx 183.17.$$

Thus, the silos will be empty in about 183 days. ▲

Now let's discuss polyhedrons. A *polyhedron* is a closed surface formed by the union of polygonal regions. Figure 9.39 illustrates some polyhedrons.

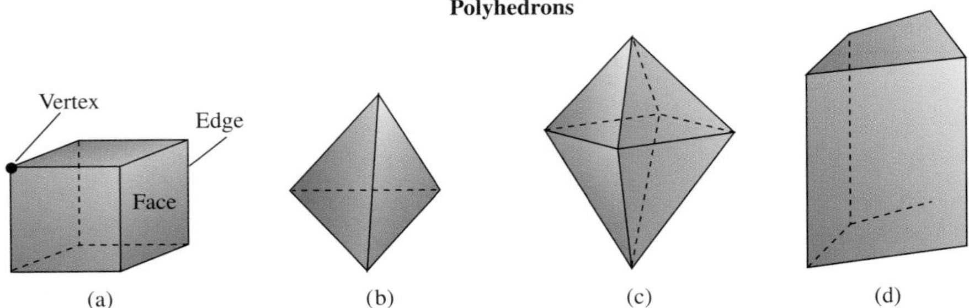

Figure 9.39

Each polygonal region is called a *face* of the polyhedron. The line segment formed by the intersection of two faces is called an *edge*. The point at which two or more edges intersect is called a *vertex*. In Fig. 9.39(a) there are 6 faces, 12 edges, and 8 vertices. Note that

$$\begin{aligned}
\text{Number of vertices} - \text{number of edges} + \text{number of faces} &= 2 \\
8 - 12 + 6 &= 2
\end{aligned}$$

This formula, credited to Leonhard Euler, is true for any polyhedron.

Euler's Polyhedron Formula

$$\text{Number of vertices} - \text{number of edges} + \text{number of faces} = 2$$

We suggest that you verify this formula holds for Fig. 9.39(b), (c), and (d).

EXAMPLE 3 *Using Euler's Polyhedron Formula*

A certain polyhedron has 6 vertices and 9 edges. Determine the number of faces on this polyhedron.

SOLUTION: Since we are seeking the number of faces, we will let x represent the number of faces on the polyhedron. Next, we will use Euler's polyhedron formula to set up an equation:

$$\begin{aligned} \text{Number of vertices} - \text{number of edges} + \text{number of faces} &= 2 \\ 6 - 9 + x &= 2 \\ -3 + x &= 2 \\ x &= 5 \end{aligned}$$

Therefore, the polyhedron has 5 faces.

A *regular polyhedron* is one whose faces are all regular polygons of the same size and shape. Figure 9.39(a) and (b) on page 510 are regular polyhedrons.

A *prism* is a special type of polyhedron whose bases are congruent polygons and whose sides are parallelograms. These parallelogram regions are called the *lateral faces* of the prism. If all the lateral faces are rectangles, the prism is said to be a *right prism*. Some right prisms are illustrated in Fig. 9.40. In this book, whenever we use the word *prism* we are referring to a right prism.

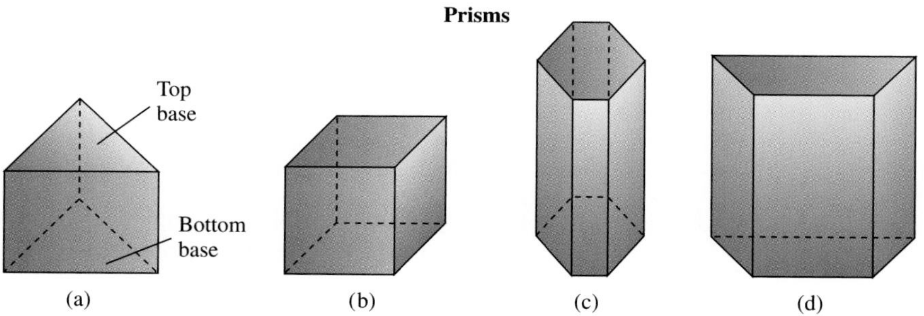

Figure 9.40

The volume of any prism can be found by multiplying the area of the base, B, by the height, h, of the prism.

Volume of a Prism

$$V = Bh$$

where B is the area of a base and h is the height.

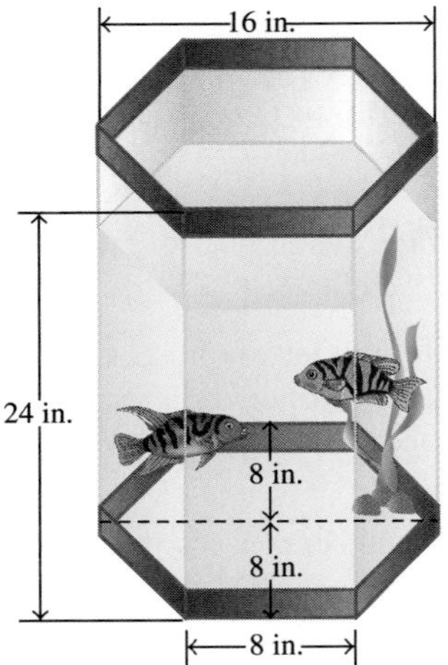

Figure 9.41

EXAMPLE 4 *Volume of a Hexagonal Prism Fish Tank*

Frank Nicolzaao's fish tank is in the shape of a hexagonal prism as shown in Fig. 9.41. Use the dimensions shown in the figure and the fact that 1 gal $=$ 231 in.^{3} to

a) determine the volume of the fish tank in cubic inches.

b) determine the volume of the fish tank in gallons (round your answer to the nearest gallon).

SOLUTION:

a) First we will need to calculate the area of the hexagonal base of the fish tank. Notice from Fig. 9.41 that by drawing a diagonal as indicated, the base can be divided into two identical trapezoids. To find the area of the hexagonal base, we will calculate the area of one of these trapezoids and then multiply by 2.

$$\text{Area of one trapezoid} = \tfrac{1}{2}h(b_1 + b_2)$$
$$= \tfrac{1}{2}(8)(16 + 8) = 96 \text{ in.}^2$$
$$\text{Area of the hexagonal base} = 2(96) = 192 \text{ in.}^2$$

Now to determine the volume of the fish tank, we will use the formula for the volume of a prism, $V = Bh$. We already determined that the area of the base, B, is 192 in.^{2}

$$V = B \cdot h = 192 \cdot 24 = 4608 \text{ in.}^3$$

In the above calculation, the area of the base, B, was measured in square inches, and the height was measured in inches. The product of in.^{2} and in. is cubic inches, or in.^{3}.

b) To determine the volume of the fish tank in gallons, we will divide the volume of the fish tank in cubic inches by 231.

$$V = \frac{4608}{231} \approx 19.95 \text{ gal}$$

Thus, the volume of the fish tank is approximately 20 gal. ▲

EXAMPLE 5 *Volumes Involving Prisms*

Find the volume of the remaining solid after the cylinder, triangular prism, and square prism have been cut from the solid (Fig. 9.42).

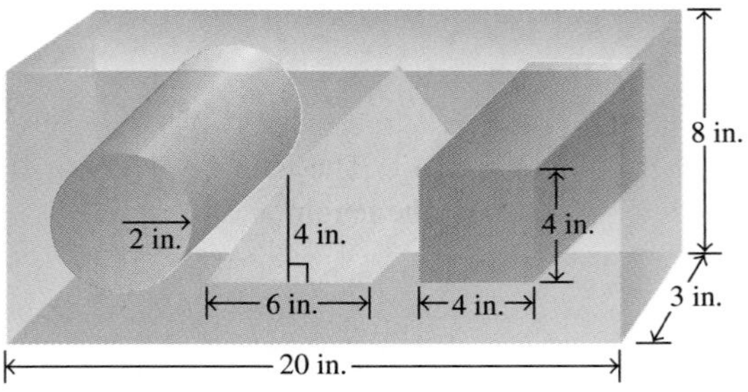

Figure 9.42

SOLUTION: To find the volume of the remaining solid, first find the volume of the rectangular solid. Then subtract the volume of the two prisms and the cylinder that were cut out.

$$\begin{aligned}
\text{Volume of rectangular solid} &= l \cdot w \cdot h \\
&= 20 \cdot 3 \cdot 8 = 480 \text{ in.}^3 \\
\text{Volume of circular cylinder} &= \pi r^2 h \\
&\approx (3.14)(2^2)(3) \\
&\approx (3.14)(4)(3) \approx 37.68 \text{ in.}^3 \\
\text{Volume of triangular prism} &= \text{area of the base} \cdot \text{height} \\
&= \tfrac{1}{2}(6)(4)(3) = 36 \text{ in.}^3 \\
\text{Volume of square prism} &= s^2 \cdot h \\
&= 4^2 \cdot 3 = 48 \text{ in.}^3 \\
\text{Volume of solid} &\approx 480 - 37.68 - 36 - 48 \\
&\approx 358.32 \text{ in.}^3
\end{aligned}$$

Another special category of polyhedrons is the *pyramid*. Unlike prisms, pyramids have only one base. Some pyramids are illustrated in Fig. 9.43. Note that all but one face of a pyramid intersect at a common vertex.

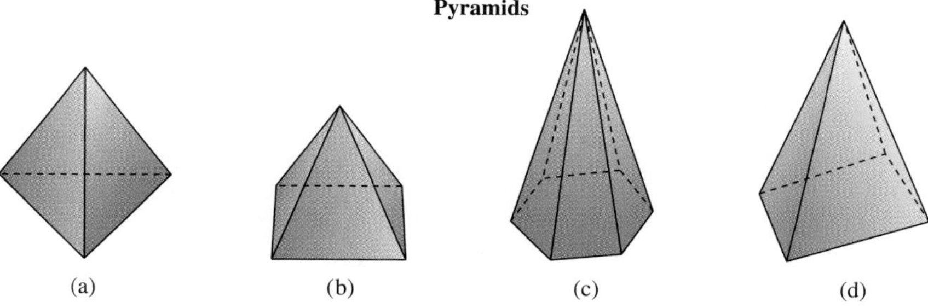

Figure 9.43

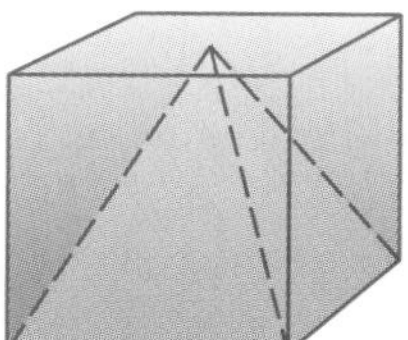

Figure 9.44

If a pyramid is drawn inside a prism, as shown in Fig. 9.44, the volume of the pyramid is less than that of the prism. In fact, the volume of the pyramid is one-third the volume of the prism.

Volume of a Pyramid

$$V = \tfrac{1}{3}Bh$$

where B is the area of the base and h is the height.

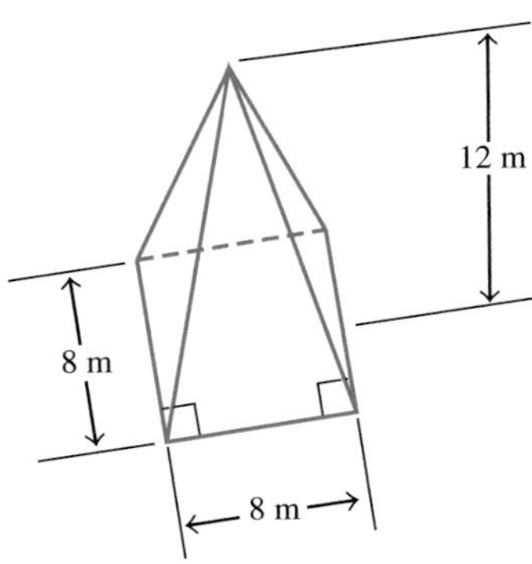

Figure 9.45

EXAMPLE 6 *Volume of a Pyramid*

Find the volume of the pyramid shown in Fig. 9.45.

SOLUTION: First find the area of the base of the pyramid. Since the base of the pyramid is a square,

$$\text{Area of base} = s^2 = 8^2 = 64 \text{ m}^2$$

Now use this information to find the volume of the pyramid.

$$\begin{aligned} V &= \tfrac{1}{3} \cdot B \cdot h \\ &= \tfrac{1}{3} \cdot 64 \cdot 12 \\ &= 256 \text{ m}^3 \end{aligned}$$

▲

In certain situations converting volume from one cubic unit to a different cubic unit might be necessary. For example, when purchasing topsoil you might have to change the amount of topsoil from cubic feet to cubic yards prior to placing your order. Example 7 shows how this may be done.

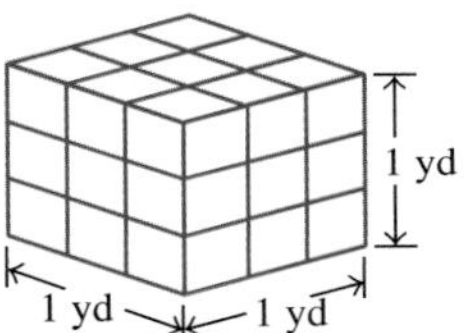

Figure 9.46

EXAMPLE 7 *Cubic Yards and Cubic Feet*

a) Convert 1 yd^3 to cubic feet. (See Fig. 9.46.)
b) Convert 8.3 yd^3 to cubic feet.

SOLUTION:
a) We know that 1 yd = 3ft. Thus,
$1 \text{ yd}^3 = 3 \text{ ft} \cdot 3 \text{ ft} \cdot 3 \text{ ft} = 27 \text{ ft}^3$.
b) In part a) we learned that $1 \text{ yd}^3 = 27\text{ft}^3$. Thus,
$8.3 \text{ yd}^3 = 8.3 \times 27 = 224.1 \text{ ft}^3$.

▲

EXAMPLE 8 *Filling in a Swimming Pool*

Julianne Peterson recently purchased a home with a rectangular swimming pool. The pool is 30 ft long, 15 ft wide, and has a uniform depth of 4.5 ft. Julianne lives in a cold climate and so she plans to fill the pool in with dirt to make a flower garden. How many cubic yards of dirt will Julianne have to purchase to fill in the swimming pool?

SOLUTION: To find the amount of dirt, we will use the formula for the volume of a rectangular solid:

$$\begin{aligned} V &= lwh \\ &= (30)(15)(4.5) \\ &= 2025 \text{ ft}^3 \end{aligned}$$

Now, we must convert this volume from cubic feet to cubic yards. In Example 7, we learned that $1 \text{ yd}^3 = 27 \text{ ft}^3$. Therefore, $2025 \text{ ft}^3 = \frac{2025}{27} = 75 \text{ yd}^3$. Thus, Julianne needs to purchase 75 yd^3 of dirt to fill in her swimming pool.

▲

SECTION 9.4 EXERCISES

Concept/Writing Exercises

1. In your own words, define *volume.*
2. What is solid geometry?
3. What is the difference between a polyhedron and a regular polyhedron?
4. What is the difference between a prism and a right prism?
5. In your own words, explain the difference between a prism and a pyramid.
6. In your own words, state Euler's polyhedron formula.

Practice the Skills

In Exercises 7–20, find the volume of the solid. When necessary, round your answer to hundredths.

7.

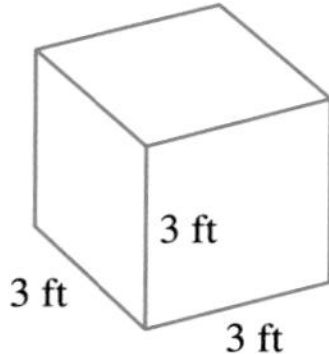

8.

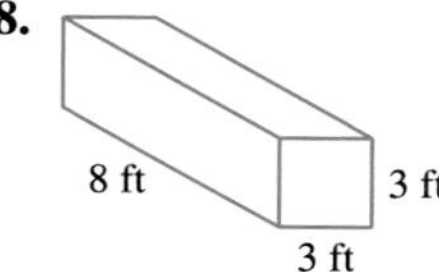

9.

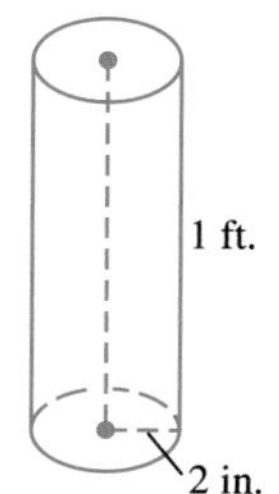

10.

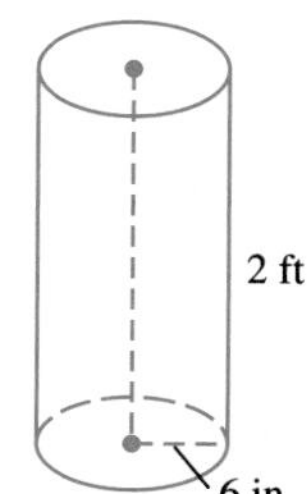

11.

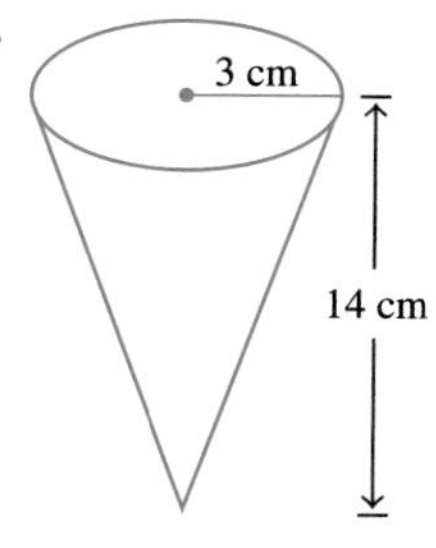

12.

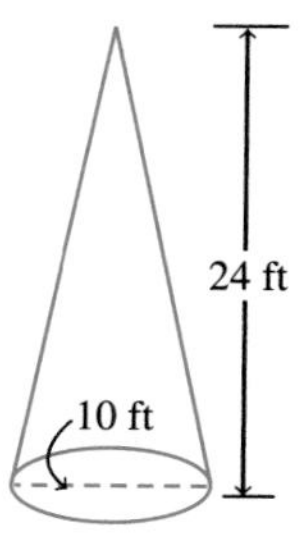

13.

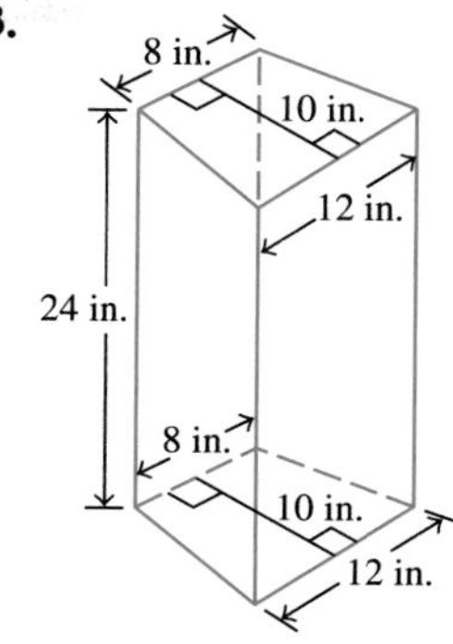

14.

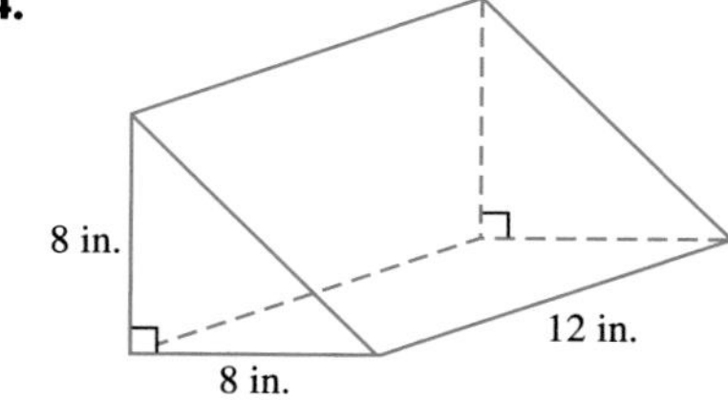

15.

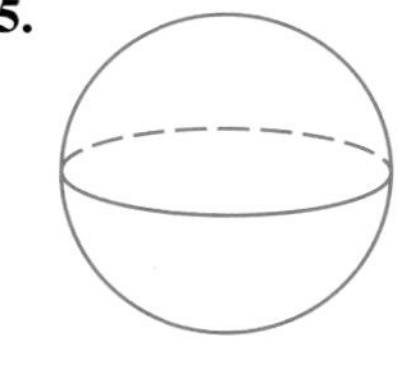

16.

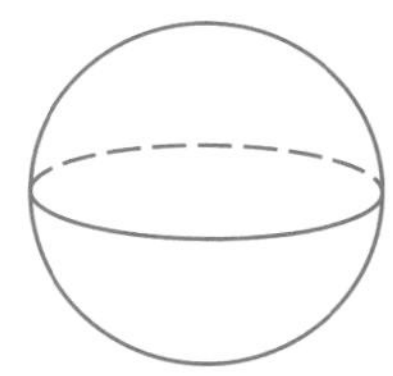

17.

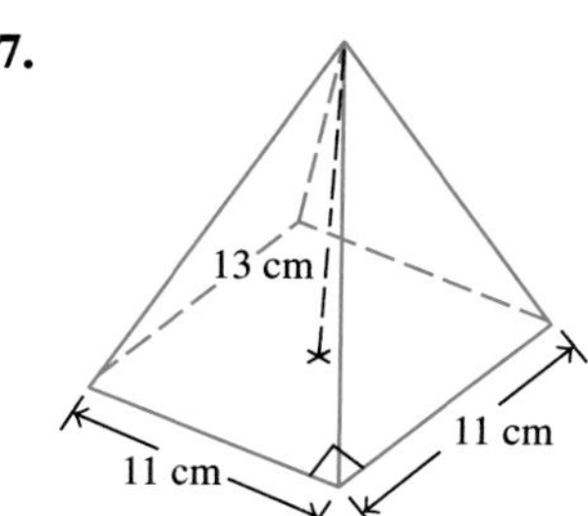

18.

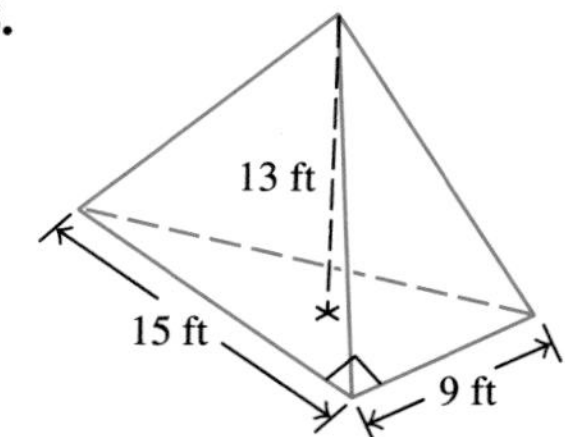

19.

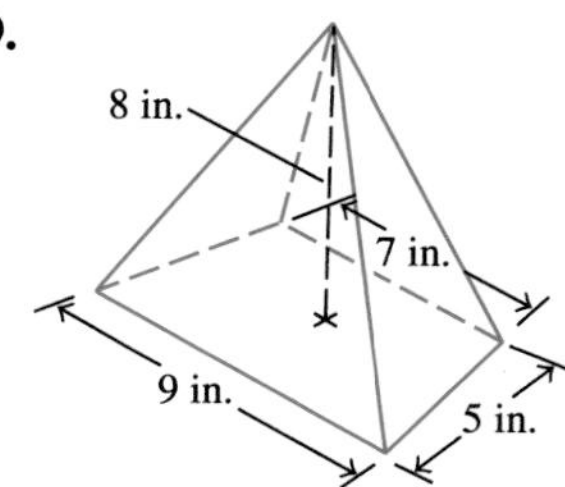

20.

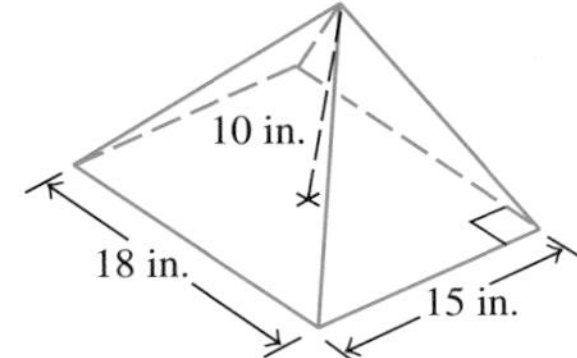

Problem Solving

In Exercises 21–28, find the volume of the shaded area. Round your answers to hundredths.

21.

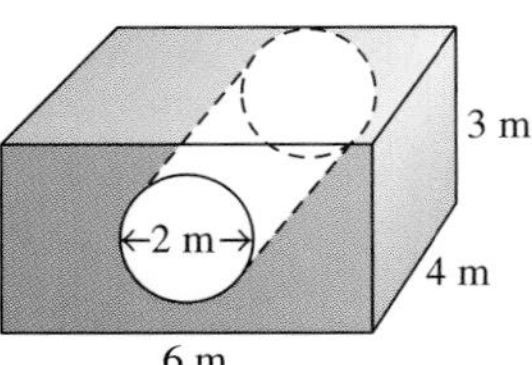

22.

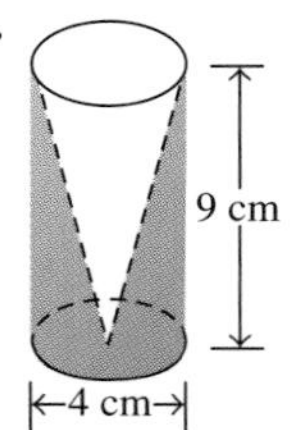

23.

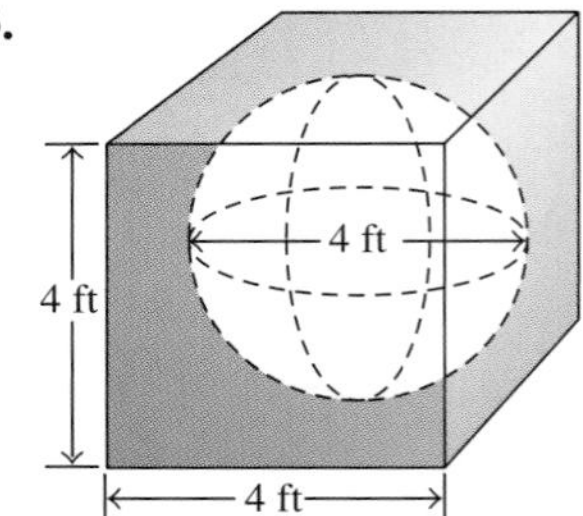

24.

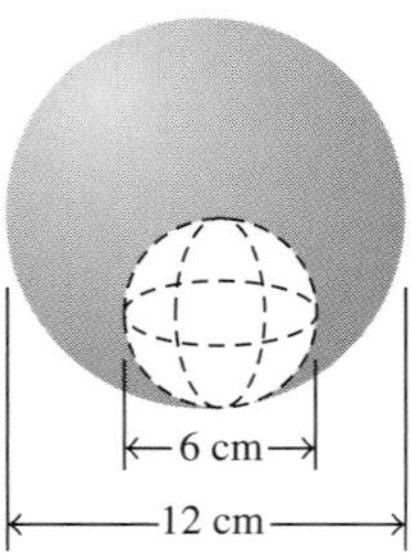

25.

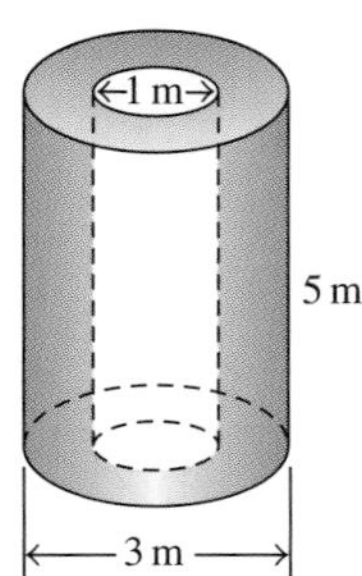

26.

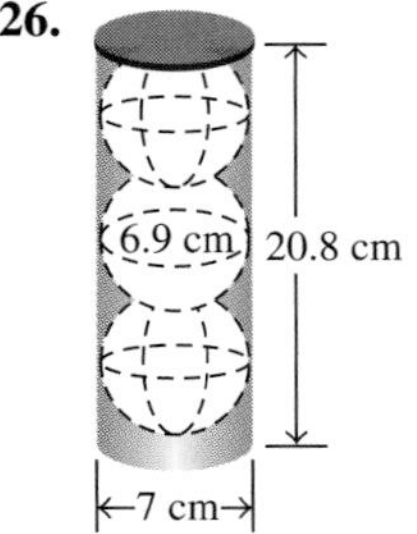

27.

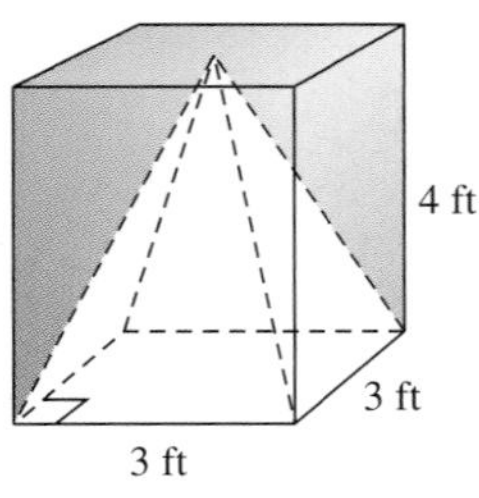

28.

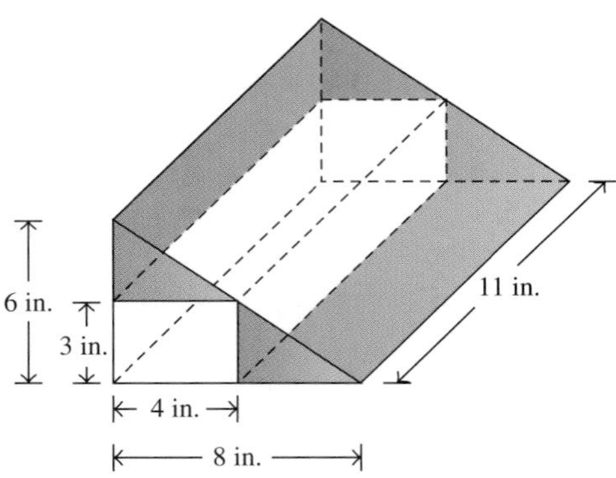

In Exercises 29–32, use the fact that 1 yd^3 equals 27 ft^3 to make the conversion.

29. 7 yd^3 to cubic feet

30. 3.8 yd^3 to cubic feet

31. 153 ft^3 to cubic yards

32. 2457 ft^3 to cubic yards

In Exercises 33–36, use the fact that 1 m^3 equals 1,000,000 cm^3 to make the conversion.

33. 5.9 m^3 to cubic centimeters

34. 17.6 m^3 to cubic centimeters

35. 3,000,000 cm^3 to cubic meters

36. 7,300,000 cm^3 to cubic meters

37. *Volume of a Freezer* The dimensions of the interior of an upright freezer are height 46 in., width 25 in., and depth 25 in. Determine the volume of the freezer
a) in cubic inches. **b)** in cubic feet.

38. *Ice Cream Comparison* The Louisburg Creamery packages its homemade ice cream in tubs and in boxes. The tubs are in the shape of a right cylinder with a radius of 3 in. and height of 5 in. The boxes are in the shape of a cube with each side measuring 5 in. Determine the volume of each container.

39. *Volume of a Bread Pan* A bread pan is 12 in. $\times$ 4 in. $\times$ 3 in. How many quarts does it hold, if 1 in.^{3} $\approx$ 0.01736 qt?

40. *Hamburger Comparison* The dimensions of a square Wendy's hamburger are length and width 4 in. and thickness $\frac{3}{16}$ in. The dimensions of a Magic Burger circular hamburger are diameter $4\frac{1}{2}$ in. and thickness $\frac{1}{4}$ in. Which hamburger has the greater volume? What is the difference in their volumes?

41. *Gasoline Containers* Mark Russo has two right cylindrical containers for storing gasoline. One has a diameter of 10 in. and a height of 12 in. The other has a diameter of 12 in. and a height of 10 in.

a) Which container holds the greater amount of gasoline, the taller one or the one with the greater diameter?

b) What is the difference in volume?

42. *A Swimming Pool*

a) What is the volume of water in a rectangular swimming pool that is 15 m long and 9 m wide and has an average depth of 2 m? Give your answer in cubic meters.

b) If $1\ m^3 = 1\ k\ell$, how many kiloliters of water will the pool hold?

43. *The Pyramid of Cheops* The Pyramid of Cheops in Egypt has a square base measuring 720 ft on a side. Its height is 480 ft. What is its volume?

44. *A Fish Tank*

a) How many cubic centimeters of water will a rectangular fish tank hold if the tank is 80 cm long, 50 cm wide, and 30 cm high?

b) If $1\ cm^3$ holds $1\ m\ell$ of liquid, how many milliliters will the tank hold?

c) If $1\ell = 1000\ m\ell$, how many liters will the tank hold?

45. *Engine Capacity* The engine in a 1957 Chevrolet Corvette has eight cylinders. Each cylinder is a right cylinder with a bore (diameter) of 3.875 in. and a stroke (height) of 3 in. Determine the total displacement (volume) of this engine.

46. *Rose Garden Topsoil* Marisa Raffaele wishes to plant a rose garden in her backyard. The rose garden will be in the shape of a 9 ft by 18 ft rectangle. Marisa wishes to add a 4 in. layer of organic topsoil on top of the rectangular area. The topsoil sells for \$32.95 per cubic yard. Determine

a) how many cubic yards of topsoil Marisa will need.

b) how much the topsoil will cost.

47. *Pool Toys* A Wacky Noodle Pool Toy, frequently referred to as a "noodle," is a cylindrical flotation device made from cell foam (see photo). One style of noodle is a cylinder that has a diameter of 2.5 in. and a length of 5.5 ft. Determine the volume of this style of noodle in

a) cubic inches.

b) cubic feet.

48. *Keeping Soda Cold* Tobi Moore and Tacinto Lopez and their friends are at a picnic at the town park. They have brought a children's wading pool in the shape of a right circular cylinder with a radius of 2 ft and a height of 1 ft into which they will put cold water to keep the soda cold. See figure (a). They carry the water from the faucet to the pool in a bucket that is also in the shape of a right circular cylinder, with a diameter of 1 ft and a height of 1 ft. See figure (b).

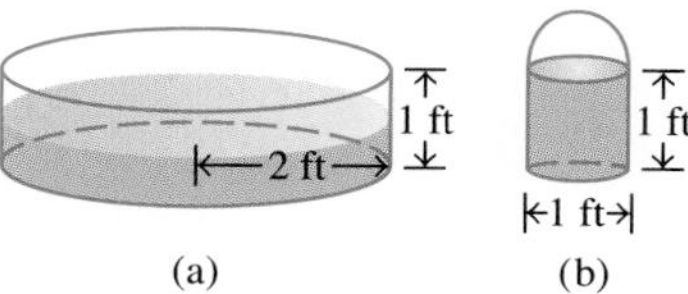

a) How many buckets of water are needed to fill the pool to a height of $\frac{1}{2}$ ft?

b) If $1\ ft^3$ of water weighs 62.5 lb, what is the weight of the water in the pool?

c) If there are 7.5 gal of water per cubic foot, how many gallons of water are in the pool?

49. *Comparing Cake Pans* When baking a cake you can choose between a round pan with a 9 in. diameter and a 7 in. × 9 in. rectangular pan.

a) Determine the area of the base of each pan.

b) If both pans are 2 in. deep, determine the volume of each pan.

c) Which pan has the larger volume?

50. *Cake Icing* A bag used to apply icing to a cake is in the shape of a cone with a diameter of 3 in. and a height of 6 in. How much icing will this bag hold when full?

51. *Flower Box* A flower box is 4 ft long, and its ends are in the shape of a trapezoid. The upper and lower bases of the trapezoid measure 12 in. and 8 in., respectively, and the height is 9 in. Find the volume of the flower box

a) in cubic inches.

b) in cubic feet.

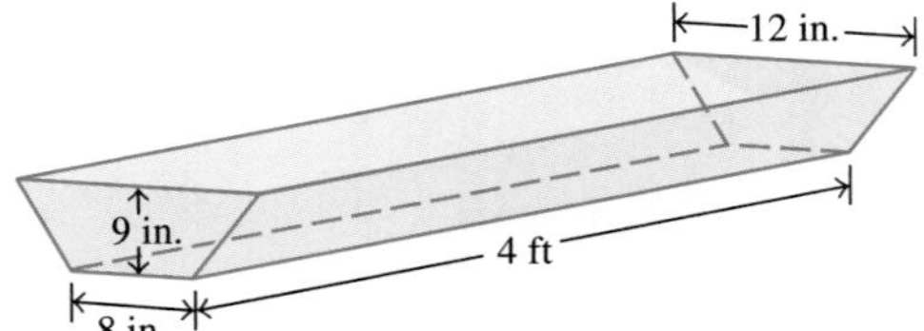

52. *The Leaning Tower of Pisa* The Leaning Tower of Pisa was designed to be a vertical bell tower for a cathedral. If the tower were vertical, it would be 60 meters high with a diameter of about 19.6 meters roughly in the shape of a cylinder. Use this information to find

a) the circumference of the tower.

b) the volume of tower.

In Exercises 53–58, find the missing value indicated by the question mark. Use the following formula.

$$\begin{pmatrix}\text{Number of}\\ \text{vertices}\end{pmatrix} - \begin{pmatrix}\text{number of}\\ \text{edges}\end{pmatrix} + \begin{pmatrix}\text{number}\\ \text{of faces}\end{pmatrix} = 2$$

	Number of Vertices	Number of Edges	Number of Faces
53.	8	?	3
54.	12	16	?
55.	?	8	4
56.	7	12	?
57.	11	?	5
58.	?	10	4

Challenge Problems/Group Activities

59. *Packing Orange Juice* A box is packed with six cans of orange juice. The cans are touching each other and the sides of the box, as shown. What percent of the volume of the interior of the box is not occupied by the cans?

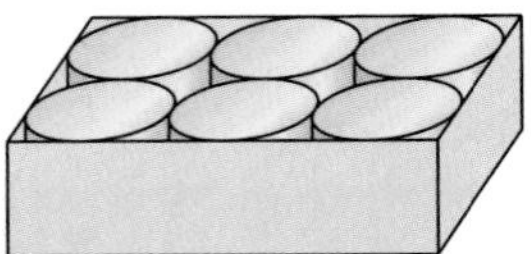

60. *Doubling the Edges of a Cube* In this exercise we will explore what happens to the volume of a cube if we double the length of each edge of the cube.

a) Choose a number between 1 and 10 and call this number s.

b) Calculate the volume of a cube with the length of each edge equal to s.

c) Now double s and call this number t.

d) Calculate the volume of a cube with the length of each edge equal to t.

e) Repeat parts (a) through (d) for a different value of s.

f) Compare the results from part (b) to the results from part (d) and explain what happens to the volume of a cube if we double the length of each edge.

61. *Doubling the Radius of a Sphere* In this exercise we will explore what happens to the volume of a sphere if we double the radius of the sphere.

a) Choose a number between 1 and 10 and call this number r.

b) Calculate the volume of a sphere with radius r (use the $\boxed{\pi}$ key on your calculator).

c) Now double r and call this number t.

d) Calculate the volume of a sphere with radius t.

e) Repeat parts (a) through (d) for a different value of r.

f) Compare the results from part (b) to the results from part (d) and explain what happens to the volume of a sphere if we double the radius.

62. *Cost of a Dripping Faucet* Leah Quintero has a faucet in her home that drips at a rate of 42 drops per minute. There are approximately 20 drops in 1 mℓ, 1000 mℓ in 1 ℓ, and approximately 3.79 ℓ in 1 gal. Assume water costs about $0.11 per gallon.

a) Determine the number of drops of water wasted over a 1-year period.

b) Determine the volume of water wasted over a 1-year period in milliliters, liters, and gallons.

c) Estimate the cost of water wasted over a 1-year period.

63. **a)** Explain how to demonstrate, using the cube shown on page 519, that

$$(a + b)^3 = a^3 + 3a^2b + 3ab^2 + b^3$$

b) What is the volume in terms of a and b of each numbered piece in the figure?

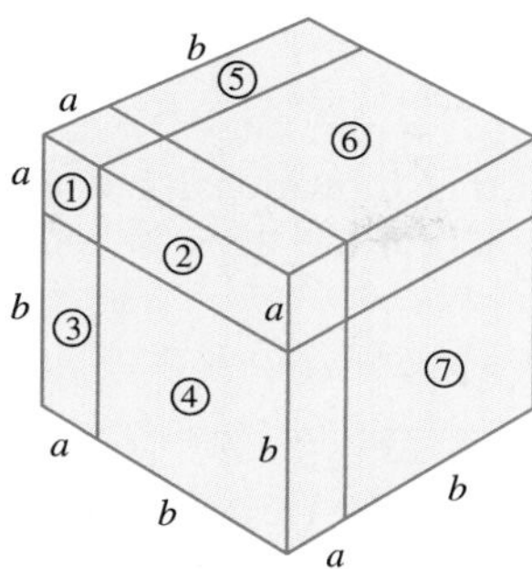

c) An eighth piece is not illustrated. What is its volume?

Recreational Mathematics

64. *More Pool Toys* Wacky Noodle Pool Toys (see Exercise 47) come in many different shapes and sizes.

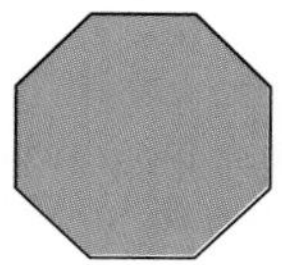
Base for part (a)

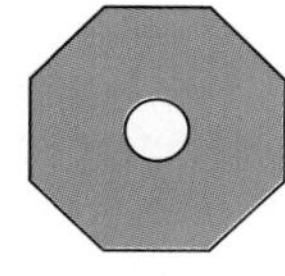
Base for part (b)

a) Determine the volume, in cubic inches, of a noodle that is in the shape of a 5.5-ft-long solid octagonal prism whose base has an area of 5 in.^{2}

b) Determine the volume, in cubic inches, of a hollow noodle that has the same shape as the noodle described in part (a) except that a right circular cylinder of diameter 0.75 in. has been removed from the center.

Internet/Research Activities

65. *Air Conditioner Selection* Calculate the volume of the room in which you sleep or study. Go to a store that sells room air conditioners and find out how many cubic feet can be cooled by the different models available. Describe the model that would be the proper size for your room. What is the initial cost? How much does that model cost to operate? If you moved to a room that had twice the amount of floor space and the same height, would the air conditioner you selected still be adequate? Explain.

66. Pappus of Alexandria (ca. A.D. 350) was the last of the well-known ancient Greek mathematicians. Write a paper on his life and his contributions to mathematics.

9.5 TRANSFORMATIONAL GEOMETRY, SYMMETRY, AND TESSELLATIONS

In our study of geometry, we have thus far focused on definitions, axioms, and theorems that are used in the study of *Euclidean geometry*. We will now introduce a second type of geometry called *transformational geometry*. In *transformational geometry*, we study various ways to move a geometric figure without altering the shape or size of the figure. When discussing transformational geometry we often use the term *rigid motion*.

The act of moving a geometric figure from some starting position to some ending position without altering its shape or size is called a **rigid motion** (or **transformation**).

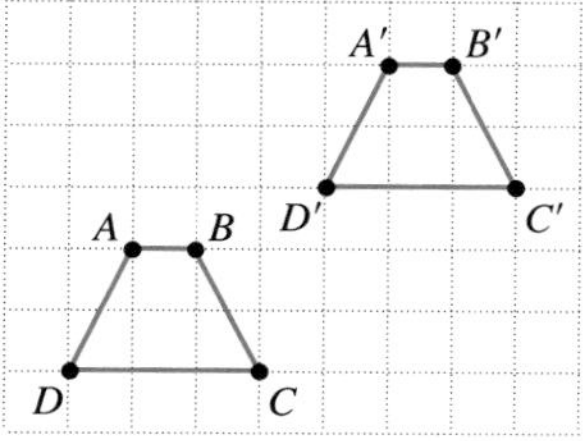

Figure 9.47

Consider trapezoid $ABCD$ in Figure 9.47. If we move each point on this trapezoid 4 units to the right and 3 units up, the trapezoid is in the location specified by trapezoid $A'B'C'D'$. This figure illustrates one type of rigid motion. When studying rigid motions we are only concerned about the starting and ending positions of the figure and not what happens in between. When discussing rigid motions of two-dimensional figures, there are four types of rigid motions: reflections, rotations, translations, and glide reflections. We call these four types of rigid motions the *basic rigid motions in a plane*. After we discuss the four rigid motions we will discuss symmetry of geometric figures and tessellations.

Reflections

The first rigid motion we will study is *reflection*. In our everyday life we are quite familiar with the concept of reflection. In transformational geometry a reflection is an image of a geometric figure that appears on the opposite side of a designated line.

A **reflection** is a rigid motion that moves a geometric figure to a new position such that the figure in the new position is a mirror image of the figure in the starting position. In two dimensions the figure and its mirror image are equidistant from a line called the **reflection line** or the **axis of reflection.**

Figure 9.48 below shows trapezoid $ABCD$, a reflection line l, and the reflected trapezoid $A'B'C'D'$. Notice that vertex A is 6 units to the *left* of reflection line l and that vertex A' is 6 units to the *right* of reflection line l. Next notice that vertex B is 2 units to the *left* of l and that vertex B' is 2 units to the right of l. A similar relationship holds true for vertices C and C' and for vertices D and D'. It is important to see that the trapezoid is not simply *moved* to the other side of the reflection line, but instead it is *reflected*. Notice in the trapezoid $ABCD$ that the longer base BC is on the right side of the trapezoid, but in the reflected trapezoid $A'B'C'D'$ the longer base $B'C'$ is on the left side of the trapezoid. Finally, notice the colors of the sides of the two trapezoids. Side AB in trapezoid $ABCD$ and side $A'B'$ in the reflected trapezoid are both blue. Sides BC and sides $B'C'$ are both red, sides CD and $C'D'$ are both gold, and sides DA and $D'A'$ are both green. In this section we will occasionally use such color coding to help you visualize the effect of a rigid transformation on a figure.

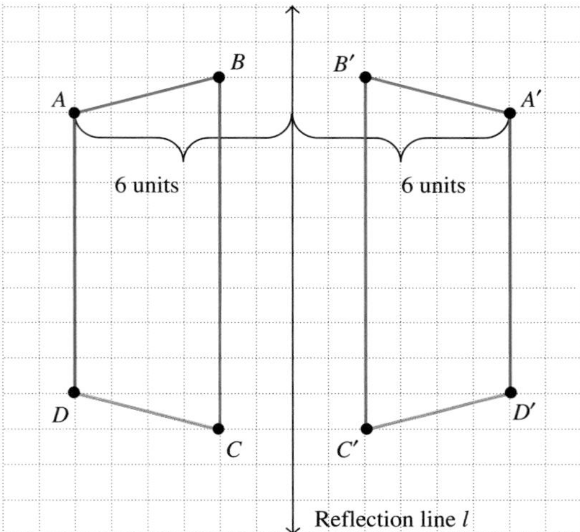

Figure 9.48

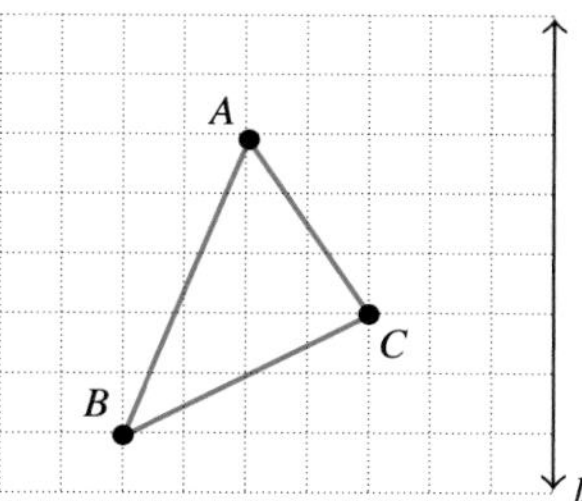

Figure 9.49

EXAMPLE 1 *Reflection of a Triangle*

Construct the reflection of triangle ABC, shown in Fig. 9.49, about reflection line l.

SOLUTION: The reflection of triangle ABC will be called $A'B'C'$. To determine the position of the reflection, we first examine vertex A in Fig. 9.49. Notice that vertex A is 5 units to the *left* of reflection line l. Thus, in the reflected triangle $A'B'C'$, vertex A' must also be 5 units away from, but to the *right* of, reflection line l. Next, notice that vertex B is 7 units to the left of l and that vertex C is 3 units to the left of l. Thus, in the reflection, vertex B' must be 7 units to the right of l and vertex C' must be 3 units to the right of l. Fig. 9.50 shows vertices A', B', and C'. Finally, we draw line segments between vertices A', B', and C' to form the sides of the reflection, triangle $A'B'C'$, as illustrated in Fig. 9.50.

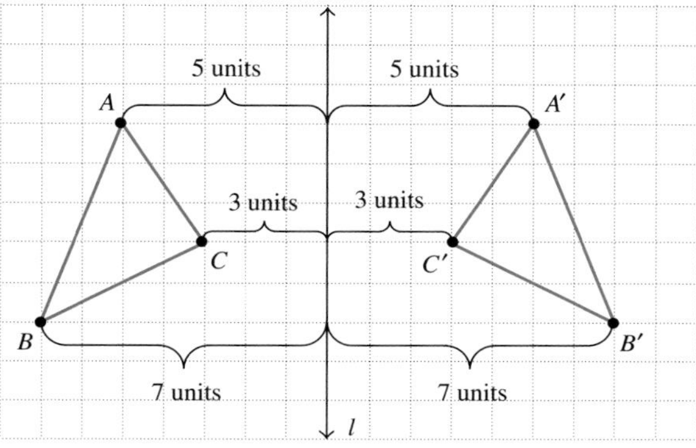

Figure 9.50

In Example 1, the reflection line did not intersect the figure being reflected. We now will study an example where the reflection line goes directly through the figure to be reflected.

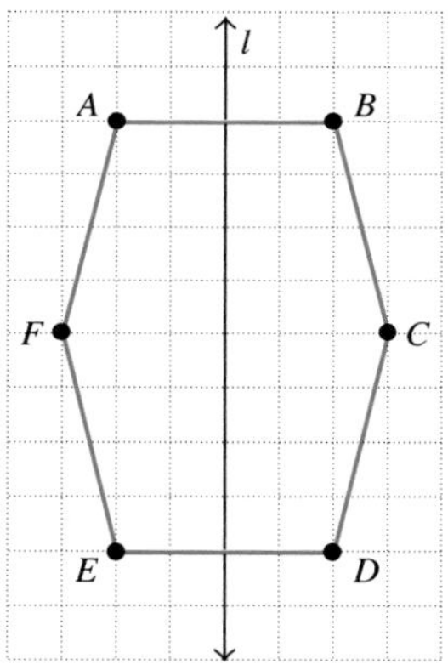

Figure 9.51

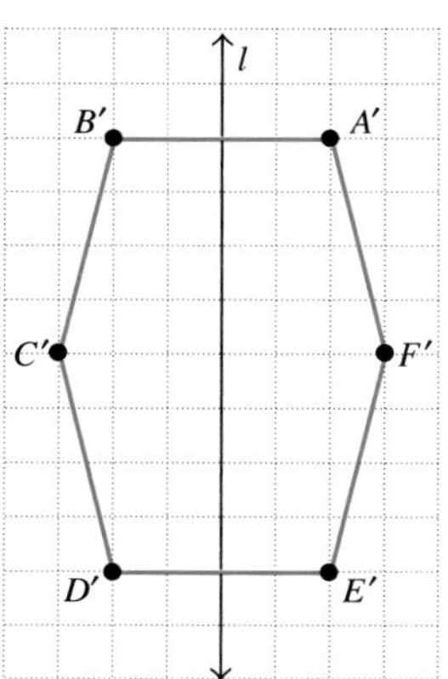

Figure 9.52

EXAMPLE 2 *Reflection of a Hexagon*

Construct the reflection of hexagon $ABCDEF$, shown in Fig. 9.51, about reflection line l.

SOLUTION: From Figure 9.51 we see that vertex A in hexagon $ABCDEF$ is 2 units to the left of reflection line l. Thus, vertex A' in the reflected hexagon will be 2 units to the right of l (see Figure 9.52). Notice that vertex A' of the reflected hexagon is in the same location as vertex B of hexagon $ABCDEF$ in Figure 9.51.

We next see that vertex B in hexagon $ABCDEF$ is 2 units to the right of l. Thus, vertex B' in the reflected hexagon will be 2 units to the left of l. Notice that vertex B' of the reflected hexagon is in the same location as vertex A of hexagon $ABCDEF$. We continue this process to determine the locations of vertices C', D', E', and F' of the reflected hexagon. Notice once again that each vertex of the reflected hexagon is in the same location as a vertex of hexagon $ABCDEF$. Finally, we draw the line segments to complete the reflected hexagon $A'B'C'D'E'F'$ (see Figure 9.52). For this example, we see that other than the vertex labels, the positions of the hexagon before and after the reflection are identical.

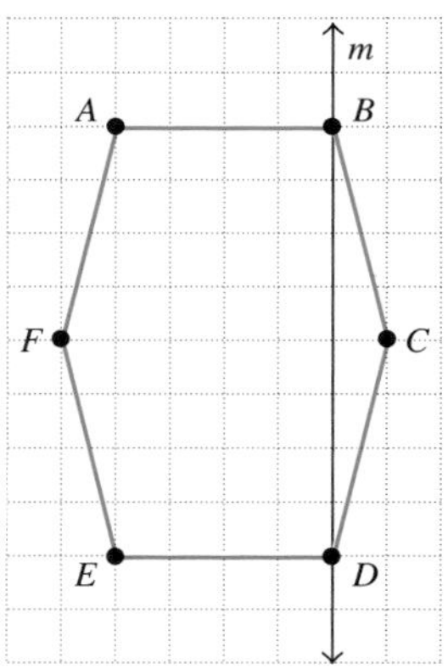

Figure 9.53

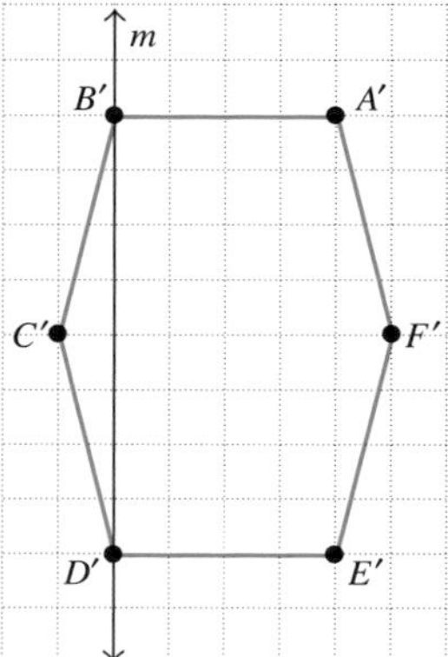

Figure 9.54

In Example 2 the reflection line was in the center of the hexagon in the original position. As a result, the reflection line was also in the center of the reflected hexagon. In this particular case the reflected hexagon lies directly on top of the hexagon in its original position. We will revisit reflections such as those in Example 2 again when we discuss *reflective symmetry* later in this section.

Now consider hexagon *ABCDEF* in Fig. 9.53 and its reflection about line *m*, hexagon *A′B′C′D′E′F′* in Fig. 9.54. Notice that the positions of the hexagon before and after the reflection, relative to line *m*, are not the same. Furthermore, if we line up reflection line *m* in Fig. 9.53 and Fig. 9.54, we would see that hexagon *ABCDEF* and hexagon *A′B′C′D′E′F′* are in different positions.

Translations

The next rigid motion we will discuss is the *translation*. In a translation we simply move a figure along a straight line to a new position.

> A **translation** (or **glide**) is a rigid motion that moves a geometric figure by sliding it along a straight line segment in the plane. The direction and length of the line segment completely determine the translation.

After conducting a translation, we say the figure was *translated* to a new position.

A concise way to indicate the direction and the distance that a figure is moved during a translation is with a *translation vector*. In mathematics, vectors are typically represented with boldface letters. For example, in Fig. 9.55 we see trapezoid *ABCD* and a translation vector, **v**, which is pointing to the right and upward. This translation vector indicates a translation of 9 units to the right and 4 units upward. Note that in Fig. 9.55 the translated vector appears on the right side of the polygon. The placement of the translation vector does not matter. Therefore, the translation vector could have been placed to the left, above, or below the polygon, and the translation would not change. When trapezoid *ABCD* is translated using **v**, every point on trapezoid *ABCD* is moved 9 units to the right and 4 units upward. This movement is demonstrated for vertex *A* in Fig. 9.56(a) on page 523. Figure 9.56(b) shows trapezoid *ABCD* and the translated trapezoid *A′B′C′D′*. Note in Fig. 9.56(b) that every point on trapezoid *A′B′C′D′* is 9 units to the right and 4 units up from its corresponding point on trapezoid *ABCD.*

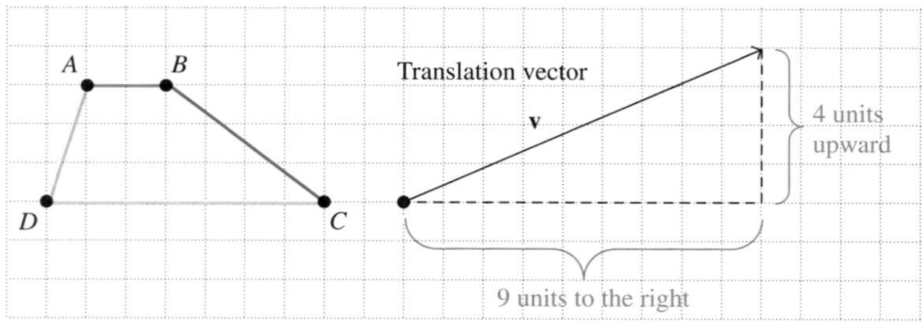

Figure 9.55

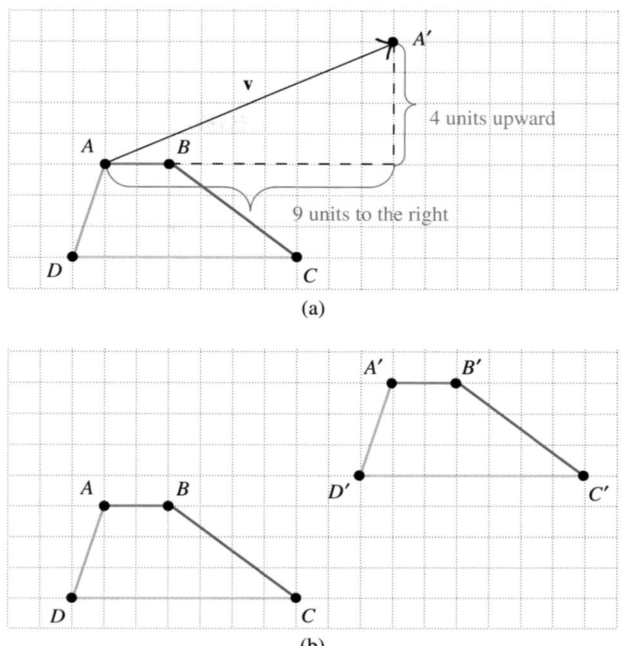

Figure 9.56

EXAMPLE 3 *A Translated Parallelogram*

Given parallelogram *ABCD* and translation vector, **v**, shown in Fig. 9.57, construct the translated parallelogram $A'B'C'D'$.

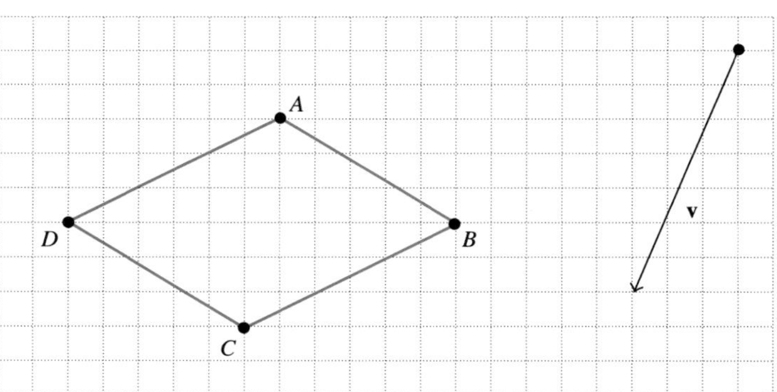

Figure 9.57

SOLUTION: The translated figure will be a parallelogram of the same size and shape as parallelogram *ABCD*. We notice that the translation vector, **v**, points 7 units downward and 3 units to the left. We next examine vertex *A*. To determine the location of vertex A' of the translated parallelogram start at vertex *A* of parallelogram *ABCD* and move down 7 units and to the left 3 units. We label this vertex A' (see Fig. 9.58a on page 524). We determine vertices B', C', and D' in a similar manner by moving down 7 units and to the left 3 units from vertices *B, C*, and *D*, respectively. Figure 9.58(b) shows parallelogram *ABCD* and the translated parallelogram $A'B'C'D'$. Note in Fig. 9.58(b) that every point on parallelogram $A'B'C'D'$ is 7 units down and 3 units to the left of its corresponding point on parallelogram *ABCD*.

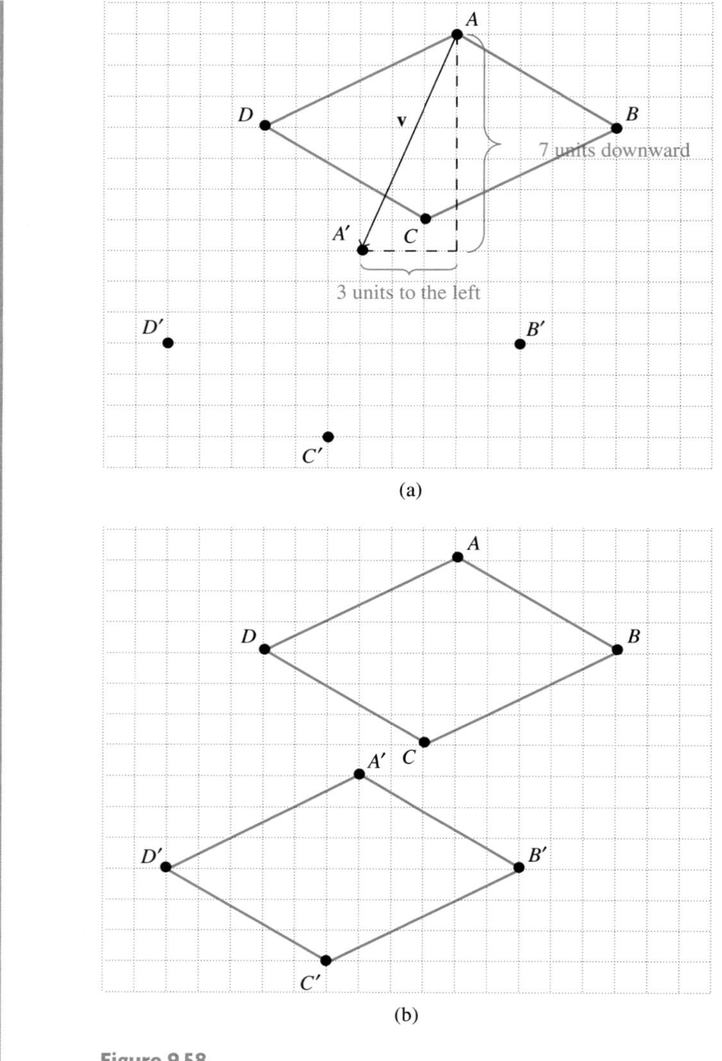

Figure 9.58

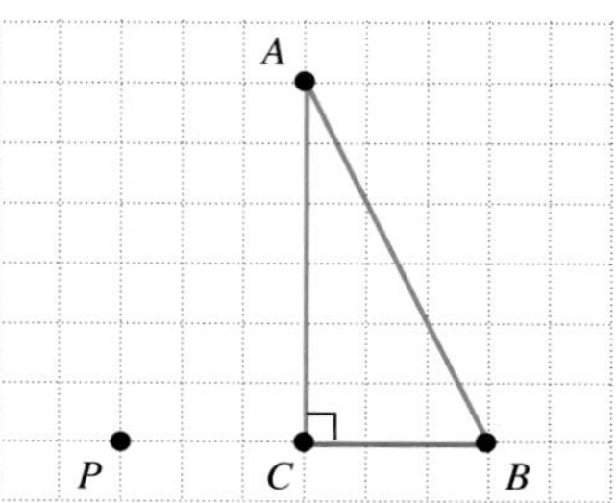

Figure 9.59

Rotations

The next rigid motion we will discuss is *rotation*. To help visualize a rotation examine Fig. 9.59, which shows right triangle ABC and point P about which right triangle ABC is to be rotated.

Imagine that this page was removed from this book and attached to a bulletin board with a single pin through point P. Next imagine rotating the page 90° in the *counterclockwise* direction. The triangle would now appear as triangle $A'B'C'$ shown in Fig. 9.60. Next, imagine rotating the original triangle 180° in a counterclockwise direction. The triangle would now appear as triangle $A''B''C''$ shown in Fig. 9.61.

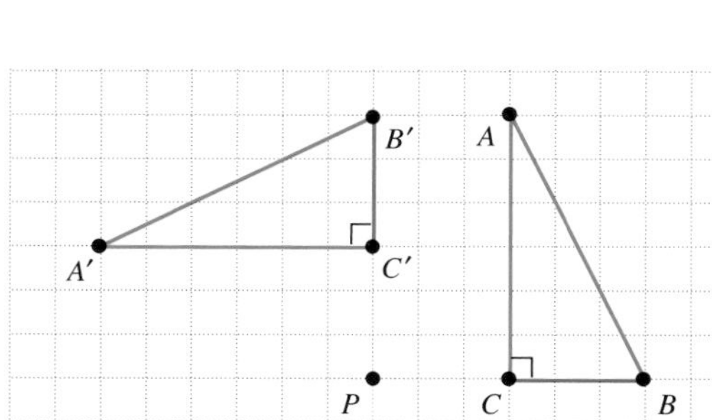

Figure 9.60

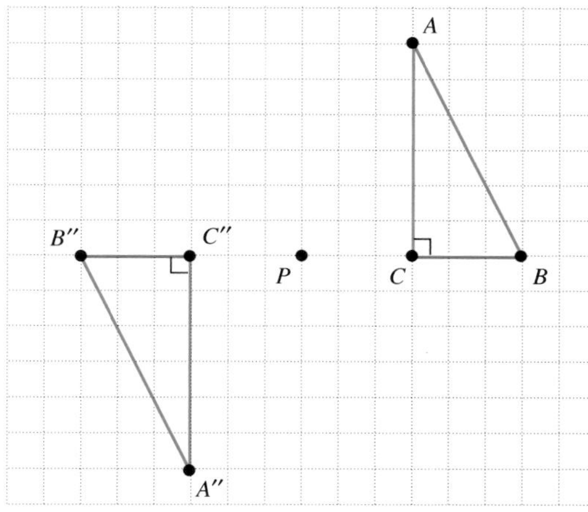

Figure 9.61

Now that we have an intuitive idea of how to determine a rotation, we give the definition of rotation.

A **rotation** is a rigid motion performed by rotating a geometric figure in the plane about a specific point, called the **rotation point** or the **center of rotation.** The angle through which the object is rotated is called the **angle of rotation.**

We will measure angles of rotation using degrees. In mathematics, generally, *counterclockwise angles have positive degree measures and clockwise angles have negative degree measures.*

EXAMPLE 4 *A Rotated Rectangle*

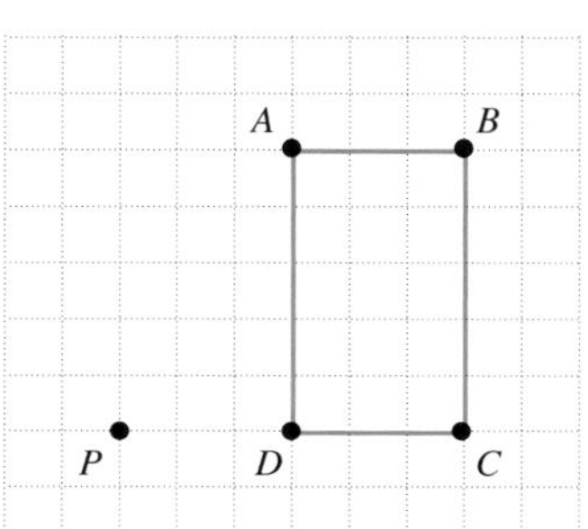

Figure 9.62

Given rectangle *ABCD* and rotation point *P*, shown in Figure 9.62, construct rectangles that result from rotations through

a) 90°. b) 180°. c) 270°.

SOLUTION:

a) First, since 90 is a *positive* number, we will rotate the figure in a counterclockwise direction. We also note that the rotated rectangle will be the same size and shape as rectangle *ABCD*. To get an idea of what the rotated rectangle will look like, pick up this book and rotate it counterclockwise 90°. Fig. 9.63 on page 526 shows rectangle *ABCD* and the rectangle rotated 90°, *A′B′C′D′*, about point *P*. Notice how line segment *AB* in rectangle *ABCD* is horizontal, but in the rotated rectangle in Fig. 9.63 line segment *A′B′* is vertical. Also notice that in rectangle *ABCD* vertex *D* is 3 units to the *right* of rotation point *P*, but in the rotated rectangle vertex *D′* is 3 units *above* rotation point *P*.

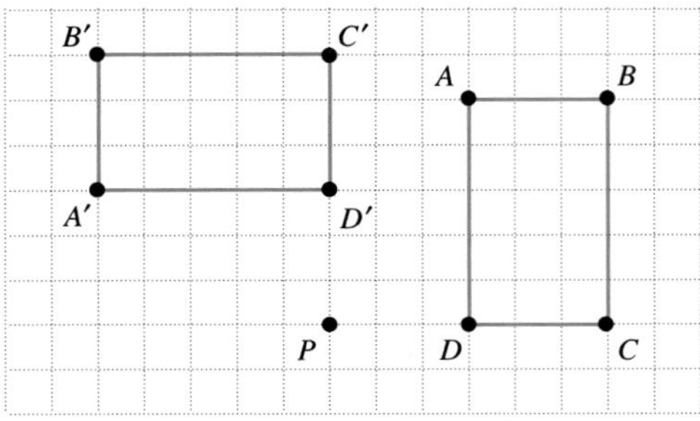

Figure 9.63

b) To gain some perspective on a 180° rotation, again pick up this book, but this time rotate the book 180° in the counterclockwise direction. The rotated rectangle $A''B''C''D''$ is shown along with the rectangle $ABCD$ in Fig. 9.64.

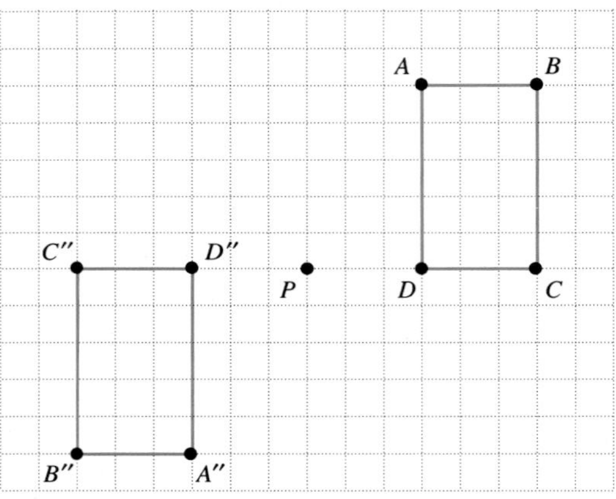

Figure 9.64

c) To gain some perspective on a 270° rotation, rotate this book 270° in the counterclockwise direction. The rotated rectangle $A'''B'''C'''D'''$ is shown along with rectangle $ABCD$ in Fig. 9.65. ▲

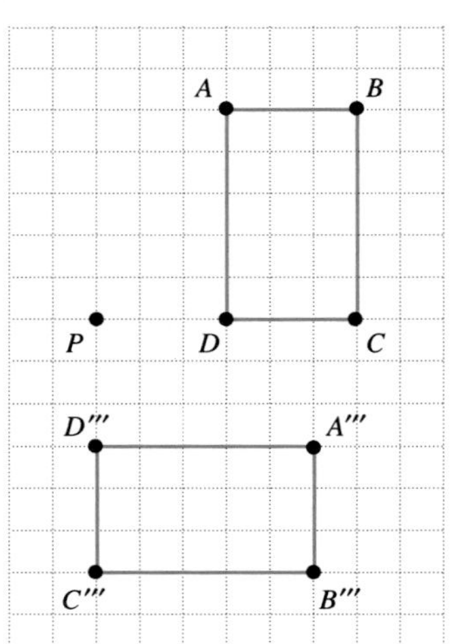

Figure 9.65

Thus far, in our examples of rotations, the rotation point was outside the figure being rotated. We now will study an example where the rotation point is inside the figure to be rotated.

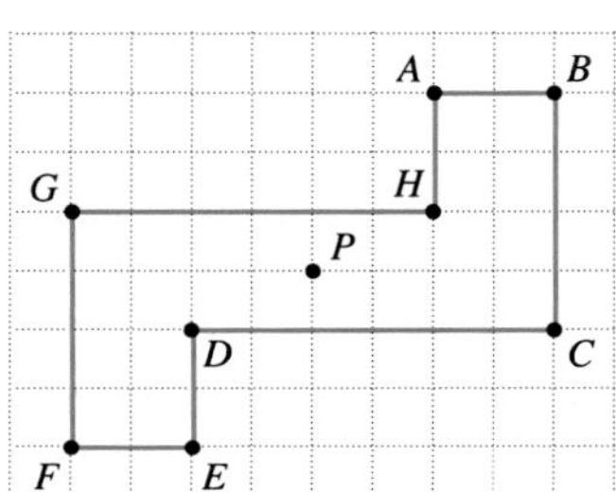

Figure 9.66

EXAMPLE 5 *A Rotation Point Inside a Polygon*

Given polygon $ABCDEFGH$ and rotation point P, shown in Fig. 9.66, construct polygons that result from rotations through

a) 90°. b) 180°.

SOLUTION:

a) We will rotate the polygon 90° in a counterclockwise direction. The resulting polygon will be the same size and shape as polygon $ABCDEFGH$. To visualize

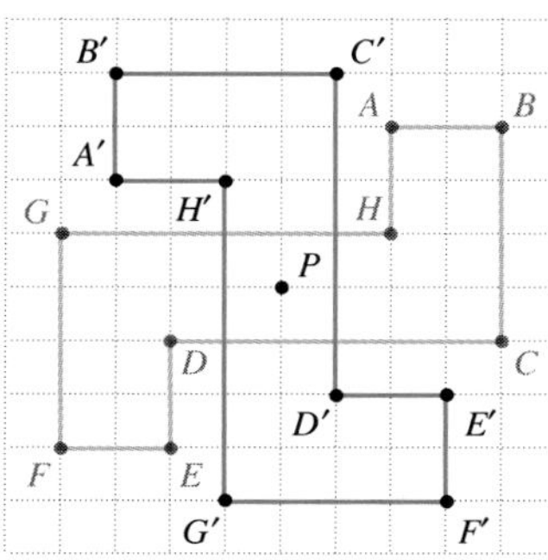

Figure 9.67

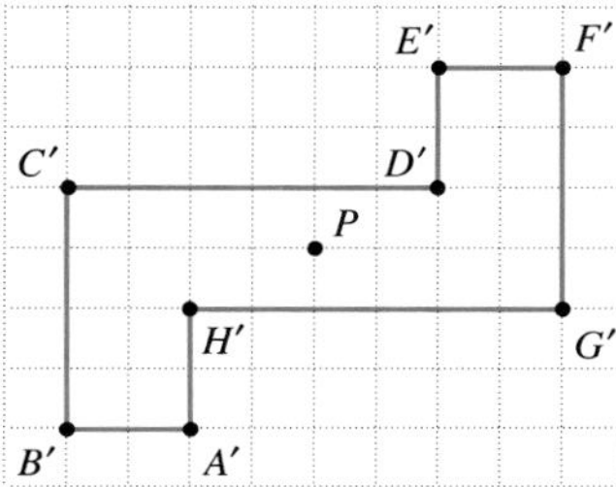

Figure 9.68

what the rotated polygon will look like pick up this book and rotate it counterclockwise 90°. Figure 9.67 shows the polygon *ABCDEFGH,* in pale blue, and the rotated polygon $A'B'C'D'E'F'G'H'$ in deeper blue. Notice how line segments *AB, CD, EF,* and *GH* in polygon *ABCDEFGH* are *horizontal*, but in the rotated polygon $A'B'C'D'E'F'G'H'$ line segments $A'B'$, $C'D'$, $E'F'$, and $G'H'$ are *vertical*. Also notice in polygon *ABCDEFGH* that line segment *GH* is 1 unit *above* rotation point *P*, but in the rotated polygon line segment $G'H'$ is 1 unit to the *left* of rotation point *P*.

b) To visualize the polygon obtained through a 180° rotation, we can pick up this book and rotate it 180° in the counterclockwise direction. Notice from Fig. 9.68 that vertex A' of the rotated polygon is in the same position as vertex *E* of polygon *ABCDEFGH*. Also notice from Fig. 9.68 that vertex B' of the rotated polygon is in the same position as vertex *F* of polygon *ABCDEFGH*. In fact, each of the vertices in the rotated polygon is in the same position as a different vertex in polygon *ABCDEFGH*. From Fig. 9.68 we see that, other than vertex labels, the position of rotated polygon $A''B''C''D''E''F''G''H''$ is the same as the position of polygon *ABCDEFGH*. ▲

The polygon used in Example 5 will be discussed again later when we discuss *rotational symmetry*. The three rigid motions we have discussed thus far are reflection, translation, and rotation. Now we will discuss the fourth rigid motion, *glide reflection*.

A **glide reflection** is a rigid motion formed by performing a *translation* (or *glide*) followed by a *reflection*.

A glide reflection, as its name suggests, is a translation (or glide) followed by a reflection. Both translations and reflections were discussed earlier in this section. Consider triangle *ABC* (shown in blue), translation vector **v**, and reflection line *l* in Fig. 9.69. The translation of triangle *ABC,* obtained using translation vector **v**, is triangle $A'B'C'$ (shown in red). The reflection of triangle $A'B'C'$ about reflection line *l* is triangle $A''B''C''$ (shown in green). Thus, triangle $A''B''C''$ is the glide reflection of triangle *ABC* using translation vector **v** and reflection line *l*.

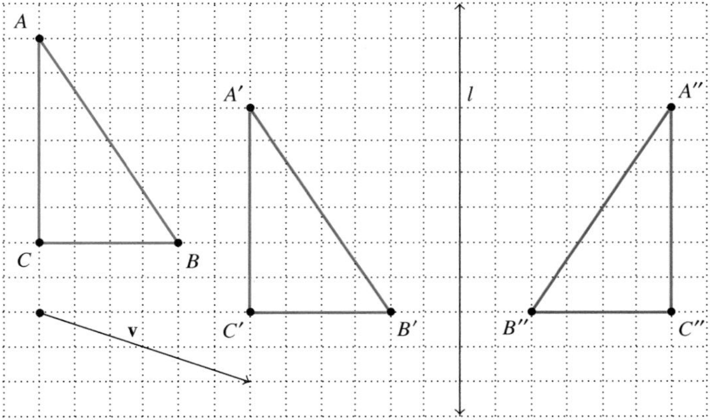

Figure 9.69

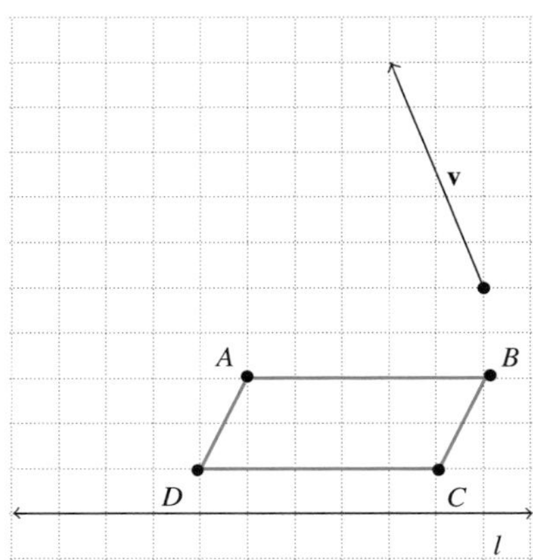

Figure 9.70

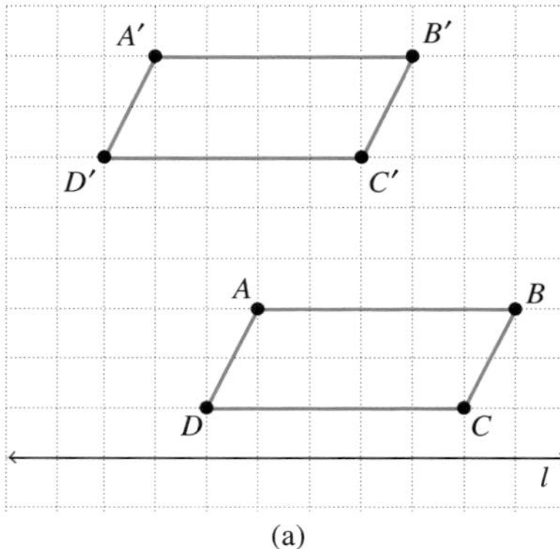

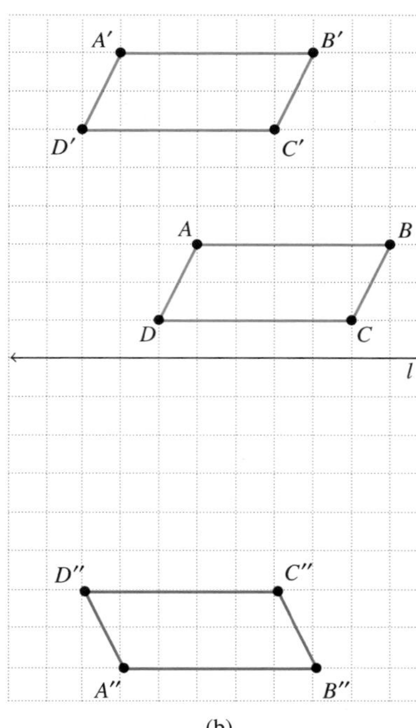

Figure 9.71

EXAMPLE 6 *A Glide Reflection of a Parallelogram*

Construct a glide reflection of parallelogram *ABCD*, shown in Fig. 9.70, using translation vector **v** and reflection line *l*.

SOLUTION: To construct the glide reflection of parallelogram *ABCD*, first translate the parallelogram 2 units to the left and 5 units up as indicated by translation vector **v**. This translated parallelogram is labeled $A'B'C'D'$, shown in red in Fig. 9.71(a). Next, we will reflect parallelogram $A'B'C'D'$ about reflection line *l*. Parallelogram $A'B'C'D'$, shown in red, and the reflected parallelogram, labeled $A''B''C''D''$, shown in green, are shown in Fig. 9.71(b). The glide reflection of the parallelogram *ABCD* is parallelogram $A''B''C''D''$.

Symmetry

We are now ready to discuss symmetry. Our discussion of symmetry involves a rigid motion of an object.

A **symmetry** of a geometric figure is a rigid motion that moves the figure back onto itself. That is, the beginning position and ending position of the figure must be identical.

Suppose we start with a figure in a specific position and perform a rigid motion on this figure. If the position of the figure after the rigid motion is identical to the position of the figure before the rigid motion (if the beginning and ending positions of the figure coincide), then the rigid motion is a symmetry and we say that the figure has symmetry. For a two-dimensional figure there are four types of symmetries: reflective symmetry, rotational symmetry, translational symmetry, and glide reflective symmetry. In this textbook, however, we will only discuss reflective symmetry and rotational symmetry.

Consider the polygon and reflection line *l* shown in Fig. 9.72(a). If we use the rigid motion of reflection and reflect the polygon *ABCDEFGH* about line *l*, we get polygon $A'B'C'D'E'F'G'H'$. Note that the ending position of the polygon is identical to the starting position as shown in Fig. 9.72(b). Compare Fig. 9.72(a) with Fig. 9.72(b). Although the vertex labels are different, the reflected polygon is in the same position as the polygon in the original position. Thus, we say that the polygon has **reflective symmetry** about line *l*. We refer to line *l* as a **line of symmetry.**

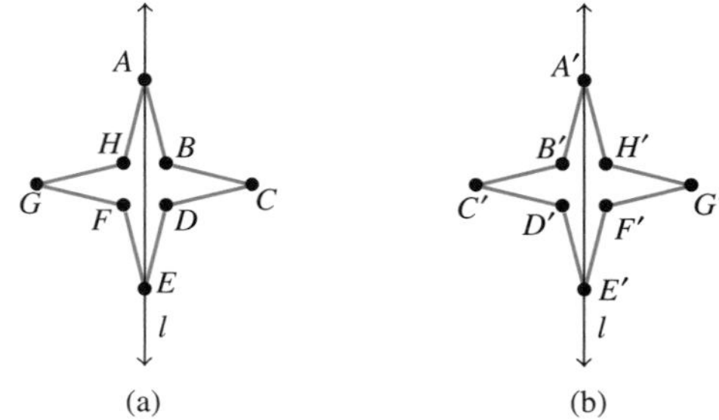

Figure 9.72

Recall Example 2 on page 521 in which hexagon *ABCDEF* was reflected about reflection line *l*. Examine the hexagon in the original position (Fig. 9.51) and the hexagon in the final position after being reflected about line *l* (Fig. 9.52). Other than the labels of the vertices, the beginning and ending positions of the hexagon are identical. Therefore, hexagon *ABCDEF* has reflective symmetry about line *l*.

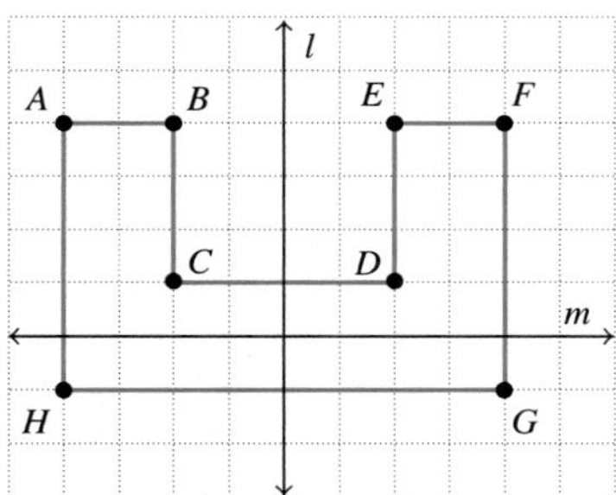

Figure 9.73

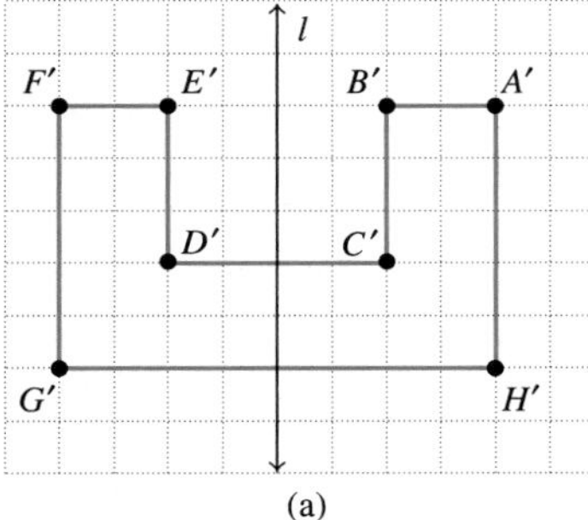

(a)

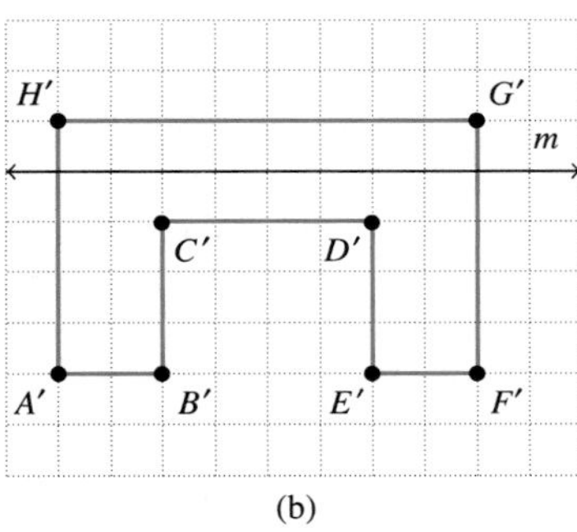

(b)

Figure 9.74

EXAMPLE 7 *Reflective Symmetries of Polygons*

Determine whether the polygon shown in Fig. 9.73 has reflective symmetry about each of the following lines.

a) Line l b) Line m

SOLUTION:

a) Examine the reflection of the polygon about line l as seen in Fig. 9.74(a). Notice that other than the vertex labels, the beginning and ending positions of the polygon are identical. Thus, the polygon has reflective symmetry about line l.

b) Examine the reflection of the polygon about line m as seen in Fig. 9.74(b). Notice that the position of the reflected polygon is different from the original position of the polygon. Thus, the polygon does not have reflective symmetry about line m. ▲

We will now discuss a second type of symmetry, rotational symmetry. Consider the polygon and rotation point P shown in Fig. 9.75(a). The rigid motion of rotation of polygon $ABCDEFGH$ through a 90° angle about point P gives polygon $A'B'C'D'E'F'G'H'$ shown in Fig. 9.75(b). Compare Fig. 9.75(a) to Fig. 9.75(b). Although the vertex labels are different, the position of the polygon before and after the rotation is identical. Thus, we say that the polygon has 90° **rotational symmetry** about point P. We refer to point P as the **point of symmetry.**

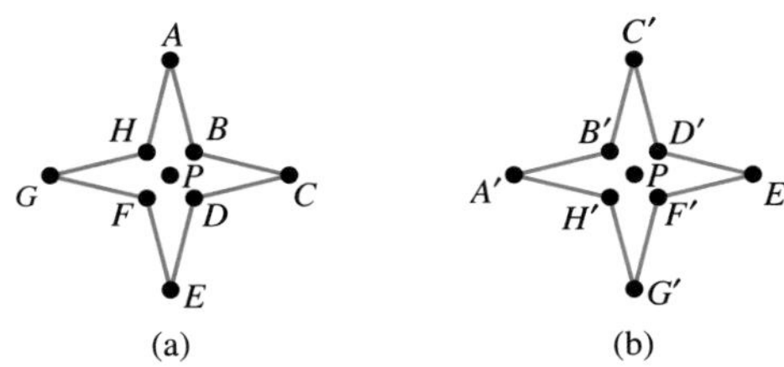

Figure 9.75

Recall Example 5 on page 526 in which polygon $ABCDEFGH$ was rotated 90° about point P in part (a) and 180° in part (b). First examine the polygon in the original position in Fig. 9.66 on page 526 and the 90° rotated polygon in Fig. 9.67 on page 527. Notice the position of the polygon after the 90° rotation is different from the original position of the polygon. Therefore, polygon $ABCDEFGH$ in Fig. 9.67 does not have 90° rotational symmetry about point P. Now examine the 180° rotated polygon in Fig. 9.68 on page 527. Notice that other than the vertex labels, the position of the two polygons $ABCDEFGH$ and $A'B'C'D'E'F'G'H'$ is identical with regard to rotation about point P. Therefore, polygon $ABCDEFGH$ in Figure 9.66 has 180° rotational symmetry about point P.

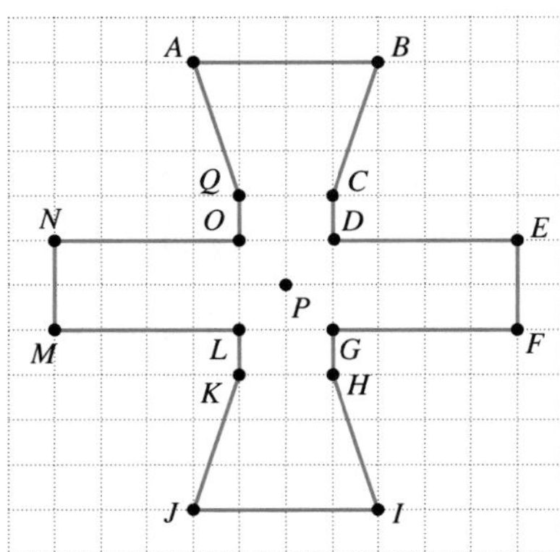

Figure 9.76

EXAMPLE 8 *Rotational Symmetries*

Determine whether the polygon shown in Fig. 9.76 has rotational symmetry about point P for rotations through each of the following angles.

a) 90° b) 180°

SOLUTION:

a) To determine whether the polygon has 90° counterclockwise rotational symmetry about point P we rotate the polygon 90° as shown in Fig. 9.77(a) on page 530. Compare Fig. 9.77(a) with Fig. 9.76. Notice that the position of the polygon after

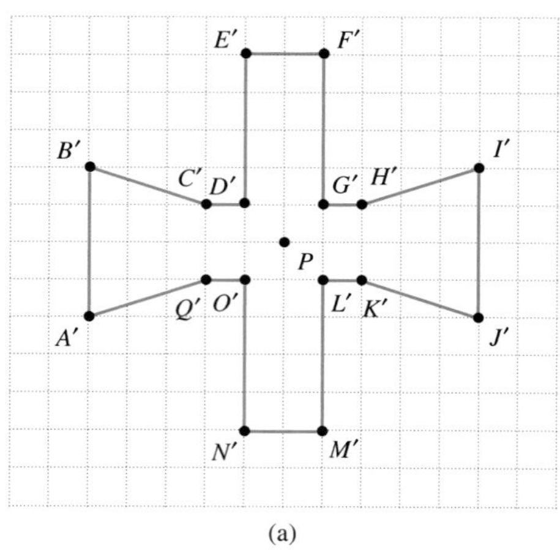

(a)

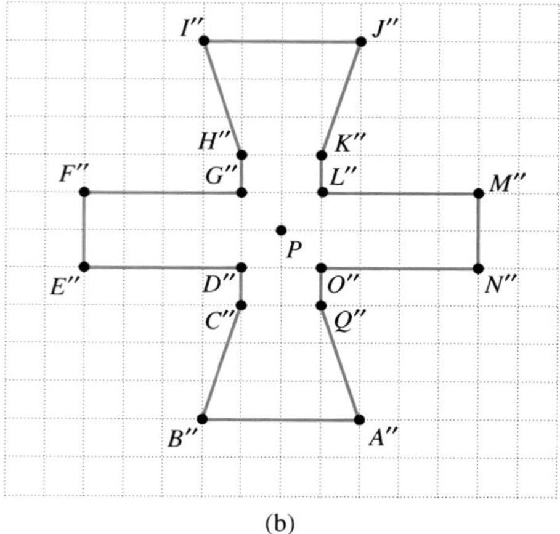

(b)

Figure 9.77

the rotation in Fig. 9.77(a) is different than the original position of the polygon (Fig. 9.76). Therefore, the polygon does not have 90° rotational symmetry.

b) To determine whether the polygon has 180° counterclockwise rotational symmetry about the point P we rotate the polygon 180° as shown in Fig. 9.77(b). Compare Fig. 9.77(b) with Fig. 9.76. Notice that other than vertex labels, the position of the polygon after the rotation in Fig. 9.77(b) is identical to the position of the polygon before the rotation (Fig. 9.76). Therefore, the polygon has 180° rotational symmetry. ▲

Tessellations

A fascinating application of transformational geometry is the creation of *tessellations*.

A **tessellation** (or **tiling**) is a pattern consisting of the repeated use of the same geometric figures to entirely cover a plane, leaving no gaps. The geometric figures used are called the **tessellating shapes** of the tessellation.

Figure 9.78 shows an example of a tessellation from ancient Egypt. Perhaps the most famous person to incorporate tessellations into his work is M. C. Escher (see Profiles in Mathematics on the next page).

Figure 9.78

PROFILE IN MATHEMATICS

MAURITS CORNELIUS ESCHER

In addition to being wonderfully engaging art, the work of M. C. Escher (1898–1972) also displays some of the more beautiful and intricate aspects of mathematics. Escher's work involves Euclidean, non-Euclidean (to be studied shortly), and transformational geometries. Amazingly, Escher had no formal training in higher mathematics.

In 1936 Escher visited the Alhambra Palace in Granada, Spain, and became fascinated with Moorish tilings. Thereafter, Escher became obsessed with creating art that used objects to cover a plane so as to leave no gaps. Escher's brother recognized the mathematics depicted in this art and gave Escher a list of mathematics articles he felt would be of assistance to the artist. That was Escher's first exposure to formal mathematics. Among the mathematicians whose work influenced Escher were George Polya and Donald Coxeter. Polya's work with symmetry became a cornerstone of Escher's famous tessellations. Coxeter's work with non-Euclidean geometry was key to Escher's later work involving infinity, multiple dimensions, and hyperbolic and spherical tessellations. In 1995 Coxeter published a paper in which he proved that the mathematics Escher displayed in the etching *Circle Limit III* was indeed perfectly consistent with mathematical theory.

Escher's *Circle Limit III*

Another testimonial of Escher's genius is found in a notebook in which he kept background information for his artwork. In this notebook Escher categorized all possible combinations of shapes, colors, and symmetrical properties of polygons in the plane. By doing so, Escher had unwittingly developed areas of a branch of mathematics called *crystallography*, years before any mathematician had done so.

The simplest tessellations use one single regular polygon as the tessellating shape. Recall that a *regular polygon* is one whose sides are all the same length and whose interior angles all have the same measure. A tessellation that uses one single regular polygon as the tessellating shape is called a *regular tessellation*. It can be shown that only three regular tessellations exist: those that use an equilateral triangle, a square, or a regular hexagon as the tessellating shape. Figure 9.79 shows each of these regular tessellations. Notice that each of these tessellations can be obtained from a single tessellating shape through the use of reflections, translations, or rotations.

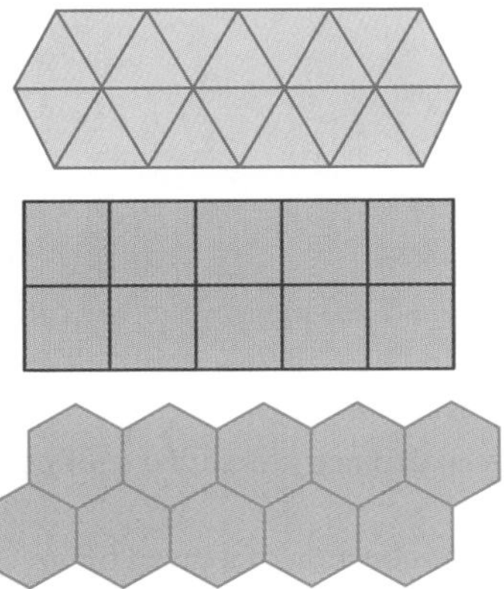

Figure 9.79

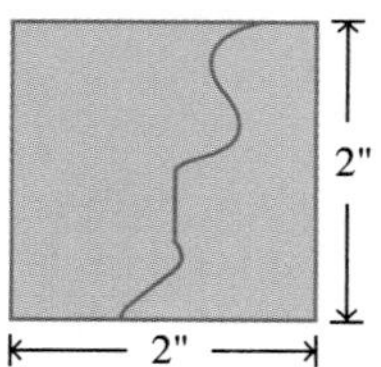

Figure 9.80

We will now learn how to create unique tessellations. We will do so by constructing a unique tessellating shape from a square. We could also construct other tessellating shapes using an equilateral triangle or a regular hexagon. If you wish to follow along with our construction, you will need some lightweight cardboard, a ruler, cellophane tape, and a pair of scissors. We will start by measuring and cutting out a square 2 in. by 2 in. from the cardboard. We next cut the square into two parts by cutting it from top to bottom using any kind of cut. One example is shown in Fig. 9.80. We then rearrange the pieces and tape the two vertical edges together as shown in Fig. 9.81. Next we cut this new shape into two parts by cutting it from left to right using any kind of cut as shown in Fig. 9.82. We then rearrange the pieces and tape the two horizontal edges together as shown in Fig. 9.83. This completes our tessellating shape.

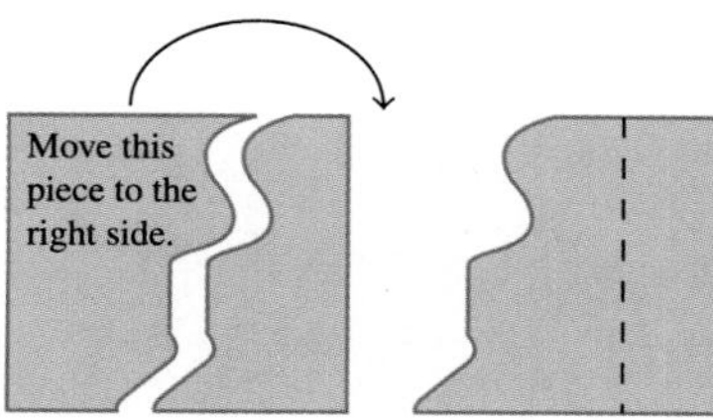

Figure 9.81

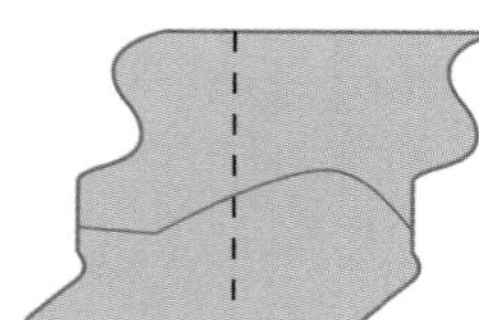

Figure 9.82

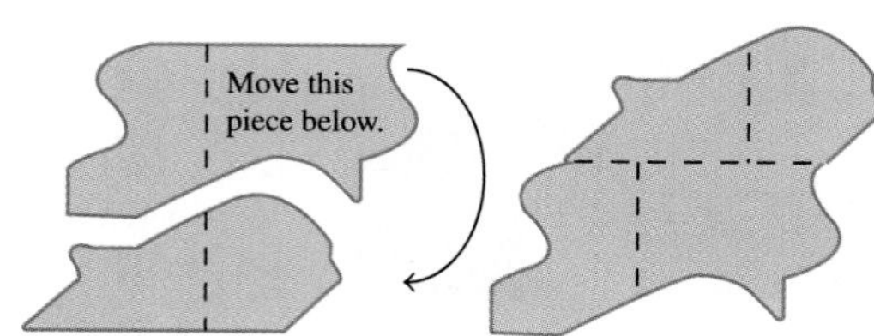

Figure 9.83

Figure 9.84

We now set the cardboard tessellating shape in the middle of a blank piece of paper (the tessellating shape can be rotated to any position as a starting point) and trace the outline of the shape onto the paper. Next move the tessellating shape so that it lines up with the figure already drawn and trace the outline again. Continue to do that until the page is completely covered. Once the page is covered with the tessellation, we can add some interesting colors or even some unique sketches to the tessellation. Figure 9.84 shows one tessellation created using the tessellation shape in Fig. 9.83. In Fig. 9.84 the tessellation shape was rotated about 45° counterclockwise.

An infinite number of different tessellations can be created using the method described by altering the cuts made. We could also create different tessellations using an equilateral triangle, a regular hexagon, or other types of polygons. There are also other, more complicated ways to create the tessellating shape. The Internet has many sites devoted to the creation of tessellations by hand. Many computer programs that generate tessellations are also available.

SECTION 9.5 EXERCISES

Concept/Writing Exercises

1. In the study of transformational geometry, what is a rigid motion? List the four rigid motions studied in this section.
2. What is transformational geometry?
3. In terms of transformational geometry, describe a reflection.
4. Describe how to construct a reflection of a given figure about a given line.
5. In terms of transformational geometry, describe a translation.
6. Describe how to construct a translation of a given figure using a translation vector.
7. In terms of transformational geometry, describe a rotation.
8. Describe how to construct a rotation of a given figure, about a given point, through a given angle.
9. In terms of transformational geometry, describe a glide reflection.

10. Describe how to construct a glide reflection of a given figure using a given translation vector and a given reflection line.
11. Describe what it means for a figure to have reflective symmetry about a given line.
12. Describe what it means for a figure to have rotational symmetry about a given point.
13. What is a tessellation?
14. Describe one way to make a unique tessellation from a 2-in. by 2-in. cardboard square.

Practice the Skills/Problem Solving

In Exercises 15–22, use the given figure and lines of reflection to construct the indicated reflections. Show the figure in the positions both before and after the reflection.

In Exercises 15 and 16, use the following figure. Construct

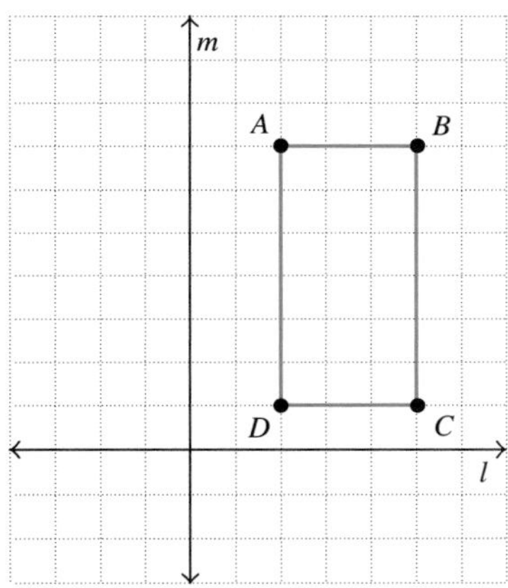

15. the reflection of rectangle *ABCD* about line *m*.

16. the reflection of rectangle *ABCD* about line *l*.

In Exercises 17 and 18, use the following figure. Construct

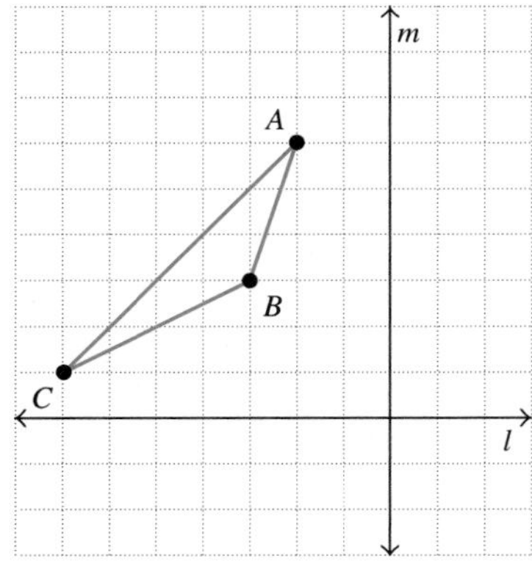

17. the reflection of triangle *ABC* about line *l*.

18. the reflection of triangle *ABC* about line *m*.

In Exercises 19 and 20, use the following figure. Construct

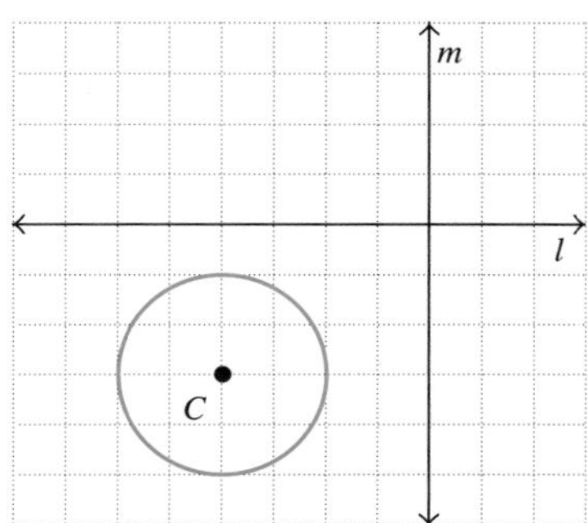

19. the reflection of circle *C* about line *l*.

20. the reflection of circle *C* about line *m*.

In Exercises 21 and 22, use the following figure. Construct

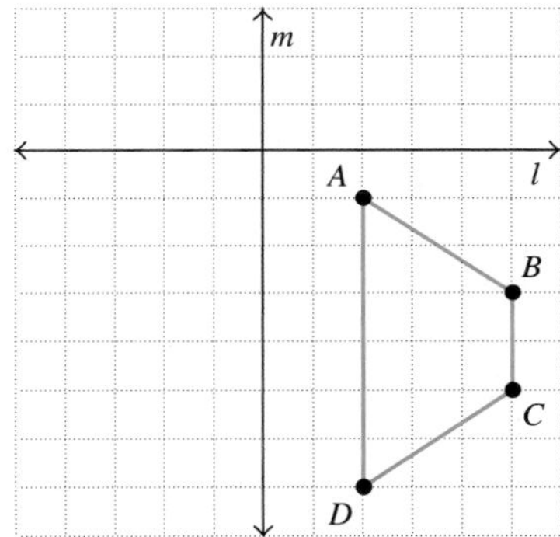

21. the reflection of trapezoid *ABCD* about line *m*.

22. the reflection of the trapezoid *ABCD* about line *l*.

In Exercises 23–30, use the translation vectors, **v** *and* **w** *shown below, to construct the translations indicated in the exercises. Show the figure in the positions both before and after the translation.*

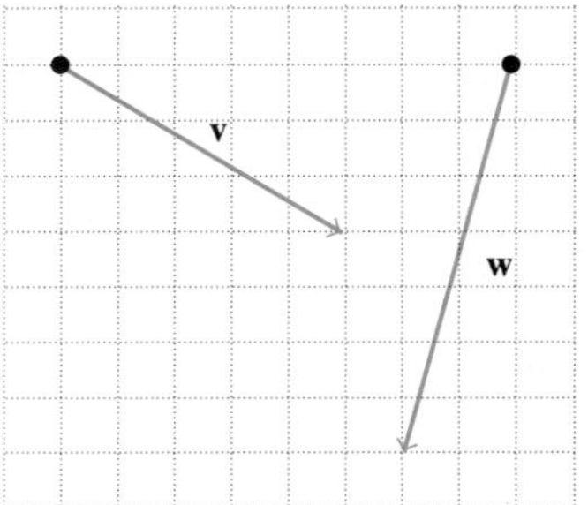

In Exercises 23 and 24, use the following figure. Construct

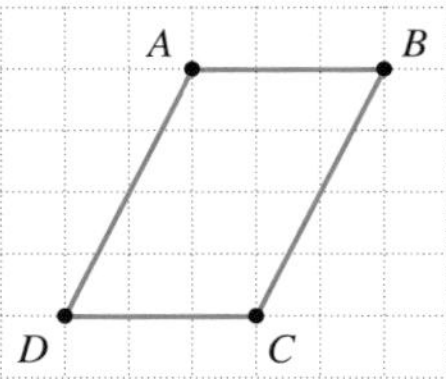

23. the translation of parallelogram *ABCD* using translation vector **v** (shown on page 533).

24. the translation of parallelogram *ABCD* using translation vector **w** (shown on page 533).

In Exercises 25 and 26, use the following figure. Construct

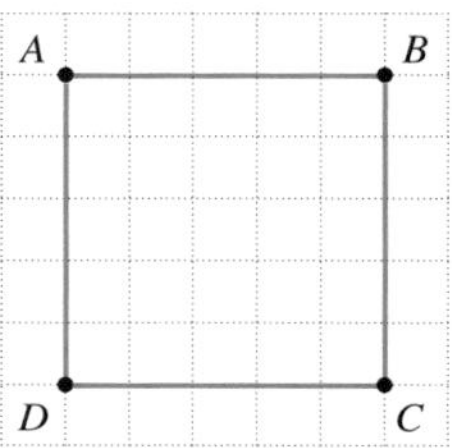

25. the translation of square *ABCD* using translation vector **w** (shown on page 533).

26. the translation of square *ABCD* using translation vector **v** (shown on page 533).

In Exercises 27 and 28, use the following figure. Construct

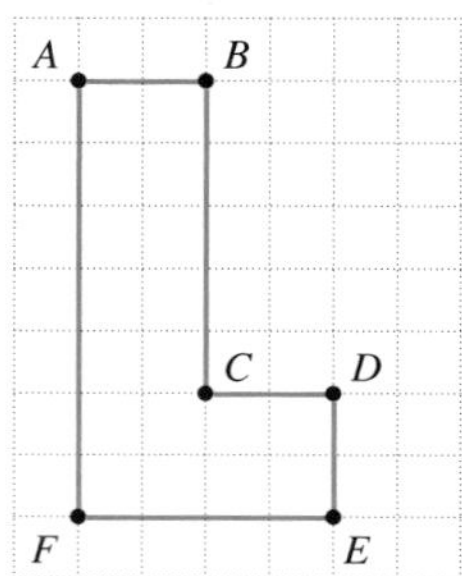

27. the translation of polygon *ABCDEF* using translation vector **v**.

28. the translation of polygon *ABCDEF* using translation vector **w**.

In Exercises 29 and 30, use the following figure. Construct

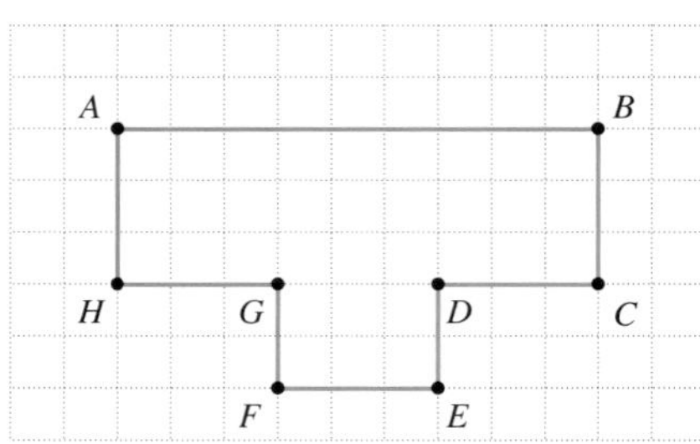

29. the translation of polygon *ABCDEFGH* using translation vector **w**.

30. the translation of polygon *ABCDEFGH* using translation vector **v**.

In Exercises 31–38, use the given figure and rotation point P to construct the indicated rotations. Show the figure in the positions both before and after the rotation.

In Exercises 31 and 32, use the following figure. Construct

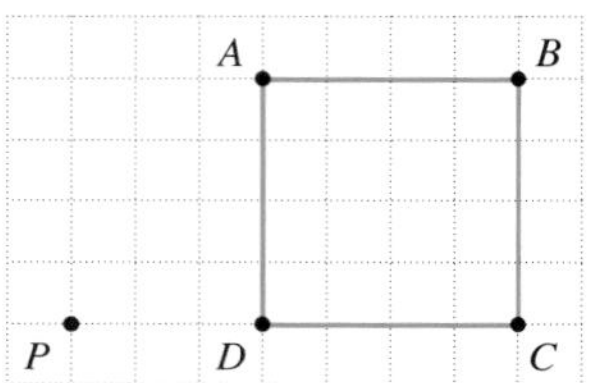

31. a 90° rotation of square *ABCD* about point *P*.

32. a 180° rotation of square *ABCD* about point *P*.

In Exercises 33 and 34, use the following figure. Construct

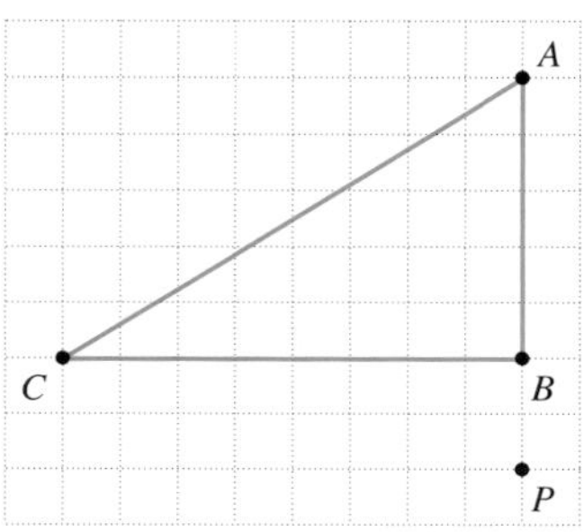

33. a 180° rotation of triangle *ABC* about point *P*.

34. a 270° rotation of triangle *ABC* about point *P*.

In Exercises 35 and 36, use the following figure. Construct

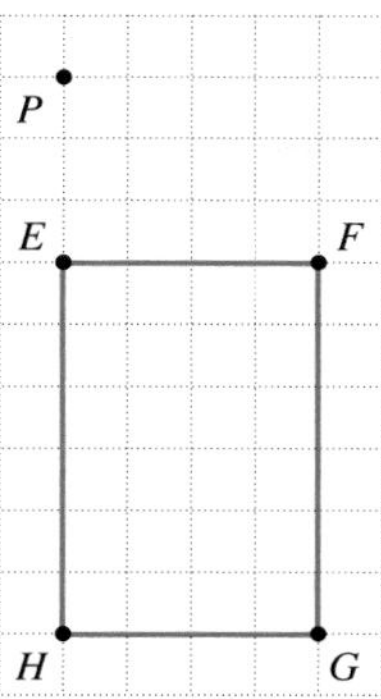

35. a 270° rotation of rectangle *EFGH* about point *P*.

36. a 180° rotation of rectangle *EFGH* about point *P*.

In Exercises 37 and 38, use the following figure. Construct

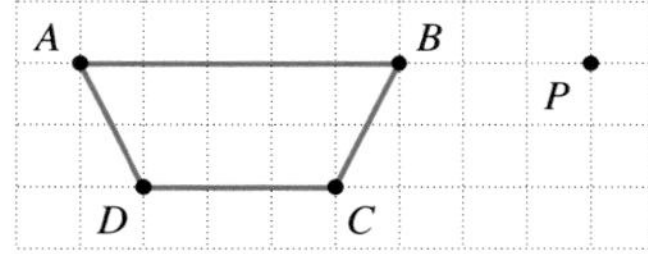

37. a 90° rotation of trapezoid *ABCD* about point *P*.

38. a 270° rotation of trapezoid *ABCD* about point *P*.

In Exercises 39–46, use the given figure, translation vectors **v** *and* **w**, *and reflection lines l and m to construct the indicated glide reflections. Show the figure in the positions before and after the glide reflection.*

In Exercises 39 and 40, use the following figure. Construct

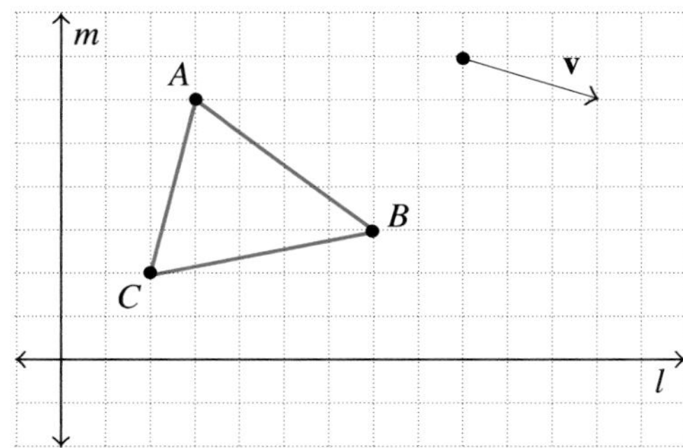

39. a glide reflection of triangle *ABC* using vector **v** and reflection line *l*.

40. a glide reflection of triangle *ABC* using vector **v** and reflection line *m*.

In Exercises 41 and 42, use the following figure. Construct

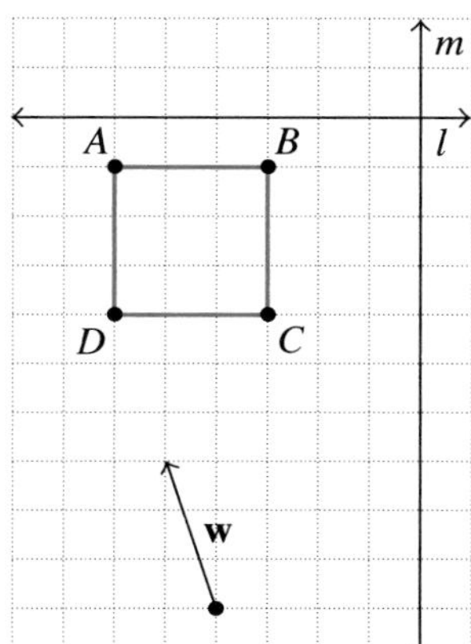

41. a glide reflection of square *ABCD* using vector **w** and reflection line *l*.

42. a glide reflection of square *ABCD* using vector **w** and reflection line *m*.

In Exercises 43 and 44, use the following figure. Construct

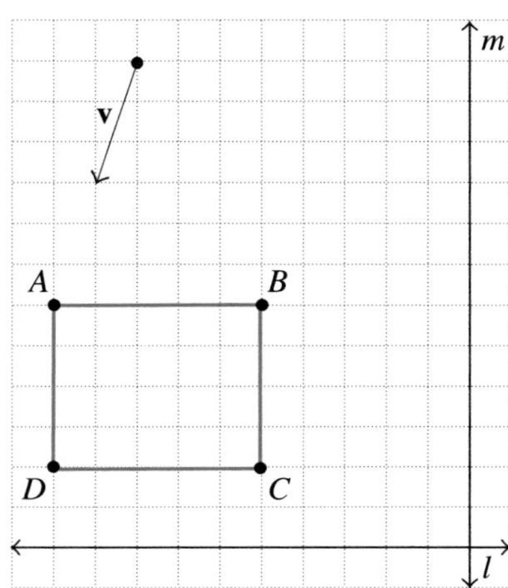

43. a glide reflection of rectangle *ABCD* using vector **v** and reflection line *l*.

44. a glide reflection of rectangle *ABCD* using vector **v** and reflection line *m*.

In Exercises 45 and 46, use the following figure. Construct

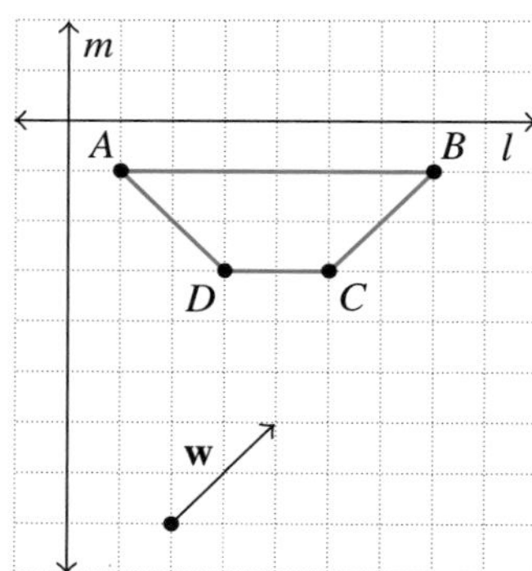

45. a glide reflection of trapezoid *ABCD* using vector **w** and reflection line *l*.

46. a glide reflection of trapezoid *ABCD* using vector **w** and reflection line *m*.

47. a) Reflect triangle *ABC*, shown below, about line *l*. Label the reflected triangle $A'B'C'$.

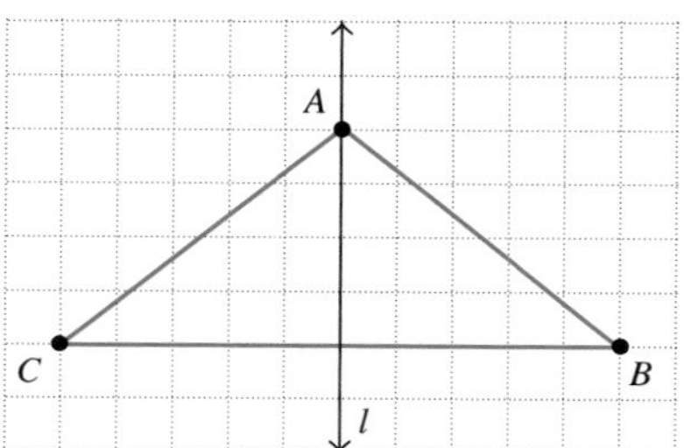

b) Other than vertex labels, is the position of triangle $A'B'C'$ identical to the position of triangle *ABC*?

c) Does triangle *ABC* have reflective symmetry about line *l*?

48. a) Reflect trapezoid *ABCD*, shown below, about line *l*. Label the reflected trapezoid $A'B'C'D'$.

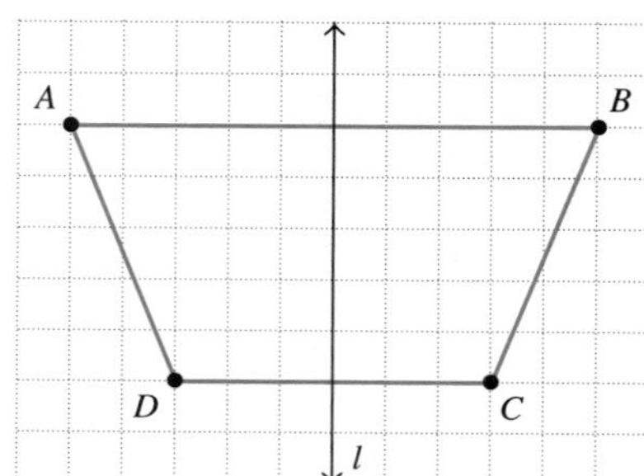

b) Other than vertex labels, is the position of trapezoid $A'B'C'D'$ identical to the position of trapezoid *ABCD*?

c) Does trapezoid *ABCD* have reflective symmetry about line *l*?

49. a) Reflect parallelogram *ABCD*, shown below, about line *l*. Label the reflected parallelogram $A'B'C'D'$.

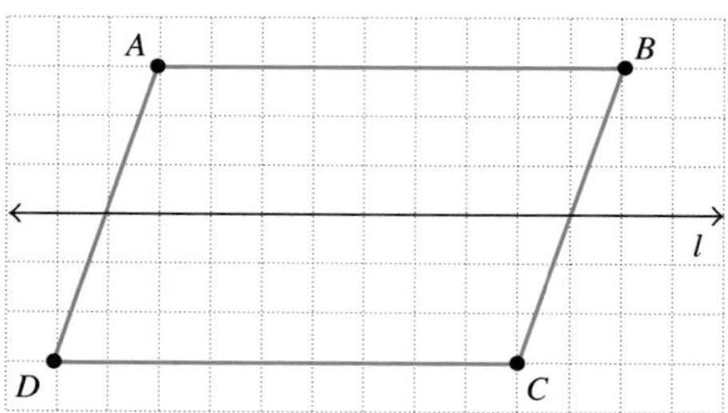

b) Other than vertex labels, is the position of parallelogram $A'B'C'D'$ identical to the position of parallelogram *ABCD*?

c) Does parallelogram *ABCD* have reflective symmetry about line *l*?

50. a) Reflect square *ABCD*, shown below, about line *l*. Label the reflected square $A'B'C'D'$.

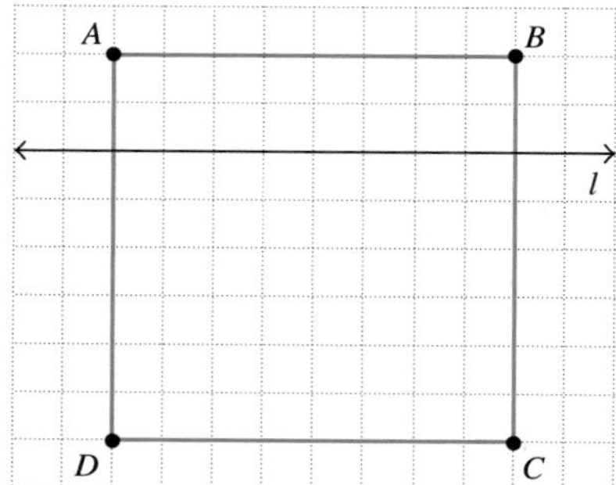

b) Other than vertex labels, is the position of square $A'B'C'D'$ identical to the position of square *ABCD*?

c) Does square *ABCD* have reflective symmetry about line *l*?

51. a) Rotate rectangle *ABCD*, shown below, 90° about point *P*. Label the rotated rectangle $A'B'C'D'$.

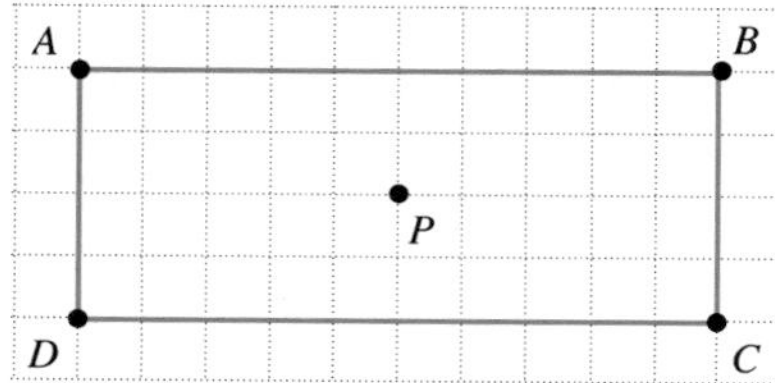

b) Other than vertex labels, is the position of rectangle $A'B'C'D'$ identical to the position of rectangle *ABCD*?

c) Does rectangle *ABCD* have 90° rotational symmetry about point *P*?

d) Now, rotate the rectangle in the original position, rectangle *ABCD*, 180° about point *P*. Label the rotated rectangle $A''B''C''D''$.

e) Other than vertex labels, is the position of rectangle $A''B''C''D''$ identical to the position of rectangle $ABCD$?

f) Does rectangle $ABCD$ have 180° rotational symmetry about point P?

52. a) Rotate parallelogram $ABCD$, shown below, 90° about point P. Label the rotated parallelogram $A'B'C'D'$.

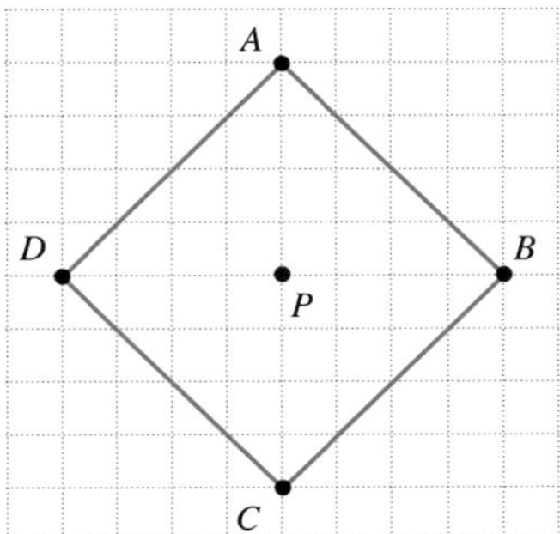

b) Other than vertex labels, is the position of parallelogram $A'B'C'D'$ identical to the position of parallelogram $ABCD$?

c) Does parallelogram $ABCD$ have 90° rotational symmetry about point P?

d) Now rotate the parallelogram in the original position, parallelogram $ABCD$, shown above, 180° about point P. Label the rotated parallelogram $A''B''C''D''$.

e) Other than vertex labels, is the position of parallelogram $A''B''C''D''$ identical to the position of parallelogram $ABCD$?

f) Does parallelogram $ABCD$ have 180° rotational symmetry about point P?

53. Consider the following figure.

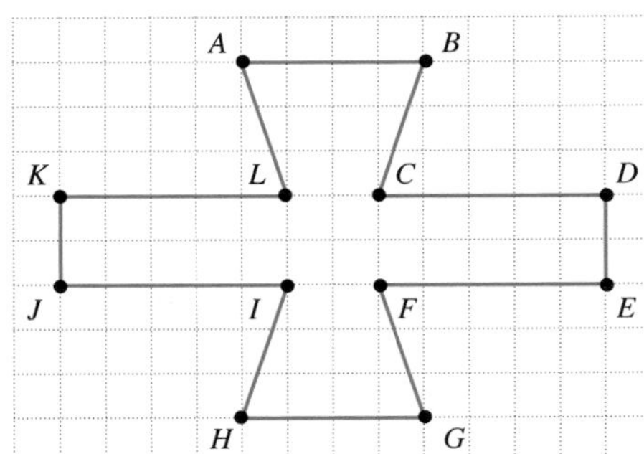

a) Insert a vertical line m through the figure so the figure has reflective symmetry about line m.

b) Insert a horizontal line l through the figure so the figure has reflective symmetry about line l.

c) Insert a point P within the figure so the figure has 180° rotational symmetry about point P.

d) Is it possible to insert a point P within the figure so the figure has 90° rotational symmetry about point P? Explain your answer.

54. Consider the following figure.

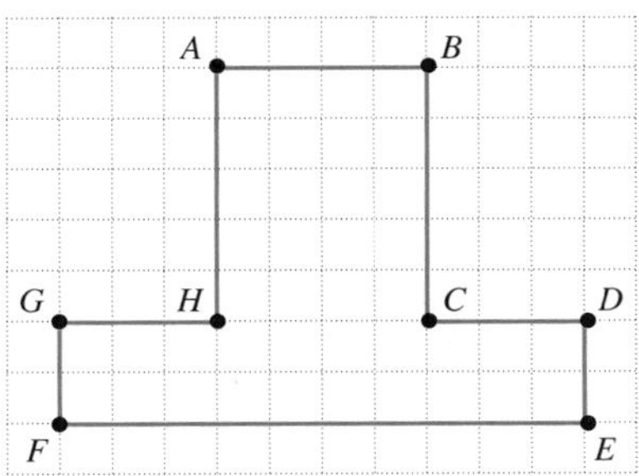

a) Insert a vertical line m through the figure so the figure has reflective symmetry about line m.

b) Is it possible to insert a horizontal line l through the figure so the figure has reflective symmetry about line l? Explain your answer.

c) Is it possible to insert a point P within the figure so the figure has 90° rotational symmetry about point P? Explain your answer.

d) Is it possible to insert a point P within the figure so the figure has 180° rotational symmetry about point P? Explain your answer.

Challenge Problems/Group Activities

55. *Glide Reflection, Order* Examine the figure below and then do the following:

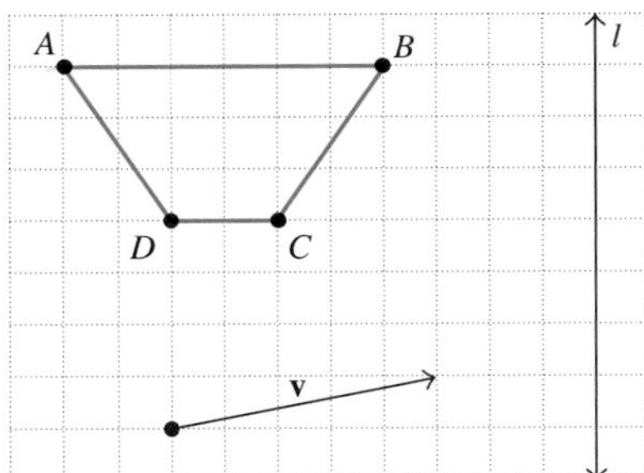

a) Determine a glide reflection of trapezoid $ABCD$ by first applying translation vector **v** and then reflecting about the line l. Label the glide reflection $A'B'C'D'$.

b) In this step we will reverse the order of the translation and the reflection. First reflect trapezoid $ABCD$ about the line l and then translate the reflection using vector **v**. Label the resulting figure $A''B''C''D''$.

c) Is figure $A'B'C'D'$ in the same position as figure $A''B''C''D''$?

d) What can be said about the order of the translation and the reflection used in a glide reflection? Is the figure obtained in part (a) or part (b) the glide reflection?

56. *Tessellation with a Square* Create a unique tessellation from a square piece of cardboard by using the method described on page 532 of the text. Be creative using color and sketches to complete your tessellation.

57. *Tessellation with a Hexagon* Using the method described on page 532, create a unique tessellation using a regular hexagon like the one shown below. Be creative using color and sketches to complete your tessellation.

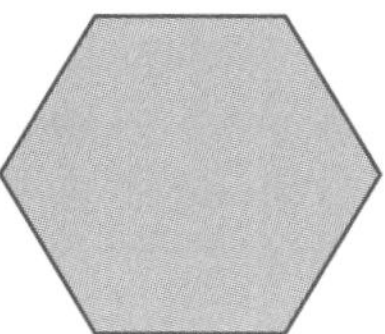

58. *Tessellation with an Octagon?* **a)** Trace the regular octagon, shown below, onto a separate piece of paper.

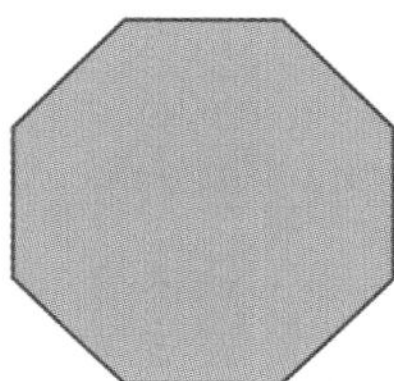

b) Try to create a regular tessellation by tracing this octagon repeatedly. Attempt to cover the entire piece of paper where no two octagons overlap each other. What conclusion can you draw about using a regular octagon as a tessellating shape?

59. *Tessellation with a Pentagon?* Repeat Exercise 58 using the regular pentagon below instead of a regular octagon.

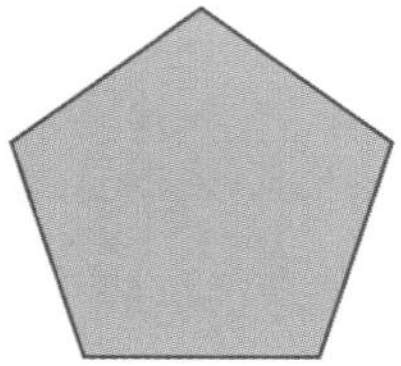

Recreational Mathematics

60. Examine each capital letter in the alphabet and determine which letters have reflective symmetry about a horizontal line through the center of the letter.

61. Examine each capital letter in the alphabet and determine which letters have reflective symmetry about a vertical line through the center of the letter.

62. Examine each capital letter in the alphabet and determine which letters have 180° rotational symmetry about a point in the center of the letter.

Internet/Research Activities

63. In the study of biology reflective symmetry is called *bilateral symmetry* and rotational symmetry is called *radial symmetry*. Do research and write a report on the role symmetry plays in the study of biology.

64. Write a paper on the mathematics displayed in the artwork of M. C. Escher. Include such topics as tessellations, optical illusions, perspective, and non-Euclidean geometry.

9.6 THE MÖBIUS STRIP, KLEIN BOTTLE, AND MAPS

The branch of mathematics called *topology* is sometimes referred to as "rubber sheet geometry" because it deals with bending and stretching of geometric figures.

One of the first pioneers of topology was the German astronomer and mathematician August Ferdinand Möbius (1790–1866). A student of Gauss, Möbius was the director of the University of Leipzig's observatory. He spent a great deal of time studying geometry and he played an essential part in the systematic development of projective geometry. He is best known for his studies of the properties of one-sided surfaces, including the one called the Möbius strip.

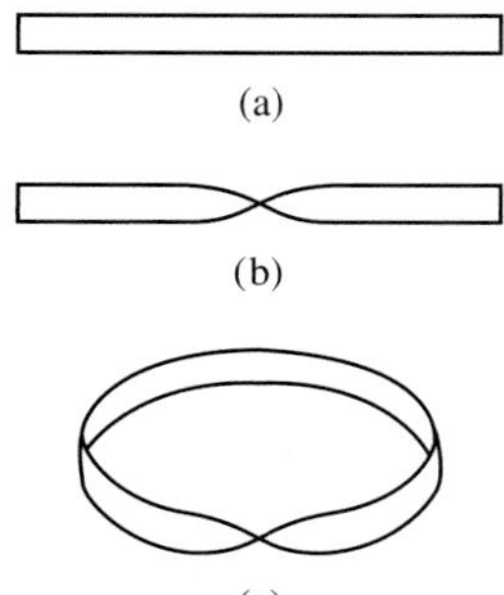

Figure 9.85

Möbius Strip

If you place a pencil on one surface of a sheet of paper and do not remove it from the sheet, you must cross the edge to get to the other surface. Thus, a sheet of paper has one edge and two surfaces. The sheet retains these properties even when crumpled into a ball. The *Möbius strip*, also called a *Möbius band*, is a one-sided, one-edged surface. You can construct one, as shown in Figure 9.85, by (a) taking a strip of paper, (b) giving one end a half twist, and (c) taping the ends together.

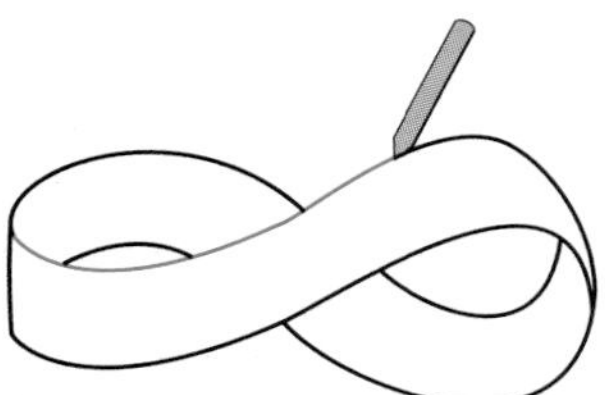

Figure 9.86

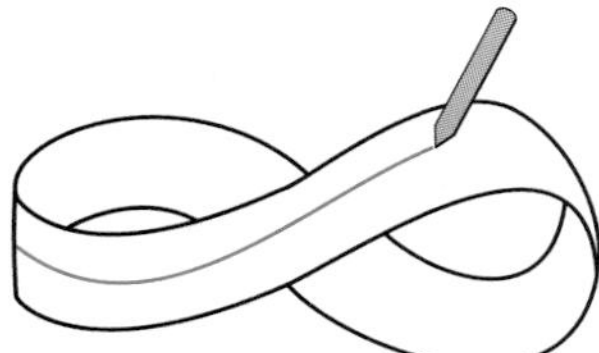

Figure 9.87

Figure 9.88

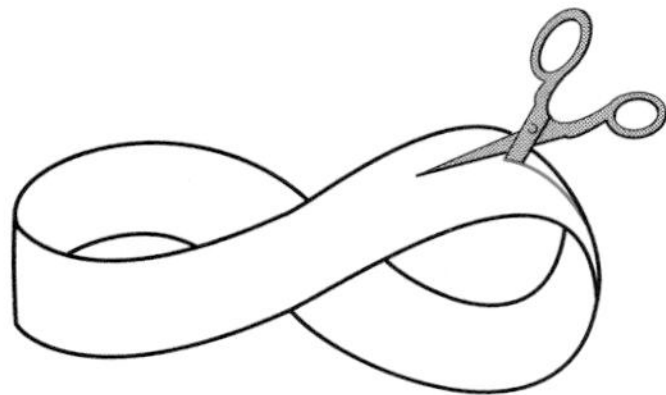

Figure 9.89

The Möbius strip has some very interesting properties. To better understand these properties, perform the following experiments.

Experiment 1 Make a Möbius strip using a strip of paper and tape as illustrated in Fig. 9.85. Place the point of a felt-tip pen on the edge of the strip (Fig. 9.86). Pull the strip slowly so that the pen marks the edge; do not remove the pen from the edge. Continue pulling the strip and observe what happens.

Experiment 2 Make a Möbius strip. Place the tip of a felt-tip pen on the surface of the strip (Fig. 9.87). Pull the strip slowly so that the pen marks the surface. Continue and observe what happens.

Experiment 3 Make a Möbius strip. Use scissors to make a small slit in the middle of the strip. Starting at the slit, cut along the strip, keeping the scissors in the middle of the strip (Fig. 9.88). Continue cutting and observe what happens.

Experiment 4 Make a Möbius strip. Make a small slit at a point about one-third of the width of the strip. Cut along the strip, keeping the scissors the same distance from the edge (Fig. 9.89). Continue cutting and observe what happens.

If you give a strip of paper several twists, you get variations on the Möbius strip. To a topologist the important distinction is between an odd number of twists, which leads to a one-sided surface, and an even number of twists, which leads to a two-sided surface. All strips with an odd number of twists are topologically the same as a Möbius strip, and all strips with an even number of twists are topologically the same as an ordinary cylinder, which has no twists.

Klein Bottle

Another topological object is the punctured *Klein bottle*; see Fig. 9.90. This object, named after Felix Klein (1849–1925), resembles a bottle but only has one side.

A punctured Klein bottle can be made by stretching a hollow piece of glass tubing. The neck is then passed through a hole and joined to the base.

Look closely at the model of the Klein bottle shown in Fig. 9.90. The punctured Klein bottle has only one edge and no outside or inside because it has just one side. Figure 9.91 shows a Klein bottle blown in glass by Alan Bennett of Bedford, England.

Limericks from unknown writers:

"A mathematician confided
That a Möbius band is one-sided,
And you'll get quite a laugh
If you cut one in half
For it stays in one piece when divided."

"A mathematician named Klein
Thought the Möbius band was divine.
He said, 'If you glue
the edges of two
You'll get a weird bottle like mine.' "

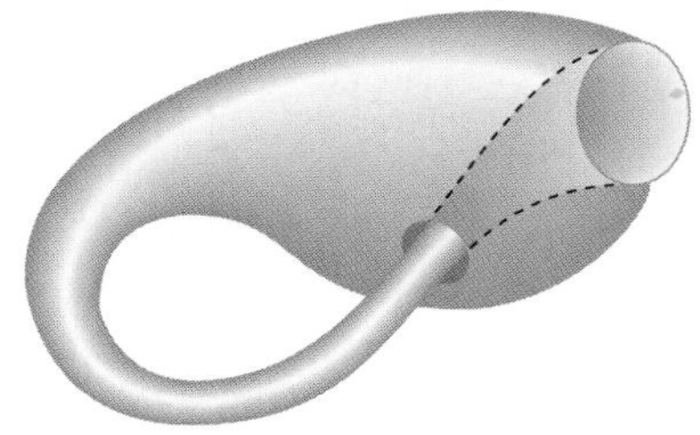

Figure 9.90

Figure 9.91 *Klein bottle,* a one-sided surface, blown in glass by Alan Bennett.

DID YOU KNOW

His Vision Came in a Dream

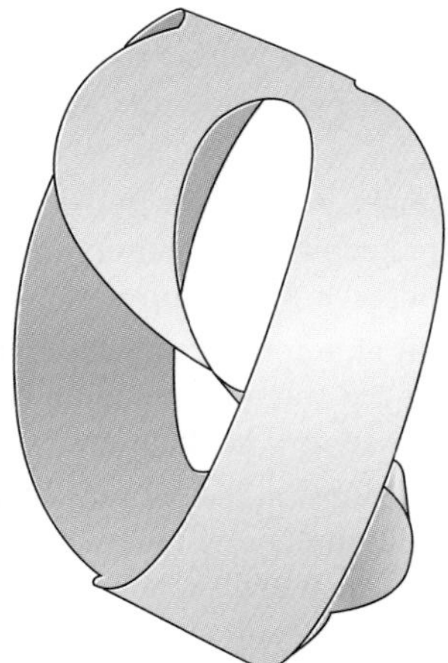

Paper-strip Klein bottle

The Life, the Times, and the Art of Branson Graves Stevenson, by Herbert C. Anderson Jr. (Janher Publishing, 1979), reports that "in response to a challenge from his son, Branson made his first Klein bottle. . . . He failed in his first try, until the famous English potter, Wedgwood, came to Branson in a dream and showed him how to make the Klein bottle." That was around 50 years ago. Branson's study of claywork and pottery eventually led to the formation of the Archie Bray Foundation in Helena, Montana.

People have made Klein bottles from all kinds of materials. There is a knitting pattern for a woolly Klein bottle and even a paper Klein bottle with a hole.

Imagine trying to paint a Klein bottle. You start on the "outside" of the large part and work your way down the narrowing neck. When you cross the self-intersection, you have to pretend temporarily that it is not there, so you continue to follow the neck, which is now inside the bulb. As the neck opens up, to rejoin the bulb, you find that you are now painting the inside of the bulb! What appear to be the inside and outside of a Klein bottle connect together seamlessly since it is one-sided.

It is interesting to note that if a Klein bottle is cut along a curve, the results are two (one-twist) Möbius strips, see Fig. 9.92. Thus, a Klein bottle could also be made by gluing together two Möbius strips along the edges.

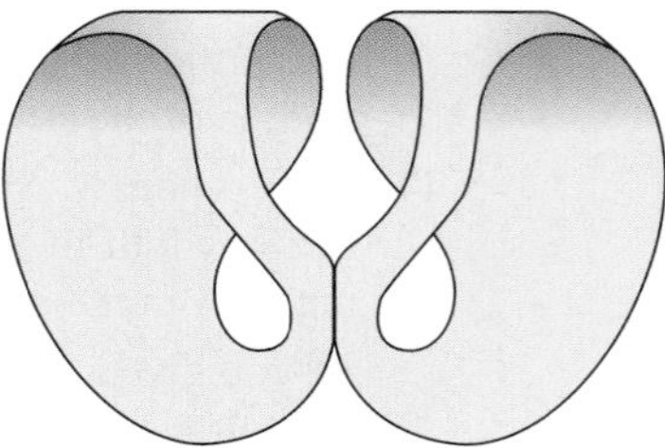

Figure 9.92 Two Möbius strips result from cutting a Klein bottle along a curve.

Maps

Maps have fascinated topologists for years because of the many challenging problems they present. Mapmakers have known for a long time that regardless of the complexity of the map and whether it is drawn on a flat surface or a sphere, only four colors are needed to differentiate each country (or state) from its immediate neighbors. Thus, every map can be drawn by using only four colors, and no two countries with a common border will have the same color. Regions that meet at only one point (such as the states of Arizona, Colorado, Utah, and New Mexico) are not considered to have a common border. In Fig. 9.93(a) no two states with a common border are marked with the same color.

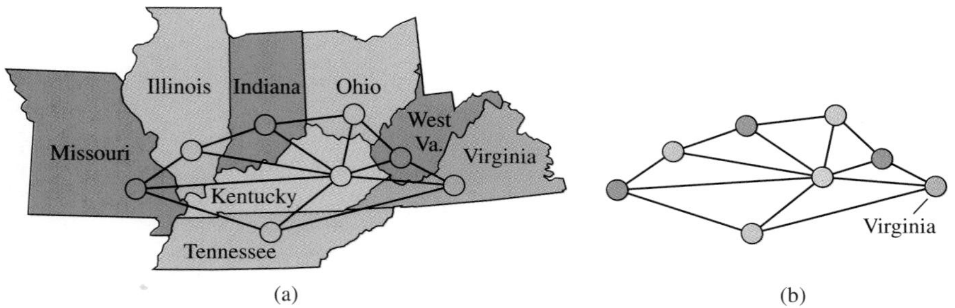

Figure 9.93

The "four-color" problem was first suggested by a student of Augustus DeMorgan in 1852. In 1976 Kenneth Appel and Wolfgang Haken of the University of Illinois—using their ingenuity, logic, and 1200 hours of computer time—succeeded in proving that only four colors are needed to draw a map. They solved the four-color map problem by reducing any map to a series of points and connecting line segments.

They replaced each country with a point. They connected two countries having a common border with a straight line; see Fig. 9.93(b). They then showed that the points of any graph in the plane could be colored by using only four colors in such a way that no two points connected by the same line were the same color.

Mathematicians have shown that, on different surfaces, more than four colors may be needed to draw a map. For example, a map drawn on a Möbius strip requires a maximum of six colors as in Fig. 9.94(a). A map drawn on a torus (the shape of a doughnut) requires a maximum of seven colors as in Fig. 9.94(b).

Figure 9.94

Jordan Curves

A *Jordan curve* is a topological object that can be thought of as a circle twisted out of shape; see Fig. 9.95 (a)–(d). Like a circle, it has an inside and an outside. To get from one side to the other at least one line must be crossed. Consider the Jordan curve in Fig. 9.95(d). Are points A and B inside or outside the curve?

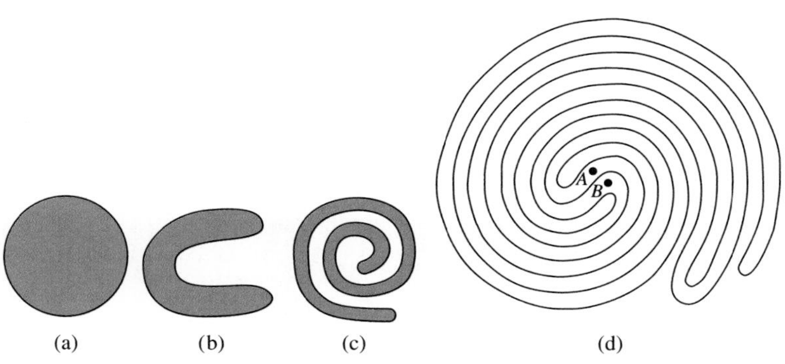

Figure 9.95

A quick way to tell whether the two dots are inside or outside the curve is to draw a straight line from each dot to a point that is clearly outside the curve. If the straight line crosses the curve an even number of times, the dot is outside. If the straight line crosses the curve an odd number of times, the dot is inside the curve. Can you explain why this procedure works? Determine whether point A and point B are inside or outside the curve (see Exercises 21 and 22 at the end of this section).

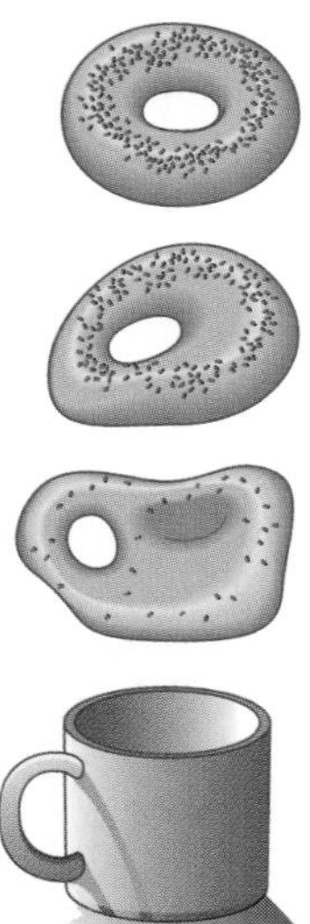

Figure 9.96

Topological Equivalence

Someone once said that a topologist is a person who does not know the difference between a doughnut and a coffee cup. Two geometric figures are said to be *topologically equivalent* if one figure can be elastically twisted, stretched, bent, or shrunk into the other figure without puncturing or ripping the original figure. If a doughnut is made of elastic material, it can be stretched, twisted, bent, shrunk, and distorted until it resembles a coffee cup with a handle, as shown in Fig. 9.96. Thus, the doughnut and coffee cup are topologically equivalent.

In topology figures are classified according to their *genus*. The *genus* of an object is determined by the number of holes in the object. A cup and a doughnut each have one hole and are of genus 1. A kettle and scissors each have two holes and are of genus 2. Figure 9.97 illustrates this type of classification.

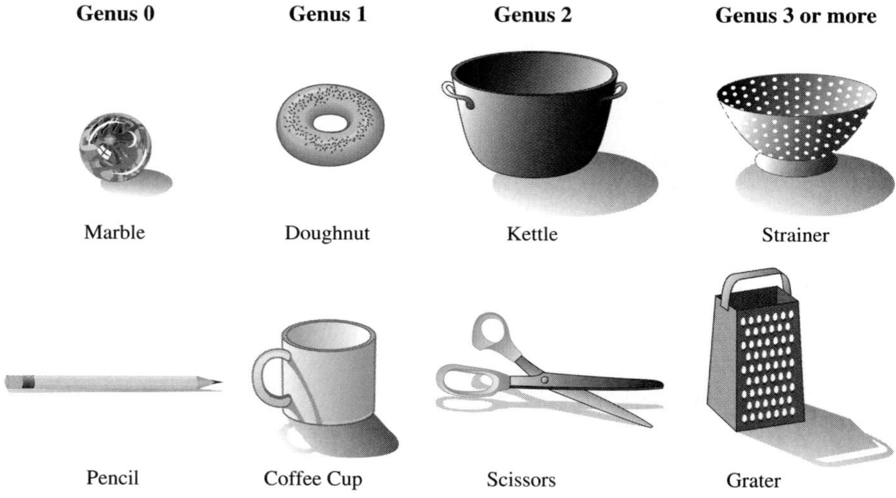

Figure 9.97

SECTION 9.6 EXERCISES

Concept/Writing Exercises

1. Explain why topology is sometimes referred to as "rubber sheet geometry."
2. What is a Möbius strip?
3. Explain how to make a Möbius strip.
4. What is a Klein bottle?
5. What is the maximum number of colors needed to create a map on a flat surface if no two regions colored the same are to share a common border?
6. What is the maximum number of colors needed to create a map if no two regions colored the same are to share a common border if the surface is a
   a) Möbius strip?
   b) torus?
7. What is a Jordan curve?
8. When testing to determine whether a point is inside or outside a Jordan curve, explain why if you count an odd number of lines, the point is inside the curve, and if you count an even number of lines, the point is outside the curve.

9. How is the genus of a figure determined?
10. When are two figures topologically equivalent?

Practice the Skills

In Exercises 11–16, color the map by using a maximum of four colors so that no two regions with a common border have the same color.

11.

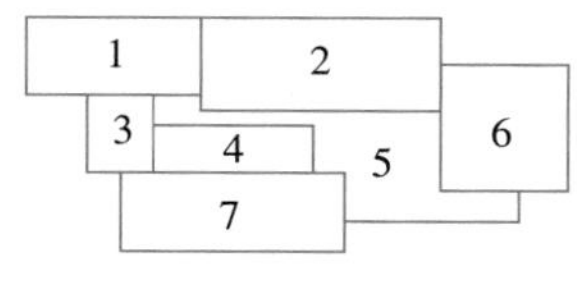

12.

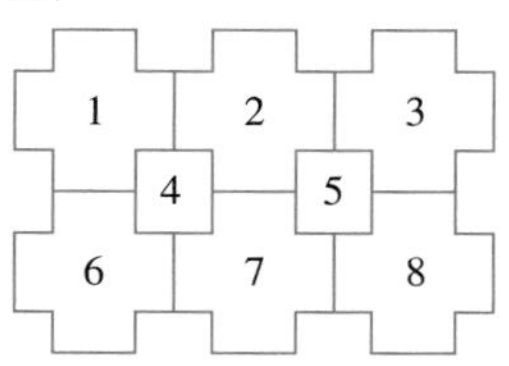

13.

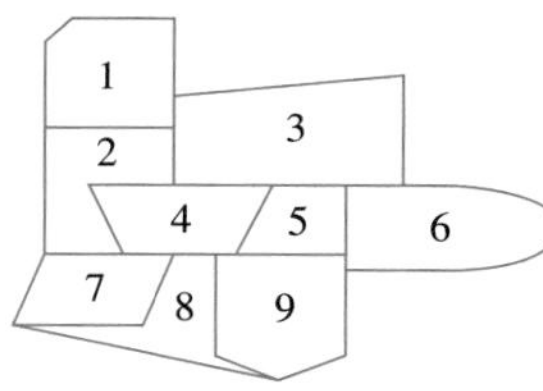

14.

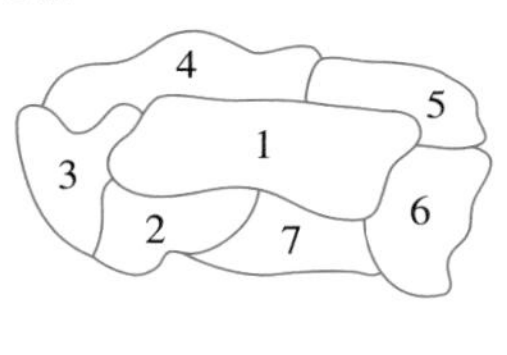

15.

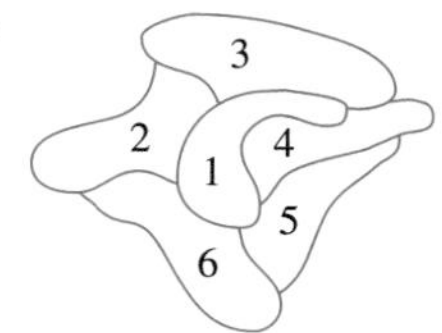

16.

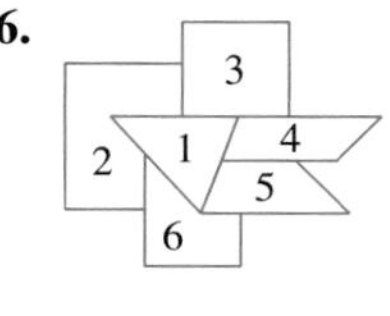

Using the Four-Color Theorem In Exercises 17–20, maps show certain areas of the United States, Canada, and Mexico. Shade in the states (or provinces) using a maximum of four colors so that no two states (or provinces) with a common border have the same color.

17.

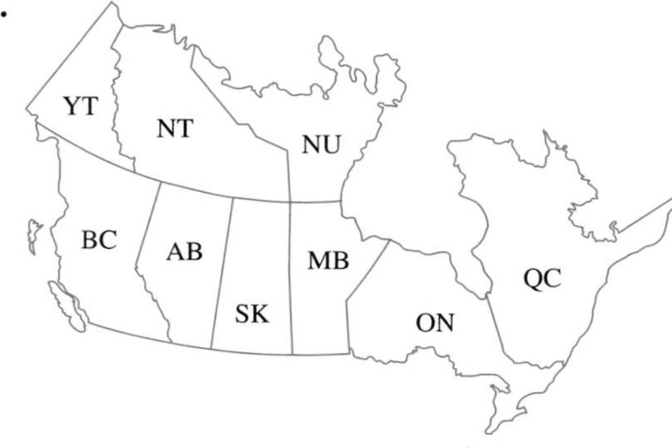

18.

19.

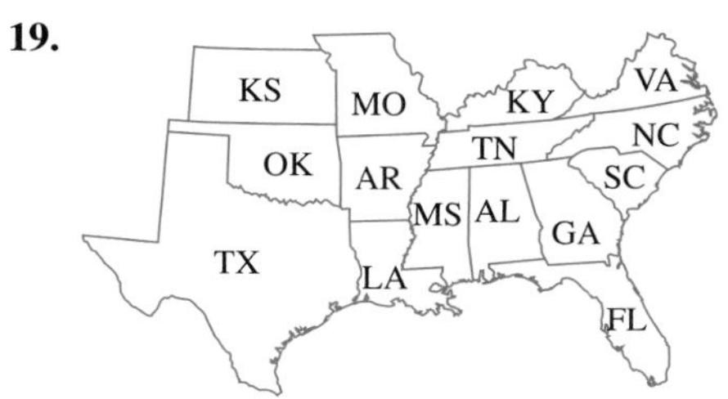

20.

21. Determine whether point A in Fig. 9.95(d) on page 541 is inside or outside the Jordan curve.
22. Determine whether point B in Fig. 9.95(d) is inside or outside the Jordan curve.

At right is a Jordan curve. In Exercises 23–28, determine if the point is inside or outside of the curve.

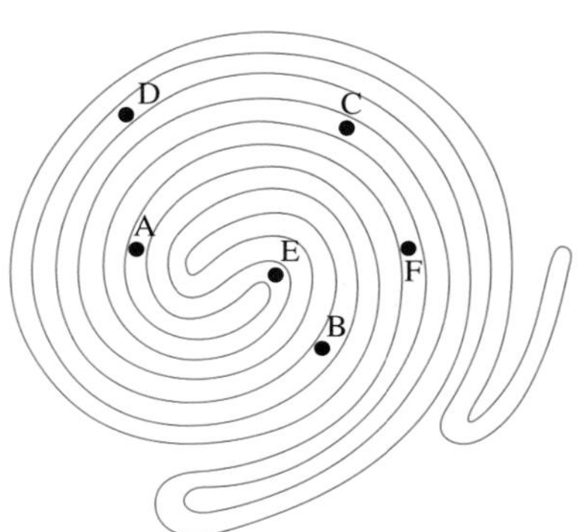

23. Point A
24. Point B
25. Point C
26. Point D
27. Point E
28. Point F

In Exercises 29–40, give the genus of the object. If the object has a genus larger than 5, write "larger than 5."

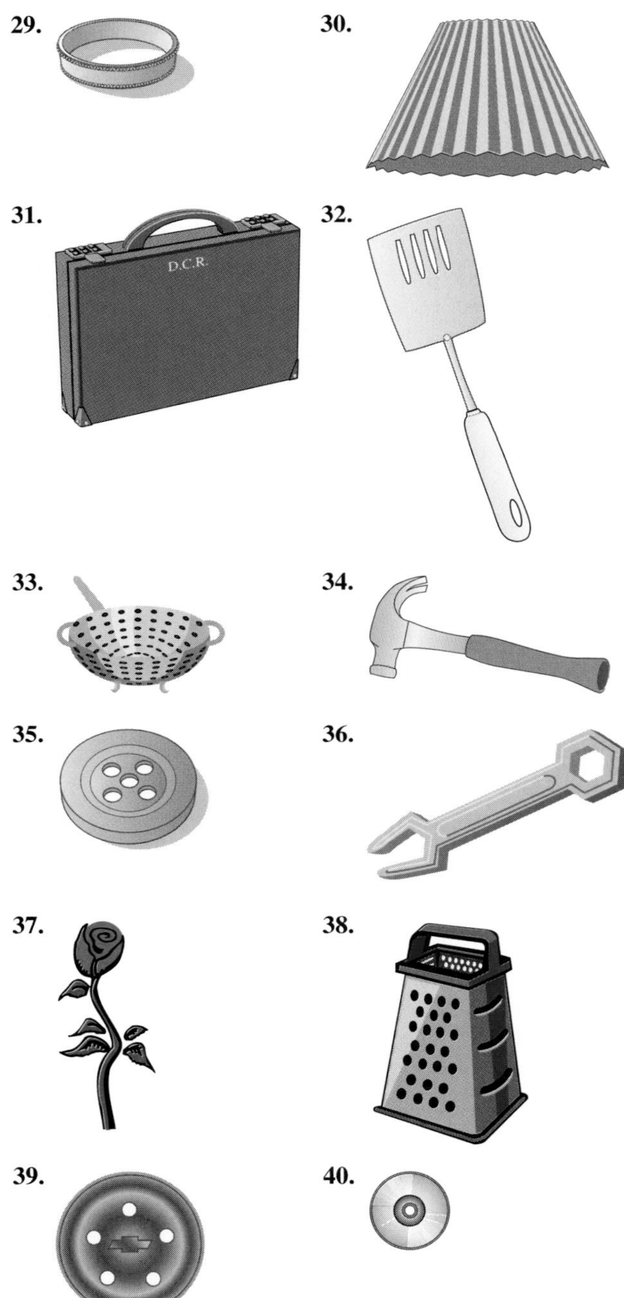

41. Name at least three objects not mentioned in this section that have
a) genus 0.
b) genus 1.
c) genus 2.
d) genus 3 or more.

42. Use the result of Experiment 1 on page 539 to find the number of edges on a Möbius strip.

43. Use the result of Experiment 2 on page 539 to find the number of surfaces on a Möbius strip.

44. How many separate strips are obtained in Experiment 3 on page 539?

45. How many separate strips are obtained in Experiment 4 on page 539?

46. **a)** Take a strip of paper, give it one full twist, and connect the ends. Is the result a Möbius strip with only one side? Explain.
b) Determine the number of edges, as in Experiment 1.
c) Determine the number of surfaces, as in Experiment 2.
d) Cut the strip down the middle. What is the result?

47. Make a Möbius strip. Cut it one-third of the way from the edge, as in Experiment 4. You should get two loops, one going through the other. Determine whether either (or both) of these loops is itself a Möbius strip.

48. Take a strip of paper, make one whole twist and another half twist, and then tape the ends together. Test by a method of your choice to determine whether this has the same properties as a Möbius strip.

Challenge Problems/Group Activities

49. Can you see any advantage in a Möbius conveyor belt? Explain.

50. Using clay (or glazing compound) make a doughnut. Without puncturing or tearing the doughnut, reshape it into a topologically equivalent figure, a cup with a handle.

51. Using at most four colors, color the map of South America. Do not use the same color for any two countries that share a common border.

52. Using at most four colors, color the following map of the counties of Arizona. Do not use the same color for any two counties that share a common border.

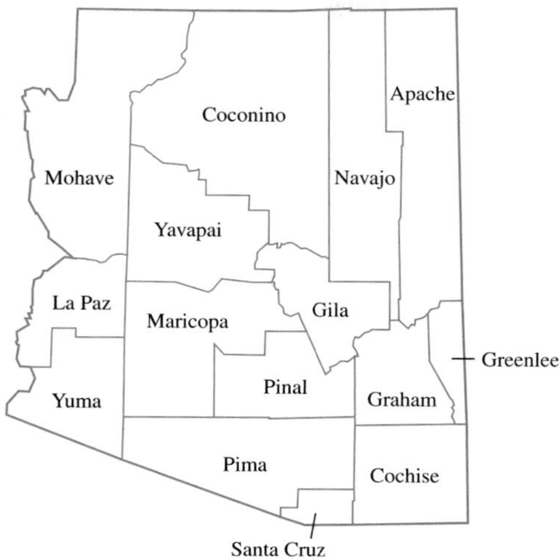

Recreational Mathematics

53. *An Interesting Surface* Construct the following surface using two strips of paper, scissors, and tape, then answer the following questions.

a) How many sides does this surface have?
b) How many edges does this surface have?
c) Attempt to cut the surface "in half" by making a small slit in the middle of the paper surface. Then cut along the surface (see dashed line in the figure), keeping the scissors the same distance from the edge. In your own words, describe what happens.

It is interesting to note that although the surface shown in the figure above shares some of the same traits as a Möbius strip, this surface is not topologically equivalent to the Möbius strip.

Internet/Research Activity

54. Use the Internet to find a map of your state that shows the outline of all the counties within your state. Print this map and, using at most four colors, color it. Do not use the same color for any two counties that share a common border.

9.7 NON-EUCLIDEAN GEOMETRY AND FRACTAL GEOMETRY

Non-Euclidean Geometry

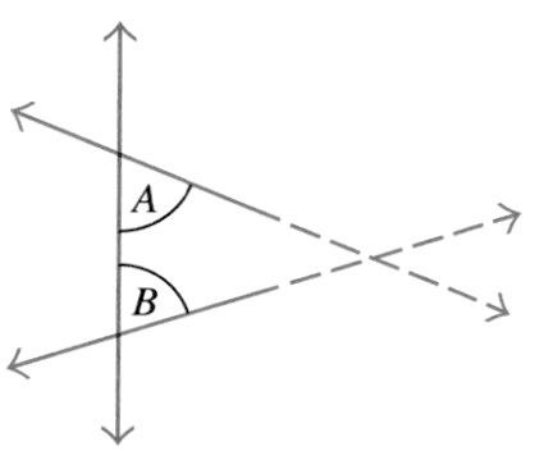

Figure 9.98

In Section 9.1 we stated postulates or axioms are statements to be accepted as true. In his book *Elements,* Euclid's fifth postulate was, "If a straight line falling on two straight lines makes the interior angles on the same side less than two right angles, the two straight lines, if produced indefinitely, meet on that side on which the angles are less than the two right angles."

Euclid's fifth axiom may be better understood by observing Fig. 9.98. The sum of angles A and B is less than the sum of two right angles (180°). Therefore, the two lines will meet if extended.

John Playfair (1748–1819), a Scottish physicist and mathematician, wrote a geometry book that was published in 1795. In his book Playfair gave a logically equivalent interpretation of Euclid's fifth postulate. This version is often referred to as Playfair's postulate or the Euclidean parallel postulate.

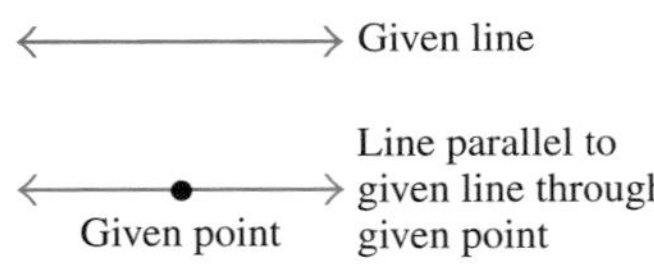

Figure 9.99

The Euclidean Parallel Postulate

Given a line and a point not on the line, one and only one line can be drawn through the given point parallel to the given line (Fig. 9.99)

The Euclidean parallel postulate may be better understood by looking at Fig. 9.99. Many mathematicians after Euclid believed that this postulate was not as self-evident as the other nine. Others believed this postulate could be proved from the other nine postulates and therefore was not needed at all. Of the many attempts to prove that the fifth postulate was not needed, the most noteworthy one was presented by Girolamo Saccheri (1667–1733), a Jesuit priest in Italy. In the course of his elaborate chain of deductions, Saccheri proved many of the theorems of what is now called hyperbolic geometry. However, Saccheri did not realize what he had done. He believed that Euclid's geometry was the only "true" geometry and concluded that his own work was in error. Thus, Saccheri narrowly missed receiving credit for a great achievement: the founding of *non-Euclidean geometry*.

Over time, geometers became more and more frustrated at their inability to prove Euclid's fifth postulate. One of them, a Hungarian named Farkos Bolyai, in a letter to his son, Janos Bolyai, wrote, "I entreat you leave the science of parallels alone. . . . I have traveled past all reefs of this infernal dead sea and have always come back with a broken mast and torn sail." The son, refusing to heed his father's advice, continued to think about parallels until, in 1823, he saw the whole truth and enthusiastically declared, "I have created a new universe from nothing." He recognized that geometry branches in two directions, depending on whether Euclid's fifth postulate is applied. He recognized two different geometries and published his discovery as a 24-page appendix to a textbook written by his father. The famous mathematician George Bruce Halsted called it "the most extraordinary two dozen pages in the whole history of thought." Farkos Bolyai proudly presented a copy of his son's work to his friend Carl Friedrich Gauss, then Germany's greatest mathematician, whose reply to the father had a devastating effect on the son. Gauss wrote, "I am unable to praise this work. . . . To praise it would be to praise myself. Indeed, the whole content of the work, the path taken by your son, the results to which he is led, coincides almost entirely with my meditations which occupied my mind partly for the last thirty or thirty-five years." We now know from his earlier correspondence that Gauss had indeed been familiar with *hyperbolic geometry* even before Janos was born. In his letter Gauss also indicated that it was his intention not to let his theory be published during his lifetime, but to record it so that the theory would not perish with him. It is believed that the reason Gauss did not publish his work was that he feared being ridiculed by other prominent mathematicians of his time.

Nikolay Ivanovich Lobachevsky

At about the same time as Bolyai's publication, Nikolay Ivanovich Lobachevsky, a Russian, published a paper that was remarkably like Bolyai's, although it was quite independent of it. Lobachevsky made a deeper investigation and wrote several books. In marked contrast to Bolyai, who received no recognition during his lifetime, Lobachevsky received great praise and became a professor at the University of Kazan.

G. F. Bernhard Riemann

After the initial discovery, little attention was paid to the subject until 1854, when G. F. Bernhard Riemann (1826–1866), a student of Gauss, suggested a second type of non-Euclidean geometry, which is now called *spherical*, *elliptical*, or *Riemannian geometry*. The hyperbolic geometry of his predecessors was synthetic; that is, it was not based on or related to any concrete model when it was developed. Riemann's

MATHEMATICS *Everywhere*

Mapping The Brain

Medical researchers and mathematicians are currently attempting to capture an image of the three-dimensional human brain on a two-dimensional map. In some respects the task is like capturing the image of Earth on a two-dimensional map, but because of the many folds and fissures on the surface of the brain, the task is much more complex. Points of the brain that are at different depths can appear too close in a flat image. Therefore, to develop an accurate mapping, researchers use topology, hyperbolic geometry, and elliptical geometry to create an image known as a *conformal mapping*. Researchers use conformal mappings to precisely identify the parts of the brain that correspond to specific functions.

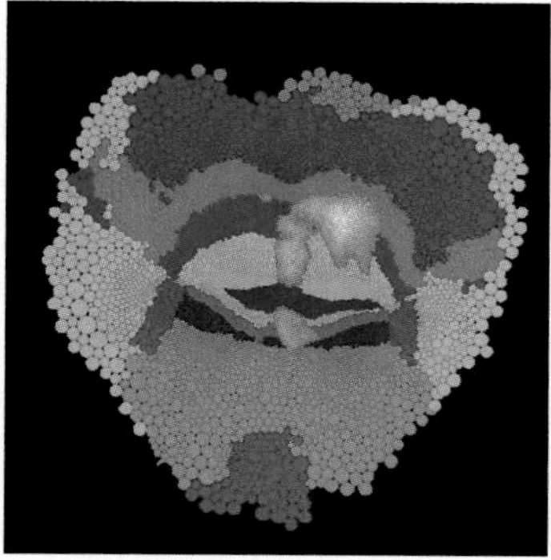

Photograph courtesy of Dr. Monica K. Hurdal (mhurdal@math.fsu.edu) Dept. of Mathematics, Florida State University

geometry was closely related to the theory of surfaces. A *model* may be considered a physical interpretation of the undefined terms that satisfies the axioms. A model may be a picture or an actual physical object.

The two types of non-Euclidean geometries we have mentioned are elliptical geometry and hyperbolic geometry. The major difference among the three geometries lies in the fifth axiom. The fifth axiom of the three geometries is summarized here.

The Fifth Axiom of Geometry

Euclidean	Elliptical	Hyperbolic
Given a line and a point not on the line, one and only one line can be drawn parallel to the given line through the given point.	Given a line and a point not on the line, no line can be drawn through the given point parallel to the given line.	Given a line and a point not on the line, two or more lines can be drawn through the given point parallel to the given line.

To understand the fifth axiom of the two non-Euclidean geometries, remember that the term *line* is undefined. Thus, a line can be interpreted differently in different geometries. A model for Euclidean geometry is a plane, such as a blackboard (Fig. 9.100a). A model for elliptical geometry is a sphere (Fig. 9.100b). A model for hyperbolic geometry is a pseudosphere (Fig. 9.100c). A pseudosphere is similar to two trumpets placed bell to bell. Obviously, a line on a plane cannot be the same as a line on either of the other two figures. The curved red lines in Fig. 9.101(a) and both colored lines in Fig. 9.102, on page 548, are examples of lines in elliptical and hyperbolic geometry, respectively.

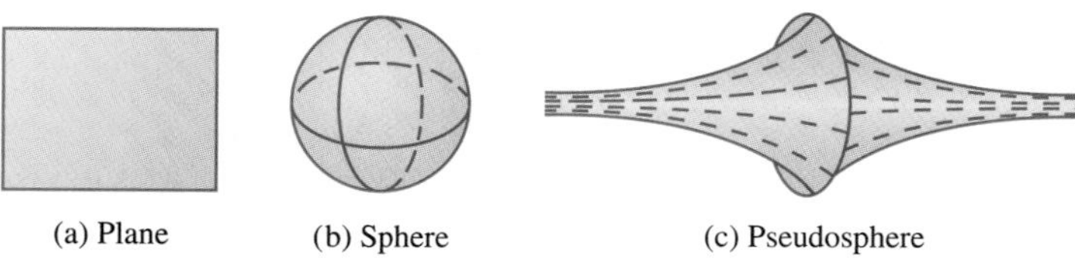

(a) Plane (b) Sphere (c) Pseudosphere

Figure 9.100

Elliptical Geometry

A circle on the surface of a sphere is called a great circle if it divides the sphere into two equal parts. If we were to cut through a sphere along a great circle, we would have two identical pieces. If we interpret a line to be a great circle, we can see the fifth axiom of elliptical geometry is true. Two great circles on a sphere must intersect; hence, there can be no parallel lines (Fig. 9.101a).

If we were to construct a triangle on a sphere, the sum of its angles would be greater than 180° (Fig. 9.101b). The theorem, "The sum of the measures of the angles of a triangle is greater than 180°," has been proven by means of the axioms of elliptical geometry. The sum of the measures of the angles varies with the area of the triangle and gets closer to 180° as the area decreases.

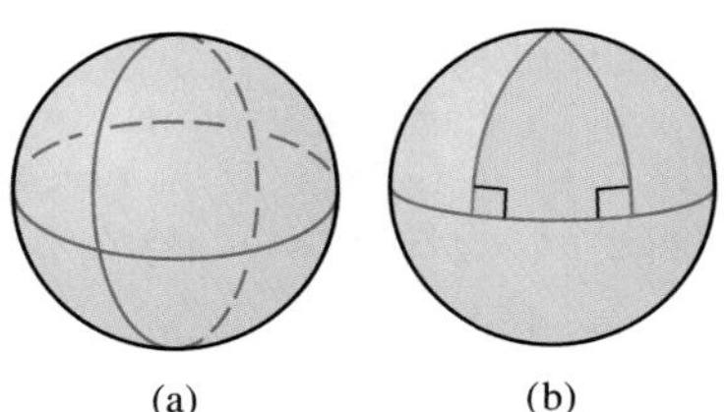

(a) (b)

Figure 9.101

Hyperbolic Geometry

The lines in hyperbolic geometry are represented by geodesics on the surface of the pseudosphere. A *geodesic* is the shortest and least-curved arc between two points on a surface. Figure 9.102 illustrates two lines on the surface of a pseudosphere.

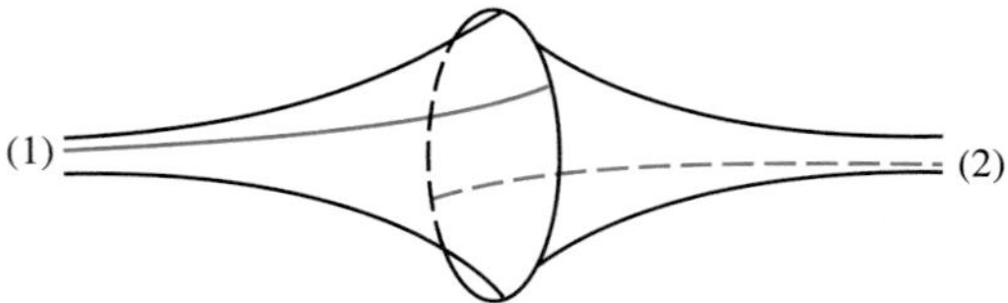

Figure 9.102

Figure 9.103(a) illustrates the fifth axiom of hyperbolic geometry.* Note that, through the given point, two lines are drawn parallel to the given line. If we were to construct a triangle on a pseudosphere, the sum of the measures of the angles would be less than 180° (Fig. 9.103b). The theorem, "The sum of the measures of the angles of a triangle is less than 180°," has been proven by means of the axioms of hyperbolic geometry.

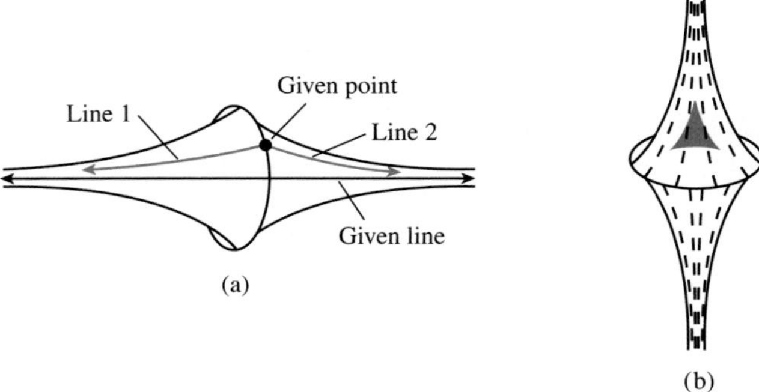

Figure 9.103

We have stated that the sum of the measures of the angles of a triangle is 180°, is greater than 180°, and is less than 180°. Which statement is correct? Each statement is correct *in its own system.* Many theorems hold true for all three geometries; vertical angles still have the same measure, we can uniquely bisect a line segment with a straightedge and compass alone, and so on.

The many theorems based on the fifth postulate may differ in each geometry. It is important for you to realize each theorem proved is true *in its own system* because each is logically deduced from the given set of axioms of the system. No one system is the "best" system. Euclidean geometry may appear to be the one to use in the classroom, where the blackboard is flat. In discussions involving Earth as a whole, however, elliptical geometry may be the most useful, since Earth is a sphere. If the object under consideration has the shape of a saddle or pseudosphere, hyperbolic geometry may be the most useful.

---

*A formal discussion of hyperbolic geometry is beyond the scope of this text.

DID YOU KNOW

It's All Relative

Albert Einstein's general theory of relativity, published in 1916, approached space and time differently from our everyday understanding of them. Einstein's theory unites the three dimensions of space with one of time in a four-dimensional space, time continuum. His theory dealt with the path that light and objects take while moving through space under the force of gravity. Einstein conjectured that mass (such as stars and planets) caused space to be curved. The greater the mass, the greater the curvature. Also, in the region nearer to the mass, the curvature of the space is greater.

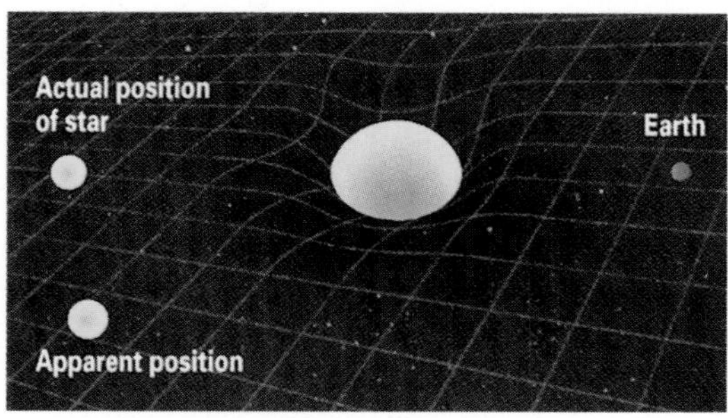

You can visualize Einstein's theory by thinking of space as a rubber sheet pulled taut on which a mass is placed, causing the rubber sheet to bend.

To prove his conjecture, Einstein exposed himself to Riemann's non-Euclidean geometry. Einstein believed that the trajectory of a particle in space represents not a straight line but the straightest curve possible, a geodesic.

Space–time is now thought to be a combination of three different types of curvature: spherical (described by Riemannian geometry), flat (described by Euclidean geometry), and saddle-shaped (described by hyperbolic geometry).

Einstein's theory was confirmed by the solar eclipses of 1919 and 1922.

"The Great Architect of the universe now appears to be a great mathematician."
British physicist Sir James Jeans

Fractal Geometry

We are familiar with one-, two-, and three-dimensional figures. Many objects, however, are difficult to categorize as one-, two-, or three-dimensional. For example, how would you classify the irregular shapes we see in nature such as a coastline, or the bark on a tree, or a mountain, or a path followed by lightning? For a long time mathematicians assumed that making realistic geometric models of natural shapes and figures was almost impossible, but the development of *fractal geometry* now makes it possible. Both color photos on this page were made by using fractal geometry. The discovery and study of fractal geometry has been one of the most popular mathematical topics in recent times.

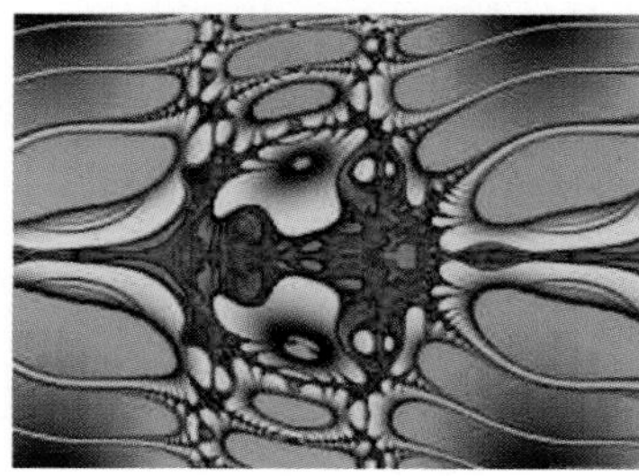
Fractal images

The word *fractal* (from the Latin word *fractus,* "broken up, fragmented") was first used in the mid-1970s by mathematician Benoit Mandelbrot to describe shapes that had several common characteristics, including some form of "self-similarity," as will be seen shortly in the Koch snowflake.

Typical fractals are extremely irregular curves or surfaces that "wiggle" enough so that they are not considered one-dimensional. Fractals do not have integer dimensions; their dimensions are between 1 and 2. For example, a fractal may have a dimension of 1.26. Fractals are developed by applying the same rule over and over again, with the end point of each simple step becoming the starting point for the next step, in a process called *recursion*.

Using the recursive process, we will develop a famous fractal called the *Koch snowflake* named after Helga von Koch, a Swedish mathematician who first discovered its remarkable characteristics. The Koch snowflake illustrates a property of all fractals called *self-similarity*; that is, each smaller piece of the curve resembles the whole curve.

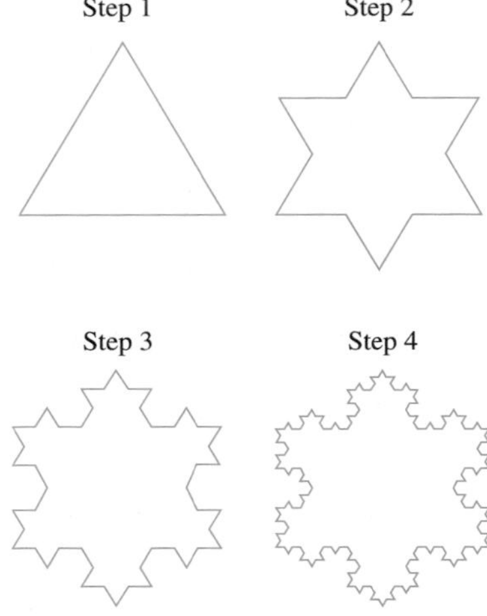

Figure 9.104

To develop the Koch snowflake:

1. Start with an equilateral triangle (step 1, Fig. 9.104).
2. Whenever you see an edge —— replace it with _/_ (steps 2–4).

What is the perimeter of the snowflake in Fig. 9.104, and what is its area? A portion of the boundary of the Koch snowflake known as the Koch curve or the snowflake curve is represented in Fig. 9.105.

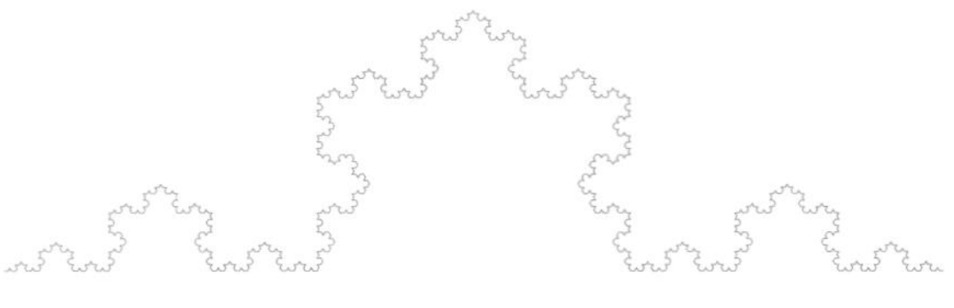

Figure 9.105

Because the Koch curve consists of infinitely many pieces of the form _/_, the perimeter is also infinite. The area of the Koch snowflake is 1.6 times the area of the starting equilateral triangle. Thus, the area of the snowflake is finite. The Koch snowflake has a finite area enclosed by an infinite boundary! This fact may seem difficult to accept, but it is true. However, the Koch snowflake, like other fractals, is not an everyday run-of-the-mill geometric shape.

Let us look at a few more fractals made using the recursive process. We will now construct what is known as a *fractal tree.* Start with a tree trunk (Fig. 9.106a on page 551). Draw two branches, each one a bit smaller than the trunk (Fig. 9.106b). Draw two branches from each of those branches, and continue; see Figs. 9.106c and Fig. 9.106d. Ideally we continue the process forever.

DID YOU KNOW

Fractal Antennas

Fractal triangle can act as a miniaturized antenna.

Recently researchers have begun to use antennas made using fractal designs. Many antennas that look like a simple unit, including most radar antennas, are actually arrays of up to thousands of small antennas. Scientists have discovered that a fractal arangement can result in antennas being as powerful as traditional antennas, using only a quarter of the number of elements. Dwight Jasserd of the University of Pennsylvania says, "Fractals bridge the gap; they have short-range disorder and long-range order." Fractal antennas are 25% more efficient than the rubbery "stubby" antennas found on most phones. Why do fractal antennas work so well? It has been proven mathematically that for an antenna to work equally well at all frequencies, it must be self-similar, having the same basic appearance at every scale. That is, it has to be fractal!

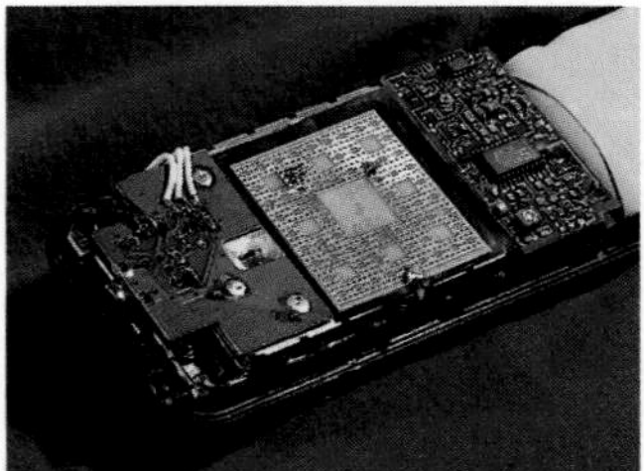

Hidden inside a cordless phone, a square fractal antenna (center board) replaces the usual rubbery stalk.

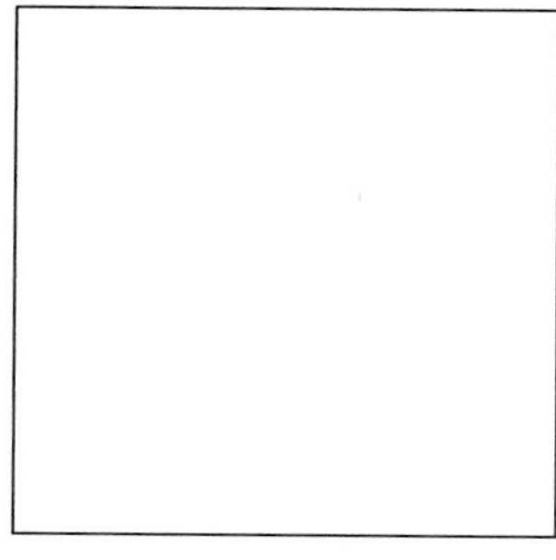
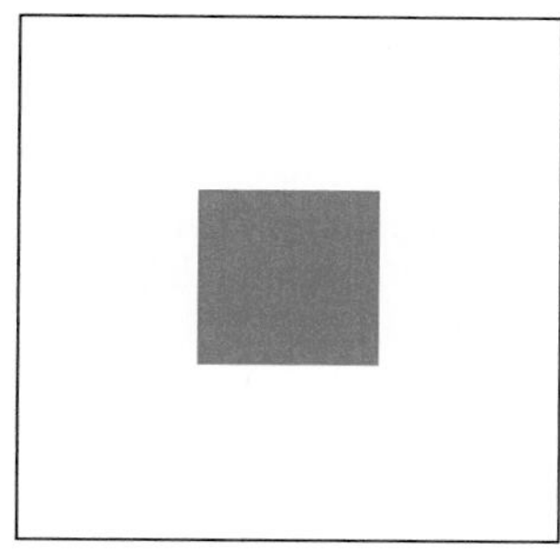
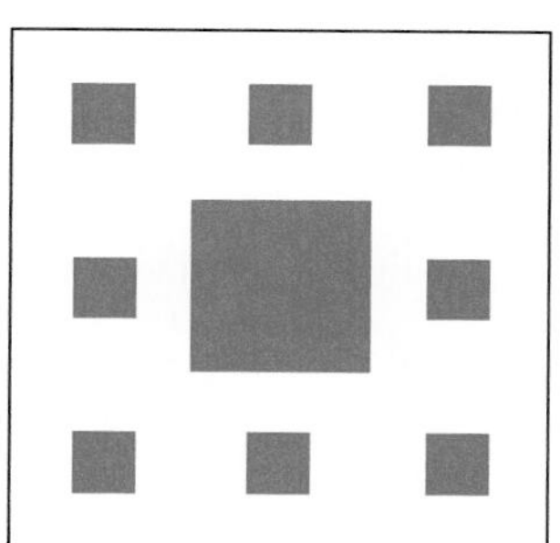
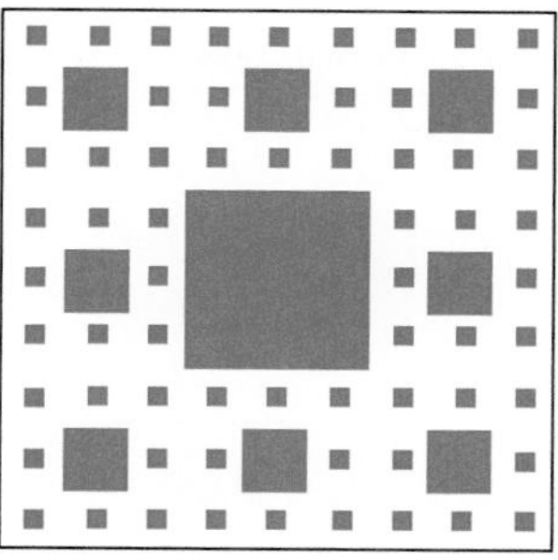
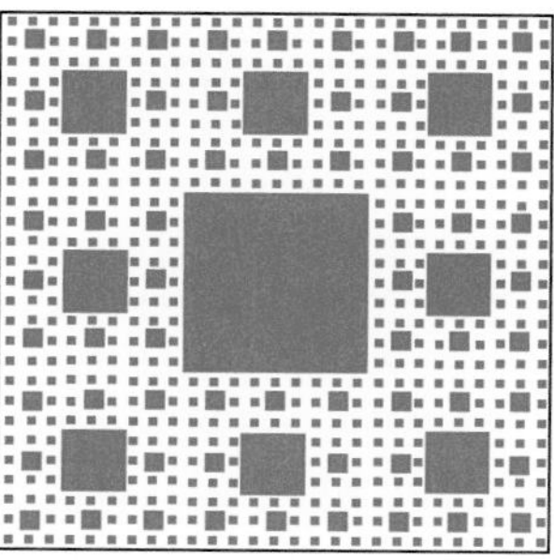

Figure 9.108 Sierpinski carpet

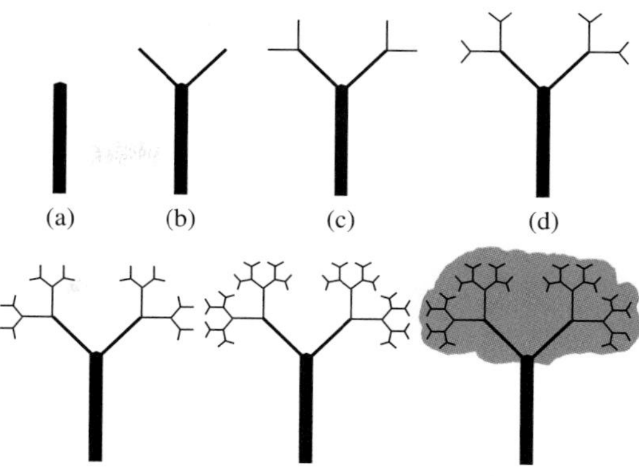

Figure 9.106 The fractal tree

If you take a little piece of any branch and zoom in on it, it will look exactly like the original tree. Fractals are *scale independent,* which means that you cannot really tell whether you are looking at something very big or something very small because the fractal looks the same whether you are close to it or far from it.

In Figs. 9.107 and 9.108 we develop two other fractals through the process of recursion. Figure 9.107 shows a fractal called the Sierpinski triangle, and Fig. 9.108 shows a fractal called the Sierpinski carpet. Both fractals are named after Waclaw Sierpinski, a Polish mathematician who is best known for his work with fractals and space-filling curves.

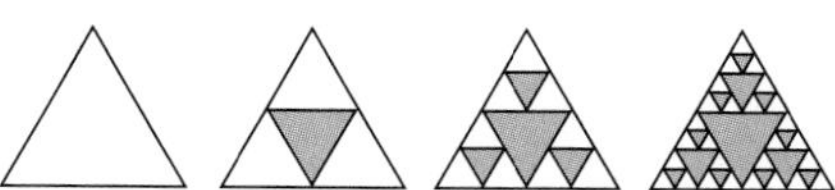

Figure 9.107 Sierpinski triangle

Fractals provide a way to study natural forms such as coastlines, trees, mountains, galaxies, polymers, rivers, weather patterns, brains, lungs, and blood supply. Fractals also help explain that which appears chaotic. The blood supply in the body is one example. The branching of arteries and veins appear chaotic, but closer inspection reveals the same type of branching occurs for smaller and smaller blood vessels, down to the capillaries. Thus, fractal geometry provides a geometric structure for chaotic processes in nature. The study of chaotic processes is called *chaos theory.*

Fractals nowadays have a potentially important role to play in characterizing weather systems and in providing insight into various physical processes such as the occurrence of earthquakes or the formation of deposits that shorten battery life. Some scientists view fractal statistics as a doorway to unifying theories of medicine, offering a powerful glimpse of what it means to be healthy.

Fractals lie at the heart of current efforts to understand complex natural phenomena. Unraveling their intricacies could reveal the basic design principles at work in our world. Until recently, there was no way to describe fractals. Today, we are beginning to see such features everywhere. Tomorrow, we may look at the entire universe through a fractal lens.

SECTION 9.7 EXERCISES

Concept/Writing Exercises

In Exercises 1–6, describe the accomplishments of the mathematician.

1. Girolamo Saccheri
2. Janos Bolyai
3. Carl Friedrich Gauss
4. Nikolay Ivanovich Lobachevsky
5. G. F. Bernhard Riemann
6. Benoit Mandelbrot
7. State the fifth axiom of
   a) Euclidean geometry. **b)** hyperbolic geometry.
   c) elliptical geometry.
8. State the theorem concerning the sum of the measures of the angles of a triangle in
   a) Euclidean geometry. **b)** hyperbolic geometry.
   c) elliptical geometry.
9. What model is often used in describing and explaining Euclidean geometry?
10. What model is often used in describing and explaining elliptical geometry?
11. What model is often used in describing and explaining hyperbolic geometry?
12. What do we mean when we say that no one axiomatic system of geometry is "best"?
13. List the three types of curvature of space and the types of geometry that correspond to them.
14. List at least five natural forms that appear chaotic that we can study using fractals.

Practice the Skills

In the following we show a fractal-like figure made using a recursive process with the letter "M." Use this fractal-like figure as a guide in constructing fractal-like figures with the letter given in Exercises 15–18. Show three steps, as is done here.

15. I **16.** W **17.** V **18.** H

19. **a)** Develop a fractal by beginning with a square and replacing each side —— with a ⎍ . Repeat this process twice.
    b) If you continue this process, will the fractal's perimeter be finite or infinite? Explain.
    c) Will the fractal's area be finite or infinite? Explain.

Problem Solving/Group Activity

20. In forming the Koch snowflake in Figure 9.104 on page 550 the perimeter becomes greater at each step in the process. If each side of the original triangle is 1 unit, a general formula for the perimeter, L, of the snowflake at any step, n, may be found by the formula

$$L = 3\left(\frac{4}{3}\right)^{n-1}$$

For example, at the first step when $n = 1$, the perimeter is 3 units, which can be verified by the formula as follows:

$$L = 3\left(\frac{4}{3}\right)^{1-1} = 3\left(\frac{4}{3}\right)^{0} = 3 \cdot 1 = 3$$

At the second step, when $n = 2$, we find the perimeter as follows:

$$L = 3\left(\frac{4}{3}\right)^{2-1} = 3\left(\frac{4}{3}\right) = 4$$

Thus, at the second step the perimeter of the snowflake is 4 units.

a) Use the formula to complete the following table.

Step	Perimeter
1	
2	
3	
4	
5	
6	

b) Use the results of your calculations to explain why the perimeter of the Koch snowflake is infinite.
c) Explain how the Koch snowflake can have an infinite perimeter, but a finite area.

Internet/Research Activities

In Exercises 21–23, references include the Internet, books on art, encyclopedias, and history of mathematics books.

21. To complete his masterpiece *Circle Limit III*, (see page 531) M. C. Escher studied a model of hyperbolic geometry

called the *Poincaré disk*. Write a paper on the Poincaré disk and how it was used in Escher's art. Include representations of *infinity* and the concepts of *point* and *line* in hyperbolic geometry.

22. To transfer his two-dimensional tiling known as *Symmetry Work 45* to a sphere, M. C. Escher used the spherical geometry of Bernhard Riemann. Write a paper on Escher's use of geometry to complete this masterpiece (see the figures below and to the right).

Symmetry Work 45

23. Go to the website *Fantastic Fractals* and study the information about fractals given there. Print copies, in color if a color printer is available, of the Mandlebrot set and the Julia set.

CHAPTER 9 SUMMARY

IMPORTANT FACTS

The sum of the measures of the angles of a triangle is 180°.

The sum of the measures of the angles of a quadrilateral is 360°.

The sum of the measures of the interior angles of an n-sided polygon is $(n - 2)180°$.

Triangle

$A = \frac{1}{2}bh$

$p = s_1 + s_2 + s_3$

Square

$A = s^2$

$p = 4s$

Rectangle

$A = lw$

$p = 2l + 2w$

Parallelogram

$A = bh$

$p = 2b + 2w$

Trapezoid

$A = \frac{1}{2}h(b_1 + b_2)$

$p = s_1 + s_2 + b_1 + b_2$

Pythagorean theorem

$a^2 + b^2 = c^2$

Circle

$A = \pi r^2$; $C = 2\pi r$ or $C = \pi d$

Cube

$V = s^3$

Rectangular solid

$V = lwh$

Cylinder

$V = \pi r^2 h$

Cone

$V = \frac{1}{3}\pi r^2 h$

Sphere

$V = \frac{4}{3}\pi r^3$

Prism

$V = Bh$, where B is the area of the base

Pyramid

$V = \frac{1}{3}Bh$, where B is the area of the base

Fifth postulate in Euclidean geometry

Given a line and a point not on the line, only one line can be drawn through the given point parallel to the given line.

Fifth postulate in elliptical geometry

Given a line and a point not on the line, no line can be drawn through the given point parallel to the given line.

Fifth postulate in hyperbolic geometry

Given a line and a point not on the line, two or more lines can be drawn through the given point parallel to the given line.

CHAPTER 9 REVIEW EXERCISES

9.1

In Exercises 1–6, use the figure shown to determine the following.

1. $\measuredangle EFI \cap \measuredangle BFC$
2. $\overline{BF} \cup \overline{FC} \cup \overline{BC}$
3. $\overleftrightarrow{AB} \cup \overline{BC}$
4. $\overrightarrow{BH} \cup \overrightarrow{HB}$ (open-ended rays)
5. $\overleftrightarrow{HI} \cap \overleftrightarrow{EG}$
6. $\overrightarrow{CF} \cap \overrightarrow{CG}$ (open-ended rays)

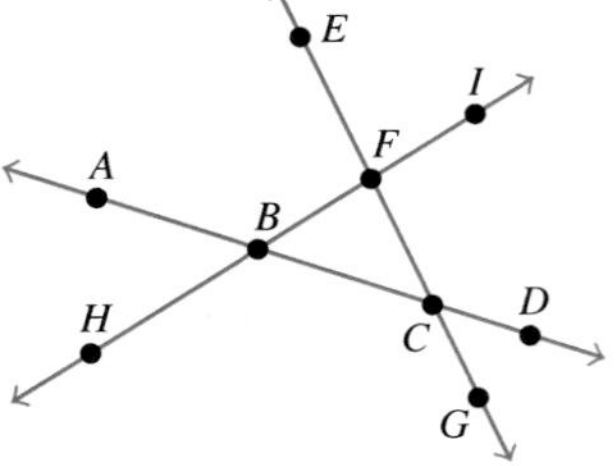

7. $m\measuredangle A = 51.2°$. Determine the measure of the complement of $\measuredangle A$.
8. $m\measuredangle B = 124.7°$. Determine the measure of the supplement $\measuredangle B$.

9.2

In Exercises 9–12, use the similar triangles ABC and A′B′C shown to determine the following.

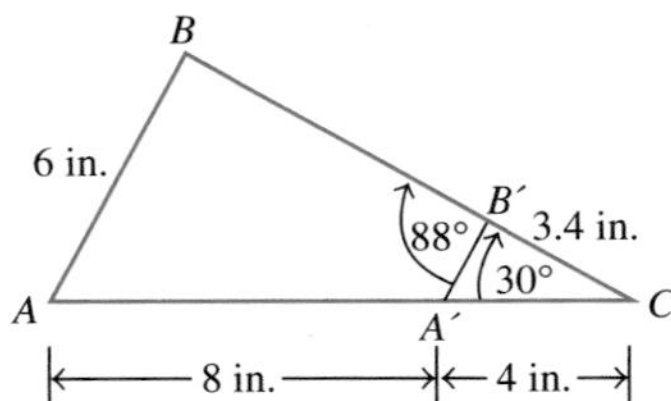

9. The length of $\overline{BC}$
10. The length of $\overline{A'B'}$
11. $m\measuredangle BAC$
12. $m\measuredangle ABC$
13. In the following figure, l_1 and l_2 are parallel lines. Determine $m\measuredangle 1$ through $m\measuredangle 6$.

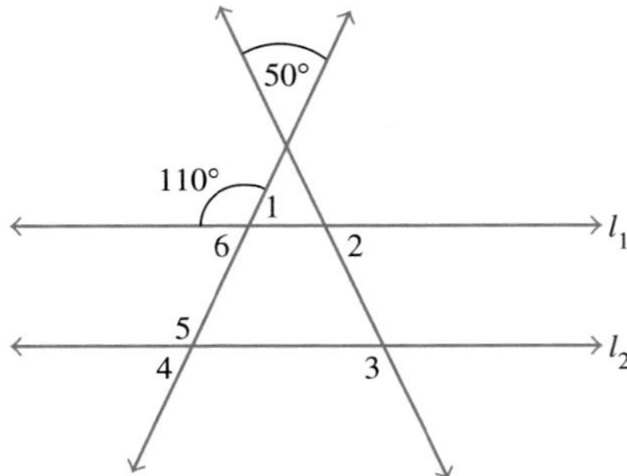

14. Determine the sum of the measures of the interior angles of a hexagon.

9.3

In Exercises 15–19, determine the area of the figure.

15.

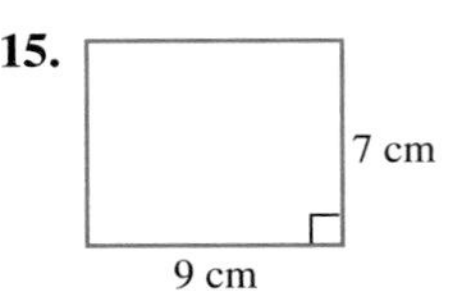

16.

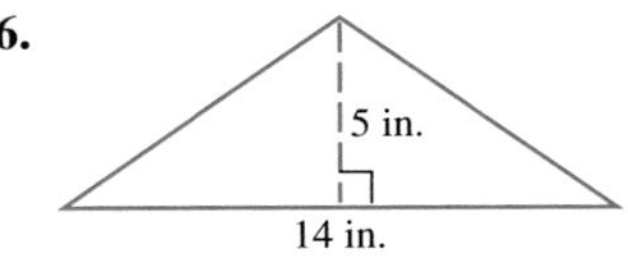

17.

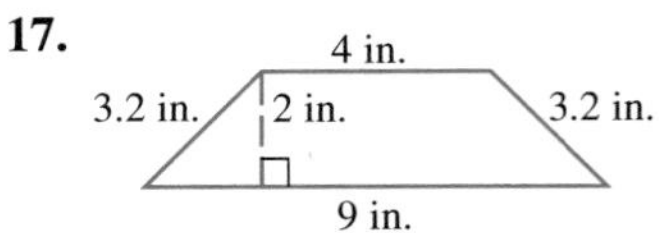

18.

19.

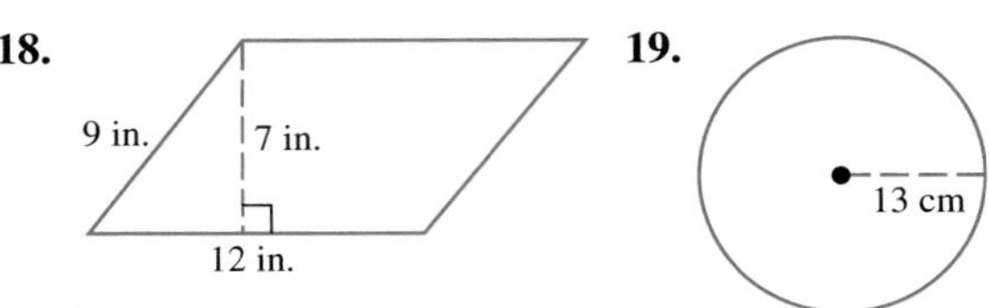

20. *Cost of Kitchen Tile* Determine the total cost of covering a 14 ft by 16 ft kitchen floor with ceramic tile. The cost of the tile selected is \$2.75 per square foot.

9.4

In Exercises 21–26, determine the volume of the figure. Round your answers to hundredths.

21.

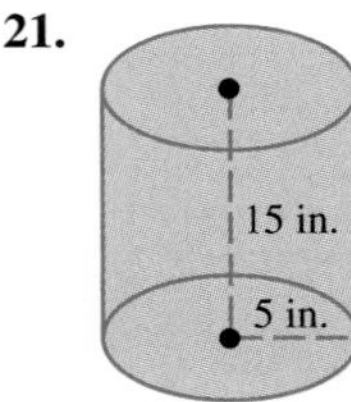

22.

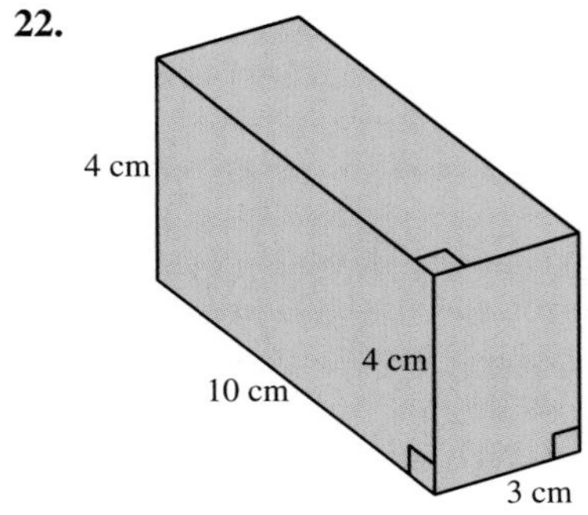

23.

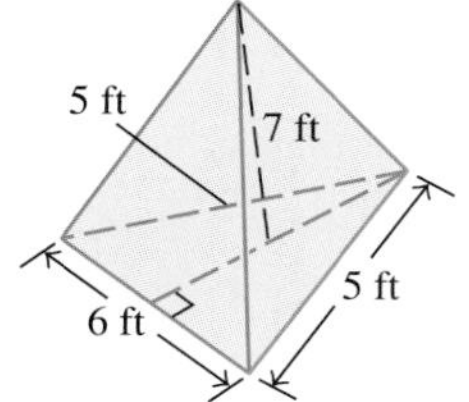

24.

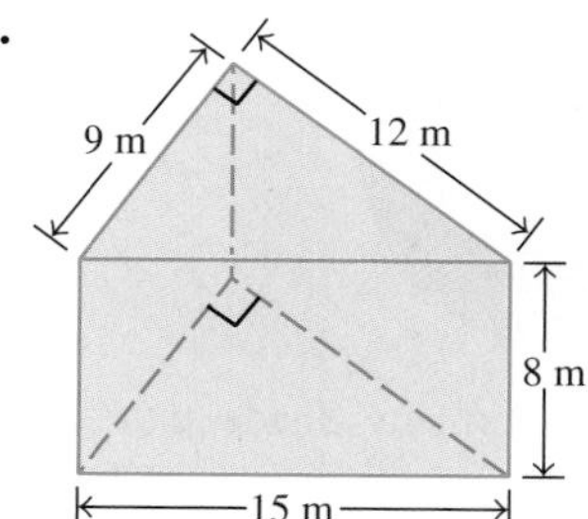

25.

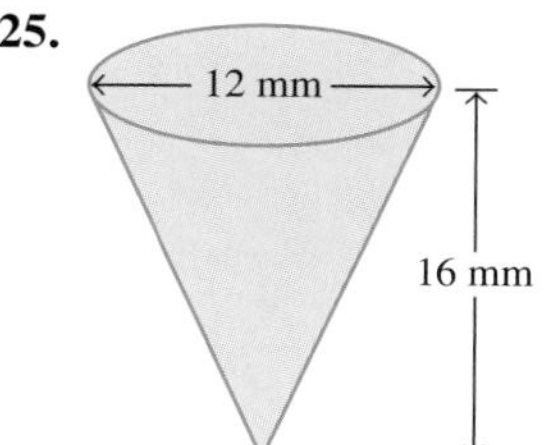

26.

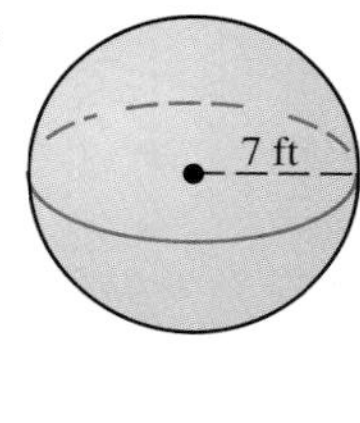

27. *Water Trough* Steven Dale has a water trough whose ends are trapezoids and whose sides are rectangles, as illustrated. He is afraid that the base it is sitting on will not support the weight of the trough when it is filled with water. He knows that the base will support 4800 lb.

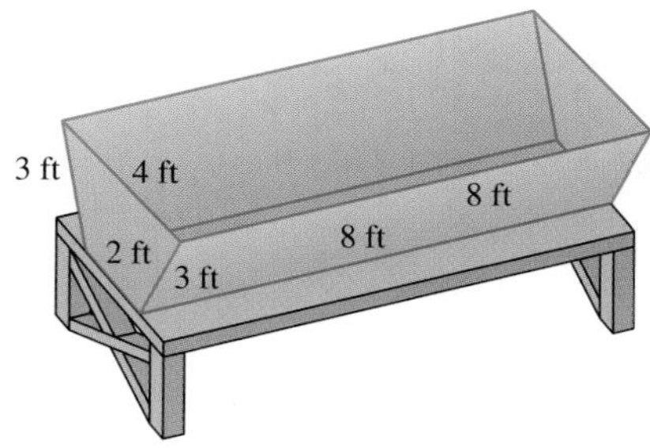

a) Find the number of cubic feet of water contained in the trough.
b) Find the total weight, assuming that the trough weighs 375 lb and the water weighs 62.5 lb per cubic foot. Is the base strong enough to support the trough filled with water?
c) If 1 gal of water weighs 8.3 lb, how many gallons of water will the trough hold?

9.5

In Exercises 28 and 29, use the given triangle and lines of reflection to construct the indicated reflections. Show the triangle in the positions both before and after the reflection. Construct

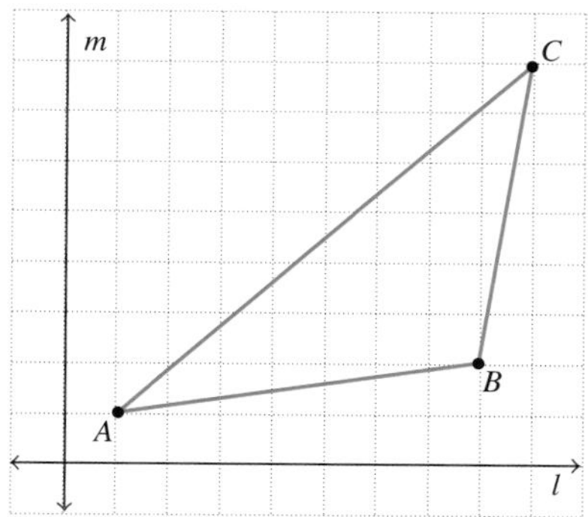

28. the reflection of triangle *ABC* about line *l*.

29. the reflection of triangle *ABC* about line *m*.

In Exercises 30 and 31, use translation vectors **v** *and* **w** *to construct the indicated translations. Show the parallelogram in the positions both before and after the translation. Construct*

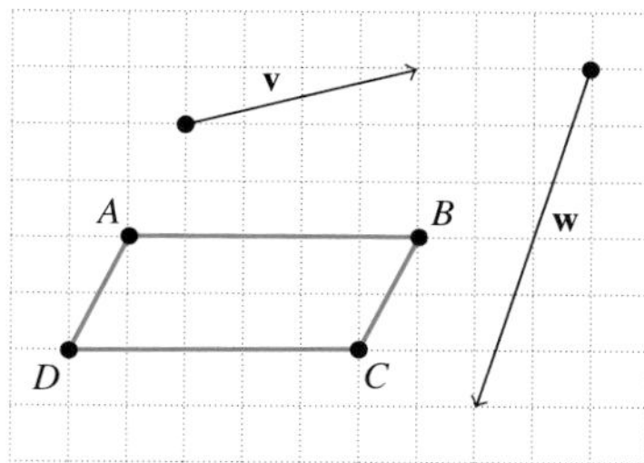

30. the translation of parallelogram *ABCD* using translation vector **v**.

31. the translation of parallelogram *ABCD* using translation vector **w**.

In Exercises 32–34, use the given figure and rotation point P to construct the indicated rotations. Show the trapezoid in the positions both before and after the rotation. Construct

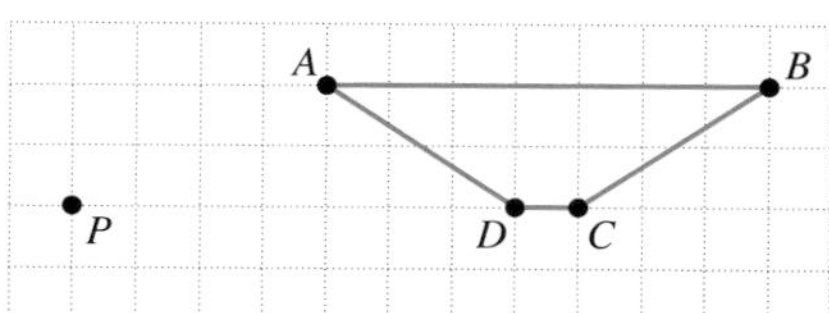

32. a 90° rotation of trapezoid *ABCD* about point *P*.

33. a 180° rotation of trapezoid *ABCD* about point *P*.

34. a 270° rotation of trapezoid *ABCD* about point *P*.

In Exercises 35 and 36, use the given figure, translation vector **v**, *and reflection lines l and m to construct the indicated glide reflections. Show the triangle in the positions both before and after the glide reflection. Construct*

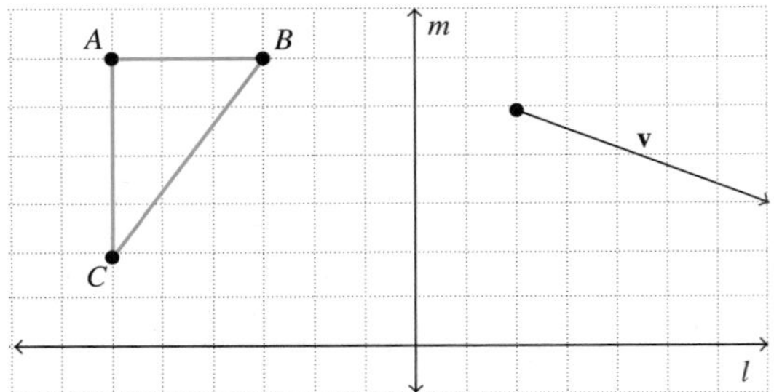

35. a glide reflection of triangle ABC using vector **v** and reflection line l.

36. a glide reflection of triangle ABC using vector **v** and reflection line m.

In Exercises 37 and 38, use the following figure to answer the following questions.

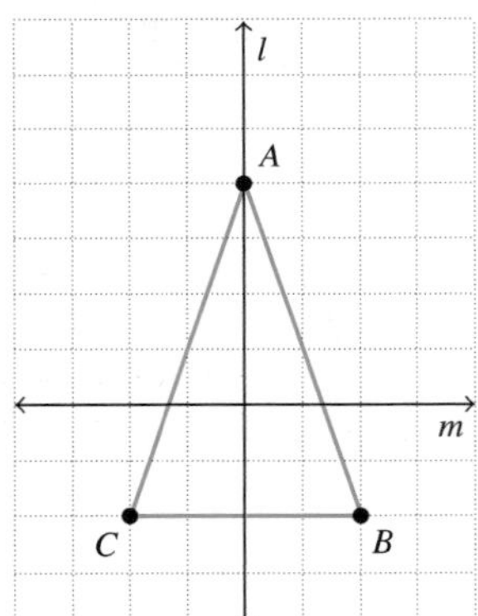

37. Does triangle ABC have reflective symmetry about line l? Explain.

38. Does triangle ABC have reflective symmetry about line m? Explain.

In Exercises 39 and 40, use the following figure to answer the following questions.

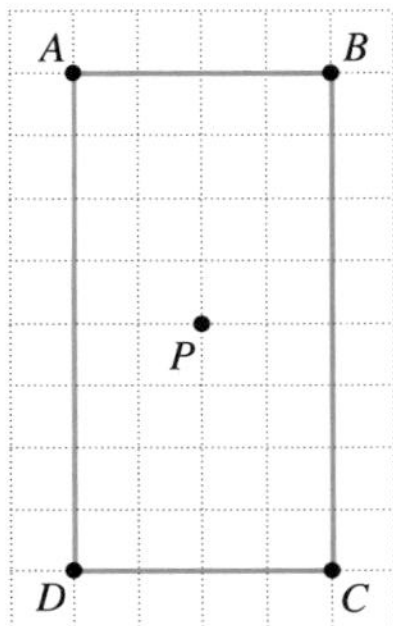

39. Does rectangle $ABCD$ have 90° rotational symmetry about point P? Explain.

40. Does rectangle $ABCD$ have 180° rotational symmetry about point P? Explain.

9.6

41. Give the genus of the object:

42. The map shows the states of Germany. Shade in the states using a maximum of four colors so that no two states with a common border have the same color.

43. Determine whether point A is inside or outside the Jordan curve.

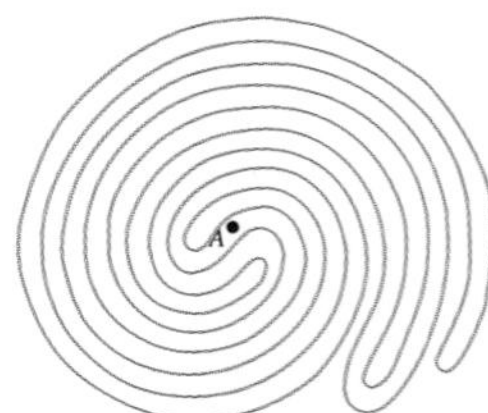

9.7

44. State the fifth axiom of Euclidean, elliptical, and hyperbolic geometry.

45. Develop a fractal by beginning with a square and replacing each side —— with a ⎍ . Repeat this process twice.

CHAPTER 9 TEST

In Exercises 1–4, use the figure to describe the following sets of points:

1. $\overleftrightarrow{AF} \cap \overleftrightarrow{EF}$
2. $\overline{BC} \cup \overline{CD} \cup \overline{BD}$
3. $\measuredangle EDF \cap \measuredangle BDC$
4. $\overrightarrow{AC} \cup \overrightarrow{BA}$

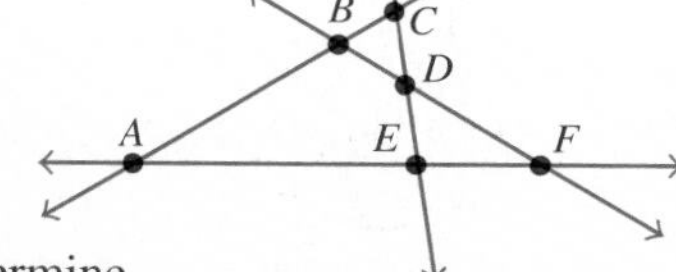

5. $m\measuredangle A = 36.9°$. Determine the measure of the complement of $\measuredangle A$.
6. $m\measuredangle B = 101.5°$. Determine the measure of the supplement of $\measuredangle B$.
7. In the figure, determine the measure of $\measuredangle x$.

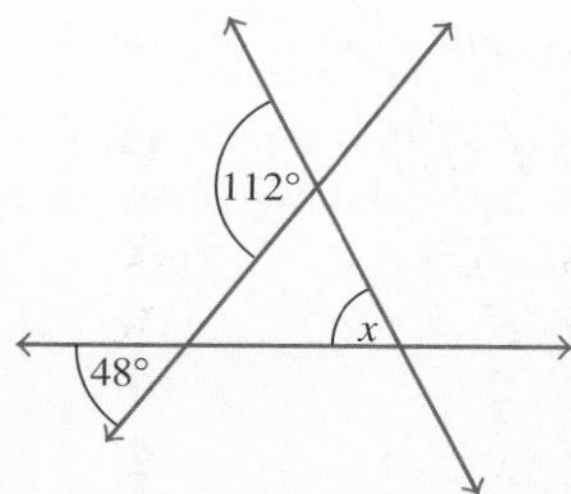

8. Determine the sum of the measures of the interior angles of an octagon.
9. Triangles ABC and $A'B'C'$ are similar figures. Determine the length of side $\overline{B'C'}$.

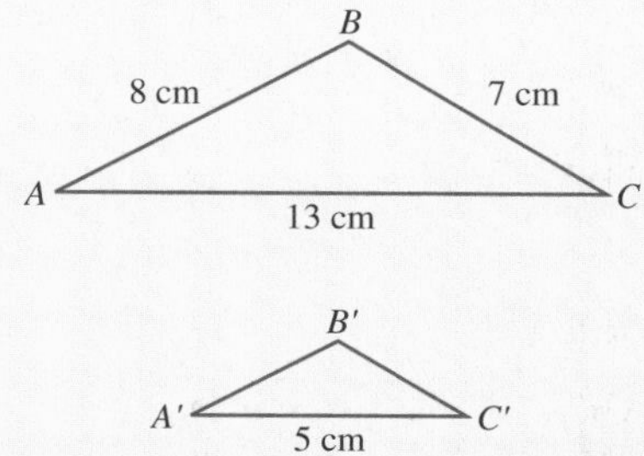

10. Right triangle ABC has one leg of length 5 in. and a hypotenuse of length 13 in.
    a) Determine the length of the other leg.
    b) Determine the perimeter of the triangle.
    c) Determine the area of the triangle.

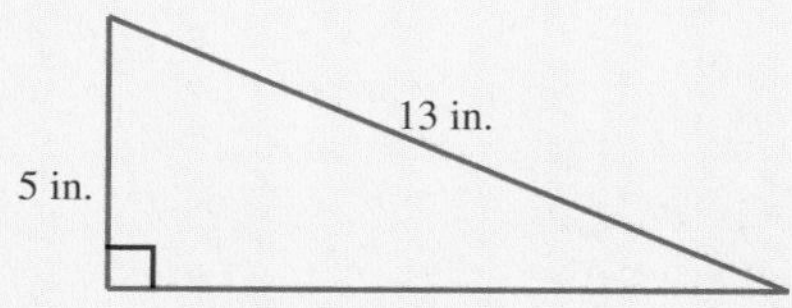

11. Determine the volume of a sphere of diameter 16 cm.
12. ***Building a Pier*** The sketch shows the dimensions of the base of a pier for a bridge with semicircular ends. How many cubic yards of concrete are needed to build a pier 6 ft high?

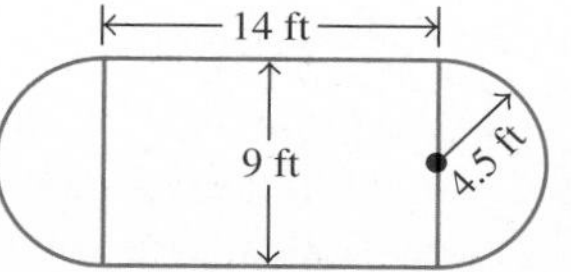

13. Determine the volume of the pyramid.

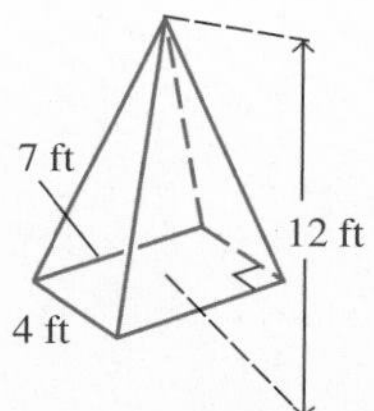

14. Construct a reflection of square $ABCD$, shown below, about line l. Show the square in the positions both before and after the reflection.

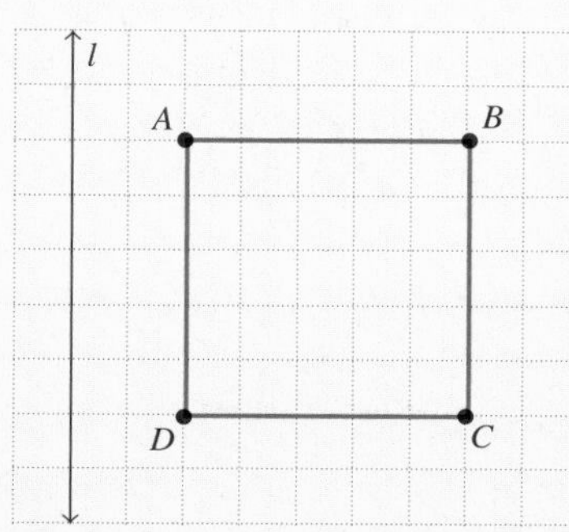

15. Construct a translation of quadrilateral $ABCD$, shown below, using translation vector **v**. Show the quadrilateral in the positions both before and after the translation.

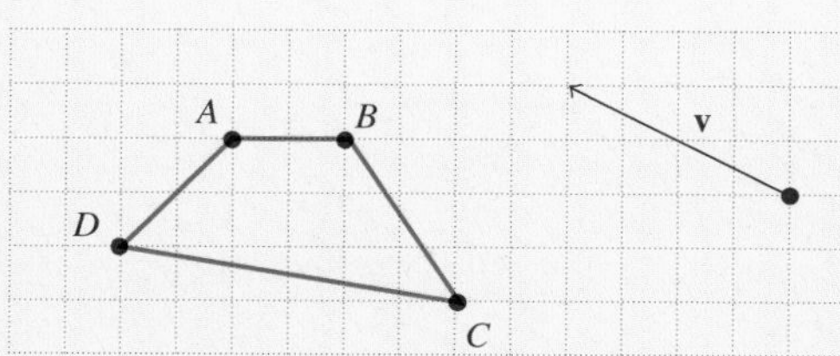

16. Construct a 180° rotation of triangle ABC, shown below, about rotation point P. Show the triangle in the positions both before and after the rotation.

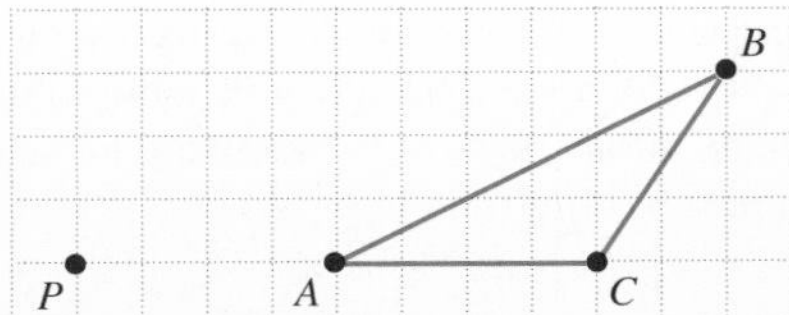

17. Construct a glide reflection of rectangle $ABCD$, shown below, using translation vector **v** and reflection line l. Show the rectangle in the positions both before and after the glide reflection.

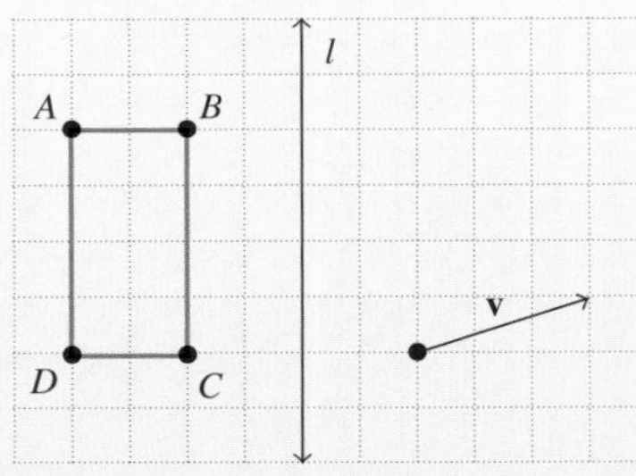

18. Use the figure below to answer the following questions.

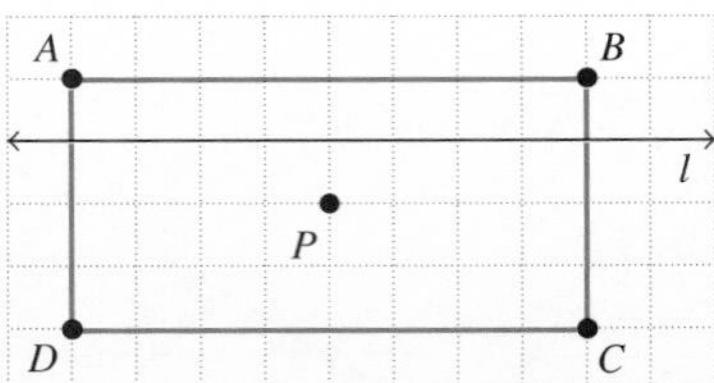

a) Does rectangle $ABCD$ have reflective symmetry about line l? Explain.
b) Does rectangle $ABCD$ have 180° rotational symmetry about point P? Explain.

19. What is a Möbius strip?

20. a) Sketch an object of genus 1.
b) Sketch an object of genus 2.

21. Explain how the fifth axiom in Euclidean geometry, elliptical geometry, and hyperbolic geometry differ.

GROUP PROJECTS

Supporting a Jacuzzi

1. Samantha Saraniti is thinking of buying a circular hot tub 12 ft in diameter, 4 ft deep, and weighing 475 lb. She wants to place the hot tub on a deck built to support 30,000 lb.
   a) Determine the volume of the water in the hot tub in cubic feet.
   b) Determine the number of gallons of water the hot tub will hold. Note 1 $ft^3 \approx 7.5$ gal.
   c) Determine the weight of the water in the hot tub. (*Hint:* Fresh water weighs about 52.4 lb/ft^3.)
   d) Will the deck support the weight of the hot tub and water?
   e) Will the deck support the weight of the hot tub, water, and four people, whose average weight is 115 lb?

Designing a Ramp

2. David and Sandra Jessee are planning to build a ramp so that their front entrance is wheelchair accessible. The ramp will be 36 in. wide. It will rise 2 in. for each foot of length of horizontal distance. Where the ramp meets the porch, the ramp must be 2 ft high. To provide stability for the ramp, the Jessees will install a slab of concrete 4 in. thick and 6 in. longer and wider than the ramp (see accompanying figure). The top of the slab will be level with the ground. The ramp may be constructed of concrete or pressure-treated lumber. You are to estimate the cost of materials for constructing the slab, the ramp of concrete, and the ramp of pressure-treated lumber.

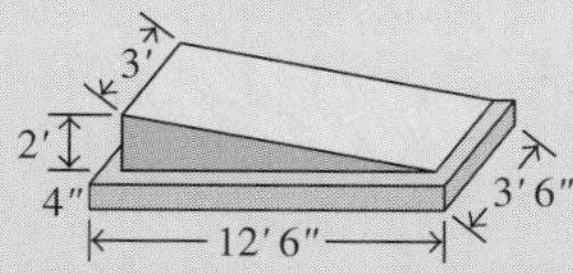

Slab

a) Determine the length of the base of the ramp.
b) Determine the dimensions of the concrete slab on which the ramp will set.
c) Determine the volume of the concrete in cubic yards needed to construct the slab.
d) If ready-mix concrete costs \$45 per cubic yard, determine the cost of the concrete needed to construct the slab.

Concrete Ramp

e) To build the ramp of concrete a form in the shape of the ramp must be framed. The two sides of the form are triangular, and the shape of the end, which is against the porch is rectangular. The form will be framed from $\frac{3}{4}$ in. plywood, which comes in 4 ft × 8 ft sheets. Determine the number of sheets of plywood needed. Assume that the entire sheet(s) will be used to make the sides and the end of the form and that there is no waste.
f) If the plywood costs \$18.95 for a 4 ft × 8 ft sheet, determine the cost of the plywood.
g) To brace the form, the Jessees will need two boards 2 in. × 4 in. × 8 ft (referred to as 8 ft 2 × 4's) and six pieces of lumber 2 in. × 4 in. × 3 ft. These six pieces of lumber will be cut from 8 ft 2 × 4 boards. Determine the number of 8 ft 2 × 4 boards needed.
h) Determine the cost of the 8 ft 2 × 4 boards needed in part (g) if one board costs \$2.14.
i) Determine the volume, in cubic yards, of concrete needed to fill the form.
j) Determine the cost of the concrete needed to fill the form.
k) Determine the total cost of materials for building the ramp of concrete by adding the results in parts (d), (f), (h), and (j).

Wooden Ramp

l) Determine the length of the top of the ramp.
m) The top of the ramp will be constructed of $\frac{5}{4}$ in. × 6 in. × 10 ft pressure-treated lumber. The boards will be butted end to end to make the necessary length and will be supported from underneath by a wooden frame. Determine the number of boards needed to cover the top of the ramp. The boards are laid lengthwise on the ramp.
n) Determine the cost of the boards to cover the top of the ramp if the price of a 10 ft length is \$6.47.
o) To support the top of the ramp, the Jessees will need 10 pieces of 8 ft 2 × 4's. The price of a pressure-treated 8 ft 2 × 4 is \$2.44. Determine the cost of the supports.
p) Determine the cost of the materials for building a wooden ramp by adding the amounts from parts (d), (n), and (o).
q) Are the materials for constructing a concrete ramp or a wooden ramp less expensive?

CHAPTER 10

Despite the seemingly endless variety of patterns the human imagination can devise, group theory can be used to catalog and define patterns by the way in which the design elements are transformed and positioned.

MATHEMATICAL SYSTEMS

In the last 200 years, much of the focus of scientific study of the fundamental laws of nature has shifted from what things are to how they change in space and time. The mathematics used for this purpose belongs to a branch of mathematics known as group theory. A group is a collection of fundamentally basic elements along with an operation that acts upon these elements. The elements in a group could be numbers, or elementary particles of physics, or a pattern of repeating geometric designs. The way in which the elements change or remain the same when acted upon by some operation or transformation defines membership in the group.

To appreciate why group theory is so useful to scientists, consider the physicists who are trying to piece together the history of the universe. They have a clear sense of what the universe is like today, but what about 15 billion years ago when some scientists believe the universe exploded into existence in what is called the Big Bang? Group theory may enable them to derive what the initial conditions of the universe may have been from knowledge of what matter is like today. Group theory can also be used to describe many other items, including number systems, and the basic building blocks of crystalline solids. It can even be used to define the symmetries that appear in the artifacts of a given culture. No wonder the physicist Sir Arthur Stanley Eddington called group theory the "super-mathematics."

10.1 GROUPS

We begin our discussion by introducing a mathematical system. As you will learn shortly, you already know and use many mathematical systems.

A **mathematical system** consists of a set of elements and at least one binary operation.

DID YOU KNOW

1 Color + 1 Color = 1 Color

The mixing of two paints is a binary operation. When two colors are mixed—for example, yellow and blue—a single color results, in this case green.

In the above definition we mention binary operation. A *binary operation* is an operation, or rule, that can be performed on two and only two elements of a set. The result is a single element. When we add *two* integers, the sum is *one* integer. When we multiply *two* integers, the product is *one* integer. Thus, addition and multiplication are both binary operations. Is finding the reciprocal of a number a binary operation? No, it is an operation on a single element of a set.

When you learned how to add integers, you were introduced to a mathematical system. The set of elements is the set of integers, and the binary operation is addition. When you learned how to multiply integers, you became familiar with a second mathematical system. The set of integers with the operation of subtraction and the set of integers with the operation of division are two other examples of mathematical systems since subtraction and division are also binary operations.

Some systems are used in solving everyday problems, such as planning work schedules. Others are more abstract and are used primarily in research, chemistry, physical structure, matter, the nature of genes, and other scientific fields.

Commutative and Associative Properties

Once a mathematical system is defined, its structure may display certain properties. Consider the set of integers:

$$I = \{\dots, -3, -2, -1, 0, 1, 2, 3, \dots\}$$

Recall that the ellipsis, the three dots at each end of the set, indicates that the set continues in the same manner.

The set of integers can be studied with the operations of addition, subtraction, multiplication, and division as separate mathematical systems. For example, when we study the set of integers under the operations of addition or multiplication, we see that the commutative and associative properties hold. The general forms of the properties are shown here.

For Any Elements a, b, and c	***Addition***	***Multiplication***
Commutative property	$a + b = b + a$	$a \cdot b = b \cdot a$
Associative property	$(a + b) + c = a + (b + c)$	$(a \cdot b) \cdot c = a \cdot (b \cdot c)$

The integers *are commutative* under the operations of *addition and multiplication*. For example,

Addition		***Multiplication***
$2 + 4 = 4 + 2$	and	$2 \cdot 4 = 4 \cdot 2$
$6 = 6$		$8 = 8$

The integers, however, *are not commutative* under the operations of *subtraction and division*. For example,

Subtraction		***Division***
$4 - 2 \neq 2 - 4$	and	$4 \div 2 \neq 2 \div 4$
$2 \neq -2$		$2 \neq \frac{1}{2}$

The integers *are associative* under the operations of *addition and multiplication*. For example,

Addition		***Multiplication***
$(1 + 2) + 3 = 1 + (2 + 3)$	and	$(1 \cdot 2) \cdot 3 = 1 \cdot (2 \cdot 3)$
$3 + 3 = 1 + 5$		$2 \cdot 3 = 1 \cdot 6$
$6 = 6$		$6 = 6$

The integers, however, *are not associative* under the operations of *subtraction and division*. See Exercises 21 and 22 at the end of this section.

To say that a set of elements is commutative under a given operation means that the commutative property holds for *any* elements a and b in the set. Similarly, to say that a set of elements is associative under a given operation means that the associative property holds for *any* elements a, b, and c in the set.

Consider the mathematical system consisting of the set of integers under the operation of addition. Because the set of integers is infinite, this mathematical system is an example of an *infinite mathematical system*. We will study certain properties of this mathematical system. The first property that we examine is closure.

Closure

The sum of any two integers is an integer. Therefore, the set of integers is said to be *closed,* or to satisfy the *closure property,* under the operation of addition.

> If a binary operation is performed on any two elements of a set and the result is an element of the set, then that set is **closed** (or has **closure**) under the given binary operation.

Is the set of integers closed under the operation of multiplication? The answer is yes. When any two integers are multiplied, the product will be an integer.

Is the set of integers closed under the operation of subtraction? Again, the answer is yes. The difference of any two integers is an integer.

Is the set of integers closed under the operation of division? The answer is no because two integers may have a quotient that is not an integer. For example, if we select the integers 2 and 3, the quotient of 2 divided by 3 is $\frac{2}{3}$, which is not an integer. Thus, the integers are not closed under the operation of division.

We showed that the set of integers was not closed under the operation of division by finding two integers whose quotient was not an integer. A specific example illustrating that a specific property is not true is called a *counterexample*. Mathematicians and scientists often try to find a counterexample to confirm that a specific property is not always true.

DID YOU KNOW

Secrets of the Universe

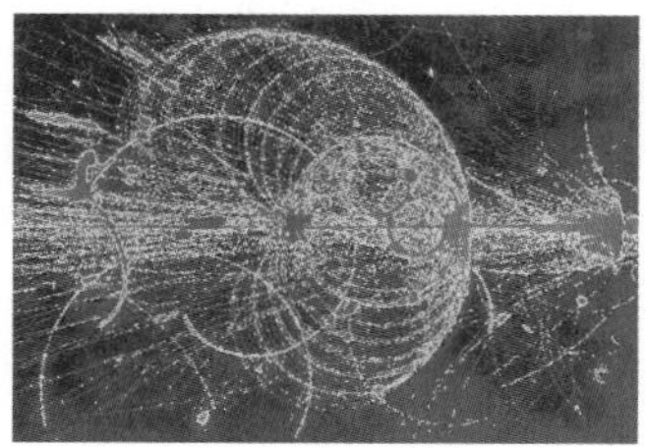

A computer simulation of protons and neutrons being smashed together.

Physicists use group theory in their search for the elementary particles that make up all matter. The operations are *transformations:* the coming together or splitting apart of elementary particles as they form larger particles such as protons and neutrons. Some transformations bring two particles together to form a different particle (binary operation), some leave a particle intact (identity), and some follow the rules of associativity and commutativity. In fact, understanding there were distinct "families" of particles related to one another as groups enabled physicists to predict the existence of certain particles before any experimental evidence was available.

Identity Element

Now we will discuss the identity element for the set of integers under the operation of addition. Is there an element in the set that, when added to any given integer, results in a sum that is the given integer? The answer is yes. The sum of 0 and any integer is the given integer. For example, $1 + 0 = 0 + 1 = 1$, $-4 + 0 = 0 + (-4) = -4$, and so on. For this reason, we call 0 the ***additive identity element*** for the set of integers. Note that for any integer a, $a + 0 = 0 + a = a$.

An **identity element** is an element in a set such that when a binary operation is performed on it and any given element in the set, the result is the given element.

Is there an identity element for the set of integers under the operation of multiplication? The answer is yes; it is the number 1. Note that $2 \cdot 1 = 1 \cdot 2 = 2$, $3 \cdot 1 = 1 \cdot 3 = 3$, and so on. For any integer a, $a \cdot 1 = 1 \cdot a = a$. For this reason, 1 is called the ***multiplicative identity element*** for the set of integers.

Inverses

What integer, when added to 4, gives a sum of 0; that is, $4 + \square = 0$? The shaded area is to be filled in with the integer -4: $4 + (-4) = 0$. We say that -4 is the additive inverse of 4 and that 4 is the additive inverse of -4. Note that the sum of the element and its additive inverse gives the additive identity element 0. What is the additive inverse of 12? Since $12 + (-12) = 0$, -12 is the additive inverse of 12.

Other examples of integers and their additive inverses are

Element	+	Additive Inverse	=	Identity Element
0	+	0	=	0
2	+	(−2)	=	0
−5	+	5	=	0

Note that for the operation of addition, every integer a has a unique inverse, $-a$, such that $a + (-a) = -a + a = 0$.

When a binary operation is performed on two elements in a set and the result is the identity element for the binary operation, each element is said to be the **inverse** of the other.

Does every integer have an inverse under the operation of multiplication? For multiplication the product of an integer and its inverse must yield the multiplicative identity element, 1. What is the multiplicative inverse of 2? That is, 2 times what number gives 1?

$$2 \cdot ? = 1 \qquad 2 \cdot \tfrac{1}{2} = 1$$

However, since $\frac{1}{2}$ is not an integer, 2 does not have a multiplicative inverse in the set of integers.

PROFILES IN MATHEMATICS

NIELS ABEL (1802–1829), ÉVARISTE GALOIS (1811–1832)

Niels Abel Évariste Galois

Untimely ends: Important contributions to the development of group theory were made by two young men who would not live to see their work gain acceptance: Niels Abel, a Norwegian, and Évariste Galois, a Frenchman. Although their work showed brilliance, neither was noticed in his lifetime by the mathematics community. Abel lived in poverty and died of malnutrition and tuberculosis at the age of 26, just 2 days before a letter arrived with the offer of a teaching position. Galois died of complications from a gunshot wound he received in a duel over a love affair. The 20-year-old man had spent the night before summarizing his theories in the hope that eventually someone would find it "profitable to decipher this mess." After their deaths, both men had important elements of group theory named after them.

Group

Let's review what we have learned about the mathematical system consisting of the set of integers under the operation of addition.

1. The set of integers is *closed* under the operation of addition.
2. The set of integers has an *identity element* under the operation of addition.
3. Each element in the set of integers has an *inverse* under the operation of addition.
4. The *associative property* holds for the set of integers under the operation of addition.

The set of integers under the operation of addition is an example of a *group*. The properties of a group can be summarized as follows.

Properties of a Group

Any mathematical system that meets the following four requirements is called a **group.**

1. The set of elements is *closed* under the given operation.
2. An *identity element* exists for the set under the given operation.
3. Every element in the set has an *inverse* under the given operation.
4. The set of elements is *associative* under the given operation.

TIMELY TIP Note that the set of elements need *not* be commutative for the mathematical system to be a group. Also note that *every* element in the set must have an inverse for the mathematical system to be a group.

It is often very time consuming to show that the associative property holds for all cases. In many of the examples that follow, we will state that the associative property holds for the given set of elements under the given operation.

Commutative Group

The commutative property does not need to hold for a mathematical system to be a group. However, if a mathematical system meets the four requirements of a group and is also commutative under the given operation, the mathematical system is a *commutative* (or *abelian*) *group*. The abelian group is named after Niels Abel (see the Profiles in Mathematics).

A group that satisfies the commutative property is called a **commutative group** (or **abelian group**)

Because the commutative property holds for the set of integers under the operation of addition, the set of integers under the operation of addition is not only a group, but it is a commutative group.

DID YOU KNOW

The Enormous Theorem

Mathematicians are attempting to make the world's longest proof shorter. It is called the *enormous theorem.* More than 100 mathematicians toiled for more than 30 years to produce the proof known formally as the classification of finite, simple groups. Completed in the 1980s, it consists of some 500 publications totaling 15,000 pages. The theorem established that all finite, simple groups can fall into 17 infinite families, or 26 so-called sporadic forms. Now an attempt is being made to whittle the theorem down to a mere 3000 pages or so. Those working on it hope that shortening the proof can be completed in a decade or so.

Properties of a Commutative Group

A mathematical system is a **commutative group** if all five conditions hold.

1. The set of elements is *closed* under the given operation.
2. An *identity element* exists for the set.
3. Every element in the set has an *inverse.*
4. The set of elements is *associative* under the given operation.
5. The set of elements is *commutative* under the given operation.

To determine whether a mathematical system is a group under a given operation, *check, in the following order,* to determine whether (a) the system is closed under the given operation, (b) there is an identity element in the set for the given operation, (c) every element in the set has an inverse under the given operation, and (d) the associative property holds under the given operation. If *any* of these four requirements is *not* met, stop and state the mathematical system is not a group. If asked to determine whether the mathematical system is a commutative group, you also need to check to determine whether the commutative property holds for the given operation.

EXAMPLE 1 *Is It a Group?*

Determine whether the set of rational numbers under the operation of multiplication forms a group.

SOLUTION: Recall from Chapter 5 that the rational numbers are the set of numbers of the form p/q where p and q are integers and $q \neq 0$. All fractions and integers are rational numbers.

1. *Closure:* The product of any two rational numbers is a rational number. Therefore, the rational numbers are closed under the operation of multiplication.
2. *Identity element:* The multiplicative identity element for the set of rational numbers is 1. Note, for example, that $3 \cdot 1 = 1 \cdot 3 = 3$, and $\frac{3}{8} \cdot 1 = 1 \cdot \frac{3}{8} = \frac{3}{8}$. For any rational number a, $a \cdot 1 = 1 \cdot a = a$.
3. *Inverse elements:* For the mathematical system to be a group under the operation of multiplication, *each and every* rational number must have a multiplicative inverse in the set of rational numbers. Remember that for the operation of multiplication, the product of a number and its inverse must give the multiplicative identity element, 1. Let's check a few rational numbers:

Rational number	·	Inverse	=	Identity element
3	·	$\frac{1}{3}$	=	1
$\frac{2}{3}$	·	$\frac{3}{2}$	=	1
$-\frac{1}{5}$	·	-5	=	1

Looking at these examples you might deduce that each rational number does have an inverse. However, one rational number, 0, does not have an inverse.

$$0 \cdot ? = 1$$

Because there is no rational number that, when multiplied by 0, gives 1, 0 does not have a multiplicative inverse. Since *every* rational number does not have an inverse, this mathematical system is not a group.

There is no need at this point to check the associative property because we have already shown that the mathematical system of rational numbers under the operation of multiplication is not a group.

MATHEMATICS *Everywhere*

The Universe

Woody Allen and Diane Keaton in *Annie Hall*

In the chapter opening material, we mentioned that group theory is helpful to scientists in determining what the universe was like 15 billion years ago. Group theory plays an important role in many areas of higher mathematics and science. We often see articles where the information provided comes in part from group theory. For example, the June 25, 2001, issue of *Time* featured an article on space. The article mentions that "the Milky Way is a huge, whirling pinwheel made of 100 billion or more stars; that tens of billions of other galaxies lie beyond its edges; and most astonishing of all, that these galaxies are rushing headlong away from one another in the aftermath of an explosive cataclysm known as the Big Bang."

The article also discusses and illustrates how Albert Einstein believed that the universe is curved. He believed that parts of the universe have a positive curvature, parts are flat, and parts have a negative curvature. The curvature of our universe is discussed briefly in Section 9.7, non-Euclidean Geometry.

How and when will the universe end? In the Woody Allen movie *Annie Hall*, a boy (Woody Allen), having just read that the universe is expanding, becomes so worried that he can't do his homework. "Someday it will break apart," he tells his psychiatrist, "and that will be the end of everything." But his mother snaps, "You're here in Brooklyn! Brooklyn is not expanding!"

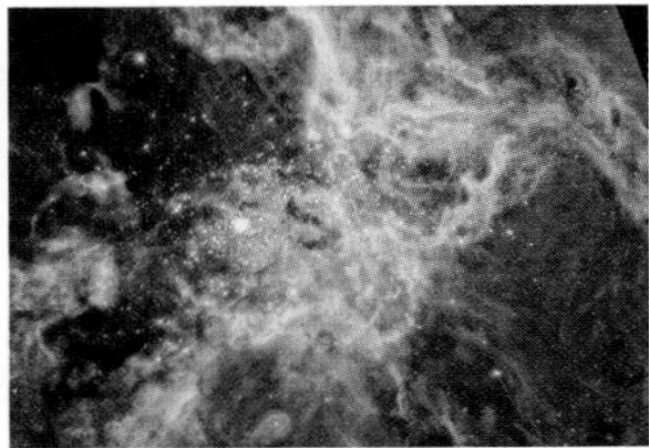

Doradus Nebula, a satellite galaxy of the Milky Way

The article indicates that Allen may have been on to something. The universe could break apart. However, this will not be for trillions of years, and life as we know it on Earth will cease long before that. Our sun, halfway through its estimated 10-billion-year lifetime, is slowly brightening. In about 1 billion years, its energy output will have increased by at least 10 percent, turning Earth into a Venus-like hothouse incapable of supporting life as we know it. The article goes on to mention that long before that, humans may make plans to escape to Mars, which is farther from the sun and hence cooler. Although quite some time away, scientists are already considering various scenarios to make the surface of Mars a livable habitat.

SECTION 10.1 EXERCISES

Concept/Writing Exercises

1. What is a binary operation?
2. What does a mathematical system consist of?
3. Explain why each of the following is a binary operation. Give an example to illustrate each binary operation.
   a) Addition **b)** Subtraction
   c) Multiplication **d)** Division
4. What properties are required for a mathematical system to be a group?
5. What properties are required for a mathematical system to be a commutative group?
6. What is another name for a commutative group?
7. Explain the closure property. Give an example of the property.
8. What is an identity element? Give the additive and multiplicative identity elements for the set of integers.
9. What is an inverse element? Give the additive and multiplicative inverse of the number 2 for the set of rational numbers.
10. What is a counterexample?
11. Is it possible that a mathematical system is a commutative group but not a group? Explain.

12. Is it possible that a mathematical system is a group but not a commutative group? Explain.
13. Which of the following properties is not required for a mathematical system to be a group?
    a) Closure
    b) An identity element
    c) Every element must have an inverse.
    d) The commutative property must apply.
    e) The associative property must apply.
14. For the set of integers, list two operations that are not binary. Explain.

Practice the Skills

15. Give the associative property of addition and illustrate the property with an example.
16. Give the associative property of multiplication and illustrate the property with an example.
17. Give the commutative property of multiplication and illustrate the property with an example.
18. Give the commutative property of addition and illustrate the property with an example.
19. Give an example to show that the commutative property does not hold for the set of integers under the operation of division.
20. Give an example to show that the commutative property does not hold for the set of integers under the operation of subtraction.
21. Give an example to show that the associative property does not hold for the set of integers under the operation of subtraction.
22. Give an example to show that the associative property does not hold for the set of integers under the operation of division.

Problem Solving

In Exercises 23–38, explain your answer.

23. Is the set of positive integers a group under the operation of addition?
24. Is the set of positive integers a commutative group under the operation of addition?
25. Is the set of integers a commutative group under the operation of addition?
26. Is the set of integers a group under the operation of addition?
27. Is the set of positive integers a group under the operation of subtraction?
28. Is the set of negative integers a group under the operation of division?
29. Is the set of negative integers a commutative group under the operation of addition?
30. Is the set of integers a group under the operation of multiplication?
31. Is the set of positive integers a commutative group under the operation of multiplication?
32. Is the set of negative integers a group under the operation of multiplication?
33. Is the set of rational numbers a group under the operation of addition?
34. Is the set of rational numbers a commutative group under the operation of multiplication?
35. Is the set of rational numbers a commutative group under the operation of division?
36. Is the set of rational numbers a group under the operation of subtraction?
37. Is the set of rational numbers a commutative group under the operation of subtraction?
38. Is the set of positive integers a commutative group under the operation of division?

Challenge Problems/Group Activities

In Exercises 39–42, explain your answer.

39. Is the set of irrational numbers a group under the operation of addition?
40. Is the set of irrational numbers a group under the operation of multiplication?
41. Is the set of real numbers a group under the operation of addition?
42. Is the set of real numbers a group under the operation of multiplication?
43. Create a mathematical system with two binary operations. Select a set of elements and two binary operations so that one binary operation with the set of elements meets the requirements for a group and the other binary operation will not. Explain why the one binary operation with the set of elements is a group. For the other binary operation and the set of elements find counterexamples to show that it is not a group.

Recreational Mathematics

44. *The Greatest Number* Find the greatest number that can be written with three digits.
45. *Sign Maker* There are 100 buildings along a street. A sign maker is asked to number the buildings from 1 to 100. How many "9's" will the sign maker need?

Internet/Research Activity

46. There are other classifications of mathematical systems besides groups. For example, there are *rings* and *fields*. Do research to determine the requirements that must be met for a mathematical system to be (a) a ring and (b) a field. (c) Is the set of real numbers, under the operations of addition and multiplication, a field? Ask your instructor for references to use.

10.2 FINITE MATHEMATICAL SYSTEMS

In the preceding section we presented infinite mathematical systems. In this section we present some finite mathematical systems. A *finite mathematical system* is one whose set contains a finite number of elements.

Clock Arithmetic

Figure 10.1

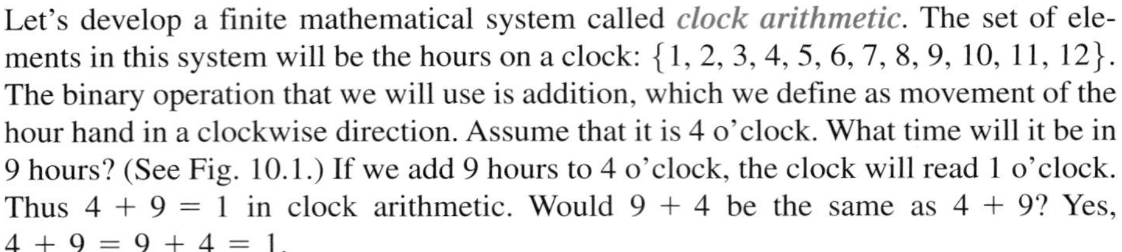

Let's develop a finite mathematical system called *clock arithmetic*. The set of elements in this system will be the hours on a clock: $\{1, 2, 3, 4, 5, 6, 7, 8, 9, 10, 11, 12\}$. The binary operation that we will use is addition, which we define as movement of the hour hand in a clockwise direction. Assume that it is 4 o'clock. What time will it be in 9 hours? (See Fig. 10.1.) If we add 9 hours to 4 o'clock, the clock will read 1 o'clock. Thus $4 + 9 = 1$ in clock arithmetic. Would $9 + 4$ be the same as $4 + 9$? Yes, $4 + 9 = 9 + 4 = 1$.

Table 10.1 is the addition table for clock arithmetic. Its elements are based on the definition of addition as previously illustrated. For example, the sum of 4 and 9 is 1, so we put a 1 in the table where the row to the right of the 4 intersects the column below the 9. Likewise, the sum of 11 and 10 is 9, so we put a 9 in the table where the row to the right of the 11 intersects the column below the 10.

The binary operation of this system is defined by the table. It is denoted by the symbol +. To determine the value of $a + b$, where a and b are any two numbers in the set, find a in the left-hand column and find b along the top row. Assume that there is a horizontal line through a and a vertical line through b; the point of intersection of these two lines is where you find the value of $a + b$. For example, $10 + 4 = 2$ has been circled in Table 10.1. Note that $4 + 10$ also equals 2, but this result will not necessarily hold for all examples in this chapter.

TABLE 10.1 Clock 12 Arithmetic

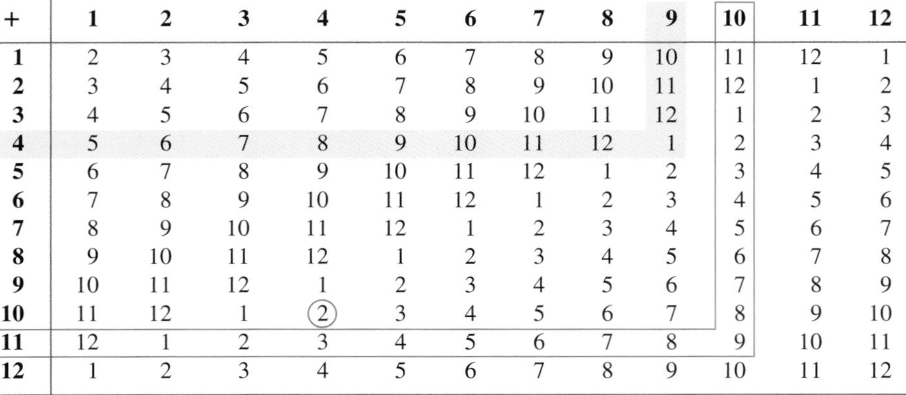

+	**1**	**2**	**3**	**4**	**5**	**6**	**7**	**8**	**9**	**10**	**11**	**12**
1	2	3	4	5	6	7	8	9	10	11	12	1
2	3	4	5	6	7	8	9	10	11	12	1	2
3	4	5	6	7	8	9	10	11	12	1	2	3
4	5	6	7	8	9	10	11	12	1	2	3	4
5	6	7	8	9	10	11	12	1	2	3	4	5
6	7	8	9	10	11	12	1	2	3	4	5	6
7	8	9	10	11	12	1	2	3	4	5	6	7
8	9	10	11	12	1	2	3	4	5	6	7	8
9	10	11	12	1	2	3	4	5	6	7	8	9
10	11	12	1	(2)	3	4	5	6	7	8	9	10
11	12	1	2	3	4	5	6	7	8	9	10	11
12	1	2	3	4	5	6	7	8	9	10	11	12

EXAMPLE 1 *A Commutative Group?*

Determine whether the clock arithmetic system under the operation of addition is a commutative group.

SOLUTION: Check the five requirements that must be satisfied for a commutative group.

1. *Closure:* Is the set of elements in clock arithmetic closed under the operation of addition? Yes, since Table 10.1 contains only the elements in the set

DID YOU KNOW

The Changing Times

Venice clock

This photo shows the famous clock at the Cathedral of St. Marks (called the Basilica) in Piazza San Marco (St. Mark's Plaza) in Venice, Italy. Construction of the basilica began in 830, but it has been rebuilt a number of times due to fires. The clock tower was built in the fifteenth century. Note that the clock displays all 24 hours in Roman numerals. Also note the position of the hours. Noon, or XII, is where we expect 9 o'clock to be.

In Florence, Italy, the oldest building is probably the Duomo (The Cathedral House) in the Piazza del Duomo in the city's center. Florentines believe it was constructed in about 1000. The large clock in the cathedral also displays 24 hours using Roman numerals.

On both clocks 4 o'clock is indicated by IIII, not IV, and 9 o'clock is indicated by VIIII, not IX.

$\{1, 2, 3, 4, 5, 6, 7, 8, 9, 10, 11, 12\}$. If Table 10.1 had contained an element other than the numbers 1 through 12, the set would not have been closed under addition.

2. *Identity element:* Is there an identity element for clock arithmetic? If the time is currently 4 o'clock, how many hours have to pass before it is 4 o'clock again? Twelve hours: $4 + 12 = 12 + 4 = 4$. In fact, given any hour, in 12 hours the clock will return to the starting point. Therefore, 12 is the additive identity element in clock arithmetic.

   In examining Table 10.1 we see that the row of numbers next to the 12 in the left-hand column is identical to the row of numbers along the top. We also see that the column of numbers under the 12 in the top row is identical to the column of numbers on the left. The search for such a column and row is one technique for determining whether an identity element exists for a system defined by a table.

3. *Inverse elements:* Is there an inverse for the number 4 in clock arithmetic for the operation of addition? Recall that the identity element in clock arithmetic is 12. What number when added to 4 gives 12; that is, $4 + \square = 12$? Table 10.1 shows that $4 + 8 = 12$ and also that $8 + 4 = 12$. Thus, 8 is the additive inverse of 4, and 4 is the additive inverse of 8.

   To find the additive inverse of 7, find 7 in the left-hand column of Table 10.1. Look to the right of the 7 until you come to the identity element 12. Determine the number at the top of this column. The number is 5. Since $7 + 5 = 5 + 7 = 12$, 5 is the inverse of 7, and 7 is the inverse of 5. The other inverses can be found in the same way. Table 10.2 shows each element in clock 12 arithmetic and its inverse. Note that each element in the set has an *inverse.*

TABLE 10.2 Clock 12 Inverses

Element	+	Inverse	=	Identity Element
1	+	11	=	12
2	+	10	=	12
3	+	9	=	12
4	+	8	=	12
5	+	7	=	12
6	+	6	=	12
7	+	5	=	12
8	+	4	=	12
9	+	3	=	12
10	+	2	=	12
11	+	1	=	12
12	+	12	=	12

4. *Associative property:* Now consider the associative property. Does $(a + b) + c = a + (b + c)$ for all values a, b, and c of the set? Remember to always evaluate the values within the parentheses first. Let's select some values for a, b, and c. Let $a = 2$, $b = 6$, and $c = 8$. Then

$$\begin{aligned}(2 + 6) + 8 &= 2 + (6 + 8)\\ 8 + 8 &= 2 + 2\\ 4 &= 4 \quad \text{True}\end{aligned}$$

   Let $a = 5$, $b = 12$, and $c = 9$. Then

$$\begin{aligned}(5 + 12) + 9 &= 5 + (12 + 9)\\ 5 + 9 &= 5 + 9\\ 2 &= 2 \quad \text{True}\end{aligned}$$

Randomly selecting *any* elements a, b, and c of the set reveals $(a + b) + c = a + (b + c)$. Thus, the system of clock arithmetic is associative under the operation of addition. Note that if there is just one set of values a, b, and c such that $(a + b) + c \neq a + (b + c)$, the system is not associative. Normally you will not be asked to check every case to determine whether the associative property holds. *If every element in the set does not appear in every row and column of the table, however, you need to check the associative property carefully.*

5. *Commutative property:* Does the commutative property hold under the given operation? Does $a + b = b + a$ for all elements a and b of the set? Let's randomly select some values for a and b to determine whether the commutative property appears to hold. Let $a = 5$ and $b = 8$; then Table 10.1 shows that

$$5 + 8 = 8 + 5$$
$$1 = 1 \quad \text{True}$$

Let $a = 9$ and $b = 6$; then

$$9 + 6 = 6 + 9$$
$$3 = 3 \quad \text{True}$$

The commutative property holds for these two specific cases. In fact, if we were to select *any* values for a and b, we would find that $a + b = b + a$. Thus, the commutative property of addition is true in clock arithmetic. Note that if there is just one set of values a and b such that $a + b \neq b + a$, the system is not commutative.

This system satisfies the five properties required for a mathematical system to be a commutative group. Thus, clock arithmetic under the operation of addition is a commutative or abelian group.

TABLE 10.3 Symmetry about the Main Diagonal

+	0	1	2	3	4
0	0	1	2	3	4
1	1	2	3	4	0
2	2	3	4	0	1
3	3	4	0	1	2
4	4	0	1	2	3

One method that can be used to determine whether a system defined by a table is commutative under the given operation is to determine whether the elements in the table are symmetric about the main diagonal. The main diagonal is the diagonal from the upper left-hand corner to the lower right-hand corner of the table. In Table 10.3 the main diagonal is shaded in color.

If the elements are symmetric about the main diagonal, then the system is commutative. If the elements are not symmetric about the main diagonal, then the system is not commutative. If you examine the system in Table 10.3, you see that its elements are symmetric about the main diagonal because the same numbers appear in the same relative positions on opposite sides of the main diagonal. Therefore, this mathematical system is commutative.

It is possible to have groups that are not commutative. Such groups are called *noncommutative* or *nonabelian groups*. However, a *noncommutative group defined by a table must be at least a six-element by six-element table*. Nonabelian groups are illustrated in Exercises 83 through 85 at the end of this section.

Now we will look at another finite mathematical system.

TABLE 10.4 Four-Element System

$\odot$	1	3	5	7
1	5	7	1	3
3	7	1	3	5
5	1	3	5	7
7	3	5	7	1

EXAMPLE 2 *A Finite System*

Consider the mathematical system defined by Table 10.4. Assume that the associative property holds for the given operation.

a) List the elements in the set of this mathematical system.

b) Identify the binary operation.

c) Determine whether this mathematical system is a commutative group.

SOLUTION:

a) The set of elements for this mathematical system consists of the elements found on the top (or left-hand side) of the table: $\{1, 3, 5, 7\}$.

b) The binary operation is $\odot$

c) We must determine whether the five requirements for a commutative group are satisfied.

1. *Closure:* All the elements in the table are in the original set of elements, $\{1, 3, 5, 7\}$, so the system is closed.
2. *Identity element:* The identity element is 5. Note that the row of elements to the right of the 5 is identical to the top row *and* the column of elements under the 5 is identical to the left-hand column.
3. *Inverse elements:* When an element operates on its inverse, the result is the identity element. For this example the identity element is 5. To determine the inverse of 1, find the element to replace the question mark:

$$1 \odot ? = 5$$

Since $1 \odot 1 = 5$, 1 is the inverse of 1. Thus, 1 is its own inverse.

To find the inverse of 3, find the element to replace the question mark:

$$3 \odot ? = 5$$

Since $3 \odot 7 = 7 \odot 3 = 5$, 7 is the inverse of 3 (and 3 is the inverse of 7). The elements and their inverses are shown in Table 10.5. Every element has a unique inverse.

TABLE 10.5 Inverses under $\odot$

Element	$\odot$	Inverse	=	Identity Element
1	$\odot$	1	=	5
3	$\odot$	7	=	5
5	$\odot$	5	=	5
7	$\odot$	3	=	5

4. *Associative property:* It is given that the associative property holds for this operation. One example of associative property is

$$\begin{aligned}(7 \odot 3) \odot 1 &= 7 \odot (3 \odot 1)\\ 5 \odot 1 &= 7 \odot 7\\ 1 &= 1 \quad \text{True}\end{aligned}$$

5. *Commutative property:* The elements in Table 10.4 are symmetric about the main diagonal, so the commutative property holds for the operation of $\odot$. One example of the commutative property is

$$\begin{aligned}3 \odot 5 &= 5 \odot 3\\ 3 &= 3 \quad \text{True}\end{aligned}$$

The five necessary properties hold. Thus, the mathematical system is a commutative group. ▲

Mathematical Systems Without Numbers

Thus far, all the systems we have discussed have been based on sets of numbers. Example 3 illustrates a mathematical system of symbols rather than numbers.

DID YOU KNOW

Group Theory Can Be Fun

Group theory can be fun as well as educational, at least so thought Erno Rubik, a Hungarian teacher of architecture and design. He presented the world with a puzzle in the form of a cube with six faces, each of a different color. Each face is divided into nine squares; each row and column of each face can rotate so that in no time at all six faces are transformed into a mix of colors. Your job, should you choose to accept it, is to get the cube back to its original condition.

EXAMPLE 3 *Investigating a System of Symbols*

Use the mathematical system defined by Table 10.6 and determine

TABLE 10.6 A System of Symbols

⊡	@	*P*	*W*
@	*P*	*W*	@
P	*W*	@	*P*
W	@	*P*	*W*

a) the set of elements.
b) the binary operation.
c) closure or nonclosure of the system.
d) the identity element.
e) the inverse of @.
f) $W \boxdot P$ and $P \boxdot @$.
g) $(@ \boxdot P) \boxdot P$ and $@ \boxdot (P \boxdot P)$.

SOLUTION:

a) The set of elements of this mathematical system is $\{@, P, W\}$.
b) The binary operation is ⊡.
c) Because the table does not contain any symbols other than @, *P*, and *W*, the system is closed under ⊡.
d) The identity element is *W*. Note that the row next to *W* in the left-hand column is the same as the top row and that the column under *W* is identical to the left-hand column. We see that

$$@ \boxdot W = W \boxdot @ = @$$
$$P \boxdot W = W \boxdot P = P$$
$$W \boxdot W = W$$

e) We know that

$$\text{element} \boxdot \text{inverse element} = \text{identity element},$$

and since *W* is the identity element, to find the inverse of @ we write

$$@ \boxdot ? = W$$

To find the inverse of @, we must determine the element to replace the question mark. Since $@ \boxdot P = W$ and $P \boxdot @ = W$, *P* is the inverse of @.

f) $W \boxdot P = P$ and $P \boxdot @ = W$

g) We first evaluate the information within parentheses.

$$(@ \boxdot P) \boxdot P = W \boxdot P = P \quad \text{and} \quad @ \boxdot (P \boxdot P) = @ \boxdot @ = P$$

▲

TABLE 10.7

*	*A*	*B*	*C*	*D*
A	*D*	*A*	*B*	*C*
B	*A*	*B*	*C*	*D*
C	*B*	*C*	*D*	*A*
D	*C*	*D*	*A*	*B*

EXAMPLE 4 *Is the System a Commutative Group?*

Determine whether the mathematical system in Table 10.7 is a commutative group. Assume that the associative property holds for the given operation.

TABLE 10.8

Element	*	Inverse	=	Identity element
A	*	C	=	B
B	*	B	=	B
C	*	A	=	B
D	*	D	=	B

SOLUTION:

1. *Closure:* The system is closed.
2. *Identity element:* The identity element is B.
3. *Inverse elements:* Each element has an inverse as illustrated in Table 10.8.
4. *Associative property:* It is given that the associative property holds. An example illustrating the associative property is

$$\begin{aligned}(D*A)*C &= D*(A*C)\\ C*C &= D*B\\ D &= D \quad \text{True}\end{aligned}$$

5. *Commutative property:* By examining the table we can see that it is symmetric about the main diagonal. Thus, the system is commutative under the given operation. One example of the commutative property is

$$\begin{aligned}D*C &= C*D\\ A &= A \quad \text{True}\end{aligned}$$

All five properties are satisfied. Thus, the system is a commutative group. ▲

EXAMPLE 5 *Another System to Study*

Determine whether the mathematical system in Table 10.9 is a commutative group under the operation of ☺.

TABLE 10.9

☺	x	y	z
x	x	z	y
y	z	y	x
z	y	x	z

SOLUTION:

1. The system is closed.
2. No row is identical to the top row, so there is no identity element. Therefore, this mathematical system is *not a group*. There is no need to go any further, but for practice, let's look at a few more items.
3. Since there is no identity element, there can be no inverses.
4. The associative property does not hold. The following counterexample illustrates the associative property does not hold for every case.

$$\begin{aligned}(x ☺ y) ☺ z &\neq x ☺ (y ☺ z)\\ z ☺ z &\neq x ☺ x\\ z &\neq x\end{aligned}$$

5. The table is symmetric about the main diagonal. Therefore, the commutative property does hold for the operation of ☺.

Note that the associative property does not hold even though the commutative property does hold. This outcome can occur when there is no identity element and every element does not have an inverse, as in this example. ▲

EXAMPLE 6 *Is the System a Commutative Group?*

Determine whether the mathematical system in Table 10.10 is a commutative group under the operation of $*$.

TABLE 10.10

*	□	b	c
□	□	b	c
b	b	b	□
c	c	□	c

SOLUTION:

1. The system is closed.
2. There is an identity element, □.

3. Each element has an inverse; $\square$ is the inverse of $\square$, b is the inverse of c, and c is the inverse of b.
4. Every element in the set does not appear in every row and every column of the table, so we need to check the associative property carefully. There are many specific cases where the associative property does hold. However, the following counterexample illustrates that the associative property does not hold for every case.

$$\begin{aligned} (b*b)*c &\neq b*(b*c) \\ b*c &\neq b*\square \\ \square &\neq b \end{aligned}$$

5. The commutative property holds because there is symmetry about the main diagonal.

Since we have shown that the associative property does not hold under the operation of $*$, this system is not a group. Therefore, it cannot be a commutative group.

DID YOU KNOW

Creating Patterns by Design

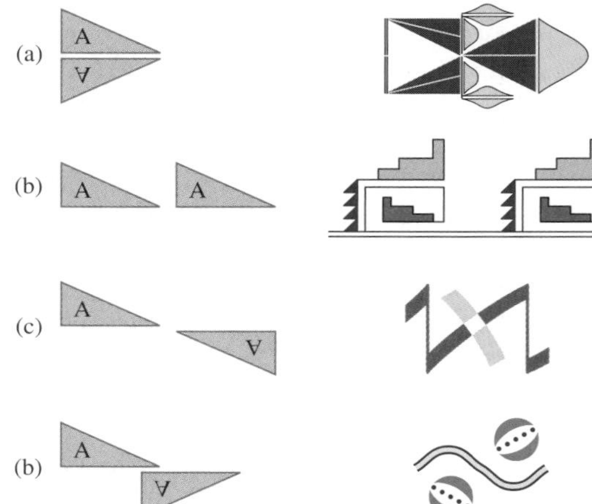

Patterns from *Symmetries of Culture* by Dorothy K. Washburn and Donald W. Crowe (University of Washington Press, 1988)

What makes group theory such a powerful tool is that it can be used to reveal the underlying structure of just about any physical phenomenon that involves symmetry and patterning, such as wallpaper or quilt patterns. Interest in the formal study of symmetry in design came out of the Industrial Revolution in the late nineteenth century. The new machines of the Industrial Revolution could vary any given pattern almost indefinitely. Designers needed a way to describe and manipulate patterns systematically. At the same time, explorers were discovering artifacts of other cultures, which stimulated interest in categorizing patterns. Shown here are the four geometric motions that generate all two-dimensional patterns: (a) reflection, (b) translation, (c) rotation, and (d) glide reflection. How these motions are applied, or not applied, is the basis of pattern analysis. The geometric motions, called *rigid motions,* were discussed in Section 9.5.

SECTION 10.2 EXERCISES

Concept/Writing Exercises

1. Explain how the clock 12 addition table is formed.
2. What is $12 + 12$ in clock 12 arithmetic? Explain how you obtained your answer.
3. **a)** Explain how to add the numbers $(4 + 10) + 3$ in clock 12 arithmetic using the addition table.
 b) What is $(4 + 10) + 3$ in clock 12 arithmetic?
4. **a)** Explain how to determine a difference of two numbers in clock arithmetic by using the face of a clock.
 b) Determine $4 - 7$ in clock 12 arithmetic using the method explained in part (a).
5. **a)** Explain how to find $5 - 9$ in clock 12 arithmetic by adding the number 12 to one of the numbers.
 b) Determine $5 - 9$ in clock 12 arithmetic using the method explained in part (a).
 c) Explain why the procedure you give in part (a) works.
6. Explain one method of determining whether a system defined by a table is commutative under the given operation.

7. Is clock 12 arithmetic closed under the operation of addition? Explain.

8. Is there an identity element for addition in clock 12 arithmetic? If so, what is it?

9. Does each element in clock 12 arithmetic have an inverse? If so, give each element and its corresponding inverse.

10. Give an example to illustrate the associative property of addition in clock 12 arithmetic.

11. Is clock 12 arithmetic commutative? Give an example to verify your answer.

12. Is clock 12 arithmetic under the operation of addition a commutative group? Explain.

13. Consider clock 5 arithmetic under the operation of addition. The set of elements in such a mathematical system is $\{1, 2, 3, 4, 5\}$.
   a) Which of these elements is the additive identity element?
   b) What is the additive inverse of 2? Explain your answer.

14. Consider clock 8 arithmetic under the operation of addition. The set of elements in such a mathematical system is $\{1, 2, 3, 4, 5, 6, 7, 8\}$.
   a) Which of these elements is the additive identity element?
   b) What is the additive inverse of 3? Explain your answer.

In Exercises 15 and 16, determine if the system is commutative. Explain how you determined your answer.

15.

⊟	A	⊗	W
A	⊗	W	A
⊗	W	A	⊗
W	A	⊗	⊗

16.

I	P	A	L
P	L	P	A
A	P	L	A
L	A	L	P

In Exercises 17 and 18, determine if the system has an identity element. If so, list the identity element. Explain how you determined your answer.

17.

W	A	B	C
A	C	B	A
B	B	C	B
C	A	B	C

18.

⊖	□	⊙	△
□	△	□	⊙
⊙	⊙	△	□
△	□	⊙	△

In Exercises 19 and 20, the identity element is C. Determine the inverse of A. Explain how you determined your answer.

19.

Ⓨ	A	B	C
A	B	C	A
B	C	A	B
C	A	B	C

20.

☮	C	A	B
C	C	A	B
A	A	C	B
B	B	B	C

Practice the Skills

In Exercises 21–32, use Table 10.1 on page 568 to determine the sum in clock 12 arithmetic.

21. $4 + 7$
22. $8 + 7$
23. $9 + 8$
24. $10 + 4$
25. $4 + 12$
26. $12 + 12$
27. $3 + (8 + 9)$
28. $(8 + 7) + 6$
29. $(6 + 4) + 8$
30. $(6 + 10) + 12$
31. $(7 + 8) + (9 + 6)$
32. $(7 + 11) + (9 + 5)$

In Exercises 33–44, determine the difference in clock 12 arithmetic by starting at the first number and counting counterclockwise on the clock the number of units given by the second number.

33. $7 - 4$
34. $11 - 8$
35. $4 - 12$
36. $3 - 9$
37. $5 - 10$
38. $3 - 10$
39. $1 - 12$
40. $6 - 10$
41. $5 - 5$
42. $8 - 8$
43. $12 - 12$
44. $5 - 8$

45. Use the following figure to develop an addition table for clock 6 arithmetic. The figure will also be used in Exercises 46–54.

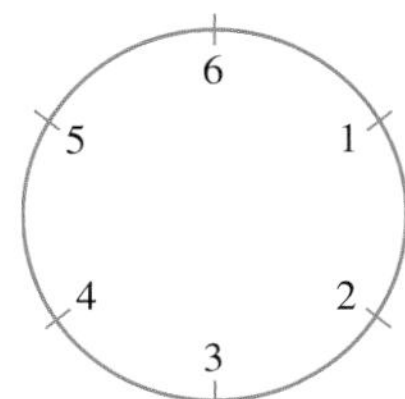

In Exercises 46–54, determine the sum or difference in clock 6 arithmetic.

46. $4 + 5$
47. $1 + 6$
48. $6 + 4$
49. $5 - 2$
50. $4 - 5$
51. $2 - 6$
52. $3 - 4$
53. $(3 - 5) - 6$
54. $2 + (1 - 3)$

55. Use the following figure to develop an addition table for clock 7 arithmetic. The figure will also be used in Exercises 56–64.

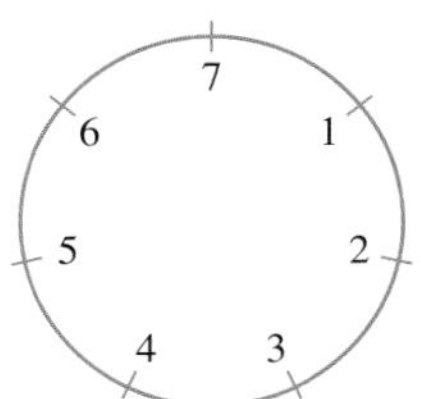

In Exercises 56–64, determine the sum or difference in clock 7 arithmetic.

56. $5 + 4$
57. $6 + 5$
58. $4 + 4$
59. $7 + 6$
60. $2 - 3$
61. $3 - 6$
62. $2 - 4$
63. $(4 - 5) - 6$
64. $3 - (2 - 6)$

65. Determine whether clock 7 arithmetic under the operation of addition is a commutative group. Explain.

66. A mathematical system is defined by a three-element by three-element table where every element in the set appears in each row and each column. Must the mathematical system be a commutative group? Explain.

67. Consider the mathematical system indicated by the following table. Assume that the associative property holds for the given operation.

	0	**1**	**2**	**3**
0	0	1	2	3
1	1	2	3	0
2	2	3	0	1
3	3	0	1	2

a) What are the elements of the set in this mathematical system?
b) What is the binary operation?
c) Is the system closed? Explain.
d) Is there an identity element for the system under the given operation? If so, what is it?
e) Does every element in the system have an inverse? If so, give each element and its corresponding inverse.
f) Give an example to illustrate the associative property.
g) Is the system commutative? Give an example to verify your answer.
h) Is the mathematical system a commutative group? Explain.

In Exercises 68–70, repeat parts (a)–(h) of Exercise 67 for the mathematical system in the table. Assume that the associative property holds for the given operation.

68.

	*****	**5**	***L***
*****	5	*L*	*
5	*L*	*	5
L	*	5	*L*

69.

	r	***s***	***t***	***u***
r	*t*	*u*	*r*	*s*
s	*u*	*r*	*s*	*t*
t	*r*	*s*	*t*	*u*
u	*s*	*t*	*u*	*r*

70.

	3	**5**	**8**	**4**
3	5	8	4	3
5	8	4	3	5
8	4	3	5	8
4	3	5	8	4

71. For the mathematical system

	f	***r***	***o***	***m***
f	*f*	*r*	*o*	*m*
r	*r*	*o*	*m*	*f*
o	*o*	*m*	*f*	*r*
m	*m*	*f*	*r*	*o*

determine
a) the elements in the set.
b) the binary operation.
c) closure or nonclosure of the system.
d) (*r* *o*) *f*
e) (*f* *r*) *m*
f) the identity element.
g) the inverse of *r*.
h) the inverse of *m*.

72. **a)** Is the following mathematical system a group? Explain your answer.
b) Find an example showing that the associative property does not hold for the given set of elements.

	1	**2**	**3**	**4**
1	2	3	4	1
2	3	4	1	2
3	4	1	2	3
4	1	4	3	2

In Exercises 73–78, for the mathematical system given, determine which of the five properties of a commutative group do not hold.

73.

⊗	⊡	*M*	🔔
⊡	⊡	*M*	🔔
M	*M*	⊡	🔔
🔔	🔔	*M*	⊡

74.

∧	*w*	*x*	*y*
w	*w*	*y*	*x*
x	*y*	*x*	*w*
y	*x*	*a*	*y*

75.

⊡	⊙	*	?	*T*	*P*
⊙	*T*	*P*	⊙	*	*
*	*P*	⊙	*	⊙	*T*
?	⊙	*	?	*T*	*P*
T	*	⊙	*T*	*P*	?
P	*	*T*	*P*	?	⊙

76.

☺	*a*	*b*	*π*	0	Δ
a	Δ	*a*	*b*	*π*	0
b	*a*	*b*	*π*	0	Δ
π	*b*	*π*	0	Δ	*a*
0	*π*	0	Δ	*a*	*b*
Δ	0	Δ	*π*	*b*	*a*

77.

⇔	*a*	*b*	*c*	*d*	*e*
a	*c*	*d*	*e*	*a*	*b*
b	*d*	*e*	*a*	*b*	*c*
c	*e*	*a*	*b*	*c*	*d*
d	*a*	*b*	*c*	*e*	*d*
e	*b*	*c*	*d*	*e*	*a*

78.

▽	0	1	2	3	4	5
0	0	0	0	0	0	0
1	0	1	2	3	4	5
2	0	2	4	0	2	4
3	0	3	0	3	0	3
4	0	4	2	0	4	2
5	0	5	4	3	2	1

Problem Solving Exercises

79. **a)** Consider the set consisting of two elements $\{E, O\}$, where E stands for an even number and O stands for an odd number. For the operation of addition, complete the table.

+	*E*	*O*
E		
O		

b) Determine whether this mathematical system forms a commutative group under the operation of addition. Explain your answer.

80. **a)** Let E and O represent even numbers and odd numbers, respectively, as in Exercise 79. Complete the table for the operation of multiplication.

×	*E*	*O*
E		
O		

b) Determine whether this mathematical system forms a commutative group under the operation of multiplication. Explain your answer.

In Exercises 81 and 82, make up your own mathematical system that is a group. List the identity element and the inverses of each element. Do so with sets containing

81. three elements.

82. four elements.

83. The tables in Exercises 83 and 84 are examples of noncommutative or nonabelian groups. For each exercise, do the following.

a) Show that the system under the given operation is a group. (It would be very time consuming to prove that the associative property holds, but you can give some examples to show that it appears to hold.)

b) Find a counterexample to show that the commutative property does not hold.

∞	1	2	3	4	5	6
1	5	3	4	2	6	1
2	4	6	5	1	3	2
3	2	1	6	5	4	3
4	3	5	1	6	2	4
5	6	4	2	3	1	5
6	1	2	3	4	5	6

84.

⊖	*A*	*B*	*C*	*D*	*E*	*F*
A	*E*	*C*	*D*	*B*	*F*	*A*
B	*D*	*F*	*E*	*A*	*C*	*B*
C	*B*	*A*	*F*	*E*	*D*	*C*
D	*C*	*E*	*A*	*F*	*B*	*D*
E	*F*	*D*	*B*	*C*	*A*	*E*
F	*A*	*B*	*C*	*D*	*E*	*F*

Challenge Problems/Group Activities

85. *Book Arrangements* Suppose that three books numbered 1, 2, and 3 are placed next to one another on a shelf. If we remove volume 3 and place it before volume 1, the new

order of books is 3, 1, 2. Let's call this replacement R. We can write

$$R = \begin{pmatrix} 1 & 2 & 3 \\ 3 & 1 & 2 \end{pmatrix}$$

which indicates the books were switched in order from 1, 2, 3 to 3, 1, 2. Other possible replacements are S, T, U, V, and I, as indicated.

$$S = \begin{pmatrix} 1 & 2 & 3 \\ 2 & 1 & 3 \end{pmatrix} \quad T = \begin{pmatrix} 1 & 2 & 3 \\ 3 & 2 & 1 \end{pmatrix} \quad U = \begin{pmatrix} 1 & 2 & 3 \\ 1 & 3 & 2 \end{pmatrix}$$

$$V = \begin{pmatrix} 1 & 2 & 3 \\ 2 & 3 & 1 \end{pmatrix} \quad I = \begin{pmatrix} 1 & 2 & 3 \\ 1 & 2 & 3 \end{pmatrix}$$

Replacement set I indicates that the books were removed from the shelves and placed back in their original order. Consider the mathematical system with the set of elements, R, S, T, U, V, I, with the operation $*$.

To evaluate $R * S$, write

$$R * S = \overset{R}{\begin{pmatrix} 1 & 2 & 3 \\ 3 & 1 & 2 \end{pmatrix}} * \overset{S}{\begin{pmatrix} 1 & 2 & 3 \\ 2 & 1 & 3 \end{pmatrix}}.$$

As shown in Fig. 10.2, R replaces 1 with 3 and S replaces 3 with 3 (no change), so $R * S$ replaces 1 with 3. R replaces 2 with 1 and S replaces 1 with 2, so $R * S$ replaces 2 with 2 (no change). R replaces 3 with 2 and S replaces 2 with 1, so $R * S$ replaces 3 with 1. $R * S$ replaces 1 with 3, 2 with 2, and 3 with 1.

$$\begin{array}{ccccc} & R & & S & \\ 1 & \longrightarrow & 3 & \longrightarrow & 3 \\ 2 & \longrightarrow & 1 & \longrightarrow & 2 \\ 3 & \longrightarrow & 2 & \longrightarrow & 1 \\ \uparrow & R & * & S & \uparrow \end{array}$$

$$R * S = \begin{pmatrix} 1 & 2 & 3 \\ 3 & 2 & 1 \end{pmatrix} = T$$

Figure 10.2

Since this result is the same as replacement set T, we write $R * S = T$.

a) Complete the table for the operation using the procedure outlined.

$*$	R	S	T	U	V	I
R		T				
S						
T						
U						
V						
I						

b) Is this mathematical system a group? Explain.

c) Is this mathematical system a commutative group? Explain.

86. If a mathematical system is defined by a four-element by four-element table, how many specific cases must be illustrated to prove the set of elements is associative under the given operation?

Recreational Mathematics

The following recreational exercises will help prepare you for the next section.

87. A table is shown below.

+	0	1	2	3	4
0	0		2		4
1	1		3		
2			4		1
3	3		0	1	2
4		0	1		

Fill in the blank areas of the table by performing the following operations.

1. Add the number in the left-hand column and the number in the top row.
2. Divide this sum by 5.
3. Place the *remainder* in the cell where the number in the left-hand column and the number in the top row would meet.

   For example, in the table, the number 2 is in red. This 2 is found as follows:

$$(3 + 4) \div 5 = 1, \text{ remainder } 2$$

88. Using Exercise 87 as a guide, complete the following table by dividing the sum of the numbers by 6. Place the remainders in the table as explained in Exercise 87.

+	0	1	2	3	4	5
0						
1						
2						
3						
4						
5						

89. Consider the mathematical system defined by the following table.

+	0	1	2	3
0	0	1	2	3
1	1	2	3	0
2	2	3	0	1
3	3	0	1	2

After working Exercises 87 and 88, can you explain how this table was formed?

Internet/Research Activities

90. In section 7.3 we introduced matrices. Show that 2×2 matrices under the operation of addition form a commutative group.

91. Show that 2×2 matrices under the operation of multiplication do not form a commutative group.

10.3 MODULAR ARITHMETIC

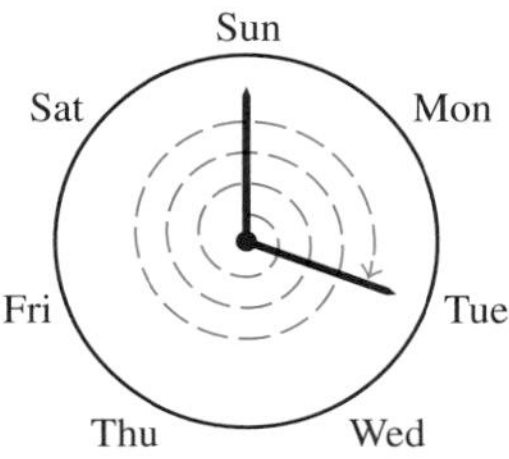

Figure 10.3

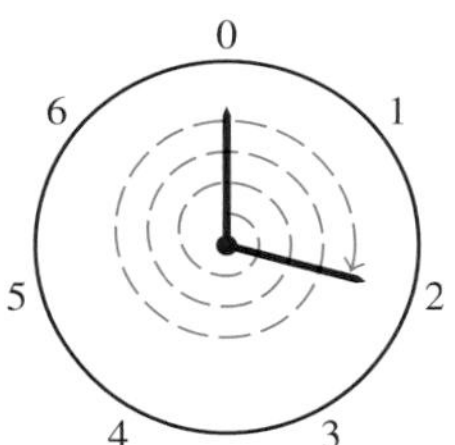

Figure 10.4

TABLE 10.11 Modular 7 Addition

+	**0**	**1**	**2**	**3**	**4**	**5**	**6**
0	0	1	2	3	4	5	6
1	1	2	3	4	5	6	0
2	2	3	4	5	6	0	1
3	3	4	5	6	0	1	2
4	4	5	6	0	1	2	③
5	5	6	0	1	2	3	4
6	6	0	1	2	3	4	5

Figure 10.5

The clock arithmetic we discussed in the previous section is similar to modular arithmetic. The set of elements $\{0, 1, 2, 3, 4, 5, 6, 7, 8, 9, 10, 11\}$ together with the operation of addition is called a modulo 12 or mod 12 system. There is one difference in notation between clock 12 arithmetic and modulo 12 arithmetic. In the modulo 12 system the symbol 12 is replaced with the symbol 0.

A *modulo m system* consists of m elements, 0 through $m - 1$, and a binary operation. In this section we will discuss modular arithmetic systems and their properties.

If today is Sunday, what day of the week will it be in 23 days? The answer, Tuesday, is arrived at by dividing 23 by 7 and observing the remainder of 2. Twenty-three days represent 3 weeks plus 2 days. Since we are interested only in the day of the week on which the twenty-third day will fall, the 3-week segment is unimportant to the answer. The remainder of 2 indicates the answer will be 2 days later than Sunday, which is Tuesday.

If we place the days of the week on a clock face as shown in Fig. 10.3, then in 23 days the hand would have made three complete revolutions and end on Tuesday. If we replace the days of the week with numbers, then a modulo 7 arithmetic system will result. See Fig. 10.4: Sunday = 0, Monday = 1, Tuesday = 2, and so on. If we start at 0 and move the hand 23 places, we will end at 2. Table 10.11 shows a modulo 7 addition table.

If we start at 4 and add 6, we end at 3 on the clock in Fig. 10.5. This number is circled in Table 10.11. The other numbers can be obtained in the same way.

A second method of determining the sum of $4 + 6$ in modulo 7 arithmetic is to divide the sum, 10, by 7 and observe the remainder.

$$10 \div 7 = 1, \quad \text{remainder } 3$$

The remainder, 3, is the sum of $4 + 6$ in a modulo 7 arithmetic system.

The concept of congruence is important in modular arithmetic.

> a is **congruent** to b modulo m, written, $a \equiv b \pmod{m}$, if a and b have the same remainder when divided by m.

We can show, for example, that $10 \equiv 3 \pmod{7}$ by dividing both 10 and 3 by 7 and observing that we obtain the same remainder in each case.

$$10 \div 7 = 1, \quad \text{remainder } 3 \qquad \text{and} \qquad 3 \div 7 = 0, \quad \text{remainder } 3$$

Since the remainders are the same, 3 in each case, 10 is congruent to 3 modulo 7, and we may write $10 \equiv 3 \pmod{7}$.

TABLE 10.12 Modulo 7 Classes

0	1	2	3	4	5	6
0	1	2	3	4	5	6
7	8	9	10	11	12	13
14	15	16	17	18	19	20
21	22	23	24	25	26	27
28	29	30	31	32	33	34
⋮	⋮	⋮	⋮	⋮	⋮	⋮

Now consider $37 \equiv 5 \pmod 8$. If we divide both 37 and 5 by 8, each has the same remainder, 5.

In any modulo system we can develop a set of ***modulo classes*** by placing all numbers with the same remainder in the appropriate modulo class. In a modulo 7 system every number must have a remainder of either 0, 1, 2, 3, 4, 5, or 6. Thus, a modulo 7 system has seven modulo classes. The seven classes are presented in Table 10.12.

Every number is congruent to a number from 0 to 6 in modulo 7. For example, $24 \equiv 3 \pmod 7$ because 24 is in the same modulo class as 3.

The solution to a problem in modular arithmetic, if it exists, will always be a number from 0 through $m - 1$, where m is the ***modulus*** of the system. For example, in a modulo 7 system, because 7 is the modulus, the solution will be a number from 0 through 6.

EXAMPLE 1 *Congruence Modulo 7*

Determine which number, from 0 through 6, the following numbers are congruent to in modulo 7.

a) 62 b) 53 c) 105

SOLUTION: We could determine the answer by listing more entries in Table 10.12. Another method of finding the answer is to divide the given number by 7 and observe the remainder. In the solutions we will use a question mark, ?, as a place holder. The ? will represent a number from 0 through 6. When we determine the answer, we will replace the ? with the answer.

a) $62 \equiv ? \pmod 7$

To determine the value that 62 is congruent to in mod 7, divide 62 by 7 and find the remainder.

$$\begin{array}{r} 8 \\ 7\overline{)62} \\ \underline{56} \\ 6 \leftarrow \text{remainder} \end{array}$$

$$62 \div 7 = 8, \quad \text{remainder } 6$$

Thus, $62 \equiv 6 \pmod 7$.

b) $53 \equiv ? \pmod 7$

$$53 \div 7 = 7, \quad \text{remainder } 4$$

Thus, $53 \equiv 4 \pmod 7$.

c) $105 \equiv ? \pmod 7$

$$105 \div 7 = 15, \quad \text{remainder } 0$$

Thus, $105 \equiv 0 \pmod 7$. ▲

EXAMPLE 2 *Congruence Modulo 5*

Evaluate each of the following in mod 5.

a) $3 + 4$ b) $4 - 3$ c) $4 \cdot 2$

DID YOU KNOW

Modular Arithmetic

January

S	M	T	W	T	F	S
				1	2	3
4	5	6	7	8	9	10
11	12	13	14	15	16	17
18	19	20	21	22	23	24
25	26	27	28	29	30	31

Car odometer

0 1 4 6 0 0

As you will see when you work the exercises, we can use modular arithmetic to solve everyday real-life situations. The hours of a clock and the months of a year form modular 12 systems. The odometer in your car is probably a modular 1,000,000 system. That is, a car that has traveled 14,600 miles would have an odometer reading of 014,600. If a car were able to travel 1,014,600 miles, the odometer would show the same reading, 014,600.

PROFILE IN MATHEMATICS

AMALIE EMMY NOETHER (1882–1935)

Professor Amalie Emmy Noether was described by Albert Einstein as the "most significant creative mathematical genius thus far produced since the higher education of women began." Because she was a woman, she was never to be granted a regular professorship in her native Germany. In 1933, with the rise of the Nazis to power, she and many other mathematicians and scientists emigrated to the United States. That brought her to Bryn Mawr College, at the age of 51, to her first full-time faculty appointment. She was to die just 2 years later.

Her special interest lay in abstract algebra, which includes such topics as groups, rings, and fields. She developed a number of theories that helped draw together many important mathematical concepts. Her innovations in higher algebra gained her recognition as one of the most creative abstract algebraists of modern times.

SOLUTION: In each part, because we are working in modulo 5, the answer will be a number from 0 through 4.

a) $3 + 4 \equiv ? \pmod 5$
$7 \equiv ? \pmod 5$

$$7 \div 5 = 1, \quad \text{remainder } 2$$

Therefore, $4 + 3 \equiv 2 \pmod 5$.

b) $4 - 3 \equiv ? \pmod 5$
$1 \equiv ? \pmod 5$
$1 \equiv 1 \pmod 5$

Remember that we want to replace the question mark with a number between 0 and 4, inclusive. Thus, $4 - 3 \equiv 1 \pmod 5$.

c) $4 \cdot 2 \equiv ? \pmod 5$
$8 \equiv ? \pmod 5$

Since $8 \div 5 = 1$, remainder 3, $8 \equiv 3 \pmod 5$. Thus, $4 \cdot 2 \equiv 3 \pmod 5$. ▲

Note in Table 10.12 that every number in the same modulo class differs by a multiple of the modulo, in this case a multiple of 7. Adding (or subtracting) a multiple of the modulo number to (or from) a given number does not change the modulo class or congruence of the given number. For example, $3, 3 + 1(7), 3 + 2(7), 3 + 3(7), \ldots, 3 + n(7)$ are all in the same modulo class, namely, 3. We use this fact in the solution to Example 3.

EXAMPLE 3 *Using Modulo Classes in Subtraction*

Find the replacement for the question mark that makes each of the following true.

a) $3 - 5 \equiv ? \pmod 7$ b) $? - 4 \equiv 3 \pmod 5$ c) $5 - ? \equiv 7 \pmod 8$

SOLUTION: In each part we wish to replace the question mark with a number less than the modular number. Therefore, in part (a) we wish to replace the ? with a number less than 7. In part (b) we wish to replace the ? with a number less than 5. In part (c) we wish to replace the ? with a number less than 8.

a) In mod 7, adding 7, or a multiple of 7, to a number results in a sum that is in the same modulo class. Thus, if we add 7, 14, 21, . . . to 3, the result will be a number in the same modulo class. We want to replace 3 with an equivalent mod 7 number that is greater than 5. Adding 7 to 3 yields a sum of 10, which is greater than 5.

$$\begin{aligned} 3 - 5 &\equiv ? \pmod 7 \\ (3 + 7) - 5 &\equiv ? \pmod 7 \\ 10 - 5 &\equiv ? \pmod 7 \\ 5 &\equiv ? \pmod 7 \\ 5 &\equiv 5 \pmod 7 \end{aligned}$$

Therefore, $? = 5$ and $3 - 5 \equiv 5 \pmod 7$.

b) We wish to replace the ? with a number less than 5. We know that $7 - 4 \equiv 3$ mod 5 because $3 \equiv 3$ mod 5. Therefore, we need to determine what number, less than 5, the number 7 is congruent to in mod 5. If we subtract the

DID YOU KNOW

The Enigma

A cipher, or secret code, actually has group properties: A well-defined operation turns plain text into a cipher and an inverse operation allows it to be deciphered. During World War II, the Germans used an encrypting device based on a modulo 26 system (for the 26 letters of the alphabet) known as the enigma. It has four rotors, each with the 26 letters. The rotors were wired to one another so that if an A were typed into the machine, the first rotor would contact a different letter on the second rotor, which contacted a different letter on the third rotor, and so on. The French secret service obtained the wiring instructions, and Polish and British cryptoanalysts were able to crack the ciphers using group theory. This breakthrough helped to keep the Allies abreast of the deployment of the German navy.

modulo, 5, from 7, we obtain 2. Thus, 2 and 7 are in the same modular class. Therefore, ? = 2.

$$\begin{aligned} ? - 4 &\equiv 3 \pmod 5 \\ 7 - 4 &\equiv 3 \pmod 5 \\ 2 - 4 &\equiv 3 \pmod 5 \end{aligned}$$

Notice in the last equivalence that if we add 5 to 2, we get $7 - 4 \equiv 3 \pmod 5$, which is a true statement.

c) $5 - ? \equiv 7 \pmod 8$

In mod 8, adding 8, or a multiple of 8, to a number results in a sum that is in the same modulo class. Thus, we can add 8 to 5 so that the statement becomes

$$\begin{aligned} (8 + 5) - ? &\equiv 7 \pmod 8 \\ 13 - ? &\equiv 7 \pmod 8 \end{aligned}$$

We can see that $13 - 6 = 7$. Therefore, ? = 6 and $5 - 6 \equiv 7 \pmod 8$. ▲

EXAMPLE 4 *Using Modulo Classes in Multiplication*

Find all replacements for the question mark that make the statements true.

a) $4 \cdot ? \equiv 3 \pmod 5$ b) $3 \cdot ? \equiv 0 \pmod 6$ c) $3 \cdot ? \equiv 2 \pmod 6$

SOLUTION:

a) One method of determining the solution is to replace the question mark with the numbers 0–4 and then find the equivalent modulo class of the product. We use the numbers 0–4 because we are working in modulo 5.

$$\begin{aligned} 4 \cdot ? &\equiv 3 \pmod 5 \\ 4 \cdot 0 &\equiv 0 \pmod 5 \\ 4 \cdot 1 &\equiv 4 \pmod 5 \\ 4 \cdot 2 &\equiv 3 \pmod 5 \\ 4 \cdot 3 &\equiv 2 \pmod 5 \\ 4 \cdot 4 &\equiv 1 \pmod 5 \end{aligned}$$

Therefore, ? = 2 since $4 \cdot 2 \equiv 3 \pmod 5$.

b) Since we are working in modulo 6, replace the question mark with the numbers 0–5 and follow the procedure used in part (a).

$$\begin{aligned} 3 \cdot ? &\equiv 0 \pmod 6 \\ 3 \cdot 0 &\equiv 0 \pmod 6 \\ 3 \cdot 1 &\equiv 3 \pmod 6 \\ 3 \cdot 2 &\equiv 0 \pmod 6 \\ 3 \cdot 3 &\equiv 3 \pmod 6 \\ 3 \cdot 4 &\equiv 0 \pmod 6 \\ 3 \cdot 5 &\equiv 3 \pmod 6 \end{aligned}$$

Therefore, replacing the question mark with 0, 2, or 4 results in true statements. The answers are 0, 2, and 4.

c) $3 \cdot ? \equiv 2 \pmod 6$

Examining the products in part (b) shows there are no values that satisfy the statement. The answer is "no solution." ▲

Modular arithmetic systems under the operation of addition are commutative groups, as illustrated in Example 5.

EXAMPLE 5 *A Commutative Group*

Construct a mod 5 addition table and show that the mathematical system is a commutative group. Assume that the associative property holds for the given operation.

SOLUTION: The set of elements in modulo 5 arithmetic is $\{0, 1, 2, 3, 4\}$; the binary operation is $+$.

+	**0**	**1**	**2**	**3**	**4**
0	0	1	2	3	4
1	1	2	3	4	0
2	2	3	4	0	1
3	3	4	0	1	2
4	4	0	1	2	3

For this system to be a commutative group, it must satisfy the five properties of a commutative group.

1. *Closure:* Every entry in the table is a member of the set $\{0, 1, 2, 3, 4\}$, so the system is closed under addition.
2. *Identity element:* An easy way to determine whether there is an identity element is to look for a row in the table that is identical to the elements at the top of the table. Note that the row next to 0 is identical to the top of the table, which indicates that 0 *might be* the identity element. Now look at the column under the 0 at the top of the table. If this column is identical to the left-hand column, then 0 is the identity element. Since the column under 0 is the same as the left-hand column, 0 is the additive identity element in modulo 5 arithmetic.

Element	+	Identity	=	Element
0	+	0	=	0
1	+	0	=	1
2	+	0	=	2
3	+	0	=	3
4	+	0	=	4

3. *Inverse elements:* Does every element have an inverse? Recall an element plus its inverse must equal the identity element. In this example, the identity element is 0. Therefore, for each of the given elements 0, 1, 2, 3, and 4, we must find the element that when added to it results in a sum of zero. These elements will be the inverses.

Element	+	Inverse	=	Identity	
0	+	?	=	0	Since $0 + 0 = 0$, 0 is its own inverse.
1	+	?	=	0	Since $1 + 4 = 0$, 4 is the inverse of 1.
2	+	?	=	0	Since $2 + 3 = 0$, 3 is the inverse of 2.
3	+	?	=	0	Since $3 + 2 = 0$, 2 is the inverse of 3.
4	+	?	=	0	Since $4 + 1 = 0$, 1 is the inverse of 4.

Note that each element has an inverse.

4. *Associative property:* It is given that the associative property holds. One example that illustrates the associative property is

$$(2 + 3) + 4 = 2 + (3 + 4)$$
$$0 + 4 = 2 + 2$$
$$4 = 4 \quad \text{True}$$

5. *Commutative property:* Is $a + b = b + a$ for *all* elements a and b of the given set? The table shows that the system is commutative because the elements are symmetric about the main diagonal. We will give one example to illustrate the commutative property.

$$4 + 2 = 2 + 4$$
$$1 = 1 \quad \text{True}$$

All five properties are satisfied. Thus, modulo 5 arithmetic under the operation of addition is a commutative group. ▲

Whenever a process is repetitive, modular arithmetic may be helpful in answering some questions about the process. Now let's look at an application of modular arithmetic.

EXAMPLE 6 *Work Schedule*

Ellis drives a bus for a living. His working schedule is to drive for 6 days, and then he gets 2 days off. If today is the third day that Ellis has been driving, determine the following.

a) Will he be driving 60 days from today?
b) Will he be driving 82 days from today?
c) Was he driving 124 days ago?

SOLUTION:

a) Since Ellis drives for 6 days and then gets 2 days off, his working schedule may be considered a modular 8 system. That is, 8, 16, 24, . . . days from today will be just like today, the third day of the 8-day cycle.

If we divide 60 by 8, we obtain

$$\begin{array}{r} 7 \\ 8\overline{)60} \\ \underline{56} \\ 4 \end{array} \leftarrow \text{remainder}$$

Therefore, in 60 days Ellis will go through 7 complete cycles and be 4 days further into the next cycle. If we let D represent a driving day and N represent a not driving day, then Ellis's cycle may be represented as follows:

$$D\ D\ D\ D\ D\ D\ N\ N$$

↑ today (third D) ↑ 4 days from today (first N)

Notice that 4 days from today will be his first nonworking day. Therefore, Ellis will not be driving 60 days from today.

b) We work this part in the same way we worked part (a). Divide 82 by 8 and determine the remainder.

$$\begin{array}{r} 10 \\ 8\overline{)82} \\ \underline{80} \\ 2 \leftarrow \text{remainder} \end{array}$$

Thus, in 82 days it will be 2 days later in the cycle than it is today. Because he is currently in day 3 of his cycle, Ellis will be in day 5 of his cycle and will be driving the bus.

c) This part is worked in the same way as parts (a) and (b), but once we find the remainders we must move backward in the cycle.

$$\begin{array}{r} 15 \\ 8\overline{)124} \\ \underline{120} \\ 4 \leftarrow \text{remainder} \end{array}$$

Thus, 124 days ago was 4 days earlier in the cycle. Marking day 3 of the cycle (indicated by word today) and then moving 4 days backwards brings us to the first nondriving day. The two N's shown at the beginning of the letters below are actually the end days of the previous cycle.

N N D D D D D D N N

↑ 4 days earlier than today ↑ today

Therefore, 124 days ago Ellis was not driving.

TIMELY TIP A knowledge of modular arithmetic may prove useful throughout life. Answers to problems like those presented in Example 6 and Exercises 67–74 can often be found using modular arithmetic.

SECTION 10.3 EXERCISES

Concept/Writing Exercises

1. What does a *modulo m* system consist of?
2. **a)** Explain the meaning of the statement, "*a* is congruent to *b* modulo *m*."
   b) Explain the meaning of $13 \equiv 3 \pmod 5$.
3. In a modulo 5 system, how many modulo classes will there be? Present a table similar to Table 10.12 on page 580 showing elements from each class.
4. In general, for a modulo *m* system, how are modulo classes developed?

5. In a modulo 12 system, how many modulo classes will there be? Explain.
6. In a modulo n system, how many modulo classes will there be? Explain.
7. Consider $27 \equiv ? \pmod{5}$. Which of the following values could replace the question mark and result in a true statement? Explain.
   a) 20 **b)** 2 **c)** 12 **d)** 107
8. Consider $106 \equiv ? \pmod{7}$. Which of the following values could replace the question mark and result in a true statement? Explain.
   a) 83 **b)** 71 **c)** 7 **d)** 22

Practice the Skills

In Exercises 9–16, assume that Sunday is represented as day 0, Monday is represented by day 1, and so on. If today is Thursday (day 4), determine the day of the week it will be at the end of each period. Assume no leap years.

9. 30 days **10.** 161 days **11.** 365 days
12. 2 years **13.** 3 years, 34 days **14.** 463 days
15. 728 days **16.** 3 years, 27 days

In Exercises 17–24, consider 12 months to be a modulo 12 system. If it is currently October, determine the month it will be in the specified number of months.

17. 9 months **18.** 36 months
19. 3 years, 5 months **20.** 4 years, 8 months
21. 83 months **22.** 7 years
23. 105 months **24.** 5 years, 9 months

In Exercises 25–36, determine what number the sum, difference, or product is congruent to in mod 5.

25. $8 + 6$ **26.** $5 + 10$
27. $1 + 9 + 12$ **28.** $9 - 3$
29. $5 - 12$ **30.** $7 \cdot 4$
31. $8 \cdot 9$ **32.** $10 - 15$
33. $4 - 8$ **34.** $3 - 7$
35. $(15 \cdot 4) - 8$ **36.** $(4 - 9) \cdot 7$

In Exercises 37–50, find the modulo class to which each number belongs for the indicated modulo system.

37. 15, mod 5 **38.** 23, mod 7 **39.** 84, mod 12
40. 43, mod 6 **41.** 60, mod 9 **42.** 75, mod 8
43. 30, mod 7 **44.** 53, mod 4 **45.** −5, mod 7
46. −7, mod 4 **47.** −13, mod 11 **48.** −11, mod 13
49. 135, mod 10 **50.** −12, mod 4

In Exercises 51–66, find all replacements (less than the modulus) for the question mark that make the statement true.

51. $3 + 4 \equiv ? \pmod{6}$ **52.** $? + 5 \equiv 3 \pmod{8}$
53. $2 + ? \equiv 4 \pmod{5}$ **54.** $4 + ? \equiv 3 \pmod{6}$
55. $4 - ? \equiv 5 \pmod{6}$ **56.** $4 \cdot 5 \equiv ? \pmod{7}$
57. $5 \cdot ? \equiv 7 \pmod{9}$ **58.** $3 \cdot ? \equiv 5 \pmod{6}$
59. $3 \cdot ? \equiv 1 \pmod{6}$ **60.** $3 \cdot ? \equiv 3 \pmod{12}$
61. $4 \cdot ? \equiv 4 \pmod{10}$ **62.** $? - 6 \equiv 4 \pmod{8}$
63. $? - 7 \equiv 9 \pmod{12}$ **64.** $6 - ? \equiv 8 \pmod{9}$
65. $3 \cdot ? \equiv 0 \pmod{10}$ **66.** $4 \cdot ? \equiv 5 \pmod{8}$

Problem Solving

67. *Presidential Elections* The upcoming presidential election years are 2004, 2008, 2012, ...
   a) List the next five presidential election years after 2012.
   b) What will be the first election year after the year 3000?
   c) List the election years between the years 2550 and 2575.

68. *Flight Schedules* A pilot is scheduled to fly for 5 consecutive days and rest for 3 consecutive days. If today is the second day of her rest shift, determine whether she will be flying or resting
   a) 60 days from today.
   b) 90 days from today.
   c) 240 days from today.
   d) Was she flying 6 days ago?
   e) Was she flying 20 days ago?

69. *Workout Schedule* A tennis pro's workout schedule is to have both morning and afternoon practice for 3 days, rest for 1 day, have only morning practice for 2 days, rest for 2 days, and then start the cycle again. If the tennis pro is on her rest for 1 day part of the schedule, determine what she will be doing
   a) 28 days from today.
   b) 60 days from today.
   c) 127 days from today.
   d) Will the tennis pro have a day off 82 days from today?

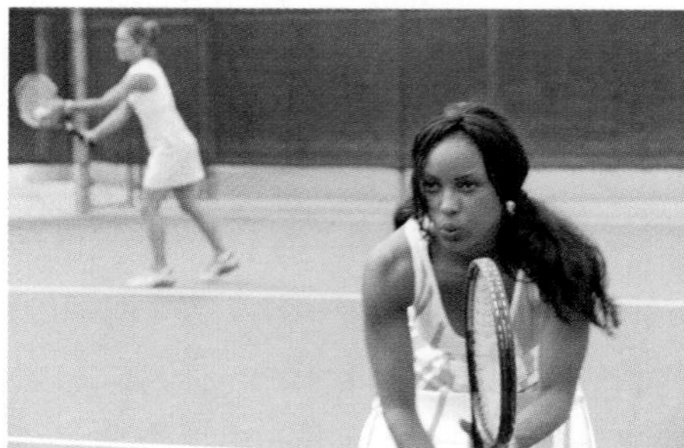

70. *Physical Therapy* A man has an Achilles' tendon injury and is receiving physical therapy. He must have physical therapy twice a day for 5 days, physical therapy once a day for 3 days, 2 days off, and then the cycle begins again. If he is in his second day of his twice a day therapy cycle, determine what he will be doing

a) 20 days from today.
b) 49 days from today.
c) 103 days from today.
d) Will the man have a day off 78 days from today?

71. *The Weekend Off* A manager of a theater has both Saturday and Sunday off every 7 weeks. This is week 2 of the 7 weeks.

a) Determine the number of weeks before she will have both Saturday and Sunday off.
b) Will she have both Saturday and Sunday off 25 weeks from this week?
c) What is the first week, after 50 weeks from this week, that she will have both Saturday and Sunday off?

72. *Nursing Shifts* A nurse's work pattern at Community Hospital consists of working the 7 A.M.–3 P.M. shift for 3 weeks and then the 3 P.M.–11 P.M. shift for 2 weeks.

a) If this is the third week of the pattern, what shift will the nurse be working 6 weeks from now?
b) If this is the fourth week of the pattern, what shift will the nurse be working 7 weeks from now?
c) If this is the first week of the pattern, what shift will the nurse be working 11 weeks from now?

73. *Restaurant Rotation* A waiter at a restaurant works both daytime and evening shifts. He works daytime for 5 consecutive days, then evenings for 3 consecutive days, then daytime for 4 consecutive days, then evenings for 2 consecutive days. Then the rotation starts again. If this is day 2 of the 5-day consecutive daytime shift, determine whether he will be working the daytime or evening shift

a) 20 days from today.
b) 52 days from today.
c) 365 days from today.

74. *A Truck Driver's Schedule* A truck driver's routine is as follows: Drive 3 days from New York to Chicago, rest 1 day in Chicago, drive 3 days from Chicago to Los Angeles, rest 2 days in Los Angeles, drive 5 days to return to New York, rest 3 days in New York. Then the cycle begins again. If the truck driver is starting his trip to Chicago today, what will he be doing

a) 30 days from today?
b) 70 days from today?
c) 2 years from today?

75. a) Construct a modulo 4 addition table.

b) Is the system closed? Explain.
c) Is there an identity element for the system? If so, what is it?
d) Does every element in the system have an inverse? If so, list the elements and their inverses.
e) The associative property holds for the system. Give an example.
f) Does the commutative property hold for the system? Give an example.
g) Is the system a commutative group?
h) Will every modulo system under the operation of addition be a commutative group? Explain.

76. Construct a modulo 8 addition table. Repeat parts (b)–(h) in Exercise 75.

77. a) Construct a modulo 4 multiplication table.

b) Is the system closed under the operation of multiplication?
c) Is there an identity element in the system? If so, what is it?
d) Does every element in the system have an inverse? Make a list showing the elements that have a multiplicative inverse and list the inverses.
e) The associative property holds for the system. Give an example.
f) Does the commutative property hold for the system? Give an example.
g) Is this mathematical system a commutative group? Explain.

78. Construct a modulo 7 multiplication table. Repeat parts (b)–(g) in Exercise 77.

Challenge Problems/Group Activities

We have not discussed division in modular arithmetic. With what number or numbers, if any, can you replace the question marks to make the statement true? The question mark must be a number less than the modulo. Hint: Use the fact that $\frac{a}{b} \equiv c \pmod{m}$ *means* $a \equiv b \cdot c \pmod{m}$ *and use trial and error to obtain your answer.*

79. $5 \div 7 \equiv ? \pmod{9}$

80. $? \div 5 \equiv 5 \pmod{9}$

81. $? \div ? \equiv 1 \pmod{4}$

82. $1 \div 2 \equiv ? \pmod{5}$

In Exercises 83–85, solve for x where k is any counting number.

83. $5k \equiv x \pmod{5}$

84. $5k + 4 \equiv x \pmod{5}$

85. $4k - 2 \equiv x \pmod{4}$

86. Find the smallest positive number divisible by 5 to which 2 is congruent in modulo 6.

Recreational Mathematics

87. *Rolling Wheel* The wheel shown below is to be rolled. Before the wheel is rolled, it is resting on number 0. The wheel will be rolled at a uniform rate of one complete roll every 4 minutes. In exactly 1 year (not a leap year), what number will be at the bottom of the wheel?

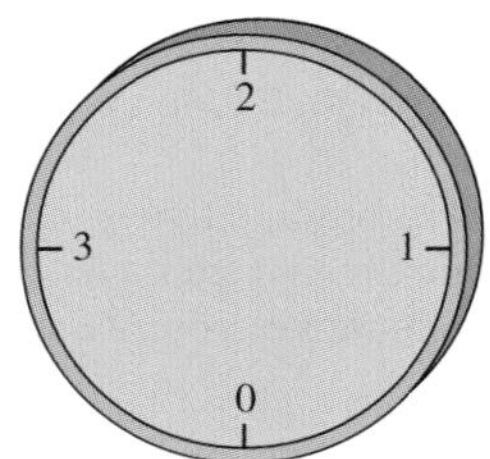

88. *Climbing a Mountain Range* A person climbs a uniform mountain range like the one shown below. Assume that the mountain range continues indefinitely.

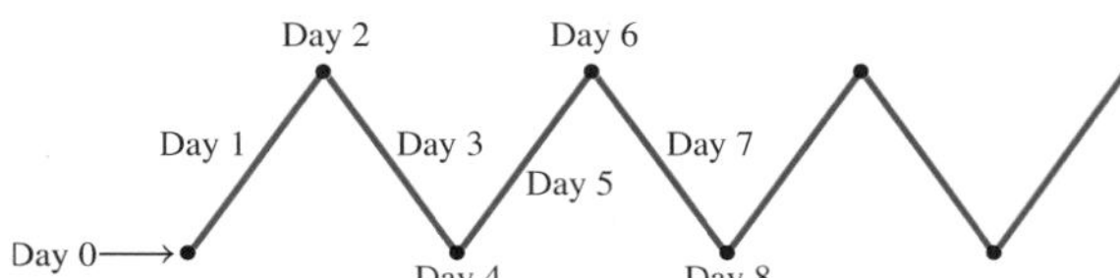

On day 0, the person rests at the bottom of a mountain in the mountain range. On day 1, the person climbs the mountain and reaches halfway up the mountain. On day 2, the person reaches the top and rests for the balance of the day. On day 3, the person starts down the mountain, and reaches halfway down the mountain. On day 4, the person reaches the bottom of the mountain and rests for the balance of the day. Then the process starts again on the next mountain in the mountain range. If today is day 4, where in the mountain range (bottom of a mountain, halfway up a mountain, halfway down a mountain, or the top of a mountain) will the person be after 1 year 21 days?

89. *Deciphering a Code* One important use of modular arithmetic is in coding. One type of coding circle is given in Fig. 10.6. To use it, the person you are sending the message to must know the code key to decipher the code. The code key to this message is *j*. Can you decipher this code? (*Hint:* Subtract the code key from the code numbers.)

23 11 3 18 10 19 2 10 16 4 24

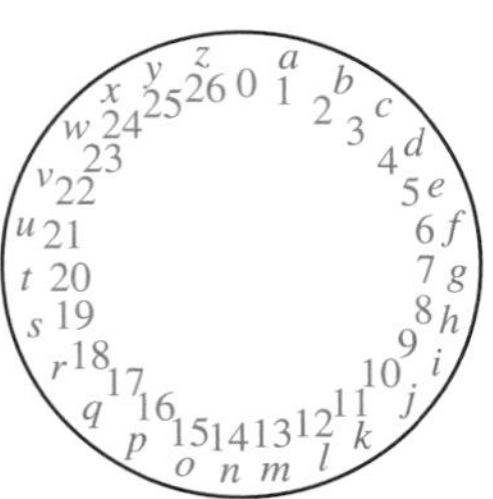

Figure 10.6

Internet/Research Activity

90. The concept and notation for modular systems were introduced by Carl Friedrich Gauss in 1801. Write a paper on Gauss's contribution to modular systems.

CHAPTER 10 SUMMARY

IMPORTANT FACTS

If the elements in a table are symmetric about the main diagonal, then the system is commutative.

	Addition	Multiplication
Commutative property	$a + b = b + a$	$a \cdot b = b \cdot a$
Associative property	$(a + b) + c = a + (b + c)$	$(a \cdot b) \cdot c = a \cdot (b \cdot c)$

a is **congruent** to b modulo m, written $a \equiv b \pmod{m}$, if a and b have the same remainder when divided by m.

Properties of a group and a commutative group

A mathematical system is a group if the first four conditions hold and a commutative group if all five conditions hold.

Commutative Group { Group {

1. The set of elements is *closed* under the given operation.
2. An *identity element* exists for the set.
3. Every element in the set has an *inverse.*
4. The set of elements is *associative* under the given operation.

5. The set of elements is *commutative* under the given operation.

CHAPTER 10 REVIEW EXERCISES

10.1, 10.2

1. List the parts of a mathematical system.
2. What is a binary operation?
3. Are the integers closed under the operation of addition? Explain.
4. Are the natural numbers closed under the operation of subtraction? Explain.

Determine the sum or difference in clock 12 arithmetic.

5. 9 + 10 **6.** 5 + 12 **7.** 8 − 10

8. 4 + 7 + 9 **9.** 7 − 4 + 6 **10.** 2 − 8 − 7

11. List the properties of a group, and explain what each property means.
12. What is an abelian group?

In Exercises 13 and 14, explain your answer.

13. Determine whether the set of positive integers under the operation of addition forms a group.
14. Determine whether the set of integers under the operation of multiplication forms a group.
15. Determine whether the set of rational numbers under the operation of addition forms a group.
16. Determine whether the set of rational numbers under the operation of multiplication forms a group.

In Exercises 17–19, for the mathematical system, determine which of the five properties of a commutative group do not hold.

17.

●	*R*	*S*	⊗
R	*R*	⊗	*S*
S	⊗	*S*	*R*
⊗	*S*	*R*	⊗

18.

□	!	?	Δ	*p*
!	?	Δ	!	*p*
?	Δ	*p*	?	!
Δ	!	?	Δ	*p*
p	*p*	!	*p*	Δ

19.

?	4	#	L	P
4	P	4	#	L
#	4	#	L	P
L	#	L	P	4
P	L	P	4	L

20. Consider the following mathematical system in which the operation is associative.

⊥	≻	⊙	?	Δ
≻	≻	⊙	?	Δ
⊙	⊙	?	Δ	≻
?	?	Δ	≻	⊙
Δ	Δ	≻	⊙	?

a) What are the elements of the set in this mathematical system?
b) What is the binary operation?
c) Is the system closed? Explain.
d) Is there an identity element for the system under the given operation?
e) Does every element in the system have an inverse? If so, give each element and its corresponding inverse.
f) Give an example to illustrate the associative property.
g) Is the system commutative? Give an example.
h) Is this mathematical system a commutative group? Explain.

10.3

In Exercises 21–30, find the modulo class to which the number belongs for the indicated modulo system.

21. 21, mod 3 **22.** 31, mod 8

23. 31, mod 6 **24.** 59, mod 8

25. 82, mod 13 **26.** 54, mod 4

27. 52, mod 12 **28.** 54, mod 14

29. 97, mod 11 **30.** 42, mod 11

In Exercises 31–40, find all replacements (less than the modulus) for the question mark that make the statement true.

31. 5 + 8 ≡ ? (mod 9) **32.** ? − 3 ≡ 0 (mod 5)

33. 4 · ? ≡ 3 (mod 6) **34.** 6 − ? ≡ 5 (mod 7)

35. ? · 4 ≡ 0 (mod 8) **36.** 10 · 7 ≡ ? (mod 12)

37. 3 − 5 ≡ ? (mod 7) **38.** ? · 7 ≡ 3 (mod 10)

39. 5 · ? ≡ 3 (mod 8) **40.** 7 · ? ≡ 2 (mod 9)

41. Construct a modulo 6 addition table. Then determine whether the modulo 6 system forms a commutative group under the operation of addition.
42. Construct a modulo 4 multiplication table. Then determine whether the modulo 4 system forms a commutative group under the operation of multiplication.

43. *Work Pattern* Toni Ranier's work pattern at the fast-food restaurant is as follows: She works 3 evenings, has 2 evenings off, then works 2 evenings, and then has 3 evenings off; then the pattern repeats. If today is the first day of the work pattern,

a) will Toni be working 18 days from today?
b) will Toni have the evening off for a party that is being held in 38 days?

CHAPTER 10 TEST

1. What is a mathematical system?
2. List the requirements needed for a mathematical system to be a commutative group.
3. Is the set of the whole numbers a commutative group under the operation of addition? Explain your answer completely.
4. Develop a clock 5 arithmetic addition table.
5. Is clock 5 arithmetic under the operation of addition a commutative group? Assume that the associative property holds. Explain your answer completely.

Determine the following in clock 5 arithmetic.

6. $4 + 3 + 2$

7. $2 - 5$

8. Consider the mathematical system

□	*W*	*S*	*T*	*R*
W	*T*	*R*	*W*	*S*
S	*R*	*W*	*S*	*T*
T	*W*	*S*	*T*	*R*
R	*S*	*T*	*R*	*W*

a) What is the binary operation?
b) Is this system closed? Explain.
c) Is there an identity element for this system under the given operation? Explain.
d) What is the inverse of the element R?
e) What is $(T \,\square\, R) \,\square\, W$?

In Exercises 9 and 10, determine whether the mathematical system is a commutative group. Explain your answer completely.

9.

*	*a*	*b*	*c*
a	*a*	*b*	*c*
b	*b*	*b*	*a*
c	*c*	*a*	*d*

10.

?	**1**	**2**	**3**
1	3	1	2
2	1	2	3
3	2	3	1

11. Determine whether the mathematical system is a commutative group. Assume that the associative property holds. Explain your answer.

○	@	$	&	%
@	%	@	$	&
$	@	$	&	%
&	$	&	%	@
%	&	%	@	$

In Exercises 12 and 13, determine the modulo class to which the number belongs for the indicated modulo system.

12. 64, mod 9

13. 58, mod 11

In Exercises 14–18, find all replacements for the question mark, less than the modulus, that make the statement true.

14. $7 + 7 \equiv ? \pmod 8$

15. $? - 3 \equiv 4 \pmod 5$

16. $3 - ? \equiv 7 \pmod 9$

17. $4 \cdot 2 \equiv ? \pmod 6$

18. $3 \cdot ? \equiv 2 \pmod 6$

19. To what number is 103 congruent to in modulo 7?

20. **a)** Construct a modulo 5 multiplication table.
b) Is this mathematical system a commutative group? Explain your answer completely.

GROUP PROJECTS

1. *Rotating a Square* The square *ABCD* below has a pin through it at point *P* so that the square can be rotated in a clockwise direction. The mathematical system consists of the set $\{A, B, C, D\}$ and the operation ♣. The elements of the set represent different rotations of the square clockwise, as follows.

   A = rotate about the point P clockwise 90°

   B = rotate about the point P clockwise 180°

   C = rotate about the point P clockwise 270°

   D = rotate about the point P clockwise 360°

   The operation ♣ means *followed by*. For example, A ♣ B means a clockwise rotation of 90° followed by a clockwise rotation of 180° or a clockwise rotation of 270°. Since C represents a clockwise rotation of 270°, we write A ♣ $B = C$. This and two other results are shown in the following table.

♣	*A*	*B*	*C*	*D*
A		*C*		
B			*A*	
C	*D*			
D				

   Complete the table and determine if the mathematical system is a commutative group. Explain.

2. *Product of Zero* In arithmetic and algebra the statement, "If $a \cdot b = 0$, then $a = 0$ or $b = 0$" is true. That is, for the product of two numbers to be 0, at least one of the factors must be 0. Can the product of two nonzero numbers equal 0 in a specific modulo system? If so, in what type of modulo systems can this result occur?
   a) Construct multiplication tables for modulo systems 3–9.
   b) Which, if any, of the multiplication tables in part (a) have products equal to 0 when neither factor is 0?
   c) Which, if any, of the multiplication tables in part (a) have products equal to 0 only when at least one factor is 0?
   d) Using the results in parts (b) and (c), can you write a conjecture as to which modulo systems have a product of 0 when neither factor is 0?

3. *Conjecture about Multiplication Inverses* Are there certain modulo systems where all numbers have multiplicative inverses?
   a) If you have not worked Group Projects Exercise 2, construct multiplication tables for modulo systems 3–9.
   b) Which of the multiplication tables in part (a) contain multiplicative inverses for all nonzero numbers?
   c) Which, if any, of the multiplication tables in part (a) do not contain multiplicative inverses for all nonzero numbers?
   d) Using the results in parts (b) and (c), can you write a conjecture as to which modulo systems contain multiplicative inverses for all nonzero numbers?

To some extent, money is a central aspect of everyone's life. Being knowledgeable about consumer mathematics can help you reach your financial goals.

CONSUMER MATHEMATICS

Managing your money takes much thought and planning. Your daily needs, such as food and transportation, as well as your monthly and yearly needs, such as car payments, rent or mortgage, and utility bills, must all be paid. At the same time, you should consider your long-term goals such as saving enough for large purchases—a new car or a new house—as well as for unexpected emergencies and for retirement. The decisions you make on a daily basis affect your ability to reach your long-term goals. This chapter will provide information on several areas of consumer mathematics, including loans, interest rates, and mortgages. A solid understanding of these and other consumer mathematics topics can help you achieve your financial goals.

11.1 PERCENT

The study of mathematics is crucial to understanding how to make better financial decisions. A basic topic necessary for understanding the material in this chapter is percent. This section will give you a better understanding of the meaning of percent and its use in real-life situations.

The word *percent* comes from the Latin *per centum,* meaning "per hundred." A *percent* is simply a ratio of some number to 100. Thus, $\frac{15}{100} = 15\%$, and $\frac{x}{100} = x\%$.

Percents are useful in making comparisons. Consider Ross, who took two psychology tests. On the first test Ross answered 18 of the 20 questions correctly, and on the second test he answered 23 of 25 questions correctly. On which test did he have the higher score? One way to compare the results is to write ratios of the number of correct answers to the number of questions on the test, and then convert the ratios to percents. We can find the grades in percent for each test by (a) writing a ratio of the number of correct answers to the total number of questions, (b) rewriting these ratios with a denominator of 100, and (c) expressing the ratios as percents.

Test 1 (a) (b) (c)

$$\frac{\text{Number of correct answers}}{\text{Number of questions on the test}} = \frac{18}{20} = \frac{18 \times 5}{20 \times 5} = \frac{90}{100} = 90\%$$

Test 2 (a) (b) (c)

$$\frac{\text{Number of correct answers}}{\text{Number of questions on the test}} = \frac{23}{25} = \frac{23 \times 4}{25 \times 4} = \frac{92}{100} = 92\%$$

By changing the results of both tests to percents, we have a common standard for comparison. The results show that Ross scored 90% on the first test and 92% on the second test. Thus, he had a higher score on the second test.

Another procedure to change a fraction to a percent follows.

Procedure to Change a Fraction to a Percent

1. Divide the numerator by the denominator.
2. Multiply the quotient by 100 (which has the effect of moving the decimal point two places to the right).
3. Add a percent sign.

Note that steps 2 and 3 together are the equivalent of multiplying by 100%. Since 100% = 100/100 = 1, we are not changing the *value* of the number, we are simply changing the number to a percent.

DID YOU KNOW

Is It Pure?

99 and 44/100% Pure: It is a slogan that has been around since 1881, ever since Procter & Gamble decided to market its laundry soap, Ivory Soap, as a cosmetic soap. "Pure what?" you may ask. Soap, as opposed to detergent, is a mixture of fatty acids and alkali, so the makeup of Ivory Soap is 99.44% soap and only 0.56% impurities. A semanticist, who studies language and its meaning, might say that's pure baloney. That which is described as pure, by definition, cannot be a mixture at all, and it is either 100% pure or not pure at all.

EXAMPLE 1 *Converting a Fraction to a Percent*

Change $\frac{17}{20}$ to a percent.

SOLUTION: Follow the steps in the procedure box.

1. $17 \div 20 = 0.85$ 2. $0.85 \times 100 = 85$ 3. 85%

Thus, $\frac{17}{20} = 85\%$ ▲

Procedure to Change a Decimal Number to a Percent

1. Multiply the decimal number by 100.
2. Add a percent sign.

The procedure for changing a decimal number to a percent is equivalent to moving the decimal point two places to the right and adding a percent sign.

EXAMPLE 2 *Converting a Decimal Number to a Percent*

Change 0.235 to percent.

SOLUTION: $0.235 = (0.235 \times 100)\% = 23.5\%$ ▲

To change a number given as a percent to a decimal number, follow this procedure.

Procedure to Change a Percent to a Decimal Number

1. Divide the number by 100.
2. Remove the percent sign.

The procedure for changing a percent to a decimal number is equivalent to moving the decimal point two places to the left and removing the percent sign. Another way to remember this is: percent means per hundred (or to *divide by 100*).

EXAMPLE 3 *Converting a Percent to a Decimal Number*

a) Change 35% to a decimal number.

b) Change $\frac{1}{2}\%$ to a decimal number.

SOLUTION:

a) $35\% = \frac{35}{100} = 0.35$. Thus, $35\% = 0.35$.

b) $\frac{1}{2}\% = 0.5\% = \frac{0.5}{100} = 0.005$. Thus, $\frac{1}{2}\% = 0.005$. ▲

EXAMPLE 4 *How Old Is Old?*

A Yahoo! Question of the Week asked, "At what age do you consider someone old?" Out of the 3496 people who responded, 1017 said age 80 is old. What percent of respondents feel that age 80 is old?

SOLUTION: To find the percent that feel age 80 is old, divide the number of those who responded that age 80 is old by the total number of respondents. Then move the decimal point two places to the right and add a percent sign:

$$\text{Percent who feel age 80 is old} = \frac{1017}{3496} \approx 0.2909 = 29.09\%$$

Thus, about 29.1% (to the nearest tenth of a percent) feel that age 80 is old. ▲

All answers in this section will be given to the nearest tenth of a percent. When rounding answers to the nearest tenth of a percent, we carry the division to four places after the decimal point (ten-thousandths) and then round to the nearest thousandths position.

EXAMPLE 5 *SARS Cases*

According to the World Health Organization, on May 1, 2003, there were 5865 reported worldwide cases of severe acute respiratory syndrome (SARS). Of these cases, 5238 were in China, 201 were in Singapore, 147 were in Canada, and 279 were in other countries.* Determine the percent of SARS cases in each of the following countries: China, Singapore, Canada, and other countries.

SOLUTION: To find each percent, divide the number of cases in each country by the total number of cases.

$$\text{Percent of cases in China} = \frac{5238}{5865} \approx 0.8931 \approx 89.3\%$$

$$\text{Percent of cases in Singapore} = \frac{201}{5865} \approx 0.0343 \approx 3.4\%$$

$$\text{Percent of cases in Canada} = \frac{147}{5865} \approx 0.0251 \approx 2.5\%$$

$$\text{Percent of cases in other countries} = \frac{279}{5865} \approx 0.0476 \approx 4.8\%$$

The sum of the percents, 89.3% + 3.4% + 2.5% + 4.8%, equals 100%. ▲

The percent increase or decrease, or percent change, over a period of time is found by the following formula:

$$\textbf{Percent change} = \frac{\begin{pmatrix}\text{amount in} \\ \text{latest period}\end{pmatrix} - \begin{pmatrix}\text{amount in} \\ \text{previous period}\end{pmatrix}}{\text{amount in previous period}} \times 100$$

If the amount in the latest period is greater than the amount in the previous period, the answer will be positive and will indicate a percent increase. If the amount in the latest period is smaller than the amount in the previous period, the answer will be negative and will indicate a percent decrease.

EXAMPLE 6 *Most Improved Baseball Record*

In 2002, the Major League baseball team with the most improved record was the Anaheim Angels. In 2001, the Angels won 75 games. In 2002, the Angels won 99 games. Find the percent increase in games won from 2001 to 2002.

*On May 1, 2003, there were 54 reported cases of SARS in the United States.

SOLUTION: The previous period is 2001 and the latest period is 2002.

$$\begin{aligned}\text{Percent change} &= \frac{\begin{pmatrix}\text{amount in}\\ \text{latest period}\end{pmatrix} - \begin{pmatrix}\text{amount in}\\ \text{previous period}\end{pmatrix}}{\text{amount in previous period}} \times 100\\ &= \frac{99 - 75}{75} \times 100\\ &= \frac{24}{75} \times 100\\ &= 0.32 \times 100\\ &= 32\end{aligned}$$

Therefore, there was a 32% increase in the number of games won by the Angels from 2001 to 2002.

EXAMPLE 7 *Labor Union Membership*

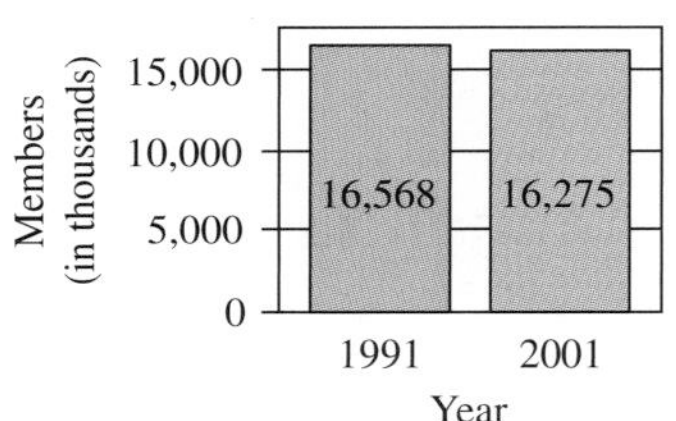

Source: U.S. Bureau of Labor Statistics

In 1991, there were approximately 16,568,000 labor union members in the United States. By 2001, this number had dropped to 16,275,000 (see the figure in the margin). Find the percent change in labor union membership from 1991 to 2001.

SOLUTION: The previous period is 1991 and the latest period is 2001.

$$\begin{aligned}\text{Percent change} &= \frac{16{,}275{,}000 - 16{,}568{,}000}{16{,}568{,}000} \times 100\\ &= \frac{-293{,}000}{16{,}568{,}000} \times 100\\ &\approx -0.0177 \times 100\\ &\approx -1.8\end{aligned}$$

Thus, union membership decreased by about 1.8% over this period.

A similar formula is used to calculate percent markup or markdown on cost. A positive answer indicates a markup and a negative answer indicates a markdown.

$$\textbf{Percent markup on cost} = \frac{\text{selling price} - \text{dealer's cost}}{\text{dealer's cost}} \times 100$$

EXAMPLE 8 *Determining Percent Markup*

Holdren Hardware stores pay \$48.76 for glass fireplace screens. They regularly sell them for \$79.88. At a sale they sell them for \$69.99. Find

a) the percent markup on the regular price.

b) the percent markup on the sale price.

c) the percent decrease of the sale price from the regular price.

DID YOU KNOW

To Insure Promptness

The practice of tipping began in an English coffeehouse in the eighteenth century. Legend has it that a hurried businessman dropped a few pence on the table as a friendly incentive to the waiter. In fact, the coffeehouses eventually installed small collection boxes marked **T**o **I**nsure **P**romptness. The tip, as it became known, was paid before, not after, the meal. Today, there are many books and Internet web sites devoted to the guidelines of tipping. For example, according to the Original Tipping Page (*www.tipping.org*), it is proper to tip waiters and waitresses 15–20% of the total restaurant bill. Hair stylists should be tipped 15% of the total bill and pizza delivery persons should receive \$1–\$3, depending on the distance driven. Skycaps at the airport should be tipped at least \$1 per bag and newspaper delivery persons should receive \$15–\$25 at holiday time.

SOLUTION:

a) We determine the percent markup on the regular price as follows.

$$\begin{aligned}\text{Percent markup} &= \frac{\text{selling price} - \text{dealer's cost}}{\text{dealer's cost}} \times 100 \\ &= \frac{\$79.88 - \$48.76}{\$48.76} \times 100 \\ &\approx 0.6382 \times 100 \\ &\approx 63.8\end{aligned}$$

Thus, the percent markup on the regular price was about 63.8%.

b) We determine the percent markup on the sale price as follows.

$$\begin{aligned}\text{Percent markup} &= \frac{\$69.99 - \$48.76}{\$48.76} \times 100 \\ &\approx 0.4354 \times 100 \\ &\approx 43.5\end{aligned}$$

Thus, the percent markup on the sale price was about 43.5%.

c) Based on the regular price, we determine the percent decrease of the sale price.

$$\begin{aligned}\text{Percent decrease} &= \frac{\$69.99 - \$79.88}{\$79.88} \times 100 \\ &\approx -0.1238 \times 100 \\ &\approx -12.4\end{aligned}$$

The sale price is about 12.4% lower than the regular price. ▲

In daily life we may need to know how to solve any one of the following three types of problems involving percent:

1. What is a 15% tip on a restaurant bill of \$24.66? The problem can be stated as

   15% of \$24.66 is what number?

2. If Nancy Johnson made a sale of \$500 and received a commission of \$25, what percent of the sale is the commission? The problem can be stated as

   What percent of \$500 is \$25?

3. If the price of a jacket was reduced by 25% or \$12.50, what was the original price of the jacket? The problem can be stated as

   25% of what number is 12.50?

To answer these questions we will write each problem as an equation. The word *is* means "is equal to," or =. In each problem we will represent the unknown quantity with the letter x. Therefore, the preceding problems can be represented as

1. 15% of $\$24.66 = x$ 2. $x\%$ of $\$500 = \$25$ 3. 25% of $x = \$12.50$

The word *of* in such problems indicates multiplication. To solve each problem, change the percent to a decimal number and express the problem as an equation; then solve the equation for the variable x. The solutions follow.

1. $$15\% \text{ of } \$24.66 = x$$
$$0.15(24.66) = x$$ 15% is written as 0.15 in decimal form.
$$3.699 = x$$

Since 15% of \$24.66 is \$3.699, the tip would be \$3.70.

2. $$x\% \text{ of } \$500 = \$25$$
$$(0.01x)500 = 25$$ x% is written as 0.01x in decimal form.
$$5x = 25$$
$$\frac{5x}{5} = \frac{25}{5}$$
$$x = 5$$

Since 5% of \$500 is \$25, the commission is 5% of the sale.

3. $$25\% \text{ of } x = \$12.50$$
$$0.25(x) = 12.50$$ 25% is written as 0.25 in decimal form.
$$\frac{0.25x}{0.25} = \frac{12.50}{0.25}$$
$$x = 50$$

Since 25% of \$50 is \$12.50, the original price of the jacket was \$50.00.

EXAMPLE 9 *Down Payment on a House*

Melissa Bell wishes to buy a house for \$87,000. To obtain a mortgage, she needs to pay 20% of the selling price as a down payment. Determine the amount of Melissa's down payment.

SOLUTION: We want to find the amount of the down payment. Let $x =$ the down payment. Then

$$\begin{aligned} x &= 20\% \text{ of the selling price} \\ &= 20\% \text{ of } \$87{,}000 \\ &= 0.20(87{,}000) \\ &= 17{,}400. \end{aligned}$$

Melissa will have a down payment of \$17,400. ▲

EXAMPLE 10 *Chess Tournaments*

In 2002, about 50,000 out of the 88,000 U.S. Chess Federation (USCF) members competed in USCF tournaments. What percent of USCF members competed in USCF tournaments?

SOLUTION: We need to determine what percent of 88,000 is 50,000. Let $x =$ percent of USCF members who competed in USCF tournaments. Then

$$\begin{aligned} x\% \text{ of } 88{,}000 &= 50{,}000 \\ 0.01x(88{,}000) &= 50{,}000 \\ 880x &= 50{,}000 \\ x &= \frac{50{,}000}{880} \\ x &\approx 56.8 \end{aligned}$$

Therefore, about 56.8% of USCF members competed in USCF tournaments in 2002.

Mexico City, Mexico

EXAMPLE 11 *Population of Mexico*

About 35,640,000, or 36%, of Mexico's population is younger than 15 years old. What is the population of Mexico?

SOLUTION: This problem can be stated as, 36% of what number is 35,640,000? Let x = the population of Mexico. Then

$$\begin{aligned} 36\% \text{ of } x &= 35{,}640{,}000 \\ 0.36x &= 35{,}640{,}000 \\ x &= \frac{35{,}640{,}000}{0.36} \\ x &= 99{,}000{,}000 \end{aligned}$$

Therefore, the population of Mexico is about 99,000,000 people.

SECTION 11.1 EXERCISES

Concept/Writing Exercises

1. What is a percent?
2. Explain how to change a percent to a decimal number.
3. Explain how to change a fraction to a percent.
4. Explain how to change a decimal number to a percent.
5. Explain how to determine percent change.
6. Explain how to determine percent markup on cost.

Practice the Skills

In Exercises 7–14, change the number to a percent. Express your answer to the nearest tenth of a percent.

7. $\frac{1}{2}$ **8.** $\frac{1}{4}$ **9.** $\frac{2}{5}$

10. $\frac{7}{8}$ **11.** 0.007654 **12.** 0.5688

13. 3.78 **14.** 13.678

In Exercises 15–24, change the percent to a decimal number.

15. 4% **16.** 6.9% **17.** 1.34% **18.** 0.0005%

19. $\frac{1}{4}\%$ **20.** $\frac{3}{8}\%$ **21.** $\frac{1}{5}\%$ **22.** 135.9%

23. 1% **24.** 0.50%

Problem Solving

For Exercises 25–46, round answers to the nearest tenth of a percent.

25. *Potassium* The U.S. Department of Agriculture's recommended daily allowance (USRDA) of potassium for adults is 3500 mg. One serving of Cheerios provides 95 mg of potassium. What percent of the USRDA of potassium does one serving of Cheerios provide?

26. *Disney Tickets* As of April 18, 2003, tickets for one day at Disney World are \$50 for adults and \$40 for children. In addition, if the tickets are purchased at Disney World, visitors must pay a 6% sales tax and a 6% tourist tax. However, if the tickets are purchased at any AAA office in Florida, visitors only pay the 6% sales tax. How much can Clarence and Joy Greenhalgh and their two children Martin and Thompson save by purchasing their tickets at a Florida AAA office?

27. *Reduced Fat Milk* One serving of whole milk contains 8 g of fat. Reduced-fat milk contains 41.25% less fat per serving than whole milk. How many grams of fat does one serving of reduced-fat milk contain?

28. *River Pollution* In a study of U.S. rivers, 693,905 miles of river were studied. According to the U.S. Environmental Protection Agency, it was found that 36% of the total miles of rivers studied had impaired water quality. How many miles of rivers had impaired quality?

Kentucky Lottery In Exercises 29–32, use the circle graph to answer the questions. In 2001, the Kentucky Lottery had total sales of $591 million. The areas where this money was used along with the corresponding percents are represented in the circle graph below.

2001 Kentucky Lottery Expenditures

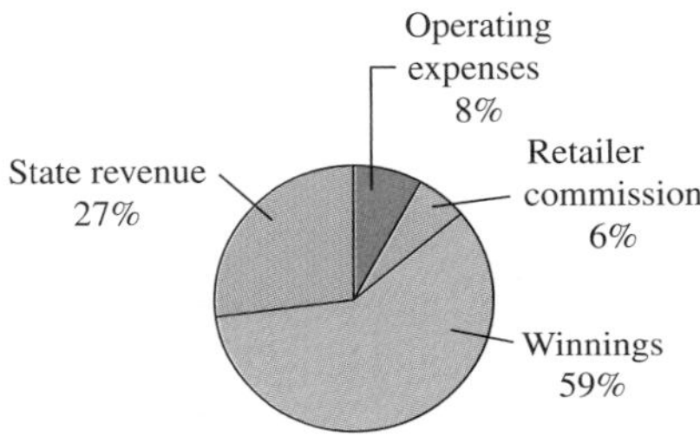

Source: Kentucky Lottery Internet Website

29. Determine the amount used on operating expenses.

30. Determine the amount used on retailer commissions.

31. Determine the amount used on state revenue.

32. Determine the amount used on winnings.

Inventory Shrinkage In Exercises 33–36, use the circle graph to answer the questions. In 2002, U.S. companies lost $32.3 billion due to "inventory shrinkage." Each sector of the circle graph shows the percent of this total due to each of four sources.

Inventory Shrinkage

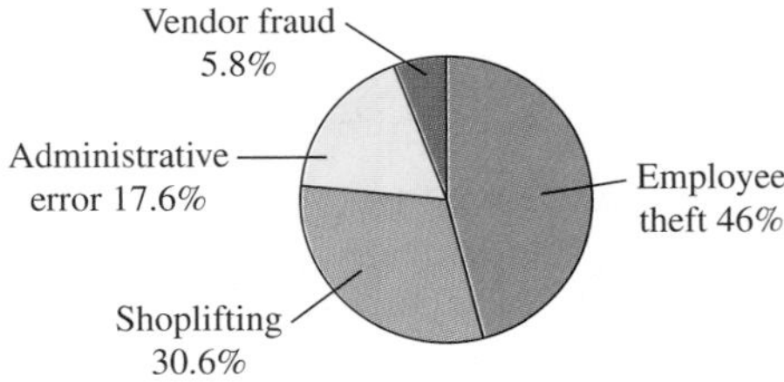

Source: Center for Retailing Education, University of Florida

33. Determine the amount lost to administrative error.

34. Determine the amount lost to vendor fraud.

35. Determine the amount lost to shoplifting.

36. Determine the amount lost to employee theft.

Top Five Advertisers In Exercises 37–40, use the circle graph to answer the questions. The top five advertisers spent $8105 million in 2001.

Top Five Advertisers in 2001 (in millions of dollars)

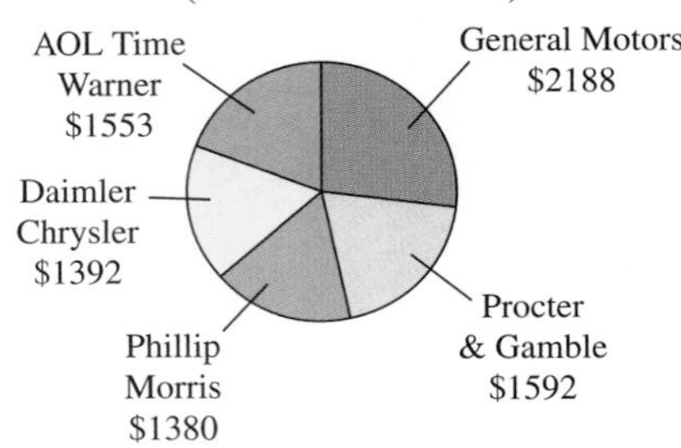

Source: Competitive Media Reporting

37. What percent of the $8105 million did AOL Time Warner spend?

38. What percent of the $8105 million did DaimlerChrysler spend?

39. What percent of the $8105 million did Procter & Gamble spend?

40. What percent of the $8105 million did General Motors spend?

41. *Decreasing Population* In 1990, the population of Pittsfield, Massachusetts, was 48,622. In 2000, the population had decreased to 45,793. Determine the percent decrease in Pittsfield's population from 1990 to 2000.

42. *Increasing Population* The population of the United States rose from approximately 248.7 million in 1990 to approximately 288.4 million in 2002.

a) Determine the percent increase in population from 1990 to 2002.

b) What will the population be in the year 2014 if it increases at the same percent as it did from 1990 to 2002?

43. *California Milk Production* The graph shows milk production in California for the years 1960, 1970, 1980, 1990, and 2000.

California Milk Production

Millions of pounds: 0, 10,000, 20,000, 30,000, 40,000
Year: 1960, 1970, 1980, 1990, 2000
8059
9457
13,577
20,947
32,240

Source: U.S. Department of Agriculture

a) Determine the percent increase in California milk production from 1960 to 1970.
b) Determine the percent increase in California milk production from 1970 to 1980.
c) Determine the percent increase in California milk production from 1980 to 1990.
d) Determine the percent increase in California milk production from 1990 to 2000.

44. *IRS Audits* The graph shows the number of individual audits conducted by the Internal Revenue Service (IRS) during the years 1996, 1997, 1998, 1999, and 2000.

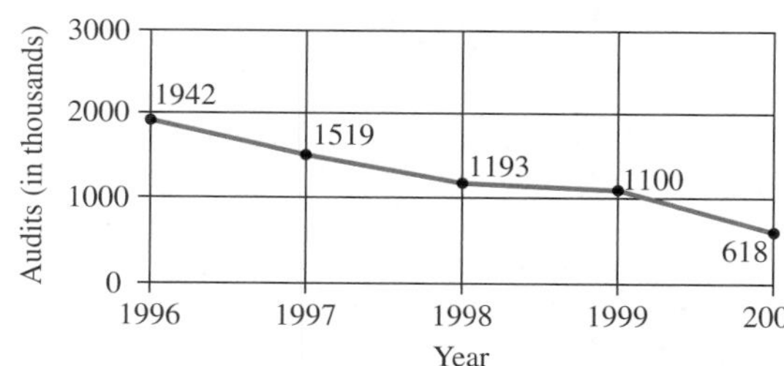

Source: Internal Revenue Service

a) Determine the percent decrease in the number of individual audits from 1996 to 1997.
b) Determine the percent decrease in the number of individual audits from 1997 to 1998.
c) Determine the percent decrease in the number of individual audits from 1998 to 1999.
d) Determine the percent decrease in the number of individual audits from 1999 to 2000.

45. *Dow Jones 2002* The following graph shows the closing Dow Jones Industrial Average (DJIA) for the months January through October 2002.

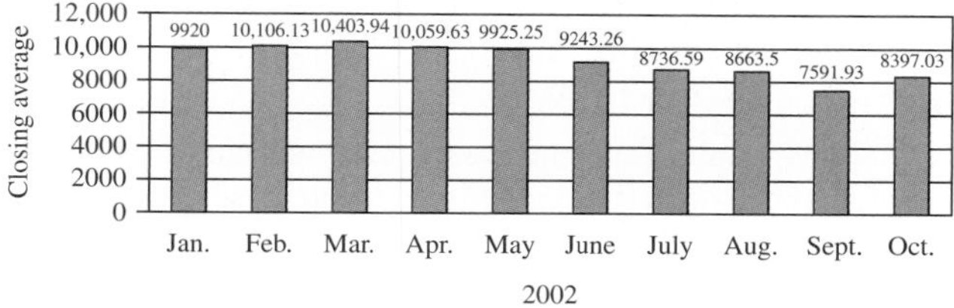

Source: *New York Times*

a) Determine the percent increase in the DJIA from January 2002 through March 2002.
b) Determine the percent decrease in the DJIA from March 2002 through September 2002.
c) Determine the percent decrease in the DJIA from January 2002 through September 2002.
d) Determine the percent increase in the DJIA from September 2002 through October 2002.

46. *Presidential Salary* The following graph shows the annual salary for the president of the United States for selected years from 1789 to 2001.

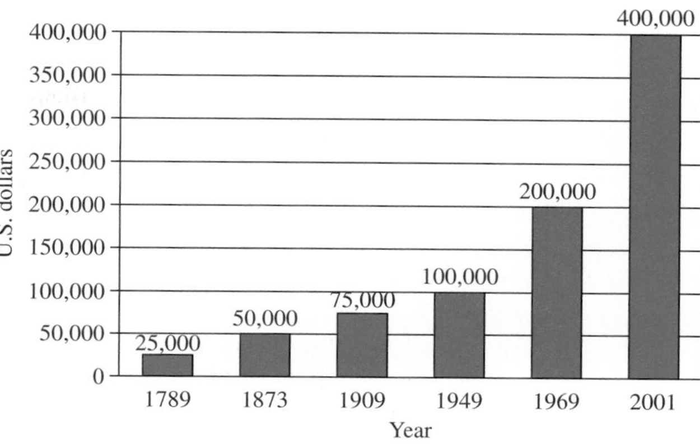

Source: *Congressional Quarterly*

a) Determine the percent increase in annual salary from 1789 to 1873.
b) Determine the percent increase in annual salary from 1909 to 1949.
c) Determine the percent increase in annual salary from 1969 to 2001.
d) Determine the percent increase in annual salary from 1789 to 2001.

In Exercises 47–52, determine the answer to the question.

47. What is 15% of $45.00?

48. What is 6.5% of $150.00?

49. What percent of 96 is 24?

50. What percent of 75 is 15?

51. Five percent of what number is 15?

52. Ten percent of what number is 75?

53. *Tax and Tip* According to the Original Tipping Page (*www.tipping.org*), it is proper to tip waiters and waitresses 15–20% of the total restaurant bill, *including the tax.* Mary and Keith's dinner costs $43.50 before tax, and the tax rate is 6%.
a) What is the tax on Mary and Keith's dinner?
b) What is the total bill, including tax, before the tip?
c) If Mary and Keith decide to tip 15% of the total bill, how much is the tip?
d) What is the total cost of the dinner including tax and tip?

54. *Fishing* The Gordon and Stallard Charter Fishing Boat Company has recently increased the number of crewmembers by 25%, or 10 crewmembers. What was the original number of crewmembers?

55. *Percentage of A's* In a mathematics class, eighteen students received an A on the third test, which is 150% of the students who received an A on the second test. How many students received an A on the second test?

56. *Employee Increase* The Fastlock Company hired 57 new employees, which increased its staff by 30%. What was the original number of employees?

57. *Salary Increase* Qami Brown's present salary is $36,500. He is getting an increase of 7% in his salary next year. What will his new salary be?

58. *Salad Dressing Preference* In a survey of 300 people, 17% prefer ranch dressing on their salad. How many people in the surveyed group prefer ranch dressing?

59. *Vacuum Cleaner Sales* A Kirby vacuum cleaner dealership sold 430 units in 2003 and 407 units in 2004. Find the percent increase or decrease in the number of units sold.

60. *Vacuum Cleaner Markup* If the Kirby dealership in Exercise 59 pays $320 for each unit and sells it for $699, what is the percent markup?

61. *Poverty in the United States* In the United States in 1993, there were 39.3 million people classified as living in poverty. In 2000, there were 31.1 million people classified as living in poverty. Find the percent decrease in the number of people living in poverty from 1993 to 2000.

62. *More Grandchildren* The Ngs had eight grandchildren in 2002. In 2003 they had 12 grandchildren. Find the percent increase in the number of grandchildren from 2002 to 2003.

63. *Television Sale* The regular price of a Phillips color TV is $539.62. During a sale, Hill TV is selling the TV for $439. Find the percent decrease in the price of this TV.

64. *Restaurant Markup* The cost of a fish dinner to the owner of the Golden Wharf restaurant is $7.95. The fish dinner is sold for $11.95. Find the percent markup.

65. *Truck Sale Profit* Bonnie James sold a truck and made a profit of $675. Her profit was 18% of the sale price. What was the sale price?

66. *Furniture Sale* Kane's Furniture Store advertised a table at a 15% discount. The original price was $115, and the sale price was $100. Was the sale price consistent with the ad? Explain.

67. *Reselling a Car* Quincy Carter purchased a used car for $1000. He decided to sell the car for 10% above his purchase price. Quincy could not sell the car so he reduced his asking price by 10%. If he sells the car at the reduced price, will he have a profit or a loss or will he break even? Explain how you arrived at your answer.

Challenge Problems/Group Activities

68. *Comparing Markdowns*
   a) A coat is marked down 10%, and the customer is given a second discount of 15%. Is that the same as a single discount of 25%? Explain.
   b) The regular price of a chair is $189.99. Determine the sale price of the chair if the regular price is reduced by 10% and this price is then reduced another 15%.
   c) Determine the sale price of the chair if the regular price of $189.99 is reduced by 25%.
   d) Examine the answers obtained in parts (b) and (c). Does your answer to part (a) appear to be correct? Explain.

69. *Selling Ties* The Tie Shoppe paid $5901.79 for a shipment of 500 ties and wants to make a profit of 40% of the cost on the whole shipment. The store is having two special sales. At the first sale it plans to sell 100 ties for $9.00 each, and at the second sale it plans to sell 150 ties for $12.50 each. What should be the selling price of the other 250 ties for the Tie Shoppe to make a 40% profit on the whole shipment?

Recreational Mathematics

70. In parts a) through d) determine which is the greater amount, and by how much.
   a) $100 increased by 25% or $200 decreased by 25%.
   b) $100 increased by 50% or $200 decreased by 50%.
   c) $100 increased by 100% or $200 decreased by 100%.

Internet/Research Activity

71. Find two circle graphs in newspapers, magazines, or on the Internet whose data are not given in percents. Redraw the graphs and label them with percents.

11.2 PERSONAL LOANS AND SIMPLE INTEREST

Often consumers want to buy clothing, appliances, or furniture but do not have the cash to do so. They then have to determine whether the cost of borrowing the money is worth the convenience or pleasure of having the item today versus waiting several months or years and paying cash. If you are faced with this choice, you may decide that you must have the item today and may choose to borrow the money from a bank or other lending institution. The money a bank is willing to lend you is called the amount of *credit* extended or the *principal of the loan*.

DID YOU KNOW

Seasonal Loan

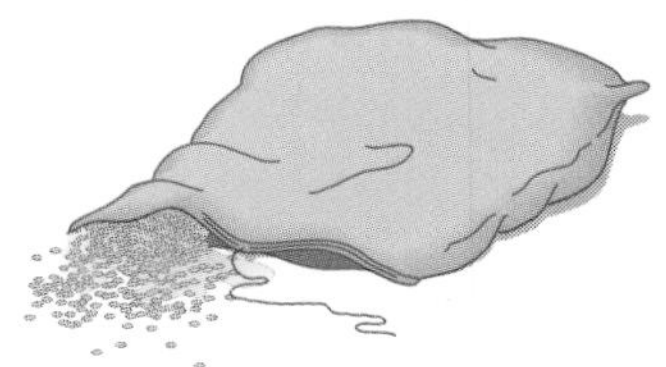

The most common type of loan is the short-term (seasonal) loan. Originally, the loans were agricultural loans provided to farmers to plant their crops and see them through the period between planting and harvesting. In fact, "seed" loans were first recorded in ancient Babylonia.

The amount of credit and the interest rate that you may obtain depend on the assurance that you can give the lender that you will be able to repay the loan. Your credit is determined by your business reputation for honesty, by your earning power, and by what you can pledge as security to cover the loan. *Security* (or *collateral*) is anything of value pledged by the borrower that the lender may sell or keep if the borrower does not repay the loan. Acceptable security may be a business, a mortgage on a property, the title to an automobile, savings accounts, or stocks or bonds. The more marketable the security, the easier it is to obtain the loan, and in some cases marketability may help in getting a lower interest rate.

Bankers sometimes grant loans without security, but they require the signature of one or more other persons, called ***cosigners***, who guarantee the loan will be repaid. For either of the two types of loans, the secured loan or the cosigner loan, the borrower (and cosigner, if there is one) must sign an agreement called a ***personal note***. This document states the terms and conditions of the loan.

The most common way for individuals to borrow money is through an installment loan or using a credit card. (Installment loans and credit cards are discussed in Section 11.4.)

The concept of simple interest is essential to the understanding of installment buying. *Interest* is the money the borrower pays for the use of the lender's money. One type of interest is called simple interest. *Simple interest* is based on the entire amount of the loan for the total period of the loan. The formula used to find simple interest follows.

Simple Interest Formula

$$\text{Interest} = \text{principal} \times \text{rate} \times \text{time}$$
$$i = prt$$

In the simple interest formula the ***principal***, p, is the amount of money lent, the ***rate***, r, is the rate of interest expressed as a percent, and the ***time***, t, is the number of days, months, or years for which the money will be lent. Time is expressed in the same period as the rate. For example, if the rate is 2% per month, the time must be expressed in months. Typically, rate means the annual rate unless otherwise stated. Principal and interest are expressed in dollars in the United States.

Ordinary Interest

The most common type of simple interest is called *ordinary interest.* For computing ordinary interest, each month has 30 days and a year has 12 months or 360 days. On the due date of a *simple interest note* the borrower must repay the principal plus the interest. (*Note:* Simple interest will mean ordinary interest unless stated otherwise.)

EXAMPLE 1 ***Calculating Interest and Payback Amount***

Chi Tran needs to borrow $1600 to have corrective eye surgery. From her credit union, she obtains a 9-month loan with an annual simple interest rate of 6.5%.

a) Calculate the simple interest on the loan.

b) Determine the amount (principal + interest) that Chi will pay the credit union at the end of the 9 months.

DID YOU KNOW

Pawn Loans

Since ancient times, pawnbroking has been a source of credit for people who cannot obtain loans from other financial institutions. A typical pawn transaction involves the customer presenting an item of value to a pawnbroker. The pawnbroker makes a loan to the customer based on a percentage of the value of the item. The pawnbroker keeps the item as collateral until the customer repays the loan plus the interest. If the customer defaults on the loan, the pawnbroker keeps the item and usually offers it for sale in his or her pawnshop. Due to the 15 to 20% default rate, as well as the need to pay for retail space, pawn loans usually have a significantly higher rate of interest than other financial institutions (see Example 3).

SOLUTION:

a) To find the interest on the loan, we use the formula $i = prt$. We know that $p = \$1600$, $r = 6.5\%$ (or converted to a decimal, 0.065), and t in years $= \frac{9}{12} = 0.75$. We substitute the appropriate values in the formula

$$\begin{aligned} i &= p \times r \times t \\ &= \$1600 \times 0.065 \times 0.75 \\ &= \$78 \end{aligned}$$

The simple interest on \$1600 at 6.5% for 9 months is \$78.

b) The amount to be repaid is equal to the principal plus interest, or

$$\begin{aligned} A &= p + i \\ &= \$1600 + \$78 \\ &= \$1678 \end{aligned}$$

To pay off her loan, Chi will pay the credit union \$1678 at the end of 9 months.

EXAMPLE 2 *Determining the Annual Rate of Interest*

Patricia Allaire lent her friend Dan Marcum \$300 to help him pay his income taxes. Six months later Dan repaid the original \$300 plus \$15.00 interest. What annual rate of interest did Patricia receive?

SOLUTION: We need to solve for the interest rate, r. Since the time is 6 months, the time in years is $\frac{6}{12}$ or 0.5. Using the formula $i = prt$, we get

$$\begin{aligned} \$15 &= \$300 \times r \times 0.50 \\ 15 &= 150r \\ \frac{15}{150} &= r \\ 0.1 &= r \end{aligned}$$

The annual rate of interest paid is 10%.

EXAMPLE 3 *A Pawn Loan*

To obtain money for new eyeglasses, Gilbert French decides to pawn his trumpet. Gilbert borrows \$240 and after 30 days he gets his trumpet back by paying the pawnbroker \$288. What annual rate of interest did Gilbert pay?

SOLUTION: Gilbert paid $\$288 - \$240 = \$48$ in interest, and the length of the loan is for one month or $\frac{1}{12}$ of a year.

Using the formula $i = prt$, we get

$$\begin{aligned} \$48 &= \$240 \times r \times \frac{1}{12} \\ 48 &= 20r \\ \frac{48}{20} &= r \\ 2.4 &= r \end{aligned}$$

DID YOU KNOW

Origin of Banks

In ancient Babylonia, as early as 2000 B.C., temples were considered safe depositories for assets. It was believed that these sacred places enjoyed the special protection of the gods and were not likely to be robbed. It is no coincidence that many of the bank buildings built during the early twentieth century resembled ancient temples. In medieval times, monies were kept in vaults protected by armies of the ruling nobility. The word *bank* is derived from the Italian word *banca,* meaning "board." It refers to the counting boards used by merchants. Dishonest moneychangers in the marketplace had their boards smashed to prevent them from continuing in business. The word *bankrupt* is literally "a broken (ruptured) board."

The annual rate of interest as a decimal number is 2.4. To change this number to a percent, multiply the number by 100 and add a percent sign. Thus, the annual rate of interest paid is 240%. ▲

In Examples 1, 2, and 3, we illustrated a simple interest loan, for which the interest and principal are paid on the due date of the note. There is another type of loan, the *discount note*, for which the interest is paid at the time the borrower receives the loan. The interest charged in advance is called the *bank discount*. A Federal Reserve Treasury bill is a bank discount note issued by the U.S. government. Example 4 illustrates a discount note.

EXAMPLE 4 *True Interest Rate of a Discount Note*

Siegrid Cook borrowed $500 on a 10% discount note for a period of 3 months. Find

a) the interest she must pay to the bank on the date she receives the loan.
b) the net amount of money she receives from the bank.
c) the actual rate of interest for the loan.

SOLUTION:

a) To find the interest, use the simple interest formula.

$$\begin{aligned} i &= prt \\ &= \$500 \times 0.10 \times \frac{3}{12} \\ &= \$12.50 \end{aligned}$$

b) Since Siegrid must pay $12.50 interest when she first receives the loan, the net amount she receives is $500 − $12.50 or $487.50.

c) We calculate the actual rate of interest charged using the simple interest formula. In the formula for the principal, we use the amount Siegrid received from the bank. For the interest we use the interest calculated in part a).

$$\begin{aligned} i &= prt \\ \$12.50 &= \$487.50 \times r \times \frac{3}{12} \\ 12.50 &= 121.875 \times r \\ \frac{12.50}{121.875} &= r \\ 0.1026 &\approx r \end{aligned}$$

Thus, the actual rate of interest is about 10.3% rather than the quoted 10%. ▲

EXAMPLE 5 *Partial Payments*

Ken Hurley wishes to purchase a new racing bicycle but does not have the $2000 purchase price. Luckily, the bike shop has two payment options. With option 1, Ken can pay $1000 as a down payment and then pay $1150 in 6 months. With option 2, Ken can pay $500 as a down payment and then pay $1700 in 6 months. Which payment option has a higher annual simple interest rate?

SOLUTION: Option 1: To determine the principal of the loan, subtract the down payment from the purchase price of the bicycle. Therefore, $p = \$2000 - \$1000 = \$1000$.

DID YOU KNOW

Tallying Accounts

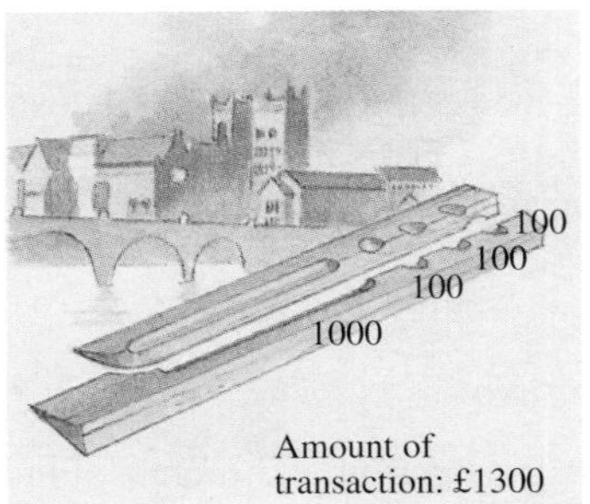

Until 1826, British banks used tally sticks to record the amounts of loans and deposits. The tallies were flat pieces of wood about 8 in. long. The amount was recorded as notches, different values represented by different widths. A hand's width signified £1000, a thumb's width £100. The stick was split in two; the borrower got "the short end of the stick" and the money lender held the larger stock ("the stockholder"). Transactions could be verified by putting the sticks together to show the accounts "tallied."

In 1834, as the tallies were being burned in the furnaces of the House of Lords, the fire got out of control and burned the old House of Parliament.

To determine the interest charged, subtract the purchase price of the bicycle from the total amount paid. Therefore, $i = (\$1000 + \$1150) - \$2000 = \$150$. Then

$$\begin{aligned} i &= p \times r \times t \\ 150 &= 1000 \times r \times \frac{1}{2} \\ 150 &= 500r \\ \frac{150}{500} &= r \\ 0.3 &= r \end{aligned}$$

Option 1 has a 30% annual simple interest rate.

Option 2: The principal is $p = \$2000 - \$500 = \$1500$, and interest is $i = (\$500 + \$1700) - \$2000 = \$200$. Then

$$\begin{aligned} i &= p \times r \times t \\ 200 &= 1500 \times r \times \frac{1}{2} \\ 200 &= 750r \\ \frac{200}{750} &= r \\ 0.2667 &\approx r \end{aligned}$$

Option 2 has about 26.7% annual simple interest rate. Therefore, option 1 charges a higher annual simple interest rate than option 2. ▲

The United States Rule

A loan has a date of maturity, at which time the principal and interest are due. It is possible to make payments on a loan before the date of maturity. A Supreme Court decision specified the method by which these payments are credited. The procedure is called the *United States rule*.

The United States rule states that if a partial payment is made on the loan, interest is computed on the principal from the first day of the loan until the date of the partial payment. The partial payment is used to pay the interest first; then the rest of the payment is used to reduce the principal. The next time a partial payment is made, interest is calculated on the unpaid principal from the date of the previous date of payment. Again, the payment goes first to pay the interest, with the rest of the payment used to reduce the principal. An individual can make as many partial payments as he or she wishes; the procedure is repeated for each payment. The balance due on the date of maturity is found by computing interest due since the last partial payment and adding this interest to the unpaid principal.

The *Banker's rule* is used to calculate simple interest when applying the United States rule. The Banker's rule considers a year to have 360 days, and any fractional part of a year is the exact number of days of the loan.

To determine the exact number of days in a period, we can use Table 11.1. Example 6 illustrates how to use the table.

DID YOU KNOW

Selling Time

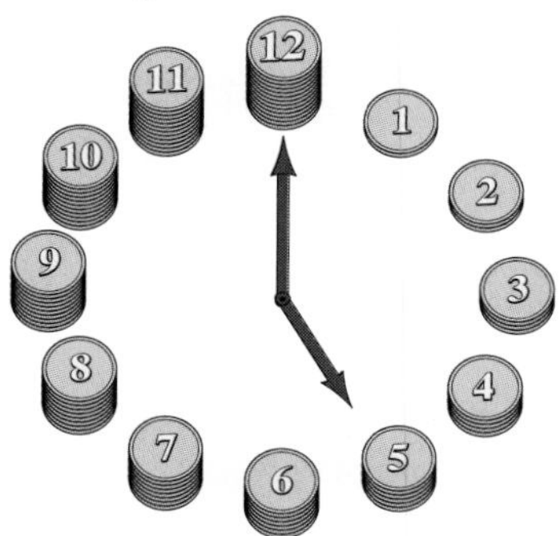

For most of human history the practice of charging interest on money borrowed, or *usury*, was considered not only immoral, but a crime. This attitude arose in ancient times, in part because a peasant or farmer could literally be enslaved if payments could not be made. Charging interest over time was considered "selling time," something not ours to sell. What was especially objectionable was the idea of "pure interest," a fixed payment set at the time of the loan, even when there was no risk to the lender. Much has changed in the past 200 years. Borrowing money has become a fact of life for individuals, businesses, and governments alike. In 2002, the U.S. government paid about $332.5 billion in interest alone on a debt of more than $6.3 trillion.

Today the word *usury* means lending money at a rate of interest that is excessive or unlawfully high. A person who engages in such loans is often referred to as a "loan shark."

TABLE 11.1

	Days in Each Month											
	31	28	31	30	31	30	31	31	30	31	30	31
Day of Month	Jan.	Feb.	Mar.	Apr.	May	June	July	Aug.	Sept.	Oct.	Nov.	Dec.
Day 1	1	32	60	91	121	152	182	213	244	274	305	335
Day 2	2	33	61	92	122	153	183	214	245	275	306	336
Day 3	3	34	62	93	123	154	184	215	246	276	307	337
Day 4	4	35	63	94	124	155	185	216	247	277	308	338
Day 5	5	36	64	95	125	156	186	217	248	278	309	339
Day 6	6	37	65	96	126	157	187	218	249	279	310	340
Day 7	7	38	66	97	127	158	188	219	250	280	311	341
Day 8	8	39	67	98	128	159	189	220	251	281	312	342
Day 9	9	40	68	99	129	160	190	221	252	282	313	343
Day 10	10	41	69	100	130	161	191	222	253	283	314	344
Day 11	11	42	70	101	131	162	192	223	254	284	315	345
Day 12	12	43	71	102	132	163	193	224	255	285	316	346
Day 13	13	44	72	103	133	164	194	225	256	286	317	347
Day 14	14	45	73	104	134	165	195	226	257	287	318	348
Day 15	15	46	74	105	135	166	196	227	258	288	319	349
Day 16	16	47	75	106	136	167	197	228	259	289	320	350
Day 17	17	48	76	107	137	168	198	229	260	290	321	351
Day 18	18	49	77	108	138	169	199	230	261	291	322	352
Day 19	19	50	78	109	139	170	200	231	262	292	323	353
Day 20	20	51	79	110	140	171	201	232	263	293	324	354
Day 21	21	52	80	111	141	172	202	233	264	294	325	355
Day 22	22	53	81	112	142	173	203	234	265	295	326	356
Day 23	23	54	82	113	143	174	204	235	266	296	327	357
Day 24	24	55	83	114	144	175	205	236	267	297	328	358
Day 25	25	56	84	115	145	176	206	237	268	298	329	359
Day 26	26	57	85	116	146	177	207	238	269	299	330	360
Day 27	27	58	86	117	147	178	208	239	270	300	331	361
Day 28	28	59	87	118	148	179	209	240	271	301	332	362
Day 29	29		88	119	149	180	210	241	272	302	333	363
Day 30	30		89	120	150	181	211	242	273	303	334	364
Day 31	31		90		151		212	243		304		365

Add 1 day for leap year if February 29 falls between the two dates under consideration.

EXAMPLE 6 *Determining the Due Date of a Note*

Use Table 11.1 to find (a) the due date of a loan made on March 15 for 120 days and (b) the number of days from April 18 to July 31.

SOLUTION:

a) To determine the due date of the loan, do the following. In Table 11.1 find Day 15 in the left column (with heading Day of Month), and then move three columns to the right (heading at the top of the column is March) to find the number 74 (circled in red). Thus, March 15 is the 74th day of the year. Add 120 to 74, since the loan will be due 120 days after March 15:

$$74 + 120 = 194$$

Thus, the due date of the note is the 194th day of the year. Find 194 (circled in blue) in Table 11.1 in the column headed July. The number in the same row as 194 in the left column is Day 13. Thus, the due date of the note is July 13.

b) To determine the number of days from April 18 to July 31, use Table 11.1 to find that April 18 is the 108th day of the year and that July 31 is the 212th day of the year. Then find the difference: 212 − 108 = 104. Thus, the number of days from April 18 to July 31 is 104 days. ▲

EXAMPLE 7 *Using the Banker's Rule*

Determine the simple interest that will be paid on a \$300 loan at an interest rate of 5% for the period March 3 to May 3 using the Banker's rule.

SOLUTION: The exact number of days from March 3 to May 3 is 61. The period of time in years is 61/360. Substituting in the simple interest formula gives

$$\begin{aligned} i &= prt \\ &= \$300 \times 0.05 \times \frac{61}{360} \\ &\approx \$2.54 \end{aligned}$$

The interest is \$2.54. ▲

DID YOU KNOW

Investing in Stocks

When the owners of a company wish to raise money for expanding their company, they often decide to sell part of the company to investors. When an investor purchases a portion of a company, the investor is said to own *stock* in the company. The unit of measure of the stock is called a *share*. By selling shares of stock, a company is selling ownership of the company. By buying shares of stock, an investor is becoming a part owner, or *shareholder*, of the company. The shares of many companies' stock are often bought and sold, or *traded*, by investors through a *stock exchange*. There are many stock exchanges located throughout the world, but the largest is the New York Stock Exchange. Shares of stock may also be bought and sold through the National Association of Securities Dealers Automated Quotation system (commonly referred to as the NASDAQ).

Large companies may have many millions or even billions of shares of stock available for trading to the general public. For example, in 2002, there were over 2 billion shares of the Coca-Cola Company owned by over 300,000 investors. On April 11, 2003, each of these shares was worth \$41.43.

Companies will often distribute their profits to their shareholders in the form of *dividends*. After the company declares the dividend per share, the amount the investor receives depends on the number of shares owned by the investor. Shareholders may also make money when they sell their shares if they sell them at a higher price than the price at which they bought the shares. Such an increase is referred to as a *capital gain*. Because companies do not always distribute dividends and the price of stocks may go down as well as go up, investing in stocks involves some risk of losing all or part of your investment. Investing in stocks over long periods of time, however, is usually a good investment. Since 1926, shareholders of large established companies have seen an average annual increase in their investment of 11.3%. This return exceeds the average annual increase in bonds, savings accounts, or certificates of deposit.

The next example illustrates how a partial payment is credited under the United States rule. Making partial payments reduces the amount of interest paid and the cost of the loan.

EXAMPLE 8 *Using the United States Rule*

Cathy Panik is a mathematics teacher and she plans to attend a national conference. To pay for her airfare, on November 1, 2002, Cathy takes out a 120-day loan for

$400 at an interest rate of 12.5%. Cathy uses some birthday gift money to make a partial payment of $150 on January 5, 2003. She makes a second partial payment of $100 on February 2, 2003.

a) Determine the due date of the loan.

b) Determine the interest and the amount credited to the principal on January 5.

c) Determine the interest and the amount credited to the principal on February 2.

d) Determine the amount that Cathy must pay on the due date.

SOLUTION:

a) Using Table 11.1, we see that November 1 is the 305th day of the year. Next, we note that the sum of 305 and 120 is 425. Since this due date will extend into the next year, we subtract 365 from 425 to get 60. From Table 11.1, we see that the 60th day of the year is March 1. Therefore, the loan due date is March 1, 2003. Had 2003 been a leap year, the due date would have been February 29, 2003.

b) Using Table 11.1, January 5 is the 5th day of the year and November 1 is the 305th day of the year. The number of days from November 1 to January 5 can be computed as follows: $(365 - 305) + 5 = 65$. Then using $i = prt$, and the Banker's rule, we get:

$$\begin{aligned} i &= \$400 \times 0.125 \times \frac{65}{360} \\ &\approx \$9.03 \end{aligned}$$

The interest of $9.03 that is due January 5, 2003, is deducted from the payment of $150. The remaining payment of $150 − $9.03 or $140.97 is then credited to the principal. Therefore, the adjusted principal is now $400 − $140.97, or $259.03.

Note that, if this had been the only partial payment made, then the balance due on March 1,2003, would be calculated by determining the interest on the balance, $259.03, for the remainder of the loan, 55 days, and adding this interest to the principal of $259.03. If this payment was the only partial payment made, then the balance due on March 1 would be

$$\begin{aligned} \text{Balance due} &= \text{principal} + \text{interest} \\ &\approx 259.03 + \left(259.03 \times 0.125 \times \frac{55}{360}\right) \\ &\approx 259.03 + 4.95 \\ &\approx 263.98 \end{aligned}$$

c) Since there was a second partial payment made, we use the Banker's rule to calculate the interest on the unpaid principal for the period from January 5 to February 2. According to Table 11.1, the number of days from January 5 to February 2 is 33 − 5, or 28 days.

$$\begin{aligned} i &= \$259.03 \times 0.125 \times \frac{28}{360} \\ &\approx \$2.52 \end{aligned}$$

The interest of $2.52 that is due February 2, 2003, is deducted from the payment of $100. The remaining payment of $100 − $2.52, or $97.48, is then credited to the principal. Therefore, the new adjusted principal is now $259.03 − $97.48, or $161.55.

d) The due date of the loan is March 1. Using Table 11.1, we see that there are

60 − 33 or 27 days from February 2 to March 1. The interest is computed on the remaining balance of \$161.51 by using the simple interest formula.

$$i = \$161.55 \times 0.125 \times \frac{27}{360}$$
$$\approx \$1.51$$

Therefore, the balance due on the maturity date of the loan is the sum of the principal and the interest, \$161.55 + \$1.51, or \$163.06. *Note:* The sum of the days in the three calculations, 65 + 28 + 27, equals the total number of days in the loan, 120.

SECTION 11.2 EXERCISES

Concept/Writing Exercises

1. What is interest?
2. What is credit?
3. What is security (or collateral)?
4. What is a cosigner?
5. Explain what each letter in the simple interest formula, $i = prt$, represents.
6. What is a personal note?
7. Explain the United States rule.
8. Explain the difference between ordinary interest and interest under the Banker's rule.

In Exercises 9–18, determine the simple interest. (The rate is an annual rate unless otherwise noted. Assume 360 days in a year.)

Practice the Skills

9. $p = \$300, r = 4\%, t = 5$ years

10. $p = \$450, r = 5.5\%, t = 2$ years

11. $p = \$900, r = 3.75\%, t = 30$ days

12. $p = \$365.45, r = 11\frac{1}{2}\%, t = 8$ months

13. $p = \$587, r = 0.045\%$ per day, $t = 2$ months

14. $p = \$6742.75, r = 6.05\%, t = 90$ days

15. $p = \$2,756.78, r = 10.15\%, t = 103$ days

16. $p = \$550.31, r = 8.9\%, t = 67$ days

17. $p = \$1372.11, r = 1\frac{3}{8}\%$ per month, $t = 6$ months

18. $p = \$41,864, r = 0.0375\%$ per day, $t = 60$ days

In Exercises 19–24, use the simple interest formula to determine the missing value.

19. $p = \$1500, r = ?, t = 3$ years, $i = \$450$

20. $p = ?, r = 3\%, t = 90$ days, $i = \$600$

21. $p = ?, r = 8\%, t = 3$ months, $i = \$12.00$

22. $p = \$800.00, r = 6\%, t = ?, i = \$64.00$

23. $p = \$957.62, r = 6.5\%, t = ?, i = \$124.49$

24. $p = \$1650.00, r = ?, t = 6.5$ years, $i = \$343.20$

Problem Solving

25. *School Loan* To encourage employees to take courses at the local community college, DCR Industries offers loans of \$1000 at an annual simple interest rate of 3.0%. If Darren Sharper gets one such loan, determine how much Darren needs to repay his employer after 6 months.

26. *Credit Union Loan* Steve and Laurie Carah borrowed \$4500 from their credit union to remodel their kitchen. The simple interest rate was 4.75% and the length of the loan was 3 years.

a) How much did the Carahs pay for the use of the money?
b) Determine the amount the Carahs paid to the credit union on the date of maturity of the loan.

27. *Bank Personal Note* Kelly Droessler borrowed \$3500 from the bank for 6 months. Her friend Ms. Harris was cosigner of Kelly's personal note. The bank collected $7\frac{1}{2}\%$ simple interest on the date of maturity.
a) How much did Kelly pay for the use of the money?
b) Find the amount she repaid to the bank on the due date of the note.

28. *Bank Discount Note* Kwame Adebele borrowed \$2500 for 5 months from his bank, using U.S. government bonds as security. The bank discounted the loan at 8%.
a) How much interest did Kwame pay the bank for the use of its money?

b) How much did he receive from the bank?
c) What was the actual rate of interest he paid?

29. *Bank Discount Note* Julie Jansen borrowed \$3650 from her bank for 8 months. The bank discounted the loan at 7.5%.
a) How much interest did Julie pay the bank for the use of its money?
b) How much did she receive from the bank?
c) What was the actual rate of interest she paid?

30. *Credit Union Loan* Enrico Montoyo wants to borrow \$350 for 6 months from his credit union, using his savings account as security. The credit union's policy is that the maximum amount a person can borrow is 80% of the amount in the person's savings account. The interest rate is 2% higher than the interest rate being paid on the savings account. The current rate on the savings account is $3\frac{1}{4}\%$.
a) How much money must Enrico have in his account in order to borrow \$350?
b) What is the rate of interest the credit union will charge for the loan?
c) Find the amount Enrico must repay in 6 months.

31. *Investing Tuition Payments* Sand Ridge School is requiring parents to pay half of the yearly tuition at the time of registration and half on the date classes begin. Registration is held 5 months prior to the beginning of school, and administrators expect 470 students to register. If annual tuition is \$4500 and if the money paid at the time of registration is placed in an account paying 5.4% simple interest, how much interest will Sand Ridge School earn by the time school begins?

32. *A Pawn Loan* Jeffrey Kowalski wants to take his mother out for dinner on her birthday, but he doesn't get paid until the following week. To borrow money, Jeffrey pawns his watch. Based on the value of the watch, the pawnbroker loans Jeffrey \$75. Fourteen days later Jeffrey gets his watch back by paying the pawnbroker \$80.25. What annual simple interest rate did the pawnbroker charge Jeffrey?

In Exercises 33–38, determine the exact time from the first date to the second date. Use Table 11.1. Assume the year is not a leap year unless otherwise indicated.

33. January 17 to July 4

34. February 12 to June 19 (the loan is due in a leap year)

35. March 17 to December 8

36. June 14 to January 24

37. August 24 to May 15

38. December 21 to April 28

In Exercises 39–42, determine the due date of the loan, using the exact time, if the loan is made on the given date for the given number of days.

39. April 15 for 60 days

40. May 18 for 180 days

41. November 25 for 120 days (the loan is due in a leap year)

42. July 5 for 210 days

In Exercises 43–52, a partial payment is made on the date(s) indicated. Use the United States rule to determine the balance due on the note at the date of maturity. (The Effective Date is the date the note was written.) Assume the year is not a leap year.

	Principal	Rate	Effective date	Partial payment(s) Amount	Partial payment(s) Date(s)	Maturity date
43.	\$2000	5%	Mar. 1	\$400	Apr. 1	May 1
44.	\$4500	3%	Jan. 15	\$2000	Mar. 1	Apr. 15
45.	\$7000	5.75%	Aug. 1	\$3500	Nov. 15	Dec. 15
46.	\$7500	12%	April 15	\$1000	Aug. 1	Oct. 1
47.	\$9000	6%	July 15	\$4000	Dec. 27	Feb. 1
48.	\$1000	12.5%	Jan. 1	\$300	Jan. 15	Feb. 15
49.	\$1800	15%	Aug. 1	\$500	Sept. 1	Nov. 1
				\$500	Oct. 1	
50.	\$5000	14%	Oct. 15	\$800	Nov. 15	Jan. 1
				\$800	Dec. 15	
51.	\$11,600	6%	Mar. 1	\$2000	Aug. 1	Dec. 1
				\$4000	Nov. 15	
52.	\$21,000	$4\frac{3}{8}\%$	July 12	\$8000	Oct. 10	Jan. 30
				\$6000	Dec. 8	

53. *Company Loan* On March 1 the Zwick Balloon Company signed a \$6500 note with simple interest of $10\frac{1}{2}\%$ for 180 days. The company made payments of \$1750 on May 1 and \$2350 on July 1. How much will the company owe on the date of maturity?

54. *Restaurant Loan* The Sweet Tooth Restaurant borrowed \$3000 on a note dated May 15 with simple interest of 11%. The maturity date of the loan is September 1. The restaurant made partial payments of \$875 on June 15 and \$940 on August 1. Find the amount due on the maturity date of the loan.

55. *U.S. Treasury Bills* The U.S. government borrows money by selling Treasury bills. Treasury bills are discounted notes issued by the U.S. government. On May 5, 2004, Kris Greenhalgh purchased a 182-day, \$1000 U.S. Treasury bill at a 4.34% discount. On the date of maturity Kris will receive \$1000.

a) What is the date of maturity of the Treasury bill?
b) How much did Kris actually pay for the Treasury bill?
c) How much interest did the U.S. goverment pay Kris on the date of maturity?
d) What is the actual rate of interest of the Treasury bill? (Round the answer to the nearest hundredth of a percent.)

56. *U.S. Treasury Bills* On August 31, 2003, Trinity Lopez purchased a 364-day, \$6000 U.S. Treasury bill at a 4.4% discount. (See Exercise 55.)

a) What is the date of maturity of the Treasury bill (2004 is a leap year)?
b) How much did Trinity actually pay for the Treasury bill?
c) How much interest did the U.S. government pay Trinity on the date of maturity?
d) What is the actual rate of interest of the Treasury bill?

57. *Tax Preparation Loan* Many tax preparation organizations will prepay customers' tax refunds if they pay a one time finance charge. In essence, the customer is borrowing the money (the refund minus the finance charge) from the tax preparer, prepaying the interest (as in a discount note) and then repaying the loan with the tax refund. This procedure allows customers access to their tax refund money without having to wait. Joy Stallard had a tax refund of \$743.21 due. She was able to get her tax refund immediately by paying a finance charge of \$39.95. What annual simple interest rate is Joy paying for this loan assuming

a) the tax refund check would be available in 5 days?
b) the tax refund check would be available in 10 days?
c) the tax refund check would be available in 20 days?

58. *Prime Interest Rate* Nick St. Louis borrowed \$600 for 3 months. The banker said that Nick must repay the loan at the rate of \$200 per month plus interest. The bank was charging a rate of 2% above the prime interest rate. The *prime interest rate* is the rate charged to preferred customers of the bank. During the first month the prime interest rate was 4.75%, during the second month it was 5%, and during the third month it was 5.25%.

a) Find the amount Nick paid the bank at the end of the first month, at the end of the second month, and at the end of the third month.
b) What was the total amount of interest Nick paid the bank?

Challenge Problem/Group Activity

59. *U.S. Treasury Bills* Mark Beiley purchased a 52-week U.S. Treasury bill (see Exercises 55 and 56) for \$93,337. The par value (the value of the bill upon maturity) was \$100,000.

a) What was the discount rate?
b) What was the actual interest rate Mark received?
c) Since U.S. Treasury bills are sold through auctions at Federal Reserve Banks, Mark did not know the purchase price of his Treasury bill until after he was notified by mail. If Mark had sent the Federal Reserve Bank a check for \$100,000 to purchase the Treasury bill, how much would the Federal Reserve Bank have to rebate him upon notice of his purchase?
d) On the day he received the rebate that was discussed in part (c) he invested it in a 1-year certificate of deposit yielding 5% interest. What is the total amount of interest he will receive from both investments?

Recreational Mathematics

60. *Columbus Investment* On August 3, 1492 Christopher Columbus set sail on a voyage that would eventually lead him to the Americas. If on this day Columbus had invested \$1 in a 5% simple interest account, determine the amount of interest the account would have earned by the following dates. Use a scientific calculator and disregard leap years in your calculations.

a) December 11, 1620 (Pilgrims land on Plymouth Rock)
b) July 4, 1776 (Declaration of Independence)
c) December 7, 1941 (U.S. enters World War II)
d) Today's date

Internet/Research Activities

61. *Banking Practices* Three types of simple interest may be calculated with the simple interest formula. They are ordinary, Banker's rule, and exact time.

a) Visit a local lending institution to determine how exact time is used in the simple interest formula.
b) Compare the results for each method on a loan for \$500 at 8% for the period January 2, 2004, to April 2, 2004.

62. *Loan Sources* Consider the following places where a loan may be obtained: banks, savings and loans, credit unions, and pawn shops. Write a report that includes the following information:

a) Describe the ownership of each.
b) Historically, what need is each fulfilling?
c) What are the advantages and disadvantages of obtaining a loan from each of those listed?

11.3 COMPOUND INTEREST

Albert Einstein was once asked to name the greatest discovery of man. His reply was, "Compound interest."

In this section we discuss some ways to put your money to work for your benefit.

An *investment* is the use of money or capital for income or profit. We can divide investments into two classes: fixed investments and variable investments. In a *fixed investment*, the amount invested as principal is guaranteed and the interest is computed at a fixed rate. *Guaranteed* means that the exact amount invested will be paid back together with any accumulated interest. Examples of a fixed investment are savings accounts and certificates of deposit. Another fixed investment is a government savings bond. In a *variable investment*, neither the principal nor the interest is guaranteed. Examples of variable investments are stocks, mutual funds, and commercial bonds.

Simple interest, introduced earlier in the chapter, is calculated once for the period of a loan using the formula $i = prt$. The interest paid on savings accounts at most banks is compound interest. A bank computes the interest periodically (for example, daily or quarterly) and adds this interest to the original principal. The interest for the following period is computed by using the new principal (original principal plus interest). In effect, the bank is computing interest on interest, which is called compound interest.

Interest that is computed on the principal and any accumulated interest is called **compound interest.**

DID YOU KNOW

A Penny Saved...

"Money makes money and the money that money makes makes more money." So said Benjamin Franklin, who in his will left 1000 pounds sterling (£) to the city of Boston, where he was born. The money was to be lent to young apprentices who would pay $\frac{1}{10}$ of the principal each year plus 5% interest compounded yearly. The money earned was to be used to benefit the inhabitants of Boston. By 1894, the fund's value had increased to 90,000 £. Part of this legacy became the Franklin Institute of Boston, a 2-year technical college founded in 1908.

EXAMPLE 1 *Computing Compound Interest*

Marjorie Thrall recently won the \$1000 first prize in a raffle contest. Marjorie deposits the \$1000 in a 1-year certificate of deposit paying 2.0% compounded quarterly. Find the amount, A, to which the \$1000 will grow in 1 year.

SOLUTION: Compute the interest for the first quarter using the simple interest formula. Add this interest to the principal to find the amount at the end of the first quarter. In our calculation, time is $\frac{1}{4}$ of a year, or $t = 0.25$.

$$\begin{aligned} i &= prt \\ &= \$1000 \times 0.02 \times 0.25 = \$5.00 \\ A &= \$1000 + \$5 = \$1005 \end{aligned}$$

Now repeat the process for the second quarter, this time using a principal of \$1005.

$$\begin{aligned} i &= \$1005 \times 0.02 \times 0.25 \approx \$5.03 \\ A &= \$1005 + \$5.03 = \$1010.03 \end{aligned}$$

For the third quarter, use a principal of \$1010.03.

$$\begin{aligned} i &= \$1010.03 \times 0.02 \times 0.25 \approx \$5.05 \\ A &= \$1010.03 + \$5.05 = \$1015.08 \end{aligned}$$

For the fourth quarter, use a principal of \$1015.08.

$$\begin{aligned} i &= \$1015.08 \times 0.02 \times 0.25 \approx \$5.08 \\ A &= \$1015.08 + \$5.08 = \$1020.16 \end{aligned}$$

Hence, the \$1000 grows to a final value of \$1020.16 over the 1-year period. ▲

DID YOU KNOW

Cost of Manhattan

In 1626 Peter Du Minuit, representing the Dutch West India Company, traded beads and blankets to the native American inhabitants of Manhattan Island for the island. The trade was valued at 60 Dutch guilders, about \$24. If at that time the \$24 had been invested at 6% interest compounded annually, the investment would be worth about \$88,273,086,435 in 2004. As a comparison, had the \$24 been invested for the 378 years in a 6% simple interest account, the investment would be worth only \$568.32. Had the \$24 been invested in a continuously compounded interest account at 6%, the investment would be worth about \$169,828,311,335. These results dramatically demonstrate the power of compounding interest over time.

This example shows the effect of earning interest on interest, or compounding interest. In 1 year, the amount of \$1000 has grown to \$1020.16, compared with \$1020 that would have been obtained with a simple interest rate of 2%. Thus, in 1 year alone the gain was \$0.16 more with compound interest than with simple interest.

A simpler and less time-consuming way to calculate compound interest is to use the compound interest formula and a calculator.

Compound Interest Formula

$$A = p\left(1 + \frac{r}{n}\right)^{nt}$$

In this formula, A is the amount, p is the principal, r is the annual rate of interest, t is the time in years, and n is the number of compounding periods per year.

EXAMPLE 2 *Using the Compound Interest Formula*

When Alexander was born, he received several gifts of cash from his relatives and his parents' friends. His father invested this money in a money market account that had a rate of 3% compounded monthly. If the amount invested was \$3200, determine the amount in the account after 5 years.

SOLUTION: We will use the formula for compound interest, $A = p\left(1 + \frac{r}{n}\right)^{nt}$.

Since the interest is compounded monthly, there are 12 periods per year. Thus, $n = 12$. Since the money is invested for 5 years, $t = 5$.

$$\begin{aligned} A &= 3200\left(1 + \frac{0.03}{12}\right)^{(12)(5)} \\ &= 3200(1 + 0.0025)^{60} \\ &= 3200(1.0025)^{60} \\ &\approx 3200(1.1616168) \\ &\approx 3717.17 \end{aligned}$$

Thus, the amount in the account after 5 years would be \$3717.17. ▲

In Example 2, to find the value of $(1.0025)^{60}$ using a scientific calculator, you can use the $\boxed{y^x}$ key. The keystrokes are

$$1.0025 \boxed{y^x} 60 \boxed{=} 1.161617$$

Note: Some scientific calculators have an $\boxed{x^y}$ or a $\boxed{\wedge}$ key in place of a $\boxed{y^x}$ key. In either case, the procedure works the same: Enter the base then the key ($\boxed{y^x}$, $\boxed{x^y}$ or $\boxed{\wedge}$), the exponent, and then the $\boxed{=}$ key.

EXAMPLE 3 *Calculating Compound Interest*

Calculate the interest on \$650 at 8% compounded semiannually for 3 years, using the compound interest formula.

SOLUTION: Since interest is compounded semiannually, there are two periods per year. Thus, $n = 2$, $r = 0.08$, and $t = 3$. Substituting into the formula, we find the amount, A.

$$\begin{aligned} A &= p\left(1 + \frac{r}{n}\right)^{nt} \\ &= 650\left(1 + \frac{0.08}{2}\right)^{(2)(3)} \\ &= 650(1.04)^6 \\ &\approx 650(1.2653190) \\ &\approx \$822.46 \end{aligned}$$

Since the total amount is \$822.46 and the original principal is \$650, the interest must be \$822.46 − \$650, or \$172.46. ▲

In Example 3, the interest rate is stated as an annual rate of 8%, but the number of compounding periods per year is two. In applying the compound interest formula, the rate for one period is $r \div n$, which was 8% ÷ 2, or 4% per period.

Now we will calculate the amount, and interest, on \$1 invested at 8% compounded semiannually for 1 year. The result is

$$A = 1\left(1 + \frac{0.08}{2}\right)^{(2)(1)} = 1.0816$$

$$\begin{aligned} \text{Interest} &= \text{amount} - \text{principal} \\ i &= 1.0816 - 1 \\ &= 0.0816 \end{aligned}$$

The interest for 1 year is 0.0816. This amount written as a percent, 8.16%, is called the *effective annual yield. Most financial institutions refer to the effective annual yield as the annual percentage yield (APY).* If \$1 was invested at a simple interest rate of 8.16% and \$1 was invested at 8% interest compounded semiannually (equivalent to an effective yield of 8.16%), the interest from both investments would be the same.

The **effective annual yield** or **annual percentage yield (APY)** is the simple interest rate that gives the same amount of interest as a compound rate over the same period of time.

Suncoast Schools Credit Union

CD Rates

Type	Rate	APY*
12 mo	3.25%	3.30%
24 mo	3.83%	3.90%
36 mo	4.21%	4.30%
48 mo	4.59%	4.70%

* Annual Percentage Yield

Many banks compound interest daily. When computing the effective annual yield, they use 360 for the number of periods in a year. To find the effective annual yield for any interest rate, calculate the amount using the compound interest formula where p is \$1. Then subtract \$1 from that amount. The difference, written as a percent, is the effective annual yield, as illustrated in Example 4.

The sign in the margin shows interest rates and the corresponding annual percentage yields (APY) that were available for certificates of deposit (CDs) from Suncoast Schools Credit Union on November 2, 2002. To determine the effective annual yield (or the APY) for the 48-month CD, calculate the amount of interest earned on $1 for 1 year.

$$A = \left(1 + \frac{0.0459}{360}\right)^{360} \approx 1.0469666 \approx 1.0470$$

From this result we subtract 1 to get $1.0470 - 1 = 0.0470$, or 4.70%. This result confirms that the effective annual yield for a rate of 4.59% is 4.70%. Confirm that the other effective yields shown on the sign are correct now.

DID YOU KNOW

Federal Reserve Banks

Federal Reserve Banks are institutions owned by commecial banks in the district they serve. The many functions performed by these banks include the following: clearing checks, issuing new currency, retiring damaged currency, making discount loans to banks within their districts, examining member state banks, collecting data on local business conditions, and conducting research to help establish monetary policy, including the setting of the discount interest rate used when banks borrow money from each other. Currently, there are Federal Reserve Banks in Boston, New York, Philadelphia, Richmond, Cleveland, Atlanta, Chicago, St. Louis, Minneapolis, Kansas City, Dallas, and San Francisco. Take a look at any dollar bill and you will see a circular symbol with the location of the Federal Reserve Bank that printed the bill.

EXAMPLE 4 *Determining Annual Percentage Yield*

Determine the annual percentage yield or the effective annual yield for $1 invested for 1 year at

a) 8% compounded daily.

b) 6% compounded quarterly.

SOLUTION:

a) With daily compounding $n = 360$.

$$\begin{aligned} A &= p\left(1 + \frac{r}{n}\right)^{nt} \\ &= 1\left(1 + \frac{0.08}{360}\right)^{(360)(1)} \\ &\approx 1.0832774 \approx 1.0833 \\ i &= A - 1 \\ &\approx 1.0833 - 1 \\ &\approx 0.0833 \end{aligned}$$

Thus, when the interest is 8% compounded daily, the annual percentage yield or the effective annual yield is about 8.33%.

b) With quarterly compounding, $n = 4$.

$$\begin{aligned} A &= p\left(1 + \frac{r}{n}\right)^{nt} \\ &= 1\left(1 + \frac{0.06}{4}\right)^{(4)(1)} \\ &\approx 1.06136355 \approx 1.0614 \\ i &= A - 1 \\ &\approx 1.0614 - 1 \\ &\approx 0.0614 \end{aligned}$$

Thus, when the interest is 6% compounded quarterly, the annual percentage yield or the effective annual yield is about 6.14%. ▲

There are numerous types of savings accounts. Many savings institutions compound interest daily. Some pay interest from the day of deposit to the day of withdrawal, and others pay interest from the first of the month on all deposits made before the tenth of the month. In each of these accounts in which interest is compounded daily, interest is

DID YOU KNOW

Investing in Bonds

A *bond* is a type of loan. When government agencies or corporations need money they often borrow money from investors by selling, or *issuing*, bonds. When an investor purchases a bond, the investor is actually lending money to the bond's issuer. The issuer agrees to pay the investor a certain interest rate over a stated period of time, usually from 1 to 30 years. The date on which the issuer repays the loan is called the *date of maturity*. Although bonds are generally considered safer investments than stocks, they do have some risks. On rare occasions, issuers may fail to make interest payments or may fail to return the investment entirely. A more common risk is that the value of a bond may decrease if interest rates increase. Such a decrease may cause investors to lose some of their investment if they decide to sell their bond before the date of maturity.

Most bonds fit into one of three distinct categories: Treasury, corporate, or municipal bonds. The U.S. government issues *Treasury bonds*. Because the U.S. government backs them, Treasury bonds are considered extremely safe investments. Corporations issue *corporate bonds*. The amount of risk to the investor of corporate bonds varies widely, depending on the issuer. State and local governments or other public agencies issue *municipal bonds*. Like corporate bonds, municipal bonds vary greatly in risk to the investor. Investors in municipal bonds may gain an additional benefit because the interest earned may be free of state and local taxes. In general, bonds offer a very stable investment that usually provide a higher return on investment than savings accounts or certificates of deposit without many of the risks associated with investing in stocks.

entered into the depositor's account only once each quarter. Some savings institutions will not pay any interest on a day-to-day account if the balance falls below a set amount.

Present Value

You may wonder about what amount of money you must deposit in an account today to have a certain amount of money in the future. For example, how much must you deposit in an account today at a given rate of interest so it will accumulate to $25,000 to pay your child's college costs in 4 years? The principal, p, which would have to be invested now is called the *present value*. Following is a formula for determining the present value.

Present Value Formula

$$p = \frac{A}{\left(1 + \frac{r}{n}\right)^{nt}}$$

In the formula, A represents the amount to be accumulated in n years. Note that the present value formula is a variation of the compound interest formula.

EXAMPLE 5 *Savings for College*

Nicholas Christos is currently in the eighth grade and intends to attend a state college when he finishes high school. Nicholas's parents currently have some money invested in mutual funds, but would like to invest this money in a more secure investment now to pay for college in 4 years. If they will need \$25,000, how much do Nicholas's parents have to invest now in a 48-month CD that has a rate of 4.59% compounded monthly?

SOLUTION: To answer this question we will use the present value formula.

$$p = \frac{A}{\left(1 + \frac{r}{n}\right)^{nt}}$$

$$= \frac{25{,}000}{\left(1 + \frac{0.0459}{12}\right)^{(12)(4)}}$$

$$= \frac{25{,}000}{(1.003825)^{48}}$$

$$\approx \frac{25{,}000}{1.2011}$$

$$\approx 20{,}814.25$$

Nicholas Christos's parents need to invest approximately \$20,814.25 now to have \$25,000 in 4 years.

In Example 5, when we evaluated $(1.003825)^{48}$ we rounded the answer to four decimal places to obtain 1.2011. Using the value 1.2011 we obtained an answer of \$20,814.25. Had we not rounded the answer to $(1.003825)^{48}$ and instead used the value given by a calculator for $(1.003825)^{48}$, we would have obtained an answer of \$20,814.00, rounded to the nearest cent.

SECTION 11.3 EXERCISES

Concept/Writing Exercises

1. What is an investment?
2. What is a fixed investment?
3. What is a variable investment?
4. What is compound interest?
5. **a)** What is effective annual yield?
   b) What is another name for effective annual yield?
6. What is meant by present value in the present value formula?

Practice the Skills

In Exercises 7–16, use the compound interest formula

$$A = p\left(1 + \frac{r}{n}\right)^{nt}$$

to compute

a) the total amount.
b) the interest earned on each investment.

7. \$2000 for 3 years at 2.00% compounded annually
8. \$2000 for 3 years at 2.00% compounded semiannually
9. \$3500 for 4 years at 3.00% compounded semiannually
10. \$3500 for 4 years at 3.00% compounded annually
11. \$1500 for 3 years at 4.75% compounded quarterly
12. \$1500 for 4 years at 4.75% compounded quarterly
13. \$2500 for 2 years at 6.25% compounded monthly
14. \$3000 for 2 years at 6.25% compounded monthly
15. \$4000 for 4 years at 4.59% compounded daily (use $n = 360$)
16. \$4000 for 8 years at 4.59% compounded daily (use $n = 360$)

Problem Solving

17. *Textbook Advance* Lynae Sakshaug, a textbook author, deposited her \$7500 advance in a money market account

that pays 2.66% interest compounded semiannually. How much money will Lynae have in this account after 4 years?

18. *Automobile Sale* Bill Palow sells his vintage 1974 Plymouth Road Runner for $9500. He uses this money to invest in a 36-month CD that pays 4.12% interest compounded quarterly. How much money will Bill receive when he cashes in the CD at the end of the 36 months?

19. *Class Trip* To help pay for a class trip at the end of their senior year, the sophomore class at Cortez High School invests $1500 from fund-raisers in a 30-month CD paying 3.9% interest compounded monthly. Determine the amount the class will receive when it cashes in the CD after 30 months.

20. *Investing Prize Winnings* Marcella Laddon wins third prize in the Clearinghouse Sweepstakes and receives a check for $250,000. After spending $10,000 on a vacation, she decides to invest the rest in a money market account that pays 1.5% interest compounded monthly. How much money will be in the account after 10 years?

21. *Investing Gifts and Scholarships* Cliff Morris just graduated from high school and has received $800 in gifts of cash from friends and relatives. Additionally, Cliff received three scholarships in the amounts of $150, $300, and $1000. If Cliff takes all his gift and scholarship money and invests it in a 24-month CD paying 2% interest compounded daily, how much will he have when he cashes in the CD at the end of the 24 months?

22. *Investing a Signing Bonus* Joe Gallegos just started a new job and has received a $5000 signing bonus. Joe decides to invest this money now so that he can buy a new car in 5 years. If Joe invests in a CD paying 3.35% interest compounded quarterly, how much money will he receive from his CD in 5 years?

23. *Savings Account Investment* When Richard Zucker was born, his father deposited $2000 in his name in a savings account. The account was paying 5% interest compounded semiannually.

a) If the rate did not change, what was the value of the account after 15 years?

b) If the money had been invested at 5% compounded quarterly, what would the value of the account have been after 15 years?

24. *Savings and Loan Investment* When Lois Martin was born, her father deposited $2000 in a savings account in her name at a savings and loan association. At the time, the savings and loan was paying 6% interest compounded semiannually on savings accounts. After 10 years, the savings and loan association changed to an interest rate of 6% compounded quarterly. How much had the $2000 amounted to after 18 years when the money was withdrawn for Lois to use to help pay her college expenses?

25. *Personal Loan* Brent Pickett borrowed $3000 from his brother Dave. He agreed to repay the money at the end of 2 years, giving Dave the same amount of interest that he would have received if the money had been invested at 1.75% compounded quarterly. How much money did Brent repay his brother?

26. *Forgoing Interest* Rikki Blair borrowed $6000 from her daughter, Lynette. She repaid the $6000 at the end of 2 years. If Lynette had left the money in a bank account that paid an interest rate of $5\frac{1}{4}\%$ compounded monthly, how much interest would she have accumulated?

27. *Saving for a Down Payment* Jean Woody invested $6000 at 8% compounded quarterly. Three years later she withdrew the full amount and used it for the down payment on a house. How much money did she put down on the house?

28. *Investing a Salary* After Karen Estes began her job as a waitress, she invested in a money market account paying 5.6% interest compounded daily. What was the effective annual yield of this account?

29. *Doubling the Rate* Determine the total amount and the interest paid on $1000 with interest compounded semiannually for 2 years at

a) 2%. **b)** 4%. **c)** 8%.

d) Is there a predictable outcome in either the amount or the interest when the rate is doubled? Explain.

30. *Doubling the Principal* Compute the total amount and the interest paid at 12% compounded monthly for 2 years for the following principals.

a) $100 **b)** $200 **c)** $400

d) Is there a predictable outcome in the interest when the principal is doubled? Explain.

31. *Doubling the Time* Compute the total amount and the interest paid on $1000 at 6% compounded semiannually for

a) 2 years. **b)** 4 years. **c)** 8 years.

d) Is there a predictable outcome in the amount when the time is doubled? Explain.

32. *Doubling the Compounding Period* Compute the total amount and the interest paid after 1 year on $1000 at 4% interest compounded
    a) annually. **b)** semiannually. **c)** quarterly.
    d) Is there a predictable outcome in the amount when the compounding period is doubled? Explain.

33. *Determining Effective Annual Yield* Determine the effective annual yield for $1 invested for 1 year at 3.5% compounded semiannually.

34. *Determining Effective Annual Yield* Determine the effective annual yield for $1 invested for 1 year at 4.75% compounded monthly.

35. *Verifying APY* Suppose at your local bank you saw a sign that said, "2.4% rate compounded monthly—2.6% Annual Percentage Yield (APY)." Is there anything wrong with the sign? Explain.

36. *Verifying APY* Suppose you saw an advertisement in the newspaper for a financial planner who was recommending a certificate of deposit that paid 4.5% interest compounded quarterly. In the fine print at the bottom of the advertisement it stated that the APY on the CD was 4.58%. Was this advertisement accurate? Explain.

37. *Comparing Investments* Dave Dudley won a photography contest and received a $1000 cash prize. Will he earn more interest in 1 year if he invests his winnings in a simple interest account that pays 5% or in an account that pays 4.75% interest compounded monthly?

38. *Comparing Loan Sources* Tom Angelo needs to borrow $1500 to expand his farm implement maintenance business. He learns that the local bank will lend him the money for 2 years at a rate of 10% compounded quarterly. After hearing this, Tom's grandfather offers to lend him the money for 2 years with a simple interest rate of 7%. How much money will Tom save by borrowing the money from his grandfather?

39. *A New Water Tower* The village of Kieler completes an exploratory study and finds that the current village water tower will need replacement in 10 years at a cost of $290,000. To finance this amount, the village board will at this time assess its 958 homeowners with a one-time surcharge and then invest this amount in a 10-year CD paying 8.25% interest compounded semiannually.

    a) How much will the village of Kieler need to invest at this time in this CD to raise the $290,000 in 10 years?
    b) What amount should each homeowner pay as a surcharge?

40. *Water Tower Surcharge* (See Exercise 39.) After seeing its neighboring village raise the money to invest in a new water tower, the city board of East Dubuque decides to adopt a similar plan. However, since East Dubuque is a much larger community, they will need to raise $783,000 to build three new water towers in 15 years. At this time the city board plans to assess its 2682 homeowners with a one-time surcharge and then invest the money received in a money market account paying 9% interest compounded monthly.
    a) How much money will the city board need to raise at this time to meet the city's water tower needs at the end of 15 years?
    b) Before applying the surcharge, the city board decides to use a $50,000 benefactor gift toward the water tower investment. Taking this gift into account, how much should the surcharge be on each homeowner?

41. *Saving for a Tractor* Jim Roznowski wants to invest some money now to buy a new tractor in the future. If he wants to have $30,000 available in 5 years, how much does he need to invest now in a CD paying 5.15% interest compounded monthly?

42. *Investing for Retirement* The Pearsons are planning to retire in 20 years and believe that they will need $200,000 in addition to income from their retirement plans. How much must they invest today at 7.5% compounded quarterly to accomplish their goal?

43. *Investment for a Newborn* How much money should parents invest at the birth of their child to provide their child with $50,000 at age 18? Assume that the money earns interest at 8% compounded quarterly.

44. *Future Value* How much money must Harry Kim invest today to have \$20,000 in 15 years? Assume that the money earns interest at 7% compounded quarterly.

Challenge Problems/Group Activities

45. *A Loaf of Bread* If the cost of a loaf of bread was \$1.35 in 2004 and the annual average inflation rate is $2\frac{1}{2}\%$, what will be the cost of a loaf of bread in 2009?

46. *Finding the Interest Rate* For a total accumulated amount of \$3586.58, a principal of \$2000, and a time period of 5 years, use the compound interest formula to find r if interest is compounded monthly.

47. *Rule of 72* A simple formula can help you estimate the number of years required to double your money. It's called the *rule of 72*. You simply divide 72 by the interest rate (without the percent sign). For example, with an interest rate of 4%, your money would double in approximately $72 \div 4$ or 18 years. In (a)–(d), determine the approximate number of years it will take for \$1000 to double at the given interest rate.
    a) 3% **b)** 6% **c)** 8% **d)** 12%
    e) If \$120 doubles in approximately 22 years, estimate the rate of interest.

48. *Finding the Interest Rate* Richard Maruszewski borrowed \$2000 from Linda Tonolli. The terms of the loan are as follows: The period of the loan is 3 years, and the rate of interest is 8% compounded semiannually. What rate of simple interest would be equivalent to the rate Linda charged Richard?

49. *Investing with an Annuity* The question may be asked, If I deposit a fixed sum of money monthly, quarterly, semiannually, or annually at a fixed rate of interest, how much money will I have accumulated in x years? This scenario is called an *annuity*. The formula for determining the value of the annuity is

$$S = \frac{R\left[\left(1 + \frac{r}{n}\right)^{nt} - 1\right]}{\frac{r}{n}}$$

where S is the value of the annuity in t years, R is the amount invested each period, r is the annual rate of interest, n is the number of payments per year, and t is the number of years. After Denisse Brown's birth, her parents invested \$500 semiannually (every 6 months) at an annual rate of 5.5% compounded semiannually. What is the value of the annuity after 17 years?

50. *A Retirement Annuity* To supplement her retirement income, Chris Dunn is investing \$50 each quarter at 8% compounded quarterly. How many dollars will she accumulate in 30 years? (See Exercise 49.)

51. *Investing in Annuities* Rodney is saving money to send his sons to college by investing in annuities. (See Exercise 49.) For his oldest son, Jacob, Rodney is investing \$150 per month in an annuity that pays 5.6% interest compounded monthly. For his second oldest son, Justin, he is investing \$900 twice a year in an annuity that pays 5.8% interest compounded semiannually.
    a) If Rodney started Jacob's annuity when he was born, how much would the annuity be worth when Jacob turns 18 years old?
    b) If Rodney started Justin's annuity when he was born, how much would the annuity be worth when Justin turns 18 years old?

Recreational Mathematics

52. *Interest Comparison* You are given a choice of taking the simple interest on \$100,000 invested for 4 years at a rate of 5% or the interest on \$100,000 invested for 4 years at an interest rate of 5% compounded daily. Which would you select? Explain your answer and give the difference in the two investments.

Internet/Research Activities

53. Imagine you have \$4000 to invest and that you need this money to grow to \$5000 by investing in a CD. Contact a local bank, a savings and loan, and a credit union to obtain CD information: the interest rate, the length of the term, and the number of times per year the CD is compounded. Find out how long it would take you to reach your goal with the institution selected. Write a report summarizing your findings.

54. Write a paper on the history of simple interest and compound interest. Answer the questions: When was simple interest first charged on loans? When was compound interest first given on investments?

11.4 INSTALLMENT BUYING

In Section 11.2, we discussed personal notes and discounted notes. When borrowing money by either of these methods, the borrower normally repays the loan as a single payment at the end of the specified time period. There may be circumstances under which it is more convenient for the borrower to repay the loan on a weekly or monthly basis or to use some other convenient time period. One method of doing so is to borrow money on an *installment plan*.

DID YOU KNOW

Rent-to-Own Transactions

The rent-to-own industry has mushroomed in recent years to a $4.4 billion per year industry. Rent-to-own retailers allow you to rent furniture, appliances, electronics and the like over a period of time and, at a certain point, obtain ownership of the item. Is such a transaction an example of installment buying? It depends. If the consumer returns the item to the retailer prior to obtaining ownership, then the transaction is truly a rental agreement. If, however, the consumer obtains ownership, then the item was actually purchased on an installment plan. Since retailers classify the transaction as a rental agreement, they are not always subject to state mandated limits on the interest they can charge. The result is that items are often purchased at rates higher than even credit cards are allowed to charge. This controversy has come to the attention of lawmakers, and several states have enacted legislation controlling the amount of interest retailers may charge on rent-to-own transactions.

There are two types of installment loans: open-end and fixed payment. An *open-end installment loan* is a loan on which you can make variable payments each month. Credit cards, such as MasterCard, Visa, and Discover, are actually open-end installment loans, used to purchase items such as clothing, textbooks, and meals. A *fixed installment loan* is one on which you pay a fixed amount of money for a set number of payments. Examples of items purchased with fixed-payment installment loans are college tuition loans and loans for cars, boats, appliances, and furniture. These loans are generally repaid in 24, 36, 48, or 60 equal monthly payments.

Lenders give any individual wishing to borrow money or purchase goods or services on the installment plan a credit rating to determine if the borrower is likely to repay the loan. The lending institution determines whether the applicant is a good "credit risk" by examining the individual's income, assets, liabilities, and history of repaying debts.

The advantage of installment buying is that the buyer has the use of an article while paying for it. If the article is essential, installment buying may serve a real need. A disadvantage is that some people buy more on the installment plan than they can afford. Another disadvantage is the interest the borrower pays for the loan. The method of determining the interest charged on an installment plan may vary with different lenders.

To provide the borrower with a way to compare interest charged, Congress passed the Truth in Lending Act in 1969. The law requires that the lending institution tell the borrower two things: the annual percentage rate and the finance charge. The *annual percentage rate (APR)* is the true rate of interest charged for the loan. The APR is calculated by using a complex formula, so we use a table to determine the APR. The technique of using a table to find the APR is illustrated in Example 1. The total *finance charge* is the total amount of money the borrower must pay for its use. The finance charge includes the interest plus any additional fees charged. The additional fees may include service charges, credit investigation fees, mandatory insurance premiums, and so on.

The finance charge a consumer pays when purchasing goods or services on an installment plan is the difference between the total installment price and the cash price. The *total installment price* is the sum of all the monthly payments and the down payment, if any.

Fixed Installment Loan

In Example 1, we learn how to determine the finance charge and the monthly payment on a fixed installment loan.

EXAMPLE 1 *High-Definition Plasma Television*

Johnny Lott wishes to buy a high-definition plasma display television for $4500. The electronics store has an advertised finance option of no down payment and 7% APR for 24 months.

a) Determine the finance charge.

b) Determine the monthly payment.

SOLUTION:

a) Table 11.2 on page 623 gives the finance charge per $100 of the amount financed. The table shows that the finance charge per $100 of the amount financed for 24 months at 7% is $7.45 (circled in red). Since Johnny is financing $4500, the number of hundreds of dollars financed is $\frac{4500}{100} = 45$. To determine the total finace charge we multiply the finance charge per $100 by the number of hundreds of dollars financed.

$$\text{Total finance charge} = \$7.45 \times 45 = \$335.25$$

Therefore, Johnny will pay a total finance charge of $335.25.

b) To determine the monthly payments, we first calculate the total installment price by adding the finance charge to the purchase price:

$$\text{Total installment price} = \$4500 + \$335.25 = \$4835.25$$

Next, to determine the monthly payment we divide the total installment price by the number of payments:

$$\text{Monthly payment} = \frac{\$4835.25}{24} \approx \$201.47$$

Johnny will have 24 monthly payments of $201.47. ▲

TABLE 11.2 Annual Percentage Rate Table for Monthly Payment Plans

Number of Payments	Annual Percentage Rate												
	3.0%	**3.5%**	**4.0%**	**4.5%**	**5.0%**	**5.5%**	**6.0%**	**6.5%**	**7.0%**	**7.5%**	**8.0%**	**8.5%**	**9.0%**
	(Finance charge per $100 of amount financed)												
6	0.88	1.02	1.17	1.32	1.46	1.61	1.76	1.90	2.05	2.20	2.35	2.49	2.64
12	1.63	1.91	2.18	2.45	2.73	3.00	3.28	3.56	3.83	4.11	4.39	4.66	4.94
18	2.39	2.79	3.20	3.60	4.00	4.41	4.82	5.22	5.63	6.04	6.45	6.86	7.28
24	3.15	3.69	4.22	4.75	5.29	5.83	6.37	6.91	7.45	8.00	8.54	9.09	9.64
30	3.92	4.58	5.25	5.92	6.59	7.26	7.94	8.61	9.30	9.98	10.66	11.35	12.04
36	4.69	5.49	6.29	7.09	7.90	8.71	9.52	10.34	11.16	11.98	12.81	13.64	14.48
48	6.24	7.31	8.38	9.46	10.54	11.63	12.73	13.83	14.94	16.06	17.18	18.31	19.45
60	7.81	9.15	10.50	11.86	13.23	14.61	16.00	17.40	18.81	20.23	21.66	23.10	24.55

Our next example shows how to determine the annual percentage rate on a loan that has a schedule for repaying a fixed amount each period.

EXAMPLE 2 *Determining the APR*

Al and Albina Cannavach are purchasing a piano for $5000, including taxes. They decide to make a $1000 down payment and finance the balance, $4000, through their credit union. The loan officer informs them that their payment will be $121 per month for 36 months.

a) Determine the finance charge. b) Determine the APR.

SOLUTION:

a) The total installment price is the down payment plus the total monthly installment payments.

$$\begin{aligned}\text{Total installment price} &= \$1000 + (36 \times \$121)\\ &= \$1000 + \$4356 = \$5356\end{aligned}$$

The finance charge is the total installment price minus the cash price.

$$\text{Finance charge} = \$5356 - \$5000 = \$356$$

b) To determine the annual percentage rate, use Table 11.2. First divide the finance charge by the amount financed and multiply the quotient by 100. The result is the finance charge per $100 of the amount financed.

$$\frac{\text{Finance charge}}{\text{Amount financed}} \times 100 = \frac{356}{4000} \times 100 = 0.089 \times 100 = 8.9$$

Thus, the Cannavaciolos pay $8.90 for each $100 being financed. To use Table 11.2 look for 36 in the left column under the heading Number of Payments. Then move across to the right until you find the value closest to $8.90. The value closest to $8.90 is $8.71 (circled in blue). At the top of this column is the value 5.5%. Therefore, the annual percentage rate is approximately 5.5%. ▲

Much more complete APR tables similar to Table 11.2 are available at your local lending institution or on the Internet.

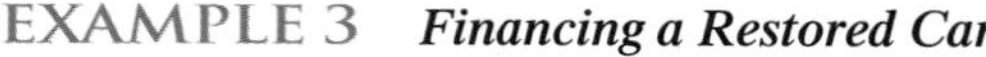

EXAMPLE 3 *Financing a Restored Car*

Tino Garcia borrowed $9800 to purchase a classic 1965 Ford Mustang. He does not recall the APR of the loan but remembers that there are 48 payments of $237. If he did not make a down payment on the car, determine the APR.

SOLUTION: First determine the finance charge by subtracting the cash price from the total amount paid.

$$\begin{aligned} \text{Finance charge} &= (237 \times 48) - 9800 \\ &= 11{,}376 - 9800 \\ &= \$1576 \end{aligned}$$

Next divide the finance charge by the amount of the loan and multiply this quotient by 100.

$$\frac{1576}{9800} \times 100 \approx 16.08$$

Next find 48 payments in the left column of Table 11.2. Move to the right until you find the value that is closest to 16.08. The value closest to 16.08 is 16.06 (circled in green). At the top of the column is the APR of 7.5%. ▲

In Example 3, if Tino made all 48 payments his finance charge would be $1576. If he decides to repay the loan after making 30 payments, must he pay the total finance charge? The answer is no. Two methods are used to determine the finance charge when you repay an installment loan early: the *actuarial method* and the *rule of 78s*. The actuarial method uses the APR tables, whereas the less frequently used method, the rule of 78s, does not. On top of page 625 we give the formulas used to calculate the unearned interest using both the actuarial method and the rule of 78s. The *unearned interest* is the interest saved by paying off the loan early.

DID YOU KNOW

Actuarial Method versus Rule of 78s

When making automobile or other installment loans, consumers should be aware of the "fine print" described in the loan agreement. If a consumer repays a loan early, the lender is required by law to rebate a portion of the original finance charge or interest. Two methods exist for determining this rebate. The actuarial method is the primary method used by financial institutions today. This method provides the consumer with an accurate interest rebate for repaying the loan early. The second method, the rule of 78s, however, is still used by some lenders. The rule of 78s charges the consumer more interest during the first few months of the loan. The result is a lower interest rebate for the consumer. The rule of 78s originated as an easy way for lenders to compute interest rebates. It has been phased out gradually as calculators and computers become more available. Today the rule of 78s is most likely to be applied by used automobile dealers who specialize in offering loans to borrowers with a poor credit history.

In 1992, the U.S. Congress outlawed the use of the rule of 78s for loans longer than 61 months. In 2001, federal legislation was introduced to outlaw it altogether. At present, 17 states outlaw the use of the rule of 78s for all loans.

Actuarial method

$$u = \frac{n \cdot P \cdot V}{100 + V}$$

u = unearned interest
n = number of remaining monthly payments (excluding current payment)
P = monthly payment
V = the value from the APR table that corresponds to the annual percentage rate for the number of remaining payments (excluding current payment)

Rule of 78s

$$u = \frac{f \cdot k(k + 1)}{n(n + 1)}$$

u = unearned interest
f = original finance charge
k = number of remaining monthly payments (excluding current payment)
n = original number of payments

Example 4 illustrates the actuarial method used for calculating the unearned interest, and Example 5 illustrates the rule of 78s used for calculating the unearned interest.

EXAMPLE 4 *Using the Actuarial Method*

In Example 3, we determined the APR of Tino's loan to be 7.5%. Instead of making his 30th payment of his 48-payment loan, Tino wishes to pay his remaining balance and terminate the loan.

a) Use the actuarial method to determine how much interest Tino will save (the unearned interest, u) by repaying the loan early.
b) What is the total amount due to pay off the loan early on the day he makes his final payment?

SOLUTION:

a) Recall from Example 3 that Tino's monthly payments are \$237.00. After 30 payments have been made, 18 payments remain. Thus, $n = 18$ and $P = \$237$. To determine V, use the APR table (Table 11.2). In the Number of Payments column find the number of remaining payments, 18, and then look to the right until you reach the column headed by 7.5%, the APR. This row and column intersect at 6.04. Thus, $V = 6.04$. Now use the actuarial method formula to determine the unearned interest, u.

$$u = \frac{n \cdot P \cdot V}{100 + V} = \frac{(18)(237)(6.04)}{100 + 6.04} \approx 242.99$$

Tino will save \$242.99 in interest by the actuarial method.

b) Because the remaining payments total $18(\$237) = \$4266$, Tino's remaining balance is

\$4266.00	Total of remaining payments (which includes interest)
− 242.99	Interest saved (unearned interest)
\$4023.01	Balance due

A payment of \$4023.01 plus the 30th monthly payment of \$237 will terminate Tino's installment loan. The total amount due is \$4023.01 + \$237 = \$4260.01. ▲

EXAMPLE 5 *Using the Rule of 78s*

In Example 3, we determine the APR of Tino's loan to be 7.5%. Instead of making his 30th payment of his 48 payment loan, Tino decides to pay his remaining balance and terminate the loan.

a) Use the rule of 78s to determine how much interest Tino will save by repaying the loan early.

b) What is the total amount due to pay off the loan early on the day he makes his final payment?

SOLUTION:

a) After the 30 payments have been made, 18 payments remain. Thus $k = 18$ and $n = 48$. From Example 3 we know the original finance charge, f, is \$1576.

$$u = \frac{f \cdot k(k + 1)}{n(n + 1)}$$

$$= \frac{1576 \cdot 18(18 + 1)}{48(48 + 1)}$$

$$= \$229.16$$

Tino saves \$229.16 in interest by the rule of 78s.

b) Tino's balance is computed in a manner similar to the method in Example 4.

\$4266.00	Total of remaining payments
− 229.16	Interest saved
\$4036.84	Balance due

A payment of \$4036.84 plus the 30th monthly payment of \$237 will terminate Tino's installment loan. The total amount due is \$4036.84 + \$237 = \$4273.84. ▲

Open-End Installment Loan

A credit card is a popular way of making purchases or borrowing money. Use of a credit card is an example of an open-end installment loan. A typical charge account with a bank or store may have the terms in Table 11.3 on page 627.

Typically, credit card monthly statements contain the following information: balance at the beginning of the period, balance at the end of the period (or new balance), the transactions for the period, statement closing date (or billing date), payment due date, and the minimum payment due. For *purchases* there is no finance or interest charge if there is no previous balance due and you pay the entire new balance by the payment due date. However, if you borrow money (*cash advances*) through this account, a finance charge is applied from the date you borrowed the money until the date you repay the money. When you make purchases or borrow money, the minimum monthly payment is sometimes determined by dividing the balance due by 36 and rounding the answer up to the nearest whole dollar, thus ensuring repayment in 36 months. However, if the balance due for any month is less than \$360, the minimum monthly payment is typically \$10. These general guidelines may vary by bank and store.

TABLE 11.3 Credit Card Terms

Type of Charge	Daily Periodic Rate*	Annual Percentage Rate*
Purchases	0.03490%	12.74%
Cash advances	0.05477%	19.99%

**These rates vary with different charge accounts and localities.*

EXAMPLE 6 *Holiday Shopping Charges*

While doing her holiday shopping in November, Jan Reckard charged the following items to her Discover card: a set of tools for her husband ($250), an original print from a local artist for her daughter ($155), a set of classical music compact discs for her best friend ($100), and a gift certificate to a bookstore for her boss ($15). Her balance on December 1 is $250 + $155 + $100 + $15, or $520. The bank requires repayment within 36 months and charges an interest rate of 0.03490% per day.

a) Determine the minimum payment due on December 1.

b) On December 1, Jan makes a payment of $100. Determine the balance due on January 1, assuming that there are no additional charges or cash advances.

SOLUTION:

a) To determine the minimum payment, we divide the balance due on December 1 by 36. $520 ÷ 36 = $14.44. Rounding up to the nearest dollar, we determine that the minimum payment due on December 1 is $15.

b) With no additional purchases or cash advances and paying $100 on December 1, the balance on January 1 is $520 − $100 or $420. In addition, she must pay interest on the outstanding balance of $420 for the month of December. December has 31 days. The interest is

$$\begin{aligned} i &= prt \\ &= \$420(0.0003490)(31) \\ &= \$4.54 \end{aligned}$$

Therefore, the balance due on January 1 is $420 + $4.54, or $424.54. ▲

In Example 6, there were no additional transactions in the account for the period. When additional charges are made during the period, the finance charges on open-end installment loans or credit cards are generally calculated in one of two ways: the *unpaid balance method* or the *average daily balance method.* Example 7 illustrates the unpaid balance method and Example 8 illustrates the average daily balance method.

With the *unpaid balance method*, the borrower is charged interest or a finance charge on the unpaid balance from the previous charge period.

EXAMPLE 7 *Finance Charges Using the Unpaid Balance Method*

Ed Laughbaum charged all the supplies for his Halloween party to his Visa card. On November 5, the billing date, Ed had a balance due of $275. From November 5 through December 4, he did some shopping and charged items totaling $320, and he also made a payment of $145.

a) Find the finance charge due on December 5, using the unpaid balance method. Assume that the interest rate charged is 1.3% per month.

b) Find the new account balance on December 5.

DID YOU KNOW

"Plastic"

Market Share 2001
(in billions of dollars)

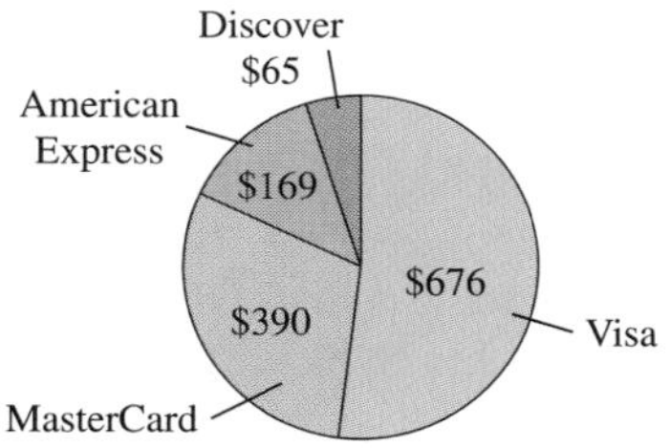

In 2001, Americans used their credit cards or "plastic" to charge $1.3 trillion at a rate of about $2.5 million per minute. The four most commonly used cards are Visa, MasterCard, American Express, and Discover. Unlike American Express and Discover, each Visa and MasterCard is associated with one of thousands of different lending institutions such as banks, savings and loans, or credit unions. The institutions that issue the cards are members of the Visa or MasterCard consortium, a system of institutions that use the name and logo and share the electronic database. The individual institutions are responsible for the money charged and reap the benefit of the interest applied. In 2001, the largest Visa and MasterCard issuers in the United States were Citigroup, MBNA America, Bank One, and Capital One.

SOLUTION:

a) The finance charge is based on the $275 balance due on November 5. To find the finance charge due on December 5 we used the simple interest formula with a time of 1 month.

$$i = \$275 \times 0.013 \times 1 \approx \$3.58$$

The finance charge on December 5 is $3.58.

b) The balance due on December 5 is found by adding the costs of the new purchases and the calculated interest to the balance due on November 5, and then subtracting the payment made from this sum.

$$\$275 + \$320 + \$3.58 - \$145 = \$453.58$$

The balance due on December 5 is $453.58. The finance charge on January 5 is based on $453.58. ▲

Many lending institutions use the *average daily balance method* of calculating the finance charge because they believe that it is fairer to the customer.

With the average daily balance method, a balance is determined each day of the billing period for which there is a transaction in the account. The average daily balance method is illustrated in Example 8.

EXAMPLE 8 *Finance Charges Using the Average Daily Balance Method*

The balance on Min Zeng's credit card account on July 1, the billing date, was $375.80. The following transactions occurred during the month of July.

July 5	Payment	$150.00
July 10	Charge: Toy store	74.35
July 18	Charge: Garage	123.50
July 28	Charge: Restaurant	42.50

a) Find the average daily balance for the billing period.

b) Find the finance charge to be paid on August 1. Assume that the interest rate is 1.3% per month.

c) Find the balance due on August 1.

SOLUTION:

a) To determine the average daily balance, we do the following. (i) Find the balance due for each transaction date.

July 1	$375.80				
July 5	$375.80	−	$150	=	$225.80
July 10	$225.80	+	$74.35	=	$300.15
July 18	$300.15	+	$123.50	=	$423.65
July 28	$423.65	+	$42.50	=	$466.15

(ii) Find the number of days that the balance did not change between each transaction. Count the first day in the period but not the last day. Note that from July 28 through August 1, the beginning of the next billing cycle, is 4 days.
(iii) Multiply the balance due by the number of days the balance did not change.
(iv) Find the sum of the products.

DID YOU KNOW

Give Credit Where Credit Is Due

The use of credit cards originated in the United States during the 1920s. The first credit cards were issued by individual firms such as oil companies, department stores, and hotels for use by only the firm that issued them.

The first universal credit card, which could be used at a variety of establishments, was introduced by Diner's Club, Inc., in 1950. The first national plan was Bank-Americard, begun on a statewide basis by the Bank of America in California in 1959 and licensed in other states beginning in 1966. Cooperating merchants throughout the world pay a service charge of 4–7% of the total credit card billings to use a bank credit card system.

Date	(i) Balance due	(ii) Number of days balance did not change	(iii) (Balance)(Days)	
July 1	\$375.80	4	(\$375.80)(4) =	\$1503.20
July 5	\$225.80	5	(\$225.80)(5) =	\$1129.00
July 10	\$300.15	8	(\$300.15)(8) =	\$2401.20
July 18	\$423.65	10	(\$423.65)(10) =	\$4236.50
July 28	\$466.15	4	(\$466.15)(4) =	\$1864.60
		31	(iv) Sum =	\$11,134.50

(v) Divide this sum by the number of days in the billing cycle (in the month). The number of days may be found by adding the days in column (ii).

$$\frac{\$11{,}134.50}{31} = \$359.18$$

Thus, the average daily balance is \$359.18.

b) The finance charge for the month is found using the simple interest formula with the average daily balance as the principal.

$$i = \$359.18 \times 0.013 \times 1 = \$4.67$$

c) Since the finance charge for the month is \$4.67, the balance owed on August 1 is \$466.15 + \$4.67 or \$470.82. ▲

The calculations in Example 8 are tedious. These calculations, however, are made almost instantaneously with computers.

Example 9 illustrates how a credit card may be used to borrow money.

EXAMPLE 9 *Using a Credit Card for a Cash Advance*

To obtain money to buy a stereo system at a garage sale Bobby Bueker obtained a cash advance of \$1500 from his credit card. He borrowed the money on July 10 and repaid it on July 31. If Bobby is charged an interest rate of 0.05477% per day, how much did Bobby pay the credit card company on July 31?

SOLUTION: The amount Bobby pays is the original principal plus any accrued interest. Interest on cash advances is generally calculated for the exact number of days of the loan, starting with the day the money is obtained. The time of the loan in this case is 21 days. Using the simple interest formula we get the following:

$$\begin{aligned} i &= prt \\ &= \$1500 \times 0.0005477 \times 21 \\ &= \$17.25 \end{aligned}$$

Therefore, Bobby must repay the credit card company \$1500 + \$17.25, or \$1517.25. ▲

Anyone purchasing a car or other costly items should consider a number of different sources for a loan. Example 10 on page 630 illustrates one method of making a comparison.

DID YOU KNOW

No More PINs

An emeging technology may make personal identification numbers (PINs) at automatic teller machines (ATMs) a thing of the past. Iris-based identification systems use a series of cameras to identify people by the characteristics of the irises in their eyes. No two irises are identical, not even an individual's left and right irises. A representation of one of your irises would be obtained and stored electronically when you open an account. Each time you used your ATM card, instead of entering your PIN, the system would scan your eye and compare this to the representation on file. Similar systems have been successfully used at highly classified military installations throughout the United States. The system has been successfully tested at ATMs in London and may soon be tested in the United States. The technology also has huge potential to reduce credit card, debit card, and telephone calling card fraud.

EXAMPLE 10 *Comparing Loan Sources*

Franz Helfenstein purchased carpeting costing $2400 with his credit card. When the bill comes due on February 1, Franz realizes he can pay $350 per month until the debt is paid off. His credit card charges 1.5% interest per month.

a) Assuming that Franz makes no other purchases with this credit card, how many payments are necessary to retire this debt?
b) What is the total interest Franz will pay?
c) How much money could Franz have saved by obtaining a fixed installment loan of $2400 with an annual percentage rate of 6% interest with 6 equal monthly payments?

SOLUTION:

a) Franz would make his first monthly payment of $350 on February 1, resulting in a new balance of $2400 − $350 = $2050. His next bill reflects the $2050 balance plus the monthly interest. He continues to make payments until the debt is retired. For each date indicated the amount on the far right represents the amount due on that date.

February 1	$2400 − $350 = $2050
March 1	$2050 + 0.015($2050) = $2080.75; $2080.75 − $350 = $1730.75
April 1	$1730.75 + 0.015($1730.75) = $1756.71; $1756.71 − $350 = $1406.71
May 1	$1406.71 + 0.015($1406.71) = $1427.81; $1427.81 − $350 = $1077.81
June 1	$1077.81 + 0.015($1077.81) = $1093.98; $1093.98 − $350 = $743.98
July 1	$743.98 + 0.015($743.98) = $755.14; $755.14 − $350 = $405.14
August 1	$405.14 + 0.015($405.14) = $411.22; $411.22 − $350 = $61.22
September 1	$61.22 + 0.015($61.22) = $62.14

After eight payments—seven for $350 and one for $62.14—Franz has paid off his credit card bill for his carpeting.

b) To calculate the total interest paid, we add up all his payments and then subtract the cost of the carpeting:

$$\begin{aligned}\text{Total of all payments} &= 7(\$350) + \$62.14\\ &= \$2512.14\\ \text{Total interest} &= \$2512.14 - \$2400\\ &= \$112.14\end{aligned}$$

c) To determine the interest that Franz pays, we will use Table 11.2 on page 623. From Table 11.2, we see that a fixed installment loan with an APR of 6% for 6 months corresponds to a finance charge of $1.76 per $100 of the amount financed. So the interest, or finance charge, is

$$\begin{aligned}\text{Finance charge} &= 1.76\left(\frac{2400}{100}\right)\\ &= 1.76(24)\\ &= \$42.24\end{aligned}$$

Therefore, Franz would save $112.14 − $42.24 = $69.90 in interest by using an installment loan instead of a credit card.

SECTION 11.4 EXERCISES

Concept/Writing Exercises

1. Explain the difference between an open-end installment loan and a fixed installment loan.
2. Explain how an installment plan differs from a personal note.
3. What is an annual percentage rate (APR)?
4. What is a finance charge?
5. What is a total installment price?
6. Name the two methods used to determine the unearned interest when a loan is repaid early.
7. Name the two methods used to determine the finance charge on an open-end installment loan.
8. What is a cash advance?

Practice the Skills/Problem Solving

9. *Financing a New Boat* Pablo Silonto purchased a new boat for $43,000. He paid 15% as a down payment and financed the balance of the purchase with a 60-month fixed installment loan with an APR of 5.5%.
   a) Determine Pablo's total finance charge.
   b) Determine Pablo's monthly payment.
10. *New Appliances* Becky Kubiac wishes to purchase new appliances for her home. The total cost for the appliances is $2900. To finance the purchase, Becky must pay 20% down, with the balance being financed with a 24-month installment loan with an APR of 8.5%.
   a) Determine Becky's total finance charge.
   b) Determine Becky's monthly payment.

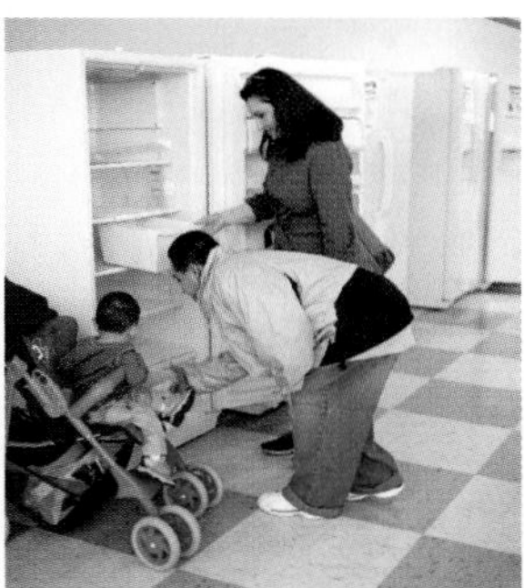

11. *Financing Student Loans* Joni Gile has a total of $4000 in student loans that will be paid with a 60-month installment loan with an APR of 7.5%.
   a) Determine Joni's total finance charge.
   b) Determine Joni's monthly payment.
12. *Financing Eye Surgery* Peg Pankowski plans to have laser eye surgery. The total cost of $2500 will be paid with a 48-month installment loan with an APR of 4.5%.
   a) Determine Peg's total finance charge.
   b) Determine Peg's monthly payment.
13. *Financing a New Business* Cheryl Sisson is a hair designer and wishes to convert her garage into a hair salon to use for her own business. The entire project would have a cash price of $3200. She decides to finance the project by paying 20% down, with the balance paid in 60 monthly payments of $53.14.
   a) What finance charge will Cheryl pay?
   b) What is the APR to the nearest half percent?
14. *Financing a Computer* Ilga Ross purchased a new laptop computer on a monthly purchase plan. The computer sold for $1495. Ilga paid 5% down and $64 a month for 24 months.
   a) What finance charge did Ilga pay?
   b) What is the APR to the nearest half percent?
15. *Financing a Used Car* Jack Keating wishes to purchase a used car that has a cash price of $12,000. The installment terms include a down payment of $3,000 and 48 monthly payments of $224.
   a) What finance charge will Jack pay?
   b) What is the APR to the nearest half percent?
16. *Financing Furniture* Mr. and Mrs. Chan want to buy furniture that has a cash price of $3450. On the installment plan they must pay 25% of the cash price as a down payment and make six monthly payments of $437.
   a) What finance charge will the Chans pay?
   b) What is the APR to the nearest half percent?
17. *Early Repayment Using the Actuarial Method* Ray Flagg took out a 60-month fixed installment loan of $12,000 to open a new pet store. He paid no money down and began making monthly payments of $232. Ray's business does better than expected and instead of making his 24th payment, Ray wishes to repay his loan in full.
   a) Determine the APR of the installment loan.
   b) How much interest will Ray save (use the actuarial method)?
   c) What is the total amount due to pay off the loan?
18. *Early Repayment Using the Actuarial Method* Coba Ling has a 48-month installment loan, with a fixed monthly payment of $167.67. The amount borrowed was $7500. Instead of making her 18th payment, Coba is paying the remaining balance on the loan.
   a) Determine the APR of the installment loan.

b) How much interest will Coba save (use the actuarial method)?
c) What is the total amount due to pay off the loan?

19. *Early Repayment Using the Actuarial Method* Nina Abu buys a new sport utility vehicle for $32,000. She trades in her old truck and receives $10,000, which she uses as a down payment. She finances the balance at 8% APR over 36 months. Before making her 24th payment, she decides to pay off the loan.
a) Use Table 11.2 to determine the total interest Nina would pay if all 36 payments were made.
b) What were Nina's monthly payments?
c) How much interest will Nina save (use the actuarial method)?
d) What is the total amount due to pay off the loan?

20. *Early Repayment Using the Actuarial Method* The cash price for furniture for Kathy Mowers's apartment was $6520. She made a down payment of $3962 and financed the balance on a 24-month fixed payment installment loan. The monthly payments are $110.52. Instead of making her 12th payment, Kathy decides to pay off the loan.
a) Determine the APR on the installment loan.
b) How much interest will Kathy save (use the actuarial method)?
c) What is the total amount due to pay off the loan?

21. *Early Repayment Using the Rule of 78s* Robert Malena wishes to purchase new equipment worth $7345, including taxes, for his landscaping business. Robert is able to secure a no-money-down, 48-month, 8.5% APR fixed installment loan from his local bank. Before making his 12th payment, Robert decides to pay off the loan with some of his yearly profits.
a) What was Robert's original finance charge?
b) What was Robert's original monthly payment?
c) How much interest will Robert save (use the rule of 78s)?
d) What is the total amount due to pay off the loan?

22. *Early Repayment Using the Rule of 78s* To pay for remodeling their kitchen the Leesebergs obtained a no-money down, 36-month, 8.5% fixed installment loan for the amount of $3600. Before making their 12th payment, the Leesebergs decide to pay off the loan.
a) What was the Leesebergs' original finance charge?
b) What was the Leesebergs' original monthly payment?
c) How much interest will the Leesebergs save (use the rule of 78s)?
d) What is the total amount due to pay off the loan?

23. *Early Repayment Using the Rule of 78s* Tony Gambino is buying a $3000 sound system by making a down payment of $500 and 18 monthly payments of $151.39. Instead of making his 12th payment, Tony decides to pay off the loan.
a) How much interest will Tony save (use the rule of 78s)?
b) What is the total amount due to pay off the loan?

24. *Early Repayment Using the Rule of 78s* Roger Golden purchased woodworking tools for $2375. He made a down payment of $850 and financed the balance with a 12-month fixed payment installment loan. Instead of making his sixth monthly payment of 134.71, he decides to pay off the loan.
a) How much interest will Roger save (use the rule of 78s)?
b) What is the total amount due to pay off the loan?

25. *Travel Expenses* To pay for his trip to Portland, Oregon for a teaching conference in November, David Dean charged the following expenses to his credit card: airfare ($365), hotel ($180), conference fee ($195), and meals ($84). David had a previous balance of zero, he bought no other items with this credit card, and on December 1 he made a payment of $200. The bank that issued the card requires repayment within 48 months and charges an interest rate of 1.1% per month.
a) What is the minimum payment due on December 1?
b) What is the balance due on January 1?

Portland, Oregon

26. *College Expenses* Brian Hickey uses his credit card in August to purchase the following college supplies: books ($425), yearlong bus pass ($175), food service meal ticket ($450), and season tickets to the basketball games ($125). On September 1, he used $650 of his financial aid check to reduce the balance. The issuing bank charges 1.2% interest per month and requires full payment within 36 months. Brian had a previous balance of zero and he makes no other purchases with this card.
a) What is the minimum payment due September 1?
b) What is the balance due on October 1?

27. *Business Expenses* In February, Denny Droessler used his credit card to pay for the following business expenses: truck repair ($423), lunch for himself and three of his

clients ($36), laundering of business uniforms ($145), and maintenance of equipment ($491). On March 1, he received payment from a client in the amount of $548 which he used as payment on his credit card. The issuing credit union charges 1.1% interest per month and requires full payment within 36 months. Denny had a previous balance of zero and he makes no other purchases with this card.

a) What is the minimum payment due March 1?

b) What is the balance due on April 1?

28. *Vacation Expenses* In June, while on vacation, the Greenbergs charged the following expenses to their credit card: airfare ($512), car rental ($172), meals ($190), and hotel ($350). On July 1, the Greenbergs paid $500 to reduce the balance. The issuing savings and loan charges 1.3% interest per month and requires full payment within 48 months. The Greenbergs make no other purchases with this card.

a) What is the minimum payment due on July 1?

b) What is the balance due on August 1?

29. *Unpaid Balance Method* On the April 5 billing date, Michaelle Chappell had a balance due of $1097.86 on her credit card. From April 5 through May 4, Michaelle charged an additional $425.79 and makes a payment of $800.

a) Find the finance charge on May 5, using the unpaid balance method. Assume that the interest rate is 1.8% per month.

b) Find the new balance on May 5.

30. *Unpaid Balance Method* On September 5, the billing date, Verna Brown had a balance due of $567.20 on her credit card. The transactions during the following month were

September 8	Payment	$275.00
September 21	Charge: Airline ticket	330.00
September 27	Charge: Hotel bill	190.80
October 2	Charge: Clothing	84.75

a) Find the finance charge on October 5, using the unpaid balance method. Assume that the interest rate is 1.1% per month.

b) Find the new balance on October 5.

31. *Unpaid Balance Method* On February 3, the billing date, Carol Ann Bluesky had a balance due of $124.78 on her credit card. Her bank charges an interest rate of 1.25% per month. She made the following transactions during the month:

February 8	Charge: Art supplies	$25.64
February 12	Payment	100.00
February 14	Charge: Flowers delivered	67.23
February 25	Charge: Music CD	13.90

a) Find the finance charge on March 3, using the unpaid balance method.

b) Find the new balance on March 3.

32. *Unpaid Balance Method* On April 15, the billing date, Gabrielle Michaelis had a balance due of $57.88 on her credit card. She is redecorating her apartment and has the following transactions.

April 16	Charge: Paint	$64.75
April 20	Payment	45.00
May 3	Charge: Curtains	72.85
May 10	Charge: Chair	135.50

a) Find the finance charge on May 15, using the unpaid balance method. Assume that the interest rate is 1.35% per month.

b) Find the new balance on May 15.

33. *Average Daily Balance Method* The balance on the Razazada's credit card on May 12, their billing date, was $378.50. For the period ending June 12, they had the following transactions.

May 13	Charge: Toys	$129.79
May 15	Payment	50.00
June 1	Charge: Clothing	135.85
June 8	Charge: Housewares	37.63

a) Find the average daily balance for the billing period.

b) Find the finance charge to be paid on June 12. Assume an interest rate of 1.3% per month.

c) Find the balance due on June 12.

34. *Average Daily Balance Method* The Levy's credit card statement shows a balance due of $1578.25 on March 23, the billing date. For the period ending April 23, they had the following transactions.

March 26	Charge: Party supplies	$79.98
March 30	Charge: Restaurant meal	52.76
April 3	Payment	250.00
April 15	Charge: Clothing	190.52
April 22	Charge: Car repairs	190.85

a) Find the average daily balance for the billing period.

b) Find the finance charge to be paid on April 23. Assume an interest rate of 1.3% per month.

c) Find the balance due on April 23.

35. *Average Daily Balance Method* Refer to Exercise 31. Instead of the unpaid balance method, suppose that Carol Ann's bank uses the average daily balance method.

a) Find Carol Ann's average daily balance for the billing period from February 3 to March 3.

b) Find the finance charge to be paid on March 3.

c) Find the balance due on March 3.

d) Compare these answers with those in Exercise 31.

36. *Average Daily Balance Method* Refer to Exercise 30. Instead of the unpaid balance method, suppose Verna's bank uses the average daily balance method.

a) Find Verna's average daily balance for the billing period from September 5 to October 5.

b) Find the finance charge to be paid on October 5.

c) Find the balance due on October 5.

d) Compare these answers with those in Exercise 30.

37. *A Cash Advance* Travis Thompson uses his credit card to obtain a cash advance of \$600 to pay for his textbooks in medical school. The interest rate charged for the loan is 0.05477% per day. Travis repays the money plus the interest after 27 days.

a) Determine the interest charged for the cash advance.
b) When he repaid the loan, how much did he pay the credit card company?

38. *A Cash Advance* John Richards borrowed \$875 against his charge account on September 12 and repaid the loan on October 14 (32 days later). Assume that the interest rate is 0.04273% per day.

a) How much interest did John pay on the loan?
b) What amount did he pay the bank when he repaid the loan?

39. *Comparing Loan Sources* Grisha Stewart needs to borrow \$1000 for an automobile repair. She finds that State National Bank charges 5% simple interest on the amount borrowed for the duration of the loan and requires the loan to be repaid in 6 equal monthly payments. Consumer's Credit Union offers loans of \$1000 to be repaid in 12 monthly payments of \$86.30.

a) How much interest is charged by the State National Bank?
b) How much interest is charged by the Consumer's Credit Union?
c) What is the APR, to the nearest half percent, on the State National Bank loan?
d) What is the APR, to the nearest half percent, on the Consumer Credit Union loan?

40. *Comparing Loan Options* Sara Lin wants to purchase a new television set. The purchase price is \$890. If she purchases the set today and pays cash, she must take money out of her savings account. Another option is to charge the TV on her credit card, take the set home today, and pay next month. Next month she will have cash and can pay her credit card balance without paying any interest. The simple interest rate on her savings account is $5\frac{1}{4}\%$. How much is she saving by using the credit card instead of taking the money out of her savings account?

Challenge Problems/Group Activity

41. *Comparing Loans* Suppose the Chans in Exercise 16 use a credit card rather than an installment plan. Assume that they make the same down payment, have no finance charge the first month, make no additional purchases on their credit card, and pay \$432 per month plus the finance charge starting with the second month. If the interest rate is 1.3% per month

a) How many months will it take them to repay the loan?
b) How much interest will they pay on the loan?
c) Which method of borrowing will cost the Chans the least amount of interest, the installment loan in Exercise 16 or the credit card?

42. *Determining Purchase Price* Ken Tucker bought a new car, but now he cannot remember the original purchase price. His payments are \$379.50 per month for 36 months. He remembers that the salesperson said the simple interest rate for the period of the loan was 6%. He also recalls he was allowed \$2500 on his old car. Find the original purchase price.

43. *Repayment Comparisons* Joscelyn Jarrett obtained a new sport utility vehicle that had a cash price of \$35,000 by paying 15% down and financing the balance with a 60-month fixed installment loan. The APR on the loan was 8.5%. Before making the 24th payment, Joscelyn decides to pay off the loan.

a) Determine the original finance charge on the 60-month loan.
b) Determine Joscelyn's monthly payment.
c) If the actuarial method is used, determine the amount of interest Joscelyn will save by paying the loan off early.
d) If the rule of 78s is used, determine the amount of interest Joscelyn will save by paying the loan off early.

44. *Repayment Comparisons* Christine Biko obtained a new speedboat that had a cash price of \$23,000 by paying 10% down and financing the balance with a 48-month fixed installment loan. The APR on the loan was 6.0%. Before making the 12th payment, Christine decides to pay off the loan.

a) Determine the original finance charge on the 48-month loan.
b) Determine Christine's monthly payment.
c) If the actuarial method is used, determine the amount of interest Christine will save by paying the loan off early.
d) If the rule of 78s is used, determine the amount of interest Christine will save by paying the loan off early.

Recreational Mathematics

45. *Borrowing Money Interest Free* Martina Saul wants to buy a camera in time for a June 30 family gathering. She knows that she will not have the money to pay for the camera until August 5. The billing date on her credit card is the 25th of the month and she has a 20-day grace period from the billing date to pay the bill with no finance charge. Explain how she can buy the camera before June 30, pay for it after August 5, and pay no interest.

Internet/Research Activities

46. Write a brief report giving the advantages and disadvantages of leasing a car. Determine all the individual costs involved with leasing a car. Indicate why you would prefer to lease or purchase a car at the present time.

47. Assume that you are married and have a child. You don't own a washer and dryer and have no money to buy the appliances. Would it be cheaper to borrow money on an installment loan and buy the appliances or to continue to go to the local coin-operated laundry for 5 years until you have saved enough to pay cash for a washer and dryer?

    With the aid of parents or friends, establish how many loads of laundry you would be doing each week. Then determine the cost of doing that number of loads at a coin-operated laundry. (Don't forget the cost of transportation to the laundry.) Shop around for a washer and dryer, and determine the total cost on an installment plan. Don't forget to include the cost of gas, electricity, and water. This information can be obtained from a local gas and electric company. With this information, you should be able to make a decision about whether to buy now or wait for 5 years.

48. Do research on the features offered by MasterCard, Visa, Discover, and American Express. Include discussions of regular, gold, and platinum cards if they exist. For example, features might include life insurance when traveling by a common carrier, miles toward air travel, discounts on automobiles, cash back at the end of the year, insurance on rental cars, and the like. Determine which card or cards have the most appropriate features for you and explain why you arrived at that conclusion.

11.5 BUYING A HOUSE WITH A MORTGAGE

DID YOU KNOW

Sears Sold Houses

Courtesy of Sears, Roebuck and Co.

Sears, Roebuck and Co. started its mail-order business in the late 1880s. Its catalog, or "Consumer's Guide," was one of the most eagerly awaited publications of the time. You could order just about anything from the 1908 "Wish Book," including a line of "modern homes." The Langston, a modest two-story home, was offered for $1630. The houses were shipped in pieces by rail, assembly required.

Buying a house is the largest purchase of a lifetime for most people. The purchaser will normally be committed to 10, 15, 20, 25, or 30 years of mortgage payments. Before selecting the "dream house," the buyer should consider the following questions: "Can I afford it?" and "Does it suit my needs?"

The question "Can I afford the house?" must be answered carefully and accurately. If a family buys a house beyond its means, it will have a difficult time living within its income. When deciding whether to purchase a particular house, the purchaser must also consider crucial questions such as "Do I have enough cash for the down payment and closing costs?" and "Can I afford the monthly payments with my current income?" These items, down payment, closing costs, and mortgage payments over time, constitute the buyer's total cost of buying a house.

Buyers usually seek a *mortgage* from a bank or other lending institution. Before approving a mortgage, which is a long-term loan, the bank will require the buyer to have a specified minimum amount for the down payment. The *down payment* is the amount of cash the buyer must pay to the seller before the lending institution will grant the buyer a mortgage. If the buyer has the down payment and meets the other criteria for the mortgage, the lending institution prepares a written agreement called the mortgage, stating the terms of the loan. The loan specifies the repayment schedule, the duration of the loan, whether the loan can be assumed by another party, and the penalty if payments are late. The party borrowing the money accepts the terms of this agreement and gives the lending institution the title or deed to the property as security.

Homeowner's Mortgage
A long-term loan in which the property is pledged as security for payment of the difference between the down payment and the sale price.

The two most popular types of mortgage loans available today are the *adjustable-rate loan* (or *variable-rate loan*) and the *conventional loan*. The major difference between the two is that the interest rate for a conventional loan is fixed for the duration of the loan, whereas the interest rate for the variable-rate loan may change every period, as specified in the loan. We will first discuss the requirements that are the same for both types of loans.

The size of the down payment required depends on who is lending the money, how old the property is, and whether or not it is easy to borrow money at that particular time. The down payment required by the lending institution can vary from 5% to 50% of the purchase price. A larger down payment is required when money is "tight," that is, when it is difficult to borrow money. Furthermore, most lending institutions tend to require larger down payments on older homes and smaller down payments on newer homes.

Most lending institutions may require the buyer to pay one or more *points* for their loan at the time of the *closing* (the final step in the sale process). *One point* amounts to 1% of the mortgage money (the amount being borrowed). By charging points, the bank reduces the rate of interest on the mortgage, thus reducing the size of the monthly payments and enabling more people to purchase houses. However, because they charge points, the rate of interest that banks state is not the APR (annual percentage rate) for the loan. The APR would be determined by adding the amount paid for points to the total interest paid and then using an APR table.

Conventional Loans

Example 1 illustrates purchasing a house with a conventional mortgage loan.

EXAMPLE 1 *Calculating Down Payment and Points*

Chris and Daryl Cahill want to purchase a house selling for $125,000. Their bank requires a 15% down payment and a payment of 1 point at the time of closing.

a) Determine the Cahill's down payment.
b) With a 15% down payment, determine the Cahill's mortgage.
c) What is the cost of the point paid by the Cahills on their mortgage.

SOLUTION:

a) The down payment is 15% of $125,000, or

$$0.15 \times \$125{,}000 = \$18{,}750$$

b) The mortgage on the Cahill's new home is the selling price minus the down payment.

$$\$125{,}000 - \$18{,}750 = \$106{,}250$$

c) One point equals 1% of the mortgage amount.

$$0.01 \times \$106{,}250 = \$1062.50$$

At the closing, the Cahills will pay the down payment of $18,750 to the seller and the 1 point, or $1062.50, to their bank. ▲

Banks use a formula to determine the maximum monthly payment that they believe is within the purchaser's ability to pay. A mortgage loan officer first determines the buyer's *adjusted monthly income* by subtracting from the gross monthly income

DID YOU KNOW

The Benefits of Home Ownership

Most of us are aware that a big part of the "American dream" is to own your own home. What we may not realize is that in addition to having a place to call your own, several important financial benefits occur when you own your own home. First, instead of paying rent to someone else, you make a mortgage payment that builds the equity in your home. *Equity* is the difference between the appraised value of your home and your loan balance, and it increases with each payment you make. As years go by, this equity may also help you qualify for other loans such as college and car loans. Second, the interest and real estate taxes you pay (in most cases) are deductible on your federal income tax. These deductions can add up to significant savings each year, and may result in a larger tax refund. Finally, over time you can typically expect your home to increase in value. Thus, your home not only becomes your place of dwelling; in most cases, it also serves as a wise financial investment.

(total income before any deductions) any fixed monthly payments with more than 10 months remaining (such as for a student loan, a car, furniture, or a television). The loan officer then multiplies the adjusted monthly income by 28%. (This percent, and the maximum number of payments remaining on other fixed loans, may vary in different locations.) In general, this product is the maximum monthly house payment the lending institution believes the purchaser can afford to pay. This payment must cover principal, interest, property taxes, and insurance. Taxes and insurance are not necessarily paid to the bank; they may be paid directly to the tax collector and the insurance company. Example 2 shows how a bank uses the formula to determine whether a prospective buyer qualifies for a mortgage.

EXAMPLE 2 *Qualifying for a Mortgage*

Suppose that the Cahill's (see Example 1) gross monthly income is $4200 and that they have 15 remaining payments of $185 per month on their car loan and 14 remaining payments of $35 per month on a loan used to purchase a new washer and dryer. The taxes on the house they want to purchase are $135 per month and the insurance is $38 per month.

a) What maximum monthly payment does the bank's loan officer think the Cahills can afford?

b) The Cahills want a 30-year $106,250 mortgage. If the interest rate is 6.5%, determine whether the Cahills qualify for the mortgage.

SOLUTION:

a) To find the maximum monthly payment the bank's loan officer believes the Cahills can afford first determine their adjusted monthly income.

$4200	Gross income
−220	Monthly payments (car and appliance loans)
$3980	Adjusted monthly income

Next find 28% of the adjusted monthly income.

$$0.28 \times \$3980 = \$1114.40$$

The loan officer determines that the Cahills can afford a maximum monthly payment—including principal, interest, taxes, and insurance—of $1114.40.

b) To determine whether the Cahills qualify for a 30-year conventional mortgage with their current income calculate the total monthly payment the Cahills would have to pay, including principal, interest, property taxes, and insurance. Then compare the calculated total monthly payment with the maximum payment the bank's loan officer thinks the Cahills can afford, which was calculated in part (a).

Lending institutions and lawyers use computer programs or calculators to determine monthly mortgage payments, per thousand dollars, for a specific number of years at a specific rate. This information can also obtained from one of many mortgage calculator webpages on the Internet. Table 11.4 on page 638 gives monthly mortgage payments, including principal and interest, per $1000 of mortgage. With an interest rate of 6.5%, a 30-year loan would have a monthly mortgage payment of $6.32 (circled in blue) per thousand dollars of mortgage.

To determine the Cahill's monthly mortgage payment of principal and interest first divide the mortgage by $1000. This will give the number of thousands of dollars of the mortgage.

$$\frac{106{,}250}{1000} = 106.25$$

Then find the monthly mortgage payment by multiplying the number of thousands of dollars of mortgage, 106.25, by the value found in Table 11.4, $6.32.

$$\$106.25 \times \$6.32 = \$671.50$$

The monthly payment for principal and interest is $671.50. To the $671.50, add the monthly cost of real estate taxes, $135, and insurance of $38, for a total cost of $844.50. Since $844.50 is less than $1114.40, the maximum monthly payment the loan officer determined the Cahills can afford, the Cahills will most likely be granted the loan. ▲

TABLE 11.4 Monthly Payment per $1000 of Mortgage, Including Principal and Interest

	Number of Years				
Rate %	**10**	**15**	**20**	**25**	**30**
4	$10.12	$7.40	$6.06	$5.28	$4.77
4.5	10.36	7.65	6.33	5.56	5.07
5	10.61	7.91	6.60	5.85	5.37
5.5	10.85	8.17	6.88	6.14	5.68
6	11.10	8.44	7.16	6.44	6.00
6.5	11.35	8.71	7.46	6.75	6.32
7	11.61	8.99	7.75	7.07	6.65
7.5	11.87	9.27	8.06	7.39	6.99
8	12.13	9.56	8.36	7.72	7.34
8.5	12.40	9.85	8.68	8.05	7.69
9	12.67	10.14	9.00	8.40	8.05
9.5	12.94	10.44	9.33	8.74	8.41
10	13.22	10.75	9.66	9.09	8.70
10.5	13.49	11.05	9.98	9.44	9.15
11	13.78	11.37	10.32	9.80	9.52
11.5	14.06	11.68	10.66	10.16	9.90
12	14.35	12.00	11.01	10.53	10.29

What is the effect on the monthly payments when only the period of time has been changed? The total monthly payments for the Cahills in Example 2 would have been $1378.94 for 10 years, $1098.44 for 15 years, $965.63 for 20 years, $890.19 for 25 years, and $844.50 for 30 years. (You should verify these numbers yourself.) Increasing the length of time decreases the monthly payment but increases the total amount of interest paid because the borrower is paying for a longer period of time. The longer the term of a mortgage, the more expensive the total cost of the house.

EXAMPLE 3 *The Total Cost of a House*

The Cahills of Examples 1 and 2 obtained a house selling for $125,000. They made a 15% down payment and obtained a 30-year conventional mortgage for $106,250 at 6.5%. They also paid 1 point at closing. Their monthly mortgage payment of principal and interest is $671.50.

a) Determine the total amount including principal, interest, the down payment, and the points that the Cahills will pay the bank for their house over 30 years.

b) How much of the cost will be interest?

c) How much of the first payment on the mortgage is applied to the principal?

SOLUTION:

a) To find the total amount the Cahills will pay for their house perform the following computation:

$671.50	Mortgage payment for 1 month
× 12	Number of months in a year
$8058.00	Mortgage payments for 1 year
× 30	Number of years in the mortgage
$241,740.00	Total mortgage payments
+ 18,750.00	Down payment
+ 1,062.50	1 point
$261,552.50	Total cost of the house

Note: The result might not be the exact cost of the house, since the final payment on the mortgage might be slightly more or less than the regular monthly payment.

b) To determine the amount of interest paid over 30 years subtract the purchase price of the house and the cost of the points from the total cost.

$261,552.50	Total cost
−125,000.00	Purchase price
$136,552.50	Total interest including one point
− 1,062.50	1 point
$135,490.00	Total interest on mortgage payments

c) To find the amount of the first payment that is applied to the principal subtract the amount of interest on the first payment from the monthly mortgage payment. Use the simple interest formula to find the interest on the first payment.

$$\begin{aligned} i &= prt \\ &= \$106{,}250 \times 0.065 \times \frac{1}{12} \\ &\approx \$575.52 \end{aligned}$$

Now subtract the interest for the first month from the monthly mortgage payment. The difference will be the amount paid on the principal for the first month.

$671.50	Monthly mortgage payment
−575.52	Interest paid for the first month
$95.98	Principal paid for the first month

Thus, the first payment of $671.50 consists of $575.52 in interest and $95.98 in principal. The $95.98 is applied to reduce the loan. Thus, the balance due after the first payment is $106,250 − $95.98, or $106,154.02. ▲

By repeatedly using the simple interest formula month-to-month on the unpaid balance, you could calculate the principal and the interest for all the payments—a tedious task. However, a list containing the payment number, payment on the interest, payment on the principal, and balance of the loan, can be prepared using a computer. Such a list is called a loan *amortization schedule*. One way to obtain an amortization schedule is by using a computer spreadsheet program. Another way is to access an amortization "calculator" program on the Internet. A part of the amortization schedule for the Cahill's loan in Example 3 is given in Table 11.5. This schedule was generated from an Internet site called *Monthly Mortgage Payment Calculator* (*www.hsh.com*). Note that the monthly payment in Table 11.5 ($671.57) is slightly more than the monthly payment we calculated using Table 11.4 ($671.50). This difference is due to a rounding error that occurs when estimating the monthly principal and interest payment from Table 11.4.

TABLE 11.5 Amortization Schedule

Annual % Rate: 6.5		**Monthly Payment: $671.57**	
Loan: $106,250		**Term: Years 30, Months 0**	
Periods: 360			
Payment Number	**Interest**	**Principal**	**Balance of Loan**
1	$575.52	$96.05	$106,153.95
2	$575.00	$96.57	$106,057.38
3	$574.48	$97.09	$105,960.28
4	$573.95	$97.62	$105,862.66
11	$570.19	$101.38	$105,164.35
12	$569.64	$101.93	$105,062.42
119	$489.88	$181.69	$90,257.31
120	$488.89	$182.68	$90,074.63
239	$324.14	$347.43	$59,493.68
240	$322.26	$349.31	$59,144.36
359	$7.22	$664.36	$667.95
360	$3.62	$667.95	$0.00

Adjustable-Rate Mortgages

Now let's consider *adjustable-rate mortgages*, ARMs (also called *variable-rate mortgages*). The rules for ARMs vary from state to state and from bank to bank, so the material presented on ARMs may not apply in your state or at your local lending institution. Generally, with adjustable-rate mortgages, the monthly mortgage payment remains the same for a 1-, 2-, or 5-year period even though the interest rate of the mortgage may change every 3 months, 6 months, or some other predetermined period. The interest rate for an adjustable-rate mortgage may be based on an index that is determined by the Federal Home Loan Bank Association or it may be based on the interest rate of a 3-month, 6-month, or 1-year Treasury bill. The interest rate of a 3-month Treasury bill may change every 3 months, the interest rate on a 6-month Treasury bill may change every 6 months, and so on. When the base is a Treasury bill, the actual interest rate charged for the mortgage is often determined by adding 3% to $3\frac{1}{2}\%$, called the *add on rate* or *margin*, to the rate of the Treasury bill. Thus, if the rate of the Treasury bill is 6% and the add on rate is 3%, the interest rate charged is 9%.

EXAMPLE 4 *An Adjustable-Rate Mortgage*

Tony and Keisha Torrence purchased a house for \$115,000 with a down payment of \$23,100. They obtained a 30-year adjustable-rate mortgage with the following terms. The interest rate is based on a 6-month Treasury bill. The interest rate charged is 3% above the interest rate of the 6-month Treasury bill (3% is the add on rate). The interest rate is adjusted every 6 months on the date of adjustment. The interest rate will not change more than 1% (up or down) when the interest rate is adjusted. The maximum interest rate for the duration of the loan is 12%. There is no lower limit on the interest rate. The initial mortgage interest rate is 5.5%, and the monthly payments (including principal and interest) are adjusted every 5 years.

a) Determine the initial monthly payment.

b) Determine the adjusted interest rate in 6 months if the interest rate on the Treasury bill at that time is 2%.

SOLUTION:

a) To determine the initial monthly payment of interest and principal divide the amount of the loan, \$115,000 − \$23,100 = \$91,900, by \$1000. The result is 91.9. Now multiply the number of thousands of dollars of mortgage, 91.9, by the value found in Table 11.4 with $r = 5.5\%$ for 30 years. The value found in the table is \$5.68.

$$\$5.68 \times 91.9 \approx \$521.99$$

Thus, the initial monthly payment for principal and interest is \$521.99. This amount will not change for the first 5 years of the mortgage.

b) The adjusted interest rate in 6 months will be the Treasury bill rate plus the add on rate.

$$2\% + 3\% = 5\%$$

In Example 4(b), note that the rate after 6 months, 5%, is lower than the initial rate of 5.5%. Since the monthly payment remains the same, the additional money paid the bank is applied to reduce the principal. The monthly interest and principal payment of \$521.99 would pay off the loan in 30 years if the interest remained constant at 5.5%. What happens if the interest rate drops and stays lower than the initial 5.5% for the length of the loan? In this case, at the end of each 5-year period the bank reduces the monthly payment so that the loan will be paid off in 30 years. What happens if the interest rate increases above the initial 5.5% rate? In this case, part of, or if necessary, all of the mortgage payment that would normally go toward repaying the principal would be used to meet the interest obligation. At the end of the 5-year period, the bank will increase the monthly payment so that the loan can be repaid by the end of the 30-year period. Or the bank may increase the time period of the loan beyond 30 years so that the monthly payment is affordable.

To prevent rapid increases in interest rates, some banks have a rate cap. A *rate cap* limits the maximum amount the interest rate may change. A *periodic rate cap* limits the amount the interest rate may increase in any one period. For example, your mortgage could provide that, even if the index increases by 2% in 1 year, your rate can only go up 1% per year. An *aggregate rate cap* limits the interest rate increase and decrease over the entire life of the loan. If the initial interest rate is 6% and the aggregate rate cap is 2%, the interest rate could go no higher than 8% and no lower than 4% over the life of the mortgage. A *payment cap* limits the amount the monthly payment may change but does not limit changes in interest rates. If interest rates increase rapidly on a loan with a payment cap, the monthly payment may not be large enough to pay the monthly principal and interest on the loan. If that happens, the borrower could end up paying interest on interest.

Other Types of Mortgages

Conventional mortgages and adjustable-rate mortgages are not the only methods of financing the purchase of a house. Next, we briefly describe four other methods, and we briefly discuss home equity loans.

FHA Mortgage

A house can be purchased with a smaller down payment than with a conventional mortgage if the individual qualifies for a Federal Housing Administration (FHA) loan. The loan application is made through a local bank. The bank's loan officer determines the maximum monthly payment a loan applicant can afford by taking 29% of the adjusted monthly income. The bank provides the money, but the FHA insures the loan. The down payment for an FHA loan is as low as 2.5% of the purchase price, rather than the standard 5–50%. Another advantage is FHA loans can be assumed at the original rate of interest on the loan. For example, if you purchase a home today that already has a 6% FHA mortgage, you, the new buyer, can assume that 6% mortgage regardless of the current interest rates. However, to be able to assume a mortgage, the purchaser must be able to make a down payment equal to the difference between the purchase price of the house and the balance due on the original mortgage. The government sets the maximum interest rate the lender may charge.

One drawback to FHA loans is that the borrower must pay an FHA insurance premium as part of the monthly mortgage payment. The insurance premium is calculated at a rate of one-half percent (0.5%) of the unpaid balance of the loan on the anniversary date of the loan. Thus, even though the insurance premium decreases each year, it adds to the monthly payments.

VA Mortgage

A veteran certified by the Department of Veterans Affairs (VA) who wants to purchase a house applies for a mortgage with a bank or lending institution. The individual must meet the requirements set by the bank or lending institution for a mortgage. The VA guarantees repayment of a certain percentage of the loan obtained by the individual should the individual default on the loan. For example, if the loan is less than $45,000, the VA guarantees 50% of the loan. For loans from $45,000 to $144,000, the VA guarantees the lesser of 40% of the loan or $36,000. Since the bank has this guarantee from the VA, qualified veterans often do not have to make a down payment. A veteran who can make the monthly mortgage payments may therefore obtain a certain mortgage without a down payment.

The government sets the maximum interest rate that the lender may charge. A VA loan is always assumable. With a VA loan, the seller may be asked to pay points, but there is no monthly insurance premium.

Graduated Payment Mortgage (GPM)

A GPM mortgage is designed so that for the first 5 to 10 years the size of the mortgage payment is smaller than the payments for the remaining time of the mortgage. After the first 5- to 10-year period, the mortgage payments are increased. The mortgage payments then remain constant for the duration of the mortgage. Depending on the size of the mortgage, the lender might find that the monthly payments made for the first few years may actually be less than the interest owed on the loan for those few years. The interest not paid during the first few years of the mortgage is then added to the original loan. This type of loan is strictly for those who are confident their annual incomes will increase as rapidly as the mortgage payments do.

Balloon-Payment Mortgage (BPM)

The BPM type of loan could be for the person who needs time to find a permanent loan. It may work this way: The individual pays the interest for 3 to 8 years, the period of the loan. At the end of the 3- to 8-year period, the buyer must repay the entire principal unless the lender agrees to a loan extension. Balloon-payment loans generally offer lower rates than conventional mortgages. This type of loan may be advantageous if the buyer plans to sell the house before the maturity date of the balloon-payment mortgage.

Home Equity Loans

As you make monthly payments and pay off the principal you owe on your home, you are said to be gaining *equity* in your home. **Equity** is the difference between the appraised value of your home and your loan balance. This equity can be used as collateral in obtaining a loan. Such a loan is referred to as a **home equity loan** or a **second mortgage.** One advantage of home equity loans over other types of loans (such as installment loans) is that the interest charged on a home equity loan is often tax deductible on federal income taxes. Home equity loans are commonly used for home improvements, for bill consolidation, or to pay for college education expenses.

For further information about the types of mortgages or loans discussed in this section, consult a loan officer at a local bank, a savings and loan, or a credit union. Additional information on buying a house may be obtained from the U.S. Government Printing Office in Pueblo, CO 81009. This information is also available on the Internet at *http://www.access.gpo.gov*.

DID YOU KNOW

Investing in Mutual Funds

A *mutual fund* is an investment tool that enables investors to indirectly own a wide variety of stocks, bonds, or other investments. When investors purchase shares in a mutual fund, they are actually placing their money in a pool along with many other investors. An investment company that invests in many different stocks, bonds, or other investments manages this pool of money. The investments within a mutual fund are called the mutual fund's *portfolio*. The investors of a mutual fund share the gains and losses from the investments within the portfolio. There are some distinct advantages to investing in mutual funds rather than investing in individual stocks and bonds. First, investors in mutual funds have their money managed by full-time professionals who research and evaluate hundreds of stocks and bonds every day. Such management may help investors avoid poor investments. Second, because large sums of money are managed within a mutual fund, costs related to investing, known as *commissions,* are generally lower than they are for purchasing individual stocks and bonds. Third, when investors purchase shares in a mutual fund, they are indirectly purchasing shares in a multitude of stocks or bonds. This diversification can greatly help to reduce some of the risks of investing. Finally, investors can easily buy and sell shares in a mutual fund through the mail, over the telephone, or on the Internet, often without having to pay commissions.

One disadvantage of mutual fund investing is the potential to miss out on a large return on investment. For example, if on October 1, 2002, an investor had invested $1000 in stock shares of Dobson Communications, by January 1, 2003, this investment would have been worth $6620.70. Meanwhile, despite this remarkable gain, several mutual funds that included Dobson Communications in their portfolios actually lost money over the same period because of poor performance by other stocks in their portfolios. In general, though, investing in mutual funds is considered an excellent way to begin investing and to maintain diverse ownership in a variety of investments.

SECTION 11.5 EXERCISES

Concept/Writing Exercises

1. What is a mortgage?
2. What is a down payment?
3. What is the difference between a variable-rate mortgage and a conventional mortgage?
4. **a)** What are points in a mortgage agreement?
   b) Explain how to determine the cost of x points.
5. Explain how to determine a buyer's adjusted monthly income.
6. What is an add on rate, or margin?
7. What is an amortization schedule?
8. Who insures an FHA loan? Who provides the money for an FHA loan?
9. What is equity?
10. What is a home equity loan?

Problem Solving

11. *Buying a House* Sally Jacobs wishes to buy a house selling for $250,000. Her credit union requires her to make a 15% down payment. The current mortgage rate is 4.5%.
a) Determine the amount of the required down payment.
b) Determine the monthly mortgage payment for a 15-year loan with a 15% down payment.

12. *Down Payment and Mortgage Payment* Thomas Osler is buying a townhouse selling for $175,000. His bank is requiring a minimum down payment of 20%. The current mortgage rate is 5.5%.
a) Determine the amount of the required down payment.
b) Determine the monthly mortgage payment for a 30-year loan with the minimum down payment.

13. *Monthly Payment on a Condominium* Mary Beth and Ken Henkel are buying a new condominium for $210,000. Their bank is requiring a minimum down payment of 10%. The current mortgage rate is 5%.
a) Determine the amount of the required down payment.
b) Determine the monthly mortgage payment for a 20-year loan with the minimum down payment.

14. *Buying a First Home* Sandra Coleman's family is purchasing their first home, which is selling for $95,000. The credit union is requiring a minimum down payment of 5%. The current mortgage rate is 7%.
a) Determine the amount of the required down payment.
b) Determine the monthly mortgage payment for a 30-year loan with the minimum down payment.

15. *Paying Points* Martha Cutler is buying a house selling for $195,000. The bank is requiring a minimum down payment of 20%. To obtain a 20-year mortgage at 6% interest she must pay 2 points at the time of closing.
a) What is the required down payment?
b) With the 20% down payment, what is the amount of the mortgage?
c) What is the cost of the 2 points?

16. *Down Payment and Points* The Nicols are buying a house selling for $245,000. They pay a down payment of $45,000 from the sale of their current house. To obtain a 15-year mortgage at 4.5% interest the Nicols must pay 1.5 points at the time of closing.
a) What is the amount of the mortgage?
b) What is the cost of the 1.5 points?

17. *Qualifying for a Mortgage* Pietr and Helga Guenther's gross monthly income is $3200. They have 25 remaining car payments of $335. The Guenthers are applying for a 15-year, $150,000 mortgage at 5% interest to buy a new house. The taxes and insurance on the house are $225 per month.
a) Determine the Guenther's adjusted monthly income.
b) Determine the maximum monthly payment a lender feels the Guenthers can afford.
c) Determine the monthly mortgage payment plus taxes and insurance.
d) Do the Guenthers qualify for this mortgage?

18. *Qualifying for a Mortgage* Ting-Fang and Su-hua Zheng's gross monthly income is $4100. They have 18 remaining boat payments of $505. The Zhengs are applying for a 20-year, $275,000 mortgage at 9% interest to buy a new house. The taxes and insurance on the house are $425 per month.
a) Determine the Zheng's adjusted monthly income.
b) Determine the maximum monthly payment a lender feels the Zhengs can afford.

c) Determine the monthly mortgage payment plus taxes and insurance.
d) Do the Zhengs qualify for this mortgage?

19. *A 30-Year Conventional Mortgage* Ingrid Holzner obtains a 30-year, $63,750 conventional mortgage at 8.5% on a house selling for $75,000. Her monthly payment, including principal and interest, is $490.24.
    a) Determine the total amount Ingrid will pay for her house.
    b) How much of the cost will be interest?
    c) How much of the first payment on the mortgage is applied to the principal?

20. *A 25-Year Conventional Mortgage* Mr. and Mrs. Alan Bell obtain a 25-year, $110,000 conventional mortgage at 10.5% on a house selling for $160,000. Their monthly mortgage payment, including principal and interest, is $1038.40.
    a) Determine the total amount the Bells will pay for their house.
    b) How much of the cost will be interest?
    c) How much of the first payment on the mortgage is applied to the principal?

21. *Evaluating a Loan Request* The Rosens found a house selling for $113,500. The taxes on the house are $1200 per year, and insurance is $320 per year. They are requesting a conventional loan from the local bank. The bank is currently requiring a 28% down payment and 3 points, and the interest rate is 10%. The Rosen's monthly income is $4750. They have more than 10 monthly payments remaining on a car, a boat, and furniture. The total monthly payments for these items is $420.
    a) Determine the required down payment.
    b) Determine the cost of the 3 points.
    c) Determine their adjusted monthly income.
    d) Determine the maximum monthly payment the bank's loan officer believes they can afford.
    e) Determine the monthly payments of principal and interest for a 20-year loan.
    f) Determine their total monthly payment, including insurance and taxes.
    g) Determine whether the Rosens qualify for the 20-year loan.
    h) Determine how much of the first payment on the loan is applied to the principal.

22. *Evaluating a Loan Request* Kathy Fields wants to buy a condominium selling for $95,000. The taxes on the property are $1500 per year, and insurance is $336 per year. Kathy's gross monthly income is $4000. She has 15 monthly payments of $135 remaining on her van. The bank is requiring 20% down and is charging 9.5% interest.
    a) Determine the required down payment.
    b) Determine the maximum monthly payment the bank's loan officer believes Kathy can afford.
    c) Determine the monthly payment of principal and interest for a 25-year loan.
    d) Determine her total monthly payment, including insurance and taxes.
    e) Does Kathy qualify for the loan?
    f) Determine how much of the first payment on the mortgage is applied to the principal.
    g) Determine the total amount she pays for the condominium with a 25-year conventional loan. (Do not include taxes or insurance.)
    h) Determine the total interest paid for the 25-year loan.

23. *Comparing Loans* The Riveras are negotiating with two banks for a mortgage to buy a house selling for $105,000. The terms at bank A are a 10% down payment, an interest rate of 10%, a 30-year conventional mortgage, and 3 points to be paid at the time of closing. The terms at bank B are a 20% down payment, an interest rate of 11.5%, a 25-year conventional mortgage, and no points. Which loan should the Riveras select in order for the total cost of the house to be less?

24. *Comparing Loans* Paul Westerberg is negotiating with two credit unions for a mortgage to buy a condominium selling for $525,000. The terms at Grant County Teacher's Credit Union are a 20% down payment, an interest rate of 7.5%, a 15-year mortgage, and 1 point to be paid at the time of closing. The terms at Sinnipee Consumer's Credit Union are a 15% down payment, an interest rate of 8.5%, a 20-year mortgage, and no points. Which loan should Paul select for the total cost of the down payment, points, and total mortgage payments of the house to be less?

Challenge Problems/Group Activities

25. *An Adjustable-Rate Mortgage* The Simpsons purchased a house for $105,000 with a down payment of $5000. They obtained a 30-year adjustable-rate mortgage. The terms of the mortgage are as follows: The interest rate is based on a 3-month Treasury bill, the interest rate charged is 3.25% above the rate of the Treasury bill on the date of adjustment, the interest rate is adjusted every 3 months, the interest rate will not change more than 1% (up or down) when the interest rate is adjusted, the maximum interest rate that

can be charged for the duration of the loan is 16%, there is no lower limit on the interest rate, the initial mortgage interest rate is 9%, and the monthly payment of interest and principal is adjusted semiannually.

a) Determine the initial monthly payment for interest and principal.
b) Determine an amortization schedule for months 1–3.
c) Determine the interest rate for months 4–6 if the interest rate on the Treasury bill at the time is 6.13%.
d) Determine an amortization schedule for months 4–6.
e) Determine the interest rate for months 7–9 if the interest rate on the Treasury bill at the time is 6.21%.

26. *An Adjustable-Rate Mortgage* The Bretz family purchased a house for \$95,000 with a down payment of \$13,000. They obtained a 30-year adjustable-rate mortgage. The terms of the mortgage are as follows: The interest rate is based on a 3-month Treasury bill, the interest rate charged is 3.25% above the rate of the Treasury bill on the date of adjustment, the interest rate is adjusted every 3 months, the interest rate will not change more than 1% (up or down) when the interest rate is adjusted, the maximum interest rate that can be charged for the duration of the loan is 16%, there is no lower limit on the interest rate, the initial mortgage interest rate is 8.5%, and the monthly payment of interest and principal is adjusted annually.

a) Determine the initial monthly payment for interest and principal.
b) Determine the interest rate in 3 months if the interest rate on the Treasury bill at the time is 5.65%.
c) Determine the interest rate in 6 months if the interest rate on the Treasury bill at the time is 4.85%.

27. *How Much House Can They Afford?* A bank's loan officer determines that the Pappys can afford to make a \$950 monthly mortgage payment. If the bank will give them a 25-year conventional mortgage at 9% and requires a 25% down payment, what is

a) the maximum mortgage the bank will grant the Pappys?
b) the highest-priced house they can afford?

28. *Comparing Mortgages* The Hassads are applying for a \$90,000 mortgage. They can choose between a conventional mortgage and a variable-rate mortgage. The interest rate on a 30-year conventional mortgage is 9.5%. The terms of the variable-rate mortgage are 6.5% interest rate the first year, an annual cap of 1%, and an aggregate cap of 6%. The interest rates and the mortgage payments are adjusted annually. Assume that the interest rates for the variable-rate mortgage increase by the maximum amount each year. Then the monthly mortgage payments for the variable-rate mortgage for years 1–6 are \$568.86, \$628.05, \$688.29, \$749.35, \$811.02, and \$873.11, respectively.

a) Knowing that they will be in the house for only 6 years, which mortgage, the conventional mortgage or the variable-rate mortgage, will be the least expensive for that period?
b) How much will they save by choosing the less expensive mortgage?

Internet/Research Activities

29. *Finding Your Dream Home* Examine a local newspaper to find your "dream home" and note the asking price. Next contact a loan officer from your local bank, savings and loan, or credit union. Assuming you can make a 20% down payment, determine the interest rates for a 15-year and a 30-year mortgage. Use an amortization calculator (see page 640) with the data you obtained to print amortizations schedules. Compare the monthly payments with the 15-year mortgage to those of the 30-year mortgage. Compare the total interest costs of the 15-year mortgage with those of the 30-year mortgage. Write a report summarizing your findings.

30. *Closing Costs* An important part of buying a house is the closing. The exact procedures for the closing differ with individual cases and in different parts of the country. In any closing, however, both the buyer and the seller have certain expenses. To determine what is involved in the closing of a property in your community, contact a lawyer, a real estate agent, or a banker. Explain that you are a student and that your objective is to understand the procedure for closing a real estate purchase and the costs to both buyer and seller. Select a specific piece of property that is for sale. Use the asking price to determine the total closing costs to both buyer and seller. The following is a partial list of the most common costs. Consider them in your research.

a) Fee for title search and title insurance
b) Credit report on buyer
c) Fees to the lender for services in granting the loan
d) Fee for property survey
e) Fee for recording of the deed
f) Appraisal fee
g) Lawyer's fee
h) Escrow accounts (taxes, insurance)
i) Mortgage assumption fee

CHAPTER 11 SUMMARY

IMPORTANT FACTS

Ordinary interest

When computing ordinary interest, each month is considered to have 30 days and a year is considered to have 360 days.

United States rule

If a partial payment is made on a loan, interest is computed on the principal from the first day of the loan until the date of the partial payment. The partial payment is used to pay the interest first; then the rest of the payment is used to reduce the principal.

Banker's rule

When computing interest with the Banker's rule, a year is considered to have 360 days and any fractional part of a year is the exact number of days.

Simple interest formula

Interest = principal × rate × time or $i = prt$

Compound interest formula

$$A = p\left(1 + \frac{r}{n}\right)^{nt}$$

Present value formula

$$p = \frac{A}{\left(1 + \frac{r}{n}\right)^{nt}}$$

Actuarial method

$$u = \frac{u \cdot P \cdot V}{100 + V}$$

Rule of 78s

$$u = \frac{f \cdot k(k + 1)}{n(n + 1)}$$

$$\textbf{Percent change} = \frac{\text{amount in latest period} - \text{amount in previous period}}{\text{amount in previous period}} \times 100$$

$$\textbf{Percent markup on cost} = \frac{\text{selling price} - \text{dealer's cost}}{\text{dealer's cost}} \times 100$$

CHAPTER 11 REVIEW EXERCISES

11.1

Change the number to a percent. Express your answer to the nearest tenth of a percent.

1. $\frac{3}{5}$ **2.** $\frac{2}{3}$ **3.** $\frac{5}{8}$

4. 0.041 **5.** 0.0098 **6.** 3.141

Change the percent to a decimal number.

7. 3% **8.** 12.1% **9.** 123%

10. $\frac{1}{4}\%$ **11.** $\frac{5}{6}\%$ **12.** 0.00045%

13. *Lambeau Field* Before undergoing renovations, the seating capacity at Lambeau Field (in Green Bay, Wisconsin) was 60,790. After the renovations were complete the seating capacity was 71,500. Determine the percent increase (to the nearest tenth of a percent) in the seating capacity of Lambeau Field.

Lambeau Field

14. *Salary Increase* Charlotte Newsom had a salary of \$46,200 in 2003 and a salary of \$51,300 in 2004. Determine the percent increase in Charlotte's salary from 2003 to 2004.

In Exercises 15–17, solve for the unknown quantity.

15. What percent of 80 is 25?

16. Forty-four is 16% of what number?

17. What is 17% of 540?

18. *Tipping* At Empress Garden Restaurant, Vishnu and Krishna's bill comes to \$42.79, including tax. If they wish to leave a 15% tip on the total bill, how much should they tip the waiter?

19. *Increased Membership* If the number of people in your chess club increased by 20%, or 8 people, what was the original number of people in the club?

20. *Increased Membership* The Sarasota Wheelers skateboard club had 75 members and increased the number of members to 95. What is the percent increase in the number of members?

11.2

In Exercises 21–24, find the missing quantity by using the simple interest formula.

21. $p = \$2500, r = 4\%, t = 60$ days, $i = ?$

22. $p = \$1575, r = ?, t = 100$ days, $i = \$41.56$

23. $p = ?, r = 8\frac{1}{2}\%, t = 3$ years, $i = \$114.75$

24. $p = \$5500, r = 11\frac{1}{2}\%, t = ?, i = \$316.25$

25. *Roof Replacement Loan* Chris Sharek borrowed \$5300 from his father to replace the roof on his house. The loan was for 36 months and had a simple interest rate of 5.75%. Determine the amount Chris paid his father on the date of maturity of the loan.

26. *A Bank Loan* Lori Holdren borrowed \$3000 from her bank for 240 days at a simple interest rate of 8.1%.

a) How much interest did she pay for the use of the money?

b) How much did she pay the bank on the date of maturity?

27. *A Bank Loan* Nikos Pappas borrowed \$6000 for 24 months from the bank, using stock as security. The bank discounted the loan at $11\frac{1}{2}\%$.

a) How much interest did Nikos pay the bank for the use of the money?

b) How much did he receive from the bank?

c) What was the actual rate of interest?

28. *Savings as Security* Golda Frankl borrowed \$800 for 6 months from her bank, using her savings account as security. A bank rule limits the amount that can be borrowed in this manner to 85% of the amount in the borrower's savings account. The rate of interest is 2% higher than the interest rate being paid on the savings account. The current rate on the savings account is $5\frac{1}{2}\%$.

a) What rate of interest will the bank charge for the loan?

b) Find the amount that Golda must repay in 6 months.

c) How much money must she have in her account in order to borrow \$800?

11.3

29. *Comparing Compounding Periods* Determine the amount and the interest when \$1000 is invested for 5 years at 10%

a) compounded annually.

b) compounded semi-annually.

c) compounded quarterly.

d) compounded monthly.

e) compounded daily (use $n = 360$).

30. *Total Amount* Choi deposited \$2500 in a savings account that pays 4.75% interest compounded quarterly. What will be the total amount of money in the account 15 years from the day of deposit?

31. *Effective Annual Yield* Determine the effective annual yield of an investment if the interest is compounded daily at an annual rate of 5.6%.

32. *Present Value* How many dollars must you invest today to have \$40,000 in 20 years? Assume that the money earns 5.5% interest compounded quarterly.

11.4

33. *Actuarial Method* Bill Jordan has a 48-month installment loan, with a fixed monthly payment of \$176.14. The amount borrowed was $7500. Instead of making his 24th payment, Bill is paying the remaining balance on the loan.

a) Determine the APR of the installment loan.

b) How much interest will Bill save, computed by the actuarial method?

c) What is the total amount due on that day?

34. *Rule of 78s* Carter Fenton is buying a book collection that costs $4000. He is making a down payment of $500 and 24 monthly payments of $163.33. Instead of making his 12th payment, Carter decides to pay the total remaining balance and terminate the loan.

a) How much interest will Carter save, computed by the rule of 78s?

b) What is the total amount due on that day?

35. *Installment Loan* Dara Holliday's cost for a new wardrobe was $3420. She made a down payment of $860 and financed the balance on a 24-month fixed payment installment loan. The monthly payments are $111.73. Instead of making her 12th payment, Dara decides to pay the total remaining balance and terminate the loan.

a) Determine the APR of the installment loan.

b) How much interest will Dara save, computed by the actuarial method?

c) What is the total amount due on that day?

36. *Finance Charge Comparison* On June 1, the billing date, Krishna Muhundan had a balance due of $485.75 on his credit card. The transactions during the month of June were

June 4	Payment	$375.00
June 8	Charge: Car repair	370.00
June 21	Charge: Airline ticket	175.80
June 28	Charge: Clothing	184.75

a) Find the finance charge on July 1 by using the unpaid balance method. Assume that the interest rate is 1.3% per month.

b) Find the new account balance on July 1 using the finance charge found in part (a).

c) Find the average daily balance for the period.

d) Find the finance charge on July 1 by using the average daily balance method. Assume that the interest rate is 1.3% per month.

e) Find the new account balance on July 1 using the finance charge found in part (d).

37. *Finance Charge Comparison* On August 5, the billing date, Pat Schaefer had a balance due of $185.72 on her credit card. The transactions during the month of August were

August 8	Charge: Shoes	$85.75
August 10	Payment	75.00
August 15	Charge: Dry cleaning	72.85
August 21	Charge: Textbooks	275.00

a) Find the finance charge on September 5 by using the unpaid balance method. Assume that the interest rate is 1.4% per month.

b) Find the new account balance on September 5 using the finance charge found in part (a).

c) Find the average daily balance for the period.

d) Find the finance charge on September 5 by using the average daily balance method. Assume that the interest rate is 1.4% per month.

e) Find the new account balance on September 5 using the finance charge found in part (d).

38. *Financing a Corvette* David Snodgress bought a new Chevrolet Corvette for $52,000. He made a 20% down payment and financed the balance with the dealer on a 48-month payment plan. The monthly payments were $930.02.

a) Determine the down payment.

b) Determine the amount to be financed.

c) Determine the total finance charge.

d) Determine the APR.

39. *Financing a Ski Outfit* Lucille Groenke can buy a cross-country skiing outfit for $275. The store is offering the following terms: $50 down and 12 monthly payments of $19.62.

a) Find the interest paid.

b) Find the APR.

11.5

40. *Building a House* The Freemans have decided to build a new house. The contractor quoted them a price of $135,700. The taxes on the house will be $3450 per year, and insurance will be $350 per year. They have applied for a conventional loan from a local bank. The bank is requiring a 25% down payment, and the interest rate on the loan

is 9.5%. The Freemans' annual income is $64,000. They have more than 10 monthly payments remaining on each of the following: $218 on a car, $120 on new furniture, and $190 on a camper. Determine
a) the required down payment.
b) their adjusted monthly income.
c) the maximum monthly payment the bank's loan officer believes they can afford.
d) the monthly payment of principal and interest for a 30-year loan.
e) their total monthly payment, including insurance and taxes.
f) Do the Freemans qualify for the mortgage?

41. *Thirty-Year Mortgage* James Whitehead purchased a home selling for $89,900 with a 15% down payment. The period of the mortgage is 30 years, and the interest rate is 11.5%. Determine the
a) amount of the down payment.
b) monthly mortgage payment.
c) amount of the first payment applied to the principal.
d) total cost of the house.
e) total interest paid.

42. *Adjustable-Rate Mortgage* The Nguyens purchased a house for $105,000 with a down payment of $26,250. They obtained a 30-year adjustable-rate mortgage. The terms of the mortgage are as follows: The interest rate is based on the 6-month Treasury bill, the interest rate charged is 3.00% above the rate of the Treasury bill on the date of adjustment, the interest rate is adjusted every 6 months, the interest rate will not change more than 1% (up or down) when the interest rate is adjusted, the maximum interest rate that can be charged for the duration of the loan is 16%, there is no lower limit on the interest rate, the initial mortgage interest rate is 7.5%, and the monthly payment of interest and principal is adjusted annually. Determine the
a) initial monthly payment for principal and interest.
b) interest rate in 6 months if the interest rate on the Treasury bill at the time is 5.00%.
c) interest rate in 6 months if the interest rate on the Treasury bill at the time is 4.75%.

CHAPTER 11 TEST

In Exercises 1 and 2, find the missing quantity by using the simple interest formula.

1. $i = ?$, $p = \$2000$, $r = 4\%$ per year, $t = 6$ months
2. $i = \$288$, $p = \$1200$, $r = 8\%$ per year, $t = ?$

In Exercises 3 and 4, Greg Wright borrowed $5000 from a bank for 18 months. The rate of simple interest charged is 8.5%.

3. How much interest did he pay for the use of the money?
4. What is the amount he repaid to the bank on the due date of the loan?

In Exercises 5 and 6, Yolanda Fernandez received a $5400 loan with interest at 12.5% for 90 days on August 1. Yolanda made a payment of $3000 on September 15.

5. How much did she owe the bank on the date of maturity?
6. What total amount of interest did she pay on the loan?

In Exercises 7 and 8, compute the amount and the compound interest.

	Principal	Time	Rate	Compounded
7.	$7500	2 years	3%	Quarterly
8.	$2500	3 years	6.5%	Monthly

A New Printer In Exercises 9–11, a new laser color printer sells for $2350. To finance the laser printer through a bank the bank will require a down payment of 15% and monthly payments of $90.79 for 24 months.

9. How much money will the purchaser borrow from the bank?

10. What finance charge will the individual pay the bank?

11. What is the APR?

12. *Rule of 78s* Sandi Abramowicz purchased a used fishing boat for $6750. She made a down payment of $1550 and financed the balance with a 12-month fixed-payment installment loan. Instead of making the sixth monthly payment of $465.85, she decides to pay off the loan.

a) How much interest will Sandi save (use the rule of 78s)?

b) What is the total amount due to pay off the loan?

13. *Actuarial Method* Gino Sedillo borrowed $7500. To repay the loan he was scheduled to make 36 monthly installment payments of $223.10. Instead of making his 24th payment, Gino decides to pay off the loan.

a) Determine the APR of the installment loan.

b) How much interest will Gino save (use the actuarial method)?

c) What is the total amount due to pay off the loan?

14. *Unpaid Balance Method* Michael Murphy's credit card statement shows a balance due of $878.25 on March 23, the billing date. For the period ending on April 23, he had the following transactions.

March 26	Charge: Groceries	$ 95.89
March 30	Charge: Restaurant bill	68.76
April 3	Payment	450.00
April 15	Charge: Clothing	90.52
April 22	Charge: Eyeglasses	450.85

a) Find the finance charge on March 23 by using the unpaid balance method. Assume that the interest rate is 1.4% per month.

b) Find the new account balance on April 23 using the finance charge found in part (a).

c) Find the average daily balance for the period.

d) Find the finance charge on March 23 by using the average daily balance method. Assume that the interest rate is 1.4% per month.

e) Find the new account balance on April 23 using the finance charge found in part (d).

Building a House In Exercises 15–21, the Leungs decided to build a new house. The contractor quoted them a price of $144,500, including the lot. The taxes on the house would be $3200 per year, and insurance would cost $450 per year. They have applied for a conventional loan from a bank. The bank is requiring a 15% down payment, and the interest rate is $10\frac{1}{2}$%. The Leung's annual income is $86,500. They have more than 10 monthly payments remaining on each of the following: $220 for a car, $175 for new furniture, and $210 on a college education loan.

15. What is the required down payment?

16. Determine their adjusted monthly income.

17. What is the maximum monthly payment the bank's loan officer believes the Leungs can afford?

18. Determine the monthly payments of principal and interest for a 30-year loan.

19. Determine their total monthly payments, including insurance and taxes.

20. Does the bank's loan officer believe that the Leungs meet the requirements for the mortgage?

21. a) Find the total cost of the house (excluding insurance and taxes) after 30 years.

b) How much of the total cost is interest?

GROUP PROJECTS

Mortgage Loan

1. The Young family is purchasing a $130,000 house with a VA mortgage. The bank is offering them a 25-year mortgage with an interest rate of 9.5%. They have $20,000 invested that could be used for a down payment. Since they do not need a down payment, Mr. Young wants to keep the money invested. Mrs. Young believes that they should make a down payment of $20,000.
   a) Determine the total cost of the house with no down payment.
   b) Determine the total cost of the house if they make a down payment of $20,000.
   c) Mr. Young believes that the $20,000 investment will have an annual rate of return of 10% compounded quarterly. Assuming that Mr. Young is right, calculate the value of the investment in 25 years. (See Section 11.3.)
   d) If the Youngs use the $20,000 as a down payment, their monthly payments will decrease. Determine the difference of the monthly payments in parts (a) and (b).
   e) Assume that the difference in monthly payments, part (d), is invested each month at a rate of 6% compounded monthly for 25 years. Determine the value of the investment in 25 years. (See Exercise 49 in Section 11.3.)
   f) Use the information from parts (a)–(e) to analyze the problem. Would you recommend that the Youngs make the down payment of $20,000 and invest the difference in their monthly payments as in part (e) or that they do not make the down payment and keep the $20,000 invested as in part (c)? Explain.

Credit Card Terms

2. With each credit card comes a credit agreement (or security agreement) the cardholder must sign. Select two members of your group who have a major credit card (MasterCard, Visa, Discover, or American Express). If possible, one of the credit cards should be a gold or platinum card. Obtain a copy of the credit agreement signed by each cardholder and answer questions (a)–(p).
   a) What are the cardholder's responsibilities?
   b) What is the cardholder's maximum line of credit?
   c) What restrictions apply to the use of the credit card?
   d) How many days after the billing date does the cardholder have to make a payment without being charged interest?
   e) What is the minimum monthly payment required, and how is it determined?
   f) What is the interest rate charged on purchases?
   g) How does the bank determine when to start charging interest on purchases?
   h) Is there an annual fee for the credit card? If so, what is it?
   i) What late charge applies if payments are not made on time?
   j) What information is given on the monthly statement?
   k) How is the finance charge computed?
   l) What other fees, if any, may the bank charge you?
   m) If the card is lost, what is the responsibility of the cardholder?
   n) If the card is lost, what is the liability of the cardholder?
   o) What are the advantages of a gold or platinum card over a regular card?
   p) In your opinion, which of the two cards is more desirable? Explain.

Retirement Plans

3. Working men and women often have the benefit of contributing to various retirement plans. Such plans include traditional and Roth IRAs, 403(b) plans, 401(k) plans, IRA–SEP plans, and Keogh plans. Research each of these plans and answer questions (a)–(d).
   a) What type of employees are eligible to use each type of plan?
   b) What is the maximum annual contribution that can be made to each plan?
   c) What type of investments can be made through each of these plans?
   d) What are the tax advantages of each type of plan?

The laws of probability have applications in the science of genetics. The larger and more diverse the population, the greater is the probability that the species will have the characteristics necessary to adapt to changes in the environment. The cheetah, the world's fastest-running mammal, faces extinction because it lacks the genetic diversity necessary to survive disease. Once found worldwide, the species now lives wild in only a few areas of Africa.

PROBABILITY

Millions of Americans each year play state lotteries or play bingo. If you play the lottery, your hope is to beat the odds and be the person with the winning numbers. Mathematicians of the sixteenth, seventeenth, and eighteenth centuries, not satisfied with leaving things to chance, invented the study of probability to use mathematics to determine the likelihood of an event (choosing the winning lottery numbers, for example) occurring.

Although the rules of probability were first applied to gaming, they have many other applications. The quality of the food you eat, the pedigree of your cat or dog, and the cost of your car insurance involve probability. In the business of insurance underwriting, when determining the likelihood of an event such as an automobile accident, the age, gender, and location of the driver are facts used in setting the cost of the insurance.

12.1 THE NATURE OF PROBABILITY

PROFILE IN MATHEMATICS

JACOB BERNOULLI (1654–1705)

Swiss mathematician Jacob Bernoulli was considered a pioneer of probability theory. He was part of a famous family of mathematicians and scientists. In *Ars Conjectand,* published posthumously in 1713, he proposed that an increased degree of accuracy can be obtained by increasing the number of trials of an experiment. This theorem, called Bernoulli's theorem (of probability), is also known as the law of large numbers. Bernoulli's theorem of fluid dynamics, used in aircraft wing design, was developed by Daniel Bernoulli (1700–1782), Jacob's second son.

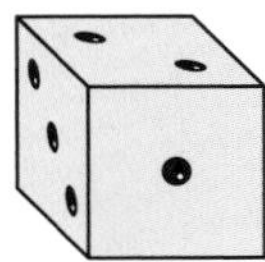

A die (one of a pair of dice) contains six surfaces, called faces. Each face contains a unique number of dots, from 1 to 6. The sum of the dots on opposite surfaces is 7.

History

Probability is used in many areas, including public finance, medicine, insurance, elections, manufacturing, educational tests and measurements, genetics, weather forecasting, investments, opinion polls, the natural sciences, and games of chance. The study of probability originated from the study of games of chance. Archaeologists have found artifacts used in games of chance in Egypt dating from about 3000 B.C.

Mathematical problems relating to games of chance were studied by a number of mathematicians of the Renaissance. Italy's Girolamo Cardano (1501–1576) in his *Liber de Ludo Aleae* (book on the games of chance) presents one of the first systematic computations of probabilities. Although it is basically a gambler's manual, many consider it the first book ever written on probability. A short time later, two French mathematicians, Blaise Pascal (1623–1662) and Pierre de Fermat (1601–1665), worked together studying "the geometry of the die." In 1657, Dutch mathematician Christian Huygens (1629–1695) published *De Ratiociniis in Luno Aleae* (on ratiocination in dice games), which contained the first documented reference to the concept of mathematical expectation (see Section 12.4). Swiss mathematician Jacob Bernoulli (1654–1705), whom many consider the founder of probability theory, is said to have fused pure mathematics with the empirical methods used in statistical experiments. The works of Pierre-Simon de Laplace (1749–1827) dominated probability throughout the nineteenth century.

The Nature of Probability

Before we discuss the meaning of the word *probability* and learn how to calculate probabilities, we must introduce a few definitions.

An **experiment** is a controlled operation that yields a set of results.

The process by which medical researchers administer experimental drugs to patients to determine their reaction is one type of experiment.

The possible results of an experiment are called its **outcomes.**

For example, the possible outcomes from administering an experimental drug may be a favorable reaction, no reaction, or an adverse reaction.

An **event** is a subcollection of the outcomes of an experiment.

For example, when a die is rolled, the event of rolling a number greater than 2 can be satisfied by any one of four outcomes: 3, 4, 5, or 6. The event of rolling a 5 can be satisfied by only one outcome, the 5 itself. The event of rolling an even number can be satisfied by any of three outcomes: 2, 4, or 6.

Probability is classified as either *empirical* (experimental) or *theoretical* (mathematical). *Empirical probability* is the relative frequency of occurrence of an event and is determined by actual observations of an experiment. *Theoretical probability* is determined through a study of the possible *outcomes* that can occur for the given experiment. We will indicate the probability of an event E by $P(E)$, which is read "P of E."

In this section, we will briefly discuss empirical probability. The emphasis in the remaining sections is on theoretical probability. Following is the formula for computing empirical probability, or relative frequency.

Empirical Probability (Relative Frequency)

$$P(E) = \frac{\text{number of times event } E \text{ has occurred}}{\text{total number of times the experiment has been performed}}$$

The probability of an event, whether empirical or theoretical, is always a number between 0 and 1, inclusive, and may be expressed as a decimal number or a fraction. An empirical probability of 0 indicates that the event has never occurred. An empirical probability of 1 indicates that the event has always occurred.

EXAMPLE 1 *Heads Up!*

In 100 tosses of a fair coin, 44 landed heads up. Find the empirical probability of the coin landing heads up.

SOLUTION: Let E be the event that the coin lands heads up. Then

$$P(E) = \frac{44}{100} = 0.44$$

▲

EXAMPLE 2 *Weight Reduction*

A pharmaceutical company is testing a new drug that is supposed to help with weight reduction. The drug is given to 500 individuals with the following outcomes.

Weight reduced	Weight unchanged	Weight increased
279	92	129

If this drug is given to an individual, find the empirical probability that the person's weight is (a) reduced, (b) unchanged, (c) increased.

DID YOU KNOW

One Blue Eye, One Brown Eye

Seeing a dog or a person with different-colored eyes is unusual, but not that unusual. Our features, such as eye color, are determined genetically, as are certain illnesses. Below is a chart listing some genetic disorders and their rates of occurrence. Genetics, which has its foundation in probability, is used in explaining the cause of these illnesses, and it plays an important role in research to find cures for genetic disorders.

Disorder/Incidence

Cystic fibrosis:
1 out of 2500 Caucasian births

Down syndrome:
1 out of 800–1000 births

Duchenne muscular dystrophy:
1 out of 3300 male births

Fragile X syndrome:
1 out of 1500 male births
1 out of 2500 female births

Hemophilia A:
1 out of 8500 male births

Polycystic kidney disease:
1 out of 3000 births

Prader–Willi syndrome:
1 out of 12,000 births

Sickle-cell anemia:
1 out of 400–600 African-American births

Tay-Sachs disease:
1 out of 3600 Ashkenazi Jewish births

SOLUTION:

a) Let E be the event that the weight is reduced.

$$P(E) = \frac{279}{500} = 0.558$$

b) Let E be the event that the weight is unchanged.

$$P(E) = \frac{92}{500} = 0.184$$

c) Let E be the event that the weight is increased.

$$P(E) = \frac{129}{500} = 0.258$$

Empirical probability is used when probabilities cannot be theoretically calculated. For example, life insurance companies use empirical probabilities to determine the chance of an individual in a certain profession, with certain risk factors, living to age 65.

Empirical Probability in Genetics

Using empirical probability, Gregor Mendel (1822–1884) developed the laws of heredity by crossbreeding different types of "pure" pea plants and observing the relative frequencies of the resulting offspring. These laws became the foundation for the study of genetics. For example, when he crossbred a pure yellow pea plant and a pure green pea plant, the resulting offspring (the first generation) were always yellow; see Fig. 12.1(a). When he crossbred a pure round-seeded pea plant and a pure wrinkled-seeded pea plant, the resulting offspring (the first generation) were always round; see Fig. 12.1(b).

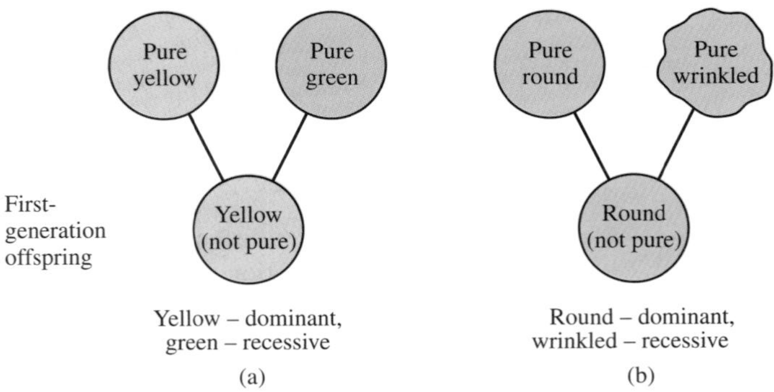

Figure 12.1

Mendel called traits such as yellow color and round seeds *dominant* because they overcame or "dominated" the other trait. He labeled the green and wrinkled traits *recessive*.

Mendel then crossbred the offspring of the first generation. The resulting second generation offspring had both the dominant and the recessive traits of their grandparents; see Fig. 12.2(a) and (b). What's more, these traits always appeared in approximately a 3 to 1 ratio of dominant to recessive.

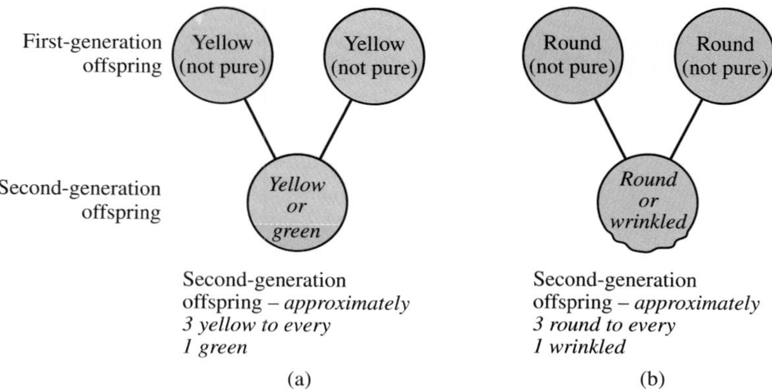

Figure 12.2

Table 12.1 lists some of the actual results of Mendel's experiments with pea plants. Note that the ratio of dominant trait to recessive trait in the second-generation offspring is about 3 to 1 for each of the experiments. The empirical probability of the dominant trait has also been calculated. How would you find the empirical probability of the recessive trait?

TABLE 12.1 Second-Generation Offspring

Dominant Trait	Number with Dominant Trait	Recessive Trait	Number with Recessive Trait	Ratio of Dominant to Recessive	*P* (Dominant Trait)
Yellow seeds	6022	Green seeds	2001	3.01 to 1	$\frac{6022}{8023} \approx 0.75$
Round seeds	5474	Wrinkled seeds	1850	2.96 to 1	$\frac{5474}{7324} \approx 0.75$

From his work Mendel concluded that the sex cells (now called gametes) of the pure yellow (dominant) pea plant carried some factor that caused the offspring to be yellow, and the gametes of the green variety had a variant factor that "induced the development of green plants." In 1909, Danish geneticist W. Johannsen called these factors "genes." Mendel's work led to the understanding that each pea plant contains two genes for color, one that comes from the mother and the other from the father. If the two genes are alike—for instance, both for yellow plants or both for green plants—the plant will be that color. If the genes for color are different, the plant will grow the color of the dominant gene. Thus, if one parent contributes a gene for the plant to be yellow (dominant) and the other parent contributes a gene for the plant to be green (recessive), the plant will be yellow.

DID YOU KNOW

The Royal Disease

The effect of genetic inheritance is dramatically demonstrated by the occurrence of hemophilia in some of the royal families of Europe. Great Britain's Queen Victoria (1819–1901) was the initial carrier. The disease was subsequently introduced into the royal lines of Prussia, Russia, and Spain through the marriages of her children.

Hemophilia is a disease that keeps blood from clotting. As a result, even a minor bruise or cut can be dangerous. It is also a recessive sex-linked disease. Females have a second gene that enables the blood to clot, which blocks the effects of the recessive carrier gene; males do not. So even though both males and females are carriers, the disease afflicts only males.

Queen Victoria had 9 children, 26 grandchildren, and 34 great-grandchildren. Among them 1 son and 9 grandsons were hemophiliacs, and 2 daughters and 4 granddaughters were carriers of the gene for hemophilia. The genetic line of the present-day British royal family is free of the disease.

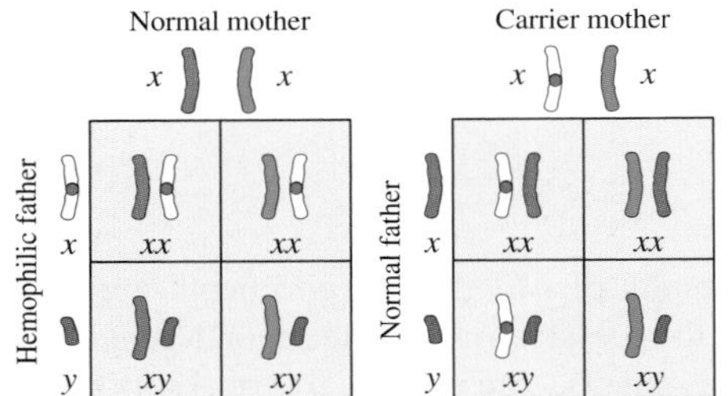

In humans, genes are located on 23 pairs of *chromosomes*. Each parent contributes one member of each pair to a child. The gene that affects blood clotting is carried on the x chromosome. Females have two x chromosomes; males have one x chromosome and one y chromosome. Thus, on the diagram on the left, xx represents a female and xy represents a male. A female with the defective gene, symbolized by [chromosome symbol], will be a carrier, whereas a male with the defective gene will have hemophilia.

The Law of Large Numbers

"The laws of probability, so true in general, so fallacious in particular."
Edward Gibbon, 1796

Most of us accept that if a "fair coin" is tossed many, many times, it will land heads up approximately half of the time. Intuitively, we can guess that the probability that a fair coin will land heads up is $\frac{1}{2}$. Does this mean that if a coin is tossed twice, it will land heads up exactly once? If a fair coin is tossed 10 times, will there necessarily be five heads? The answer is clearly no. What then does it mean when we state that the probability that a fair coin will land heads up is $\frac{1}{2}$? To answer this question, let's examine Table 12.2, which shows what may occur when a fair coin is tossed a given number of times.

TABLE 12.2

Number of Tosses	Expected Number of Heads	Actual Number of Heads Observed	Relative Frequency of Heads
10	5	4	$\frac{4}{10} = 0.4$
100	50	45	$\frac{45}{100} = 0.45$
1000	500	546	$\frac{546}{1,000} = 0.546$
10,000	5000	4852	$\frac{4852}{10,000} = 0.4852$
100,000	50,000	49,770	$\frac{49,770}{100,000} = 0.49770$

The last column of Table 12.2, the relative frequency of heads, is a ratio of the number of heads observed to the total number of tosses of the coin. The relative frequency is the empirical probability, as defined earlier. Note as the number of tosses increases, the relative frequency of heads gets closer and closer to $\frac{1}{2}$, or 0.5, which is what we expect.

The nature of probability is summarized by the law of large numbers.

> The **law of large numbers** states that probability statements apply in practice to a large number of trials, not to a single trial. It is the relative frequency over the long run that is accurately predictable, not individual events or precise totals.

What does it mean to say that the probability of rolling a 2 on a die is $\frac{1}{6}$? It means that over the long run, on the average, one of every six rolls will result in a 2.

DID YOU KNOW

Batting Averages

If Barry Bonds of the San Francisco Giants gets three hits in his first three at bats of the season, he is batting a thousand (1.000). But over the course of the 162 games of the season (with three or four at bats per game), however, his batting average will fall closer to .370 (his 2002 major league leading batting average). In 2002, out of 403 at bats Bonds had 149 hits, an average above all other players' but much less than 1.000. His batting average is a relative frequency (or empirical probability) of hits to at bats. It is only the long-term average that we take seriously because it is based on the law of large numbers.

SECTION 12.1 EXERCISES

Concept/Writing Exercises

1. What is an experiment?
2. **a)** What are outcomes of an experiment?
   b) What is an event?
3. What is empirical probability and how is empirical probability determined?
4. What are theoretical probabilities based on?
5. Explain in your own words the law of large numbers.
6. Explain in your own words why empirical probabilities are used in determining premiums for life insurance policies.
7. The theoretical probability of a coin landing heads up is $\frac{1}{2}$. Does this probability mean that if a coin is flipped two times, one flip will land heads up? If not, what does it mean?

8. The theoretical probability of rolling a 4 on a die is $\frac{1}{6}$. Does this probability mean that, if a die is rolled six times, one 4 will appear? If not, what does it mean?

9. To determine premiums, life insurance companies must compute the probable date of death. On the basis of a great deal of research Mr. Duncan, age 36, is expected to live another 43.21 years. Does this determination mean that Mr. Duncan will live until he is 79.21 years old? If not, what does it mean?

10. **a)** Explain how you would find the empirical probability of rolling a 5 on a die.
b) What do you believe is the empirical probability of rolling a 5?
c) Determine the empirical probability of rolling a 5 by rolling a die 40 times.

Practice the Skills

11. *Flip a Coin* Flip a coin 50 times and record the results. Determine the empirical probability of tossing
a) a head.
b) a tail.
c) Does the probability of tossing a head appear to be the same as tossing a tail?

12. *Roll a Die* Roll a die 50 times and record the results. Determine the empirical probability of rolling
a) a 1.
b) a 6.
c) Does the probability of rolling a 1 appear to be the same as the probability of rolling a 6? Explain.

13. *Pair of Dice* Roll a pair of dice 60 times and record the sums. Determine the empirical probability of rolling a sum of
a) 2.
b) 7.
c) Does the probability of rolling a sum of 2 appear to be the same as the probability of rolling a sum of 7?

14. *Two Coins* Toss two coins 50 times and record the number of times exactly one head was obtained. Determine the empirical probability of tossing exactly one head.

Problem Solving

15. *Birds at a Feeder* The last 30 birds that fed at the Haines' bird feeder were 14 finches, 10 cardinals, and 6 blue jays. Use this information to determine the empirical probability that the next bird to feed from the feeder is
a) a finch.
b) a cardinal.
c) a blue jay.

See Exercise 15

16. *Music Purchases* At the Virgin Music store in Times Square, 60 people entering the store were selected at random and were asked to state their favorite type of music. Of the 60, 24 selected rock, 16 selected country, 8 selected classical, and 12 said something other than rock, country, or classical. Determine the empirical probability that the next person entering the store favors
a) rock music.
b) country music.
c) something other than rock, country, or classical music.

17. *Veterinarian* In a given week, a veterinarian treated the following animals.

Animal	Number Treated
Dog	40
Cat	35
Bird	15
Iguana	5

Determine the empirical probability that the next animal she treats is
a) a dog.
b) a cat.
c) an iguana.

18. *Prader–Willi Syndrome* In a sample of 50,000 first-born babies, 5 were found to have Prader–Willi syndrome. Find the empirical probability that a family's first child will be born with this syndrome.

19. *Favorite Fruit* At the produce department of a grocery store, 900 people were asked to name their favorite fruit. The following graph indicates their response.

Favorite Fruit

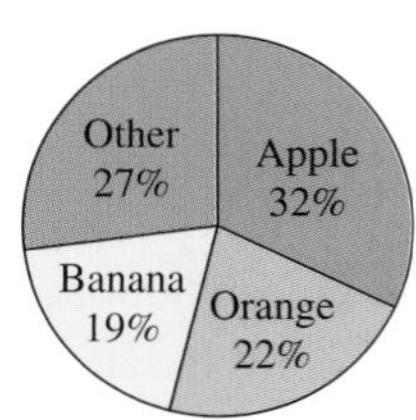

a) Explain why this graph illustrates empirical probabilities.

If one person is selected at random from the 900 people sampled, determine the empirical probability that the person's favorite fruit is

b) an apple.
c) an orange.
d) a banana.

20. *Tax Returns* In 2001 about 40,244 million tax returns were filed electronically out of a total of about 131,100 million tax returns that were filed. Determine the empirical probability that a person selected at random who filed a tax return in 2001 filed it

a) electronically.
b) nonelectronically.

21. *Dow Jones Gains* The following table shows the gain made by the Dow Jones Industrial Average (DJIA) in each year that ends in a 5 since records have been kept.

Years Ending in 5

Year	DJIA Return
1885	27.7%
1895	2.3%
1905	38.2%
1915	86.5%
1925	30.0%
1935	38.5%
1945	26.6%
1955	20.8%
1965	10.9%
1975	38.3%
1985	27.7%
1995	36.8%
2005	?

a) What is the empirical probability that the DJIA will increase in a year ending in 5?
b) Is it possible that the DJIA could have a loss in the year 2005? Explain.

22. *Grade Distribution* Mr. Doole's grade distribution over the past 3 years for a course in college algebra is shown in the chart below.

Grade	Number
A	43
B	182
C	260
D	90
F	62
I	8

If Sue Gilligan plans to take college algebra with Mr. Doole, determine the empirical probability she receives a grade of

a) A.
b) C.
c) D or higher.

23. *Election* In an election for student council president at Russell Sage College, a sample of 80 students were polled and asked for whom they planned to vote. The table shows the results of the poll.

Candidate	Votes
Allison	22
Emily	18
Kimberly	20
Johanna	14
Other	6

If one student from the sample was selected at random, determine the empirical probability the person planned to vote for

a) Allison.
b) Emily.
c) Kimberly.
d) Johanna.
e) Someone other than the four people listed above.

24. *Housing Prices* The following illustration shows how house prices have changed in the United States, by state, for the period March 1, 1997, through March 1, 2002.*

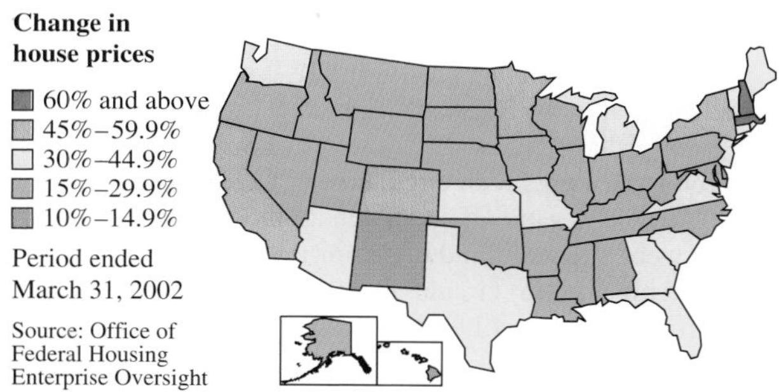

If one state is selected at random, determine the empirical probability that the state's average house price increased by

a) 60% and above.
b) 45%–59.9%.
c) 30%–44.9%.
d) 15%–29.9%.
e) 10%–14.9%.

25. *Hitting a Bull's-Eye* The pattern of hits shown on the target resulted from a marksman firing 20 rounds. For a single shot

a) find the empirical probability that the marksman hits the 50-point bull's-eye (the center of the target).

*For information purposes, the District of Columbia's change in housing prices was 60% and above (do not consider this information in answering the exercises).

b) find the empirical probability that the marksman does not hit the bull's-eye.
c) find the empirical probability that the marksman scores at least 20 points.
d) find the empirical probability that the marksman does not score any points (the area outside the large circle).

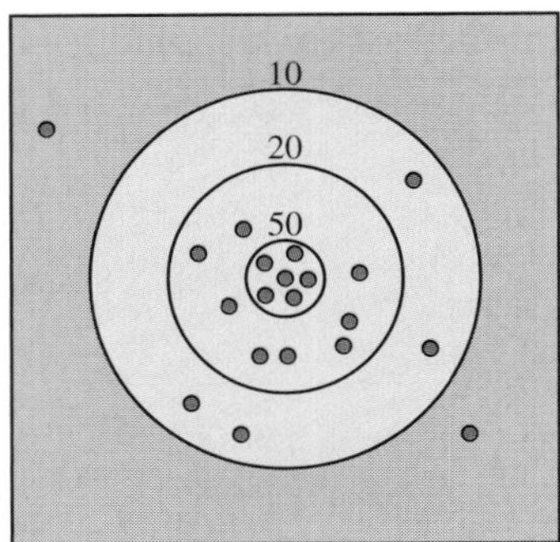

26. *Rock Toss* Jim Handy finds an irregularly shaped five-sided rock. He labels each side and tosses the rock 100 times. The results of his tosses are shown in the table. Find the empirical probability that the rock will land on side 4 if tossed again.

Side	1	2	3	4	5
Frequency	32	18	15	13	22

27. *Cell Biology Experiment* An experimental serum was injected into 500 guinea pigs. Initially, 150 of the guinea pigs had circular cells, 250 had elliptical cells, and 100 had irregularly shaped cells. After the serum was injected, none of the guinea pigs with circular cells were affected, 50 with elliptical cells were affected, and all those with irregular cells were affected. Find the empirical probability that a guinea pig with (a) circular cells, (b) elliptical cells, and (c) irregular cells will be affected by injection of the serum.

28. *Baby Gender* In the United States, more male babies are born than female. In 2000, 2,076,969 males were born and 1,981,845 females were born. Find the empirical probability of an individual being born
a) male.
b) female.

29. *Mendel's Experiment* In one of Mendel's experiments (see pages 656–657), he crossbred nonpure purple flower pea plants. These purple pea plants had two traits for flowers, purple (dominant) and white (recessive). The result of this crossbreeding was 705 second-generation plants with purple flowers and 224 second-generation plants with white flowers. Find the empirical probability of a second-generation plant having
a) white flowers.
b) purple flowers.

30. *Second-Generation Offspring* In another experiment, Mendel crossbred nonpure tall pea plants. As a result, the second-generation offspring were 787 tall plants and 277 short plants. Find the empirical probability of a second-generation plant being
a) tall.
b) short.

Challenge Problems/Group Activities

31. **a)** *Design an Experiment* Which do you believe is used more frequently in a book or newspaper article, the word *a* or the word *the*?
b) Design an experiment to determine the empirical probabilities (or relative frequencies) of the words *a* and *the* appearing in a book or newspaper article.
c) Perform the experiment in part (b) and determine the empirical probabilities.
d) Which word appears to occur more frequently?

Recreational Mathematics

32. *Cola Preference* Can people selected at random distinguish Coke from Pepsi? Which do they prefer?
a) Design an experiment to determine the empirical probability that a person selected at random can select Coke when given samples of both Coke and Pepsi.
b) Perform the experiment in part (a) and determine the empirical probability.
c) Determine the empirical probability that a person selected at random will prefer Coke over Pepsi.

Internet/Research Activities

33. Write a paper on how insurance companies use empirical probabilities in determining insurance premiums. An insurance agent may be able to direct you to a source of information.

34. Write a paper on how Gregor Mendel's use of empirical probability led to the development of the science of genetics. You may want to check with a biology professor to determine references to use.

12.2 THEORETICAL PROBABILITY

Should you spend the 37 cents for a stamp to return a sweepstakes ticket? What are your chances of winning a lottery? If you go to a carnival, bazaar, or casino, which games provide the greatest chance of winning? These and similar questions can be answered once you have an understanding of theoretical probability. *In the remainder of this chapter, the word probability will refer to theoretical probability.*

TIMELY TIP To be able to do the problems in this section and the remainder of the chapter you must have a thorough understanding of fractions. If you have forgotten how to work with fractions, we strongly suggest that you review Section 5.3 before beginning this section.

Recall from Section 12.1 that the results of an experiment are called outcomes. When you roll a die and observe the number of points that face up, the possible outcomes are 1, 2, 3, 4, 5, and 6. It is equally likely that you will roll any one of the possible numbers.

If each outcome of an experiment has the same chance of occurring as any other outcome, they are said to be **equally likely outcomes.**

Can you think of a second set of equally likely outcomes when a die is rolled? An odd number is as likely to be rolled as an even number. Therefore, odd and even are another set of equally likely outcomes.

If an event E has *equally likely outcomes,* the probability of event E, symbolized by $P(E)$, may be calculated with the following formula.

Probability

$$P(E) = \frac{\text{number of outcomes favorable to } E}{\text{total number of possible outcomes}}$$

Example 1 illustrates how to use this formula.

EXAMPLE 1 *Finding Probabilities*

A die is rolled. Find the probability of rolling

a) a 2. b) an even number. c) a number greater than 4.

d) a 7. e) a number less than 7.

SOLUTION:

a) There are six possible equally likely outcomes: 1, 2, 3, 4, 5, and 6. The event of rolling a 2 can occur in only one way.

$$P(2) = \frac{\text{number of outcomes that will result in a 2}}{\text{total number of possible outcomes}} = \frac{1}{6}$$

b) The event of rolling an even number can occur in three ways: 2, 4, or 6.

$$P(\text{even number}) = \frac{\text{number of outcomes that result in an even number}}{\text{total number of possible outcomes}} = \frac{3}{6} = \frac{1}{2}$$

c) Two numbers are greater than 4, namely, 5 and 6.

$$P(\text{number greater than 4}) = \frac{2}{6} = \frac{1}{3}$$

d) No outcomes will result in a 7. Thus, the event cannot occur and the probability is 0.

$$P(7) = \frac{0}{6} = 0$$

e) All the outcomes 1 through 6 are less than 7. Thus, the event must occur and the probability is 1.

$$P(\text{number less than 7}) = \frac{6}{6} = 1$$

Four important facts about probability follow.

Important Facts

1. The probability of an event that cannot occur is 0.
2. The probability of an event that must occur is 1.
3. Every probability is a number between 0 and 1 inclusive; that is, $0 \le P(E) \le 1$.
4. The sum of the probabilities of all possible outcomes of an experiment is 1.

Northern Cardinal

EXAMPLE 2 *Choosing One Bird from a List*

The names of 15 birds and their food preferences are listed in Table 12.3 on page 665. Each of the 15 birds' names is listed on a slip of paper, and the 15 slips are placed in a bag. One slip is to be selected at random from the bag. Find the probability that the slip contains the name of

a) a sparrow (any type listed).

b) a bird that has a high attractiveness to peanut kernels.

c) a bird that has a low attractiveness to peanut kernels, *and* a low attractiveness to cracked corn, *and* a high attractiveness to black striped sunflower seeds.

d) a bird that has a high attractiveness to either peanut kernels *or* cracked corn (or both).

TABLE 12.3 Birds and Their Food Preferences

Bird	Peanut Kernels	Cracked Corn	Black Striped Sunflower Seeds
American goldfinch	L	L	H
Blue jay	H	M	H
Chickadee	M	L	H
Common grackle	M	H	H
Evening grosbeak	L	L	H
House finch	M	L	H
House sparrow	L	M	M
Mourning dove	L	M	M
Northern cardinal	L	L	H
Purple finch	L	L	H
Scrub jay	H	L	H
Song sparrow	L	L	M
Tufted titmouse	H	L	H
White-crowned sparrow	H	M	H
White-throated sparrow	H	H	H

Note: H = high attractiveness; M = medium attractiveness; L = low attractiveness.

Source: How to Attract Birds (Ortho Books)

SOLUTION:

a) Four of the 15 birds listed are sparrows (house sparrow, song sparrow, white-crowned sparrow, and white-throated sparrow).

$$P(\text{sparrow}) = \frac{4}{15}$$

b) Five of the 15 birds listed have a high attractiveness to peanut kernels (blue jay, scrub jay, tufted titmouse, white-crowned sparrow, and white-throated sparrow).

$$P(\text{high attractiveness to peanut kernels}) = \frac{5}{15} = \frac{1}{3}$$

c) Reading across the rows reveals that 4 birds have a low attractiveness to peanut kernels and cracked corn and a high attractiveness to black striped sunflower seeds (American goldfinch, evening grosbeak, northern cardinal, and purple finch).

$$P\begin{pmatrix}\text{low attractiveness to peanuts, and low} \\ \text{to corn, and high to black sunflower seeds}\end{pmatrix} = \frac{4}{15}$$

d) Six birds have a high attractiveness to either peanut kernels or to cracked corn (or both). They are the blue jay, common grackle, scrub jay, tufted titmouse, white-crowned sparrow, and white-throated sparrow.

$$P(\text{high attractiveness to peanut kernels or cracked corn}) = \frac{6}{15} = \frac{2}{5}$$ ▲

In any experiment, an event must either occur or not occur. *The sum of the probability that an event will occur and the probability that it will not occur is 1.* Thus, for any event A we conclude that

$$P(A) + P(\text{not } A) = 1$$

or

$$P(\text{not } A) = 1 - P(A)$$

For example, if the probability that event A will occur is $\frac{5}{12}$, the probability that event A will not occur is $1 - \frac{5}{12}$, or $\frac{7}{12}$. Similarly, if the probability that event A will not occur is 0.3, the probability that event A will occur is $1 - 0.3 = 0.7$ or $\frac{7}{10}$. We make use of this concept in Example 3.

EXAMPLE 3 *Selecting One Card from a Deck*

A standard deck of 52 playing cards is shown in Figure 12.3. The deck consists of four suits: hearts, clubs, diamonds, and spades. Each suit has 13 cards, including numbered cards ace (1) through 10 and three picture (or face) cards, the jack, the queen, and the king. Hearts and diamonds are red cards; clubs and spades are black cards. There are 12 picture cards, consisting of 4 jacks, 4 queens, and 4 kings. One card is to be selected at random from the deck of cards. Find the probability that the card selected is

a) a 9.
b) not a 9.
c) a diamond.
d) a jack *or* queen *or* king (a picture card).
e) a heart *and* a club.
f) a card greater than 6 *and* less than 9.

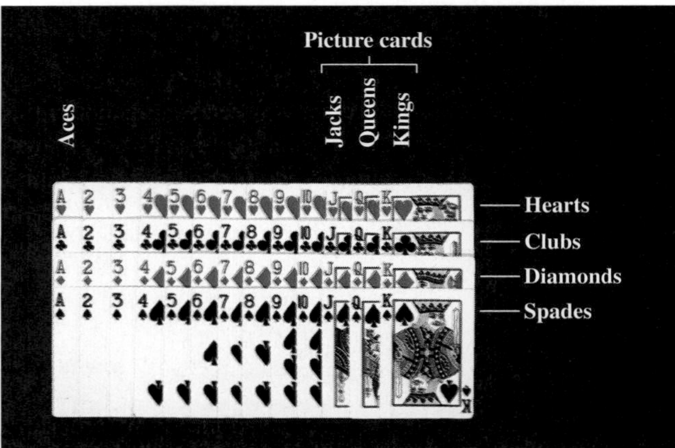

Figure 12.3

SOLUTION:

a) There are four 9's in a deck of 52 cards.

$$P(9) = \frac{4}{52} = \frac{1}{13}$$

b) $P(\text{not a } 9) = 1 - P(9) = 1 - \frac{1}{13} = \frac{12}{13}$

This probability could also have been found by noting that there are 48 cards that are not 9's in a deck of 52 cards.

$$P(\text{not a } 9) = \frac{48}{52} = \frac{12}{13}$$

c) There are 13 diamonds in the deck.

$$P(\text{diamond}) = \frac{13}{52} = \frac{1}{4}$$

d) There are 4 jacks, 4 queens, and 4 kings, or a total of 12 picture cards.

$$P(\text{jack } or \text{ queen } or \text{ king}) = \frac{12}{52} = \frac{3}{13}$$

e) The word *and* means that *both* events must occur. Since it is not possible to select one card that is both a heart and a club, the probability is 0.

$$P(\text{heart and club}) = \frac{0}{52} = 0$$

f) The cards that are both greater than 6 and less than 9 are 7's and 8's. There are four 7's and four 8's, or a total of eight cards.

$$P(\text{greater than 6 } and \text{ less than 9}) = \frac{8}{52} = \frac{2}{13}$$

SECTION 12.2 EXERCISES

Concept/Writing Exercises

1. What are equally likely outcomes?
2. Explain in your own words how to find the theoretical probability of an event.
3. State the relationship that exists for $P(A)$ and $P(\text{not } A)$.
4. If the probability that an event occurs is $\frac{4}{9}$, determine the probability that the event does not occur.
5. If the probability that an event occurs is 0.3, determine the probability that the event does not occur.
6. If the probability that an event does not occur is 0.25, determine the probability that the event occurs.
7. If the probability that an event does not occur is $\frac{5}{12}$, determine the probability that the event occurs.
8. How many of each of the following are there in a standard deck of cards?
   a) Total cards
   b) Hearts
   c) Red cards
   d) Fives
   e) Black cards
   f) Picture cards
   g) Aces
   h) Queens
9. Using the definition of probability, explain in your own words why the probability of an event that cannot occur is 0.
10. Using the definition of probability, explain in your own words why the probability of an event that must occur is 1.
11. Between what two numbers (inclusively) will all probabilities lie?
12. What is the sum of all the probabilities of all possible outcomes of an experiment?

Practice the Skills

13. *Multiple-Choice Test* A multiple-choice test has five possible answers for each question.
    a) If you guess at an answer, what is the probability that you select the correct answer for one particular question?

b) If you eliminate one of the five possible answers and guess from the remaining possibilities, what is the probability that you select the correct answer to that question?

14. *Remote Control* A TV remote has keys for channels 0 through 9. If you select one key at random,
   a) what is the probability that you press channel 6?
   b) what is the probability that you press a key for an even number?
   c) what is the probability that you press a key for a number less than 7?

15. *Raffle* In a raffle where one number is chosen, determine the probability you would win if you have a choice of 50 numbers to choose from. Explain your answer.

16. *Raffle* In a raffle where one number is chosen, determine the probability you would win if you have a choice of 46 numbers to choose from. Explain your answer.

Select a Card In Exercises 17–26, one card is selected at random from a deck of cards. Find the probability that the card selected is

17. a 7.
18. a 7 or a 9.
19. not a 7.
20. the five of diamonds.
21. a black card.
22. a heart.
23. a red card or a black card.
24. a red card and a black card.
25. a card greater than 4 and less than 9.
26. a jack and a heart.

Spin the Spinner In Exercises 27–30, assume that the spinner cannot land on a line. Find the probability that the spinner lands on (a) red, (b) green, (c) yellow, (d) blue.

27. 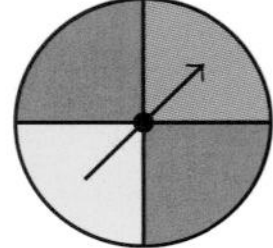

28. 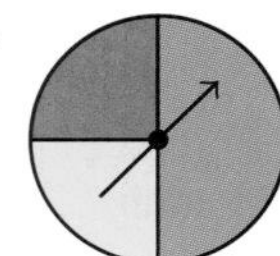

29. 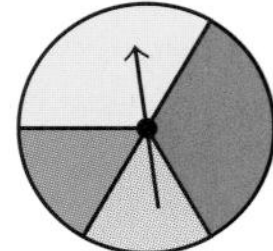

30. 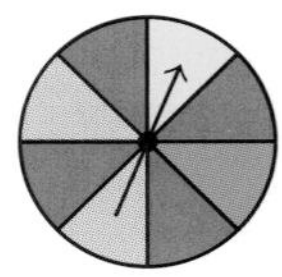

Picnic In Exercises 31–34, a bin at a picnic contains 100 cans of soda covered by ice. There are 30 cans of cola, 40 cans of orange soda, 10 cans of ginger ale, and 20 cans of root beer. The cans are all the same size and shape. If one can is selected at random from the bin, find the probability that the soda selected is

31. orange soda.
32. cola or orange soda.
33. cola, root beer, or orange soda.
34. ginger ale.

Wheel of Fortune In Exercises 35–38, use the small replica of the Wheel of Fortune.

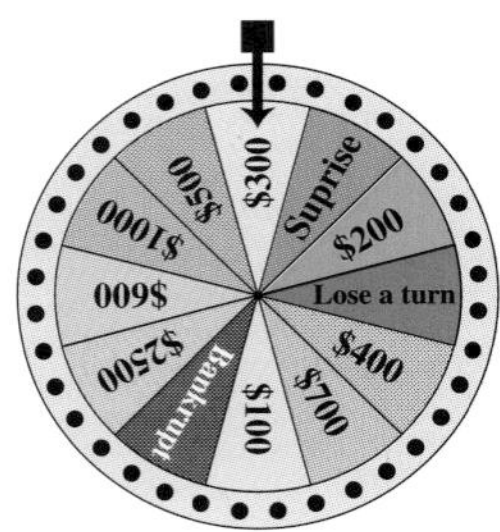

If the wheel is spun at random, find the probability of the sector indicated stopping under the pointer.

35. \$600
36. A number greater than \$400
37. Lose a turn or Bankrupt
38. \$2500 or Surprise

Tennis Balls In Exercises 39–42, 50 tennis balls including 23 Wilson, 17 Penn, and 10 other brand-name balls are on a tennis court. Barry Wood closes his eyes and arbitrarily picks up a ball from the court. Determine the probability the ball selected is

39. a Wilson.
40. a Penn.
41. not a Penn.
42. a Wilson or a Penn.

Traffic Light In Exercises 43–46, a traffic light is red for 25 sec, yellow for 5 sec, and green for 55 sec. What is the probability that when you reach the light,

43. the light is green.

44. the light is yellow.

45. the light is not red.

46. the light is not green.

Mississippi In Exercises 47–52, each individual letter of the word Mississippi is placed on a piece of paper, and all 11 pieces of paper are placed in a hat. If one letter is selected at random from the hat, find the probability that

47. the letter *s* is selected.

48. the letter *s* is not selected.

49. a vowel is selected.

50. the letter *i* or *p* is selected.

51. the letter *v* is not selected.

52. the letter *w* is selected.

Manatees In Exercises 53–56, use the following chart, which shows information about manatee deaths from 1991 through 2001.

Year	Boat Deaths	Totals Deaths*
1991	53	174
1992	38	163
1993	35	145
1994	49	193
1995	42	201
1996	60	415
1997	54	242
1998	66	231
1999	82	268
2000	78	273
2001	81	325

* Manatees are an endangered species. A major cause of manatee deaths is related to boating accidents. The year 2001 was the second worst year ever recorded for manatee deaths.

Source: St. Petersburg Times, January 5, 2002

If one year from 1991 up to and including 2001 is selected at random, determine the probability that in the year selected

53. exactly 60 manatee deaths that year were caused by boating accidents.

54. the total number of manatee deaths exceeded 250.

55. the number of manatee deaths by boating accidents exceeded 50 and the number of total manatee deaths exceeded 250.

56. the number of manatee deaths by boating accidents was less than or equal to 40 but the total number of manatee deaths was greater than or equal to 163.

Dart Board In Exercises 57–60, a dart is thrown randomly and sticks on the circular dart board with 26 partitions, as shown.

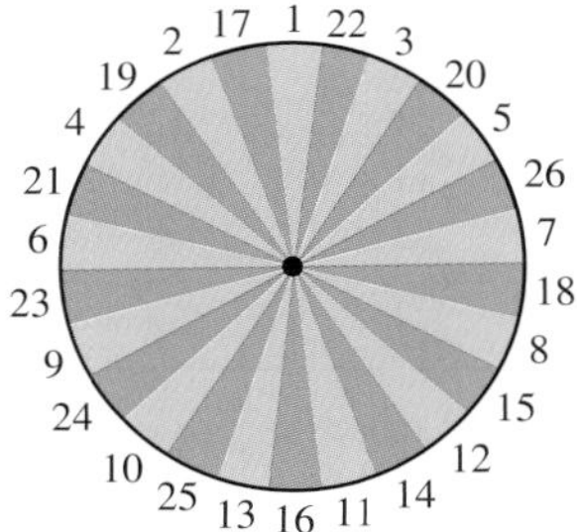

Assuming that the dart cannot land on the black area or on a border between colors, find the probability that the dart lands on

57. the area marked 15.

58. an orange area.

59. an area marked with a number greater than or equal to 22.

60. an area marked with a number greater than 6 and less than or equal to 9.

Car Manufacturer In Exercises 61–66, refer to the following table, which shows the results of a survey regarding the manufacturer of the cars driven by the people who were interviewed. In the table, GM represents General Motors.

	GM, Ford, DaimlerChrysler	Other Manufacturer	Total
Men	260	85	345
Women	273	97	370
Total	533	182	715

If one person who completed the survey is selected at random, determine the probability that the person selected

61. is a man.

62. is a woman.

63. drives a car manufactured by GM, Ford or DaimlerChrysler.

64. drives a car manufactured by a company other than GM, Ford, or DaimlerChrysler.

65. is a woman who drives a car manufactured by a company other than GM, Ford, or DaimlerChrysler.

66. is a man who drives a car manufactured by GM, Ford, or DaimlerChrysler.

Peanut Butter Preference *In Exercises 67–72, refer to the following table, which contains information about a sample of shoppers selecting various brands of peanut butter at a grocery store. Assume that each shopper purchased exactly one jar of peanut butter.*

Brand	Smooth	Chunky	Total
Peter Pan	30	23	53
Jiff	28	22	50
Skippy	23	16	39
Other	12	5	17
Total	93	66	159

If one shopper from the sample is selected at random, determine the probability the shopper selected

67. Jif peanut butter.

68. Skippy peanut butter.

69. a chunky peanut butter.

70. a smooth peanut butter.

71. Peter Pan chunky peanut butter.

72. Jiff smooth peanut butter.

Bean Bag Toss *In Exercises 73–77, a bean bag is randomly thrown onto the square table and does not touch a line.*

Find the probability that the bean bag lands on

73. a red area.

74. a green area.

75. a yellow area.

76. a red or green area.

77. a yellow or green area.

78. a red or yellow area.

Challenge Problems/Group Activities

Before working Exercises 79 and 80, reread the material on genetics in Section 12.1.

79. Genetics Cystic fibrosis is an inherited disease that occurs in about 1 in every 2500 Caucasian births in North America and in about 1 in every 250,000 non-Caucasian births in North America. Let's denote the cystic fibrosis gene as c and a disease-free gene as C. Since the disease-free gene is dominant, only a person with cc genes will have the disease. A person who has Cc genes is a carrier of cystic fibrosis but does not actually have the disease. If one parent has CC genes and the other parent has cc genes, find the probability that
 a) an offspring will inherit cystic fibrosis, that is, cc genes.
 b) an offspring will be a carrier of cystic fibrosis but not contract the disease.

80. Genetics Sickle-cell anemia is an inherited disease that occurs in about 1 in every 500 African-American births and about 1 in every 160,000 non-African-American births. Unlike cystic fibrosis, in which the cystic fibrosis gene is recessive, sickle-cell anemia is *codominant.* In other words, a person inheriting two sickle-cell genes will have sickle-cell anemia, whereas a person inheriting only one of the sickle-cell genes will have a mild version of sickle-cell anemia, called *sickle-cell trait.* Let's call the disease-free genes s_1 and the sickle cell gene s_2. If both parents have $s_1 s_2$ genes, determine the probability that
 a) an offspring will have sickle-cell anemia.
 b) an offspring will have the sickle-cell trait.
 c) an offspring will have neither sickle-cell anemia nor the sickle cell trait.

In Exercises 81 and 82, the solutions involve material that we will discuss in later sections of the chapter. Try to solve them before reading ahead.

81. Marbles A bottle contains two red and two green marbles, and a second bottle also contains two red and two green marbles. If you select one marble at random from each bottle, find the probability (to be discussed in Section 12.6) that you obtain
 a) two red marbles.
 b) two green marbles.
 c) a red marble from the first bottle and a green marble from the second bottle.

82. Birds Consider Table 12.3 on page 665. Suppose you are told that one bird's name was selected from the birds listed and the bird selected has a low attractiveness to peanut kernels. Find the probability (to be discussed in Section 12.7) that
 a) the bird is a sparrow.
 b) the bird has a high attractiveness to cracked corn.
 c) the bird has a high attractiveness to black striped sunflower seeds.

Recreational Exercise

83. *Dice* On a die, the sum of the dots on the opposite faces is seven. Two six-sided dice are placed together on top of one another, on a table, as shown in the figure below. The top and bottom faces of the bottom die, and the bottom face of the top die cannot be seen. If you walk around the table, what is the sum of all the dots on all the visible faces of the dice?

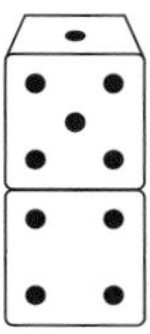

Internet/Research Activity

84. On page 654 we briefly discuss Jacob Bernoulli. The Bernoulli family produced several prominent mathematicians, including Jacob I, Johann I, and Daniel. Write a paper on the Bernoulli family, indicating some of the accomplishments of each of the three Bernoullis named and their relationship to each other. Indicate which Bernoulli the Bernoulli numbers are named after, which Bernoulli the Bernoulli theorem in statistics is named after, and which Bernoulli the Bernoulli theorem of fluid dynamics is named after.

12.3 ODDS

The odds against winning a lottery are 7 million to 1; the odds against being audited by the IRS this year are 47 to 1. We see the word *odds* daily in newspapers and magazines and often use it ourselves. Yet there is a widespread misunderstanding of its meaning. In this section, we will explain the meaning of odds.

The odds given at horse races, at craps, and at all gambling games in Las Vegas and other casinos throughout the world are always *odds against* unless they are otherwise specified. The *odds against* an event is a ratio of the probability that the event will fail to occur (failure) to the probability the event will occur (success). Thus, *to find odds you must first know or determine the probability of success and the probability of failure.*

$$\text{Odds against event} = \frac{P(\text{event fails to occur})}{P(\text{event occurs})} = \frac{P(\text{failure})}{P(\text{success})}$$

DID YOU KNOW

Gambling in Ancient Times

Archaeologists have found evidence of gambling in all cultures, from the Stone Age Australian aborigines to the ancient Egyptians, and across cultures touched by the Roman Empire. There is an equally long history of moral and legal opposition to gambling.

EXAMPLE 1 *Rolling a 4*

Find the odds against rolling a 4 on one roll of a die.

SOLUTION: Before we can determine the odds, we must first determine the probability of rolling a 4 (success) and the probability of not rolling a 4 (failure). When a die is rolled there are six possible outcomes: 1, 2, 3, 4, 5, and 6.

$$P(\text{rolling a 4}) = \frac{1}{6} \qquad P(\text{failure to roll a 4}) = \frac{5}{6}$$

Now that we know the probabilities of success and failure, we can determine the odds against rolling a 4.

$$\text{Odds against rolling a 4} = \frac{P(\text{failure to roll a 4})}{P(\text{rolling a 4})}$$

$$= \frac{\frac{5}{6}}{\frac{1}{6}} = \frac{5}{\cancel{6}_1} \cdot \frac{\cancel{6}^1}{1} = \frac{5}{1}$$

The ratio $\frac{5}{1}$ is commonly written as 5:1 and is read "5 to 1." Thus, the odds against rolling a 4 are 5 to 1. ▲

TIMELY TIP The denominators of the probabilities in an odds problem will always divide out, as was shown in Example 1.

In Example 1, consider the possible outcomes of the die: 1, 2, 3, 4, 5, 6. Over the long run, one of every six rolls will result in a 4, and five of every six rolls will result in a number other than 4. Therefore, if a person was gambling, for each dollar bet in favor of the rolling of a 4, \$5 should be bet against the rolling of a 4 if the person is to break even. The person betting in favor of the rolling of a 4 will either lose \$1 (if a number other than a 4 is rolled) or win \$5 (if a 4 is rolled). The person betting against the rolling of a 4 will either win \$1 (if a number other than a 4 is rolled) or lose \$5 (if a 4 is rolled). If this game is played for a long enough period, each player theoretically will break even.

EXAMPLE 2 *Tax Returns*

In 2002, about 5 of every 23 tax returns are filed electronically. If one tax return that was filed is selected at random, what are the odds against that tax return being filed electronically?

SOLUTION: The probability that a tax return selected at random was filed electronically is $\frac{5}{23}$. Therefore, the probability that a tax return selected at random *is not* filed electronically is $1 - \frac{5}{23}$, or $\frac{18}{23}$.

$$\begin{array}{c}\text{Odds against return}\\ \text{being filed electronically}\end{array} = \frac{P(\text{return not filed electronically})}{P(\text{return filed electronically})}$$

$$= \frac{18/23}{5/23} = \frac{18}{\cancel{23}} \cdot \frac{\cancel{23}}{5} = \frac{18}{5} \text{ or } 18:5$$

Thus, the odds against the tax return being filed electronically are 18:5. ▲

DID YOU KNOW

Flying Is Safer

Many people fear the risks of air travel, but the odds in favor of dying in an airplane crash are relatively low: approximately 1 to 354,319. The odds in favor of dying in a car accident are approximately 1 to 18,585. Thus, the odds in favor of dying in a car accident are much greater than the odds in favor of dying in an airplane accident.

Although odds are generally given against an event, at times they may be given in favor of an event. The *odds in favor of* an event are expressed as a ratio of the probability that the event will occur to the probability that the event will fail to occur.

$$\text{Odds in favor of event} = \frac{P(\text{event occurs})}{P(\text{event fails to occur})} = \frac{P(\text{success})}{P(\text{failure})}$$

DID YOU KNOW

What Were the Odds on This One?

In the year 1905, only two automobiles were registered in the entire state of Missouri.

In the year 1905, Missouri's only two registered automobiles were involved in a head-on collision!

If the odds *against* an event are $a{:}b$, the odds *in favor of* the event are $b{:}a$.

Example 3 involves a circle graph that contains percents; see Fig. 12.4. Before we discuss Example 3, let us briefly discuss percents. Recall that probabilities are numbers between 0 and 1, inclusive. We can change a percent between 0% and 100% to a probability by writing the percent as a fraction or a decimal number. In Fig. 12.4, we see 38% in one of the sectors (or areas) of the circle. To change 38% to a probability we can write $\frac{38}{100}$ or 0.38. Notice that both the fraction and the decimal number are numbers between 0 and 1, inclusive.

EXAMPLE 3 *Attending the Least Expensive Colleges*

The following circle graph shows that most college students in the United States attend the least expensive colleges.

Most Attend Least Expensive Schools
Percent of full-time undergraduates attending schools by cost of tuition and fees, 2002–2003: Numbers are rounded to the nearest percent.

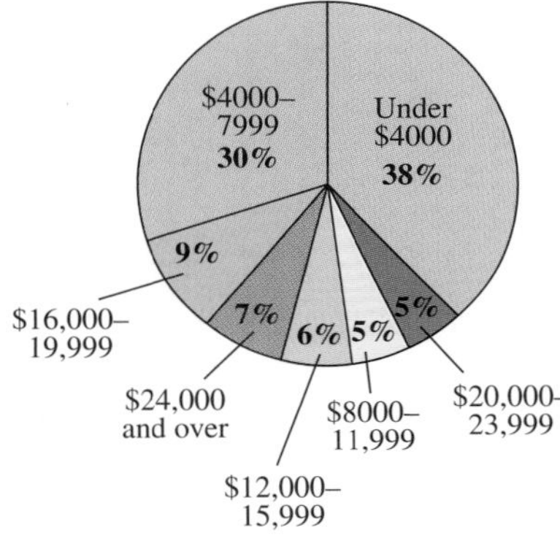

Figure 12.4 Source: The College Board.

Use the graph to determine

a) the odds against a college student selected at random attending a college that costs \$16,000–\$19,999.

b) the odds in favor of a college student selected at random attending a college that costs \$16,000–\$19,999.

SOLUTION:

a) The graph shows that 9%, or $\frac{9}{100}$, of all college students attend a college whose costs are \$16,000–\$19,999. Thus, the probability that a student attends a college whose costs are \$16,000–\$19,999 is $\frac{9}{100}$. The probability the student's college costs are not \$16,000–\$19,999 is therefore $1 - \frac{9}{100} = \frac{91}{100}$.

$$\text{Odds against college whose costs are \$16,000–\$19,999} = \frac{P(\text{college costs not \$16,000–\$19,999})}{P(\text{college costs \$16,000–\$19,999})}$$

$$= \frac{\frac{91}{100}}{\frac{9}{100}} = \frac{91}{\cancel{100}} \cdot \frac{\cancel{100}}{9} = \frac{91}{9} \text{ or } 91:9$$

Thus, the odds against a college student selected at random attending a college that costs \$16,000–\$19,999 are 91 : 9.

b) The odds in favor of a college student selected at random attending a college that costs \$16,000–\$19,999 are 9 : 91.

Finding Probabilities from Odds

When odds are given, either in favor of or against a particular event, it is possible to determine the probability that the event occurs and the probability that the event does not occur. The denominators of the probabilities are found by adding the numbers in the odds statement. The numerators of the probabilities are the numbers given in the odds statements.

EXAMPLE 4 *Determining Probabilities from Odds*

The odds against Robin Murphy being admitted to the college of her choice are 9:2. Find the probability that (a) Robin is admitted and (b) Robin is not admitted.

SOLUTION:

a) We have been given odds against and have been asked to find probabilities.

$$\text{Odds against being admitted} = \frac{P(\text{fails to be admitted})}{P(\text{is admitted})}$$

Since the odds statement is 9:2, the denominators of both the probability of success and the probability of failure must be $9 + 2$ or 11. To get the odds ratio of 9:2 the probabilities must be $\frac{9}{11}$ and $\frac{2}{11}$. Since odds against is a ratio of failure to success, the $\frac{9}{11}$ and $\frac{2}{11}$ represent the probabilities of failure and success, respectively. Thus, the probability that Robin is admitted (success) is $\frac{2}{11}$.

b) The probability that Robin is not admitted (failure) is $\frac{9}{11}$.

Odds and probability statements are sometimes stated incorrectly. For example, consider the statement, "The odds of being selected to represent the district are 1 in 5." Odds are given using the word *to,* not *in*. Thus, there is a mistake in this statement. The correct statement might be, "The odds of being selected to represent the district are 1 to 5" or "The probability of being selected to represent the district is 1 in 5." Without additional information, it is not possible to tell which is the correct interpretation.

SECTION 12.3 EXERCISES

Concept/Writing Exercises

1. Explain in your own words how to determine the odds against an event.

2. Explain in your own words how to determine the odds in favor of an event.

3. Which odds are generally quoted, odds against or odds in favor?

4. Explain how to determine probabilities when you are given an odds statement.

5. The odds in favor of winning the door prize are 5 to 9. Find the odds against winning the door prize.

6. The odds against Thor's Lightning winning the horse race are 7:3. Find the odds in favor of Thor's Lightning winning.

7. If the odds against an event are 1:1, what is the probability the event will
   a) occur.
   b) fail to occur.
   Explain your answer.

8. If the probability an event will occur is $\frac{1}{2}$, determine
   a) the probability the event will fail to occur.
   b) the odds against the event occurring.
   c) the odds in favor of the event occurring.
   Explain your answer.

Practice the Skills/Problem Solving

9. *Dressing Up* Lalo Jaquez is going to wear a blue sportcoat and is trying to decide what tie he should wear to work. In his closet, he has 27 ties, 8 of which he feels go well with the sportcoat. If Lalo selects one tie at random, determine
a) the probability it goes well with the sportcoat.
b) the probability it does not go well with the sportcoat.
c) the odds against it going well with the sportcoat.
d) the odds in favor of it going well with the sportcoat.

10. *Making a Donation* In her wallet, Anne Kelly has 14 bills. Seven are $1 bills, two are $5 bills, four are $10 bills, and one is a $20 bill. She passes a volunteer seeking donations for the Salvation Army and decides to select one bill at random from her wallet and give it to the Salvation Army. Determine
a) the probability she selects a $5 bill.
b) the probability she does not select a $5 bill.
c) the odds in favor of her selecting a $5 bill.
d) the odds against her selecting a $5 bill.

Toss a Die In Exercises 11–14, a die is tossed. Find the odds against rolling

11. a 3.

12. an even number.

13. a number less than 3.

14. a number greater than 4.

Deck of Cards In Exercises 15–18, a card is picked from a deck of cards. Find the odds against and the odds in favor of selecting

15. a queen.

16. a heart.

17. a picture card.

18. a card greater than 5 (ace is low).

Spin the Spinner In Exercises 19–22, assume that the spinner cannot land on a line. Find the odds against the spinner landing on the color red.

19.

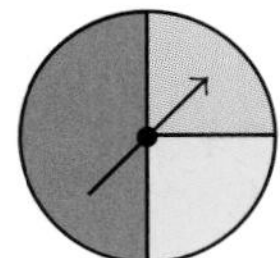

20.

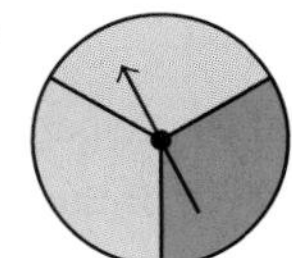

21.

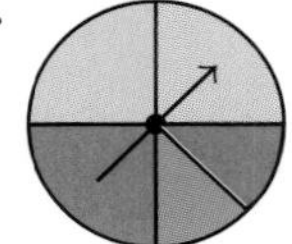

22.

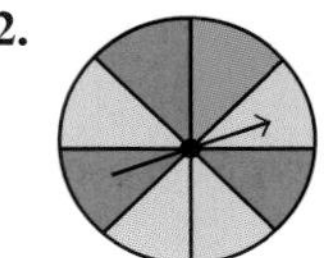

23. *Students* One person is selected at random from a class of 16 men and 14 women. Find the odds against selecting
a) a woman.
b) a man.

24. *Lottery* One million tickets are sold for a lottery where a single prize will be awarded.
a) If you purchase a ticket, find your odds against winning.
b) If you purchase 10 tickets, find your odds against winning.

Billiard Balls In Exercises 25–30, use the rack of 15 billiard balls shown.

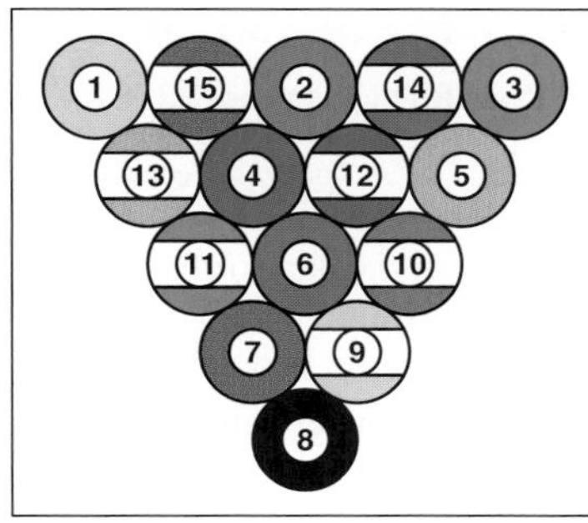

25. If one ball is selected at random, find the odds against it containing a stripe. (Balls numbered 9 through 15 contain stripes.)

26. If one ball is selected at random, find the odds in favor of it being an even-numbered ball.

27. If one ball is selected at random, find the odds in favor of it being a ball other than the 8 ball.

28. If one ball is selected at random, find the odds against it containing any red coloring (solid or striped).

29. If one ball is selected at random, find the odds against it containing a number greater than or equal to 9.

30. If one ball is selected at random, find the odds in favor of it containing two digits.

31. *Bowl Games* The number of college football bowl games has grown in recent years. The table below indicates the nine bowl games with the greatest per-team payout in 2003.

Bowl Game	Per-team Payout (millions)
Tostitos Fiesta	13.5
Fed Ex Orange	13.5
Rose	13.5
Nokia Sugar	13.5
Capital One	5.125
SBC Cotton	3
Outback	2.55
Chick-Fil A Peach	2
Pacific Life Holiday	2

If one of the bowl games listed in the chart is selected at random, determine
a) the probability the per-team payout is greater than $5 million.

b) the odds against the per-team payout being greater than \$5 million.

The Rose Bowl in Pasadena, CA

32. *Rolling a Special Die* A special die used in a game contains one dot on one side, two dots on two sides, and three dots on three sides. If the die is rolled, determine
a) the probability of rolling two dots.
b) the odds against rolling two dots.

33. *Medical Tests* The results of a medical test show that of 76 people selected at random who were given the test, 72 tested negative and 4 tested positive. Determine the odds against a person selected at random testing negative on the test. Explain how you determined your answer.

34. *A Red Marble* A box contains 9 red and 2 blue marbles. If you select one marble at random from the box, determine the odds against selecting a red marble. Explain how you determined your answer.

35. *Teaching Award* The odds in favor of June White winning the teaching award are 8:5. Find the probability that
a) June wins.
b) June does not win.

36. *Hot Dog Contest* The odds in favor of Boris Penzed winning the hot dog eating contest are 2:7. Find the probability that Boris will
a) win the contest.
b) not win the contest.

37. *Getting Promoted* The odds against Jason Judd getting promoted are 4:11. Find the probability that Jason gets promoted.

38. *Winning a Race* The odds against Paul Phillips winning the 100 yard dash are 5:2. Find the probability that
a) Paul wins.
b) Paul loses.

Playing Bingo When playing bingo, 75 balls are placed in a bin and balls are selected at random. Each ball is marked with a letter and number as indicated in the following chart.

B	I	N	G	O
1–15	16–30	31–45	46–60	61–75

For example, there are balls marked B1, B2, up to B15; I16, I17, up to I30; and so on. In Exercises 39–44, assuming one bingo ball is selected at random, determine

39. the probability it contains the letter G.

40. the probability it does not contain the letter G.

41. the odds in favor of it containing the letter G.

42. the odds against it containing the letter G.

43. the odds against it being $B9$.

44. the odds in favor of it being $B9$.

Blood Types In Exercises 45–50, the following circle graph shows the percent of Americans with the various types of blood.

Blood Types of Americans

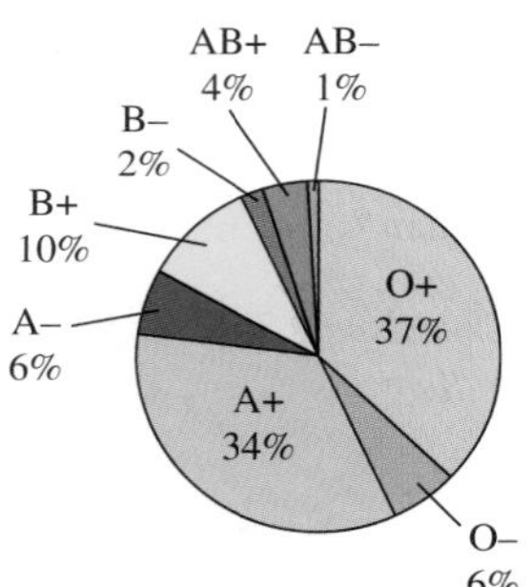

Source: 2003 Time Almanac

If one American is selected at random, use the graph to determine

45. the probability the person has A+ blood.

46. the probability the person has B− blood.

47. the odds against the person having A+ blood.

48. the odds in favor of the person having B− blood.

49. the odds in favor of the person having either O+ or O− blood.

50. the odds against the person having either A+ or O+ blood.

51. *Rock Concert* Suppose that the probability that a rock concert sells out is 0.9. Determine the odds against the concert selling out.

52. *Working Overtime* Suppose that the probability that you are asked to work overtime this week is 3/8. Determine the odds in favor of your being asked to work overtime.

53. *Bookcase Assembly* Suppose that the probability that all the parts needed to assemble a bookcase are included in the carton is $\frac{7}{8}$. Determine the odds in favor of the carton including all the needed parts.

54. *IRS Audit* One in 42 individuals whose salaries range between \$10,000 and \$40,000 will be randomly selected to have his or her income tax returns audited. Mr. Frank is in this income tax range. Find
  a) the probability that Mr. Frank will be audited.
  b) the odds against Mr. Frank being audited.

55. *Arthritis* Gout, a form of arthritis, is much less common in women than in men. In general, 20 out of 21 people with gout are men.
  a) If J. Douglas has gout, what is the probability that J. Douglas is a man?
  b) If J. Douglas has gout, what are the odds against J. Douglas's being a woman?

Challenge Problems/Group Activities

56. *Odds Against* Find the odds against an even number or a number greater than 3 being rolled on a die.

57. *Horse Racing* Racetracks quote the approximate odds against each horse winning on a large board called a *tote board.* The odds quoted on a tote board for a race with five horses is as follows.

Horse Number	Odds
1	7:2
2	2:1
3	15:1
4	7:5
5	1:1

Find the probability of each horse winning the race. (Do not be concerned that the sum of the probabilities is not 1.)

58. *Roulette* Turn to the roulette wheel illustrated on page 687. If the wheel is spun, find
  a) the probability that the ball lands on red.
  b) the odds against the ball landing on red.
  c) the probability that the ball lands on 0 or 00.
  d) the odds in favor of the ball landing on 0 or 00.

Recreational Exercise

59. *Multiple Births* Multiple births make up about 3% of births a year in the United States. The following illustrates the number and type of multiple births in 2000.

Multiple Births in the United States in 2000

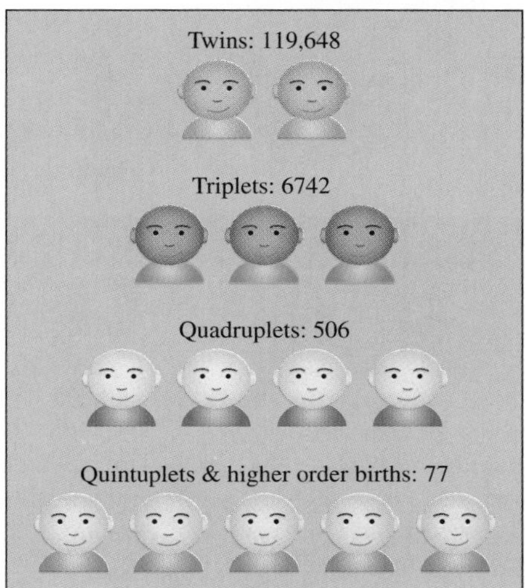

Source: National Center for Health Statistics

Using the above information, determine an estimate for the odds against a birth being a multiple birth in 2000.

Internet/Research Activities

60. *State Lottery* Determine whether your state has a lottery. If so, do research and write a paper indicating
  a) the probability of winning the grand prize.
  b) the odds against winning the grand prize.
  c) Explain, using real objects such as pennies or table tennis balls, what these odds really mean.

61. *Casino Advantages* There are many types of games of chance to choose from at casinos. The house has the advantage in each game, but the advantages differ according to the game.
  a) List the games available at a typical casino.
  b) List those for which the house has the smallest advantage of winning.
  c) List those for which the house has the greatest advantage of winning.

12.4 EXPECTED VALUE (EXPECTATION)

DID YOU KNOW

Tough Decision

The concept of expected value can be used to help evaluate the consequences of many decisions. You use this concept when you consider whether to double-park your car for a few minutes. The probability of being caught may be low, but the penalty, a parking ticket, may be high. You weigh these factors when you decide whether to spend time looking for a legal parking spot.

Expected value, also called *expectation,* is often used to determine the expected results of an experiment or business venture *over the long run.* Expectation is used to make important decisions in many different areas. In business, for example, expectation is used to predict future profits of a new product. In the insurance industry, expectation is used to determine how much each insurance policy should cost for the company to make an overall profit. Expectation is also used to predict the expected gain or loss in games of chance such as the lottery, roulette, craps, and slot machines.

Consider the following: Tim tells Barbara that he will give her \$1 if she can roll an even number on a single die. If she fails to roll an even number, she must give Tim \$1. Who would win money in the long run if this game were played many times? We would expect in the long run that half the time Tim would win \$1 and half the time he would lose \$1; therefore, Tim would break even. Mathematically, we could find Tim's expected gain or loss by the following procedure:

$$\text{Tim's expected gain or loss} = P\begin{pmatrix}\text{Tim}\\\text{wins}\end{pmatrix}\cdot\begin{pmatrix}\text{amount}\\\text{Tim wins}\end{pmatrix} + P\begin{pmatrix}\text{Tim}\\\text{loses}\end{pmatrix}\cdot\begin{pmatrix}\text{amount}\\\text{Tim loses}\end{pmatrix}$$

$$= \frac{1}{2}(\$1) + \frac{1}{2}(-\$1) = \$0$$

Note that the loss is written as a negative number. This procedure indicates that Tim has an expected gain or loss (or expected value) of \$0. The expected value of zero indicates that he would indeed break even, as we had anticipated. Thus, the game is a *fair game.* If his expected value were positive, it would indicate a gain; if negative, a loss.

The expected value, E, is calculated by multiplying the probability of an event occurring by the *net* amount gained or lost if the event occurs. If there are a number of different events and amounts to be considered, we use the following formula.

Expected Value

$$E = P_1 \cdot A_1 + P_2 \cdot A_2 + P_3 \cdot A_3 + \cdots + P_n \cdot A_n$$

The symbol P_1 represents the probability that the first event will occur, and A_1 represents the net amount won or lost if the first event occurs. P_2 is the probability of the second event, and A_2 is the net amount won or lost if the second event occurs, and so on. The sum of these products of the probabilities and their respective amounts is the expected value. The expected value is the average (or mean) result that would be obtained if the experiment were performed a great many times.

EXAMPLE 1 *A New Business Venture*

Southwest Airlines is considering adding a route to the city of Cedar Rapids, Iowa. Before the company makes their decision as to whether or not to service Cedar Rapids, it needs to consider many factors, including their potential profits and losses. Factors that may affect the company's profits and losses include the number

of competing airlines, the potential number of customers, the overhead costs, and fees they must pay. After considerable research, the company estimates that if it serves Cedar Rapids, there is a 60% chance of making an \$800,000 profit, a 10% chance of breaking even, and a 30% chance of losing \$1,200,000. How much can Southwest Airlines "expect" to make on this new route?

SOLUTION: The three amounts to be considered are a gain of \$800,000, breaking even at \$0, and a loss of \$1,200,000. The probability of gaining \$800,000 is 0.6, the probability of breaking even is 0.1, and the probability of losing \$1,200,000 is 0.3.

$$\begin{aligned}\text{Southwest's expectation} &= \overbrace{P_1 \cdot A_1}^{\text{Gain}} + \overbrace{P_2 \cdot A_2}^{\text{Break even}} + \overbrace{P_3 \cdot A_3}^{\text{Loss}} \\ &= (0.6)(\$800{,}000) + (0.1)(\$0) + (0.3)(-\$1{,}200{,}000) \\ &= \$480{,}000 + \$0 - \$360{,}000 \\ &= \$120{,}000\end{aligned}$$

Southwest Airlines has an expectation, or expected average gain, of \$120,000 for adding this particular service. Thus, if the company opened routes like this one, with these particular probabilities and amounts, in the long run it would have an average gain of \$120,000 per route. However, you must remember, that there is a 30% chance that Southwest will lose \$1,200,000 on this *particular* route (or any particular route with these probabilities and amounts.) ▲

EXAMPLE 2 *Test-Taking Strategy*

Maria is taking a multiple-choice exam in which there are five possible answers for each question. The instructions indicate that she will be awarded 2 points for each correct response, that she will lose $\frac{1}{2}$ point for each incorrect response, and that no points will be added or subtracted for answers left blank.

a) If Maria does not know the correct answer to a question, is it to her advantage or disadvantage to guess at an answer?

b) If she can eliminate one of the possible choices, is it to her advantage or disadvantage to guess at the answer?

SOLUTION:

a) Let's determine the expected value if Maria guesses at an answer. Only one of five possible answers is correct.

$$P(\text{guesses correctly}) = \frac{1}{5} \qquad P(\text{guesses incorrectly}) = \frac{4}{5}$$

$$\begin{aligned}\text{Maria's expectation} &= \overbrace{P_1 \cdot A_1}^{\text{Guesses correctly}} + \overbrace{P_2 \cdot A_2}^{\text{Guesses incorrectly}} \\ &= \frac{1}{5}(2) + \frac{4}{5}\left(-\frac{1}{2}\right) \\ &= \frac{2}{5} - \frac{2}{5} = 0\end{aligned}$$

Thus, Maria's expectation is zero when she guesses. Therefore, over the long run she will neither gain nor lose points by guessing.

b) If Maria can eliminate one possible choice, one of four answers will be correct.

$$P(\text{guesses correctly}) = \frac{1}{4} \qquad P(\text{guesses incorrectly}) = \frac{3}{4}$$

$$\begin{aligned}\text{Maria's expectation} &= \overbrace{P_1 \cdot A_1}^{\text{Guesses correctly}} + \overbrace{P_2 \cdot A_2}^{\text{Guesses incorrectly}} \\ &= \frac{1}{4}(2) + \frac{3}{4}\left(-\frac{1}{2}\right) \\ &= \frac{2}{4} - \frac{3}{8} = \frac{4}{8} - \frac{3}{8} = \frac{1}{8}\end{aligned}$$

Since the expectation is a positive $\frac{1}{8}$, Maria will, on average, gain $\frac{1}{8}$ point each time she guesses when she can eliminate one possible choice. ▲

EXAMPLE 3 *Pothole Repairs*

A highway crew repairs 30 potholes a day in dry weather and 12 potholes a day in wet weather. If the weather in this region is wet 40% of the time, find the expected (average) number of potholes that can be repaired per day.

SOLUTION: The amounts in this problem are the number of potholes repaired. Since the weather is wet 40% of the time, it will be dry 100% − 40% = 60% of the time. When written as probabilities, 60% and 40% are 0.60 and 0.40, respectively.

$$\begin{aligned}E &= P(\text{dry}) \cdot (\text{amount repaired}) + P(\text{wet}) \cdot (\text{amount repaired}) \\ &= 0.60(30) + 0.40(12) = 18.0 + 4.8 = 22.8\end{aligned}$$

Thus, the average, or expected, number of potholes repaired per day is 22.8. ▲

When we gave the expectation formula we indicated that the amounts were the **net amounts,** which are the actual amounts gained or lost. Examples 4 and 5 illustrate how net amounts are used in two applications of expected value.

EXAMPLE 4 *Winning a Door Prize*

When Josh Rosenberg attends a charity event, he is given a free ticket for the \$50 door prize. A total of 100 tickets will be given out. Determine his expectation of winning the door prize.

SOLUTION: The probability of winning the door prize is $\frac{1}{100}$ since Josh has 1 of 100 tickets. If he wins, his net or actual winnings will be \$50 since he did not pay for the ticket. The probability that Josh loses is $\frac{99}{100}$. If Josh loses, the amount he loses is \$0 because he did not pay for the ticket.

$$\begin{aligned}\text{Expectation} &= P(\text{Josh wins}) \cdot (\text{amount won}) + P(\text{Josh loses}) \cdot (\text{amount lost}) \\ &= \frac{1}{100}(50) + \frac{99}{100}(0) = \frac{50}{100} = 0.50\end{aligned}$$

Thus, Josh's expectation is \$0.50, or 50 cents. ▲

DID YOU KNOW

It Pays to Be Original

Commonly Selected Numbers

7-14-21-28-35-42
1-2-3-4-5-6
5-10-15-20-25-30
3-8-13-18-23-28

The numbers you select when picking lottery numbers have no effect on your probability of winning. However, your expectation (expected winnings) varies greatly with the numbers you select. Because the jackpot is divided among all the winners, the fewer the number of winners the more each winner receives. There are some groups of numbers that are commonly selected. Some are illustrated above. In a lottery with a large jackpot, there may be as many as 10,000 people who select the numbers 7-14-21-28-35-42. If the jackpot was $40 million and these numbers were selected, each winner would receive about $4000. That is quite a difference from the $40 million a single winner would receive. Many people use birthdays or other dates when selecting lottery numbers. Therefore, there may be fewer people selecting numbers greater than 31.

TIMELY TIP In Example 4 and in all the previous examples of expectation, the sum of the probabilities of the events has always been 1. This should always be the case, that is, the sum of the probabilities in any expectation problem should always be 1.

Now we will consider a problem similar to Example 4, but this time we will assume that Josh must purchase the ticket for the door prize.

EXAMPLE 5 *Winning a Door Prize*

When Josh Rosenberg attends a charity event, he is given the opportunity to purchase a ticket for the $50 door prize. The cost of the ticket is $2, and 100 tickets will be sold. Determine Josh's expectation if he purchases one ticket.

SOLUTION: As in Example 4, Josh's probability of winning is $\frac{1}{100}$. However, if he does win, his actual or net winnings will be $48. The $48 is obtained by subtracting the cost of the ticket, $2, from the amount of the door prize, $50. There is also a probability of $\frac{99}{100}$ that Josh will not win the door prize. If he does not win the door prize, he has lost the $2 that he paid for the ticket. Therefore, there are two amounts that we must consider when we determine Josh's expectation, winning $48 and losing $2.

$$\begin{aligned}\text{Expectation} &= P(\text{Josh wins})\cdot(\text{amount won}) + P(\text{Josh loses})\cdot(\text{amount lost})\\ &= \frac{1}{100}(48) + \frac{99}{100}(-2)\\ &= \frac{48}{100} - \frac{198}{100} = -\frac{150}{100} = -1.50\end{aligned}$$

Josh's expectation is $-\$1.50$ when he purchases one ticket. ▲

In Example 5, we determined that Josh's expectation was $-\$1.50$ when he purchased one ticket. If he purchased two tickets, his expectation would be $2(-\$1.50)$, or $-\$3.00$. We could also compute Josh's expectation if he purchased two tickets as follows

$$E = \frac{2}{100}(46) + \frac{98}{100}(-4) = -3.00$$

This answer, $-\$3.00$, checks with the answer obtained by multiplying the expectation for a single ticket by 2.

Let's look at one more example where a person must pay for a chance to win a prize. In the following example, there will be more than two amounts to consider.

EXAMPLE 6 *Raffle Tickets*

One thousand raffle tickets are sold for $1 each. One grand prize of $500 and two consolation prizes of $100 will be awarded. The tickets are placed in a bin. The winning tickets will be selected from the bin. Assuming that each ticket selected for a prize is returned to the bin before the next ticket is selected, determine

a) Irene Drew's expectation if she purchases one ticket.
b) Irene's expectation if she purchases five tickets.

SOLUTION:

a) Three amounts are to be considered: the net gain in winning the grand prize, the net gain in winning one of the consolation prizes, and the loss of the cost of the ticket. If Irene wins the grand prize, her net gain is \$499 (\$500 minus \$1 spent for the ticket). If Irene wins one of the consolation prizes, her net gain is \$99 (\$100 minus \$1). We are told to assume that each winning ticket is replaced in the bin after being selected. The probability that Irene wins the grand prize is $\frac{1}{1000}$. Since two consolation prizes will be awarded, the probability that she wins a consolation prize is $\frac{2}{1000}$. The probability that she does not win a prize is $1 - \frac{1}{1000} - \frac{2}{1000} = \frac{997}{1000}$.

$$E = P_1 \cdot A_1 + P_2 \cdot A_2 + P_3 \cdot A_3$$
$$= \frac{1}{1000}(\$499) + \frac{2}{1000}(\$99) + \frac{997}{1000}(-\$1)$$
$$= \frac{499}{1000} + \frac{198}{1000} - \frac{997}{1000} = -\frac{300}{1000} = -0.30$$

Thus, Irene's expectation is −\$0.30 per ticket purchased.

b) On average, Irene loses 30 cents on each ticket purchased. On five tickets her expectation is (−\$0.30)(5), or −\$1.50.

In Example 5, we determined that Josh's expectation was −\$1.50. Now let's determine how to find out how much should have been charged for a ticket so that his expectation would be \$0. If Josh's expectation were to be \$0, he could be expected to break even over the long run. Suppose that Josh paid 50 cents, or \$0.50, for the ticket. His expectation, if paying \$0.50 for the ticket, would be calculated as shown below.

$$\text{Expectation} = P(\text{Josh wins}) \cdot (\text{amount won}) + P(\text{Josh loses}) \cdot (\text{amount lost})$$
$$= \frac{1}{100}(49.50) + \frac{99}{100}(-\$0.50)$$
$$= \frac{49.50}{100} - \frac{49.50}{100} = 0$$

Thus, if Josh paid 50 cents per ticket, his expectation would be \$0. The 50 cents, in this case, is called the fair price of the ticket. The *fair price* is the amount to be paid that will result in an expected value of \$0. The fair price may be found by adding the *cost to play* to the *expected value*.

Fair price = expected value + cost to play

In Example 5, the cost to play was \$2 and the expected value was determined to be −\$1.50. The fair price for a ticket in Example 5 may be found as follows.

$$\text{Fair price} = \text{expected value} + \text{cost to play}$$
$$= -1.50 + 2.00 = 0.50$$

We obtained a fair price of \$0.50. If the tickets were sold for the fair price of \$0.50 each, Josh's expectation would be \$0, as shown above. Can you now find the fair price that Irene would pay for a raffle ticket in Example 6? In Example 6, the cost of a ticket was \$1 and we determined that the expected value was −\$0.30.

$$\text{Fair price} = \text{expected value} + \text{cost to play}$$
$$= -\$0.30 + \$1.00 = \$0.70$$

Thus, the fair price for a ticket in Example 6 is \$0.70, or 70 cents. Verify for yourself now that if the tickets were sold for \$0.70, the expectation would be \$0.00.

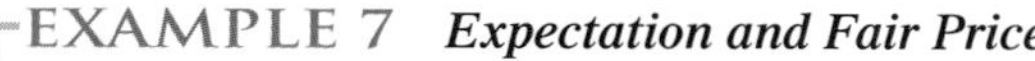

EXAMPLE 7 *Expectation and Fair Price*

Suppose that you are playing a game in which you spin the pointer shown in the figure in the margin, and you are awarded the amount shown under the pointer. If it costs \$8 to play the game, determine

a) the expectation of a person who plays the game.

b) the fair price to play the game.

SOLUTION:

a) There are four numbers on which the pointer can land: 1, 5, 10, and 20. The following chart shows the probability of the pointer landing on each number and the actual amount won or lost if the pointer lands on that number. The probabilities are obtained using the areas of the circle. The amounts won or lost are determined by subtracting the cost to play, \$8, from each indicated amount.

Amount Shown on Wheel	**\$1**	**\$5**	**\$10**	**\$20**
Probability	$\frac{1}{4}$	$\frac{3}{8}$	$\frac{1}{4}$	$\frac{1}{8}$
Amount Won or Lost	−\$7	−\$3	\$2	\$12

Notice that the sum of the probabilities is 1, which shows that all possible outcomes have been considered.

Now let's find the expectation. There are four amounts to consider.

$$\begin{aligned}\text{Expectation} &= P(\text{lands on } \$1)\cdot(\text{amount}) + P(\text{lands on } \$5)\cdot(\text{amount}) \\ &\quad + P(\text{lands on } \$10)\cdot(\text{amount}) + P(\text{lands on } \$20)\cdot(\text{amount}) \\ &= \frac{1}{4}(-7) + \frac{3}{8}(-3) + \frac{1}{4}(2) + \frac{1}{8}(12) \\ &= -\frac{7}{4} - \frac{9}{8} + \frac{2}{4} + \frac{12}{8} \\ &= -\frac{14}{8} - \frac{9}{8} + \frac{4}{8} + \frac{12}{8} = -\frac{7}{8} = -\$0.875\end{aligned}$$

Thus, the expectation is −\$0.875.

b) Fair price = expectation + cost to play

$$= -\$0.875 + \$8 = \$7.125$$

Thus, the fair price is about \$7.13. ▲

SECTION 12.4 EXERCISES

Concept/Writing Exercises

1. What does the expected value of an experiment or business venture represent?

2. What does an expected value of 0 mean?

3. What is meant by the fair price of a game of chance?

4. Write the formula used to find the expected value of an experiment with

a) two possible outcomes.

b) three possible outcomes.

5. If the expected value and cost to play are known for a particular game of chance, explain how you can determine the fair price to pay to play that game of chance.
6. Is the fair price to pay for a game of chance the same as the expected value of that game of chance? Explain your answer.
7. If a particular game costs \$1.50 to play and the expectation for the game is −\$1.00, what is the fair price to pay to play the game? Explain how you determined your answer.
8. If a particular game cost \$3.00 to play and the expectation for the game is −\$2.00, what is the fair price to pay to play the game? Explain how you determined your answer.

Practice the Skills/Problem Solving

9. *Three Tickets* On a \$1 lottery ticket, Marty Smith's expected value is −\$0.40. What is Marty's expected value if he purchases three lottery tickets?
10. *Expected Value* If on a \$1 bet, Paul Goldstein's expected value is \$0.20. What is Paul's expected value on a \$5 bet?
11. *Expected Attendance* For a showing of a specific movie, an AMC theater estimates that 120 people will attend if it is not raining. If it is raining, the theater estimates that 200 people will attend. The meteorologist predicts a 70% chance of rain tomorrow. Determine the expected number of people who will attend the movie.
12. *A New Business* In a proposed business venture, Stephanie Morrison estimates that there is a 60% chance she will make \$80,000 and a 40% chance she will lose \$20,000. Determine Stephanie's expected value.
13. *Basketball* Diana Taurasi is a star player for the University of Connecticut Huskies women's basketball team. She has injured her ankle and it is doubtful if she will be able to play in an upcoming game. If she can play, the coach estimates that the Huskies will score 78 points. If she is not able to play, the coach estimates that they will score 62 points. The team doctor estimates that there is a 50% chance Diana will play. Determine the number of points the team can expect to score.

Shareese Grant (left) and Diana Taurasi

14. *Seminar Attendance* At an investment tax seminar, Judy Johnson estimates that 20 people will attend if it does not rain and 12 people will attend if it rains. The weather forecast indicates a 40% chance it will not rain and a 60% chance it will rain on the day of the seminar. Determine the expected number of people who will attend the seminar.
15. *TV Shows* The NBC television network is scheduling its fall lineup of shows. For the Thursday night 8 P.M. slot, NBC has selected the show *The West Wing*. If its rival network CBS schedules the show *CSI—Crime Scene Investigation* during the same time slot, NBC estimates that *The West Wing* will get 1.2 million viewers. However, if CBS schedules the show *Judging Amy* during that time slot, NBC estimates that *The West Wing* will get 1.6 million viewers. NBC believes that the probability that CBS will show *CSI* is 0.4 and the probability that CBS will show *Judging Amy* is 0.6. Determine the expected number of viewers for the show *The West Wing*.
16. *Seattle Greenery* In July in Seattle, the grass grows $\frac{1}{2}$ in. a day on a sunny day and $\frac{1}{4}$ in. a day on a cloudy day. In Seattle in July, 75% of the days are sunny and 25% are cloudy.
    a) Find the expected amount of grass growth on a typical day in July in Seattle.
    b) Find the expected total grass growth in the month of July in Seattle.
17. *Investment Club* The Triple L investment club is considering purchasing a certain stock. After considerable research the club members determine that there is a 60% chance of making \$10,000, a 10% chance of breaking even, and a 30% chance of losing \$7200. Find the expectation of this purchase.
18. *Clothing Sale* At a special clothing sale at the Crescent Oaks Country Club, after the cashier rings up your purchase, you select a slip of paper from a box. The slip of paper indicates the dollar amount, either \$5 or \$10, that is deducted from your purchase price. The probability of selecting a slip indicating \$5 is $\frac{7}{10}$ and the probability of selecting a slip indicating \$10 is $\frac{3}{10}$. If your original purchase before you select the slip of paper is \$100, determine
    a) the expected dollar amount to be deducted from your purchase.
    b) the expected dollar amount you will pay for your purchase.
19. *Fortune Cookies* At the Royal Dragon Chinese restaurant, a slip in the fortune cookies indicates a dollar amount that will be subtracted from your total bill. A bag of 10 fortune cookies is given to you from which you will select 1. If seven fortune cookies contain "\$1 off," two contain "\$2 off," and one contains "\$5 off," find the expectation of a selection.
20. *Pick a Card* Mike and Dave play the following game: Mike picks a card from a deck of cards. If he selects a heart, Dave gives him \$5. If not, he gives Dave \$2.
    a) Find Mike's expectation.
    b) Find Dave's expectation.
21. *Roll a Die* Cortney and Kelly play the following game. Kelly rolls a die. If she rolls a number greater than 4, Cortney gives Kelly \$8. If Kelly does not roll a number greater than 4, she gives Cortney \$5.
    a) Determine Kelly's expectation.
    b) Determine Cortney's expectation.

22. *Blue Chips and Red Chips* A bag contains 3 blue chips and 2 red chips. Chi and Dolly play the following game. Chi selects one chip at random from the bag. If Chi selects a blue chip, Dolly gives Chi $5. If Chi selects a red chip, Chi gives Dolly $8.
a) Determine Chi's expectation.
b) Determine Dolly's expectation.

23. *Multiple-Choice Test* A multiple-choice exam has five possible answers for each question. For each correct answer, you are awarded 5 points. For each incorrect answer, 1 point is subtracted from your score. For answers left blank, no points are added or subtracted.
a) If you do not know the correct answer to a particular question, is it to your advantage to guess? Explain.
b) If you do not know the correct answer but can eliminate one possible choice, is it to your advantage to guess? Explain.

24. *Multiple-Choice Test* A multiple-choice exam has four possible answers for each question. For each correct answer, you are awarded 5 points. For each incorrect answer, 2 points are subtracted from your score. For answers left blank, no points are added or subtracted.
a) If you do not know the correct answer to a particular question, is it to your advantage to guess? Explain.
b) If you do not know the correct answer but can eliminate one possible choice, is it to your advantage to guess? Explain.

Exercises 25–28 deal with raffle drawings. Assume that after each ticket is drawn, the ticket that was drawn is mixed in with the other tickets before the next selection. Therefore, the selections are being made with replacement.

25. *Raffle Tickets* Five hundred raffle tickets are sold for $2 each. One prize of $400 is to be awarded.
a) Raul Mondesi purchases one ticket. Find his expected value.
b) Determine the fair price of a ticket.

26. *Raffle Tickets* One thousand raffle tickets are sold for $1 each. One prize of $800 is to be awarded.
a) Rena Condos purchases one ticket. Find her expected value.
b) Determine the fair price of a ticket.

27. *Raffle Tickets* Two thousand raffle tickets are sold for $3.00 each. Three prizes will be awarded: one for $1000 and two for $500. Jeremy Sharp purchases one of these tickets.
a) Determine his expected value.
b) Determine the fair price of a ticket.

28. *Raffle Tickets* Ten thousand raffle tickets are sold for $5 each. Four prizes will be awarded: one for $10,000, one for $5000, and two for $1000. Sidhardt purchases one of these tickets.
a) Determine his expected value.
b) Determine the fair price of a ticket.

In Exercises 29 and 30, assume that a person spins the pointer and is awarded the amount indicated by the pointer. Determine the person's expectation.

29.

$1 $10

30.

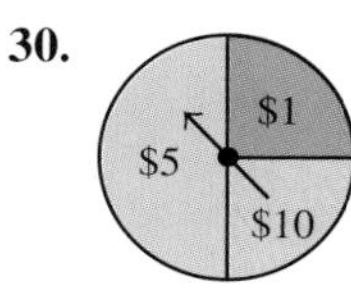

In Exercises 31 and 32, assume that a person spins the pointer and is awarded the amount indicated if the pointer points to a positive number but must pay the amount indicated if the pointer points to a negative number. Determine the person's expectation if the person plays the game.

31.

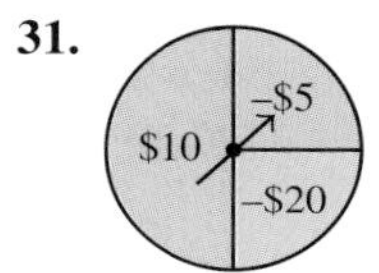

32.

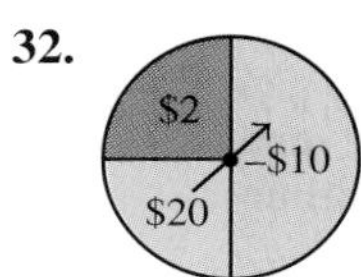

In Exercises 33–36, assume that a person spins the pointer and is awarded the amount indicated by the pointer. If it costs $2 to play the game, determine
a) *the expectation of a person who plays the game.*
b) *the fair price to play the game.*

33.

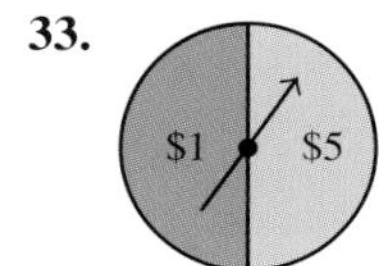

34.

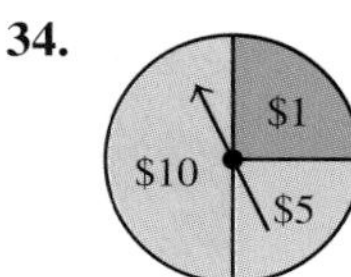

35.

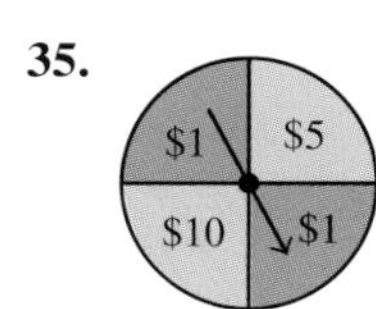

36.

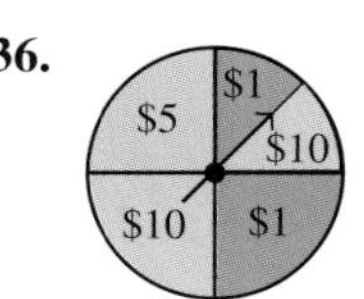

37. *Reaching Base Safely* Based on past history, Jim Devias has a 17% chance of reaching first base safely, a 10% chance of hitting a double, a 2% chance of hitting a triple, an 8% chance of hitting a home run, and a 63% chance of making an out at his next at bat. Determine Jim's expected number of bases for his next at bat.

38. *Life Insurance* According to Bristol Mutual Life Insurance's mortality table, the probability that a 20-year-old woman will survive 1 year is 0.994 and the probability that she will die within 1 year is 0.006. If she buys a $10,000 1-year policy for $100, what is Bristol Mutual's expected gain or loss?

39. *Choosing a Colored Chip* In a box there are a total of 10 chips. The chips are orange, green, or yellow, as shown below.

If you select an orange chip you get 4 points, a green chip 3 points, and a yellow chip 1 point. If you select one chip at random, determine the expected number of points you will get.

40. *Choosing a Colored Chip* Repeat Exercise 39 but assume that an orange chip is worth 5 points, a green chip 2 points, and a yellow chip -3 points (3 points are taken away).

41. *Airline Hiring* American Airlines has requested new routes. If the new routes are granted, American will hire 850 new employees. If the new routes are not granted, American will hire only 140 new employees. If the probability that the new routes will be approved is 0.34, what is the expected number of new employees to be hired by American Airlines?

42. *Salary Negotiating* The Tampa Bay Buccaneers are negotiating salary with quarterback Brad Johnson. If the Buccaneers reach the Super Bowl, his salary will be \$2.3 million per year. If the Bucs do not reach the Super Bowl, his salary will be \$1.7 million per year. There is a 62% chance that the Bucs will reach the Super Bowl and a 38% chance that the Bucs will not reach the Super Bowl. Determine the expected salary for Brad Johnson.

Brad Johnson

43. *Spam* Spam (or junk e-mail) is a growing nuisance. In January 2003, about 11% of computer users spent about 10 minutes reviewing and deleting spam, 65% of computer users spent about 15 minutes reviewing and deleting spam, and 24% of computer users spent about 20 minutes reviewing and deleting spam. If a computer user is selected at random, determine the expected time he or she will spend reviewing and deleting spam.

44. *Rolltop Desk* The owner of an antique store estimates that there is a 40% chance she will make \$1000 when she sells an antique rolltop desk, a 50% chance she will make \$500 when she sells the desk, and a 10% chance she will break even when she sells the desk. Determine the expected amount she will make when she sells the desk.

45. *Rolling a Die* A die is rolled many times, and the points facing up are recorded. Find the expected (average) number of points facing up over the long run.

46. *Lawsuit* Don Vello is considering bringing a lawsuit against the Dummote Chemical Company. His lawyer estimates that there is a 70% chance Don will make \$40,000, a 10% chance Don will break even, and a 20% chance they will lose the case and Don will need to pay \$30,000 in legal fees. Estimate Don's expected gain or loss if he proceeds with the lawsuit.

47. *Road Service* On a clear day in Boston, the Automobile Association of American (AAA) makes an average of 110 service calls for motorist assistance, on a rainy day it makes an average of 160 service calls, and on a snowy day it makes an average of 210 service calls. If the weather in Boston is clear 200 days of the year, rainy 100 days of the year, and snowy 65 days of the year, find the expected number of service calls made by the AAA in a given day.

48. *Real Estate* The expenses for Jorge Estrada, a real estate agent, to list, advertise, and attempt to sell a house are \$1000. If Jorge succeeds in selling the house, he will receive a commission of 6% of the sales price. If an agent with a different company sells the house, Jorge still receives 3% of the sales price. If the house is unsold after 3 months, Jorge loses the listing and receives nothing. Suppose that the probability that he sells a \$100,000 house is 0.2, the probability that another agent sells the house is 0.5, and the probability that the house is unsold after 3 months is 0.3. Find Jorge's expectation if he accepts this house for listing. Should Jorge list the house? Explain.

In Exercises 49 and 50, assume that you are blindfolded and throw a dart at the dart board shown. Assuming your dart sticks in the dart board,

a) *determine the probabilities that the dart lands on \$1, \$10, \$20, and \$100, respectively.*

b) *If you win the amount of money indicated by the section of the board where the dart lands, find your expectation when you throw the dart.*

c) *If the game is to be fair, how much should you pay to play?*

49.

$1
$20
$1
$100
$10

50. 

Challenge Problems/Group Activities

51. *Term Life Insurance* An insurance company will pay the face value of a term life insurance policy if the insured person dies during the term of the policy. For how much should an insurance company sell a 10-year term policy with a face value of $40,000 to a 30-year-old man for the company to make a profit? The probability of a 30-year-old man living to age 40 is 0.97. Explain your answer. Remember the customer pays for the insurance before the policy becomes effective.

52. *Lottery Ticket* Is it possible to determine your expectation when you purchase a lottery ticket? Explain.

Roulette In Exercises 53 and 54, use the roulette wheel illustrated. A roulette wheel typically contains slots with numbers 1–36 and slots marked 0 and 00. A ball is spun on the wheel and comes to rest in one of the 38 slots. Eighteen numbers are colored red, and 18 numbers are colored black. The 0 and 00 are colored green. If you bet on one particular number and the ball lands on that number, the house pays off odds of 35 to 1. If you bet on a red number or black number and win, the house pays 1 to 1 (even money).

53. Find the expected value of betting $1 on a particular number.

54. Find the expected value of betting $1 on red.

Recreational Exercise

55. *Wheel of Fortune* The following is a miniature version of the Wheel of Fortune. When Dave Salem spins the wheel, he is awarded the amount on the wheel indicated by the pointer. If the wheel points to Bankrupt, he loses the total amount he has accumulated and also loses his turn. Assume that the wheel stops on a position at random and that each position is equally likely to occur.

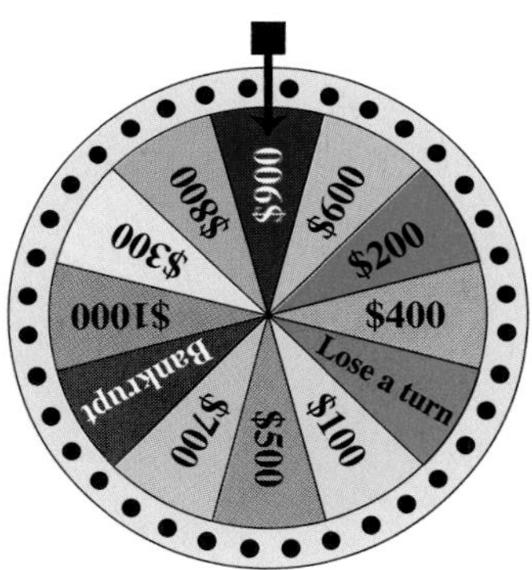

a) Find Mr. Salem's expectation when he spins the wheel at the start of the game (he has no money to lose if he lands on Bankrupt).

b) If Mr. Salem presently has a balance of $1800, find his expectation when he spins the wheel.

12.5 TREE DIAGRAMS

We stated earlier that the possible results of an experiment are called its outcomes. To solve more difficult probability problems we must first be able to determine all the possible outcomes of an experiment. The counting principle can be used to determine the number of outcomes of an experiment and is helpful in constructing tree diagrams.

DID YOU KNOW

Mix and Match

Fashion experts often suggest that you select your wardrobe from pieces of clothing that can be mixed and matched. If, for example, you have 7 shirts, 3 sweaters, and 4 pairs of pants, you actually have 84 possible outfits. If you wear only a sweater or a shirt, but not both, the possibilities go down to 40.

Counting Principle

If a first experiment can be performed in M distinct ways and a second experiment can be performed in N distinct ways, then the two experiments in that specific order can be performed in $M \cdot N$ distinct ways.

If we wanted to find the number of possible outcomes when a coin is tossed and a die is rolled, we could reason that the coin has two possible outcomes, heads and tails. The die has six possible outcomes: 1, 2, 3, 4, 5, and 6. Thus, the two experiments together have $2 \cdot 6$, or 12, possible outcomes.

A list of all the possible outcomes of an experiment is called a *sample space*. Each individual outcome in the sample space is called a *sample point*. *Tree diagrams* are helpful in determining sample spaces.

A tree diagram illustrating all the possible outcomes when a coin is tossed and a die is rolled (see Fig. 12.5) has two initial branches, one for each of the possible outcomes of the coin. Each of these branches will have six branches emerging from them, one for each of the possible outcomes of the die. That will give a total of 12 branches, the same number of possible outcomes found by using the counting principle. We can obtain the sample space by listing all the possible combinations of branches. Note that this sample space consists of 12 sample points.

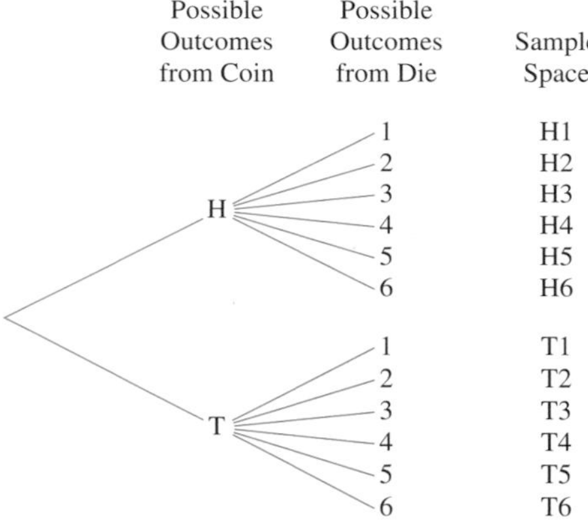

Figure 12.5

Example 1 uses the phrase "without replacement." This phrase tells us that once an item is selected, it cannot be selected again, making it impossible to select the same item twice.

EXAMPLE 1 *Selecting Balls without Replacement*

Two balls are to be selected *without replacement* from a bag that contains one red, one blue, one green, and one orange ball (see Fig. 12.6).

a) Use the counting principle to determine the number of points in the sample space.
b) Construct a tree diagram and list the sample space.
c) Find the probability that one green ball is selected.
d) Find the probability that an orange ball followed by a red ball is selected.

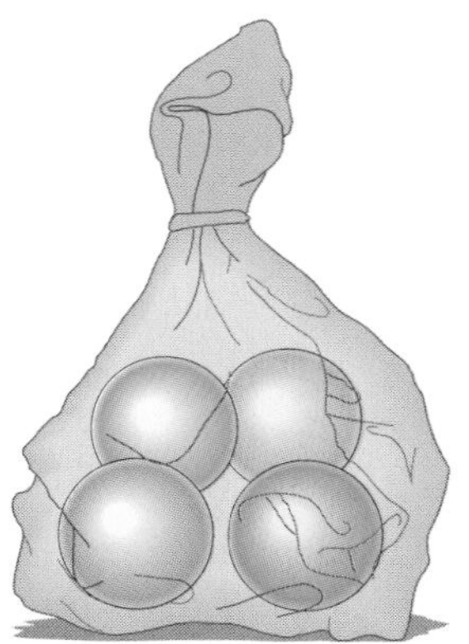

Figure 12.6

SOLUTION:

a) The first selection may be any one of the four balls. Once the first ball is selected, only three balls remain for the second selection. Thus, there are $4 \cdot 3$, or 12, sample points in the sample space.

b) The first ball selected can be red, blue, green, or orange. Since this experiment is done without replacement, the same colored ball cannot be selected twice. For example, if the first ball selected is red, the second ball selected must be either blue, green, or orange. The tree diagram and sample space are shown in Fig. 12.7. The sample space contains 12 points. That result checks with the answer obtained in part (a) using the counting principle.

First Selection	Second Selection	Sample Space
R	B	RB
	G	RG
	O	RO
B	R	BR
	G	BG
	O	BO
G	R	GR
	B	GB
	O	GO
O	R	OR
	B	OB
	G	OG

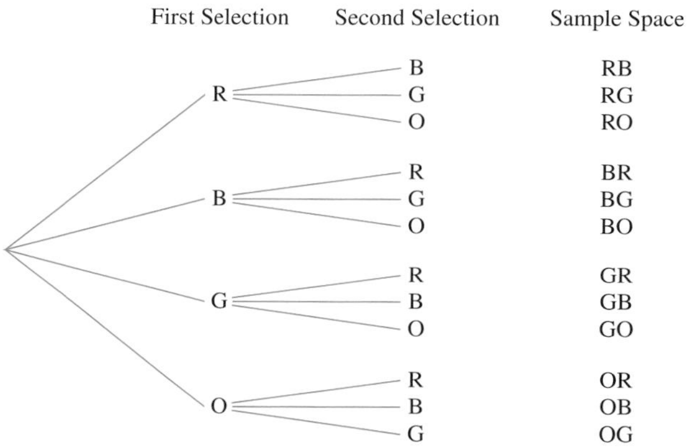

Figure 12.7

c) If we know the sample space, we can compute probabilities using the formula

$$P(E) = \frac{\text{number of outcomes favorable to } E}{\text{total number of outcomes}}$$

The total number of outcomes will be the number of points in the sample space. From Fig. 12.7 we determine that there are 12 possible outcomes. Six outcomes have one green ball: RG, BG, GR, GB, GO, and OG.

$$P(\text{one green ball is selected}) = \frac{6}{12} = \frac{1}{2}$$

d) One possible outcome meets the criteria of an orange ball followed by a red ball: OR.

$$P(\text{orange followed by red}) = \frac{1}{12}$$

The counting principle can be extended to any number of experiments, as illustrated in Example 2.

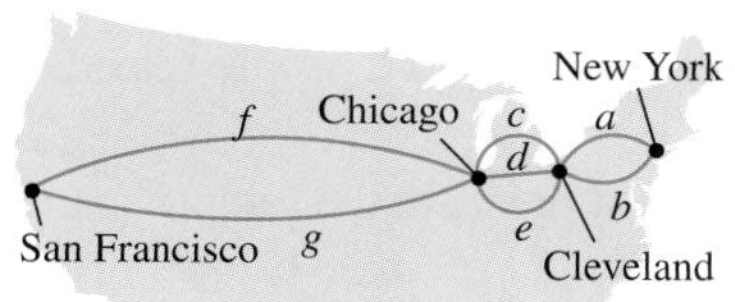

Figure 12.8

EXAMPLE 2 *Using the Counting Principle*

The Gilligans are driving from New York to San Francisco and wish to stop in Cleveland and Chicago. They are considering two highways from New York to Cleveland, three highways from Cleveland to Chicago, and two highways from Chicago to San Francisco, as illustrated in Fig. 12.8.

a) Use the counting principle to determine the number of different routes the Gilligans can take from New York to San Francisco.

b) Use a tree diagram to determine the routes.

c) If a route from New York to San Francisco is selected at random and all routes are considered equally likely, find the probability that both routes a and g are used.

d) Find the probability that neither routes d nor f are used.

SOLUTION:

a) Using the counting principle, we can determine that there are $2 \cdot 3 \cdot 2$, or 12, different routes the Gilligans can take from New York to San Francisco.

b) The tree diagram illustrating the 12 possibilities is given in Fig. 12.9.

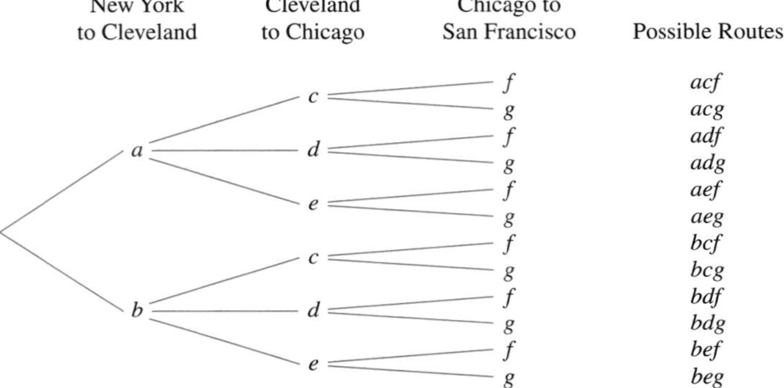

Figure 12.9

c) Of the 12 possible routes, 3 use both a and g (acg, adg, aeg).

$$P(\text{routes } a \text{ and } g \text{ are both used}) = \frac{3}{12} = \frac{1}{4}$$

d) Of the 12 possible routes, 4 use neither d nor f (acg, aeg, bcg, beg).

$$P(\text{neither } d \text{ nor } f \text{ is used}) = \frac{4}{12} = \frac{1}{3}$$

▲

EXAMPLE 3 *Selecting Ticket Winners*

A radio station has two tickets to give away to a Britney Spears concert. It held a contest and narrowed the possible recipients down to four people: Christine (C), Mike Hammer (MH), Mike Levine (ML), and Phyllis (P). The names of two of these four people will be selected at random from a hat and the two people selected will be awarded the tickets.

a) Use the counting principle to determine the number of points in the sample space.

b) Construct a tree diagram and list the sample space.

c) Determine the probability that Christine is selected.

d) Determine the probability that neither Mike Hammer nor Mike Levine is selected.

e) Determine the probability that at least one Mike is selected.

SOLUTION:

a) The first selection may be any one of the four people; see Fig. 12.10. Once the first person is selected, only three people remain for the second selection. Thus, there are 4 · 3 or 12 sample points in the sample space.

b)

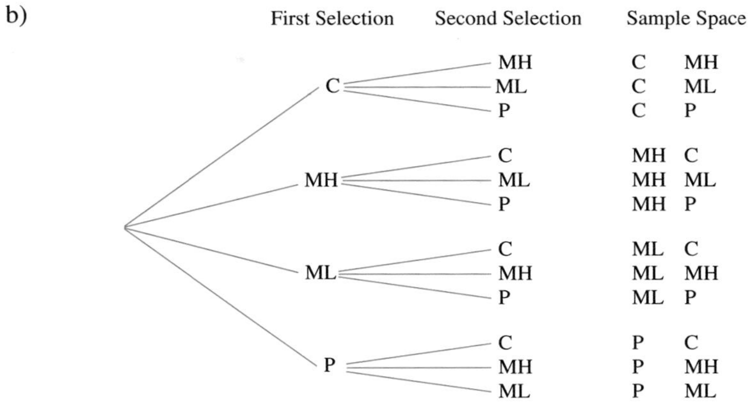

Figure 12.10

c) Of the 12 points in the sample space, 6 have Christine. They are C MH, C ML, C P, MH C, ML C, and P C.

$$P(\text{Christine is selected}) = \frac{6}{12} = \frac{1}{2}$$

d) Of the 12 points in the sample space, two have neither Mike. They are C P (Christine Phyllis) and P C (Phyllis Christine).

$$P(\text{neither Mike selected}) = \frac{2}{12} = \frac{1}{6}$$

e) At least one Mike means that one or more Mikes are selected. There are 10 points in the sample space with at least one Mike (all those except Christine Phyllis and Phyllis Christine).

$$P(\text{at least one Mike is selected}) = \frac{10}{12} = \frac{5}{6}$$

▲

In Example 3, if you add the probability of no Mike being selected with the probability of at least one Mike being selected, you get $\frac{1}{6} + \frac{5}{6}$, or 1. In any probability problem, if E is a specific event, then either E happens at least one time or it does not happen at all. Thus, $P(E$ happening at least once$) + P(E$ does not happen$) = 1$.This leads to the following rule.

$$P(\text{event happening at least once}) = 1 - P(\text{event does not happen})$$

For example, suppose that the probability of not getting any red flowers from 3 seeds that are planted is $\frac{2}{7}$. Then the probability of getting at least one red flower from the 3 seeds that are planted is $1 - \frac{2}{7} = \frac{5}{7}$. We will use this rule in later sections.

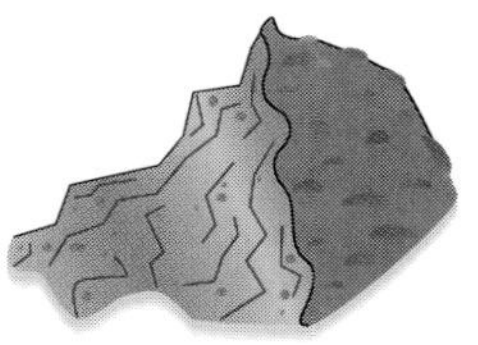

Figure 12.11

In all the tree diagrams in this section, the outcomes were always equally likely; that is, each outcome had the same probability of occurrence. Consider a rock that has 4 faces such that each face has a different surface area and the rock is not uniform in density (see Figure 12.11). When the rock is dropped, the probability that the rock lands on face 1 will not be the same as the probability that the rock lands on face 2. In fact, the probability that the rock lands on face 1, face 2, face 3, and face 4 may all be different. Therefore, the outcomes of the rock landing on face 1, face 2, face 3, and face 4 are not equally likely outcomes. Because the outcomes are not equally likely and we are not given additional information, we cannot determine the theoretical probability of the rock landing on each individual face. However, we can still determine the sample space indicating the faces that the rock may land on when the rock is dropped twice. The tree diagram and sample space is shown in Figure 12.12.

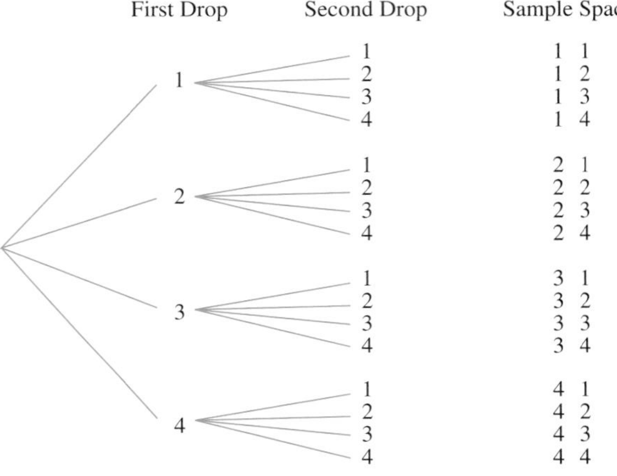

Figure 12.12

Since the outcomes are not equally likely, the probability of each of the sixteen sample points in the sample space occurring cannot be determined. If the outcomes were equally likely, then each of the sixteen points in the sample space would have a probability of $\frac{1}{16}$. See Exercises 29 and 30 which deal with outcomes that are not equally likely.

SECTION 12.5 EXERCISES

Concept/Writing Exercises

1. Explain the counting principle.
2. **a)** What is a sample space?
   b) What is a sample point?
3. If a first experiment can be performed in two distinct ways and a second experiment can be performed in seven distinct ways, how many possible ways can the two experiments be performed? Explain your answer.
4. In your own words, describe how to construct a tree diagram.
5. A problem states that two selections are made "without replacement." Explain what that means.
6. One experiment has five equally likely outcomes, and a second experiment has two equally likely outcomes. How many sample points will be in the sample space when the two experiments are performed one after the other?

Practice the Skills

7. *Selecting States* If two states are selected at random from the 50 states, use the counting principle to determine the number of possible outcomes if the states are selected
   a) with replacement.
   b) without replacement.
8. *Selecting Dates* If two dates are selected at random from the 365 days of the year, use the counting principle to determine the number of possible outcomes if the dates are selected
   a) with replacement.
   b) without replacement.
9. *Selecting Batteries* A bag contains six batteries, all of which are the same size and are equally likely to be selected. Each battery is a different brand. If you select three batteries at random, use the counting principle to deter-

mine how many points will be in the sample space if the batteries are selected
a) with replacement.
b) without replacement.

10. *Remote Control* Your television remote control has buttons for digits 0–9. If you press two buttons, how many numbers are possible if
a) the same button may be pressed twice.
b) the same button may not be pressed twice.

In Exercises 11–28, use the counting principle to determine the answer to part (a). Assume that each event is equally likely to occur.

11. *Coin Toss* Two coins are tossed.
a) Determine the number of points in the sample space.
b) Construct a tree diagram and list the sample space.
Find the probability that
c) no heads are tossed.
d) exactly one head is tossed.
e) two heads are tossed.

12. *Boys and Girls* A couple plans to have two children.
a) Determine the number of points in the sample space of the possible arrangements of boys and girls.
b) Construct a tree diagram and list the sample space.
Find the probability that the couple has
c) two girls.
d) at least one girl.
e) a girl and then a boy.

13. *Cards* A box contains three cards. On one card there is a circle. On another card there is a square, and on the third card there is a triangle.

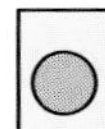 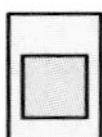 

Two cards are to be selected at random with replacement.
a) Determine the number of points in the sample space.
b) Construct a tree diagram and determine the sample space.
Find the probability that
c) two circles are selected.
d) a square and then a triangle are selected.
e) at least one circle is selected.

14. *Cards* Repeat Exercise 13 but assume the cards are drawn without replacement.

15. *Marble Selection* A hat contains four marbles: 1 yellow, 1 red, 1 blue, and 1 green. Two marbles are to be selected at random without replacement from the hat.

a) Determine the number of points in the sample space.
b) Construct a tree diagram and list the sample space.
Find the probability of selecting
c) exactly 1 red marble.
d) at least 1 marble that is not red.
e) no green marbles.

16. *Three Coins* Three coins are tossed.
a) Determine the number of points in the sample space.
b) Construct a tree diagram and list the sample space.
Find the probability that
c) no heads are tossed.
d) exactly one head is tossed.
e) three heads are tossed.

17. *Three Children* A couple plans to have three children.
a) Determine the number of points in the sample space of possible arrangements of boys and girls.
b) Construct a tree diagram and list the sample space.
Find the probability that the couple has
c) no boys.
d) at least one girl.
e) either exactly two boys or two girls.
f) two boys first, then one girl.

18. *Pet Shop* A pet shop is selling a calico cat, a Siamese cat, a Persian cat, and a Himalayan cat. The Chens are going to select two cats to bring home as pets.
a) Determine the number of points in the sample space.
b) Construct a tree diagram and list the sample space.
Find the probability they select
c) the Persian cat.
d) the Persian cat and the calico cat.
e) cats other than the Persian cat.

Problem Solving

19. *Rolling Dice* Two dice are rolled.
a) Determine the number of points in the sample space.
b) Construct a tree diagram and list the sample space.
Find the probability that
c) a double (a 1, 1 or 2, 2, etc.) is rolled.
d) a sum of 7 is rolled.
e) a sum of 2 is rolled.
f) Are you as likely to roll a sum of 2 as you are of rolling a sum of 7? Explain your answer.

20. *Voting* At a homeowners association meeting, a board member can vote yes, no, or abstain on a motion. There are three motions the board is voting on.
a) Determine the number of points in the sample space.
b) Construct a tree diagram and determine the sample space.
Find the probability that a board member votes
c) no, yes, no in that order.
d) yes on exactly two of the motions.
e) yes on at least one motion.

21. *Gift Certificates* Three different people are to be selected at random, and each will be given one gift certificate. There is one certificate from Home Depot, one from Sears, and one from Outback Steakhouse. The first person selected gets to choose one of the certificates. The second person

selected gets to choose between the two remaining certificates. The third person selected gets the third certificate.

a) Determine the number of points in the sample space.

b) Construct a tree diagram and determine the sample space.

Find the probability that

c) Sears is selected first.

d) Home Depot is selected first and Outback Steakhouse is selected last.

e) The certificates are selected in this order: Sears, Outback Steakhouse, Home Depot.

22. *Shopping* Susan Forman has to purchase a box of cereal, a bottle of soda, and a can of vegetables. The types of cereal, soda, and vegetables she is considering are shown below.

Cereal: Rice Krispies, Frosted Flakes, Honeycomb
Soda: orange, black cherry, ginger ale
Vegetables: peas, carrots

a) Determine the number of points in the sample space.

b) Construct a tree diagram and determine the sample space.

Find the probability that she selects

c) Honeycomb for the cereal.

d) Rice Krispies and ginger ale.

e) a soda other than black cherry.

23. *Vacationing in Florida* Peter and Carol Collinge are vacationing in the Orlando, Florida, area. They have listed what they consider the major attractions in the area in two groups: Disney attractions and non-Disney attractions. The four Disney attractions are the Magic Kingdom, Epcot Center, MGM Studios, and Animal Kingdom. The four non-Disney attractions are Sea World, Universal Studios, Islands of Adventure, and Busch Gardens (Tampa). They decided to first visit one Disney and then one non-Disney attraction. They will select the name of the one Disney attraction from one hat and the name of one non-Disney attraction from a second hat.

a) Determime the number of points in the sample space.

b) Construct a tree diagram and determine the sample space.

Find the probability that they select

c) the Magic Kingdom or Epcot Center.

d) MGM Studios or Universal Studios.

e) the Magic Kingdom and either Sea World or Busch Gardens.

Epcot Center

24. *A New Computer* You visit Computer City to purchase a new computer system. You are going to purchase a computer, printer, and monitor from among the following brands.

Computer	Printer	Monitor
Compaq	Hewlett-Packard	Omega
IBM	Epson	Toshiba
Apple		
Dell		

a) Determine the number of points in the sample space.

b) Construct a tree diagram and determine the sample space.

Find the probability of selecting

c) an Apple computer.

d) a Hewlett-Packard printer.

e) an Apple computer and a Hewlett-Packard printer.

25. *Buying Appliances* Mr. and Mrs. J. B. Davis just moved into a new home and need to purchase kitchen appliances. The chart below shows the brands of appliances they are considering.

Refrigerator	Stove	Dishwasher
General Electric	General Electric	General Electric
Kenmore	Frigidaire	KitchenAid
Maytag	Roper	Whirlpool

The Davises are going to purchase a refrigerator, a stove, and a dishwasher.

a) Determine the number of points in the sample space.

b) Construct a tree diagram and determine the sample space.

Find the probability that they select

c) General Electric for all three appliances.

d) no General Electric appliances.

e) at least one General Electric appliance.

26. *Summer School* You decide to take three courses during summer school: an English course, a mathematics course, and a science course. The available courses that you can take are illustrated below.

English	Mathematics	Science
English composition	College algebra	Biology
English literature	Statistics	Geology
	Calculus	Chemistry
		Physics

a) Determine the number of points in the sample space.

b) Construct a tree diagram and determine the sample space.

Find the probability that

c) geology is selected.

d) either geology or chemistry is selected.

e) calculus is not selected.

27. *Personal Characteristics* An individual can be classified as male or female with red, brown, black, or blonde hair and with brown, blue, or green eyes.

a) How many different classifications are possible (for example, male, red-headed, blue-eyed)?
b) Construct a tree diagram to determine the sample space.
c) If each outcome is equally likely, find the probability that the individual will be a male with black hair and blue eyes.
d) Find the probability that the individual will be a female with blonde hair.

28. *Mendel Revisited* A pea plant must have exactly one of each of the following pairs of traits: short (s) or tall (t); round (r) or wrinkled (w) seeds; yellow (y) or green (g) peas; and white (wh) or purple (p) flowers (for example, short, wrinkled, green pea with white flowers).

a) How many different classifications of pea plants are possible?
b) Use a tree diagram to determine all the classifications possible.
c) If each characteristic is equally likely, find the probability that the pea plant will have round peas.
d) Find the probability that the pea plant will be short, have wrinkled seeds, have yellow seeds, and have purple flowers.

Challenge Problems/Group Activities

29. *Three Chips* Suppose that a bag contains one white chip and two red chips. Two chips are going to be selected at random from the bag *with replacement*.
a) What is the probability of selecting a white chip from the bag on the first selection?
b) What is the probability of selecting a red chip from the bag on the first selection?
c) Are the outcomes of selecting a white chip and selecting a red chip on the first selection equally likely? Explain.
d) The sample space when two chips are selected from the bag with replacement is ww, wr, rw, rr. Do you believe that the probability of selecting ww is greater than, equal to, or less than the probability of selecting rr? Explain.

30. A thumbtack is dropped on a concrete floor. Assume the thumbtack can only land point up (u) and point down (d), as shown in the figure below.

If two thumbtacks are dropped, one after the other, the tree diagram below can be used to show the possible outcomes.

First Thumbtack	Second Thumbtack	Sample Space
u	u	$u\,u$
	d	$u\,d$
d	u	$d\,u$
	d	$d\,d$

a) Do you believe the outcomes of the thumbtack landing point up and the thumbtack landing point down are equally likely? Explain.
b) List the sample points in the sample space of this experiment.
c) Do you believe the probability that both thumbtacks land point up (uu) is the same as the probability that both thumbtacks land point down (dd)? Explain.
d) Can you compute the theoretical probability of the thumbtack landing point up and the theoretical probability of the thumbtack landing point down? Explain.
e) Obtain a box of thumbtacks and drop the thumbtacks out of the box with care. Determine the empirical probability of a thumbtack landing point up when dropped and the empirical probability of a thumbtack landing point down when dropped.

Recreational Exercise

31. *Ties* All my ties are red except two. All my ties are blue except two. All my ties are brown except two. How many ties do I have?

32. *Rock Faces* An experiment consists of 3 parts: flipping a coin, tossing a rock, and rolling a die. If the sample space consists of 60 sample points, determine the number of faces on the rock.

12.6 *OR* AND *AND* PROBLEMS

In Section 12.5, we showed how to work probability problems by constructing sample spaces. Often it is inconvenient or too time consuming to solve a problem by first constructing a sample space. For example, if an experiment consists of selecting two cards with replacement from a deck of 52 cards, there would be $52 \cdot 52$ or 2704 points in the sample space. Trying to list all these sample points could take hours. In this section, we learn how to solve *compound probability* problems that contain the words *and* or *or* without constructing a sample space.

Or Problems

The *or probability problem* requires obtaining a "successful" outcome for *at least one* of the given events. For example, suppose that we roll one die and we are interested in finding the probability of rolling an even number *or* a number greater than 4. For this situation rolling either a 2, 4, or 6 (an even number) or a 5 or 6 (a number greater than 4) would be considered successful. Note that the number 6 satisfies both criteria. Since 4 out of 6 of the numbers meet the criteria (the 2, 4, 5, and 6), the probability of rolling an even number *or* a number greater than 4 is $\frac{4}{6}$ or $\frac{2}{3}$.

A formula for finding the probability of event A or event B, symbolized $P(A \text{ or } B)$, follows.

$$P(A \text{ or } B) = P(A) + P(B) - P(A \text{ and } B)$$

Since we add (and subtract) probabilities to find $P(A \text{ or } B)$, this formula is sometimes referred to as the *addition formula.* We explain the use of the *or* formula in Example 1.

EXAMPLE 1 *Using the Addition Formula*

Each of the numbers 1, 2, 3, 4, 5, 6, 7, 8, 9, and 10 is written on a separate piece of paper. The 10 pieces of paper are then placed in a hat, and one piece is randomly selected. Find the probability that the piece of paper selected contains an even number or a number greater than 6.

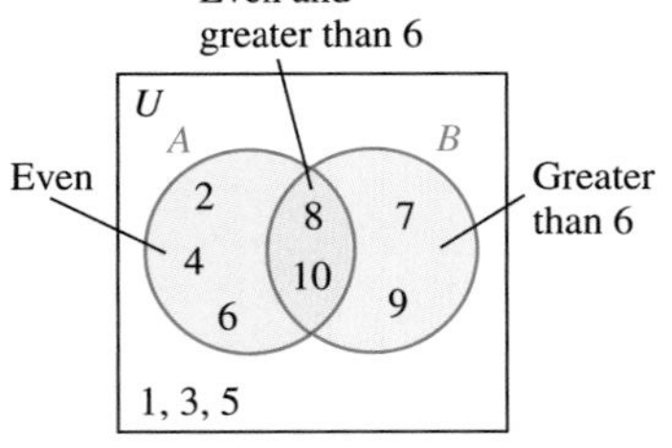

Figure 12.13

SOLUTION: We are asked to find the probability the number selected *is even* or *greater than 6.* Let's use set A to represent the statement, "the number is even," and set B to represent the statement, "the number is greater than 6." Figure 12.13 is a Venn diagram, as introduced in Chapter 2, with sets A (even) and B (greater than 6). There are a total of 10 numbers, of which five are even (2, 4, 6, 8, and 10). Thus, the probability of selecting an even number is $\frac{5}{10}$. Four numbers are greater than 6: the 7, 8, 9, and 10. Thus, the probability of selecting a number greater than 6 is $\frac{4}{10}$. Two numbers are both even and greater than 6: the 8 and 10. Thus, the probability of selecting a number that is both even and greater than 6 is $\frac{2}{10}$.

If we substitute the appropriate statements for A and B in the formula, we obtain

$$\begin{aligned} P(A \text{ or } B) &= P(A) + P(B) - P(A \text{ and } B) \\ P\begin{pmatrix}\text{even or} \\ \text{greater than 6}\end{pmatrix} &= P(\text{even}) + P\begin{pmatrix}\text{greater} \\ \text{than 6}\end{pmatrix} - P\begin{pmatrix}\text{even and} \\ \text{greater than 6}\end{pmatrix} \\ &= \frac{5}{10} + \frac{4}{10} - \frac{2}{10} \\ &= \frac{7}{10} \end{aligned}$$

Thus, the probability of selecting an even number or a number greater than 6 is $\frac{7}{10}$. ▲

Example 1 illustrates that when finding the probability of A or B, we add the probabilities of events A and B and then subtract the probability of both events occurring simultaneously.

EXAMPLE 2 *Using the Addition Formula*

Consider the same sample space, the numbers 1 through 10, as in Example 1. If one piece of paper is selected, find the probability that it contains a number less than 4 or a number greater than 6.

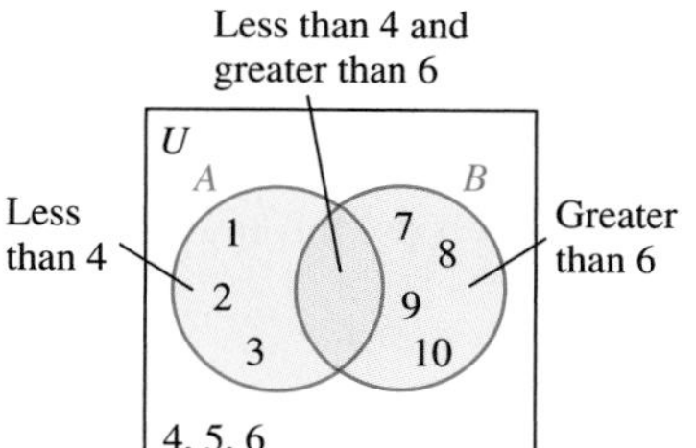

Figure 12.14

SOLUTION: Let A represent the statement, "the number is less than 4," and let B represent the statement, "the number is greater than 6." A Venn diagram illustrating these statements is shown in Fig. 12.14.

$$P(\text{number is less than 4}) = \frac{3}{10}$$

$$P(\text{number is greater than 6}) = \frac{4}{10}$$

Since there are no numbers that are *both* less than 4 and greater than 6, $P(\text{number is less than 4 and greater than 6}) = 0$. Therefore,

$$P\begin{pmatrix}\text{number is}\\ \text{less than 4}\\ \text{or greater}\\ \text{than 6}\end{pmatrix} = P\begin{pmatrix}\text{number is}\\ \text{less than 4}\end{pmatrix} + P\begin{pmatrix}\text{number is}\\ \text{greater than 6}\end{pmatrix} - P\begin{pmatrix}\text{number is}\\ \text{less than 4}\\ \text{and greater}\\ \text{than 6}\end{pmatrix}$$

$$= \frac{3}{10} + \frac{4}{10} - 0 = \frac{7}{10}$$

Thus, the probability of selecting a number less than 4 or greater than 6 is $\frac{7}{10}$. ▲

In Example 2, it is impossible to select a number that is both less than 4 *and* greater than 6 when only one number is to be selected. Events such as these are said to be *mutually exclusive.*

> Two events A and B are **mutually exclusive** if it is impossible for both events to occur simultaneously.

If events A and B are mutually exclusive, then $P(A \text{ and } B) = 0$, and the addition formula simplifies to $P(A \text{ or } B) = P(A) + P(B)$.

EXAMPLE 3 *Probability of A or B*

One card is selected from a standard deck of playing cards. Determine whether the following pairs of events are mutually exclusive, and find $P(A \text{ or } B)$.

a) A = an ace, B = a king

b) A = an ace, B = a heart

c) A = a red card, B = a black card

d) A = a picture card, B = a red card

The ace of hearts is both an ace and a heart.

SOLUTION:

a) There are four aces and four kings in a standard deck of 52 cards. It is impossible to select both an ace and a king when only one card is selected. Therefore, these events are mutually exclusive.

$$P(\text{ace or king}) = P(\text{ace}) + P(\text{king}) = \frac{4}{52} + \frac{4}{52} = \frac{8}{52} = \frac{2}{13}$$

b) There are 4 aces and 13 hearts in a standard deck of 52 cards. One card, the ace of hearts, is both an ace and a heart. Therefore, these events are not mutually exclusive.

$$P(\text{ace}) = \frac{4}{52} \qquad P(\text{heart}) = \frac{13}{52} \qquad P(\text{ace and heart}) = \frac{1}{52}$$

$$\begin{aligned} P(\text{ace or heart}) &= P(\text{ace}) + P(\text{heart}) - P(\text{ace and heart}) \\ &= \frac{4}{52} + \frac{13}{52} - \frac{1}{52} \\ &= \frac{16}{52} = \frac{4}{13} \end{aligned}$$

c) There are 26 red cards and 26 black cards in a standard deck of 52 cards. It is impossible to select one card that is both a red card and a black card. Therefore, the events are mutually exclusive.

$$\begin{aligned} P(\text{red or black}) &= P(\text{red}) + P(\text{black}) \\ &= \frac{26}{52} + \frac{26}{52} = \frac{52}{52} = 1 \end{aligned}$$

Therefore, a red card or a black card must be selected.

d) There are 12 picture cards in a standard deck of 52 cards. Six of the 12 picture cards are red (the jacks, queens, and kings of hearts and diamonds). Thus, selecting a picture card and a red card are not mutually exclusive.

$$\begin{aligned} P\begin{pmatrix}\text{picture card}\\ \text{or red card}\end{pmatrix} &= P\begin{pmatrix}\text{picture}\\ \text{card}\end{pmatrix} + P\begin{pmatrix}\text{red}\\ \text{card}\end{pmatrix} - P\begin{pmatrix}\text{picture card}\\ \text{and red card}\end{pmatrix} \\ &= \frac{12}{52} + \frac{26}{52} - \frac{6}{52} \\ &= \frac{32}{52} = \frac{8}{13} \end{aligned}$$

And Problems

A second type of probability problem is the *and probability problem*, which requires obtaining a favorable outcome in *each* of the given events. For example, suppose that *two* cards are to be selected from a deck of cards and we are interested in the probability of selecting two aces (one ace *and* then a second ace). Only if *both* cards selected are aces would this experiment be considered successful. A formula for finding the probability of events A and B, symbolized $P(A \text{ and } B)$, follows.

DID YOU KNOW

They're All Girls

The probability that a husband and wife have nine children where all are girls is $(\frac{1}{2})^9$ or $\frac{1}{512}$. The odds against a family having nine girls in a row is 511 to 1. Thus, for every 512 families with nine children, on average, one family will have all girls and 511 will not. However, should another child be born to that one family, the probability of the tenth child being a girl is still $\frac{1}{2}$.

$$P(A \text{ and } B) = P(A) \cdot P(B)$$

Since we multiply to find $P(A \text{ and } B)$, this formula is sometimes referred to as the *multiplication formula.* When using the multiplication formula, we **always assume that event A has occurred when calculating $P(B)$** because we are determining the probability of obtaining a favorable outcome in both of the given events.*

Unless we specify otherwise, $P(A \text{ and } B)$ indicates that we are determining the probability that event A occurs *and then* event B occurs (in that order). Consider a bag that contains three chips: 1 red chip, 1 blue chip, and 1 green chip. Suppose that two chips are selected from the bag with replacement. The tree diagram and sample space for the experiment are shown in Fig. 12.15. There are nine possible outcomes for the two selections, as indicated in the sample space.

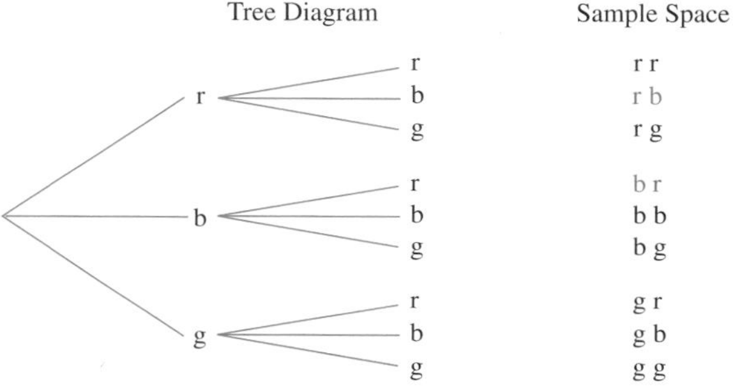

Figure 12.15

Notice that the probability of selecting a red chip followed by a blue chip (rb), indicated by $P(\text{red and blue})$, is $\frac{1}{9}$. The probability of selecting a red chip and a blue chip, in any order (rb or br), is $\frac{2}{9}$. In this section, when we ask for $P(A \text{ and } B)$, it means the probability of event A occurring *and then* event B occurring, in that order.

EXAMPLE 4 *An Experiment with Replacement*

Two cards are to be selected *with replacement* from a deck of cards. Find the probability that two aces will be selected.

SOLUTION: Since the deck of 52 cards contains four aces, the probability of selecting an ace on the first draw is $\frac{4}{52}$. The card selected is then returned to the deck. Therefore, the probability of selecting an ace on the second draw remains $\frac{4}{52}$.

If we let A represent the selection of the first ace and B represent the selection of the second ace, the formula may be written

$$
\begin{aligned}
P(A \text{ and } B) &= P(A) \cdot P(B) \\
P(2 \text{ aces}) = P(\text{ace 1 } \textit{and} \text{ ace 2}) &= P(\text{ace 1}) \cdot P(\text{ace 2}) \\
&= \frac{4}{52} \cdot \frac{4}{52} \\
&= \frac{1}{13} \cdot \frac{1}{13} = \frac{1}{169}
\end{aligned}
$$

*$P(B)$, assuming that event A has occurred, may be denoted $P(B \mid A)$, which is read "the probability of B, given A." We will discuss this type of probability (conditional probability) further in Section 12.7.

DID YOU KNOW

The Birthday Problem

Among 24 people chosen at random, what would you guess is the probability that at least 2 of them have the same birthday? It might surprise you to learn it is greater than 0.5. There are 365 days on which the first person selected can have a birthday. That person has a 365/365 chance of having a birthday on one of those days. The probability that the second person's birthday is on any other day is 364/365. The probability that the third person's birthday is on a day different from the first two is 363/365, and so on. The probability that the 24th person has a birthday on any other day than the first 23 people is 342/365. Thus, the probability, P, that of 24 people, no 2 have the same birthday is $(365/365) \times (364/365) \times (363/365) \times \cdots \times (342/365) = 0.462$. Then the probability of at least 2 people of 24 having the same birthday is $1 - P = 1 - 0.462 = 0.538$, or slightly larger than $\frac{1}{2}$.

EXAMPLE 5 *An Experiment without Replacement*

Two cards are to be selected *without replacement* from a deck of cards. Find the probability that two aces will be selected.

SOLUTION: This example is similar to Example 4. However, this time we are doing the experiment without replacing the first card selected to the deck before selecting the second card.

The probability of selecting an ace on the first draw is $\frac{4}{52}$. When calculating the probability of selecting the second ace, we must assume that the first ace has been selected. Once this first ace has been selected, only 51 cards, including 3 aces, remain in the deck. The probability of selecting an ace on the second draw becomes $\frac{3}{51}$. The probability of selecting two aces without replacement is

$$\begin{aligned} P(2 \text{ aces}) &= P(\text{ace } 1) \cdot P(\text{ace } 2) \\ &= \frac{4}{52} \cdot \frac{3}{51} \\ &= \frac{1}{13} \cdot \frac{1}{17} = \frac{1}{221} \end{aligned}$$

▲

Now we introduce *independent events*.

Event A and event B are **independent events** if the occurrence of either event in no way affects the probability of occurrence of the other event.

Rolling dice and tossing coins are examples of independent events. In Example 4, the events are independent since the first card was returned to the deck. The probability of selecting an ace on the second draw was not affected by the first selection. The events in Example 5 are not independent since the probability of the selection of the second ace was affected by removing the first ace selected from the deck. Such events are called *dependent events*. *Experiments done with replacement will result in independent events, and those done without replacement will result in dependent events.*

EXAMPLE 6 *Independent or Dependent Events?*

One hundred people attended a dinner to recognize the accomplishments of a local citizen. Three people in attendance will be selected at random without replacement, and each will be awarded one door prize. Are the events of selecting the three people who will be awarded the door prize independent or dependent events?

SOLUTION: The events are dependent since each time one person is selected, it changes the probability of the next person being selected. In the first selection, the probability that a specific individual is selected is $\frac{1}{100}$. If that person is not selected first, the probability that the specific person is selected second changes to $\frac{1}{99}$. In general, in any experiment in which two or more items are selected *without replacement*, the events will be dependent.

▲

The multiplication formula may be extended to more than two events, as illustrated in Example 7.

EXAMPLE 7 *Flower Probabilities*

A package of 25 zinnia seeds contain 12 seeds for red flowers, 5 seeds for white flowers, and 8 seeds for yellow flowers. Three seeds are randomly selected and planted. Find the probability of each of the following.

a) All three seeds will produce yellow flowers.

b) The first seed selected will produce a yellow flower, the second seed will produce a red flower, and the third seed will produce a yellow flower.

c) None of the seeds will produce yellow flowers.

d) At least one will produce yellow flowers.

SOLUTION: Each time a seed is selected and planted, the number of seeds remaining decreases by one.

a) The probability that the first seed selected produces a yellow flower is $\frac{8}{25}$. If the first seed selected is yellow, only 7 yellow seeds in 24 are left. The probability of selecting a second yellow seed is $\frac{7}{24}$. If the second seed selected is yellow, only 6 yellow seeds in 23 are left. The probability of selecting a third yellow seed is $\frac{6}{23}$.

$$P(3 \text{ yellow seeds}) = P(\text{yellow seed } 1) \cdot P(\text{yellow seed } 2) \cdot P(\text{yellow seed } 3)$$
$$= \frac{8}{25} \cdot \frac{7}{24} \cdot \frac{6}{23} = \frac{14}{575}$$

b) The probability that the first seed selected produces a yellow flower is $\frac{8}{25}$. Once a seed for a yellow flower is selected, only 24 seeds are left. Twelve of the remaining 24 will produce red flowers. Thus, the probability that the second seed selected will produce a red flower is $\frac{12}{24}$. After the second seed has been selected, there are 23 seeds left, 7 of which will produce yellow flowers. The probability that the third seed produces a yellow flower is therefore $\frac{7}{23}$.

$$P\begin{pmatrix}\text{first yellow and second} \\ \text{red and third yellow}\end{pmatrix} = P\begin{pmatrix}\text{first} \\ \text{yellow}\end{pmatrix} \cdot P\begin{pmatrix}\text{second} \\ \text{red}\end{pmatrix} \cdot P\begin{pmatrix}\text{third} \\ \text{yellow}\end{pmatrix}$$
$$= \frac{8}{25} \cdot \frac{12}{24} \cdot \frac{7}{23} = \frac{28}{575}$$

c) If no flowers are to be yellow, they must either be red or white. Seventeen seeds will not produce yellow flowers (12 for red and 5 for white). The probability that the first seed does not produce a yellow flower is $\frac{17}{25}$. After the first seed has been selected, 16 of the remaining 24 seeds will not produce yellow flowers. After the second seed has been selected, 15 of the remaining 23 seeds will not produce yellow flowers.

$$P(\text{none yellow}) = P\begin{pmatrix}\text{first not} \\ \text{yellow}\end{pmatrix} \cdot P\begin{pmatrix}\text{second not} \\ \text{yellow}\end{pmatrix} \cdot P\begin{pmatrix}\text{third not} \\ \text{yellow}\end{pmatrix}$$
$$= \frac{17}{25} \cdot \frac{16}{24} \cdot \frac{15}{23} = \frac{34}{115}$$

d) In Section 12.5, we learned that

$$P(\text{event happening at least once}) = 1 - P(\text{event does not happen})$$

In part (c), we found that the probability of selecting no yellow flowers is $\frac{34}{115}$. Therefore, the probability that at least one of the seeds will produce yellow flowers can be found as follows:

$$\begin{aligned} P(\text{at least one yellow flower}) &= 1 - P(\text{no yellow flowers}) \\ &= 1 - \frac{34}{115} = \frac{115}{115} - \frac{34}{115} = \frac{81}{115} \end{aligned}$$

TIMELY TIP

Which formula to use

It is sometimes difficult to determine when to use the *or* formula and when to use the *and* formula. The following information may be helpful in deciding which formula to use.

***Or* formula**

Or problems will almost always contain the word *or* in the statement of the problem. For example, find the probability of selecting a heart *or* a 6. *Or* problems in this book generally involve only *one* selection. For example, "one card is selected" or "one die is rolled."

***And* formula**

And problems often do *not* use the word *and* in the statement of the problem. For example, "find the probability that both cards selected are red" or "find the probability that none of those selected is a banana" are both *and*-type problems. *And* problems in this book will generally involve *more than one* selection. For example, the problem may read "two cards are selected" or "three coins are flipped."

DID YOU KNOW

Slot Machines

You pull the handle, or push the button, to activate the slot machine. You are hoping that each of the three reels stops on Jackpot. The first reel shows Jackpot, the second reel shows Jackpot, and you hold your breath hoping the third reel shows Jackpot. When the third reel comes to rest, though, the Jackpot appears either one row above or one row below the row containing the other two Jackpots. You say to yourself, "I came so close" and try again. In reality, however, you probably did not come close to winning the Jackpot at all. The instant you pull the slot machine's arm, or push the button, the outcome is decided by a computer inside the slot machine. The computer uses stop motors to turn each reel and stop it at predetermined points. A random number generator within the computer is used to determine the outcomes. Although each reel generally has 22 stops, each stop is not equally likely. The computer assigns only a few random numbers to the stop that places the Jackpots on all three reels at the same time and many more random numbers to stop the reels on stops that are not the Jackpot. If you do happen to hit the Jackpot, it is simply because the random number generator happened to generate the right sequence of numbers the instant you activated the machine. For more detailed information check *www.howstuffworks.com.* Also see the Did You Know on page 714.

SECTION 12.6 EXERCISES

Concept/Writing Exercises

1. a) In $P(A \text{ or } B)$, what does the word *or* indicate?
   b) In $P(A \text{ and } B)$, what does the word *and* indicate?
2. a) Give the formula for $P(A \text{ or } B)$.
   b) In your own words, explain how to determine $P(A \text{ or } B)$ with the formula.
3. a) What are mutually exclusive events? Give an example.
   b) How do you calculate $P(A \text{ or } B)$ when A and B are mutually exclusive?
4. a) Give the formula for $P(A \text{ and } B)$.
   b) In your own words, explain how to determine $P(A \text{ and } B)$ with the formula.
5. When finding $P(B)$ using the formula $P(A \text{ and } B)$, what do we always assume?
6. What are independent events? Give an example.
7. What are dependent events? Give an example.
8. A family is selected at random. Let event A be the mother likes classical music. Let event B be the daughter likes classical music.
   a) Are events A and B mutually exclusive? Explain.
   b) Are they independent events? Explain.
9. A family is selected at random. Let event A be the father likes to swim. Let event B be the mother likes chocolate chip cookies.
   a) Are events A and B mutually exclusive? Explain.
   b) Are they independent events? Explain.
10. An individual is selected at random. Let event A be the individual owns a computer. Let event B be the individual owns a digital camera.
    a) Are events A and B mutually exclusive? Explain.
    b) Are they independent events? Explain.
11. If events A and B are mutually exclusive, explain why the formula $P(A \text{ or } B) = P(A) + P(B) - P(A \text{ and } B)$ can be simplified to $P(A \text{ or } B) = P(A) + P(B)$.
12. a) Write a problem that you would use the *or formula* to solve. Solve the problem and give the answer.
    b) Write a problem that you would use the *and formula* to solve. Solve the problem and give the answer.

Practice the Skills

In Exercises 13–16, find the indicated probability.

13. If $P(A) = 0.6$, $P(B) = 0.4$, and $P(A \text{ and } B) = 0.3$, find $P(A \text{ or } B)$.

14. If $P(A \text{ or } B) = 0.9$, $P(A) = 0.5$, and $P(B) = 0.6$, find $P(A \text{ and } B)$.

15. If $P(A \text{ or } B) = 0.8$, $P(A) = 0.4$, and $P(A \text{ and } B) = 0.1$, find $P(B)$.

16. If $P(A \text{ or } B) = 0.6$, $P(B) = 0.3$, and $P(A \text{ and } B) = 0.1$, find $P(A)$.

Roll a Die In Exercises 17–20, a single die is rolled one time. Find the probability of rolling

17. a 2 or 5.

18. an odd number or a number greater than 2.

19. a number greater than 4 or less than 2.

20. a number greater than 3 or less than 5.

Select One Card In Exercises 21–26, one card is selected from a deck of playing cards. Find the probability of selecting

21. an ace or a 2.

22. a jack or a diamond.

23. a picture card or a red card.

24. a club or a red card.

25. a card less than 7 or a club. (*Note:* The ace is considered a low card.)

26. a card greater than 9 or a black card.

Problem Solving

Select Two Cards In Exercises 27–34, a board game uses the deck of 20 cards shown.

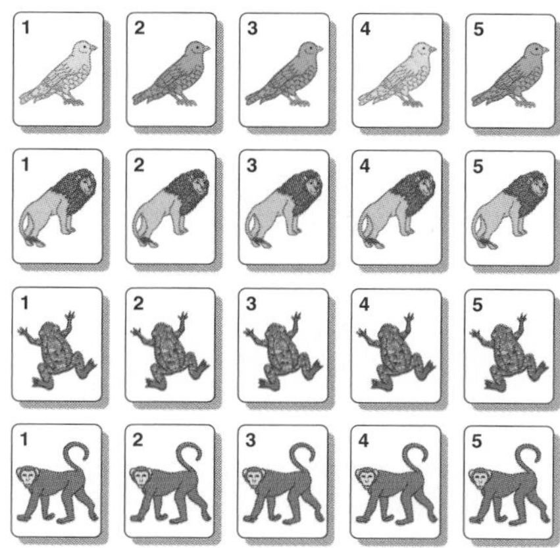

Two cards are selected at random from this deck. Find the probability of the following
 a) *with replacement.*
 b) *without replacement.*

27. They both show frogs.

28. They both show the number 3.

29. The first shows a lion, and the second shows a bird.

30. The first shows a 2, and the second shows a 4.

31. The first shows a red bird, and the second shows a monkey.

32. They both show even numbers.

33. Neither shows an even number.

34. The first shows a lion, and the second shows a red bird.

Select One Card *In the deck of cards used in Exercises 27–34, if one card is drawn, find the probability that the card shows*

35. a monkey or an even number.

36. a yellow bird or a number greater than 4.

37. a lion or a 2.

38. a red bird or an even number.

Two Spins *In Exercises 39–48, assume that the pointer cannot land on the line and that each spin is independent.*

If the pointer in Fig. 12.16 is spun twice, find the probability that the pointer lands on

39. red on both spins.

40. red and then yellow.

Figure 12.16

Two Spins *If the pointer in Fig. 12.17 is spun twice, find the probability that the pointer lands on*

41. green and then red.

42. red on both spins.

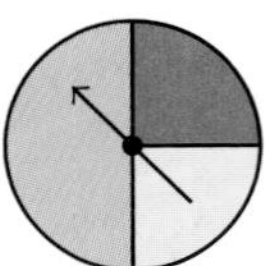

Figure 12.17

Two Spins *If the pointer in Fig. 12.18 is spun twice, find the probability that the pointer lands on*

43. red on both spins.

44. a color other than yellow on both spins.

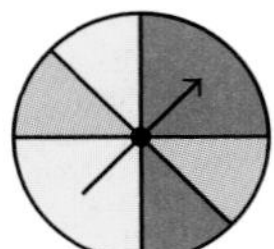

Figure 12.18

Two Spins *In Exercises 45–48, assume that the pointer in Fig. 12.16 is spun and then the pointer in Fig. 12.17 is spun. Find the probability of the pointers landing on*

45. red on both spins.

46. red on the first spin and yellow on the second spin.

47. a color other than yellow on both spins.

48. yellow on the first spin and a color other than yellow on the second spin.

Having a Family *In Exercises 49–52, a couple has three children. Assuming independence and the probability of a boy is $\frac{1}{2}$, find the probability that*

49. all three children are girls.

50. all three children are boys.

51. the youngest child is a boy and the older children are girls.

52. the youngest child is a girl, the middle child is a boy, and the oldest child is a girl.

53. a) Five Children The Martinos plan to have four children. Find the probability that all their children will be boys. (Assume that $P(\text{boy}) = \frac{1}{2}$, and assume independence.)

b) If their first four children are boys and Mrs. Martino is expecting another child, what is the probability that the fifth child will be a boy?

54. a) The Probability of a Girl The Bronsons plan to have seven children. Find the probability that all their children will be girls. (Assume that $P(\text{girl}) = \frac{1}{2}$, and assume independence.)

b) If their first seven children are girls and Mrs. Bronson is expecting another child, what is the probability that the eighth child will be a girl?

Golf Balls *Angel Sanchez has seven golf balls in one pocket of his golf bag: 4 Titleist balls, 2 Top Flite balls, and 1 Pinnacle ball. In Exercises 55–58, two balls will be selected at random. Find the probability of selecting each of the following*

a) *with replacement.*

b) *without replacement.*

55. a Titleist ball and then a Pinnacle ball

56. no Top Flite balls

57. at least one Top Flite ball

58. two Pinnacle balls

Health Insurance *A sample of 40 people yielded the following information about their health insurance.*

Number of People	Type of Insurance
22	Managed care plan
14	Traditional insurance
4	No insurance

Two people who provided information for the table were selected at random, without replacement. Find the probability that

59. neither had traditional insurance.

60. they both had a managed care plan.

61. at least one had traditional insurance.

62. the first had traditional insurance and the second had a managed care plan.

Home Builder *A sample of 30 women who recently had a home built yielded the following information about their builder.*

Number of Women	Would You Recommend Your Home Builder to a Friend
19	Yes
6	No
5	Not sure

Three women who provided information for the table were selected at random. Find the probability that

63. they would all recommend their home builder.

64. the first would not recommend her home builder but the second and third would recommend their home builder.

65. the first two would not recommend their home builder and the third is not sure if she would recommend her home builder.

66. the first would recommend her home builder but the second and third would not recommend their home builder.

An Experimental Drug *In Exercises 67–70, an experimental drug was given to a sample of 100 hospital patients with an unknown sickness. Of the total, 70 patients reacted favorably, 10 reacted unfavorably, and 20 were unaffected by the drug. Assume that this sample is representative of the entire population. If this drug is given to Mr. and Mrs. Rivera and their son Carlos, what is the probability of each of the following? (Assume independence.)*

67. Mrs. Rivera reacts favorably.

68. Mr. and Mrs. Rivera react favorably, and Carlos is unaffected.

69. All three react favorably.

70. None reacts favorably.

Multiple-Choice Exam *In Exercises 71–76, each question of a five-question multiple-choice exam has four possible answers. Gurshawn Salk picks an answer at random for each question. Find the probability that he selects the correct answer on*

71. any one question.

72. only the first question.

73. only the third and fourth questions.

74. all five questions.

75. none of the questions.

76. at least one of the questions.

A Slot Machine *In Exercises 77–80, consider a slot machine.*

Most people who play slot machines end up losing money because the machines are designed to favor the casino (the house). There are 22 positions on each reel. Assume that the following is a list of the number of symbols of each type on each of the three reels and that each symbol has the same chance of occurring (which is not the case; see the Did You Know on page 702).

Pictures on Reels	Reels 1	2	3
Cherries	2	5	4
Oranges	5	4	5
Plums	6	4	4
Bells	3	4	4
Melons	3	2	3
Bars BAR	2	2	1
7s 7	1	1	1

For this slot machine, assuming that the wheels are independent, find the probability of obtaining

77. a bell on the first reel.

78. oranges on all three reels.

79. no bars.

80. three 7's.

Two Wheels In Exercises 81–84, the following double wheel is spun.

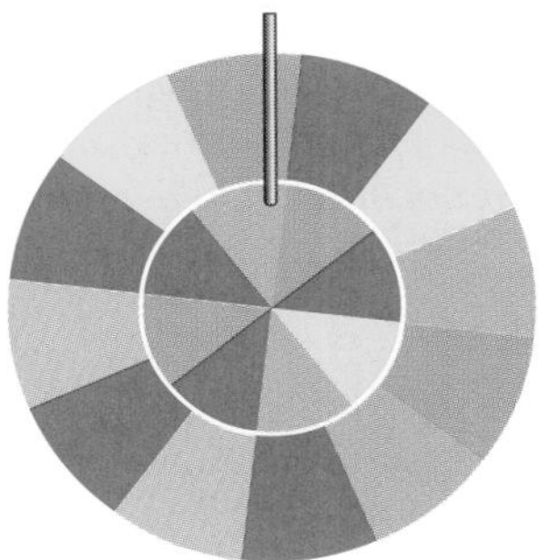

Assuming that the wheels are independent and each outcome is equally likely, find the probability of obtaining

81. yellow on both wheels.

82. red on the outer wheel and blue on the inner wheel.

83. red on neither wheel.

84. red on at least one wheel.

Hitting a Target In Exercises 85–88, the probability that a heat-seeking torpedo will hit its target is 0.4. If the first torpedo hits its target, the probability that the second torpedo will hit the target increases to 0.9 because of the extra heat generated by the first explosion. If two heat-seeking torpedoes are fired at a target, find the probability that

85. neither hits the target.

86. the first hits the target and the second misses the target.

87. both hit the target.

88. the first misses the target and the second hits the target.

89. *Polygenetic Afflictions* Certain birth defects and syndromes are *polygenetic* in nature. Typically, the chance that an offspring will be born with a polygenetic affliction is small. However, once an offspring is born with the affliction, the probability that future offspring of the same parents will be born with the same affliction increases. Let's assume that the probability of a child being born with affliction A is 0.001. If a child is born with this affliction, the probability of a future child being born with the same affliction becomes 0.04.

a) Are the events of the births of two children in the same family with affliction A independent? Explain.

b) A couple plans to have one child. Find the probability that the child will be born with this affliction.

A couple plans to have two children. Use the information provided to determine the probability that

c) both children will be born with the affliction.

d) the first will be born with the affliction and the second will not.

e) the first will not be born with the affliction and the second will.

f) neither will be born with the affliction.

90. *Lottery Ticket* In a bin are an equal number of balls marked with the digits 0, 1, 2, 3, . . . , 9. Three balls are to be selected from the bin, one after the other, at random with replacement to make the winning three digit lottery number. Ms. Jones has a lottery ticket with a three-digit number in the range 000 to 999. Find the probability that Ms. Jones's number is the winning number.

Chance of an Audit In Exercises 91–94, assume that 32 in every 1000 people in the \$14,000–\$56,800 income bracket are audited yearly. Assuming that the returns to be audited are selected at random and that each year's selections are independent of the previous year's selections, find the probability that a person in this income bracket will be audited

91. this year.

92. the next two years in succession.

93. this year but not next year.

94. neither this year nor next year.

Challenge Problems/Group Activities

95. *Picking Chips* A bag contains five red chips, three blue chips, and two yellow chips. Two chips are selected from the bag without replacement. Find the probability that two chips of the same color are selected.

96. *Ten Yen Coins* Ron has ten coins from Japan: three 1-yen coins, one 10-yen coin, two 20-yen coins, one 50-yen coin, and three 100-yen coins. He selects two coins at random without replacement. Assuming that each coin is equally likely to be selected, find the probability that Ron selects at least one 1-yen coin.

97. *A Fair Game?* Two playing cards are dealt to you from a well-shuffled standard deck of 52 cards. If either card is a diamond or if both are diamonds, you win; otherwise, you lose. Determine whether this game favors you, is fair, or favors the dealer. Explain your answer.

98. *Picture Card Probability* You have three cards: an ace, a king, and a queen. A friend shuffles the cards, selects two of them at random, and discards the third. You ask your friend to show you a picture card, and she turns over the king. What is the probability that she also has the queen?

Recreational Exercises

A Different Die For Exercises 99–102, consider a die that has 1 dot on one side, 2 dots on two sides, and 3 dots on three sides.

If the die is rolled twice, find the probability of rolling

99. two 2's. **100.** two 3's.

If the die is rolled only once, determine the probability of rolling

101. an even number or a number less than 3.

102. an odd number or a number greater than 1.

Internet/Research Activity

103. Girolamo Cardano (1501–1576) wrote *Liber de Ludo Aleae,* which is considered to be the first book on probability. Cardano had a number of different vocations. Do research and write a paper on the life and accomplishments of Girolamo Cardano.

12.7 CONDITIONAL PROBABILITY

In Section 12.6, we indicated that events are independent when the outcome of either event has no effect on the outcome of the other event. For example, selecting two cards from the deck of cards *with replacement* represents independent events. However, not all events with two selections are independent events. Consider this problem: Find the probability of selecting two aces from a deck of cards *without replacement.* The probability of selecting the first ace is $\frac{4}{52}$. The probability that the second ace is selected becomes $\frac{3}{51}$, since we assume that an ace was removed from the deck with the first selection. Since the probability of the second ace being selected is affected by the first ace being selected, these two events are *dependent.* Probability problems involving dependent events can be solved by using conditional probability.

Conditional Probability

In general, the probability of event E_2 occurring, given that an event E_1 has happened (or will happen; the time relationship does not matter) is called a **conditional probability** and is written $P(E_2 \mid E_1)$.

The symbol $P(E_2 \mid E_1)$, read "the probability of E_2, given E_1," represents the probability of E_2 occurring, assuming that E_1 has already occurred (or will occur).

DID YOU KNOW

Dice-Y Music

During the eighteenth century, composers were fascinated with the idea of creating compositions by rolling dice. A musical piece attributed by some to Wolfgang Amadeus Mozart, titled "Musical Dice Game," consists of 176 numbered musical fragments of three-quarter beats each. Two charts show the performer how to order the fragments given a particular roll of two dice. Modern composers can create music by using the same random technique and a computer; as in Mozart's time, however, the music created by random selection of notes or bars is still considered by musicologists to be aesthetically inferior to the work of a creative composer.

EXAMPLE 1 *Using Conditional Probability*

A single card is selected from a deck of cards. Find the probability it is a heart, given that it is red.

SOLUTION: We are told that the card is red. Thus, only 26 cards are possible, of which 13 are hearts. Therefore,

$$P(\text{heart} \mid \text{red}) \text{ or } P(\text{H} \mid \text{R}) = \frac{13}{26} = \frac{1}{2}$$

▲

EXAMPLE 2 *Girls in a Family*

Given a family with two children, and assuming that boys and girls are equally likely, find the probability that the family has

a) two girls.

b) two girls if you know that at least one of the children is a girl.

c) two girls given that the older child is a girl.

DID YOU KNOW

Chance of Showers

At one time or another, you probably have been caught in a downpour on a day the weather forecaster had predicted sunny skies. Short-term (24-hour) weather forecasts are correct nearly 85% of the time, a level of accuracy achieved through the use of conditional probability. Computer models are used to analyze data taken on the ground and in the air and make predictions of atmospheric pressures at some future time, say 10 minutes ahead. Based on these predicted conditions, another forecast is then computed. This process is repeated until a weather map has been generated for the next 12, 24, 36, and 48 hours. Since each new prediction relies on the previous prediction being correct, the margin of error increases as the forecast extends farther into the future.

SOLUTION:

a) To find the probability that the family has two girls we can determine the sample space of a family with two children. Then, from the sample space we can determine the probability that both children are girls. The sample space of two children can be determined by a tree diagram (see Fig. 12.19).

1st Child	2nd Child	Sample Space
B	B	BB
	G	BG
G	B	GB
	G	GG

Figure 12.19

There are four possible equally likely outcomes, BB, BG, GB, and GG. Only one of the outcomes has two girls, GG. Thus,

$$P(2\text{ girls}) = \frac{1}{4}$$

b) We are given that at least one of the children is a girl. Therefore, for this problem the sample space is BG, GB, GG. Since there are three possibilities, of which only one has two girls, GG,

$$P(\text{both girls} \mid \text{at least one is a girl}) = \frac{1}{3}$$

c) If the older child is a girl, the sample space reduces to GB, GG. Thus,

$$P(\text{both girls} \mid \text{older child is a girl}) = \frac{1}{2}$$

There are a number of formulas that can be used to find conditional probabilities. The one we will use follows.

Conditional Probability

For any two events, E_1 and E_2,

$$P(E_2 \mid E_1) = \frac{n(E_1 \text{ and } E_2)}{n(E_1)}$$

In the formula, $n(E_1 \text{ and } E_2)$ represents the number of sample points common to both event 1 and event 2, and $n(E_1)$ is the number of sample points in event E_1, the given event. Since the intersection of E_1 and E_2, symbolized $E_1 \cap E_2$, represents the sample points common to both E_1 and E_2, the formula can also be expressed as

$$P(E_2 \mid E_1) = \frac{n(E_1 \cap E_2)}{n(E_1)}$$

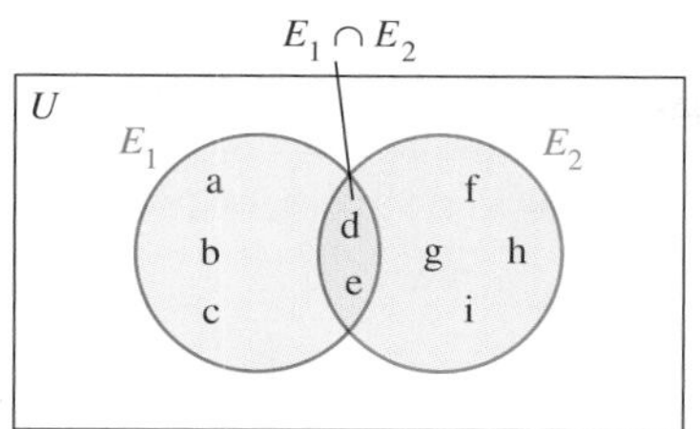

Figure 12.20

Figure 12.20 is helpful in explaining conditional probability.

Here, the number of elements in E_1 is five, the number of elements in E_2 is six, and the number of elements in both E_1 and E_2, or $E_1 \cap E_2$, is two.

$$P(E_2 \mid E_1) = \frac{n(E_1 \text{ and } E_2)}{n(E_1)} = \frac{2}{5}$$

Thus, for this situation, the probability of selecting an element from E_2, given that the element is in E_1, is $\frac{2}{5}$.

EXAMPLE 3 *Using the Conditional Probability Formula*

Each person in a sample of 200 residents in Lincoln County was asked whether he or she favored having a single countywide police department. The county consists of one large city and a number of small townships. The response of those sampled, with their place of residence specified, is given in the following table.

Residence	Favor	Oppose	Total
Live in city	80	50	130
Live outside city	60	10	70
Total	140	60	200

If one person from the sample is selected at random, find the probability that the person

a) favors a single countywide police department.
b) favors a single countywide police department, given that the person lives in the city.
c) opposes a single countywide police department, given that the person lives outside the city.
d) lives outside the city, given that the person opposes a single countywide police department.

SOLUTION:

a) The total number of respondents is 200, of which 140 favor a single countywide police department.

$$P(\text{favors a single countywide police department}) = \frac{140}{200} = \frac{7}{10}$$

b) We are given that the person lives in the city. Thus, this is a conditional probability problem. Let E_1 be the given information "the person lives in the city." Let E_2 be "the person favors a single countywide police department." We are being asked to find $P(E_2 \mid E_1)$. The number of people who live in the city, $n(E_1)$, is 130. The number of people who live in the city and favor a single countywide police department, $n(E_1 \text{ and } E_2)$, is 80. Thus,

$$P(E_2 \mid E_1) = \frac{n(E_1 \text{ and } E_2)}{n(E_1)} = \frac{80}{130} = \frac{8}{13}$$

c) We are given that the person lives outside the city. Thus, this is a conditional probability problem. Let E_1 be the given information "the person lives outside the city." Let E_2 be "the person opposes a single countywide police department." We are asked to find $P(E_2 \mid E_1)$. The number of people who live outside the city, $n(E_1)$, is 70. The number of people who live outside the city and oppose the countywide police department, $n(E_1 \text{ and } E_2)$, is 10. Thus,

$$P(E_2 \mid E_1) = \frac{n(E_1 \text{ and } E_2)}{n(E_1)} = \frac{10}{70} = \frac{1}{7}$$

d) We are given that the person opposes a single countywide police department. Thus, this is a conditional probability problem. Let E_1 be the given information "the person opposes a single countywide police department." Let E_2 be "the person lives outside the city." We are asked to find $P(E_2 \mid E_1)$. The number of people who oppose a single countywide police department, $n(E_1)$, is 60. The number of people who oppose a single countywide police department and live outside the city, $n(E_1 \text{ and } E_2)$, is 10. Thus,

$$P(E_2 \mid E_1) = \frac{n(E_1 \text{ and } E_2)}{n(E_1)} = \frac{10}{60} = \frac{1}{6}$$

In many of the examples, we used the words *given that*. Other words may be used instead. For example, in Example 3(b), the question could have been worded, "favors a single countywide police department *if* the person lives in the city."

SECTION 12.7 EXERCISES

Concept/Writing Exercises

1. What does the notation $P(E_2 \mid E_1)$ mean?
2. Give the formula for $P(E_2 \mid E_1)$.
3. If $n(E_1 \cap E_2) = 4$ and $n(E_1) = 12$, find $P(E_2 \mid E_1)$.
4. If $n(E_1 \cap E_2) = 5$ and $n(E_1) = 22$, find $P(E_2 \mid E_1)$.

Practice the Skills

Select a Circle *In Exercises 5–10, consider the circles shown.*

Assume that one circle is selected at random and each circle is equally likely to be selected. Find the probability of selecting

5. a 5, given that the circle is orange.
6. a 3, given that the circle is yellow.
7. an even number, given that the circle is not orange.
8. a number less than 2, given that the number is less than 5.
9. a red number, given that the circle is orange.
10. a number greater than 3, given that the circle is yellow.

Select a Number *In Exercises 11–16, consider the following figures.*

Assume that one number from 1 to 7 is equally likely to be selected at random. Each number corresponds to one of the seven figures shown. Find the probability of selecting

11. a circle, given that an odd number is selected.
12. a circle, given that a number greater than or equal to 5 is selected.
13. a red figure, given that an even number is selected.
14. a red or blue figure, given that an even number is selected.
15. a circle or square, given that a number less than 4 is selected.
16. a circle, given that an even number is selected.

Spin the Wheel In Exercises 17–24, consider the following wheel.

If the wheel is spun and each section is equally likely to stop under the pointer, find the probability that the pointer lands on

17. a five, given that the color is red.

18. an even number, given that the color is red.

19. purple, given that the number is odd.

20. a number greater than 4, given that the color is red.

21. a number greater than 4, given that the color is purple.

22. an even number, given that the color is red or purple.

23. gold, given that the number is greater than 5.

24. gold, given that the number is greater than 10.

Money from a Hat In Exercises 25–28, assume that a hat contains four bills: a \$1 bill, a \$5 bill, a \$10 bill, and a \$20 bill. Two bills are to be selected at random with replacement. Construct a sample space as was done in Example 2 and find the probability that

25. both bills are \$1 bills.

26. both bills are \$1 bills if the first selected is a \$1 bill.

27. both bills are \$5 bills if at least one of the bills is a \$5 bill.

28. both bills have a value greater than a \$5 bill if the second bill is a \$10 bill.

Two Dice In Exercises 29–34, two dice are rolled one after the other. Construct a sample space and find the probability that the sum of the dots on the dice total

29. 6.

30. 6 if the first die is a 1.

31. 6 if the first die is a 3.

32. an even number if the second die is a 2.

33. a number greater than 7 if the second die is a 5.

34. a 7 or 11 if the first die is a 5.

Problem Solving

Taste Test In Exercises 35–40, use the following results of a taste test given at a local mall.

	Prefers Coke	Prefers Pepsi	Total
Men	60	45	105
Women	50	62	112
Total	110	107	217

If one person from the sample is selected at random, find the probability the person selected

35. prefers Pepsi.

36. is a woman.

37. prefers Coke, given that a woman is selected.

38. prefers Pepsi, given that a man was selected.

39. is a man, given that the person prefers Coke.

40. is a woman, given that the person prefers Pepsi.

Lion or Elephant Use the following information for Exercises 41–46. At a zoo, a sampling of children was asked if the zoo were to get one additional animal, would they prefer a lion or an elephant. The results of the survey follow.

	Lion	Elephant	Total
Boys	90	110	200
Girls	75	85	160
Total	165	195	360

If one child who was in the survey is selected at random, find the probability that

41. the child is a girl.

42. the child selected the lion.

43. the child selected the elephant, given that the child was a boy.

44. the child selected the lion, given that the child was a girl.

45. the child selected is a boy, given that the child preferred the elephant.

46. the child selected is a girl, given that the child preferred the lion.

Videotapes and DVDs Use the following information for Exercises 47–52. At a Blockbuster video store, 300 people were surveyed to determine whether they rented DVDs, videotapes, or both. The results of the survey follow.

Age	Only DVDs	Only Videotapes	Both	Total
Under 30	60	39	21	120
30 or Older	64	94	22	180
Total	124	133	43	300

If one person from this survey is selected at random, determine the probability that the person

47. rented only videotapes.

48. was 30 or older.

49. rented only DVDs, given that the person was under 30.

50. rented both videotapes and DVDs, given that the person was 30 or older.

51. was under 30, given that the person rented both DVDs and videotapes.

52. was 30 or older, given that the person rented only videotapes.

Military Trials Use the following information concerning military trials for Exercises 53–58. Use the following data for the years indicated.

	Convictions	Acquittals	Total Cases
Air Force (1992–2001)	8166	667	8833
Army (1997–2001)	5024	434	5458
Navy/Marine Corps (1997–2001)	12,866	473	13,339
Total cases	26,056	1574	27,630

Source: Department of Defense.
Note: Statistics for the Navy and Marine Corps are maintained jointly.

Assuming that one person who was on trial was selected at random, determine the probability (as a decimal number rounded to four decimal places) that

53. the person was from the Air Force.

54. the person was acquitted.

55. the person was acquitted, given that the person was from the Army.

56. the person was convicted, given that the person was from the Navy/Marine Corps.

57. the person was in the Army, given that the person was convicted.

58. the person was in the Air Force, given that the person was acquitted.

Quality Control In Exercises 59–64, Sally Horsefall, a quality control inspector, is checking a sample of light bulbs for defects. The following table summarizes her findings.

Wattage	Good	Defective	Total
20	80	15	95
50	100	5	105
100	120	10	130
Total	300	30	330

If one of these light bulbs is selected at random, find the probability that the light bulb is

59. good.

60. good, given that it is 50 watts.

61. defective, given that it is 20 watts.

62. good, given that it is 100 watts.

63. good, given that it is 50 or 100 watts.

64. defective, given that it is not 50 watts.

News Survey In Exercises 65–70, 270 individuals are asked which evening news they watch most often. The results are summarized as follows.

Viewers	ABC	NBC	CBS	Other	Total
Men	30	20	40	55	145
Women	50	10	20	45	125
Total	80	30	60	100	270

If one of these individuals is selected at random, find the probability that the person watches

65. ABC or NBC.

66. ABC, given that the individual is a woman.

67. ABC or NBC, given that the individual is a man.

68. a station other than CBS, given that the individual is a woman.

69. a station other than ABC, NBC, or CBS, given that the individual is a man.

70. NBC or CBS, given that the individual is a woman.

Brian Williams, MSNBC. See Exercises 65–70 on page 712.

Challenge Problems/Group Activities

Mutual Fund Holdings Use the following information in Exercises 71–74. Mutual funds often hold many stocks. Each stock may be classified as a value stock, a growth stock, or a blend of the two. The stock may also be categorized by how large the company is. It may be classified as a large company stock, medium company stock, or small company stock. A selected mutual fund contains 200 stocks as illustrated in the following chart.

	Value	Blend	Growth
Large	28	23	42
Medium	19	15	18
Small	26	12	17

Equity Investment Style

If one stock is selected at random from the mutual fund, find the probability that it is

71. a large company stock.

72. a value stock.

73. a blend, given that it is a medium company stock.

74. a large company stock, given that it is a blend stock.

75. Consider the Venn diagram above and to the right. The numbers in the regions of the circle indicate the number of items that belong to that region. For example, 60 items are in set A but not in set B.

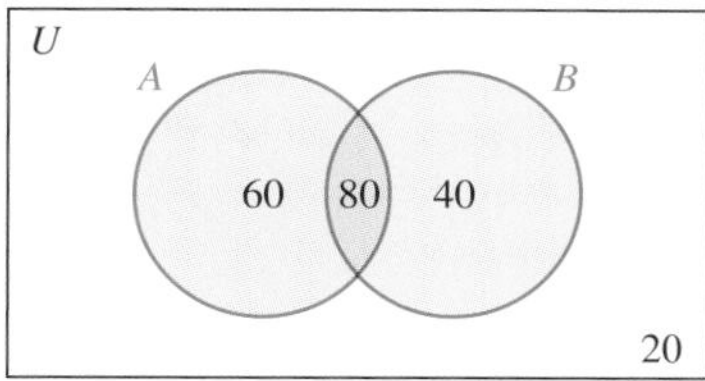

Find **a)** $n(A)$ **b)** $n(B)$ **c)** $P(A)$ **d)** $P(B)$
Use the formula on page 708 to find:
e) $P(A \mid B)$ **f)** $P(B \mid A)$
g) Explain why $P(A \mid B) \neq P(A) \cdot P(B)$.

76. A formula we gave for conditional probability is

$$P(E_2 \mid E_1) = \frac{n(E_1 \text{ and } E_2)}{n(E_1)}$$

This formula may be derived from the formula

$$P(E_2 \mid E_1) = \frac{P(E_1 \text{ and } E_2)}{P(E_1)}$$

Can you explain why? [*Hint:* Consider what happens to the denominators of $P(E_1 \text{ and } E_2)$ and $P(E_1)$ when they are expressed as fractions and the fractions are divided out.]

77. Given that $P(A) = 0.3$, $P(B) = 0.4$, and $P(A \text{ and } B) = 0.12$, use the formula

$$P(E_2 \mid E_1) = \frac{P(E_1 \text{ and } E_2)}{P(E_1)}$$

to find
a) $P(A \mid B)$.
b) $P(B \mid A)$.
c) Are A and B independent? Explain.

Recreational Exercises

In Exercises 78–83, suppose that each circle is equally likely to be selected. One circle is selected at random.

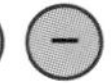

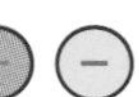

Find the probability indicated.

78. P(green circle | + obtained)

79. P(+ | orange circle obtained)

80. P(yellow circle | − obtained)

81. P(green + | + obtained)

82. P(green or orange circle | green + obtained)

83. P(orange circle with green + | + obtained)

12.8 THE COUNTING PRINCIPLE AND PERMUTATIONS

In Section 12.5, we introduced the counting principle, which is repeated here for your convenience.

Counting Principle
If a first experiment can be performed in M distinct ways and a second experiment can be performed in N distinct ways, then the two experiments in that specific order can be performed in $M \cdot N$ distinct ways.

The counting principle is illustrated in Examples 1 and 2.

EXAMPLE 1 *Counting Principle: Passwords*

A password used to gain access to a computer account is to consist of two lowercase letters followed by four digits. Determine how many different passwords are possible if

a) repetition of letter and digits is permitted.
b) repetition of letters and digits is not permitted.
c) the first letter must be a vowel (a, e, i, o, u) and the first digit cannot be a 0, and repetition of letters and digits is not permitted.

SOLUTION: There are 26 letters and 10 digits (0–9). We have six positions to fill, as indicated.

$$\overline{L}\,\overline{L}\,\overline{D}\,\overline{D}\,\overline{D}\,\overline{D}$$

a) Since repetition is permitted, there are 26 possible choices for both the first and second positions. There are 10 possible choices for the third, fourth, fifth, and sixth positions.

$$\underset{L}{\underline{26}}\ \underset{L}{\underline{26}}\ \underset{D}{\underline{10}}\ \underset{D}{\underline{10}}\ \underset{D}{\underline{10}}\ \underset{D}{\underline{10}}$$

Since $26 \cdot 26 \cdot 10 \cdot 10 \cdot 10 \cdot 10 = 6{,}760{,}000$, there are 6,760,000 different possible arrangements.

b) There are 26 possibilities for the first position. Since repetition of letters is not permitted, there are only 25 possibilities for the second position. The same reasoning is used when determining the number of digits for positions 3 through 6.

$$\underset{L}{\underline{26}}\ \underset{L}{\underline{25}}\ \underset{D}{\underline{10}}\ \underset{D}{\underline{9}}\ \underset{D}{\underline{8}}\ \underset{D}{\underline{7}}$$

Since $26 \cdot 25 \cdot 10 \cdot 9 \cdot 8 \cdot 7 = 3{,}276{,}000$, there are 3,276,000 different possible arrangements.

c) Since the first letter must be an a, e, i, o, or u, there are five possible choices for the first position. The second position can be filled by any of the letters except

DID YOU KNOW

Slot Machines

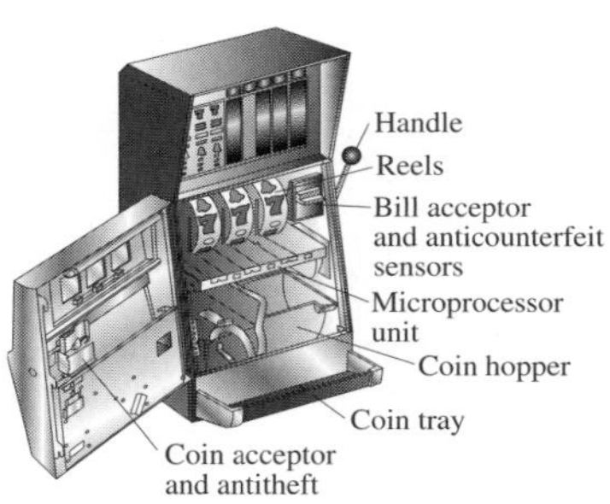

Source: Scientific American

Slot machines may qualify as the world's most user-friendly computer. Nowadays, slot machines are a conglomeration of microprocessors and memory chips. Each of three reels can stop at 22 positions (each one displaying various symbols or a blank space) for a total of 10,648 different arrangements ($22 \times 22 \times 22 = 10{,}648$). The heart of a contemporary slot machine is a microprocessor programmed to generate random numbers that can be assigned to any arrangement of the reels. In effect, the microprocessor generates random numbers that dictates the machine's display and payout. Because many or few random numbers can be assigned to any given arrangement, makers of slot machines can vary the odds as desired. For the top jackpot, the odds of obtaining the single correct arrangement may be set at 10 million to 1.

for the vowel selected for the first position. Therefore, there are 25 possibilities for the second position.

Since the first digit cannot be a 0, there are nine possibilities for the third position. The fourth position can be filled by any digit except the one selected for the third position. Thus, there are nine possibilities for the fourth position. Since the fifth position cannot be filled by any of the two digits previously used, there are eight possibilities for the fifth position. The last position can be filled by any of the seven remaining digits.

$$\underset{L}{\underline{5}}\ \underset{L}{\underline{25}}\ \underset{D}{\underline{9}}\ \underset{D}{\underline{9}}\ \underset{D}{\underline{8}}\ \underset{D}{\underline{7}}$$

Since $5 \cdot 25 \cdot 9 \cdot 9 \cdot 8 \cdot 7 = 567{,}000$, there are 567,000 different arrangements that meet the conditions specified. ▲

EXAMPLE 2 *Counting Principle: Computer Colors*

At Best Buy, they have just received a supply of Apple iMac computers. The computers come in the following colors: tangerine, strawberry, blueberry, grape, and lime. Sadie Bragg, the floor manager, decides to display one of each color computer.

a) In how many different ways can she display the five different color computers on a shelf?

b) If she wants to place the strawberry computer in the middle, in how many different ways can she arrange the computers?

c) If Sadie wants the blueberry computer to be the first computer and the strawberry computer to be the last computer, in how many different ways can she arrange the computers?

SOLUTION:

a) There are five positions to fill, using the five colors. In the first position, on the left, she can use any one of the five colors. In the second position, she can use any of the four remaining colors. In the third position, she can use any of the three remaining colors, and so on. The number of distinct possible arrangements is

$$\underline{5} \cdot \underline{4} \cdot \underline{3} \cdot \underline{2} \cdot \underline{1} = 120$$

b) We begin by satisfying the specified requirements stated. In this case, the strawberry computer must be placed in the middle. Therefore, there is only one possibility for the middle position.

$$\underline{\ }\ \underline{\ }\ \underline{1}\ \underline{\ }\ \underline{\ }$$

For the first position, there are now four possibilities. For the second position, there will be three possibilities. For the fourth position, there will be two possibilities. Finally, in the last position, there is only one possibility.

$$\underline{4} \cdot \underline{3} \cdot \underline{1} \cdot \underline{2} \cdot \underline{1} = 24$$

Thus, under the condition stated, there are 24 different possible arrangements.

c) For the first computer, there is only 1 possible computer, the blueberry computer. For the last computer, there is only 1 possible computer, the strawberry computer. This gives

$$\underline{1}\ \underline{\ \ }\ \underline{\ \ }\ \underline{\ \ }\ \underline{1}$$

The second position can be filled by any of the three remaining computers. The third position can be filled by any of the two remaining computers. There is only one computer left for the fourth position. Thus, the number of possible arrangements is

$$\underline{1} \cdot \underline{3} \cdot \underline{2} \cdot \underline{1} \cdot \underline{1} = 6$$

There are only six possible arrangements that satisfy the given conditions. ▲

Permutations

Now we introduce the definition of a permutation.

A **permutation** is any *ordered arrangement* of a given set of objects.

Curly Moe Larry

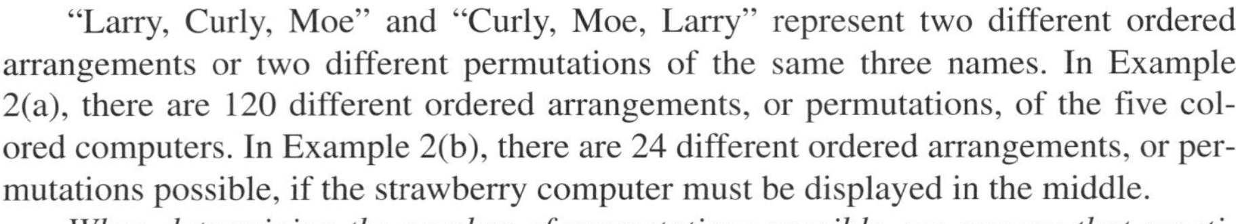

"Larry, Curly, Moe" and "Curly, Moe, Larry" represent two different ordered arrangements or two different permutations of the same three names. In Example 2(a), there are 120 different ordered arrangements, or permutations, of the five colored computers. In Example 2(b), there are 24 different ordered arrangements, or permutations possible, if the strawberry computer must be displayed in the middle.

When determining the number of permutations possible, we assume that repetition of an item is not permitted. To help you understand and visualize permutations, we illustrate the various permutations possible when a triangle, rectangle, and circle are to be placed in a line, see Figure 12.21.

Six Permutations

Figure 12.21

Note for this set of three shapes that six different arrangements, or six permutations, are possible. We can obtain the number of permutations by using the counting principle. For the first position, there are three choices. There are then two choices for the second position, and only one choice is left for the third position.

$$\text{Number of permutations} = 3 \cdot 2 \cdot 1 = 6$$

The product $3 \cdot 2 \cdot 1$ is referred to as 3 factorial, and is written 3!. Thus,

$$3! = 3 \cdot 2 \cdot 1 = 6$$

Number of Permutations

The number of permutations of n distinct items is n factorial, symbolized $n!$, where

$$n! = n(n-1)(n-2)\cdots(3)(2)(1)$$

It is important to note that 0! is defined to be 1. Many calculators have the ability to determine factorials. Often to determine factorials you need to press the 2nd or INV key. Read your calculator manual to determine how to find factorials on your calculator.

EXAMPLE 3 *Children in Line*

In how many different ways can seven children be arranged in a line?

SOLUTION: Since there are seven children, the number of permutations is 7!

$$7! = 7 \cdot 6 \cdot 5 \cdot 4 \cdot 3 \cdot 2 \cdot 1 = 5040$$

The seven children can be arranged in 5040 different ways. ▲

Example 4 illustrates how to use the counting principle to determine the number of permutations possible when only a part of the total number of items is to be selected and arranged.

Some of the many permutations of 3 of the 5 letters

c d e

Figure 12.22

EXAMPLE 4 *Permutations of Three Out of Five Letters*

Consider the five letters a, b, c, d, e. In how many distinct ways can three letters be selected and arranged if repetition is not allowed?

SOLUTION: We are asked to select and arrange only three of the five possible letters. Figure 12.22 shows some possibilities. Using the counting principle, we find that there are five possible letters for the first choice, four possible letters for the second choice, and three possible letters for the third choice:

$$5 \cdot 4 \cdot 3 = 60$$

Thus, there are 60 different possible ordered arrangements, or permutations. ▲

In Example 4, we determined the number of different ways in which we could select and arrange three of the five items. We can indicate that by using the notation ${}_5P_3$. The notation ${}_5P_3$ is read "the number of permutations of five items taken three at a time." The notation ${}_nP_r$ is read "the number of permutations of n items taken r at a time."

We use the counting principle below to evaluate ${}_8P_4$, ${}_9P_3$, and ${}_{10}P_5$. Notice the relationship between the number preceding the P, the number following the P, and the last number in the product.

One more than $8 - 4$

$${}_8P_4 = 8 \cdot 7 \cdot 6 \cdot 5$$

One more than $9 - 3$

$${}_9P_3 = 9 \cdot 8 \cdot 7$$

One more than $10 - 5$

$${}_{10}P_5 = 10 \cdot 9 \cdot 8 \cdot 7 \cdot 6$$

Notice that to evaluate ${}_nP_r$ we begin with n and form a product of r consecutive descending factors. For example, to evaluate ${}_{10}P_5$, we start with 10 and form a product of 5 consecutive descending factors (see the preceding illustration).

In general, the number of permutations of n items taken r at a time, ${}_nP_r$, may be found by the formula

One more than $n - r$

$${}_nP_r = n(n - 1)(n - 2)\cdots\overbrace{(n - r + 1)}$$

Therefore, when evaluating ${}_{20}P_{15}$, we would find the product of consecutive decreasing integers from 20 to $(20 - 15 + 1)$ or 6, which is written as $20 \cdot 19 \cdot 18 \cdot 17 \cdot \cdots \cdot 6$.

Now let's develop an alternative formula that we can use to find the number of permutations possible when r objects are selected from n objects:

$${}_nP_r = n(n - 1)(n - 2)\cdots(n - r + 1)$$

Now multiply the expression on the right side of the equals sign by $\dfrac{(n - r)!}{(n - r)!}$ which is equivalent to multiplying the expression by 1.

$${}_nP_r = n(n - 1)(n - 2)\cdots(n - r + 1) \times \frac{(n - r)!}{(n - r)!}.$$

For example,

$${}_{10}P_5 = 10 \cdot 9 \cdot \cdots \cdot 6 \times \frac{5!}{5!}$$

or

$${}_{10}P_5 = \frac{10 \cdot 9 \cdot \cdots \cdot 6 \times 5!}{5!}$$

Since $(n - r)!$ means $(n - r)(n - r - 1)\cdots(3)(2)(1)$, the expression for ${}_nP_r$ can be rewritten as

$${}_nP_r = \frac{n(n - 1)(n - 2)\cdots(n - r + 1)\overbrace{(n - r)(n - r - 1)\cdots(3)(2)(1)}^{(n - r)!}}{(n - r)!}$$

Since the numerator of this expression is $n!$, we can write

$$_nP_r = \frac{n!}{(n-r)!}$$

For example,

$$_{10}P_5 = \frac{10!}{(10-5)!}$$

> The number of permutations possible when r objects are selected from n objects is found by the **permutation formula**
>
> $$_nP_r = \frac{n!}{(n-r)!}$$

In Example 4, we found that when selecting three of five letters, there were 60 permutations. We can obtain the same result using the permutation formula:

$$_5P_3 = \frac{5!}{(5-3)!} = \frac{5!}{2!} = \frac{5 \cdot 4 \cdot 3 \cdot \cancel{2 \cdot 1}}{\cancel{2 \cdot 1}} = 60$$

EXAMPLE 5 *Using the Permutation Formula*

You are among nine people forming a hiking club. Collectively, you decide to put each person's name in a hat and to randomly select a president, a vice president, and a secretary. How many different arrangements or permutations of officers are possible?

SOLUTION: Since there are nine people, $n = 9$, of which three are to be selected; thus $r = 3$.

$$_9P_3 = \frac{9!}{(9-3)!} = \frac{9!}{6!} = \frac{9 \cdot 8 \cdot 7 \cdot \cancel{6 \cdot 5 \cdot 4 \cdot 3 \cdot 2 \cdot 1}}{\cancel{6 \cdot 5 \cdot 4 \cdot 3 \cdot 2 \cdot 1}} = 504$$

Thus, with nine people there can be 504 different arrangements for president, vice president, and secretary. ▲

In Example 5, the fraction

$$\frac{9 \cdot 8 \cdot 7 \cdot \cancel{6 \cdot 5 \cdot 4 \cdot 3 \cdot 2 \cdot 1}}{\cancel{6 \cdot 5 \cdot 4 \cdot 3 \cdot 2 \cdot 1}} = 504$$

can be also expressed as

$$\frac{9 \cdot 8 \cdot 7 \cdot \cancel{6!}}{\cancel{6!}} = 504$$

The solution to Example 5, like other permutation problems, can also be obtained using the counting principle.

EXAMPLE 6

The Prince George County bicycle club has 10 different routes members wish to travel exactly once, but they only have 6 specific dates for their trips. In how many ways can the different routes be assigned to the dates scheduled for their trips?

SOLUTION: There are 10 possibile routes but only 6 specific dates scheduled for the trips. Since traveling route A on day 1 and traveling route B on day 2 is different than traveling route B on day 1 and traveling route A on day 2, we have a permutation problem. There are 10 possible routes; thus, $n = 10$. There are 6 routes that are going to be selected and assigned to different days; thus, $r = 6$. Now we calculate the number of different permutations of selecting and arranging the dates for 6 out of 10 possible routes.

$${}_{10}P_6 = \frac{10!}{(10-6)!} = \frac{10!}{4!} = \frac{10 \cdot 9 \cdot 8 \cdot 7 \cdot 6 \cdot 5 \cdot \cancel{4!}}{\cancel{4!}} = 151{,}200$$

There are 151,200 different ways that 6 routes can be selected and scheduled from the 10 possible routes. ▲

Example 6 could also be worked using the counting principle because we are discussing an *ordered arrangement* (a permutation) that is done *without replacement*. For the first date scheduled, there are 10 possible outcomes. For the second date selected, there are 9 possible outcomes. By continuing this process we would determine that the number of possible outcomes for the 6 different trips is $10 \cdot 9 \cdot 8 \cdot 7 \cdot 6 \cdot 5 = 151{,}200$.

We have worked permutation problems (selecting and arranging, without replacement, r items out of n *distinct* items) by using the counting principle and using the permutation formula. When you are given a permutation problem, unless specified by your instructor, you may use either technique to determine its solution.

Permutations of Duplicate Items

So far, all the examples we have discussed in this section have involved arrangements with distinct items. Now we will consider permutation problems in which some of the items to be arranged are duplicates. For example, the name BOB contains three letters, of which the two Bs are duplicates. How many permutations of the letters in the name BOB are possible? If the two Bs were distinguishable (one red and the other blue), there would be six permutations.

BOB BBO OBB

BOB BBO OBB

However, if the Bs are not distinguishable (replacing all colored Bs with black print), we see there are only three permutations.

BOB BBO OBB

The number of permutations of the letters in BOB can be computed as

$$\frac{3!}{2!} = \frac{3 \cdot \cancel{2 \cdot 1}}{\cancel{2 \cdot 1}} = 3$$

where 3! represents the number of permutations of three letters, assuming that none are duplicates, and 2! represents the number of ways the two items that are duplicates can be arranged (BB or BB). In general, we have the following rule.

Permutations of Duplicate Objects

The number of distinct permutations of n objects where n_1 of the objects are identical, n_2 of the objects are identical, . . . , n_r of the objects are identical is found by the formula

$$\frac{n!}{n_1!n_2!\cdots n_r!}$$

EXAMPLE 7 *Duplicate Letters*

In how many different ways can the letters of the word "MISSISSIPPI" be arranged?

SOLUTION: Of the 11 letters, four are I's, four are S's, and two are P's. The number of possible arrangements is

$$\frac{11!}{4!4!2!} = \frac{11\cdot 10\cdot 9\cdot 8\cdot 7\cdot 6\cdot 5\cdot \cancel{4\cdot 3\cdot 2\cdot 1}}{\cancel{4\cdot 3\cdot 2\cdot 1}\cdot 4\cdot 3\cdot 2\cdot 1\cdot 2\cdot 1} = 11\cdot 10\cdot 9\cdot 7\cdot 5 = 34{,}650$$

There are 34,650 different possible arrangements of the letters in the word "MISSISSIPPI."

SECTION 12.8 EXERCISES

Concept/Writing Exercises

1. In your own words, state the counting principle.
2. In your own words, describe a permutation.
3. Give the formula for the number of permutations of n distinct items.
4. In your own words, explain how to find $n!$ for any whole number n.
5. How do you read ${}_nP_r$? When you evaluate ${}_nP_r$, what does the outcome represent?
6. Give the formula for the number of permutations of n objects when $n_1, n_2, \ldots, n_r$, of the objects are identical.
7. Give the formula for the number of permutations when r objects are selected from n objects.
8. Does ${}_1P_1 = {}_1P_0$? Explain.

Practice the Skills

In Exercises 9–20, evaluate the expression.

9. 6!	**10.** 8!	**11.** ${}_6P_2$	**12.** ${}_5P_2$
13. 0!	**14.** ${}_6P_4$	**15.** ${}_8P_0$	**16.** ${}_5P_0$
17. ${}_9P_4$	**18.** ${}_4P_4$	**19.** ${}_8P_5$	**20.** ${}_{10}P_6$

Problem solving

21. *ATM Codes* To use an automated teller machine, you generally must enter a four-digit code, using the digits 0–9. How many four-digit codes are possible if repetition of digits is permitted?
22. *Daily Double* The daily double at most racetracks consists of selecting the winning horse in both the first and second races. If the first race has seven entries and the second race

has eight entries, how many daily double tickets must you purchase to guarantee a win?

23. *Passwords* Assume that a password to log onto a computer account is to consist of three letters followed by three digits. Determine the number of possible passwords if
a) repetition is not permitted.
b) repetition is permitted.

24. *Passwords* Assume that a password to log onto a computer account is to consist of any four digits or letters (repetition is permitted). Determine the number of passwords possible if
a) the letters are not case sensitive (that is, a lowercase letter is treated the same as an uppercase letter).
b) the letters are case sensitive (that is, an uppercase letter is considered different than the same lowercase letter).

25. *Car Door Locks* Some doors on cars can be opened by pressing the correct sequence of buttons. A display of the five buttons by the door handle of a car follows.*

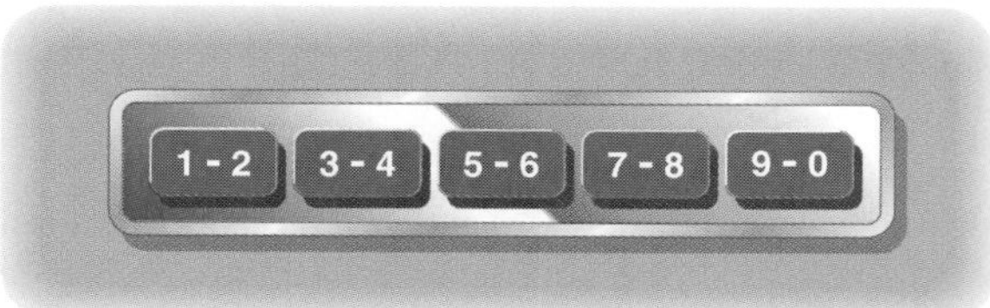

The correct sequence of five buttons must be pressed to unlock the door. If the same button may be pressed consecutively,
a) how many possible ways can the five buttons be pressed (repetition is permitted)?
b) If five buttons are pressed at random, find the probability that a sequence that unlocks the door will be entered.

26. *Social Security Numbers* A social security number consists of nine digits. How many different social security numbers are possible if repetition of digits is permitted?

27. *License Plate* A license plate is to have five uppercase letters or digits. Determine the number of license plates possible if repetition is permitted and if any position can contain either a letter or digit with the exception that the first position cannot contain the letter *O* or the number 0.

28. *Winning the Trifecta* The trifecta at most racetracks consists of selecting the first-, second-, and third-place finishers in a particular race in their proper order. If there are seven entries in the trifecta race, how many tickets must you purchase to guarantee a win?

29. *Advertising* The operator of the Sound Great Stereo store is planning a grand opening. He wishes to advertise that the store has many different sound systems available. It stocks 8 different CD players, 10 different receivers, and 9 different sets of speakers. Assuming that a sound system will consist of one CD player, one receiver, and one set of speakers and that all pieces are compatible, how many different sound systems can they advertise?

30. *Geometric Shapes* Consider the five figures shown.

In how many different ways can the figures be arranged
a) from left to right?
b) from top to bottom if placed one under the other?
c) from left to right if the triangle is to be placed on the far right?
d) from left to right if the circle is to be placed on the far left and the triangle is to be placed on the far right?

31. *Arranging Pictures* The six pictures shown are to be placed side by side along a wall.

In how many ways can they be arranged from left to right if
a) they can be arranged in any order?
b) the bird must be on the far left?
c) the bird must be on the far left and the giraffe must be next to the bird?
d) a four-legged animal must be on the far right?

32. *Car Rental* At a local car rental agency 8 identical midsize cars are available and 3 customers want midsize cars. If each customer can choose his or her own car, how many different ways can the cars be selected?

33. *Club Officers* If a club consists of ten members, how many different arrangements of president, vice-president, and secretary are possible?

34. *ISBN Codes* Each book registered in the Library of Congress must have an ISBN code number. For an ISBN number of the form D-DD-DDDDDD-D, where D represents a digit from 0–9, how many different ISBN numbers are possible if repetition of digits is allowed? (See research activity Exercise 70.)

35. *Wedding Reception* At the reception line of a wedding, the bride, the groom, the best man, the maid of honor, the four ushers, and the four bridesmaids must line up to receive the guests.

*On most cars, although each key lists two numbers, the key acts as a single number. Therefore, if your code is 1, 6, 8, 5, 3, the code 2, 5, 7, 6, 4 will also open the lock.

a) If these individuals can line up in any order, how many arrangements are possible?
b) If the groom must be the last in line and the bride must be next to the groom, and the others can line up in any order, how many arrangements are possible?
c) If the groom is to be last in line, the bride next to the groom, and males and females are to alternate, how many arrangements are possible?

36. *Disc Jockey* A disc jockey has 8 songs to play. Three are slow songs and 5 are fast songs. Each song is to be played only once. In how many ways can the disc jockey play the 8 songs if
a) the songs can be played in any order.
b) the first song must be a slow song and the last song must be a slow song.
c) the first two songs must be fast songs.

Letter Codes In Exercises 37–40, an identification code is to consist of two letters followed by four digits. How many different codes are possible if

37. repetition is not permitted?

38. repetition is permitted?

39. the first two entries must both be the same letter and repetition of the digits is not permitted?

40. the first letter must be a A, B, C, or D and repetition is not permitted?

License Plates In Exercises 41–44, a license plate is to consist of three digits followed by two uppercase letters. Determine the number of different license plates possible if

41. repetition of numbers and letters is permitted.

42. repetition of numbers and letters is not permitted.

43. the first and second digits must be odd, and repetition is not permitted.

44. the first digit cannot be zero, and repetition is not permitted.

45. *Possible Phone Numbers* A telephone number consists of seven digits with the restriction that the first digit cannot be 0 or 1.
a) How many distinct telephone numbers are possible?
b) How many distinct telephone numbers are possible with three-digit area codes preceding the seven-digit number, where the first digit of the area code is not 0 or 1?
c) With the increasing use of cellular phones and paging systems, our society is beginning to run out of usable phone numbers. Various phone companies are developing phone numbers that use 11 digits instead of 7. How many distinct phone numbers can be made with 11 digits, assuming that the area code remains three digits and the first digit of the area code and the phone number cannot be 0 or 1?

46. *Selecting Cereal* Mrs. Williams and her 3 children go shopping at a local grocery store. Each of the children will be allowed to select one box of cereal for their own. On the store's shelf there are only 12 boxes of cereal, and each box contains a different type of cereal. In how many ways can the selections be made?

47. *Swimming Event* A swimming meet has 15 participants for the 100 meter freestyle event. The 6 participants with the fastest speeds will be listed, in the order of their speed, on the tote board. How many different ways are there for the names to be listed?

48. *History Test* In one question of a history test the student is asked to match 10 dates with 10 events; each date can only be matched with 1 event. In how many different ways can this question be answered?

49. *Color Permutations* Find the number of permutations of the colors in the spectrum that follows.

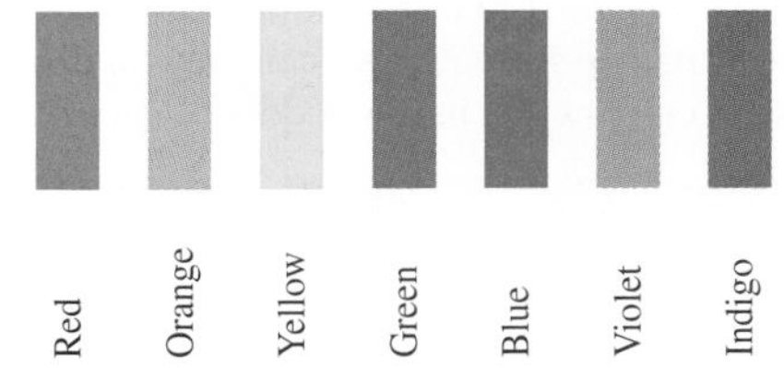

50. *Drive-Through at a Bank* A bank has three drive-through stations. Assuming that each is equally likely to be selected by customers, how many different ways can the next six drivers select a station?

51. *Computer Systems* At a computer store, a customer is considering 5 different computers, 4 different monitors, 7 different printers, and 2 different scanners. Assuming that each of the components is compatible with one another and that one of each is to be selected, determine the number of different computer systems possible.

52. *Selecting Furniture* The Johnsons just moved into their new home and are selecting furniture for the family room. They are considering 5 different sofas, 2 different chairs, and 6 different tables. They plan to select one item from each category. Determine the number of different ways they can select the furniture.

53. Determine the number of permutations of the letters of the word "EDUCATION."

54. Determine the number of permutations of the letters of the word "GOVERNMENT."

55. In how many ways can the letters in the word "INDEPENDENCE" be arranged?

56. In how many ways can the letters in the word "TENNESSEE" be arranged?

57. In how many ways can the digits in the number 1,324,324 be arranged?

58. In how many ways can the digits of the number 2,142,332 be arranged?

59. *Flag Messages* Five different colored flags will be placed on a pole, one beneath another. The arrangement of the colors indicates the message. How many messages are possible if five flags are to be selected from eight different colored flags?

60. *Multiple-Choice Test* Keri Kershaw is taking a 10-question multiple-choice exam. Each question has three possible answers, (a), (b), and (c). In how many possible ways can Keri answer the questions?

61. *Batting Order* In how many ways can the manager of a National League baseball team arrange his batting order of nine players if

a) the pitcher must bat last?

b) there are no restrictions?

62. *Painting Exhibit* Five Monet paintings are to be displayed in a museum.

a) In how many different ways can they be arranged if they must be next to one another?

b) In how many different ways can they be displayed if a specific one is to be in the middle?

The Water Lily Pond, Pink Harmony, 1900 by Claude Monet

Challenge Problems/Group Activities

63. *Car Keys* Door keys for a certain automobile are made from a blank key on which five cuts are made. Each cut may be one of five different depths.

a) How many different keys can be made?

b) If 400,000 of these automobiles are made such that each of the keys determined in part a) opens the same number of cars, find how many cars can be opened by a specific key?

c) If one of these cars is selected at random, what is the probability that the key selected at random will unlock the door?

64. *Voting* On a ballot, each committee member is asked to rank three of seven candidates for recommendation for promotion, giving their first, second, and third choices (no ties). What is the minimum number of ballots that must be cast to guarantee that at least two ballots are the same?

65. *Scrabble* Nancy Lin, who is playing Scrabble with Dale Grey, has seven different letters. She decides to test each five-letter permutation before her next move. If each permutation takes 5 sec, how long will it take Nancy to check all the permutations?

66. *Scrabble* In Exercise 65, assume, of Nancy's seven letters, that three are identical and two are identical. How long will it take Nancy to try all different permutations of her seven letters?

67. Does ${}_nP_r = {}_nP_{(n-r)}$ for all whole numbers, where $n \geq r$? Explain.

Recreational Exercises

68. *Stations* There are eight bus stations from town A to town B. How many different single tickets must be printed so that a passenger may purchase a ticket from any station to any other station?

A o—o—o—o—o—o—o—o B

69. *Bus Loop* How many tickets with different points of origin and destination can be sold on a bus line that travels a loop with 25 stops?

Internet/Research Activity

70. When a book is published it is assigned a 10-digit code number called the International Standard Book Number (ISBN). Do research and write a report on how this coding system works.

12.9 COMBINATIONS

When the order of the selection of the items is important to the final outcome, the problem is a permutation problem. When the order of the selection of the items is unimportant to the final outcome, the problem is a *combination* problem.

Recall from Section 12.8 that permutations are *ordered* arrangements. Thus, for example, a, b, c and b, c, a are two different permutations because the ordering of the three letters is different. The letters a, b, c and b, c, a represent the same combination

DID YOU KNOW

Powerball

On Thursday, December 26, 2002, Andrew "Jack" Whittaker Jr., 55, of Scott Depot, South Carolina, won the largest Powerball jackpot in history, a \$314.9 million prize. It was the largest jackpot won worldwide by one person up until that date. Rather than taking 30 annual payments of about \$10.5 million or about \$6.9 million after taxes, Whittaker chose to take a smaller lump sum of \$170.5 million, which comes to about \$111.6 million after taxes.

As of this writing Powerball tickets are sold in 23 states, the District of Columbia, and the U.S. Virgin Islands. Winning the Powerball requires selecting 5 matching numbers of 53 as well as the Powerball number from 1 to 42. The winning numbers were 5-14-16-29-53, and the Powerball number was 7. The probability of winning is about 1 in 120,526,770. Verify this probability yourself.

The largest lottery jackpot in U.S. history was a Big Game prize of \$363 million, won in 2000 by two ticket holders in Michigan and Illinois. The largest lottery in the world is Spain's annual contest known as *El Gordo,* or "the Fat One," which in 2002 paid out a total of \$1.7 billion. Each year El Gordo has about 10,000 winners, ranging from \$20 to \$200,000 each.

of letters because the *same letters* are used in each set. However, the letters a, b, c and a, b, d represent two different combinations of letters because the letters contained in each set are different.

A **combination** is a distinct group (or set) of objects without regard to their arrangement.

EXAMPLE 1 *Permutation or Combination*

Determine whether the situation represents a permutation or combination problem.

a) A group of five friends, Arline, Inez, Judy, Dan, and Eunice, are forming a club. The group will elect a president and a treasurer. In how many different ways can the president and treasurer be selected?

b) Of the five individuals named, two will be attending a meeting together. In how many different ways can they do so?

SOLUTION:

a) Since the president's position is different from the treasurer's position, we have a permutation problem. Judy as president with Dan as treasurer is different from Dan as president with Judy as treasurer. The order of the selection is important.

b) Since the order in which the two individuals selected to attend the meeting is not important, we have a combination problem. There is no difference if Judy is selected and then Dan is selected, or if Dan is selected and then Judy is selected. ▲

In Section 12.8, you learned that ${}_nP_r$ represents the number of permutations when r items are selected from n distinct items. *Similarly,* ${}_nC_r$ *represents the number of combinations when r items are selected from n distinct items.*

Consider the set of elements $\{a, b, c, d, e\}$. The number of permutations of two letters from the set is represented as ${}_5P_2$, and the number of combinations of two letters from the set is represented as ${}_5C_2$. Twenty permutations of two letters and 10 combinations of two letters are possible from these five letters. Thus, ${}_5P_2 = 20$ and ${}_5C_2 = 10$, as shown.

Permutations		**Combinations**	
$ab, ba, ac, ca, ad, da, ae, ea, bc, cb,$ $bd, db, be, eb, cd, dc, ce, ec, de, ed$	20	$ab, ac, ad, ae, bc,$ bd, be, cd, ce, de	10

When discussing both combination and permutation problems, we always assume that the experiment is performed without replacement. That is why duplicate letters such as *aa* or *bb* are not included in the preceding example.

Note that from one combination of two letters two permutations can be formed. For example, the combination *ab* gives the permutations *ab* and *ba*, or twice as many permutations as combinations. Thus, for this example we may write

$${}_5P_2 = 2 \cdot ({}_5C_2)$$

Since $2 = 2!$ we may write

$${}_5P_2 = 2!({}_5C_2)$$

DID YOU KNOW

Poker versus Bridge

A nice bridge hand

In popular card games, there is such a variety of possible combinations of cards that a player rarely gets the same hand twice. The total number of different 5-card poker hands using a standard deck of 52 cards is 2,598,960, and for 13-card bridge hands this number increases to 635,013,559,600.

If we repeated this same process for comparing the number of permutations in ${}_nP_r$ with the number of combinations in ${}_nC_r$, we would find that

$${}_nP_r = r!({}_nC_r)$$

Dividing both sides of the equation by $r!$ gives

$${}_nC_r = \frac{{}_nP_r}{r!}$$

Since ${}_nP_r = \dfrac{n!}{(n-r)!}$, the combination formula may be expressed as

$${}_nC_r = \frac{n!/(n-r)!}{r!} = \frac{n!}{(n-r)!r!}$$

The number of combinations possible when r objects are selected from n objects is found by the **combination formula**

$${}_nC_r = \frac{n!}{(n-r)!r!}$$

EXAMPLE 2 *Exam Question Selection*

An exam consists of six questions. Any four may be selected for answering. In how many ways can this selection be made?

SOLUTION: This problem is a combination problem because the order in which the four questions are answered does not matter.

$${}_6C_4 = \frac{6!}{(6-4)!4!} = \frac{6!}{2!4!} = \frac{\overset{3}{\cancel{6}}\cdot 5\cdot \cancel{4\cdot 3\cdot 2\cdot 1}}{\cancel{2}\cdot 1\cdot \cancel{4\cdot 3\cdot 2\cdot 1}} = 15$$

There are 15 different ways that four of the six questions can be selected. ▲

EXAMPLE 3 *Airline Seat Availability*

A commercial airline is seating passengers. With a few minutes to go before take-off, there are 9 passengers who are flying standby and wish to get seats, but only 5 seats are available. Determine the number of different ways a group of 5 people can be selected from the 9 to board the plane.

SOLUTION: This problem is a combination problem because the order in which the five people are selected is unimportant. There are a total of nine people, so $n = 9$. Five are to be selected, so $r = 5$.

$${}_9C_5 = \frac{9!}{(9-5)!5!} = \frac{9!}{4!5!} = \frac{9\cdot \overset{2}{\cancel{8}}\cdot 7\cdot \cancel{6}\cdot \cancel{5\cdot 4\cdot 3\cdot 2\cdot 1}}{\cancel{4}\cdot \cancel{3}\cdot \cancel{2}\cdot 1\cdot \cancel{5\cdot 4\cdot 3\cdot 2\cdot 1}} = 126$$

Thus, 126 different combinations are possible when five from a group of nine are to be selected. ▲

EXAMPLE 4 *Dinner Combinations*

At the Royal Dynasty Chinese restaurant, dinner for eight people consists of 3 items from column A, 4 items from column B and 3 items from column C. If columns A, B, and C have 5, 7, and 6 items, respectively, how many different dinner combinations are possible?

SOLUTION: For column A, 3 of 5 items must be selected, which can be represented as ${}_5C_3$. For column B, 4 of 7 items must be selected, which can be represented as ${}_7C_4$. For column C, 3 of 6 items must be selected, or ${}_6C_3$.

$${}_5C_3 = 10 \qquad {}_7C_4 = 35 \quad \text{and} \quad {}_6C_3 = 20$$

Using the counting principle, we can determine the total number of dinner combinations by multiplying the number of choices from columns A, B, and C:

$$\begin{aligned}\text{Total number of dinner choices} &= {}_5C_3 \cdot {}_7C_4 \cdot {}_6C_3 \\ &= 10 \cdot 35 \cdot 20 = 7000\end{aligned}$$

Therefore, 7000 different combinations are possible under these conditions. ▲

We have presented various counting methods, including the counting principle, permutations, and combinations. You often need to decide which method to use to solve a problem. Table 12.4 may help you in selecting the procedure to use.

TABLE 12.4 Summary of Counting Methods

	Determining the number of ways of selecting *r* items from *n* items. Repetition not permitted.	
	Permutations	**Combinations**
Counting Principle: If a first experiment can be performed in M distinct ways and a second experiment can be performed in N distinct ways, then the two experiments in that specific order can be performed in $M \cdot N$ distinct ways. The counting principle may be used with or without repetition of items. It is used when determining the number of different ways that two or more experiments can occur. It is also used when there are specific placement requirements, such as the first digit must be a 0 or 1.	Permutations are used when order is important. For example, a, b, c and b, c, a are two different permutations of the same three letters. $${}_nP_r = \frac{n!}{(n-r)!}$$ Problems solved with the permutation formula may also be solved by using the counting principle.	Combinations are used when order is not important. For example, a, b, c and b, c, a are the same combination of three letters. But a, b, c, and a, b, d are two different combinations of three letters. $${}_nC_r = \frac{n!}{(n-r)!r!}$$

SECTION 12.9 EXERCISES

Concept/Writing Exercises

1. In your own words, explain what is meant by a combination.
2. What does ${}_nC_r$ mean?
3. Give the formula for finding ${}_nC_r$.
4. What is the relationship between ${}_nC_r$ and ${}_nP_r$?
5. In your own words, explain the difference between a permutation and a combination.
6. Assume that you have six different objects and that you are going to select four without replacement. Will there be more combinations or more permutations of the four items? Explain.

Practice the Skills

In Exercises 7–20, evaluate the expression.

7. ${}_5C_3$

8. ${}_7C_2$

9. a) ${}_6C_4$ b) ${}_6P_4$

10. a) ${}_8C_2$ b) ${}_8P_2$

11. a) ${}_8C_0$ b) ${}_8P_0$

12. a) ${}_{12}C_8$ b) ${}_{12}P_8$

13. a) ${}_{10}C_3$ b) ${}_{10}P_3$

14. a) ${}_5C_5$ b) ${}_5P_5$

15. $\dfrac{{}_5C_3}{{}_5P_3}$

16. $\dfrac{{}_6C_2}{{}_6P_2}$

17. $\dfrac{{}_8C_5}{{}_8C_2}$

18. $\dfrac{{}_6C_6}{{}_8C_0}$

19. $\dfrac{{}_9P_5}{{}_{10}C_4}$

20. $\dfrac{{}_7P_0}{{}_7C_0}$

Problem Solving

21. *Car Rental* At a car rental agency, the agent has 9 identical midsize cars on her lot and 6 people have reserved midsize cars. In how many different ways can the 6 cars to be used be selected?

22. *Banana Split* An ice cream parlor has 20 different flavors. Cynthia orders a banana split and has to select 3 different flavors. How many different selections are possible?

23. *Test Essays* A student must select and answer four of five essay questions on a test. In how many ways can she do so?

24. *Book Selection* A textbook search committee is considering eight books for possible adoption. The committee has decided to select three of the eight for further consideration. In how many ways can it do so?

25. *Dinner Party* Cliff Michael is having a dinner party. He has 10 different bottles of wine on his wine rack and wants to select and bring out 4 bottles. In how many ways can he do so?

26. *Attending Plays* While visiting New York City, the Nygents want to attend 3 plays out of 10 plays they would like to see. In how many ways can they do so?

Theater district, New York City

27. *Taxi Ride* A group of 9 people wants to use taxis to go to a local restaurant. When the first taxi arrives, the group decides that 5 people should get into the taxi. In how many ways can this be done?

28. *Plants* Mary Robinson purchased a package of 24 plants, but she only needed 20 plants for planting. In how many ways can she select the 20 plants from the package to be planted?

29. *Entertainers* Ruth Eckerd Hall must select 8 of 12 possible entertainers for their summer schedule. In how many ways can that be done?

30. *CD Purchase* Neo Anderson wants to purchase six different CDs but only has enough money to purchase four. In how many ways can he select four of six CDs for purchase?

31. *Painting* Paula Dunst has 10 framed paintings she would like to mount on her wall in a straight line. However, her wall is only wide enough to hold 8 of her paintings. In how many ways can Paula select the 8 paintings to mount on the wall?

32. *Printers and Keyboards* Office Depot has nine different printers in stock and six different cordless keyboards in stock. The manager wants to place three of the nine printers and two of the six cordless keyboards on sale. In how many ways can the manager select the items to be listed as sale items?

33. *Quinella Bet* A quinella bet consists of selecting the first- and second-place winners, in any order, in a particular event. For example, suppose you select a 2–5 quinella. If 2 wins and 5 finishes second, or if 5 wins and 2 finishes second, you win. Mr. Smith goes to a jai alai match. In the match, 8 jai alai teams compete. How many quinella tickets must Mr. Smith purchase to guarantee a win?

Jai alai game

34. *Test Question* On an English test, Tito Ramirez must write an essay for three of the five questions in Part 1 and four of the six questions in Part 2. How many different combinations of questions can he answer?

35. *Big-Screen TVs* A television/stereo store has 12 different big-screen televisions and 8 stereo systems in stock. The store's manager wishes to place 3 big-screen TVs and 2 stereo systems on sale. In how many ways can that be done?

36. *Medical Research* At a medical research center an experimental drug is to be given to 12 people, 6 men and 6 women. If 10 men and 9 women have volunteered to be given the drug, in how many ways can the researcher choose the 12 people to be given the drug?

37. *An Editor's Choice* An editor has eight manuscripts for mathematics books and five manuscripts for computer science books. If he is to select five mathematics and three computer science manuscripts for publication, how many different choices does he have?

38. *Selecting Soda* Michael Miller is sent to the store to get 5 different bottles of regular soda and 3 different bottles of diet soda. If there are 10 different types of regular sodas and 7 different types of diet sodas to choose from, how many different choices does Michael have?

39. *Forming a Committee* How many different committees can be formed from 6 teachers and 50 students if the committee is to consist of 2 teachers and 3 students?

40. *Constructing a Test* A teacher is constructing a mathematics test consisting of 10 questions. She has a pool of 28 questions, which are classified by level of difficulty as follows: 6 difficult questions, 10 average questions, and 12 easy questions. How many different 10-question tests can she construct from the pool of 28 questions if her test is to have 3 difficult, 4 average, and 3 easy questions?

41. *Selecting Mutual Funds* George Holloway recently graduated from college and now has a job that provides a retirement investment plan. George wants to diversify his investments, so he wants to invest in three stock mutual funds and two bond mutual funds. If he has a choice of eight stock mutual funds and five bond mutual funds, how many different selections of mutual funds does he have?

42. *Door Prize* As part of a door prize, Mary McCarty won three tickets to a baseball game and three tickets to a theater performance. She decided to give all the tickets to friends. For the baseball game she is considering six different friends, and for the theater she is considering eight different friends. In how many ways can she distribute the tickets?

43. *New Breakfast Cereals* General Mills is testing 6 oat cereals, 5 wheat cereals, and 4 rice cereals. If it plans to market 3 of the oat cereals, 2 of the wheat cereals, and 2 of the rice cereals, how many different combinations are possible?

44. *Catering Service* A catering service is making up trays of hors d'oeuvres. The hors d'oeuvres are categorized as inexpensive, average, and expensive. If the client must select three of the seven inexpensive, five of the eight average, and two of the four expensive hors d'oeuvres, how many different choices are possible?

Challenge Problems/Group Activities

45. *Test Answers* Consider a 10-question test in which each question can be answered either correctly or incorrectly.
   a) How many different ways are there to answer the questions so that eight are correct and two are incorrect?
   b) How many different ways are there to answer the questions so that at least eight are correct?

46. a) *A Dinner Toast* Four people at dinner make a toast. If each person is to tap glasses with each other person one at a time, how many taps will take place?
   b) Repeat part (a) with five people.
   c) How many taps will there be if there are n people at the dinner table?

47. *Pascal's Triangle* The notation ${}_nC_r$ may be written $\binom{n}{r}$.
   a) Use this notation to evaluate each of the combinations in the following array. Form a triangle of the results, similar to the one given, by placing the answer to each combination in the same relative position in the triangle.

$$\binom{0}{0}$$
$$\binom{1}{0} \quad \binom{1}{1}$$
$$\binom{2}{0} \quad \binom{2}{1} \quad \binom{2}{2}$$
$$\binom{3}{0} \quad \binom{3}{1} \quad \binom{3}{2} \quad \binom{3}{3}$$
$$\binom{4}{0} \quad \binom{4}{1} \quad \binom{4}{2} \quad \binom{4}{3} \quad \binom{4}{4}$$

   b) Using the number pattern in part (a), find the next row of numbers of the triangle (known as *Pascal's triangle*).

48. *Lottery Combinations* Determine the number of combinations possible in a state lottery where you must select
   a) 6 of 46 numbers.
   b) 6 of 47 numbers.
   c) 6 of 48 numbers.
   d) 6 of 49 numbers.
   e) Does the number of combinations increase by the same amount going from part (a) to part (b) as from part (b) to part (c)?

49. a) *Table Seating Arrangements* How many distinct ways can four people be seated in a row?
   b) How many distinct ways can four people be seated at a circular table?

50. Show that ${}_nC_r = {}_nC_{(n-r)}$.

Recreational Exercise

51. a) *Combination Lock* To open a combination lock, you must know the lock's three-number sequence in its proper order. Repetition of numbers is permitted. Why is this lock more like a permutation lock than a combination lock? Why is it not a true permutation problem?
    b) Assuming that a combination lock has 40 numbers, determine how many different three-number arrangements are possible if repetition of numbers is allowed.
    c) Answer the question in part (b) if repetition is not allowed.

12.10 SOLVING PROBABILITY PROBLEMS BY USING COMBINATIONS

In Section 12.9, we discussed combination problems. Now we will use combinations to solve probability problems.

Suppose that we want to find the probability of selecting two picture cards (jacks, queens, or kings) when two cards are selected, without replacement, from a standard deck of 52 cards. Using the *and* probability formula discussed in Section 12.6, we could reason as follows.

$$\begin{aligned} P(2 \text{ picture cards}) &= P(1\text{st picture card}) \cdot P(2\text{nd picture card}) \\ &= \frac{12}{52} \cdot \frac{11}{51} = \frac{132}{2652}, \quad \text{or} \quad \frac{11}{221} \end{aligned}$$

Since the order of the two picture cards selected is not important to the final answer, this problem can be considered a combination probability problem.

We can also find the probability of selecting two picture cards, using combinations, by finding the number of possible successful outcomes (selecting two picture cards) and dividing that answer by the total number of possible outcomes (selecting any two cards).

The number of ways in which two picture cards can be selected from the 12 picture cards in a deck is ${}_{12}C_2$, or

$${}_{12}C_2 = \frac{12!}{(12-2)!2!} = \frac{\overset{6}{\cancel{12}} \cdot 11 \cdot \cancel{10!}}{\cancel{10!} \cdot \cancel{2} \cdot 1} = 66$$

The number of ways in which two cards can be selected from a deck of 52 cards is ${}_{52}C_2$, or

$${}_{52}C_2 = \frac{52!}{(52-2)!2!} = \frac{\overset{26}{\cancel{52}} \cdot 51 \cdot \cancel{50!}}{\cancel{50!} \cdot \cancel{2} \cdot 1} = 1326$$

Thus,

$$P(\text{selecting 2 picture cards}) = \frac{{}_{12}C_2}{{}_{52}C_2} = \frac{66}{1326} = \frac{11}{221}$$

DID YOU KNOW

The Dead Man's Hand

James Butler "Wild Bill" Hickok (1837–1876), known as "the fastest gun in the West," was a famous scout and federal marshal. Hickok was shot in the back and mortally wounded by Jack McCall while playing poker in a saloon in Deadwood, Dakota Territory (now South Dakota). McCall was hanged for his deed. The cards Hickok was holding when shot are shown here. Since that time, the two pair, aces and eights, have become known as "the dead man's hand" (see Exercise 43). Hickok is buried next to Martha Canary, better known as "Calamity Jane," in a cemetery overlooking the town of Deadwood.

Note that the same answer is obtained with either method. To give you more exposure to counting techniques, we will work the problems in this section using combinations.

EXAMPLE 1 *Committee of Three Women*

A club consists of four men and five women. Three members are to be selected at random to form a committee. What is the probability that the committee will consist of three women?

SOLUTION: The order in which the three members are selected is not important. Therefore, we may work this problem using combinations.

$$P\begin{pmatrix}\text{committee consists}\\ \text{of 3 women}\end{pmatrix} = \frac{\text{number of possible committees with 3 women}}{\text{total number of possible 3-member committees}}$$

Since there are a total of 5 women, the number of possible committees with three women is ${}_5C_3 = 10$. Since there are a total of 9 people, the total number of possible three-member committees is ${}_9C_3 = 84$.

$$P(\text{committee consists of 3 women}) = \frac{10}{84} = \frac{5}{42}$$

The probability of randomly selecting a committee with three women is $\frac{5}{42}$.

EXAMPLE 2 *A Heart Flush*

A flush in the game of poker is five cards of the same suit (5 hearts, 5 diamonds, 5 clubs, or 5 spades). If you are dealt a five-card hand, find the probability that you will be dealt a heart flush.

SOLUTION: The order in which the five hearts are dealt is not important. Therefore, we may work this problem using combinations.

$$P(\text{heart flush}) = \frac{\text{number of possible 5-card heart flushes}}{\text{total number of possible 5-card hands}}$$

Since there are 13 hearts in a deck of cards, the number of possible five-card heart flush hands is ${}_{13}C_5 = 1287$. The total number of possible five-card hands in a deck of 52 cards is ${}_{52}C_5 = 2{,}598{,}960$.

$$P(\text{heart flush}) = \frac{{}_{13}C_5}{{}_{52}C_5} = \frac{1287}{2{,}598{,}960} = \frac{33}{66{,}640}$$

The probability of being dealt a heart flush is $\frac{33}{66{,}640}$, or ≈ 0.000495.

EXAMPLE 3 *Employment Assignments*

A temporary employment agency has six men and five women who wish to be assigned for the day. One employer has requested four employees for security positions, and the second employer has requested three employees for moving furniture in an office building. If we assume that each of the potential employees has the same chance of being selected and being assigned at random and that only seven employees will be assigned, find the probability that

a) three men will be selected for moving furniture.
b) three men will be selected for moving furniture and four women will be selected for security positions.

SOLUTION:

a) $$P\begin{pmatrix}\text{3 men selected}\\ \text{for moving furniture}\end{pmatrix} = \frac{\begin{pmatrix}\text{number of possible combinations}\\ \text{of 3 men selected}\end{pmatrix}}{\begin{pmatrix}\text{total number of possible combinations}\\ \text{for selecting 3 people}\end{pmatrix}}$$

The number of possible combinations with 3 men is ${}_6C_3$. The total number of possible selections of three people is ${}_{11}C_3$.

$$P\begin{pmatrix}\text{3 men selected}\\ \text{for moving furniture}\end{pmatrix} = \frac{{}_6C_3}{{}_{11}C_3} = \frac{20}{165} = \frac{4}{33}$$

Thus, the probability that 3 men are selected is $\frac{4}{33}$.

b) The number of ways of selecting 3 men out of 6 is ${}_6C_3$ and the number of ways of selecting 4 women out of 5 is ${}_5C_4$. The total number of possible selections when 7 people are selected from 11 is ${}_{11}C_7$. Since both the 3 men *and* the 4 women must be selected, the probability is calculated as follows:

$$P\begin{pmatrix}\text{3 men and}\\ \text{4 women selected}\end{pmatrix} = \frac{\begin{pmatrix}\text{number of combinations}\\ \text{of 3 men selected}\end{pmatrix}\cdot\begin{pmatrix}\text{number of combinations}\\ \text{of 4 women selected}\end{pmatrix}}{\text{total number of possible combinations of 7 people}}$$

$$= \frac{{}_6C_3 \cdot {}_5C_4}{{}_{11}C_7} = \frac{20 \cdot 5}{330} = \frac{100}{330} = \frac{10}{33}$$

Thus, the probability is $\frac{10}{33}$.

EXAMPLE 4 *New Breakfast Cereals*

Kellogg's is testing 12 new cereals for possible production. They are testing 3 oat cereals, 4 wheat cereals, and 5 rice cereals. If we assume that each of the 12 cereals has the same chance of being selected and that 4 new cereals will be produced, find the probability that

a) no wheat cereals are selected.
b) at least 1 wheat cereal is selected.
c) 2 wheat cereals and 2 rice cereals are selected.

SOLUTION:

a) If no wheat cereals are to be selected, then only oat and rice cereals must be selected. A total of 8 cereals are oat or rice. Thus, the number of ways that 4 oat or rice cereals may be selected from the 8 possible oat or rice cereals is ${}_8C_4$. The total number of possible selections is ${}_{12}C_4$.

$$P(\text{no wheat cereals}) = \frac{{}_8C_4}{{}_{12}C_4} = \frac{70}{495} = \frac{14}{99}$$

b) When 4 cereals are selected, the choice must contain either no wheat cereal or at least 1 wheat cereal. Since one of these outcomes must occur, the sum of the probabilities must be 1, or

$$P(\text{no wheat cereal}) + P(\text{at least 1 wheat cereal}) = 1$$

Therefore,

$$P(\text{at least 1 wheat cereal}) = 1 - P(\text{no wheat cereal})$$
$$= 1 - \frac{14}{99} = \frac{99}{99} - \frac{14}{99} = \frac{85}{99}$$

Note that the probability of selecting no wheat cereals, $\frac{14}{99}$, was found in part (a).

c) The number of ways of selecting 2 wheat cereals out of 4 wheat cereals is ${}_4C_2$, which equals 6. The number of ways of selecting 2 rice cereals out of 5 rice cereals is ${}_5C_2$, which equals 10. The total number of possible selections when 4 cereals are selected from the 12 choices is ${}_{12}C_4$. Since both the 2 wheat *and* the 2 rice cereals must be selected, the probability is calculated as follows.

$$P(\text{2 wheat and 2 rice}) = \frac{{}_4C_2 \cdot {}_5C_2}{{}_{12}C_4} = \frac{6 \cdot 10}{495} = \frac{60}{495} = \frac{4}{33}$$

▲

EXAMPLE 5 *Baseball Cards*

Derek Brock has 18 valuable baseball cards, including 6 Mickey Mantle cards, 4 Ken Griffey Jr. cards, and 5 Cal Ripken Jr. cards. He plans to sell 7 of his cards to finance part of his college education. If he selects the cards at random, what is the probability that 3 Mickey Mantle cards, 2 Ken Griffey Jr. cards, and 2 Cal Ripken Jr. cards are selected?

Honus Wagner (left) generally considered the most valuable baseball card. The Mickey Mantle rookie card is on the right.

SOLUTION: The number of ways that Derek can select 3 out of 6 Mickey Mantle cards is ${}_6C_3$. The number of ways he can select 2 out of 4 Ken Griffey Jr. cards is ${}_4C_2$. The number of ways he can select 2 out of 5 Cal Ripken Jr. cards is ${}_5C_2$. He will select 7 from a total of 18 baseball cards. The number of ways he can do so is ${}_{18}C_7$. The probability that Derek selects 3 Mantle, 2 Griffey Jr., and 2 Ripken Jr. cards is calculated as follows.

$$P(\text{3 Mantle, 2 Griffey Jr., and 2 Ripken Jr.}) = \frac{{}_6C_3 \cdot {}_4C_2 \cdot {}_5C_2}{{}_{18}C_7}$$
$$= \frac{20 \cdot 6 \cdot 10}{31{,}824} = \frac{1200}{31{,}824} = \frac{25}{663}$$

▲

SECTION 12.10 EXERCISES

Concept/Writing Exercises

In Exercises 1–8, set up the problem as if it were to be solved, but do not solve. Assume that each problem is to be done without replacement. Explain why you set up the exercises as you did.

1. Six red balls and four blue balls are in a bag. If four balls from the bag are to be selected at random, determine the probability of selecting four red balls.
2. A class consists of 19 girls and 15 boys. If 12 of the students are to be selected at random, determine the probability that they are all girls.

3. Three letters are to be selected at random from the English alphabet of 26 letters. Determine the probability that 3 vowels (a, e, i, o, u) are selected.
4. Determine the probability of being dealt 3 aces from a standard deck of 52 cards when 3 cards are dealt.
5. On a horse farm there are 18 horses, of which 10 are palaminos. If 5 horses are selected at random, determine the probability they are all palaminos.

6. Of 80 people attending a dance 28 have a college degree. If 4 people at the dance are selected at random, find the probability that each of the 4 has a college degree.
7. In a small area of a forest there are 30 trees, of which 16 are oak trees. Nine trees are to be selected at random. Find the probability that *none* of those selected are oak trees.
8. A class of 16 people contains 4 people whose birthday is in October. If 3 people from the class are selected at random, find the probability that *none* of those selected has an October birthday.

Practice the Skills/Problem Solving

In Exercises 9–18, the problems are to be done without replacement. Use combinations to determine probabilities.

9. *Green and Red Balls* A bag contains four red balls and five green balls. You plan to draw three balls at random. Find the probability of selecting three green balls.

10. *Drawing from a Hat* Each of the numbers 1–6 is written on a piece of paper, and the six pieces of paper are placed in a hat. If two numbers are selected at random, find the probability that both numbers selected are even.

11. *Flu Serum* A doctor has five doses of flu protection serum left. He has six women and eight men who want the medication. If the names of five of these people are selected at random, find the probability that five men's names are selected.

12. *Bills of Four Denominations* Duc Tran's wallet contains 8 bills of the following denominations: four \$5 bills, two \$10 bills, one \$20 bill, and one \$50 bill. If he selects two bills at random, determine the probability that he selects two \$5 bills.

13. *Selecting Digits* Each of the digits 0–9 is written on a slip of paper, and the slips are placed in a hat. If three slips of paper are selected at random, find the probability that the three numbers selected are greater than 4.

14. *Bike Riding* A bicycle club has 10 members. Six members ride Huffy bicycles, two members ride Roadmaster bicycles, and two members ride American Flier bicycles. If four of the members are selected at random, determine the probability that they all ride Huffy bicycles.

15. *Gift Certificates* The sales department at Atwell Studios consists of three people, the manufacturing department consists of six people, and the accounting department consists of two people. Three people will be selected at random from these people and will be given gift certificates to Sweet Tomatoes, a local restaurant. Find the probability that all those selected will be from the manufacturing department.

16. *Faculty-Student Committee* A committee of four is to be randomly selected from a group of seven teachers and eight students. Find the probability that the committee will consist of four students.

17. *Winning the Grand Prize* A lottery consists of 46 numbers. You select 6 numbers and if they match the 6 numbers selected by the lottery commission, you win the grand prize. Find the probability of winning the grand prize.

18. *Red Cards* You are dealt 5 cards from a standard deck of 52 cards. Find the probability that you are dealt 5 red cards.

TV Game Show In Exercises 19–22, a television game show has five doors, of which the contestant must pick two. Behind two of the doors are expensive cars, and behind the other three doors are consolation prizes. The contestant gets to keep the items behind the two doors she selects. Find the probability that the contestant wins

19. no cars.
20. both cars.
21. at least one car.
22. exactly one car.

Baseball In Exercises 23–26, assume that a particular professional baseball team has 10 pitchers, 6 infielders, and 9 other players. If 3 players' names are selected at random, determine the probability that

23. all 3 are infielders.
24. none of the three is a pitcher.
25. 2 are pitchers and 1 is an infielder.

26. 1 is a pitcher and 2 are players other than pitchers and infielders.

Jury Selection In Exercises 27–30, a jury pool has 17 men and 22 women, from which 12 will be selected. Assuming that each person is equally likely to be selected and that the jury is selected at random, find the probability the jury consists of

27. all women.

28. 8 women and 4 men.

29. 6 men and 6 women.

30. at least 1 man.

Airline Routes In Exercises 31–34, an airline is given permission to fly 5 new routes of its choice. The airline is considering 15 new routes: 4 routes in Florida, 5 routes in Kentucky, and 6 routes in Virginia. If the airline selects the 5 new routes at random from the 15 possibilities, find the probability that

31. 3 are in Florida and 2 are in Virginia.

32. 4 are in Kentucky and 1 is in Florida.

33. 1 is in Florida, 2 are in Kentucky, and 2 are in Virginia.

34. at least one is in Virginia.

Theater In Exercises 35–38, five men and six women are going to be assigned to a specific row of seats in a theater. If the 11 tickets for the numbered seats are given out at random, find the probability that

35. five women are given the first five seats next to the center aisle.

36. at least one woman is in one of the first five seats.

37. exactly one woman is in one of the first five seats.

38. three women are seated in the first three seats and two men are seated in the next two seats.

39. *Work Shift* Among 24 employees who work at Wendy's, three are brothers. If 6 of the 24 are selected at random to work a late shift, find the probability that the three brothers are selected.

40. *Poker Probability* A full house in poker consists of three of one kind and two of another kind in a five-card hand. For example, if a hand contains three kings and two 5's, it is a full house. If 5 cards are dealt at random from a standard deck of 52 cards, without replacement, find the probability of getting three kings and two 5's.

41. *A Royal Flush* A royal flush consists of the ace, king, queen, jack, and 10 all in the same suit. If 7 cards are dealt at random from a standard deck of 52 cards, find the probability of getting a

a) royal flush in spades.

b) royal flush in any suit.

42. *Restaurant Staff* The staff of a restaurant consists of 25 people, including 8 waiters, 12 waitresses, and 5 cooks. For Mother's Day a total of 9 people will need to be selected to work. If the selections are made at random, determine the probability that 3 waiters, 4 waitresses, and 2 cooks will be selected.

43. *"Dead Man's Hand"* A pair of aces and a pair of 8's is often known as the "dead man's hand." (See the Did You Know on page 731.)

a) Determine the probability of being dealt the dead man's hand (any two aces, any two eights, and one other card that is not an ace or an eight) when 5 cards are dealt, without replacement, from a standard deck of 52 cards.

b) The actual cards "Wild Bill" Hickok was holding when he was shot were the aces of spades and clubs, the 8's of spades and clubs, and the 9 of diamonds. If you are dealt five cards without replacement, determine the probability of being dealt this exact hand.

Challenge Problems/Group Activities

44. *Alternate Seating* If three men and three women are to be assigned at random to six seats in a row at a theater, find the probability they will alternate by gender.

45. *Selecting Officers* A club consists of 15 people including Ali, Kendra, Ted, Alice, Marie, Dan, Linda, and Frank. From the 15 members a president, vice president, and treasurer will be selected at random, and an advisory committee of 5 other individuals will also be selected at random.

a) Find the probability that Ali is selected president, Kendra is selected vice president, Ted is selected treasurer, and the other 5 individuals named form the advisory committee.

b) Find the probability that 3 of the 8 individuals named are selected for the three officers' positions and the other 5 are selected for the advisory board.

46. *A Marked Deck* A number is written with a magic marker on each card of a deck of 52 cards. The number 1 is put on the first card, 2 on the second, and so on. The cards are then shuffled and cut. What is the probability that the top 4 cards will be in ascending order? (For example, the top card is 12, the second 22, the third 41, and the fourth 51.)

Recreational Exercise

47. *Hair* When the Isle of Flume took its most recent census, the population was 100,002 people. Nobody on the isle has more than 100,001 hairs on his or her head. Determine the probability that at least two people have exactly the same number of hairs on their head.

12.11 BINOMIAL PROBABILITY FORMULA

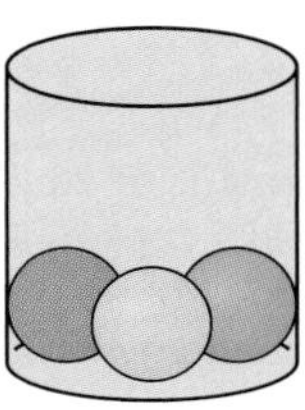

Figure 12.23

Suppose that a basket contains three identical balls, except for their color. One is red, one is blue, and one is yellow (Fig. 12.23). Suppose further that we are going to select three balls *with replacement* from the basket. We can determine specific probabilities by examining the tree diagram shown in Fig. 12.24. Note that 27 different selections are possible, as indicated in the sample space.

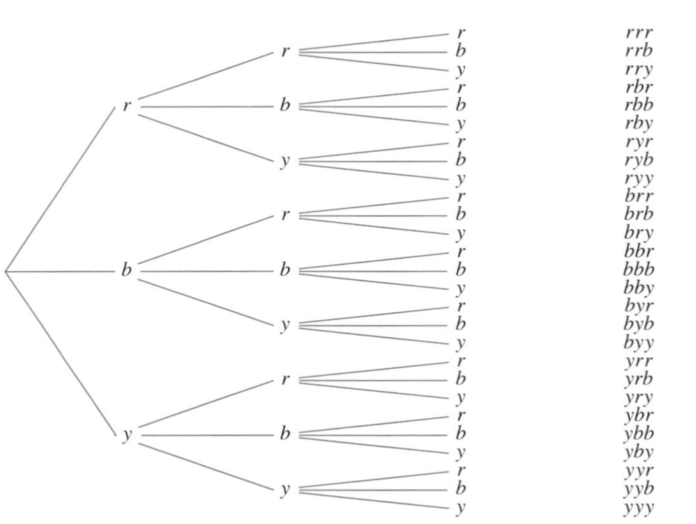

Figure 12.24

Our three selections may yield 0, 1, 2, or 3 red balls. We can determine the probability of selecting exactly 0, 1, 2, or 3 red balls by using the sample space. To determine the probability of selecting 0 red balls, we count those outcomes that do not contain a red ball. There are 8 of them (*bbb, bby, byb, byy, ybb, yby, yyb, yyy*). Thus, the probability of obtaining exactly 0 red balls is 8/27. We determine the probability of selecting exactly 1 red ball by counting the sample points that contain exactly 1 red ball. There are 12 of them. Thus, the probability is 12/27, or 4/9.

We can determine the probability of selecting exactly 2 red balls and exactly 3 red balls in a similar manner. The probabilities of selecting exactly 0, 1, 2, and 3 red balls are illustrated in Table 12.5.

TABLE 12.5 A Probability Distribution for Three Balls Selected with Replacement

Number of Red Balls Selected, (x)	**Probability of Selecting the Number of Red Balls, $P(x)$**
0	$\frac{8}{27}$
1	$\frac{12}{27}$
2	$\frac{6}{27}$
3	$\frac{1}{27}$
	Sum $= \frac{27}{27} = 1$

Note that the sum of the probabilities is 1. This table is an example of a *probability distribution*, which shows the probabilities associated with each specific outcome of an experiment. *In a probability distribution, every possible outcome must be listed, and the sum of the probabilities must be 1.*

Let us specifically consider the probability of selecting 1 red ball in 3 selections. We see from Table 12.5 that this probability is $\frac{12}{27}$ or $\frac{4}{9}$. Can we determine this probability without developing a tree diagram? The answer is yes.

Suppose we consider selecting a red ball success, S, and a non–red ball failure, F. Furthermore, suppose we let p represent the probability of success and q the probability of failure on any trial. Then $p = \frac{1}{3}$ and $q = \frac{2}{3}$. We can obtain 1 success in three selections in the following ways:

SFF FSF FFS

We can compute the probabilities of each of these outcomes using the multiplication formula because each of the selections is independent.

$$P(\text{SFF}) = P(\text{S}) \cdot P(\text{F}) \cdot P(\text{F}) = p \cdot q \cdot q = pq^2 = \frac{1}{3}\left(\frac{2}{3}\right)^2 = \frac{4}{27}$$

$$P(\text{FSF}) = P(\text{F}) \cdot P(\text{S}) \cdot P(\text{F}) = q \cdot p \cdot q = pq^2 = \frac{1}{3}\left(\frac{2}{3}\right)^2 = \frac{4}{27}$$

$$P(\text{FFS}) = P(\text{F}) \cdot P(\text{F}) \cdot P(\text{S}) = q \cdot q \cdot p = pq^2 = \frac{1}{3}\left(\frac{2}{3}\right)^2 = \frac{4}{27}$$

$$\text{Sum} = \frac{12}{27} = \frac{4}{9}$$

We obtained an answer of 4/9, the same answer that was obtained using the tree diagram. Note that each of the 3 sets of outcomes above has one success and two failures. Rather than listing all the possibilities containing 1 success and 2 failures, we can use the combination formula to determine the number of possible combinations of 1 success in 3 trials. To do so, evaluate ${}_3C_1$.

$$ {}_3C_1 = \frac{3!}{(3-1)!1!} = \frac{3 \cdot 2 \cdot 1}{2 \cdot 1 \cdot 1} = 3 $$

Number of trials (3); Number of successes (1)

Thus, we see that there are 3 ways the 1 success could occur in 3 trials. To compute the probability of 1 success in 3 trials we can multiply the probability of success in any one trial, $p \cdot q^2$, by the number of ways the 1 success can be arranged among the 3 trials, ${}_3C_1$. Thus, the probability of selecting 1 red ball, $P(1)$, in 3 trials may be found as follows,

$$ P(1) = ({}_3C_1)p^1q^2 = 3\left(\frac{1}{3}\right)\left(\frac{2}{3}\right)^2 = \frac{12}{27} = \frac{4}{9} $$

The binomial probability formula, which we introduce shortly, explains how to obtain expressions like $P(1) = ({}_3C_1)p^1q^2$ and is very useful in finding certain types of probabilities.

To use the binomial probability formula the following three conditions must hold.

To Use the Binomial Probability Formula

1. There are n repeated independent trials.
2. Each trial has two possible outcomes, *success* and *failure*.
3. For each trial, the probability of success (and failure) remains the same.

Before going further, let's discuss why we can use the binomial probability formula to find the probability of selecting a specific number of red balls when three balls are selected with replacement. First, since each trial is performed *with replacement,* the three trials are independent of each other. Second, we may consider selecting a red ball as success and selecting any ball of another color as failure. Third, for each selection, the probability of success (selecting a red ball) is $\frac{1}{3}$, and the probability of failure (selecting a ball of another color) is $\frac{2}{3}$. Now let's discuss the binomial probability formula.

Binomial Probability Formula

The probability of obtaining exactly x successes, $P(x)$, in n independent trials is given by

$$ P(x) = ({}_nC_x)p^xq^{n-x} $$

where p is the probability of success on a single trial and q is the probability of failure on a single trial.

In the formula, p will be a number between 0 and 1, inclusive, and $q = 1 - p$. Therefore, if $p = 0.2$, then $q = 1 - 0.2 = 0.8$. If $p = 3/5$, then $q = 1 - 3/5 = 2/5$. Note that $p + q = 1$ and the values of p and q remain the same for each independent trial. The combination ${}_nC_x$ is called the *binomial coefficient.*

In Example 1, we use the binomial probability formula to solve the same problem we recently solved by using a tree diagram.

EXAMPLE 1 *Selecting Colored Balls with Replacement*

A basket contains 3 balls: 1 red, 1 blue, and 1 yellow. Three balls are going to be selected with replacement from the basket. Find the probability that

a) no red balls are selected.

b) exactly 1 red ball is selected.

c) exactly 2 red balls are selected.

d) exactly 3 red balls are selected.

SOLUTION:

a) We will consider selecting a red ball a success and selecting a ball of any other color a failure. Since only 1 of the 3 balls is red, the probability of success on any single trial, p, is 1/3. The probability of failure on any single trial, q, is $1 - 1/3 = 2/3$. We are finding the probability of selecting 0 red balls, or 0 successes. Since x represents the number of successes, we let $x = 0$. There are 3 independent selections (or trials), so $n = 3$. In our calculations we will need to evaluate $(\frac{1}{3})^0$. Note that any nonzero number raised to a power of 0 is 1. Thus, $(\frac{1}{3})^0 = 1$. We determine the probability of 0 successes, or $P(0)$, as follows.

$$\begin{aligned} P(x) &= ({}_nC_x)p^xq^{n-x} \\ P(0) &= ({}_3C_0)\left(\frac{1}{3}\right)^0\left(\frac{2}{3}\right)^{3-0} \\ &= (1)(1)\left(\frac{2}{3}\right)^3 \\ &= \left(\frac{2}{3}\right)^3 = \frac{8}{27} \end{aligned}$$

b) We are finding the probability of obtaining exactly 1 red ball or exactly 1 success in 3 independent selections. Thus, $x = 1$ and $n = 3$. We find the probability of exactly 1 success, or $P(1)$, as follows.

$$\begin{aligned} P(x) &= ({}_nC_x)p^xq^{n-x} \\ P(1) &= ({}_3C_1)\left(\frac{1}{3}\right)^1\left(\frac{2}{3}\right)^{3-1} \\ &= 3\left(\frac{1}{3}\right)\left(\frac{2}{3}\right)^2 \\ &= 3\left(\frac{1}{3}\right)\left(\frac{4}{9}\right) = \frac{4}{9} \end{aligned}$$

c) We are finding the probability of selecting exactly 2 red balls in 3 independent trials. Thus, $x = 2$ and $n = 3$. We find $P(2)$ as follows.

$$\begin{aligned} P(x) &= ({}_nC_x)p^xq^{n-x} \\ P(2) &= {}_3C_2\left(\frac{1}{3}\right)^2\left(\frac{2}{3}\right)^{3-2} \\ &= 3\left(\frac{1}{3}\right)^2\left(\frac{2}{3}\right)^1 \\ &= 3\left(\frac{1}{9}\right)\left(\frac{2}{3}\right) = \frac{2}{9} \end{aligned}$$

d) We are finding the probability of selecting exactly 3 red balls in 3 independent trials. Thus, $x = 3$ and $n = 3$. We find $P(3)$ as follows.

$$P(x) = ({}_nC_x)p^x q^{n-x}$$
$$P(3) = ({}_3C_3)\left(\frac{1}{3}\right)^3\left(\frac{2}{3}\right)^{3-3}$$
$$= 1\left(\frac{1}{3}\right)^3\left(\frac{2}{3}\right)^0$$
$$= 1\left(\frac{1}{27}\right)(1) = \frac{1}{27}$$

All the probabilities obtained in Example 1 agree with the answers obtained by using the tree diagram. Whenever you obtain a value for $P(x)$, you should obtain a value between 0 and 1, inclusive. If you obtain a value greater than 1, you have made a mistake.

EXAMPLE 2 *Quality Control for Flashlights*

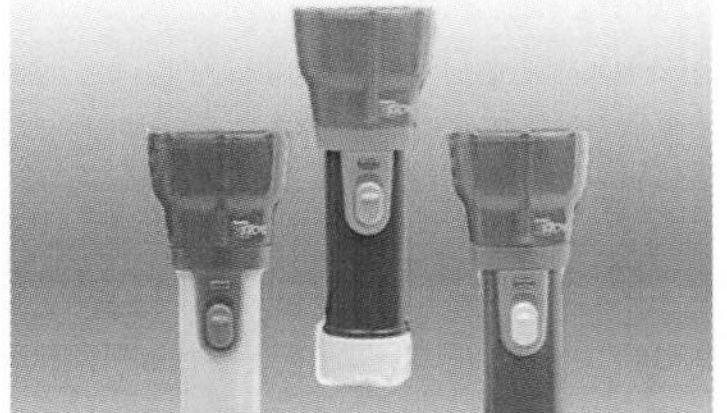

A manufacturer of flashlights knows that 0.5% of the flashlights produced by the company are defective.

a) Write the binomial probability formula that would be used to determine the probability that exactly x out of n flashlights produced are defective.

b) Write the binomial probability formula that would be used to find the probability that exactly 3 flashlights of 50 produced will be defective.

SOLUTION:

a) We want to find the probability that exactly x flashlights are defective where selecting a defective flashlight is considered success. The probability, P, that an individual flashlight is defective is 0.5%, or 0.005 in decimal form. The probability that a flashlight is not defective, q, is $1 - 0.005$, or 0.995. The general formula for finding the probability that exactly x out of n flashlights produced are defective is

$$P(x) = ({}_nC_x)p^x q^{n-x}$$

Substituting 0.005 for p and 0.995 for q, we obtain the formula

$$P(x) = ({}_nC_x)(0.005)^x(0.995)^{n-x}$$

b) We want to determine the probability that exactly 3 flashlights out of 50 produced are defective. Thus, $x = 3$ and $n = 50$. Substituting these values into the formula in part (a) gives

$$P(3) = ({}_{50}C_3)(0.005)^3(0.995)^{50-3}$$
$$= ({}_{50}C_3)(0.005)^3(0.995)^{47}$$

The answer may be obtained using a scientific calculator.

EXAMPLE 3 *Weather Forecast Accuracy*

The local weatherperson has been accurate in her temperature forecast 80% of the time. Find the probability that she is accurate

a) exactly 3 of the next 5 days.

b) exactly 4 of the next 4 days.

SOLUTION:

a) We want to find the probability that the forecaster is successful (or accurate) exactly 3 of the next 5 days. Thus, $x = 3$ and $n = 5$. The probability of success on any one day, p, is 80%, or 0.8. The probability of failure, q, is $1 - 0.8 = 0.2$. Substituting these values into the binomial probability formula yields

$$\begin{aligned} P(x) &= ({}_nC_x)p^x q^{n-x} \\ P(3) &= ({}_5C_3)(0.8)^3(0.2)^{5-3} \\ &= 10(0.8)^3(0.2)^2 \\ &= 10(0.512)(0.04) \\ &= 0.2048 \end{aligned}$$

Thus, the probability she is accurate in exactly 3 of the next 5 days is 0.2048.

b) We want to find the probability that she is accurate in each of the next 4 days. Thus, $x = 4$ and $n = 4$. We wish to find $P(4)$.

$$\begin{aligned} P(x) &= ({}_nC_x)p^x q^{n-x} \\ P(4) &= ({}_4C_4)(0.8)^4(0.2)^{4-4} \\ &= 1(0.8)^4(0.2)^0 \\ &= 1(0.4096)(1) \\ &= 0.4096 \end{aligned}$$

▲

TIMELY TIP Now would be a good time to spend a few minutes learning how to evaluate expressions such as $(0.8)^3(0.2)^2$ using your calculator. Read the manual that comes with your calculator to learn the procedure to follow to evaluate exponential expressions. Many scientific calculators have keys for evaluating factorials, and some can be used to evaluate permutations and combinations. Check to see if your calculator can be used to evaluate factorials, permutations, and combinations. Also, check with your instructor to see if these features on your calculator can be used on exams.

EXAMPLE 4 *Blue Eyes*

The probability that an individual selected at random has blue eyes is 0.4. Find the probability that

a) none of four people selected at random has blue eyes.

b) at least one of four people selected at random has blue eyes.

SOLUTION:

a) Success is selecting a person with blue eyes. Thus, $p = 0.4$ and $q = 1 - 0.4 = 0.6$. We want to find the probability of 0 successes in 4 trials. Thus, $x = 0$ and $n = 4$. We find the probability of 0 successes, or $P(0)$, as follows.

$$\begin{aligned} P(x) &= ({}_nC_x)p^x q^{n-x} \\ P(0) &= ({}_4C_0)(0.4)^0(0.6)^{4-0} \\ &= 1(1)(0.6)^4 \\ &= 1(1)(0.1296) \\ &= 0.1296 \end{aligned}$$

b) The probability that at least one person of the four has blue eyes can be found by subtracting from 1 the probability none of the people has blue eyes. We worked problems of this type in earlier sections of the chapter.

In part (a), we determined the probability that none of the people has blue eyes is 0.1296. Thus,

$$\begin{aligned} P(\text{at least 1 has blue eyes}) &= 1 - P(\text{none has blue eyes}) \\ &= 1 - 0.1296 \\ &= 0.8704 \end{aligned}$$

SECTION 12.11 EXERCISES

Concept/Writing Exercises

1. What is a probability distribution?
2. What are the three requirements that must be met to use the binomial probability formula?
3. Write the binomial probability formula.
4. In the binomial probability formula what do p and q represent?

Practice the Skills

In Exercises 5–10, assume that each of the n *trials is independent and that p is the probability of success on a given trial. Use the binomial probability formula to find P(x).*

5. $n = 4, x = 2, p = 0.3$

6. $n = 3, x = 2, p = 0.6$

7. $n = 5, x = 2, p = 0.4$

8. $n = 3, x = 3, p = 0.9$

9. $n = 6, x = 0, p = 0.5$

10. $n = 5, x = 3, p = 0.4$

Problem Solving

11. *A Dozen Eggs* An egg distributor determines that the probability that any individual egg has a crack is 0.14.
  a) Write the binomial probability formula to determine the probability that exactly x of n eggs are cracked.
  b) Write the binomial probability formula to determine the probability that exactly 2 in a one-dozen egg carton are cracked.

12. *Getting Audited* The probability that a family selected at random will be audited by the Internal Revenue Service (IRS) is 0.0237.
  a) Write the binomial probability formula to determine the probability that exactly x out of n families selected at random will be audited by the IRS.
  b) Write the binomial probability formula to determine the probability that exactly 5 of 20 families selected at random will be audited by the IRS.

In Exercises 13–21, use the binomial probability formula to answer the question. Round answers to five decimal places.

13. *Cell Phones* Thirty percent of the cell phone users in Texas use a Motorola phone. Determine the probability that if six cell phone users in Texas are selected at random, exactly four of them use a Motorola cell phone.

14. *Traffic Tickets* In Georgia, the probability that a driver is actually given a ticket when he or she is pulled over for a traffic infraction by a member of the Georgia State Police is 0.6. If eight people who were pulled over are selected at random, determine the probability that exactly five of them were given tickets.

15. *Vision Surgery* Ninety-six percent of Dr. William's LASIK surgery patients end up with 20–30, or better, vision. Find the probability that exactly 2 of her next 3 patients end up with 20–30, or better, vision.

16. *Basketball* Jean Woody makes 80% of her free throws in a basketball game. Find the probability that she makes exactly four of the next six free throws.

17. *Dolphin Drug Care* When treated with the antibiotic resonocyllin, 92% of all dolphins are cured of a particular bacterial infection. If six dolphins with the particular bacterial infection are treated with resonocyllin, find the probability that exactly four are cured.

18. *Manufacturing Light bulbs* A quality control engineer at a GE light bulb plant finds that 1% of its bulbs are defective. Find the probability that exactly two of the next four bulbs made are defective.

19. *Water Heaters* The probability that a specific brand of water heater produces the water temperature it is set to produce is $\frac{4}{5}$. Find the probability that if five of these water heaters are selected at random, exactly four of them will produce the water temperature it is set to produce.

20. *TV Purchases* At a Circuit City store, $\frac{1}{4}$ of those purchasing color televisions purchase a large-screen TV. Find the probability that

a) none of the next four people who purchase a color television at Circuit City purchases a large-screen TV.

b) at least one of the next four people who purchase a color television at Circuit City purchases a large-screen TV.

21. *Supporting a Candidate* Sixty percent of the eligible voting residents of a certain community support Ms. Stein, the incumbent candidate. If five of the residents are selected at random, find the probability that

a) none supports Ms. Stein.

b) at least one supports Ms. Stein.

Challenge Problems/Group Activities

22. *Transportation to Work* In a random sample of 80 working mothers in Duluth, Minnesota, the following data indicating how they get to work were obtained.

Mode of Transportation	Number of Mothers
Car	40
Bus	20
Bike	16
Other	4

If this sample is representative of all working mothers in Duluth, find the probability that exactly three of five working mothers selected at random

a) take a car to work.

b) take a bus to work.

23. *Selecting 6 Cards* Six cards are selected from a deck of playing cards with replacement. Find the probability that

a) exactly three picture cards are obtained.

b) exactly two spades are obtained.

24. *Office Visit* The probability that a person visiting Dr. Guillermo Suarez's office is over 60 years old is 0.7. Find the probability that

a) exactly three of the next five people visiting the office are over 60 years old.

b) at least three of the next five people visiting the office are over 60 years old.

Recreational Exercise

25. *Aruba* The island of Aruba is well known for its beaches and predictable warm sunny weather. In fact, Aruba's weather is so predictable that the daily newspapers don't even bother to print a forecast. Strangely enough, however, on New Year's Eve, as the islanders were counting down the last 10 sec of 2002, it began to rain. What is the probability, from 0 to 1, that 72 hr later the sun will be shining?

CHAPTER 12 SUMMARY

IMPORTANT FACTS

Empirical probability

$$P(E) = \frac{\text{number of times event } E \text{ has occurred}}{\left(\begin{array}{c}\text{total number of times the}\\ \text{experiment has been performed}\end{array}\right)}$$

The law of large numbers

Probability statements apply in practice to a large number of trials, not to a single trial. It is the relative frequency over the long run that is accurately predictable, not individual events or precise totals.

Theoretical probability

$$P(E) = \frac{\text{number of outcomes favorable to } E}{\text{total number of possible outcomes}}$$

The probability of an event that cannot occur is 0. The probability of an event that must occur is 1. Every probability must be a number between 0 and 1 inclusively; that is

$$0 \leq P(E) \leq 1$$

The sum of the probabilities of all possible outcomes of an event is 1.

$$P(A) + P(\text{not } A) = 1$$

Odds against an event

$$\text{Odds against} = \frac{P(\text{event fails to occur})}{P(\text{event occurs})} = \frac{P(\text{failure})}{P(\text{success})}$$

Odds in favor of an event

$$\text{Odds in favor} = \frac{P(\text{event occurs})}{P(\text{event fails to occur})} = \frac{P(\text{success})}{P(\text{failure})}$$

Expected value

$$E = P_1A_1 + P_2A_2 + P_3A_3 + \cdots + P_nA_n$$

Fair price = expected value + cost to play

Counting principle

If a first experiment can be performed in M distinct ways and a second experiment can be performed in N distinct ways, then the two experiments in that specific order can be performed in $M \cdot N$ distinct ways.

Or and And problems

$$P(A \text{ or } B) = P(A) + P(B) - P(A \text{ and } B)$$

$$P(A \text{ and } B) = P(A) \cdot P(B)$$

Conditional probability

$$P(E_2 \mid E_1) = \frac{n(E_1 \text{ and } E_2)}{n(E_1)}$$

The **number of permutations** of n items is $n!$.

$$n! = n(n-1)(n-2)\cdots(3)(2)(1)$$

Permutation formula

$${}_nP_r = \frac{n!}{(n-r)!}$$

The number of different permutations of n objects where $n_1, n_2, \ldots, n_r$ of the objects are identical is

$$\frac{n!}{n_1!n_2!\cdots n_r!}$$

Combination formula

$${}_nC_r = \frac{n!}{(n-r)!r!}$$

Binomial probability formula

$$P(x) = ({}_nC_x)p^x q^{n-x}$$

CHAPTER 12 REVIEW EXERCISES

12.1–12.11

1. In your own words, explain the law of large numbers.
2. Explain how empirical probability can be used to determine whether a die is "loaded" (not a fair die).
3. *Bicycles* Of 40 people who purchase a bike at a bike shop, 8 purchased a mountain bike. Find the empirical probability that the next person who purchases a bike from that bike shop purchases a mountain bike.
4. *Cards* Select a card from a deck of cards 40 times with replacement and compute the empirical probability of selecting a heart.
5. *Television News* In a small town, 200 people were asked whether they watched ABC, CBS, NBC, or Fox news. The results are indicated below.

Network	Number of People
ABC	80
CBS	30
NBC	55
Fox	35

Find the empirical probability that the next person selected at random from the town watches ABC news.

Digits In Exercises 6–9, each of the digits 0, 1, 2, 3, 4, 5, 6, 7, 8, 9 is written on a piece of paper, and all the pieces of paper are placed in a hat. One number is selected at random. Find the probability that the number selected is

6. even.
7. odd or greater than 5.
8. greater than 2 or less than 5.
9. even and greater than 4.

National Parks In Exercises 10–13, assume that 240 people are selected at random and are asked to name their favorite national park. The results are summarized in the following chart.

National Park	Number of People
Grand Canyon	50
Yosemite	40
Rocky Mountain	35
Great Smoky Mountains	45
Other park	70

If one person from those surveyed is selected at random, find the probability that the person chose

10. Grand Canyon National Park.
11. Yosemite National Park.

Yosemite National Park, CA

12. either Rocky Mountain National Park or Great Smoky Mountains National Park.
13. a park other than Grand Canyon National Park.
14. *Gold Star* Planters Peanuts is having a contest. They indicate that 1 in 10 jars will have a gold star under the cover. Jason Dwyer purchases one jar. Find the odds
    a) against his jar having a star.
    b) in favor of his jar having a star.
15. *Vegetable Mix-up* Nicholas Delaney, a mischievous little boy, has removed labels on the eight cans of vegetables in the cabinet. Nicholas's father knows that there are three cans of corn, three cans of beans, and two cans of carrots. If the father selects and opens one can at random, find the odds against his selecting a can of corn.
16. *Horseracing* The odds against Buttermilk winning the Triple Crown in horse racing are 82:3. Find the probability that Buttermilk wins the Triple Crown.
17. *Jitterbug Contest* The probability that Bob and Sue Nenno will win the jitterbug contest is 0.7. Find the odds in favor of them winning.
18. *Raffle Tickets* A thousand raffle tickets are sold at $2 each. Three prizes of $200 and two prizes of $100 will be awarded. Assume that the tickets are replaced after each selection.
    a) Find the expectation of a person who purchases a ticket.
    b) Find the expectation of a person who purchases three tickets.
    c) Find the fair price to pay for a ticket.
19. *Expectation of a Card* If Cameron selects a picture card from a deck of cards, Lindsey will give him $9. If Cameron does not select a picture card, he must give Lindsey $3.

a) Find Cameron's expectation.
b) Find Lindsey's expectation.
c) If Cameron plays this game 100 times, how much can he expect to lose or gain?

20. *Expected Attendance* If the day is sunny, 1000 people will attend the baseball game. If the day is cloudy, only 500 people will attend. If it rains, only 100 people will attend. The local meteorologist states that the probability of a sunny day is 0.4, of a cloudy day is 0.5, and of a rainy day is 0.1. Find the number of people that are expected to attend.

21. *Club Officers* Tina, Jake, Gina, and Carla form a club. They plan to select a president and a vice president.
a) Construct a tree diagram showing all the possible outcomes.
b) List the sample space.
c) Find the probability that Gina is selected president and Jake is selected vice president.

22. *A Coin and a Number* A coin is flipped and then a number from 1 through 4 is selected at random from a bag.
a) Construct a tree diagram showing all the possible outcomes.
b) List the sample space.
c) Find the probability that a head is flipped and an odd number is selected.
d) Find the probability that a head is flipped or an odd number is selected.

Spinning Two Wheels In Exercises 23–28, the outer and inner wheels are spun.

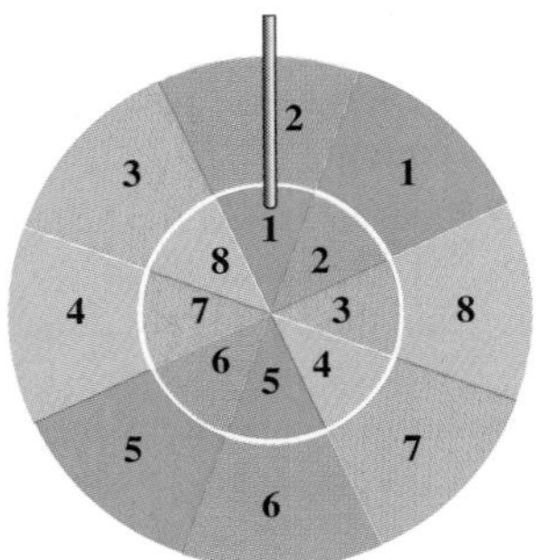

Assuming that the wheels are independent and the outcomes are equally likely, find the probability of obtaining

23. even numbers on both wheels.

24. numbers greater than 5 on both wheels.

25. an odd number on the outer wheel and a number less than 6 on the inner wheel.

26. an even number or a number less than 6 on the outer wheel.

27. an even number or a color other than green on the inner wheel.

28. gold on the outer wheel and a color other than gold on the inner wheel.

Candy Selections In Exercises 29–32, assume that the Pollingers are going to purchase three bags of candy. The store has only 12 different bags of candies, including 5 different varieties made by Hershey, 4 different varieties made by Nestlé, and 3 different varieties made by H.B. Reese Candy Company. If Mrs. Pollinger selects 3 of these 12 varieties at random, find the probability she selects

29. 3 varieties of Hershey candies.

30. no varieties of Nestlé candies.

31. at least 1 variety of Nestlé candy.

32. varieties of Hershey, Hershey, Reese in this order.

Spinner Probabilities In Exercises 33–36, assume that the spinner cannot land on a line.

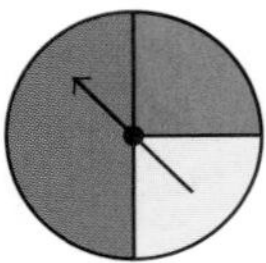

If spun once, find

33. the probability that the spinner lands on yellow.

34. the odds against and the odds in favor of the spinner landing on yellow.

35. You are awarded \$5 if the spinner lands on red, \$10 if it lands on yellow, and \$20 if it lands on green. Find your expected value.

36. If the spinner is spun twice, find the probability that it lands on red and then green (assume independence).

Spinner Probabilities In Exercises 37–40, assume that the spinner cannot land on a line.

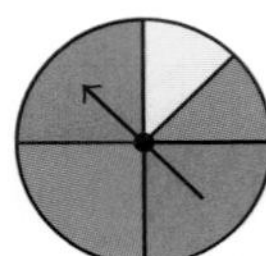

If spun once, find

37. the probability that the spinner does not land on green.

38. the odds in favor of and the odds against the spinner landing on green.

39. A person wins \$10 if the spinner lands on green, wins \$5 if the spinner lands on red, and loses \$20 if the spinner lands on yellow. Find the expectation of a person who plays this game.

40. If the spinner is spun three times, find the probability that at least one spin lands on red.

Automobile Quality Control In Exercises 41–44, a sample of 180 new cars was checked for defects. The following table shows the results of the survey.

Car	Fewer than Six Defects	Six or More Defects	Total
American built	89	17	106
Foreign built	55	19	74
Total	144	36	180

Find the probability that if one car is selected from this sample, the car has

41. fewer than six defects, given that it is American built.

42. fewer than six defects, given that it is foreign built.

43. six or more defects, given that it is foreign built.

44. six or more defects, given that it is American built.

Neuroscience In Exercises 45–48, assume that in a neuroscience course the students perform an experiment. Tests are given to determine if people are right brained, left brained, or have no predominance. It is also recorded whether they are right handed or left handed. The following chart shows the results obtained.

	Right Brained	Left Brained	No Predominance	Total
Right handed	40	130	60	230
Left handed	120	30	20	170
Total	160	160	80	400

If one person who completed the survey is selected at random, find the probability the person selected is

45. right handed.

46. left brained, given that the person is left handed.

47. right handed, given that the person has no predominance.

48. right brained, given that the person is left handed.

49. *Television Show* Four contestants are on a television show. There are four different-colored rubber balls in a box, and each contestant gets to pick one from the box. Inside each ball is a slip of paper indicating the amount the contestant has won. The amounts are \$10,000, \$5000, \$2000, and \$1000.

a) In how many different ways can the contestants select the balls?

b) What is the expectation of a contestant?

50. *Spelling Bee* Five finalists remain in a high school spelling bee. Two will receive \$50 each, two will receive \$100 each, and one will receive \$500. How many different arrangements of prizes are possible?

51. *Candy Selection* Mrs. Williams takes her 3 children shopping. Each of her children gets to select a different type of candy that only that child will eat. At the store there are only 10 boxes of candy left, and each is a different type. In how many ways can the 3 children select the candy?

52. *Astronaut Selection* Three of nine astronauts must be selected for a mission. One will be the captain, one will be the navigator, and one will perform scientific experiments. In how many ways can a three-person crew be selected so that each person has a different assignment?

53. *Medicine* Dr. Goldberg has three doses of serum for influenza type A. Six patients in the office require the serum. In how many different ways could Dr. Goldberg dispense the serum?

54. *Dogsled*

a) Ten of 15 huskies are to be selected to pull a dogsled. In how many ways can this selection be made?

b) How many different arrangements of the 10 huskies on a dogsled are possible?

55. *Mega Millions* The Big Game Mega Millions is a multi-state lottery game offered in Georgia, Illinois, Maryland, Massachusetts, Michigan, New Jersey, New York, Ohio, Texas, Virginia, and Washington. To play, you select 5 numbers from 1 through 52 and 1 Big Money Ball number from 1 through 52. If you win the Big Game by matching all 6 numbers, your guaranteed minimum payoff is $10 million. If you match the 5 numbers but do not match the Big Money number, your guaranteed payoff is $175,000.

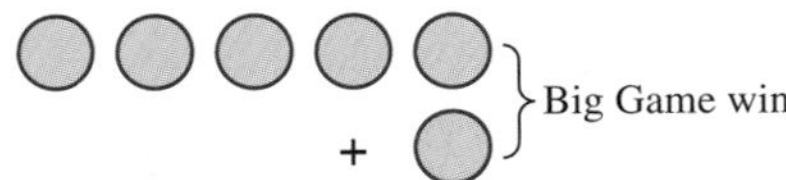

a) What is the probability you match the 5 numbers?
b) What is the probability you have a Big Game win?

56. *Parent-Teacher Committee* A committee of 6 is to be formed from 8 parents and 10 teachers. If the committee is to consist of 2 parents and 4 teachers, how many different combinations are possible?

57. *Selecting Test Subjects* In a psychology research laboratory, one room contains eight men and another room contains five women. Three men and two women are to be selected at random to be given a psychological test. How many different combinations of these people are possible?

58. *Choosing Two Aces* Two cards are selected at random, without replacement, from a deck of 52 cards. Find the probability that two aces are selected (use combinations).

Color Chips In Exercises 59–62, a bag contains five red chips, three white chips, and two blue chips. Three chips are to be selected at random, without replacement. Find the probability that

59. all are red.

60. the first two are red and the third is blue.

61. the first is red, the second is white, and the third is blue.

62. at least one is red.

Magazines In Exercises 63–66, on a table in a doctor's office are six Newsweek *magazines, five* U.S. News and World Report *magazines, and three* Time *magazines. If Ramona Cleary randomly selects three magazines, find the probability that*

63. three *U.S. News and World Report* magazines were selected.

64. two *Newsweek* magazines and one *Time* magazine were selected.

65. no *Newsweek* magazines were selected.

66. at least one *Newsweek* magazine was selected.

67. *New Homes* In the community of Spring Hill, 60% of the homes purchased cost more than $125,000.
 a) Write the binomial probability formula to determine the probability that exactly x of the next n homes purchased in Spring Hill cost more than $125,000.
 b) Write the binomial probability formula to determine the probability that exactly 75 of the next 100 home purchases cost more than $125,000.

68. *Long-Stemmed Roses* At the Floyd's Flower Shop, $\frac{1}{5}$ of those ordering flowers select long-stemmed roses. Find the probability that exactly 3 of the next 5 customers ordering flowers select long-stemmed roses.

69. *Taking a Math Course* During any semester at City College, 60% of the students are taking a mathematics course. Find the probability that of four students selected at random
 a) none is taking a mathematics course this semester.
 b) at least one is taking a mathematics course this semester.

CHAPTER 12 TEST

1. *Fishing* Of 30 people who went fishing in Lake Mead, 22 were fishing for bass. Find the empirical probability that the next person who goes fishing in Lake Mead will be fishing for bass.

One Sheet of Paper In Exercises 2–5, each of the numbers 1–9 is written on a sheet of paper, and the nine sheets of paper are placed in a hat. If one sheet of paper is selected at random from the hat, find the probability that the number selected is

2. greater than 7.

3. odd.

4. even or greater than 4.

5. odd and greater than 4.

Two Sheets of Paper In Exercises 6–9, if 2 of the same 9 sheets of paper mentioned above are selected, without replacement, from the hat, find the probability that

6. both numbers are greater than 5.

7. both numbers are even.

8. the first number is odd and the second number is even.

9. neither of the numbers is greater than 6.

10. One card is selected at random from a deck of cards. Find the probability that the card selected is a red card or a picture card.

One Chip and One Die In Exercises 11–15, one colored chip—red, blue, or green—is selected at random, and a die is rolled.

11. Use the counting principle to determine the number of sample points in the sample space.

12. Construct a tree diagram illustrating all the possible outcomes, and list the sample space.

In Exercises 13–15, by observing the sample space of the chips and die, determine the probability of obtaining

13. the color blue and the number 1.

14. the color blue or the number 1.

15. a color other than red or an odd number.

16. *Passwords* A personal password for an Internet brokerage account is to consist of a digit, followed by two letters, followed by two digits. Find the number of personal codes possible if the first digit cannot be zero and repetition is permitted.

17. *Puppies* A litter of collie puppies consists of four males and five females. If one of the puppies is selected at random, find the odds

a) against the puppy being male.
b) in favor of the puppy being female.

18. *Tennis Odds* The odds against Aimee Calhoun winning the Saddlebrook Tennis Tournament are 5:2. Find the probability that Aimee wins the tournament.

19. *Pick a Card* You get to select one card at random from a deck of cards. If you pick a club, you win $8. If you pick a heart, you win $4. If you pick any other suit, you lose $6. Find your expectation for this game.

20. *Cars and SUVs* The number of cars and the number of sport utility vehicles (SUVs) going through the toll gates of two bridges is recorded. The results are shown below.

Bridge	Cars	SUVs	Total
George Washington	120	106	226
Golden Gate	94	136	230
Total	214	242	456

The toll booths at Golden Gate bridge

If one of these vehicles going over the bridges is selected at random, find the probability that
a) it is a car.
b) it is going over the Golden Gate Bridge.
c) it is an SUV, given that it is going over the Golden Gate Bridge.
d) it is going over the George Washington Bridge, given that it is a car.

21. *Awarding Prizes* Three of six people are to be selected and given small prizes. One will be given a book, one will be given a calculator, and one will be given a $10 bill. In how many different ways can these prizes be awarded?

Quality Control In Exercises 22 and 23, a bin contains a total of 20 batteries, of which 8 are defective. If you select 2 at random, without replacement, find the probability that

22. none of the batteries is good.

23. at least one battery is good.

24. *Apples from a Bucket* Five green apples and seven red apples are in a bucket. Five apples are to be selected at random, without replacement. Find the probability that two green apples and three red apples are selected.

25. *University Admission* The probability that a person is accepted for admission to a specific university is 0.1. Find the probability that exactly three of the next five people who apply to the university get accepted.

GROUP PROJECTS

The Probability of an Exact Measured Value

1. Your car's speedometer indicates that you are traveling at 65 mph. What is the probability that you are traveling at *exactly* 65 mph? Explain your answer.

Taking an Exam

2. A 10-question multiple-choice exam is given, and each question has five possible answers. Pascal Gonyo takes this exam and guesses at every question. Use the binomial probability formula to find the probability (to 5 decimal places) that
   a) he gets exactly 2 questions correct.
   b) he gets no questions correct.
   c) he gets at least 1 question correct (use the information from part (b) to answer this part).
   d) he gets at least 9 questions correct.
   e) Without using the binomial probability formula, determine the probability that he gets exactly 2 questions correct.
   f) Compare your answers to parts (a) and (e). If they are not the same explain why.

Keyless Entry

3. Many cars have keyless entry. To open the lock you may press a 5-digit code on a set of buttons like that illustrated. The code may include repeated digits like 11433 or 55512.

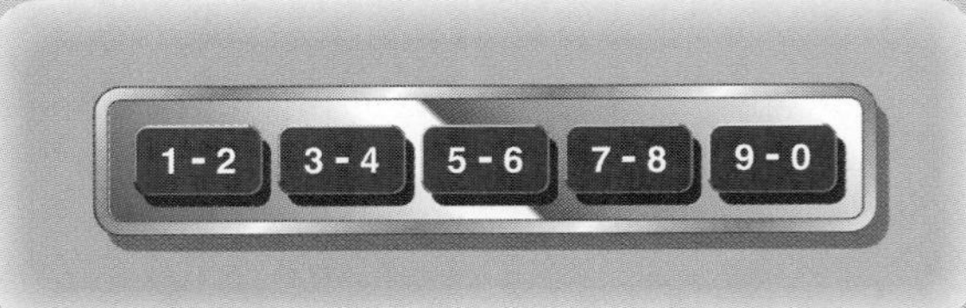

   a) How many different 5-digit codes can be made using the 10 digits if repetition is permitted?
   b) How many different ways are there of pressing 5 buttons if repetition is allowed?
   c) A burglar is going to press 5 buttons at random, with repetition allowed. Find the probability that the burglar hits the sequence to open the door.
   d) Suppose that each button had only one number associated with it as illustrated below. How many different 5-digit codes can be made with the 5 digits if repetition is permitted?

   e) Using the buttons labeled 1–5, how many different ways are there to press 5 buttons if repetition is allowed?
   f) A burglar is going to press 5 buttons of those labeled 1–5 at random with repetition allowed. Find the probability that the burglar hits the sequence to open the door.
   g) Is a burglar more likely, is he or she less likely, or does he or she have the same likelihood of pressing 5 buttons and opening the car door if the buttons are labeled as in the first illustration or as in the second illustration? Explain your answer.
   h) Can you see any advantages in labeling the buttons as in the first illustration? Explain.*

*In actuality, in most cars that have key pads like that shown on the bottom left, each key acts as if it contains a single digit. For example, if your code is 7, 9, 5, 1, 3 the code 8, 0, 6, 2, 4 will unlock the door. The extra numbers, in effect, give the owner a false sense of security.

Now in its 122nd edition and weighing over 2 pounds, the *Statistical Abstract of the United States* is a compilation of facts and figures taken from the U.S. census. In it you'll find many items, including what Americans owe and what kinds of pets Americans have (in 2001 36.1% of households had dogs, 31.6% had cats, and 4.6% had birds).

STATISTICS

Benjamin Disraeli (1804–1881), once Prime Minister of Britain, said that there are three kinds of lies: "Lies, damned lies, and statistics." Do numbers lie? Numbers are, after all, the foundation of all statistical information. The "lie" comes in when, either intentionally or carelessly, a number is used in such a way as to lead us to a conclusion that is unjustified or incorrect.

The first large-scale survey was commissioned in 1086 by William the Conqueror of England to provide a basis for taxation. The first modern census, for use as a basis for government representation, was taken in the United States in August 1790. A census has been taken in the United States every 10 years since then, and it is now used for many purposes.

Today there are methods by which statistics obtained from a small sample are used to represent a much larger population. In fact, a sample as small as 1600 may be used to predict the outcome of a national election. Information gathering is conducted by a variety of people, including medical researchers, scientists, advertisers, and political pollsters. You are likely to find examples of statistics every day on the front page of your newspaper.

When evaluating statistical information, remember that you need to judge the sampling methods as well as the numbers given. Ask yourself who conducted the study and whether they have a bias, how large the sample was and whether it was representative, and where the study appeared and whether there was an opposing side. Numbers may not lie, but they can be manipulated and misinterpreted.

13.1 SAMPLING TECHNIQUES

DID YOU KNOW

Tune in Tomorrow

The A. C. Nielsen Company, which has been measuring the viewing population of TV shows for more than 50 years, uses a sample of 5100 households in the United States to draw conclusions about more than 93 million viewers. An electronic measurement system, called the People Meter, is placed on each TV in the sample household. Nielsen uses the People Meter to measure what program is being tuned in and who is watching. Each household member is assigned a personal viewing button on the People Meter to keep track of which channels he or she watches and for how long. Nielsen then computes the rating of the show, using the data obtained from their sample. The People Meter is used to collect audience information for broadcast and cable networks, nationally syndicated programs, Spanish-language networks, and satellite distributors. Due in part to the claim by ABC, CBS, and NBC that its statistics are unreliable, Nielsen may soon begin using a new video recorder to monitor viewing habits.

Statistics is the art and science of gathering, analyzing, and making inferences (predictions) from numerical information obtained in an experiment. This numerical information is referred to as *data*. The use of statistics, originally associated with numbers gathered for governments, has grown significantly and is now applied in all walks of life.

Governments use statistics to estimate the amount of unemployment and the cost of living. Thus, statistics has become an indispensable tool in attempting to regulate the economy. In psychology and education, the statistical theory of tests and measurements has been developed to compare achievements of individuals from diverse places and backgrounds. Another use of statistics with which we are all familiar is the public opinion poll. Newspapers and magazines carry the results of different polls on topics ranging from the president's popularity to the number of cans of soda consumed. In recent years, these polls have attained a high degree of accuracy. The A. C. Nielsen rating is a public opinion poll that determines the country's most and least watched TV shows. Statistics is used in scores of other professions; in fact, it is difficult to find one that does not depend on some aspect of statistics.

Statistics is divided into two main branches: descriptive and inferential. *Descriptive statistics* is concerned with the collection, organization, and analysis of data. *Inferential statistics* is concerned with making generalizations or predictions from the data collected.

Probability and statistics are closely related. Someone in the field of probability is interested in computing the chance of occurrence of a particular event when all the possible outcomes are known. A statistician's interest lies in drawing conclusions about possible outcomes through observations of only a few particular events.

If a probability expert and a statistician find identical boxes, the probability expert might open the box, observe the contents, replace the cover, and proceed to compute the probability of randomly selecting a specific object from the box. The statistician might select a few items from the box without looking at the contents and make a prediction as to the total contents of the box.

The entire contents of the box constitute the *population*. A population consists of all items or people of interest. The statistician often uses a subset of the population, called a *sample*, to make predictions concerning the population. It is important to understand the difference between a population and a sample. A population includes *all* items of interest. A sample includes *some* of the items in the population.

When a statistician draws a conclusion from a sample, there is always the possibility that the conclusion is incorrect. For example, suppose that a jar contains 90 blue marbles and 10 red marbles, as shown in Fig. 13.1. If the statistician selects a random sample of five marbles from the jar and all are blue, he or she may wrongly conclude

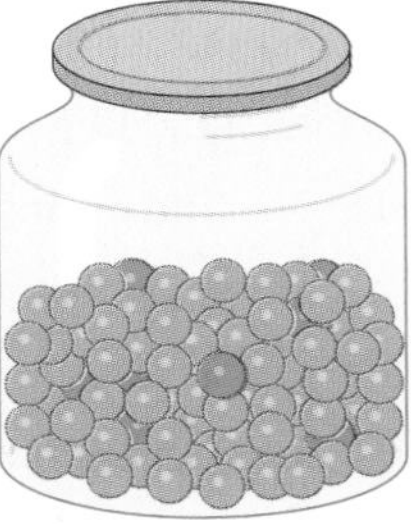

Figure 13.1

that the jar contains all blue marbles. If the statistician takes a larger sample, say, 15 marbles, he or she is likely to select some red marbles. At that point, the statistician may make a prediction about the contents of the jar based on the sample selected. Of course, the most accurate result would occur if every object in the jar, the entire population, were observed. However, in most statistical experiments, observing the entire population is not practical.

Statisticians use samples instead of the entire population for two reasons: (a) it is often impossible to obtain data on an entire population, and (b) sampling is less expensive because collecting the data takes less time and effort. For example, suppose that you wanted to determine the number of each species of all the fish in a lake. To do so would be almost impossible without using a sample. If you did try to obtain this information from the entire population, the cost would be astronomical. Or suppose that you wanted to test soup cans for spoilage. If every can produced by the company was opened and tested, the company wouldn't have any product left to sell. Instead of testing the entire population of soup cans, a sample is selected. The results obtained from the sample of soup cans selected are used to make conclusions about the entire population of soup cans.

Later in this chapter we will discuss statistical measures such as the *mean* and the *standard deviation*. When statisticians calculate the mean and the standard deviation of the entire population they use different symbols and formulas than when they calculate the mean and standard deviation of a sample. The following chart shows the symbols used to represent the mean and standard deviation of a sample and of a population. Note that the mean and standard deviation of a population are symbolized by Greek letters.

Measure	Sample	Population
Mean	$\bar{x}$ (read "x bar")	μ (mu)
Standard deviation	s	σ (sigma)

Unless otherwise indicated, in this book we will always assume that we are working with a sample and so we will use $\bar{x}$ and s. If you take a course in statistics, you will use all four symbols and different formulas for a sample and for a population.

Consider the task of determining the political strength of a certain candidate running in a national election. It is not possible for pollsters to ask each of the approximately 190 million eligible voters his or her preference for a candidate. Thus, pollsters must select and use a sample of the population to obtain their information. How large a sample do you think they use to make predictions about an upcoming national election? You might be surprised to learn that pollsters use only about 1600 registered voters in their national sample. How can a pollster using such a small percentage of the population make an accurate prediction?

The answer is that, when pollsters select a sample, they use sophisticated statistical techniques to obtain an unbiased sample. An *unbiased sample* is one that is a small replica of the entire population with regard to income, education, gender, race, religion, political affiliation, age, and so on. The procedures statisticians use to obtain unbiased samples are quite complex. The following sampling techniques will give you a brief idea of how statisticians obtain unbiased samples.

Random Sampling

If a sample is drawn in such a way that each time an item is selected each item in the population has an equal chance of being drawn, the sample is said to be a *random sample*. When using a random sample, one combination of a specified number of items has

DID YOU KNOW

The Birth of Inferential Statistics

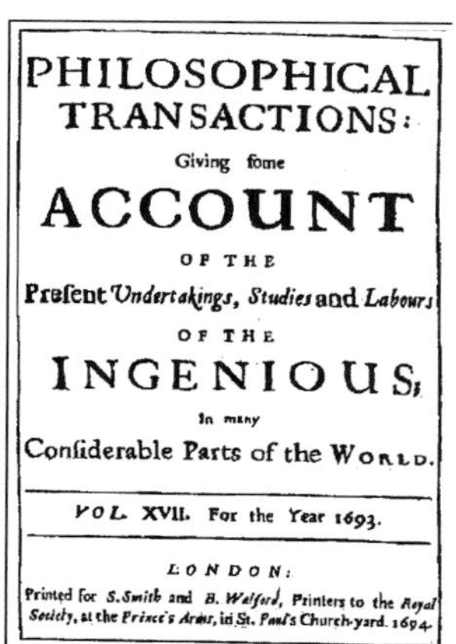

PHILOSOPHICAL
TRANSACTIONS:
Giving fome
ACCOUNT
OF THE
Prefent Undertakings, Studies and Labours
OF THE
INGENIOUS,
In many
Confiderable Parts of the WORLD.

VOL. XVII. For the Year 1693.

LONDON:
Printed for S. Smith and B. Walford, Printers to the Royal Society, at the Prince's Arms, in St. Paul's Church-yard. 1694.

John Gaunt, a London merchant, is credited with being the first person to make statistical predictions, or inferences, from a set of data rather than basing the predictions simply on the laws of chance. He studied the vital statistics (births, deaths, marriages) contained in the Bills of Mortality published during the years of the Great Plague. He observed that more males were born than females and that women lived longer than men. From these observations, he made predictions about life expectancies. The keeping of mortality statistics was stimulated considerably by the growth of the insurance industry.

DID YOU KNOW

Don't Count Your Votes Until They're Cast

A classic instance of faulty sampling occurred in the 1936 presidential election. On the basis of the responses of 2,300,000 voters, selected from automobile owners and telephone subscribers, the *Literary Digest* confidently predicted that the Republican candidate, Alf Landon, would be elected. As it turned out, Franklin D. Roosevelt, the Democratic candidate, won by a large margin. The erroneous prediction occurred because the voters used in the sample were not representative of the general voting population. In 1936, telephones and automobiles were unaffordable to the average voter.

the same probability of being selected as any other combination. When all the items in the population are similar with regard to the specific characteristic we are interested in, a random sample can be expected to produce satisfactory results. For example, consider a large container holding 300 tennis balls that are identical except for color. One-third of the balls are yellow, one-third are white, and one-third are green. If the balls can be thoroughly mixed between each draw of a tennis ball so that each ball has an equally likely chance of being selected, randomness is not difficult to achieve. However, if the objects or items are not all the same size, shape, or texture, it might be impossible to obtain a random sample by reaching into a container and selecting an object.

The best procedure for selecting a random sample is to use a random number generator or a table of random numbers. A random number generator is a device, usually a calculator or computer program, that produces a list of random numbers. A random number table is a collection of random digits in which each digit has an equal chance of appearing. To select a random sample first assign a number to each element in the population. Numbers are usually assigned in order. Then select the number of random numbers needed, which is determined by the sample size. Each numbered element from the population that corresponds to a selected random number becomes part of the sample.

Systematic Sampling

When a sample is obtained by drawing every *n*th item on a list or production line, the sample is a *systematic sample*. The first item should be determined by using a random number.

It is important that the list from which a systematic sample is chosen include the entire population being studied. See the Did You Know called "Don't Count Your Votes Until They're Cast." Another problem that must be avoided when this method of sampling is used is the constantly recurring characteristic. For example, on an assembly line, every 10th item could be the work of robot X. If only every 10th item is checked, the work of other robots doing the same job may not be checked and may be defective.

Cluster Sampling

A *cluster sample* is sometimes referred to as an *area sample* because it is frequently applied on a geographical basis. Essentially, the sampling consists of a random selection of groups of units. To select a cluster sample we divide a geographic area into sections. Then we randomly select the sections or clusters. Either each member of the selected cluster is included in the sample or a random sample of the members of each cluster is used. For example, geographically we might randomly select city blocks to use as a sample unit. Then either every member of each selected city block would be used or a random sample from each selected city block would be used. Another example would be to select x boxes of screws from a whole order, count the number of defective screws in the x boxes, and use this number to determine the expected number of defective screws in the whole order.

Stratified Sampling

When a population is divided into parts, called strata, for the purpose of drawing a sample, the procedure is known as *stratified sampling*. Stratified sampling involves dividing the population by characteristics called *stratifying factors* such as gender, race, religion, or income. When a population has varied characteristics, it is desirable to separate the population into classes with similar characteristics and then take a random sample from each stratum (or class). For example, we could separate the population of undergraduate college students into strata called freshmen, sophomores, juniors, and seniors.

The use of stratified sampling requires some knowledge of the population. For example, to obtain a cross section of voters in a city, we must know where various groups are located and the approximate numbers in each location.

Convenience Sampling

A *convenience sample* uses data that are easily or readily obtained. Occasionally, data that are conveniently obtained may be all that is available. In some cases, some information is better than no information at all. Nevertheless, convenience sampling can be extremely biased. For example, suppose that a town wants to raise taxes to build a new elementary school. The local newspaper wants to obtain the opinion of some of the residents and sends a reporter to a senior citizens center. The first 10 people who exit the building are asked if they are in favor of raising taxes to build a new school. This sample could be biased against raising taxes for the new school. Most senior citizens would not have school-age children and may not be interested in paying increased taxes to build a new school. Although a convenience sample may be very easy to select, one must be very cautious when using the results obtained from this method.

EXAMPLE 1 *Identifying Sampling Techniques*

Identify the sampling technique used to obtain a sample in the following. Explain your answer.

a) Every 20th calculator coming off an assembly line is checked for defects.

b) A $50 gift certificate is given away at the Annual Bankers Convention. Tickets are placed in a bin and the tickets are mixed up. Then the winning ticket is selected by a blindfolded person.

c) Children in a large city are classified based on the neighborhood school they attend. A random sample of five schools is selected. All the children from each selected school are included in the sample.

d) The first 30 people entering an opera house are asked if they support an increase in funding for the arts.

e) Students at Portland State University are classified according to their major. Then a random sample of 15 students from each major is selected.

SOLUTION:

a) Systematic sampling. The sample is obtained by drawing every nth item. In this example, every 20th item on an assembly line is selected.

b) Random sampling. Every ticket has an equal chance of being selected.

c) Cluster sampling. A random sample of geographic areas is selected.

d) Convenience sampling. The sample is selected by picking people that are easily obtained.

e) Stratified sampling. The students are divided into strata based on their majors. Then random samples are selected from each strata. ▲

SECTION 13.1 EXERCISES

Concept/Writing Exercises

1. Define *statistics* in your own words.
2. Explain the difference between descriptive and inferential statistics.
3. When you hear the word *statistics,* what specific words or ideas come to mind?
4. Attempt to list at least two professions in which no aspect of statistics is used.
5. Name five areas other than those mentioned in this section in which statistics is used.
6. Explain the difference between probability and statistics.
7. **a)** What is a population?
   b) What is a sample?
8. **a)** What is a systematic sample?
   b) How might a systematic sample be selected?
9. **a)** What is a random sample?
   b) How might a random sample be selected?
10. **a)** What is a cluster sample?
    b) How might a cluster sample be selected?
11. **a)** What is a stratified sample?
    b) How might a stratified sample be selected?
12. **a)** What is a convenience sample?
    b) How might a convenience sample be selected?
13. What is an unbiased sample?
14. *Family Size* The principal of an elementary school wishes to determine the "average" family size of the children who attend the school. To obtain a sample, the principal visits each room and selects the four students closest to each corner of the room. The principal asks each of these students how many people are in his or her family.
    a) Will this technique result in an unbiased sample? Explain your answer.
    b) If the sample is biased, will the average be greater than or less than the true family size? Explain.

Practice the Skills

Sampling Techniques In Exercises 15–24, identify the sampling technique used to obtain a sample. Explain your answer.

15. A group of people are classified according to age and then random samples of people from each group are taken.
16. Every 15th CD player coming off an assembly line is checked for defects.
17. A state is divided into regions using zip codes. A random sample of 20 zip code areas is selected.
18. A door prize is given away at a teachers' convention. Tickets are placed in a bin and the tickets are mixed up. Then a ticket is selected by a blindfolded person.
19. Every 17th person in line to buy tickets for a rock concert is asked his or her age.
20. The businesses in Iowa City are grouped according to type: medical, service, retail, manufacturing, financial, construction, restaurant, hotel, tourism, and other. A random sample of 10 businesses from each type is selected.
21. The first 25 students leaving the cafeteria are asked how much money they spent on textbooks for the semester.
22. The Food and Drug Administration randomly selects five stores from each of four randomly selected sections of a large city and checks food items for freshness. These stores are used as a representative sample of the entire city.
23. Bingo balls in a bin are shaken and then balls are selected from the bin.

24. The Student Senate at the University of New Orleans is electing a new president. The first 25 people leaving the library are asked for whom they will vote.

Challenge Problems/Group Activities

25. **a)** *Random Sampling* Select a topic and population of interest to which a random sampling technique can be applied to obtain data.
    b) Explain how you or your group would obtain a random sample for your population of interest.
    c) Actually obtain the sample by the procedure stated in part (b).
26. *Data from Questionnaire* Some subscribers of *Consumer Reports* respond to an annual questionnaire regarding their satisfaction with new appliances, cars, and other items. The information obtained from these questionnaires is then used as a sample from which frequency of repairs and other ratings are made by the magazine. Are the data obtained from these returned questionnaires representative of the entire population or are they biased? Explain your answer.

Recreational Mathematics

27. Statistically speaking, what is the most dangerous job in the United States?

28. Refer to the Did You Know on page 700. Select a random sample of 30 people and see how many of the 30 people have the same birthday. (*Hint:* The probability of at least 2 of the 30 sharing the same birthday is greater than 0.5).

Internet/Research Problem

29. We have briefly introduced sampling techniques. Using statistics books and Internet websites as references, select one type of sampling technique (it may be one that we have not discussed in this section) and write a report on how statisticians obtain that type of sample. Also indicate when that type of sampling technique may be preferred. List two examples of when the sampling technique may be used.

13.2 THE MISUSES OF STATISTICS

Statistics, when used properly, is a valuable tool to society. However, many individuals, businesses, and advertising firms misuse statistics to their own advantage. You should examine statistical statements very carefully before accepting them as fact. Two questions you should ask yourself are: Was the sample used to gather the statistical data unbiased and of sufficient size? Is the statistical statement ambiguous; that is, can it be interpreted in more than one way?

Let's examine two advertisements. "Four out of five dentists recommend sugarless gum for their patients who chew gum." In this advertisement, we do not know the sample size and the number of times the experiment was performed to obtain the desired results. The advertisement does not mention that possibly only 1 of 100 dentists recommended gum at all.

In a golf ball commercial, a "type A" ball is hit, and a second ball is hit in the same manner. The type A ball travels farther. We are supposed to conclude that the type A is the better ball. The advertisement does not mention the number of times the experiment was previously performed or the results of the earlier experiments. Possible sources of bias include (1) wind speed and direction, (2) that no two swings are identical, and (3) that the ball may land on a rough or smooth surface.

Vague or ambiguous words also lead to statistical misuses or misinterpretations. The word *average* is one such culprit. There are at least four different "averages," some of which are discussed in Section 13.5. Each is calculated differently, and each may have a different value for the same sample. During contract negotiations, it is not uncommon for an employer to state publicly that the average salary of its employees is $35,000, whereas the employees' union states that the average is $30,000. Who is lying? Actually, both sides may be telling the truth. Each side will use the average that best suits its needs to present its case. Advertisers also use the average that most enhances their products. Consumers often misinterpret this average as the one with which they are most familiar.

Another vague word is *largest*. For example, ABC claims that it is the largest department store in the United States. Does that mean largest profit, largest sales, largest building, largest staff, largest acreage, or largest number of outlets?

Still another deceptive technique used in advertising is to state a claim from which the public may draw irrelevant conclusions. For example, a disinfectant manufacturer claims that its product killed 40,760 germs in a laboratory in 5 seconds. "To prevent colds, use disinfectant A." It may well be that the germs killed in the laboratory were not related to any type of cold germ. In another example, company C claims that its paper towels are heavier than its competition's towels. Therefore, they will hold more water. Is weight a measure of absorbency? A rock is heavier than a sponge, yet a sponge is more absorbent.

DID YOU KNOW

Creative Displays

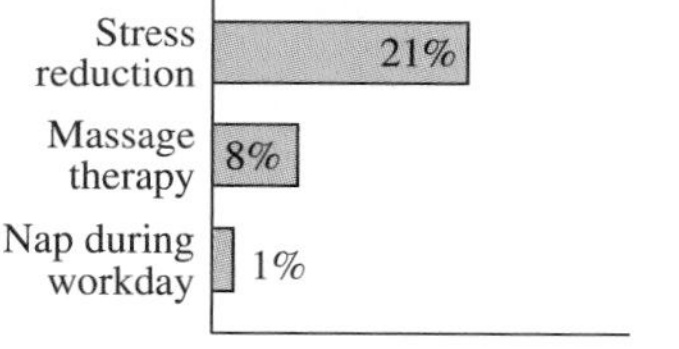

Visual graphics are often used to "dress up" what might otherwise be considered boring statistics. Although visually appealing, such creative displays of numerical data can be misleading. The graph above, shows the percentage of employers that offer "perks" such as stress reduction, massage therapy, or a nap during the workday to make workers happy. This graph is misleading because the lengths of the bars are not proportional to one another as they should be to accurately reflect the percent of employers offering each of the named perks. For example, the bar for massage therapy should be eight times as long as the bar for nap during workday instead of being approximately four times as long, as the graph shows.

An insurance advertisement claims that in Duluth, Minnesota, 212 people switched to insurance company Z. One may conclude that this company is offering something special to attract these people. What may have been omitted from the advertisement is that 415 people in Duluth, Minnesota, dropped insurance company Z during the same period.

A foreign car manufacturer claims that 9 of every 10 of a popular model car it sold in the United States during the previous 10 years were still on the road. From this statement the public is to conclude that this foreign car is well manufactured and would last for many years. The commercial neglects to state that this model has been selling in the United States for only a few years. The manufacturer could just as well have stated that 9 of every 10 of these cars sold in the United States in the previous 100 years were still on the road.

Charts and graphs can also be misleading or deceptive. In Fig. 13.2, the two graphs show the performance of two stocks over a 6-month period. Based on the graphs, which stock would you purchase? Actually, the two graphs present identical information; the only difference is that the vertical scale of the graph for stock B has been exaggerated.

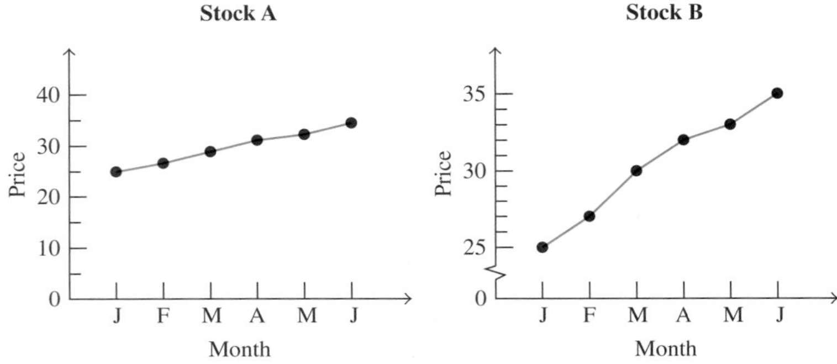

Figure 13.2

The two graphs in Fig. 13.3 show the same change. However, the graph in part (a) appears to show a greater increase than the graph in part (b), again because of a different scale.

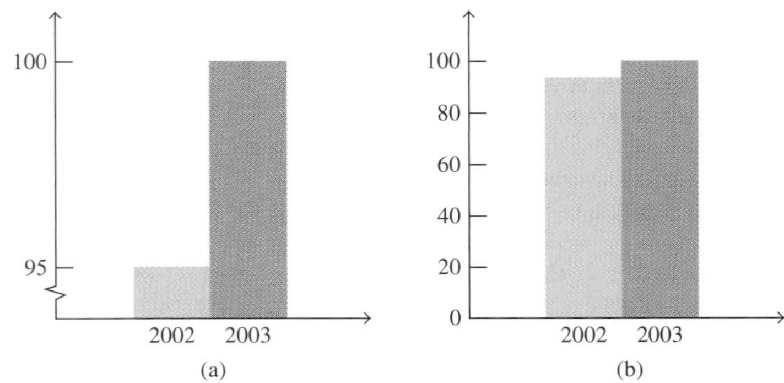

Figure 13.3

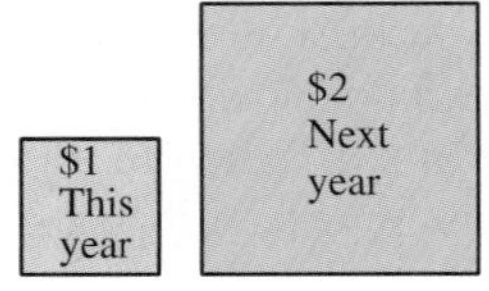

Figure 13.4

Consider a claim that if you invest $1, by next year you will have $2. This type of claim is sometimes misrepresented, as in Fig. 13.4. Actually, your investment has only doubled, but the area of the square on the right is four times that of the square on the left. By expressing the amounts as cubes (Fig. 13.5), you increase the volume eightfold.

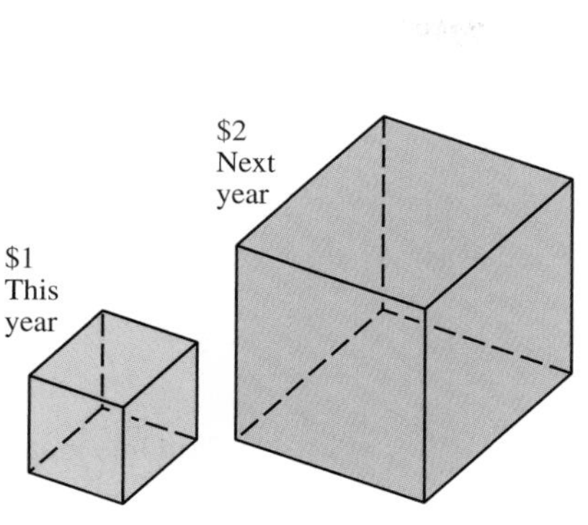

Figure 13.5

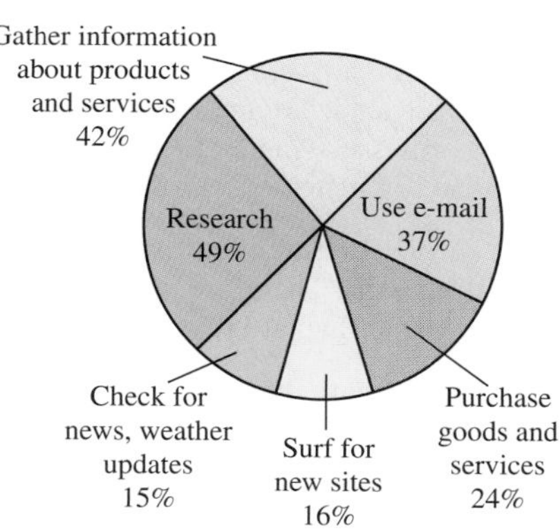

Figure 13.6

The graph in Fig. 13.6 is an example of a circle graph. We will discuss how to construct circle graphs in Section 13.4. In a circle graph, the total circle represents 100%. Therefore, the sum of the parts should add up to 100%. This graph is misleading since the sum of its parts is 183%. A graph other than a circle graph should have been used to display the top six reasons Americans say they use the Internet.

Despite the examples presented in this section, you should not be left with the impression that statistics is used solely for the purpose of misleading or cheating the consumer. As stated earlier, there are many important and necessary uses of statistics. Most statistical reports are accurate and useful. You should realize, however, the importance of being an aware consumer.

SECTION 13.2 EXERCISES

Concept/Writing Exercises

1. Find five advertisements or commercials that may be statistically misleading. Explain why each may be misleading.
2. A sample of 300 people leaving a restaurant was asked the following question: "On which of the following special occasions are you likely to dine out: your birthday, Mother's Day, Valentine's Day, or New Year's Eve?" The following circle graph shows the percent of responses for each of the above special occasions. Is the graph misleading? Explain.

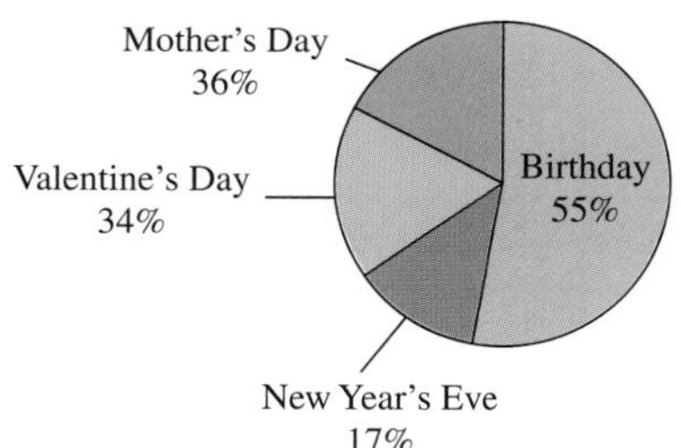

Practice the Skills

Misinterpretations of Statistics In Exercises 3–16, discuss the statement and tell what possible misuses or misinterpretations may exist.

3. In 2003, there were more car thefts in Baltimore, Maryland, than in Reno, Nevada. Therefore, you are more likely to have your car stolen in Baltimore than in Reno.
4. There are more empty spaces in the parking lot of Mama Mia's Italian restaurant than at Shanghi Chinese restaurant. Therefore, more people prefer Chinese food than Italian food.
5. Healthy Snacks cookies are fat free. So eat as many as you like and you will not gain weight.
6. Morgan's is the largest department store in New York. So shop at Morgan's and save money.
7. Most accidents occur on Saturday night. This means that people do not drive carefully on Saturday night.
8. Eighty percent of all automobile accidents occur within 10 miles of the driver's home. Therefore, it is safer to take long trips.

9. Arizona has the highest death rate for asthma in the country. Therefore, it is unsafe to go to Arizona if you have asthma.

10. Thirty students said they would recommend Professor Malone to a friend. Twenty students said they would recommend Professor Wagner to a friend. Therefore, Professor Malone is a better teacher than Professor Wagner.

11. Milk is cheaper at Star Food Markets than at Price Chopper Food Markets. Therefore, groceries at Star Food Markets are cheaper than at Price Chopper Food Markets.

12. Treadware Tires are the most expensive tires. Therefore, they will last the longest.

13. The average depth of the pond is only 3 ft, so it is safe to go wading.

14. More men than women are involved in automobile accidents. Therefore, women are better drivers.

15. At West High School, half the students are below average in mathematics. Therefore, the school should receive more federal aid to raise student scores.

16. More women than men applied for admission to the 2003 class of Mesa Community College. Therefore, in 2003, more women than men will attend Mesa Community College.

17. *Hospital Care Expenditures* The following table shows the percent of national health expenditures spent on hospital care for selected years.

Year	Percent
1985	39.1
1990	36.5
1995	34.7
1997	33.6
1998	33.0
1999	32.3
2000	31.7

Source: National Center for Health Statistics, U.S. Department of Health and Human Services.

Draw a line graph that makes the decrease in the percent of national health expenditures spent on hospital care from 1985 through 2000 appear to be
a) small.
b) large.

18. *Infant Mortality Rate* The table above and to the right shows the United States infant mortality rate, per 1000 births from 1994 to 2000.

Year	Rate
1994	8.0
1995	7.6
1996	7.3
1997	7.2
1998	7.2
1999	7.1
2000	6.9

Source: National Center for Health Statistics, U.S. Department of Health and Human Services.

Draw a line graph that makes the decrease in the U.S. infant mortality rate from 1994 through 2000 appear to be
a) small.
b) large.

First Marriage In Exercises 19 and 20, use the following table.

Median Age at First Marriage

Male		Female	
Year	**Age**	**Year**	**Age**
1970	23.2	1970	20.8
1980	24.7	1980	22.0
1990	26.1	1990	23.9
2000	26.9	2000	25.1

Source: U.S. Census Bureau.

19. a) Draw a bar graph that appears to show a small increase in the median age at first marriage for males.
b) Draw a bar graph that appears to show a large increase in the median age at first marriage for males.

20. a) Draw a bar graph that appears to show a small increase in the median age at first marriage for females.
b) Draw a bar graph that appears to show a large increase in the median age at first marriage for females.

21. *Online Purchasing* The following graph shows the percent of males and the percent of females surveyed that purchased clothing accessories online during the months from November 2000 to January 2001.
a) Draw a bar graph that shows the entire scale from 0 to 6.
b) Does the new graph give a different impression? Explain.

Percent of Survey Respondents Who Purchased Clothing Accessories Online, Nov. 2000–Jan. 2001

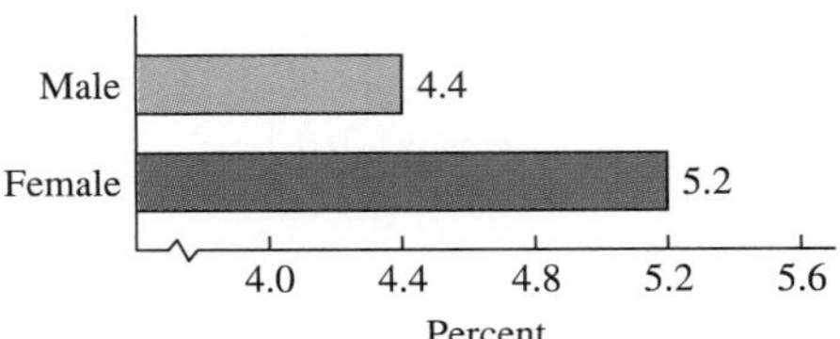

Source: Forrester Research

Challenge Problem/Group Activity

22. Consider the following graph, which shows the U.S. population in 2000 and the projected U.S. population in 2050.

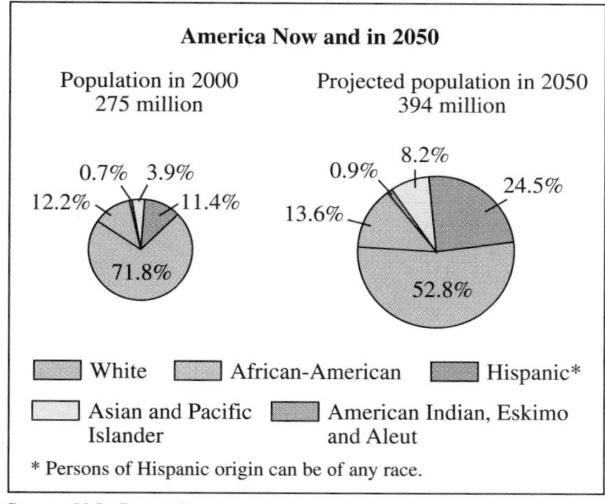

Source: U.S. Census Bureau

a) Compute the projected percent increase in population from 2000 to 2050 by using the formula given on page 595.
b) Measure the radius and then compute the area of the circle representing 2000. Use $A = \pi r^2$.
c) Repeat part (b) for the circle representing 2050.
d) Compute the percent increase in the size of the area of the circle from 2000 to 2050.
e) Are the circle graphs misleading? Explain your answer.

Recreational Mathematics

23. What mathematical symbol can you place between 1 and 2 to obtain a number greater than 1 but less than 2?

Internet/Research Activity

24. Read the book *How to Lie with Statistics* by Darrell Huff and write a book report on it. Select three illustrations from the book that show how people manipulate statistics.

13.3 FREQUENCY DISTRIBUTIONS

It is not uncommon for statisticians and others to have to analyze thousands of pieces of data. A *piece of data* is a single response to an experiment. When the amount of data is large, it is usually advantageous to construct a frequency distribution. A *frequency distribution* is a listing of the observed values and the corresponding frequency of occurrence of each value.

"Statistical thinking will one day be as necessary for efficient citizenship as the ability to read and write."
H. G. Wells

EXAMPLE 1 *Frequency Distribution*

The number of children per family is recorded for 64 families surveyed. Construct a frequency distribution of the following data:

0	1	1	2	2	3	4	5
0	1	1	2	2	3	4	5
0	1	1	2	2	3	4	6
0	1	2	2	2	3	4	6
0	1	2	2	2	3	4	7
0	1	2	2	3	3	4	8
0	1	2	2	3	3	5	8
0	1	2	2	3	3	5	9

DID YOU KNOW

Can You Count the F's?

Statistical errors often result from careless observations. To see how such errors can occur, consider the statement below. How many F's do you count in the statement? You can find the answer in the answer section (Section 13.3, Exercise 3.2).

FINISHED FILES ARE THE RESULT OF YEARS OF SCIENTIFIC STUDY COMBINED WITH THE EXPERIENCE OF YEARS.

SOLUTION: Listing the number of children (observed values) and the number of families (frequency) gives the following frequency distribution.

Number of Children (Observed Values)	Number of Families (Frequency)
0	8
1	11
2	18
3	11
4	6
5	4
6	2
7	1
8	2
9	1
	64

Eight families had no children, 11 families had one child, 18 families had two children, and so on. Note that the sum of the frequencies is equal to the original number of pieces of data, 64. ▲

Often data are grouped in classes to provide information about the distribution that would be difficult to observe if the data were ungrouped. Graphs called *histograms* and *frequency polygons* can be made of grouped data, as will be explained in Section 13.4. These graphs also provide a great deal of useful information.

When data are grouped in classes, certain rules should be followed.

Rules for Data Grouped by Classes

1. The classes should be of the same "width."
2. The classes should not overlap.
3. Each piece of data should belong to only one class.

In addition, it is often suggested that a frequency distribution should be constructed with 5 to 12 classes. If there are too few or too many classes, the distribution may become difficult to interpret. For example, if you use fewer than 5 classes, you risk losing too much information. If you use more than 12 classes, you may gain more detail but you risk losing clarity. Let the spread of the data be a guide in deciding the number of classes to use.

To understand these rules, let's consider a set of observed values that go from a low of 0 to a high of 26. Let's assume that the first class is arbitrarily selected to go from 0 through 4. Thus, any of the data with values of 0, 1, 2, 3, 4 would belong in this class. We say that the *class width* is 5, since there are five integral values that belong to the class. This first class ended with 4, so the second class must start with 5. If this class is to have a width of 5, at what value must it end? The answer is 9 (5, 6, 7, 8, 9). The second class is 5–9. Continuing in the same manner, we obtain the following set of classes.

TABLE 13.1

Newspaper	Circulation (thousands)
USA Today	2150
Wall Street Journal	1781
New York Times	1109
Los Angeles Times	944
Washington Post	760
New York Daily News	734
Chicago Tribune	676
Long Island Newsday	577
Houston Chronicle	552
New York Post	534
San Francisco Chronicle	512
Dallas Morning News	495
Chicago Sun-Times	481
Boston Globe	471
Phoenix Arizona Republic	451
Newark Star-Ledger	411
Atlanta Journal-Constitution	396
Detroit Free Press	371
Philadelphia Inquirer	365
Cleveland Plain Dealer	360
San Diego Union-Tribune	352
Portland Oregonian	351
Minneapolis Star Tribune	340
St. Petersburg Times	332
Orange County Register	324
Miami Herald	318
Denver Rocky Mountain News	310
Baltimore Sun	306
Denver Post	306
St. Louis Post-Dispatch	291
Sacramento Bee	286
Investor's Business Daily	281
San Jose Mercury News	269
Kansas City Star	260
Boston Herald	259
Milwaukee Journal Sentinel	255
Orlando Sentinel	255
New Orleans Times-Picayune	255
Indianapolis Star	252
Fort Lauderdale Sun-Sentinel	252
Columbus Dispatch	244
Detroit News	243
Pittsburgh Post-Gazette	242
Charlotte Observer	235
Louisville Courier-Journal	222
Seattle Times	219
Buffalo News	219
Fort Worth Star-Telegram	214
Tampa Tribune	213
San Antonio Express-News	209

Source: *2002 Editor & Publisher International Yearbook.*

Classes

Lower class limits { 0–4, 5–9, 10–14, 15–19, 20–24, 25–29 } Upper class limits

We need not go beyond the 25–29 class because the largest value we are considering is 26. The classes meet our three criteria: They have the same width, there is no overlap among the classes, and each of the values from a low of 0 to a high of 26 belongs to one and only one class.

The choice of the first class, 0–4, was arbitrary. If we wanted to have more classes or fewer classes, we would make the class widths smaller or larger, respectively.

The numbers 0, 5, 10, 15, 20, 25 are called the *lower class limits*, and the numbers 4, 9, 14, 19, 24, 29 are called the *upper class limits*. Each class has a width of 5. Note that the class width, 5, can be obtained by subtracting the first lower class limit from the second lower class limit: $5 - 0 = 5$. The difference between any two consecutive lower class or upper class limits is also 5.

EXAMPLE 2 *A Frequency Distribution of Daily Newspaper Circulation*

Table 13.1 in the margin shows the circulation for the 50 leading U.S. daily newspapers, in 2001. The circulation is rounded to the nearest thousand. Construct a frequency distribution of the data, letting the first class be 209–402.

SOLUTION: Fifty pieces of data are given in *descending order* from highest to lowest. We are given that the first class is 209–402. The second class must therefore start at 403. To find the class width we subtract 209 (the lower class limit of the first class) from 403 (the lower class limit of the second class) to obtain a class width of 194. The upper class limit of the second class is found by adding the class width, 194, to the upper class limit of the first class, 402. Therefore, the upper class limit of the second class is $402 + 194 = 596$. Thus,

$$209 - 402 = \text{first class}$$
$$403 - 596 = \text{second class}$$

The other classes are found using a similar technique. The other classes are 597–790, 791–984, 985–1178, 1179–1372, 1373–1566, 1567–1760, 1761–1954, 1955–2148, and 2149–2342. Since the highest value in the data is 2150, there is no need to go any further. Note that each two consecutive lower class limits differ by 194, as do each two consecutive upper class limits. There are 34 pieces of data in the 209–402 class. There are 9 pieces of data in the 403–596 class, 3 in the 597–790 class, 1 in the 791–984 class, 1 in the 985–1178 class, 0 in the 1179–1372 class, 0 in the 1373–1566 class, 0 in the 1567–1760 class, 1 in the 1761–1954 class, 0 in the 1955–2148 class, and 1 in the 2149–2342 class. The complete frequency distribution of the 11 classes is given on page 764. The number of newspapers totals 50, so we have included each piece of data.

Circulation	Number of Newspapers
209–402	34
403–596	9
597–790	3
791–984	1
985–1178	1
1179–1372	0
1373–1566	0
1567–1760	0
1761–1954	1
1955–2148	0
2149–2342	1
	50

The *modal class* of a frequency distribution is the class with the greatest frequency. In Example 2, the modal class is 209–402. The *midpoint of a class*, also called the *class mark*, is found by adding the lower and upper class limits and dividing the sum by 2. The midpoint of the first class in Example 2 is

$$\frac{209 + 402}{2} = \frac{611}{2} = 305.5$$

Note that the difference between successive class marks is the class width. The class mark of the second class can therefore be obtained by adding the class width, 194, to the class mark of the first class, 305.5. The sum is $305.5 + 194 = 499.5$. Note that $\frac{403 + 596}{2} = \frac{999}{2} = 499.5$ which checks with the class mark obtained by adding the class width to the first class mark.

EXAMPLE 3 *A Frequency Distribution of Family Income*

The following set of data represents the family income (in thousands of dollars, rounded to the nearest hundred) of 15 randomly selected families.

31.5	16.8	30.8	29.7	25.9
50.2	37.4	29.6	38.7	33.8
20.5	25.3	24.8	41.3	35.7

Construct a frequency distribution with a first class of 16.5–22.6.

SOLUTION: First rearrange the data from lowest to highest so that the data will be easier to categorize.

16.8	25.3	29.7	33.8	38.7
20.5	25.9	30.8	35.7	41.3
24.8	29.6	31.5	37.4	50.2

The first class goes from 16.5 to 22.6. Since the data are in tenths, the class limits will also be given in tenths. The first class ends with 22.6; therefore, the second class must start with 22.7. The class width of the first class is $22.7 - 16.5$, or 6.2. The upper class limit of the second class must therefore be $22.6 + 6.2$, or 28.8. The frequency distribution is as follows.

Income ($1000)	Number of Families
16.5–22.6	2
22.7–28.8	3
28.9–35.0	5
35.1–41.2	3
41.3–47.4	1
47.5–53.6	1
	15

Note in Example 3 that the class width is 6.2, the modal class is 28.9 − 35.0, and the class mark of the first class is $(16.5 + 22.6)/2$, or 19.55.

DID YOU KNOW

Cyberspace Is the Place to Be

Portland, Oregon, the "most wired" city.

The Internet, first developed as a communications tool for the government and for research universities, is now being used for everything from research to entertainment to shopping to chatting with friends. Internet usage continues to increase dramatically. In 2001, 56.5% of households in the United States had computers, and 50.5% of households in the United States were online. Thirty-nine percent of Internet users made online purchases and 35% of Internet users searched for health information. Consumers are also using the Internet to provide themselves with automotive information prior to purchasing a new vehicle. According to a study done by J.D. Power and Associates, 62% of all new vehicle buyers in 2001 used the Internet for automotive information before purchasing their vehicle. Internet usage was growing at a rate of 2 million new users per month in 2001. According to Nielsen/NetRatings, as of March 2001, Portland, Oregon, was the overall "most wired" city in the United States, with almost 70% of households having Internet access from a personal computer at home.

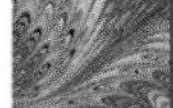

SECTION 13.3 EXERCISES

Concept/Writing Exercises

1. What is a frequency distribution?

2. How can a class width be determined using class limits?

3. Suppose that the first class of a frequency distribution is 9–15.
   a) What is the width of this class?
   b) What is the second class?
   c) What is the lower class limit of the second class?
   d) What is the upper class limit of the second class?

4. Repeat Exercise 3 for a frequency distribution whose first class is 12–20.

5. What is the modal class of a frequency distribution?

6. What is another name for the midpoint of a class? How is the midpoint of a class determined?

Practice the Skills/Problem Solving

In Exercises 7 and 8, use the frequency distribution to determine
   a) *the total number of observations.*
   b) *the width of each class.*
   c) *the midpoint of the second class.*
   d) *the modal class (or classes).*
   e) *the class limits of the next class if an additional class were to be added.*

7.

Class	Frequency
9–15	3
16–22	6
23–29	1
30–36	0
37–43	3
44–50	5

8.

Class	Frequency
40–49	7
50–59	5
60–69	3
70–79	2
80–89	7
90–99	1

9. *Sales* A car dealership is interested in the number of cars sold daily. A sample is taken over 40 days to obtain the following data regarding the number of cars sold daily. Construct a frequency distribution, letting each class have a width of 1 (as in Example 1).

0	1	1	3	4	5	7	8
0	1	2	3	5	5	7	8
0	1	2	3	5	5	7	9
1	1	2	3	5	6	8	10
1	1	3	4	5	6	8	10

10. *Park Visits by Families* The town of Brighton is planning to improve the local park. The responses of 32 families who were asked how many times per year they visit the park are shown below. Construct a frequency distribution letting each class have a width of 1.

20	21	24	25	26	27	29	32
20	23	24	25	26	27	30	32
20	23	24	26	26	28	31	33
21	23	24	26	26	28	31	34

Note: No one visited the park 22 times per year. However, it is customary to include a missing value as an observed value and assign to it a frequency of 0.

IQ Scores In Exercises 11–14, use the following data, which show the result of 50 sixth-grade I.Q. scores.

80	89	92	95	97	100	102	106	110	120
81	89	93	95	98	100	103	108	113	120
87	90	94	97	99	100	103	108	114	122
88	91	94	97	100	100	103	108	114	128
89	92	94	97	100	101	104	109	119	135

Use this data to construct a frequency distribution with a first class of

11. 78–86.

12. 80–88.

13. 80–90.

14. 80–92.

Placement Test Scores In Exercises 15–18, use the following data, which represent the English placement test scores of a sample of 30 students.

559	482	490	520	514
498	472	490	523	491
480	490	562	486	491
498	543	506	539	576
508	509	499	515	501
593	512	510	577	533

Use this data to construct a frequency distribution with a first class of

15. 472–492.

16. 470–486.

17. 472–487.

18. 472–496.

Newspaper Circulation In Exercises 19–22, use the data in Example 2 on page 763 to construct a frequency distribution with a first class (in thousands) of

19. 209–458.

20. 205–414.

21. 209–408.

22. 209–358.

County Population In Exercises 23–26, use the following data, which represent the 2000 population of the 25 largest counties in the United States, in millions of people (rounded to the nearest 100,000).

9.5	2.8	2.1	1.5	1.4
5.4	2.5	1.7	1.5	1.4
3.4	2.3	1.7	1.5	1.4
3.1	2.3	1.7	1.5	1.4
2.8	2.2	1.6	1.4	1.4

Use this data to construct a frequency distribution with a first class of

23. 1.4–2.1.

24. 1.0–2.7.

25. 1.0–2.5.

26. 1.4–2.9.

Price of Eggs In Exercises 27–30, use the data in the following table.

Average Price per Dozen Eggs for Selected States in 2000

State	Price ($)	State	Price ($)
AL	1.31	MT	0.46
AR	1.06	NE	0.38
CA	0.45	NH	0.86
CO	0.70	NJ	0.53
CT	0.56	NY	0.56
DE	0.67	NC	1.07
FL	0.48	OH	0.50
GA	0.87	OK	0.84
HI	0.89	OR	0.48
ID	0.61	PA	0.55
IL	0.47	RI	0.63
IN	0.52	SC	0.64
IA	0.38	SD	0.35
KS	0.39	TN	1.24
KY	0.90	TX	0.70
LA	0.81	UT	0.43
ME	0.60	VT	0.60
MD	0.60	VA	0.96
MA	0.63	WA	0.55
MI	0.42	WV	1.46
MS	1.18	WI	0.48
MO	0.52		

Source: National Agricultural Statistics Service, U.S. Dept. of Agriculture.

Construct a frequency distribution with a first class of

27. 0.35–0.44.

28. 0.35–0.45.

29. 0.35–0.54.

30. 0.35–0.48.

Recreational Mathematics

31. In what month do people take the least number of daily vitamins?

32. a) Count the number of F's in the sentence at the bottom of the Did You Know on page 762.

b) Can you explain why so many people count the number of F's incorrectly?

13.4 STATISTICAL GRAPHS

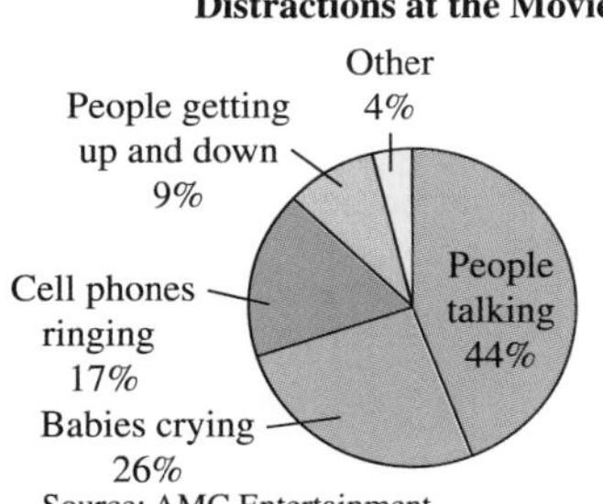

Figure 13.7

Now we will consider four types of graphs: the circle graph, the histogram, the frequency polygon, and the stem-and-leaf display.

Circle graphs (also known as pie charts) are often used to compare parts of one or more components of the whole to the whole. The circle graph in Fig. 13.7 shows what moviegoers say is the most annoying distraction during a movie. Since the total circle represents 100%, the sum of the percents of the sectors should be 100%, and it is.

In the next example, we will discuss how to construct a circle graph given a set of data.

EXAMPLE 1 *Labor Day Travel*

According to the American Automobile Association (AAA), 27.7 million Americans traveled by car during Labor Day weekend in 2001. The following table indicates the destinations of these travelers.

Destination	Number of People (millions)
Major cities	6.4
Oceans and beaches	5.5
Towns and rural areas	5.3
Mountains	3.9
Other	6.6
	27.7

Use this information to construct a circle graph illustrating the percent of people who traveled to major cities, oceans and beaches, towns and rural areas, the mountains, and other places during Labor Day weekend in 2001.

SOLUTION: Determine the measure of the corresponding central angle, as illustrated in the following table.

Destination	Number of People (millions)	Percent of Total (to the nearest tenth of a percent)	Measure of Central Angle (degrees)
Major cities	6.4	$\frac{6.4}{27.7} \times 100 = 23.1\%$	$0.231 \times 360 = 83.2°$
Oceans and beaches	5.5	$\frac{5.5}{27.7} \times 100 = 19.9\%$	$0.199 \times 360 = 71.6°$
Towns and rural areas	5.3	$\frac{5.3}{27.7} \times 100 = 19.1\%$	$0.191 \times 360 = 68.8°$
Mountains	3.9	$\frac{3.9}{27.7} \times 100 = 14.1\%$	$0.141 \times 360 = 50.8°$
Other	6.6	$\frac{6.6}{27.7} \times 100 = 23.8\%$	$0.238 \times 360 = 85.7°$
Total	27.7	100.0%	360.1° *

*Due to rounding we get 360.1°, not exactly 360°. If the measure of the central angle were rounded to hundredths, the sum would be exactly 360°.

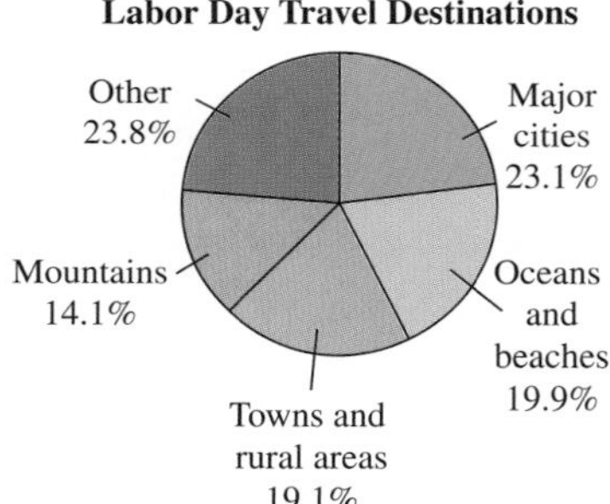

Figure 13.8

Now use a protractor (See Section 9.1, page 479.) to construct a circle graph and label it properly, as illustrated in Fig. 13.8. The measure of the central angle for major cities is about 83.2°, for oceans and beaches it is about 71.6°, for towns and rural areas it is about 68.8°, for mountains it is about 50.8°, for other areas it is about 85.7°. ▲

TIMELY TIP When constructing circle graphs remember the following information:

If you have to round your percent in the percent of total column, the sum of the percents may not be exactly 100%. Due to rounding the percents, the sum may be either slightly below 100% or slightly above 100%.

When calculating the measure of a central angle, if you have to round the central angle measure, the sum of the angles may not be exactly 360°. Due to rounding measurements, the sum of the angles may be slightly above 360° or slightly below 360°.

DID YOU KNOW

Polls

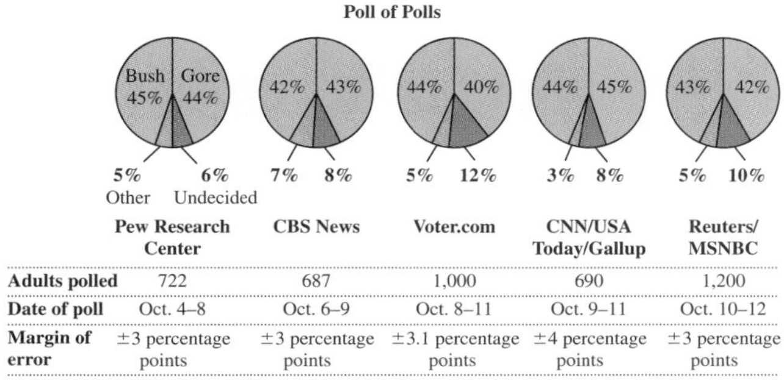

	Pew Research Center	CBS News	Voter.com	CNN/USA Today/Gallup	Reuters/ MSNBC
Adults polled	722	687	1,000	690	1,200
Date of poll	Oct. 4–8	Oct. 6–9	Oct. 8–11	Oct. 9–11	Oct. 10–12
Margin of error	±3 percentage points	±3 percentage points	±3.1 percentage points	±4 percentage points	±3 percentage points

During every major election we are told the results of many polls. The following circle graphs, shown in the October 23, 2000, issue of *U.S. News and World Report*, show the results of various polls taken less than a month before the 2000 U.S. presidential election. The polls were repeated at various times closer to election time. Notice that no two polling services had exactly the same results. Also notice that the margin of error varied between ±3 points to ±4 points. The margin of error is determined by the size of the sample used and by other items. Many people believe that exit polls, taken right after people vote, should not be allowed to be broadcast until after the election results are finalized, because these polls may influence the people who have not yet voted. For example, since the people on the East Coast vote about 3 hours earlier than people on the West Coast, polls showing voters' preferences in eastern states could affect voters' choices in western states.

Histograms and frequency polygons are statistical graphs used to illustrate frequency distributions. A *histogram* is a graph with observed values on its horizontal scale and frequencies on its vertical scale. A bar is constructed above each observed value (or class when classes are used), indicating the frequency of that value. The horizontal scale need not start at zero, and the calibrations on the horizontal and vertical scales do not have to be the same. The vertical scale must start at zero. To accommodate large frequencies on the vertical scale, it may be necessary to break the scale. Because histograms and other bar graphs are easy to interpret visually, they are used a great deal in newspapers and magazines.

EXAMPLE 2 *Construct a Histogram*

The frequency distribution developed in Example 1, Section 13.3, is repeated here. Construct a histogram of this frequency distribution.

Number of Children (Observed Values)	Number of Families (Frequency)
0	8
1	11
2	18
3	11
4	6
5	4
6	2
7	1
8	2
9	1

SOLUTION: The vertical scale must extend at least to the number 18, since that is the greatest recorded frequency (see Fig. 13.9 on page 770). The horizontal scale must include the numbers 0–9, the number of children observed. Eight families have no children. We indicate this by constructing a bar above the number 0, centered at 0, on the horizontal scale extended up to 8 on the vertical scale. Eleven families have one child, so we construct a bar extending to 11 above the number 1, centered at 1, on the horizontal scale. We continue this procedure for each observed value. Both the horizontal and vertical scales should be labeled, the bars should be the same width and centered at the observed value, and the histogram should have a title. In a histogram, the bars should always touch.

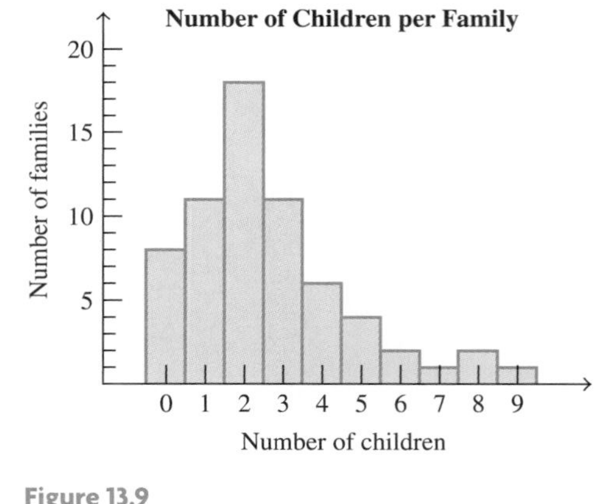

Figure 13.9

Frequency polygons are line graphs with scales the same as those of the histogram; that is, the horizontal scale indicates observed values and the vertical scale indicates frequency. To construct a frequency polygon, place a dot at the corresponding frequency above each of the observed values. Then connect the dots with straight-line segments. When constructing frequency polygons, always put in two additional class marks, one at the lower end and one at the upper end on the horizontal scale (values for these added class marks are not needed on the frequency polygon). Since the frequency at these added class marks is 0, the end points of the frequency polygon will always be on the horizontal scale.

EXAMPLE 3 *Construct a Frequency Polygon*

Construct a frequency polygon of the frequency distribution in Example 2.

SOLUTION: Since eight families have no children, place a mark above the 0 at 8 on the vertical scale, as shown in Fig. 13.10. Because there are 11 families with one

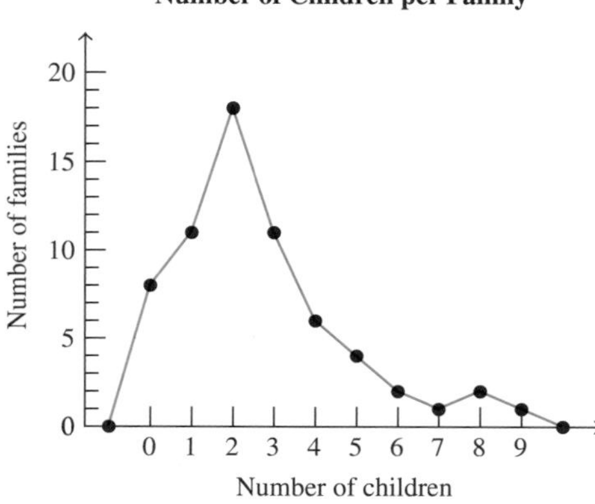

Figure 13.10

child, place a mark above the 1 on the horizontal scale at the 11 on the vertical scale, and so on. Connect the dots with straight-line segments, and bring the end points of the graph down to the horizontal scale, as shown. ▲

TIMELY TIP When constructing a histogram or frequency polygon, be sure to label both scales of the graph.

EXAMPLE 4 *Gas Mileage*

The frequency distribution of average gas mileage for selected 2003 automobiles is listed in Table 13.2. Construct a histogram and then construct a frequency polygon on the histogram.

TABLE 13.2

Mileage (mpg)	Number of Cars
10–14	5
15–19	31
20–24	36
25–29	8
30–34	5
35–39	4
40–44	1

SOLUTION: The histogram can be constructed with either class limits or class marks (class midpoints) on the horizontal scale. Frequency polygons are constructed with class marks on the horizontal scale. Since we will construct a frequency polygon on the histogram, we will use class marks. Recall that class marks are found by adding the lower class limit and upper class limit and dividing the sum by 2. For the first class, the class mark is $(10 + 14)/2$, or 12. Since the class widths are five units, the class marks will also differ by five units (see Fig. 13.11).

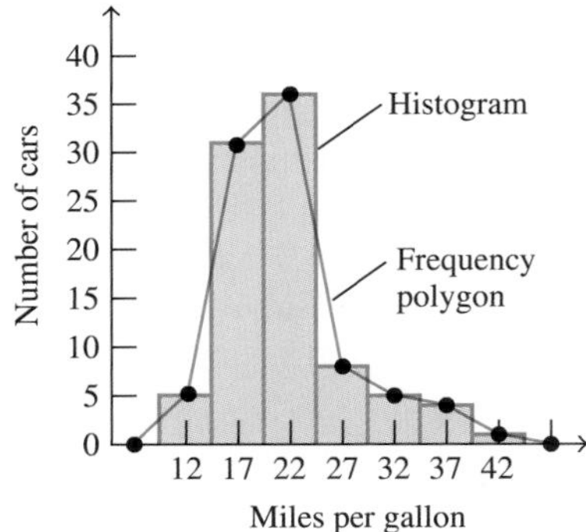

Figure 13.11

▲

EXAMPLE 5 *Carry-on Luggage Weights*

The histogram in Fig. 13.12 on page 772 shows the weights of selected pieces of carry-on luggage at an airport. Construct the frequency distribution from the histogram in Fig. 13.12.

TABLE 13.3

Weight (pounds)	Number of Pieces of Luggage
1–5	8
6–10	10
11–15	7
16–20	5
21–25	6
26–30	3
31–35	1
36–40	2

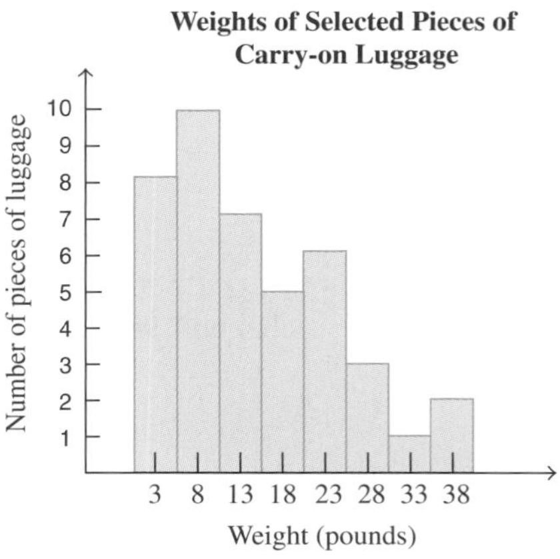

Figure 13.12

SOLUTION: There are five units between class midpoints, so each class width must also be five units. Since three is the midpoint of the first class, there must be two units below and two above it. The first class must be 1–5. The second class must therefore be 6–10. The frequency distribution is given in Table 13.3.

Frequency distributions and histograms provide very useful tools to organize and summarize data. However, if the data are grouped, we cannot identify specific data values in a frequency distribution and in a histogram. For example, in Example 5, we know that there are eight pieces of luggage in the class of 1 to 5 pounds, but we don't know the specific weights of those eight pieces of luggage.

A *stem-and-leaf display* is a tool that organizes and groups the data while allowing us to see the actual values that make up the data. To construct a stem-and-leaf display each value is represented with two different groups. The left group of digits is called the *stem.* The remaining group of digits on the right is called the *leaf.* There is no rule for the number of digits to be included in the stem. Usually the units digit is the leaf and the remaining digits are the stem. For example, the number 53 would be broken up into 5 and 3. The 5 would be the stem and the 3 would be the leaf. The number 417 would be broken up into 41 and 7. The 41 would be the stem and the 7 would be the leaf. The number 6, which can be represented as 06, would be broken up into 0 and 6. The stem would be the 0 and the leaf would be the 6. With a stem-and-leaf display, the stems are listed, in order, to the left of a vertical line. Then we place each leaf to the right of its corresponding stem, to the right of the vertical line.* Example 6 illustrates this procedure.

EXAMPLE 6 *Stem-and-Leaf Display*

The table below indicates the ages of a sample of 20 guests who stayed at Captain Fairfield House Bed and Breakfast. Construct a stem-and-leaf display.

29	31	39	43	56
60	62	59	58	32
47	27	50	28	71
72	44	45	44	68

*In stem-and-leaf displays, the leaves are sometimes listed from lowest digit to greatest digit, but this is not necessary.

SOLUTION: By quickly glancing at the data, we can see the ages consist of two digit numbers. Let's use the first digit, the tens digit, as our stem and the second digit, the units digit, as the leaf. For example, for an age of 62, the stem is 6, and the leaf is 2. Our values are numbers in the 20s, 30s, 40s, 50s, 60s, and 70s. Therefore, the stems will be 2, 3, 4, 5, 6, 7 as shown below.

```
2|
3|
4|
5|
6|
7|
```

Next we place each leaf on its stem. We will do so by placing the second digit of each value next to its stem, to the right of the vertical line. Our first value is 29. The 2 is the stem and the 9 is the leaf. Therefore, we place a 9 next to the stem of 2 and to the right of the vertical line.

```
2|9
```

The next value is 31. We will place a leaf of 1 next to the stem of 3.

```
2|9
3|1
```

The next value is 39. Therefore, we will place a leaf of 9 after the leaf of 1 that is next to the stem of 3.

```
2|9
3|1 9
```

We continue this process until we have listed all the leaves on the display. The diagram below shows the stem-and-leaf display for the ages of the guests. In our display, we will also include a legend to indicate the values represented by the stems and leaves. For example, 5 | 6 represents 56.

```
     5|6 represents 56
Stem  Leaves
     2|9 7 8
     3|1 9 2
     4|3 7 4 5 4
     5|6 9 8 0
     6|0 2 8
     7|1 2
```

▲

Every piece of the original data can be seen in a stem-and-leaf display. From the above diagram, we can see that five of the guests' ages were in the 40s. Only two guests were older than 70. Note that the stem-and-leaf display gives the same visual impression as a sideways histogram.

SECTION 13.4 EXERCISES

Concept/Writing Exercises

1. In your own words, explain how to construct a circle graph from a table of values.

2. a) What is listed on the horizontal axis of a histogram and frequency polygon?
   b) What is listed on the vertical axis of a histogram and frequency polygon?

3. In your own words, explain how to construct a frequency polygon from a set of data.

4. In your own words, explain how to construct a histogram from a set of data.

5. a) In your own words, explain how to construct a frequency polygon from a histogram.
   b) Construct a frequency polygon from the histogram below.

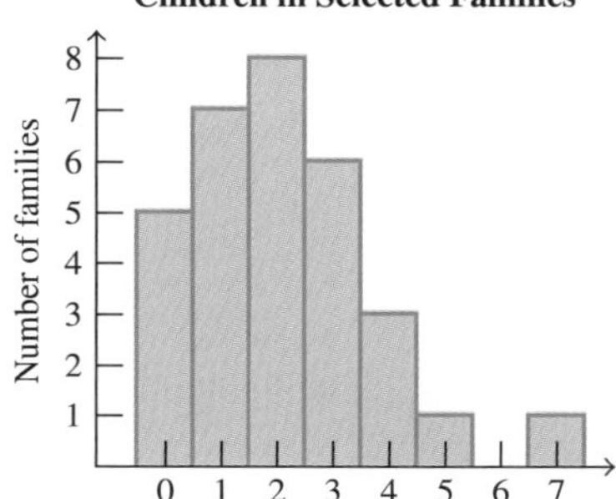

6. a) In your own words, explain how to construct a histogram from a frequency polygon.
   b) Construct a histogram from the frequency polygon below.

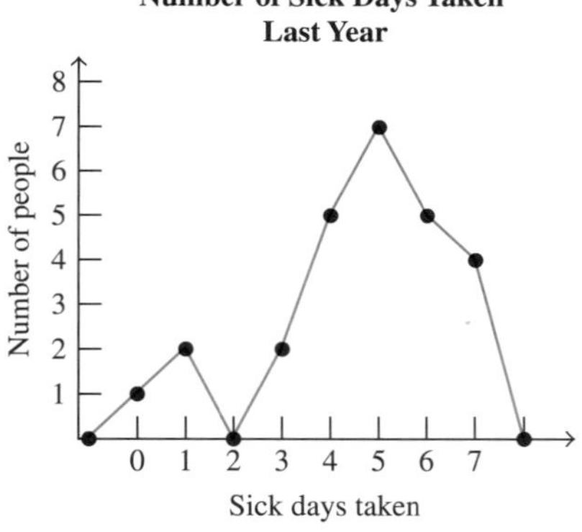

7. a) In your own words, explain how to construct a stem-and-leaf display.
   b) Construct a frequency distribution, letting each class have a width of 1, from the stem-and-leaf display above and to the right.

4 | 5 represents 45

Stem	Leaf
4	5 5 5 7 9
5	0 1 1

8. Construct a frequency distribution, letting each class have a width of 1, from the following stem-and-leaf display.

2 | 3 represents 23

Stem	Leaf
1	7 8 7 9 6
2	3 1 2 2 5 5 4

Practice the Skills

9. *Bringing Home Leftovers* ConAgra foods, along with the American Dietetic Association, conducted a survey asking the question, "How often do you bring home leftovers from restaurants?" The following circle graph shows the percent of respondents who answered occasionally, most times, every time, or never. If 500 people were surveyed, determine the number of people in each category.

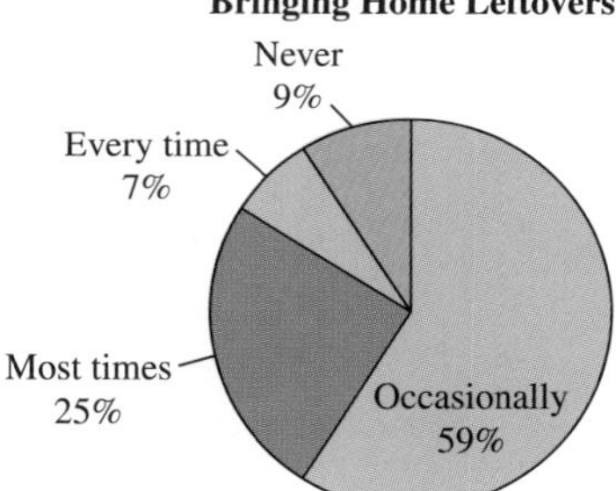

10. *Where Teens Work* The National Academy Press surveyed teens to determine where they work. The following circle graph shows the percent of 15- to 17-year-old teens surveyed who answered retail, services, or other as their type of employment. If 700 teens were surveyed, determine the number of teens working in each category. Round answers to nearest person.

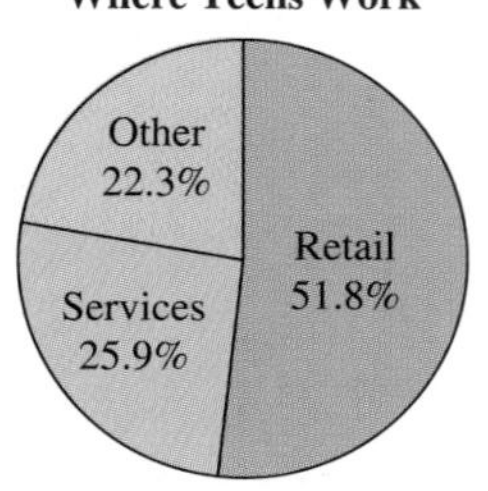

Source: National Academy Press

11. *Online Travel Websites* A sample of 500 travelers who used the Internet to book a trip were asked which travel website they used. Their responses are given in the table below. Construct a circle graph, with sectors given in percent, which illustrates this information.

Online Travel Website	Number of Bookings
Travelocity	175
Expedia	125
Priceline	85
Other	115

12. *Housing Permits* A sample of 600 housing permits for new houses was randomly selected. The number of bedrooms listed in the permits was recorded, as indicated in the following table. Construct a circle graph, with sectors given in percent, which illustrates this information. Round percents to tenths.

Number of Bedrooms	Number of Permits
2	182
3	230
4	100
5 or more	88

13. *Heights* The frequency distribution indicates the heights of 45 male high school seniors.

Height (in.)	Number of Males
64	2
65	6
66	7
67	9
68	10
69	6
70	3
71	0
72	2

a) Construct a histogram of the frequency distribution.
b) Construct a frequency polygon of the frequency distribution.

14. *Jazz Concert* The frequency distribution indicates the ages of a group of 45 people attending a jazz concert.

Age	Number of People
17	2
18	5
19	7
20	8
21	0
22	10
23	5
24	8

a) Construct a histogram of the frequency distribution.
b) Construct a frequency polygon of the frequency distribution.

15. *DVDs* The frequency distribution indicates the number of DVDs owned by a sample of 40 people.

Number of DVDs	Number of People
6–13	4
14–21	5
22–29	10
30–37	11
38–45	6
46–53	3
54–61	1

a) Construct a histogram of the frequency distribution.
b) Construct a frequency polygon of the frequency distribution.

16. *Annual Salaries* The frequency distribution illustrates the annual salaries, in thousands of dollars, of the people in management positions at the X-Chek Corporation.

Salary (in $1000)	Number of People
30–35	4
36–41	7
42–47	8
48–53	9
54–59	7
60–65	5
66–71	3

a) Construct a histogram of the frequency distribution.
b) Construct a frequency polygon of the frequency distribution.

Problem Solving

17. *Number of Soft Drinks Purchased* Use the histogram below to answer the following questions.

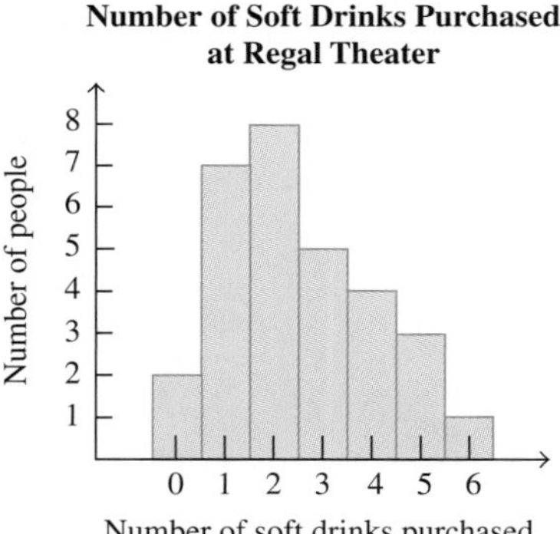

a) How many people were surveyed?
b) How many people purchased four soft drinks?
c) What is the modal class?
d) How many soft drinks were purchased?
e) Construct a frequency distribution from this histogram.

18. *Car Insurance* Use the histogram below to answer the following questions.

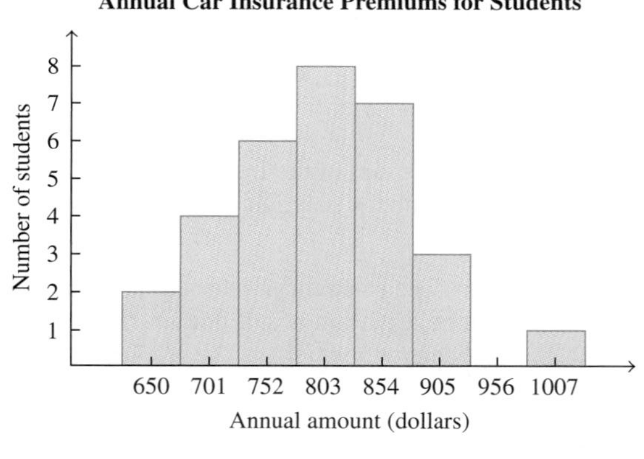

a) How many students were surveyed?
b) What are the lower and upper class limits of the first and second classes?
c) How many students have an annual car insurance premium in the class with a class mark of $752?
d) What is the class mark of the modal class?
e) Construct a frequency distribution from this histogram. Use a first class of 625–675.

19. *Response Time* Use the frequency polygon below to answer the following questions.

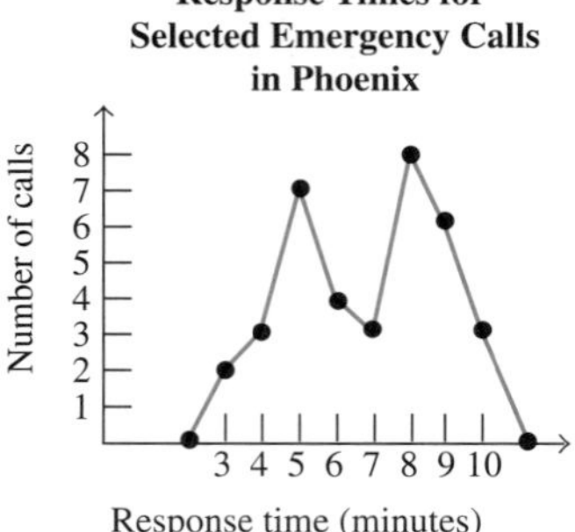

a) How many calls were responded to in 5 minutes?
b) How many calls were responded to in 6 minutes or less?
c) How many calls were included in the survey?
d) Construct a frequency distribution from the frequency polygon.
e) Construct a histogram from the frequency distribution in part (d).

20. *San Diego Zoo* Use the frequency polygon below to answer the following questions.

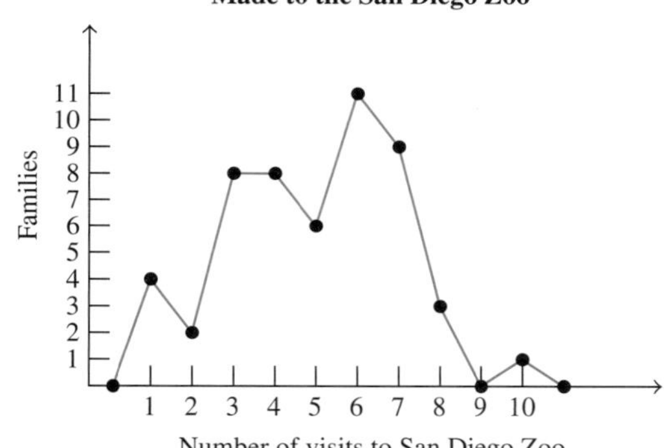

a) How many families visited the San Diego Zoo four times?
b) How many families visited the San Diego Zoo at least six times?
c) How many families were surveyed?
d) Construct a frequency distribution from the frequency polygon.
e) Construct a histogram from the frequency distribution in part (d).

21. Construct a histogram and a frequency polygon from the frequency distribution given in Exercise 7 of Section 13.3. See page 765.

22. Construct a histogram and a frequency polygon from the frequency distribution given in Exercise 8 of Section 13.3. See page 765.

23. *College Credits* Eighteen students in a geology class were asked how many college credits they had earned. The responses are as follows. Construct a stem-and-leaf display.

10	15	24	36	48	45
42	53	60	17	24	30
33	45	48	62	54	60

24. *Distance to Work* Twenty workers at a small company were asked how many miles they drive to work, one way. The responses are as follows. Construct a stem-and-leaf display. For single-digit data, use a stem of 0.

12	18	3	8	12	25	21
33	15	2	5	27	41	22
19	13	23	34	17	16	

25. *Starting Salaries* Starting salaries (in thousands of dollars) for social workers with a bachelor of science degree and no experience are shown for a random sample of 25 different social workers.

27	28	29	31	33
28	28	29	31	33
28	28	30	32	33
28	29	30	32	34
28	29	30	32	34

a) Construct a frequency distribution. Let each class have a width of one.
b) Construct a histogram.
c) Construct a frequency polygon.
d) Construct a stem-and-leaf display.

26. *Broadway Shows* The ages of a random sample of people attending a Broadway show are

20	23	25	30	32	35	39	44
21	23	26	30	33	35	40	45
21	24	27	30	34	35	40	45
22	24	28	31	34	37	40	46
23	25	28	31	34	38	42	47

Broadway in New York City

a) Construct a frequency distribution with a first class of 20–24.
b) Construct a histogram.
c) Construct a frequency polygon.
d) Construct a stem-and-leaf display.

27. *Advertising* The following table shows the 50 leading companies in terms of dollars spent for advertising in 2001, in the United States, rounded to the nearest million dollars.

Company	Advertising Spending (millions)
General Motors Corp.	$3374
Procter & Gamble Co.	$2541
Ford Motor Co.	$2408
PepsiCo	$2210
Pfizer	$2190
DaimlerChrysler	$1985
AOL Time Warner	$1885
Philip Morris Cos.	$1816
Walt Disney Co.	$1757
Johnson & Johnson	$1618
Unilever	$1484
Sears, Roebuck & Co.	$1480
Verizon Communications	$1462
Toyota Motor Corp.	$1399
AT&T Corp.	$1372
Sony Corp.	$1310
Viacom	$1283
McDonald's Corp.	$1195
Diageo	$1181
Sprint Corp.	$1160
Merck & Co.	$1137
Honda Motor Co.	$1103
J.C. Penney Corp.	$1086
U.S. Government	$1057
L'Oreal	$1041
IBM Corp.	$994
Bristol-Myers Squibb Co.	$974
Nestlé	$967
SBC Communications	$943
Target Corp.	$926
Microsoft Corp.	$920
Coca-Cola Co.	$904
Hewlett-Packard Co.	$899
AT&T Wireless	$888
General Mills	$884
GlaxoSmithKline	$881
WorldCom	$840
Sara Lee Corp.	$812
Home Depot	$778
Nissan Motor Co.	$775

(continued on next page)

Company	Advertising Spending (millions)
Wyeth	$771
Estee Lauder Cos.	$766
Federated Department Stores	$746
Yum Brands	$677
News Corp.	$670
ConAgra	$668
General Electric Co.	$664
Anheuser-Busch Cos.	$656
Mars Inc.	$615
Kmart Corp.	$597

Source: AdAge.com

a) Construct a frequency distribution with the first class $597 million–$905 million.
b) Construct a histogram.
c) Construct a frequency polygon.

28. *U.S. Ambassadors* The ages of a random sample of U.S. ambassadors are

40	43	45	50	52	55	59	64
41	43	46	50	53	55	60	65
41	44	47	50	54	55	60	65
42	44	48	51	54	57	60	66
43	45	48	51	54	58	62	67

a) Construct a frequency distribution with the first class 40–44.
b) Construct a histogram.
c) Construct a frequency polygon.

Challenge Problems/Group Activities

29. **a)** *Birthdays* What do you believe a histogram of the months in which the students in your class were born (January is month 1 and December is month 12) would look like? Explain.
b) By asking, determine the month in which the students in your class were born (include yourself).
c) Construct a frequency distribution containing 12 classes.
d) Construct a histogram from the frequency distribution in part (c).
e) Construct a frequency polygon of the frequency distribution in part (c).

30. *Social Security Numbers* Repeat Exercise 29 for the last digit of the students' social security numbers. Include classes for the digits 0–9.

Internet/Research Activity

31. Over the years many changes have been made in the U.S. Social Security System.
a) Do research and determine the number of people receiving social security benefits for the years 1945, 1950, 1955, 1960, . . . , 2000. Then construct a frequency distribution and histogram of the data.
b) Determine the maximum amount that self-employed individuals had to pay into social security (the FICA tax) for the years 1945, 1950, 1955, 1960, . . . , 2000. Then construct a frequency distribution and a histogram of the data.

13.5 MEASURES OF CENTRAL TENDENCY

Most people have an intuitive idea of what is meant by an "average." The term is used daily in many familiar ways: "This car averages 19 miles per gallon." "The average test grade was 78." "The average height of adult males is 5 feet 9 inches."

An *average* is a number that is representative of a group of data. There are at least four different averages: the mean, the median, the mode, and the midrange. Each is calculated differently and may yield different results for the same set of data. Each will result in a number near the center of the data; for this reason, averages are commonly referred to as *measures of central tendency*.

The *arithmetic mean*, or simply the *mean*, is symbolized either by $\bar{x}$ (read "*x* bar") or by the Greek letter mu, μ. The symbol $\bar{x}$ is used when the mean of a *sample* of the population is calculated. The symbol μ is used when the mean of the *entire population* is calculated. Unless otherwise indicated, we will assume that the data featured in this book represent samples; therefore, we will use $\bar{x}$ for the mean.

The Greek letter sigma, Σ, is used to indicate "summation." The notation Σx, read "the sum of *x*," is used to indicate the sum of all the data. For example, if there are five pieces of data, 4, 6, 1, 0, 5, then $\Sigma x = 4 + 6 + 1 + 0 + 5 = 16$.

DID YOU KNOW

Buying the American Dream

San Francisco, California

One of the biggest dreams for most people is to own their own home. Yet depending on where you live, the American dream may be hard to achieve. In 2002, San Francisco had the highest median home price for major metropolitan housing markets, \$482,300. Beaumont/Port Arthur had the lowest median home price for major metropolitan housing markets, \$76,800.

Beaumont/Port Arthur, Texas
Source: National Association of Realtors, Quarterly Housing Affordability Index

Now we can discuss the procedure for determining the mean of a set of data.

The mean, $\bar{x}$, is the sum of the data divided by the number of pieces of data. The formula for calculating the mean is

$$\bar{x} = \frac{\Sigma x}{n}$$

where Σx represents the sum of all the data and n represents the number of pieces of data.

The most common use of the word *average* is the mean.

EXAMPLE 1 *Find the Mean*

Find the mean age of a group of volunteers at an American Red Cross office if the ages of the individuals are 27, 18, 48, 34, and 48.

SOLUTION:

$$\bar{x} = \frac{\Sigma x}{n} = \frac{27 + 18 + 48 + 34 + 48}{5} = \frac{175}{5} = 35$$

Therefore, the mean, $\bar{x}$, is 35 years. ▲

The mean represents "the balancing point" of a set of data. For example, if a seesaw were pivoted at the mean and uniform weights were placed at points corresponding to the ages in Example 1, the seesaw would balance. Figure 13.13 shows the five ages given in Example 1 and the calculated mean.

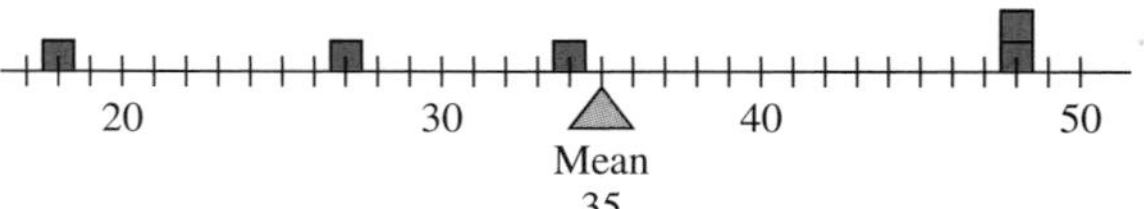

Figure 13.13

A second average is the *median.* To find the median of a set of data, *rank the data* from smallest to largest, or largest to smallest, and determine the value in the middle of the set of *ranked data.* This value will be the median.

The **median** is the value in the middle of a set of *ranked data.*

EXAMPLE 2 *Find the Median*

Determine the median of the volunteers' ages in Example 1.

SOLUTION: Ranking the data from smallest to largest gives 18, 27, 34, 48, and 48. Since 34 is the value in the middle of this set of ranked data (two pieces of data above it and two pieces below it) 34 years is the median. ▲

If there are an even number of pieces of data, the median will be halfway between the two middle pieces. In this case, to find the median, add the two middle pieces and divide this sum by 2.

EXAMPLE 3 *Find the Median of an Even Number of Pieces of Data*

Determine the median of the following sets of data.

a) 7, 12, 14, 15, 9, 14, 9, 10

b) 7, 8, 8, 8, 9, 10

SOLUTION:

a) Ranking the data gives 7, 9, 9, 10, 12, 14, 14, 15. There are eight pieces of data. Therefore, the median will lie halfway between the two middle pieces, the 10 and the 12. The median is $\frac{10 + 12}{2}$ or $\frac{22}{2}$ or 11.

b) There are six pieces of data and they are already ranked. Therefore, the median lies halfway between the two middle pieces. Both middle pieces are 8's. The median is $\frac{8 + 8}{2}$, or $\frac{16}{2}$, or 8. ▲

TIMELY TIP Data must be ranked before finding the median. A common error made when finding the median is neglecting to arrange the data in ascending (increasing) or in descending (decreasing) order.

A third average is the *mode*.

The **mode** is the piece of data that occurs most frequently.

EXAMPLE 4 *Find the Mode*

Determine the mode of the volunteers' ages in Example 1.

SOLUTION: The ages are 27, 18, 48, 34, and 48. The age 48 is the mode because it occurs twice and the other values occur only once. ▲

If each piece of data occurs only once, the set of data has no mode. For example, the set of data 1, 2, 3, 4, 5 has no mode. If two values in a set of data occur more often than all the other data, we consider both these values as modes and say the data is **bimodal*** (which means two modes). For example, the set of data 1, 1, 2, 3, 3, 5 has two modes, 1 and 3.

The last average that we will discuss is the midrange. The *midrange* is the value halfway between the lowest (L) and highest (H) values in a set of data. It is found by adding the lowest and highest values and dividing the sum by 2. A formula for finding the midrange follows.

$$\textbf{Midrange} = \frac{\text{lowest value} + \text{highest value}}{2}$$

*Some textbooks say that sets of data such as 1, 1, 2, 3, 3, 5 have no mode.

DID YOU KNOW

When Babies' Eyes Are Smiling

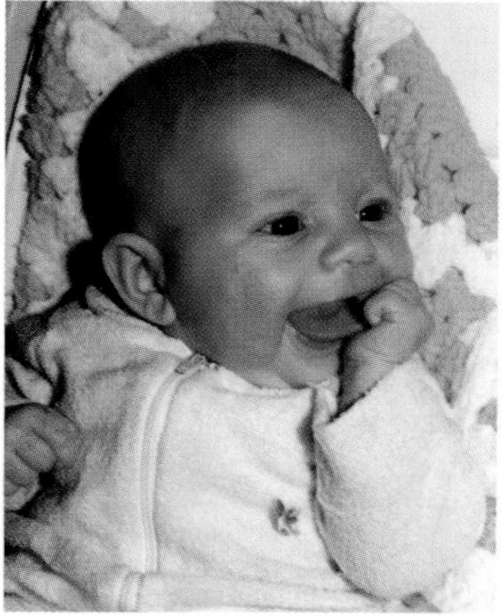

Experimental psychologists formulate hypotheses about human behavior, design experiments to test them, make observations, and draw conclusions from their data. They use statistical concepts at each stage to help ensure that their conclusions are valid. In one experiment, researchers observed that 2-month-old infants who learned to move their heads so as to make a mobile turn began to smile as soon as the mobile turned. Babies in the control group did not smile as often when the mobile moved independently of their head turning. The researchers concluded that it was not the movement of the mobile that made the infant smile; rather, the infants smiled at their own achievement.

EXAMPLE 5 *Find the Midrange*

Determine the midrange of the volunteers' ages given in Example 1.

SOLUTION: The ages of the volunteers are 27, 18, 48, 34, and 48. The lowest age is 18 and the highest age is 48.

$$\text{Midrange} = \frac{\text{lowest} + \text{highest}}{2} = \frac{18 + 48}{2} = \frac{66}{2} = 33 \text{ years}$$

▲

The "average" of the ages 27, 18, 48, 34, 48 can be considered any one of the following values: 35 (mean), 34 (median), 48 (mode), or 33 (midrange). Which average do you feel is most representative of the ages? We will discuss this question later in this section.

EXAMPLE 6 *Measures of Central Tendency*

The salaries of eight selected teachers rounded to the nearest thousand dollars are 40, 25, 28, 35, 42, 60, 60, and 73. For this set of data, determine the (a) mean, (b) median, (c) mode, and (d) midrange, and then (e) rank the measures of central tendency from lowest to highest.

SOLUTION:

a) $\bar{x} = \dfrac{\Sigma x}{n} = \dfrac{40 + 25 + 28 + 35 + 42 + 60 + 60 + 73}{8} = \dfrac{363}{8} = 45.375$

b) Ranking the data from the smallest to largest gives

25, 28, 35, 40, 42, 60, 60, 73

Since there are an even number of pieces of data, the median is halfway between 40 and 42. The median $= (40 + 42)/2 = 82/2 = 41$.

c) The mode is the piece of data that occurs most frequently. The mode is 60.

d) The midrange $= (\text{L} + \text{H})/2 = (25 + 73)/2 = 98/2 = 49$.

e) The averages from lowest to highest are the median, mean, midrange, and mode. Their values are 41, 45.375, 49, and 60, respectively. ▲

At this point you should be able to calculate the four measures of central tendency: mean, median, mode, and midrange. Now let's examine the circumstances in which each is used.

The mean is used when each piece of data is to be considered and "weighed" equally. It is the most commonly used average. It is the only average that can be affected by *any* change in the set of data; for this reason, it is the most sensitive of all the measures of central tendency (see Exercise 23).

Occasionally, one or more pieces of data may be much greater or much smaller than the rest of the data. When this situation occurs, these "extreme" values have the effect of increasing or decreasing the mean significantly so that the mean will not be representative of the set of data. Under these circumstances, the median should be used instead of the mean. The median is often used in describing average family incomes because a relatively small number of families have extremely large incomes. These few incomes would inflate the mean income, making it nonrepresentative of the millions of families in the population.

Consider a set of exam scores from a mathematics class: 0, 16, 19, 65, 65, 65, 68, 69, 70, 72, 73, 73, 75, 78, 80, 85, 88, 92. Which average would best represent these grades? The mean is 64.06. The median is 71. Since only 3 of the 18 scores fall below the mean, the mean would not be considered a good representative score. The median of 71 probably would be the better average to use.

The mode is the piece of data, if any, that occurs most frequently. Builders planning houses are interested in the most common family size. Retailers ordering shirts are interested in the most common shirt size. An individual purchasing a thermometer might choose one, from those on display, whose temperature reading is the most common reading among those on display. These examples illustrate how the mode may be used.

The midrange is sometimes used as the average when the item being studied is constantly fluctuating. Average daily temperature, used to compare temperatures in different areas, is calculated by adding the lowest and highest temperatures for the day and dividing the sum by 2. The midrange is actually the mean of the high value and the low value of a set of data. Occasionally, the midrange is used to estimate the mean, since it is much easier to calculate.

Sometimes an average itself is of little value, and care must be taken in interpreting its meaning. For example, Jim is told that the average depth of Willow Pond is only 3 feet. He is not a good swimmer but decides it is safe to go out a short distance in this shallow pond. After he is rescued, he exclaims, "I thought this pond was only 3 feet deep." Jim didn't realize an average does not indicate extreme values or the spread of the values. The spread of data is discussed in Section 13.6.

Measures of Position

Measures of position are used to describe the position of a piece of data in relation to the rest of the data. If you took the Scholastic Aptitude Test (SAT) before applying to college, your score was described as a measure of position rather than a measure of central tendency. *Measures of position* are often used to make comparisons, such as comparing the scores of individuals from different populations, and are generally used when the amount of data is large.

Two measures of position are *percentiles* and *quartiles*. There are 99 percentiles dividing a set of data into 100 equal parts; see Fig. 13.14. For example, suppose that you scored 490 on the math half of the SAT, and the score of 490 was reported to be in the 78th percentile of high school students. This wording *does not* mean that 78% of your answers were correct; it *does* mean that you outperformed about 78% of all those taking the exam. In general, a score in the nth percentile means that you outperformed about $n\%$ of the population who took the test and that $(100 - n)\%$ of the people taking the test performed better than you did.

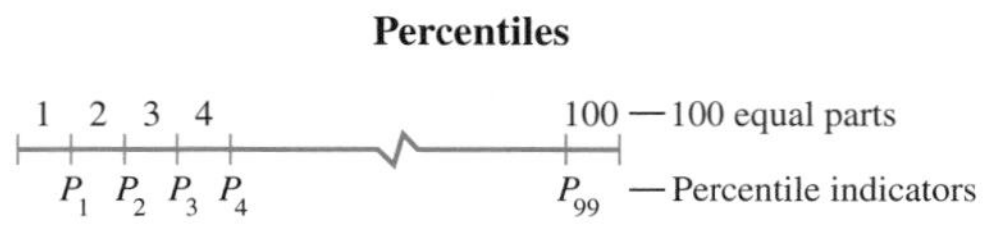

Percentiles

1 2 3 4 ... 100 — 100 equal parts

P_1 P_2 P_3 P_4 ... P_{99} — Percentile indicators

Figure 13.14

EXAMPLE 7 *English Achievement Test*

Kara Hopkins took an English achievement test to obtain college credit by exam for freshman English. Her score was at the 81st percentile. Explain what that means.

SOLUTION: If a score is at the 81st percentile, it means that about 81% of the scores are below that score. Therefore, Kara scored better than about 81% of the students taking the exam. Also, about 19% of all students taking the exam scored higher than she did.

Quartiles are another measure of position. Quartiles divide data into four equal parts: The first quartile is the value that is higher than about 1/4 or 25% of the population. It is the same as the 25th percentile. The second quartile is the value that is higher than about 1/2 the population and is the same as the 50th percentile, or the median. The third quartile is the value that is higher than about 3/4 of the population and is the same as the 75th percentile; see Fig. 13.15.

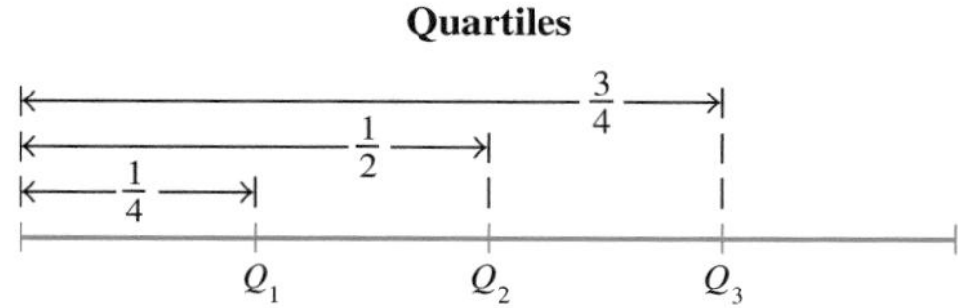

Figure 13.15

To Find the Quartiles of a Set of Data:

1. Order the data from smallest to largest.
2. Find the median, or the 2nd quartile, of the set of data. If there are an odd number of pieces of data, the median is the middle value. If there are an even number of pieces of data, the median will be halfway between the two middle pieces of data.
3. The first quartile, Q_1, is the median of the lower half of the data; that is, Q_1 is the median of the data less than Q_2.
4. The third quartile, Q_3, is the median of the upper half of the data; that is, Q_2 is the median of the data greater than Q_2.

EXAMPLE 8 *Finding Quartiles*

Electronics World is concerned about the high turnover of its sales staff. A survey was done to determine how long (in months) their sales staff had been in their current positions. The responses of 27 sales staff follow. Determine Q_1, Q_2, and Q_3.

25	3	7	15	31	36	17	21	2
11	42	16	23	19	21	9	20	5
8	12	27	14	39	24	18	6	10

SOLUTION: First we order the data from smallest to largest.

2	3	5	6	7	8	9	10	11
12	14	15	16	17	18	19	20	21
21	23	24	25	27	31	36	39	42

Next we find the median. Since there are 27 pieces of data, an odd number, the median will be the middle value. The middle value is 17, with 13 pieces of data less than 17 and 13 pieces of data greater than 17. Therefore, the median, Q_2, is 17, shown in red.

To find Q_1, the median of the lower half of the data, we need to find the median of the 13 pieces of data that are less than Q_2. The middle value of the lower half of the data is 9. There are 6 pieces of data less than 9 and 6 pieces of data greater than 9. Therefore, Q_1 is 9, shown in blue.

To find Q_3, the median of the upper half of the data, we need to find the median of the 13 pieces of data that are greater than 17, or Q_2. The middle value of the upper half of the data is 24. There are 6 pieces of data greater than 17 but less than 24 and 6 pieces of data greater than 24. Therefore, Q_3 is 24, shown in blue. ▲

SECTION 13.5 EXERCISES

Concept/Writing Exercises

1. What is a set of *ranked data*?
2. Describe the mean of a set of data and explain how to find it.
3. Describe the median of a set of data and explain how to find it.
4. Describe the midrange of a set of data and explain how to find it.
5. Describe the mode of a set of data and explain how to find it.
6. When might the mode be the preferred average to use? Give an example.
7. When might the median be the preferred average to use? Give an example.
8. When might the midrange be the preferred average to use? Give an example.
9. When might the mean be the preferred average to use? Give an example.

10. a) What symbol is used for the sample mean?
b) What symbol is used for the population mean?

Practice the Skills

In Exercises 11–20, determine the mean, median, mode, and midrange of the set of data. Round your answer to the nearest tenth.

11. 5, 6, 6, 9, 10, 10, 10, 20, 23

12. 9, 10, 15, 17, 15, 14, 370, 45, 42, 13

13. 66, 72, 84, 45, 90, 42, 86

14. 7, 5, 8, 8, 8, 10, 12

15. 1, 3, 5, 7, 9, 11, 13, 15

16. 40, 50, 30, 60, 90, 100, 140

17. 1, 7, 11, 27, 36, 14, 12, 9, 1

18. 1, 1, 1, 1, 4, 4, 4, 4, 6, 8, 10, 12, 15, 21

19. 6, 8, 12, 13, 11, 13, 15, 17

20. 5, 15, 5, 15, 5, 15

21. *Best-Seller List* The number of weeks the top 10 hardcover fiction novels were on the best-seller list as of July 28, 2002, is 2, 3, 3, 5, 19, 7, 4, 5, 11, 6. Find the mean, median, mode, and midrange.

22. *Daily Tips* The amount of money Amy Striegel collected in tips as a waitress in each of seven days is \$25, \$60, \$37, \$48, \$100, \$140, \$59. Find the mean, median, mode, and midrange.

23. *Change in the Data* The mean is the "most sensitive" average because it is affected by any change in the data.
a) Determine the mean, median, mode, and midrange for 1, 2, 3, 5, 5, 7, 11.
b) Change the 7 to a 10 in part (a). Find the mean, median, mode, and midrange.
c) Which averages were affected by changing the 7 to a 10?
d) Which averages will be affected by changing the 11 to a 10 in part (a)?

24. *Life Expectancy* In 2000, the National Center for Health Statistics indicated a new record "average life expectancy" of 76.9 years for the total U.S. population. The average life expectancy for men was 74.1 years, and for women it was 79.5 years. Which "average" do you think the National Center for Health is using? Explain your answer.

25. *A Grade of B* To get a grade of B, a student must have a mean average of 80 or greater. Jim Condor has a mean average of 79 for 10 quizzes. He approaches his teacher and

asks for a B, reasoning that he missed a B by only one point. What is wrong with Jim's reasoning?

26. *Employee Salaries* The salaries of 10 employees of a small company follow.

$28,000	$64,000
25,000	24,000
31,000	27,000
26,000	81,000
26,000	29,000

Calculate the
a) mean.
b) median.
c) mode.
d) midrange.
e) If the employees wanted to demonstrate the need for a raise, which average would they use to show they are being underpaid: the mean or the median? Explain.
f) If the management did not want to give the employees a raise, which average would they use: the mean or the median? Explain.

27. *National Parks* The 10 national parks and recreation areas with the most visitors in 2001 are listed below.

Park	Number of Visitors (millions of people)
Blue Ridge Parkway	19.7
Golden Gate National Recreation Area	13.4
Great Smoky Mountains National Park	9.5
Lake Mead National Recreation Area	8.8
Gateway National Recreation Area	8.2
George Washington Memorial Parkway	7.8
Natchez Trace Parkway	5.5
Statue of Liberty National Monument	5.4
Delaware Water Gap National Recreation Area	4.8
Castle Clinton	4.6

Source: National Park Service (www.nps.gov.)

Rounding your answers to the nearest tenth, determine the
a) mean.
b) median.
c) mode.
d) midrange.

28. *Living Expenses* Bob Exler's monthly living expenses for 1 year are as follows:

$1000	$850	$1370	$1400
1900	850	1350	1250
1600	900	1110	1230

When appropriate, round your answer to the nearest cent. Determine the
a) mean.
b) median.
c) mode.
d) midrange.

29. *Costliest Hurricanes* The damage caused, in billions of dollars, by the 11 costliest hurricanes in the United States is listed below.

Hurricane	Year	Cost (billions)
Andrew	1992	$26.5
Hugo	1989	7.0
Floyd	1999	4.5
Fran	1996	3.2
Opal	1995	3.0
Georges	1998	2.3
Frederic	1979	2.3
Agnes	1972	2.1
Alicia	1983	2.0
Bob	1991	1.5
Juan	1985	1.5

Source: National Oceanic and Atmospheric Administration

Rounding your answers to the nearest tenth, determine the
a) mean.
b) median.
c) mode.
d) midrange.
e) Which average is the best measure of central tendency for this set of data? Explain.

30. *Exam Average* Malcolm Sander's mean average on five exams is 76. Find the sum of his scores.

31. *Exam Average* Jeremy Urban's mean average on six exams is 85. Find the sum of his scores.

32. *Creating a Data Set* Construct a set of five pieces of data in which the mode has a lower value than the median and the median has a lower value than the mean.

33. *Creating a Data Set* Construct a set of six pieces of data with a mean, median, and midrange of 75 and where no two pieces of data are the same.

34. *Creating a Data Set* Construct a set of six pieces of data with a mean of 88 and where no two pieces of data are the same.

35. *Water Park* For the 2003 season, 24,000 people visited the Blue Lagoon Water Park. The park was open 120 days for water activities. The highest number of visitors on a single day was 500. The lowest number of visitors on a single day was 50. Determine whether it is possible to find the following with the given information:

a) the mean number of visitors per day.
b) the median number of visitors per day.
c) the mode number of visitors per day.
d) the midrange number of visitors per day.
e) Find all the measures of central tendency that can be found with the information and explain why the others cannot be found.

36. *Determine a Necessary Grade* A mean average of 80 or greater for five exams is needed for a final grade of B. Jorge Rivera's first four exam grades are 73, 69, 85, and 80. What grade does Jorge need on the fifth exam to get a B in the course?

37. *Grading Methods* A mean average of 60 on seven exams is needed to pass a course. On her first six exams, Sheryl Ward received grades of 49, 72, 80, 60, 57, and 69.

a) What grade must she receive on her last exam to pass the course?
b) An average of 70 is needed to get a C in the course. Is it possible for Sheryl to get a C? If so, what grade must she receive on the seventh exam?
c) If her lowest grade of the exams already taken is to be dropped, what grade must she receive on her last exam to pass the course?
d) If her lowest grade of the exams already taken is to be dropped, what grade must she receive on her last exam to get a C in the course?

38. *Central Tendencies* Which of the measures of central tendency *must* be an actual piece of data in the distribution? Explain.

39. *Creating a Data Set* Construct a set of six pieces of data such that if only one piece of data is changed, the mean, median, and mode will all change.

40. *Changing One Piece of Data* Consider the set of data 1, 1, 1, 2, 2, 2. If one 2 is changed to a 3, which of the following will change: mean, median, mode, midrange? Explain.

41. *Changing One Piece of Data* Is it possible to construct a set of six different pieces of data such that by changing only one piece of data you cause the mean, median, mode, and midrange to change? Explain.

42. *Grocery Expenses* The Taylor's have recorded their weekly grocery expenses for the past 12 weeks and determined the mean weekly expense was $85.20. Later Mrs. Taylor discovered 1 week's expense of $74 was incorrectly recorded as $47. What is the correct mean?

43. *Percentiles* For any set of data, what must be done to the data before percentiles can be determined?

44. *Percentiles* Josie Waverly scored in the 73rd percentile on the verbal part of her College Board test. What does that mean?

45. *Percentiles* When a national sample of heights of kindergarten children was taken, Kevin was told that he was in the 35th percentile. Explain what that means.

46. *Percentiles* A union leader is told that, when all workers' salaries are considered, the first quartile is $20,750. Explain what that means.

47. *Quartiles* The prices of the 21 top-rated 27-inch direct-view television sets, as rated in the March 2002 issue of *Consumer Reports*, are as follows:

$290	$330	$350	$400	$450	$650	$700
300	350	350	430	500	650	750
300	350	350	450	600	700	800

Determine
a) Q_2. **b)** Q_1. **c)** Q_3.

48. *Quartiles* The cost, in cents, per half-cup serving of the top 20 rated brands of vanilla ice cream, as reported in the July 2002 issue of *Consumer Reports*, are as follows:

17	21	27	27	28	33	80
17	24	27	28	28	38	81
20	25	27	28	31	74	

Determine
a) Q_2. **b)** Q_1. **c)** Q_3.

49. *The 50th Percentile* Give the names of two other statistics that have the same value as the 50th percentile.

50. *College Admissions* Jonathan took an admission test for the University of California and scored in the 85th percentile. The following year Jonathan's sister Kendra took a similar admission test for the University of California and scored in the 90th percentile.

a) Is it possible to determine which of the two answered the higher percent of questions correctly on their respective exams? Explain your answer.

b) Is it possible to determine which of the two was in a better relative position with regard to their respective populations? Explain.

51. *Employee Salaries* The following statistics represent weekly salaries at the Midtown Construction Company:

Mean	\$510	First quartile	\$470
Median	\$500	Third quartile	\$535
Mode	\$490	83rd percentile	\$575

a) What is the most common salary?
b) What salary did half the employees' salaries surpass?
c) About what percent of employees' salaries surpassed \$535?
d) About what percent of employees' salaries were less than \$470?
e) About what percent of employees' salaries surpassed \$575?
f) If the company has 100 employees, what is the total weekly salary of all employees?

Challenge Problems/Group Activities

52. *The Mean of the Means* Consider the following five sets of values.

i) 5 6 7 7 8 9 14
ii) 3 6 8 9
iii) 1 1 1 2 5
iv) 6 8 9 12 15
v) 50 51 55 60 80 100

a) Compute the mean of each of the five sets of data.
b) Compute the mean of the five means in part (a).
c) Find the mean of the 27 pieces of data.
d) Compare your answer in part (b) to your answer in part (c). Are the values the same? Does your answer make sense? Explain.

53. *Ruth versus Mantle* The tables to the right compare the batting performances for selected years for two well-known former baseball players, Babe Ruth and Mickey Mantle.

Babe Ruth
Boston Red Sox 1914–1919
New York Yankees 1920–1934

Year	At Bats	Hits	Pct.
1925	359	104	
1930	518	186	
1933	459	138	
1916	136	37	
1922	406	128	
Total	1878	593	

Mickey Mantle
New York Yankees 1951–1968

Year	At Bats	Hits	Pct.
1954	543	163	
1957	474	173	
1958	519	158	
1960	527	145	
1962	377	121	
Total	2440	760	

a) For each player, compute the batting average percent (pct.) for each year by dividing the number of hits by the number of at bats. Round off to the nearest thousandth. Place the answers in the pct. column.
b) Going across each of the five horizontal lines (for example Ruth, 1925, vs. Mantle, 1954), compare the percents (pct.) and determine which is greater in each case.
c) For each player, compute the mean batting average percent for the 5 given years by dividing the total hits by the total at bats. Which is greater, Ruth's or Mantle's?
d) Based on your answer in part (b), does your answer in part (c) make sense? Explain.
e) Find the mean percent for each player by adding the five pcts. and dividing by 5. Which is greater, Ruth's or Mantle's?
f) Why do the answers obtained in parts (c) and (e) differ? Explain.
g) Who would you say has the better batting average percent for the 5 years selected? Explain.

54. *Employee Salaries* The following table gives the annual salary distribution for employees at Kulzer's Home Improvement.

Annual Salary	Number Receiving Salary
$100,000	1
85,000	2
24,000	6
21,000	4
18,000	5
17,000	7

Using the information provided in the table, find the
a) mean annual salary.
b) median annual salary.
c) mode annual salary.
d) midrange annual salary.
e) Which is the best measure of central tendency for this set of data? Explain your answer.

Weighted Average Sometimes when we wish to find an average, we may wish to assign more importance, or weight, to some of the pieces of data. To calculate a ***weighted average,*** *we use the formula: weighted average* $= \frac{\Sigma xw}{\Sigma w}$, *where w is the weight of the piece of data, x;* Σxw *is the sum of the products of each piece of data multiplied by its weight; and* Σw *is the sum of the weights. For example, suppose that students in a class need to submit a report that counts 20% of their grade, they need to take a midterm exam that counts 30% of their grade, and they need to take a final exam that counts 50% of their grade. Suppose that a student got a 72 on the report, an 85 on the midterm exam, and a 93 on the final exam. To determine this student's weighted average, first find* Σxw: $\Sigma xw = 72(0.20) + 85(0.30) + 93(0.50) = 86.4$. *Next find* Σw, *the sum of the weights:* $\Sigma w = 0.20 + 0.30 + 0.50 = 1.00$. *Now determine the weighted average as follows*

$$\text{weighted average} = \frac{\Sigma xw}{\Sigma w} = \frac{86.4}{1.00} = 86.4$$

Thus, the weighted average is 86.4. Note that Σw *does not always have to be 1.00. In Exercise 55 and 56, use the weighted average formula.*

55. *Course Average* Suppose that your final grade for a course is determined by a midterm exam and a final exam. The midterm exam is worth 40% of your grade and the final exam is worth 60%. If your midterm exam grade is 84 and your final exam grade is 94, calculate your final average.

56. *Grade Point Average* In a four-point grade system, an A corresponds to 4.0 points, a B corresponds to a 3.0 points, a C corresponds to a 2.0 points, and a D corresponds to a 1.0 points. No points are awarded for an F. Last semester Tanya Reeves received a B in a four-credit hour course, an A in a three-credit hour course, a C in a three-credit hour course, and an A in another three-credit hour course. Grade point average (GPA) is calculated as a weighted average using the credit hours as weights and the number of points corresponding to the grade as pieces of data. Calculate Tanya's GPA for the previous semester. (Round your answer to the nearest hundredth.)

Recreational Mathematics

57. *Your Exam Average* **a)** Calculate the mean, median, mode, and midrange of your exam grades in your mathematics course.
b) Which measure of central tendency best represents your average grade?
c) Which measure of central tendency would you rather use as your average grade?

58. *Purchases* Matthew Abbott purchased some items at Staples each day for five days. The mode of the number of items Matthew purchased is higher than the median of the number of items he purchased. The median of the number of items Matthew purchased is higher than the mean of the number of items he purchased. Each day he purchased at least two items but no more than seven items.
a) How many items did Matthew purchase each day? (*Note:* There is more than one correct answer.)
b) Determine the mean, median, and mode for your answer to part (a).

Internet/Research Activity

59. Two other measures of location that we did not mention in this section are *stanines* and *deciles.* Use statistics books, books on educational testing and measurements, and Internet websites to write a report on what stanines and deciles are and when percentiles, quartiles, stanines, and deciles are used.

13.6 MEASURES OF DISPERSION

The measures of central tendency by themselves do not always give sufficient information to analyze a situation and make decisions. As an example, two manufacturers of airplane engines are being considered for a contract. Manufacturer A's engines have an average (mean) life of 1000 hours of flying time before they must be rebuilt. Manufacturer B's engines have an average life of 950 hours of flying time before they must be rebuilt. If you assume that both cost the same, which engines should be purchased? The average engine life may not be the most important factor. The fact that manufacturer A's engines have an average life of 1000 hours could mean that half will last about 500 hours and the other half will last about 1500 hours. If in fact all manufacturer B's engines have a life span of between 900 and 1000 hours, then all of manufacturer B's engines are more consistent and reliable. If A's engines were purchased, they would all have to be rebuilt every 300 hours or so because it would be impossible to determine which ones would fail first. If B's engines were purchased, they could go much longer before having to be rebuilt. This example is of course an exaggeration used to illustrate the importance of knowing something about the *spread,* or *variability,* of the data.

Measures of dispersion are used to indicate the spread of the data. The range and standard deviation* are the measures of dispersion that will be discussed in this book.

The *range* is the difference between the highest and lowest values; it indicates the total spread of the data.

Range = highest value − lowest value

EXAMPLE 1 *Find the Range*

Twelve different foods were selected and the amount of potassium, in milligrams, in each was recorded. Find the range of the following amounts of potassium.

900, 789, 400, 408, 860, 780, 451, 502, 496, 503, 555, 566

SOLUTION: Range = highest value − lowest value = 900 − 400 = 500. The range of the amounts of potassium is 500 mg. ▲

The second measure of dispersion, the *standard deviation*, measures how much the data *differ from the mean.* It is symbolized either by the letter s or by the Greek letter sigma, σ.† The s is used when the standard deviation of a *sample* is calculated. The σ is used when the standard deviation of the entire *population* is calculated. Since we are assuming that all data presented in this section are for samples, we use s to represent the standard deviation (note, however, that on the doctors' charts on page 794, σ is used. Also, we will use σ in the next section when we find standard scores.) The larger the spread of the data about the mean, the larger is the standard deviation. Consider the following two sets of data.

5, 8, 9, 10, 12, 13   8, 9, 9, 10, 10, 11

Both have a mean of 9.5. Which set of values on the whole do you believe differs less from the mean of 9.5? Figure 13.16 may make the answer more apparent. The scores in the second set of data are closer to the mean and therefore have a smaller standard deviation. You will soon be able to verify such relationships yourself.

Sometimes only a very small standard deviation is desirable or acceptable. Consider a cereal box that is to contain 8 oz of cereal. If the amount of cereal put into the

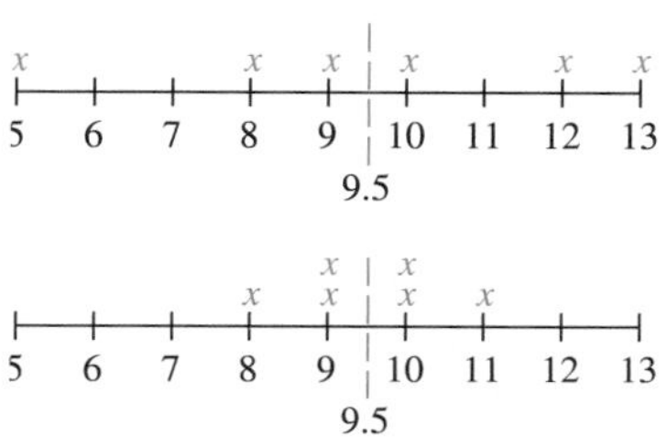

Figure 13.16

Variance, another measure of dispersion, is the square of the standard deviation.

†Our alphabet uses both uppercase and lowercase letters, for example, A and a. The Greek alphabet also uses both uppercase and lowercase letters. The symbol Σ is the capital Greek letter sigma, and σ is the lowercase Greek letter sigma.

boxes varies too much—sometimes underfilling, sometimes overfilling—the manufacturer will soon be in trouble with consumer groups and government agencies.

At other times, a larger spread of data is desirable or expected. For example, intelligence quotients (IQs) are expected to exhibit a considerable spread about the mean, since everyone is different. The following procedure explains how we determine the standard deviation of a set of data.

To Find the Standard Deviation of a Set of Data:

1. Find the mean of the set of data.
2. Make a chart having three columns:

$$\text{Data} \qquad \text{Data} - \text{Mean} \qquad (\text{Data} - \text{Mean})^2$$

3. List the data vertically under the column marked Data.
4. Subtract the mean from each piece of data and place the difference in the Data − Mean column.
5. Square the values obtained in the Data − Mean column and record these values in the $(\text{Data} - \text{Mean})^2$ column.
6. Determine the sum of the values in the $(\text{Data} - \text{Mean})^2$ column.
7. Divide the sum obtained in step 6 by $n - 1$, where n is the number of pieces of data.*
8. Determine the square root of the number obtained in step 7. This number is the standard deviation of the set of data.

Example 2 illustrates the procedure to follow to find the standard deviation of a set of data.

EXAMPLE 2 *Find the Standard Deviation*

A veterinarian in an animal hospital recorded the following life spans of selected Labrador retrievers (to the nearest year):

$$7, 9, 11, 15, 18, 12$$

Find the standard deviation of the ages.

SOLUTION: First determine the mean:

$$\bar{x} = \frac{\Sigma x}{n} = \frac{7 + 9 + 11 + 15 + 18 + 12}{6} = \frac{72}{6} = 12$$

Next construct a table with three columns, as illustrated in Table 13.4, and list the data in the first column (it is often helpful to list the data in ascending or descending order). Complete the second column by subtracting the mean, 12 in this case, from each piece of data in the first column.

*To find the standard deviation of a sample divide the sum of $(\text{Data} - \text{Mean})^2$ column by $n - 1$. To find the standard deviation of a population divide the sum by n. In this book, we assume that the set of data represents a sample and divide by $n - 1$. The quotient obtained in step 7 represents a measure of dispersion called the *variance*.

DID YOU KNOW

Statistics and Opera Houses

Sydney Opera House in Sydney, Austrailia

Architects have developed a mathematical rule based on statistics to help them construct opera houses with exceptional acoustics. The rule was first developed by having conductors rate the overall sound quality in 23 opera houses. Then acoustical engineers measured several acoustical properties in those 23 buildings. By using statistical analysis, the engineers were able to determine which combination of properties produced exceptional sound and which of the acoustic characteristics were most important. This mathematical rule is now used in the development of new opera houses.

TABLE 13.4

Data	Data − Mean	(Data − Mean)^2
7	7 − 12 = −5	
9	9 − 12 = −3	
11	11 − 12 = −1	
12	12 − 12 = 0	
15	15 − 12 = 3	
18	18 − 12 = 6	
	0	

The sum of the values in the Data − Mean column should always be zero; if not, you have made an error. (If the mean is a decimal number, there may be a slight rounding error.)

Next square the values in the second column and place the squares in the third column (Table 13.5).

TABLE 13.5

Data	Data − Mean	(Data − Mean)^2
7	−5	$(-5)^2 = (-5)(-5) = 25$
9	−3	$(-3)^2 = (-3)(-3) = 9$
11	−1	$(-1)^2 = (-1)(-1) = 1$
12	0	$(0)^2 = (0)(0) = 0$
15	3	$(3)^2 = (3)(3) = 9$
18	6	$(6)^2 = (6)(6) = 36$
	0	80

Add the squares in the third column. In this case, the sum is 80. Divide this sum by one less than the number of pieces of data $(n - 1)$. In this case, the number of pieces of data is 6. Therefore, we divide by 5 and get

$$\frac{80}{5} = 16^*$$

Finally, take the square root of this number. Since $\sqrt{16} = 4$, the standard deviation, symbolized s, is 4.

Now we will develop a formula for finding the standard deviation of a set of data. If we call the individual data x and the mean $\bar{x}$, we could write the three column heads Data, Data − Mean, and (Data − Mean)^2 in Table 13.4 as

$$x \qquad x - \bar{x} \qquad (x - \bar{x})^2$$

Let's follow the procedure we used to obtain the standard deviation in Example 2. We found the sum of the (Data − Mean)^2 column, which is the same as the sum of the $(x - \bar{x})^2$ column. We can represent the sum of the $(x - \bar{x})^2$ column by using the summation notation, $\Sigma(x - \bar{x})^2$. Thus, in Table 13.5, $\Sigma(x - \bar{x})^2 = 80$. We then divided this number by 1 less than the number of pieces of data, $n - 1$. Thus, we have

$$\frac{\Sigma(x - \bar{x})^2}{n - 1}$$

*16 is the variance, symbolized s^2, of this set of data.

Finally, we took the square root of this value to obtain the standard deviation.

Standard Deviation

$$s = \sqrt{\frac{\Sigma(x - \bar{x})^2}{n - 1}}$$

EXAMPLE 3 *Find the Standard Deviation of Stock Prices*

The following are the prices of nine stocks on the New York Stock Exchange. Find the standard deviation of the prices.

$15, $28, $32, $36, $50, $52, $68, $74, $104

SOLUTION: The mean, $\bar{x}$, is

$$\bar{x} = \frac{\Sigma x}{n} = \frac{15 + 28 + 32 + 36 + 50 + 52 + 68 + 74 + 104}{9} = \frac{459}{9} = 51$$

The mean is $51.

TABLE 13.6

x	$x - \bar{x}$	$(x - \bar{x})^2$
15	−36	1296
28	−23	529
32	−19	361
36	−15	225
50	−1	1
52	1	1
68	17	289
74	23	529
104	53	2809
	0	6040

Table 13.6 shows us that $\Sigma(x - \bar{x})^2 = 6040$. Since there are nine pieces of data, $n - 1 = 9 - 1$, or 8.

$$s = \sqrt{\frac{\Sigma(x - \bar{x})^2}{n - 1}} = \sqrt{\frac{6040}{8}} = \sqrt{755} \approx 27.5$$

The standard deviation, to the nearest tenth, is $27.5. ▲

Standard deviation will be used in Section 13.7 to find the percent of data between any two values in a normal curve. Standard deviations are also often used in determining norms for a population (see Exercise 31).

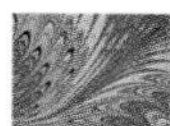

SECTION 13.6 EXERCISES

Concept/Writing Exercises

1. Explain how to find the range of a set of data.
2. What does the standard deviation of a set of data measure?
3. Explain how to find the standard deviation of a set of data.
4. What is the standard deviation of a set of data in which all the data values are the same? Explain.
5. Why is measuring dispersion in observed data important?
6. What symbol is used to represent the sample standard deviation?
7. What symbol is used to represent the population standard deviation?
8. Can you think of any situations in which a large standard deviation may be desirable? Explain.
9. Can you think of any situations in which a small standard deviation may be desirable? Explain.
10. Without actually doing the calculations, decide which of the following two sets of data will have the greater standard deviation. Explain why.

    13, 16, 17, 18, 20, 24   16, 17, 17, 18, 18, 19

11. Without actually doing the calculations, decide which, if either, of the following two sets of data will have the greater standard deviation. Explain why.

    2, 4, 6, 8, 10   102, 104, 106, 108, 110

12. By studying the standard deviation formula, explain why the standard deviation of a set of data will always be greater than or equal to 0.
13. Patricia Wolff has two statistics classes, one in the morning and the other in the evening. On the midterm exam, the morning class had a mean of 75.2 and a standard deviation of 5.7. The evening class had a mean of 75.2 and a standard deviation of 12.5.
    a) How do the means compare?
    b) If we compare the set of scores from the first class with those in the second class, how will the distributions of the two sets of scores compare? Explain.
14. Explain why the standard deviation is usually a better measure of dispersion than the range.

Practice the Skills

In Exercises 15–22, determine the range and standard deviation of the set of data. When appropriate, round standard deviations to the nearest hundredths.

15. 7, 5, 2, 8, 13

16. 10, 10, 14, 16, 8, 8

17. 120, 121, 122, 123, 124, 125, 126

18. 3, 7, 8, 12, 0, 9, 11, 12, 6, 2

19. 4, 8, 9, 11, 13, 15

20. 9, 9, 9, 9, 9, 9, 9

21. 7, 9, 7, 9, 9, 10, 12

22. 52, 50, 54, 59, 40, 43, 64, 62

23. *Computer Games* Find the range and standard deviation of the following prices of selected computer games: \$28, \$28, \$50, \$45, \$30, \$45, \$48, \$18, \$45, \$23.

24. *Years until Retirement* Seven employees at a large company were asked the number of additional years they planned to work before retirement. Their responses were 10, 23, 28, 4, 1, 6, 12. Find the range and standard deviation of the number of years.

25. *Fishing Poles* Find the range and standard deviation of the following prices of selected fishing poles: \$50, \$120, \$130, \$60, \$55, \$75, \$200, \$110, \$125, \$175.

26. *Holiday Gifts* The amount of money seven college students planned to spend on gifts during the holiday season are as follows: \$60, \$100, \$85, \$35, \$250, \$150, \$300. Find the range and standard deviation of the amounts.

Problem Solving

27. *Count Your Money* Six people were asked to determine the amount of money they were carrying, to the nearest dollar. The results were

    \$32, \$60, \$14, \$25, \$5, \$68

    a) Determine the range and standard deviation of the amounts.
    b) Add \$10 to each of the six amounts. How do you expect the range and standard deviation of the new set of data to change? Explain your answer.
    c) Determine the range and standard deviation of the new set of data. Do the results agree with your answer to part (b)? If not, explain why.

28. **a)** *Adding to or Subtracting from Each Number* Pick any five numbers. Compute the mean and the standard deviation of this distribution.

b) Add 20 to each of the numbers in your original distribution and compute the mean and the standard deviation of this new distribution.

c) Subtract 5 from each number in your original distribution and compute the mean and standard deviation of this new distribution.

d) What conclusions can you draw about changes in the mean and the standard deviation when the same number is added to or subtracted from each piece of data in a distribution?

e) How will the mean and standard deviation of the numbers 6, 7, 8, 9, 10, 11, 12 differ from the mean and standard deviation of the numbers 596, 597, 598, 599, 600, 601, 602? Find the mean and standard deviation of both sets of numbers.

29. **a)** *Multiplying Each Number* Pick any five numbers. Compute the mean and standard deviation of this distribution.

b) Multiply each number in your distribution by 3 and compute the mean and the standard deviation of this new distribution.

c) Multiply each number in your original distribution by 9 and compute the mean and the standard deviation of this new distribution.

d) What conclusions can you draw about changes in the mean and the standard deviation when each value in a distribution is multiplied by the same number?

e) The mean and standard deviation of the distribution 1, 3, 4, 4, 5, 7 are 4 and 2, respectively. Use the conclusion drawn in part (d) to determine the mean and standard deviation of the distribution

5, 15, 20, 20, 25, 35

30. *Waiting in Line* Consider the following illustrations of two bank-customer waiting systems.

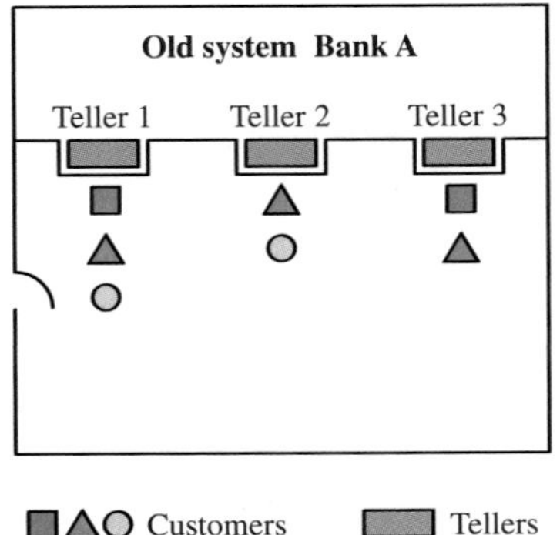

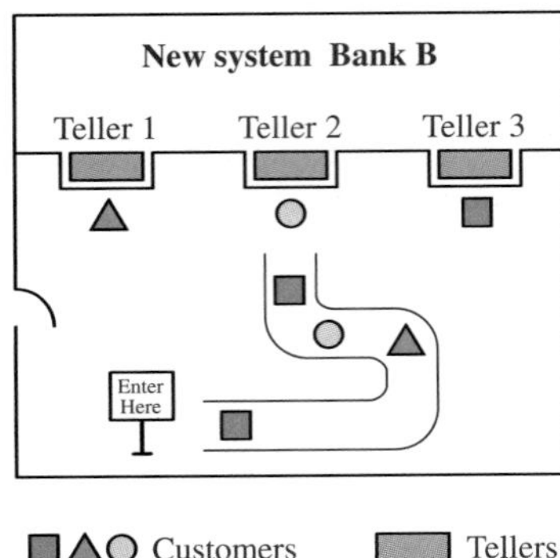

a) How would you expect the mean waiting time in Bank A to compare with the mean waiting time in Bank B? Explain your answer.

b) How would you expect the standard deviation of waiting times in Bank A to compare with the standard deviation of waiting times in Bank B? Explain your answer.

31. *Height and Weight Distribution* The chart shown below uses the symbol σ to represent the standard deviation. Note that 2σ represents the value that is two standard deviations above the mean; -2σ represents the value that is two standard deviations below the mean. The unshaded areas, from two standard deviations below the mean to two standard deviations above the mean, are considered the normal range. For example, the average (mean) 8-year-old boy has a height of about 50 inches, but any heights between approximately 45 inches and 55 inches are considered normal for 8-year-old boys. Refer to the chart below to answer the following questions.

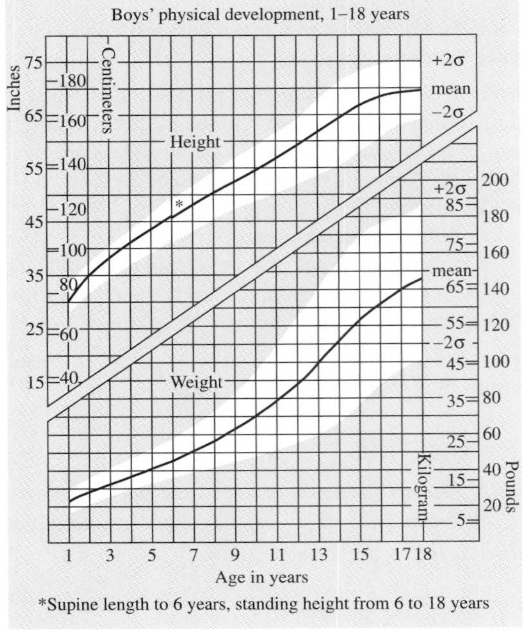

a) What happens to the standard deviation for weights of boys as the age of boys increases? What is the significance of this fact?
b) At age 16, what is the mean weight, in pounds, of boys?
c) What is the approximate standard deviation of boys' weights at age 16?
d) Find the mean weight and normal range for boys at age 13.
e) Find the mean height and normal range for boys at age 13.
f) Assuming that this chart was constructed so that approximately 95% of all boys are always in the normal range, determine what percentage of boys are not in the normal range.

Challenge Problems/Group Activities

32. *Athletes' Salaries* The following table lists the 10 highest paid athletes in Major League Baseball and in the National Football League.

Major League Baseball (2003 Season)

Player	Salary (millions)
1. Alex Rodriguez	$22
2. Manny Ramirez	20
3. Carlos Delgado	18.7
4. Mo Vaughn	17.2
5. Sammy Sosa	16
6. Kevin Brown	15.7
7. Shawn Green	15.7
8. Derek Jeter	15.6
9. Mike Piazza	15.6
10. Barry Bonds	15.5

Source: Major League Players Association.

National Football League (2002 Season)

Player	Salary (millions)
1. Donovan McNabb	$15.4
2. Curtis Martin	13.3
3. Larry Allen	13.0
4. David Carr	12.0
5. Rod Smith	11.7
6. Jeff Garcia	11.7
7. Michael Strahan	11.4
8. Aaron Glenn	11.3
9. Tarik Glenn	11.3
10. Ray Lewis	10.5

Source: National Football League Players Association.

a) Without doing any calculations, which do you believe is greater, the mean salary of the 10 baseball players or the mean salary of the 10 football players? Explain.
b) Without doing any calculations, which do you believe is greater, the standard deviation of the salary of the 10 baseball players or the standard deviation of the salary of the 10 football players? Explain.
c) Compute the mean salary of the 10 baseball players and the mean salary of the 10 football players and determine whether your answer in part (a) was correct.
d) Compute the standard deviation of the salary of the 10 baseball players and the standard deviation of the salary of the 10 football players and determine whether your answer in part (b) is correct. Round each mean to the nearest tenth to determine the standard deviation.

33. *Oil Change* Jiffy Lube has franchises in two different parts of the city. The number of oil changes made daily, for 25 days, is given below.

East Store					West Store				
33	59	27	30	42	38	46	38	38	30
19	42	25	22	32	38	38	37	39	31
43	27	57	37	52	39	36	40	37	47
40	67	38	44	43	30	34	42	45	29
15	31	49	41	35	31	46	28	45	48

a) Construct a frequency distribution for each store with a first class of 15–20.
b) Draw a histogram for each store.
c) Using the histogram, determine which store appears to have a greater mean, or do the means appear about the same? Explain.
d) Using the histogram, determine which store appears to have the greater standard deviation? Explain.
e) Calculate the mean for each store and determine whether your answer in part (c) was correct.
f) Calculate the standard deviation for each store and determine whether your answer in part (d) was correct.

Recreational Mathematics

34. Calculate the range and standard deviation of your exam grades in this mathematics course. Round the mean to the nearest tenth to calculate the standard deviation.

35. Construct a set of 5 pieces of data with a mean, median, mode, and midrange of 6 and a standard deviation of 0.

Internet/Research Problem

36. Use a calculator with statistical function keys to find the mean and standard deviation of the salaries of the 10 Major League Baseball players and the 10 National Football League players in Exercise 32.

13.7 THE NORMAL CURVE

When examining data using a histogram, we can refer to the overall appearance of the histogram as the *shape* of the distribution of the data. Certain shapes of distributions of data are more common than others. In this section, we will illustrate and discuss a few of the more common ones. In each case, the vertical scale is the frequency and the horizontal scale is the observed values.

In a *rectangular distribution* (Fig. 13.17), all the observed values occur with the same frequency. If a die is rolled many times, we would expect the numbers 1–6 to occur with about the same frequency. The distribution representing the outcomes of the die is rectangular.

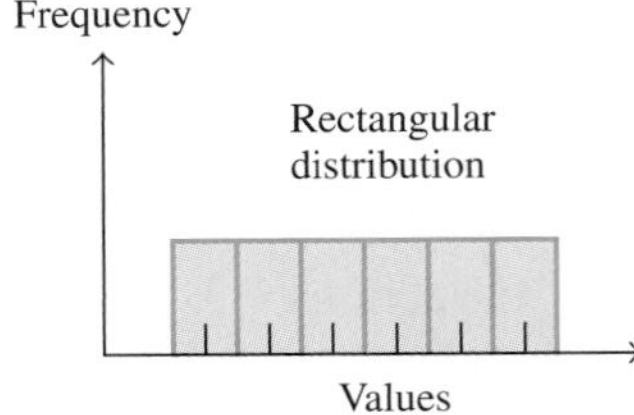

Figure 13.17

In *J-shaped distributions*, the frequency is either constantly increasing (Fig. 13.18a) or constantly decreasing (Fig. 13.18b). The number of hours studied per week by students may have a distribution like that in Fig. 13.18(b). The bars might represent (from left to right) 0–5 hours, 6–10 hours, 11–15 hours, and so on.

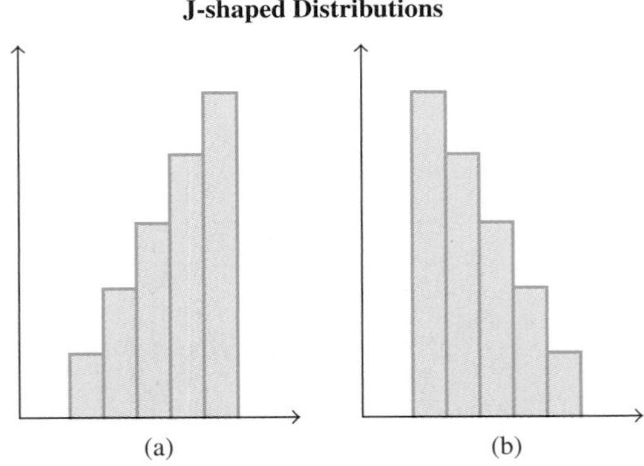

Figure 13.18

A *bimodal distribution* (Fig. 13.19) is one in which two nonadjacent values occur more frequently than any other values in a set of data. For example, if an equal number of men and women were weighed, the distribution of their weights would probably be bimodal, with one mode for the women's weights and the second for the men's weights. For a distribution to be considered bimodal, both modes need not have the same frequency but they must both have a frequency greater than the frequency of each of the other values in the distribution.

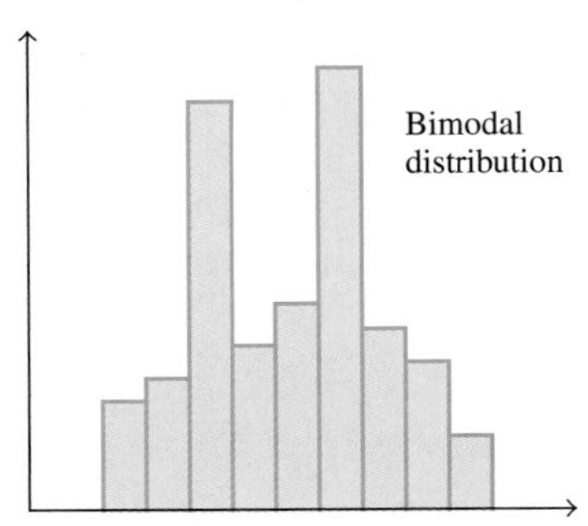

Figure 13.19

The life expectancy of light bulbs has a bimodal distribution: a small peak very near 0 hours of life, resulting from the bulbs that burned out very quickly because of a manufacturing defect, and a much broader peak representing the nondefective bulbs. A bimodal frequency distribution generally means that you are dealing with two distinct populations, in this case, defective and nondefective bulbs.

Another distribution, called a *skewed distribution*, has more of a "tail" on one side than the other. A skewed distribution with a tail on the right (Fig. 13.20a) is said to be skewed to the right. If the tail is on the left (Fig. 13.20b), the distribution is referred to as skewed to the left.

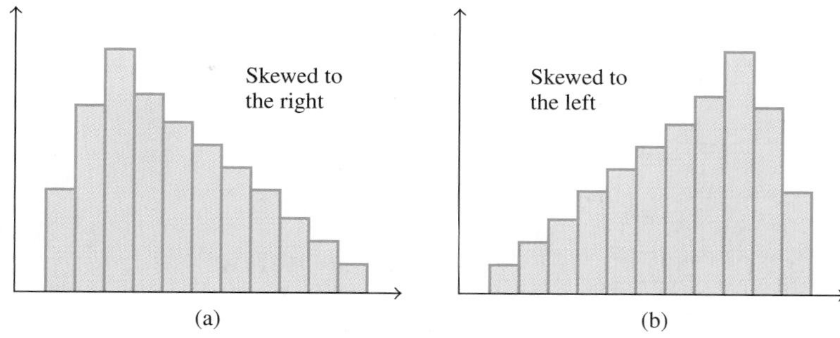

Figure 13.20

The number of children per family might be a distribution skewed to the right. Some families have no children, more families may have one child, the greatest percentage may have two children, fewer may have three children, still fewer may have four children, and so on.

Since few families have high incomes, distributions of family incomes might be skewed to the right.

Smoothing the histograms of the skewed distributions shown in Fig. 13.20 to form curves gives the curves illustrated in Fig. 13.21.

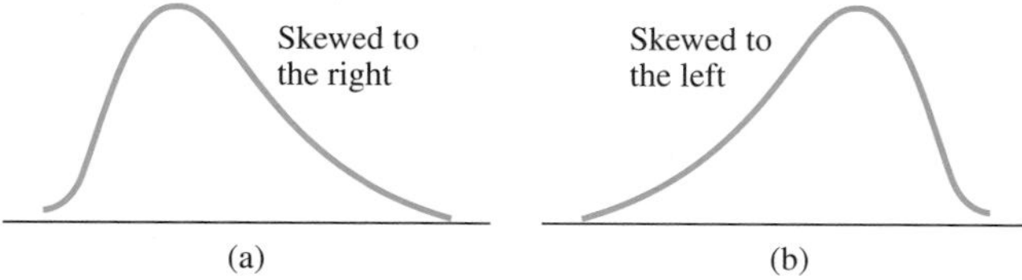

Figure 13.21

In Fig. 13.21(a), the greatest frequency appears on the left side of the curve, and the frequency decreases from left to right. Since the mode is the value with the greatest frequency, the mode would appear on the left side of the curve.

Every value in the set of data is considered in determining the mean. The values on the far right side of the curve in Fig. 13.21(a) would tend to increase the value of the mean. Thus, the value of the mean would be farther to the right than the mode. The median would be between the mode and the mean. The relationship between the mean, median, and mode for curves that are skewed to the right and left is given in Fig. 13.22.

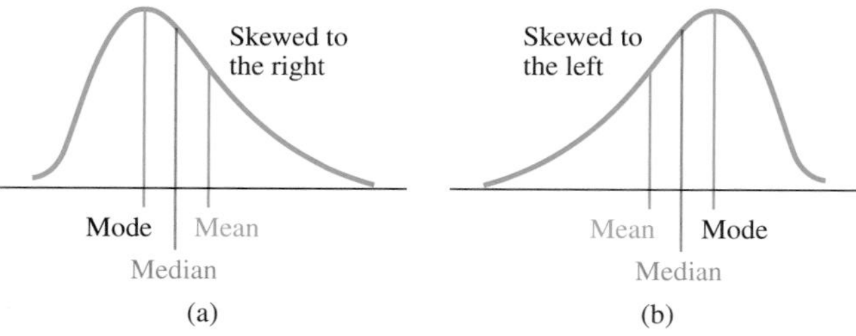

Figure 13.22

DID YOU KNOW

What Conclusions Can You Draw?

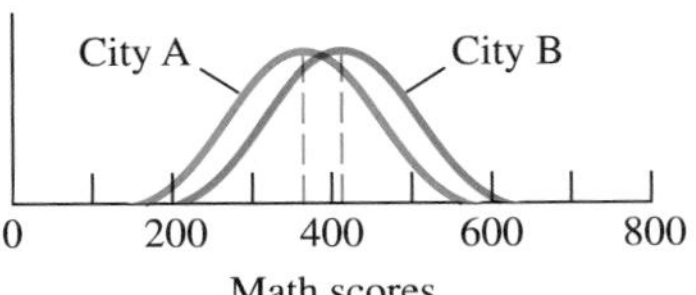

Based on the figure, which shows the distribution of scores on the mathematics part of the SAT test for two different cities, can we say that any given person selected at random from city B has outperformed any given person selected at random from city A? Both distributions appear normal, and the mean of city A is slightly smaller than the mean of city B. Consider, however, two randomly selected students who took this test: Sally from city A and Kendra from city B. The graph shows that many students in city A outperformed students from city B, so we cannot conclude that Sally scored higher than Kendra or that Kendra scored higher than Sally.

Each of these distributions is useful in describing sets of data. However, the most important distribution is the *normal* or *Gaussian distribution*, named for the German mathematician Carl Friedrich Gauss. The histogram of a normal distribution is illustrated in Fig. 13.23.

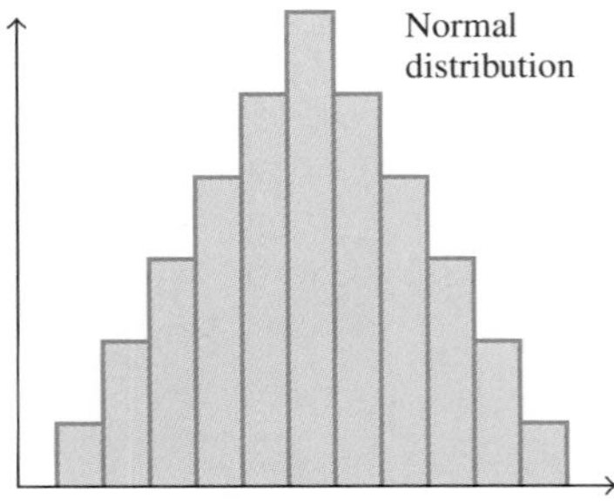

Figure 13.23

The normal distribution is important because many sets of data are normally distributed, or they closely resemble a normal distribution. Such distributions include intelligence quotients, heights and weights of males, heights and weights of females, lengths of full-grown boa constrictors, weights of watermelons, wearout mileage of automobile brakes, and life spans of refrigerators, to name just a few.

The normal distribution is symmetric about the mean. If you were to fold the histogram down the middle, the left side would fit the right side exactly. **In a normal distribution, the mean, median, and mode all have the same value.**

When the histogram of a normal distribution is smoothed to form a curve, the curve is bell-shaped. The bell may be high and narrow or short and wide. Each of the three curves in Fig. 13.24 represents a normal curve. Curve 13.24(a) has the smallest standard deviation (spread from the mean); curve 13.24(c) has the largest.

When we work with a distribution, we are working with an entire population. Therefore, when we discuss the normal distribution, we use μ for the mean and σ for the standard deviation.

Since the curve is symmetric, 50% of the data always falls above (to the right of) the mean and 50% of the data falls below (to the left of) the mean. In addition, every normal distribution has approximately 68% of the data between the value that is one standard deviation below the mean, $\mu - 1\sigma$, and the value that is one standard deviation above the mean, $\mu + 1\sigma$; see Fig. 13.25. Approximately 95% of the data falls between the value that is two standard deviations below the mean, $\mu - 2\sigma$, and the value that is two standard deviations above the mean, $\mu + 2\sigma$.

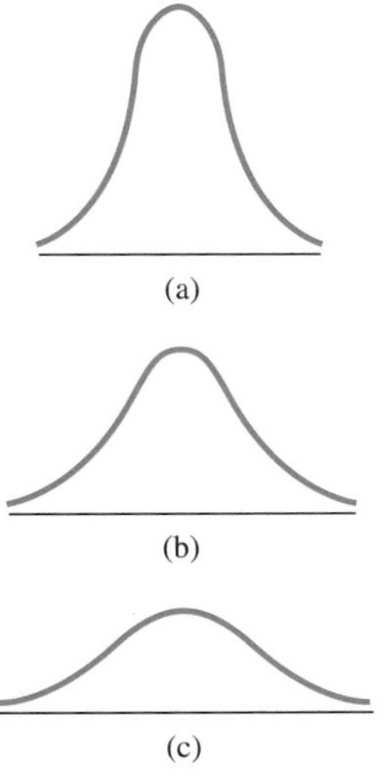

Figure 13.24

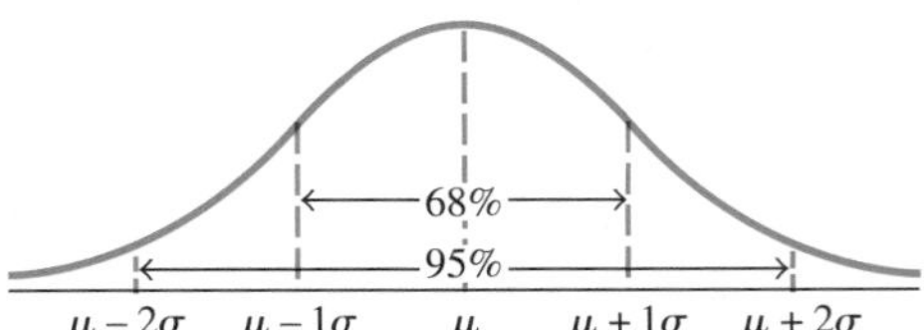

Figure 13.25

Thus, if a normal distribution has a mean of 100 and a standard deviation of 10, then approximately 68% of all the data falls between 100 − 10 and 100 + 10, or be-

PROFILE IN MATHEMATICS

DAVID BLACKWELL (1919–)

David H. Blackwell, (1919–), professor of statistics, is the author of more than 90 publications on statistics, probability, game theory, set theory, dynamic programming, and information theory. Blackwell, past president of the American Statistical Society, was the first African-American elected to the National Academy of Sciences.

His first interests in mathematics were in Geometry as a schoolboy growing up in southern Illinois. When he enrolled at the University of Illinois at the age of 16, he planned to be an elementary school teacher. When he received his Ph.D. in mathematics from the University of Illinois in 1941, he was just the sixth African-American to receive a doctorate in mathematics in the country.

After spending 2 years as a postdoctoral fellow at the Institute for Advanced Study at Princeton, he taught mathematics for 10 years at Howard University. While at Howard, Blackwell built a strong national reputation as a gifted teacher and creative researcher.

In 1954, he joined the Department of Statistics at the University of California, Berkeley. Blackwell, who has taught a wide variety of mathematics courses, says, "Basically, I'm not interested in doing research and I never have been. I'm interested in understanding, which is quite a different thing."

tween 90 and 110. Approximately 95% of the data falls between $100 - 20$ and $100 + 20$, or between 80 and 120. In fact, given any normal distribution with a known standard deviation and mean, it is possible through the use of Table 13.7 page 801 (the z-table) to determine the percent of data between any two given values.

We use *z-scores* (or *standard scores*) to determine how far, in terms of standard deviations, a given score is from the mean of the distribution. For example, a score that has a z-value of 1.5 indicates the score is 1.5 standard deviations above the mean. The standard or z-score is calculated as follows.

The formula for finding z-scores or standard scores is

$$z = \frac{\text{value of the piece of data} - \text{mean}}{\text{standard deviation}}$$

If we let x represent the value of the given piece of data, μ represent the mean, and σ represent the standard deviation, we can symbolize the z-score formula as

$$z = \frac{x - \mu}{\sigma}$$

In this book, the notation z_x represents the z-score, or standard score, of the value x. For example, if a normal distribution has a mean of 86 with a standard deviation of 12, a score of 110 has a standard or z-score of

$$z_{110} = \frac{110 - 86}{12} = \frac{24}{12} = 2$$

Therefore, a value of 110 in this distribution has a z-score of 2. The score of 110 is two standard deviations above the mean.

Data below the mean will always have negative z-scores; data above the mean will always have positive z-scores. The mean will always have a z-score of 0.

EXAMPLE 1 *Finding z-Scores*

A normal distribution has a mean of 100 and a standard deviation of 10. Find z-scores for the following values.

a) 110 b) 115 c) 100 d) 84

SOLUTION:

a)

$$z = \frac{\text{value} - \text{mean}}{\text{standard deviation}}$$

$$z_{110} = \frac{110 - 100}{10} = \frac{10}{10} = 1$$

A score of 110 is one standard deviation above the mean.

b)

$$z_{115} = \frac{115 - 100}{10} = \frac{15}{10} = 1.5$$

A score of 115 is 1.5 standard deviations above the mean.

DID YOU KNOW

Six Sigma

Many companies use a process called Six Sigma, a quality control strategy, to help the company improve quality and reduce errors. Six Sigma refers to an interval in a normal distribution from six standard deviations below the mean to six standard deviations above the mean. As 99.9997% of a normal distribution is within six standard deviations of the mean, Six Sigma means the company's goal is to produce error-free products 99.9997% of the time. Companies such as General Electric (GE), Whirlpool, and Motorola have all reported success after implementing Six Sigma. GE's former CEO, Jack Welch, a major advocate for Six Sigma, reported that Six Sigma saved the company more than $2 billion in 1999.

c) $$z_{100} = \frac{100 - 100}{10} = \frac{0}{10} = 0$$

The mean always has a z-score of 0.

d) $$z_{84} = \frac{84 - 100}{10} = \frac{-16}{10} = -1.6$$

A score of 84 is 1.6 standard deviations below the mean. 

Let's now consider finding areas under the normal curve. The total area under the normal curve is 1.00. Table 13.7 on page 801 will be used to determine the area under the normal curve between any two given points (the values in the table have been rounded). **Table 13.7 gives the area under the normal curve from the mean (a z-value of 0) to a z-value to the right of the mean.**

For example, between the mean and $z = 2.00$, the table shows a value of 0.477. Thus, there is 0.477 of the total area under the curve between the mean and $z = 2.00$, see Fig. 13.26. To change this area of 0.477 to a percent, simply multiply by 100%: $0.477 \times 100\%$ is 47.7%. Thus, 47.7% of all scores will be between the mean and the score that is two standard deviations above the mean.

When you are finding the area under the normal curve, it is often helpful to draw a picture such as the one in Fig. 13.26, indicating the area or percent to be found.

The normal curve is symmetric about the mean. Thus, the same percent of data is between the mean and a positive z-score as between the mean and the corresponding negative z-score. For example, there is the same area under the normal curve between a z of 1.60 and the mean as between a z of -1.60 and the mean. By looking up a z-score of 1.60 in Table 13.7, we see that both have an area of 0.445 (Fig. 13.27). Since an area of 0.445 corresponds to 44.5%, we can reason that 44.5% + 44.5% or 89.0% of the data is between z-scores of -1.60 and 1.60.

You now have the necessary knowledge to find the percent of data between any two values in a normal distribution.

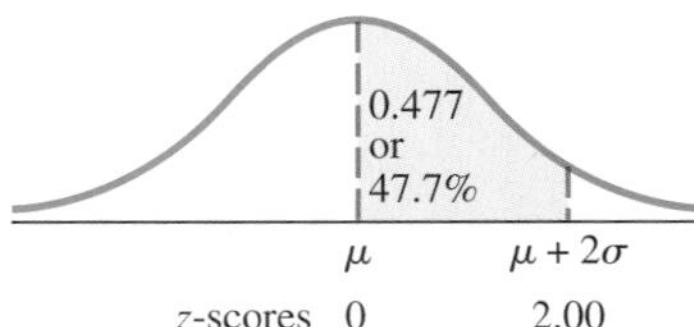

Figure 13.26

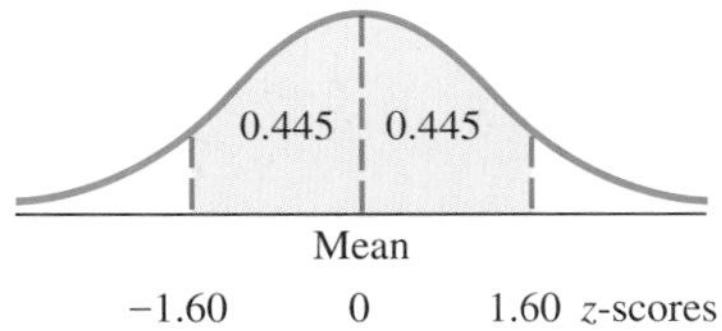

Figure 13.27

To Find the Percent of Data Between any Two Values:

1. Draw a diagram of the normal curve, indicating the area or percent to be determined.
2. Use the formula $z = \dfrac{x - \mu}{\sigma}$ to convert the given values to z-scores. Indicate these z-scores on the diagram.
3. Look up the percent that corresponds to each z-score in Table 13.7.
4. **a)** When finding the percent of data between two z-scores on the opposite side of the mean (when one z-score is positive and the other is negative), you find the sum of the individual percents.
   b) When finding the percent of data between two z-scores on the same side of the mean (when both z-scores are positive or both are negative), subtract the smaller percent from the larger percent.
   c) When finding the percent of data to the right of a positive z-score or to the left of a negative z-score, subtract the percent of data between 0 and z from 50%.
   d) When finding the percent of data to the left of a positive z-score or to the right of a negative z-score, add the percent of data between 0 and z to 50%.

TABLE 13.7 Areas under the Standard Normal Curve (the z-table)

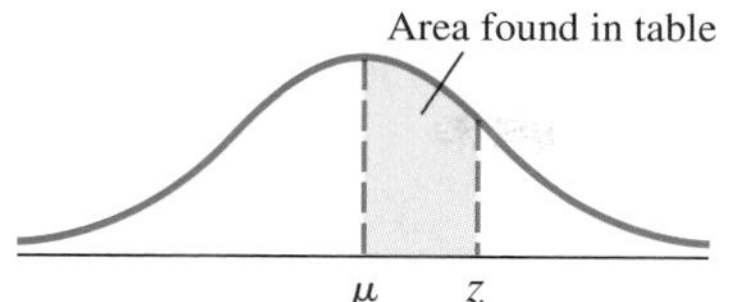

The column under A gives the area under the entire curve that is between $z = 0$ (or the mean) and a positive value of z

z	A	z	A	z	A	z	A	z	A	z	A	z	A	z	A	z	A
.00	.000	.37	.144	.74	.270	1.11	.367	1.48	.431	1.85	.468	2.22	.487	2.59	.495	2.96	.499
.01	.004	.38	.148	.75	.273	1.12	.369	1.49	.432	1.86	.469	2.23	.487	2.60	.495	2.97	.499
.02	.008	.39	.152	.76	.276	1.13	.371	1.50	.433	1.87	.469	2.24	.488	2.61	.496	2.98	.499
.03	.012	.40	.155	.77	.279	1.14	.373	1.51	.435	1.88	.470	2.25	.488	2.62	.496	2.99	.499
.04	.016	.41	.159	.78	.282	1.15	.375	1.52	.436	1.89	.471	2.26	.488	2.63	.496	3.00	.499
.05	.020	.42	.163	.79	.285	1.16	.377	1.53	.437	1.90	.471	2.27	.488	2.64	.496	3.01	.499
.06	.024	.43	.166	.80	.288	1.17	.379	1.54	.438	1.91	.472	2.28	.489	2.65	.496	3.02	.499
.07	.028	.44	.170	.81	.291	1.18	.381	1.55	.439	1.92	.473	2.29	.489	2.66	.496	3.03	.499
.08	.032	.45	.174	.82	.294	1.19	.383	1.56	.441	1.93	.473	2.30	.489	2.67	.496	3.04	.499
.09	.036	.46	.177	.83	.297	1.20	.385	1.57	.442	1.94	.474	2.31	.490	2.68	.496	3.05	.499
.10	.040	.47	.181	.84	.300	1.21	.387	1.58	.443	1.95	.474	2.32	.490	2.69	.496	3.06	.499
.11	.044	.48	.184	.85	.302	1.22	.389	1.59	.444	1.96	.475	2.33	.490	2.70	.497	3.07	.499
.12	.048	.49	.188	.86	.305	1.23	.391	1.60	.445	1.97	.476	2.34	.490	2.71	.497	3.08	.499
.13	.052	.50	.192	.87	.308	1.24	.393	1.61	.446	1.98	.476	2.35	.491	2.72	.497	3.09	.499
.14	.056	.51	.195	.88	.311	1.25	.394	1.62	.447	1.99	.477	2.36	.491	2.73	.497	3.10	.499
.15	.060	.52	.199	.89	.313	1.26	.396	1.63	.449	2.00	.477	2.37	.491	2.74	.497	3.11	.499
.16	.064	.53	.202	.90	.316	1.27	.398	1.64	.450	2.01	.478	2.38	.491	2.75	.497	3.12	.499
.17	.068	.54	.205	.91	.319	1.28	.400	1.65	.451	2.02	.478	2.39	.492	2.76	.497	3.13	.499
.18	.071	.55	.209	.92	.321	1.29	.402	1.66	.452	2.03	.479	2.40	.492	2.77	.497	3.14	.499
.19	.075	.56	.212	.93	.324	1.30	.403	1.67	.453	2.04	.479	2.41	.492	2.78	.497	3.15	.499
.20	.079	.57	.216	.94	.326	1.31	.405	1.68	.454	2.05	.480	2.42	.492	2.79	.497	3.16	.499
.21	.083	.58	.219	.95	.329	1.32	.407	1.69	.455	2.06	.480	2.43	.493	2.80	.497	3.17	.499
.22	.087	.59	.222	.96	.332	1.33	.408	1.70	.455	2.07	.481	2.44	.493	2.81	.498	3.18	.499
.23	.091	.60	.226	.97	.334	1.34	.410	1.71	.456	2.08	.481	2.45	.493	2.82	.498	3.19	.499
.24	.095	.61	.229	.98	.337	1.35	.412	1.72	.457	2.09	.482	2.46	.493	2.83	.498	3.20	.499
.25	.099	.62	.232	.99	.339	1.36	.413	1.73	.458	2.10	.482	2.47	.493	2.84	.498	3.21	.499
.26	.103	.63	.236	1.00	.341	1.37	.415	1.74	.459	2.11	.483	2.48	.493	2.85	.498	3.22	.499
.27	.106	.64	.239	1.01	.344	1.38	.416	1.75	.460	2.12	.483	2.49	.494	2.86	.498	3.23	.499
.28	.110	.65	.242	1.02	.346	1.39	.418	1.76	.461	2.13	.483	2.50	.494	2.87	.498	3.24	.499
.29	.114	.66	.245	1.03	.349	1.40	.419	1.77	.462	2.14	.484	2.51	.494	2.88	.498	3.25	.499
.30	.118	.67	.249	1.04	.351	1.41	.421	1.78	.463	2.15	.484	2.52	.494	2.89	.498	3.26	.499
.31	.122	.68	.252	1.05	.353	1.42	.422	1.79	.463	2.16	.485	2.53	.494	2.90	.498	3.27	.500
.32	.126	.69	.255	1.06	.355	1.43	.424	1.80	.464	2.17	.485	2.54	.495	2.91	.498	3.28	.500
.33	.129	.70	.258	1.07	.358	1.44	.425	1.81	.465	2.18	.485	2.55	.495	2.92	.498	3.29	.500
.34	.133	.71	.261	1.08	.360	1.45	.427	1.82	.466	2.19	.486	2.56	.495	2.93	.498	3.30	.500
.35	.137	.72	.264	1.09	.362	1.46	.428	1.83	.466	2.20	.486	2.57	.495	2.94	.498	3.31	.500
.36	.141	.73	.267	1.10	.364	1.47	.429	1.84	.467	2.21	.487	2.58	.495	2.95	.498	3.32	.500

EXAMPLE 2 *IQ Scores*

Intelligence quotients (IQs) are normally distributed with a mean of 100 and a standard deviation of 15. Find the percent of individuals with IQs

a) between 100 and 115.
b) between 70 and 100.
c) between 70 and 115.
d) between 115 and 130.
e) below 130.
f) above 122.5.

SOLUTION:

a) We want to find the area under the normal curve between the values of 100 and 115, as illustrated in Fig. 13.28(a). Converting 100 to a z-score yields a z-score of 0.

$$z_{100} = \frac{100 - 100}{15} = \frac{0}{15} = 0$$

Converting 115 to a z-score yields a z-score of 1.00.

$$z_{115} = \frac{115 - 100}{15} = \frac{15}{15} = 1.00$$

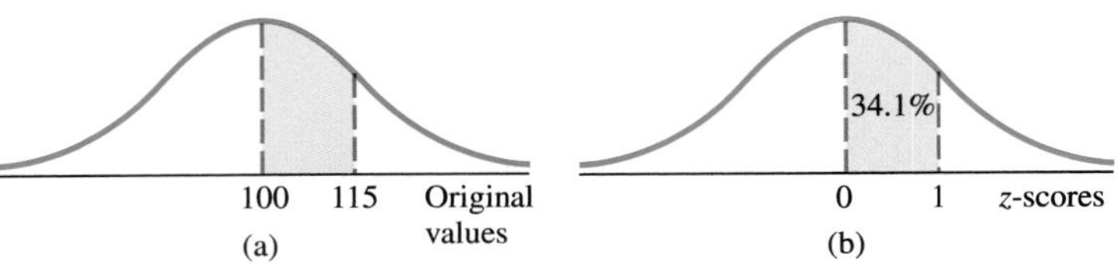

Figure 13.28

The percent of individuals with IQs between 100 and 115 is the same as the percent of data between z-scores of 0 and 1 (Fig. 13.28b).

From Table 13.7, we determine that 0.341 of the area, or 34.1% of all the data, is between z-scores of 0 and 1.00. Therefore, 34.1% of individuals have IQs between 100 and 115.

b) Begin by finding z-scores for 70 and 100.

$$z_{70} = \frac{70 - 100}{15} = \frac{-30}{15} = -2.00$$

$$z_{100} = 0 \text{ (from part a)}$$

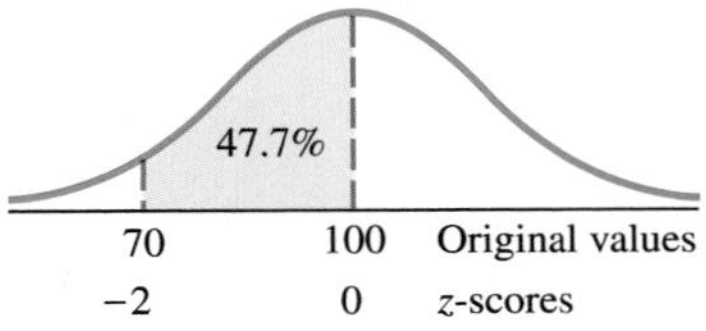

47.7%
70 100 Original values
−2 0 z-scores

Figure 13.29

The percent of data between scores of 70 and 100 is the same as the percent between $z = -2$ and $z = 0$ (Fig. 13.29). The percent of data between the mean and two standard deviations below the mean is the same as the percent of data between the mean and two standard deviations above the mean. From Table 13.7, we determine that 47.7% of the data is between $z = 0$ and $z = 2$. Thus, 47.7% of the data is also between $z = -2.00$ and $z = 0$. Therefore, 47.7% of all individuals have IQs between 70 and 100.

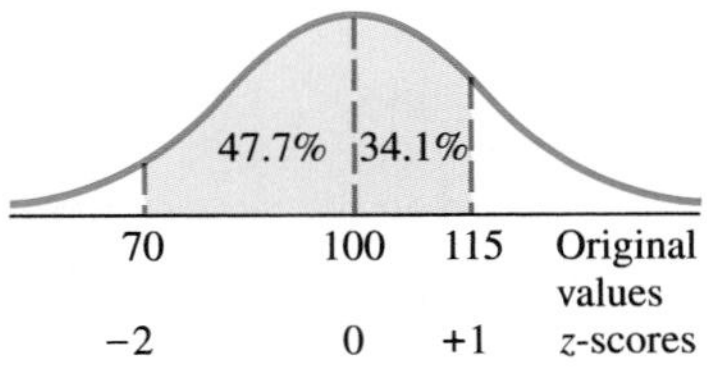

Figure 13.30

c) In parts (a) and (b), we determined that $z_{115} = 1.00$ and $z_{70} = -2.00$. Since the values are on opposite sides of the mean, the percent of data between the two values is found by adding the individual percents: 34.1% + 47.7% = 81.8% (Fig. 13.30). Thus, 81.8% of the IQs are between 70 and 115.

d) Begin by finding z-scores for 115 and 130.

$$z_{115} = 1.00 \text{ (from part a)}$$

$$z_{130} = \frac{130 - 100}{15} = \frac{30}{15} = 2.00$$

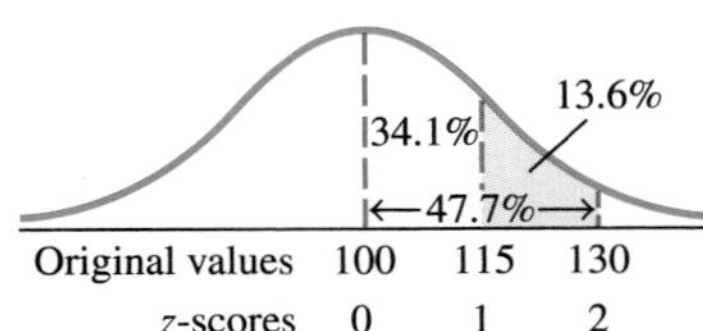

Figure 13.31

Since both values are on the same side of the mean (Fig. 13.31), the smaller percent must be subtracted from the larger percent to obtain the percent of data in the shaded area: 47.7% − 34.1% is 13.6%. Thus, 13.6% of all the individuals have IQs between 115 and 130.

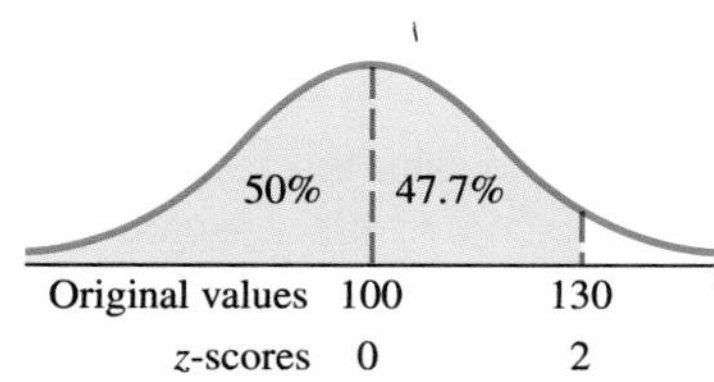

Figure 13.32

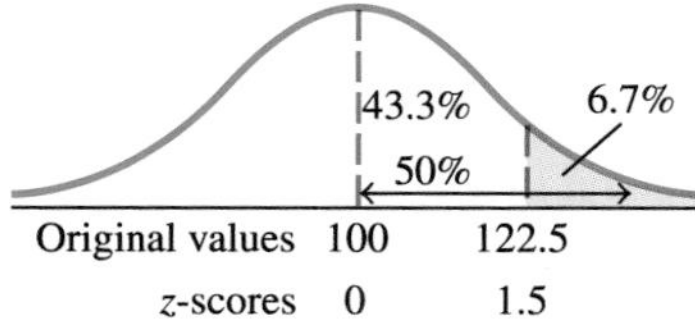

Figure 13.33

e) The percent of IQs below 130 is the same as the percent of data below a z-score of 2. Between $z = 2$ and the mean is 47.7% of the data (Fig. 13.32). To this 47.7% we add the 50% of the data below the mean to give 97.7%. Thus, 50% + 47.7%, or 97.7%, of all IQs are below 130.

f)

$$z_{122.5} = \frac{122.5 - 100}{15} = \frac{22.5}{15} = 1.50$$

The percent of IQs above 122.5 is the same as the percent of data above $z = 1.5$ (Fig. 13.33). Fifty percent of the data is to the right of the mean. Since 43.3% of the data is between the mean and $z = 1.5$, 50% − 43.3%, or 6.7%, of the data is greater than $z = 1.5$. Thus, 6.7% of all IQs are greater than 122.5. ▲

EXAMPLE 3 *Waiting Time at a Restaurant*

Assume that the waiting times for customers at a popular restaurant before being seated for lunch are normally distributed with a mean of 16 min and standard deviation of 4 min.

a) Find the percent of customers who wait for at least 16 min before being seated.
b) Find the percent of customers who wait between 12 and 24 min before being seated.
c) Find the percent of customers who wait at least 21 min before being seated.
d) Find the percent of customers who wait less than 9 min before being seated.
e) In a random sample of 500 customers, how many wait at least 21 min before being seated?

SOLUTION:

a) In a normal distribution, half the data are always above the mean. Since 16 min is the mean, half, or 50%, of customers wait at least 16 min before being seated.
b) Convert 12 min and 24 min to z-scores.

$$z_{12} = \frac{12 - 16}{4} = -1.00$$

$$z_{24} = \frac{24 - 16}{4} = 2.00$$

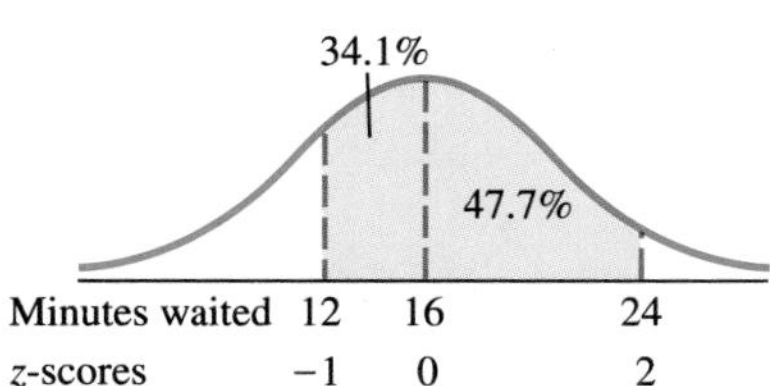

Figure 13.34

Now look up the areas in Table 13.7. The percent of customers who wait between 12 and 24 min before being seated is 34.1% + 47.7% or 81.8% (Fig. 13.34).

c) Convert 21 min to a z-score.

$$z_{21} = \frac{21 - 16}{4} = 1.25$$

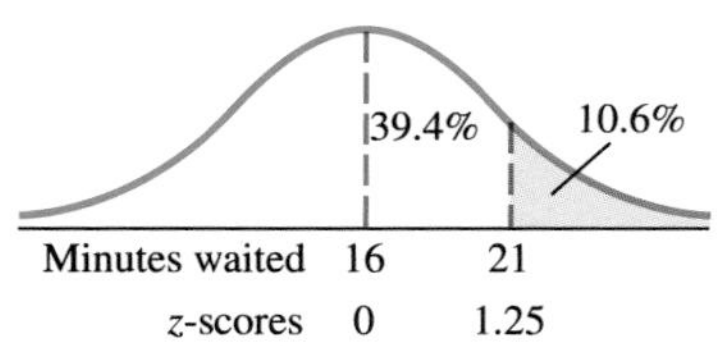

Figure 13.35

Look up the area in Table 13.7. Figure 13.35 shows 39.4% of the data is between the mean and $z = 1.25$. Therefore, the percent of data above $z = 1.25$ is 50% − 39.4% = 10.6%. Thus, 10.6% of customers wait at least 21 min before being seated.

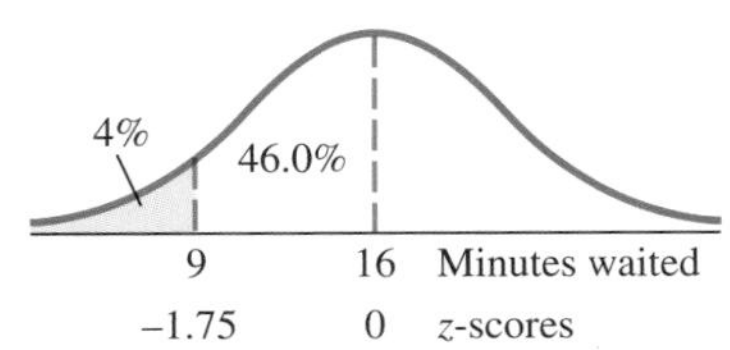

Figure 13.36

d) Convert 9 min to a z-score.

$$z_9 = \frac{9 - 16}{4} = -1.75$$

Look up the area in Table 13.7. Figure 13.36 shows that 46.0% of the data is between the mean and $z = -1.75$. The percent of data to the left of $z = -1.75$ is found by subtracting 46.0% from 50.0% to obtain 4.0%. Thus, 4% of customers wait less than 9 min before being seated.

e) In part (c), we determined that 10.6% of all customers wait at least 21 min before being seated. We now multiply 0.106 times 500 to determine the number of customers who wait at least 21 min before being seated. There are $0.106 \times 500 = 53$ customers who wait at least 21 min before being seated. 

TIMELY TIP Remember that area cannot be negative. A negative z-score indicates that the corresponding value in the original distribution is below the mean.

SECTION 13.7 EXERCISES

Concept/Writing Exercises

In Exercises 1–6, describe

1. a rectangular distribution.
2. a J-shaped distribution.
3. a bimodal distribution.
4. a distribution that is skewed to the right.
5. a distribution that is skewed to the left.
6. a normal distribution.

7. Consider the following normal curve, representing a normal distribution, with points A, B, and C. One of these points corresponds to μ, one point corresponds to $\mu + \sigma$, and one point corresponds to $\mu - 2\sigma$.

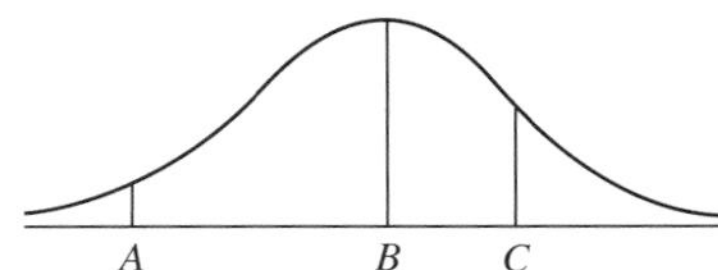

a) Which point corresponds to μ?
b) Which point corresponds to $\mu + \sigma$?
c) Which point corresponds to $\mu - 2\sigma$?

8. Consider the following normal curves.

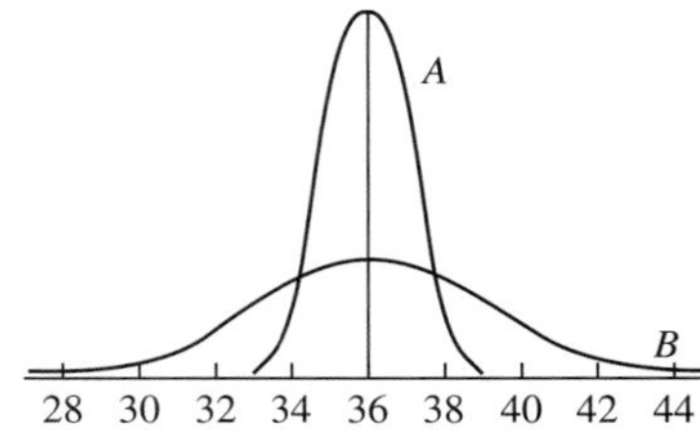

a) Do these distributions have the same mean? If so, what is the mean?
b) One of these curves corresponds to a normal distribution with $\sigma = 1$. The other curve corresponds to a normal distribution with $\sigma = 3$. Which curve, A or B, has $\sigma = 3$? Explain.

In Exercises 9–12, give an example of the type of distribution.

9. Rectangular
10. Skewed
11. J-shaped
12. Bimodal

For the distributions in Exercises 13–16, state whether you think the distribution would be normal, J-shaped, bimodal, rectangular, skewed left, or skewed right. Explain your answers.

13. The life expectancy of a sample of microwave ovens

14. The numbers resulting from tossing a die many times
15. The salaries of teachers at Roosevelt High School, where there are many newly hired teachers
16. The heights of a sample of high school seniors, where there are an equal number of males and females
17. In a distribution that is skewed to the right, which has the greatest value: the mean, median, or mode? Which has the smallest value? Explain.
18. In a distribution skewed to the left, which has the greatest value: the mean, median, or mode? Which has the smallest value? Explain.
19. List three populations other than those given in the text that may be normally distributed.
20. List three populations other than those given in the text that may not be normally distributed.
21. In a normal distribution, what is the relationship between the mean, median, and mode?
22. What does the z, or standard score, measure?
23. When will a z-score be negative?
24. Explain in your own words how to find the z-score of a particular piece of data.
25. What is the value of the z-score of the mean of a set of data?
26. In a normal distribution, approximately what percent of the data is between
    a) one standard deviation below the mean to one standard deviation above the mean?
    b) two standard deviations below the mean to two standard deviations above the mean?

Practice the Skills

In Exercises 27–38, use Table 13.7 on page 801 to find the specified area.

27. Above the mean
28. Below the mean
29. Between two standard deviations below the mean and one standard deviation above the mean
30. Between 1.10 and 1.70 standard deviations above the mean
31. To the right of $z = 1.82$
32. To the left of $z = 1.19$
33. To the left of $z = -1.78$
34. To the right of $z = -1.78$
35. To the right of $z = 2.08$
36. To the left of $z = 1.96$
37. To the left of $z = -1.62$
38. To the left of $z = -0.90$

In Exercises 39–48, use Table 13.7 to determine the percent of data specified.

39. Between $z = 0$ and $z = 0.71$
40. Between $z = -0.15$ and $z = -0.82$
41. Between $z = -1.34$ and $z = 2.24$
42. Less than $z = -1.90$
43. Greater than $z = -1.90$
44. Greater than $z = 2.66$
45. Less than $z = 1.96$
46. Between $z = 0.72$ and $z = 2.14$
47. Between $z = -1.53$ and $z = -1.82$
48. Between $z = -2.15$ and $z = 3.31$

Problem Solving

Fitness Test Scores *In Exercises 49 and 50, suppose that the results on a fitness test are normally distributed. The z-scores for some participants are shown below.*

Jake	1.3	Marie	0.0	Justin	−1.9	Kevin	0.0
Sarah	1.7	Omar	−2.1	Carol	0.8	Kim	−1.2

49. a) Which of these participants scored above the mean?
    b) Which of these participants scored at the mean?
    c) Which of these participants scored below the mean?
50. a) Which participant had the highest score?
    b) Which participant had the lowest score?

Hours Worked by College Students *In Exercises 51–54, assume that the number of hours college students spend working per week is normally distributed with a mean of 18 hours and standard deviation of 4 hours.*

51. Find the percent of college students who work at least 18 hours per week.
52. Find the percent of college students who work between 14 and 26 hours per week.

53. Determine the percent of college students who work at least 23 hours per week.

54. In a random sample of 500 college students, how many work at least 23 hours per week?

Watching Television *In Exercises 55–60, assume that the amount of time children spend watching television per year is normally distributed with a mean of 1600 hours and a standard deviation of 100 hours.*

55. What percent of children watch television less than 1650 hours per year?

56. What percent of children watch television more than 1750 hours per year?

57. What percent of children watch television between 1650 and 1750 hours per year?

58. What percent of children watch television less than 1400 hours per year?

59. What percent of children watch television between 1500 and 1625 hours per year?

60. What percent of children watch television more than 1480 hours per year?

Vending Machine *In Exercises 61–64, a vending machine is designed to dispense a mean of 7.6 oz of coffee into an 8 oz cup. If the standard deviation of the amount of coffee dispensed is 0.4 oz and the amount is normally distributed, find the percent of times the machine will*

61. dispense from 7.4 oz to 7.7 oz.

62. dispense less than 7.0 oz.

63. dispense less than 7.7 oz.

64. result in the cup overflowing (therefore dispense more than 8 oz).

Cholesterol Levels *In Exercises 65–70, assume that the cholesterol levels for females are normally distributed with a mean of 206 and a standard deviation of 12.*

65. What percent of females have a cholesterol level greater than 206?

66. What percent of females have a cholesterol level between 197 and 215?

67. What percent of females have a cholesterol level less than 191?

68. What percent of females have a cholesterol level greater than 224?

69. If 200 women are selected at random, how many will have a cholesterol level less than 191?

70. If 200 women are selected at random, how many will have a cholesterol level greater than 224?

Tire Mileage *In Exercises 71–74, the wearout mileage of a certain tire is normally distributed with a mean of 35,000 miles and standard deviation of 2500 miles.*

71. Determine the percent of tires that will last between 30,750 miles and 38,300 miles.

72. Determine the percent of tires that will last at least 39,000 miles.

73. If the manufacturer guarantees the tires to last at least 30,750 miles, what percent of tires will fail to live up to the guarantee?

74. If 200,000 tires are produced, how many will last at least 39,000 miles?

Ages of Children at Day Care *In Exercises 75–80, assume that the ages of children at Happy Times Day Care are normally distributed with a mean of 3.7 years and a standard deviation of 1.2 years. Find the percent of children at Happy Times Day Care who are*

75. Older than 3.1 years.

76. Between 2.5 and 4.3 years.

77. Older than 6.7 years.

78. Younger than 6.7 years.

79. If 120 children are enrolled at Happy Times Day Care, how many of them are older than 3.1 years?

80. If 120 children are enrolled at Happy Times Day Care, how many of them are between 2.5 and 4.3 years?

81. Weight Loss A weight-loss clinic guarantees that its new customers will lose at least 5 lb by the end of their first month of participation or their money will be refunded. If the loss of weight of customers at the end of their first month is normally distributed, with a mean of 6.7 lb and a standard deviation of 0.81 lb, find the percent of customers who will be able to claim a refund.

82. *Appliance Warranty* The warranty on the motor of a dishwasher is 8 yr. If the breakdown times of this motor are normally distributed, with a mean of 10.2 yr and a standard deviation of 1.8 yr, find the percent of motors that can be expected to require repair or replacement under warranty.

83. *Coffee Machine* A vending machine that dispenses coffee does not appear to be working correctly. The machine rarely gives the proper amount of coffee. Some of the time the cup is underfilled, and some of the time the cup overflows. Does this variation indicate that the mean number of ounces dispensed has to be adjusted or that the standard deviation of the amount of coffee dispensed by the machine is too large? Explain your answer.

84. *Grading on a Normal Curve* Mr. Sanderson marks his class on a normal curve. Those with z-scores above 1.8 will receive an A, those between 1.8 and 1.1 will receive a B, those between 1.1 and -1.2 will receive a C, those between -1.2 and -1.9 will receive a D, and those under -1.9 will receive an F. Find the percent of grades that will be A, B, C, D, and F.

Challenge Problems/Group Activities

85. *Salesperson Promotion* The owner at Kim's Home Interiors is reviewing the sales records of two managers who are up for promotion, Katie and Stella, who work in different stores. At Katie's store, the mean sales have been \$23,200 per month, with a standard deviation of \$2170. At Stella's store, the mean sales have been \$25,600 per month, with a standard deviation of \$2300. Last month Katie's store sales were \$28,408 and Stella's store sales were \$29,510. At both stores, the distribution of monthly sales is normal.

a) Convert last month's sales for Katie's store and for Stella's store to z-scores.

b) If one of the two were to be promoted based solely on the increase in sales last month, who should be promoted? Explain.

86. *Chebyshev's Theorem* How can you determine whether a distribution is approximately normal? A statistical theorem called *Chebyshev's theorem* states that the *minimum percent* of data between plus and minus K standard deviations from the mean $(K > 1)$ in *any distribution* can be found by the formula

$$\text{Minimum percent} = 1 - \frac{1}{K^2}$$

Thus, for example, between ± 2 standard deviations from the mean, there will always be a minimum of 75% of data. This minimum percent is true for any distribution. For $K = 2$,

$$\text{Minimum percent} = 1 - \frac{1}{2^2}$$
$$= 1 - \frac{1}{4} = \frac{3}{4}, \quad \text{or} \quad 75\%$$

Likewise, between ± 3 standard deviations from the mean, there will always be a minimum of 89% of the data. For $K = 3$,

$$\text{Minimum percent} = 1 - \frac{1}{3^2}$$
$$= 1 - \frac{1}{9} = \frac{8}{9}, \quad \text{or} \quad 89\%$$

The following table lists the minimum percent of data in *any distribution* and the actual percent of data in *the normal distribution* between ± 1.1, ± 1.5, ± 2.0, and ± 2.5 standard deviations from the mean. The minimum percents of data in any distribution were calculated by using Chebyshev's theorem. The actual percents of data for the normal distribution were calculated by using the area given in the standard normal, or z, table.

	$K = 1.1$	$K = 1.5$	$K = 2$	$K = 2.5$
Minimum (for any distribution)	17.4%	55.6%	75%	84%
Normal distribution	72.8%	86.6%	95.4%	99.8%
Given distribution				

The third row of the chart has been left blank for you to fill in the percents when you reach part (e).

Consider the following 30 pieces of data obtained from a quiz.

1, 1, 1, 1, 2, 2, 2, 2, 3, 3, 4, 4, 4, 5, 6,
6, 6, 7, 7, 7, 7, 8, 8, 8, 8, 9, 9, 9, 10, 10

a) Find the mean of the set of scores.

b) Find the standard deviation of the set of scores.

c) Determine the values that correspond to 1.1, 1.5, 2, and 2.5 standard deviations above the mean. (For example, the value that corresponds to 1.5 standard deviations above the mean is $\mu + 1.5\sigma$.)

Then determine the values that correspond to 1.1, 1.5, 2, and 2.5 standard deviations below the mean. (For example, the value that corresponds to 1.5 standard deviations below the mean is $\mu - 1.5\sigma$.)

d) By observing the 30 pieces of data, determine the actual percent of quiz scores between

± 1.1 standard deviations from the mean.

± 1.5 standard deviations from the mean.

± 2 standard deviations from the mean.

± 2.5 standard deviations from the mean.

e) Place the percents found in part (d) in the third row of the chart.

f) Compare the percents in the third row of the chart with the minimum percents in the first row and the normal percents in the second row, and then make a judgment as to whether this set of 30 scores is approximately normally distributed. Explain your answer.

87. *Using Data from Your Class* Obtain a set of test scores from your teacher.
   a) Find the mean, median, mode, and midrange of the test scores.
   b) Find the range and standard deviation of the set of scores. (You may round the mean to the nearest tenth when finding the standard deviation.)
   c) Construct a frequency distribution of the set of scores. Select your first class so that there will be between 5 and 12 classes.
   d) Construct a histogram and frequency polygon of the frequency distribution in part (c).
   e) Does the histogram in part (d) appear to represent a normal distribution? Explain.
   f) Use the procedure explained in Exercise 86 to determine whether the set of scores approximates a normal distribution. Explain.

88. Find a value of z such that $z \geq 0$ and 47.5% of the standard normal curve lies between 0 and the z-value.

89. Find a value of z such that $z \leq 0$ and 38.1% of the standard normal curve lies between 0 and the z-value.

Recreational Mathematics

90. Ask your instructor for the class mean and class standard deviation for one of the exams taken by your class. For that exam, calculate the z-score for your exam grade. How many standard deviations is your exam grade away from the mean?

91. If the mean score on a math quiz is 12.0 and 77% of the students in your class scored between 9.6 and 14.4, determine the standard deviation of the quiz scores.

Internet/Research Activity

92. In this project, you actually become the statistician.
   a) Select a project of interest to you in which data must be collected.
   b) Write a proposal and submit it to your instructor for approval. In the proposal, discuss the aims of your project and how you plan to gather the data to make your sample unbiased.
   c) After your proposal has been approved, gather 50 pieces of data by the method you proposed.
   d) Rank the data from smallest to largest.
   e) Compute the mean, median, mode, and midrange.
   f) Determine the range and standard deviation of the data. You may round the mean to the nearest tenth when computing the standard deviation.
   g) Construct a frequency distribution, histogram, frequency polygon, and stem-and-leaf display of your data. Select your first class so that there will be between 5 and 12 classes. Be sure to label your histogram and frequency polygon.
   h) Does your distribution appear to be normal? Explain your answer. Does it appear to be another type of distribution discussed? Explain.
   i) Determine whether your distribution is approximately normal by using the technique discussed in Exercise 86.

13.8 LINEAR CORRELATION AND REGRESSION

In this section, we discuss two important statistical topics: correlation and regression. *Correlation* is used to determine whether there is a relationship between two quantities and, if so, how strong that relationship is. *Regression* is used to determine the equation that relates the two quantities. Although there are other types of correlation and regression, in this section we discuss only linear correlation and linear regression. We begin by discussing linear correlation.

Linear Correlation

Do you believe that there is a relationship between

a) the time a person studied for an exam and the exam grade received?
b) the age of a car and the value of the car?
c) the height and weight of adult males?
d) a person's IQ and income?

Correlation is used to answer questions of this type. The *linear correlation coefficient*, r, is a unitless measure that describes the strength of the linear relationship between two variables. A positive value of r, or a positive correlation, means that as one variable increases, the other variable also increases. A negative value of r, or a negative correlation, means that as one variable increases, the other variable decreases. The correlation coefficient, r, will always be a value between -1 and 1 inclusive. A value of 1 indicates the strongest possible positive correlation, a value of -1 indicates the strongest possible negative correlation, and a value of 0 indicates no correlation (Fig. 13.37).

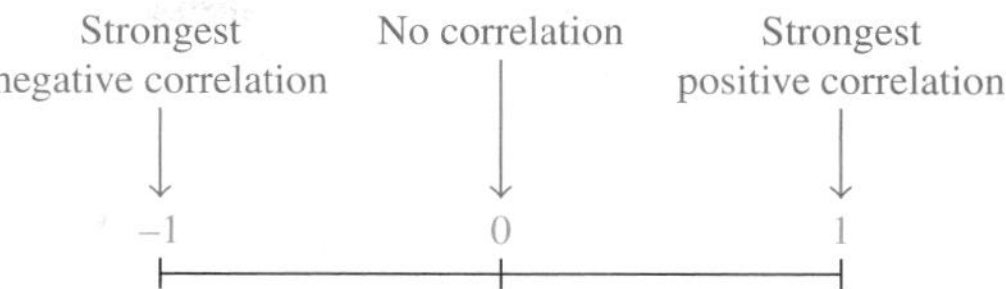

Figure 13.37

A visual aid used with correlation is the *scatter diagram*, a plot of data points. To help understand how to construct a scatter diagram, consider the following data from Egan Electronics. During a 6-day period, Egan Electronics kept daily records of the number of assembly line workers absent and the number of defective parts produced. The information is provided in the following chart.

Day	**1**	**2**	**3**	**4**	**5**	**6**
Number of workers absent	3	5	0	1	2	6
Number of defective parts	15	22	7	12	20	30

For each of the 6 days, two pieces of data are provided: number of workers absent and number of defective parts. We call the set of data *bivariate data*. Often when we have a set of bivariate data, we can control one of the quantities. We generally denote the quantity that can be controlled, the *independent variable*, x. The other variable, the *dependent variable*, is denoted as y. In this problem, we will assume the number of defective parts produced is affected by the number of workers absent. Therefore, we will call the number of workers absent x and the number of defective parts produced y. When we plot bivariate data, the independent variable is marked on the horizontal axis and the dependent variable is marked on the vertical axis. Therefore, for this example, number of workers absent is marked on the horizontal axis and number of defective parts is marked on the vertical axis. If we plot the six pieces of bivariate data in the Cartesian coordinate system, we get a scatter diagram, as shown in Fig. 13.38.

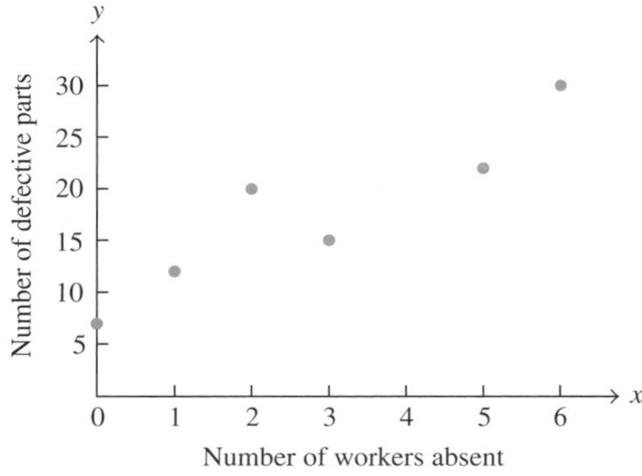

Figure 13.38

Figure 13.38 shows that, generally, the more workers that are absent, the more defective parts are produced.

In Fig. 13.39, we show some scatter diagrams and indicate the corresponding strength of correlation between the quantities on the horizontal and vertical axes.

Earlier we mentioned that r will always be a value between -1 and 1 inclusive. A value of $r = 1$ is obtained only when every point of the bivariate data on a scatter diagram lies in a straight line and the line is increasing from left to right (see Fig. 13.39a). In other words, the line has a positive slope, as discussed in Section 6.6.

A value of $r = -1$ will be obtained only when every point of the bivariate data on a scatter diagram lies in a straight line and the line is decreasing from left to right (see Fig. 13.39e). In other words, the line has a negative slope.

The value of r is a measure of how far a set of points varies from a straight line. The greater the spread, the weaker the correlation and the closer the value of r is to 0. Figure 13.39 shows that the more the dots diverge from a straight line, the weaker the correlation becomes.

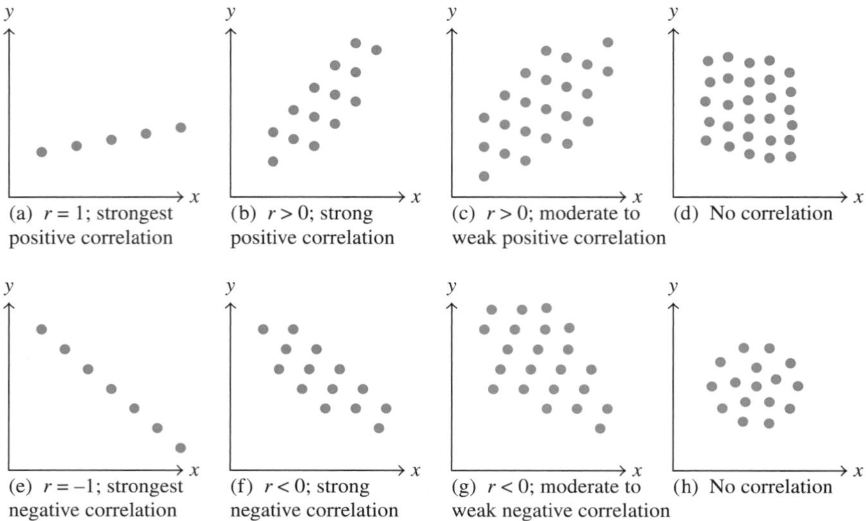

Figure 13.39

The following formula is used to calculate r.

Linear Correlation Coefficient

The formula to calculate the **correlation coefficient, *r*,** is as follows.

$$r = \frac{n(\Sigma xy) - (\Sigma x)(\Sigma y)}{\sqrt{n(\Sigma x^2) - (\Sigma x)^2}\sqrt{n(\Sigma y^2) - (\Sigma y)^2}}$$

To determine the correlation coefficient, r, and the equation of the line of best fit (to be discussed shortly), a statistical calculator may be used. On the calculator you enter the ordered pairs, (x, y), and press the appropriate keys.

In Example 1, we show how to determine r for a set of bivariate data without the use of a statistical calculator. We will use the same set of bivariate data given on page 809 that was used to make the scatter diagram in Figure 13.38.

EXAMPLE 1 ***Number of Absences versus Number of Defective Parts***

Egan Electronics provided the following daily records about the number of assembly line workers absent and the number of defective parts produced for 6 days. Determine the correlation coefficient between the number of workers absent and the number of defective parts produced.

Day	**1**	**2**	**3**	**4**	**5**	**6**
Number of workers absent	3	5	0	1	2	6
Number of defective parts	15	22	7	12	20	30

SOLUTION: We plotted this set of data on the scatter diagram in Figure 13.38. We will call the number of workers absent x. We will call the number of defective parts produced y. We list the values of x and y and calculate the necessary sums: Σx, Σy, Σxy, Σx^2, Σy^2. We determine the values in the column labeled x^2 by squaring the x's (multiplying the x's by themselves). We determine the values in the column labeled y^2 by squaring the y's. We determine the values in the column labeled xy by multiplying each x value by its corresponding y value.

Number of Workers Absent	**Number of Defective Parts**			
x	y	x^2	y^2	xy
3	15	9	225	45
5	22	25	484	110
0	7	0	49	0
1	12	1	144	12
2	20	4	400	40
6	30	36	900	180
17	106	75	2202	387

Thus, $\Sigma x = 17$, $\Sigma y = 106$, $\Sigma x^2 = 75$, $\Sigma y^2 = 2202$, and $\Sigma xy = 387$. In the formula for r, we use both $(\Sigma x)^2$ and Σx^2. Note that $(\Sigma x)^2 = (17)^2 = 289$ and that $\Sigma x^2 = 75$. Similarly, $(\Sigma y)^2 = (106)^2 = 11{,}236$ and $\Sigma y^2 = 2202$.

The n in the formula represents the number of pieces of bivariate data. Here $n = 6$. Now let's determine r.

$$
\begin{aligned}
r &= \frac{n(\Sigma xy) - (\Sigma x)(\Sigma y)}{\sqrt{n(\Sigma x^2) - (\Sigma x)^2}\sqrt{n(\Sigma y^2) - (\Sigma y)^2}} \\
&= \frac{6(387) - (17)(106)}{\sqrt{6(75) - (17)^2}\sqrt{6(2202) - (106)^2}} \\
&= \frac{2322 - 1802}{\sqrt{6(75) - 289}\sqrt{6(2202) - 11{,}236}} \\
&= \frac{520}{\sqrt{450 - 289}\sqrt{13{,}212 - 11{,}236}} \\
&= \frac{520}{\sqrt{161}\sqrt{1976}} \approx 0.92
\end{aligned}
$$

TABLE 13.8 Correlation Coefficient, r

n	$\alpha = 0.05$	$\alpha = 0.01$
4	0.950	0.990
5	0.878	0.959
6	0.811	0.917
7	0.754	0.875
8	0.707	0.834
9	0.666	0.798
10	0.632	0.765
11	0.602	0.735
12	0.576	0.708
13	0.553	0.684
14	0.532	0.661
15	0.514	0.641
16	0.497	0.623
17	0.482	0.606
18	0.468	0.590
19	0.456	0.575
20	0.444	0.561
22	0.423	0.537
27	0.381	0.487
32	0.349	0.449
37	0.325	0.418
42	0.304	0.393
47	0.288	0.372
52	0.273	0.354
62	0.250	0.325
72	0.232	0.302
82	0.217	0.283
92	0.205	0.267
102	0.195	0.254

The derivation of this table is beyond the scope of this text. It shows the critical values of the Pearson correlation coefficient.

Since the maximum possible value for r is 1.00, a correlation coefficient of 0.92 is a strong, positive correlation. This result implies that, generally, the more assembly line workers absent, the more defective parts produced. ▲

In Example 1, had we found r to be a value greater than 1 or less than -1, it would have indicated that we had made an error. Also, from the scatter diagram, we should realize that r should be a positive value and not negative.

In Example 1, there appears to be a cause–effect relationship. That is, the more assembly line workers who are absent, the more defective parts are produced. *However, a correlation does not necessarily indicate a cause–effect relationship.* For example, there is a positive correlation between police officers' salaries and the cost of medical insurance over the past 10 years (both have increased), but that does not mean that the increase in police officers' salaries caused the increase in the cost of medical insurance.

Suppose in Example 1 that r had been 0.53. Would this value have indicated a correlation? What is the minimum value of r needed to assume that a correlation exists between the variables? To answer this question, we introduce the term *level of significance*. The *level of significance*, denoted α (alpha), is used to identify the cutoff between results attributed to chance and results attributed to an actual relationship between the two variables. Table 13.8 gives *critical values** (or cutoff values) that are sometimes used for determining whether two variable are related. The table indicates two different levels of significance: $\alpha = 0.05$ and $\alpha = 0.01$. A level of significance of 5%, written $\alpha = 0.05$, means that there is a 5% chance that, when you say the variables are related, they actually are *not* related. Similarly, a level of significance of 1%, or $\alpha = 0.01$, means that there is a 1% chance that, when you say the variables are related, they actually are *not* related. More complete critical value tables are available in statistics books.

To explain the use of the table, we use *absolute value*, symbolized $|\ |$. The absolute value of a nonzero number is the positive value of the number and the absolute value of 0 is 0. Therefore,

$$|3| = 3, \qquad |-3| = 3, \qquad |5| = 5, \qquad |-5| = 5, \qquad \text{and} \qquad |0| = 0$$

If the absolute value of r, written $|r|$, is *greater than* the value given in the table under the specified α and appropriate sample size n, we assume that a correlation does exist between the variables. If $|r|$ is less than the table value, we assume that no correlation exists.

Returning to Example 1, if we want to determine whether there is a correlation at a 5% level of significance, we find the critical value (or cutoff value) that corresponds to $n = 6$ (there are 6 pieces of bivariate data) and $\alpha = 0.05$. The value to the right of $n = 6$ and under the $\alpha = 0.05$ column is the critical value 0.811. From the formula, we had obtained $r = 0.92$. Since $|0.92| > 0.811$, or $0.92 > 0.811$, we assume that a correlation between the variables exists.

Note in Table 13.8 that the larger the sample size, the smaller is the value of r needed for a significant correlation.

EXAMPLE 2 *Amount of Drug Remaining in the Bloodstream*

To test the length of time that an infection-fighting drug stays in a person's bloodstream, a doctor gives 300 milligrams of the drug to 10 patients labeled 1–10 in the table on page 813. Once each hour, for 8 hours, one of the 10 patients is selected at

*This table of values may be used only under certain conditions. If you take a statistics course, you will learn more about which critical values to use to determine whether a linear correlation exists.

random and that person's blood is tested to determine the amount of the drug remaining in the bloodstream. The results are as follows.

Patient	1	2	3	4	5	6	7	8	9	10
Time (hr)	1	2	3	4	5	6	7	8	9	10
Drug remaining (mg)	250	230	200	210	140	120	210	100	90	85

Determine at a level of significance of 5% whether a correlation exists between the time elapsed and the amount of drug remaining.

SOLUTION: Let time be represented by x and the amount of drug remaining by y. We first draw a scatter diagram (Fig. 13.40).

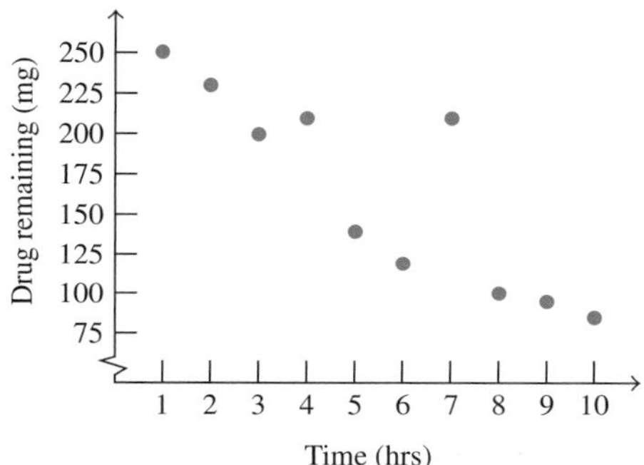

Figure 13.40

The scatter diagram suggests that, if a correlation exists, it will be negative. We now construct a table of values and calculate r.

x	y	x^2	y^2	xy
1	250	1	62,500	280
2	230	4	52,900	460
3	200	9	40,000	600
4	210	16	44,100	840
5	140	25	19,600	700
6	120	36	14,400	720
7	210	49	44,100	1470
8	100	64	10,000	800
9	90	81	8100	810
10	85	100	7225	850
55	1635	385	302,925	7500

$$
\begin{aligned}
r &= \frac{n(\Sigma xy) - (\Sigma x)(\Sigma y)}{\sqrt{n(\Sigma x^2) - (\Sigma x)^2}\sqrt{n(\Sigma y^2) - (\Sigma y)^2}} \\
&= \frac{10(7500) - (55)(1635)}{\sqrt{10(385) - (55)^2}\sqrt{10(302{,}925) - (1635)^2}} \\
&= \frac{-14{,}925}{\sqrt{825}\sqrt{355{,}025}} \approx \frac{-14{,}925}{17{,}114.19} \\
&\approx -0.872
\end{aligned}
$$

From Table 13.8, for $n = 10$ and $\alpha = 0.05$, we get 0.632. Since $|-0.872| = 0.872$ and $0.872 > 0.632$, a correlation exists. The correlation is negative, which indicates that the longer the time period, the smaller is the amount of drug remaining. ▲

Linear Regression

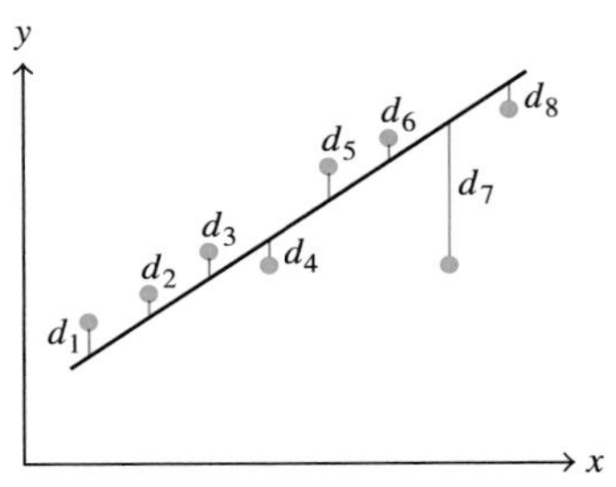

Figure 13.41

Let's now turn to regression. *Linear regression* is the process of determining the linear relationship between two variables. Recall from Section 6.6 that the slope–intercept form of a straight line is $y = mx + b$, where m is the slope and b is the y-intercept.

Using the set of bivariate data, we will determine the equation of *the line of best fit*. The line of best fit is also called *the regression line*, or *the least squares line*. The *line of best fit* is the line such that the sum of the vertical distances from the line to the data points (on the scatter diagram) is a minimum, as shown in Fig. 13.41. In Fig. 13.41, the line of best fit minimizes the sum of d_1 through d_8. To determine the equation of the line of best fit, $y = mx + b$ * we must find m and then b. The formulas for finding m and b are as follows.

The equation of the line of best fit is

$$y = mx + b,$$

$$\text{where} \quad m = \frac{n(\Sigma xy) - (\Sigma x)(\Sigma y)}{n(\Sigma x^2) - (\Sigma x)^2}, \quad \text{and} \quad b = \frac{\Sigma y - m(\Sigma x)}{n}$$

Note that the numerator of the fraction used to find m is identical to the numerator used to find r. Therefore, if you have previously found r, you do not need to repeat this calculation. Also, the denominator of the fraction used to find m is identical to the radicand of the first square root in the denominator of the fraction used to find r.

EXAMPLE 3 *The Line of Best Fit*

a) Use the data in Example 1 to find the equation of the line of best fit that relates the number of workers absent on an assembly line and the number of defective parts produced.

b) Graph the equation of the line of best fit on a scatter diagram that illustrates the set of bivariate points.

SOLUTION:

a) In Example 1, we found $n(\Sigma xy) - (\Sigma x)(\Sigma y) = 520$ and $n(\Sigma x^2) - (\Sigma x)^2 = 161$. Thus,

$$m = \frac{n(\Sigma xy) - (\Sigma x)(\Sigma y)}{n(\Sigma x^2) - (\Sigma x)^2} = \frac{520}{161} \approx 3.23$$

*Some statistics books use $y = ax + b$, $y = b_0 + b_1x$, or something similar for the equation of the line of best fit. In any case, the letter next to the variable x represents the slope of the line of best fit, and the other letter represents the y-intercept of the graph.

Now we find the y-intercept, b. In Example 1, we found $n = 6$, $\Sigma x = 17$, and $\Sigma y = 106$.

$$b = \frac{\Sigma y - m(\Sigma x)}{n}$$

$$\approx \frac{106 - 3.23(17)}{6} \approx \frac{51.09}{6} \approx 8.52$$

Therefore, the equation of the line of best fit is

$$y = mx + b$$
$$y = 3.23x + 8.52$$

where x represents the number of workers absent and y represents the predicted number of defective parts produced.

b) To graph $y = 3.23x + 8.52$, we need to plot at least two points. We will plot three points and then draw the graph.

	$y = 3.23x + 8.52$
$x = 2$	$y = 3.23(2) + 8.52 = 14.98$
$x = 4$	$y = 3.23(4) + 8.52 = 21.44$
$x = 6$	$y = 3.23(6) + 8.52 = 27.90$

x	y
2	14.98
4	21.44
6	27.90

These three calculations indicate that if 2 assembly line workers are absent on the assembly line, the predicted number of defective parts produced is about 15. If 4 assembly line workers are absent, the predicted number of defective parts produced is about 21, and if 6 assembly line workers are absent, the predicted number of defective parts produced is about 28. Plot the three points (the three red points in Figure 13.42) and then draw a straight line through the three points. The scatter diagram and graph of the equation of the line of best fit are plotted in Fig. 13.42.

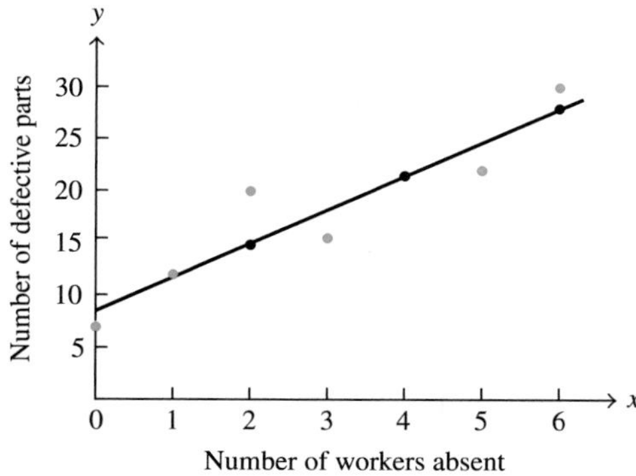

Figure 13.42

In Example 3, the line of best fit intersects the y-axis at 8.52, the value we determined for b in part (a).

EXAMPLE 4 *Line of Best Fit for Example 2*

a) Determine the equation of the line of best fit between the time elapsed and the amount of drug remaining in a person's bloodstream in Example 2 on pages 812 and 813.

b) If the average person is given 300 mg of the drug, how much will remain in the person's bloodstream after 5 hr?

SOLUTION:

a) From the scatter diagram on page 813 we see that the slope of the line of best fit, m, will be negative. In Example 2, we found that $n(\Sigma xy) - (\Sigma x)(\Sigma y) = -14{,}925$ and that $n(\Sigma x^2) - (\Sigma x)^2 = 825$. Thus,

$$m = \frac{n(\Sigma xy) - (\Sigma x)(\Sigma y)}{n(\Sigma x^2) - (\Sigma x)^2} = \frac{-14{,}925}{825} \approx -18.1$$

From Example 2, $n = 10$, $\Sigma x = 55$, and $\Sigma y = 1635$.

$$b = \frac{\Sigma y - m\Sigma x}{n} = \frac{1635 - (-18.1)(55)}{10} \approx 263.1$$

Thus, the equation of the line of best fit is

$$y = mx + b$$
$$y = -18.1x + 263.1$$

where x is the elapsed time and y is the amount of drug remaining.

b) We evaluate $y = -18.1x + 263.1$ at $x = 5$.

$$y = -18.1x + 263.1$$
$$y = -18.1(5) + 263.1 = 172.6$$

Thus, after 5 hr, about 173 mg of the drug remains in the average person's bloodstream. ▲

SECTION 13.8 EXERCISES

Concept/Writing Exercises

1. What does the correlation coefficient measure?
2. What is the purpose of regression?
3. What value of r represents the maximum positive correlation?
4. What value of r represents the maximum negative correlation?
5. What value of r represents no correlation between the variables?
6. What does a negative correlation between two variables indicate?

7. What does a positive correlation between two variables indicate?

8. What does the line of best fit represent?

9. What does the level of significance signify?

10. What is a scatter diagram?

In Exercises 11–14, indicate if you believe that a correlation exists between the quantities on the horizontal and vertical axis. If so, indicate if you believe that the correlation is a strong positive correlation, a strong negative correlation, a weak positive correlation, a weak negative correlation, or no correlation. Explain your answer.

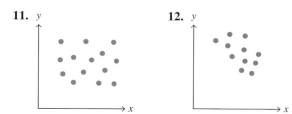

11. **12.**

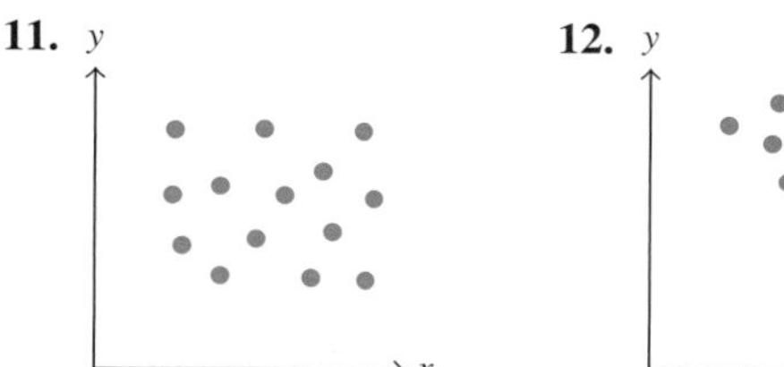

13. **14.**

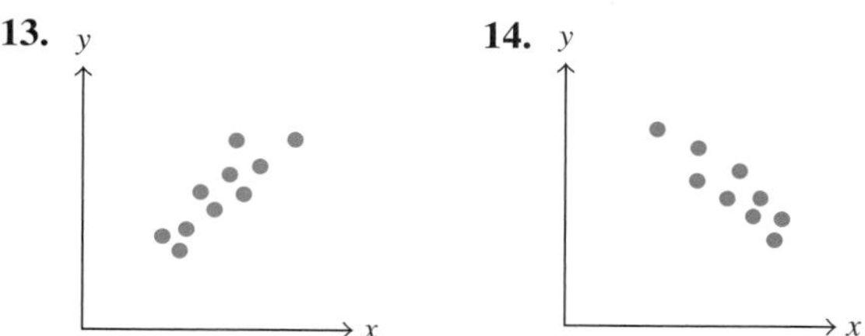

Practice the Skills

In Exercises 15–22, assume that a sample of bivariate data yields the correlation coefficient, r, indicated. Use Table 13.8 on page 812 for the specified sample size and level of significance to determine whether a linear correlation exists.

15. $r = 0.76$ when $n = 13$ at $\alpha = 0.01$

16. $r = 0.43$ when $n = 22$ at $\alpha = 0.01$

17. $r = -0.73$ when $n = 8$ at $\alpha = 0.05$

18. $r = -0.49$ when $n = 11$ at $\alpha = 0.05$

19. $r = -0.23$ when $n = 102$ at $\alpha = 0.01$

20. $r = -0.49$ when $n = 18$ at $\alpha = 0.01$

21. $r = 0.82$ when $n = 6$ at $\alpha = 0.01$

22. $r = 0.96$ when $n = 5$ at $\alpha = 0.01$

In Exercises 23–30, (a) draw a scatter diagram; (b) find the value of r, rounded to the nearest thousandth; (c) determine whether a correlation exists at $\alpha = 0.05$; and (d) determine whether a correlation exists at $\alpha = 0.01$.

23.

x	y
4	6
5	9
6	11
7	11
10	13

24.

x	y
6	12
8	11
11	9
14	10
17	8

25.

x	y
23	29
35	37
31	26
43	20
49	39

26.

x	y
90	3
80	4
60	6
60	5
40	5
20	7

27.

x	y
5.3	10.3
4.7	9.6
8.4	12.5
12.7	16.2
4.9	9.8

28.

x	y
12	15
16	19
13	45
24	30
100	60
50	28

29.

x	y
100	2
80	3
60	5
60	6
40	6
20	8

30.

x	y
90	90
70	70
65	65
60	60
50	50
40	40
15	15

In Exercises 31–38, determine the equation of the line of best fit from the data in the exercise indicated. Round both the slope and the y-intercept to the nearest tenth.

31. Exercise 23

32. Exercise 24

33. Exercise 25

34. Exercise 26

35. Exercise 27

36. Exercise 28

37. Exercise 29

38. Exercise 30

Problem Solving

39. *Commuting Time* Six students provided the following data about the distance (in miles) from their home to their college and the time (in minutes) required to commute from their home to their college.

Distance (miles)	8	20	9	15	16	2
Time (minutes)	15	28	20	25	28	5

a) Determine the correlation coefficient between the distance and the time to commute.

b) Determine whether a correlation exists at $\alpha = 0.05$.

c) Find the equation of the line of best fit for the distance and the time to commute.

40. *Amount of Fat in Pizza* The January 2002 issue of *Consumer Reports* reported the following information regarding the number of calories and the number of grams of fat in a 5 oz slice of pizza for the top-rated pizza chains.

Calories	321	380	350	358	378	391
Fat (grams)	13	23	16	14	19	19

a) Determine the correlation coefficient between the number of calories and the number of grams of fat.
b) Determine whether a correlation exists at $\alpha = 0.05$.
c) Find the equation of the line of best fit for the number of calories and the number of grams of fat.

41. *Amount of Time Spent Studying* Six students provided the following data about the lengths of time they studied for a psychology exam and the grades they received on the exam.

Time studied (minutes)	20	40	50	60	80	100
Grade received (percent)	40	45	70	76	92	95

a) Determine the correlation coefficient between the length of time studied and the grade received.
b) Determine whether a correlation exists at $\alpha = 0.01$.
c) Find the equation of the line of best fit for the length of time studied and the grade received.

42. *Hiking* The following table shows the number of hiking permits issued for a specific trail at Yellowstone National Park for selected years and the corresponding number of mountain lions sighted by the hikers on that trail.

Hiking permits	765	926	1145	842	1485	1702
Mountain lions	119	127	150	119	153	156

a) Determine the correlation coefficient between the number of hiking permits issued and the number of mountain lions sighted by hunters.
b) Determine whether a correlation exists at $\alpha = 0.05$.
c) Determine the equation of the line of best fit for the number of hiking permits issued and the number of mountain lions sighted by hunters.
d) Use the equation in part (c) to estimate the number of mountain lions sighted by hunters if 1500 hiking permits were issued.

43. *City Muggings* In a certain section of a city, muggings have been a problem. The number of police officers patrolling that section of the city has varied. The following chart shows the number of police officers and the number of muggings for 10 successive days.

Police officers	20	12	18	15	22	10	20	12
Muggings	8	10	12	9	6	15	7	18

a) Determine the correlation coefficient for number of police officers and number of muggings.
b) Determine whether a correlation exists at $\alpha = 0.05$.
c) Find the equation of the line of best fit for number of police officers and number of muggings.
d) Use the equation in part (c) to estimate the average number of muggings when 14 police officers are patrolling that section of the city.

44. *Higher Education Enrollment* The following table shows the projected enrollment in higher education, in millions, for the years 2000–2005.

Year	2000	2001	2002	2003	2004	2005
Enrollment (in millions)	15.0	15.3	15.5	15.8	16.1	16.3

Source: U.S. Center for Educational Statistics

a) Determine the correlation coefficient between the year and the projected enrollment in higher education.
b) Determine whether a correlation exists at $\alpha = 0.05$.
c) Find the equation of the line of best fit for the year and the projected enrollment in higher education.
d) Use the equation in part (c) to estimate the projected enrollment in higher education in 2008.

45. *Selling Popcorn at the Movies* The number of movie tickets sold and the number of units of popcorn sold at Regal Cinema for 8 days is shown below.

Ticket sales	89	110	125	92	100	95	108	97
Units of popcorn	22	28	30	26	22	21	28	25

a) Determine the correlation coefficient between ticket sales and units of popcorn sold.
b) Determine whether a correlation exists at $\alpha = 0.05$.
c) Determine the equation of the line of best fit for tickets sold and units of popcorn sold.
d) Use the equation in part (c) to estimate the units of popcorn sold if 115 tickets are sold.

46. *Movie Ratings* A popular newspaper rates movies from one to four stars (four stars is the highest rating). The following table shows the ratings of 10 movies selected at random and the gross earnings of each movie.

Rating (stars)	4	4	3	2	1	3	4	2	4	1
Earnings (millions of dollars)	100	67	80	120	40	90	60	60	90	100

a) Determine the correlation coefficient between number of stars and the movies' earnings.
b) Determine whether a correlation exists at $\alpha = 0.05$.
c) Determine the equation of the line of best fit for the number of stars and the movies' earnings.

47. *Chlorine in a Swimming Pool* A gallon of chlorine is put into a swimming pool. Each hour later for the following 6 hr the percent of chlorine that remains in the pool is measured. The following information is obtained.

Time	1	2	3	4	5	6
Chlorine remaining (percent)	80.0	76.2	68.7	50.1	30.2	20.8

a) Determine the correlation coefficient for time and percent of chlorine remaining.
b) Determine whether a correlation exists at $\alpha = 0.01$.
c) Determine the equation of the line of best fit for time and amount of chlorine remaining.
d) Use the equation in part (c) to estimate the average amount of chlorine remaining after 4.5 hr.

48. *Social Security Numbers* **a)** Match the first 9 digits in your phone number (including area code) with the 9 digits in your social security number. To do so, match the first digit in your phone number with the first digit in your social security number to get one ordered pair. Match the second digits to get a second ordered pair. Continue this process until you get a total of nine ordered pairs.
b) Do you believe that this set of bivariate data has a positive correlation, a negative correlation, or no correlation? Explain your answer.
c) Construct a scatter diagram for the nine ordered pairs.
d) Calculate the correlation coefficient, r.
e) Is there a correlation at $\alpha = 0.05$? Explain.
f) Calculate the equation of the line of best fit.
g) Use the equation in part (f) to estimate the digit in a social security number that corresponds with a 7 in a telephone number.

49. *Hitting the Brakes* **a)** Examine the art below. Do you believe that there is a positive correlation, a negative correlation, or no correlation between speed of a car and stopping distance when the brakes are applied? Explain.
b) Do you believe that there is a stronger correlation between speed of a car and stopping distance on wet or dry roads? Explain.
c) Use the figure to construct two scatter diagrams, one for dry pavement and the other for wet pavement. Place the speed of the car on the horizontal axis.
d) Compute the correlation coefficient for speed of the car and stopping distance for dry pavement.
e) Repeat part (d) for wet pavement.
f) Were your answers to parts (a) and (b) correct? Explain.
g) Determine the equation of the line of best fit for dry pavement.
h) Repeat part (g) for wet pavement.

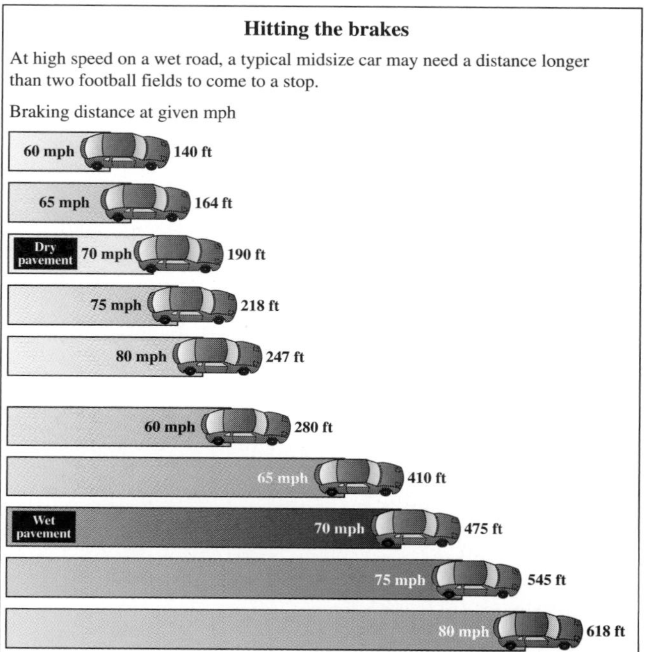

Source: Car and Driver, American Automobile Association

i) Use the equations in parts (g) and (h) to estimate the stopping distance of a car going 77 mph on both dry and wet pavements.

Challenge Problems/Group Activities

50. a) Assume that a set of bivariate data yields a specific correlation coefficient. If the x and y values are interchanged and the correlation coefficient is recalculated, will the correlation coefficient change? Explain.
b) Make up a table of five pieces of bivariate data and determine r using the data. Then switch the values of the x's and y's and recompute the correlation coefficient. Has the value of r changed?

51. a) Do you believe that a correlation exists between a person's height and the length of a person's forearm? Explain.
b) Select 10 people from your class and measure (in inches) their heights and the lengths of their forearms.
c) Plot the 10 ordered pairs on a scatter diagram.
d) Calculate the correlation coefficient, r.
e) Determine the equation of the line of best fit.
f) Estimate the length of the forearm of a person who is 58 in. tall.

52. a) Have your group select a category of bivariate data that it thinks has a strong positive correlation. Designate the independent variable and the dependent variable. Explain why your group believes that the bivariate data have a strong positive correlation.
b) Collect at least 10 pieces of bivariate data that can be used to determine the correlation coefficient. Explain how your group chose these data.
c) Plot a scatter diagram.
d) Calculate the correlation coefficient.
e) Does there appear to be a strong positive correlation? Explain your answer.
f) Calculate the equation of the line of best fit.
g) Explain how the equation in part (f) may be used.

53. Use the following table. CPI represents consumer price index.

Year	1996	1997	1998	1999	2000	2001
CPI	157	161	163	167	172	177

a) Calculate r.
b) If 1996 is subtracted from each year, the table obtained becomes:

Year	0	1	2	3	4	5
CPI	157	161	163	167	172	177

If r is calculated from these values, how will it compare with the r determined in part (a)? Explain.
c) Calculate r from the values in part (b) and compare the results with the value of r found in part (a). Are they the same? If not, explain why.

54. a) There are equivalent formulas that can be used to find the correlation coefficient and the equation of the line of best fit. A formula used in some statistics books to find the correlation coefficient is

$$r = \frac{SS(xy)}{\sqrt{SS(x)SS(y)}}$$

where

$$SS(x) = \Sigma x^2 - \frac{(\Sigma x)^2}{n}$$

$$SS(y) = \Sigma y^2 - \frac{(\Sigma y)^2}{n}$$

and $$SS(xy) = \Sigma xy - \frac{(\Sigma x)(\Sigma y)}{n}$$

Use this formula to find the correlation coefficient of the set of bivariate data given in Example 1 on page 811.
b) Compare your answer with the answer obtained in Example 1.

Internet/Research Activities

55. a) Obtain a set of bivariate data from a newspaper or magazine.
b) Plot the information on a scatter diagram.
c) Indicate whether you believe that the data show a positive correlation, a negative correlation, or no correlation. Explain your answer.
d) Calculate r and determine whether your answer to part (c) was correct.
e) Determine the equation of the line of best fit for the bivariate data.

56. Find a scatter diagram in a newspaper or magazine and write a paper on what the diagram indicates. Indicate whether you believe that the bivariate data show a positive correlation, a negative correlation, or no correlation and explain why.

CHAPTER 13 SUMMARY

IMPORTANT FACTS

Rules for data grouped by classes

1. The classes should be the same width.
2. The classes should not overlap.
3. Each piece of data should belong to only one class.

Measures of central tendency

The **mean** is the sum of the data divided by the number of pieces of data: $\bar{x} = \dfrac{\Sigma x}{n}$.

The **median** is the value in the middle of a set of ranked data.

The **mode** is the piece of data that occurs most frequently (if there is one).

The **midrange** is the value halfway between the lowest and highest values: midrange $= \dfrac{L + H}{2}$.

Statistical graphs

Circle graph

Histogram

Frequency polygon

Stem-and-leaf display

Measures of dispersion

The **range** is the difference between the highest value and lowest value in a set of data.

The **standard deviation,** s, is a measure of the spread of a set of data about the mean: $s = \sqrt{\dfrac{\Sigma(x - \bar{x})^2}{n - 1}}$.

z-scores

$$z = \frac{x - \mu}{\sigma}$$

Chebyshev's theorem

$$\text{Minimum percentage} = 1 - \frac{1}{K^2},\ K > 1$$

Linear correlation and regression

Linear correlation coefficient, r, is

$$r = \frac{n(\Sigma xy) - (\Sigma x)(\Sigma y)}{\sqrt{n(\Sigma x^2) - (\Sigma x)^2}\sqrt{n(\Sigma y^2) - (\Sigma y)^2}}$$

Equation of the line of the best fit

$y = mx + b$, where

$$m = \frac{n(\Sigma xy) - (\Sigma x)(\Sigma y)}{n(\Sigma x^2) - (\Sigma x)^2}, \text{ and}$$

$$b = \frac{\Sigma y - m(\Sigma x)}{n}$$

CHAPTER 13 REVIEW EXERCISES

13.1

1. **a)** What is a population?
   b) What is a sample?
2. What is a random sample?

13.2

In Exercises 3 and 4, tell what possible misuses or misinterpretations may exist in the statements.

3. The Stay Healthy Candy Bar indicates on its label that it has no cholesterol. Therefore, it is safe to eat as many of these candy bars as you want.
4. More copies of *Time* magazine are sold than are copies of *Money* magazine. Therefore, *Time* is a more profitable magazine than *Money*.
5. Network News For the week of July 15–21, 2002, ABC's *World News Tonight* had 9.1 million viewers and NBC's

Nightly News had 8.9 million viewers. Draw a graph that appears to show
a) small difference in the number of viewers.
b) a large difference in the number of viewers.

13.3, 13.4

6. Consider the following set of data.

35	37	38	41	43
36	37	38	41	43
36	37	39	41	43
36	37	39	41	44
37	37	39	42	45

a) Construct a frequency distribution letting each class have a width of 1.
b) Construct a histogram.
c) Construct a frequency polygon.

7. *Average Daily High Temperature* Consider the following average daily high temperature in January for 40 selected cities.

87	41	54	34	89	83	65	80
69	75	47	43	68	67	48	35
86	30	56	44	50	33	77	77
42	84	75	54	81	86	79	66
48	42	73	83	36	88	66	55

a) Construct a frequency distribution. Let the first class be 30–39.
b) Construct a histogram of the frequency distribution.
c) Construct a frequency polygon of the frequency distribution.
d) Construct a stem-and-leaf display.

13.5, 13.6

In Exercises 8–13, for the following test scores 63, 76, 79, 83, 86, 93, determine the

8. mean.
9. median.
10. mode.
11. midrange.
12. range.
13. standard deviation.

In Exercises 14–19, for the set of data 4, 5, 12, 14, 19, 7, 12, 23, 7, 17, 15, 21, determine the

14. mean.
15. median.
16. mode.
17. midrange.
18. range.
19. standard deviation.

13.7

Anthropology In Exercises 20–24, assume that anthropologists have determined that a certain type of primitive animal had a mean head circumference of 42 cm with a standard deviation of 5 cm. Given that head sizes were normally distributed, determine the percent of heads

20. between 37 and 47 cm.
21. between 32 and 52 cm.
22. less than 50 cm.
23. greater than 50 cm.
24. greater than 39 cm.

Pizza Delivery In Exercises 25–28, assume that the amount of time to prepare and deliver a pizza from Dominos is normally distributed with a mean of 20 min and standard deviation of 5 min. Find the percent of pizzas that were prepared and delivered

25. between 20 and 25 min.
26. in less than 18 min.
27. between 22 and 28 min.
28. If Dominos advertises that the pizza is free if it takes more than 30 min to deliver, what percent of the pizza will be free?

13.8

In Exercises 29 and 30, use the following table that shows both the U.S. sales of flowers and plants and the number of greenhouse owners in New York State, where the column labeled as Year refers to the number of years since 1997. Thus 1997 would correspond to year 0.

Year	U.S. Flower and Plant Sales (billions)	Greenhouse Owners in New York State
0	$3.9	897
1	3.9	800
2	4.1	770
3	4.6	760
4	4.8	735
5	4.9	663

29. a) *Flower and Plant Sales* Construct a scatter diagram for the U.S. sales of flowers and plants with the years, from 0 through 5, on the horizontal axis.
b) Use the scatter diagram in part (a) to determine whether you believe a correlation exists between the year and U.S. sales of flowers and plants. If so, is it a positive or a negative correlation? Explain.
c) Calculate the correlation coefficient between the year and the U.S. sales of flowers and plants.
d) Determine whether a correlation exists at $\alpha = 0.05$.
e) Determine the equation of the line of best fit between the year and U.S. sales of flowers and plants.

30. *Greenhouse Owners in New York State*
a) Construct a scatter diagram for the number of greenhouse owners in New York State with the years, from 0 through 5, on the horizontal axis.

b) Use the scatter diagram in part (a) to determine whether you believe a correlation exists between the year and the number of greenhouse owners in New York State. If so, is it a positive or a negative correlation? Explain.
c) Calculate the correlation coefficient between the year and the number of greenhouse owners in New York State.
d) Determine whether a correlation exists at $\alpha = 0.05$.
e) Determine the equation of the line of best fit between the year and the number of greenhouse owners in New York State.

31. *Daily Sales* Ace Hardware recorded the number of a particular item sold per week for 6 weeks and the corresponding weekly price, in dollars, of the item as shown in the table below.

Price ($)	0.75	1.00	1.25	1.50	1.75	2.00
Number sold	200	160	140	120	110	95

a) Construct a scatter diagram with price on the horizontal axis.
b) Use the scatter diagram in part (a) to determine whether you believe that a correlation exists between the price of the item and number sold. If so, it is a positive or a negative correlation? Explain.
c) Determine the correlation coefficient between the price and the number sold.
d) Determine whether a correlation exists at $\alpha = 0.05$. Explain how you arrived at your answer.
e) Determine the equation of the line of best fit for the price and the number sold.
f) Use the equation in part (e) to estimate the number sold if the price is $1.60.

13.5–13.7

Men's Weight In Exercises 32–39, use the following data obtained from a study of the weights of adult men.

Mean	187 lb	First quartile	173 lb
Median	180 lb	Third quartile	227 lb
Mode	175 lb	86th percentile	234 lb
Standard deviation	23 lb		

32. What is the most common weight?

33. What weight did half of those surveyed exceed?

34. About what percent of those surveyed weighed more than 227 lb?

35. About what percent of those surveyed weighed less than 173 lb?

36. About what percent of those surveyed weighed more than 234 lb?

37. If 100 men were surveyed, what is the total weight of all men?

38. What weight represents two standard deviations above the mean?

39. What weight represents 1.8 standard deviations below the mean?

13.2–13.7

Presidential Children The following list shows the names of the 42 U.S. presidents and the number of children in their families.

Washington	0	Cleveland	5
J. Adams	5	B. Harrison	3
Jefferson	6	McKinley	2
Madison	0	T. Roosevelt	6
Monroe	2	Taft	3
J. Q. Adams	4	Wilson	3
Jackson	0	Harding	0
Van Buren	4	Coolidge	2
W. H. Harrison	10	Hoover	2
Tyler	14	F. D. Roosevelt	6
Polk	0	Truman	1
Taylor	6	Eisenhower	2
Fillmore	2	Kennedy	3
Pierce	3	L. B. Johnson	2
Buchanan	0	Nixon	2
Lincoln	4	Ford	4
A. Johnson	5	Carter	4
Grant	4	Reagan	4
Hayes	8	G. Bush	6
Garfield	7	Clinton	1
Arthur	3	G. W. Bush	2

In Exercises 40–51, use the data above to determine the following.

40. Mean number of children

41. Mode

42. Median

43. Midrange

44. Range

45. Standard deviation (round the mean to the nearest tenth)

46. Construct a frequency distribution; let the first class be 0–1.

47. Construct a histogram.

48. Construct a frequency polygon.

49. Does this distribution appear to be normal? Explain.

50. On the basis of this sample, do you think the number of children per family in the United States is a normal distribution? Explain.

51. Do you believe that this sample is representative of the population? Explain.

CHAPTER 13 TEST

In Exercises 1–6, for the set of data 21, 37, 37, 39, 46, determine the

1. mean.
2. median.
3. mode.
4. midrange.
5. range.
6. standard deviation.

In Exercises 7–9, use the set of data

26	28	35	46	49	56
26	30	36	46	49	58
26	32	40	47	50	58
26	32	44	47	52	62
27	35	46	47	54	66

to construct

7. a frequency distribution; let the first class be 25–30.

8. a histogram of the frequency distribution.

9. a frequency polygon of the frequency distribution.

Statistics on Salaries In Exercises 10–16, use the following data on weekly salaries at Maxwell Mechanical Contractors.

Mean	\$700	First quartile	\$650
Median	\$670	Third quartile	\$705
Mode	\$695	79th percentile	\$712
Standard deviation	\$40		

10. What is the most common salary?

11. What salary did half the employees exceed?

12. About what percent of employees' salaries exceeded \$650?

13. About what percent of employees' salaries was less than \$712?

14. If the company has 100 employees, what is the total weekly salary of all employees?

15. What salary represents one standard deviation above the mean?

16. What salary represents 1.5 standard deviations below the mean?

Mileage of 5-Year-Old Cars In Exercises 17–20, the mileage of 5-year-old cars is normally distributed with a mean of 75,000 and a standard deviation of 12,000 miles.

17. What percent of 5-year-old cars have mileage between 50,000 and 70,000 miles?

18. What percent of 5-year-old cars have mileage greater than 60,000 miles?

19. What percent of 5-year-old cars have mileage greater than 90,000 miles?

20. If a random sample of 300 five-year-old cars is selected, how many would have mileage between 60,000 and 70,000 miles?

21. *The Elderly U.S. Population* The following chart shows the percent of the U.S. population that was age 65 and over for the years 1970, 1980, 1990, 1995, and 2000, where the column labeled Year refers to the number of years since 1970.

Percent of U.S. Population Age 65 and Over

Year	**Percent**
0	9.8
10	11.3
20	12.5
25	12.8
30	12.4

Source: U.S. Bureau of the Census, U.S. Dept. of Commerce.

a) Construct a scatter diagram placing the year on the horizontal axis.
b) Use the scatter diagram in part (a) to determine whether you believe that a correlation exists between the year and the percent of the U.S. population age 65 and over. Explain.
c) Determine the correlation coefficient between the year and the percent of the U.S. population age 65 and over.
d) Determine whether a correlation exists at $\alpha = 0.05$.
e) Assuming that this trend continues, determine the equation of the line of best fit between the year and the percent of the U.S. population age 65 and over.
f) Use the equation in part (e) to predict the percent of the U.S. population age 65 and over in 2010, or 40 years after 1970.

GROUP PROJECTS

Watching TV

1. Do you think that men or women, aged 17–20, watch more hours of TV weekly, or do you think that they watch the same number of hours?
   a) Write a procedure to use to determine the answer to that question. In your procedure, use a sample of 30 men and 30 women. State how you will obtain an unbiased sample.
   b) Collect 30 pieces of data from men aged 17–20 and 30 pieces of data from women aged 17–20. Round answers to the nearest 0.5 hr. Follow the procedure developed in part (a) to obtain your unbiased sample.
   c) Compute the mean for your two groups of data to the nearest tenth.
   d) Using the means obtained in part (c), answer the question asked at the beginning of the problem.
   e) Is it possible that your conclusion in part (d) is wrong? Explain.
   f) Compute the standard deviation for each group to the nearest tenth. How do the standard deviations compare?
   g) Do you believe that the distribution of data from either or both groups resembles a normal distribution? Explain.
   h) Add the two groups of data to get one group of 60 pieces of data. If these 60 pieces of data are added and divided by 60, will you obtain the same mean as when you add the two means from part (c) and divide the sum by 2? Explain.
   i) Compute the mean of the 60 pieces of data by using both methods mentioned in part (h). Are they the same? If so, why? If not, why not?
   j) Do you believe that this group of 60 pieces of data represents a normal distribution? Explain.

Binomial Probability Experiment

2. **a)** Have your group select a category of bivariate data that it thinks has a strong negative correlation. Indicate the variable that you will designate as the independent variable and the variable that you will designate as the dependent variable. Explain why your group believes that the bivariate data have a strong negative correlation.
   b) Collect at least 10 pieces of bivariate data that can be used to determine the correlation coefficient. Explain how your group chose these data.
   c) Plot a scatter diagram.
   d) Calculate the correlation coefficient.
   e) Is there a negative correlation at $\alpha = 0.05$? Explain your answer.
   f) Calculate the equation of the line of best fit.
   g) Explain how the equation in part (f) may be used.

APPENDIX: GRAPH THEORY

DID YOU KNOW

Making Connections

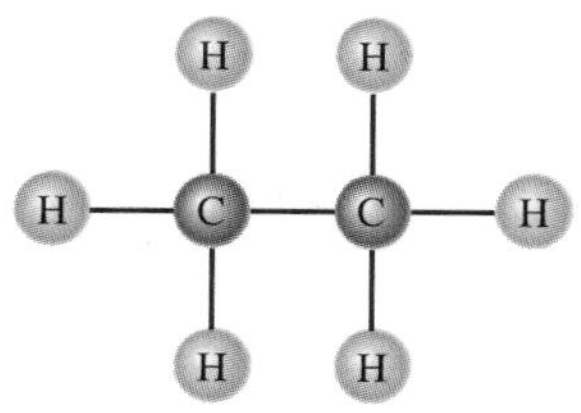

Ethane

A chemist sketches the relationship among atoms in a molecule. An engineer draws a schematic diagram of an electrical network. A genealogist charts a family tree. A sociologist maps the power structure of a large corporation. What do all of these drawings have in common? Each one is a system of points (atoms, electrical components, people, segments) connected by line segments. The German mathematician Denes Konig, who made the first systematic study of such patterns, named them "graphs."

In the eighteenth-century Prussian town of Königsberg (now the city of Kaliningrad, Russia near the Baltic Sea) seven bridges crossed the Prigel River (Fig. A.1a). Individuals in the area tried to determine whether it was possible to walk a path that would cross each of the seven bridges exactly once. They found that they ended up either not crossing one of the bridges, or crossing one of the bridges more than once. The problem was brought to the attention of the Swiss mathematician Leonhard Euler (pronounced "oiler," 1707–1783). His study of this problem, now known as the Königsberg bridge problem, laid the groundwork for a modern branch of mathematics called *graph theory,* a topic in a more general area called *topology*. To solve the problem, Euler drew figures called *graphs or networks* like that shown in red in Fig. A.1(b). Each dot represented a plot of land and each line a bridge or path. Can you find a path that crosses each bridge exactly once?

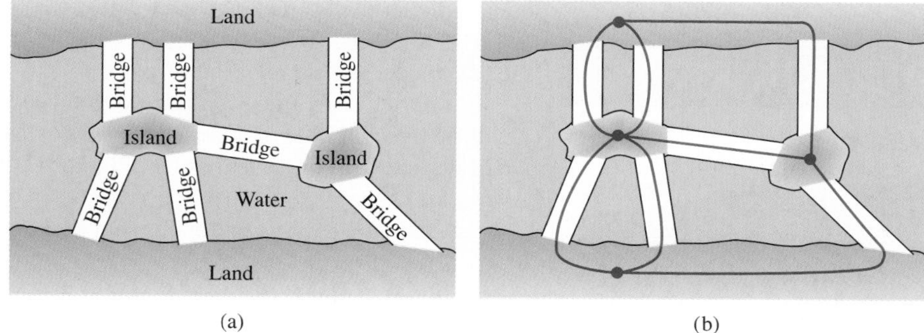

Figure A.1

Before we determine whether there is a path that will cross each bridge exactly once, let's consider Example 1.

EXAMPLE 1

In Fig. A.2 start at any point and try to trace each figure without retracing a line and without removing your pencil from the paper. If you succeed, indicate your starting point and ending point.

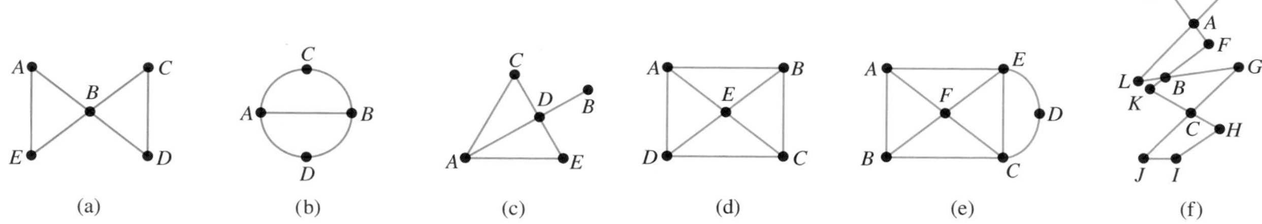

Figure A.2

SOLUTION:

a) The figure can be traced if you start at any point. You will end at the point at which you started.

b) The figure can be traced, but only if you start at point A or point B. If you start at point A, you will end at point B, and vice versa.

c) The figure can be traced, but only if you start at point A or point B. If you start at point A, you will end at point B, and vice versa.

d) The figure cannot be traced without retracing a line.

e) The figure can be traced, but only if you start at point A or point B. If you start at point A, you will end at point B, and vice versa.

f) The figure can be traced if you start at any point. You will end at the point at which you started. ▲

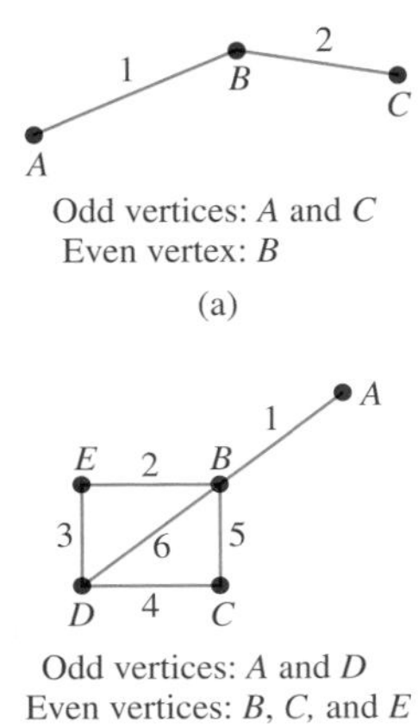

Figure A.3

In order for you to be able to answer the Königsberg bridge problem and understand why we were able to trace all but one of the figures in Example 1 without retracing a line, we must introduce some new terms and concepts.

A *vertex* is any designated point. An *edge* (or an *arc*) is any line, either straight or curved, that begins and ends at a vertex. Figure A.3(a) has three designated vertices, A, B, and C, and two edges, 1 and 2. Figure A.3(b) has five designated vertices, A, B, C, D, and E, and six edges.

A vertex with an odd number of attached edges is called an *odd vertex*. A vetex with an even number of attached edges is called an *even vertex*. Figure A.3(a) has two odd vertices and one even vertex. Figure A.3(b) has two odd vertices and three even vertices.

A *network* is any continuous (not broken) system of edges and vertices.

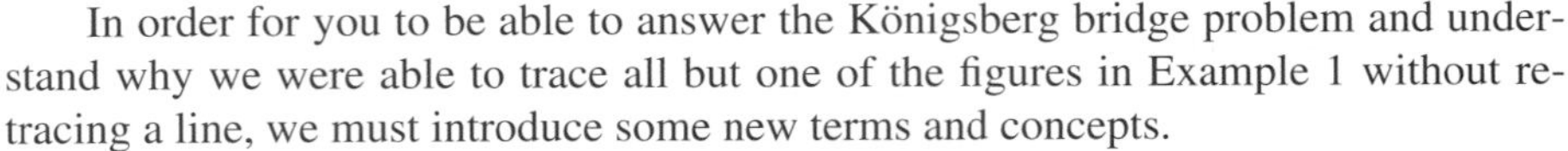
A network is said to be **traversable** if it can be traced without removing the pencil from the paper and without tracing an edge more than once.

After completing Example 1, do you have an intuitive feeling as to when a figure is traversable? Think about odd and even vertices. Leonhard Euler discovered an important scientific principle concealed in the Königsberg bridge problem. He presented his simple and ingenious solution of that problem to the Russian Academy at St. Petersburg in 1735. Euler developed the following rules of traversability in solving the problem.

Rules of Traversability

1. A network with no odd (all even) vertices is traversable; you may start from any vertex, and you will end where you began.
2. A network with exactly two odd vertices is traversable; you must start at either of the odd vertices and you will finish at the other.
3. A network with more than two odd vertices is not traversable.

The network in the Königsberg bridge problem (Fig. A.1b) has four odd vertices, so it cannot be traversed. Therefore crossing each bridge only once is impossible. Note that it is impossible for a network to contain an odd number of odd vertices. (If you don't believe this statement, try to construct such a network.)

DID YOU KNOW

Networks

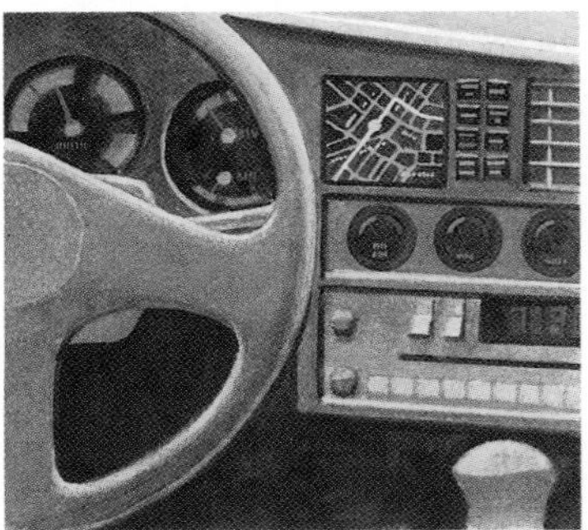

Today, graph (network) theory is a flourishing field, useful for dealing with a variety of everyday problems. It is used by the airline industry to arrange the complex network of flights connecting cities and by sports officials planning tournaments and playoff schedules. City planners use it to work out new roads and traffic patterns. Tests are underway in traffic-clogged cities such as Los Angeles to link cars, via an on-board navigational computer, to the city's central traffic-control computer. In this way motorists can be rerouted to less congested roadways.

Now go back to Example 1 and determine which figures are traversable, using these rules. Note that Figs. A.2(a) and A.2(f) are traversable from any point because they contain only even vertices. Figures A.2(b), (c), and (e) have exactly two odd vertices and can be traversed but only by starting at one of the odd vertices, either point A or point B. Figure A.2(d) contains more than two odd vertices and therefore cannot be traversed.

EXAMPLE 2

The floor plan of a six-gallery art museum is shown in Fig. A.4(a). The openings represent doors, and the letters represent galleries.

a) Determine the galleries that contain an odd number of doors; an even number of doors.
b) Each gallery can be represented as an odd or even vertex. Use this information to determine whether it is possible to walk through each gallery by using each door only once.
c) Determine a path to walk through each gallery by using each door only once.

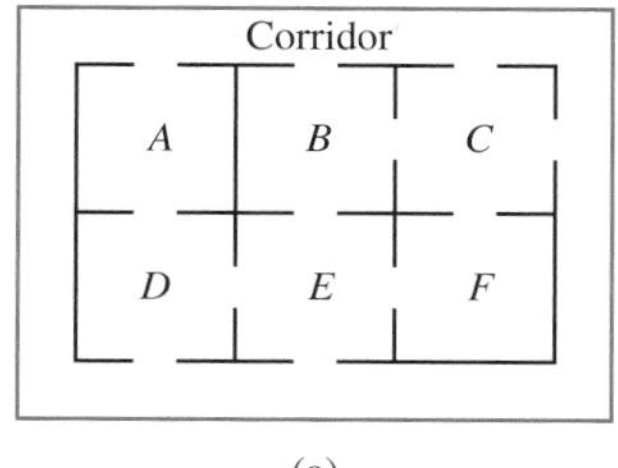

(a)

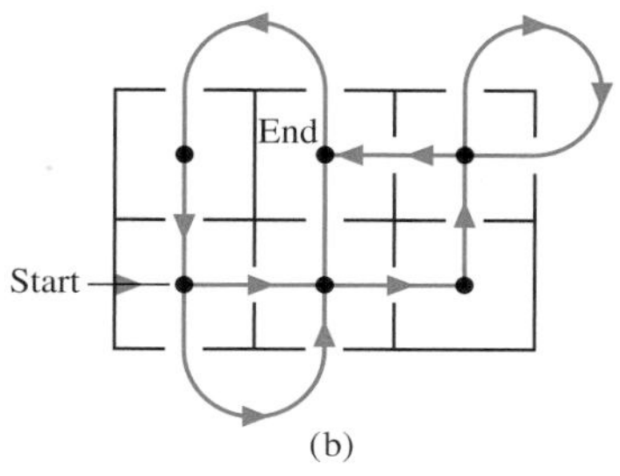

(b)

Figure A.4

SOLUTION:

a) Galleries B and D contain three doors each. Galleries A and F contain two doors each. Galleries C and E have four doors.
b) There are only two odd vertices, B and D, so the figure is traversable, and you can walk through the museum by using each door only once.
c) You must start in either Gallery B or D (see Fig. A.4b). If you start in B, you will end in D, and vice versa.. When you leave Gallery B, you can leave by any of the three doors.

The floor plan in Example 2 can be reduced to a map, where the rooms are the vertices and the doors are the paths. Construct a map of that floor plan now.

APPENDIX EXERCISES

Concept/Writing Exercises

1. What is a vertex?
2. What is an edge (or an arc)?
3. Explain how to determine whether a vertex is odd or even.
4. In your own words, explain the rules for determining whether a graph is traversable.

Practice the Skills

In Exercises 5–8, determine the number of vertices and the number of edges.

5. 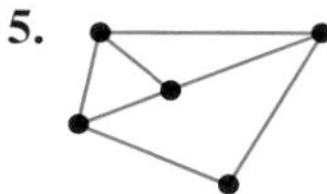

6. 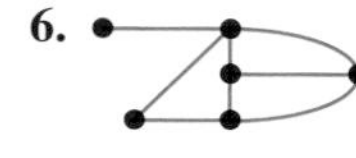

7. 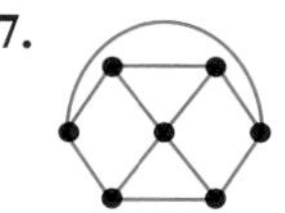

8. 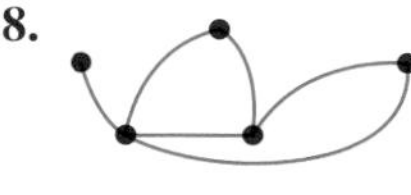

In Exercises 9 and 10, explain why these two figures represent the same graph.

9. 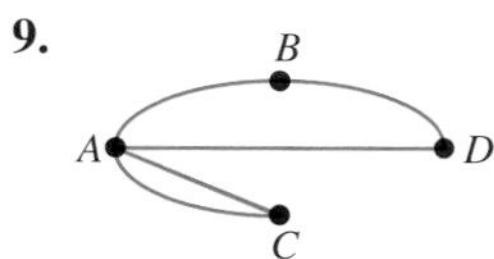

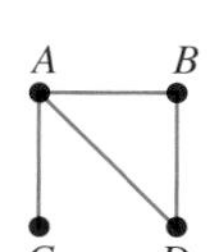

10. 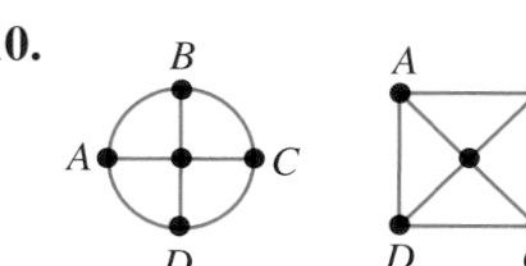

In Exercises 11 and 12, list the vertices that are odd and the vertices that are even.

11.

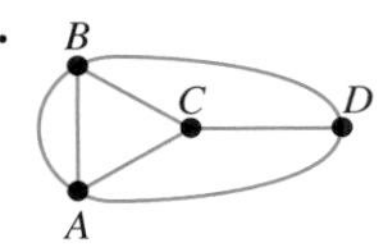

12.

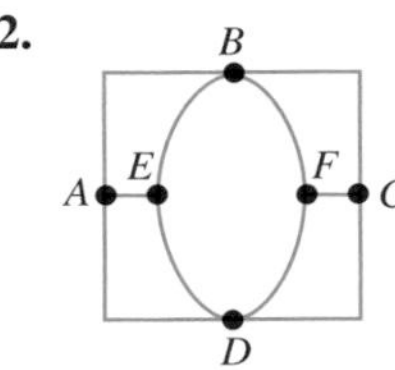

Problem Solving

In Exercises 13–20, determine whether the network is traversable. If it is, state the points from which you may start and end.

13.

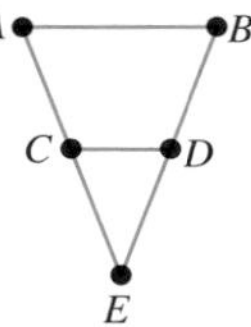

14.

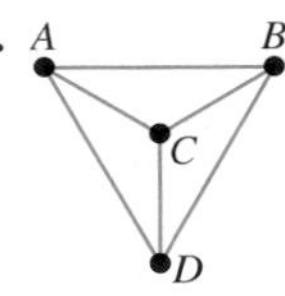

15.

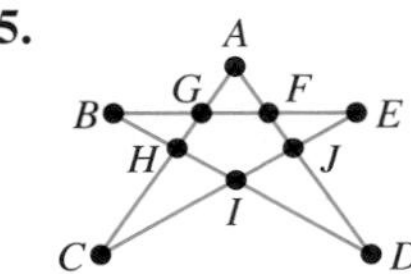

16.

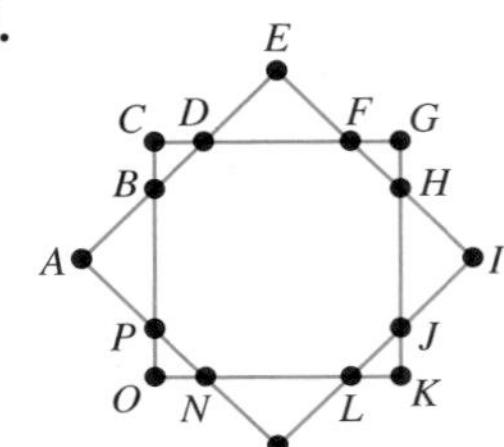

17.

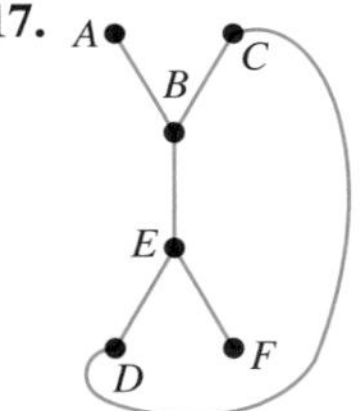

18.

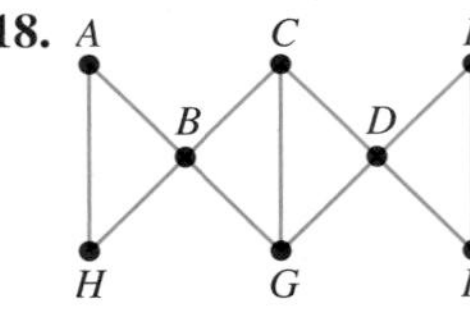

19.

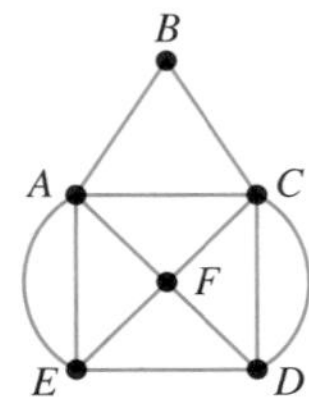

20.

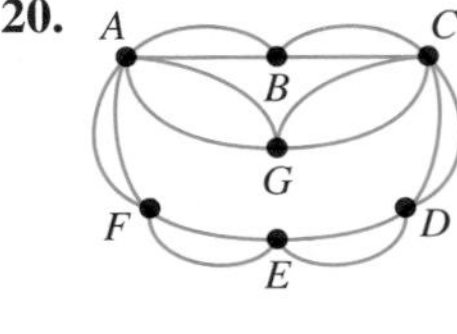

In Exercises 21–28, the floor plan of a building is shown.

a) *Determine the number of rooms that contain an odd number of doors; an even number of doors.*

b) *Use the rules of traversability to determine whether it is possible to walk through the building using each door only once.*

c) *If the answer to part (b) is yes, indicate where you can start and where you will end and describe one such path (for example, A to D to B to . . . etc.).*

21.

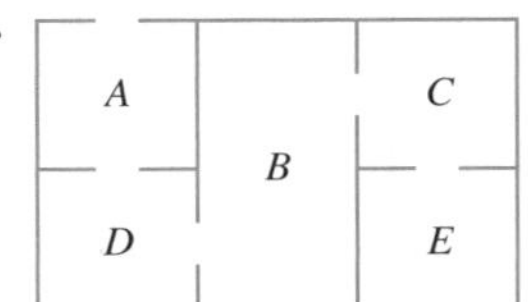

22.

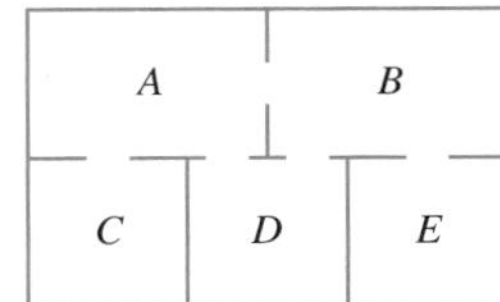

23.

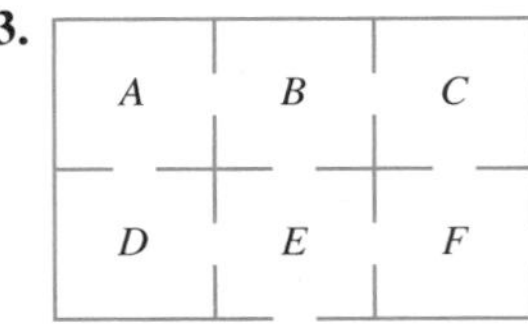

24.

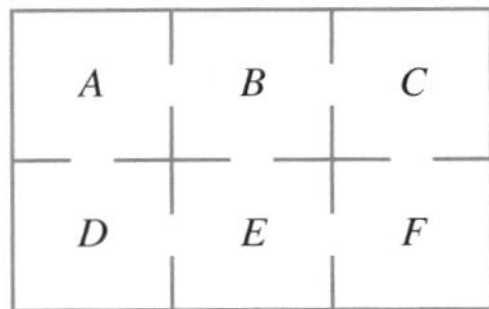

25.

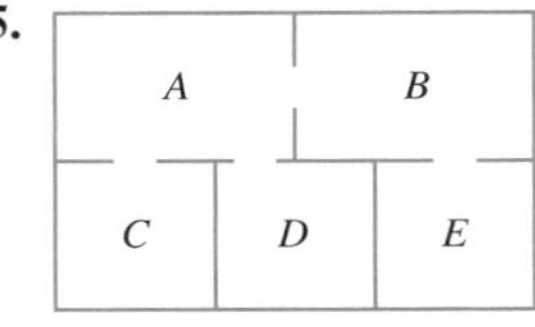

26.

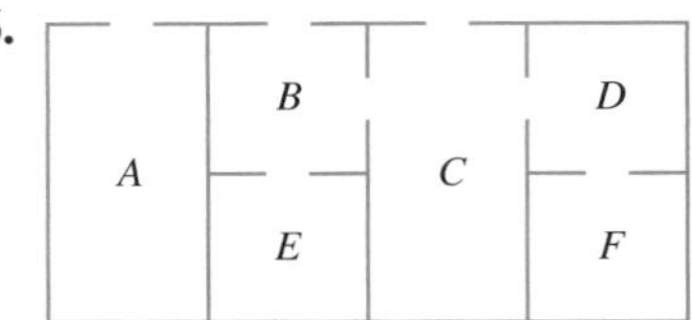

27.

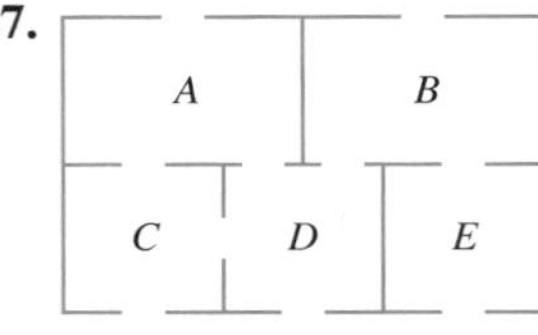

28.

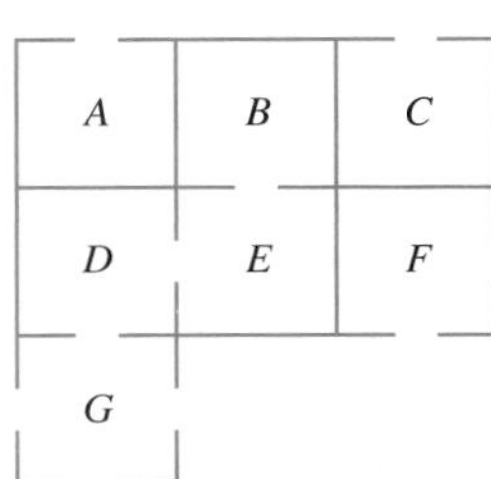

In Exercises 29 and 30, the floor plan for a suite of rooms is shown. Add an exit door in one of the rooms so that the security guard can enter through the door marked enter, pass through each door only once locking it behind him, and then exit by the door you added. Explain why there is only one possible room in which the door may be placed.

29.

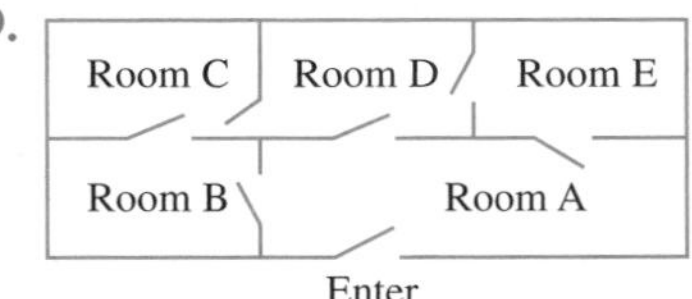

30.

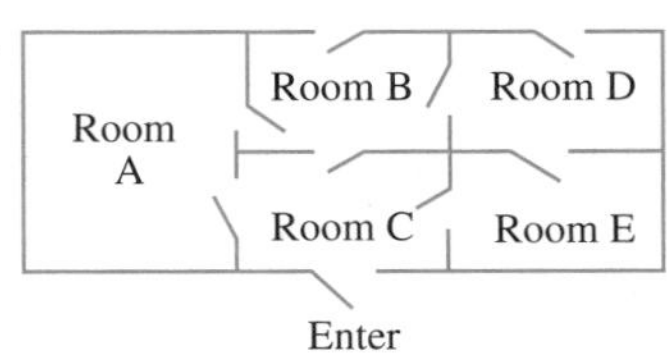

In Exercises 31 and 32, is it possible to cross each bridge exactly once? If it is possible, indicate where the person can start and where the person will finish. Explain your answer.

31.

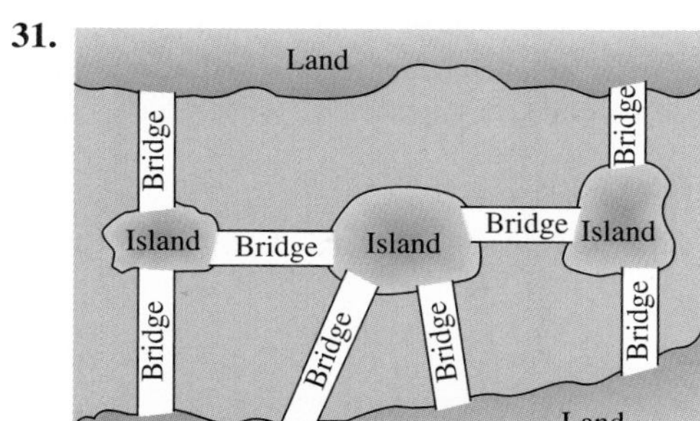

32.

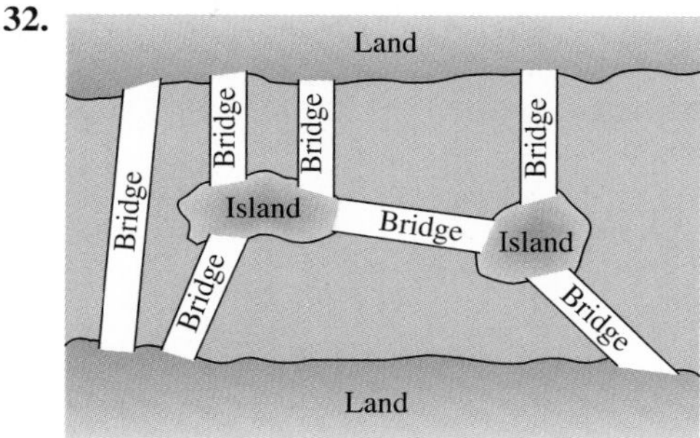

33. Draw a graph that contains four vertices and is traversable from exactly two points.

34. Draw a graph that contains five vertices that is traversable from exactly two points.

35. In the figure, lines connecting two states indicate that the two states share a common border. Which of these states share a common border with (a) Tennessee? (b) Missouri?

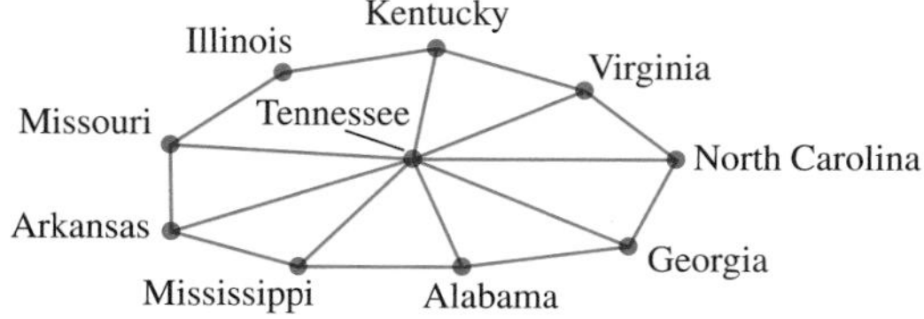

36. In the figure, a line connecting two countries indicates that they share a common border. Which of these countries share a common border with (a) Brazil? (b) Bolivia?

37. The remaining games in a soccer league are illustrated in the graph.

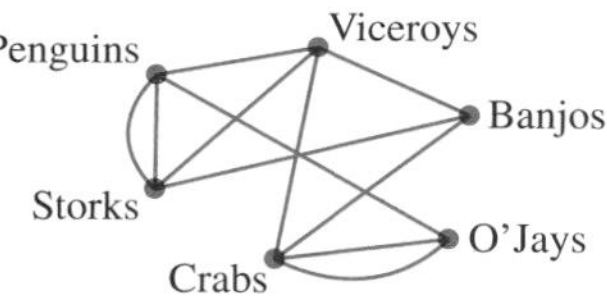

a) How many games do the Viceroys have remaining?
b) How many games do the Penguins have remaining?
c) How many games still need to be played?

38. Dawn, Jessica, Pam, Bill, Ed, and Scott go to a dance. Dawn dances with Bill and Scott. Jessica dances with all the boys and Pam dances only with Scott. Draw a graph that displays this information.

39. France has common borders with Belgium, Germany, Switzerland, Italy, and Spain. Belgium has common borders with the Netherlands, France, and Germany. Germany has common borders with Belgium, Poland, the Czech Republic, Austria, Switzerland, and France. Switzerland has common borders with France, Italy, Austria, and Germany. Draw a graph that displays this information.

40. Can you draw a graph that contains an odd number of odd vertices? If you answer yes draw such a graph.

Challenge problems/Group Activities

41. Gretchen's Delivery service, located in town C, makes deliveries every day to each town shown on the map. The map also shows the highways connecting the towns and the distances between towns. The delivery service's expenses are lowest when the driver's route is the shortest.

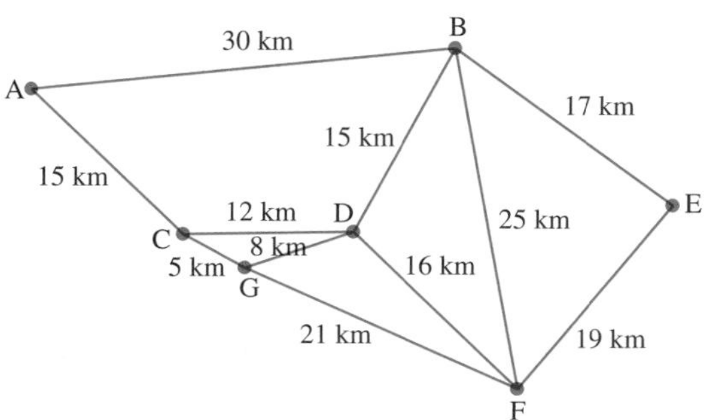

a) Is the map traversable? Explain.
b) The least number of miles that can be driven, starting at *C* and returning to *C*, that covers every city is 110 miles. Determine a path that covers all the cities and whose distance is 110 miles.

42. Networks (a) and (b) on page AA-6 illustrate a relationship between the number of edges, regions, and vertices. Network (a) has three vertices (*A*, *B*, and *C*), four edges (1, 2, 3, and 4), and three regions (*x*, *y*, and *z*). Network (b) has five vertices (*A*, *B*, *C*, *D*, and *E*), seven edges (1, 2, 3, 4, 5, 6, 7), and four regions (w, x, y, and z). Using these networks and others that you can make up yourself, develop a formula expressing the number of edges in terms of the number of vertices and the number of regions. This formula is known as *Euler's formula for networks*.

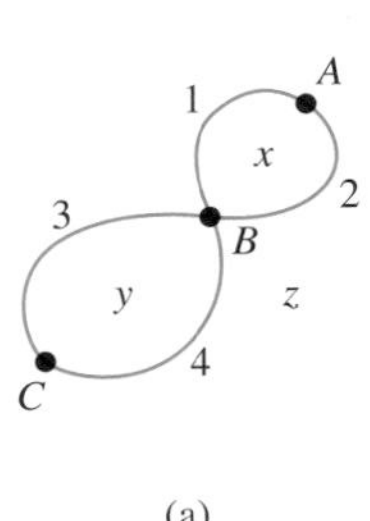

(a)

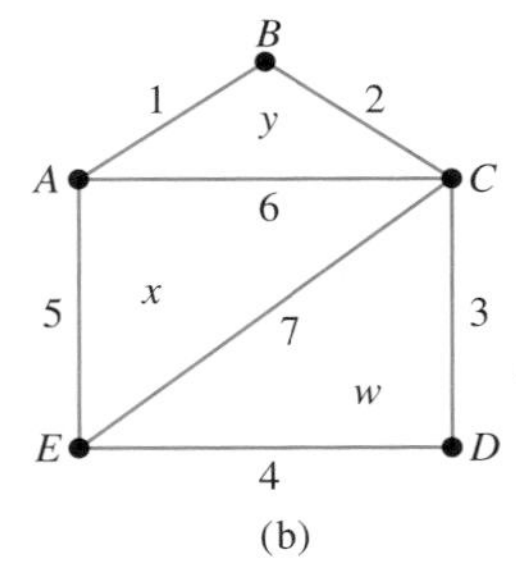

(b)

Internet/Research Activity

43. In 1856 William Rowan Hamilton (1805–1865) introduced a problem similar to the Königsberg bridge problem. Hamilton turned his problem into a game that was marketed in 1856. Write a short paper explaining the game. (References include encyclopedias and history of mathematics books and the Internet.)

ANSWERS

Chapter 1

Section 1.1, Page 5

1. a) 1, 2, 3, 4, 5, . . . **b)** Counting numbers

3. A conjecture is a belief based on specific observations that has not been proven or disproven.

5. Deductive reasoning is the process of reasoning to a specific conclusion from a general statement.

7. Inductive reasoning

9. Inductive reasoning, because a generalization was made from specific cases.

11. 1 5 10 10 5 1 **13.** $5 \times 9 = 45$

15. **17.** **19.** 15, 18, 21 **21.** $-1, 1, -1$

23. $\frac{1}{81}, \frac{1}{243}, \frac{1}{729}$ **25.** 36, 49, 64 **27.** 34, 55, 89 **29.** Y

31. a) 36, 49, 64 **b)** square 6, 7, 8, 9 and 10
c) No, 72 is between 8^2 and 9^2, so it is not a square number.

33. Blue: 1, 5, 7, 10, 12 Purple: 2, 4, 6, 9, 11 Yellow: 3, 8

35. a) ≈58 million **b)** ≈45 million
c) We are using specific cases to make a prediction.

37.

39. a) You should obtain the original number.
b) You should obtain the original number.
c) The result is the original number.
d) $n, 4n, 4n + 8, \frac{4n + 8}{4} = n + 2, n + 2 - 2 = n$

41. a) 5 **b)** You should obtain the number 5.
c) The result is always the number 5.
d) $n, n + 1, \frac{n + (n + 1) + 9}{2} = \frac{2n + 10}{2} = n + 5,$
$n + 5 - n = 5$

43. $999 \times 999 = 998{,}001$

45. $(3 + 2)/2 = 5/2$, which is not an even number.

47. $1 - 2 = -1$, which is not a counting number.

49. a) The sum of the measures of the interior angles should be 180°.
b) Yes, the sum of the measures of the interior angles should be 180°.
c) The sum of the measures of the interior angles of a triangle is 180°.

51. 129, the numbers in positions are found as follows:

$a \quad b$
$c \quad a + b + c$

53. Counterexample **54.** (c)

Section 1.2, Page 14

Answers in this section will vary depending on how you round your numbers. All answers are approximate.

1. 1170 **3.** 1,200,000,000 **5.** 8000 **7.** 100

9. 364,000,000 **11.** 1,200,000,000 **13.** $20

15. 8500 mi **17.** $85.70 **19.** $9000 **21.** 45 lb

23. 2000 mi **25.** $840 **27.** $6 **29.** $41 **31.** ≈ 375 mi

33. a) 32.5 million **b)** 558 counties
c) Answers will vary.

35. a) 4 million **b)** 98 million **c)** 64 million
d) 275 million

37. a) 83% **b)** 20% **c)** 91,771 square miles
d) No, since we are not given the area of each state.

39. 25 **41.** ≈ 90 berries **43.** 150° **45.** 10%

47. 9 square units **49.** 150 feet

51.–59. Answers will vary. **60.** 118 ridges

61. There are 336 dimples on a regulation golf ball.

62. b) 11.6 days

63. Answers will vary. The U.S. government categorized the middle class as $32,000 − $50,000 in 2001.

Section 1.3, Page 29

1. 187.5 mi **3.** 19.36 ft **5.** $4707.53 **7.** $800

9. $12.50 **11.** $70 **13.** $57,240

15. a) ≈ 122
b) Answers will vary. A close approximation can be obtained by multiplying the U.S. sizes by 2.54.

17. a) 9.2 min **b)** 62 min **c)** 40 min **d)** 150 min

19. a) 30,063,000 **b)** 97,000 **c)** 29,100

21. $82.08 **23. a)** $74.40 **b)** $264 **c)** $64

25. $34,600 **27.** 13,906 violations

29. a) Answers will vary. **b)** ≈ 267.65 million
c) ≈ 1241.2 million or 1.2412 billion

31. $990, less than initial investment

33. a) 48 rolls **b)** $198 if she purchases four 10 packs and two 4 packs

35. a) Water/milk: 3 cups; salt: $\frac{3}{8}$ tsp; Cream of Wheat: 9 tbsp (or $\frac{9}{16}$ cup)

b) Water/milk: $2\frac{7}{8}$ cups; salt: $\frac{3}{8}$ tsp; Cream of Wheat: $\frac{5}{8}$ cup (or 10 tbsp)
c) Water/milk: $2\frac{3}{4}$ cups; salt: $\frac{3}{8}$ tsp; Cream of Wheat: $\frac{9}{16}$ cup (or 9 tbsp)
d) Differences exist in water/milk because the amount for 4 servings is not twice that for 2 servings. Differences also exist in Cream of Wheat because $\frac{1}{2}$ cup is not twice 3 tbsp.

37. 144 square inches **39.** The area is 4 times as large.

41. 1 and 9 **43.** at -1 **45.** 10 birds and 12 lizards

47. a) 30 **b)** 140

49.

(4)
(3) (2)
(5) (1) (6)

51.

8	6	16
18	10	2
4	14	12

53. The sum of the four corners is 4 times the number in the center.

55. Multiply the center number by 9. **57.** 6 ways

59.

	7	
3	1	4
5	8	6
	2	

Other answers are possible, but 1 and 8 must appear in the center.

61.

1	2	3	4	5
2	3	4	5	1
3	4	5	1	2
4	5	1	2	3
5	1	2	3	4

Other answers are possible.

63. Mary is the skier. **65.** 714 square units

66. 3 ostriches

Review Exercises, Page 35

1. 23, 28, 33 **2.** 25, 36, 49 **3.** $-48, 96, -192$

4. 25, 32, 40 **5.** 15, 9, 2 **6.** $\frac{3}{8}, \frac{3}{16}, \frac{3}{32}$

7.  **8.** 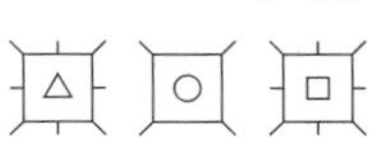 **9.** (c)

10. a) The original number and the final number are the same.
b) The original number and the final number are the same.
c) The final number is the same as the original number.
d) $n, 2n, 2n + 10, \dfrac{2n + 10}{2} = n + 5, n + 5 - 5 = n$

11. This process will always result in an answer of 3.

12. $1^2 + 2^2 = 5$

The answers to Exercises 13-25 will vary, depending upon how you round the numbers. All answers are approximate.

13. 420,000,000 **14.** 2150 **15.** 200

16. Answers will vary. **17.** \$88 **18.** \$12

19. 3 mph **20.** \$14.00 **21.** 2 mi **22.** 70% **23.** 5%

24. 13 square units **25.** Length $\approx$ 22 ft; height $\approx$ 8 ft

26. \$7.50 **27.** \$1.16

28. Berkman's is cheaper by \$20.00.

29. \$32,996 **30.** \$16.20 **31.** \$311 **32.** 7.05 mg

33. \$744.80 **34.** 6 hr 45 min **35.** July 26, 11:00 A.M.

36. a) 6.45 cm^2
b) 16.39 cm^3
c) 1 cm $\approx$ 0.39 in.

37. 201

38.

21	7	8	18
10	16	15	13
14	12	11	17
9	19	20	6

39.

23	25	15
13	21	29
27	17	19

40. 59 min 59 sec **41.** 6

42. \$25 Room
\$ 3 Men
\$ 2 Clerk
\$30

43. 140 lb

44. Yes; 3 quarters and 4 dimes, or 1 half dollar, 1 quarter and 4 dimes, or 1 quarter and 9 dimes. Other answers are possible.

45. 216 cm^3

46. Place six coins in each pan with one coin off to the side. If it balances, the heavier coin is the one on the side. If the pan does not balance, take the six coins on the heavier side and split them into two groups of three. Select the three heavier coins and weigh two coins. If the pan balances, it is the third coin. If the pan does not balance, you can identify the heavier coin.

47. 125,250 **48.** 16 blue **49.** 90

50. The fifth figure will be an octagon with sides of equal length. Inside the octagon will be a seven sided figure with each side of equal length. The figure will have one antenna.

51. 61

52. Some possible answers are shown. Others are possible.

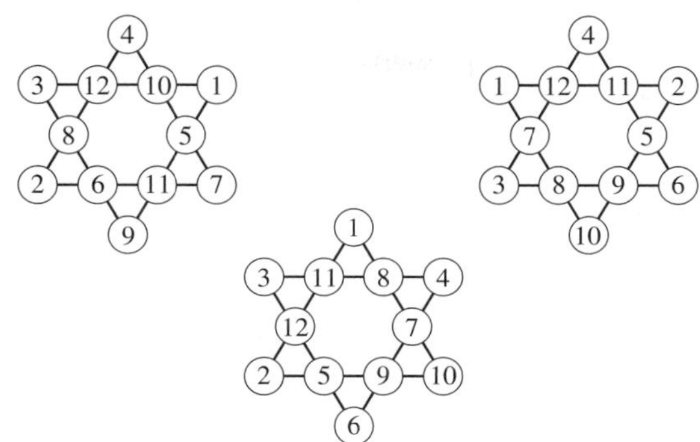

53. a) 2 **b)** 6 **c)** 24 **d)** 120
e) $n(n-1)(n-2)\ldots 1$, (or $n!$), where n = the number of people in line

Chapter Test, Page 38

1. 18, 21, 24

2. $\frac{1}{81}, \frac{1}{243}, \frac{1}{729}$

3. a) The result is the original number plus 1.
b) The result is the original number plus 1.
c) The result will always be the original number plus 1.
d) $n, 5n, 5n + 10, \frac{5n+10}{5} = n + 2,$
$n + 2 - 1 = n + 1$

The answers for Exercises 4–6 are approximate.

4. 6000 **5.** 33,000,000 **6.** 7 square units

7. a) ≈ 23.03
b) He is in the at risk range.

8. 159.25 therms **9.** 32 cans **10.** $7\frac{1}{2}$ min

11. $\approx$ 39.5 in. by 29.6 in. (The actual dimensions are 100.5 cm by 76.5 cm)

12. $49.00

13.

40	15	20
5	25	45
30	35	10

14. Less time if she had driven at 45 mph for the entire trip

15. $2 \cdot 6 \cdot 8 \cdot 9 \cdot 13$; 11 does not divide 11,232.

16. 243 jelly beans

17. a) $11.97
b) $11.81
c) Save 16 cents by using the 25% off coupon.

18. 8

Chapter 2

Section 2.1, Page 46

1. A set is a collection of objects.

3. Description, roster form, and set-builder notation; the set of counting numbers less than 7, $\{1, 2, 3, 4, 5, 6\}$, and $\{x \mid x \in N \text{ and } x < 7\}$

5. An infinite set is a set that is not finite.

7. Two sets are equivalent if they contain the same number of elements.

9. The empty set is a set that contains no elements.

11. Two sets that can be placed in a one-to-one correspondence have the same cardinal number and are equivalent.

13. Not well defined **15.** Well defined **17.** Well defined

19. Infinite **21.** Infinite **23.** Infinite

25. {Atlantic, Pacific, Arctic, Indian}

27. $\{11, 12, 13, 14, \ldots, 177\}$ **29.** $B = \{2, 4, 6, 8, \ldots\}$

31. { } or Ø **33.** $E = \{6, 7, 8, 9, \ldots, 71\}$

35. {Sony DSC-S50, Sony DSC-S70, Sony Mavica FD-90}

37. {Sony Mavica FD-73, Olympus D-360L, Sony DSC-S50, Kodak DC215, H-P Photo Smart C315}

39. {2002, 2003, 2004, 2005, 2006, 2007, 2008}

41. {2002, 2005, 2006, 2007, 2008}

43. $B = \{x \mid x \in N \text{ and } 3 < x < 11\}$ or
$B = \{x \mid x \in N \text{ and } 4 \le x \le 10\}$

45. $C = \{x \mid x \in N \text{ and } x \text{ is a multiple of } 3\}$

47. $E = \{x \mid x \in N \text{ and } x \text{ is odd}\}$

49. $C = \{x \mid x \text{ is February}\}$

51. Set A is the set of natural numbers less than or equal to 7.

53. Set V is the set of vowels in the English alphabet.

55. Set C is the set of companies that make calculators.

57. Set B is the set of members of the Beatles.

59. {St. Louis} **61.** { } or Ø

63. {1999, 2000, 2001, 2002} **65.** {1999, 2001, 2002}

67. False; $\{b\}$ is a set, and not an element of the set.

69. False; h is not an element of the set.

71. False; 3 is an element of the set.

73. True **75.** 4 **77.** 0 **79.** Both **81.** Neither

83. Equivalent

85. a) Set A is the set of natural numbers greater than 2. Set B is the set of all numbers greater than 2.
b) Set A contains only natural numbers. Set B contains other types of numbers, including fractions and decimal numbers.
c) $A = \{3, 4, 5, 6, \ldots\}$

d) No; set B cannot be written in roster form since we cannot list all the elements in set B.

87. Cardinal **89.** Ordinal **91.** Answers will vary.

93. Answers will vary.

95.

	2			7			2	
7	H	6	6	D	1	1	C	6
	8			4			3	
	8			4			3	
4	F	7	7		5	5	E	4
	3			2			8	
	3			2			8	
6	A	5	5	G	3	3	B	7
	1			6			5	

Section 2.2 Page 54

1. Set A is a subset of set B, symbolized $A \subseteq B$, if and only if all the elements of set A are also elements of set B.

3. If $A \subseteq B$, then every element of set A is an element of set B. If $A \subset B$, then every element of set A is an element of set B and set $A \neq$ set B.

5. The number of proper subsets is determined by the formula $2^n - 1$, where n is the number of elements in the set.

7. False; gold is an element of the set, not a subset.

9. True **11.** True

13. False; the set $\{\emptyset\}$ contains the element $\emptyset$.

15. True **17.** False; the set $\{0\}$ contains the element 0.

19. False **21.** True

23. False; no set is a proper subset of itself.

25. $B \subseteq A, B \subset A$ **27.** $B \subseteq A, B \subset A$ **29.** $B \subseteq A, B \subset A$

31. $A = B, A \subseteq B, B \subseteq A$ **33.** $\{ \}$

35. $\{ \}$, {pen}, {pencil}, {pen, pencil}

37. a) $\{ \}, \{a\}, \{b\}, \{c\}, \{d\}, \{a, b\}, \{a, c\}, \{a, d\}, \{b, c\}, \{b, d\}, \{c, d\}, \{a, b, c\}, \{a, b, d\}, \{a, c, d\}, \{b, c, d\}, \{a, b, c, d\}$
b) $\{a, b, c, d\}$

39. False **41.** True **43.** True **45.** True **47.** True

49. True **51.** 2^4 or 16 **53.** 2^6 or 64 **55.** $E = F$

57. a) Yes **b)** No **c)** Yes **59.** 1 **60.** Yes **61.** Yes

62. No

Section 2.3, Page 62

1.

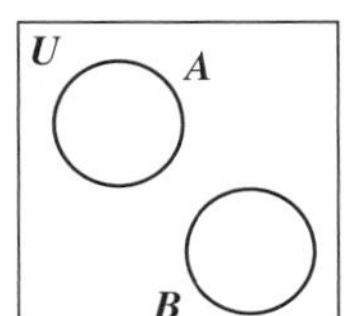

3.

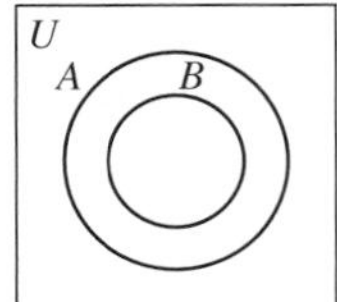

5.

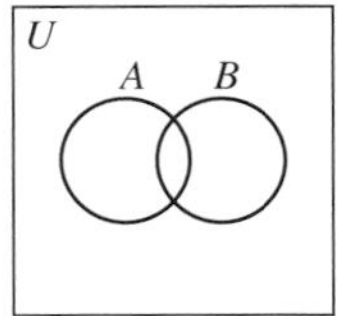

7. Combine the elements from set A and set B into one set. List any element that is contained in both sets only once.

9. Take the elements common to both set A and set B.

11. a) *Or* is generally interpreted to mean *union*.
b) *And* is generally interpreted to mean *intersection*.

13. Region II, the intersection of the two sets.

15.

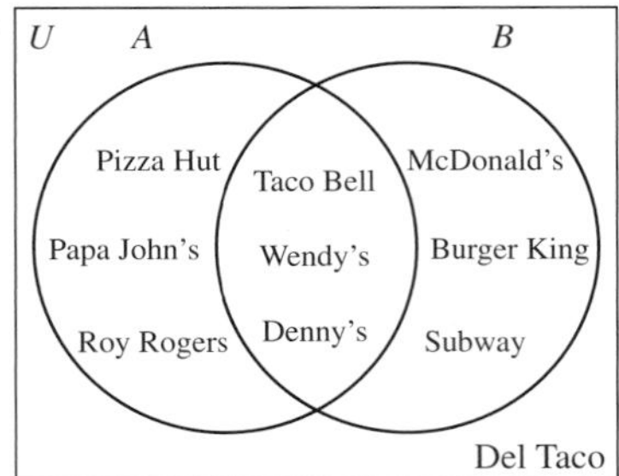

17.

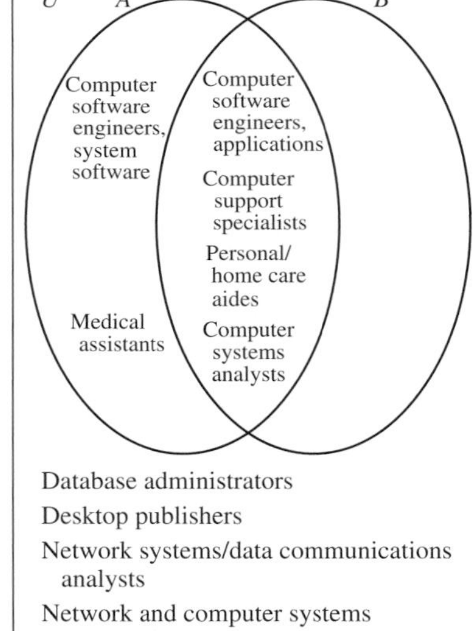

19. The set of U.S. colleges and universities that are not in the state of North Dakota

21. The set of insurance companies in the United States that do not offer life insurance

23. The set of insurance companies in the United States that offer life insurance or car insurance

25. The set of insurance companies in the United States that offer life insurance and do not offer car insurance

27. The set of U.S. corporations whose headquarters are in New York State and whose chief executive officer is a woman

29. The set of U.S. corporations whose chief executive officer is not a woman and who employ at least 100 people

31. The set of U.S. corporations whose headquarters are in New York State or whose chief executive officer is a woman or that employ at least 100 people

33. $\{b, c, t, w, a, h\}$ **35.** $\{a, h\}$

37. $\{c, w, b, t, a, h, f, g, r\}$ **39.** $\{p, m, z\}$

41. $\{L, \Delta, @, \$, *\}$

43. $\{L, \Delta, @, *, \$, R, \square, \alpha, \infty, \Sigma, Z\}$ **45.** $\{*, \$\}$

47. $\{R, \square, \alpha\}$ **49.** $\{1, 2, 3, 4, 5, 6, 8\}$

51. $\{1, 5, 7, 8\}$ **53.** $\{7\}$ **55.** $\{\ \}$ **57.** $\{7\}$

59. $\{a, e, h, i, j, k\}$ **61.** $\{a, f, i\}$

63. $\{b, c, d, e, g, h, j, k\}$ **65.** $\{a, c, d, e, f, g, h, i, j, k\}$

67. $\{a, b, c, d, e, f, g, h, i, j, k\}$, or U **69.** $\{\ \}$

71. $\{2, 4, 6, 8\}$, or B **73.** $\{7, 9\}$

75. $\{1, 3, 5, 6, 7, 8, 9\}$ **77.** $\{1, 2, 3, 4, 5, 7, 9\}$

79. $\{2, 4, 6, 8\}$, or B **81.** $\{1, 2, 3, 4, 5\}$, or C

83. A set and its complement will always be disjoint. For example, if $U = \{1, 2, 3\}$, $A = \{1, 2\}$, and $A' = \{3\}$, then $A \cap A' = \{\ \}$.

85. 49

87. a) $8 = 4 + 6 - 2$ **b)** and **c)** Answers will vary.

89. $\{1, 2, 3, 4, \ldots\}$, or A **91.** $\{4, 8, 12, 16, \ldots\}$, or B

93. $\{2, 4, 6, 8, \ldots\}$, or C **95.** $\{2, 6, 10, 14, 18, \ldots\}$

97. $\{2, 6, 10, 14, 18, \ldots\}$ **99.** U **101.** A **103.** U

105. U **107.** $B \subseteq A$ **109.** A and B are disjoint sets.

111. $A \subseteq B$ **113.** $\{e, f, h\}$ **115.** $\{d, j, k\}$ **117.** $\{13\}$

119. $\{1, 2, 3, 4, 5, 6, 7, 8, 9, 10, 11, 12, 14, 15\}$

121. $\{2, 3, 4, 5, 7, 9, 10, 11, 12, 13, 14, 15\}$

123. Complement

Section 2.4, Page 71

1. 8 **3.** II, IV, VI **5.** 8

7. a) Yes **b)** No, one specific case cannot be used as proof. **c)** No

9.

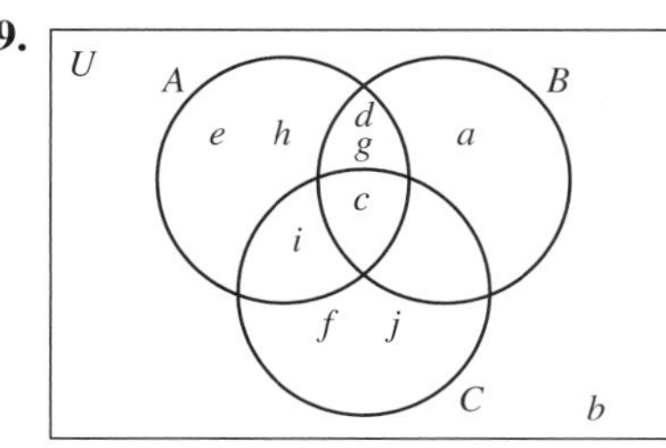

11.

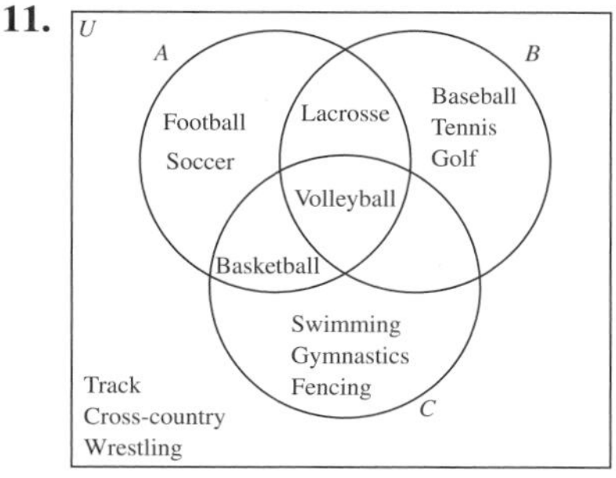

13.

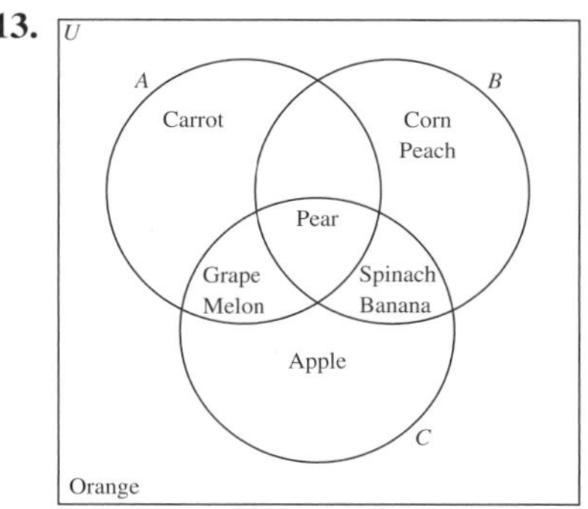

15.

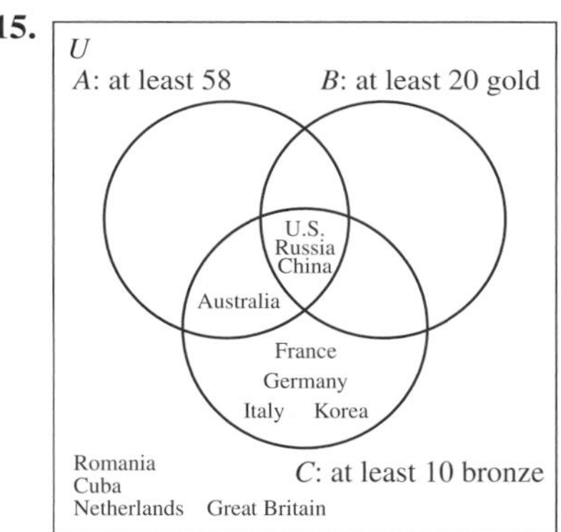

17. V **19.** VIII **21.** VI **23.** IV **25.** II **27.** VII

29. VI **31.** III **33.** III **35.** V **37.** II **39.** VII

41. I **43.** VIII **45.** VI **47.** $\{1, 2, 3, 4, 5, 6\}$

49. $\{3, 4, 5, 7, 8, 9, 12\}$ **51.** $\{3, 4, 5\}$

53. $\{1, 2, 3, 6, 9, 10, 11, 12\}$

55. $\{1, 2, 3, 4, 5, 6, 7, 8, 9, 12\}$ **57.** $\{9, 11, 12\}$

59. $\{7, 8, 9, 10, 11, 12\}$ **61.** Yes **63.** No **65.** No

67. Yes **69.** No **71.** Yes **73.** Yes **75.** Yes **77.** No

79. $(A \cup B)'$ **81.** $(A \cup B) \cap C'$

83. a) Both equal $\{6, 7\}$. **b)** Answers will vary. **c)** Both are represented by the regions IV, V, VI.

85.

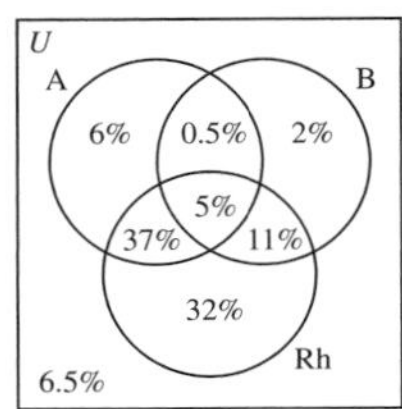

87. a)

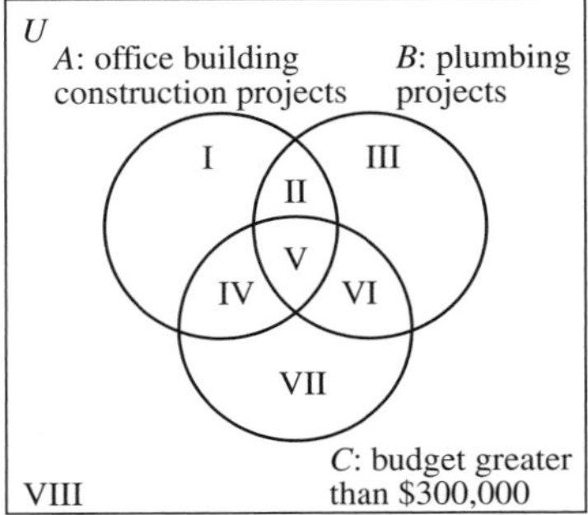

b) V; $A \cap B \cap C$
c) VI; $A' \cap B \cap C$
d) I; $A \cap B' \cap C'$

89. a)

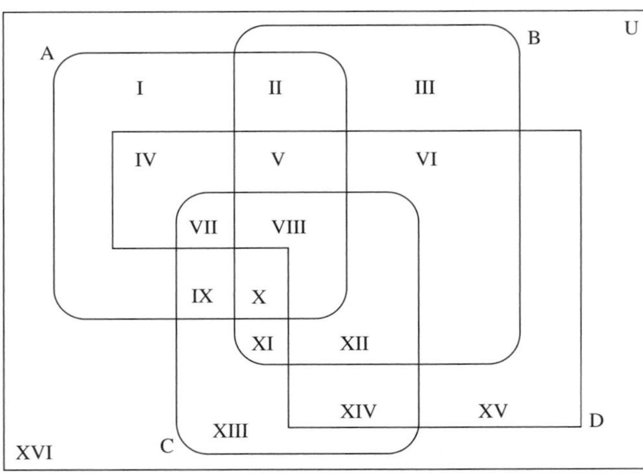

b)

Region	*Set*	*Region*	*Set*
I	$A \cap B' \cap C' \cap D'$	IX	$A \cap B' \cap C \cap D'$
II	$A \cap B \cap C' \cap D'$	X	$A \cap B \cap C \cap D'$
III	$A' \cap B \cap C' \cap D'$	XI	$A' \cap B \cap C \cap D'$
IV	$A \cap B' \cap C' \cap D$	XII	$A' \cap B \cap C \cap D$
V	$A \cap B \cap C' \cap D$	XIII	$A' \cap B' \cap C \cap D'$
VI	$A' \cap B \cap C' \cap D$	XIV	$A' \cap B' \cap C \cap D$
VII	$A \cap B' \cap C \cap D$	XV	$A' \cap B' \cap C' \cap D$
VIII	$A \cap B \cap C \cap D$	XVI	$A' \cap B' \cap C' \cap D'$

Section 2.5, Page 80

1.

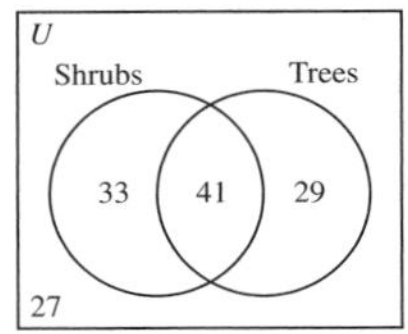

a) 33
b) 29
c) 27

3.

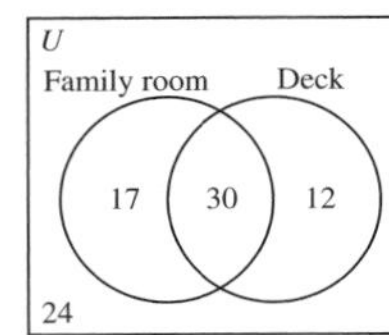

a) 17
b) 12
c) 59

5.

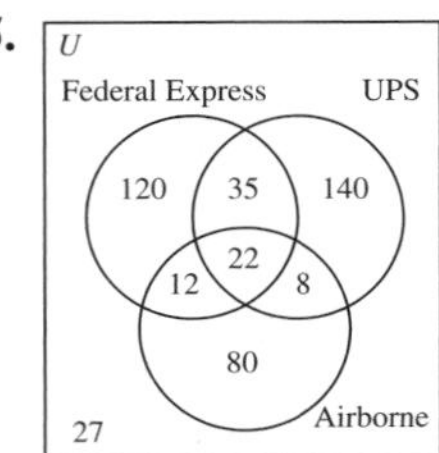

a) 27
b) 80
c) 340
d) 55
e) 337

7.

U
Mysteries Science fiction
22 11 17
2
9 3
15
6 Romances

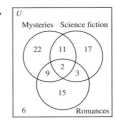

a) 22
b) 11
c) 64
d) 50
e) 23

9.

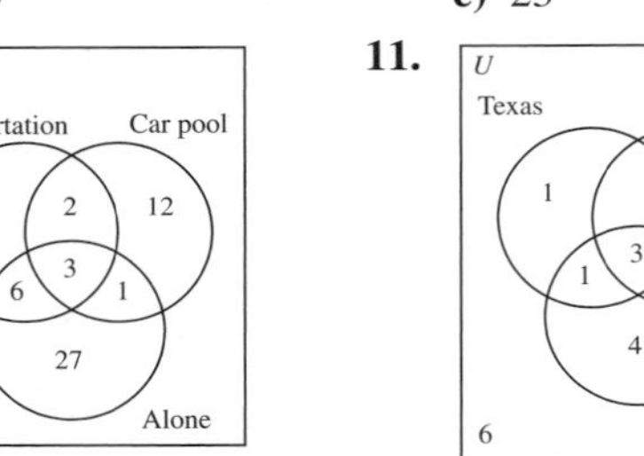

a) 17
b) 27
c) 2
d) 31
e) 2

11.

U
Texas
Greater than 750,000
1 7
3
1 3
4
6 Greater than 10%

a) 10
b) 15
c) 0
d) 6

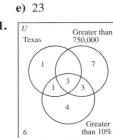

13. In a Venn diagram, regions II, IV, and V contain a total of 37 cars driven by women. This total is greater than the 35 cars driven by women, as given in the exercise.

15.

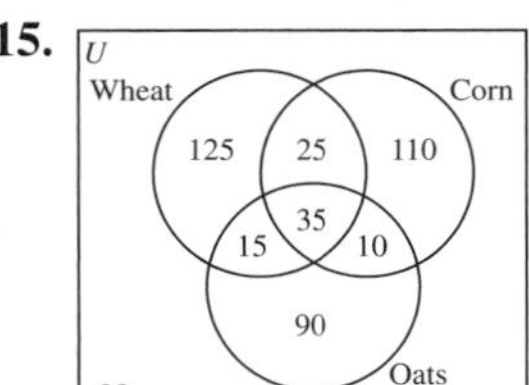

a) 410 **b)** 35 **c)** 90 **d)** 50

16. a) 10 **b)** 10 **c)** 6

Section 2.6, Page 86

1. An infinite set is a set that can be placed in a one-to-one correspondence with a proper subset of itself.

3.
$$\begin{array}{cccccccc} \{7, & 8, & 9, & 10, & 11, & \ldots, & n+6, & \ldots\} \\ \downarrow & \downarrow & \downarrow & \downarrow & \downarrow & & \downarrow & \\ \{8, & 9, & 10, & 11, & 12, & \ldots, & n+7, & \ldots\} \end{array}$$

5.
$$\begin{array}{cccccccc} \{3, & 5, & 7, & 9, & 11, & \ldots, & 2n+1, & \ldots\} \\ \downarrow & \downarrow & \downarrow & \downarrow & \downarrow & & \downarrow & \\ \{5, & 7, & 9, & 11, & 13, & \ldots, & 2n+3, & \ldots\} \end{array}$$

7.
$$\begin{array}{ccccccc} \{4, & 7, & 10, & 13, & \ldots, & 3n+1, & \ldots\} \\ \downarrow & \downarrow & \downarrow & \downarrow & & \downarrow & \\ \{7, & 10, & 13, & 16, & \ldots, & 3n+4, & \ldots\} \end{array}$$

9.
$$\begin{array}{cccccccc} \{6, & 11, & 16, & 21, & 26, & \ldots, & 5n+1, & \ldots\} \\ \downarrow & \downarrow & \downarrow & \downarrow & \downarrow & & \downarrow & \\ \{11, & 16, & 21, & 26, & 31, & \ldots, & 5n+6, & \ldots\} \end{array}$$

11.
$$\begin{array}{ccccccc} \left\{\frac{1}{2}, \right. & \frac{1}{4}, & \frac{1}{6}, & \frac{1}{8}, & \ldots, & \frac{1}{2n}, & \left. \ldots \right\} \\ \downarrow & \downarrow & \downarrow & \downarrow & & \downarrow & \\ \left\{\frac{1}{4}, \right. & \frac{1}{6}, & \frac{1}{8}, & \frac{1}{10}, & \ldots, & \frac{1}{2n+2}, & \left. \ldots \right\} \end{array}$$

13.
$$\begin{array}{ccccccc} \{1, & 2, & 3, & 4, & \ldots, & n, & \ldots\} \\ \downarrow & \downarrow & \downarrow & \downarrow & & \downarrow & \\ \{6, & 12, & 18, & 24, & \ldots, & 6n, & \ldots\} \end{array}$$

15.
$$\begin{array}{ccccccc} \{1, & 2, & 3, & 4, & \ldots, & n, & \ldots\} \\ \downarrow & \downarrow & \downarrow & \downarrow & & \downarrow & \\ \{4, & 6, & 8, & 10, & \ldots, & 2n+2, & \ldots\} \end{array}$$

17.
$$\begin{array}{ccccccc} \{1, & 2, & 3, & 4, & \ldots, & n, & \ldots\} \\ \downarrow & \downarrow & \downarrow & \downarrow & & \downarrow & \\ \{2, & 5, & 8, & 11, & \ldots, & 3n-1, & \ldots\} \end{array}$$

19.
$$\begin{array}{ccccccc} \{1, & 2, & 3, & 4, & \ldots, & n, & \ldots\} \\ \downarrow & \downarrow & \downarrow & \downarrow & & \downarrow & \\ \{5, & 8, & 11, & 14, & \ldots, & 3n+2, & \ldots\} \end{array}$$

21.
$$\begin{array}{ccccccc} \{1, & 2, & 3, & 4, & \ldots, & n, & \ldots\} \\ \downarrow & \downarrow & \downarrow & \downarrow & & \downarrow & \\ \left\{\frac{1}{3}, \right. & \frac{1}{4}, & \frac{1}{5}, & \frac{1}{6}, & \ldots, & \frac{1}{n+2}, & \left. \ldots \right\} \end{array}$$

23.
$$\begin{array}{ccccccc} \{1, & 2, & 3, & 4, & \ldots, & n, & \ldots\} \\ \downarrow & \downarrow & \downarrow & \downarrow & & \downarrow & \\ \{1, & 4, & 9, & 16, & \ldots, & n^2, & \ldots\} \end{array}$$

25.
$$\begin{array}{ccccccc} \{1, & 2, & 3, & 4, & \ldots, & n, & \ldots\} \\ \downarrow & \downarrow & \downarrow & \downarrow & & \downarrow & \\ \{3, & 9, & 27, & 81, & \ldots, & 3^n, & \ldots\} \end{array}$$

27. = **28.** = **29.** = **30.** = **31.** =

Review Exercises, Page 87

1. True
2. False; the word *best* makes the statement not well defined.
3. True **4.** False; no set is a proper subset of itself.
5. False; the elements 6, 12, 18, 24, . . . are members of both sets.
6. True
7. False; both sets do not contain exactly the same elements.
8. True **9.** True **10.** True **11.** True **12.** True
13. True **14.** True **15.** $A = \{7, 9, 11, 13, 15\}$
16. {California, Oregon, Idaho, Utah, Arizona}
17. $C = \{1, 2, 3, 4, \ldots, 296\}$
18. $D = \{9, 10, 11, 12, \ldots, 96\}$
19. $A = \{x \mid x \in N \text{ and } 52 < x < 100\}$
20. $B = \{x \mid x \in N \text{ and } x > 63\}$
21. $C = \{x \mid x \in N \text{ and } x < 3\}$
22. $D = \{x \mid x \in N \text{ and } 23 \leq x \leq 41\}$
23. *A* is the set of capital letters in the English alphabet from E through M, inclusive.
24. *B* is the set of U.S. coins with a value of less than a dollar.
25. *C* is the set of the last three lowercase letters in the English alphabet.
26. *D* is the set of numbers greater than or equal to 3 and less than 9.
27. {5, 6} **28.** {1, 2, 3, 4, 5, 6, 7, 8} **29.** {9, 10}
30. {1, 2, 4, 6, 7, 8, 10} **31.** 16 **32.** 15
33.

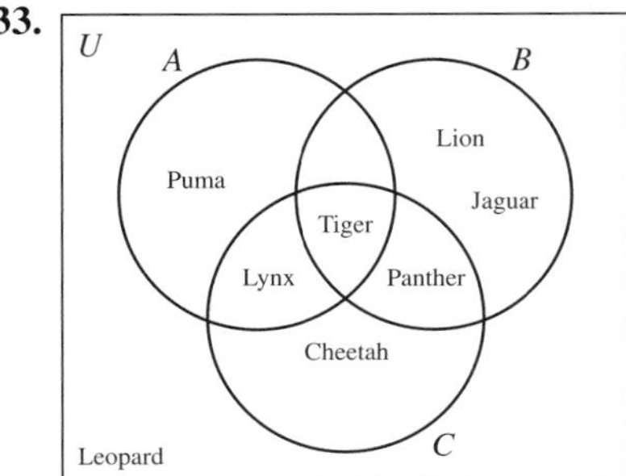

34. {b, e, g, k, c, d, f, a} **35.** {b, d}
36. {b, e, g, k, c, d, f, a, i} **37.** {f} **38.** {d, f, a}
39. {e, g, f, d, a, i} **40.** True **41.** True **42.** II
43. V **44.** VIII **45.** IV **46.** IV **47.** VII **48.** \$450

49. 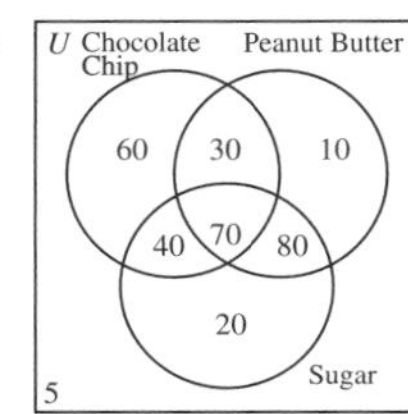

a) 315
b) 10
c) 30
d) 110

50. 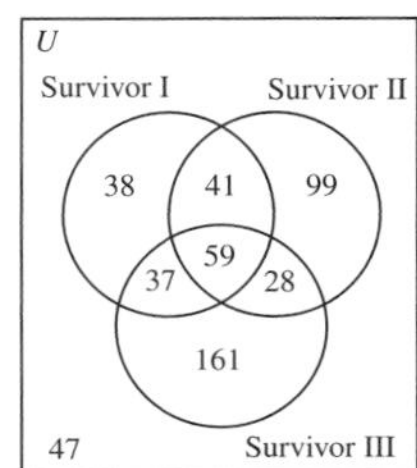

a) 38
b) 298
c) 28
d) 236
e) 106

51. $\{2, 4, 6, 8, \ldots, 2n, \ldots\}$
$\downarrow \downarrow \downarrow \downarrow \quad \downarrow$
$\{4, 6, 8, 10, \ldots, 2n + 2, \ldots\}$

52. $\{3, 5, 7, 9, \ldots, 2n + 1, \ldots\}$
$\downarrow \downarrow \downarrow \downarrow \quad \downarrow$
$\{5, 7, 9, 11, \ldots, 2n + 3, \ldots\}$

53. $\{1, 2, 3, 4, \ldots, n, \ldots\}$
$\downarrow \downarrow \downarrow \downarrow \quad \downarrow$
$\{5, 8, 11, 14, \ldots, 3n + 2, \ldots\}$

54. $\{1, 2, 3, 4, \ldots, n, \ldots\}$
$\downarrow \downarrow \downarrow \downarrow \quad \downarrow$
$\{4, 9, 14, 19, \ldots, 5n - 1, \ldots\}$

Chapter Test, Page 90

1. True

2. False; the sets do not contain exactly the same elements.

3. True

4. False; the second set has no subset that contains the element 7.

5. False; the empty set is a proper subset of every set except itself.

6. False; the set has 2^3, or 8 subsets. **7.** True

8. False; for any set A, $A \cup A' = U$, not $\{\ \}$. **9.** True

10. $A = \{1, 2, 3, 4, 5, 6, 7, 8\}$

11. Set A is the set of natural numbers less than 9.

12. $\{7, 9\}$ **13.** $\{3, 5, 7, 9, 13\}$

14. $\{3, 5, 7, 9\}$, or A **15.** 2

16. 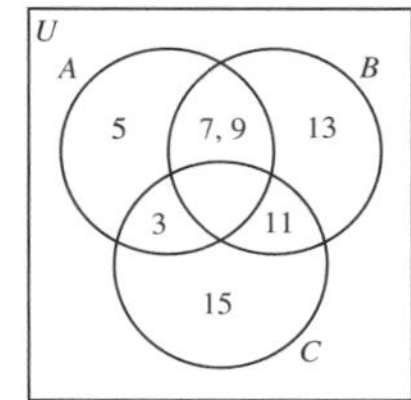

17. Equal

18. 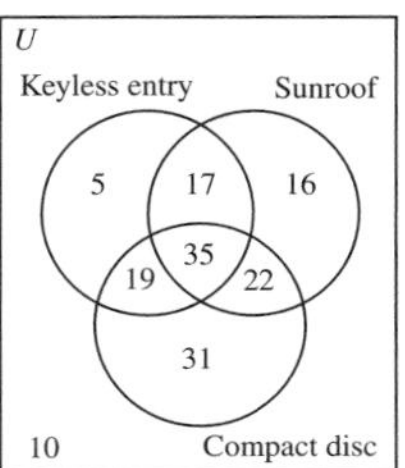

a) 52 **b)** 10 **c)** 93 **d)** 17 **e)** 38 **f)** 31

19. $\{7, 8, 9, 10, \ldots, n + 6, \ldots\}$
$\downarrow \downarrow \downarrow \downarrow \quad \downarrow$
$\{8, 9, 10, 11, \ldots, n + 7, \ldots\}$

20. $\{1, 2, 3, 4, \ldots, n, \ldots\}$
$\downarrow \downarrow \downarrow \downarrow \quad \downarrow$
$\{1, 3, 5, 7, \ldots, 2n - 1, \ldots\}$

Chapter 3

Section 3.1, Page 103

1. a) A simple statement is a statement that conveys only one idea.
b) Compound statements are statements consisting of two or more simple statements.

3. a) Some are **b)** All are
c) Some are not **d)** None are

5. a) $\rightarrow$ **b)** $\vee$ **c)** $\wedge$ **d)** $\sim$ **e)** $\leftrightarrow$

7. When two simple statements are on the same side of the comma, they are placed together in parentheses when translated into symbolic form.

9. Compound; conjunction, $\wedge$

11. Compound; biconditional, $\leftrightarrow$

13. Compound; disjunction, $\vee$

15. Simple statement **17.** Compound; negation, $\sim$

19. Compound; conjunction, $\wedge$

21. Compound; negation, $\sim$

23. No picnic tables are portable.

25. Some chickens do not fly. **27.** All turtles have claws.

29. Some bicycles have three wheels.

31. All pine trees produce pinecones.

33. No pedestrians are in the crosswalk.

35. $\sim p$ **37.** $\sim q \vee \sim p$ **39.** $\sim p \rightarrow \sim q$ **41.** $\sim q \rightarrow \sim p$

43. $\sim p \wedge \sim q$ **45.** $\sim(q \rightarrow \sim p)$

47. Firemen do not work hard.

49. Firemen wear red suspenders or firemen work hard.

51. Firemen do not work hard if and only if firemen do not wear red suspenders.

53. It is false that firemen wear red suspenders or firemen work hard.

55. Firemen do not work hard and firemen do not wear red suspenders.

57. $(p \vee \sim q) \rightarrow r$ **59.** $(p \wedge q) \vee r$ **61.** $p \rightarrow (q \vee \sim r)$

63. $(r \leftrightarrow q) \wedge p$ **65.** $q \rightarrow (p \leftrightarrow r)$

67. The water is 70° or the sun is shining, and we do not go swimming.

69. The water is not 70°, and the sun is shining or we go swimming.

71. If we do not go swimming, then the sun is shining and the water is 70°.

73. If the sun is shining then we go swimming, and the water is 70°.

75. The sun is shining if and only if the water is 70°, and we go swimming.

77. Not permissible, you cannot have both soup and salad. The *or* used on menus is the exclusive *or.*

79. Not permissible, you cannot have both potatoes and pasta. The *or* used on menus is the exclusive *or.*

81. a) $(\sim p) \rightarrow q$ **b)** Conditional

83. a) $(\sim q) \wedge (\sim r)$ **b)** Conjunction

85. a) $(p \vee q) \rightarrow r$ **b)** Conditional

87. a) $r \rightarrow (p \vee q)$ **b)** Conditional

89. a) $(\sim p) \leftrightarrow (\sim q \rightarrow r)$ **b)** Biconditional

91. a) $(r \wedge \sim q) \rightarrow (q \wedge \sim p)$ **b)** Conditional

93. a) $\sim[(p \wedge q) \leftrightarrow (p \vee r)]$ **b)** Negation

95. a) $r \wedge \sim c$ **b)** Conjunction

97. a) $\sim(b \rightarrow \sim p)$ **b)** Negation

99. a) $(f \vee v) \rightarrow h$ **b)** Conditional

101. a) $c \leftrightarrow (\sim f \vee p)$ **b)** Biconditional

103. a) $(c \leftrightarrow w) \vee s$ **b)** Disjunction

105. $[(\sim q) \rightarrow (r \vee p)] \leftrightarrow [(\sim r) \wedge q]$; Biconditional

107. a) The conjunction and disjunction have the same dominance.
b) Answers will vary. **c)** Answers will vary.

Section 3.2, Page 115

1. a) 4 **b)**

p	q
T	T
T	F
F	T
F	F

3. a)

p	q	$p \vee q$
T	T	T
T	F	T
F	T	T
F	F	F

b) Only when both p and q are false

5.	7.	9.	11.	13.	15.	17.	19.
T	F	F	T	T	T	F	T
T	T	F	F	F	F	F	T
	F	T	T	T	T	T	T
	F	F	T	T	T	F	T
				F	T	F	F
				F	F	F	F
				T	T	T	T
				T	F	T	F

21. $p \wedge q$	23. $p \wedge \sim q$
T	F
F	T
F	F
F	F

25. $\sim(p \wedge q)$	27. $p \vee (q \vee r)$	29. $p \wedge (q \vee \sim q)$
F	T	T
T	T	T
T	T	F
T	T	F
	T	
	T	
	T	
	F	

31. a) False **b)** True **33. a)** False **b)** False

35. a) True **b)** True **37. a)** False **b)** True

39. a) True **b)** True **41. a)** True **b)** True

43. True **45.** True **47.** False **49.** False

51. False **53.** True **55.** True **57.** False

59. $p \wedge \sim q$	61. $p \vee \sim q$
F	T
T	T
F	F
F	T
True in case 2	True in cases 1, 2, and 4, when p is true, or when p and q are both false.

63. $(r \vee q) \wedge p$	**65.** $q \vee (p \wedge \sim r)$
T	T
T	T
T	F
F	T
F	T
F	T
F	F
F	F

True in cases 1, 2, and 3

True in cases 1, 2, 4, 5, and 6. True except when p, q, r have truth values TFT, FFT, or FFF.

67. a) Mr. Duncan and Mrs. Tuttle qualify.
b) Mrs. Rusinek does not qualify, since their combined income is less than $46,000.

69. a) Wing Park qualifies; the other four do not.
b) Gina Vela is returning on April 2. Kara Sharo is returning on a Monday. Christos Supernaw is not staying over on a Saturday. Alex Chang is returning on a Monday.

71. T
T
T
T
F
T
F
T

73. Yes

Section 3.3, Page 125

1. a)

p	q	$p \to q$
T	T	T
T	F	F
F	T	T
F	F	T

b) The conditional is false only when the antecedent is true and the consequent is false.

3. a) Substitute the truth values for the simple statements. Then evaluate the compound statement, using the assigned truth values.
b) True

5. A self-contradiction is a compound statement that is never true.

7.	**9.**	**11.**	**13.**	**15.**	**17.**	**19.**	**21.**
T	F	F	T	F	T	F	F
F	F	T	T	T	F	T	F
T	T	T	F	T	T	F	T
T	F	F	T	T	F	F	F
					T	F	T
					F	F	T
					F	F	F
					F	T	T

23.	**25.**
T	T
T	T
T	T
T	F
T	T
F	T
F	F
F	T

27. $p \to (q \wedge r)$	**29.** $(p \leftrightarrow \sim q) \vee r$	**31.** $(\sim p \to q) \vee r$
T	T	T
F	F	T
F	T	T
F	T	T
T	T	T
T	T	T
T	T	T
T	F	F

33. Neither **35.** Self-contradiction **37.** Tautology
39. Not an implication **41.** Implication **43.** Implication
45. True **47.** True **49.** False **51.** True **53.** True
55. True **57.** True **59.** False **61.** True **63.** True
65. True **67.** False **69.** False **71.** True **73.** True
75. True

77. No, the statement only states what will occur if your sister gets straight A's. If your sister does not get straight A's, your parents may still get her a computer.

79. T
F
F
T
T
F
F
F

81. It is a tautology. The statement may be expressed as $(p \to q) \vee (\sim p \to q)$, where p: It is a head and q: I win. This statement is a tautology.

83.

Tiger	Boots	Sam	Sue
Blue	Yellow	Red	Green
Nine Lives	Whiskas	Friskies	Meow Mix

84. Katie was born last. Katie and Mary are saying the same thing.

Section 3.4, Page 139

1. Statements that have exactly the same truth values

3. The two statements are equivalent.

5. a) $q \rightarrow p$ **b)** $\sim p \rightarrow \sim q$ **c)** $\sim q \rightarrow \sim p$ **7.** $\sim p \vee q$

9. Equivalent **11.** Not equivalent **13.** Equivalent

15. Equivalent **17.** Equivalent **19.** Equivalent

21. Equivalent **23.** Equivalent **25.** Not equivalent

27. Not equivalent **29.** Equivalent

31. The Mississippi River does not run through Ohio and the Ohio River does not run through Mississippi.

33. It is false that the snowmobile was an Arctic Cat or a Ski-Do.

35. It is false that the hotel has a weight room and the conference center has an auditorium.

37. If we go to Cozumel, then it is false that we will not go snorkeling and we will go to Senior Frogs.

39. You do not drink a glass of orange juice or you will get a full day supply of folic acid.

41. If Bob the Tomato didn't visit the nursing home then he did not visit the Cub Scout meeting.

43. The plumbers meet in Kansas City and the Rainmakers will not provide the entertainment.

45. It is cloudy if and only if the front is coming through.

47. If the chemistry teacher teaches mathematics then there is a shortage of mathematics teachers and if there is a shortage of mathematics teachers then the chemistry teacher teaches mathematics.

49. *Converse:* If I finish the book in one week, then the book is interesting.
Inverse: If the book is not interesting, then I will not finish the book in one week.
Contrapositive: If I do not finish the book in one week, then the book is not interesting.

51. *Converse:* If you can watch television, then you finish your homework.
Inverse: If you did not finish your homework, then you cannot watch television.
Contrapositive: If you cannot watch television, then you did not finish your homework.

53. *Converse:* If I scream, then that annoying paper clip shows up on my computer screen.
Inverse: If that annoying paper clip does not show up on my computer screen, then I will not scream.
Contrapositive: If I do not scream, then that annoying paper clip does not show up on my screen.

55. *Converse:* If we go down to the marina and take out the sailboat, then the sun is shining.
Inverse: If the sun is not shining, then we will not go down to the marina or we will not take out the sailboat.
Contrapositive: If we do not go down to the marina or we will not take out the sailboat, then the sun is not shining.

57. If a natural number is divisible by 10, then the natural number is divisible by 5. True.

59. If a natural number is not divisible by 6, then the natural number is not divisible by 3. False.

61. If two lines are not parallel, then the two lines intersect in at least one point. True.

63. If the polygon is a quadrilateral, then the sum of the interior angles of the polygon measures 360°. True.

65. b) and **c)** are equivalent. **67. a)** and **c)** are equivalent.

69. b) and **c)** are equivalent. **71. b)** and **c)** are equivalent.

73. None are equivalent. **75.** None are equivalent.

77. a) and **c)** are equivalent. **79. a)** and **b)** are equivalent.

81. True. If $p \rightarrow q$ is false, it must be of the form $T \rightarrow F$. Therefore, the converse must be of the form $F \rightarrow T$, which is true.

83. False. A conditional statement and its contrapositive always have the same truth values.

85. Answers will vary. **87.** Answers will vary.

89. a) Conditional **b)** Biconditional **c)** Inverse **d)** Converse **e)** Contrapositive

Section 3.5, Page 150

1. The conclusion necessarily follows from the given set of premises.

3. Yes, if the conclusion necessarily follows from the set of premises, the argument is valid, even if the conclusion is false.

5. Yes, if the conclusion necessarily follows from the set of premises, the argument is valid, even if the premises themselves are false.

7. a)
$$\begin{array}{l} p \vee q \\ \underline{\sim p} \\ \therefore q \end{array}$$
b) Answers will vary.

9. a)
$$\begin{array}{l} p \rightarrow q \\ \underline{q \rightarrow r} \\ \therefore p \rightarrow r \end{array}$$
b) Answers will vary.

11. a)
$$\begin{array}{l} p \rightarrow q \\ \underline{\sim p} \\ \therefore \sim q \end{array}$$
b) Answers will vary.

13. Valid **15.** Valid **17.** Invalid **19.** Invalid

21. Invalid **23.** Valid **25.** Valid **27.** Invalid

29. Invalid **31.** Valid

33. a) $p \rightarrow q$
$\underline{\sim p}$
$\therefore \sim q$
b) Invalid

35. a) $p \rightarrow q$
$\underline{p}$
$\therefore q$
b) Valid

37. a) $p \rightarrow q$
$\underline{\sim q}$
$\therefore \sim p$
b) Valid

39. a) $p \rightarrow q$
$\underline{q}$
$\therefore p$
b) Invalid

41. a) $p \vee q$
$\underline{\sim p}$
$\therefore q$
b) Valid

43. a) $p \rightarrow q$
$\underline{q \rightarrow r}$
$\therefore p \rightarrow r$
b) Valid

45. a) $f \rightarrow w$
$\underline{w \wedge \sim t}$
$\therefore w \rightarrow \sim f$
b) Invalid

47. a) $s \wedge g$
$\underline{g \rightarrow c}$
$\therefore s \rightarrow c$
b) Valid

49. a) $h \rightarrow b$
$\underline{\sim p \rightarrow \sim b}$
$\therefore h \rightarrow p$
b) Valid

51. a) $p \rightarrow q$
$\underline{\sim q}$
$\therefore \sim p$
b) Valid

53. a) $t \vee \sim p$
$\underline{p}$
$\therefore t$
b) Valid

55. a) $t \wedge g$
$\underline{\sim t \vee \sim g}$
$\therefore \sim t$
b) Valid

57. a) $c \wedge \sim h$
$\underline{h \rightarrow c}$
$\therefore h$
b) Invalid

59. a) $f \rightarrow d$
$\underline{d \rightarrow \sim s}$
$\therefore f \rightarrow s$
b) Invalid

61. Therefore, your face will break out.

63. Therefore, a tick is an arachnid.

65. Therefore, you did not close the deal.

67. Therefore, if you do not pay off your credit card bills, then the bank makes money.

69. No. The conditional statement will always be true, and therefore it will be a tautology, and a valid argument.

Section 3.6, Page 158

1. It is an invalid argument.

3. The conclusion necessarily follows from the premises.

5. Yes, if the conclusion necessarily follows from the premises, the argument is valid.

7. Valid **9.** Valid **11.** Invalid **13.** Valid
15. Invalid **17.** Invalid **19.** Invalid **21.** Valid
23. Invalid **25.** Invalid **27.** Valid **29.** Invalid

Review Exercises, Page 161

1. No rock bands play ballads.

2. All bananas are ripe.

3. Some chickens have lips.

4. Some panthers are not endangered.

5. Some pens do not use ink.

6. Some rabbits wear glasses.

7. The coffee is Maxwell House or the coffee is hot.

8. The coffee is not hot and the coffee is strong.

9. If the coffee is hot, then the coffee is strong and the coffee is not Maxwell House.

10. The coffee is Maxwell House if and only if the coffee is not strong.

11. The coffee is not Maxwell House, if and only if the coffee is strong and the coffee is not hot.

12. The coffee is Maxwell House or the coffee is not hot, and the coffee is not strong.

13. $r \wedge q$ **14.** $p \rightarrow r$ **15.** $(r \rightarrow q) \vee \sim p$
16. $(q \leftrightarrow p) \wedge \sim r$ **17.** $(r \wedge q) \vee \sim p$ **18.** $\sim(r \wedge q)$

19.	20.	21.	22.	23.	24.
F	T	T	T	F	F
F	F	T	F	T	T
T	F	T	T	F	T
F	F	T	T	F	T
		T	F	T	T
		F	F	T	T
		F	F	T	T
		T	F	T	T

25. False **26.** True **27.** False **28.** True **29.** True
30. True **31.** True **32.** True **33.** False **34.** False
35. Not equivalent **36.** Equivalent **37.** Equivalent
38. Not equivalent

39. It is false that Johnny Cash is not in the Rock and Roll Hall of Fame or India Arie did not record *Acoustic Soul*.

40. If her foot did not fall asleep then she injured her ankle.

41. Altec Lansing does not produce only speakers and Harman Kardon does not produce only stereo receivers.

42. It is false that Travis Tritt won an Academy Award or Randy Jackson does commercials for Milk Bone Dog Biscuits.

43. The temperature is above 32° or we will go ice fishing at O'Leary's Lake.

44. a) If you enjoy life, then you hear a beautiful songbird today.
b) If you do not hear a beautiful songbird today, then you do not enjoy life.
c) If you do not enjoy life, then you do not hear a beautiful songbird today.

45. a) If the quilt has a uniform design, then you followed the correct pattern.
b) If you did not follow the correct pattern, then the quilt does not have a uniform design.

c) If the quilt does not have a uniform design, then you did not follow the correct pattern.

46. a) If Maureen Gerald is helping at the school, then she is not in attendance.
b) If Maureen Gerald is in attendance, then she is not helping at the school.
c) If Maureen Gerald is not helping at the school, then she is in attendance.

47. a) If we will not buy a desk at Miller's Furniture, then the desk is made by Winner's Only and the desk is in the Rose catalog.
b) If the desk is not made by Winner's Only or the desk is not in the Rose catalog, then we will buy a desk at Miller's Furniture.
c) If we will buy a desk at Miller's Furniture, then the desk is not made by Winner's Only or the desk is not in the Rose catalog.

48. a) If I let you attend the prom, then you will get straight A's on your report card.
b) If you do not get straight A's on your report card, then I will not let you attend the prom.
c) If I will not let you attend the prom, then you did not get straight A's on your report card.

49. a), b), and **c)** are equivalent. **50.** None are equivalent.

51. a) and **c)** are equivalent. **52.** None are equivalent.

53. Invalid **54.** Valid **55.** Invalid **56.** Valid

57. Invalid **58.** Invalid

Chapter Test, Page 163

1. $(p \wedge r) \vee \sim q$ **2.** $(r \rightarrow q) \vee \sim p$ **3.** $\sim(r \leftrightarrow \sim q)$

4. Ann is the secretary, if and only if Dick is the vice president and Elaine is the president.

5. It is false that if Ann is the secretary, then Elaine is not the president.

6.	**7.**
F	T
T	T
F	T
F	T
F	F
F	T
F	T
F	F

8. True **9.** True **10.** True **11.** True **12.** Equivalent

13. a) and **b)** are equivalent. **14. a)** and **b)** are equivalent.

15.
$$\begin{array}{l} s \rightarrow f \\ \underline{f \rightarrow p} \\ \therefore\ s \rightarrow p \end{array}$$
Valid

16. Invalid

17. Some leopards are not spotted.

18. No jacks-in-the-box are electronic.

19. *Converse:* If today is Saturday, then the garbage truck comes.
Inverse: If the garbage truck does not come, then today is not Saturday.
Contrapositive: If today is not Saturday, then the garbage truck does not come.

20. Yes

Chapter 4

Section 4.1, Page 173

1. A number is a quantity, and it answers the question "How many?" A numeral is a symbol used to represent the number.

3. A system of numeration consists of a set of numerals and a scheme or rule for combining the numerals to represent numbers.

5. The Hindu–Arabic numeration system

7. In a multiplicative system, there are numerals for each number less than the base and for powers of the base. Each numeral less than the base is multiplied by a numeral for the power of the base, and these products are added to obtain the number.

9. 142 **11.** 2423 **13.** 334,214

15. 𓍢𓍢𓍢𓍢𓍢𓍢𓎆𓎆𓎆𓏺𓏺𓏺𓏺 **17.** 𓆼𓆼𓎆𓎆𓎆𓎆𓏺𓏺𓏺𓏺𓏺

19. 𓆐𓂭𓂭𓂭𓂭𓂭𓆼𓆼𓆼𓍢𓍢𓍢𓍢𓍢𓍢𓍢𓍢𓎆𓎆𓎆𓎆𓏺𓏺𓏺𓏺𓏺

21. 19 **23.** 547 **25.** 1492 **27.** 2946 **29.** 12,666

31. 9464 **33.** LIX **35.** CXXXIV **37.** MMV

39. $\overline{\text{IV}}$DCCXCIII **41.** $\overline{\text{IX}}$CMXCIX **43.** $\overline{\text{XX}}$DCXLIV

45. 74 **47.** 4081 **49.** 8550 **51.** 4003

53. 五十三 **55.** 三百七十八 **57.** 四千二百六十 **59.** 七千零五十六

61. 341 **63.** 22,505 **65.** 9607 **67.** $\nu\theta$ **69.** ψκϛ

71. $\pi'\beta'\psi\delta$

73. Advantage: You can write some numbers more compactly.
Disadvantage: There are more numerals to memorize.

75. Advantage: You can write some numbers more compactly.
Disadvantage: There are more numerals to memorize.

77. 1936, 𓆼𓍢𓍢𓍢𓍢𓍢𓍢𓍢𓍢𓍢𓎆𓎆𓎆𓏺𓏺𓏺𓏺𓏺𓏺, α'ϡλϛ, 一千九百三十六

79. 422, 9999∩∩II, CDXXII, 四百二十二

81. ΤΤ'Q'θ' ΤΤ Q θ **83.** Turn the book upside down.

84. MM **85.** 1888, MDCCCLXXXVIII

Section 4.2, Page 180

1. A base 10 place-value system **3.** Four tens, four hundreds

5. A symbol for zero and for each counting number less than the base are required.

7. Write each digit times its corresponding positional value.

9. a) There may be confusion because numbers could be interpreted in different ways. For example, ▼ could be interpreted to be either 1 or 60.
b) ▼▼ ◀▼▼▼ for both numbers.

11. 1, 20, 18 × 20, 18 × $(20)^2$, 18 × $(20)^3$

13. (6 × 10) + (3 × 1)

15. (3 × 100) + (5 × 10) + (9 × 1)

17. (8 × 100) + (9 × 10) + (7 × 1)

19. (4 × 1000) + (3 × 100) + (8 × 10) + (7 × 1)

21. (1 × 10,000) + (6 × 1000) + (4 × 100) + (0 × 10) + (2 × 1)

23. (3 × 100,000) + (4 × 10,000) + (6 × 1000) + (8 × 100) + (6 × 10) + (1 × 1)

25. 42 **27.** 784 **29.** 4868 **31.** ▼ ◀◀◀▼▼▼

33. ▼▼▼▼ ◀◀◀◀◀▼▼▼▼▼ **35.** ▼ ▼ ◀◀▼▼▼▼▼▼

37. 92 **39.** 4321 **41.** 4000 **43.** •• ═══ **45.** •••• ═══ / •• ═══

47. • — / ⊜ / •••

49. Advantages: In general, they are more compact; large and small numbers can be written more easily; there are fewer symbols to memorize.

Disadvantage: If many of the symbols in the numeral represent zero, the place-value system may be less compact.

51. 33, ∴ ═

53. (△ × ▢²) + (□ × ▢) + (◇ × 1)

55. a) No largest number
b) ▼▼▼▼ ◀◀◀◀▼▼▼▼ ◀◀◀◀▼▼▼▼▼▼▼ ◀◀◀◀▼▼

57. ▼▼ ◀◀◀◀◀▼▼▼▼▼

59. ••• ═ / •••• ═ / ••• ═ **61.**

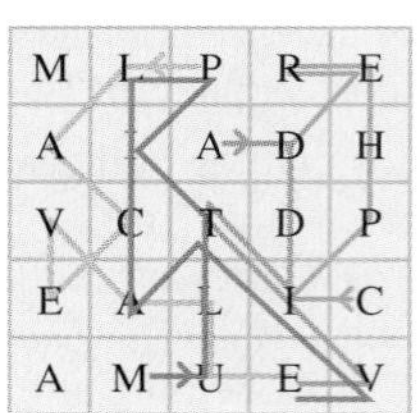

Section 4.3, Page 187

1. Answers will vary. **3.** 5 **5.** 22 **7.** 11 **9.** 100

11. 373 **13.** 1367 **15.** 867 **17.** 83 **19.** 6597

21. 1000_2 **23.** 10111_2 **25.** 2535_6 **27.** 1239_{12}

29. 1021_8 **31.** $17TE_{12}$ **33.** 1111110011_2 **35.** 4403_8

37. 1845 **39.** 447,415 **41.** $23D_{16}$ **43.** 1566_{16}

45. 11111010101_2 **47.** 31010_5 **49.** $11E1_{12}$

51. Incorrect; cannot have a 5 in base 5.

53. Written correctly

55. Incorrect; cannot have an 8 in base 7. **57.** 13

59. 73 **61.** ⊖○$_5$ **63.** ⦶⦶⦶$_5$ **65.** 7

67. 36 **69.** ●●$_4$ **71.** ●●●$_4$

73. a) Answers will vary. **b)** 10213_5 **c)** 1373_8

75. Answers will vary. **77.** $b = 6$ **78.** $d = 4$

79. a) 876 **b)** ●●●●$_4$

Section 4.4, Page 197

1. a) 1, b, b^2, b^3, b^4 **b)** 1, 6, 6^2, 6^3, 6^4

3. No, cannot have a 6 in base 5. **5.** Answers will vary.

7. 134_5 **9.** 3201_4 **11.** $9E5_{12}$ **13.** 2200_3

15. 24001_7 **17.** 10100_2 **19.** 213_4 **21.** 400_5

23. 644_{12} **25.** 11_2 **27.** 3616_7 **29.** 1011_3 **31.** 121_5

33. 2403_7 **35.** 21020_6 **37.** 12233_9 **39.** 100011_2

41. 6031_7 **43.** 110_2 **45.** 31_5 **47.** 123_4

49. 103_4 $R1_4$ **51.** 41_5 $R1_5$ **53.** 45_7 $R2_7$ **55.** ⊖○$_5$

57. ⊖⊖⦶$_5$ **59.** ●●$_4$ **61.** ●●●$_4$ **63.** ●●$_4$

65. ●●●$_4$ **67.** 2302_5, 327 **69.** 13_5

71. a) 21252_8 **b)** 306 and 29 **c)** 8874 **d)** 8874 **e)** Yes

72. 5 **73.** ● = 0, ● = 1, ● = 2, ● = 3

Section 4.5, Page 201

1. Duplation and mediation, the galley method, and Napier rods

3. a) Answers will vary. **b)** 10,498

5. 713 **7.** 1458 **9.** 8260 **11.** 8649 **13.** 2250

15. 2332 **17.** 900 **19.** 204,728 **21.** 504 **23.** 406

25. 625 **27.** 60,678 **29. a)** 46×253 **b)** 11,638
31. a) 4×382 **b)** 1528 **33.** 99∩∩∩∩∩∩∩∩∩||||||
35. Answers will vary. **37.** 12331_5 **38. a)** 1776

Review Exercises, Page 203

1. 3103 **2.** 1211 **3.** 1311 **4.** 2114 **5.** 2314
6. 2312 **7.** *bbbbbaaaaaa* **8.** *cbbaaaaa*
9. *ccbbbbbbbbbaaa* **10.** *ddaaaaa*
11. *dddddcccccccbbbbba* **12.** *ddcccbaaaa*
13. 43 **14.** 27 **15.** 749 **16.** 4068 **17.** 5648
18. 6905 **19.** *hxb* **20.** *byixe* **21.** *hyfxb*
22. *czixd* **23.** *fzd* **24.** *bza* **25.** 45 **26.** 308
27. 568 **28.** 46,883 **29.** 64,481 **30.** 60,529
31. qe **32.** upb **33.** vrc **34.** BArg **35.** ODvog
36. QFvrf **37.** 𓆼9999∩∩∩∩∩∩||
38. MCDLXII **39.** 一千四百六十二 **40.** $\alpha'\upsilon\xi\beta$

41. <<▼▼▼▼ <<▼▼ **42.** ····
·
·· **43.** 222,035 **44.** 8254

45. 685 **46.** 1991 **47.** 1277 **48.** 2690 **49.** 39
50. 5 **51.** 28 **52.** 1244 **53.** 1451 **54.** 186
55. 13033_4 **56.** 122011_3 **57.** 111001111_2 **58.** 3323_5
59. 327_{12} **60.** 717_8 **61.** 140_7 **62.** 101111_2
63. 176_{12} **64.** 1023_7 **65.** 12102_5 **66.** 12423_8
67. 3411_7 **68.** 100_2 **69.** $2E3_{12}$ **70.** 3324_5 **71.** 450_8
72. 1102_3 **73.** 212_6 **74.** 1314_5 **75.** 5656_{12}
76. 21102_3 **77.** 110111_2 **78.** 13632_8 **79.** 1011_2
80. 130_4 **81.** $23_5 R1_5$ **82.** 433_6 **83.** $411_6 R1_6$
84. $664_8 R\ 2_8$ **85.** 3408 **86.** 3408 **87.** 3408

Chapter Test, Page 204

1. A number is a quantity and answers the question "How many?" A numeral is a symbol used to represent a number.
2. 3646 **3.** 1275 **4.** 8090 **5.** 969 **6.** 122,142
7. 9999 **8.** 9999∩∩∩∩∩∩||| **9.** $\beta'\upsilon\ o\ \varsigma$
10. ···
··
≡
····
═ **11.** <<▼▼▼▼▼▼ <<<▼▼▼▼▼▼

12. MMCCCLXXVIII
13. In an additive system, the number represented by a particular set of numerals is the sum of the values of the numerals.
14. In a multiplicative system, there are numerals for each number less than the base and for powers of the base. Each numeral less than the base is multiplied by a numeral for the power of the base, and these products are added to obtain the number.
15. In a ciphered system, the number represented by a particular set of numerals is the sum of the values of the numerals. There are numerals for each number up to and including the base and multiples of the base.
16. In a place-value system, each number is multiplied by a power of the base. The position of the numeral indicates the power of the base by which it is multiplied.
17. 41 **18.** 103 **19.** 45 **20.** 305 **21.** 100100_2
22. 333_5 **23.** 1444_{12} **24.** 11365_7 **25.** 1122_5
26. 142_6 **27.** 2003_6 **28.** 220_5 **29.** 980 **30.** 8428

Chapter 5

Section 5.1, Page 216

1. Number theory is the study of numbers and their properties.
3. a) *a* divides *b* means that *b* divided by *a* has a remainder of zero.
b) *a* is divisible by *b* means that *a* divided by *b* has a remainder of zero.
5. A composite number is a natural number that is divisible by a number other than itself and 1.
7. a) The LCM of a set of natural numbers is the smallest natural number that is divisible by each number in the set.
b) Answers will vary. **c)** 80
9. Mersenne primes are prime numbers of the form $2^n - 1$, where n is a prime number.
11. Goldbach's conjecture states that every even number greater than or equal to 4 can be represented as the sum of two (not necessarily distinct) prime numbers.
13. The prime numbers between 1 and 100 are 2, 3, 5, 7, 11, 13, 17, 19, 23, 29, 31, 37, 41, 43, 47, 53, 59, 61, 67, 71, 73, 79, 83, 89, and 97.
15. True **17.** False; 21 is a multiple of 7.
19. False; 56 is divisible by 8. **21.** True
23. False; if a number is divisible by 3, then the sum of the digits of the number is divisible by 3.
25. True
27. 10,368 is divisible by 2, 3, 4, 6, 8, and 9.
29. 2,763,105 is divisible by 3 and 5.
31. 1,882,320 is divisible by 2, 3, 4, 5, 6, 8, and 10.
33. 60 (other answers are possible)
35. $45 = 3^2 \cdot 5$ **37.** $196 = 2^2 \cdot 7^2$

39. $303 = 3 \cdot 101$ **41.** $513 = 3^3 \cdot 19$

43. $1336 = 2^3 \cdot 167$ **45.** $2001 = 3 \cdot 23 \cdot 29$

47. a) 3 **b)** 30 **49. a)** 6 **b)** 432 **51. a)** 20 **b)** 1800

53. a) 4 **b)** 5088 **55. a)** 8 **b)** 384

57. 17, 19, and 29, 31

59. a) Yes **b)** No **c)** Yes **d)** Yes

61. $4 = 2 + 2, 6 = 3 + 3, 8 = 3 + 5, 10 = 3 + 7, 12 = 5 + 7, 14 = 7 + 7, 16 = 3 + 13, 18 = 5 + 13, 20 = 3 + 17$

63. 70 dolls **65.** 72 cards **67.** 180 min **69.** 30 days

71. a) 4, 5, 10, 20, 25 **b)** 25, 20, 10, 5, and 4, respectively

73. A number is divisible by 15 if both 3 and 5 divide the number.

75. 5 **77.** 36 **79.** 30 **81.** No **83.** Yes

85. a) 12 **b)** 1, 2, 3, 4, 5, 6, 10, 12, 15, 20, 30, 60

87. For any three consecutive natural numbers, one of the numbers is divisible by 2 and another number is divisible by 3. Therefore, the product of the three numbers would be divisible by 6.

89. Yes

91. $8 = 2 + 3 + 3, 9 = 3 + 3 + 3, 10 = 2 + 3 + 5,$
$11 = 2 + 2 + 7, 12 = 2 + 5 + 5, 13 = 3 + 3 + 7,$
$14 = 2 + 5 + 7, 15 = 3 + 5 + 7, 16 = 2 + 7 + 7,$
$17 = 5 + 5 + 7, 18 = 2 + 5 + 11, 19 = 3 + 5 + 11,$
$20 = 2 + 7 + 11$

92. The answer most people select is Denmark, kangaroo, and orange.

Section 5.2, Page 225

1. Begin at zero. Represent the first addend with an arrow. Draw the arrow to the right if the addend is positive, to the left if negative. From the tip of the first arrow, represent the second addend with a second arrow. The sum of the two integers is at the tip of the second arrow.

3. To rewrite a subtraction problem as an addition problem, rewrite the minus sign as a plus sign and change the second number to its opposite.

5. The quotient of two numbers with like signs is positive. The quotient of two numbers with unlike signs is negative.

7. 3 **9.** 2 **11.** −5 **13.** 2 **15.** −21 **17.** −3

19. −10 **21.** −2 **23.** −6 **25.** −2 **27.** −20

29. 144 **31.** 96 **33.** −60 **35.** −720 **37.** 2 **39.** −1

41. −7 **43.** −15 **45.** −48

In Exercises 47–55, false answers can be modified in a variety of ways. We give one possible answer.

47. True

49. False; the difference of two negative integers may be a positive integer, a negative integer, or zero.

51. True **53.** True

55. False; the sum of a positive integer and a negative integer may be a positive integer, a negative integer, or zero.

57. 6 **59.** −17 **61.** −12 **63.** −5 **65.** −6

67. −15, −10, −5, 0, 5, 10

69. −6, −5, −4, −3, −2, −1

71. 213.8°F **73.** 210 points **75.** 1769 ft

77. a) 9 hours **b)** 2 hours **79.** −1

81. $0 + 1 - 2 + 3 + 4 - 5 + 6 - 7 - 8 + 9 = 1$

Section 5.3, Page 238

1. The set of rational numbers is the set of numbers of the form p/q, where p and q are integers and $q \neq 0$.

3. a) Divide both the numerator and the denominator by their greatest common factor.
b) $\frac{5}{9}$

5. For positive mixed numbers, multiply the denominator of the fraction by the integer preceding it. Add this product to the numerator. This sum is the numerator of the improper fraction; the denominator is the same as the denominator in the mixed number. For negative mixed numbers, temporarily ignore the negative sign, perform the conversion described above, and then add the negative sign.

7. a) The reciprocal of a number is 1 divided by the number.
b) $-\frac{1}{2}$

9. a) To add or subtract two fractions with a common denominator, perform the indicated operation on the numerators. Keep the common denominator. Reduce the new fraction to lowest terms, if possible.
b) $\frac{2}{3}$ **c)** $\frac{1}{2}$

11. Answers will vary. **13.** $\frac{2}{3}$ **15.** $\frac{2}{7}$ **17.** $\frac{21}{32}$ **19.** $\frac{7}{11}$

21. $\frac{1}{11}$ **23.** $\frac{25}{7}$ **25.** $-\frac{31}{16}$ **27.** $-\frac{79}{16}$ **29.** $\frac{17}{8}$ **31.** $\frac{15}{8}$

33. $1\frac{3}{8}$ **35.** $-12\frac{1}{6}$ **37.** $-58\frac{8}{15}$ **39.** 0.6 **41.** $0.\overline{2}$

43. 0.375 **45.** $4.\overline{3}$ **47.** $5.\overline{6}$ **49.** $\frac{25}{100} = \frac{1}{4}$ **51.** $\frac{45}{1000} = \frac{9}{200}$

53. $\frac{2}{10} = \frac{1}{5}$ **55.** $\frac{452}{1000} = \frac{113}{250}$ **57.** $\frac{1}{10{,}000}$ **59.** $\frac{2}{3}$ **61.** $\frac{2}{1}$

63. $\frac{15}{11}$ **65.** $\frac{46}{45}$ **67.** $\frac{574}{165}$ **69.** $\frac{3}{22}$ **71.** $\frac{2}{5}$ **73.** $\frac{49}{64}$ **75.** $\frac{36}{35}$

77. $\frac{5}{14}$ **79.** $\frac{13}{15}$ **81.** $\frac{21}{26}$ **83.** $\frac{23}{54}$ **85.** $\frac{17}{144}$ **87.** $-\frac{109}{600}$

89. $\frac{51}{40}$ **91.** $-\frac{1}{24}$ **93.** $\frac{19}{24}$ **95.** 1 **97.** $\frac{4}{11}$ **99.** $\frac{23}{42}$ **101.** $\frac{1}{12}$

103. $120\frac{3}{4}$ in. **105. a)** $\frac{3}{8}$ cup **b)** $\frac{3}{16}$ tsp **c)** $\frac{3}{4}$ cup

107. $\frac{59}{60}$ **109.** 27 pages **111.** $58\frac{7}{8}$ in. **113.** $12\frac{7}{16}$ in.

115. $26\frac{5}{32}$ in. **117. a)** $37\frac{5}{6}$ ft **b)** 88 ft^2 **c)** $806\frac{2}{3}$ ft^3

In Exercises 119–125, an infinite number of answers are possible. We give one answer.

119. 0.105 **121.** −2.1755 **123.** 3.1234505

125. 4.8725 **127.** $\frac{1}{2}$ **129.** $\frac{11}{200}$ **131.** $\frac{9}{40}$ **133.** $\frac{11}{200}$

135. a) $1\frac{3}{8}$ cup water (or milk) and $\frac{3}{4}$ cup oatmeal
b) $1\frac{1}{2}$ cup water (or milk) and $\frac{3}{4}$ cup oatmeal

137. a) $\frac{1}{8}$ **b)** $\frac{1}{16}$ **c)** 5 **d)** 6

Section 5.4, Page 247

1. A rational number can be written as a ratio of two integers. Real numbers that cannot be written as a ratio of two integers are irrational numbers.

3. A perfect square is any number that is the square of a natural number.

5. a) To add or subtract two or more square roots with the same radicand, add or subtract their coefficients and then multiply the sum or difference by the common radical.
b) $-\sqrt{6}$

7. a) Multiply both the numerator and denominator by a radical that will result in the radicand in the denominator becoming a perfect square.
b) $\frac{7\sqrt{3}}{3}$

9. Rational **11.** Rational **13.** Irrational **15.** Rational **17.** Irrational **19.** 8 **21.** 10 **23.** -13 **25.** -15 **27.** -10 **29.** Rational number, integer, natural number

31. Rational number, integer, natural number

33. Rational number **35.** Rational number

37. Rational number **39.** $3\sqrt{2}$ **41.** $4\sqrt{3}$ **43.** $3\sqrt{7}$ **45.** $4\sqrt{5}$ **47.** $9\sqrt{2}$ **49.** $7\sqrt{6}$ **51.** $5\sqrt{3}$ **53.** $-13\sqrt{3}$ **55.** $4\sqrt{3}$ **57.** $23\sqrt{2}$ **59.** 4 **61.** $2\sqrt{15}$ **63.** $10\sqrt{2}$ **65.** $\sqrt{2}$ **67.** 3 **69.** $\frac{\sqrt{2}}{2}$ **71.** $\frac{\sqrt{21}}{7}$ **73.** $\frac{2\sqrt{15}}{3}$ **75.** $\frac{3\sqrt{2}}{2}$ **77.** $\frac{\sqrt{15}}{3}$

79. $\sqrt{7}$ is between 2 and 3 since 7 is between 4 and 9. $\sqrt{7}$ is between 2.5 and 3 since 7 is closer to 9 than to 4. $\sqrt{7} \approx 2.65$.

81. $\sqrt{107}$ is between 10 and 11 since 107 is between 100 and 121. $\sqrt{107}$ is between 10 and 10.5 since 107 is closer to 100 than to 121. $\sqrt{107} \approx 10.34$.

83. $\sqrt{170}$ is between 13 and 14 since 170 is between 169 and 196. $\sqrt{170}$ is between 13 and 13.5 since 170 is closer to 169 than to 196. $\sqrt{170} \approx 13.04$.

In Exercises 85–89, false answers can be modified in a variety of ways. We give one possible answer.

85. False. $\sqrt{p}$ is an irrational number for any prime number p.

87. True

89. False. The product of a rational number and an irrational number may be a rational number or an irrational number.

91. $\pi + (-\pi) = 0$ **93.** $\sqrt{2} \cdot \sqrt{3} = \sqrt{6}$

95. $\sqrt{5} \neq 2.236$ since $\sqrt{5}$ is irrational and 2.236 is rational.

97. No. π is irrational; therefore, it cannot equal $\frac{22}{7}$ or 3.14, both of which are rational.

99. $\sqrt{4 \cdot 9} = \sqrt{4} \cdot \sqrt{9}$, $\sqrt{36} = 2 \cdot 3$, $6 = 6$

101. a) 10 mph **b)** 20 mph **c)** 40 mph **d)** 80 mph

103. a) If the result on the calculator is a terminating or repeating decimal number, then the number is rational; if the result is not a terminating or repeating decimal number, then the number is irrational.
b) Rational. $\sqrt{0.04} = 0.2$, which is a rational number.
c) Irrational. $\sqrt{0.7} = 0.8366600265\ldots$. Since the decimal number is not a terminating or a repeating decimal number, this is an irrational number.

105. a) $(44 \div \sqrt{4}) \div \sqrt{4} = 11$ **b)** $(44 \div 4) + \sqrt{4} = 13$
c) $4 + 4 + 4 + \sqrt{4} = 14$
d) $\sqrt{4}(4 + 4) + \sqrt{4} = 18$
Other answers are possible.

Section 5.5, Page 254

1. The real numbers are the union of the rational numbers and the irrational numbers.

3. If whenever the operation is performed on two elements of a set the result is also an element of the set, then the set is closed under that operation.

5. $a + b = b + a$, the order in which two numbers are added is immaterial. One example is $4 + 5 = 5 + 4$.

7. $(a \cdot b) \cdot c = a \cdot (b \cdot c)$, when multiplying three numbers, you may place parentheses around any two adjacent numbers. One example is $(1 \times 2) \times 3 = 1 \times (2 \times 3)$.

9. Yes **11.** No **13.** Yes **15.** No **17.** Yes **19.** Yes **21.** No **23.** No **25.** Yes **27.** No

29. Commutative property of addition. The only difference between the expressions on both sides of the equal sign is the order of $(3 + 4)$ and x being added.

31. $(-3)(-4) = (-4)(-3) = 12$

33. No. $3 \div 4 \neq 4 \div 3$.

35. $[(-2)(-3)](-4) = (-2)[(-3)(-4)] = -24$

37. No. $(16 \div 8) \div 2 \neq 16 \div (8 \div 2)$.

39. No. $(81 \div 9) \div 3 \neq 81 \div (9 \div 3)$.

41. Commutative property of addition

43. Associative property of multiplication

45. Associative property of addition

47. Commutative property of multiplication

49. Distributive property

51. Commutative property of addition

53. Distributive property

55. Commutative property of addition

57. $2c + 14$ **59.** $\frac{2}{3}x - 4$ **61.** $3x + 4$ **63.** $2x - 1$

65. $15 - 3\sqrt{5}$ **67.** $2 + \sqrt{6}$

69. a) Distributive property
b) Associative property of addition

71. a) Distributive property
b) Associative property of addition
c) Commutative property of addition
d) Associative property of addition

73. a) Distributive property
b) Commutative property of addition
c) Associative property of addition
d) Commutative property of addition

75. Yes **77.** No **79.** Yes **81.** Yes **83.** Yes **85.** Yes

87. Yes **89.** Answers will vary.

91. No. $0 \div a = 0$ (when $a \neq 0$), but $a \div 0$ is undefined.

92. a) No **b)** No **c)** Answers will vary.

Section 5.6, Page 264

1. The 2 is the base and the 3 is the exponent or power.

3. a) To multiply two exponential expressions with the same base, add the exponents and use this sum as the exponent on the common base.
b) $2^3 \cdot 2^4 = 2^{3+4} = 2^7 = 128$

5. a) Any nonzero expression raised to the power of 0 equals 1.
b) $7^0 = 1$

7. a) Any base with an exponent raised to another exponent is equal to the base raised to the product of the exponents.
b) $(3^2)^4 = 3^{2 \cdot 4} = 3^8 = 6561$

9. a) -1^{500} means $-(1)^{500}$ or $-1 \cdot 1^{500}$. Since 1 raised to any power equals 1, $-1^{500} = -1 \cdot 1^{500} = -1 \cdot 1 = -1$.
b) $(-1)^{500}$ means (-1) multiplied by itself 500 times. Since 500 is even, $(-1)^{500} = 1$.
c) -1^{501} means $-(1)^{501}$ or $-1 \cdot 1^{501}$. Since 1 raised to any power equals 1, $-1^{501} = -1 \cdot 1^{501} = -1 \cdot 1 = -1$.
d) $(-1)^{501}$ means (-1) multiplied by itself 501 times. Since 501 is odd, $(-1)^{501} = -1$.

11. a) If the exponent is positive, move the decimal point in the number to the right the same number of places as the exponent, adding zeros where necessary. If the exponent is negative, move the decimal point in the number to the left the same number of places as the exponent, adding zeros where necessary.
b) 0.000576

13. 25 **15.** 16 **17.** -9 **19.** $\frac{4}{9}$ **21.** 25 **23.** 72 **25.** 25

27. $\frac{1}{49}$ **29.** 1 **31.** 81 **33.** $\frac{1}{9}$ **35.** 4096 **37.** 121

39. 16 **41.** -16 **43.** $\frac{1}{64}$ **45.** 2.31×10^5

47. 1.5×10^1 **49.** 5.6×10^{-1} **51.** 1.9×10^4

53. 1.86×10^{-4} **55.** 4.23×10^{-6} **57.** 7.11×10^2

59. 1.53×10^{-1} **61.** 2300 **63.** 0.003901

65. 0.0000862 **67.** 0.312 **69.** 9,000,000 **71.** 231

73. 35,000 **75.** 10,000 **77.** 800,000 **79.** 0.0153

81. 320 **83.** 0.0021 **85.** 20 **87.** 6.0×10^{11}

89. 4.5×10^{-7} **91.** 2.0×10^3 **93.** 2.0×10^{-7}

95. 3.0×10^8

97. 8.3×10^{-4}; 3.2×10^{-1}; 4.6; 5.8×10^5

99. 8.3×10^{-5}; 0.00079; 4.1×10^3; 40,000

101. a) $\approx$\$35,590.18 **b)** \$3.559018 $\times 10^4$

103. a) $\approx$210,109,000,000,000,000,000 sec
b) $\approx 2.1 \times 10^{20}$ sec (about 6.7 trillion years!)

105. a) 18,000 hours **b)** 1.8×10^4

107. a) 20,000,000,000,000,000 drops **b)** 2×10^{16}

109. a) 18,000 times **b)** 1.8×10^4

111. a) $\approx$ \$17,093.02 per person **b)** \$5989.17

113. a) \$720,000,000 **b)** \$300,000,000 **c)** \$120,000,000
d) \$60,000,000

115. 1000 **117.** 333,333 times

119. 230,000 sec (about 2.66 days)

121. a) About 5.87×10^{12} (5.87 trillion) mi
b) About 500 sec or 8 min 20 sec

Section 5.7, Page 273

1. A sequence is a list of numbers that are related to each other by a given rule. One example is 1, 3, 5, 7, 9, ...

3. a) An arithmetic sequence is one in which each term differs from the preceding term by a constant amount. One example is 4, 7, 10, 13, 16, ...
b) A geometric sequence is one in which the ratio of any two successive terms is a constant amount. One example is 3, 6, 12, 24, ...

5. a) a_n is the nth term or the general term.
b) a_1 is the first term.
c) d is the common difference.
d) s_n is the sum of the first n terms.

7. 3, 5, 7, 9, 11 **9.** $-5, -2, 1, 4, 7$ **11.** $5, 3, 1, -1, -3$

13. $\frac{1}{2}, 1, \frac{3}{2}, 2, \frac{5}{2}$ **15.** 17 **17.** 13 **19.** $-\frac{91}{5}$ **21.** 9

23. $a_n = n$ **25.** $a_n = 2n$ **27.** $a_n = \frac{1}{3}n - 2$

29. $a_n = \frac{3}{2}n - \frac{9}{2}$ **31.** $s_{50} = 1275$ **33.** $s_{50} = 2500$

35. $s_8 = -52$ **37.** $s_8 = 60$ **39.** 3, 6, 12, 24, 48

41. $2, -4, 8, -16, 32$ **43.** $-3, 3, -3, 3, -3$

45. $-16, 8, -4, 2, -1$ **47.** 3072 **49.** $\frac{3}{4}$ **51.** 8

53. $a_{10} = -39{,}366$

55. $a_n = 2^{n-1}$ **57.** $a_n = 3 \cdot (-1)^{n-1}$

59. $a_n = \dfrac{1}{4} \cdot (2)^{n-1}$ **61.** $a_n = 9 \cdot \left(\dfrac{1}{3}\right)^{n-1}$ **63.** 45

65. 27,305 **67.** $-620{,}011$ **69.** $s_{15} = -10{,}923$

71. 5050 **73.** 10,000 **75. a)** \$28,600 **b)** \$195,200
77. 12 in. **79.** 496 pinecones **81.** 52.4288 g
83. \$45,218 **85.** \$486,000 **87.** 161.4375 **89.** 267
91. 191.3568 ft

Section 5.8, Page 280

1. The first and second terms are one. Each term thereafter is the sum of the previous two terms.

3. a) The golden number is $\frac{\sqrt{5}+1}{2}$.
b) When a line segment *AB* is divided at a point *C*, such that the ratio of the whole, *AB*, to the larger part, *AC*, is equal to the ratio of the larger part, *AC*, to the smaller part, *CB*, then each of the two ratios *AB/AC* and *AC/CB* is known as the golden ratio.
c) The proportion made by using the two golden ratios, $AB/AC = AC/CB$, is known as the golden proportion.
d) A golden rectangle is one where the ratio of the length to the width is equal to the golden number.

5. Answers will vary. **7. a)** 1.618 **b)** 0.618 **c)** 1

9. $\frac{1}{1} = 1, \frac{2}{1} = 2, \frac{3}{2} = 1.5, \frac{5}{3} \approx 1.667, \frac{8}{5} = 1.6, \frac{13}{8} = 1.625, \frac{21}{13} \approx 1.615, \frac{34}{21} \approx 1.619, \frac{55}{34} \approx 1.6176, \frac{89}{55} \approx 1.6182$. The consecutive ratios alternate, increasing and decreasing about the golden ratio.

11. Each number in the Fibonacci sequence is either a prime number or is relatively prime with the number preceding or succeeding it in the sequence. Therefore, the GCF of any two consecutive Fibonacci numbers is 1.

13. Answers will vary. **15.** Answers will vary.
17. Answers will vary. **19.** Answers will vary.
21. Answers will vary. **23.** Yes; 29, 47 **25.** No
27. Yes; 105, 170 **29.** Yes; $-1, -1$
31. Answers will vary. **33.** Answers will vary.
35. a) 1, 3, 4, 7, 11, 18, 29, 47
b) $8 + 21 = 29$, $13 + 34 = 47$
c) It is the Fibonacci sequence.
37. Answers will vary. **39.** Answers will vary.

Review Exercises, Page 283

1. 2, 3, 4, 6, 9 **2.** 2, 3, 4, 6, 9 **3.** $2^2 \cdot 3^2 \cdot 7$
4. $5 \cdot 7 \cdot 11$ **5.** $2^3 \cdot 3 \cdot 5 \cdot 7$ **6.** $2 \cdot 3^2 \cdot 7^2$ **7.** $2^2 \cdot 3 \cdot 11^2$
8. 15; 60 **9.** 9; 756 **10.** 5; 2250 **11.** 40; 6720
12. 4; 480 **13.** 36; 432 **14.** 45 days **15.** 3 **16.** -3
17. -4 **18.** -6 **19.** -9 **20.** 3 **21.** 0 **22.** 4 **23.** 33
24. -36 **25.** -56 **26.** 5 **27.** -2 **28.** 6 **29.** 6
30. 3 **31.** 0.3 **32.** 0.6 **33.** 0.375 **34.** 3.25
35. $0.\overline{428571}$ **36.** $0.58\overline{3}$ **37.** 0.375 **38.** 0.875
39. $0.\overline{714285}$ **40.** $\frac{9}{40}$ **41.** $\frac{9}{2}$ **42.** $\frac{2}{3}$ **43.** $\frac{235}{99}$ **44.** $\frac{83}{1000}$
45. $\frac{21}{5000}$ **46.** $\frac{211}{90}$ **47.** $\frac{19}{7}$ **48.** $\frac{25}{6}$ **49.** $-\frac{13}{4}$ **50.** $-\frac{283}{8}$
51. $2\frac{1}{5}$ **52.** $1\frac{4}{5}$ **53.** $-1\frac{5}{7}$ **54.** $-27\frac{1}{5}$ **55.** $\frac{13}{10}$ **56.** $\frac{1}{8}$
57. $\frac{17}{12}$ **58.** $\frac{3}{4}$ **59.** $\frac{35}{54}$ **60.** $\frac{53}{28}$ **61.** $\frac{1}{6}$ **62.** $\frac{13}{40}$ **63.** $\frac{8}{15}$
64. $2\frac{7}{32}$ tsp **65.** $5\sqrt{2}$ **66.** $10\sqrt{2}$ **67.** $8\sqrt{5}$ **68.** $-3\sqrt{3}$
69. $8\sqrt{2}$ **70.** $-20\sqrt{3}$ **71.** $8\sqrt{3}$ **72.** $3\sqrt{2}$ **73.** $4\sqrt{3}$
74. 3 **75.** $2\sqrt{7}$ **76.** $\frac{4\sqrt{3}}{3}$ **77.** $\frac{\sqrt{15}}{5}$ **78.** $6 + 3\sqrt{7}$
79. $4\sqrt{3} + 3\sqrt{2}$ **80.** $3\sqrt{2} + 3\sqrt{5}$
81. Commutative property of addition
82. Commutative property of multiplication
83. Associative property of addition
84. Distributive property
85. Commutative property of addition
86. Commutative property of addition
87. Associative property of multiplication
88. Commutative property of multiplication
89. Distributive property
90. Commutative property of multiplication
91. Yes **92.** No **93.** No **94.** Yes **95.** No **96.** No
97. 9 **98.** $\frac{1}{9}$ **99.** 81 **100.** 125 **101.** 1 **102.** $\frac{1}{64}$
103. 64 **104.** 81 **105.** 2.3×10^8 **106.** 1.58×10^{-5}
107. 2.75×10^{-3} **108.** 4.95×10^6 **109.** 43,000,000
110. 0.000139 **111.** 0.000175 **112.** 100,000
113. 1.4×10^{-1} **114.** 1.0×10^5 **115.** 2.1×10^1
116. 3.0×10^0 **117.** 8,000,000,000 **118.** 0.7
119. 3200 **120.** 5 **121.** $\approx$388 times **122.** $\approx$ \$5555.56
123. Arithmetic; 14, 17 **124.** Geometric; 8, 16
125. Arithmetic; $-15, -18$ **126.** Geometric; $\frac{1}{32}, \frac{1}{64}$
127. Arithmetic; 16, 19 **128.** Geometric; $-2, 2$ **129.** 15
130. -34 **131.** 25 **132.** 48 **133.** $\frac{1}{4}$ **134.** -48
135. 1365 **136.** -25 **137.** 632 **138.** 57.5 **139.** 200
140. 80 **141.** 33 **142.** -21
143. Arithmetic; $a_n = -3n + 10$
144. Arithmetic; $a_n = 3n$ **145.** Arithmetic; $a_n = -\frac{3}{2}n + \frac{11}{2}$
146. Geometric; $a_n = 3(2)^{n-1}$
147. Geometric; $a_n = 2(-1)^{n-1}$
148. Geometric; $a_n = 5\left(\frac{1}{3}\right)^{n-1}$ **149.** Yes; 13, 21
150. Yes; 17, 28 **151.** No **152.** No

Chapter Test, Page 285

1. 2, 3, 5, 6, 9, 10 **2.** $2^3 \cdot 3 \cdot 5 \cdot 7$ **3.** -7 **4.** -20
5. -175 **6.** $\frac{37}{8}$ **7.** $19\frac{5}{9}$ **8.** 0.625 **9.** $\frac{129}{20}$ **10.** $\frac{121}{240}$

11. $\frac{13}{24}$ **12.** $9\sqrt{3}$ **13.** $\dfrac{\sqrt{14}}{7}$

14. Yes; the product of any two integers is an integer.

15. Associative property of addition

16. Distributive property **17.** 64 **18.** 1024 **19.** $\frac{1}{81}$

20. 8.0×10^{11} **21.** $a_n = -4n + 2$ **22.** -187 **23.** 243

24. 1023 **25.** $a_n = 3(2)^{n-1}$

26. 1, 1, 2, 3, 5, 8, 13, 21, 34, 55

Chapter 6

Section 6.1, Page 291

1. Letters of the alphabet used to represent numbers are called variables.

3. The solution to an equation is the number or numbers that replace the variable to make the equation a true statement.

5. a) The 4 is the base and the 5 is the exponent.
b) Answers will vary.

7. 12 **9.** 49 **11.** -9 **13.** 686 **15.** -3 **17.** 18

19. -27 **21.** $-\frac{10}{9}$ **23.** 7 **25.** 9 **27.** 0 **29.** No

31. No **33.** No **35.** Yes **37.** Yes **39.** \$12.25

41. \$426.25 **43.** 16,000,000 sec **45.** 1.71 in.

47. The two expressions are not equal.

Section 6.2, Page 302

1. The parts that are added or subtracted in an algebraic expression are called terms. In $3x - 2y$, the $3x$ and $-2y$ are terms.

3. The numerical part of a term is called its numerical coefficient. For the term $3x$, 3 is the numerical coefficient.

5. To simplify an expression means to combine like terms by using the commutative, associative, and distributive properties.

7. If $a = b$, then $a - c = b - c$ for all real numbers a, b, and c. If $2x + 3 = 5$, then $2x + 3 - 3 = 5 - 3$.

9. If $a = b$, then $a/c = b/c$ for all real numbers a, b, and c, where $c \neq 0$. If $4x = 8$, then $\dfrac{4x}{4} = \dfrac{8}{4}$.

11. A ratio is a quotient of two quantities. An example is $\dfrac{7}{9}$.

13. Yes. They have the same variable and the same exponent on the variable.

15. $11x$ **17.** $2x + 12$ **19.** $3x + 11y$ **21.** $-8x + 2$

23. $-5x + 3$ **25.** $13.3x - 8.3$ **27.** $-\frac{2}{15}x - 4$

29. $13x - 7y + 3$ **31.** $8s - 17$ **33.** $1.5x - 4.2$

35. $\frac{5}{12}x + \frac{3}{7}$ **37.** $4.52x - 13.5$ **39.** 5 **41.** 1 **43.** $\frac{24}{7}$

45. $\frac{2}{3}$ **47.** 3 **49.** 3 **51.** 17 **53.** -8

55. No solution **57.** All real numbers **59.** $\frac{4}{3}$ **61.** -3

63. 4 **65.** \$72.37 **67.** 59 times

69. 20,746,600 households

71. a) 1.6 kph **b)** 56.25 mph **73.** 0.3 cc

75. a) Answers will vary. **b)** -1

77. a) An equation that has no solution.
b) You will obtain a false statement.

79. a) $2:5$ **b)** $m : m + n$

Section 6.3, Page 311

1. A formula is an equation that typically has a real-life application.

3. Subscripts are numbers (or letters) placed below and to the right of variables. They are used to help clarify a formula.

5. An exponential equation is of the form $y = a^x$, $a > 0$, $a \neq 1$.

7. 20 **9.** 56 **11.** 25 **13.** 62.8 **15.** 37.1 **17.** 2

19. 3000 **21.** 8 **23.** 25 **25.** 6 **27.** 200 **29.** 7.2

31. 14 **33.** 3240 **35.** 0.5 **37.** 14

39. $y = \dfrac{10x - 13}{4}$ or $y = \dfrac{5}{2}x - \dfrac{13}{4}$

41. $y = \dfrac{-4x + 14}{7}$ or $y = -\dfrac{4}{7}x + 2$

43. $y = \dfrac{2x + 6}{3}$ or $y = \dfrac{2}{3}x + 2$

45. $y = \dfrac{2x - z + 15}{3}$ or $y = \dfrac{2}{3}x - \dfrac{1}{3}z + 5$

47. $y = \dfrac{9x + 4z - 7}{8}$ or $y = \dfrac{9}{8}x + \dfrac{1}{2}z - \dfrac{7}{8}$

49. $R = \dfrac{E}{I}$ **51.** $a = p - b - c$ **53.** $B = \dfrac{3V}{h}$

55. $r = \dfrac{C}{2\pi}$ **57.** $b = y - mx$ **59.** $w = \dfrac{P - 2l}{2}$

61. $c = 3A - a - b$ **63.** $T = \dfrac{PV}{K}$

65. $C = \frac{5}{9}(F - 32)$ **67.** $s = \dfrac{S - \pi r^2}{\pi r}$

69. a) \$12 **b)** \$612 **71.** ≈ 18.4 in.^{3}

73. 486,000 bacteria **75.** \$300,976,658,300,000

77. ≈ 1051.47 in.^{3}

Section 6.4, Page 317

1. A mathematical expression is a collection of variables, numbers, parentheses, and operation symbols. An equation is two algebraic expressions joined by an equal sign.

3. $4 + 3x$ **5.** $6r + 5$ **7.** $15 - 2r$ **9.** $2m + 9$

11. $\dfrac{18 - s}{4}$ **13.** $(5y - 6) + 3$ **15.** $x - 6 = 5$; 11

17. $x - 4 = 20$; 24 **19.** $12 + 5x = 47$; 7

21. $8x + 16 = 88$; 9 **23.** $x + 11 = 3x + 1$; 5

25. $x + 10 = 2(x + 3)$; 4

27. $x + 3x = 600$; 150 tickets to nonstudents, 450 tickets to students

29. $x + 0.116x = 34.20$; $\approx$ 30.65 million taxpayers

31. $x - 0.10x = 15.72$; \$17.47

33. $x + 3x = 12$; Samantha: 3, Josie: 9

35. $2000 + 50x = 13{,}350$; \$227

37. **a)** $x + x + 3x = 45{,}000$; 9000 ft^2, 9000 ft^2, 27,000 ft^2 **b)** Yes

39. $x + 3x + 3 = 55$; United States: 13, Italy: 42

41. $3w + 2(2w) = 140$; width: 20 ft, length: 40 ft

43. $70x = 760$; $\approx$ 11 months

45. $\frac{r}{2} + 0.07r = 257$; \$450.88

47. Deduct \$720 from Mr. McAdams's income and \$2920 from Mrs. McAdams's income.

49.
$$\begin{aligned} x + (x + 1) + (x + 2) &= 3(x + 2) - 3 \\ 3x + 3 &= 3x + 6 - 3 \\ 3x + 3 &= 3x + 3 \end{aligned}$$

51. $-40°$

Section 6.5, Page 326

1. Inverse variation: As one variable increases, the other decreases and vice versa.

3. Joint variation: One quantity varies directly as the product of two or more other quantities.

5. Direct **7.** Inverse **9.** Direct **11.** Inverse

13. Inverse **15.** Inverse **17.** Direct **19.** Direct

21. Answers will vary. **23.** **a)** $y = kx$ **b)** 15

25. **a)** $m = \frac{k}{n^2}$ **b)** 0.25 **27.** **a)** $R = \frac{k}{W}$ **b)** 0.05

29. **a)** $F = kDE$ **b)** 210 **31.** **a)** $t = \frac{kd^2}{f}$ **b)** 200

33. **a)** $Z = kWY$ **b)** 100 **35.** **a)** $H = kL$ **b)** 3

37. **a)** $A = kB^2$ **b)** 720 **39.** **a)** $F = \frac{kq_1q_2}{d^2}$ **b)** 672

41. **a)** $R = kL$ **b)** 0.32 ohm **43.** **a)** $l = \frac{k}{d^2}$ **b)** 80 dB

45. **a)** $R = \frac{kA}{P}$ **b)** 4800 tapes

47. **a)** $s = kwd^2$ **b)** 6480 pounds per square inch

49. **a)** $N = \frac{kp_1p_2}{d}$ **b)** $\approx$ 121,528 calls

51. **a)** Inversely **b)** Stays 0.3 **53.** \$132.27

Section 6.6, Page 332

1. $a < b$ means that a is less than b, $a \le b$ means that a is less than or equal to b, $a > b$ means that a is greater than b, $a \ge b$ means that a is greater than or equal to b.

3. When both sides of an inequality are multiplied or divided by a negative number, the direction of the inequality symbol must be reversed.

5. Yes, the inequality symbol points to the -3 in both cases.

7. 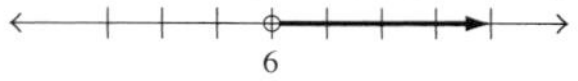 6

9.  0 3

11. 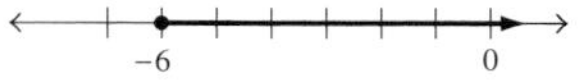 −6 0

13. 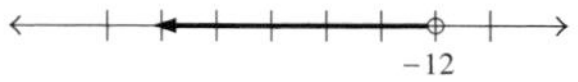 −12

15. 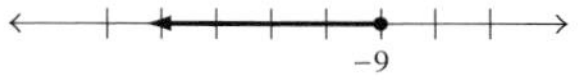 −9

17.  0 4

19. 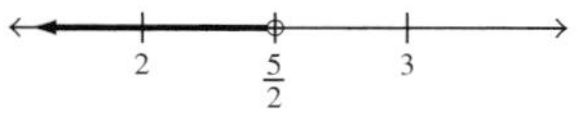 2 $\frac{5}{2}$ 3

21. 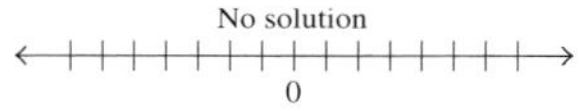 No solution 0

23. 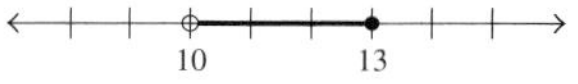 10 13

25. 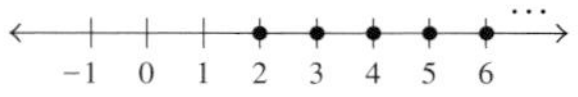 −1 0 1 2 3 4 5 6

27. 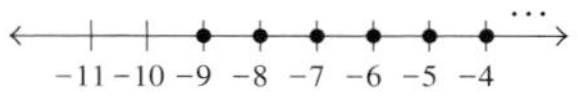 −11 −10 −9 −8 −7 −6 −5 −4

29. 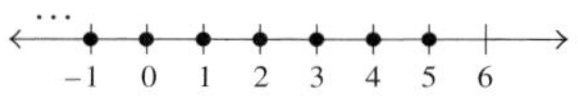 −1 0 1 2 3 4 5 6

31. 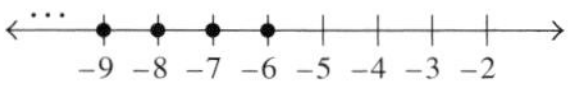 −9 −8 −7 −6 −5 −4 −3 −2

33. 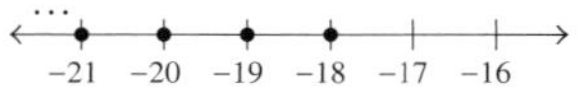 −21 −20 −19 −18 −17 −16

35. 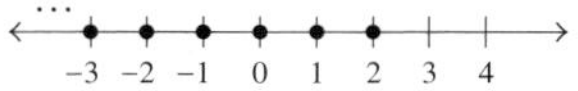 −3 −2 −1 0 1 2 3 4

37. 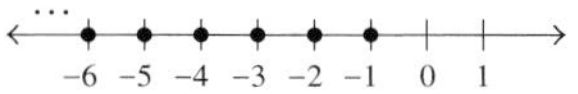 −6 −5 −4 −3 −2 −1 0 1

39. 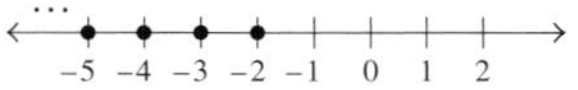 −5 −4 −3 −2 −1 0 1 2

41. 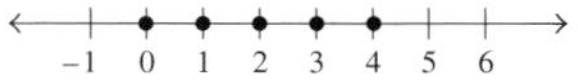 −1 0 1 2 3 4 5 6

43. 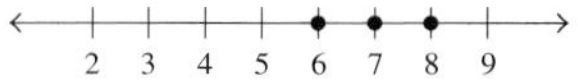 2 3 4 5 6 7 8 9

45. **a)** 2000, 2001 **b)** 1997, 1998 **c)** 1997, 1998, 1999, 2000 **d)** 1998, 1999, 2000, 2001

47. 19 videos **49.** Less than 360 mi **51.** $x \le \$15.57$

53. $0.5 < t < 1.5$

55. $94 \le x \le 100$, assuming 100 is the highest grade possible

57. $6.875 \le x \le 11$

59. The student's answer is $x \le -12$, whereas the correct answer is $x \ge -12$. Yes, -12 is in both solution sets.

Section 6.7, Page 344

1. A graph is an illustration of all the points whose coordinates satisfy an equation.

3. To find the y-intercept, set $x = 0$ and solve the equation for y.

5. a) Answers will vary. **b)** $-\frac{1}{3}$

7. a) First **b)** Second

For Exercises 9–15, see the following figure.

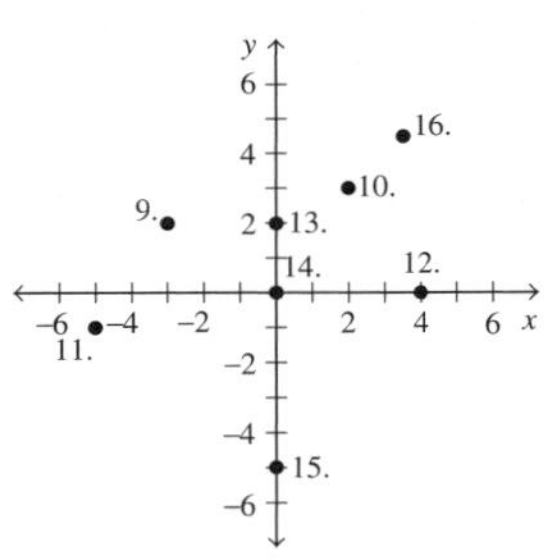

For Exercises 17–23, see the following figure.

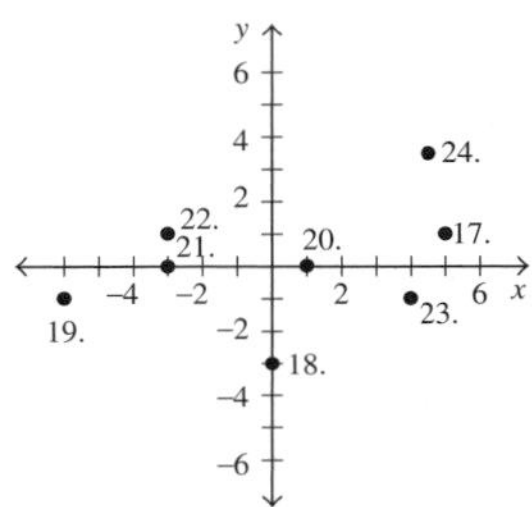

25. $(0, 2)$ **27.** $(-2, 0)$ **29.** $(-5, -3)$ **31.** $(2, -3)$
33. $(2, 2)$ **35.** $(1, 4), (-1, 10)$ **37.** $(5, 0), (0, -\frac{10}{3})$
39. $(-3, -2)$ **41.** $(0, \frac{8}{3}), (4, 0)$

43.

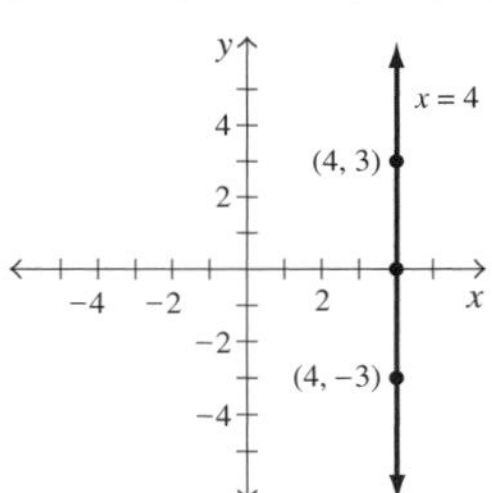

Slope: undefined

45.

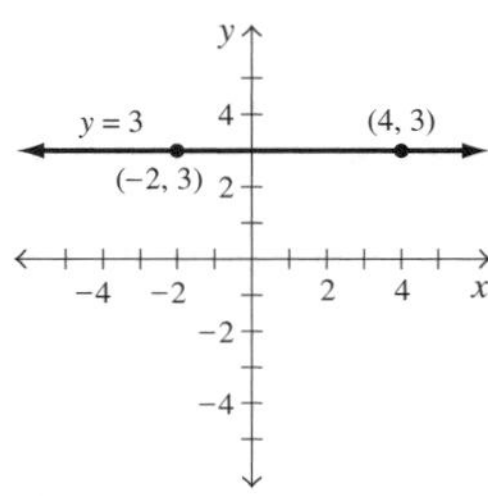

Slope is 0.

47.

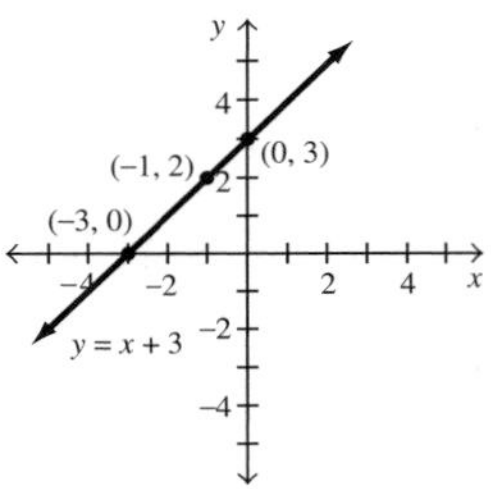

49.

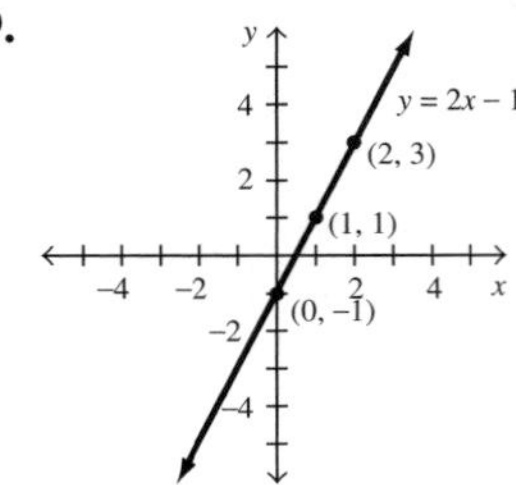

51.

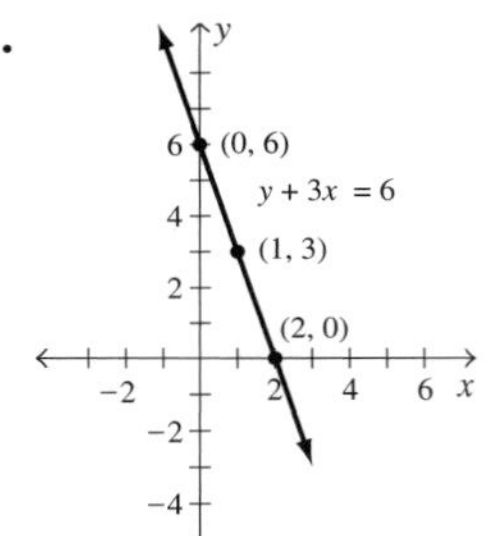

53.

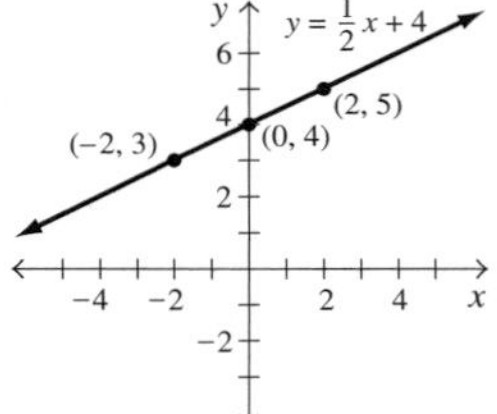

55.

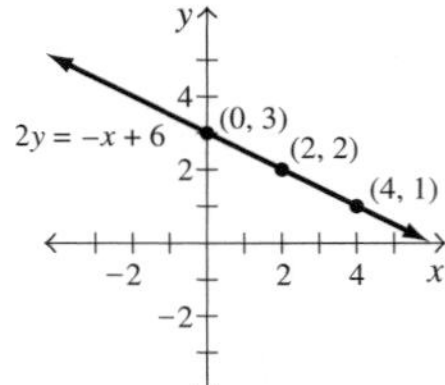

57.

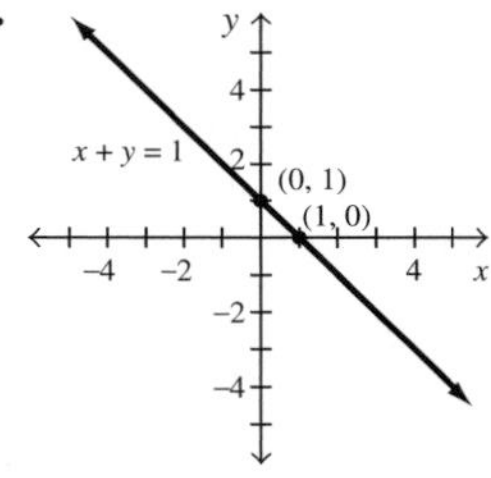

59.

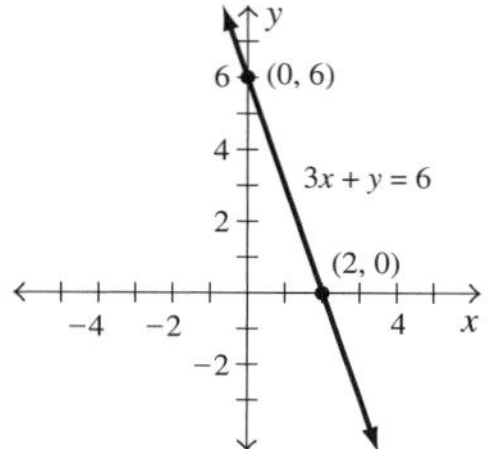

61.

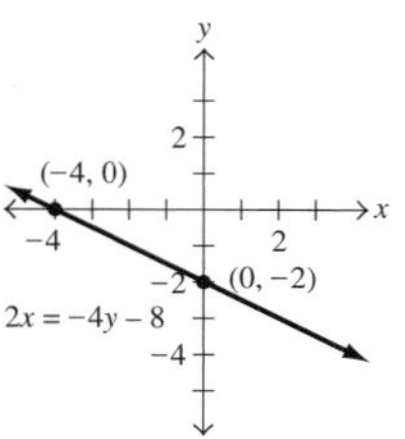

63.

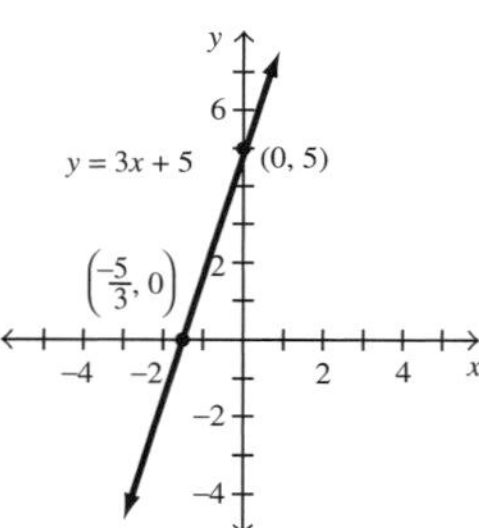

65.

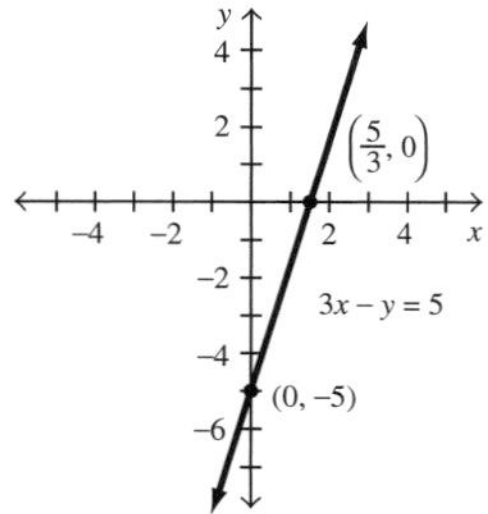

67. 2 **69.** $\frac{15}{7}$ **71.** 0 **73.** Undefined **75.** $-\frac{4}{3}$

77.

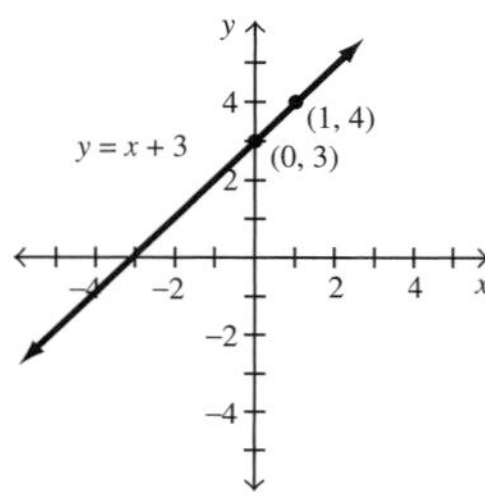

79.

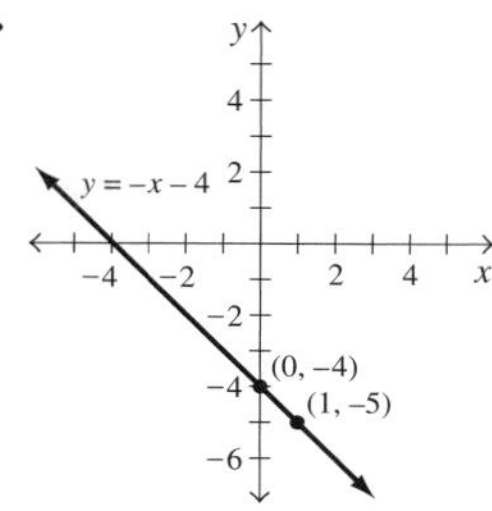

81.

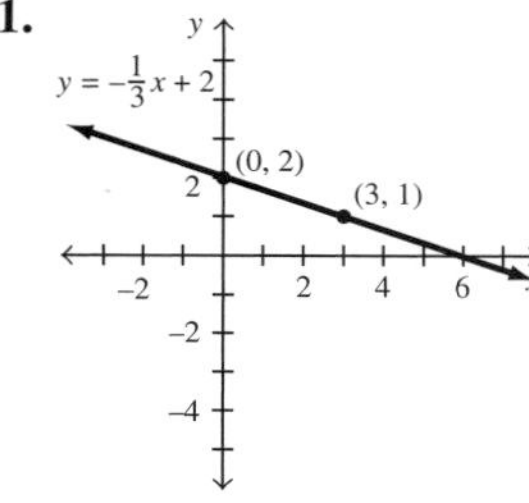

83.

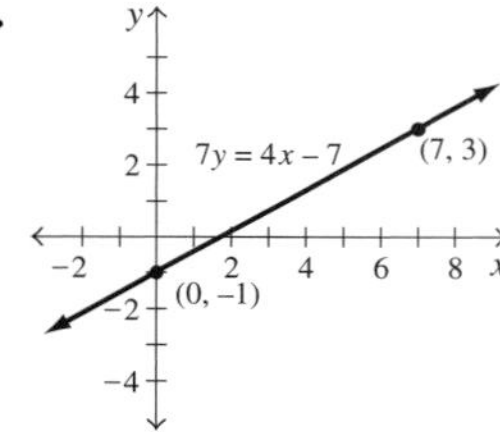

85.

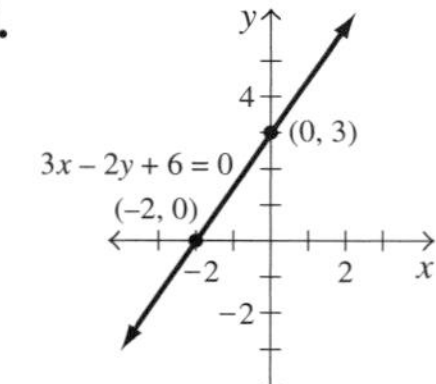

87. $y = -\frac{3}{4}x + 3$ **89.** $y = 3x + 2$

91. a) $D(-1, 2)$ **b)** $A = 10$ square units

93. (7, 2) or $(-1, 2)$ **95.** 3 **97.** 3

99. a)

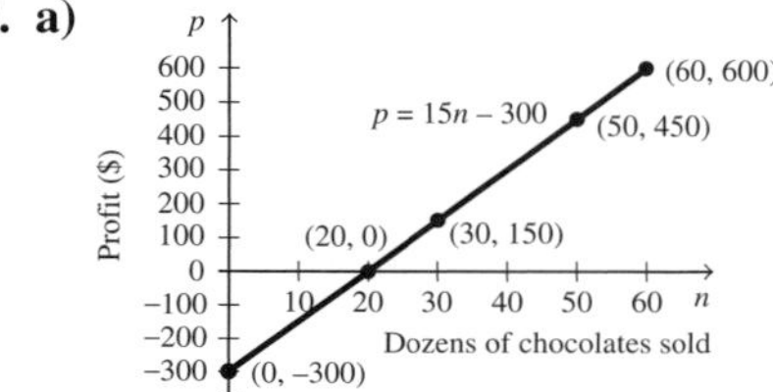

b) $300 **c)** 20 dozens of chocolates

101. a)

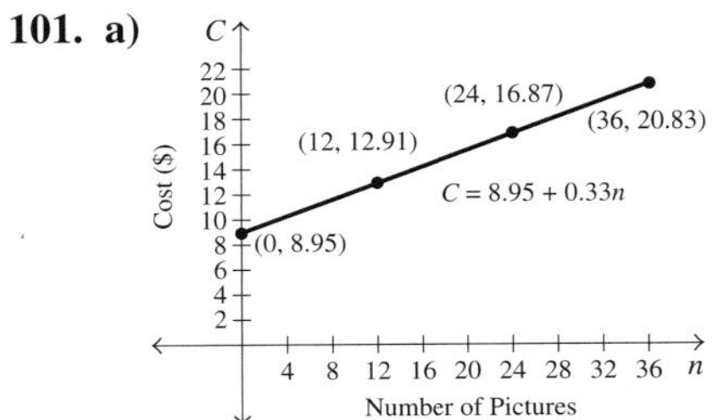

b) \$15.55 **c)** 36 pictures

103. a) 10.75 **b)** $y = 10.75x + 53$
c) 85.25 **d)** ≈ 2.5 hours

105. a) $-\frac{8}{15}$ **b)** $y = -\frac{8}{15}x + 40$
c) 32% **d)** 18.75 years after 1970 or 1988

107. a) Solve the equations for y to put them in slope–intercept form. Then compare the slopes and y-intercepts. If the slopes are equal but the y-intercepts are different, then the lines are parallel.
b) The lines are parallel.

Section 6.8, Page 350

1. (1) Mentally substitute the equal sign for the inequality sign and plot points as if you were graphing the equation. (2) If the inequality is < or >, draw a dashed line through the points. If the inequality is ≤ or ≥, draw a solid line through the points. (3) Select a test point not on the line and substitute the x- and y-coordinates into the inequality. If the substitution results in a true statement, shade in the area on the same side of the line as the test point. If the test point results in a false statement, shade in the area on the opposite side of the line as the test point.

3.

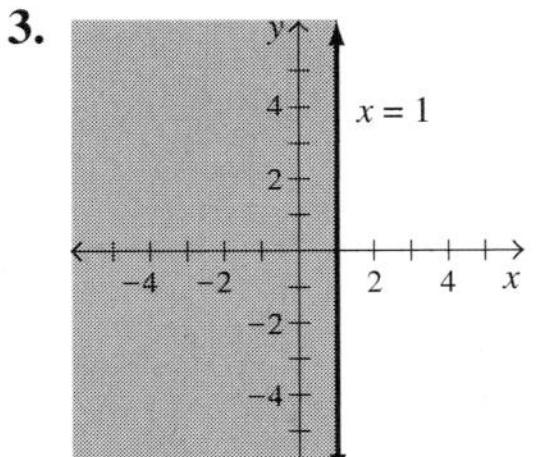

5.

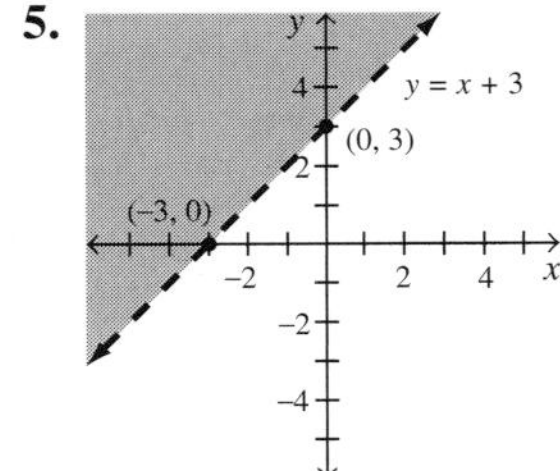

7.

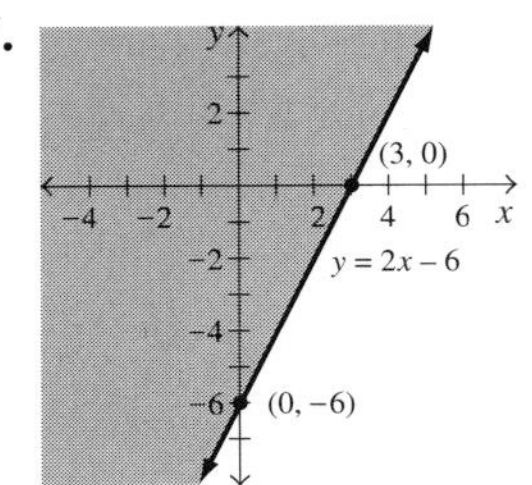

9.

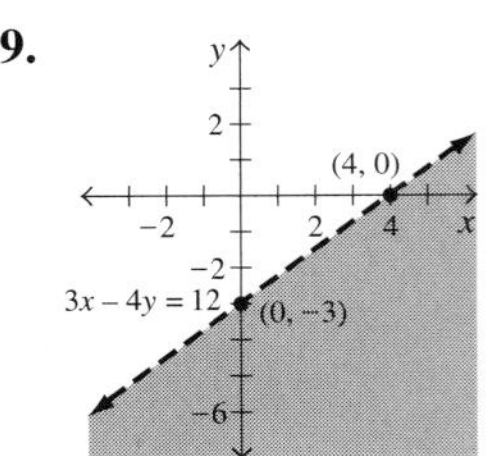

11.

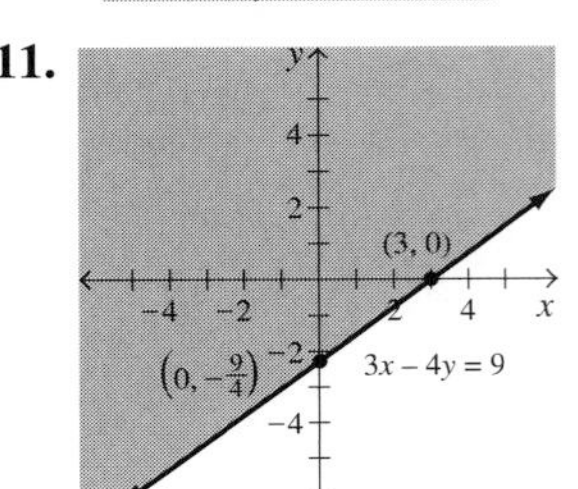

13.

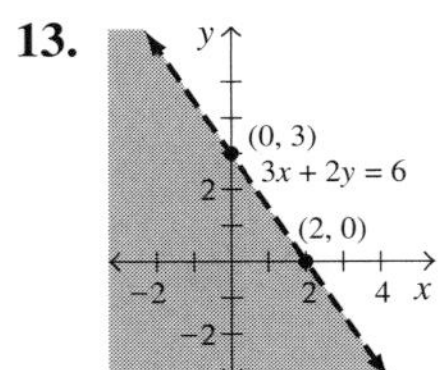

15.

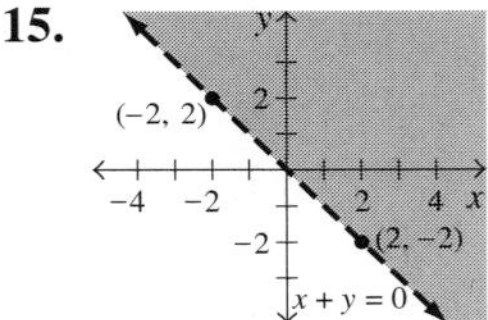

17.

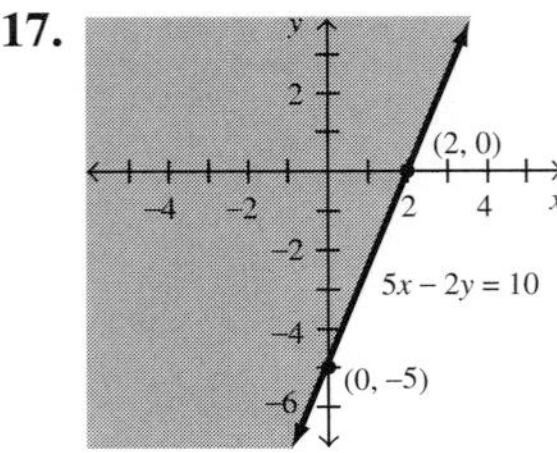

19. 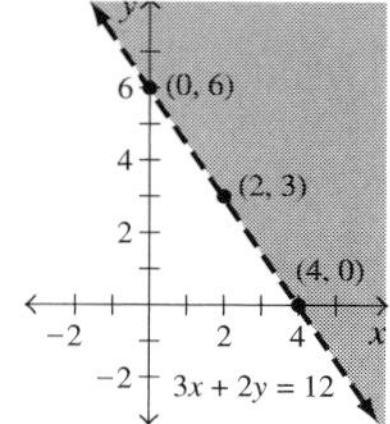

21. 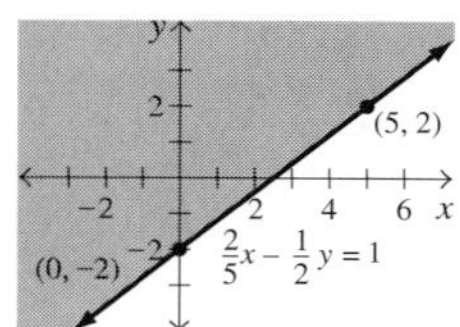

23. 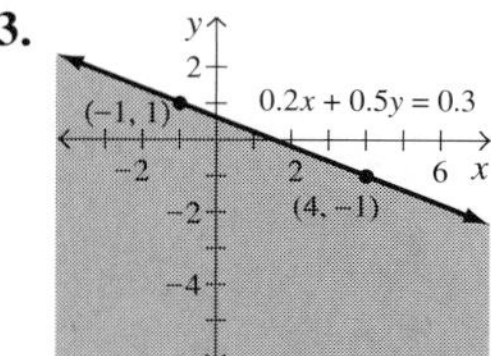

25. a) $x + y \leq 300$
b) 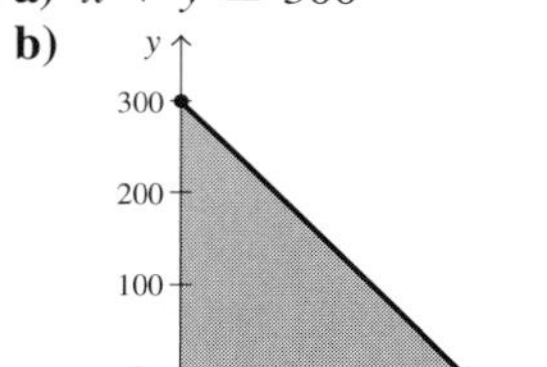

27. a) x = the number of acres of land, y = the number of square feet in the house
b) 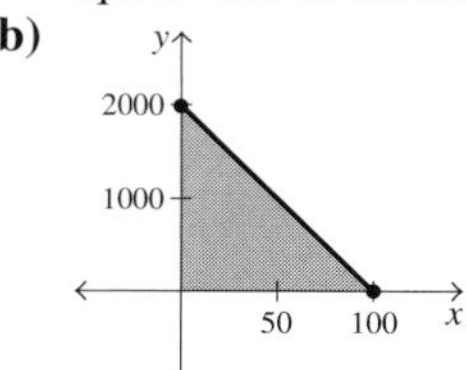
c) 2.5 acres or less
d) 1900 ft^2 or less

29. (a), (b), and (d)

Section 6.9, Page 360

1. A binomial is an expression that contains two terms in which each exponent that appears on the variable is a whole number. $2x + 3, x - 7, x^2 - 9$
3. Answers will vary.
5. $ax^2 + bx + c = 0, a \neq 0$ **7.** $(x + 6)(x + 3)$
9. $(x - 3)(x + 2)$ **11.** $(x + 6)(x - 4)$
13. $(x + 1)(x - 3)$ **15.** $(x - 7)(x - 3)$
17. $(x - 5)(x + 5)$ **19.** $(x + 7)(x - 4)$
21. $(x + 9)(x - 7)$ **23.** $(2x - 5)(x + 2)$
25. $(4x + 1)(x + 3)$ **27.** $(5x + 2)(x + 2)$
29. $(4x + 3)(x + 2)$ **31.** $(4x - 3)(x - 2)$
33. $(3x + 4)(x - 6)$ **35.** $1, -2$ **37.** $-\frac{4}{3}, \frac{1}{2}$
39. $-7, -3$ **41.** $3, 1$ **43.** $5, -3$ **45.** $3, 1$ **47.** $9, -9$
49. $-9, 4$ **51.** $\frac{2}{3}, -4$ **53.** $-\frac{1}{5}, -2$ **55.** $\frac{1}{3}, 1$ **57.** $\frac{1}{4}, 2$
59. $3, -5$ **61.** $6, -3$ **63.** $9, -1$ **65.** No real solution
67. $2 \pm \sqrt{2}$ **69.** $\dfrac{4 \pm \sqrt{13}}{3}$ **71.** $\dfrac{1 \pm \sqrt{17}}{8}$ **73.** $-1, -\frac{5}{2}$
75. $\frac{7}{3}, 1$ **77.** No real solution
79. Width = 12 m, length = 22 m
81. a) The zero-factor property cannot be used.
b) $\approx 8.37, \approx 2.63$
83. $x^2 - 2x - 3 = 0$

Section 6.10, Page 372

1. A function is a special type of relation where each value of the independent variable corresponds to a unique value of the dependent variable.
3. The domain of a function is the set of values that can be used for the independent variable.
5. If a vertical line touches more than one point on the graph, then for each value of x there is not a unique value for y and the graph does not represent a function.
7. Not a function **9.** Function, domain: $\mathbb{R}$; range: $\mathbb{R}$
11. Function, domain: $\mathbb{R}$; range: $y = 2$
13. Function, domain: $\mathbb{R}$; range: $y \geq -4$
15. Not a function
17. Function, domain: $0 \leq x < 12$; range: $y = 1, 2, 3$
19. Not a function
21. Function, domain: $\mathbb{R}$; range: $y > 0$
23. Not a function
25. Yes **27.** No **29.** Yes **31.** 5 **33.** 1 **35.** -6
37. 5 **39.** 4 **41.** -38 **43.** -17
45. 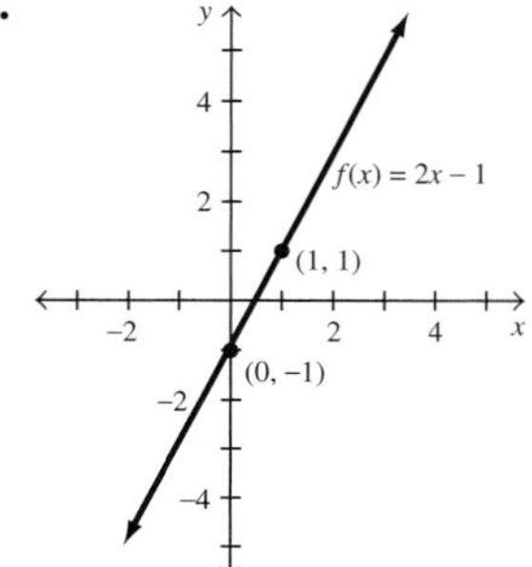

47.

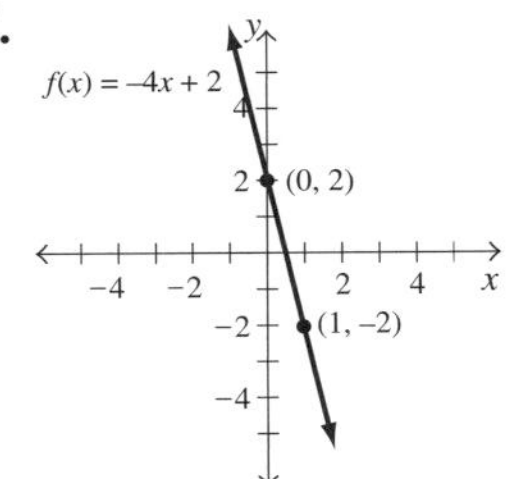

49.

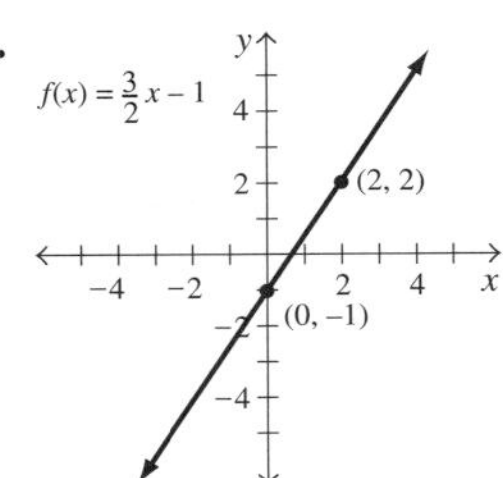

51. a) Upward **b)** $x = 0$ **c)** $(0, -16)$ **d)** $(0, -16)$
e) $(4, 0)$, $(-4, 0)$
f)

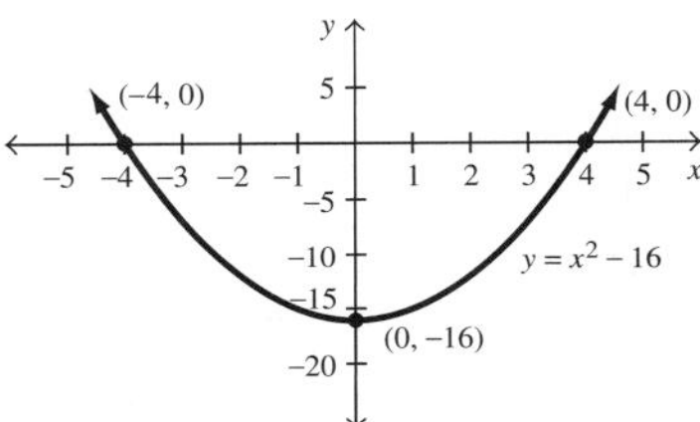

g) Domain: $\mathbb{R}$; range: $y \geq -16$

53. a) Downward **b)** $x = 0$ **c)** $(0, 4)$ **d)** $(0, 4)$
e) $(-2, 0)$, $(2, 0)$
f)

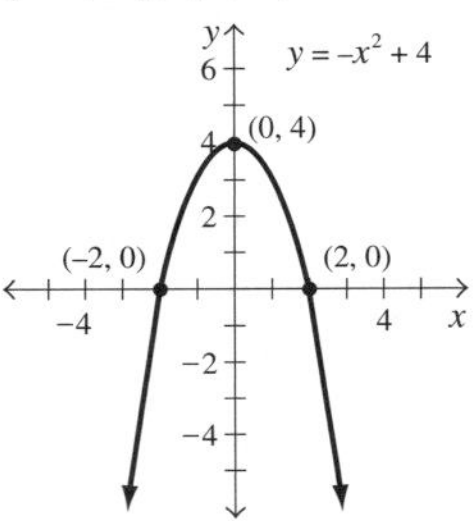

g) Domain: $\mathbb{R}$; range: $y \leq 4$

55. a) Downward **b)** $x = 0$ **c)** $(0, -4)$ **d)** $(0, -4)$
e) No x-intercepts
f)

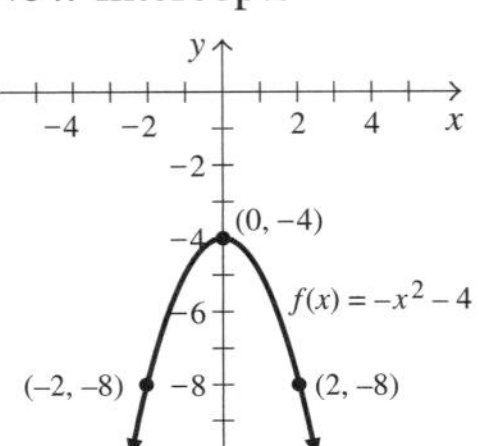

g) Domain: $\mathbb{R}$; range: $y \leq -4$

57. a) Upward **b)** $x = 0$ **c)** $(0, -3)$ **d)** $(0, -3)$
e) $(-1.22, 0)$, $(1.22, 0)$
f)

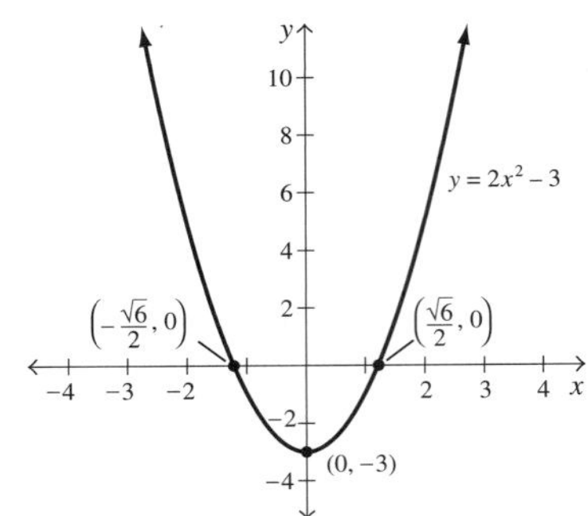

g) Domain: $\mathbb{R}$; range: $y \geq -3$

59. a) Upward **b)** $x = -1$ **c)** $(-1, 5)$ **d)** $(0, 6)$
e) No x-intercepts
f)

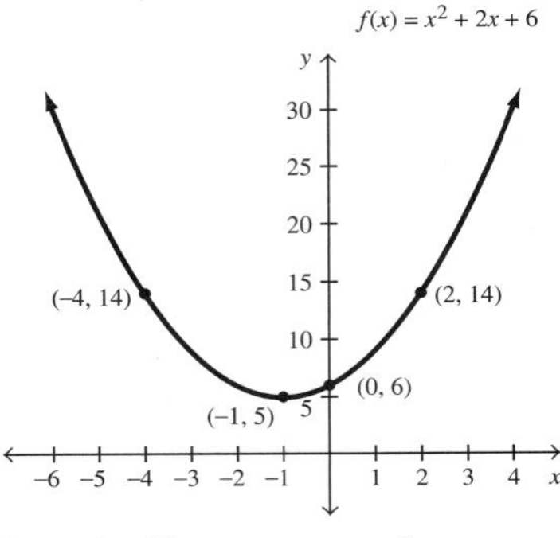

g) Domain: $\mathbb{R}$; range: $y \geq 5$

61. a) Upward **b)** $x = -\frac{5}{2}$ **c)** $(-2.5, -0.25)$ **d)** $(0, 6)$
e) $(-3, 0)$, $(-2, 0)$
f)

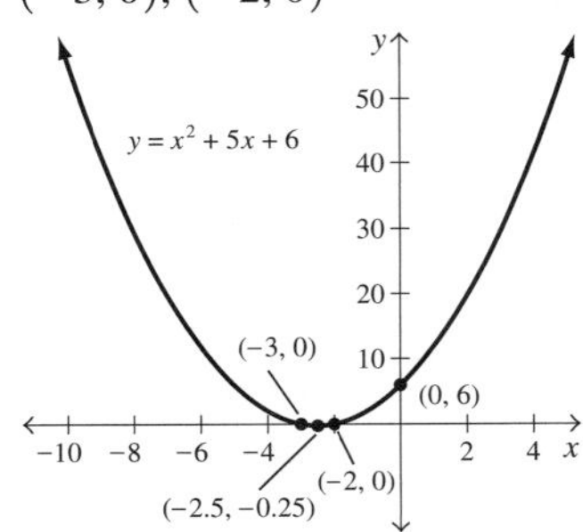

g) Domain: $\mathbb{R}$; range: $y \geq -0.25$

63. a) Downward **b)** $x = 2$ **c)** $(2, -2)$ **d)** $(0, -6)$
e) No x-intercepts
f)

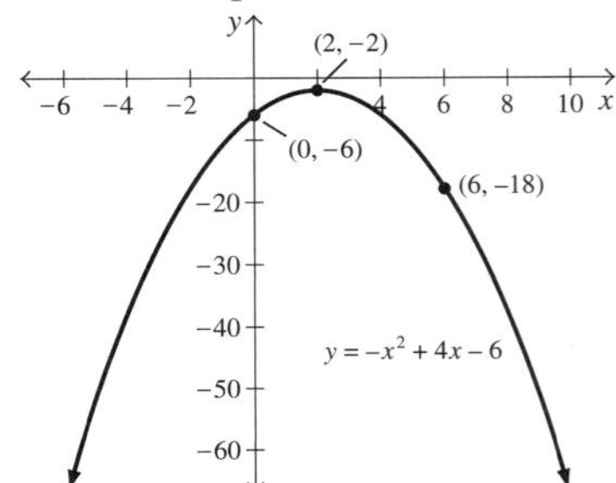

g) Domain: $\mathbb{R}$; range: $y \leq -2$

65. a) Downward **b)** $x = \frac{7}{3}$ **c)** $\left(\frac{7}{3}, \frac{25}{3}\right)$ **d)** $(0, -8)$
e) $\left(\frac{2}{3}, 0\right)$, $(4, 0)$
f)

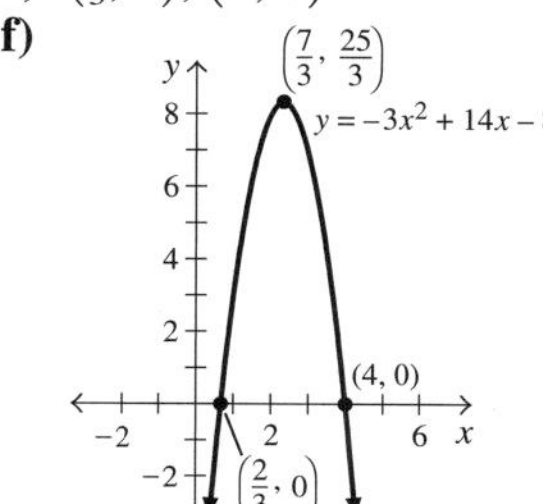

g) Domain: $\mathbb{R}$; range: $y \le \frac{25}{3}$

67. Domain: $\mathbb{R}$; range: $y > 0$

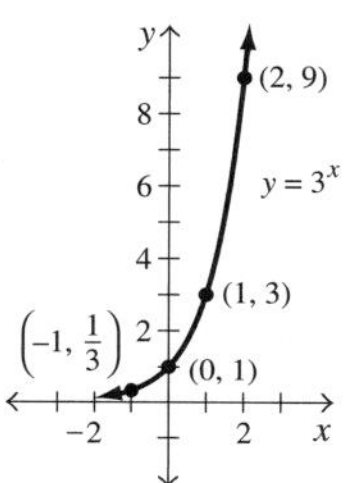

69. Domain: $\mathbb{R}$; range: $y > 0$

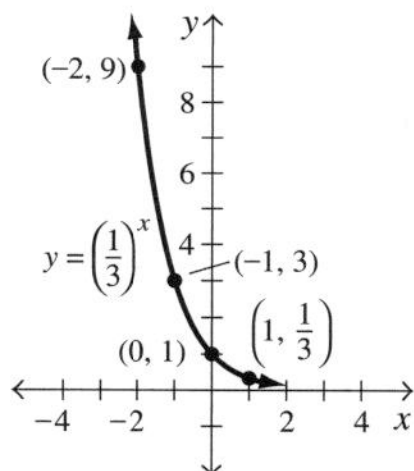

71. Domain: $\mathbb{R}$; range: $y > 1$

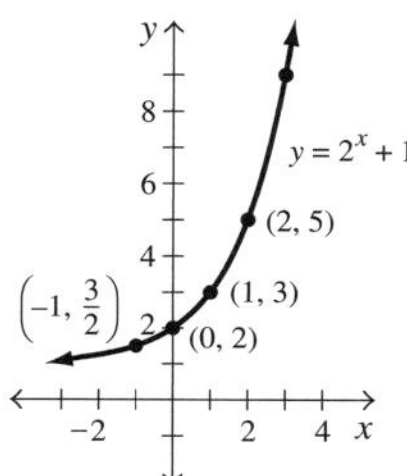

73. Domain: $\mathbb{R}$; range: $y > 1$

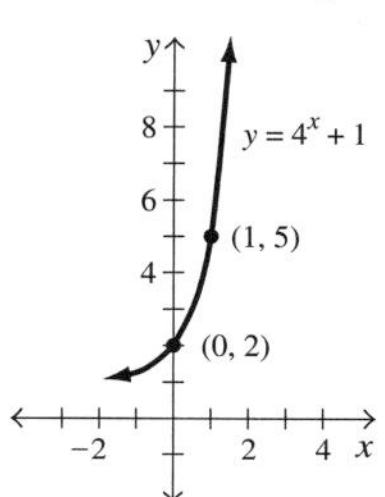

75. Domain: $\mathbb{R}$; range: $y > 0$

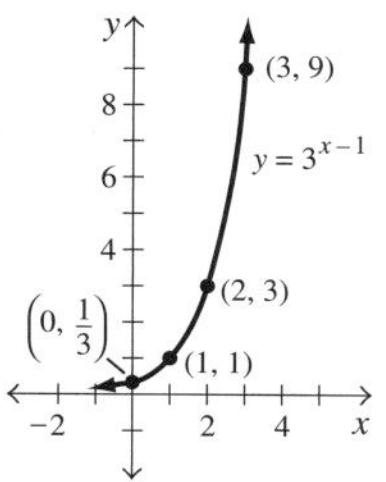

77. Domain: $\mathbb{R}$; range: $y > 0$

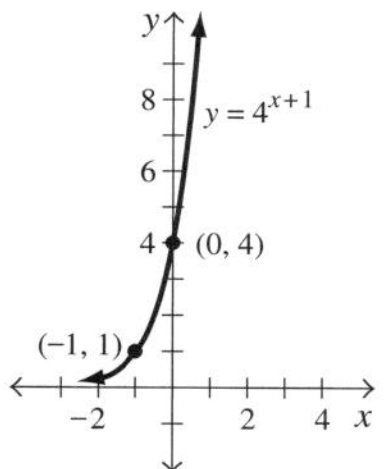

79. \$2300

81. a) 52.23% **b)** 1997 **c)** $x \approx 4.85$; $\approx 46.67\%$

83. a) 5200 people **b)** $\approx$ 14,852 people

85. a) Yes **b)** $\approx$ 6500 scooter injuries

87. a) 23.2 cm **b)** 55.2 cm **c)** 69.5 cm

89. a) 170 beats per minute
b) $\approx$ 162 beats per minute
c) $\approx$ 145 beats per minute
d) 136 beats per minute
e) 120 years of age

Review Exercises, Page 376

1. 21 **2.** −10 **3.** 17 **4.** $\frac{1}{4}$ **5.** −65 **6.** 13
7. $4x + 1$ **8.** $13x - 8$ **9.** $7x - 3$ **10.** −10
11. 7 **12.** 11 **13.** −31 **14.** $\frac{128}{5}$ **15.** $\frac{1}{2}$ cup
16. 250 min, or 4 hr 10 min **17.** 48 **18.** $\approx$ 173.1
19. 101.5 **20.** 10

21. $y = \dfrac{3x - 18}{9}$ or $y = \dfrac{1}{3}x - 2$

22. $y = \dfrac{-2x + 12}{5}$ or $y = -\dfrac{2}{5}x + \dfrac{12}{5}$

23. $y = \dfrac{2x + 22}{3}$ or $y = \dfrac{2}{3}x + \dfrac{22}{3}$

24. $y = \dfrac{-3x + 5z - 4}{4}$ or $y = -\dfrac{3}{4}x + \dfrac{5}{4}z - 1$

25. $w = \dfrac{A}{l}$

26. $w = \dfrac{P - 2l}{2}$

27. $l = \dfrac{L - 2wh}{2h}$ or $l = \dfrac{L}{2h} - w$

28. $d = \dfrac{a_n - a_1}{n - 1}$

29. $8 + 2x$ **30.** $3y - 7$ **31.** $10 + 3r$ **32.** $\dfrac{8}{q} - 11$

33. $4 + 3x = 22; x = 6$ **34.** $3x + 8 = x - 6; x = -7$

35. $5(x - 4) = 45; x = 13$

36. $10x + 14 = 8(x + 12); x = 41$

37. $x + 2x = 15{,}000$; bonds: \$5000, mutual funds: \$10,000

38. $9.50x + 15{,}000 = 95{,}000$; $\approx$ 8421 chairs

39. $x + 2x + 140 = 1130$; Philadelphia: 330, San Diego: 800

40. $x + (x + 12{,}000) = 68{,}000$; \$28,000 for B and \$40,000 for A

41. 6 **42.** 98 **43.** 20 **44.** ≈ 426.7

45. a) 150 lb **b)** 5 bags **46.** 4 in. **47.** \$119.88

48. 400 ft

49. −6

50. −1 0

51. 16

52. −4 6

53. −2 −1 0 1 2 . . .

54. −8 −7 −6 −5 −4 −3 −2 −1 . . .

55. −1 0 1 2 3 4 5 6 7 8 9 10

56. −10 −8 −6 −4 −2 0 2 4 6

For Exercises 57–60, see the following figure.

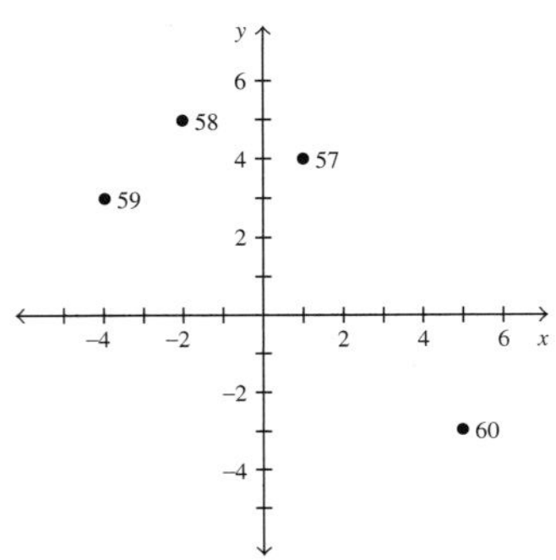

61. $D(-3, -1)$; area = 20 square units

62. $D(4, 1)$; area = 21 square units

63.

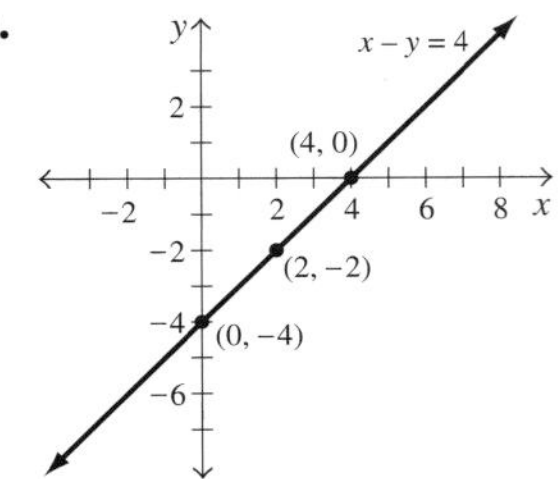

64.

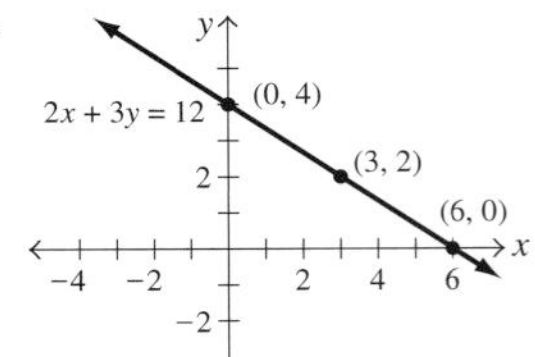

65.

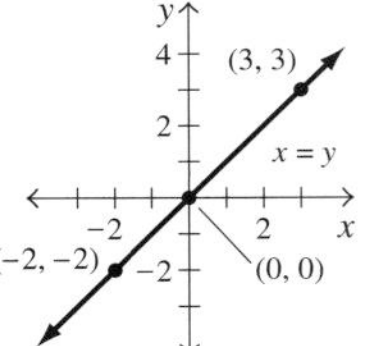

66.

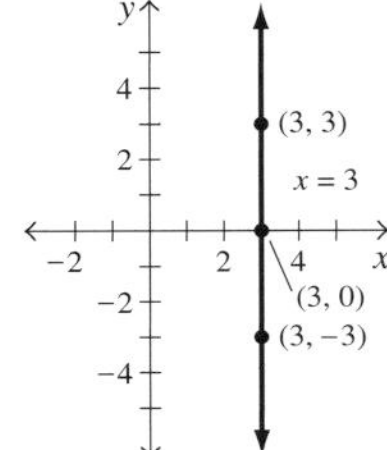

67.

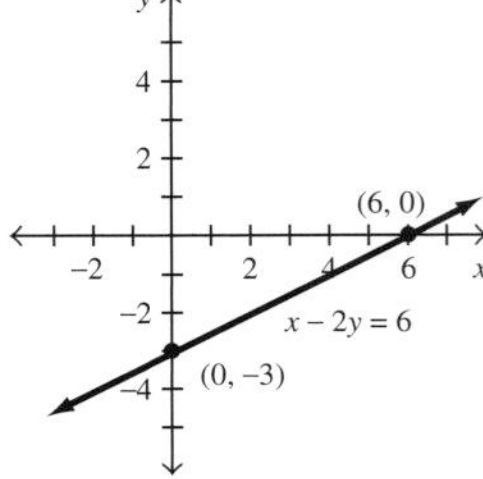

68.

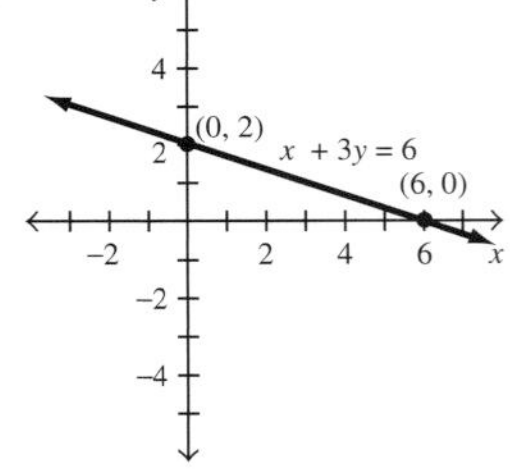

69.

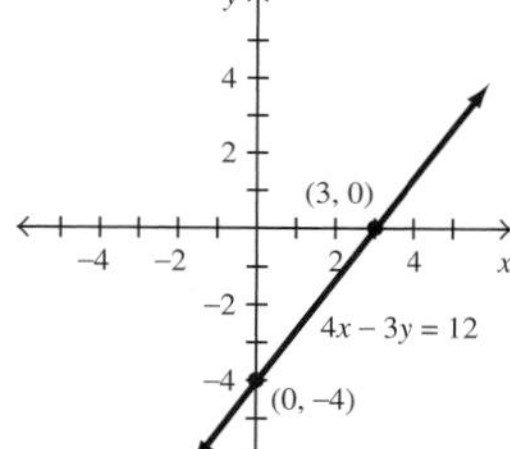

70.

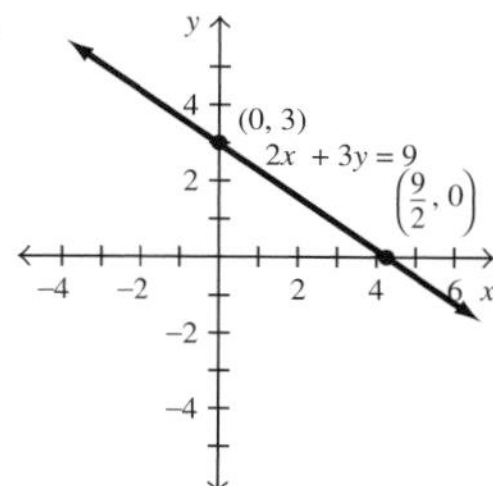

71. $\frac{2}{5}$ **72.** $-\frac{3}{2}$ **73.** $\frac{7}{3}$ **74.** Undefined

75.

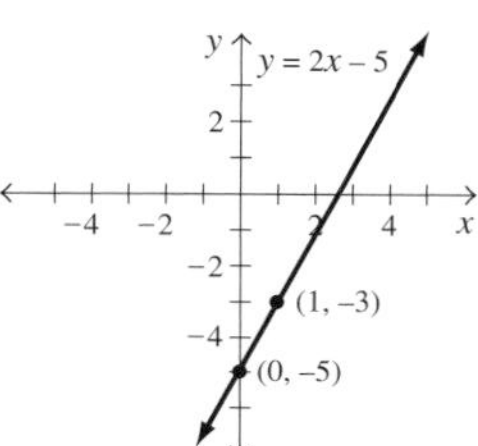

76.

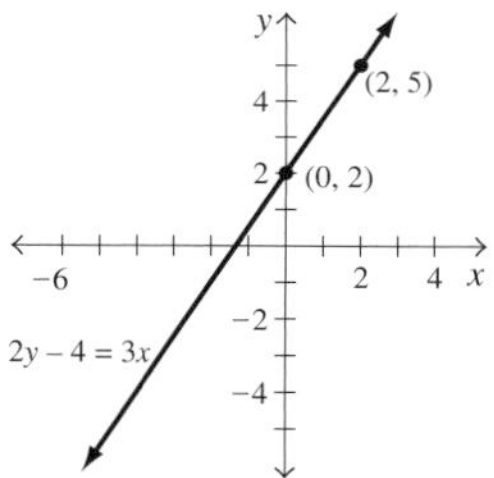

77.

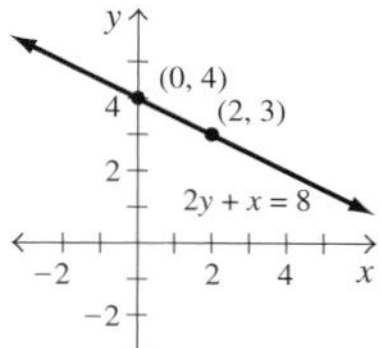

78.

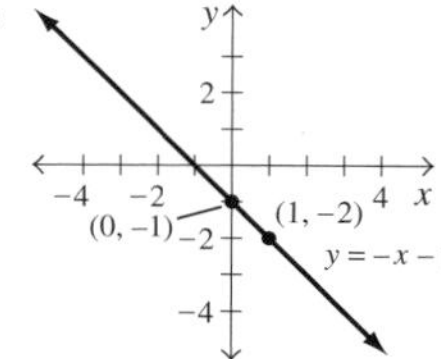

79. $y = 2x + 4$ **80.** $y = -x + 1$

81. a)

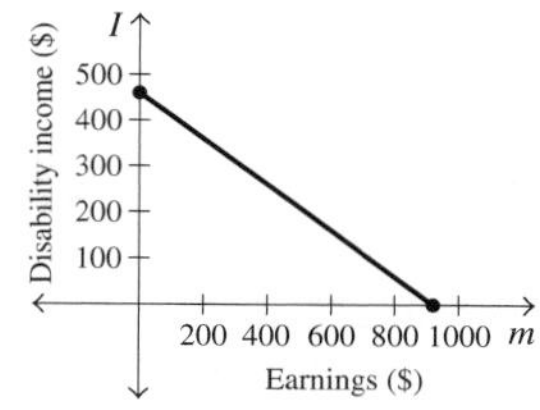

b) About \$160 **c)** About \$160

82. a)

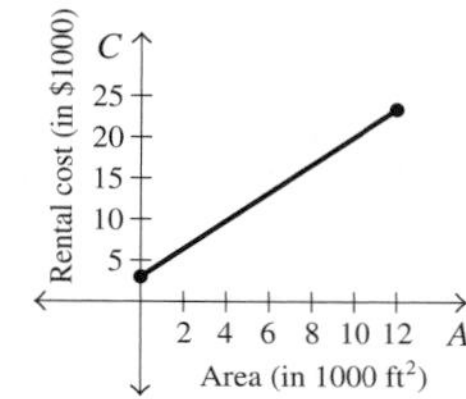

b) About \$6400 **c)** About 4120 ft^2

83.

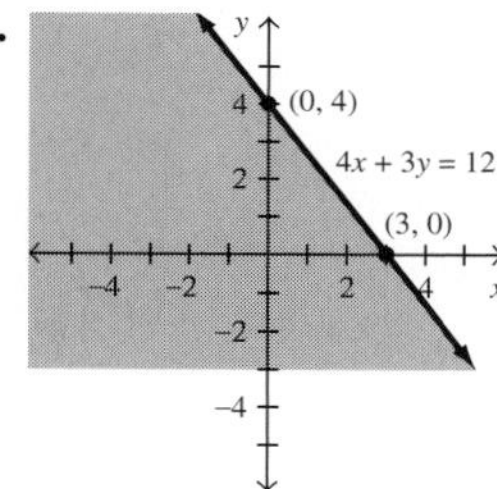

84.

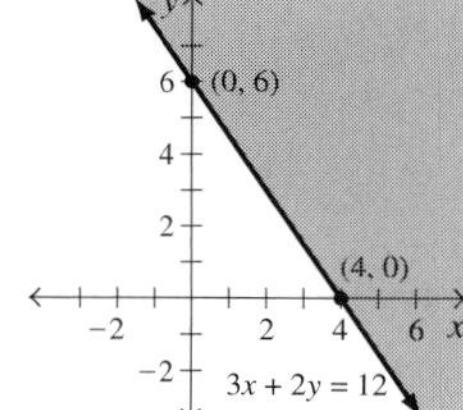

85.

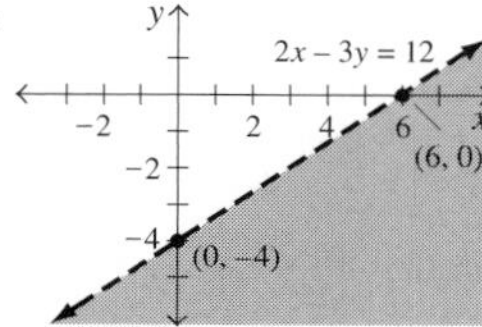

86.

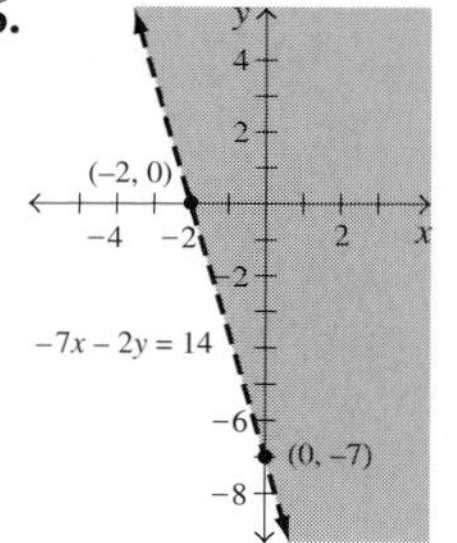

87. $(x + 3)(x + 6)$ **88.** $(x + 5)(x - 4)$
89. $(x - 6)(x - 4)$ **90.** $(x - 5)(x - 4)$

91. $(3x - 1)(2x + 3)$ **92.** $(2x - 1)(x + 7)$

93. $-1, -2$ **94.** $1, 4$ **95.** $\frac{2}{3}, 5$

96. $-2, -\frac{1}{3}$ **97.** $2 \pm \sqrt{5}$ **98.** $1, 2$

99. No real solution **100.** $-1, \frac{3}{2}$

101. Function, domain: $x = -2, -1, 2, 3$; range: $y = -1, 0, 2$

102. Not a function **103.** Not a function

104. Function, domain: $\mathbb{R}$; range: $\mathbb{R}$

105. 18 **106.** 13 **107.** 39 **108.** -27

109. a) Downward **b)** $x = -2$ **c)** $(-2, 25)$ **d)** $(0, 21)$
e) $(-7, 0), (3, 0)$
f)

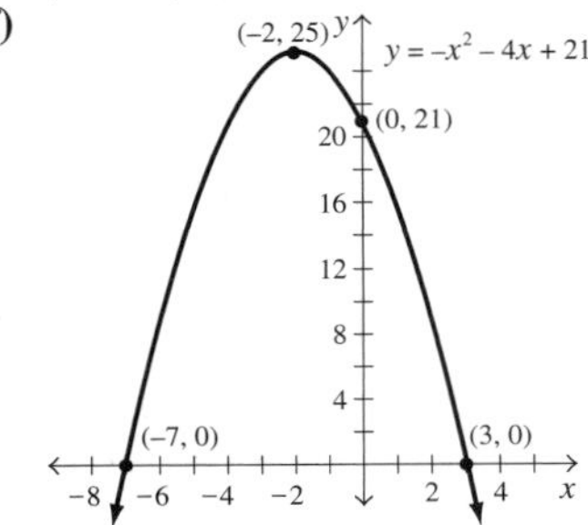

g) Domain: $\mathbb{R}$; range: $y \leq 25$

110. a) Upward **b)** $x = 2$ **c)** $(2, 2)$ **d)** $(0, 10)$
e) No x-intercepts
f)

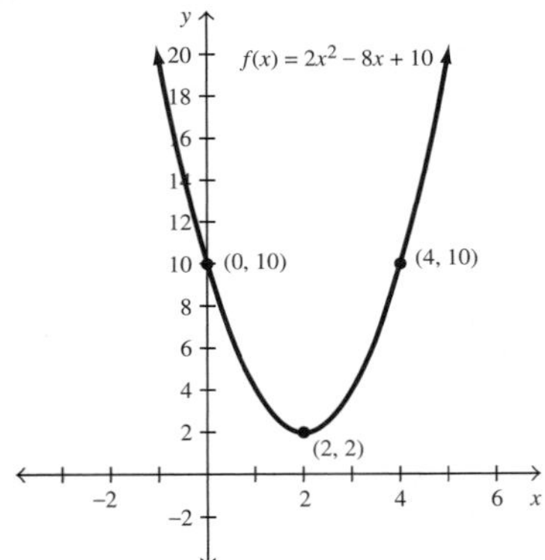

g) Domain: $\mathbb{R}$; range: $y \geq 2$

111. Domain: $\mathbb{R}$; range: $y > 0$

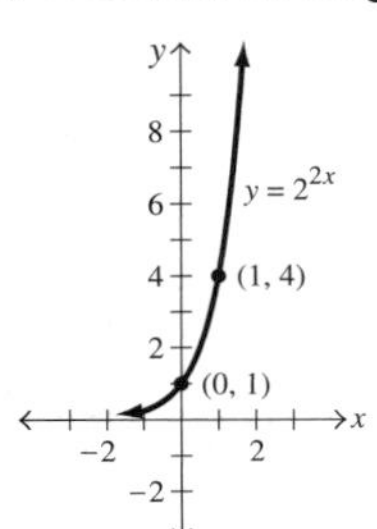

112. Domain: $\mathbb{R}$; range: $y > 0$

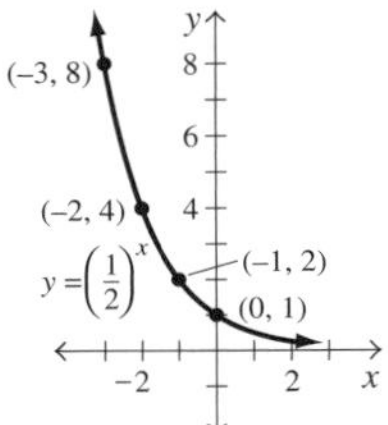

113. 22.8 mpg

114. a) 4208 **b)** 4250

115. $\approx 68.7\%$

Chapter Test, Page 380

1. 3 **2.** $\frac{19}{5}$ **3.** 18 **4.** $2x + 7 = 25$; 9

5. $x + 0.07x = 26{,}750$; \$25,000 **6.** 84

7. $y = \dfrac{-3x + 11}{5}$ or $y = -\dfrac{3}{5}x + \dfrac{11}{5}$

8. $3\frac{1}{3}$ **9.** 6.75 ft

10.

11. $\frac{7}{10}$

12.

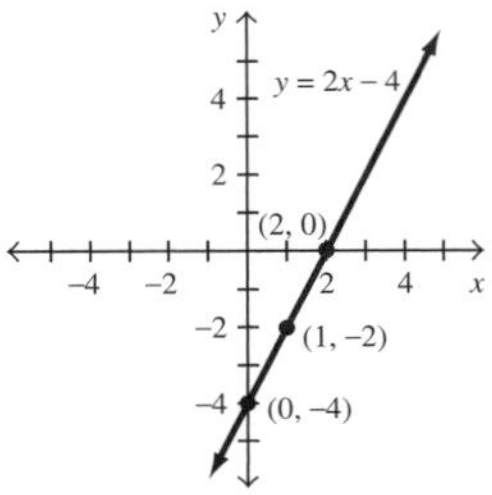

13.

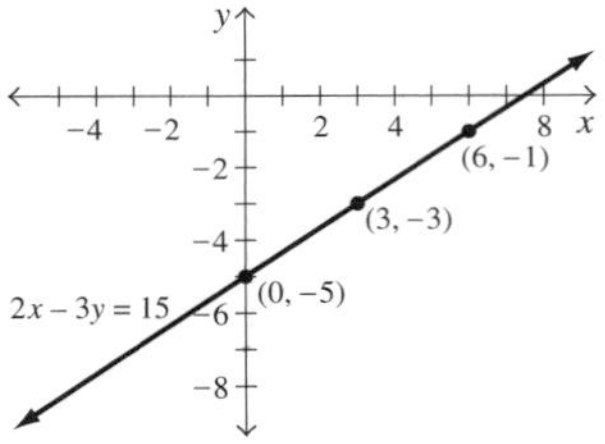

14.

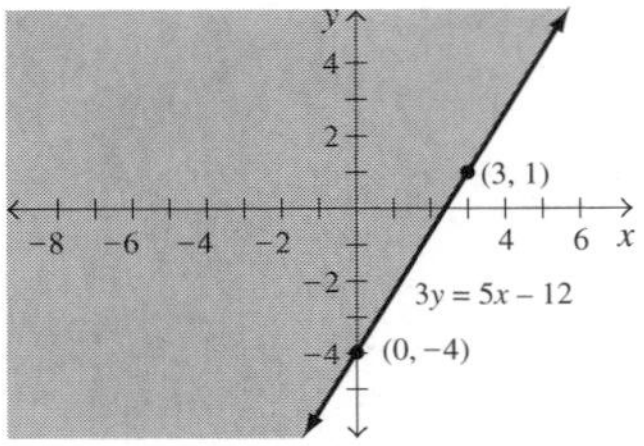

15. $7, -4$ **16.** $\frac{4}{3}, -2$

17. It is a function **18.** 11

19. a) Upward **b)** $x = 1$ **c)** $(1, 3)$ **d)** $(0, 4)$
e) No x-intercepts

f)

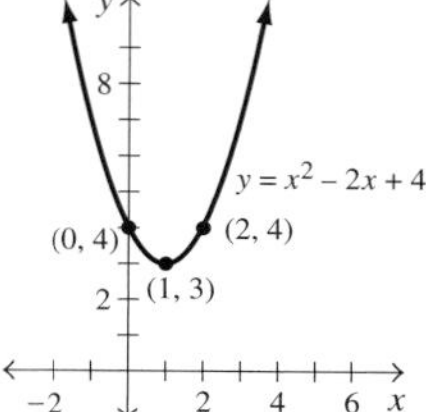

g) Domain: $\mathbb{R}$; range: $y \geq 3$

Chapter 7

Section 7.1, Page 388

1. Two or more linear equations form a system of linear equations.
3. A consistent system of equations is a system that has a solution.
5. An inconsistent system of equations is one that has no solution.
7. The graphs of the equations will be parallel.
9. The graphs of the equations will be the same line.
11. Yes: (3, 0); no: (2, −2), (1, 2)

13.

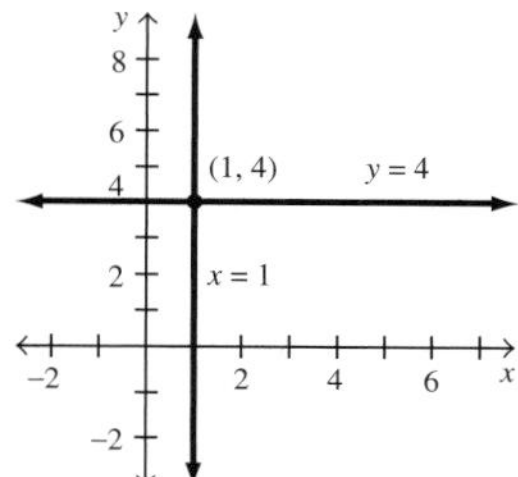

15.

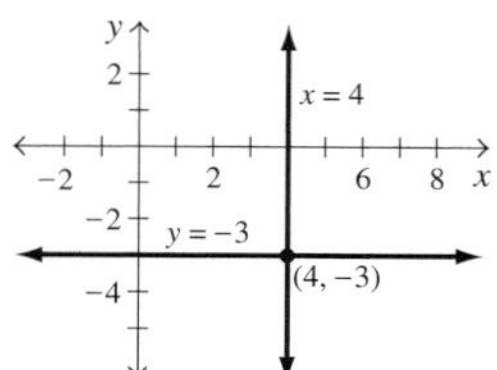

17.

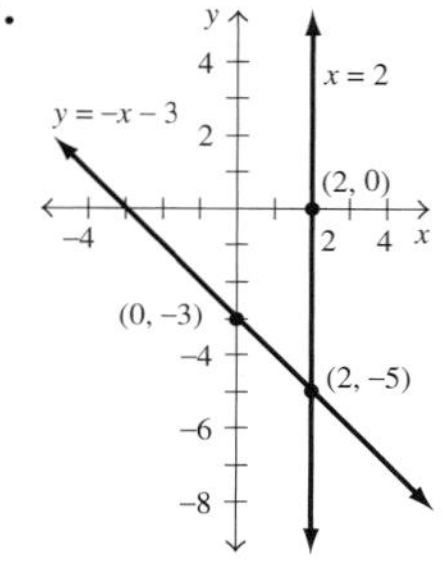

19.

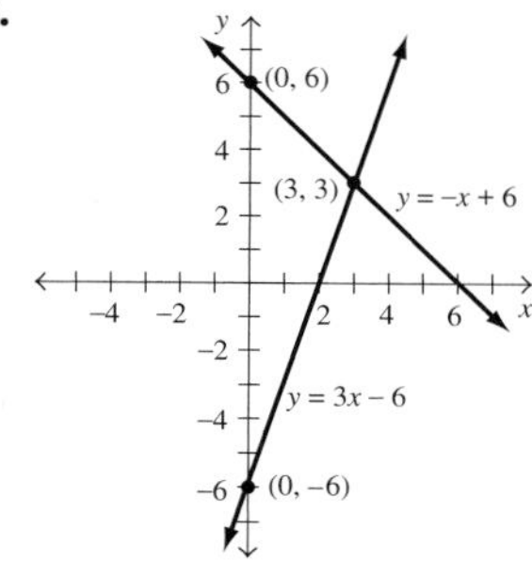

21.

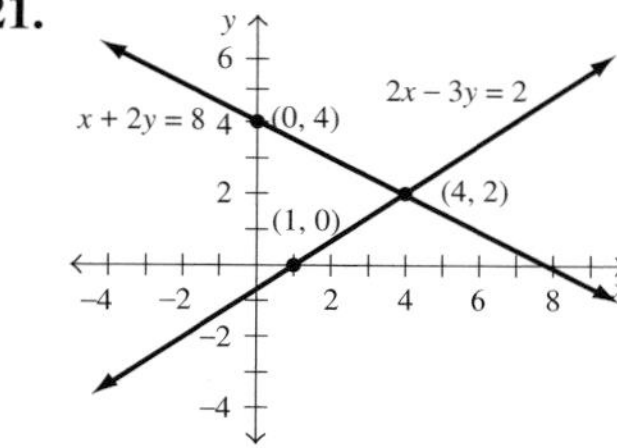

23.

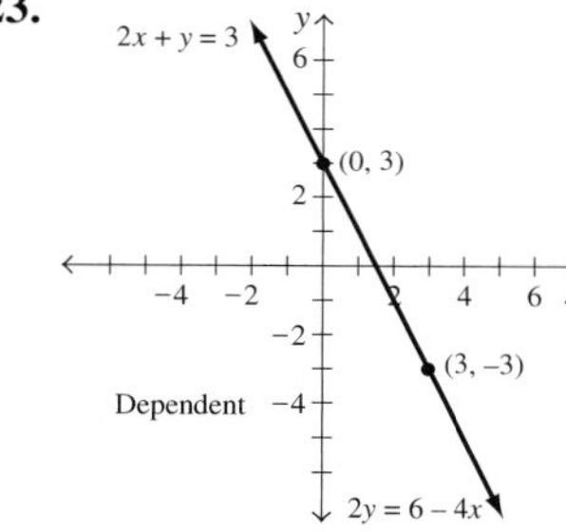

25.

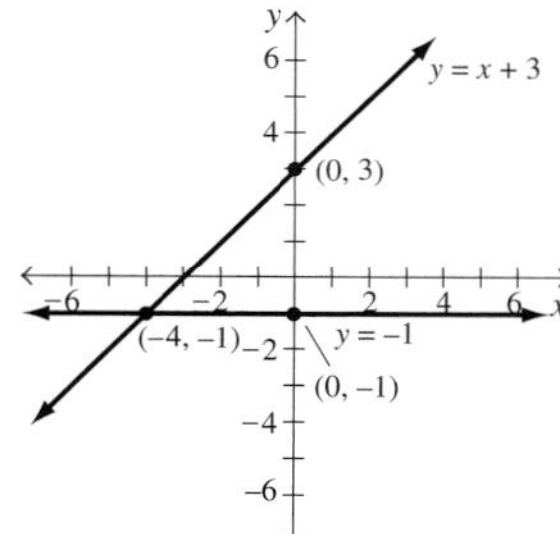

27.

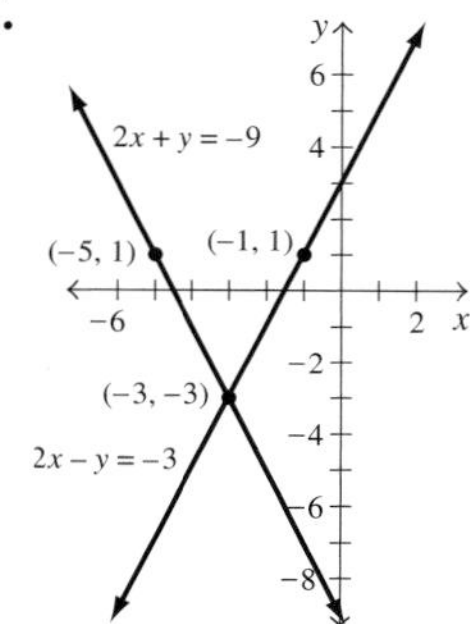

29.

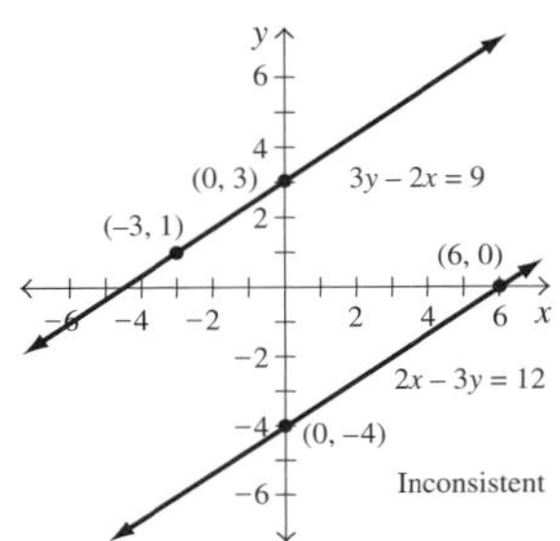

31.

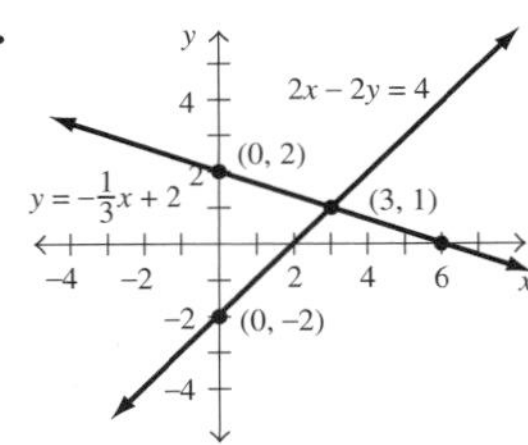

33. a) One unique solution; the lines intersect at one and only one point.
b) No solution; the lines do not intersect.
c) Infinitely many solutions; the lines coincide.

35. An infinite number of solutions **37.** One solution

39. No solution **41.** An infinite number of solutions

43. No solution **45.** One solution

47. Not perpendicular **49.** Perpendicular

51. a) Tom's Tree and Landscape: $C = 60h + 200$
Lawn Perfect: $C = 25h + 305$

b)

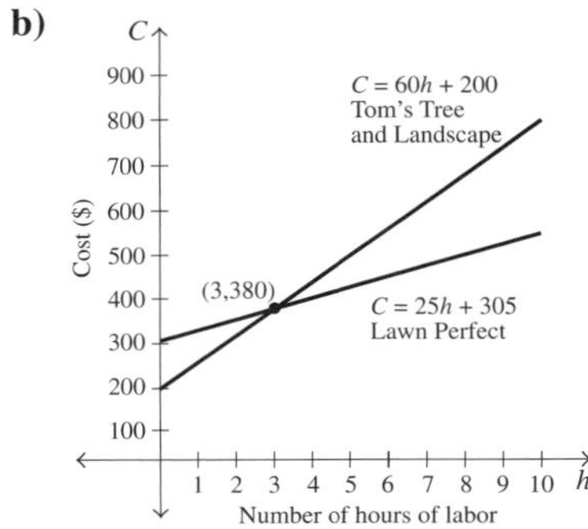

c) 3 hours

53. a) $C = 15x + 400$
$R = 25x$

b)

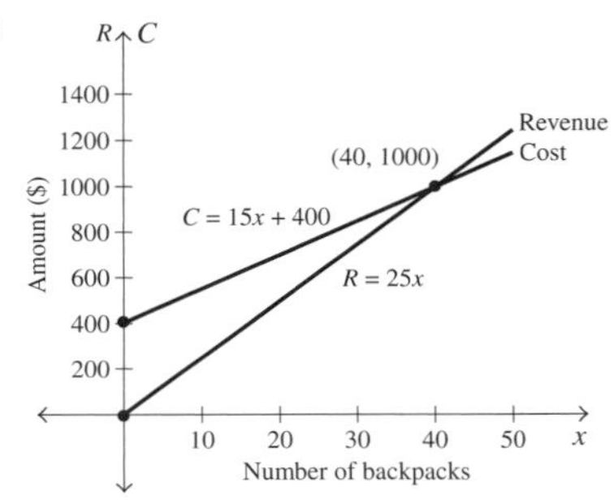

c) 40 backpacks **d)** $P = 10x - 400$
e) Loss of $100 **f)** 140 backpacks

55. a) $C = 155x + 8400,\ R = 225x$

b)

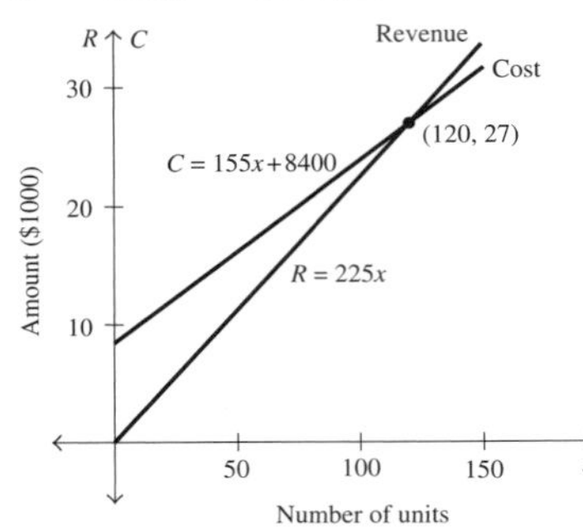

c) 120 units **d)** $P = 70x - 8400$
e) Loss of $1400 **f)** 138 units

57. a) Job 1: $s = 0.15x + 300$
Job 2: $s = 450$

b)

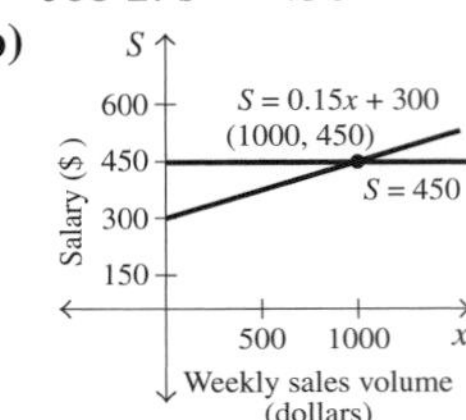

c) $1000 sales volume

59. a) One **b)** Three **c)** Six **d)** Ten
e) To find the number of points of intersection for n lines, add $n - 1$ to the number of points of intersection for $n - 1$ lines. For example, for 5 lines there were 10 points of intersection. Therefore, for 6 lines ($n = 6$) there are $10 + (6 - 1) = 15$ points of intersection.

Section 7.2, Page 399

1. Answers will vary.

3. The system is dependent if the same value is obtained on both sides of the equal sign.

5. x, in the first equation **7.** $(-1, 6)$ **9.** $(0, 2)$

11. No solution; inconsistent system **13.** $(-2, \frac{8}{3})$

15. An infinite number of solutions; dependent system

17. $(-3, -6)$ **19.** $(\frac{11}{5}, -\frac{13}{5})$ **21.** $(-\frac{1}{5}, -\frac{8}{5})$

23. No solution; inconsistent system

25. $(2, 4)$ **27.** $(6, 4)$ **29.** $(-2, 0)$ **31.** $(-4, 5)$

33. $(3, 5)$ **35.** $(1, -2)$

37. No solution; inconsistent system **39.** $(2, -1)$

41. $s = 12{,}000 + 0.15p$
$s = 27{,}000 + 0.05p$
$150,000 annual profit

43. $m + l = 50$
$10.95m + 14.95l = 663.50$
21 medium, 29 large

45. $x + y = 10$
$0.25x + 0.50y = 0.40(10)$
4ℓ of 25%, 6ℓ of 50%

47. $y = 18 + 0.02x$
$y = 24 + 0.015x$
1200 copies

49. $x + y = 20$
$3x + y = 30$
Mix 5 lb of nuts with 15 lb of pretzels.

51. $x + y = 250$
$2x + 5y = 950$
100 student tickets and 150 nonstudent tickets

53. ≈26.7 years after 1981 or in 2007

55. $(\frac{1}{2}, \frac{1}{3})$ **57.–59.** Answers will vary.

Section 7.3, page 408

1. A matrix is a rectangular array of elements.

3. A square matrix contains the same number of rows and columns.

5. 2 **7. a)** Answers will vary. **b)** $\begin{bmatrix} 4 & 9 & -7 \\ 2 & 4 & 9 \end{bmatrix}$

9. a) The number of columns of the first matrix must be the same as the number of rows of the second matrix.
b) 2×3

11. a) $I = \begin{bmatrix} 1 & 0 \\ 0 & 1 \end{bmatrix}$ **b)** $I = \begin{bmatrix} 1 & 0 & 0 \\ 0 & 1 & 0 \\ 0 & 0 & 1 \end{bmatrix}$

13. $\begin{bmatrix} -4 & 2 \\ 12 & 9 \end{bmatrix}$ **15.** $\begin{bmatrix} 0 & 4 \\ 4 & 4 \\ 5 & -1 \end{bmatrix}$

17. $\begin{bmatrix} 6 & -7 \\ -12 & 4 \end{bmatrix}$ **19.** $\begin{bmatrix} 2 & 11 \\ 16 & 13 \\ -2 & 2 \end{bmatrix}$

21. $\begin{bmatrix} 6 & 4 \\ 10 & 0 \end{bmatrix}$ **23.** $\begin{bmatrix} 0 & 13 \\ 22 & 0 \end{bmatrix}$

25. $\begin{bmatrix} 13 & 0 \\ 7 & 0 \end{bmatrix}$ **27.** $\begin{bmatrix} 4 & 12 \\ 14 & 22 \end{bmatrix}$

29. $\begin{bmatrix} 15 \\ 22 \end{bmatrix}$ **31.** $\begin{bmatrix} 4 & 7 & 6 \\ -2 & 3 & 1 \\ 5 & 1 & 2 \end{bmatrix}$

33. $A + B = \begin{bmatrix} 6 & 2 & 1 \\ 6 & -2 & 4 \end{bmatrix}$; cannot be multiplied

35. Cannot be added; $A \times B = \begin{bmatrix} 26 & 38 \\ 24 & 24 \end{bmatrix}$

37. Cannot be added; $\begin{bmatrix} 1 \\ -1 \end{bmatrix}$

39. $A + B = B + A = \begin{bmatrix} 5 & 7 \\ 8 & 4 \end{bmatrix}$

41. $A + B = B + A = \begin{bmatrix} 8 & 0 \\ 6 & -8 \end{bmatrix}$

43. $(A + B) + C = A + (B + C) = \begin{bmatrix} 7 & 10 \\ 6 & 13 \end{bmatrix}$

45. $(A + B) + C = A + (B + C) = \begin{bmatrix} 5 & 5 \\ 7 & -37 \end{bmatrix}$

47. No **49.** No **51.** Yes

53. $(A \times B) \times C = A \times (B \times C) = \begin{bmatrix} 41 & 13 \\ 56 & 16 \end{bmatrix}$

55. $(A \times B) \times C = A \times (B \times C) = \begin{bmatrix} 16 & -10 \\ -24 & 2 \end{bmatrix}$

57. $(A \times B) \times C = A \times (B \times C) = \begin{bmatrix} 17 & 0 \\ -7 & 0 \end{bmatrix}$

59. Large Small $\begin{bmatrix} 38 & 50 \\ 56 & 72 \\ 17 & 26 \\ 10 & 14 \end{bmatrix}$ **61.** [\$36.04 \$47.52]

63. Answers will vary. **65.** Yes **67.** False

69. a) \$28.70 **b)** \$60.10 **c)** $\begin{bmatrix} 28.7 & 24.6 \\ 41.3 & 35.7 \\ 69.3 & 60.1 \end{bmatrix}$

71. Yes. Answers will vary. One example is

$$A = \begin{bmatrix} 2 & 7 & 6 \\ -3 & 0 & 8 \end{bmatrix}, \quad B = \begin{bmatrix} 1 & 2 \\ 3 & 4 \\ 5 & 6 \end{bmatrix}.$$

Section 7.4, Page 417

1. a) An augmented matrix is a matrix formed with the coefficients of the variables and the constants. The coefficients of the variables are separated from the constants by a vertical bar.
b) $\left[\begin{array}{cc|c} 1 & 3 & 7 \\ 2 & -1 & 4 \end{array}\right]$

3. If you obtain an augmented matrix in which one row of numbers on the left side of the vertical line are all zeroes but a zero does not appear in the same row on the right side of the vertical line, the system is inconsistent.

5. Change the −2 to a 1 by multiplying the numbers in the second row by $-\frac{1}{2}$.

7. (3, 0) **9.** (3, 2)

11. An infinite number of solutions; dependent system

13. $(\frac{7}{2}, -1)$ **15.** $(\frac{1}{2}, -6)$

17. No solution; inconsistent system **19.** (3, 5)

21. 20 small flags and 35 large flags

23. Truck driver: $7\frac{1}{9}$ hours; laborer: $9\frac{1}{9}$ hours

25. Nonrefillable pencils: 125; refillable pencils: 75

Section 7.5, Page 420

1. The solution set of a system of linear inequalities is the set of points that satisfy all inequalities in the system.

3.

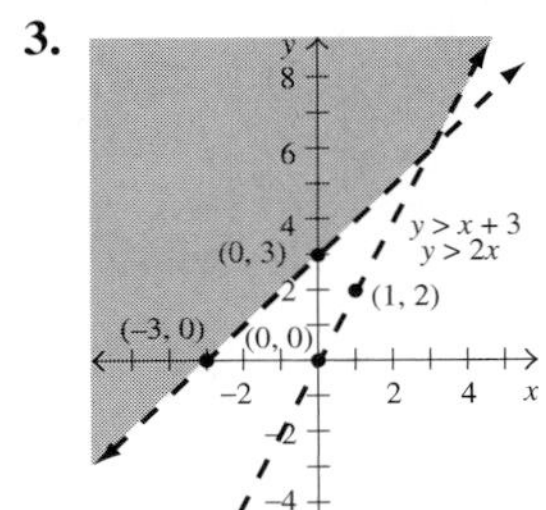

5.

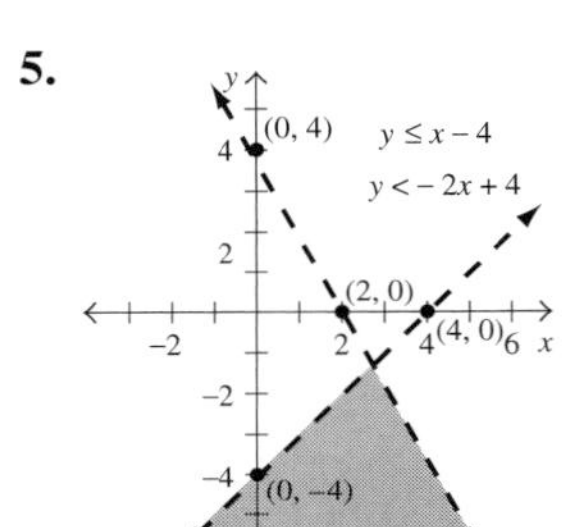

7.

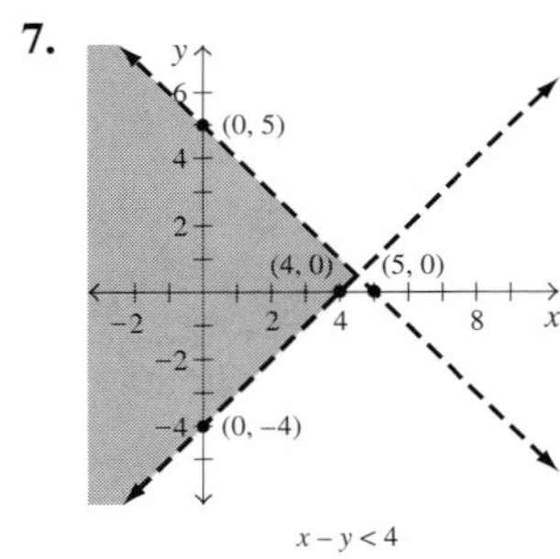

9.

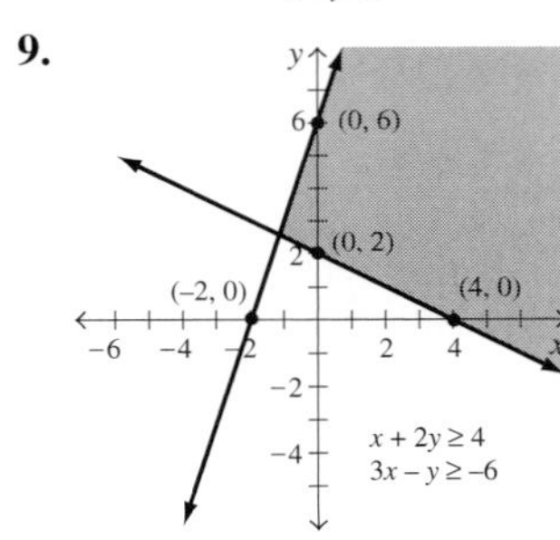

11.

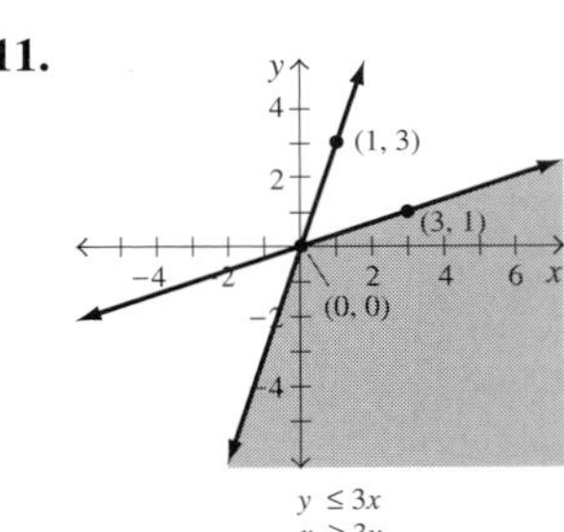

13.

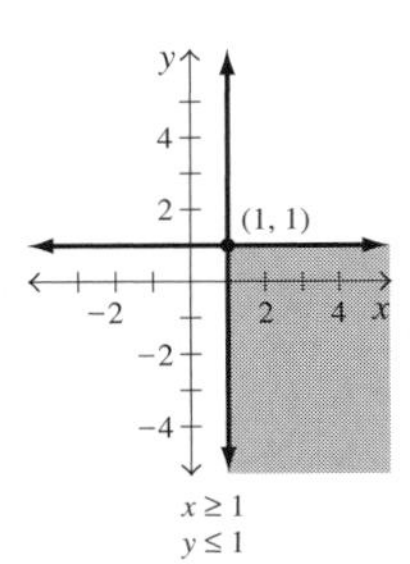

15.

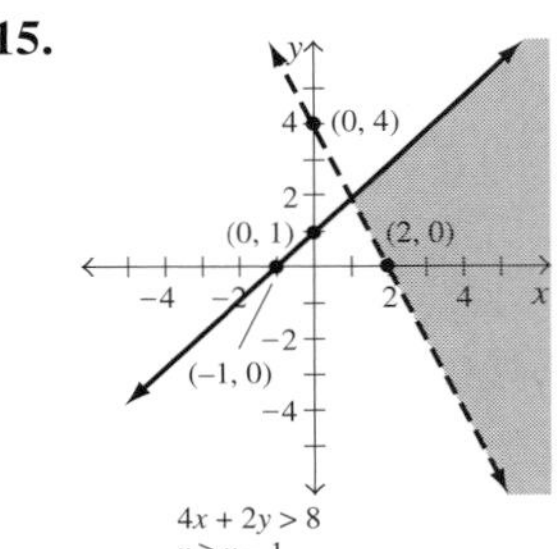

17.

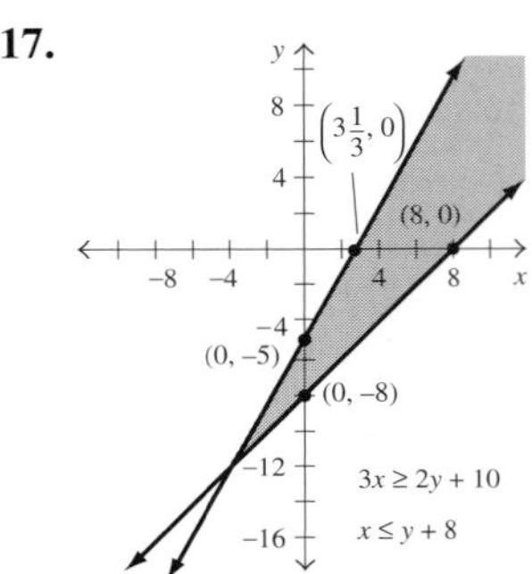

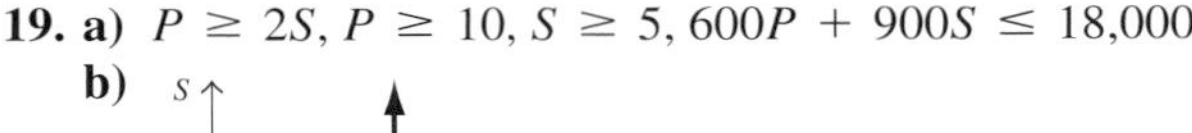

19. a) $P \geq 2S$, $P \geq 10$, $S \geq 5$, $600P + 900S \leq 18{,}000$

b)

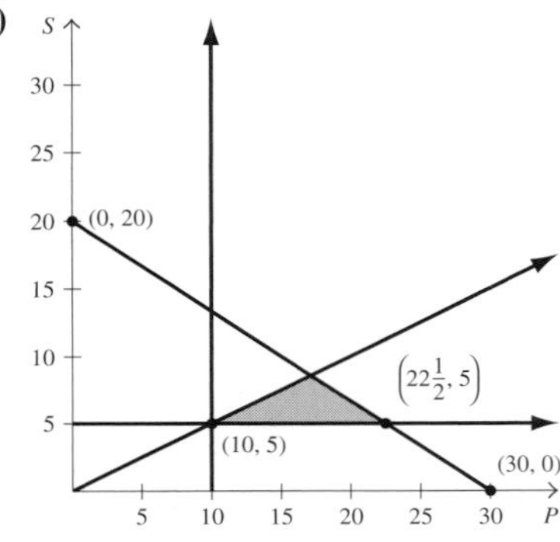

c) One example is (15, 6); \$14,400 of inventory cost

21. a) No **b)** One example: $x + y > 4$
$x + y < 1$

23. No, every line divides the plane into two half planes, only one of which can be part of the solution.

25. Answers will vary.

Section 7.6, Page 426

1. Constraints are restrictions that are represented as linear inequalities.

3. Vertices **5.** Answers will vary.

7. Maximum is 30 at (5, 0), minimum is 0 at (0, 0).

9. a)

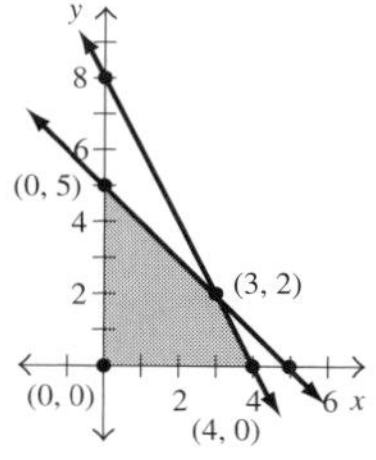

b) Maximum is 23 at (3, 2), minimum is 0 at (0, 0).

11. a)

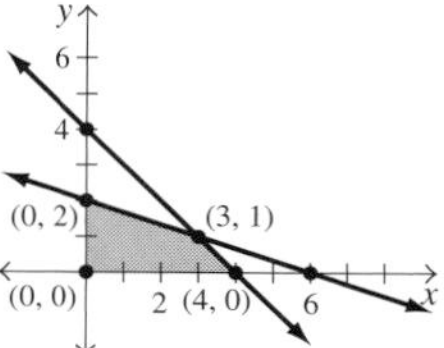

b) Maximum is 28 at (4, 0), minimum is 0 at (0, 0).

13. a)

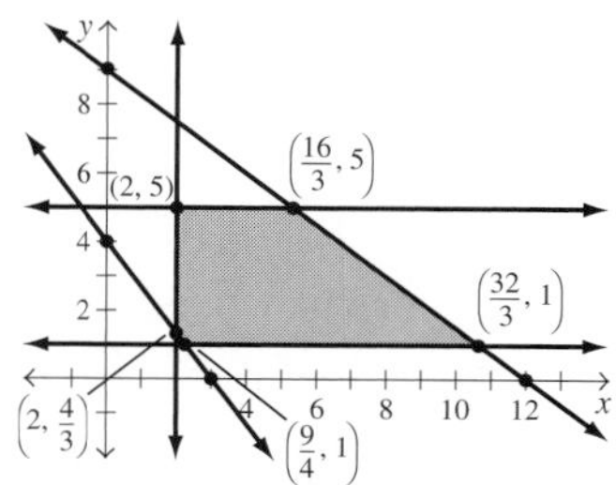

b) Maximum is ≈ 25.12 at $\left(\frac{32}{3}, 1\right)$, minimum is 6.6 at $\left(\frac{9}{4}, 1\right)$ or $\left(2, \frac{4}{3}\right)$.

15. a) $x + y \leq 24$, $x \geq 2y$, $y \geq 4$, $x \geq 0$, $y \geq 0$

b) $P = 0.35x + 0.50y$

c)

d) (8, 4), (16, 8), (20, 4)

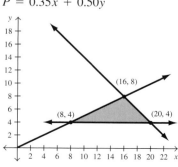

e) 16 rolls of Kodak and 8 rolls of Fuji

f) \$9.60

17. a) $3x + 4y \geq 60$, $10x + 5y \geq 100$, $x \geq 0$, $y \geq 0$

b) $C = 28x + 33y$

c)

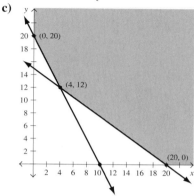

d) (0, 20), (4, 12), (20, 0)

e) 4 hours for Machine I and 12 hours for Machine II

f) \$508

19. Three car seats and seven strollers, \$320

Review Exercises, Page 428

1.

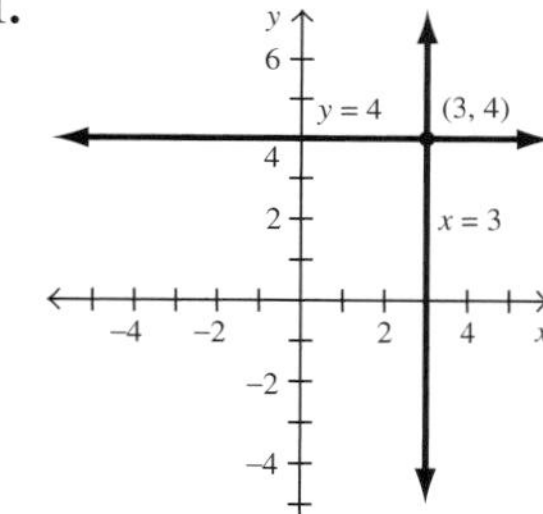

2.

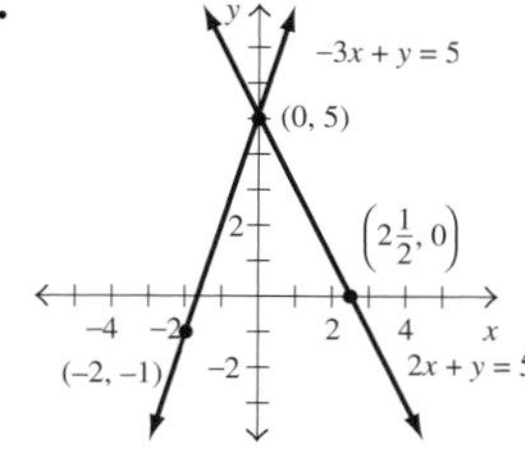

3.

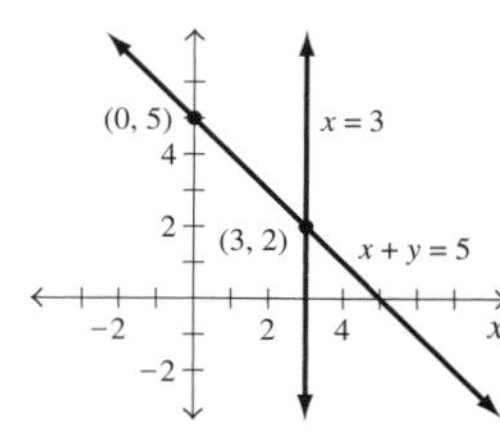

4.

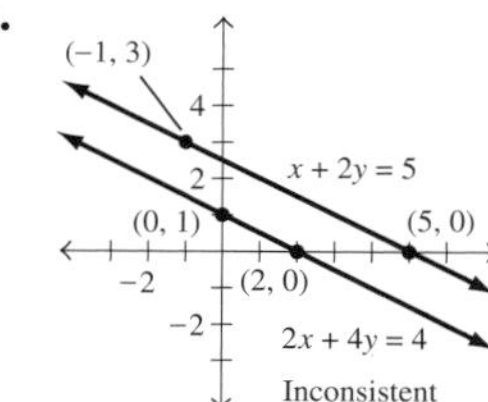

5. An infinite number of solutions **6.** No solution
7. One solution **8.** One solution **9.** $(-9, 3)$
10. $(-1, -5)$ **11.** $(-2, -8)$
12. No solution; inconsistent **13.** $(3, 1)$ **14.** $(-7, 16)$
15. $(4, -2)$ **16.** An infinite number of solutions; dependent
17. $(30, -15)$ **18.** $(2, 0)$

19. $\begin{bmatrix} -1 & -8 \\ 8 & 7 \end{bmatrix}$ **20.** $\begin{bmatrix} 3 & 2 \\ -4 & 1 \end{bmatrix}$

21. $\begin{bmatrix} 2 & -6 \\ 4 & 8 \end{bmatrix}$ **22.** $\begin{bmatrix} 8 & 9 \\ -14 & -1 \end{bmatrix}$

23. $\begin{bmatrix} -20 & -14 \\ 20 & 2 \end{bmatrix}$ **24.** $\begin{bmatrix} -12 & -14 \\ 12 & -6 \end{bmatrix}$

25. $(2, 2)$ **26.** $(-2, 2)$ **27.** $(3, -3)$
28. $(1, 0)$ **29.** $(\frac{12}{11}, \frac{7}{11})$ **30.** $(1, 2)$
31. \$350,000 at 8%, \$250,000 at 10%
32. Mix $83\frac{1}{3}\ell$ of 80% acid solution with $16\frac{2}{3}\ell$ of 50% acid solution.
33. \$500 salary, 4% commission rate
34. a) 32.5 months **b)** Model 6070B
35. a) 3 hr **b)** All-Day parking lot

36.

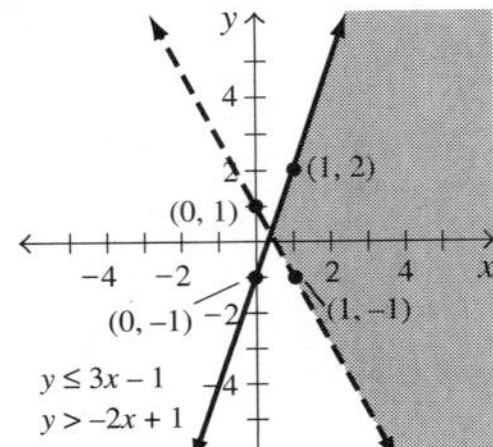

37.

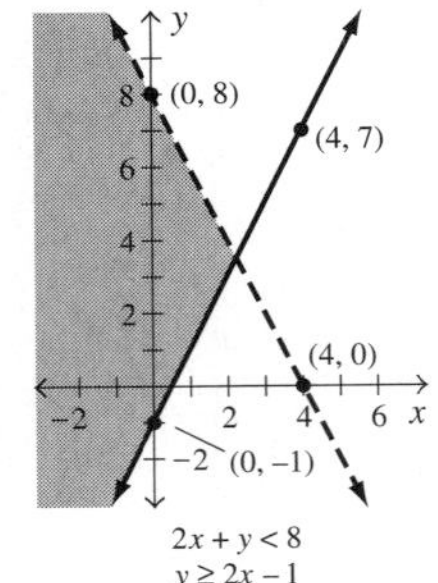

38.

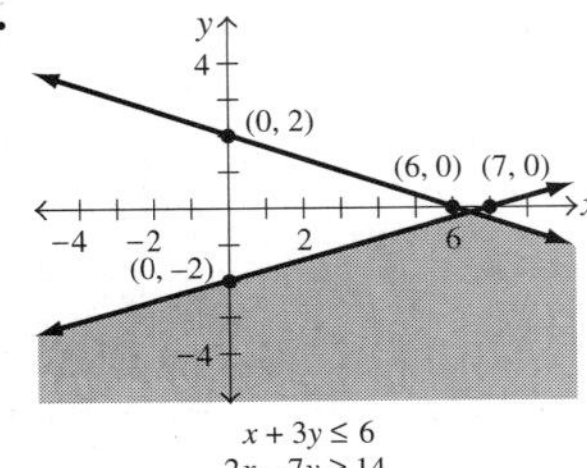

39.

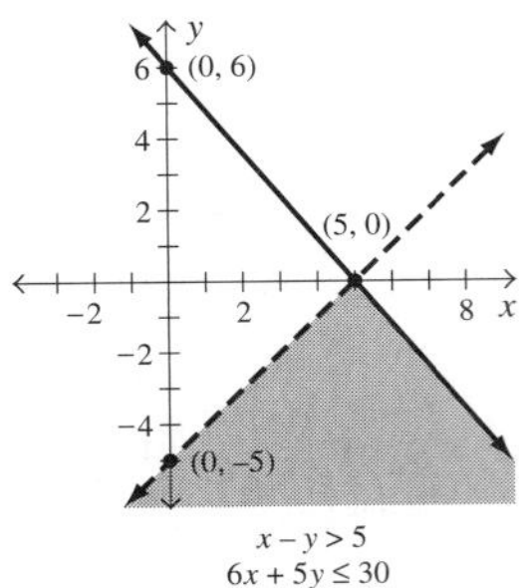

40. The maximum is 54 at $(9, 0)$.

Chapter Test, Page 429

1. If the lines do not intersect (are parallel) the system of equations is inconsistent. The system of equations is consistent if the lines intersect. If both equations represent the same line, then the system of equations is dependent.

2.

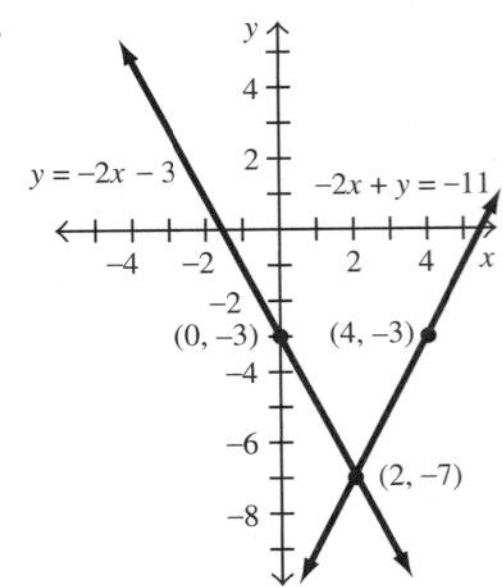

3. One solution **4.** $(2, -3)$ **5.** $(-2, -3)$
6. $(3, -1)$ **7.** $(-1, 3)$ **8.** $(2, 0)$
9. $(-2, 2)$ **10.** $\begin{bmatrix} 1 & -8 \\ 6 & 5 \end{bmatrix}$

11. $\begin{bmatrix} 7 & -12 \\ -2 & 7 \end{bmatrix}$ **12.** $\begin{bmatrix} -27 & -16 \\ 14 & 3 \end{bmatrix}$

13.

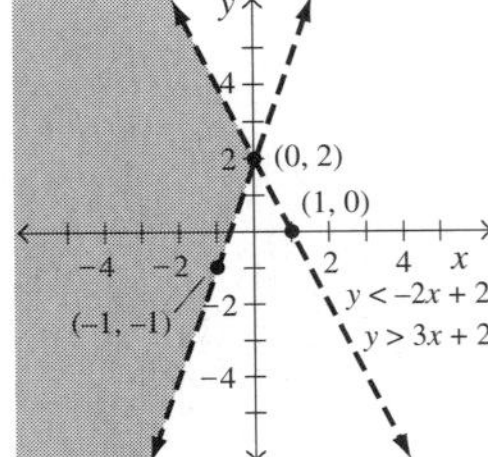

14. 20 lb at \$7.50, 10 lb at \$6.00
15. a) 40 checks **b)** Citrus Bank
16. a)

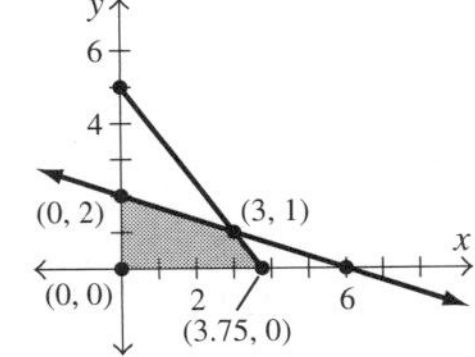

b) Maximum is 18.75 at $(3.75, 0)$, minimum is 0 at $(0, 0)$.

Chapter 8

Section 8.1, Page 437

1. The metric system

3. It is the standard of measurement accepted worldwide. There is only one basic unit of measurement for each quantity. It is based on the number 10, which makes many calculations easier than the U.S. customary system.

5. a) Answers will vary. **b)** 0.007 146 km **c)** 30 800 dm

7. Answers will vary.

9. a) 100 times greater **b)** 100 dm **c)** 0.01 dam

11. 2 **13.** 5 **15.** 22 **17.** (d) **19.** (c) **21.** (f)

23. a) 10 **b)** $\frac{1}{100}$ **c)** $\frac{1}{1000}$ **d)** $\frac{1}{10}$ **e)** 1000 **f)** 100

25. mg; $\frac{1}{1000}$ g **27.** dg; $\frac{1}{10}$ g **29.** hg; 100 g

31. 320 000 g **33.** 200 **35.** 9.5 **37.** 0.024 26

39. 0.040 36 **41.** 13 400 **43.** 92 500 g

45. 895 000 mℓ **47.** 0.024 hg **49.** 4.0302 daℓ

51. 590 cm, 5.1 dam, 0.47 km

53. 2.2 kg, 2400 g, 24 300 dg

55. 203 000 mm, 2.6 km, 52.6 hm

57. Jim, 1 m > 1 yd

59. The pump that removes 1 daℓ per minute

61. a) 346 cm **b)** 3460 mm

63. a) 108 m **b)** 0.108 km **c)** 108 000 mm **65.** 3.2 km

67. a) 2160 mℓ **b)** 2.16 ℓ **c)** $1.13 per liter

69. a) 6900 g **b)** 23 000 dg **71.** 1000

73. 1×10^{24} = 1 000 000 000 000 000 000 000 000

75. ≈ 30 eggs **77.** ≈ 4.1 cups **79.** 5 dam **81.** 6 mg

83. 2 daℓ **85.** gram **86.** decigram **87.** liter

88. dekaliter **89.** meter **90.** milliliter **91.** kilometer

92. centimeter **93.** degrees celsius **94.** hectogram

Section 8.2, Page 446

1. Volume **3.** Area **5.** Volume **7.** Volume **9.** Area

11. Length **13.–17.** Answers will vary.

19. A cubic decimeter **21.** A cubic centimeter **23.** Area

25. Centimeters **27.** Centimeters or millimeters

29. Centimeters **31.** Millimeters

33. Centimeters or millimeters **35.** Kilometers

37. (c) **39.** (c) **41.** (a) **43.** (b)

45.–49. Answers will vary. **51.** centimeter, kilometer

53. meter **55.** a centimeter **57.** Square centimeters

59. Square meters **61.** Square meters or hectares

63. Square centimeters or square millimeters

65. Square kilometers or hectares

67. (b) **69.** (a) **71.** (c) **73.** (c)

75.–79. Answers will vary. **81.** Kiloliters

83. Milliliters **85.** Liters **87.** Cubic meters

89. Liters or milliliters **91.** (c) **93.** (c) **95.** (a)

97. (a) **99. b)** 152 561 cm^3 **101.** ≈ 0.20 m^3

103. Longer side = 4 cm, shorter side = 2.2 cm, area = 8.8 cm^2

105. 2984 cm^2 **107. a)** 5.25 km^2 **b)** 525 ha

109. a) 450 m^3 **b)** 450 kℓ

111. a) 56 000 cm^3 **b)** 56 000 mℓ **c)** 56 ℓ

113. 100 times larger **115.** 1000 times larger

117. 1 000 000 **119.** 100 **121.** 0.001 **123.** 1 000 000

125. 435 **127.** 76 **129.** 600 000 dℓ

131. Answers will vary. **133.** 6700

135. a) 4,014,489,600 sq in. **b)** Answers will vary.

137. Answers will vary.

138. a) 100 cm longer **b)** 3 times longer

139. a) Answers will vary. The average use is 5150.7 ℓ/day. **b)** Answers will vary. The average use is 493.2 ℓ/day.

Section 8.3, Page 455

1. Kilogram **3.** 2 **5.–7.** Answers will vary.

9. Grams **11.** Grams **13.** Grams **15.** Kilograms

17. Grams **19.** (b) **21.** (b) **23.** (b)

25.–27. Answers will vary. **29.** (c) **31.** (b) **33.** (b)

35. (c) **37.** (b) **39.** 86°F **41.** ≈ 33.3°C **43.** ≈ 82.2°C

45. 98.6°F **47.** ≈ −10.6°C **49.** 113°F **51.** ≈ −28.9°C

53. 71.6°F **55.** 95.18°F **57.** 64.04°F–74.30°F

59. $4.34 **61.** 444 g

63. a) 2304 m^3 **b)** 2304 kℓ **c)** 2304 t

65. Yes: 78°F is about 25.6°C. **67.** 0.0042

69. 17 400 000 g **71. a)** 1200 g **b)** 1200 cm^3

73. a) 5.625 ft^3 **b)** ≈351.6 lb **c)** ≈ 42.4 gal **74.** 1500 g

75. a) −79.8°F **b)** 36.5°F **c)** 510 000 000°C

Section 8.4, Page 465

1. Dimensional analysis is a procedure used to convert from one unit of measurement to a different unit of measurement.

3. $\frac{60 \text{ seconds}}{1 \text{ minute}}$ or $\frac{1 \text{ minute}}{60 \text{ seconds}}$ **5.** $\frac{1 \text{ ft}}{30 \text{ cm}}$ **7.** $\frac{3.8\ \ell}{1 \text{ gal}}$

9. 132.08 cm **11.** 1.26 m **13.** 12 m^2 **15.** 62.4 km

17. 1687.5 acres **19.** ≈ 33.19 pints **21.** 1.52 fl oz

23. 54 kg **25.** 28 grams **27.** 0.45 kilogram

29. 2.54 centimeters, 1.6 kilometers **31.** 9 meters

33. ≈ 561.11 yd **35.** ≈ 1146.67 ft **37.** ≈ 53.13 mph
39. 43.2 m^2 **41.** ≈ 14.29 oz **43.** 240 mℓ **45.** 360 m^3
47. $0.495 per pound **49.** ≈ 9078.95 gal
51. a) ≈ 50.91 kg **b)** ≈ 113.13 lb
53. a) −8460 cm **b)** −84.6 m
55. a) 10.89 ft^2 **b)** 35.937 ft^3 **57.** 25.2 mg
59. 6840 mg, or 6.84 g **61. a)** 25 mg **b)** 900 mg
63. a) 289.2 m **b)** 76 500 t **c)** 44.8 kph
65. a) ≈ 41.1 yd **b)** 231,337.5 mi
c) 27.5 mi **d)** 2300°F
e) 209.375 mph **f)** 65,520 lb
g) 5 yd × 20 yd **h)** ≈ 45,104.21 gal/min
i) ≈ 16,733.68 gal/min **j)** 52.1 yd
k) ≈ 9.33 yd **l)** 1,406,160 lb
m) 235,871.11 lb **n)** −419.8°F
67. 7.8 lb **69.** A meter **70.** A kilogram **71.** A hectare
72. A liter **73.** A tonne **74.** A decimeter **75.** wonton
76. 1 microscope **77.** 1 kilohurtz **78.** 1 pound cake
79. 1 megaphone **80.** 2 megacycles **81.** 2 kilomockingbird
82. 1 decacards **83.** 1 decoration **84.** 1 microfiche

Review Exercises, Page 469

1. $\frac{1}{100}$ of base unit **2.** 1000 × base unit
3. $\frac{1}{1000}$ of base unit **4.** 100 × base unit
5. 10 times base unit **6.** $\frac{1}{10}$ of base unit **7.** 0.20 g
8. 320 cℓ **9.** 0.004 mm **10.** 1 kg **11.** 4620 ℓ
12. 19 260 dg **13.** 3000 mℓ, 14 630 cℓ, 2.67 kℓ
14. 0.047 km, 47 000 cm, 4700 m **15.** Centimeters
16. Grams **17.** Degrees Celsius
18. Millimeters or centimeters **19.** Square meters
20. Milliliters or cubic centimeters **21.** Millimeters
22. Kilograms or tonnes **23.** Kilometers
24. Meters or centimeters **25. a)** and **b)** Answers will vary.
26. a) and **b)** Answers will vary. **27.** (c) **28.** (b) **29.** (c)
30. (a) **31.** (a) **32.** (b) **33.** 2.5 t **34.** 6 300 000 g
35. 64.4°F **36.** 20°C **37.** ≈ −21.1°C **38.** 102.2°F
39. $l = 4$ cm, $w = 1.6$ cm, $A = 6.4$ cm^2
40. $r = 1.5$ cm, $A \approx 7.07$ cm^2
41. a) 80 m^3 **b)** 80 000 kg
42. a) 660 m^2 **b)** 0.000 66 km^2
43. a) 96 000 cm^3 **b)** 0.096 m^3
c) 96 000 mℓ **d)** 0.096 kℓ
44. 10,000 times larger **45.** ≈ 7.87 in. **46.** ≈ 233.33 lb
47. 74.7 m **48.** ≈ 111.11 yd **49.** 72 kph **50.** ≈ 42.11 qt
51. 57 ℓ **52.** ≈ 52.63 yd^3 **53.** ≈ 12.77 $in.^2$ **54.** 3.8 ℓ
55. 11.4 m^3 **56.** 99.2 km **57.** 0.9 ft **58.** 82.55 mm
59. a) 1050 kg **b)** ≈ 2333.33 lb **60.** 32.4 m^2
61. a) 190 kℓ **b)** 190 000 kg
62. a) 56 kph **b)** 56 000 meters per hour
63. a) 252 ℓ **b)** 252 kg
64. $1.58 per pound

Chapter Test, Page 471

1. 0.204 daℓ **2.** 123 000 000 mm **3.** 100 times greater
4. 2.4 km **5.** (b) **6.** (a) **7.** (c) **8.** (c) **9.** (b)
10. 10,000 times greater **11.** 1,000,000,000 times greater
12. 1148.08 cm **13.** ≈ 166.67 yd **14.** ≈ −23.33°C
15. 68°F
16. 360 cm or 365.76 cm, depending on which conversion factor you used
17. a) 3200 m^3 **b)** 3 200 000 ℓ (or 3200 kℓ)
c) 3 200 000 kg
18. $245

Chapter 9

Section 9.1, Page 482

1. a) Undefined terms, definitions, postulates (axioms), and theorems
b) First, Euclid introduced undefined terms. Second, he introduced certain definitions. Third, he stated primitive propositions called postulates about the undefined terms and definitions. Fourth, he proved, using deductive reasoning, other propositions called theorems.
3. Two lines in the same plane that do not intersect are parallel lines.
5. Two angles in the same plane are adjacent angles when they have a common vertex and a common side but no common interior points.
7. Two angles the sum of whose measure is 90° are called complementary angles.
9. An angle whose measure is greater than 90° but less than 180° is an obtuse angle.
11. An angle whose measure is 90° is a right angle.
13. Half line, $\overset{\circ\rightarrow}{AB}$ **15.** Line segment, $\overline{AB}$ **17.** Line, $\overleftrightarrow{AB}$
19. Open line segment, $\overset{\circ\!-\!\circ}{AB}$ **21.** $\overrightarrow{BD}$ **23.** $\overset{\circ\rightarrow}{BD}$
25. $\{B, F\}$ **27.** $\{C\}$ **29.** $\overline{BC}$ **31.** $\overleftrightarrow{BC}$ **33.** ∅
35. $\overrightarrow{BC}$ **37.** ∡ABE **39.** ∡EBC **41.** $\overset{\circ\rightarrow}{AC}$ **43.** $\overset{\circ\rightarrow}{BE}$
45. Obtuse **47.** Straight **49.** Right **51.** None of these
53. 71° **55.** $57\frac{1}{4}°$ **57.** 25.3° **59.** 89° **61.** 159.5°
63. $136\frac{2}{7}°$ **65.** (d) **67.** (c) **69.** (e)

71. $m\measuredangle 1 = 47°$, $m\measuredangle 2 = 43°$

73. 134° and 46°

75. Angles 3, 4, and 7 each measure 125°; angles 1, 2, 5, and 6 each measure 55°.

77. Angles 2, 5, and 6 each measure 25°; angles 1, 3, 4, and 7 each measure 155°.

79. $m\measuredangle 1 = 70°$, $m\measuredangle 2 = 20°$

81. $m\measuredangle 1 = 33°$, $m\measuredangle 2 = 57°$

83. $m\measuredangle 1 = 115°$, $m\measuredangle 2 = 65°$

85. $m\measuredangle 1 = 29°$, $m\measuredangle 2 = 151°$

87. a) An infinite number **b)** An infinite number

89. An infinite number

For Exercises 91–97, the answers given are one of many possible answers.

91. Plane ABG and plane JCD

93. $\overleftrightarrow{BG}$ and $\overleftrightarrow{DG}$

95. Plane $AGB \cap$ plane $ABC \cap$ plane $BCD = \{B\}$

97. $\overleftrightarrow{BC} \cap$ plane $ABG = \{B\}$

99. Always true. If any two lines are parallel to a third line, then they must be parallel to each other.

101. Sometimes true. Vertical angles are only complementary when each is equal to 45°.

103. Sometimes true. Alternate interior angles are only complementary when each is equal to 45°.

105. No. Line m and line n may intersect.

107.

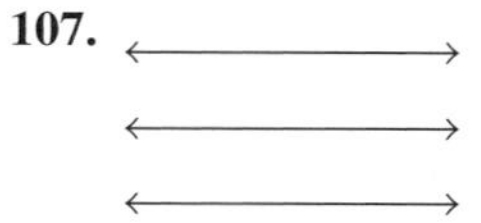

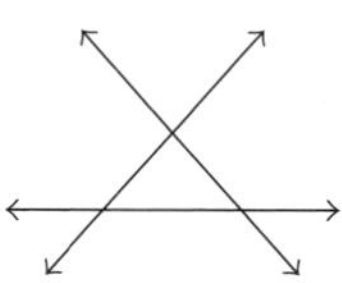

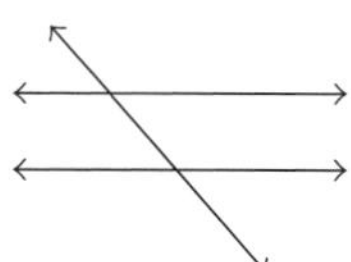

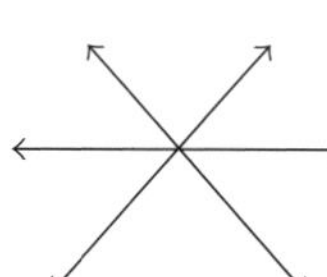

109. a)

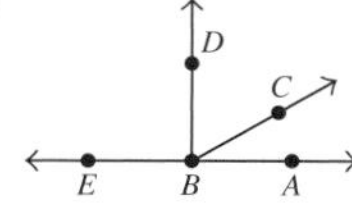

Other answers are possible.

b) 30° **c)** 60° **d)** 90°

Section 9.2, Page 491

1. A polygon is a closed figure in a plane determined by three or more straight line segments.

3. The different types of triangles are acute, obtuse, right, isosceles, equilateral, and scalene. Descriptions will vary.

5. If the corresponding sides of two similar figures are the same length, the figures are congruent figures.

7. a) Rectangle **b)** Not regular

9. a) Hexagon **b)** Regular

11. a) Rhombus **b)** Not regular

13. a) Octagon **b)** Not regular

15. a) Scalene **b)** Right

17. a) Isosceles **b)** Obtuse

19. a) Equilateral **b)** Acute

21. a) Scalene **b)** Obtuse

23. Parallelogram **25.** Rhombus **27.** Trapezoid

29. 17° **31.** 150°

33. $m\measuredangle 1 = 50°$, $m\measuredangle 2 = 63°$, $m\measuredangle 3 = 67°$, $m\measuredangle 4 = 67°$, $m\measuredangle 5 = 50°$, $m\measuredangle 6 = 113°$, $m\measuredangle 7 = 50°$, $m\measuredangle 8 = 130°$, $m\measuredangle 9 = 67°$, $m\measuredangle 10 = 113°$, $m\measuredangle 11 = 130°$, $m\measuredangle 12 = 50°$

35. 540° **37.** 720° **39.** 3240° **41. a)** 60° **b)** 120°

43. a) 135° **b)** 45° **45. a)** 150° **b)** 30°

47. $x = 6$, $y = \frac{16}{5}$ **49.** $x = \frac{12}{5}$, $y = \frac{15}{2}$

51. $x = 1.2$, $y = 0.625$ **53.** 6 **55.** $\frac{20}{3}$ **57.** 14 **59.** 28

61. 28° **63.** 8 **65.** 16 **67.** 70° **69.** 55° **71.** 35°

73. 70 ft **75. a)** 246.25 mi **b)** 270.875 mi

77. $\overline{D'E'} = 4$, $\overline{E'F'} = 5$, $\overline{D'F'} = 3$

79. a) $m\measuredangle HMF = m\measuredangle TMB$, $m\measuredangle HFM = m\measuredangle TBM$, $m\measuredangle MHF = m\measuredangle MTB$

b) 44 ft

Section 9.3, Page 503

Throughout this section, we used the π key on a scientific calculator to determine answers in calculations involving π. If you use 3.14 for π, your answers may vary slightly.

1. a) Answers will vary.

b) Answers will vary.

c) (rectangle with sides 6 and 2)

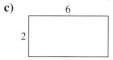

The area of this rectangle is 12 square units. The perimeter of this rectangle is 16 units.

3. a) To determine the number of square inches, multiply the number of square feet by 144.

b) To determine the number of square feet, divide the number of square inches by 144.

5. 35 in.^{2} **7.** 17.5 cm^2

9. Area = 105 ft^2; perimeter = 44 ft

11. Area = 6000 cm^2; perimeter = 654 cm

13. Area = 288 in.^{2}; perimeter = 74 in.

15. ≈ 153.94 in.^{2}, 43.98 in.

17. ≈ 63.62 ft^2, 28.27 ft
19. a) 9 in. **b)** 36 in. **c)** 54 $in.^2$
21. a) 26 cm **b)** 60 cm **c)** 120 cm^2
23. ≈ 21.99 cm^2 **25.** 8 $in.^2$ **27.** ≈ 65.73 $in.^2$
29. ≈ 114.90 ft^2 **31.** ≈ 41.20 $in.^2$ **33.** ≈ 11.89 yd^2
35. 132.3 ft^2 **37.** 234,000 cm^2 **39.** 0.1075 m^2
41. a) $3239.50 **b)** $4889.50 **43.** $1700 **45.** $2908.80
47. $38.93 **49. a)** 177.1 m^2 **b)** 0.01771 hectare
51. ≈103.94 ft **53.** 40 ft
55. a) $A = s^2$ **b)** $A = 4s^2$ **c)** Four times larger
57. 24 cm^2 **59.** Answers will vary.

Section 9.4, Page 515

Throughout this section, we used the π key on a scientific calculator to determine answers in calculations involving π. If you use 3.14 for π, your answers may vary slightly.

1. Volume is a measure of the capacity of a figure.
3. A polyhedron is a closed surface formed by the union of polygonal regions. A regular polyhedron is one whose faces are all regular polygons of the same size and shape.
5. Answers will vary. **7.** 27 ft^3 **9.** 150.80 $in.^3$
11. 131.95 cm^3 **13.** 2400 $in.^3$ **15.** 381.70 cm^3
17. 524.33 cm^3 **19.** 106.67 $in.^3$ **21.** 59.43 m^3
23. 30.49 ft^3 **25.** 31.42 m^3 **27.** 24 ft^3 **29.** 189 ft^3
31. ≈ 5.67 yd^3 **33.** 5,900,000 cm^3 **35.** 3 m^3
37. a) 28,750 $in.^3$ **b)** ≈ 16.64 ft^3 **39.** ≈ 2.50 qt
41. a) The container with the larger diameter holds more. **b)** ≈ 188.50 $in.^3$
43. 82,944,000 ft^3 **45.** ≈ 283.04 $in.^3$
47. a) ≈ 323.98 $in.^3$ **b)** ≈ 0.19 ft^3
49. a) Round pan base ≈ 63.62 $in.^2$; rectangular pan base = 63 $in.^2$
b) Round pan volume ≈ 127.24 $in.^3$; rectangular pan volume = 126 $in.^3$
c) Round pan
51. a) 4320 $in.^3$ **b)** 2.5 ft^3 **53.** Nine edges
55. Six vertices **57.** Fourteen edges **59.** ≈ 21.46%
61. a)–e) Answers will vary.
f) If we double the radius of a sphere, the new volume will be eight times the original volume.
63. a) Answers will vary.
b) $V_1 = a^3$; $V_2 = a^2b$; $V_3 = a^2b$; $V_4 = ab^2$; $V_5 = a^2b$; $V_6 = ab^2$; $V_7 = b^3$
c) ab^2
64. a) 330 $in.^3$ **b)** ≈ 300.84 $in.^3$

Section 9.5, Page 532

1. The act of moving a geometric figure from some starting position to some ending position without altering its shape or size is called rigid motion. The four main rigid motions studied in this section are reflections, translations, rotations, and glide reflections.
3. A reflection is a rigid motion that moves a figure to a new position that is a mirror image of the figure in the starting position.
5. A translation is a rigid motion that moves a figure by sliding it along a straight line segment in the plane.
7. A rotation is a rigid motion performed by rotating a figure in the plane about a specific point.
9. A glide reflection is a rigid motion formed by performing a translation (or glide) followed by a reflection.
11. A geometric figure is said to have reflective symmetry if the positions of a figure before and after a reflection are identical (except for vertex labels).
13. A tessellation is a pattern consisting of the repeated use of the same geometric figures to entirely cover a plane, leaving no gaps.

This figure contains the answers for Exercises 15 and 16.

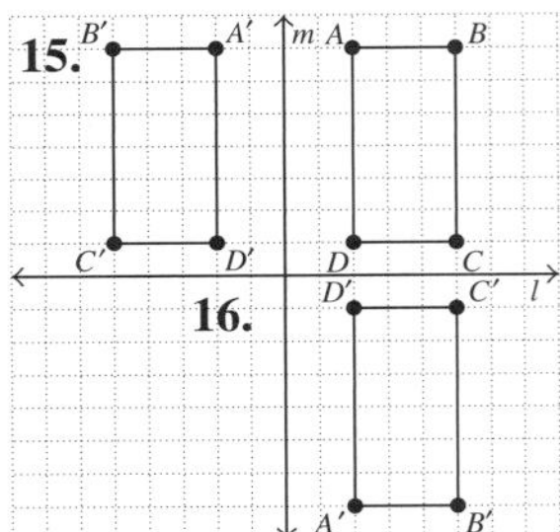

This figure contains the answers for Exercises 17 and 18.

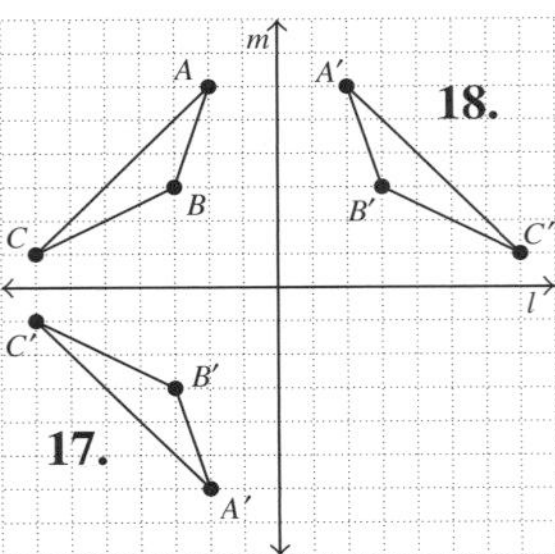

This figure contains the answers for Exercises 19 and 20.

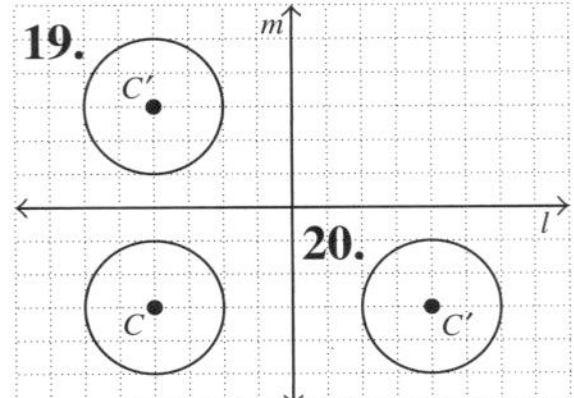

This figure contains the answers for Exercises 21 and 22.

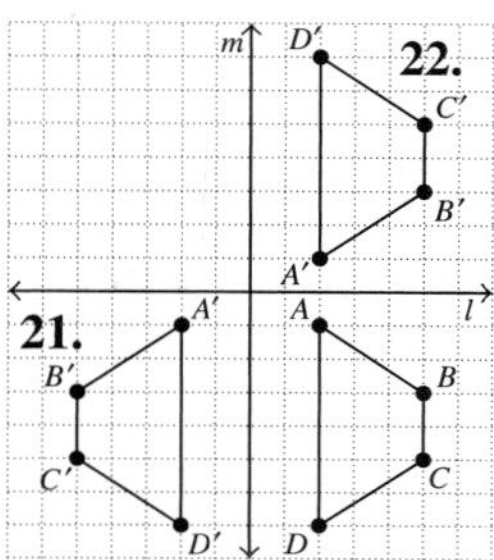

This figure contains the answers to Exercises 23 and 24.

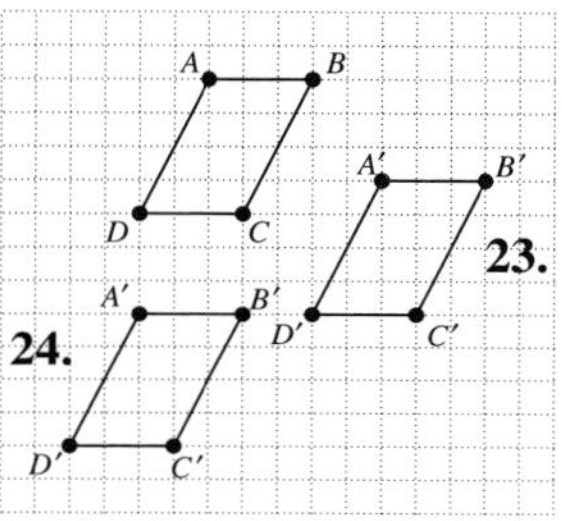

This figure contains the answers to Exercises 25 and 26.

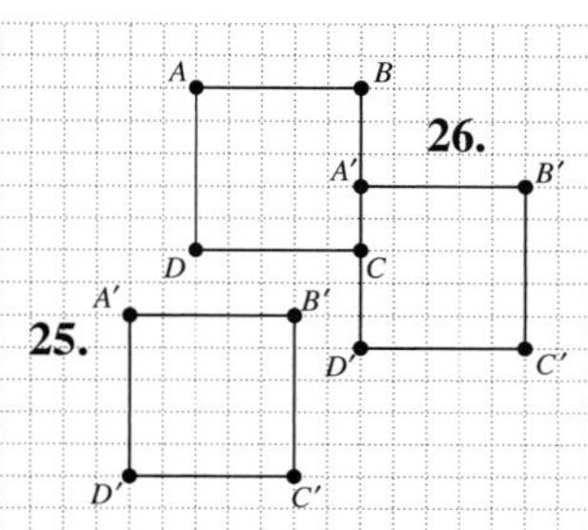

This figure contains the answers to Exercises 27 and 28.

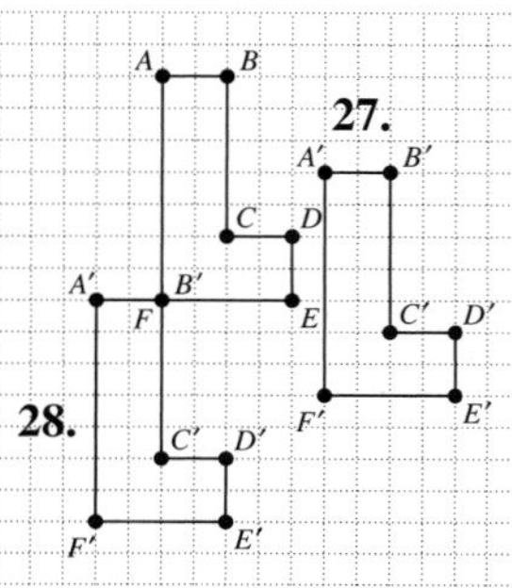

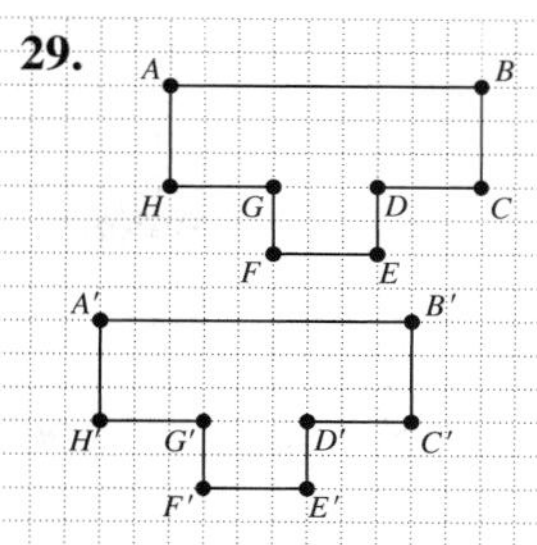

This figure contains the answers to Exercises 31 and 32.

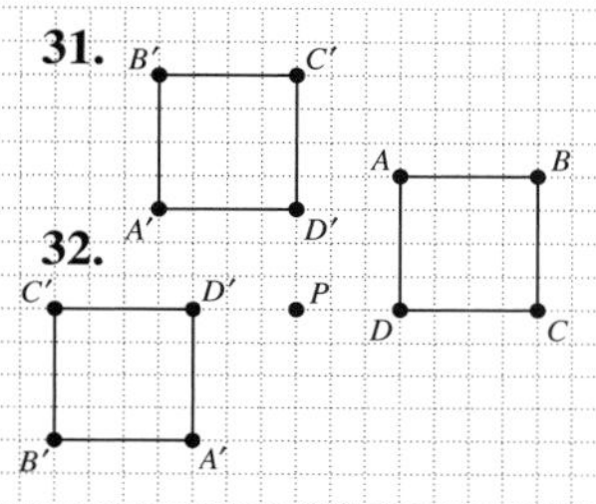

This figure contains the answers to Exercises 33 and 34.

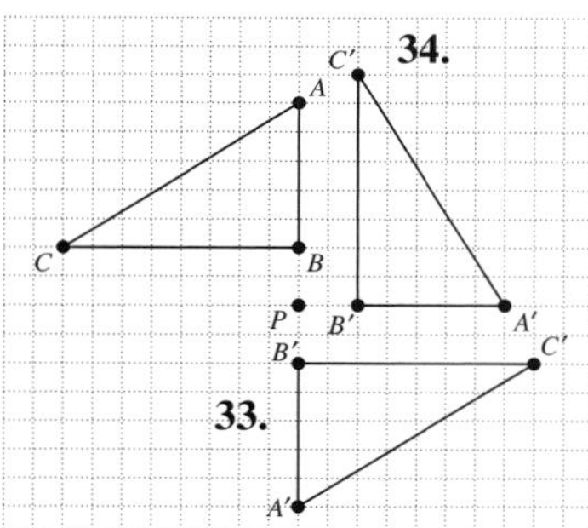

This figure contains the answers to Exercises 35 and 36.

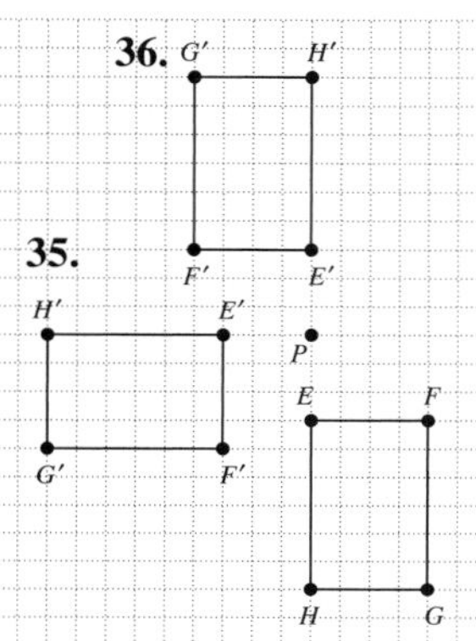

This figure contains the answers to Exercises 37 and 38.

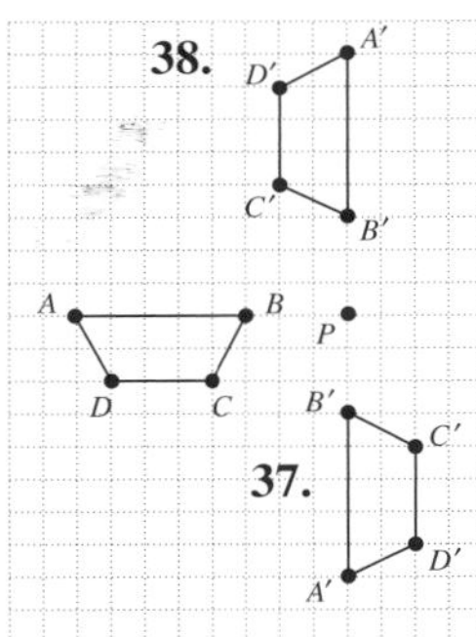

This figure contains the answers to Exercises 39 and 40.

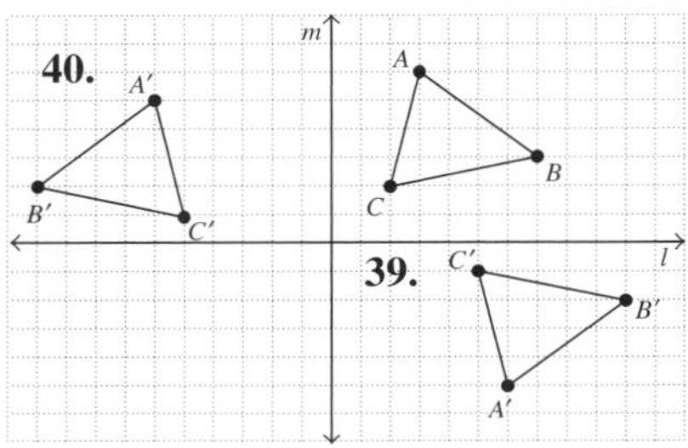

This figure contains the answers to Exercises 41 and 42.

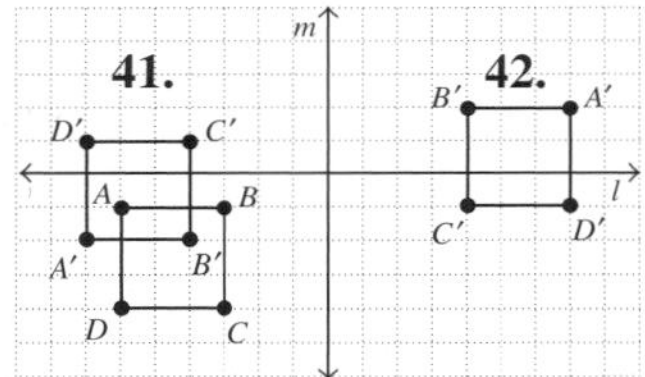

This figure contains the answers to Exercises 43 and 44.

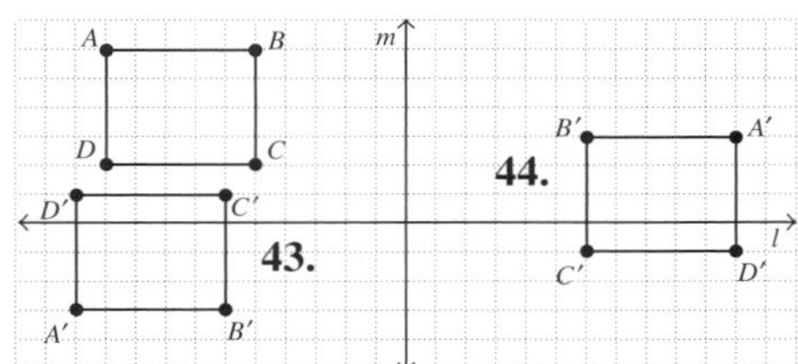

This figure contains the answers to Exercises 45 and 46.

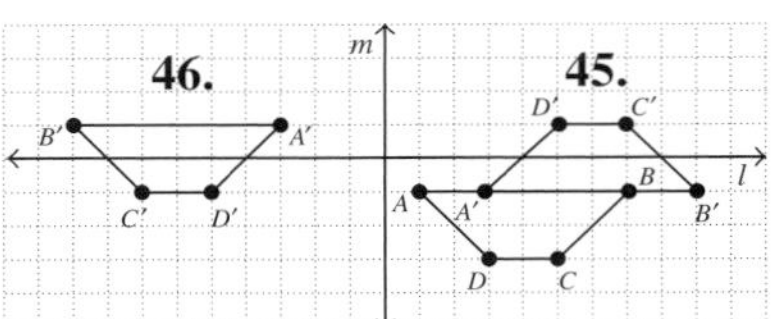

47. a)

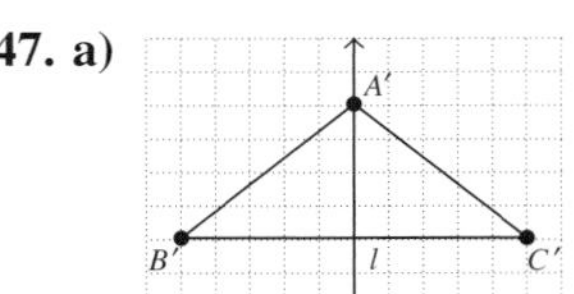

b) Yes
c) Yes

49. a)

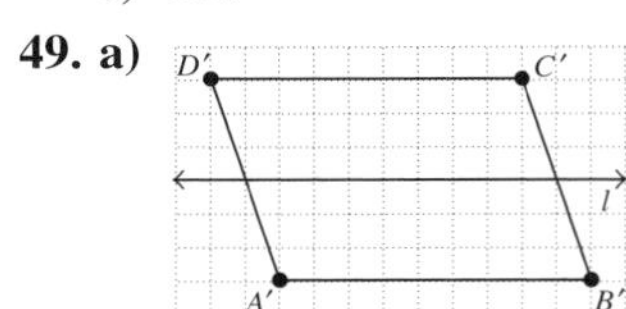

b) No
c) No

51. a)

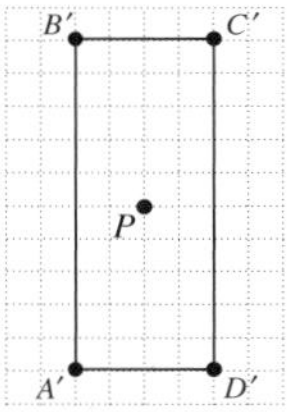

b) No
c) No
d)

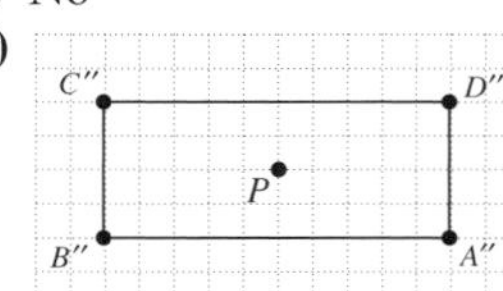

e) Yes
f) Yes

53. a)–c)

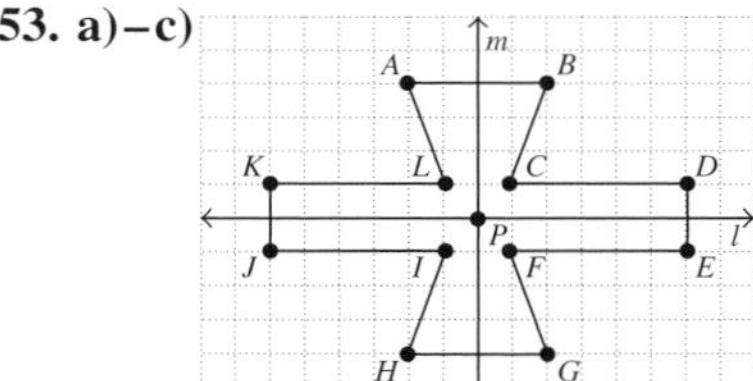

d) No. Any 90° rotation will result in the figure being in a different position than the starting position.

55. a)–b)

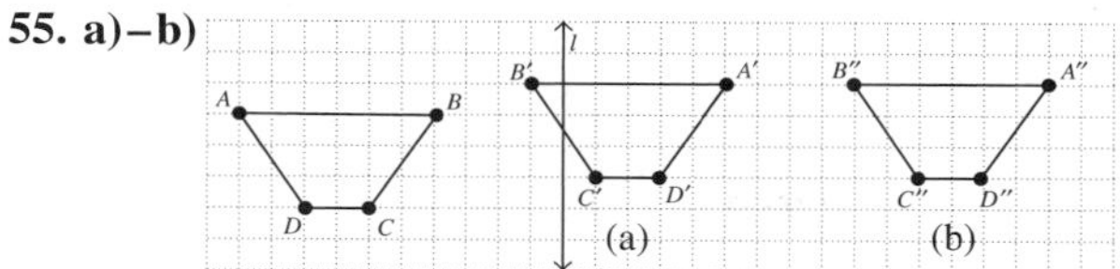

   c) No.
   d) The order in which the translation and the reflection are performed is important. The figure obtained in part (a) is the glide reflection.

57. Answers will vary.

59. a) Answers will vary.
   b) A regular pentagon cannot be used as a tessellating shape.

60. Although answers will vary depending on the font, the following capital letters have reflective symmetry about a horizontal line drawn through the center of the letter: B, C, D, E, H, I, K, O, X.

61. Although answers will vary depending on the font, the following capital letters have reflective symmetry about a vertical line drawn through the center of the letter: A, H, I, M, O, T, U, V, W, X, Y.

62. Although answers will vary depending on the font, the following capital letters have 180° rotational symmetry about a point in the center of the letter: H, I, O, S, X, Z.

Section 9.6, Page 542

1. Topology is sometimes referred to as "rubber sheet geometry" because it deals with bending and stretching of geometric figures.

3. Take a strip of paper, give one end a half twist, and tape the ends together.

5. Four

7. A Jordan curve is a topological object that can be thought of as a circle twisted out of shape.

9. The number of holes in the object determines the genus of an object.

11.–19. Answers will vary. **21.** Outside **23.** Outside

25. Outside **27.** Inside **29.** 1 **31.** 1 **33.** Larger than 5

35. 5 **37.** 0 **39.** 5 **41. a)–d)** Answers will vary.

43. One **45.** Two

47. The smaller one is a Möbius strip; the larger one is not.

49. Yes. "Both sides" of the belt experience wear.

51. Answers will vary.

53. a) 1 **b)** 1 **c)** Answers will vary.

Section 9.7, Page 552

1.–5. Answers will vary.

7. a) *Euclidean:* Given a line and a point not on the line, one and only one line can be drawn parallel to the given line through the given point.
   b) *Elliptical:* Given a line and a point not on the line, no line can be drawn through the given point parallel to the given line.
   c) *Hyperbolic:* Given a line and a point not on the line, two or more lines can be drawn through the given point parallel to the given line.

9. A plane **11.** A pseudosphere

13. Spherical: elliptical geometry; flat: Euclidean geometry; saddle-shaped: hyperbolic geometry

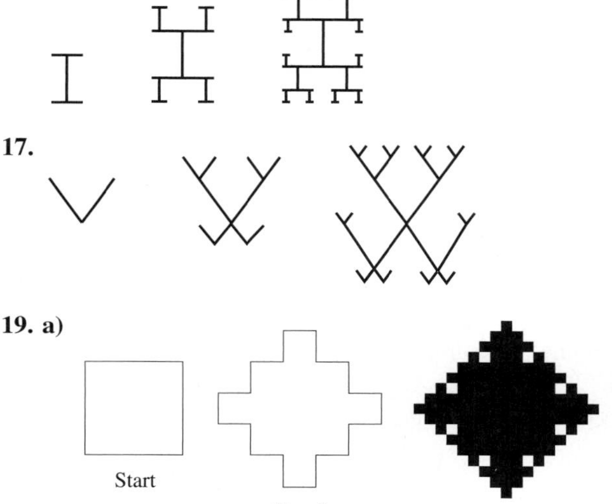

   b) Infinite
   c) Finite

Review Exercises, Page 554

In the Review Exercises and Chapter Test questions, the π key on the calculator is used to determine answers in calculations involving π. If you use 3.14 for π, your answers may vary slightly.

1. $\{F\}$ **2.** ΔBFC **3.** $\overleftrightarrow{BC}$ **4.** $\overleftrightarrow{BH}$ **5.** $\{F\}$

6. $\{\ \}$ **7.** 38.8° **8.** 55.3° **9.** 10.2 in. **10.** 2 in.

11. 58° **12.** 92°

13. $m\measuredangle 1 = 70°, m\measuredangle 2 = 60°, m\measuredangle 3 = 120°, m\measuredangle 4 = 70°, m\measuredangle 5 = 110°, m\measuredangle 6 = 70°$

14. 720° **15.** 63 cm^2 **16.** 35 in.^2 **17.** 13 in.^2

18. 84 in.^2 **19.** $\approx$ 530.93 cm^2 **20.** \$616

21. 1178.10 in.^3 **22.** 120 cm^3 **23.** 28 ft^3 **24.** 432 m^3

25. 603.19 mm^3 **26.** 1436.76 ft^3

27. a) $\approx$ 67.88 ft^3 **b)** 4617.5 lb; yes **c)** $\approx$ 511.14 gal

This figure contains the answers for Exercises 28 and 29.

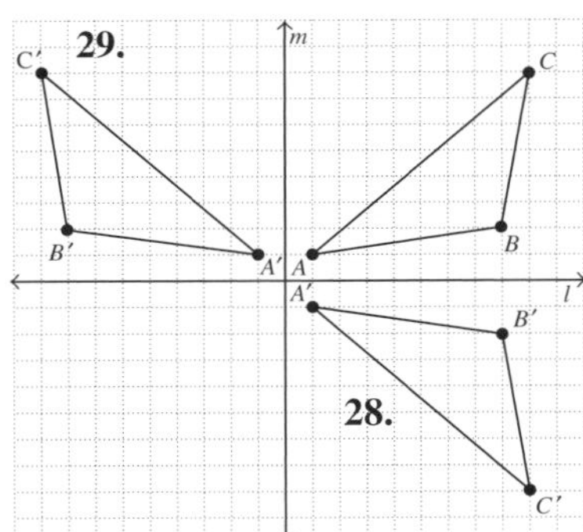

This figure contains the answers for Exercises 30 and 31.

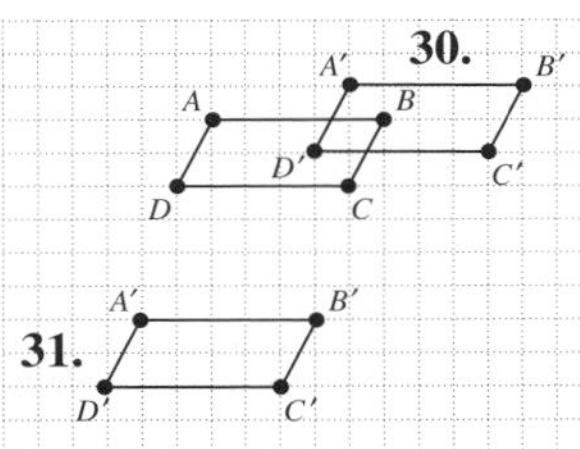

This figure contains the answers for Exercises 32–34.

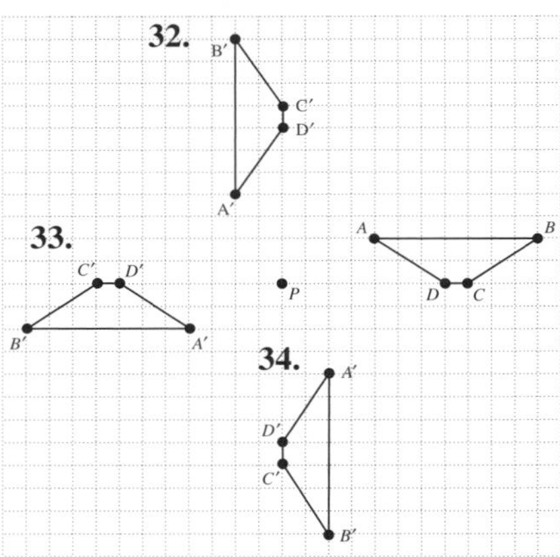

This figure contains the answers for Exercises 35 and 36.

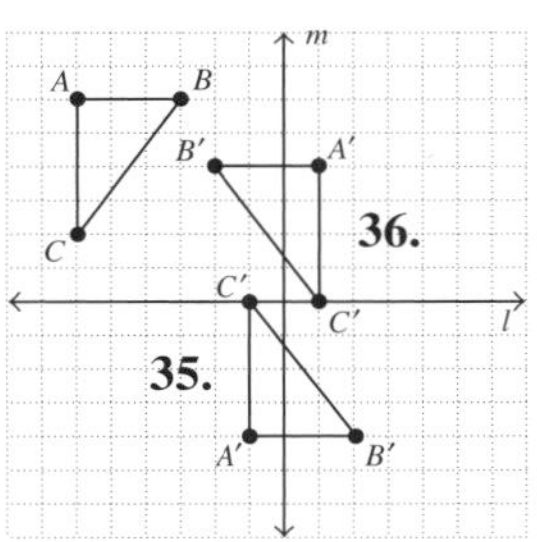

37. Yes

38. No

39. No

40. Yes

41. 1 **42.** Answers will vary. **43.** Outside

44. Euclidean: Given a line and a point not on the line, one and only one line can be drawn parallel to the given line through the given point. Elliptical: Given a line and a point not on the line, no line can be drawn through the given point parallel to the given line. Hyperbolic: Given a line and a point not on the line, two or more lines can be drawn through the given point parallel to the given line.

45.

Chapter Test, Page 557

1. $\overset{\circ}{\overrightarrow{EF}}$ **2.** ΔBCD **3.** $\{D\}$ **4.** $\overleftrightarrow{AC}$ **5.** 53.1°

6. 78.5° **7.** 64° **8.** 1080° **9.** $\approx$ 2.69 cm

10. a) 12 in. **b)** 30 in. **c)** 30 in.^{2}

11. $\approx$ 2144.66 cm^3 **12.** $\approx$ 42.14 yd^3 **13.** 112 ft^3

14.

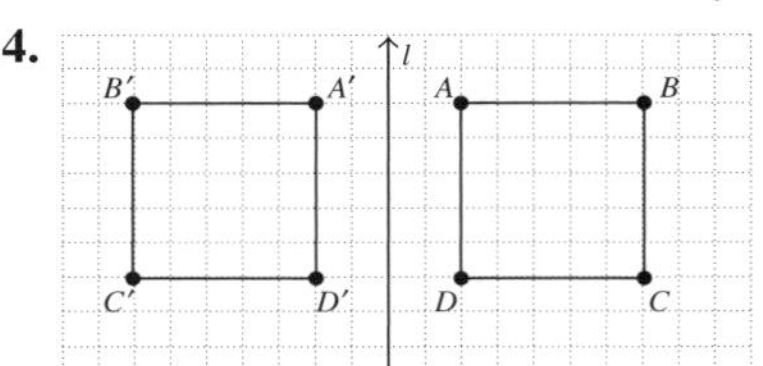

15.

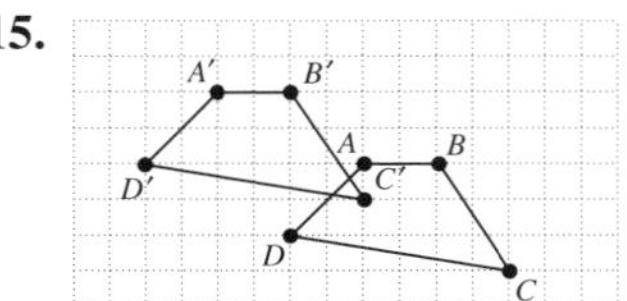

16.

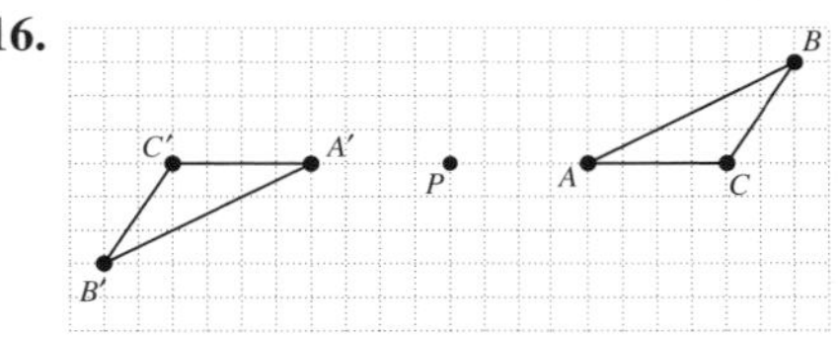

17. 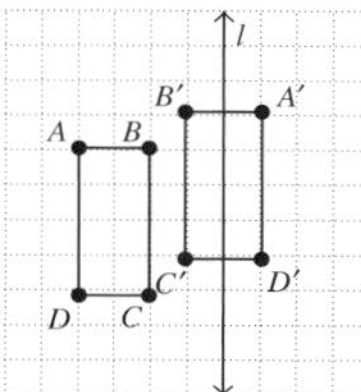

18. **a)** No **b)** Yes

19. A surface with one side and one edge

20. Answers will vary.

21. Euclidean: Given a line and a point not on the line, one and only one line can be drawn parallel to the given line through the given point. Elliptical: Given a line and a point not on the line, no line can be drawn through the given point parallel to the given line. Hyperbolic: Given a line and a point not on the line, two or more lines can be drawn through the given point parallel to the given line.

Chapter 10

Section 10.1, Page 566

1. A binary operation is an operation, or rule, that can be performed on two and only two elements of a set. The result is a single element.

3. **a)** When we add two numbers, the sum is one number: $4 + 5 = 9$.
   b) When we subtract two numbers, the difference is one number: $5 - 4 = 1$.
   c) When we multiply two numbers, the product is one number: $5 \times 4 = 20$.
   d) When we divide two numbers, the quotient is one number: $20 \div 5 = 4$.

5. A mathematical system is a commutative group if all five of the following conditions hold.
   1. The set of elements is closed under the given operation.
   2. An identity element exists for the set.
   3. Every element in the set has an inverse.
   4. The set of elements is associative under the given operation.
   5. The set of elements is commutative under the given operation.

7. If a binary operation is performed on any two elements of a set and the result is an element of the set, then that set is *closed* under the given binary operation. For all integers a and b, $a + b$ is an integer. Therefore, the set of integers is closed under the operation of addition.

9. When a binary operation is performed on two elements in a set and the result is the identity element for the binary operation, then each element is said to be the *inverse* of the other. For the set of rational numbers, the additive inverse of 2 is -2 since $2 + (-2) = 0$ and the multiplicative inverse of 2 is $\frac{1}{2}$ since $2 \times \frac{1}{2} = 1$.

11. No; every commutative group is also a group.

13. d); The commutative property need not apply.

15. $(a + b) + c = a + (b + c)$, for any elements a, b, and c; $(3 + 4) + 5 = 3 + (4 + 5)$

17. $a \cdot b = b \cdot a$ for any elements a and b; $2 \cdot 3 = 3 \cdot 2$

19. $4 \div 2 \neq 2 \div 4$

21. $$\begin{aligned}(6 - 4) - 1 &\neq 6 - (4 - 1)\\ 2 - 1 &\neq 6 - 3\\ 1 &\neq 3\end{aligned}$$

23. No; there is no identity element.

25. Yes; satisfies the five properties needed.

27. No; the system is not closed.

29. No; there is no identity element.

31. No; not all elements have inverses.

33. Yes; satisfies the four properties needed.

35. No; the system is not closed. For example, $\frac{1}{0}$ is undefined.

37. No; does not satisfy the associative property.

39. No; the system is not closed. For example, $\sqrt{2} + (-\sqrt{2}) = 0$, which is rational. There is also no identity element.

41. Yes **43.** Answers will vary. **44.** $(9^9)^9$ **45.** 20

Section 10.2, Page 574

1. Numbers are obtained by starting at the first addend (on a clock face), then moving clockwise the number of hours equal to the second addend.

3. **a)** Add $4 + 10$ to get 2, then add $2 + 3$.
   b) 5

5. **a)** Add 12 to 5 to get $17 - 9$.
   b) 8
   c) Since 12 is the identity element, you can add 12 to any number without changing the answer.

7. Yes, the sum of any two numbers in clock 12 arithmetic is a number in clock 12 arithmetic.

9. Yes; 1:11, 2:10, 3:9, 4:8, 5:7, 6:6, 7:5, 8:4, 9:3, 10:2, 11:1, and 12:12

11. Yes, $6 + 9 = 9 + 6$ since both equal 3

13. **a)** 5 **b)** 3. If you add 5 to any number, then you get the number you started with. Thus, in clock 5 arithmetic, 5 is the identity element. Three is the additive inverse of 2 since $2 + 3 = 5$.

15. Yes; the elements are symmetric about the main diagonal.

17. Yes; C is the identity element since the row next to C is identical to the top row and the column under C is identical to the left-hand column.

19. The inverse of A is B since $A \oslash B = C$ and $B \oslash A = C$.

21. 11 **23.** 5 **25.** 4 **27.** 8 **29.** 6 **31.** 6 **33.** 3 **35.** 4
37. 7 **39.** 1 **41.** 12 **43.** 12

45.

+	1	2	3	4	5	6
1	2	3	4	5	6	1
2	3	4	5	6	1	2
3	4	5	6	1	2	3
4	5	6	1	2	3	4
5	6	1	2	3	4	5
6	1	2	3	4	5	6

47. 1 **49.** 3 **51.** 2 **53.** 4

55.

+	1	2	3	4	5	6	7
1	2	3	4	5	6	7	1
2	3	4	5	6	7	1	2
3	4	5	6	7	1	2	3
4	5	6	7	1	2	3	4
5	6	7	1	2	3	4	5
6	7	1	2	3	4	5	6
7	1	2	3	4	5	6	7

57. 4 **59.** 6 **61.** 4 **63.** 7

65. Yes; it satisfies the five required properties.

67. a) $\{0, 1, 2, 3\}$
b)
c) Yes
d) Yes; 0
e) Yes; 0–0, 1–3, 2–2, 3–1
f) (2 3) 1 = 2 (3 1)
g) Yes; 2 3 = 3 2
h) Yes

69. a) $\{r, s, t, u\}$
b)
c) Yes
d) Yes; t
e) Yes; r–r, s–u, t–t, u–s
f) (r s) u = r (s u)
g) Yes; s r = r s
h) Yes

71. a) $\{f, r, o, m\}$ **b)** **c)** Closed
d) m **e)** f **f)** f **g)** m **h)** r

73. Not associative:
$(M \otimes$) $\otimes M \neq M \otimes ($ $\otimes M)$;
not commutative: $\otimes M \neq M \otimes$

75. No inverse for ⊙ or for $*$, not associative

77. No identity element, no inverses, not associative, not commutative

79. a)

+	E	O
E	E	O
O	O	E

b) Yes, it is a commutative group; it satisfies the five properties.

81. Answers will vary.

83. a) Is closed; identity element is 6; inverses: 1–5, 2–2, 3–3, 4–4, 5–1, 6–6; is associative—for example,

$$(2 \infty 5) \infty 3 = 2 \infty (5 \infty 3)$$
$$3 \infty 3 = 2 \infty 2$$
$$6 = 6$$

b) $3 \infty 1 \neq 1 \infty 3$
$2 \neq 4$

85. a)

$*$	R	S	T	U	V	I
R	V	T	U	S	I	R
S	U	I	V	R	T	S
T	S	R	I	V	U	T
U	T	V	R	I	S	U
V	I	U	S	T	R	V
I	R	S	T	U	V	I

b) Yes, the associative property will hold.
c) No, it is not commutative. For example, $R*S \neq S*R$

87.

+	0	1	2	3	4
0	0	1	2	3	4
1	1	2	3	4	0
2	2	3	4	0	1
3	3	4	0	1	2
4	4	0	1	2	3

89. Add the number in the top row and the number in the left-hand column and divide the sum by 4. The remainder is placed in the table.

Section 10.3, Page 585

1. A modulo m system consists of m elements, 0 through $m - 1$, and a binary operation.

3. 5;

0	1	2	3	4
0	1	2	3	4
5	6	7	8	9
10	11	12	13	14
.	.	.	.	.
.	.	.	.	.

5. 12 classes **7.** (b), (c) or (d) **9.** Saturday **11.** Friday
13. Saturday **15.** Thursday **17.** July **19.** March
21. September
23. July **25.** 4 **27.** 2 **29.** 3 **31.** 2 **33.** 1 **35.** 2
37. 0 **39.** 0 **41.** 6 **43.** 2 **45.** 2 **47.** 9 **49.** 5
51. 1 **53.** 2 **55.** 5 **57.** 5 **59.** { }
61. 1 and 6 **63.** 4 **65.** 0

67. a) 2016, 2020, 2024, 2028, 2032 **b)** 3004
c) 2552, 2556, 2560, 2564, 2568, 2572

69. a) Resting (for the second of two days)
b) Resting (for the second of two days)
c) Morning and afternoon practice
d) No

71. a) 5 **b)** No **c)** 54 weeks from this week

73. a) Evening **b)** Day **c)** Day

75. a)

+	0	1	2	3
0	0	1	2	3
1	1	2	3	0
2	2	3	0	1
3	3	0	1	2

b) Yes **c)** Yes, 0 **d)** Yes; 0–0, 1–3, 2–2, 3–1
e) $(1 + 2) + 3 = 1 + (2 + 3)$
f) Yes; $2 + 3 = 3 + 2$ **g)** Yes **h)** Yes

77. a)

×	0	1	2	3
0	0	0	0	0
1	0	1	2	3
2	0	2	0	2
3	0	3	2	1

b) Yes **c)** Yes, 1
d) No; no inverse for 0 or for 2, inverse of 1 is 1, inverse of 3 is 3
e) $(1 \times 2) \times 3 = 1 \times (2 \times 3)$
f) Yes, $2 \times 3 = 3 \times 2$ **g)** No

79. 2 **81.** 1, 2, 3 **83.** 0 **85.** 2 **87.** 0

88. Halfway up the mountain **89.** Math is fun.

Review Exercises, Page 589

1. A mathematical system consists of a set of elements and at least one binary operation.

2. A binary operation is an operation that can be performed on two and only two elements of a set. The result is a single element.

3. Yes; the sum of any two integers is an integer.

4. No; for example $2 - 3 = -1$ and -1 is not a natural number.

5. 7 **6.** 5 **7.** 10 **8.** 8 **9.** 9 **10.** 11

11. Closure, identity element, inverses, and associative property

12. A commutative group **13.** No; no identity element

14. No; no inverse for any integer except 1 and -1 **15.** Yes

16. No; no inverse for 0 **17.** No identity element

18. Not associative. For example, (! □ p) □ ? ≠ ! □ (p □ ?)

19. Not every element has an inverse; not associative. For example, (P ? P) ? 4 ≠ P ? (P ? 4).

20. a) {)-, ⊙, ?, △} **b)** ⎍ **c)** Yes **d)** Yes;)-
e) Yes;)- –)-, ⊙ –△, ?–?, △–⊙
f) (⊙ ⎍ ?) ⎍ △ = ⊙ ⎍ (? ⎍ △)
g) Yes; ⊙ ⎍ ? = ? ⎍ ⊙ **h)** Yes

21. 0 **22.** 7 **23.** 1 **24.** 3 **25.** 4 **26.** 2 **27.** 4
28. 12 **29.** 9 **30.** 9 **31.** 4 **32.** 3 **33.** { } **34.** 1
35. 0, 2, 4, 6 **36.** 10 **37.** 5 **38.** 9 **39.** 7 **40.** 8

41.

+	0	1	2	3	4	5
0	0	1	2	3	4	5
1	1	2	3	4	5	0
2	2	3	4	5	0	1
3	3	4	5	0	1	2
4	4	5	0	1	2	3
5	5	0	1	2	3	4

Yes, it is a commutative group.

42.

×	0	1	2	3
0	0	0	0	0
1	0	1	2	3
2	0	2	0	2
3	0	3	2	1

No; no inverse for 0 or 2

43. a) No, she will be off.
b) Yes, she will have the evening off.

Chapter Test, Page 590

1. A set of elements and a binary operation

2. Closure, identity element, inverses, associative property, commutative property

3. No, not all elements have inverses.

4.

+	1	2	3	4	5
1	2	3	4	5	1
2	3	4	5	1	2
3	4	5	1	2	3
4	5	1	2	3	4
5	1	2	3	4	5

5. Yes, it is a commutative group. **6.** 4 **7.** 2

8. a) □ **b)** Yes **c)** Yes, T **d)** S **e)** S

9. No, not closed. **10.** Yes, it is a commutative group.

11. Yes, it is a commutative group. **12.** 1 **13.** 3

14. 6 **15.** 2 **16.** 5 **17.** 2 **18.** { } **19.** 5

20. a)

×	0	1	2	3	4
0	0	0	0	0	0
1	0	1	2	3	4
2	0	2	4	1	3
3	0	3	1	4	2
4	0	4	3	2	1

b) No; no inverse for 0

Chapter 11

Section 11.1, Page 599

1. A percent is a ratio of some number to 100.
3. Divide the numerator by the denominator, multiply the quotient by 100, and add a percent sign.
5. $\text{Percent change} = \dfrac{\begin{pmatrix}\text{amount in}\\ \text{latest period}\end{pmatrix} - \begin{pmatrix}\text{amount in}\\ \text{previous period}\end{pmatrix}}{\text{amount in the previous period}} \times 100$

7. 50.0% **9.** 40.0% **11.** 0.8% **13.** 378% **15.** 0.04
17. 0.0134 **19.** 0.0025 **21.** 0.002 **23.** 0.01 **25.** ≈ 2.7%
27. 4.7 grams **29.** $47.28 million **31.** $159.57 million
33. $5.6848 billion **35.** $9.8838 billion **37.** 19.2%
39. 19.6% **41.** ≈ 5.8%
43. a) 17.3% **b)** 43.6% **c)** 54.3% **d)** 53.9%
45. a) 4.9% **b)** 27.0% **c)** 23.5% **d)** 10.6%
47. $6.75 **49.** 25% **51.** 300
53. a) $2.61 **b)** $46.11 **c)** $6.92 **d)** $53.03
55. 12 students **57.** $39,055 **59.** ≈ 5.3% decrease
61. ≈ 20.9% **63.** ≈ 18.6% decrease **65.** $3750
67. He will have a loss of $10. **69.** $21.95

Section 11.2, Page 610

1. Interest is the money the borrower pays for the use of the lender's money.
3. Security or collateral is anything of value pledged by the borrower that the lender may sell or keep if the borrower does not repay the loan.
5. *i* is the *interest, p* is the *principal, r* is the interest *rate* expressed as a percent, and *t* is the *time*.
7. The United States rule states that if a partial payment is made on a loan, interest is computed on the principal from the first day of the loan until the date of the partial payment.

9. $60.00 **11.** $2.81 **13.** $15.85 **15.** $80.06
17. $113.20 **19.** 10% **21.** $600 **23.** 2 years **25.** $1015
27. a) $131.25 **b)** $3631.25
29. a) $182.50 **b)** $3467.50 **c)** ≈ 7.9%
31. $23,793.75 **33.** 168 days **35.** 266 days **37.** 264 days
39. June 14 **41.** March 24 **43.** $1615.31 **45.** $3635.85
47. $5278.99 **49.** $850.64 **51.** $6086.82 **53.** $2646.24
55. a) November 3, 2004 **b)** $978.06 **c)** $21.94 **d)** ≈ 4.44%
57. a) ≈ 409.0% **b)** ≈ 204.5% **c)** ≈ 102.3%
59. a) 6.663% **b)** ≈ 7.139% **c)** $6663 **d)** $6996.15
60. a) $6.42 **b)** $14.20 **c)** $22.47 **d)** Answers will vary.

Section 11.3, page 618

1. An investment is the use of money or capital for income or profit.
3. A variable investment is one in which neither the principal nor the interest is guaranteed.
5. **a)** The effective annual yield is the simple interest rate that gives the same amount of interest as a compound rate over the same period of time.
   b) Another name for effective annual yield is annual percentage yield.

7. a) $2122.42 **b)** $122.42
9. a) $3942.72 **b)** $442.72
11. a) $1728.28 **b)** $228.28
13. a) $2831.95 **b)** $331.95
15. a) $4806.08 **b)** $806.08
17. $8336.15 **19.** $1653.36 **21.** $2341.82
23. a) $4195.14 **b)** $4214.36
25. $3106.62 **27.** $7609.45
29. a) $1040.60, $40.60 **b)** $1082.43, $82.43 **c)** $1169.86, $169.86 **d)** No
31. a) $1125.51, $125.51 **b)** $1266.77, $266.77 **c)** $1604.71, $604.71
d) Yes; new amount $= \dfrac{(\text{old amount})^2}{1000}$
33. ≈ 3.53% **35.** Yes, the APY should be 2.43%.
37. He will earn more interest in the account that pays the 5% simple interest
39. a) $129,210.47 **b)** $134.88
41. $23,202.23 **43.** $12,015.94 **45.** $1.53
47. a) 24 years **b)** 12 years **c)** 9 years **d)** 6 years **e)** 3.27%
49. $27,550.11 **51. a)** $55,726.01 **b)** $55,821.15

Section 11.4, page 631

1. An open-end installment loan is one with which you can make different payments each month. A fixed installment loan is one in which you pay a fixed amount each month for a set number of months.
3. The APR is the true rate of interest charged on a loan.
5. The total installment price is the sum of all the monthly payments and the down payment, if any.
7. The unpaid balance method and the average daily balance method

9. a) $5339.96 **b)** $698.17
11. a) $809.20 **b)** $80.15
13. a) $628.40 **b)** 9.0%

15. a) $1752 **b)** 9%

17. a) 6% **b)** $726.00 **c)** $7858.00

19. a) $2818.20 **b)** $689.39 **c)** $347.90 **d)** $8614.17

21. a) $1344.87 **b)** $181.04 **c)** $761.64 **d)** $5936.84

23. a) $27.63 **b)** $1032.10

25. a) $18 **b)** $630.86

27. a) $31 **b)** $553.02

29. a) $19.76 **b)** $743.41

31. a) $1.56 **b)** $133.11

33. a) $512.00 **b)** $6.66 **c)** $638.43

35. a) $121.78 **b)** $1.52 **c)** $133.07
d) The interest charged using the average daily balance method is $0.04 less than the interest charged using the unpaid balance method.

37. a) $8.87 **b)** $608.87

39. a) $25 **b)** $35.60 **c)** 8.5% **d)** 6.5%

41. a) 6 months **b)** $83.95
c) The installment loan saves them $49.45.

43. a) $6872.25 **b)** $610.37 **c)** $2637.42 **d)** $2501.05

45. Since Martina's billing date is June 25th, she can buy the camera from June 26th through June 29th and the purchase will appear on her July 25th bill. Since she has a 20-day grace period, she can pay for the camera on August 5th without paying interest.

Section 11.5, Page 644

1. A mortgage is a long-term loan in which the property is pledged as security for payment of the difference between the down payment and the sale price.

3. The major difference is that the interest rate for a conventional loan is fixed for the duration of the loan, whereas the interest rate for a variable-rate loan may change every period, as specified in the loan agreement.

5. A buyer's adjusted monthly income is found by subtracting any fixed monthly payments with more than 10 months remaining from the gross monthly income.

7. An amortization schedule lists payment dates and payment numbers. For each payment it lists the amount that goes to pay the interest and the principal. It also gives the balance remaining on the loan after each payment.

9. Equity is the difference between the appraised value of your home and the loan balance.

11. a) $37,500 **b)** $1625.63

13. a) $21,000 **b)** $1247.40

15. a) $39,000 **b)** $156,000 **c)** $3120

17. a) $2865 **b)** $802.20 **c)** $1411.50 **d)** No

19. a) $187,736.40 **b)** $112,736.40 **c)** $38.68

21. a) $31,780 **b)** $2451.60 **c)** $4330 **d)** $1212.40
e) $789.42 **f)** $916.09 **g)** Yes. **h)** $108.42

23. Bank B

25. a) 805
b)

Payment Number	Interest	Principal	Balance of Loan
1	$750.00	$55.00	$99,945.00
2	$749.59	$55.41	$99,889.59
3	$749.17	$55.83	$99,833.76

c) 9.38%
d)

Payment Number	Interest	Principal	Balance of Loan
4	$780.37	$24.63	$99,809.13
5	$780.17	$24.83	$99,784.30
6	$779.98	$25.02	$99,759.28

e) 9.46%

27. a) $113,095.24 **b)** $150,793.65

Review Exercises, page 647

1. 60.0% **2.** 66.7% **3.** 62.5% **4.** 4.1%

5. 0.98% ≈ 1.0% **6.** 314.1% **7.** 0.03 **8.** 0.121

9. 1.23 **10.** 0.0025 **11.** $0.008\overline{3}$ **12.** 0.0000045

13. ≈17.6% **14.** ≈11.0% **15.** 31.25% **16.** 275

17. 91.8 **18.** $6.42 **19.** 40 people **20.** 26.7%

21. $16.67 **22.** 9.5% **23.** $450 **24.** 0.5 year

25. $6214.25 **26. a)** $162 **b)** $3162

27. a) $1380 **b)** $4620 **c)** ≈14.9%

28. a) $7\frac{1}{2}$% **b)** $830 **c)** $941.18

29. a) $1610.51, $610.51 **b)** $1628.89, $628.89
c) $1638.62, $638.62 **d)** $1645.31, $645.31
e) $1648.61, $648.61

30. $5076.35 **31.** 5.76% **32.** $13,415.00

33. a) 6.0% **b)** 253.16 **c)** $4150.34

34. a) $109.18 **b)** $2014.11

35. a) 4.5% **b)** $32.06 **c)** $1420.43

36. a) $6.31 **b)** $847.61 **c)** $508.99 **d)** $6.62
e) $847.92

37. a) $2.60 **b)** $546.92 **c)** $382.68 **d)** $5.36
e) $549.68

38. a) $10,400 **b)** $41,600 **c)** $3040.96 **d)** 3.5%

39. a) $10.44 **b)** 8.5%

40. a) $33,925 **b)** $4805.33 **c)** $1345.49
d) $855.93 **e)** $1172.60 **f)** Yes

41. a) $13,485 **b)** $756.51 **c)** $24.20 **d)** $285,828.60
e) $195,928.60

42. a) $550.46 **b)** 8% **c)** 7.75%

Chapter Test, page 650

1. \$40 **2.** 3 years **3.** \$637.50 **4.** \$5637.50
5. \$2523.20 **6.** \$123.20 **7.** \$7961.99, \$461.99
8. \$3036.68, \$536.68 **9.** \$1997.50 **10.** \$181.46
11. 8.5% **12. a)** \$105.05 **b)** \$3155.90
13. a) 4.5% **b)** \$64.02 **c)** \$2836.28
14. a) \$12.30 **b)** \$1146.57 **c)** \$765.67 **d)** \$10.72 **e)** \$1144.99
15. \$21,675 **16.** \$6603.33 **17.** \$1848.93 **18.** \$1123.85
19. \$1428.02 **20.** Yes **21. a)** \$426,261 **b)** \$281,761

Chapter 12

Section 12.1, Page 659

1. An experiment is a controlled operation that yields a set of results.

3. Empirical probability is the relative frequency of occurrence of an event. It is determined by actual observation of an experiment.

$$P(E) = \frac{\text{number of times event has occurred}}{\text{number of times experiment was performed}}$$

5. Answers will vary.

7. No, it means that if a coin was flipped many times, about $\frac{1}{2}$ of the tosses would land heads up.

9. No, it means that the average person with traits similar to Mr. Duncan's will live another 43.21 years.

11.–13. Answers will vary.

15. a) $\frac{7}{15}$ **b)** $\frac{1}{3}$ **c)** $\frac{1}{5}$ **17. a)** $\frac{8}{19}$ **b)** $\frac{7}{19}$ **c)** $\frac{1}{19}$

19. a) The percents are relative frequencies of the events occurring.
b) 0.32 **c)** 0.22 **d)** 0.19

21. a) 1 **b)** Yes

23. a) $\frac{11}{40}$ **b)** $\frac{9}{40}$ **c)** $\frac{1}{4}$ **d)** $\frac{7}{40}$ **e)** $\frac{3}{40}$

25. a) $\frac{6}{20} = \frac{3}{10}$ **b)** $\frac{14}{20} = \frac{7}{10}$ **c)** $\frac{14}{20} = \frac{7}{10}$
d) $\frac{2}{20} = \frac{1}{10}$

27. a) 0 **b)** $\frac{50}{250} = 0.2$ **c)** 1

29. a) $\frac{224}{929} \approx 0.24$ **b)** $\frac{705}{929} \approx 0.76$

31. Answers will vary.

Section 12.2, Page 667

1. If each outcome of an experiment has the same chance of occurring as any other outcome, they are said to be equally likely outcomes.

3. $P(A) + P(\text{not } A) = 1$ **5.** 0.7 **7.** $\frac{7}{12}$

9. Answers will vary. **11.** 0 and 1 **13. a)** $\frac{1}{5}$ **b)** $\frac{1}{4}$

15. $\frac{1}{50}$ **17.** $\frac{1}{13}$ **19.** $\frac{12}{13}$ **21.** $\frac{1}{2}$ **23.** 1 **25.** $\frac{4}{13}$

27. a) $\frac{1}{2}$ **b)** $\frac{1}{4}$ **c)** $\frac{1}{4}$ **d)** 0

29. a) $\frac{1}{2}$ **b)** 0 **c)** $\frac{1}{3}$ **d)** $\frac{1}{6}$

31. $\frac{2}{5}$ **33.** $\frac{9}{10}$ **35.** $\frac{1}{12}$ **37.** $\frac{1}{6}$ **39.** $\frac{23}{50}$ **41.** $\frac{33}{50}$

43. $\frac{11}{17}$ **45.** $\frac{12}{17}$ **47.** $\frac{4}{11}$ **49.** $\frac{4}{11}$ **51.** 1 **53.** $\frac{1}{11}$

55. $\frac{4}{11}$ **57.** $\frac{1}{26}$ **59.** $\frac{5}{26}$ **61.** $\frac{345}{715} = \frac{69}{143}$ **63.** $\frac{533}{715} = \frac{41}{55}$

65. $\frac{97}{715}$ **67.** $\frac{50}{159}$ **69.** $\frac{66}{159} = \frac{22}{53}$ **71.** $\frac{23}{159}$ **73.** $\frac{13}{36}$

75. $\frac{1}{3}$ **77.** $\frac{23}{36}$ **79. a)** 0 **b)** 1 **81. a)** $\frac{1}{4}$ **b)** $\frac{1}{4}$ **c)** $\frac{1}{4}$

83. 29 dots

Section 12.3, Page 674

1. Answers will vary. **3.** Odds against

5. 9 to 5 **7. a)** $\frac{1}{2}$ **b)** $\frac{1}{2}$

9. a) $\frac{8}{27}$ **b)** $\frac{19}{27}$ **c)** 19 : 8 **d)** 8 : 19

11. 5 : 1 **13.** 4 : 2 or 2 : 1 **15.** 12 : 1, 1 : 12
17. 10 : 3, 3 : 10 **19.** 1 : 1 **21.** 5 : 3
23. a) 8 : 7 **b)** 7 : 8 **25.** 8 : 7 **27.** 14 : 1 **29.** 8 : 7

31. a) $\frac{5}{9}$ **b)** 4 : 5 **33.** 1 : 18 **35. a)** $\frac{8}{13}$ **b)** $\frac{5}{13}$

37. $\frac{11}{15}$ **39.** $\frac{1}{5}$ **41.** 1 : 4 **43.** 74 : 1 **45.** 0.34 **47.** 33 : 17

49. 43 : 57 **51.** 1 : 9 **53.** 7 : 1 **55. a)** $\frac{20}{21}$ **b)** 20 : 1

57. Horse 1, $\frac{2}{9}$; Horse 2, $\frac{1}{3}$; Horse 3, $\frac{1}{16}$; Horse 4, $\frac{5}{12}$; Horse 5, $\frac{1}{2}$

59. $\approx 97 : 3$

Section 12.4, Page 683

1. The expected value is the expected gain or loss of an experiment over the long run.

3. The fair price is the amount that should be charged for the game to be fair and result in an expectation of 0.

5. To obtain the fair price, add the cost to the expected value.
7. \$0.50. Since you would lose \$1.00 on average for each game you played, the price of the game should be \$1.00 less than the actual cost. Then the expectation would be \$0, and the game would be fair. The results could also be obtained from the fair price formula, fair price = expectation + cost to play.
9. −\$1.20 **11.** 176 people **13.** 70 points
15. 1.44 million viewers **17.** \$3840 **19.** \$1.60 off
21. a) ≈ −\$0.67 **b)** ≈ \$0.67
23. a) Yes, because you have a positive expectation of $\frac{1}{5}$
b) Yes, because you have a positive expectation of $\frac{1}{2}$
25. a) \$−1.20 **b)** \$0.80 **27. a)** \$−2.00 **b)** \$1.00
29. \$5.50 **31.** \$−1.25 **33. a)** \$1 **b)** \$3
35. a) \$2.25 **b)** \$4.25 **37.** 0.75 base **39.** 2.9 points
41. 381.4 employees **43.** ≈ 15.65 min **45.** 3.5
47. ≈141.51 service calls
49. a) $\frac{9}{16}, \frac{1}{4}, \frac{1}{8}, \frac{1}{16}$ **b)** \$11.81 **c)** \$11.81
51. An amount greater than \$1200
53. −\$0.053 or −5.3¢ **55. a)** \$458.33 **b)** \$308.33

Section 12.5, Page 692

1. If a first experiment can be performed in M distinct ways and a second experiment can be performed in N distinct ways, then the two experiments in that specific order can be performed in $M \cdot N$ distinct ways.
3. 14
5. The first selection is made. Then the second selection is made without the first selection being returned to the group of items being selected.
7. a) 2500 **b)** 2450 **9. a)** 216 **b)** 120
11. a) 4
b)

Sample Space
H — H: HH
H — T: HT
T — H: TH
T — T: TT

c) $\frac{1}{4}$ **d)** $\frac{1}{2}$ **e)** $\frac{1}{4}$

13. a) 9
b)

Sample Space
C — C, S, T: CC, CS, CT
S — C, S, T: SC, SS, ST
T — C, S, T: TC, TS, TT

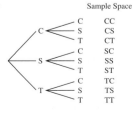

c) $\frac{1}{9}$ **d)** $\frac{1}{9}$ **e)** $\frac{5}{9}$

15. a) 12
b)

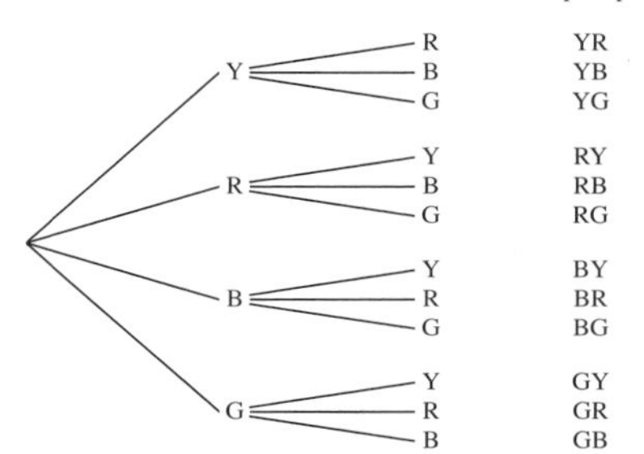

c) $\frac{1}{2}$ **d)** 1 **e)** $\frac{1}{2}$

17. a) 8
b)

Sample Space
GGG
GGB
GBG
GBB
BGG
BGB
BBG
BBB

c) $\frac{1}{8}$ **d)** $\frac{7}{8}$ **e)** $\frac{3}{4}$ **f)** $\frac{1}{8}$

19. a) 36
b)

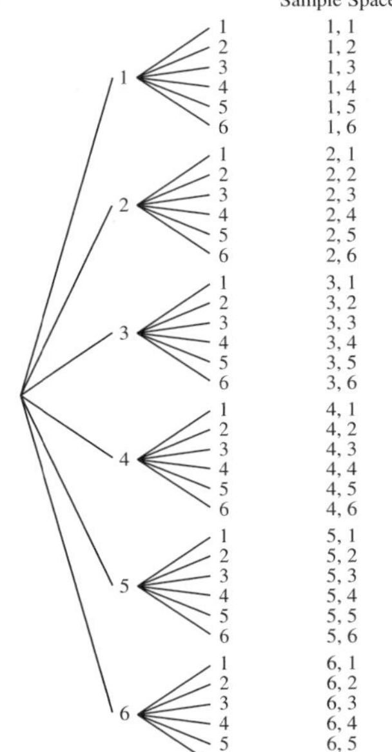

c) $\frac{1}{6}$ **d)** $\frac{1}{6}$ **e)** $\frac{1}{36}$ **f)** No

21. a) 6
b)

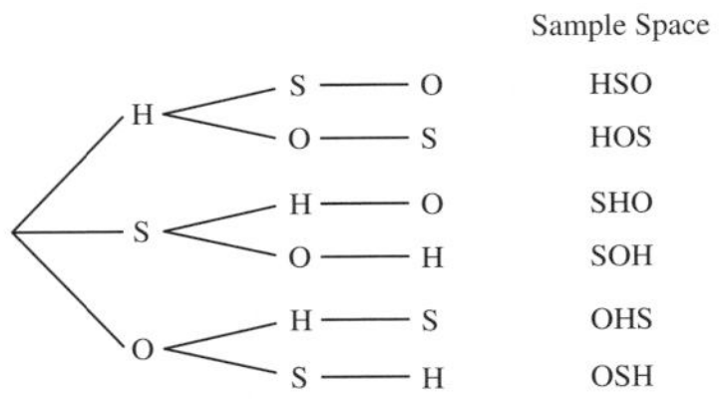

Sample Space

H — S — O HSO
H — O — S HOS
S — H — O SHO
S — O — H SOH
O — H — S OHS
O — S — H OSH

c) $\frac{1}{3}$ **d)** $\frac{1}{6}$ **e)** $\frac{1}{6}$

23. a) 16
b) In the tree diagram, K represents the Magic Kingdom and M represents MGM Studios.

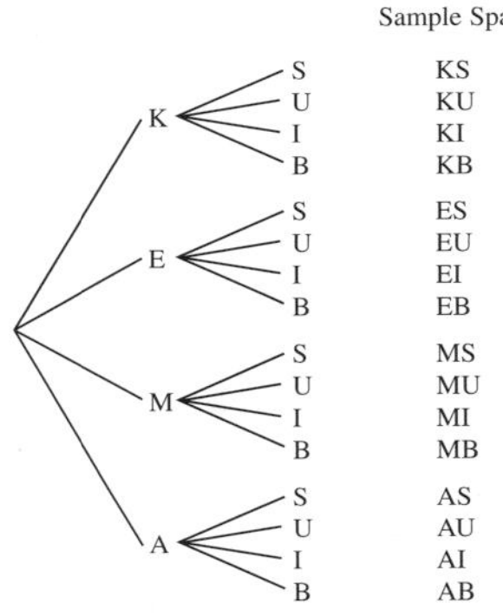

c) $\frac{1}{2}$ **d)** $\frac{7}{16}$ **e)** $\frac{1}{8}$

25. a) 27
b)

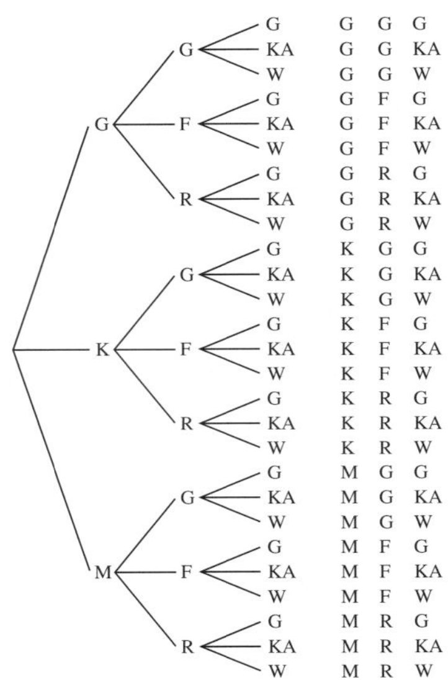

c) $\frac{1}{27}$ **d)** $\frac{8}{27}$ **e)** $\frac{19}{27}$

27. a) 24
b)

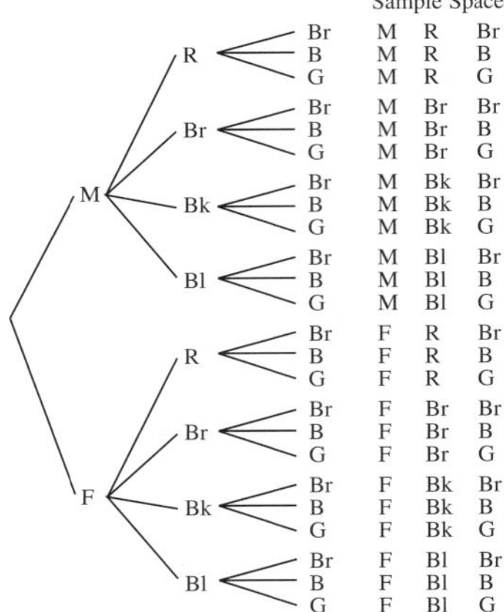

c) $\frac{1}{24}$ **d)** $\frac{1}{8}$

29. a) $\frac{1}{3}$ **b)** $\frac{2}{3}$ **c)** No, the probability of selecting a red chip is not the same as the probability of selecting a white chip. **d)** Answers will vary.

31. 3; 1 red, 1 blue, and 1 brown **32.** 5 faces

Section 12.6, Page 703

1. a) At least one event, A *or* B, must occur.
b) Both events, A *and* B, must occur.

3. a) Events that cannot happen simultaneously are mutually exclusive events.
b) $P(A \text{ or } B) = P(A) + P(B)$

5. We assume that event A has already occurred.

7. Two events are dependent when the probability of one item being selected has an effect on the probability of a second item being selected.

9. a) No, both events can occur at the same time.
b) Yes, the outcome of one event has no affect on the outcome of the other event.

11. Events A and B cannot occur at the same time. Therefore, $P(A \text{ and } B) = 0$.

13. 0.7 **15.** 0.5 **17.** $\frac{1}{3}$ **19.** $\frac{1}{2}$ **21.** $\frac{2}{13}$ **23.** $\frac{8}{13}$

25. $\frac{31}{52}$ **27. a)** $\frac{1}{16}$ **b)** $\frac{1}{19}$

29. a) $\frac{1}{16}$ **b)** $\frac{5}{76}$ **31. a)** $\frac{3}{80}$ **b)** $\frac{3}{76}$

33. a) $\frac{9}{25}$ **b)** $\frac{33}{95}$ **35.** $\frac{11}{20}$ **37.** $\frac{2}{5}$ **39.** $\frac{1}{4}$

41. $\frac{1}{8}$ **43.** $\frac{9}{64}$ **45.** $\frac{1}{8}$ **47.** $\frac{3}{8}$ **49.** $\frac{1}{8}$ **51.** $\frac{1}{8}$

53. a) $\frac{1}{16}$ **b)** $\frac{1}{2}$ **55. a)** $\frac{4}{49}$ **b)** $\frac{2}{21}$

57. **a)** $\frac{24}{49}$ **b)** $\frac{11}{21}$ **59.** $\frac{5}{12}$ **61.** $\frac{7}{12}$ **63.** $\frac{969}{4060}$

65. $\frac{5}{812}$ **67.** 0.7 **69.** 0.343 **71.** $\frac{1}{4}$ **73.** $\frac{27}{1024}$

75. $\frac{243}{1024}$ **77.** $\frac{3}{22}$ **79.** $\frac{1050}{1331}$ **81.** $\frac{1}{48}$ **83.** $\frac{5}{12}$

85. 0.36 **87.** 0.36

89. **a)** No **b)** 0.001 **c)** 0.00004 **d)** 0.00096 **e)** 0.000999 **f)** 0.998001

91. $\frac{32}{1000}$, or 0.032 **93.** $\frac{484}{15{,}625}$, or 0.030976 **95.** $\frac{14}{45}$

97. Favors dealer, the probability of at least one diamond is ≈ 0.44, which is less than 0.5

99. $\frac{1}{9}$ **100.** $\frac{1}{4}$ **101.** $\frac{1}{2}$ **102.** 1 **103.** Answers will vary.

Section 12.7, Page 710

1. The probability of E_2 given that E_1 has occurred

3. $\frac{1}{3}$ **5.** $\frac{1}{3}$ **7.** $\frac{2}{3}$ **9.** $\frac{2}{3}$ **11.** $\frac{3}{4}$ **13.** $\frac{2}{3}$ **15.** $\frac{2}{3}$ **17.** $\frac{1}{3}$

19. $\frac{1}{3}$ **21.** $\frac{3}{5}$ **23.** $\frac{1}{7}$ **25.** $\frac{1}{16}$ **27.** $\frac{1}{7}$ **29.** $\frac{5}{36}$

31. $\frac{1}{6}$ **33.** $\frac{2}{3}$ **35.** $\frac{107}{217}$ **37.** $\frac{25}{56}$ **39.** $\frac{6}{11}$ **41.** $\frac{4}{9}$

43. $\frac{11}{20}$ **45.** $\frac{22}{39}$ **47.** $\frac{133}{300}$ **49.** $\frac{1}{2}$ **51.** $\frac{21}{43}$

53. ≈ 0.3197 **55.** ≈ 0.0795 **57.** ≈ 0.1928 **59.** $\frac{10}{11}$

61. $\frac{3}{19}$ **63.** $\frac{44}{47}$ **65.** $\frac{11}{27}$ **67.** $\frac{10}{29}$ **69.** $\frac{11}{29}$ **71.** $\frac{93}{200}$

73. $\frac{15}{52}$ **75.** **a)** 140 **b)** 120 **c)** $\frac{7}{10}$ **d)** $\frac{3}{5}$
e) $\frac{2}{3}$ **f)** $\frac{4}{7}$
g) Because A and B are not independent events

77. **a)** 0.3 **b)** 0.4 **c)** Yes; $P(A \mid B) = P(A) \cdot P(B)$

78. $\frac{1}{3}$ **79.** $\frac{2}{3}$ **80.** $\frac{1}{3}$ **81.** $\frac{1}{3}$ **82.** 100 **83.** $\frac{1}{3}$

Section 12.8, Page 721

1. Answers will vary.

3. $n! = n(n-1)(n-2)\cdots(3)(2)(1)$

5. The number of permutations of n items taken r at a time.

7. ${}_nP_r = \frac{n!}{(n-r)!}$ **9.** 720 **11.** 30 **13.** 1 **15.** 1

17. 3024 **19.** 6720 **21.** 10,000

23. **a)** 11,232,000 **b)** 17,576,000

25. **a)** $5^5 = 3125$ **b)** $\frac{1}{3125} = 0.00032$

27. 57,106,944 **29.** 720 systems

31. **a)** 720 **b)** 120 **c)** 24 **d)** 600 **33.** 720

35. **a)** 479,001,600 **b)** 3,628,800 **c)** 14,400

37. 3,276,000 **39.** 131,040 **41.** 676,000 **43.** 104,000

45. **a)** 8,000,000 **b)** 6,400,000,000 **c)** 64,000,000,000,000

47. 3,603,600 **49.** 5040 **51.** 280 **53.** 362,880

55. 1,663,200 **57.** 630 **59.** 6720

61. **a)** 40,320 **b)** 362,880

63. **a)** 3125 **b)** ≈ 128 **c)** 0.00032

65. 12,600 sec, or 3.5 hr **67.** No **68.** 56 **69.** 600

Section 12.9, Page 727

1. Answers will vary.

3. ${}_nC_r = \frac{n!}{(n-r)!r!}$ **5.** Answers will vary.

7. 10 **9.** **a)** 15 **b)** 360 **11.** **a)** 1 **b)** 1

13. **a)** 120 **b)** 720 **15.** $\frac{1}{6}$ **17.** 2 **19.** 72

21. 84 **23.** 5 **25.** 210 **27.** 126 **29.** 495

31. 45 **33.** 28 **35.** 6160 **37.** 560 **39.** 294,000

41. 560 **43.** 1200 **45.** **a)** 45 **b)** 56

47. **a)** and **b)**

```
                1
             1     1
          1     2     1
       1     3     3     1
    1     4     6     4     1
 1     5    10    10     5     1
```

49. **a)** 24 **b)** 24

51. **a)** The order is important. Since the numbers may be repeated, it is not a true permutation lock.
b) 64,000 **c)** 59,280

Section 12.10, Page 733

1. $\frac{{}_6C_4}{{}_{10}C_4}$ **3.** $\frac{{}_5C_3}{{}_{26}C_3}$ **5.** $\frac{{}_{10}C_5}{{}_{18}C_5}$ **7.** $\frac{{}_{14}C_9}{{}_{30}C_9}$ **9.** $\frac{5}{42}$ **11.** $\frac{4}{143}$

13. $\frac{1}{12}$ **15.** $\frac{4}{33}$ **17.** $\frac{1}{9{,}366{,}819}$ **19.** $\frac{3}{10}$ **21.** $\frac{7}{10}$

23. $\frac{1}{115}$ **25.** $\frac{27}{230}$ **27.** $\frac{646{,}646}{3{,}910{,}797{,}436} \approx 0.0001653$

29. $\frac{923{,}410{,}488}{3{,}910{,}797{,}436} \approx 0.236$ **31.** $\frac{20}{1001}$ **33.** $\frac{200}{1001}$

35. $\frac{1}{77}$ **37.** $\frac{5}{77}$ **39.** $\frac{5}{506} \approx 0.010$

41. a) $\dfrac{1}{123{,}760}$ **b)** $\dfrac{1}{30{,}940}$

43. a) $\dfrac{33}{54{,}145}$ **b)** $\dfrac{1}{2{,}598{,}960}$

45. a) $\dfrac{1}{2{,}162{,}160}$ **b)** $\dfrac{1}{6435}$

47. 1; Since there are more hairs than people, two or more people must have the same number of hairs on their head.

Section 12.11, Page 742

1. A probability distribution shows the probability associated with each specific outcome of an experiment. In a probability distribution every possible outcome must be listed and the sum of all the probabilities must be 1.

3. $P(x) = ({}_nC_x)p^x q^{n-x}$

5. 0.2646 **7.** 0.3456 **9.** 0.015625

11. a) $P(x) = ({}_nC_x)(0.14)^x(0.86)^{n-x}$
b) $P(2) = ({}_{12}C_2)(0.14)^2(0.86)^{10}$

13. 0.05954 **15.** 0.11059 **17.** 0.06877 **19.** 0.4096

21. a) 0.01024 **b)** 0.98976

23. a) ≈ 0.1119 **b)** ≈ 0.2966

25. 0; it will be midnight.

Review Exercises, Page 745

1. Answers will vary. **2.** Answers will vary.

3. $\frac{1}{5}$ **4.** Answers will vary. **5.** $\frac{2}{5}$ **6.** $\frac{1}{2}$ **7.** $\frac{7}{10}$

8. 1 **9.** $\frac{1}{5}$ **10.** $\frac{5}{24}$ **11.** $\frac{1}{6}$ **12.** $\frac{1}{3}$ **13.** $\frac{19}{24}$

14. a) 9 : 1 **b)** 1 : 9 **15.** 5 : 3 **16.** $\frac{3}{85}$ **17.** 7 : 3

18. a) −\$1.20 **b)** −\$3.60 **c)** \$0.80

19. a) −\$0.23 **b)** \$0.23 **c)** Lose \$23.08

20. 660 people

21. a)

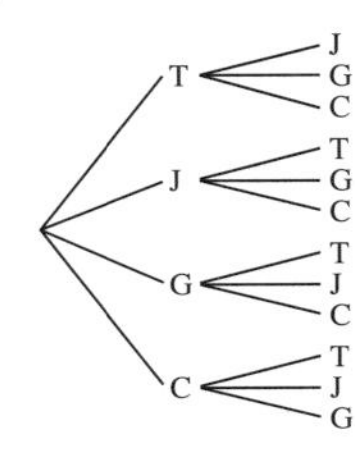

b) Sample Space **c)** $\frac{1}{12}$

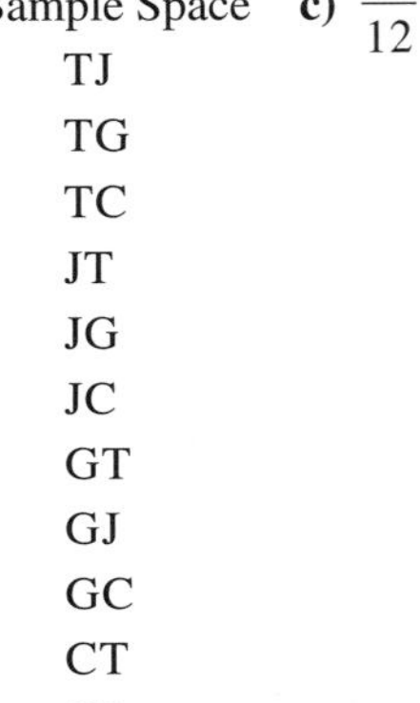

TJ
TG
TC
JT
JG
JC
GT
GJ
GC
CT
CJ
CG

22. a)

H: 1, 2, 3, 4
T: 1, 2, 3, 4

b) Sample Space **c)** $\frac{1}{4}$ **d)** $\frac{3}{4}$

H1
H2
H3
H4
T1
T2
T3
T4

23. $\frac{1}{4}$ **24.** $\frac{9}{64}$ **25.** $\frac{5}{16}$ **26.** $\frac{7}{8}$ **27.** 1 **28.** $\frac{3}{16}$

29. $\frac{1}{22}$ **30.** $\frac{14}{55}$ **31.** $\frac{41}{55}$ **32.** $\frac{1}{22}$ **33.** $\frac{1}{4}$

34. Against, 3 : 1; in favor, 1 : 3 **35.** \$13.75

36. $\frac{1}{8}$ **37.** $\frac{5}{8}$ **38.** In favor, 3 : 5; against, 5 : 3 **39.** \$3.75

40. $\frac{7}{8}$ **41.** $\frac{89}{106}$ **42.** $\frac{55}{74}$ **43.** $\frac{19}{74}$ **44.** $\frac{17}{106}$

45. $\frac{23}{40}$ **46.** $\frac{3}{17}$ **47.** $\frac{3}{4}$ **48.** $\frac{12}{17}$

49. a) 24 **b)** \$4500 **50.** 30 **51.** 720 **52.** 504

53. 20 **54. a)** 3003 **b)** 3,628,800

55. a) $\dfrac{1}{2{,}598{,}960}$ **b)** $\dfrac{1}{135{,}145{,}920}$ **56.** 5880 **57.** 560

58. $\frac{1}{221}$ **59.** $\frac{1}{12}$ **60.** $\frac{1}{18}$ **61.** $\frac{1}{24}$ **62.** $\frac{11}{12}$ **63.** $\frac{5}{182}$

64. $\frac{45}{364}$ **65.** $\frac{2}{13}$ **66.** $\frac{11}{13}$

67. a) $P(x) = ({}_nC_x)(0.6)^x(0.4)^{n-x}$
b) $P(75) = ({}_{100}C_{75})(0.6)^{75}(0.4)^{25}$

68. 0.0512 **69. a)** 0.0256 **b)** 0.9744

Chapter Test, Page 748

1. $\frac{11}{15}$ **2.** $\frac{2}{9}$ **3.** $\frac{5}{9}$ **4.** $\frac{7}{9}$ **5.** $\frac{1}{3}$ **6.** $\frac{1}{6}$ **7.** $\frac{1}{6}$

8. $\frac{5}{18}$ **9.** $\frac{5}{12}$ **10.** $\frac{8}{13}$ **11.** 18

12.

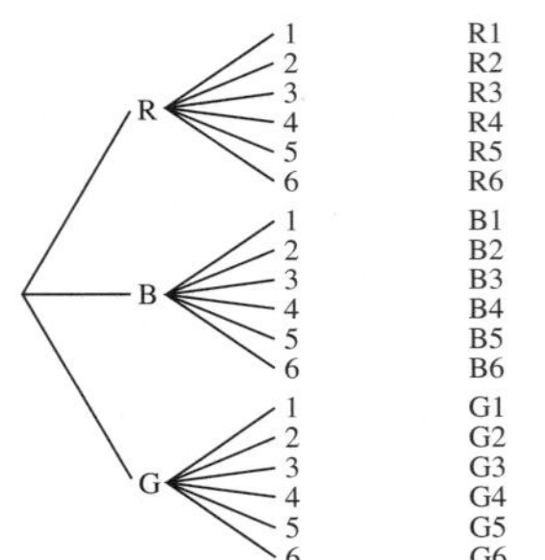

13. $\frac{1}{18}$ **14.** $\frac{4}{9}$ **15.** $\frac{5}{6}$ **16.** 608,400

17. a) 5 : 4 **b)** 5 : 4 **18.** $\frac{2}{7}$ **19.** $0

20. a) $\frac{107}{228}$ **b)** $\frac{115}{228}$ **c)** $\frac{68}{115}$ **d)** $\frac{60}{107}$

21. 120 **22.** $\frac{14}{95}$ **23.** $\frac{81}{95}$ **24.** $\frac{175}{396}$ **25.** 0.0081

Chapter 13

Section 13.1, Page 756

1. Answers will vary.

3.–5. Answers will vary.

7. a) A population is all items or people of interest.
b) A sample is a subset of the population.

9. a) A random sample is a sample drawn in such a way that each item in the population has an equal chance of being selected.
b) Number each item in the population. Write each number on a piece of paper and put each numbered piece of paper in a hat. Select pieces of paper from the hat and use the numbered items selected as your sample.

11. a) A stratified sample is one that includes items from each part (or strata) of the population.
b) First identify the strata in which you are interested. Then select a random sample from each strata.

13. An unbiased sample is one that is a small replica of the entire population with regard to income, education, gender, race, religion, political affiliation, age, and so forth.

15. Stratified sample **17.** Cluster sample

19. Systematic sample **21.** Convenience sample

23. Random sample **25. a)–c)** Answers will vary.

27. President; four out of 42 U.S. presidents have been assassinated (Lincoln, Garfield, McKinley, Kennedy).

Section 13.2, Page 759

1. Answers will vary.

3. There may have been more car thefts in Baltimore, Maryland than Reno, Nevada because many more people live in Baltimore than in Reno. But, Reno may have more car thefts per capita than Baltimore.

5. Although the cookies are fat free, they still contain calories. Eating many of them may still cause you to gain weight.

7. More people drive on Saturday evening. Thus, one might expect more accidents.

9. People with asthma may move to Arizona because of its climate. Therefore, more people with asthma may live in Arizona.

11. Although milk is less expensive at Star Food Markets than at Price Chopper Food Markets, other items may be more expensive at Star Food Markets.

13. There may be deep sections in the pond, so it may not be safe to go wading.

15. Half the students in a population are expected to be below average.

17. a)

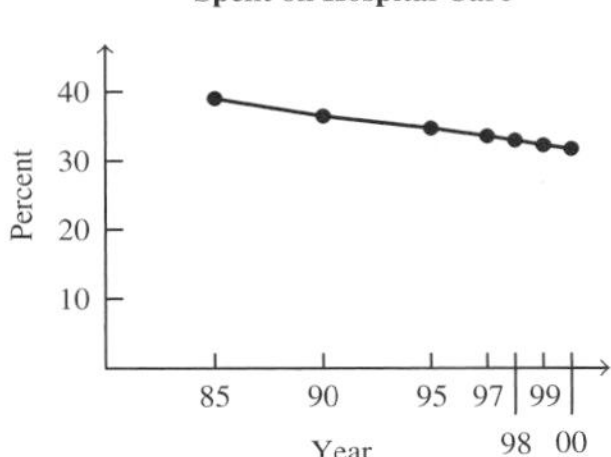

b)

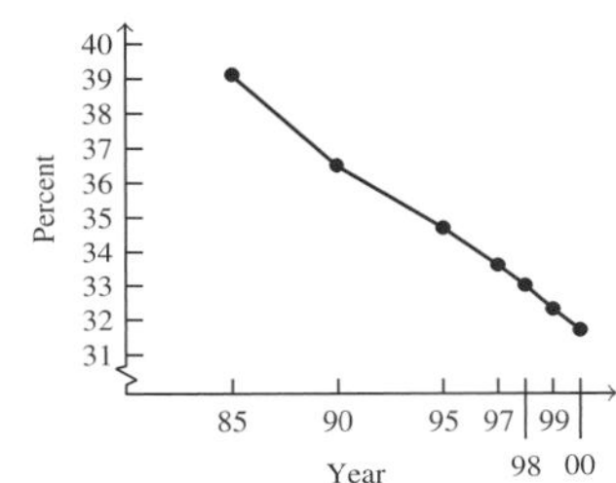

19. a)

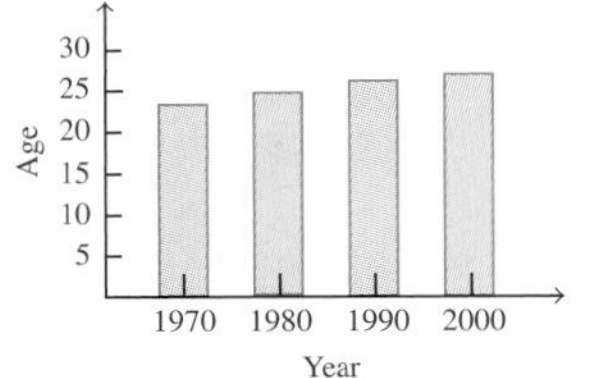

b)

21. a)

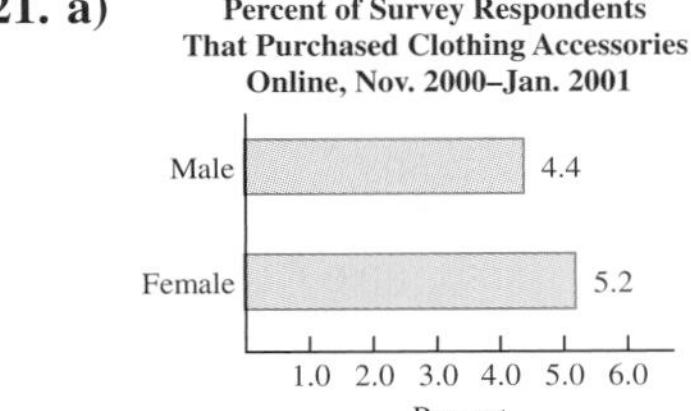

b) Yes. Answers will vary.

23. A decimal point

Section 13.3, Page 765

1. A frequency distribution is a listing of observed values and the corresponding frequency of occurrence of each value.

3. a) 7 **b)** 16–22 **c)** 16 **d)** 22

5. The modal class is the class with the greatest frequency.

7. a) 18 **b)** 7 **c)** 19 **d)** 16–22 **e)** 51–57

9.

Number Sold	Number of Days
0	3
1	8
2	3
3	5
4	2
5	7
6	2
7	3
8	4
9	1
10	2

11.

IQ	Number of Students
78–86	2
87–95	15
96–104	18
105–113	7
114–122	6
123–131	1
132–140	1

13.

IQ	Number of Students
80–90	8
91–101	22
102–112	11
113–123	7
124–134	1
135–145	1

15.

Placement Test Scores	Number of Students
472–492	9
493–513	9
514–534	5
535–555	2
556–576	3
577–597	2

17.

Placement Test Scores	Number of Students
472–487	4
488–503	9
504–519	7
520–535	3
536–551	2
552–567	2
568–583	2
584–599	1

19.

Circulation (thousands)	Number of Newspapers
209–458	36
459–708	8
709–958	3
959–1208	1
1209–1458	0
1459–1708	0
1709–1958	1
1959–2208	1

21.

Circulation (thousands)	Number of Newspapers
209–408	34
409–608	9
609–808	3
809–1008	1
1009–1208	1
1209–1408	0
1409–1608	0
1609–1808	1
1809–2008	0
2009–2208	1

23.

Population (millions)	Number of Counties
1.4–2.1	15
2.2–2.9	6
3.0–3.7	2
3.8–4.5	0
4.6–5.3	0
5.4–6.1	1
6.2–6.9	0
7.0–7.7	0
7.8–8.5	0
8.6–9.3	0
9.4–10.1	1

25.

Population (millions)	Number of Counties
1.0–2.5	19
2.6–4.1	4
4.2–5.7	1
5.8–7.3	0
7.4–8.9	0
9.0–10.5	1

27.

Price ($)	Number of States
0.35–0.44	6
0.45–0.54	10
0.55–0.64	11
0.65–0.74	3
0.75–0.84	2
0.85–0.94	4
0.95–1.04	1
1.05–1.14	2
1.15–1.24	2
1.25–1.34	1
1.35–1.44	0
1.45–1.54	1

29.

Price ($)	Number of States
0.35–0.54	16
0.55–0.74	14
0.75–0.94	6
0.95–1.14	3
1.15–1.34	3
1.35–1.54	1

31. February, since it has the fewest numbers of days

32. a) Did You Know?, page 762: There are 6 F's.
b) Answers will vary.

Section 13.4, page 774

1. Answers will vary. **3.** Answers will vary.

5. a) Answers will vary.
b)

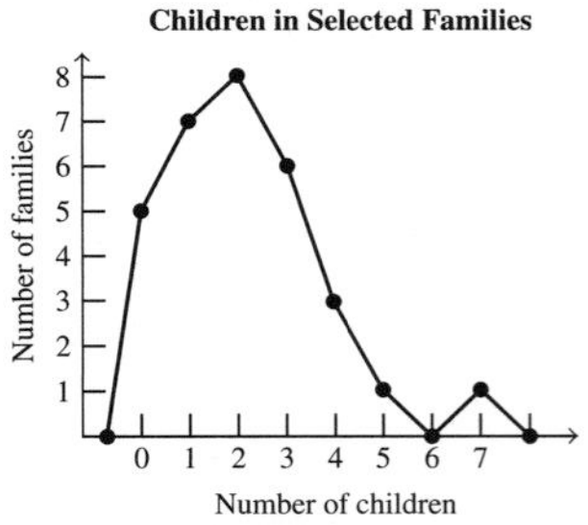

Children in Selected Families

Number of families

0 1 2 3 4 5 6 7

Number of children

7. a) Answers will vary.
b)

Observed values	Frequency
45	3
46	0
47	1
48	0
49	1
50	1
51	2

9. Occasionally: 295; most times: 125; every time: 35; never: 45

11.

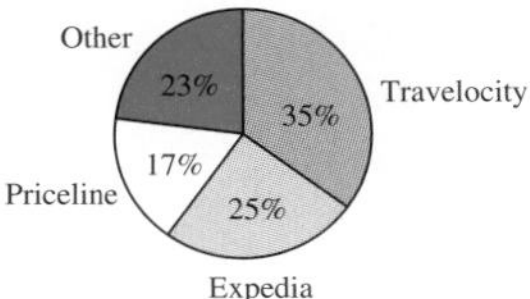

13. a) and **b)**

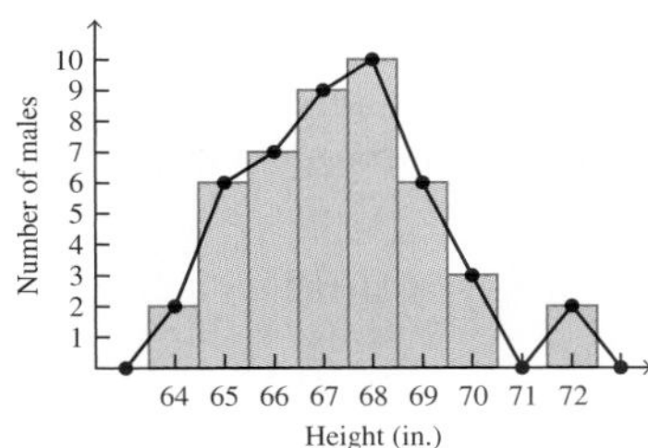

15. a) and **b)**

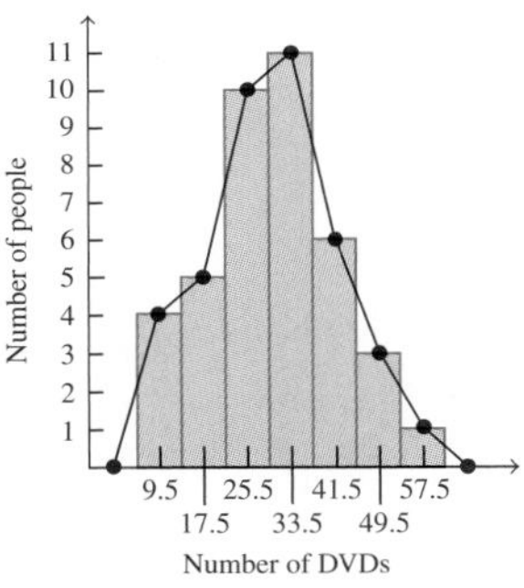

17. a) 30 **b)** 4 **c)** 2 **d)** 75

e)

Number of Soft Drinks Purchased	Number of People
0	2
1	7
2	8
3	5
4	4
5	3
6	1

19. a) 7 **b)** 16 **c)** 36

d)

Response Time (min)	Number of Calls	Response Time (min)	Number of Calls
3	2	7	3
4	3	8	8
5	7	9	6
6	4	10	3

e)

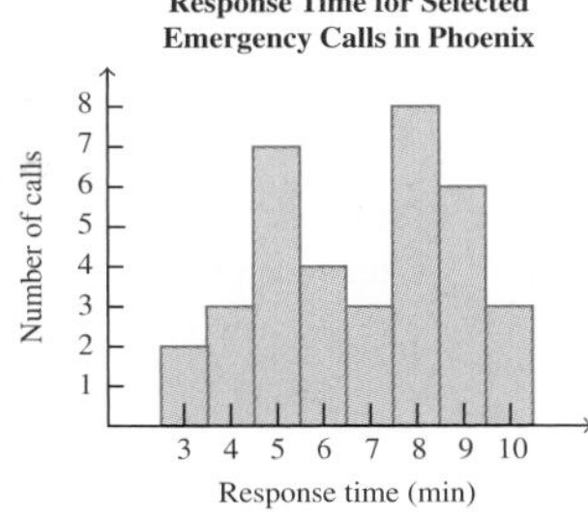

21.

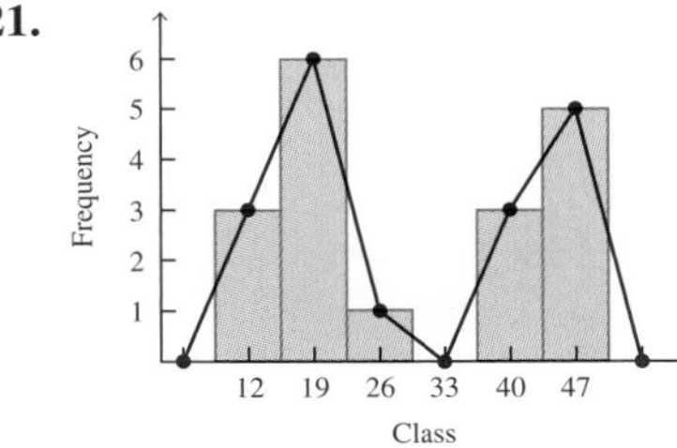

23. 1 | 5 represents 15

1	0 5 7
2	4 4
3	6 0 3
4	8 5 2 5 8
5	3 4
6	0 2 0

25. a)

Salaries (1000s of dollars)	Number of Companies
27	1
28	7
29	4
30	3
31	2
32	3
33	3
34	2

b) and **c)**

Starting Salaries for 25 Different Social Workers

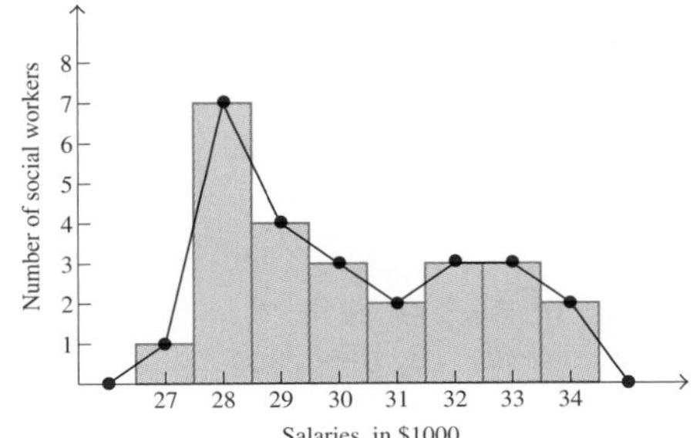

d) 2 | 3 represents 23

2 | 7 8 8 8 8 8 8 8 9 9 9 9
3 | 0 0 0 1 1 2 2 2 3 3 3 4 4

27. a)

Advertising Spending (millions of dollars)	Number of Companies
597–905	19
906–1214	14
1215–1523	7
1524–1832	3
1833–2141	2
2142–2450	3
2451–2759	1
2760–3068	0
3069–3377	1

b) and **c)**

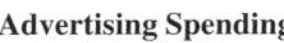

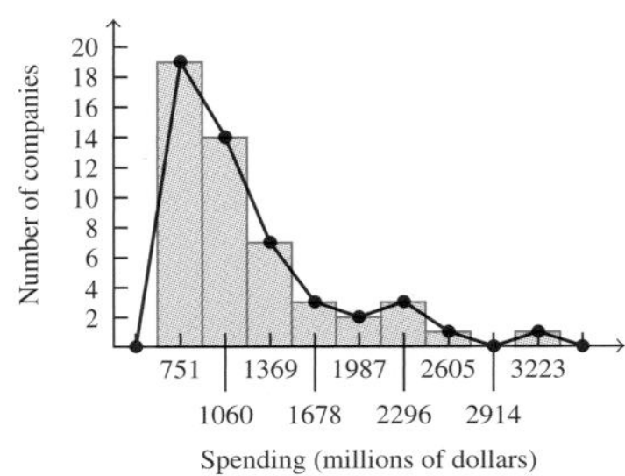

29. a)–e) Answers will vary.

Section 13.5, Page 784

1. Ranked data are data listed from the lowest value to the highest value or from the highest value to the lowest value.

3. The median is the value in the middle of a set of ranked data. To find the median rank the data and select the value in the middle.

5. The mode is the most common piece of data. The piece of data that occurs most frequently is the mode.

7. The median should be used when there are some values that differ greatly from the rest of the values in the set, for example, salaries.

9. The mean is used when each piece of data is to be considered and "weighed" equally, for example, weights of adult males.

11. 11, 10, 10, 14 **13.** 69.3, 72, none, 66

15. 8, 8, none, 8 **17.** 13.1, 11, 1, 18.5

19. 11.9, 12.5, 13, 11.5 **21.** 6.5, 5, 3 and 5, 10.5

23. a) 4.9, 5, 5, 6 **b)** 5.3, 5, 5, 6
c) Only the mean **d)** The mean and the midrange

25. A 79 mean average on 10 quizzes gives a total of 790 points. An 80 mean average on 10 quizzes requires a total of 800 points. Thus, Jim missed a B by 10 points, not 1 point.

27. a) 8.8 million **b)** 8.0 million **c)** None
d) 12.2 million

29. a) \$5.1 billion **b)** \$2.3 billion
c) \$2.3 billion and \$1.5 billion **d)** \$14 billion
e) Answers will vary.

31. 510 **33.** One example is 72, 73, 74, 76, 77, 78.

35. a) Yes **b)** No **c)** No **d)** Yes
e) Mean = 200; midrange = 275

37. a) 33 or greater
b) It is not possible if 100 is the maximum possible grade.
c) 22 or greater **d)** 82 or greater

39. One example: 1, 2, 3, 3, 4, 5, changed to 1, 2, 3, 4, 4, 5.

41. No, by changing only one piece of the 6 pieces of data you cannot alter both the median and the midrange.

43. The data must be ranked.

45. He is taller than approximately 35 percent of all kindergarten children.

47. a) \$430 **b)** \$350 **c)** \$650

49. Second quartile, median

51. a) \$490 **b)** \$500 **c)** 25%
d) 25% **e)** 17% **f)** \$51,000

53. a)

Ruth	Mantle
0.290	0.300
0.359	0.365
0.301	0.304
0.272	0.275
0.315	0.321

b) Mantle's is greater in every case.
c) Ruth: 0.316; Mantle: 0.311; Ruth's is greater.
d) Answers will vary.
e) Ruth: 0.307; Mantle: 0.313; Mantle's is greater.
f) Answers will vary. **g)** Answers will vary.

55. 90 **57. a)–c)** Answers will vary.

Section 13.6, Page 793

1. Range = highest value − lowest value

3. Answers will vary. **5.** Answers will vary. **7.** σ

9. Answers will vary.

11. They would be the same since the spread of data about each mean is the same.

13. a) The grades will be centered about the same number since the mean, 75.2, is the same for both classes.
b) The spread of the data about the mean is greater for the evening class since the standard deviation is greater for the evening class.

15. 11, $\sqrt{16.5} \approx 4.06$ **17.** 6, $\sqrt{4.67} \approx 2.16$

19. 11, $\sqrt{15.2} \approx 3.90$ **21.** 5, $\sqrt{3} \approx 1.73$

23. \$32, $\sqrt{137.78} \approx$ \$11.74 **25.** \$150, $\sqrt{2600} \approx$ \$50.99

27. a) \$63, $\sqrt{631.6} \approx$ \$25.13 **b)** Answers will vary.

c) Answers remain the same, range: \$63, standard deviation ≈ \$25.13.

29. a)–c) Answers will vary.

d) If each number in a distribution is multiplied by n, the mean and standard deviation of the new distribution will be n times that of the original distribution.

e) The mean of the second set is $4 \times 5 = 20$, and the standard deviation of the second set is $2 \times 5 = 10$.

31. a) The standard deviation increases. There is a greater spread from the mean as they get older.

b) ≈133 lb **c)** ≈21 lb

d) Mean: ≈100 lb; normal range: ≈60 to 140 lb

e) Mean: ≈62 in.; normal range: ≈53 to 68 in. **f)** 5%

33. a)

East: Number of Oil Changes Made	East: Number of Days	West: Number of Oil Changes Made	West: Number of Days
15–20	2	15–20	0
21–26	2	21–26	0
27–32	5	27–32	6
33–38	4	33–38	9
39–44	7	39–44	4
45–50	1	45–50	6
51–56	1	51–56	0
57–62	2	57–62	0
63–68	1	63–68	0

b)

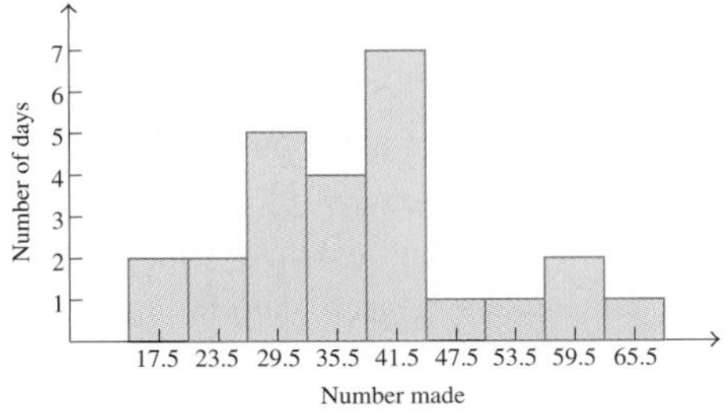

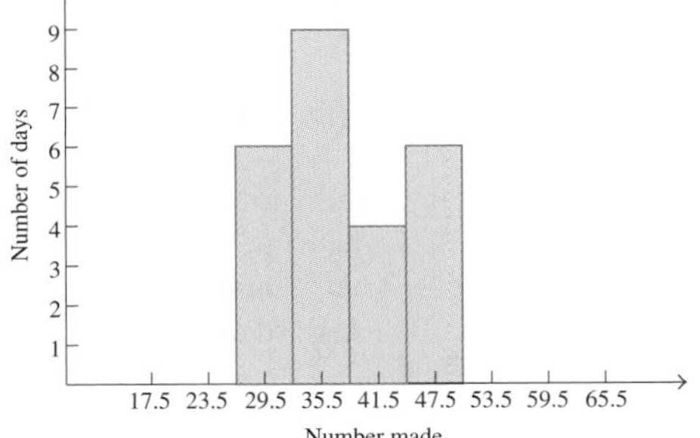

c) They appear to have about the same mean since they are both centered around 38.

d) The distribution for East is more spread out. Therefore, East has a greater standard deviation.

e) East: 38, West: 38 **f)** East: ≈12.64, West: ≈5.98

35. 6, 6, 6, 6, 6

Section 13.7, Page 804

1. A rectangular distribution is one in which all the values have the same frequency.

3. A bimodal distribution is one in which two nonadjacent values occur more frequently than any other values in a set of data.

5. A distribution skewed to the left is one that has "a tail" on its left.

7. a) *B* **b)** *C* **c)** *A* **9.–11.** Answers will vary.

13. Normal **15.** Skewed right

17. The mean is the greatest value. The median is lower than the mean. The mode is the lowest value.

19. Answers will vary. **21.** They all have the same value.

23. A z-score will be negative when the piece of data is less than the mean.

25. 0 **27.** 0.500 **29.** 0.818 **31.** 0.034 **33.** 0.037

35. 0.019 **37.** 0.053 **39.** 26.1% **41.** 89.8%

43. 97.1% **45.** 97.5% **47.** 2.9%

49. a) Jake, Sarah, Carol **b)** Marie, Kevin
c) Omar, Justin, Kim

51. 50% **53.** 10.6% **55.** 69.2% **57.** 24.1% **59.** 44.0%

61. 29.1% **63.** 59.9% **65.** 50.0% **67.** 10.6%

69. ≈ 21 women **71.** 86.2% **73.** 4.5% **75.** 69.2%

77. 0.6% **79.** ≈ 83 children **81.** 1.8%

83. The standard deviation is too large.

85. a) Katie: $z = 2.4$; Stella: $z = 1.7$

b) Katie. Her z-score is higher than Stella's z-score, which means her sales are further above the mean than Stella's sales.

87. Answers will vary. **89.** −1.18 **91.** 2

Section 13.8, Page 816

1. The correlation coefficient measures the strength of the relationship between the quantities.

3. 1 **5.** 0

7. A positive correlation indicates that as one quantity increases, the other increases.

9. The level of significance is used to identify the cutoff between results attributed to chance and results attributed to an actual relationship between the two variables.

11. –13. Answers will vary.

15. Yes **17.** Yes **19.** No **21.** No

The answers in the remainder of this section may differ slightly from your answers, depending on how your answers are rounded and which calculator you used. The answers given here were obtained from a Texas Instruments TI-36x solar calculator.

23. a)

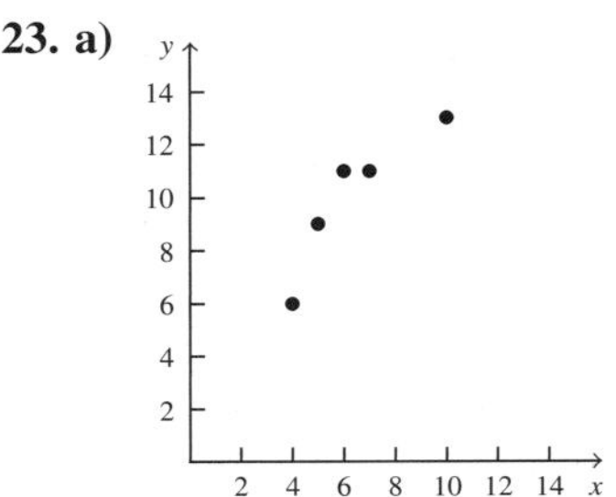

b) 0.903 **c)** Yes **d)** No

25. a)

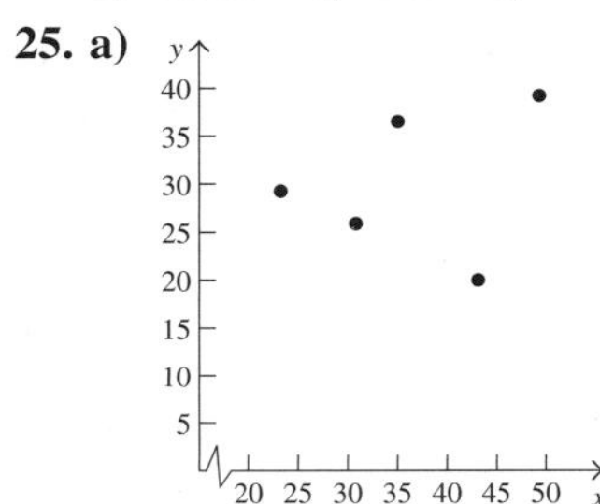

b) 0.228 **c)** No **d)** No

27. a)

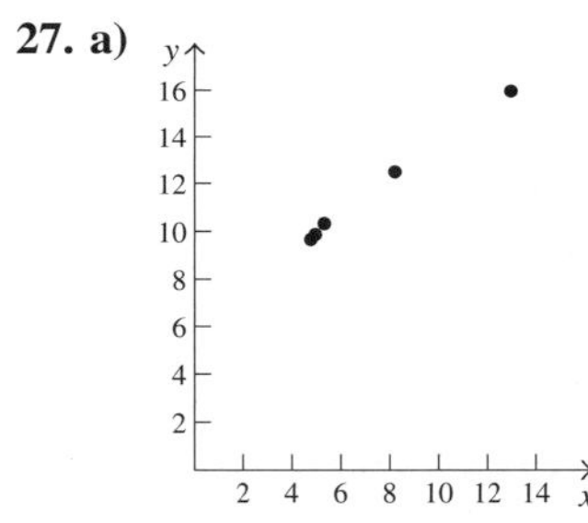

b) 0.999 **c)** Yes **d)** Yes

29. a)

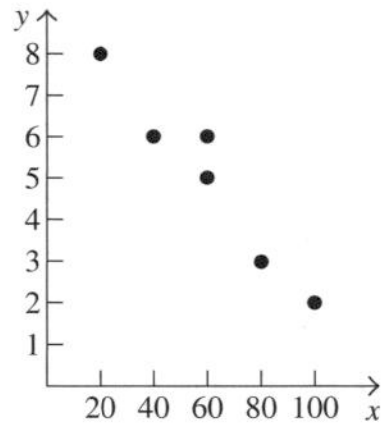

b) −0.968 **c)** Yes **d)** Yes

31. $y = 1.0x + 3.4$ **33.** $y = 0.2x + 23.8$

35. $y = 0.8x + 5.8$ **37.** $y = -0.1x + 9.5$

39. a) 0.960 **b)** Yes **c)** $y = 1.3x + 4.8$

41. a) 0.950 **b)** Yes **c)** $y = 0.8x + 24.9$

43. a) −0.782 **b)** Yes **c)** $y = -0.7x + 22.3$
d) 12.5 muggings

45. a) 0.800 **b)** Yes **c)** $y = 0.2x + 2.3$ **d)** ≈ 25 units

47. a) −0.977 **b)** Yes **c)** $y = -12.9x + 99.6$
d) 41.6%

49. a) and **b)** Answers will vary.
c)

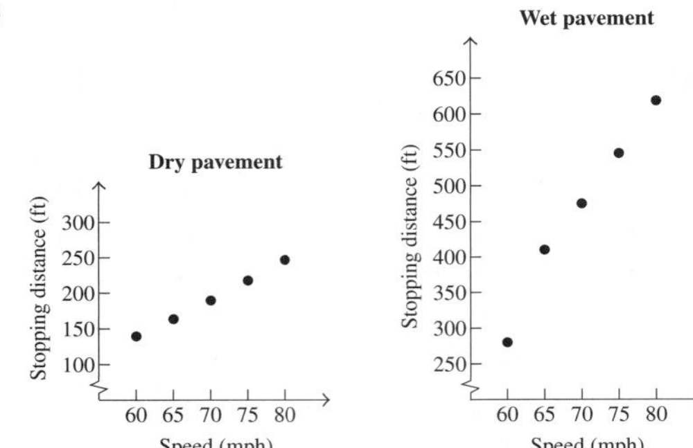

d) 0.999 **e)** 0.990 **f)** Answers will vary.
g) $y = 5.4x - 183.4$ **h)** $y = 16.2x - 669.8$
i) Dry, 232.4 ft; wet, 577.6 ft

51. Answers will vary.

53. a) 0.991 **b)** Should be the same.
c) 0.991, the values are the same.

Review Exercises, Page 821

1. a) A population consists of all items or people of interest.
b) A sample is a subset of the population.

2. A random sample is one where every item in the population has the same chance of being selected.

3. The candy bars may have lots of calories, or fat, or sodium. Therefore, it may not be healthy to eat them.

4. Sales may not necessarily be a good indicator of profit. Expenses must also be considered.

5. a)

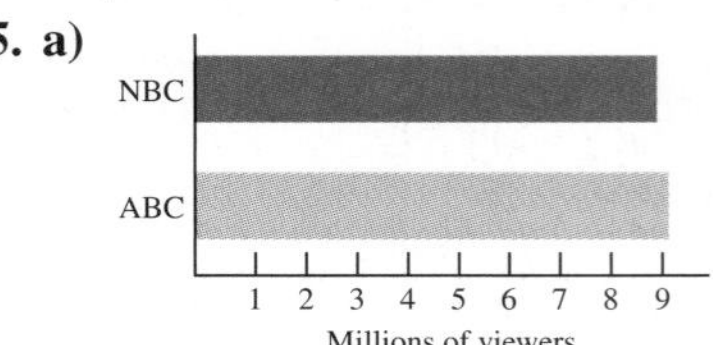

b)

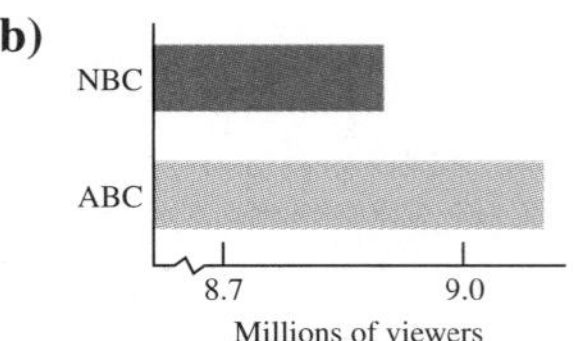

6. a)

Class	Frequency
35	1
36	3
37	6
38	2
39	3
40	0
41	4
42	1
43	3
44	1
45	1

b) and **c)**

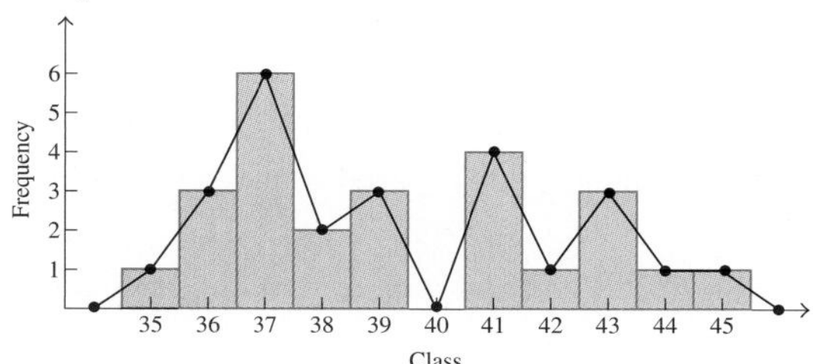

7. a)

High temperature	Number of Cities
30–39	5
40–49	8
50–59	5
60–69	6
70–79	6
80–89	10

b) and **c)**

Average Daily High Temperature in January for Selected Cities

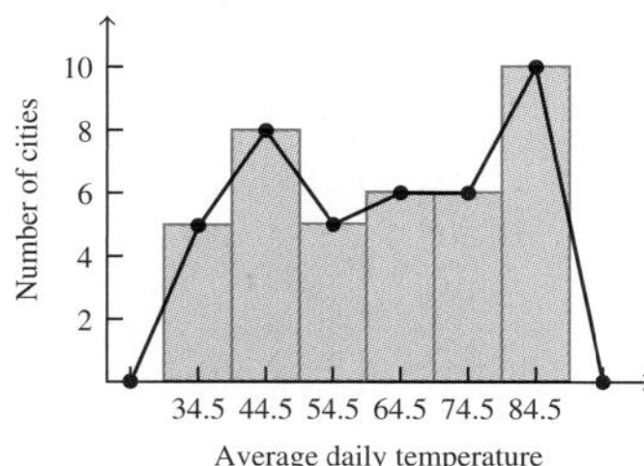

d) 3 | 6 represents 36

3 | 0 3 4 5 6
4 | 1 2 2 3 4 7 8 8
5 | 0 4 4 5 6
6 | 5 6 6 7 8 9
7 | 3 5 5 7 7 9
8 | 0 1 3 3 4 6 6 7 8 9

8. 80 **9.** 81 **10.** None **11.** 78 **12.** 30
13. $\sqrt{104} \approx 10.20$ **14.** 13 **15.** 13 **16.** 7 and 12

17. 13.5 **18.** 19 **19.** $\sqrt{40} \approx 6.32$ **20.** 68.2%
21. 95.4% **22.** 94.5% **23.** 5.5% **24.** 72.6%
25. 34.1% **26.** 34.5% **27.** 29.0% **28.** 2.3%

29. a)

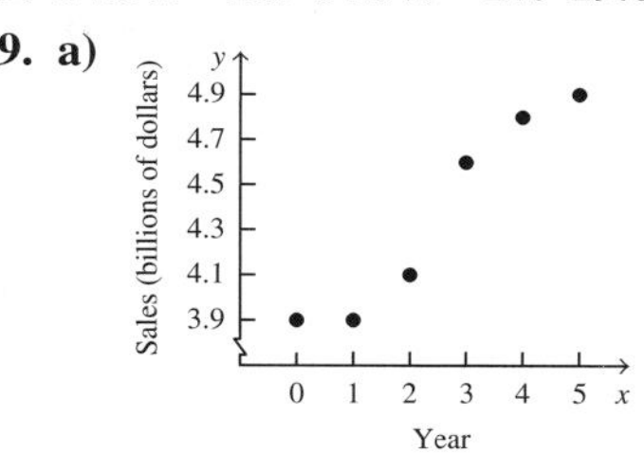

b) Yes; positive **c)** 0.964 **d)** Yes **e)** $y = 0.2x + 3.8$

30. a)

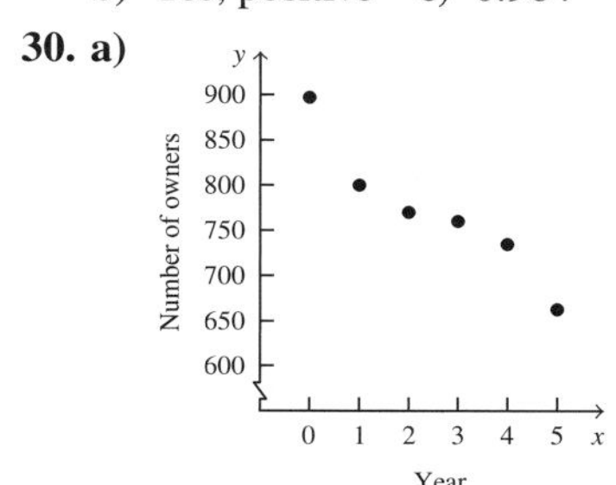

b) Yes; negative **c)** -0.952 **d)** Yes
e) $y = -39.3x + 869.0$

31. a)

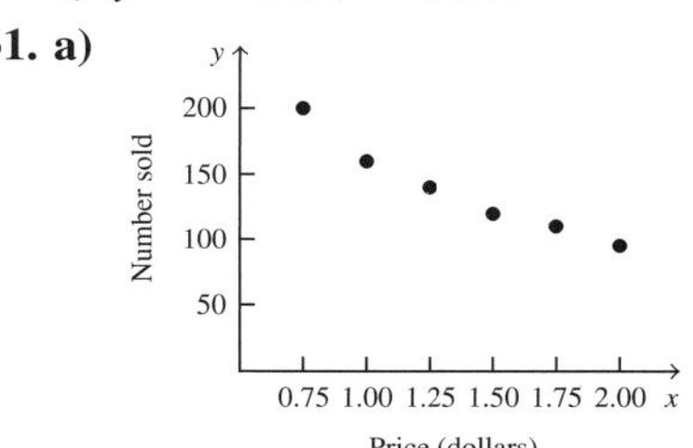

b) Yes; negative **c)** -0.973 **d)** Yes
e) $y = -79.4x + 246.7$ **f)** ≈ 120 sold

32. 175 lb **33.** 180 lb **34.** 25% **35.** 25% **36.** 14%
37. 18,700 lb **38.** 233 lb **39.** 145.6 lb **40.** ≈ 3.57
41. 2 **42.** 3 **43.** 7 **44.** 14 **45.** $\sqrt{8.105} \approx 2.85$

46.

Number of Children	Number of Presidents
0–1	8
2–3	15
4–5	10
6–7	6
8–9	1
10–11	1
12–13	0
14–15	1

47. and **48.**

Number of Children of U.S. Presidents

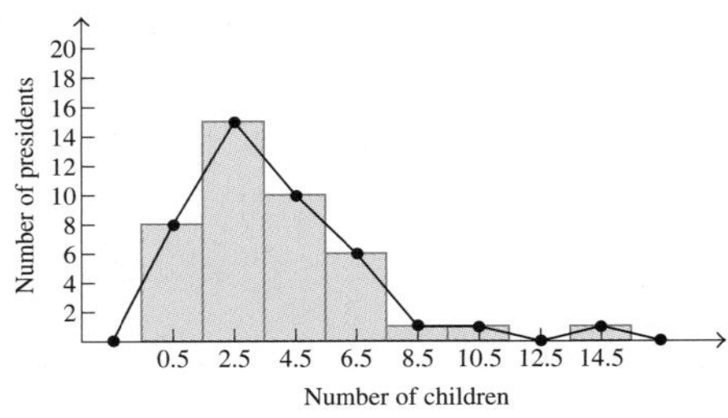

49. No, it is skewed to the right. **50.** Answers will vary.

51. Answers will vary.

Chapter Test, Page 824

1. 36 **2.** 37 **3.** 37 **4.** 33.5 **5.** 25 **6.** $\sqrt{84} \approx 9.17$

7.

Class	Frequency
25–30	7
31–36	5
37–42	1
43–48	7
49–54	5
55–60	3
61–66	2

8.

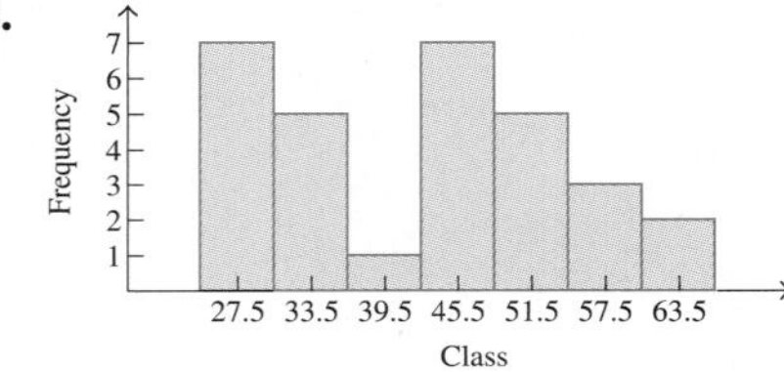

9.

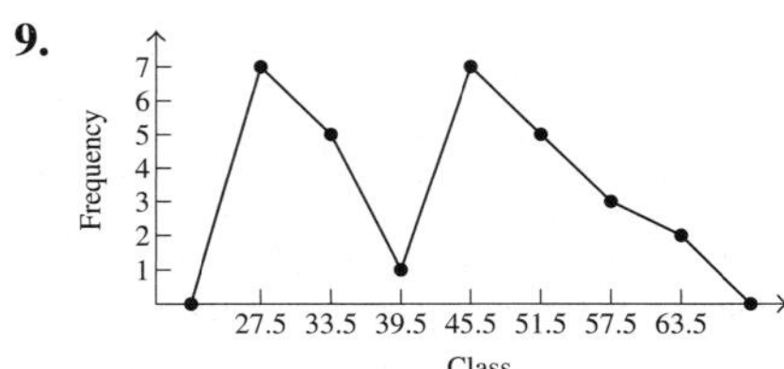

10. \$695 **11.** \$670 **12.** 75% **13.** 79%

14. \$70,000 **15.** \$740 **16.** \$640 **17.** 31.8%

18. 89.4% **19.** 10.6% **20.** ≈69 cars

21. a)

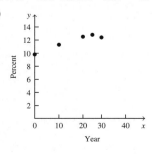

b) Yes **c)** 0.932 **d)** Yes **e)** $y = 0.1x + 10.1$
f) 14.1%

Appendix

Appendix, Page AA1

1. A vertex is a designated point.

3. To determine whether a vertex is odd or even, count the number of edges attached to the vertex. If the number of edges is odd, the vertex is odd. If the number of edges is even, the vertex is even.

5. 5 vertices, 7 edges **7.** 7 vertices, 11 edges

9. Each graph has the same number of edges from the corresponding vertices.

11. odd vertices: *C*, *D*; even vertices: *A*, *B*

13. Yes; start at *C* and end at *D*, or start at *D* and end at *C*.

15. Yes; start at any point and end where you started.

17. No.

19. Yes; start at *A* and end at *C*, or start at *C* and end at *A*.

21. a) 0 odd, 5 even **b)** Yes
c) Start in any room and end where you began. One path is *A* to *D* to *B* to *C* to *E* to *A*.
b) not traversable.

23. a) 2 odd, 4 even **b)** Yes
c) Start at *B* and end at *F*, or start at *F* and end at *B*. One path is *B* to *C* to *F* to *E* to *D* to *A* to *B* to *E* to *F*.

25. a) 4 odd, 1 even **b)** Not possible

27. a) 3 odd, 2 even **b)** Not possible

29. The door must be placed in room D. Room D is the only room with an odd number of doors.

31. Yes; there are two odd vertices. Begin at either the island on the left or on the right and end at the other island.

33.

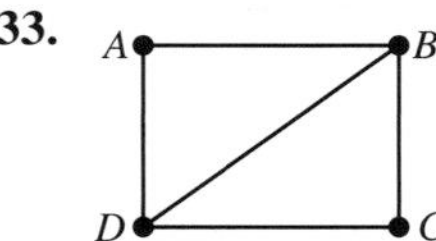

35. a) Kentucky, Virginia, North Carolina, Georgia, Alabama, Mississippi, Arkansas, Missouri
b) Illinois, Arkansas, Tennessee

37. a) 4 **b)** 4 **c)** 11

39.

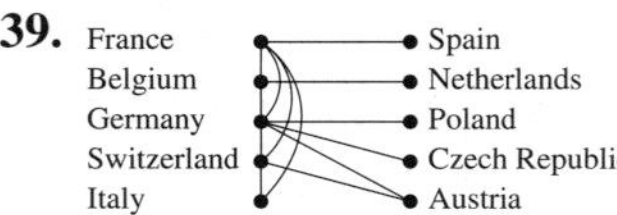

41. a) Yes, the graph has exactly 2 odd vertices.
b) One possiblity is *C*, *A*, *B*, *E*, *F*, *D*, *G*, *C*.

CREDITS

Page 1, photo of California city at night © PhotoDisc; Page 3. Photo of ATM using eye recognition ©International Business Machines Corporation; Page 4, photo of Leaning Tower of Pisa © PhotoDisc; Page 6, graph "Satellite TVs Big Gains" from *Fortune* © Time, Inc. 2/5/01. All rights reserved. Page 7, graph "Income and Tuition" © *U.S. News and World Report, L.P.* 10/30/00, p.14, Reprinted with permission; Page 8, photo of fresh donuts © PhotoDisc; Page 9, photo of a vineyard© PhotoDisc; Page 11, photo of blood sample analysis © PhotoDisc; Page 13, photo of birds © Art Wolfe/Stone/Getty Images; Page 13, photo of birds © Art Wolfe/Stone/Getty Images; Page 16, graph "Getting Older Means Getting Fatter" from Mayo Clinic Newsletter © Mayo Foundaton for Medical Education and Research. All rights reserved. Used with permission. Page 17, photo of leaves © Allen R. Angel; Page 17, photo of berries © PhotoDisc; Page 17, photo of Statue of Liberty © Allen R. Angel; Page 18, photo of person and tree © Corbis Royalty-Free; Page 18, data in table "Middle Class Perception" from *U.S. News and World Report*; Page 21, photo of George Polya © *Palo Alto Weekly*; Page 23 graph "Top Ten Irrigators" from *Scientific American* © illustration by Johnny Johnson from "Growing More Food with Less Water," by Sandra Postel. © February, 2001 Scientific American, Inc. All rights reserved. Page 25, photo of two women eating at restaurant © PhotoDisc; Page 26, photo of family vacationing in mountains © Allen R. Angel; Page 27, photo of someone spraying lawn with weed killer © Martin Barraud/Stone/Getty Images; Page 29, graph "A Wide Variety of Pump Prices" © *Rochester Democrat & Chronicle*, 8/6/01 Reprinted with permission. Page 30, graph "Demographics and Doctors" © *U.S. News and World Report, L.P.* 12/21/98 , p. 28. Reprinted with permission. Page 31, data in table "Producers of Major Greenhouse Gases" © *The New York Times*, 6/12/01; Page 39, photo of "Sunflowers" by Vincent Van Gogh © Christie's Images/Corbis; Page 41, photo of baby with blocks © Creatas; Page 42, photo of Georg Cantor © The Granger Collection; Page 45 illustration of "Creature Card 5: Jexums" © Education Development Center, Inc. 1966, 1968 All rights reserved Page 48, scene from Disney's *Snow White* © Walt Disney, 1937 The Kobal Collection; Page 48, photo of St. Louis © PhotoDisc; Page 48, graph "Yahoo Ad Revenues" © NEWSWEEK, Inc., 2/24/03, All rights reserved. Reprinted with permission. Page 49, photo of John Grisham © Edelstein Jillian/Corbis Sygma; Page 53, photo of ice cream sundae with many toppings © PhotoDisc; Page 55, photo of computer circuit board © PhotoDisc; Page 56, photo of koala and woman © Allen R. Angel; Page 58, photo of General Motors building © Peter Yates/Corbis; Page 65, photo of Hollywood Bowl © Nik Wheeler/Corbis; Page 72, photo of character from Lion King © AP/Wide World Photos/Kathy Willens; Page 76, photo of person on exercise machine © PhotoDisc; Page 79, photo of birds at sanctuary birdfeeder © Allen R. Angel; Page 80, photo of house For Sale sign © PhotoDisc; Page 81, photo of Fenway Park © Sara Anderson; Page 82, photo of people eating appetizers at restaurant © Digital Vision; Page 82, photo of wheat combine in a golden wheat field © PhotoDisc; Page 84, photo of Leopold Kronecker © The Granger Collection; Page 88, photo of a tiger © PhotoDisc; Page 92, lithograph illustration of Ptolemy © Stapleton Collections/Corbis; Page 93, photo of Abbott & Costello © Bettmann/Corbis; Page 94, photo of Brooklyn Bridge © Allen R. Angel; Page 97, photo of Geena Davis © Trapper Frank/Corbis Sygma; Page 98, photo of George Boole © The Granger Collection; Page 104, photo of a horse with a mane © EyeWire ; Page 104, photo of food on a grill © Brand X Pictures; Page 109, photo of a person studying © PhotoDisc; Page 116, photo of PowerMac G4 cube and user © Sara Anderson; Page 116, photo of Elvis Presley © Corbis; Page 116, photo of a collie © PhotoDisc Blue; Page 117, graph "Hours of Sleep" © NEWSWEEK, Inc. 8/2/99. All rights reserved. Reprinted with permission. Page 118, photo of a plane taking off © DigitalVision; Page 126, photo of a soccer game © Corbis Royalty-Free; Page 126, photo of Halle Berry © Reuters NewMedia Inc./Corbis; Page 132, photo of Charles Dodson © The Granger Collection; Page 132, photo of Alice in Wonderland © The Granger Collection; Page 133, photo of Augustus De Morgan © The Granger Collection; Page 136, photo of a sitar © PhotoDisc; Page 137, photo of boathouses at marina © Kevin Fleming/CORBIS; Page 138, photo of people posing for photo © ThinkStock; Page 139, photo of a snowmobile rider doing a stunt © PhotoDisc; Page 139, photo of people snorkeling in Cozumel © Corbis Royalty-Free; Page 140, photo of Harley Davidson motorcycles © Beth Anderson; Page 140, photo of a sailboat © Allen R. Angel; Page 141, photo of Bruce Springsteen and Clarence Clemmons on stage © Lynn Goldsmith/Corbis; Page 142, photo of a camera with adjustable lens © PhotoDisc; Page 146, photo of blindfolded scales of justice © Reza Estakhrian/Stone/Getty Images; Page 148, photo of US Constitution © Corbis Royalty-Free; Page 151, photo of Will Smith © Gerard Burkhart/AFP/Corbis; Page 152, photo of Bryce Canyon © Allen R. Angel; Page 152, photo of fireworks © PhotoDisc Blue; Page 152, photo of a football game © Allen R. Angel; Page 156, photo of a barcode and scanner © PhotoDisc; Page 158, photo of a milk cow © PhotoDisc; Page 158, photo of a mushroom © PhotoDisc; Page 158, photo of Tiger Woods © AFP/Corbis; Page 159, photo of Bonnie Raitt and guitar © Henry Diltz/Corbis; Page 161, photo of a Florida panther © PhotoDisc; Page 161, photo of Lincoln statue at Lincoln memorial © PhotoDisc; Page 162, photo of India Arie © Gary Hershorn/Reuters NewMedia Inc./Corbis; Page 163, photo of a leopard © DigitalVision; Page 165, photo of Ontario Science center logic game © courtesy Ontario Science Center; Page 166, photo of Stonehenge © Digital Vision; Page 167, photo of Denise Schmandt-Besserat © courtesy Denise Schmandt-Besserat; Page 169, photo of a clock with roman numerals © Allen R. Angel; Page 172, photo of product showing Hindu-Arabic numerals © Mark Gibson; Page 176, photo of Counting Board © Taxi/Getty Images; Page 177, photo of ship navigation © PhotoDisc; Page 179, photo of Jan Fleck's numbers © Allen R. Angel; Page 186, photo of both Arabic and Japanese or Chinese numerals © Robert Brenner / Photo Edit; Page 188, photo of kids using computers © PhotoDisc; Page 200, photo of John Napier © The Granger Collection; Page 207, photo of a doctor using flat screen monitor © Jose Luis Pelaez, Inc./Corbis; Page 211, photo of Eugene Oliver Carissan's 1914 factoring machine © Jeffrey Shallit; Page 212, photo of Srinivasa Ramanujan © The Granger Collection; Page 214, photo of a bank vault © Digital Vision; Page 215, photo of George Woltman (Mersenne GIMPS prime number) © George Woltman; Page 217, photo of Barbie dolls © Beth Anderson; Page 217, photo of O'Hare International Airport © Allen R. Angel; Page 219, photo of Roman poet Virgil © Corbis; Page 222, photo of Mount Everest © Corbis Royalty-Free; Page 236, photo of vibrating strings © Fundamental Photos; Page 240, photo of blue thistle © PhotoDisc; Page 240, photo of a silo © Digital Vision; Page 243, photo of Pythagoras of Samos © Corbis; Page 255, photo of a person in a car with seat belt fastened © PhotoDisc; Page 256, photo of cracking an egg © PhotoDisc; Page 260, photo of the galaxy © NASA Media Services; Page 265, photo of a diverse city crowd © PhotoDisc; Page 266, photo of Niagara Falls © PhotoDisc; Page 267, photo of Dr. John Allen Paulos © courtesy Temple University; Page 267, photo of minting paper money © Charles O'Rear/Corbis; Page 274, photo of Samurai sword construction © Bettmann/Corbis; Page 274, photo of a gray squirrel © DigitalVision; Page 275, photo of a card game at a casino © PhotoDisc; Page 276, photo of Fibonacci © Corbis; Page 276, photo of a sunflower center © Corbis; Page 277, photo of the Great Pyramid at Gizeh © PhotoDisc Blue; Page 278, photo of a chambered nautilus cross section © PhotoDisc; Page 278, photo of the Parthenon © Corbis Royalty-Free; Page 278, *La Parade de Cirque* by George Seurat © Metropolitan Museum of Art; Page 278, *Fibonacci's Garden* by Caryl Bryer Fallert © Caryl Bryer Fallert; Page 283, photo of a train engine © Milwaukee Road Historical Association; Page 284, photo of a roast turkey

© PhotoDisc; Page 287, photo of a person sitting in a car reading a map © PhotoDisc; Page 290, *I Saw the Figure 5 in Gold* by Charles Demuth © Metropolitan Museum of Art; Page 303, photo of Niagara Falls © Allen R. Angel; Page 309, photo of Albert Einstein © Lucien Aigner/Corbis; Page 310, photo of statue of Sophie Germaine © Stock Montage; Page 312, photo of a can of soup © Beth Anderson; Page 318, photo of a pet supply store © David Young-Wolf/PhotoEdit; Page 319, photo of San Diego © Allen R. Angel; Page 326, photo of Allen R. Angel © Allen R. Angel; Page 327, photo of workers building a deck for a house © PictureQuest; Page 333, graph "Annual Health Care Costs Per Employee, National Average" © *USA Today*, 12/1/00. Reprinted with permission. Page 333, graph "Foreign-born Population in Millions" © *U.S. News and World Report, L.P.*, 1//12/01, Reprinted with permission. Page 335, photo of Mike Myers with "Shrek" © Reuters NewMedia Inc./Corbis; Page 336, photo of illustration of René Descartes © International Business Machines Corporation; Page 339, photo of an archaeological dig © Richard T. Nowitz/Corbis; Page 341, photo of Simon Ammann in ski jumping competition © Reuters NewMedia Inc./Corbis; Page 346, photo of a CD of photos © Beth Anderson Page 347, graph "Married Househlolders with Kids" © *Newsweek*, Inc. 5/28/01. All rights reserved. Reprinted with permission. Page 363, photo of a horse-drawn carriage in a city © Allen R. Angel; Page 365, photo of a computer model of a thunderstorm © Information and Publications; Page 371, photo of an e-mail notification screen © Brand X Pictures/Getty Images; Page 374, graph "Percentage of first-time California State University freshmen entering with college-level mathematics proficiency" Adapted from *The Christian Science Monitor* © 3/5/02; Page 375, graph "U.S. Average PC Cost" © *Newsweek*, Inc. 9/20/1999; Page 377, photo of an animal at the San Diego Zoo © San Diego Zoo; Page 382, photo of telecommunications satellite testing © PhotoDisc Blue; Page 386, photo of a toy vintage convertible car © Corbis Royalty-Free; Page 389, photo of landscaping service © Allen R. Angel; Page 400, photo of the University of Tennessee women's basketball team © University of Tennessee Lady Vols; Page 400, photo of school play © Beth Anderson; Page 401, graph "Beverage Consumption in the U.S." © *Rochester Democrat & Chronicle* 9/4/01. Reprinted with permission. Page 402, photo of High Sierra Loop trail sign © Allen R. Angel; Page 406, photo of James Sylvester (matrix theory) © Stock Montage; Page 406, photo of William Rowen Hamilton © Stock Montage; Page 406, photo of Arthur Cayley © Stock Montage; Page 410, photo of cookie bakery display © Allen R. Angel; Page 414, photo of Woody from *Toy Story* © James A. Sugar/Corbis; Page 417, photo of candy store display © Creatas; Page 417, photo of the American flag © Allen R. Angel; Page 421, photo of people looking at camcorders © PhotoDisc Blue; Page 422, map of Allied invasion of Normandy © Bettmann/Corbis; Page 424, photo of a washer and dryer and people using them © Corbis Royalty-Free; Page 426, photo of in-line skaters © Allen R. Angel; Page 428, photo of chemistry class, mixing solution © Corbis Royalty-Free; Page 429, photo of premium coffee by the pound © Allen R. Angel; Page 432, photo of the *Mars Climate Orbiter* © NASA Media Services; Page 433, photo of one and two-liter bottles © Allen R. Angel; Page 433, photo of a meter stick and yard stick © Beth Anderson; Page 434, photo of a computer that can store 80GB of information © PhotoDisc; Page 434, photo of a Canadian metric road sign © Allen R. Angel; Page 436, photo of a Mexican metric road sign © Allen R. Angel; Page 438, photo of a Swiss metric aerial tram sign © Allen R. Angel; Page 438, photo of framed Picasso painting © Allen R. Angel; Page 439, photo of roast turkey © PhotoDisc; Page 440, photo of a man and guitar © Corbis Royalty-Free; Page 441, photo of Grand Canyon National Park © Allen R. Angel; Page 444, photo of Alaskan pipeline © Allen R. Angel; Page 447, photo of home plate on baseball field © Allen R. Angel; Page 447, photo of New River Gorge Bridge © Allen R. Angel; Page 447, photo of Death Valley National Park © Allen R. Angel; Page 447, photo of the American flag © Artville; Page 448, photo of Niagara falls © Allen R. Angel; Page 449, photo of a glacier © Allen R. Angel; Page 451, photo of Killer whales at Sea World © Allen R. Angel; Page 453, photo of Steven Thoburn, grocer, Sunderland, England (2/5/01) © Alan Glen Wright/First Voice; Page 456, photo of freezing rain © Allen R. Angel; Page 457, photo of Fiji market prices © Allen R. Angel; Page 458, photo of International Falls, Minnesota © Northwinds courtesy Laurel Beager; Page 461, photo of a tall man and an elevator © Allen R. Angel; Page 462, photo of land for sale in Fiji © Allen R. Angel; Page 462, photo of fruit for sale © Allen R. Angel; Page 463, photo of a Mexican road sign © Allen R. Angel; Page 465, photo of bright light bulb © Corbis Royalty-Free; Page 466, photo of a road sign © Allen R. Angel; Page 466, photo of a rice price sign © Allen R. Angel; Page 467, photo of sneakers on scale © Allen R. Angel; Page 467, photo of a Disney cruise ship © Allen R. Angel; Page 469, photo of a dolphin © Allen R. Angel; Page 470, photo of a milk tank © Allen R. Angel; Page 471, photo of a foreign Marriott sign © Allen R. Angel; Page 472, photo of a New Zealand Mobil sign © Allen R. Angel; Page 473, photo of a fractal landscape © Ann M. Burns; Page 474, photo of frontispiece of "The Elements of Euclid" from Edward R. Tufte's *Envisioning Information* © Edward R. Tufte (Cheshire, CT: Graphics Press, 1990); Page 485, photo of a dinosaur from latest Jurassic Park movie © The Kobal Collection;Page 491, photo of soccer ball hexagons © PhotoDisc Blue; Page 499, photo of Wizard of Oz scarecrow © The Kobal Collection; Page 501, photo of Dr. Andrew J. Wiles © AP/Wide World Photos; Page 509, photo of volleyball © PhotoDisc; Page 510, photo of *Les Demoiselles d'Avignon* by Pablo Picasso © Museum of Modern Art; Page 517, photo of the Pyramid of Cheops © Corbis Royalty-Free; Page 517, photo of aquarium © Reef Club; Page 517, photo of decorated cakes and icing bags © PhotoDisc Blue; Page 518, photo of the Leaning Tower of Pisa © PhotoDisc; Page 531, photo of Escher's Circle Limit III © Cordon Art; Page 539, photo of Klein bottle © Alan Bennett, Hi-Q Glass; Page 546, photo of Ivanovich Lobachevsky © The Granger Collection; Page 546, photo of G.F. Bernhard Riemann © The Granger Collection; Page 547, photo of mapping the brain © Dr. Monica K. Hurdal; Page 549, photo of Albert Einstein © Hebrew University, Jerusalem; Page 550, photo of square fractal antenna in cell phone © Andy Ryan/Andy Ryan Photography; Page 553, photo of Escher's Symmetry Work 45 © Cordon Art; Page 553, photo of Escher's Symmetry Work 45 transferred to a sphere © Cordon Art; Page 560, photo of Mongolian Girls at Ao Bao Festival © Liu Liqun/Corbis; Page 563, photo of computer simulation of protons and neutrons © Fermilab/Peter Arnold, Inc.; Page 566, photo of Woody Allen and Diane Keaton in "Annie Hall" © The Korbal Collection; Page 566, photo of Doradus Nebula © NASA Media Services; Page 569, photo of clock at Cathedral of St. Marks, Venice © Allen R. Angel; Page 574, illustration of patterns from *Symmetries of Culture* by Dorothy K. Wasburn and Donald W. Crowe © University of Washing Press, 1988. Page 581, photo of Amalie Emmy Noether © The Granger Collection; Page 584, photo of people getting on bus © PhotoDisc; Page 586, photo of female tennis pro playing tennis © Corbis Royalty-Free; Page 587, photo of people entering movie theater © Allen R. Angel; Page 592, photo of US paper currency © BrandX Pictures; Page 595, photo of Anaheim Angels player at bat © Reuters NewMedia Inc./Corbis; Page 598, photo of a young girl playing chess © Corbis Royalty-Free; Page 599, photo of Mexico City, Mexico © Digital Vision; Page 601, photo of a chartered fishing boat © Digital Vision; Page 603, photo of GTE federal Credit Union © Allen R. Angel; Page 604, photo of a pawn shop © Dana White/Photo Edit; Page 610, photo of a modern kitchen © PhotoDisc; Page 611, photo of Benny's Pawn Shop © Allen R. Angel; Page 611, photo of a hot air balloon © Allen R. Angel; Page 613, photo of Benjamin Franklin © The Granger Collection; Page 614, photo of Manhattan © Allen R. Angel; Page 617, photo of US Savings Bonds © Corbis Royalty-Free; Page 619, photo of 1974 Plymouth Roadrunner © MOPAR Nationals; Page 619, photo of a waitress in a nice restaurant © PhotoDisc; Page 620, photo of a water tower © Dale Runde; Page 620, photo of a tractor in farm field © Digital Vision; Page 620, photo of golfers © Allen R. Angel; Page 622, photo of a couple buying appliances in store © PhotoDisc; Page 624, photo of 1965 Ford Mustang © Martin Sohr; Page 627, photo of holiday shopping at busy mall © Corbis; Page 630, photo of ATM machine © PhotoDisc Blue; Page 631, photo of a woman looking at appliances © PhotoDisc; Page 632, photo of a furniture store © PhotoDisc; Page 632, photo of Portland, Oregon © PhotoDisc; Page 634, photo of a family watching television © Digital Vision; Page 636, photo of a small house © PhotoDisc Blue; Page 637, photo of a large house © PhotoDisc; Page 641, photo of a Real Estate sign © PhotoDisc; Page 644, photo of a house with Sold sign © PhotoDisc; Page 645, photo of couple applying for mortgage © PhotoDisc Page 646, photo of a house on waterfront © Digital Vision; Page 648, photo of a skateboarder © Duomo/Corbis; Page 649, photo of skiing © PhotoDisc; Page 650, photo of a house for sale with sign © PhotoDisc; Page 651, photo of someone in a fishing boat © PhotoDisc; Page 653, photo of Cheetahs © PhotoDisc Blue; Page 654, photo of Jacob Bernouilli © Corbis/Bettman; Page 656, photo of a dog © Allen R. Angel; Page 658, photo of Queen Victoria and family © Corbis/Bettman; Page 659, photo of Barry Bonds © Duomo/Corbis; Page 660, photo of birds at birdfeeder © PhotoDisc; Page 664, photo of a bird © Corbis; Page 668, photo of tennis courts © Allen R. Angel; Page 669, photo of manatees © Allen R. Angel; Page 676, photo of a bingo board © Allen R. Angel; Page 677, photo of horse race © Kevin R. Morris/Corbis; Page 677, graph Multiple Births in the United States in 2000 © *USA Today*, 1/2/03 "Stork Strain." Reprinted with permission. Page 678, photo of a man receiving a parking ticket © PhotoDisc; Page 684, photo of Diana Taurasi of the UCONN Huskies © AP Wide World Photos/Jeffrey Phelps; Page 694, photo of Epcot Center © Allen R. Angel; Page 695, photo of a flowering pea plant © PhotoDisc; Page 699, photo of family with eight girls © Corbis/Bettman; Page 700, photo of large group of people © PhotoDisc; Page 702, photo of a slot machine © Allen R. Angel; Page 704, photo of people playing golf © Allen R. Angel; Page 705, photo of a slot machine © Allen R. Angel; Page 711, photo of assortment of paper money © Beth Anderson; Page 714, illustration of Slot Machine as found in *Scientific American* 1996 © Tomo Narashima. Reprinted with permission. Page 716, photo of The Three Stooges © Bettmann/Corbis; Page 717, photo of seven children in a line © Corbis Royalty-Free; Page 719, photo of people in a hiking club © Corbis Royalty-Free; Page 720, photo of bicycle club © Corbis Royalty-Free; Page 724, Monet's *The Water Lily Pond, Pink Harmony* © Musée d'Orsay Paris/Lauros-Giraudon, Paris/SuperStock;Page 726, photo of 13-card bridge hand © Beth Anderson; Page 728, photo of theater district in NYC © Allen R. Angel; Page 728, photo of a jai alai game © Tom McCarthy/Photo Edit; Page 731, photo of Wild Bill Hickock's poker hand © Allen R. Angel; Page 731, photo of men moving furniture © Digital Vision; Page 733, photo of Mickey Mantle rookie baseball card © Allen R. Angel; Page 734, photo of a horse farm © Corbis Royalty-Free; Page 735, photo of a baseball game © Allen R. Angel; Page 743, photo of dolphins © Allen R. Angel; Page 745, photo of a national park © Allen R. Angel; Page 746, photo of a baseball field © Allen R. Angel; Page 747, photo of family riding on a dogsled © Corbis Royalty-Free; Page 748, photo of people fishing in a lake © Digital Vision; Page 749, photo of Golden Gate bridge © Allen R. Angel; Page 751, photo of a Woman with cat and dog © PhotoDisc; Page 752, photo of family watching TV © Neilsen Media Research; Page 756, photo of a bingo board © Corbis; Page 765, photo of Portland, Oregon most wired city © PhotoDisc; Page 766, photo of a car dealership © Beth Anderson; Page 767, photo of a chicken farm © PhotoDisc; Page 769, graph "Poll of Polls" © 10/23/03 *U.S. News and World Report, L.P.* Reprinted with permission. Page 772, photo of Captain Fairfield Inn Bed and Breakfast © courtesy Janet and Rick Wolf; Page 774, graph "Where Teens Work" © Rochester Democrat & Chronicle 5/8/01. Reprinted with permission. Page 775, photo of a saxophone player © BrandX Pictures; Page 777, photo of a Broadway show © Allen R. Angel; Page 779, photo of San Francisco © Allen R. Angel; Page 779, photo of Beaumont/Port Arthur, Texas © Cody Sewell; Page 781, photo of a smiling baby © Bruce Anderson; Page 784, photo of books on display at bookstore © Corbis Royalty-Free; Page 785, photo of a national park © Allen R. Angel; Page 786, photo of Blue Lagoon Water Park © Allen R. Angel; Page 787, photo of Babe Ruth © Bettmann/Corbis; Page 787, photo of Mickey Mantle © Bettmann/Corbis; Page 791, photo of Sydney opera house © Corbis Royalty-Free; Page 793, photo of someone fishing © Allen R. Angel; Page 799, photo of Blackwell; courtesy of David Blackwell, University of California at Berkeley; Page 806, photo of coffee in vending machine © Allen R. Angel; Page 806, photo of students studying © PhotoDisc Red; Page 818, photo of people eating pizza © PhotoDisc; Page 819, Photo of a movie theater © PhotoDisc; Page 819, Photo of a pool © Allen R. Angel.

INDEX

Note: Page numbers preceded by AA indicate material in the Appendix.